Ocean Wave Measurement and Analysis

Proceedings of the Second International Symposium

Honoring Professor Robert L. Wiegel

Sponsored by the

Waterways, Port, Coastal and Ocean Division of the American Society of Civil Engineers
Coastal Zone Foundation, Middletown, California
National Data Buoy Center, National Oceanic and Atmospheric Administration, Stennis, Mississippi

Edited by Orville T. Magoon and J. Michael Hemsley

New Orleans, Louisiana
July 25-28, 1993

Published by the
American Society of Civil Engineers
345 East 47th Street
New York, New York 10017-2398

ABSTRACT

This proceedings, *Ocean Wave Measurement and Analysis*, consists of papers presented at the Second International Symposium on Ocean Wave Measurement and Analysis which was held in New Orleans, Louisiana, July 25-28, 1993. It explores the major advances in wave measurement and quantification of ocean and lake waves, including technical knowledge and applications on wave theory, characteristics, design and techniques. The topics addressed in these proceedings are both national and international in scope and include practical examples and case histories on wave transformation, data analysis and reliability, wave modeling, applications, long waves, extreme wave statistics, and other topics relating to wave research over the last two decades. This proceedings will provide anyone involved with coastal technology a primary reference to the latest information in the field of wave measurement and analysis.

Library of Congress Cataloging-in-Publication Data

Ocean wave measurement and analysis: proceedings of the second international symposium, honoring professor Robert L. Wiegel, New Orleans, Louisiana, July 25-28, 1993/spon sored by Waterway, Port, Coastal, and Ocean Division of the American Society of Civil Engineers...[et al.]; edited by Orville T. Magoon and J. Michael Hemsley.
p. cm.
"The Second International Symposium on Ocean Wave Measurement and Analysis, was held in New Orleans, Louisiana, U.S.A., from July 25- 28, 1993"—Galley.
Includes index.
ISBN 0-87262-922-8
1. Ocean waves—Congresses. I. Wiegel, Robert L. II. Magoon, Orville T. III. Hemsley, J. Michael (James Michael), 1944- . IV. American Society of Civil Engineers. Waterway, Port, Coastal, and Ocean Division. V. International Symposium on Ocean Wave Measurement and Analysis (2nd: 1993: New Orleans, La.)
GC206.0255 1994 94-7190
551.47'02—dc20 CIP

Library of Congress Catalog Card No: 94-7190
ISBN 0-87262-922-8
Manufactured in the United States of America.
Abstract for Ocean Wave Measurement and Analysis

FOREWORD

WAVES 93, The Second International Symposium on Ocean Wave Measurement and Analysis—was held in New Orleans, Louisiana, U.S.A. from July 25–28, 1993.

The symposium honors Professor Robert L. Wiegel, a pioneer in the utilization of oceanographic and coastal science and engineering in finding solutions to our complex coastal problems. In addition to his professional and scientific work, Professor Wiegel has had a great inspirational impact on his many students and colleagues relative to coastal studies. Professor Wiegel is the Editor of *Shore & Beach* magazine (a publication by the American Shore & Beach Preservation Association); the past Chairman of the Coastal Engineering Research Council, American Society of Civil Engineers; Chairman of the conference on Directional Wave Spectra Applications held at the University of California, Berkeley; and the recipient of many awards and author of many technical publications.

The measurement and quantification of ocean and lake waves for verification of wave theory, understanding ocean wave characteristics, and producing economical and environmentally sensitive design is an important need in modern coastal technology.

The First International Symposium on Ocean Wave Measurement and Analysis was held in 1974 in New Orleans, and was organized by Orville T. Magoon and Dr. Billy L. Edge and produced a then state-of-the-art publication on waves. Due to major advances in wave measurement and analysis in the last two decades combined with a need to extend the scope to include the formulation of wave statistics for design, a second symposium on ocean wave measurement and analysis was essential.

WAVES 93 includes 78 papers published in this post-conference proceeding consisting of over 1,000 pages. This publication includes elements such as wave measurement theory, surf zone currents, wave measurement instruments, wave hydrodynamics, wave analysis, wave measurement project and programs, statistical methods and techniques as well as practical examples and case histories—both national and international.

To this end, **WAVES 93** was a forum for the exchange of twenty years of wave measurement and analysis research and development which will promote and improve communication, technology transfer, improved theory and design.

It is anticipated that this publication will become a primary reference in the field of wave measurement and analysis research.

The next WAVES symposium is currently being planned for 1997 to continue the exchange of ideas and data in modern coastal wave theory, measurement and analysis, and technology. For more information, please contact Orville Magoon, President, Coastal Zone Foundation, P.O. Box 279, Middletown, California 95461–0279 U.S.A.

Chairmen:

Orville T. Magoon
President
Coastal Zone Foundation
and the American Shore & Beach Preservation Association
Middletown, California U.S.A.

J. Michael Hemsley
Chief
Program Management Division
National Data Buoy Center
U.S. National Oceanic & Atmospheric Administration
Stennis Space Center, Mississippi U.S.A.

CONTENTS

SESSION 1: LONG WAVES

SESSION 2: WAVE THEORY I

SESSION 3: WAVE MEASUREMENT PROGRAMS I

SESSION 4: WAVE THEORY II

KEYNOTE SESSION; KEYNOTE ADDRESS

SESSION 5: WAVE MEASUREMENT PROGRAMS II

SESSION 6: WAVE/SEDIMENT INTERACTION

SESSION 7: WAVE MEASUREMENT HARDWARE

SESSION 9: WAVE TRANSFORMATION/DEFORMATION I

SESSION 10: DATA ANALYSIS/RELIABILITY I

SESSION 11: WAVE MODELING I

SESSION 12: WAVE TRANSFORMATION/DEFORMATION II

SESSION 13: DATA ANALYSIS/RELIABILITY II

SESSION 14: WAVE MODELING II

SESSION 15: WAVE TRANSFORMATION/DEFORMATION III

SESSION 16: DATA ANALYSIS/RELIABILITY III

SESSION 17: WAVE MODELING III

SESSION 18: WAVE TRANSFORMATION/DEFORMATION IV

SESSION 19: APPLICATIONS I

SESSION 20: WAVE MEASUREMENT TECHNIQUES

SPECIAL SESSION: EXTREME WAVE STATISTICS I— AN OVERVIEW

SESSION 21: APPLICATIONS II

ON THE GENERATION OF INFRAGRAVITY WAVES BY SHOALING MULTIDIRECTIONAL WAVES

Okey Nwogu
Hyd. Lab., Institute for Eng. in the Canadian Environment
National Research Council, Ottawa, Canada K1A 0R6

ABSTRACT

The generation of long period waves in shallow water by shoaling multidirectional waves is investigated using both numerical simulations and laboratory experiments. The numerical model is based on a new form of the Boussinesq equations that is applicable to non-breaking waves in both intermediate and shallow water depths. The model can simulate the nonlinear effects that occur in the propagation of irregular wave trains from "deep" to shallow water, including the amplification of the forced long period waves. It cannot, however, simulate free infragravity waves that are generated inside the surf zone. The results of the numerical model are compared to experimental results for the shoaling of bichromatic waves and irregular multidirectional waves on a constant slope beach.

1. INTRODUCTION

It has long been recognized that storm waves arriving at the shoreline are accompanied by infragravity waves that have periods much longer than the individual wave periods. These long waves are an important factor in determining the response of moored vessels in harbours, the oscillation of harbours, the erosion of beaches, the migration of sand bars, and wave runup on beaches. Several mechanisms for generating long period waves have been proposed by various authors. These include forced waves induced by nonlinear interactions between the primary waves (e.g. Longuet-Higgins and Stewart, 1962); free waves generated at the point of breaking (Symonds *et al.*, 1982); free waves generated due to an uneven bottom (e.g. Mei and Benmoussa, 1984); and edge waves generated by the trapping of reflected waves near the shoreline (e.g. Bowen and Guza, 1978).

Laboratory experiments on the shoaling and refraction of irregular multidirectional waves have been carried out by Briggs and Smith (1990), and Takayama *et al.* (1991). Both sets of tests found the infragravity wave energy induced in multidirectional waves to be significantly smaller than those in unidirectional waves. Theoretical investigations have also found a significant reduction of the long period energy in directional seas. Dean and Sharma (1981) derived expressions for the amplitudes of second-order forced waves in water of constant depth from the Laplace equation. Sand (1982) evaluated the expressions and found the amplitude of the long wave component of bidirectional, bichromatic waves to be smaller than those of unidirectional bichromatic waves by a factor of 5 to 10, even with small angles of separation between the primary wave components. These studies highlight the importance of including wave directionality in theoretical predictions of infragravity wave energy.

Several authors have proposed different theoretical models to simulate the nonlinear generation of infragravity waves in the nearshore zone from the offshore conditions. For unidirectional waves, Symonds *et al.* (1982) postulated that the time varying wave breaking position was primarily responsible for generating long period waves in the surf zone. The authors ignored the incident forced long waves and used the linearized form of the depth and wave period averaged continuity and momentum equations to estimate the amplitude of the radiated free long waves. The forcing term for the equations is the radiation stress gradient at the breakpoint. Schäffer and Svendsen (1988) extended the work of Symonds *et al.* to include the effect of the incident forced long waves and the propagation of wave groups into the surf zone. van Leeuwen and Battjes (1990), and List (1992), used frequency and time domain models respectively to solve the equations for irregular wave propagation. Different incident wave shoaling models were used to provide the radiation stress gradients. For multidirectional waves, Gallagher (1971) proposed a method for calculating the alongshore component of the forced long period waves. The method makes several simplifying assumptions about the change in wave height due to shoaling, wave breaking and seabed friction.

A more complete approach to simulating the nonlinear transformation of waves in shallow water is with the use of Boussinesq-type equations. The equations represent the depth integrated continuity and momentum equations and include the lowest order effects of frequency dispersion and nonlinearity. The assumption of weak frequency dispersion restricts the region of applicability of the standard form of the equations (Peregrine, 1967) to relatively shallow water. Recently, a new set of Boussinesq-type equations that is applicable to a wider range of water depths was derived by Nwogu (1993). The new set of equations is used in this paper to simulate the nonlinear generation of infragravity waves by multidirectional waves propagating from "deep" to shallow water. The results of the theoretical model are compared to experimental results for the shoaling of multidirectional waves on a constant slope beach.

2. THEORETICAL MODEL

Infragravity waves generally consist of forced waves accompanying incident short-wave groups and free waves generated from a variety of sources such as the reflection of forced waves by a sloping seabed. In this section, we shall consider forced waves induced by shoaling multidirectional waves outside the surf zone. The theoretical model is based on Boussinesq-type equations derived by Nwogu (1993). Boussinesq equations represent the depth-integrated equations of continuity and conservation of momentum for an inviscid and incompressible fluid. The equations can be considered to be a perturbation from shallow water theory to include the effect of weak frequency dispersion, and can be written in terms of the water surface elevation, $\eta(\mathbf{x},t)$, and the horizontal velocity, $\mathbf{u}_\alpha(\mathbf{x},t)$, at an arbitrary distance z_α from the still water level as (see Nwogu, 1993):

$$\left.\begin{array}{l} \eta_t + \nabla\cdot[(h+\eta)\mathbf{u}_\alpha] + \\ \nabla\cdot\left[\left(\frac{z_\alpha^2}{2}-\frac{h^2}{6}\right)h\nabla(\nabla\cdot\mathbf{u}_\alpha)+\left(z_\alpha+\frac{h}{2}\right)h\nabla[\nabla\cdot(h\mathbf{u}_\alpha)]\right] = 0 \\ \\ \mathbf{u}_{\alpha t} + g\nabla\eta + (\mathbf{u}_\alpha\cdot\nabla)\mathbf{u}_\alpha + \left[\frac{z_\alpha^2}{2}\nabla(\nabla\cdot\mathbf{u}_{\alpha t}) + z_\alpha\nabla[\nabla\cdot(h\mathbf{u}_{\alpha t})]\right] = 0 \end{array}\right\} \quad (1)$$

where $h(\mathbf{x})$ is the water depth, g is the gravitational acceleration and $\nabla = (\partial/\partial x, \partial/\partial y)$. The equations include both nonlinear and frequency dispersion terms. The frequency dispersion terms allow waves of different frequencies to travel at different speeds. The assumptions made in deriving the equations restricts the region of applicability of the equations to weakly nonlinear and mildly dispersive waves in water of varying depth, where the changes in water depth are of the same order of magnitude as the water depth. The elevation of the velocity variable z_α is a free parameter and can be chosen to minimize the differences between the linear dispersion characteristics of the Boussinesq model and Airy theory. This gives a value $z_\alpha = -0.54h$.

The nonlinear properties of Boussinesq-type equations was examined by Nwogu (1994a) who derived expressions for second-order waves induced by bidirectional, bichromatic waves in water of constant depth. The second-order forced waves occur at the sum and difference frequencies of the primary wave components and are commonly referred to as super-harmonics and sub-harmonics respectively. The sub-harmonic or set-down component travels at the velocity of the wave group and is partially responsible for the infragravity wave energy observed in the nearshore zone. Consider a bidirectional bichromatic wave train consisting of two periodic waves with frequencies, ω_1 and ω_2, amplitudes, a_1 and a_2, propagating in directions, θ_1 and θ_2. The water surface elevation is given by:

$$\eta^{(1)}(\mathbf{x},t) = a_1\cos(\mathbf{k}_1\cdot\mathbf{x}-\omega_1 t) + a_2\cos(\mathbf{k}_2\cdot\mathbf{x}-\omega_2 t), \quad (2)$$

where $\mathbf{k} = (k\cos\theta, k\sin\theta)$. The second-order forced wave at the difference frequency $\omega_- = \omega_2 - \omega_1$ can be written as:

$$\eta^{(2)}(\mathbf{x},t) = a_1 a_2 G_-(\omega_1,\omega_2,\theta_1,\theta_2)\cos(\mathbf{k}_- \cdot \mathbf{x} - \omega_- t), \tag{3}$$

where $\mathbf{k}_- = \mathbf{k}_2 - \mathbf{k}_1$, and $G_-(\omega_1,\omega_2,\theta_1,\theta_2)$ is the bidirectional quadratic transfer function for the Boussinesq equations (see Nwogu, 1994a):

$$G_-(\omega_1,\omega_2,\theta_1,\theta_2) = \frac{\omega_1\omega_2(k_-h)^2\cos\Delta\theta[1-(\alpha+\frac{1}{3})(k_-h)^2]}{2\lambda k_1' k_2' h^3} +$$

$$\frac{\omega_-[1-\alpha(k_-h)^2][\omega_1 k_2' h[k_1 h - k_2 h\cos\Delta\theta] + \omega_2 k_1' h[k_1 h\cos\Delta\theta - k_2 h]]}{2\lambda k_1' k_2' h^3}, \tag{4}$$

where $\Delta\theta = \theta_2 - \theta_1$ is the angular separation, $k_- = |\mathbf{k}_-|$, $k' = k[1-(\alpha + 1/3)(kh)^2]$, $\alpha = (z_\alpha/h)^2/2 + (z_\alpha/h)$ and

$$\lambda = \omega_-^2[1-\alpha(k_-h)^2] - gk_-^2 h[1-(\alpha+1/3)(k_-h)^2]. \tag{5}$$

$\lambda = 0$ is the dispersion relation for first-order free waves in the Boussinesq model. The quadratic transfer function is plotted against $\Delta\theta$ in Figure 1 for an example with $\omega^2 h/2\pi g = 0.05$, $\omega_-/\omega = 0.1$, where $\omega = (\omega_1+\omega_2)/2$. For unidirectional waves ($\Delta\theta = 0°$), the wavenumber of the forced wave approaches that of a free wave in shallow water with $\lambda \to 0$. This leads to a near-resonant amplification of the set-down wave in shallow water prior to wave breaking. In bidirectional seas, the wavenumber $|\mathbf{k}_2 - \mathbf{k}_1|$ of the forced long wave is significantly larger than that of the corresponding undirectional free wave. The forced wave is thus no longer close to satisfying the dispersion relation for free waves, leading to the observed significant reduction of the magnitude of the set-down wave in bidirectional seas.

When irregular waves shoal from deep to shallow water, the amplitude and phase of the forced wave changes as it interacts with the seabed as well as the primary incident waves. Free long waves are also generated due to the changing water depth. The expression for the magnitude of the forced long period wave (Eqn. 4) is no longer valid and the governing equations for water of variable depth would have to be solved. In this paper, we employ the time domain model of Nwogu (1994b) for irregular multi-directional waves. The model solves the equations using an iterative Crank-Nicolson finite difference method, with a predictor-corrector scheme used to provide the initial estimate. Details of the third-order accurate scheme are given in Nwogu (1994b). Although the Boussinesq model cannot simulate the generation of free infragravity waves in the surf zone at the present time, it implicitly includes the effect of shoaling on the forced waves, the generation of reflected free waves by the seabed as well as the interaction of the forced waves with the first-order free waves. It is anticipated that with the extension of the model to include wave breaking and runup, it should provide a comprehensive model for infragravity waves in the nearshore zone.

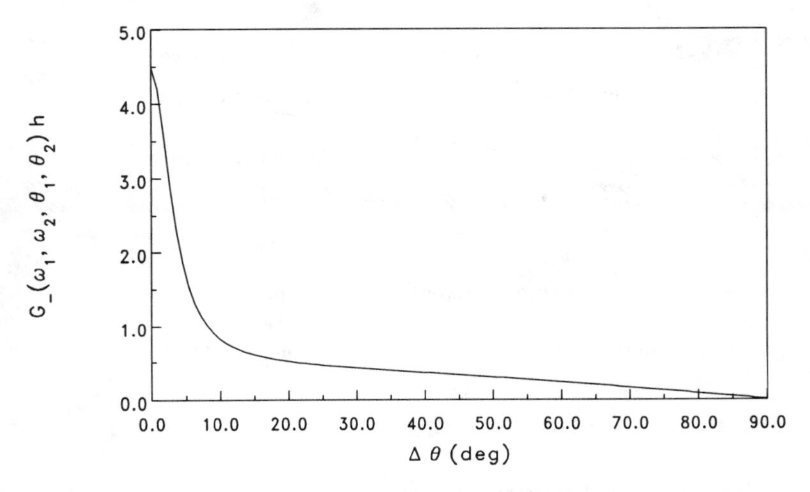

Figure 1. Variation of quadratic transfer function with angular separation.

3. EXPERIMENTS

Laboratory experiments have also been carried out to investigate the generation of infragravity waves by multidirectional waves shoaling on a constant slope beach. The experiments were conducted in the multidirectional wave basin of the Hydraulics Laboratory, National Research Council of Canada in Ottawa. The basin is equipped with a sixty-segment directional wavemaker. Perforated vertical metal sheet absorbers are placed along three sides of the basin. A 1:25 constant slope beach with an impermeable concrete cover was constructed in the basin. The toe of the beach was located 4.6 m away from the waveboard. Removable 4.6 m long solid walls were used to partially cover the side absorbers, up to the toe of the slope. The side walls were used to increase the useful working area in the basin through the application of the corner reflection method (see Funke and Miles, 1987). A constant water depth of 0.56 m was maintained at the wave generator for all the tests.

Water surface elevations along the centerline of the basin were measured with a linear array of twenty three capacitance wire wave probes. The spacing between the probes varied from 0.3 m to 1.6 m as shown in Figure 2. Different unidirectional, bidirectional and multidirectional wave conditions were generated in the basin. The irregular multidirectional sea states were synthesized from target directional wave spectra using the random phase, single direction per frequency model (e.g. Miles, 1989). The JONSWAP spectrum was used for the frequency distribution while the cosine power function was used for the

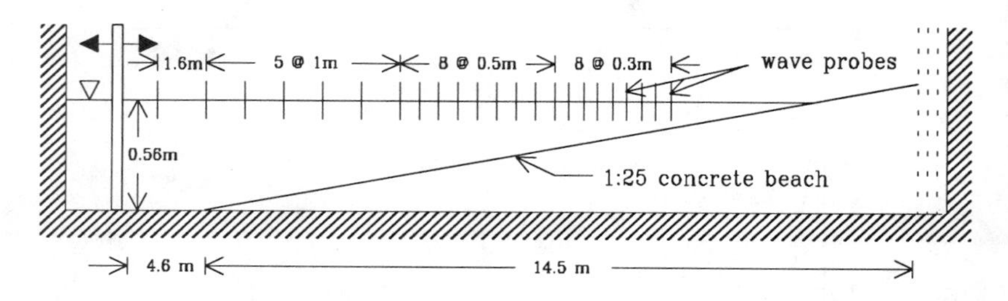

Figure 2. Experimental setup.

directional distribution. Data was sampled at a frequency of 20 Hz.

4. RESULTS AND DISCUSSION

4.1 Shoaling of Bichromatic Waves

The experimental results for bichromatic waves provide an excellent tool for investigating nonlinear wave-wave interaction effects in the shoaling zone. Let us now examine one of the tests with wave periods $T_1 = 1.65\,\mathrm{s}$, $T_2 = 1.5\,\mathrm{s}$, and heights $H_1 = 0.041\,\mathrm{m}$, $H_2 = 0.037\,\mathrm{m}$. The tests were carried for a unidirectional version with $\theta_1 = \theta_2 = 0°$, and a bidirectional version with $\theta_1 = 30°$ and $\theta_2 = -15°$. The time history of the water surface elevation at four different water depths is shown in Figure 3 for the unidirectional wave train. The depths correspond to the toe of the slope ($h = 0.56\,\mathrm{m}$), an intermediate depth in the shoaling zone ($h = 0.32\,\mathrm{m}$), the line of first breaking ($h = 0.134\,\mathrm{m}$), and a shallow depth in the surf zone ($h = 0.08\,\mathrm{m}$). The low frequency component of the surface elevation was obtained by filtering out wave components with frequencies greater than 0.25 Hz and is also shown in Figure 3.

At the toe of the slope ($h = 0.56\,\mathrm{m}$), the magnitude of the low frequency component is initially negligible. After about 25 s, which is approximately the time required for the wave group to travel to the shoreline and back to the toe of the slope, we observe a sudden increase in the long period wave energy. This is due to the arrival of reflected free waves and is consistent with early observations of surf beats by Munk (1949) and Tucker (1950). After the waves break, there is a marginal decrease in the amplitude of forced waves until the waves run up the beach and are reflected back offshore as free waves. The long wave also changes from set-down beneath groups of high waves to set-up in shallower water ($h = 0.08\,\mathrm{m}$) during the runup process. The relative significance of the forced and free waves depends on the location where the waves are measured.

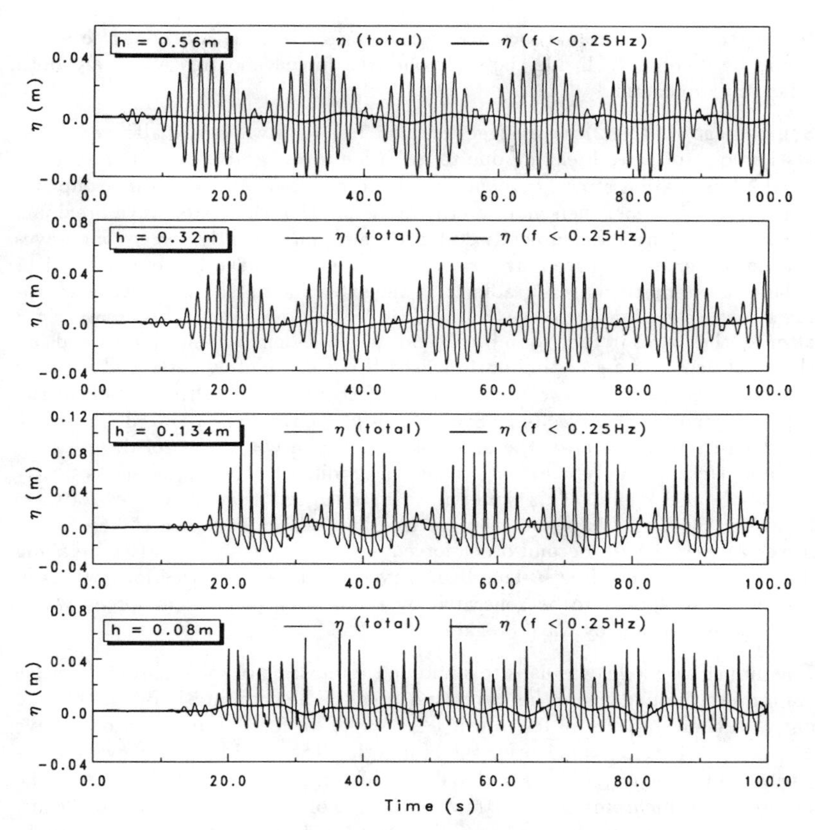

Figure 3. Measured time histories of shoaling bichromatic wave train at different depths ($T_1 = 1.65\,\text{s}$, $T_2 = 1.5\,\text{s}$).

In deeper water where the forced waves are negligible, the reflected free waves would be dominant. Inside the surf zone, the forced and free waves are both relatively important.

Symonds *et al.* (1982) have suggested that free long waves are also generated at the point of wave breaking due to radiation stress gradients. Prior to wave breaking, radiation stress gradients induce forced long waves at the group frequency. The authors then make a critical assumption that after wave breaking, the wave groupiness is completely destroyed, and consequently, forced long waves can no longer exist. Free long waves would therefore have to be generated to balance the radiation stress gradient. In the experimental data, the wave grouping structure gradually changes after the individual waves start to break. Even after all the waves in the group break and the wave height primarily depends on the local depth, the grouping pattern is still not completely destroyed as there is a quiescent period between wave groups (see Figure 3). Other experimental evidence for regular waves (e.g. Svendsen, 1984) have demonstrated that there is a transition region after the onset of breaking in which the set-down or radiation stress is nearly constant. Thus, the radiation stress gradient is nearly zero at the breakpoint. This would all seem to suggest that the process of wave breaking, by itself, does not generate free long waves or "release" the forced waves as free waves. Second-order forced long waves still exist after breaking due to the nonlinear free surface boundary conditions. The free long waves in the surf zone appear to be generated by a reflection of incident forced waves after runup and not by wave breaking.

The numerical model was used to simulate the shoaling of the bichromatic wave train up to the line where the waves first break. The computations were carried out at a time step size of 0.05 s using $\Delta x = 0.05$ m for the unidirectional wave and $\Delta x = 0.1$ m, $\Delta y = 0.5$ m for the bidirectional wave. Figure 4 shows a comparison of the measured and simulated time histories and spectral densities of the undirectional bichromatic wave train at the line of first breaking ($h = 0.134$ m), while Table 1 shows the corresponding measured and predicted wave amplitudes. The Boussinesq model is observed to reasonably predict the amplitudes of the first-order waves but not the set-down component. The numerical model predicts an amplitude of 0.003 m for the set-down wave while the measured amplitude is 0.006 m. The Boussinesq model underestimates the magnitude of the long period wave partly because it does not simulate the reflected free waves that are generated after runup. It can also be seen from Table 1 that the directional separation of the wave components ($\Delta\theta = 45°$) reduces the magnitude of the difference frequency component from 0.006 m for the unidirectional wave to virtually zero for the bidirectional wave because of the non-resonant nature of sub-harmonic interactions in directional seas.

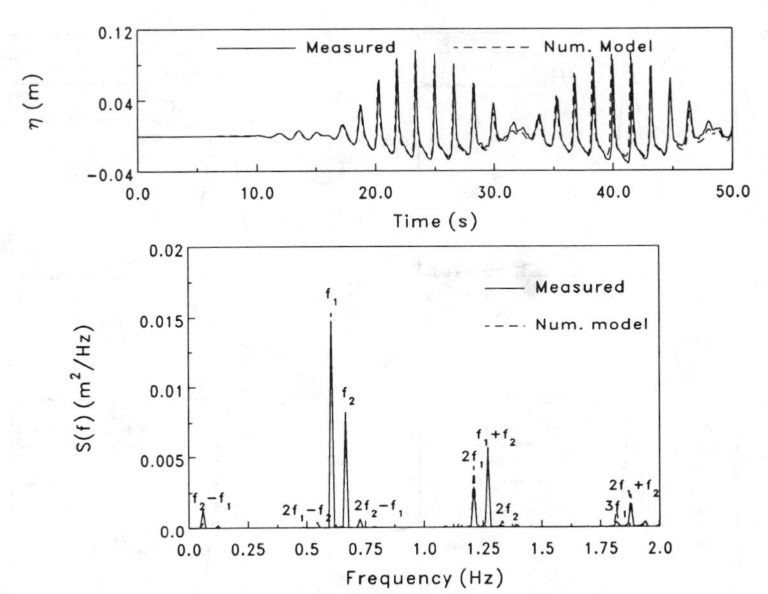

Figure 4. Measured and simulated time histories and spectral densities for bichromatic wave train ($h = 0.134$ m).

Table 1. Measured and predicted wave amplitudes (m) in shallow water ($h = 0.134$ m).

	Freq. (Hz)	Unidirectional ($\Delta\theta = 0°$)		Bidirectional ($\Delta\theta = 45°$)	
		Measured	Predicted	Measured	Predicted
f_1	0.61	0.0189	0.0193	0.0182	0.0191
f_2	0.67	0.0143	0.0133	0.0150	0.0147
$f_2 - f_1$	0.06	0.0057	0.0031	0.0007	0.0008

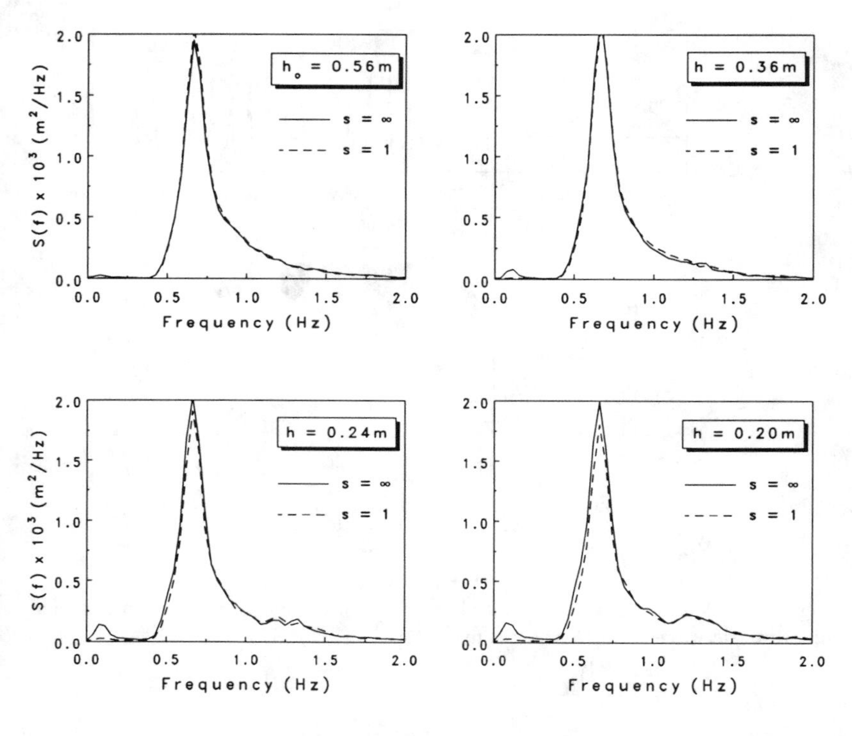

Figure 5. Surface elevation spectral densities of unidirectional ($s = \infty$) and multidirectional ($s = 1$) sea states at different water depths.

4.2 Shoaling of Irregular Waves

The experimental data for irregular waves can be used to investigate the differences between the infragravity wave energy induced by shoaling unidirectional and multidirectional waves. Figure 5 shows a comparison of the surface elevation spectral densities at different depths for a JONSWAP sea state with significant height, $H_s = 0.09\,\mathrm{m}$, peak frequency $f_p = 0.67\,\mathrm{Hz}$ and peak enhancement factor $\gamma = 3.3$. The multidirectional version of the sea state was synthesized from a cosine-squared directional distribution ($D(\theta) = 2\cos^2\theta/\pi$). The measured time series had record lengths of 819.2 s at a time interval of 0.05 s. The spectral estimates were averaged over frequency bands of width 0.04 Hz (130 dof). At all the water depths, the infragravity wave energy for the unidirectional sea state is seen to be significantly larger than that for the corresponding multidirectional sea state. At a water depth of 0.2 m, the directional spreading of wave energy

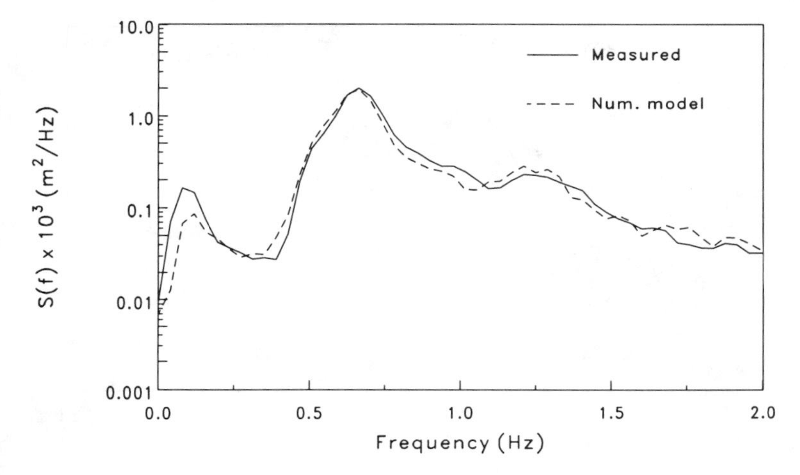

Figure 6. Measured and predicted surface elevation spectral densities for unidirectional sea state ($h = 0.18\,\mathrm{m}$).

reduces the variance of long period waves by about 78%.

The Boussinesq model was used to simulate the shoaling of the unidirectional and multidirectional sea states from deep to shallow water. The numerical simulations were carried out using $\Delta t = 0.05\,\mathrm{s}$, $\Delta x = 0.04\,\mathrm{m}$ for the undirectional wave, and $\Delta x = 0.1\,\mathrm{m}$, $\Delta y = 0.5\,\mathrm{m}$ for the multidirectional wave. The end absorbing boundary was placed at a water depth of 0.18 m. The measured and numerically predicted surface elevation spectral densities at $h = 0.18\,\mathrm{m}$ are compared in Figures 6 and 7 respectively for the unidirectional and multidirectional sea states. The numerical model underestimates the magnitude of the infragravity wave energy for both sea states. The reason for the differences between the measured and predicted infragravity wave spectrum can be further investigated by analyzing the measured data at a finer frequency resolution. Figure 8 shows the "raw" infragravity spectrum for the unidirectional sea state, in which the spectral estimates were not averaged (2 dof). It can be seen that most of the infragravity energy is concentrated around a frequency of 0.08 Hz. This frequency corresponds to a standing wave mode in the basin with an antinode at the waveboard and a node at the shoreline. The numerical model does not include wave breaking and runup at the present time and is thus unable to reproduce the observed low frequency standing wave pattern.

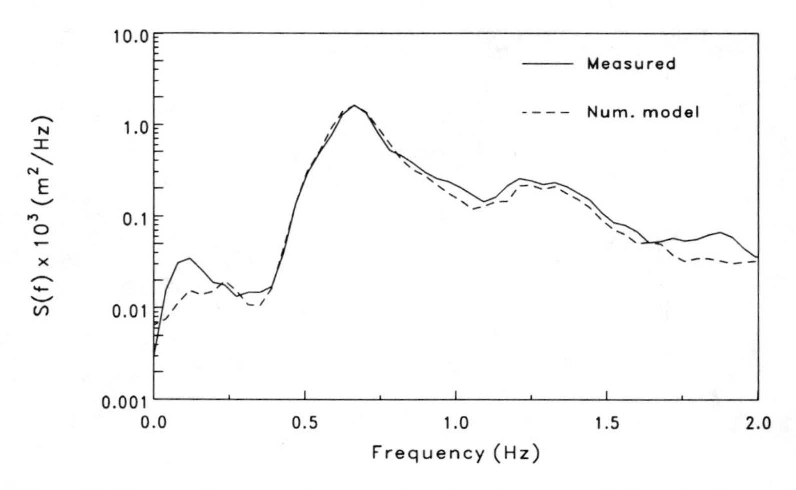

Figure 7. Measured and predicted surface elevation spectral densities for multi-directional sea state ($h = 0.18\,\mathrm{m}$).

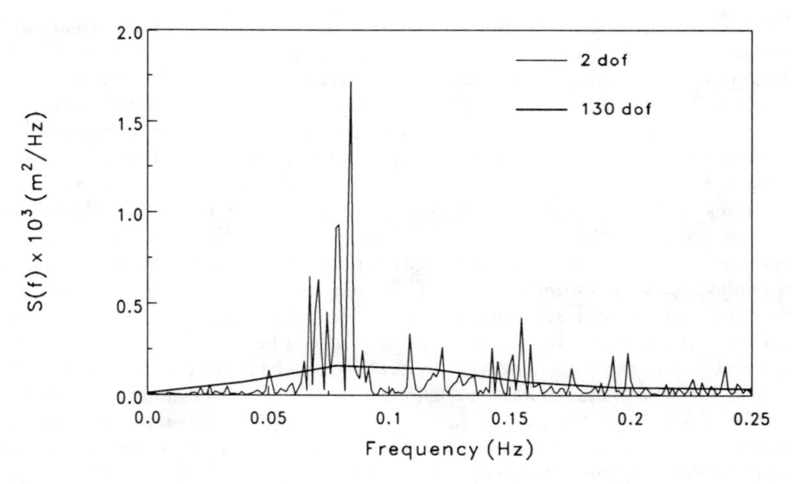

Figure 8. High frequency resolution of infragravity wave spectrum for unidirectional sea state ($h = 0.18\,\mathrm{m}$).

5. CONCLUSIONS

Numerical simulations and laboratory experiments have been used to investigate the generation of infragravity waves in shallow water by shoaling multidirectional waves. The observed infragravity waves consist of forced waves induced by second-order interactions and reflected free waves. The directional spreading of wave energy leads to a significant reduction of the induced long period waves due to the non-resonant nature of sub-harmonic interactions in directional seas. It is therefore important to include wave directionality in theoretical models for infragravity wave generation. A numerical model based on Boussinesq-type equations was used to simulate the shoaling of multidirectional waves from "deep" to shallow water. The Boussinesq model can simulate the generation of forced and free long period waves prior to breaking. By extending the model to include wave breaking and runup, we should be able to simulate the free long period waves that are generated after the reflection of the forced waves at the shoreline.

REFERENCES

1. Bowen, A.J., and Guza, R.T. 1978. Edge waves and surf beat. *J. of Geophysical Research*, **83**, C4, pp. 1913-1920.

2. Briggs, M.J., and Smith, J.M. 1990. The effect of wave directionality on nearshore waves. *Proc. 22nd Int. Conference on Coastal Engineering*, Delft, The Netherlands, pp. 267-280.

3. Dean, R.G., and Sharma, J.N. 1981. Simulation of wave systems due to nonlinear directional spectra. *Proc. Int. Symposium on Hydrodynamics in Ocean Engineering*, Trondheim, Norway, pp. 1211-1222.

4. Funke, E.R., and Miles, M.D. 1987. Multi-directional wave generation with corner reflectors. *Institute for Mechanical Engineering Technical Report* TR-HY-021, National Research Council, Canada.

5. Gallagher, B. 1971. Generation of surf beat by non-linear wave interactions. *J. of Fluid Mechanics*, **49**, pp. 1-20.

6. List, J.H. 1992. A model for the generation of two-dimensional surf beat. *J. of Geophysical Research*, **97**, C4, pp. 5623-5635.

7. Longuet-Higgins, M.S., and Stewart, R.W. 1962. Radiation stress and mass transport in gravity waves, with application to 'surf beats'. *J. of Fluid Mechanics*, **13**, pp. 481-504.

8. Mei, C.C., and Benmoussa, C. 1984. Long waves induced by short-wave groups over an uneven bottom. *J. of Fluid Mechanics*, **139**, pp. 219-235.

9. Miles, M.D. 1989. A note on directional random wave synthesis by the single-

summation method. *Proc. XXIII IAHR Congress*, Ottawa, pp. C243-250.

10. Munk, W.H. 1949. Surf beats. *Trans. American Geophysical Union*, **30**, pp. 849-854.

11. Nwogu, O. 1993. Alternative form of Boussinesq equations for nearshore wave propagation. *J. of Waterway, Port, Coastal and Ocean Engineering*, ASCE, **119**(6), in press.

12. Nwogu, O. 1994a. Second-order directional wave interactions in shallow water. *submitted to Applied Ocean Research.*

13. Nwogu, O. 1994b. Time domain simulation of shoaling multi-directional waves. *submitted to Coastal Engineering.*

14. Peregrine, D.H. 1967. Long waves on a beach. *J. of Fluid Mechanics*, **27**, pp. 815-827.

15. Sand, S.E. 1982. Long waves in directional seas. *Coastal Engineering*, **6**, pp. 195-208.

16. Schäffer, H.A., and Svendsen, I.A. 1988. Surf beat generation on a mild slope beach. *Proc. 21st Int. Conference on Coastal Engineering*, Malaga, Spain, pp. 1058-1072.

17. Svendsen, I.A. 1984. Wave heights and set-up in a surf zone. *Coastal Engineering*, **8**, pp. 303-329.

18. Symonds, G., Huntley, D.A., and Bowen, A.J. 1982. Two-dimensional surf beat: Long wave generation by a time-varying breakpoint. *J. of Geophysical Research*, **87**, C1, pp. 492-498.

19. Takayama, T., Ikeda, N., and Kosugi, Y. 1991. Hydraulic model tests on directional random wave transformation in Kamaishi bay. *Rep. of the Port and Harbour Research Institute*, Yokosuka, Japan, **30**(1), pp. 69-136.

20. Tucker, M.J. 1950. Surf beats: Sea waves of 1 to 5 min. period. *Proc. Royal Soc. London*, **A202**, pp. 565-573.

21. van Leeuwen P.J., and Battjes, J.A. 1990. A model for surf beat. *Proc. 22nd Int. Conference on Coastal Engineering*, Delft, The Netherlands, pp. 32-40.

ANALYSIS OF PROTOTYPE LONG PERIOD WAVES FOR LOS ANGELES/LONG BEACH HARBOR RESONANCE STUDIES

William C. Seabergh[1], M. ASCE

Abstract

Long period wave data have been collected at an offshore wave gage and seven harbor gages for over eight years in the Los Angeles/Long Beach Harbor region. These data are being used in studies for the design of expansion of the harbors. In particular, the data have been used to select long period spectra for reproduction in a physical model of the harbors. They included two storm conditions representative of storms from the west and a summer spectrum from the south. These long period spectra were used to program model wave generators and wave data were collected at locations in the model which corresponded to prototype gage locations. A comparison of model and prototype wave amplification indicated good model-prototype correlation. Also long wave information was examined to determine the effect of the source of the short period waves (southern hemisphere swell, hurricanes, or winter storms) on the nature of the long period component of the waves. Wave statistics were used to interpret effects of energy level on wave transformation into the harbors and correlate long and short wave spectrum.

Introduction

A physical scale model of Los Angeles and Long Beach Harbors was constructed at the US Army Engineer Waterways Experiment Station in 1973. This 1:400 horizontal, 1:100

---

[1]Research Hydraulic Engineer, Coastal Engineering Research Center, US Army Engineer Waterways Experiment Station, 3909 Halls Ferry Road, Vicksburg, MS 39180.

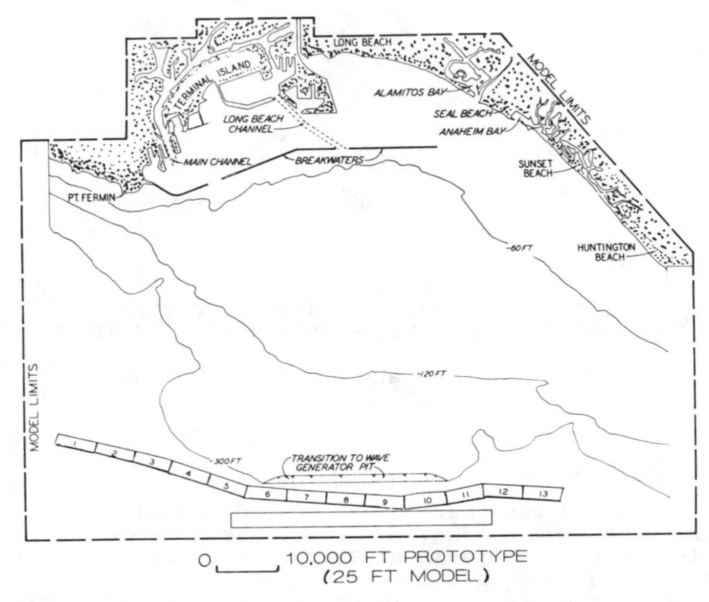

Figure 1. Los Angeles-Long Beach Harbors model layout.

vertical scale model (Figure 1) was designed to reproduce tides and waves. Since 1973, the model has been used to examine the effects of harbor expansion projects on tidal currents and harbor resonance. More recently it has been used exclusively for performing harbor resonance tests with simulation of long-period waves. These tests in the past have involved the construction of proposed projects in the model and subjecting them to a series of over 200 monochromatic wave tests, with wave periods ranging from 30 to 400 sec. Wave data are usually collected at 50 or more locations throughout the harbors at existing and proposed berths (see Seabergh (1985) for example). Using prototype long-period wave data collected offshore of the harbors, it was possible to develop long-period wave spectra which could be input to the computer-controlled model wave generators. This approach permits many periods (or frequencies) to be reproduced simultaneously over a broad range for an individual test. These results can be used to pinpoint troublesome wave period ranges, which create harbor surge conditions that may lead to difficult loading/unloading conditions and possible ship damage. Tests may then be conducted with monochromatic waves, if necessary, to better define peak amplifications. These harbor amplifications may then be related to long wave period ocean statistics to provide information on energy levels to be expected at a particular berth. This information may then be used in conjunction with a moored ship motion model to determine if troublesome conditions exist so that changes to the plan and/or mooring arrangements can be made.

Prototype Data Collection

At the initiation of this work, about 2 years of long-period wave data had been collected at an offshore oil platform located in a depth of 49 m (sensor depth 10 m), 13 kilometers south of the harbors, and at seven harbor gages (see Figure 2 for location of harbor gages). More data became available as the study progressed. Rosati and Puckette (1993) provide information on the data collection instrumentation, analysis parameters and additional gage locations deployed as the study progressed. Using these data, long-period spectra were selected for programming the model wave generators, data were collected in the model at locations of prototype gages, and these data were compared to prototype spectra. Needed adjustments to wave generator energy distribution were made and model to prototype comparisons were rechecked. This allowed base data spectra to be collected at stations throughout the harbor and used for comparison with proposed harbor expansion tests.

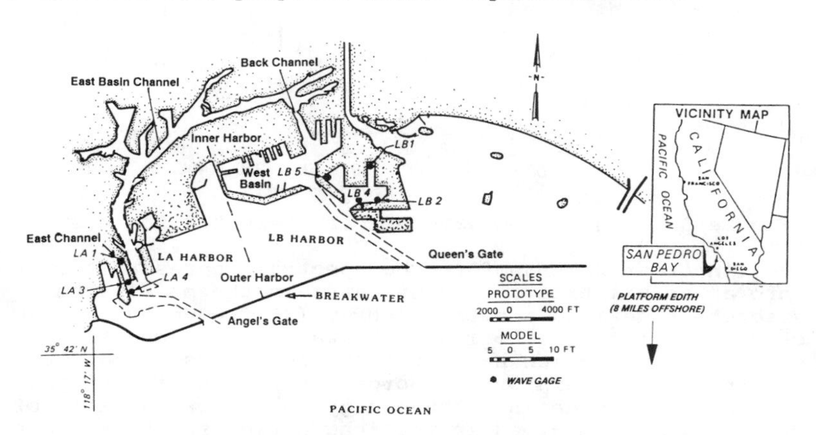

Figure 2. Locations of harbors' wave gages.

Selection of Test Conditions

An analysis of Platform Edith long-period wave data was made to determine appropriate input to the model wave generators. Two storms were outstanding in the data record as far as their impact on the harbors. The largest event recorded was the Martin Luther King Day storm of 17 January 1988. The short period portion of the wave spectrum had a significant wave height of 7.5 m (24.6 ft) during the peak period of energy measured at Platform Edith. The long-period portion of the wave spectrum contained 270 cm^2 (0.29 ft^2) of energy and was distributed as seen in Figure 3. This event caused significant damage to the Southern California coastline. The second event selected occurred on 2 February 1986 and resulted in significant harbor agitation with numerous reports of moored ship difficulties (Figure 3). The third long wave spectrum selected was based on an average or mean long-period wave spectrum condition representative of a southerly approach (Figure 3). Since the mean spectrum was nearly flat, a uniform, constant-energy spectrum was created for use in the model.

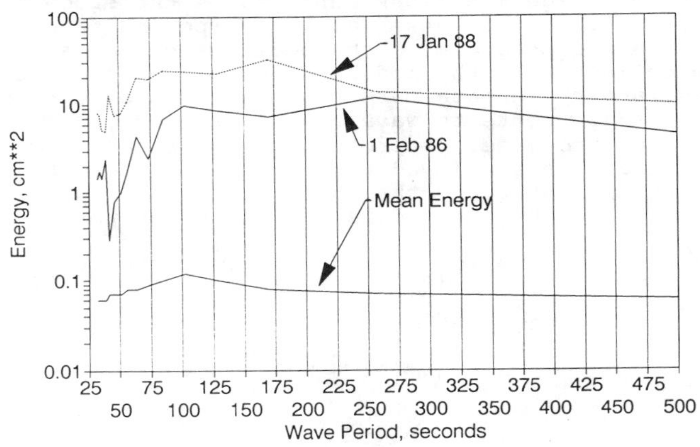

Figure 3. Long period wave spectra selected for testing

In order to transform the spectral representation into an actual time series of waves in the model, the discretely defined spectral energy (36 frequency components from 0.1 to 1.33 cps) was transformed to a control signal which has 256 frequency bands (Δf=0.00479) for the wave generator. In order to produce an analysis that accurately defines the energy in the broad range of wave periods contained in the long-period spectrum, each individual test was run for 512-sec. Runs of shorter test durations compared closely to the 512-sec test, indicating no problems with contamination of the wave records due to rereflected waves off model boundaries or the wave generator. The boundaries have multiple layers of a fibrous matrix wave absorber and the irregular ocean contours and shoreline boundary did not appear to direct significant energy back to the wave generator. The 13 individual units that make up the wave generator were operated in phase, but wave amplitude was varied along the wave front to create an appropriate energy distribution approaching the harbors. Since the two storms being run approached from a westerly quadrant, energy distribution was adjusted for that approach. The uniform-spectrum energy distribution was adjusted for a southerly wave approach, more typical of summer conditions. Ship motion observations in the prototype indicate that these two directional approaches (the west for winter storms, and the south for hurricane and Southern Hemisphere swell) create an annual bimodal distribution for significant moored ship motion events. Offshore islands and the mainland limit the primary wave approaches to the harbors for these directions.

Model wave height data were initially collected at the seven prototype gage locations seen in Figure 2. These wave heights are typically converted to a wave height amplification. Wave height amplification is traditionally defined as the ratio of the wave height at a particular location in a harbor to twice the incident wave height at the harbor mouth. This definition results from the fact that the standing wave height for a fully reflective straight coast with no harbor would be twice the incident wave height due to superposition of the incident and reflected waves. However, in the hydraulic model there is variation in wave height along the harbor boundary due to wave refraction. In this study, data were available at the ocean wave gage on Platform Edith. In order to facilitate direct comparison with prototype data, wave height data at each harbor gage were divided by wave height measured at a gage located at the analogous location of Platform Edith in the model ocean. Since the waves being studied were composed of many frequencies (or a spectrum), the digital output from the gage was analyzed by Fast Fourier Analysis (FFT) to determine an energy level that could be converted to a wave height (by taking four times the square root of the energy) for each frequency band. Water elevation data were collected at a rate of 20 readings per second at each gage location. A total of 8,192 data points were collected at each gage during a test. The data were windowed with a cosine square taper and after FFT analysis, the raw spectral estimates (Δf=0.0024414) were smoothed by averaging 8 bands, so that Δf for model data was 0.01953.

Figure 4 shows a comparison between model and prototype wave height amplification (determined by the square root of the ratio of energy for a certain frequency band at a given harbor gage to that at the ocean gage at Platform Edith) for the February 86 storm at gage LB-2 and the January 88 storm at gage LA-4. The comparison is not direct since the prototype data were analyzed with a wider frequency interval, while the model data have finer frequency (or wave period) resolution. Figure 4 indicates that the model harbor spectral response closely reproduced the prototype.

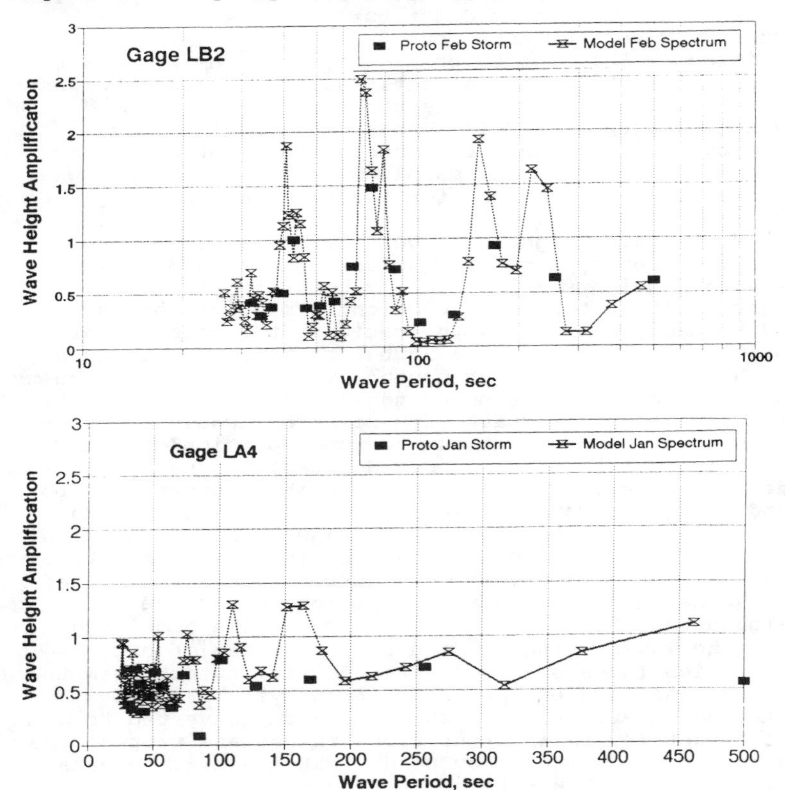

Figure 4. Prototype-to-model comparison of wave height amplifications at gages LB2 and LA4.

Comparisons of prototype and model wave height amplification for the January 1988 storm indicated the model is a little more responsive than the prototype for this extreme event. In examining many data sets, it was apparent that as ocean energy increased, harbor wave height amplification decreased. This may be caused by highly nonlinear wave motions effecting a different basin response than at lower energy conditions. There were reports of waves overtopping the outer harbor breakwater during this storm, which would produce a complex wave field in the harbor itself. Also the northwest to southeast storm track may have afforded the harbor more protection relative to the wave gage at Platform Edith, creating lower prototype wave amplifications when a wave height amplification ratio between the harbor gage and the Platform Edith gage is taken. Whatever the reason, the model results are reasonable and perhaps slightly conservative for the January 1988 storm. Figure 5 shows that when taking the ratio of the energy exceedance curves of gages LB-2 and Edith for the 73 sec energy band, the ratio curve indicates a decrease with higher energy conditions. This possibly indicates the harbors are sheltered relative to Platform Edith for higher energy conditions, which are usually from the west.

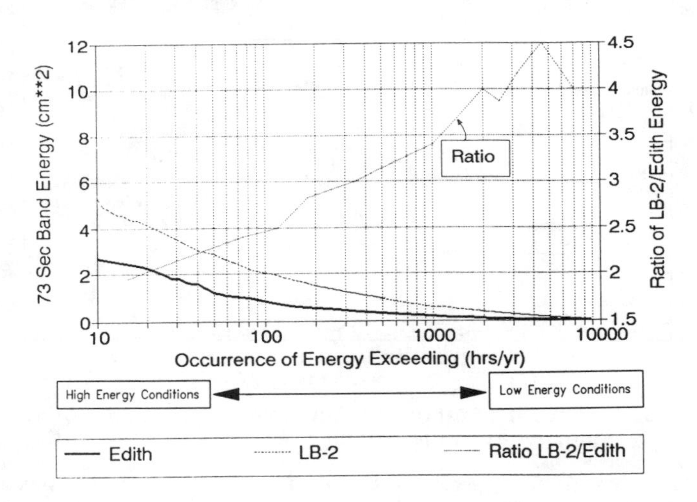

Figure 5. Variation of ratio of energy of gages LB2 and Edith from high to low energy conditions

The uniform wave spectrum, designed to typify a somewhat average wave condition with a southerly approach to the harbors, also indicated good model-prototype correlation for wave height amplifications with respect to median energy conditions.

Long Wave Energy Relationships to Short Period Waves

During the analysis of prototype data for modeling harbor resonance, correlation between the long and short period wave energy was noted, similar to others (e.g. Longuet Higgins and Stewart (1964), Bowers (1980)). For example, Figure 6 shows the low frequency energy increasing with the total wave energy at gage Edith. The events noted in Figure 6, driving the wave activity, are distant events. The waves associated with hurricanes off the coast of Mexico are typically moderate swell of 10-12 seconds once they reach the vicinity of Los Angeles. The southern hemisphere swell usually is 15-20 sec period.

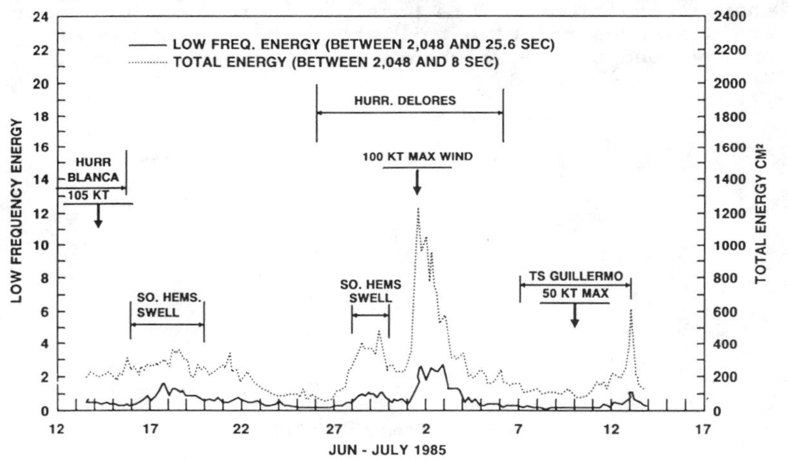

Figure 6. Correspondence of low frequency and total wave energy at gage Edith.

A more direct examination of the long-short wave correlation is shown in Figure 7, associated with a one month data set of 174 wave records. The log-linear relationship between long, or infragravity wave height and the associated wind wave height can be noted. When these data were separated into three classes, based on whether there was one primary wave (class 1), a unidirectional multiple wave (class 2), or a multiple directional wave (class 3), it was noted that long wave energy relative to short wave energy decreased for multiple direction wave spectra (discussed by Bowers (1992), as well as for multiple wave colinear spectra.

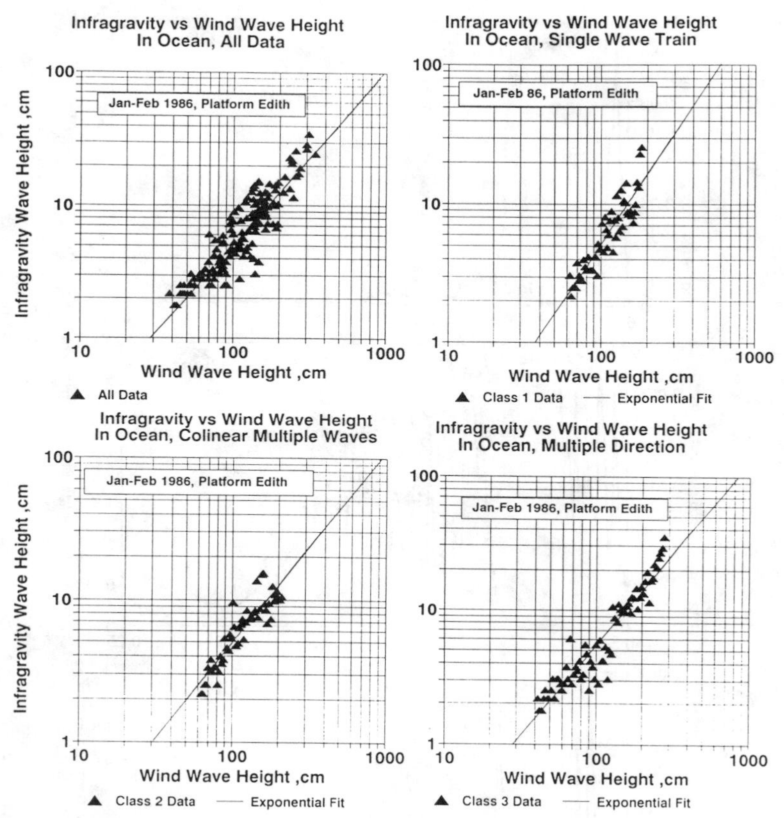

Figure 7. Infragravity wave height versus wind wave height relationships.

Also of interest was the close correlation of spectral shape of the short and long spectrum and therelationship of peak periods. Figure 8 shows three such spectra representing the three classes discussed earlier. This similarity in shape led to correlating peak periods of the long and short spectra to one another, indicating trends for shorter wind wave periods to correlate to shorter long waves and longer wind wave periods relating to longer long wave periods. Figure 9 shows this correlation for the same data set represented in Figure 7. The prototype data analysis provided data up to 512 second period band, with the next lowest band 256 sec. It was noted in correlations between the long wave energy bands at the ocean gage and the short wave

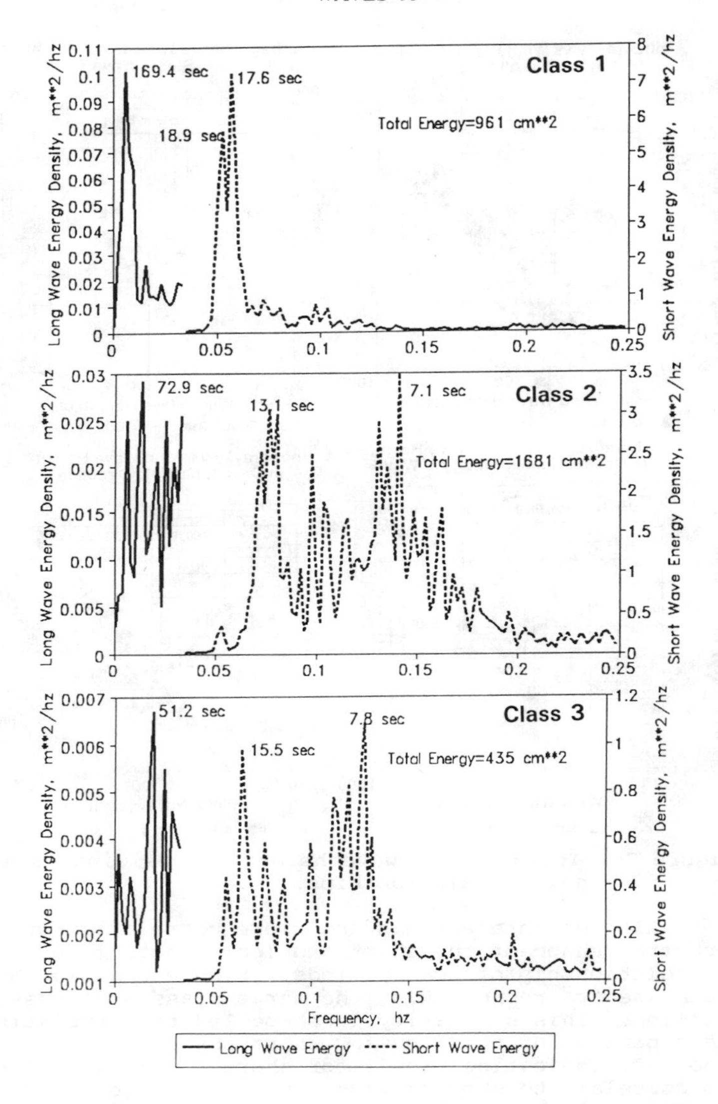

Figure 8. Long and short wave energy spectra comparison

energy there, that 256 sec band showed strong correlation with short period wave energy but the 512 sec band showed no correlation to short period wave energy. The plot of Figure 9 seems to indicate this also, except for one data point, which was for a relatively low energy level. The presence of energy at periods 500 seconds and greater is believed to be associated with shelf oscillations.

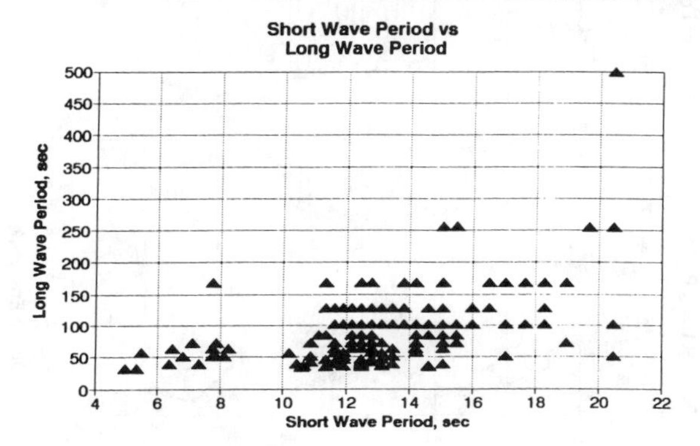

Figure 9. Short wave period versus long wave period

Identification of the response of various long period energy bands with respect to the period of the approaching short waves is illustrated in Figure 10, a plot of short period wave energy approaching the harbor with time as measured at platform Edith and the variation of percent of the total long wave energy for a given frequency band. The sources are noted on the figure. SHS indicates Southern Hemisphere swell, with periods greater than 15 seconds, and HS is hurricane swell with periods in the range of 9-12 seconds. It can be noted that longer period peak wind waves (SHS) are associated with increases of percent total energy of the longer long wave periods. In particular, the 85 and 128 sec bands respond well, while the 46 second band tends to decrease as a percentage of total long period energy. In the time period preceeding 200 hours, during the arrival of hurricane swell, the percent of energy in the 46 sec band rises, while 85 and 128 sec bands decrease as a percentage of long period energy. Finally, the 512 sec energy band tends to show little or no correlation to these swell events.

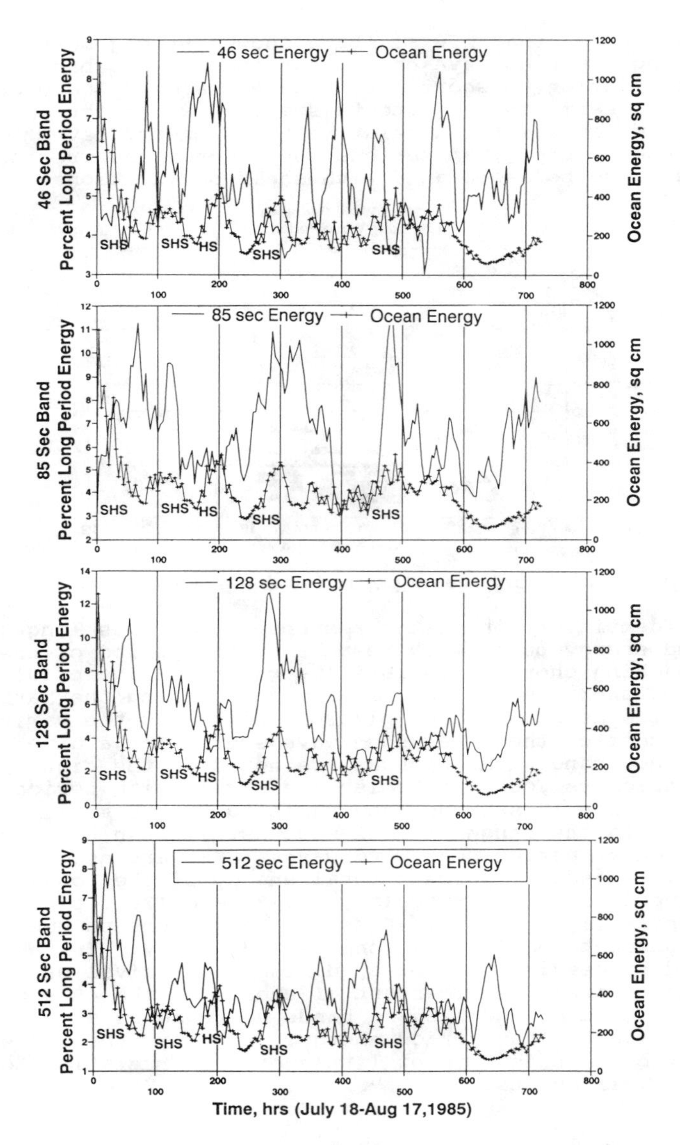

Figure 10. Variation of percent energy of long period wave energy bands with ocean swell.

Conclusions

Prototype long period wave data collected at an ocean gage were examined for use in programming the wave generators of a physical model of Los Angeles-Long Beach Harbors. Three long period wave spectra were selected for testing and were reproduced in the model. Comparison of prototype and model harbor resonance response indicated similar wave height amplifications. Further examination of the long period data indicated good correlation between the wind wave and long period wave height and spectral shape. Increases in energy in longer period swell waves correlated with increases in energy in the longer period portion of long waves and increases in energy in shorter period swell waves had associated increases in the shorter period range of the long wave spectrum.

Acknowledgment

The Office, Chief of Engineers, US Army Corps of Engineers, is gratefully acknowledged for authorizing publication of this paper. The data described and the resulting analysis presented herein, unless otherwise noted, were obtained from studies sponsored by the U.S. Army Engineer District, Los Angeles and the Ports of Los Angeles and Long Beach. Appreciation is also expressed to Ms. Leonette Thomas for compilation of data and Ms. Debbie Fulcher for preparation of manuscript.

References

Bowers, E.C. (1977). "Harbour Resonance Due to Set-Down Beneath Wave Groups." J. Fluid Mech. 79(1), pp 71-92.

Longuet Higgins, M.S. and Stewart, R.W. (1964). "Radiation Stresses in Water Waves: A Physical Discussion With Applications." Deep Sea Research, Vol. 11, p529.

Rosati, J. and Puckette, P.T. (1993). "Measurement and Analysis of Wave Data in Los Angeles and Long Beach Harbors." Proceedings, WAVES 93, The Second Symposium on Ocean Wave Measurement and Analysis.

Seabergh, W.C. 1985 "Los Angeles and Long Beach Harbors Model Study, Deep-Draft Dry Bulk Export Terminal, Alternative No. 6: Resonant Response and Tidal Circulation Studies", M.P. CERC-85-8, U.S. Army Engineer Waterways Experiment Station, Vicksburg, MS.

Instrumented Buoy Network Response to Ocean Swell

Theodore Mettlach[1] and David Gilhousen[2]

Abstract

The National Data Buoy Center has an extensive network of more than 20 moored buoys along the west coast of North America and in the Eastern North Pacific Ocean. The network ranges from the Gulf of Alaska to the Southern California Bight. All of the buoys in this network measure wave spectra and some buoys also measure directional wave spectra. The individual buoy responses to the long-period swell that were generated by intense midlatitude cyclones in the mid-Pacific during September and December 1991 are documented and described. These cyclones account for three of the four such storms since January 1988 that have generated swell that reached the west coast with a dominant wave period of 25 seconds or greater. It is shown that the characteristically coherent response by the buoy network to the arrival of the swell from these storms can be exploited using linear wave theory to forecast the arrival of such swell.

1. Introduction

Long-period swell can adversely affect a wide variety of human marine activities. The purpose of this paper is to demonstrate the utility of National Data Buoy Center (NDBC) moored buoy stations in monitoring, predicting, and studying the characteristics of ocean swell waves. The response of the NDBC west coast network of moored buoys to the long-period swell from three mid-Pacific storms has been studied. The first two storms were the extratropical remnants of tropical cyclones. The third storm was a very deep winter marine storm. Each storm reached maximum intensity near the dateline, between 40° N. and 45° N., and generated high wind and waves. The swell from each storm radiated outward from the generation area and reached the west coast two to three days later.

The general response by the west coast network was similar in each case. The propagation of the 20-second-period wave energy was accurately modelled using the

[1] Oceanographer, Applied Technology Division, Computer Sciences Corporation, John C. Stennis Space Center, MS 39529-6000

[2] Meteorologist, Data Systems Division, National Data Buoy Center, John C. Stennis Space Center, MS 39529-6000

assumption that swell radiates outward from its generation area, like a stone in a pond (Irvine 1985) under the constraints of linear wave theory.

2. The West Coast Network of NDBC Moored Buoys

In September 1991, there were 21 moored buoy stations along the west coast of North America and 1 in the Bering Sea. The various NDBC buoy stations on the west coast are listed in Table 1 by the World Meteorological Organization station location numbers with the position, hull type, and depth of the water at each location. Figure 1 shows the position of each station. Nineteen buoys provided complete or nearly complete sets of hourly data during the two September 1991 storms. Twelve buoys provided complete or nearly complete data sets of hourly data during the December 1991 event.

The wave measurements from each buoy were made with onboard instrumentation consisting of an accelerometer and data processor that produce estimates of nondirectional wave spectra. The details of the data collection and analysis procedures for NDBC wave measurement systems are best described by Steele and Earle (1979).

Stations 46025, 46042, and 46045 reported directional wave information during the September events. Station 46042 was moored offshore Monterey Bay, California, in a water depth of about 2,100 m and produced useable measurements during the December event. Stations 46025 and 46045 are located in the California Bight and were therefore sheltered from incoming swell energy.

The buoy heave, pitch angle and roll angle measurements required for directional wave estimates were measured with a Datawell Hippy 40 Mark II sensor. An NDBC Directional Wave Analyzer system computed the frequency-dependent wave direction and other directional wave parameters based on the method of Longuet-Higgins *et al.* (1963). The details of the measurement system and the computation of the directional wave spectra are given by Steele *et al.* (1985, 1990).

Figure 2 shows significant wave height and dominant wave period acquired at station 46042 for more than 4 years. It can be seen that there have been only four recorded time periods when the dominant wave period reached 25 s. This paper documents the three most recent cases.

3. Determination of Swell Generation Area

It was assumed that the long-period swell that affected the buoy network were generated a great distance away so that the source may be considered a point source in time and space. It is assumed that wave energy travels along great circle routes at the frequency-dependent group velocity given by

$$C_g(f) = \frac{g}{4\pi f} \quad , \tag{1}$$

where g is the acceleration due to gravity and f is the frequency in Hertz. It is a simple matter to model the propagation of wave energy, once the point of origin is found. Swell generated at time t_0 arrives at a location a distance d away at the time t_a given by

$$t_a = t_o + \frac{d}{C_g(f)}. \tag{2}$$

Differentiation of Equation (2) with respect to f (Earle *et al.* 1984) yields an approximate equation for the distance from the buoy platform and the wave generation area,

$$\frac{dt_a}{df} = \frac{4\pi d}{g} \approx \left(\frac{t_2 - t_1}{f_2 - f_1}\right), \tag{3}$$

where t_1 and t_2 are arrival times of energy at frequencies f_1 and f_2, respectively, at the measuring platform. Equation (3) can be used to find the distance to the wave generation area as well as the distance the swell energy of frequency f has propagated outward from the storm in time.

From the buoy-derived spectral estimates, the areas and times of wave generation were estimated using two different methods. The first method used was that of Munk *et al.* (1963), which employs time-frequency plots of spectral energy density to determine the time of swell generation. The location of the swell origin can be found from weather maps for the swell generation time. The second method used was that of Wang and Carolan (1991), which determines the location of swell origin from either equation (3) or the time-frequency-spectral energy density plots, and the wave direction inherent in the wave estimates from directional wave measuring buoys. Equations (4) and (5) describe the relationship between the distance and azimuth between two points *o* and *1*, where ϕ_o and ϕ_1 are latitudes of the points, λ_o and λ_1 are the longitudes of the points, $r_{o,1}$ is the angular distance between the two points, and $\delta_{1,0}$ is the azimuth, clockwise from north from *1* to *o*.

$$\phi_o = \sin^{-1}(\sin\phi_1 \cos r_{o,1} + \cos\phi_1 \sin r_{o,1} \cos\delta_{1,o}) \tag{4}$$

$$\lambda_o = \lambda_1 + \tan^{-1}\left(\frac{\sin r_{o,1} \sin\delta_{1,o}}{\cos\phi_1 \cos r_{o,1} - \sin\phi_1 \sin r_{o,1} \cos\delta_{1,o}}\right). \tag{5}$$

4. September 1991 Storms

During September 1991 there were two deep cyclones that generated high-height, long-period swell that travelled across the central and eastern Pacific. Both storms were the extratropical remnants of western Pacific tropical cyclones—Typhoon Ivy and Tropical Storm Luke. One death was directly attributed to the high swell from these storms when a fisherman was swept from the rocks and drowned north of Santa Cruz, California, on September 15. Both storms provided very clear spectral signals consistent with linear wave theory, unmodified by local wind waves.

4.1 Typhoon Ivy

The response of the NDBC network to the swell from the remnants of Ivy is clearly shown in Figures 3 through 6. In Figure 3, there was a marked increase in dominant wave period to 20-25 s at each station during Julian days 257 and 258. Dominant wave periods increased to 20 s at stations 46025 (Catalina Ridge) and 46045 (Long Beach) even though these stations were ostensibly sheltered from the incoming wave energy by Point Conception. This sheltering effect is obvious from the relatively low significant wave heights at these stations in Figure 4. Significant wave height reached 3 m at several of the stations. The highest waves occurred at the more northern and western stations. Figure 5 is a time series plot of the 6-h running mean of the 0.05-Hz spectral energy density. Each value corresponds to the mean of the spectral energy density for the time shown and the previous 5 h. It shows the coherent response of the buoy network to 0.05-Hz wave energy.

Figures 6(a) and 6(b) show the hourly 0.05-Hz and 0.06-Hz spectral energy densities and mean wave directions from station 46042 (Monterey Bay). During the time of maximum energy on Figure 6(a), the corresponding wave directions match quite closely. The direction from which the swell energy arrived on September 15 was approximately 300°.

Figure 7 is the time-frequency plot of spectral energy density from station 46042. The intersection of the line through the energy peaks with the time axis yields the time of generation of the swell — approximately 0000 UTC, September 11. National Weather Service surface pressure analyses for that time reveal that the remnants of Ivy was centered near 41° N., 165° E., with a central pressure of 980 hPa.

A second estimate of the swell generation point, 42° N., 164° E., was found by applying the spectral wave information acquired from station 46042 to equations (4) and (5). The distance to the wave generation area used was 6,104 km, found using equation (2) and the ridge line time in Figure 7.

4.2 Tropical Storm Luke

The response of the NDBC network, as revealed in Figures 3 through 7, for Julian days 264 through 270, was very similar to the Typhoon Ivy event. The ridge line shows that the time of swell generation was approximately 0000 UTC, September 21, 1991. National Weather Service surface pressure analyses for that time reveal that the remnants of Luke were located near 44° N., 170° E., with a central pressure of 978 hPa. A second estimate of the swell generation point, also 44° N., 170° E., was found by applying the spectral wave information acquired from station 46042 to Equations (4) and (5). The distance to the wave generation area used was 5,610 km, found using Equation (2) and the ridge line time in Figure 7. The direction from which the swell energy arrived was, again, approximately 300°.

5. December 1991 Storm

The highest 25-s waves recorded during the nearly five continuous years that station 46042 has been in operation occurred during the last week of December 1991. However, swell waves were only a partial contribution to the total wave energy present at that time. This is an interesting event because it shows that the NDBC network is capable of discerning long-period swell events within a chaotic, vigorous, wave environment. There were a series of deep weather systems in the eastern Pacific from December 25 through 31 that generated high waves all along the west coast.

Figure 8 shows the time series of dominant wave period from December 26, 1991, to January 5, 1992. The arrival of long-period swell is clearly indicated at each station. The corresponding significant wave heights at the stations, as shown on Figure 9, reveal notable increases at some but not all stations. The highest wave heights at the coastal stations were associated with local storms rather than swell waves. Nevertheless, the time series plots of 0.05-Hz spectral energy density, as shown in Figure 10, reveal a coherent pattern similar to the patterns in Figure 5.

The source of the swell was an intense cyclone centered near 46° N., 179° E., at 1200 UTC, December 28, 1991. The swell generation area was southeast of the cyclone center. Figures 11(a) and 11(b) provide sufficient information to determine the source of the swell. In Figure 11(a) the time difference between the arrival of 0.05-Hz and 0.06-Hz wave energy was approximately 13 hours. Applying Equation (3) to this suggests the swell generation point was 3,653 km from station 46042. From Figure 12(b) the azimuth to the generation point was approximately 290°. Applying this information to Equations (4) and (5) yields a generation point near 41° N., 165° W. Applying this information to Equation (2) yields a swell generation time of 1200 UTC December 28, 1991. The calculated generation point is southeast of the actual storm position. The time of swell generation derived from Equation (3) was consistent with the time found from the ridge line, as shown in Figure 12.

6. Forecasting Swell Arrival

The viability of using the known position of the generation point for long-period swell to forecast the propagation of such swell was tested using the information derived from the December 1991 case in Section 5. The known time and position of the swell origin was used with Equations (1), (2), (4), and (5) to obtain isochrones of the leading edge of 0.05-Hz wave energy, as shown in Figure 13. The basic model was that of swell waves radiating outward from a point like ripples in a pond.

The station-by-station verification of the forecast is shown in Figure 14. This shows that the model is essentially valid, but that large errors, associated with the stations closest to the derived swell generation area, where the assumption of *point* source breaks down, did occur.

7. Summary

This paper has documented and described the response of the NDBC moored buoy network to somewhat rare long-period swell from distant storms. In each case dominant wave periods reached 25 s. The source of the swell waves that affected the buoys was found using principles from classical wave theory. It was shown that it is possible to predict the arrival of long period swell waves if the source of the swell is known.

The response of the NDBC network of buoys has clearly shown the nature of swell propagation. There are several areas of further investigation that may prove fruitful. First, the great amount of energy in swell waves makes them especially dangerous when reaching coastal areas. This paper has shown that buoys located at strategic locations well offshore could be used to provide several hours warning of the impending impact of high swell at coastal locations. Swell propagation models based on classical wave theory for deep water waves are quite simple, as has been shown, and would not be difficult to implement. The nearly continuous observations of wave spectra by a network as vast as the NDBC network may be useful as initial conditions for spectral ocean wave models such as the Global Spectral Ocean Wave Model of the Fleet Numerical Oceanography Center or the Third Generation Wave Model of the European Center for Medium Range Weather forecasting. Currently, these models do not use buoy wave estimates. Future model development activities should consider the tremendous data resource available in the NDBC network.

8. References

Earle, M.D., Bush, K.A., and Hamilton, G. (1984). "High-height long-period ocean waves generated by a severe storm in the northeast Pacific Ocean during February 1983." *J. Phys. Oceano.*, 14, 1286-1299.

Hamilton, G.D. (1990). "Guide to moored buoys and other ocean data acquisition systems." WMO Report on Marine Science Affairs, No. 16, WMO—No. 750.

Irvine, D.E. (1985). "Waves across the ocean." *Johns Hopkins APL Technical Digest*, 6(4), 313-319.

Longuet-Higgins, M.S., Cartwright, D.E., and Smith, N.D. (1963). "Observations of the directional spectrum of sea waves using the motion of a floating buoy," in *Ocean Wave Spectrum*. Prentice-Hall, Englewood Cliffs, NJ, 111-136.

Munk, W.A., Miller, G.R., Snodgrass F.E., and Barber, N.F. (1963). "Directional recording of swell from distant storms." *Phil. Trans. R. Soc. London*, 1062, Vol. 255, 505-584.

Steele, K.E., and Earle, M.D. (1979). "The status of data produced by NDBO Wave Data Analyzer (WDA) systems." *Proc. Oceans '79*, San Diego, CA, Marine Technology Society and IEEE, 212-220.

Steele, K.E., Lau J., and Hsu, L. (1985). "Theory and application of calibration techniques for an NDBC directional wave measurements buoy." *IEEE J. of Oceanic Engineering*, OE-10(4), 382-396.

Steele, K.E., Wang, D., Teng, C.C., and Lang, N. (1990). "Directional wave measurements with NDBC 3-meter discus buoys." Report No. 1804-01.05, National Data Buoy Center, Stennis Space Center, MS, 35 pp.

Wang, D.W., and Carolan, R. (1991). "Estimation of swell direction by a small discus buoy in high seas." *Fifth Conference on Meteorology and Oceanography of the Coastal Zone* (preprint), May 1991, Miami, FL, 53-59.

Table 1. Station Information

Station	Latitude (° N.)	Longitude (° W.)	Hull Type	Water Depth (m)
46003	51.08	155.92	6N	4709
46001	56.47	148.18	6N	4206
46006	40.81	137.65	12D	3932
46005	41.10	130.10	6N	2853
46002	42.53	130.25	6N	3420
46041	47.42	124.53	3D	44
46040	44.78	124.30	3D	112
46027	41.84	124.40	ELB	57
46022	40.75	124.50	3D	250
46030	40.44	124.50	ELB	57
46014	39.22	123.98	10D	371
46013	38.23	123.30	10D	125
46026	37.75	122.68	LNB	33
46012	37.39	122.72	3D	92
46042	36.80	122.40	3D	2103
46028	35.76	121.87	3D	1138
46011	34.88	120.87	3D	201
46023	34.25	120.65	3D	600
46045	33.84	118.45	3D	77
46025	33.75	119.07	3D	839

Note: There are seven different hulls used by NDBC: 12D (12-m discus), 10D (10-m discus), 6N (6-m Naval Oceanographic Meteorological Automated Device-NOMAD), 3D (3-m discus), LNB (U.S. Coast Guard 12-m discus Large Navigational Buoy), ELB (U.S. Coast Guard 2.7-m Exposed Location Buoy), and 2D (2.3-m foam-discus-hulled Coastal Buoy). Descriptions of the NDBC buoy hulls are given by Hamilton (1990).

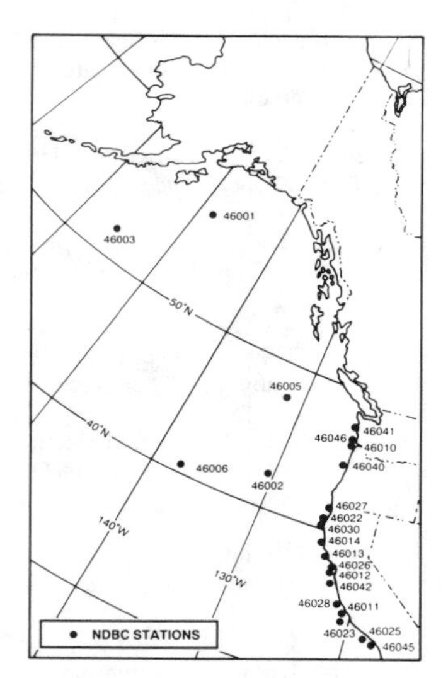

Figure 1. The National Data Buoy Center network of moored buoys.

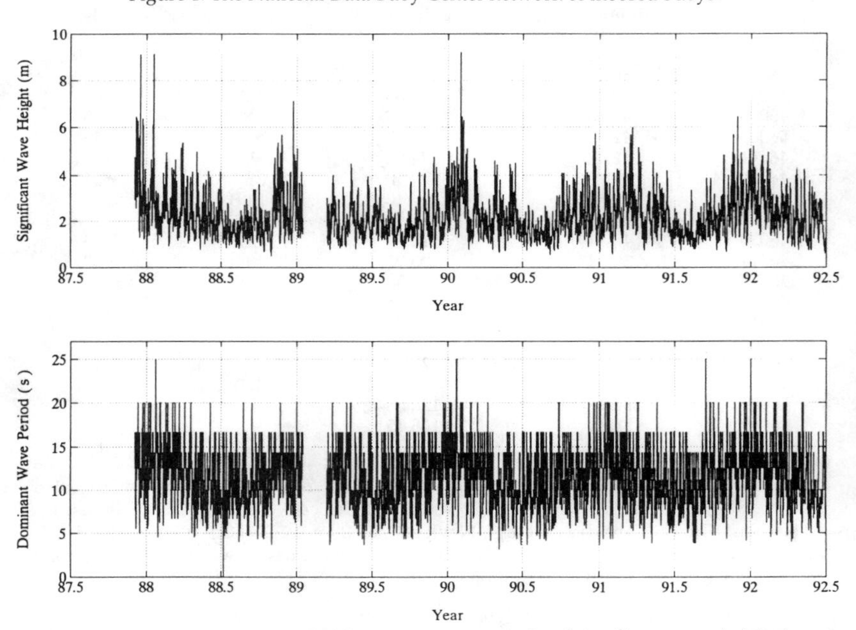

Figure 2. More than a 4-year record of dominant wave period and significant wave height from December 1987 to June 1992 at station 46042.

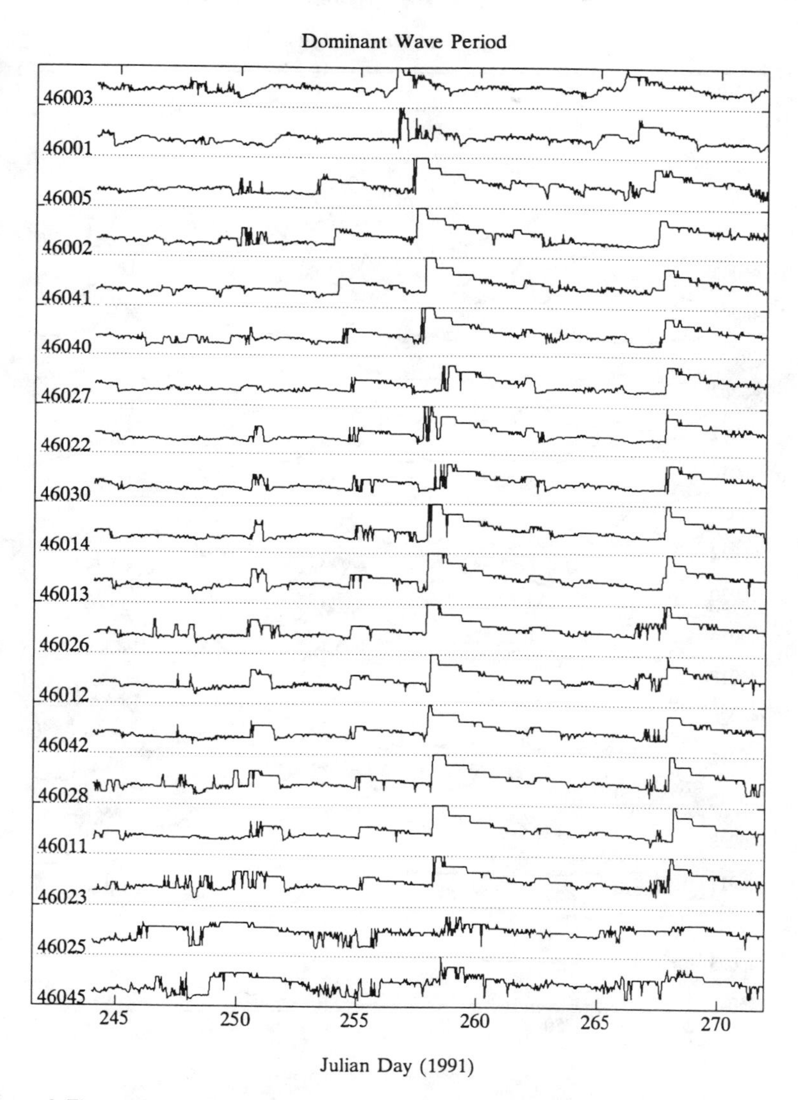

Figure 3. Time series plots of dominant wave period from 19 NDBC moored buoy stations in September 1991. Each individual ordinate covers 0 to 25 s.

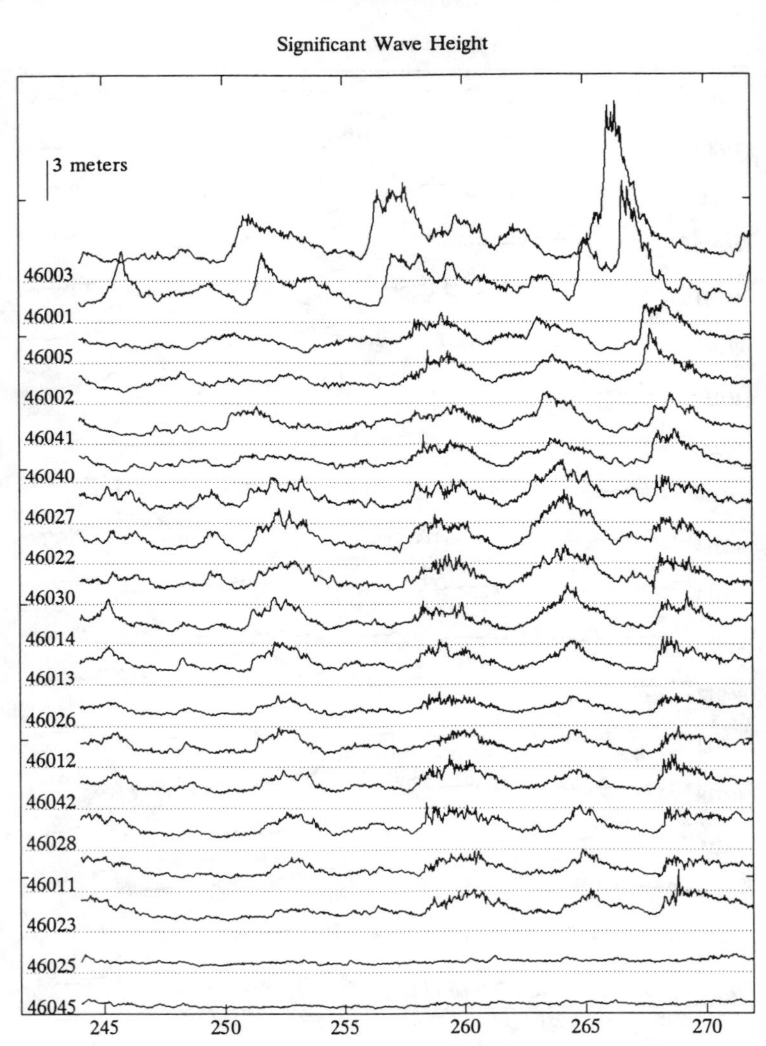

Figure 4. Time series plots of significant wave period from 19 NDBC moored buoy stations in September 1991.

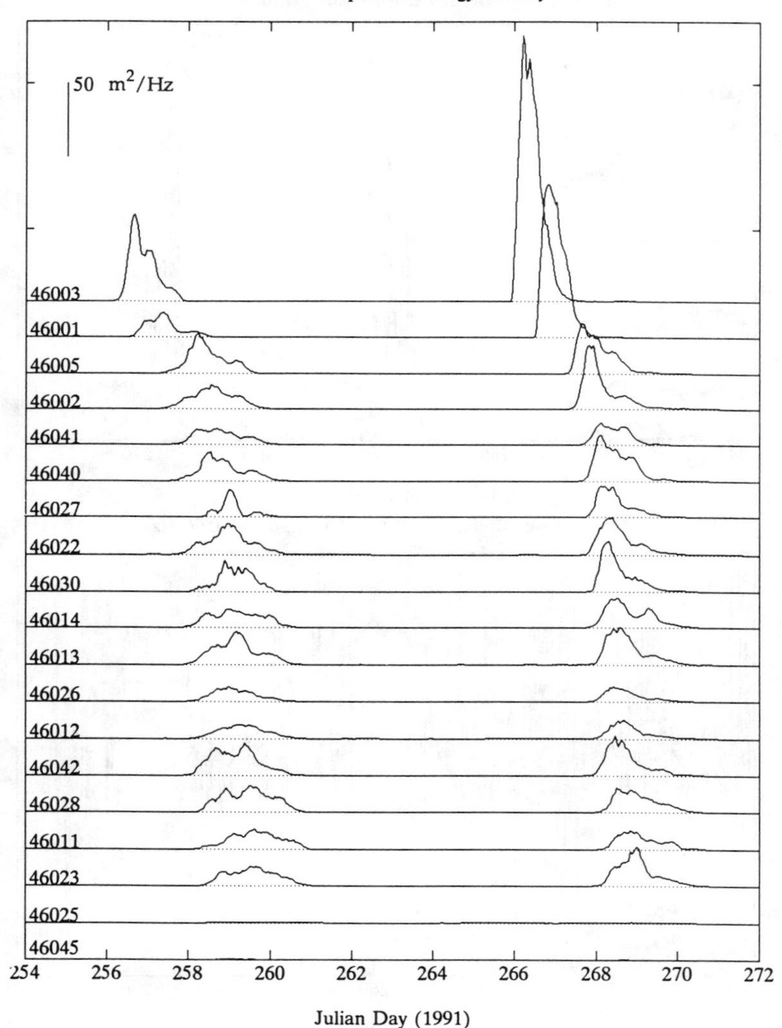

Figure 5. Time series plots of 0.05-Hz spectral energy density from 19 NDBC moored buoy stations in September 1991.

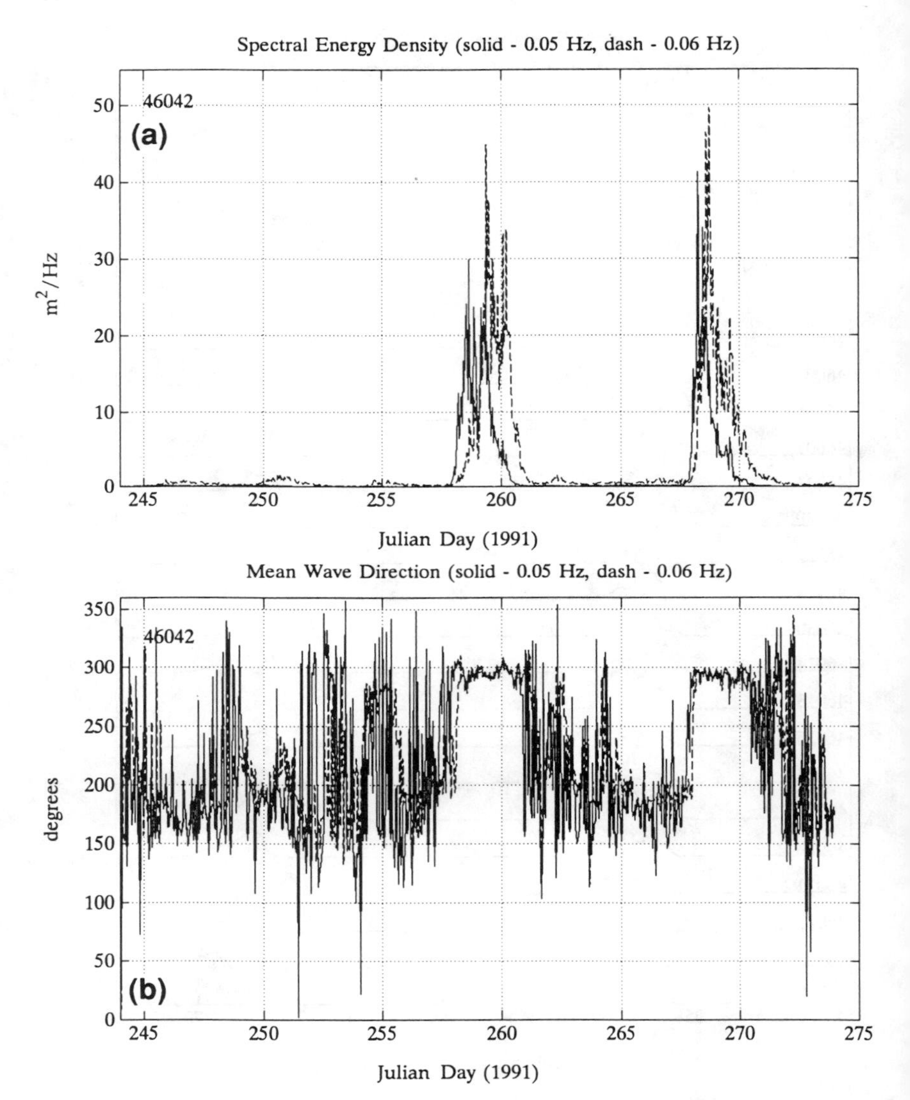

Figure 6. Time series plots of (a) 0.05- and 0.06-Hz spectral energy density, and (b) mean wave direction, at station 46042 (Monterey Bay) in September 1991.

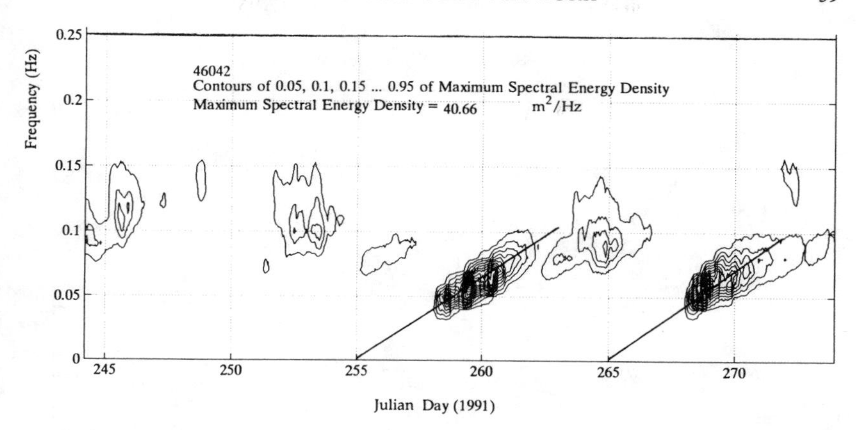

Figure 7. Time-frequency plot of spectral energy density at station 46042 (Monterey Bay) in September 1991.

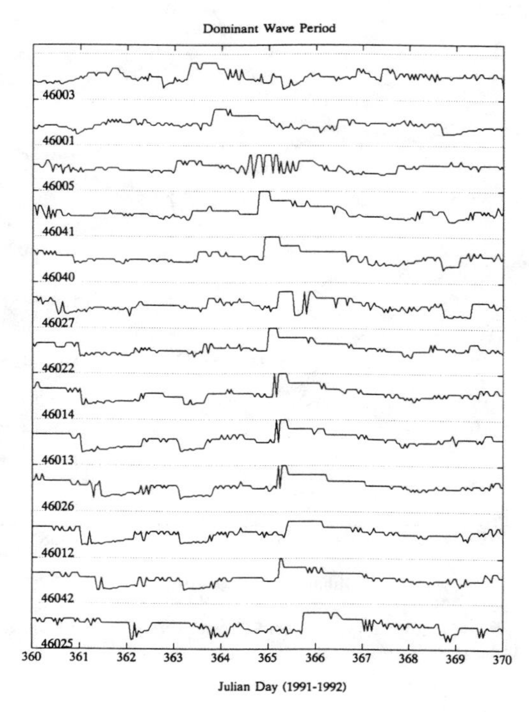

Figure 8. Time series plots of dominant wave period from 13 NDBC moored buoy stations from December 26, 1991 to January 5, 1992. Each individual ordinate covers 0 to 25 s.

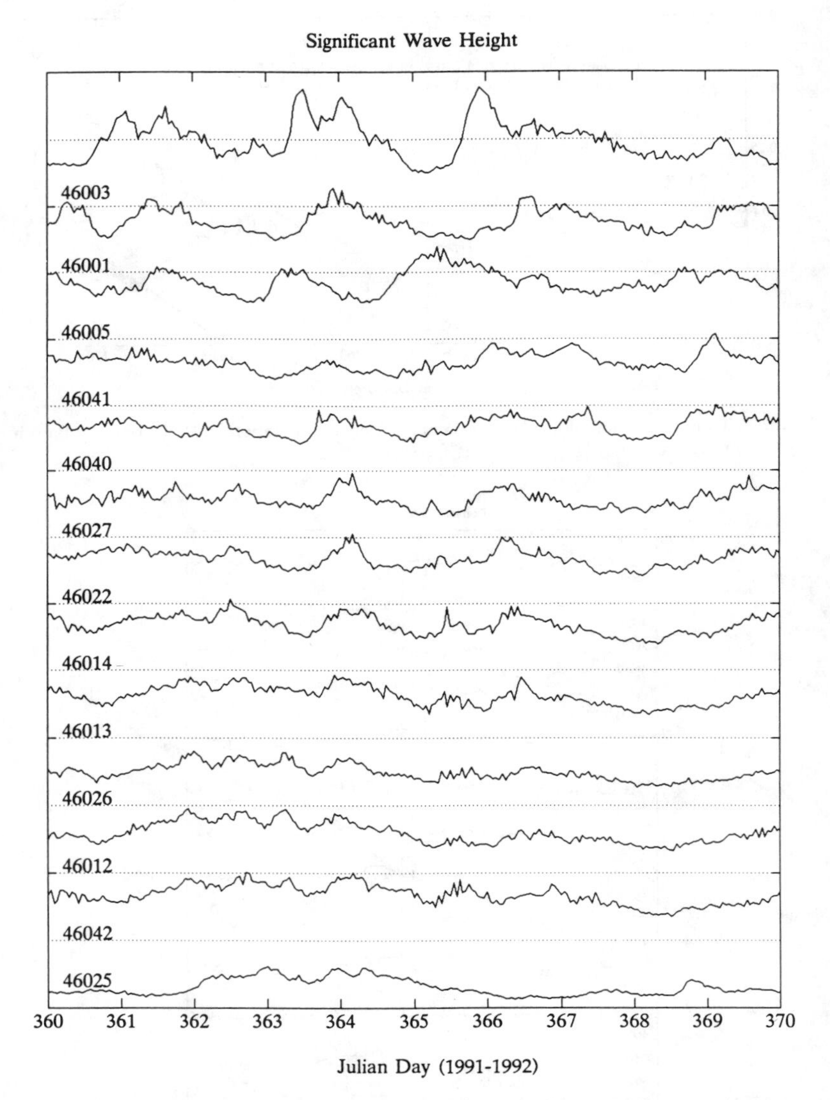

Figure 9. Time series plots of significant wave period from 13 NDBC moored buoy stations from December 26, 1991, to January 5, 1992.

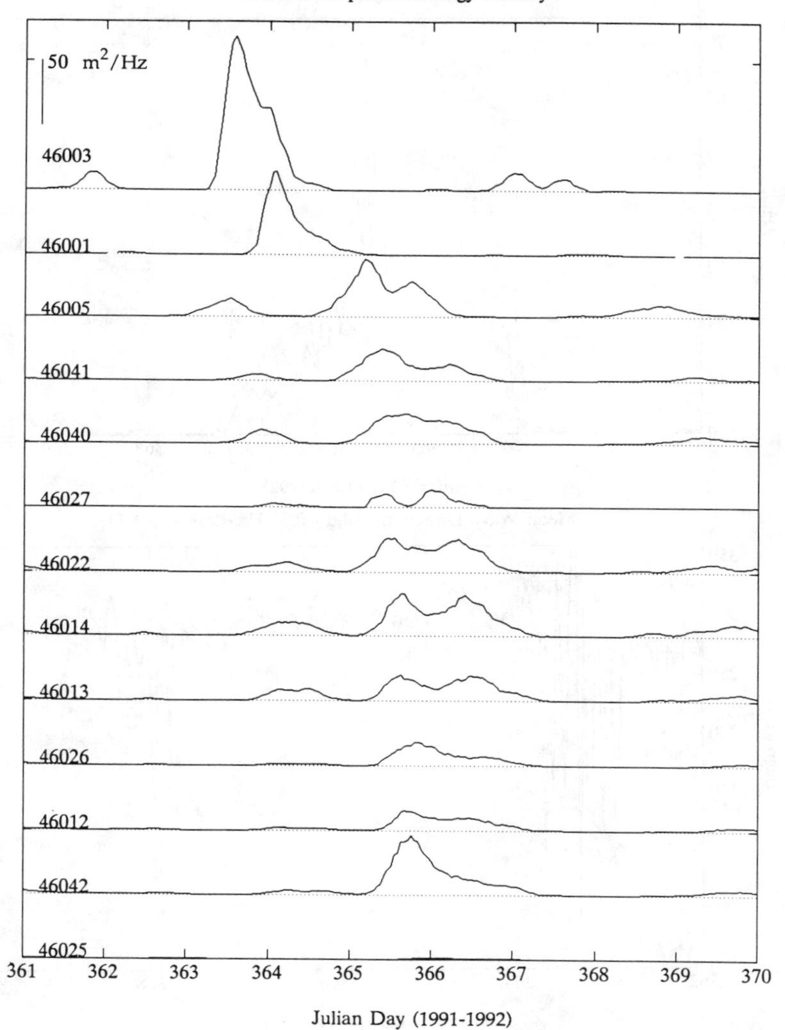

Figure 10. Time series plots of 0.05-Hz spectral energy density from 13 NDBC moored buoy stations from December 27, 1991, to January 5, 1992.

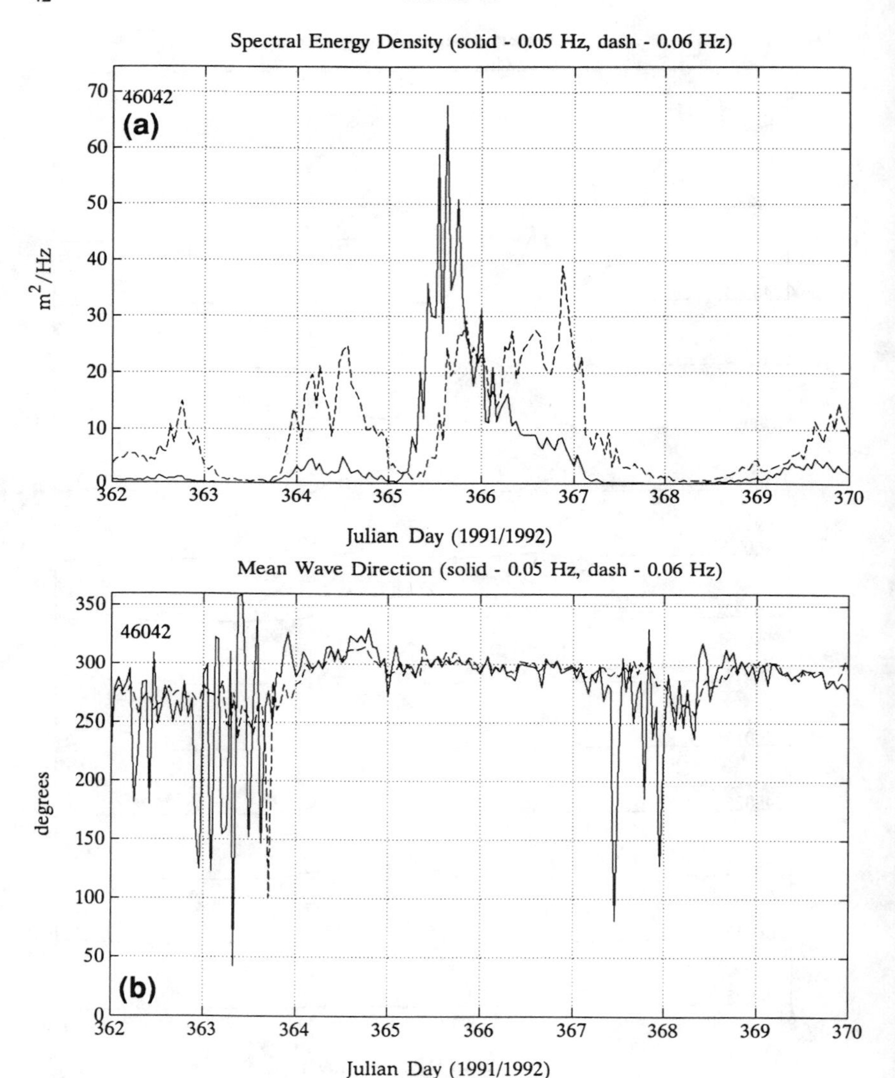

Figure 11. Time series plots of (a) 0.05- and 0.06-Hz spectral energy density, and (b) mean wave direction, at station 46042 (Monterey Bay) from December 28, 1991, to January 5, 1992.

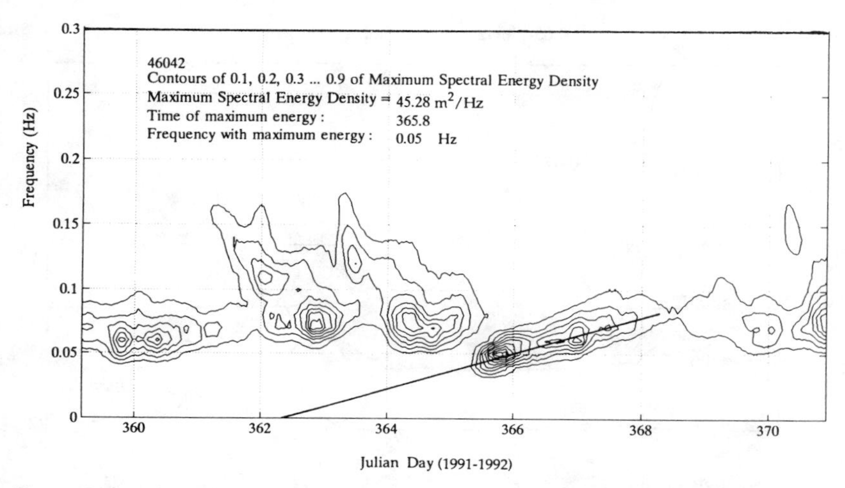

Figure 12. Time-frequency plot of spectral energy density at station 46042 (Monterey Bay) from December 25, 1991, to January 6, 1992.

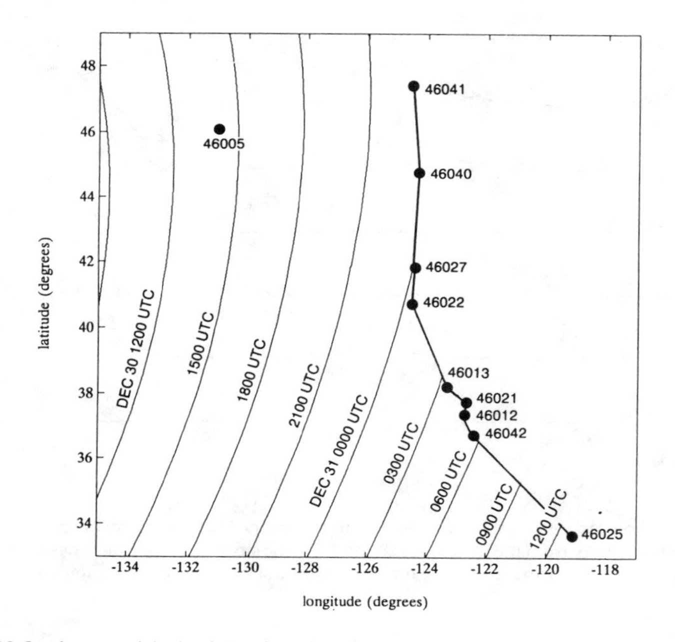

Figure 13. Isochrones of the leading edge of 0.05-Hz spectral energy density.

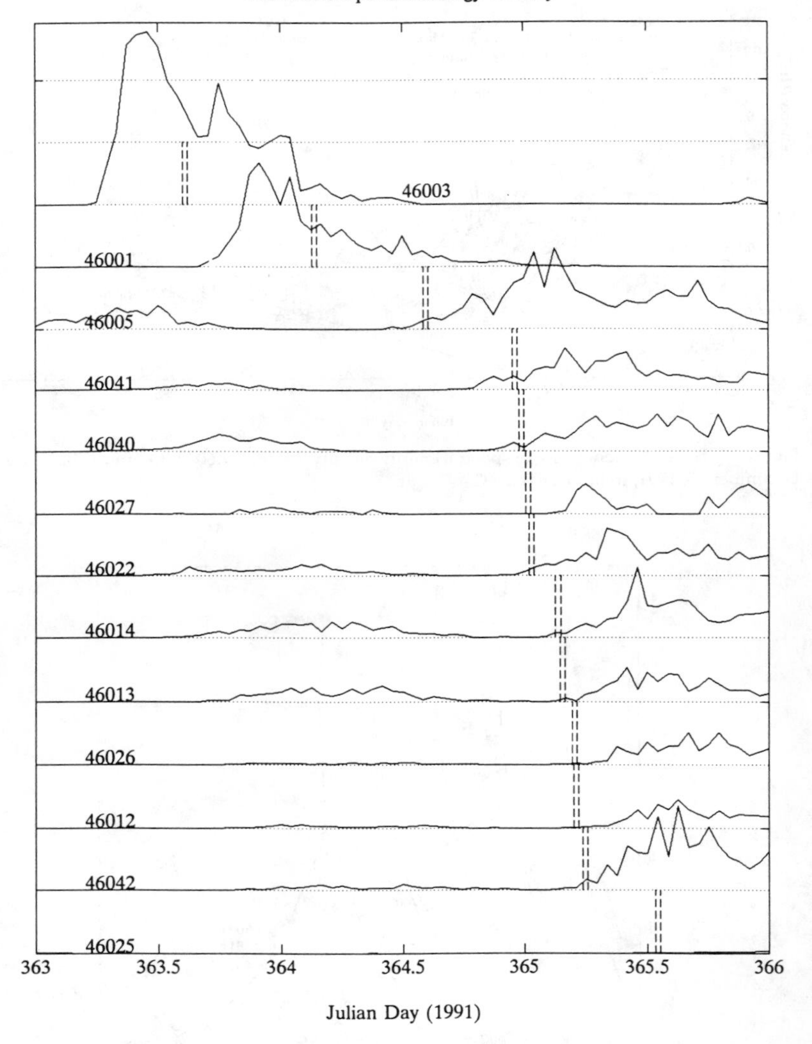

Figure 14. Forecast verification. Dashed bars represent the forecast arrival time of the first 0.05-Hz spectral energy from the December 28, 1991, Pacific storm. (Individual time series plots are the same as Figure 10, except for the time period from December 29, 1991, to January 1, 1992).

THE DAYTONA BEACH "LARGE WAVE" EVENT OF 3 JULY 1992

by
Robert J. Thieke[1], Robert G. Dean[2] and Andrew W. Garcia[3]

ABSTRACT

The fundamental mechanisms behind the anomalous large wave which struck Daytona Beach on 3 July 1992 are addressed. Although the wave was originally believed to have resulted from a submarine landslide, both physical data and anecdotal accounts are presented which link the wave to a southward propagating squall system. The wave appears to be an example of a general type of forced long wave motion which may be called a "squall-line surge", in which a resonant coupling occurs between the moving pressure front and the free surface wave, resulting in potentially large wave amplitudes. Historical precedent is found for the generation of such waves; two notable examples are compared with the wave which impacted Daytona Beach.

INTRODUCTION

At approximately 11:00 p.m. (EDT) on July 3, 1992, a highly unusual wave event occurred in the vicinity of Daytona Beach, Florida. The wave was initially believed be due to a submarine landslide and the height was reported to be as large as 18 feet. The magnitude of the wave was anecdotally based, since there were no direct measurements of the wave height at the site. However the wave clearly caused movement of automobiles parked on the beach (with many sustaining damage), and left abnormally high debris lines on the beaches. The longshore extent of the shoreline impacted by the wave was approximately 30 miles (48 km). Observations indicate that weather and sea conditions at Daytona Beach both before and after this single large wave event were relatively calm.

---

[1]Assistant Professor, Department of Coastal and Oceanographic Engineering, University of Florida, Gainesville, FL 32611, (904)-392-1436.
[2]Chairman and Graduate Research Professor, Department of Coastal and Oceanographic Engineering, University of Florida, Gainesville, FL 32611.
[3]Research Oceanographer, Coastal Engineering Research Center, U.S. Army Corps of Engineers Waterways Experiment Station, Vicksburg, MS 39180

Subsequent meteorological data and reports shifted emphasis toward a relatively rapidly moving squall line system that moved southward along the Georgia and Florida coast on the evening of July 3. The speed and location of the squall line system are consistent with the observed time of impact of the wave at Daytona Beach. Eyewitness accounts of the progress of the wave impact along the shoreline also suggest that the wave propagated from the north to the south.

EYEWITNESS ACCOUNTS

Owing to the lack of tide or wave gauging stations in the immediate proximity of Daytona Beach, a particularly notable feature of the wave event was the complete lack of any direct wave measurements. This shortcoming demanded a certain degree of reliance on anecdotal evidence provided by eyewitnesses to the wave event and related occurrences.

Wave Characteristics

A general summary of the characteristics of the wave as it impacted the Daytona Beach shoreline is presented below. This information was derived entirely from interviews with five members of the Volusia County Beach Rangers who were on patrol on the beach on the evening of 3 July. These rangers were considered to be the most reliable source of eyewitness information since 1) They were professionals with a great many years experience working on beaches, 2) Their respective locations along the shoreline were well established and 3) The timing of events as witnessed could be verified through their radio transmission logs. Their observations yielded the following information about the wave:

a) The wave progressed from north to south along Daytona Beach. The ranger stationed at the northernmost location observed the wave first (2256 EDT). The two rangers stationed immediately to the south responded to the call and were each struck in turn by the wave while driving northward along the beach in the traffic lane (slightly above MHW). In each case their 4-wheel drive vehicles were nearly overturned, swept shoreward toward the seawall during the wave runup, and finally seaward again during the backwash. Two rangers stationed at New Smyrna Beach, just south of Ponce De Leon Inlet and approximately 15 miles (24 km) south of the Daytona Beach Pier, observed the wave approximately 30 minutes later (The pier may be considered the geographical center of the Daytona Beach shoreline).

b) The wave was reported as either a single large crest or as a large primary crest followed by two smaller secondary crests. The primary crest was generally observed to be breaking.

c) The wave height of the primary crest was estimated at 8-10 ft (2.4-3.0 m) in the vicinity of the traffic lane. This estimate was based largely on the fact that the wave was observed to be significantly higher than their vehicles.

d) The duration of the event (from the initial observation to the completion of the backwash) was approximately 40-60 seconds.

Meteorological Observations

Following the wave impact the National Weather Service reported no large scale storm activity within reasonable proximity of Daytona Beach to which the wave event might be attributed. However eyewitness reports were received from various locations along the Florida shoreline (Figure 1) regarding unusual smaller scale meteorological activity which the observers believed may have some connection with the Daytona Beach wave. These accounts include the observation of large scale thunderstorm systems with several waterspouts offshore of Jacksonville, a waterspout observed on the beach just north of St. Augustine, a sudden increase in wind speed from near calm to approximately 40 mph (64 km/hr) measured by a shipboard anemometer in St. Augustine marina, and sudden changes in wind speed and temperature along Crescent Beach. In total, these observations were instrumental in drawing attention to the relatively small scale squall system which is believed to have caused the wave, particularly since there were no reports of unusual weather at Daytona Beach before or after the wave impact.

PHYSICAL DATA

Wave Runup Elevations

The wave impact left well defined and unusually high debris lines along most of Daytona Beach. The locations of maximum wave runup persisted for several days after the event and were recorded in field surveys. The elevation of wave runup is presented as a function of distance along the beach (centered at the Main Street Pier) in Figure 2. The approximate elevation of the mean water level at the time of impact is shown in the figure as 4.3 ft (1.3 m), which corresponds to the high tide elevation (the wave struck at nearly high tide). Hence the maximum wave runup occurred just north of the Main Street Pier and extended approximately 6.1 ft (1.9 m) above the local water level. Note that although the wave impacted a substantial length of coastline, the most significant runup was confined to a relatively narrow region approximately 5 miles (8 km) in width and centered roughly on the pier. In the regions north and south of the surveyed points, the maximum runup elevation could not be determined since it was sufficiently low to fall within regions of the beach which were well traveled; no evidence of the runup line was located in these areas.

Figure 1: Region of interest: North Florida and Georgia coast.

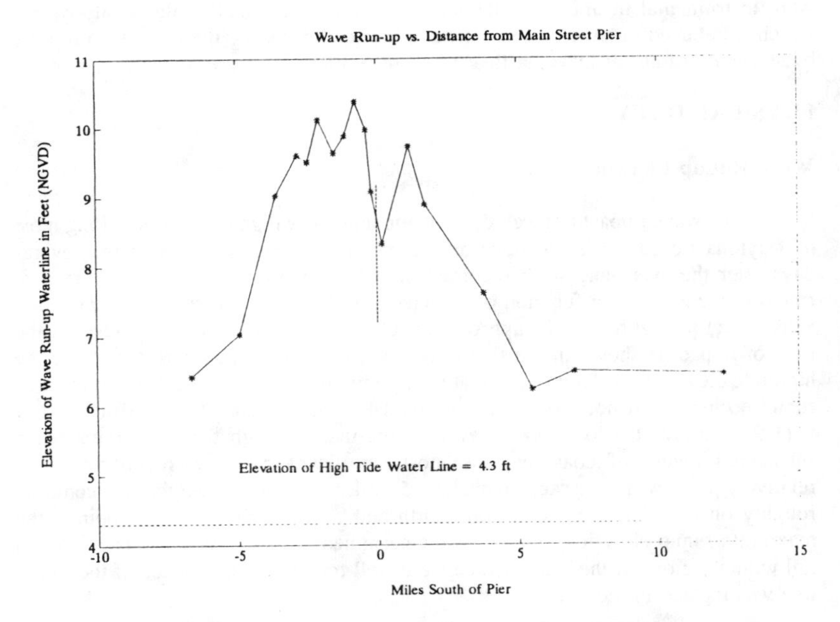

Figure 2: Wave runup elevation as a function of longshore distance, with origin at the Main Street Pier, Daytona Beach.

Tide Gauge Data

In an effort to determine if any trace of the wave had been recorded at other locations along the Atlantic Coast, tide data from four National Oceanic and Atmospheric Administration (NOAA) tide stations were analyzed: Fort Pulaski (Savannah), Georgia; Fernandina Beach (just south of the Florida-Georgia border), St. Augustine, Florida, and Haulover Pier (Miami), Florida. Only the tidal record at St. Augustine indicated any anomalous behavior (it should be noted that the Fernandina Beach Station is on the back side of the barrier island, hence it not surprising that there was no significant response at that location despite its proximity to the St. Augustine Station, which is on the open coast).

The tidal record measured at St. Augustine for the period including 3 July 1992 is shown in Figure 3a. A small anomalous wave is noted just prior to the occurrence of high tide on the evening of 3 July. The peak of this wave occurred at approximately 2100 or 9:00 p.m. EDT, two hours prior to the wave impact at Daytona Beach. Figure 3b focusses solely on this wave with the astronomical tide removed. The magnitude of this wave is seen to be only roughly 1.4 ft (0.4 m), with a period of approximately 2.2 hours. The characteristics of this disturbance are quite different from those of the wave observed at Daytona Beach; this will be discussed in a subsequent section.

Meteorological Radar Imagery

The eyewitness accounts of unusual weather patterns and the St.Augustine tidal record indicating a disturbance to the north prior to the wave impact at Daytona Beach focussed attention on a relatively rapidly moving squall line which formed over inland Georgia and South Carolina and progressed from north to south along the Georgia and Florida coastline on the evening of 3 July. The approximate positions of the squall line were derived from sequential radar images obtained from the National Weather Service in Daytona Beach. These conventional radar returns only indicate the intensity of precipitation, however this does delineate the progress of the squall front fairly accurately, if not the exact positions. Figure 4 shows a summary of the radar imagery data displayed as a series of approximate squall line positions and respective times. The concentric circles centered at Daytona Beach indicate 25 mile (40 km) increments of radial distance. The southward propagation speed of the squall line as estimated from the radar plots was approximately 30 mph (48 km/hr). The final position of the squall line is indicated at 10:35 p.m., which is 25 minutes prior to the wave impact. Interestingly, the successive radar images showed that the squall had largely dissipated by 11:30 p.m. EDT. However the position, speed and timing of the squall are all consistent with the meteorological observations and the timing and direction of the wave impact at Daytona Beach.

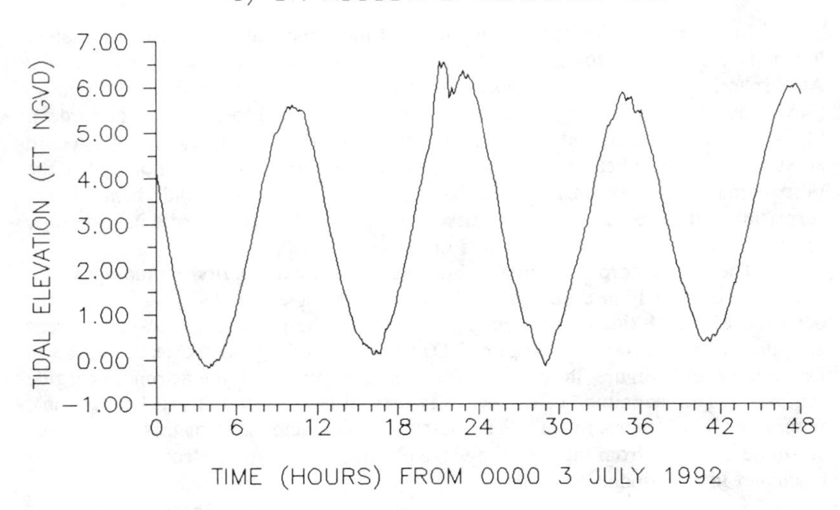

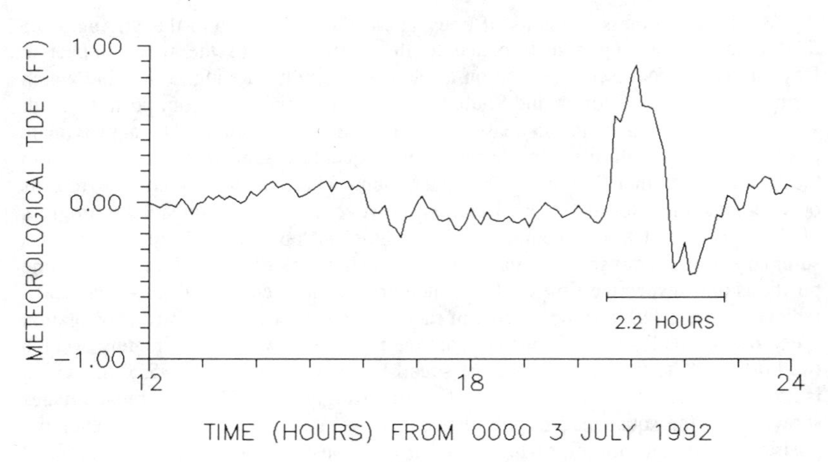

Figure 3: Water surface elevation tidal curves measured at NOAA St. Augustine tide station:
a) Total measured tide
b) Meteorological tide (astronomical tide removed)

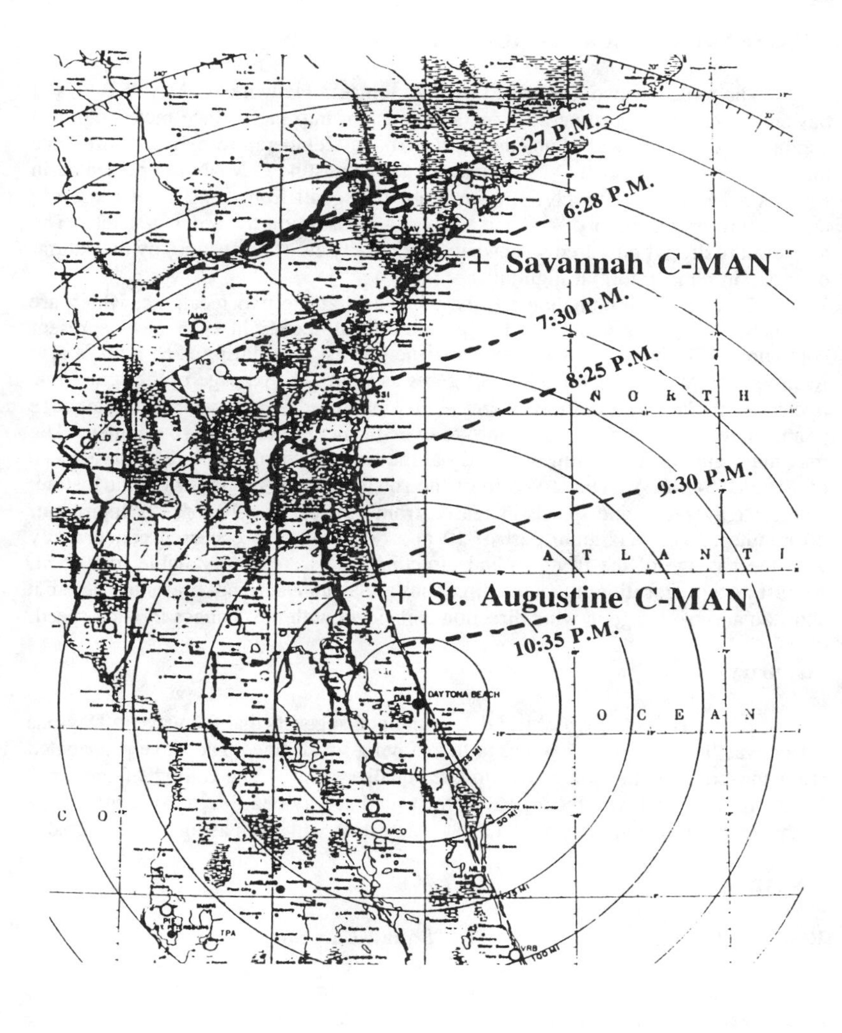

Figure 4: Approximate positions of squall line on evening of 3 July 1992 based on National Weather Service radar plots from Daytona Beach (all times EDT)

Offshore Meteorological Data (C-MAN Buoys)

The siting of two Coastal Marine Automated Network (C-MAN) Buoys at Savannah and St. Augustine afforded some insight into the surface meteorological conditions which existed along the squall front. Temperature and pressure time histories for the Savannah C-MAN and the St. Augustine C-MAN are presented in Figures 5 and 6, respectively. In each case the squall front consists of a drop in temperature on the order of 4°C and a jump in pressure on the order of 1 mb. The temperature drop is obscured somewhat (particularly at St. Augustine) by the natural decrease in temperature at nightfall.

Even more illuminating are the wind data at the two locations, which are shown in Figures 7a and 7b. The primary wind speeds in each plot represent continuous 10-minute averages, however the highest gust during each hour is also recorded and these are plotted in the figures as well. The passage of the squall line is clearly marked by an abrupt increase in wind speed. At Savannah the average winds increased from 6 m/s (13 mph) to over 14 m/s (31 mph) in 10 minutes. The maximum gust occurred almost exactly at the time of squall passage, with a speed of 24 m/s (54 mph). The structure of the squall front is similar at St. Augustine, where the average wind speeds increased from near calm to 13 m/s (29 mph) over 20 minutes. The maximum gust of 19 m/s (42 mph) also occurred concurrently with the passage of the front. Wind direction data were not available for the St. Augustine buoy location; the wind directions at Savannah are not presented here but show an abrupt change in wind direction coincident with the changes in wind speed.

Summary

Analysis of all the physical data combined suggests that the wave at Daytona Beach was linked in some manner to this moving squall line. The wave propagated from the north to the south, as did the squall system. The timing of the wave impact is consistent with the arrival of the squall at Daytona Beach, as is the timing of the anomaly on the tidal record at St. Augustine with the passage of the squall.

ANALYSIS

Resonant Coupling Mechanism: The "Squall-Line Surge"

Historical precedent exists for the generation of this type of large wave by a propagating pressure front. Ewing, Press and Donn (1954) reported on a large wave which struck the southern Lake Michigan shore on 26 June 1954, resulting in 7 fatalities along the Chicago waterfront. The wave was associated with a moving squall line in which a pressure jump of 3 mb was measured. The cause was thought to involve a resonant coupling of the traveling atmospheric disturbance to the surface wave. When the speed of the disturbance approaches that of a long wave in a given depth of water $(gh)^{1/2}$, the free wave and forced wave essentially reinforce each other, and this coupled wave can attain extremely large amplitudes.

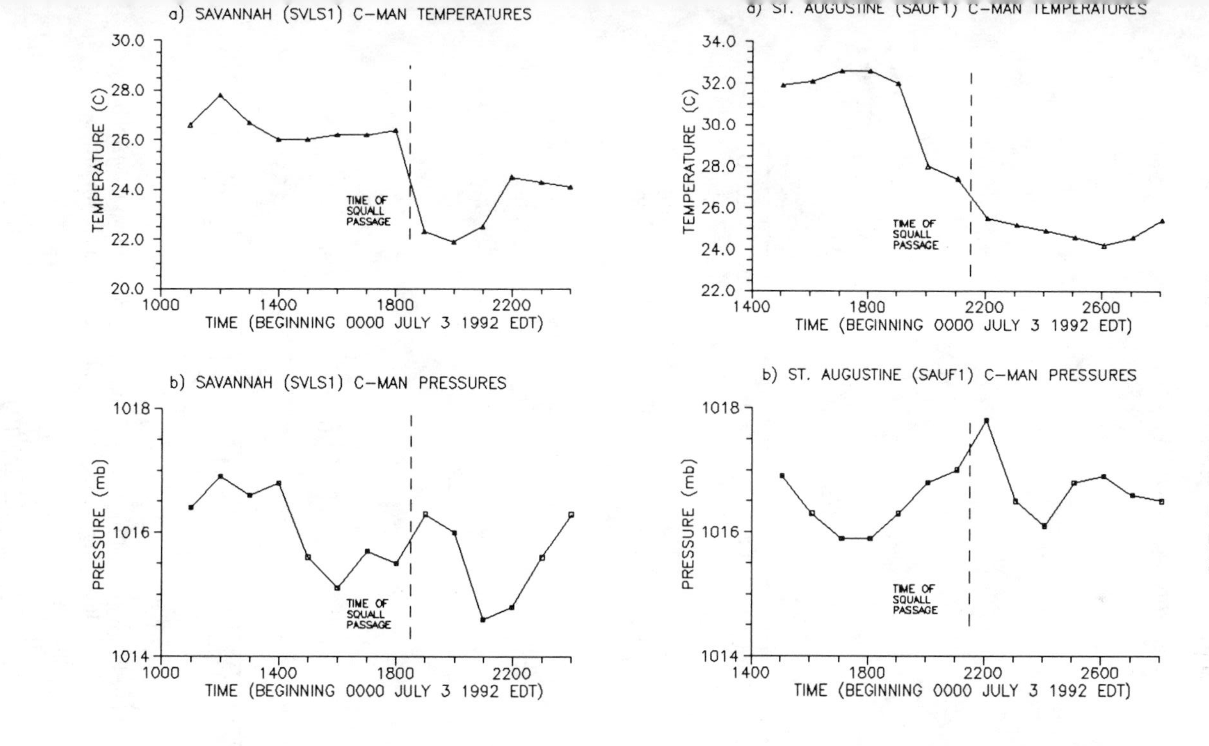

Figure 5: Time histories of meteorological parameters on 3 July 1992 at the Savannnah C-MAN buoy:
a) Air temperature
b) Barometric pressure

Figure 6: Time histories of meteorological parameters on 3 July 1992 at the St. Augustine C-MAN buoy:
a) Air temperature
b) Barometric pressure

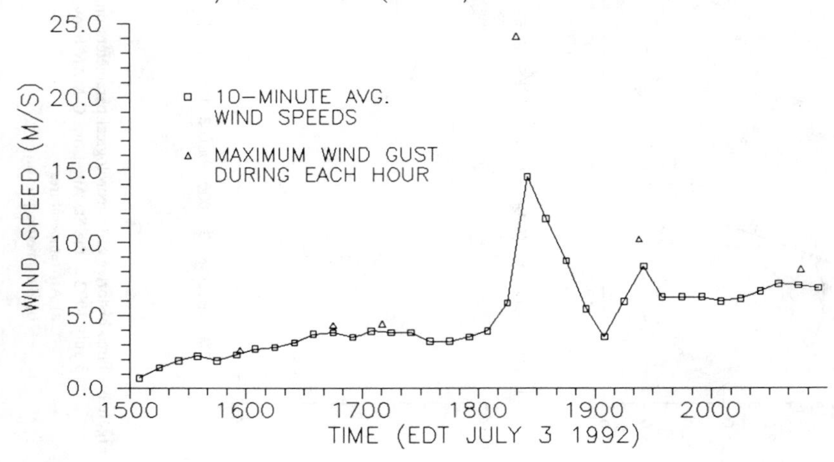

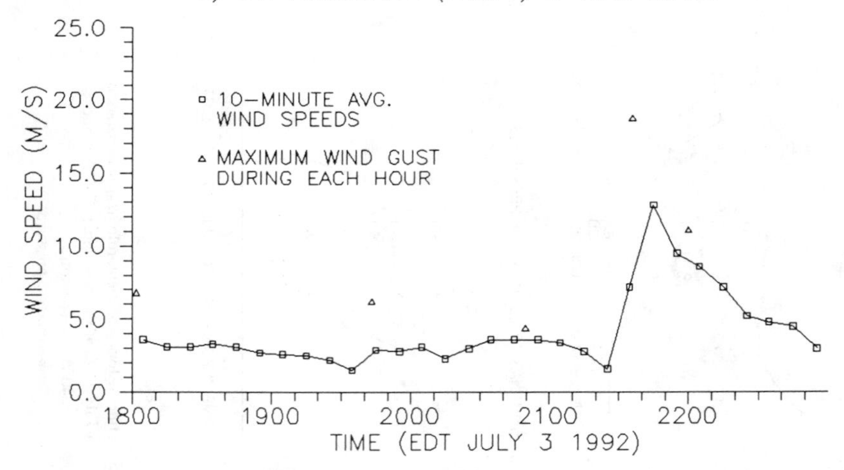

Figure 7: Time history of continuous 10-minute average winds at C-MAN buoy locations illustrating passage of squall line. Also shown are highest wind gusts within each hour.

a) Savannah (SVLS1)
b) St. Augustine (SAUF1)

Owing to the great depth of Lake Michigan, it is noted that the value of $(gh)^{1/2}$ is quite high, however the pressure jump was observed to move at a speed of 66 mph (106 km), which is consistent with a resonant condition over the southern end of the lake which the front traversed. Platzman (1958) modeled the propagation of such a wave using a finite difference model and was able to correctly reproduce the arrival times of the direct and reflected wave in the lake. Owing to complexity of the resonant response and the lake bathymetry, the shoreline amplitudes were not modelled.

Proudman (1929,1952) identified a similar mechanism for another single large wave which struck southeastern England in 1929, also resulting in fatalities. In this case a pressure jump of 1.5 mb was observed to cross the English Channel from south to north. Dean and Dalrymple (1984) offer a simple linear analytical treatment of waves forced by a moving pressure disturbance which indicates the potential for resonant amplification when the disturbance propagates at the long wave speed.

Ewing, Press and Donn (1954) contoured the depth h^* in Lake Michigan for which $(gh^*)^{1/2}$ was equal to the speed of the atmospheric disturbance and identified the resulting region as a potential "generation area" for the resonantly coupled wave. A similar analysis of the squall system in the present case shows that a speed of 30 mph (48 km/hr) yields a corresponding depth of 60 ft (18 m). Figure 8 shows a sketch of the Florida shoreline from St. Augustine to Daytona Beach with the 60 ft depth contour included. The continental shelf in this region is extremely flat, and it is clear that the squall traversed a large region with the approximately correct depth for resonant coupling.

A summary of these three events is given in Table 1. In each case the speed of the atmospheric disturbance is consistent with the long wave speed over a large portion of the traversed area. This is most particularly evident in the case of the wave at Daytona Beach, in which the depth is the most uniform of the three. Note that the magnitude of the pressure jump in the present case is the smallest of the three; the greater areal extent of the resonant generation area may well explain the ensuing wave reached a similar magnitude to those in Lake Michigan and England.

	South Sussex England 20 July 1929 Proudman (1952)	Lake Michigan Chicago 26 June 1954 Ewing et al.(1954)	Daytona Beach 3 July 1992
Pressure Disturbance	+ 1.5 mb	+ 3 mb	+ 1.0 mb
Disturbance Speed	40 mph	66 mph	30 mph
Approximate Depth, h	100 ft	270 ft	60 ft
$(gh)^{1/2}$	39 mph	64 mph	30 mph

Table 1: Comparison of approximate characteristics of the 1929 Sussex and 1954 Lake Michigan single large wave events with the present 1992 Daytona Beach event.

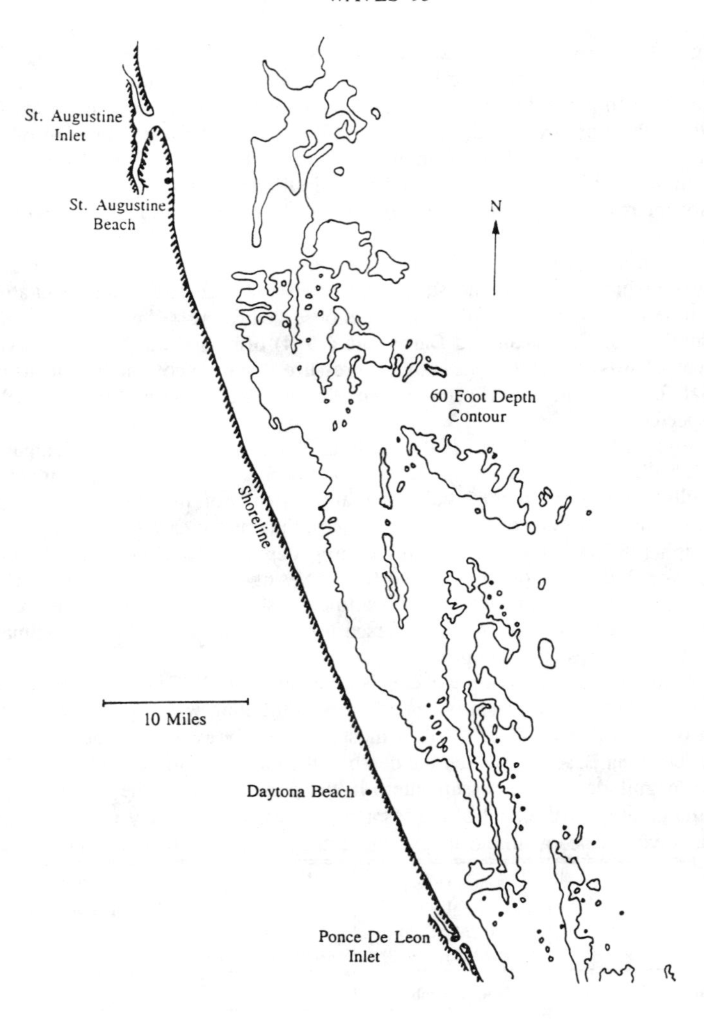

Figure 8: Florida coast between St. Augustine and Daytona Beach and areal extent of 60 ft (18 m) depth contour indicating generation area for resonant growth of wave forced by atmospheric pressure disturbance moving at a speed of 30 mph (48 km/hr).

Ruffman (1992), while cataloguing seismically generated wave events in Eastern Canada, encountered instances of tsunami-like waves which were not connected with any seismic activity but seemed to be associated with propagating pressure disturbances. These waves were described with the term "squall-line surge". The Daytona Beach wave would appear to be an example of this general type of forced long wave motion.

Edge Wave Response

As previously indicated, the characteristics of the wave response measured at the tide gauge in St. Augustine were quite different than the observed characteristics of the wave at Daytona Beach. The disturbance coincident with the squall transit past St. Augustine had a relatively smaller amplitude and much longer period (approximately 2.2 hours). It appears likely that this was the result of an excitation of an edge wave response of longer period. It is also likely that the edge wave response was present at Daytona Beach, but the small amplitude and long period effectively rendered it "invisible" relative to the large wave in the absence of any mechanical recording devices. Munk, Snodgrass and Carrier (1956) describe the propagation of edge waves due to impulsive atmospheric disturbances along the Southern California coast; an analytical treatment is provided by Greenspan (1956). Donn and Ewing (1956) show that the "zeroth" edge wave mode explained the behavior of long waves in Lake Michigan following the passage of a pressure jump parallel to the shoreline on 6 July 1954 (only 10 days after the fatal wave at Chicago).

The zero edge wave mode excited will have phase speed C equal to the speed of the moving pressure disturbance and given by

$$C = (gT\sin\beta)/2\pi \tag{1}$$

where T = wave period and $\sin\beta$ = beach slope. Inserting the values for the squall moving over the continental shelf off of Daytona Beach we have C = 30 mph and $\sin\beta = 0.001$ as an approximate measure of the average offshore slope. The edge wave period can be estimated as

$$T = \frac{2\pi C}{g\sin\beta} \approx 2.4\ hr \tag{2}$$

which is consistent with the period of 2.2 hours characterizing the wave response at St. Augustine.

Solitary Wave Runup

Given the lack of any direct measurements of the water surface profile at Daytona Beach, the height of the wave can be estimated by relating it to the measured runup elevation on the beach. The lack of any general analytic solution or measurement of the wave period leads naturally to a comparison with solitary wave runup laboratory results. For example the data of Synolakis (1987) for breaking solitary waves on a 1:20 slope provides an empirical relationship of the form $R/d = 0.918(H/d)^{0.606}$, where R = runup elevation, d = depth at "toe" of beach, and H=offshore wave height. In the present case, using the maximum observed runup, we estimate R = 6 ft (1.8 m) and d = 30 ft (9 m). Hence R/d = 0.2 and H/d = 0.08, which yields an offshore wave height of 2.4 ft (0.7 m).

To extend this result to a breaking (observed) wave height, the laboratory data of Ippen and Kulin (1955) may be used as a guide. For the given conditions, an initial amplitude to depth ration of H/d = 0.08 yield a breaking amplitude to initial amplitude ratio of approximately $H_b/H = 1.6$ (which requires extrapolation of the data). The resulting breaking wave height is $H_b = 3.8$ ft (1.2 m), which is (although of the proper order) substantially lower than any observation in the vicinity of the maximum wave runup on Daytona Beach.

However this wave height is clearly only a gross estimate since it requires extrapolation from artificially constrained laboratory conditions (which do not include effects of beach permeability and variable slope) and is predicated on the assumption that the mean water level (the datum for the runup elevation) was determined by the local tide. The long period edge wave response discussed in the previous section clearly could have produced a "local" mean water level significantly higher or lower than that due to the tide alone.

Other Related Cases

There are many other cases of similar or related phenomena that have been well documented in the literature. Donn (1959), Irish (1965) and Hughes (1965) reported on squall-line surges on the Great Lakes that were of somewhat smaller amplitude than the fatal event of 26 June 1954. In these instances, good information is available regarding both the atmospheric forcing and the lake response, owing to the presence of many lake level recorders and meteorological data from the entire lake boundary and surrounding region. This is in stark contrast to the lack of data pertaining to the present event on the open seacoast.

Wiegel, Snyder and Williams (1958) investigated the propagation of the coupled waves associated with Hurricane Carol as it propagated up the U.S. Atlantic coast in 1954. The large scale nature of the pressure disturbance allowed the response to be measured at numerous tide gaging stations along the coast, unlike the present case, in which the squall system moved over the coast, developed and dissipated all within the vicinity of only two stations. Wiegel, Snyder and Williams also performed laboratory experiments to investigate the development of resonant coupled waves induced by a moving low pressure system. Similar experiments are now being undertaken by the authors to measure the growth of the coupled wave induced by a propagating pressure jump as observed in the various cases of squall-line surge.

Draper and Bownass (1982) reported on the case of waves of unusually long period and duration which impacted the English channel coast in February 1979. It was determined that these waves resulted from an extratropical cyclone in the mid-Atlantic which moved at a speed equal to that of 18-20 second period waves and thus continued to feed energy preferentially into this longer period part of the spectrum in a resonant fashion.

CONCLUSIONS

The large wave which struck Daytona Beach on 3 July 1992 was most likely a meteorologically forced wave resulting from a resonant coupling between a propagating atmospheric pressure disturbance and free long surface gravity waves, thus falling under a general class of forced wave motions which might be denoted as "squall-line surge". The data also suggest the existence of an underlying long period edge wave response which was however not directly observed at Daytona Beach due to the lack of instrumentation and the presence of the larger amplitude, shorter period wave event.

The overall small-to-mesoscale nature of the squall system, and the fact that it propagated over an open coast environment resulted, in this instance, in the entire wave event taking place with no direct measurements of the wave at impact and limited information about the pressure front. The general lack of detailed data regarding both the meteorological forcing and the actual characteristics of the wave leaves little recourse but to attempt simple and largely conceptual analysis efforts. Thus although the general mechanisms of this event may be reasonably well understood, the details of the process will remain elusive.

ACKNOWLEDGEMENTS

The authors would like to acknowledge the following contributions of the data used in this paper:

C-MAN buoy data were provided by Mr. Sam Houston of the Hurricane Research Division (HRD) of the National Oceanic and Atmospheric Administration (NOAA) Atlantic Oceanographic and Meteorological Laboratory (AOML), Miami, Florida.

Tide gauge data were provided by NOAA Tidal Analysis Branch, Rockville, Maryland.

Squall line positions were determined from radar plots provided by the National Weather Service, Daytona Beach.

REFERENCES

Dean, R.G. and Dalrymple, R.A. (1984). *Water Wave Mechanics for Engineers and Scientists*. Prentice-Hall, Inc., New Jersey, pp.163-166.

Donn, W.L. (1959). The Great Lakes storm surge of May 5, 1952. *Journal of Geophysical Research*, vol. 64, no. 2, 191-198.

Draper, L.D. and Bownass, T.M. (1982). Unusual waves on European coasts, February 1979. *Proceedings 18th Conference on Coastal Engineering*, Cape Town, South Africa, ASCE, vol 1, 270-281.

Donn, W.L. and Ewing, M. (1956). Stokes' edge waves in Lake Michigan. *Science*, vol. 124, 1238-1242.

Ewing, M., Press, F. and Donn, W.L. (1954). An explanation of the Lake Michigan wave of 26 June 1954. *Science*, vol. 120, 684-686.

Greenspan, H.P. (1956). the generation of edge waves by moving pressure distributions. *Journal of Fluid Mechanics*, vol. 1, 574-592.

Hughes, L.A. (1965). The prediction of surges in the southern basin of Lake Michigan. Part III. The operational basis for prediction. *Monthly Weather Review*, vol. 93, no.5, 292-296.

Ippen, A.T. and Kulin, G. (1954). The shoaling and breaking of the solitary wave. *Proceedings 5th Conference on Coastal Engineering*, Berkeley, California, vol. 1, 27-49.

Irish, S.M. (1965). The prediction of surges in the southern basin of Lake Michigan. Part II. A case study of the surge of August 3, 1960. *Monthly Weather Review*, vol. 93, no.5, 282-291.

Munk, W., Snodgrass, F. and Carrier G. (1956). Edge waves on the continental shelf. *Science*, vol. 123, 127-132.

Platzman, G.W. (1958). A numerical computation of the surge of 26 June 1954 on Lake Michigan. *Geophysica*, vol. 6, no. 3-4, 407-438.

Proudman, J. (1929). The effects on the sea of changes in atmospheric pressure. *Geophysical Supplement, Monthly Notices of Royal Astronomical Society*, London, vol. 2, 197-209.

Proudman, J. (1952). *Dynamical Oceanography*, Wiley, New York, pp. 295-300.

Ruffman, A. (1992). The need for a compilation of historic Great Lakes seiches, storm surges and possible tectonically induced tsunamis. *Abstracts, International Association for Great Lakes Research 35th Conference*, University of Waterloo, Canada, p. 29.

Synolakis, C.E. (1987). The runup of solitary waves. *Journal of Fluid Mechanics*, vol. 185, 523-545.

Wiegel, R.L., Snyder, C.M. and Williams, J.E. (1958). Water gravity waves generated by a moving low pressure area. *Transactions, American Geophysical Union*, 39, 2, 224-236.

NONLINEARITY IN WAVE CREST STATISTICS

David L. Kriebel[1] and Thomas H. Dawson[2]

ABSTRACT

The statistical properties of nonlinear wave crests amplitudes are discussed for a narrow-band random sea consistent with Stokes second-order wave theory. In particular, the probability distribution for nonlinear crest amplitudes is shown to be a modified form of the Rayleigh distribution that incorporates nonlinear effects through a spectral wave steepness parameter. Predictions from this distribution are shown to be in close agreement with laboratory data and with recently published North Sea measurements for both deep water and intermediate depth conditions. This distribution is then modified to account for wave breaking in a random sea and results are again shown to be in good agreement with laboratory data.

INTRODUCTION

The probability that a wave height in a random sea will exceed a specified height has long been recognized as an important statistic in practical work. As is well known, this probability may be determined from the Rayleigh distribution for seas with relatively narrow-banded frequencies. However, in certain applications, as in the selection of the deck elevation for a fixed offshore platform for example, it is the amplitude of the wave crests rather than the wave heights that are of interest. For relatively mild sea states, the crest amplitudes are simply one-half the wave heights so that the Rayleigh distribution can continue to be used. In the case of severe sea states, however, which are usually of interest in practical considerations, the crest amplitudes are greater than one-half the wave heights and an accurate estimate of the crest statistics must include nonlinear effects.

---

[1] Associate Professor and [2] Professor, Ocean Engineering Program, United States Naval Academy, 590 Holloway Road, Annapolis, Maryland 21402

From linear theory, the probability of wave crests exceeding a given amplitude is described by the Rayleigh distribution for narrow-band seas (Longuet-Higgins, 1957) or by a modified form of this distribution for seas with finite spectral bandwidth (Cartwright and Longuet-Higgins, 1956). Modifications to account for nonlinear effects based on Stokes second-order wave theory for narrow-band random seas have been proposed by Tayfun (1980), Arhan and Plaisted (1981), and Huang et al. (1986). Empirical modifications to the Rayleigh distribution for wave crest amplitudes have also been proposed by Jahns and Wheeler (1972) and Haring et al. (1976) based on wave measurements from offshore platforms in the Gulf of Mexico.

In recent work, the writers have further examined the effect of second-order Stokes-type nonlinearity on wave crest statistics in deep water (Kriebel and Dawson, 1991, and Dawson et al., 1993). The purpose of the present paper is to review these previous developments and to consider additional measurements supporting their validity. For the first time, we compare field measurements to our theoretical distribution for nonlinear crest amplitudes. We also extend the previous results to finite water depth and present both laboratory and field measurements to support this extension of the theory. Because wave breaking eventually limits the validity of the nonlinear theory, we then review previous work which describes the effect of wave breaking on the distribution of nonlinear wave crest amplitudes.

NONLINEAR CREST STATISTICS IN DEEP WATER

The free-surface displacement at a point in a random sea with narrow-banded wave frequencies is expressible to second-order as

$$\eta(t) = a(t)\cos(\omega t+\phi(t)) + \frac{1}{2}\,k\,a^2(t)\cos(2\omega t+2\phi(t)) \tag{1}$$

where a is the modulated amplitude of linear wave components, ϕ is the slowly-varying phase, ω is the mean wave frequency, and k is the mean wavenumber given by $k = \omega^2/g$ in deep water. It is further assumed that the probability density function of linear crest amplitudes is given by the Rayleigh distribution as

$$p(a) = 16\,\frac{a}{H_s^2}\exp\left(-8\,\frac{a^2}{H_s^2}\right) \tag{2}$$

while the probability that a linear wave crest exceeds a given amplitude is given by

$$P(a) = \int_a^\infty p(a)\,da = \exp\left(-8\frac{a^2}{H_s^2}\right) \tag{3}$$

In these expressions, H_S is the significant wave height. This is defined from the zero-moment or variance of the wave spectrum as $4(m_o)^{1/2}$ as suggested by narrow-band theory. As discussed by Forristall (1978), however, this could be defined from a direct counting of wave heights as the average of the one-third largest waves in order to calibrate the Rayleigh distribution to give a better fit to a specific data set.

At the wave crest phase, the nonlinear crest amplitude, r, is determined from Stokes theory in terms of the linear amplitude, a, as

$$r = a + \frac{1}{2} k a^2 \tag{4}$$

Following Tayfun (1980), the probability density function for nonlinear crest amplitudes, $p(r)$, may be found from a transformation of random variables as

$$p(r) = p(a) \frac{da}{dr} \tag{5}$$

which requires that a be expressed as a function of r for substitution into the right-hand side. Tayfun (1980), Arhan and Plaisted (1981), and Huang et al. (1986) solved for a from the quadratic equation in (4). A simpler approach, however, involves reversion of the series in (4) which gives an approximate functional expression between a and r as

$$a \approx r - \frac{1}{2} k r^2 \tag{6}$$

Based on this relationship, it is readily found that the corresponding probability density function for the nonlinear crest amplitudes is given by equations (2) and (5) as

$$p(r) = 16 \frac{r}{H_S^2} (1 - \frac{3}{2} R \frac{r}{H_S} + \frac{1}{2} R^2 \frac{r^2}{H_S^2}) \exp(-8 \frac{r^2}{H_S^2} (1 - \frac{1}{2} R \frac{r}{H_S})^2) \tag{7}$$

Integration of equation (7) then gives the probability that the nonlinear crest amplitudes will exceed the level r as

$$P(r) = \int_r^\infty p(r)\, dr = \exp(-8 \frac{r^2}{H_S^2} (1 - \frac{1}{2} R \frac{r}{H_S})^2) \tag{8}$$

In these expressions, the parameter that represents the effect of nonlinearity is the characteristic steepness of the random sea, R, which is defined as

$$R = kH_S \tag{9}$$

Interpretation of the distributions in (7) and (8) shows that, in the absence of nonlinearity when the steepness R equals zero, the probability functions reduce to the Rayleigh distribution. The additional terms containing the steepness, R, then modify the probabilities, particularly of the highest wave crests, to reflect increased amplitudes of wave crests due to the Stokes-type nonlinearity involving the phase-locked second-order wave components.

Although equations (7) and (8) are obtained directly from the transformation of random variables, they contain terms proportional to the wave steepness squared which are seemingly inconsistent with second-order approximations. As a result, the authors have previously reported results consistent to the first power in R only as

$$p(r) \sim 16 \frac{r}{H_S^2} (1 - \frac{3}{2} R \frac{r}{H_S}) \exp(-8 \frac{r^2}{H_S^2}) \exp(+8R \frac{r^3}{H_S^3}) \tag{10}$$

and

$$P(r) = \int_r^\infty p(r)\, dr = \exp(-8 \frac{r^2}{H_S^2}) \exp(+8R \frac{r^3}{H_S^3}) \tag{11}$$

As noted in a related paper (Dawson, Kriebel, and Wallendorf, 1993), these approximate distributions must be truncated and renormalized if the steepness becomes large. As may be seen in equation (10), the probability densities become negative for amplitudes greater than $r/H_S = 2/3R$. As a practical matter, this effect is unimportant except for relatively large values of the steepness parameter of about 0.4 or greater, and such high values of spectral steepness seem to be quite rare in nature. Similar truncation is required of equation (7) for amplitudes greater than $r/H_S = 1/R$, however, this amplitude is so high that it is beyond the range of realistic crest amplitudes for almost all sea states and, in practice, equation (7) may be applied without requiring truncation.

Figures 1 and 2 show comparisons of the probability distribution for nonlinear crest amplitudes from equation (8) with wave data from laboratory tests and from field measurements respectively. Predictions from the Rayleigh distribution in equation (3) are also shown. The agreement between the measurements and the predictions from the modified nonlinear distribution can be seen to be very good for both data sets. Furthermore, the predictions with nonlinear effects included in terms of the spectral steepness are far superior to those given by the linear Rayleigh distribution, especially for the highest wave crests in the sea state. While previous laboratory data sets have supported the validity of the nonlinear distribution, the results in Figure 2 demonstrate, for the first time, the validity of the nonlinear distribution when applied to three-dimensional field measurements.

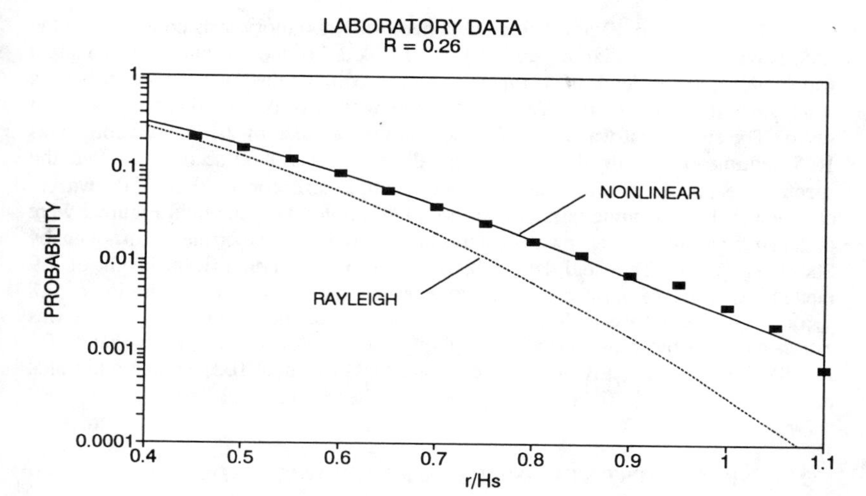

Figure 1. Comparison of laboratory data for crest amplitudes to linear and nonlinear probability distributions.

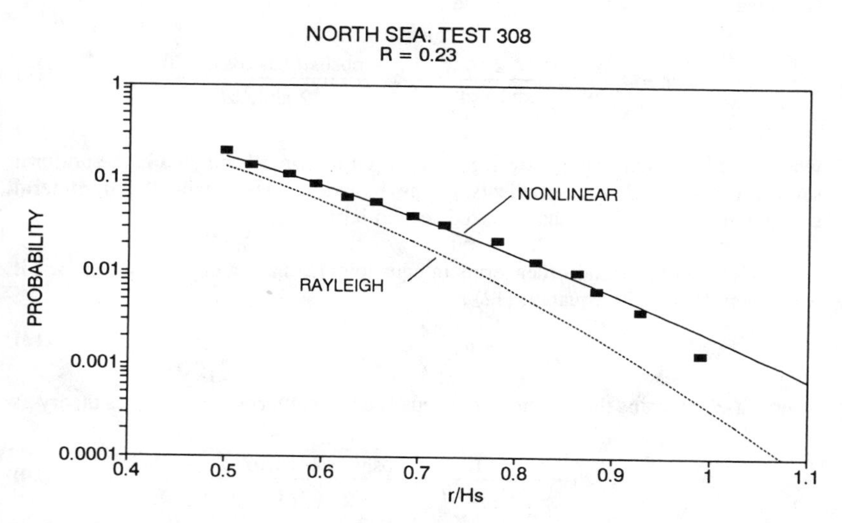

Figure 2. Comparison of North Sea field data for crest amplitudes (from Stansberg, 1991) to linear and nonlinear probability distributions.

Data shown in Figure 1 were obtained from laboratory tests conducted in the U.S. Naval Academy's large wave tank. This tank is 116 meters long and 7.9 meters wide, with a water depth of 4.9 meters. Uni-directional deep-water random waves were generated from a JONSWAP spectrum with a peak amplification factor of seven. The significant wave height, based on the variance of the wave record, was 16.5 centimeters, while the mean spectral period was 1.60 seconds so that the spectral steepness, R, was 0.26. The total sample size consisted of 1503 waves, obtained in three separate tests of 800 second duration. Data shown in Figure 2 were obtained from North Sea measurements in the WADIC experiment, as given by Stansberg (1991). The field data were reported to have a significant height of 4.9 meters based on the variance of the wave record and a peak spectral period of 9.8 seconds. In order to calculate the spectral steepness, the mean wave period was estimated from the peak period by applying a correction based on the shape of a JONSWAP spectrum, giving a value of steepness of about 0.23. Some additional details concerning the field data are given by Allender et al. (1989).

NONLINEAR CREST STATISTICS IN FINITE WATER DEPTH

The probability distributions given above are based on deep water Stokes theory. A more general and useful form of these distributions may be obtained by accounting for finite depth effects. As shown by Dean and Dalrymple (1984) or others, the nonlinear crest amplitude from Stokes second-order theory in finite depth is related to the linear crest amplitude as

$$r = a - \frac{1}{2}ka^2\frac{1}{\sinh 2kd} + \frac{1}{2}ka^2\,\frac{\cosh kd\,(2+\cosh 2kd)}{2\sinh^3 kd} \tag{12}$$

where d is the water depth. The second term on the right-hand side is the uniform set-down associated with gradients in the radiation stress while the third term contains the effect of the phase-locked second harmonic.

Since both second-order terms in equation (12) have a dependence on ka^2, it is convenient to write equation (12) as

$$r = a + \frac{1}{2}ka^2\, f_2(kd) \tag{13}$$

where $f_2(kd)$ contains the effects of the depth dependent terms from Stokes theory as

$$f_2(kd) = -\frac{1}{\sinh 2kd} + \frac{\cosh kd\,(2+\cosh 2kd)}{2\sinh^3 kd} \tag{14}$$

In finite depth, k is the wavenumber associated with the mean wave frequency based on the usual dispersion relationship from linear wave theory

$$\omega^2 = gk \tanh(kd) \tag{15}$$

Following the method used previously, we may transform random variables to obtain the probability density function for nonlinear crest amplitudes in finite depth. This was first carried out by Arhan and Plaisted (1981), but was based on solution of the quadratic equation in (13) so that the resulting distribution was difficult to interpret. Once again, a simpler and more useful result is obtained by reverting the series in (13). The resulting probability density function is then given by

$$p(r) = 16 \frac{r}{H_s^2} \left(1 - \frac{3}{2} R_* \frac{r}{H_s} + \frac{1}{2} R_*^2 \frac{r^2}{H_s^2}\right) \exp\left(-8 \frac{r^2}{H_s^2} \left(1 - \frac{1}{2} R_* \frac{r}{H_s}\right)^2\right) \tag{16}$$

Integration of equation (16) then gives the probability that a nonlinear wave crest will exceed amplitude r in finite depth as

$$P(r) = \int_r^\infty p(r)\, dr = \exp\left(-8 \frac{r^2}{H_s^2} \left(1 - \frac{1}{2} R_* \frac{r}{H_s}\right)^2\right) \tag{17}$$

The results may again be truncated to retain terms only to the first-order in wave steepness.

The probability functions for finite depth in equations (16) and (17) are identical in form to those given in equations (7) and (8) for deep water, with the exception that the spectral steepness has a different definition. In finite depth, the effective steepness, R_*, is given by

$$R_* = k\, H_s\, f_2(kd) \tag{18}$$

As a result, the steepness is influenced both by the shortening of the wavelength in finite depth and by the effect of the depth-dependent terms in Stokes wave theory. As is well known, these terms generally increase in magnitude as the depth decreases. Thus, for a given value of the significant wave height, the effective wave steepness for a random sea is larger in finite depth than in deep water.

Figures 3 and 4 show comparisons of the probability distribution for nonlinear crest amplitudes in finite water depth, given by equation (17), to wave data from laboratory tests and from field measurements. Predictions from the Rayleigh distribution in equation (3) and from the deep water form of the distribution for nonlinear crests in equation (8) are also shown. As may be seen, the modified theory accounting for finite depth effects is in general agreement with the data and provides a far better description of crest probabilities than the linear Rayleigh distribution.

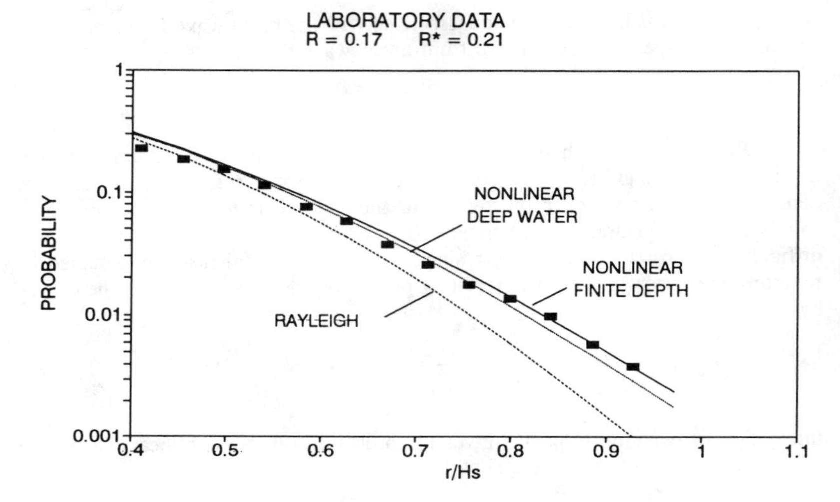

Figure 3. Comparison of laboratory data for crest amplitudes in finite depth to linear and nonlinear probability distributions.

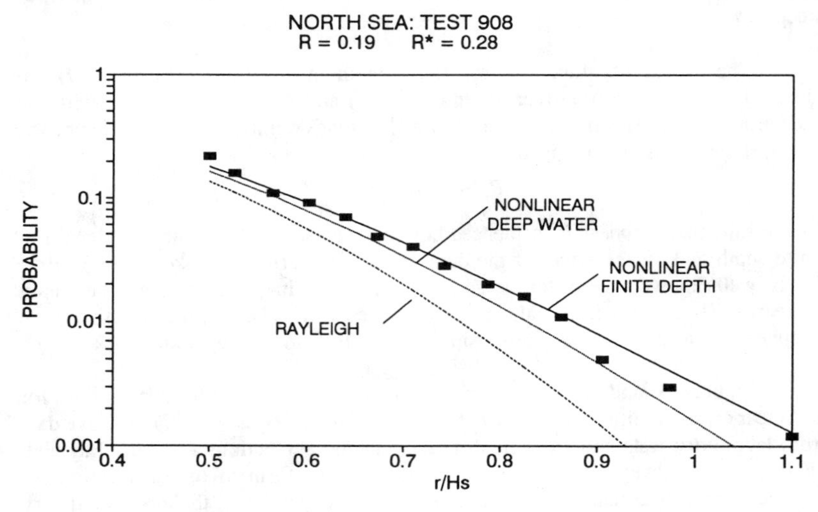

Figure 4. Comparison of North Sea field data for crest amplitudes in finite depth (from Stansberg, 1991) to linear and nonlinear probability distributions.

Data shown in Figure 3 were obtained from laboratory tests conducted in the U.S. Naval Academy's small wave tank, which is 36.5 meters long and 1.3 meters wide with a water depth of 1.55 meters, again with a uni-directional JONSWAP spectrum with an amplification factor of seven. Short wave runs of just 60 seconds were used so that reflected waves would not interfere with the incident waves at the wave gage. Data shown are based on 9 such runs, giving a total sample size of just 255 waves. The significant wave height was 14.1 centimeters and the mean spectral period was 1.83 seconds, giving a deepwater steepness, R, of 0.17 and an effective steepness based on the relative water depth, R_*, of 0.21. Due to limitations of the wavemaker, conditions departed only slightly from deep water. Despite this, the data seem to support the finite depth correction for nonlinear wave crests.

Data shown in Figure 4 were again obtained from North Sea measurements as reported by Stansberg (1991). The exact spectral form for this sea state is again unknown to the authors, however, other North Sea spectra shown by Stansberg (1991) closely resemble a JONSWAP form. The significant height was reported to be 9.7 meters and peak spectral period was 14.9 seconds. These conditions give a deepwater steepness of about 0.19 but, because of the very long wave period and the 75 meter water depth at the gage location, the effective steepness based on finite depth effects is estimated to be about 50% larger or 0.28. It can be seen that the nonlinear theory is modified more strongly by depth effects in this case than in the laboratory case. Furthermore, the nonlinear theory seems to be in good agreement with the field data.

MODIFICATIONS DUE TO WAVE BREAKING

To present a more complete description of nonlinear crest modifications, we next review some earlier results concerning the effects of wave breaking (see Dawson, Kriebel, and Wallendorf, 1993). It is assumed that breaking waves may be identified in a stochastic sense by adopting a breaking criterion linked to the downward acceleration of the nonlinear wave crest. Longuet-Higgins (1969) suggested that breaking occurs when the acceleration in the crest reaches one-half that of gravity. Experimental studies by Ochi and Tsai (1983) and Ramberg and Griffin (1987) have suggested, however, that for the inception of breaking, the limiting acceleration is actually smaller than $0.5g$.

In the previous study, the writers also found experimental evidence for a lower critical value of the downward crest acceleration of breaking waves. Starting with equation (1), it may be shown that to second-order, the downward acceleration in the wave crest for the narrow-band wave form is given by

$$\ddot{\eta}_b - a_b \, \omega^2 \tag{19}$$

where ω is interpreted to be the mean wave frequency. By setting the right-side of equation (19) equal to some fraction, β_o, of the acceleration of gravity, the average linear crest amplitude for breaking waves in a narrow-band random sea is given by

$$a_b = \beta_o \frac{g}{\omega^2} \tag{20}$$

Consequently, from equation (4), the average amplitude for nonlinear wave crests at breaking is given by

$$r_b = a_b + \frac{1}{2} k\, a_b^2 = \beta \frac{g}{\omega^2} \tag{21}$$

where β equals $\beta_o + \beta_o^2/2$. In our previous study, breaking waves were observed for a JONSWAP sea state with spectral steepness $R = 0.53$ and the appropriate values of β_o and β were determined to be 0.34 and 0.40 respectively. One critical distinction between these values and previous breaking criteria is that our values are for so-called "active" breaking conditions, where breaking was defined by any observable white-water as waves passed a fixed wave gage. Previous criteria of Ochi and Tsai (1983) or of Ramberg and Griffin (1987) pertain to the incipient breaking and do not account for the fact that waves continue to break to lower amplitudes.

Based on the mean breaking amplitude from equation (21), the overall probability of breaking in a random sea in deep water can then be estimated from integration of the probability density function in (7) or (10). Using (10) gives a simple result for the probability of breaking as

$$Q_B = \int_{r_b}^{\infty} p(r)\, dr = \exp\left(-8 \frac{\beta^2 (1-\beta)}{R^2}\right) \tag{22}$$

This result suggests that the steepness of the sea state is the dominant parameter for determining the frequency of breaking in a random sea according to narrow-band assumptions. This was tested in the Naval Academy large wave tank with a series of JONSWAP spectra, all having the same dimensionless spectral shape, but with different values of the mean wave period and significant wave height producing several distinct values of the steepness R. Waves were recorded by a fixed wave gage and breaking was determined from video-tape records of these waves passing the gage. In each case, the number of breaking waves, i.e. white-cap events, was divided by the total number of waves to determine the probability of breaking. As shown in Figure 5, breaking probabilities are related to the spectral steepness, R, as suggested by equation (22). In addition, the active breaking criterion with $\beta = 0.40$ was found to work well for all six JONSWAP sea states.

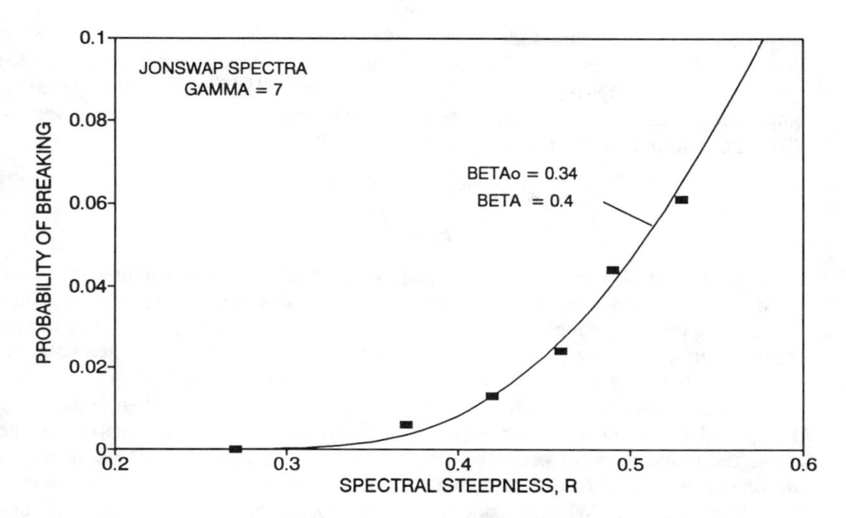

Figure 5. Probability of breaking for laboratory-generated JONSWAP seas.

Of more interest in the present paper, however, is the probability distribution for wave crest amplitudes in a random sea where a significant percentage of waves are breaking. For this purpose, we again refer to the previous paper where it was shown that for a narrow-band random sea, the cumulative distribution may be obtained from the following integration

$$P(r) = \int_r^\infty \int_{T_b(r)}^\infty p(r)\, p(T)\, dT\, dr \tag{23}$$

where $P(r)$ denotes the probability of crest amplitudes being greater than r, and where $p(r)$ is the crest probability density function given by equation (8). In addition, $p(T)$ is the wave period distribution, which is assumed to be independent of the crest-amplitude distribution for narrow-band seas. In this work, no attempt was made to describe the wave period distribution theoretically. Instead, this distribution was selected empirically from the laboratory data for JONSWAP seas. As shown by Dawson et al. (1993), the wave periods had approximately a triangular distribution centered on the mean period and extending between about 0.8 and 1.2 times the mean wave period.

A key assumption behind equation (23) is that the acceleration-based breaking criterion, given in terms of the mean wave frequency in equation (21), also holds for other wave frequencies in the random sea. A related assumption is that breaking waves are reduced in amplitude so that, on average, their amplitudes are limited to that given by this breaking criterion. Following these assumptions, equation (23) gives the probability of both breaking and non-breaking wave crests exceeding amplitude r, based on the critical breaking wave period for amplitude r, given by $T_b(r)$ and determined from equation (21) as

$$T_b = 2\pi \left(\frac{r}{g\beta}\right)^{1/2} \quad \textbf{(24)}$$

Equation (23) excludes those crests whose periods are less than $T_b(r)$, since these are breaking waves whose amplitudes are, on average, smaller than r. When considering the exceedance probability for very small amplitudes, the critical period from equation (24) may be lower than the range of wave periods in the sea state. As a result, no breaking waves exist at lower amplitudes and the distribution in equation (23) will be identical to that in equation (8). On the other hand, when considering the exceedance probability for very large amplitudes, the critical period may be above the range of wave periods in the sea state. As a result, all breaking waves must have amplitudes below that being considered and the distribution from equation (23) must go to zero. When considering the exceedance probability for an amplitude in between these extremes, some breaking waves (with periods below $T_b(r)$) have amplitudes below that being considered while other breaking and nonbroken waves (with periods greater than $T_b(r)$) have larger amplitudes. As a result, the probability of exceedance for this amplitude will be less than that given by equation (8).

Application of equation (23) to two laboratory-generated JONSWAP seas is shown in Figures 6 and 7. In each case, the results of (23) are compared to the Rayleigh distribution in equation (3) and to the deep-water form of the modified distribution for nonlinear wave crests in equation (8). Data are based on tests performed in the Naval Academy large wave tank with JONSWAP spectra using 1152 and 1121 waves respectively. In general, the data seem to follow distribution suggested by equation (23) which gives lower probabilities of exceedance for the larger wave crests due to breaking limitations.

For crests with low amplitudes, below about $r = 0.5H_S$ or so, there is little breaking in these JONSWAP seas. As a result, the probability distribution continues to follow the nonlinear distribution in equation (8). For crest amplitudes above $0.5H_S$, breaking waves exist at amplitudes both above and below the amplitude being considered. In these cases, equation (23) excludes those breaking crests below amplitude r and gives a probability of exceedance that is smaller than that given by equation (8) but which is still usually larger than that given by the Rayleigh distribution. The results of these tests suggest that, at least for narrow band sea conditions, the method yields an acceptable modification of the wave crest distribution due to breaking. In addition, it suggests that for breaking waves, the distribution of wave periods must be considered even for narrow-band seas due to the strong dependence of the breaking criterion on the wave period.

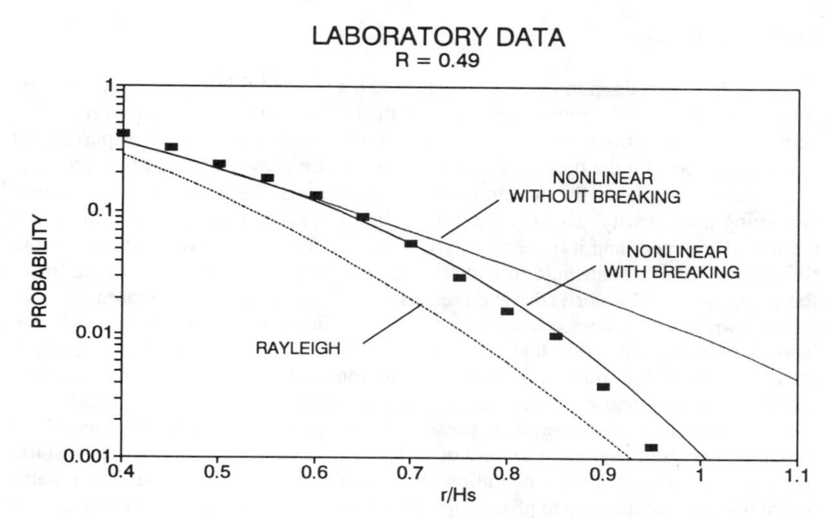

Figure 6. Comparison of the distributions of crest amplitudes in deep water random seas with breaking, $R = 0.49$.

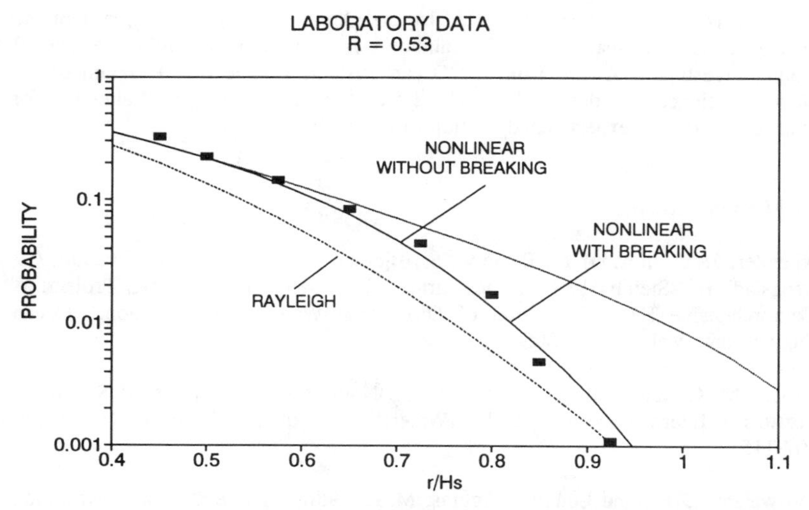

Figure 7. Comparison of the distribution of crest amplitudes in deep water random seas with breaking, $R = 0.53$.

CONCLUSIONS

It has been the purpose of this paper to discuss the effects of nonlinearity on wave crest statistics by reviewing previous theoretical developments, by presenting additional clarifications of the theory, and by presenting further experimental evidence supporting the theory. Theoretical results are presented first for deep water crest statistics, where the spectral wave steepness is the controlling parameter governing nonlinearity. Expressions for nonlinear crest probabilities in finite depth are then developed, and it is found that both the incident spectral wave steepness and finite-depth effects contribute to a departure of the crest amplitudes from the linear Rayleigh theory. The derived wave crest distributions in nonlinear random seas are then shown to be in close accord with North Sea field measurements and with new laboratory data, for both the deep water and finite depth conditions. Further modification of the nonlinear crest distribution due to wave breaking is then discussed, since breaking places an upper limit on the growth of nonlinear wave crests. Laboratory data are then presented that generally verify the theoretical limitations on nonlinear crest amplitudes imposed by wave breaking. These results, however, are for laboratory conditions with unidirectional narrow-band sea states where breaking occurs due to phase-superposition of random wave components; their applicability to field conditions is the subject of continuing research.

ACKNOWLEDGEMENTS

This work has been supported by the Ocean Technology Program of the Office of Naval Research. The first author was also supported by a Presidential Young Investigator Award from the Ocean Systems Engineering Program of the National Science Foundation. The authors are grateful to Louise A. Wallendorf for assistance with the experimental portion of the study.

REFERENCES

Allender, J., Audunson, T., Barstow, S., Bjerken, F., Borgman, L., Graham, C., Krogstad, H., Steinbakke, P., and Vartdal, L., 1989, "The WADIC Project: A Comprehensive Field Evaluation of Directional Wave Instrumentation," Ocean Engineering, Vol. 16, pp. 505-536.

Arhan, M.K., and Plaisted, R.O., 1981, ''Nonlinear Deformation of Sea-Wave Profiles in Intermediate and Shallow Water,'' Oceanol. Acta, Vol. 4, No. 2, pp. 107-115.

Cartwright, D.E., and Longuet-Higgins, M.S., 1956, ''The Statistical Distribution of the Maxima of a Random Function,'' Proc. Royal Soc. London, Ser. A, Vol. 237, pp. 212-232.

Dawson, T.H., Kriebel, D.L., and Wallendorf, L.A., 1993, ''Breaking Waves in Laboratory-Generated JONSWAP Seas,'' Applied Ocean Research, Vol. 15, pp. 85-93.

Dean, R.G., and Dalrymple, R.A., 1984, *Water Wave Mechanics for Engineers and Scientists,* Prentice-Hall, Englewood Cliffs, New Jersey.

Forristall, G.Z., 1978, "On the Statistical Distribution of Wave Heights in a Storm," J. Geophys. Research, Vol. 83, No. C5, pp. 2353-2358.

Haring, R.E., Osborne, A.R., and Spencer, L.P, 1976, "Extreme Wave Parameters Based on Continental Shelf Storm Wave Records," Proc. 15th Intl. Conf. on Coastal Engineering, pp. 151-170.

Huang, N., Bliven, S., Long, S., and Tung, C., 1986, ''An Analytic Model for Oceanic Whitecap Coverage,'' J. Phys. Ocean., Vol. 16, pp. 1597-1604.

Jahns, H.O., and Wheeler, J.D., 1972, "Long-Term Wave Probabilities Based on Hindcasting of Severe Storms," Offshore Technology Conference, Paper OTC 1590.

Kriebel, D.L., and Dawson, T.H., 1991, ''Nonlinear Effects on Wave Groups in Random Seas,'' J. Offshore Mech. and Offshore Engr., Vol. 113, No. 2, pp. 142-147.

Longuet-Higgins, M.S., 1957, ''The Statistical Analysis of a Random Moving Surface,'' Phil. Trans. Royal Soc. London, A 249, pp. 321-387.

Longuet-Higgins, M.S., 1969, ''On Wave Breaking and the Equilibrium Spectrum of Wind Generated Waves,'' Proc. Royal Soc. London, Ser. A, Vol. 310, pp. 151-159.

Ochi, M.K., and Tsai, C.H., 1983, ''Prediction of the Occurrence of Breaking Waves in Deep Water,'' J. Phys. Ocean., Vol. 13, pp. 2008-2018.

Ramberg. S.E., and Griffin, O.M., 1987,''Laboratory Study of Steep and Breaking Deep Water Waves,'' J. Waterway, Port, Coastal, and Ocean Engr., Vol. 113, pp. 493-506.

Stansberg, C.T., 1991, "Extreme Wave Asymmetry in Full Scale and Model Scale Experimental Wave Trains," Proc. 10th Intl. Conf. on Offshore Mechanics and Arctic Engineering, pp. 215-222.

Tayfun, M.A., 1980, ''Narrow-Band Nonlinear Sea Waves,'' J. Geophys. Research, Vol. 85, pp. 1548-1552.

Directional Wave Data - Measurements and Modeling
Coast of Florida Erosion and Storm Effects Study

Thomas D. Smith[1], Jon M. Hubertz[2]

Abstract

The U.S. Army Corps of Engineers, Jacksonville District, and the Florida Department of Environmental Protection are co-sponsors of a study to evaluate the coastal processes along the shores of the state. This evaluation includes measurements and modeling. Hindcast wave information is compared to measurements at two sites off the south east coast and found to accurately represent the wave climate during 1990. The wave information is used to estimate along shore sand transport and shoreline change. Net transport estimates agree with values generally accepted to characterize sand transport along this segment of coastline.

Introduction

The U.S. Army Corps of Engineers (USACE) and the State of Florida Department of Environmental Protection, Division of Beaches and Shores, are cooperating in the jointly funded Coast of Florida Erosion and Storm Effects Study (COFS). This study addresses both coastal and navigation projects with the goal of developing a comprehensive body of knowledge of coastal erosion and storm effects.

The state's shoreline is divided into five regions; Region I, the panhandle, II, the west coast, III-V, the southern, central, and northern east coast. This paper describes work done in the first region studied, Region III

[1]Coastal Engineer, U.S. Army Corps of Engineers, Jacksonville District

[2]Manager, Wave Information Study, Coastal Engineering Research Center, U.S. Army Waterways Experiment Station, Vicksburg, MS

(Dade to Palm Beach Counties).

Wave information near the coast is a critical component of any comprehensive coastal study such as COFS. Wave parameters such as height, period and direction, are some of the most important, but least known variables in studies of beach change and other coastal processes.

It would be difficult and expensive, to measure wave conditions along a coast as extensive as Florida's in order to supply wave information at all the locations and times it is needed in COFS. The accepted alternative is to provide this information with numerical models. Such models can be operated in a forecast or hindcast mode. The former uses predicted weather information to estimate future wave conditions while the latter uses known past weather data to estimate wave conditions which have occurred.

The purpose of this paper is to demonstrate that wave conditions near the coast can be hindcast accurately and used with shoreline change models to give representative estimates of beach change. Patterns of beach erosion and accretion in the past can then be studied using the 20 years (1956-1975) of hindcast wave information, (Hubertz, et. al, 1993), (Hubertz and Brooks, 1989) available along the coast of Florida.

Wave Measurements

Pressure gages and current meters were deployed by the Coastal Engineering Research Center (CERC) on the sea bed offshore of Lake Worth and Hallandale, Florida in February 1989. The depth at both sites was approximately 10 meters. The approximate locations are shown in Figure 1. The latitude and longitude of the instruments were; Lake Worth, 26.61 Deg North, 80.03 Deg West, and Hallandale, 25.97 Deg North, 80.10 Deg West. The instruments were recovered in March and February 1991 respectively.

The measurement systems consisted of a CERC internal recording P-U-V gage. The P-U-V gage is composed of a near bottom mounted pressure sensor and an electromagnetic current meter combined to give pressure (P) and components of the wave orbital velocities and ambient current (U and V). The gages operated every 4 hours for approximately 17 minutes collecting 1024 measurements at a rate of 1 sample per second (1 Hertz). Dates and time were referenced to Greenwich Mean Time (GMT).

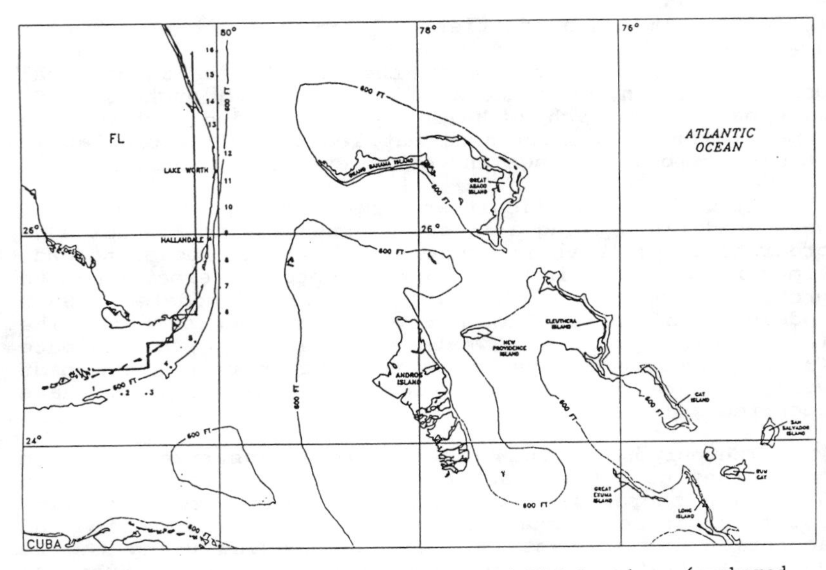

Figure 1 Location of Wave Gages and WIS Stations (numbered dots)

Wave Hindcast

The wave hindcast was conducted on three levels of resolution and regional coverage. All used a latitude, longitude grid. The level 1 grid has a spacing of one degree and covered the North Atlantic Ocean, the level two grid, a spacing of one quarter degree, and covered the U.S East coast continental Shelf, the level 3 grid a spacing of 0.5 nm covering the region between the gages and about 13 nm offshore. Deep water was assumed for all points in the level 1 grid. Depths (at Mean Low Water) from navigational charts were used at grid points in levels 2 and 3. The Bahamian Islands and shoals were represented at the resolution of level 2 (15 nm). The major configuration of the coastline of the U.S. was approximated with this 15 nm spacing so that sheltering by large features of the coast is represented by the grid.

The latest version of CERC's second generation discrete directional spectral wave model, WISWAVE 2.0, Hubertz (1992) was used for all levels. Thus, wave generation, dissipation, and the effects of depth and island (Bahamas) sheltering were included in the results saved along the coast. No other dissipative effects such as bottom friction were included. The effect of wind was included in all levels. Twenty frequency bands were used with corresponding mid band periods of 3 through 20 seconds and 22 and 24 seconds. Sixteen directional bands were used, each 22.5 degrees in width. Model results were output every 4 hours to coincide with measured data.

Windfields over the Atlantic Ocean during 1990 were obtained from the U.S. Navy's Fleet Numerical Oceanography Center (FNOC). These winds are derived from a combination of measured meteorological data and numerical weather models. Estimates of surface (19.5 meter elevation) wind speed and direction are available every six hours on a latitude longitude grid at 2.5 degree intervals. Estimates of wind speed and direction for the higher resolution grids of levels 2 and 3 were obtained by linear interpolation of the FNOC values of level 1.

Comparison of Measured and Hindcast Results

Gage data were analyzed to obtain time series of spectral wave height, peak period, and mean direction of wave energy in the peak frequency band. Similar quantities were calculated from the hindcast results. These values are compared qualitatively by plotting measured and hindcast values by month. Examples at Hallandale during January are shown in Figure 2.

No solid line is plotted in the figures when acceptable data are not available from the gages. When one or more gage data points are available within a data sparse time span, they are represented by small boxes. Hindcast results are available for the entire time period. Times are GMT and directions are coming from, in the compass sense. That is a value of 90 degrees represents waves coming from the east and going toward the west, or at both sites directly toward the shore.

Spectral wave heights are an integrated quantity. That is, they are based on the sum of energy under the discrete spectrum. Wave peak mean directions are similar, but to a lesser degree an integral quantity, in that they are the energy weighted mean of all wave directions in the peak frequency band. Wave peak periods, however are not an integrated or mean value. They are based only on the location along the frequency axis of the peak spectral energy. Thus, if the spectrum is double peaked, representing sea and swell for example, the peak frequency can shift from say 4 to 10 seconds depending on the relative magnitude of the two peaks. This tends to introduce large differences in period comparisons when, for

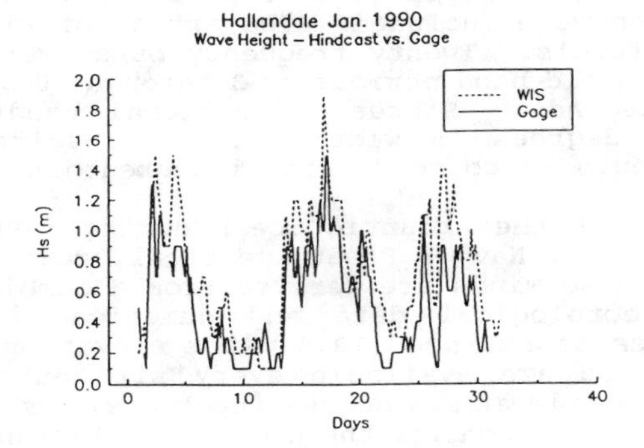

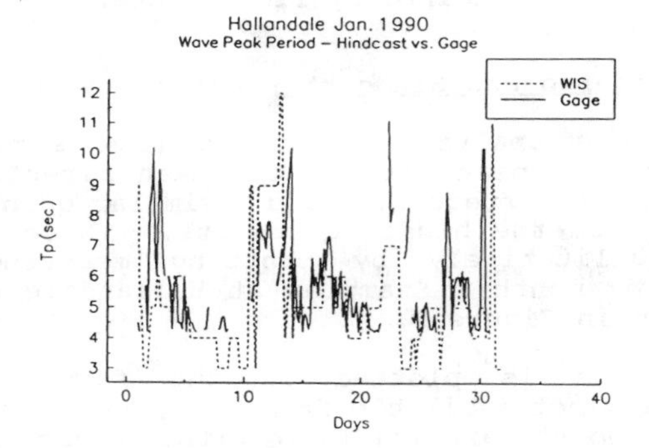

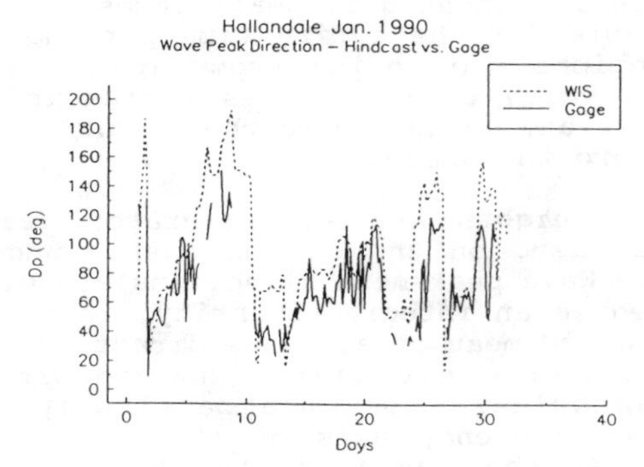

Figure 2 Comparison of Hindcast and Measured Wave Data

example, a sea peak is largest in a gage record and the swell peak is largest in a hindcast for the same time period or vice versa.

In the above example, both gage and hindcast results would be correct. There are two main components in the spectra and thus two peak periods. If two periods were calculated from the gage spectrum and two from the hindcast spectrum, the corresponding sea and swell periods from gage and hindcast would most likely be in agreement.

Peak periods in single peaked spectra can also vary in time due to the peak shifting from one frequency band to another. The amount of variation depends on the width of the frequency bands and is generally larger in the lower frequency bands. For example, for a fixed band width of 0.01 hertz, the period will vary from 14.3 to 12.5 sec for a shift from 0.07 to 0.08 hertz and from 7.1 to 6.7 for a shift from 0.14 to 0.15 hertz. This can lead to relatively large differences in peak periods for spectra whose peaks are within one or two bandwidths of each other at lower frequencies. The number and size of the frequency bandwidths can also affect the variability of peak period in time for gage or model results. Peak periods calculated from spectra with fewer frequency bands which are wider, more closely approximate a mean value than periods associated with one of many narrow bands. Spectra produced from the analysis of gage data usually have more frequency bands (typically 40) than those calculated in WIS hindcasts (typically 20). This may introduce differences on the order of seconds when comparing peak periods. Sometimes the gage spectra are band averaged, that is a number of adjacent bands are averaged together to produce one band. If the size and location in frequency space of the resulting gage spectrum is similar to the hindcast spectrum, peak periods will be comparable.

Mean periods, which are energy weighted means over all frequencies, can be compared, but for a double peaked spectrum, the mean period may be between the sea and swell periods, say 7 seconds in the example above, and thus not be representative of either sea or swell wave period. In a broad single peaked spectrum, the mean period will be one value representing the many periods present. The mean period will approach the peak period for single peaked narrow spectra. We have chosen to use peak period to characterize the distribution of energy density in frequency, since this is now the most commonly used variable from measurements and hindcast/forecast studies.

Problems similar to the variability in peak periods occur in picking one direction characteristic of an entire spectrum. The peak mean direction is the energy weighted mean of all directions at the peak frequency. The mean direction is the energy weighted mean of all directions in all frequency bands. Using the examples above for peak period, it is apparent that the mean direction of a double peaked spectrum may not be representative of either major component of the spectrum. The mean direction of a broad spectrum will be one value representing the many directions present. For coastal engineering applications in which wave direction is a critical parameter, use of mean direction may be an unacceptable parameter if the local wave climate typically has wave energy approaching simultaneously from different directions.

We have chosen to use the peak mean direction to characterize the direction in the spectrum. This value is more commonly used from measurements and hindcast/forecast studies than mean direction. Tests conducted at CERC to determine the most appropriate wave parameters to use in coastal processes modeling indicate that spectral height, peak period, and peak mean direction give the most meaningful results if one is constrained to use just three wave parameters. To preserve all the information in a directional wave spectrum, one should use the complete spectrum. However, this is not practical, at present, since most coastal engineering applications are limited to use of three wave parameters as input; height, period, and direction.

The gages deployed at Lake Worth and Hallandale are expected to have the following accuracies under good environmental conditions and proper functioning of the instruments;

Hm0 +/- 0.1 m,
Tp +/- 0.5 sec,
Dp +/- 5.0 deg.

The time series from the hindcast model at the level 3 stations were compared to available data in the measured time series. Measured values were available every four hours. The bias and root mean square difference (RMSD) were calculated by month for each site. These results are shown in Tables 1 and 2 for Lake Worth and Hallandale respectively.

Tables 1 and 2 indicate that hindcast values of wave height are high on the average with respect to the measurements on an annual basis by about 0.35 m for spectral wave height. Monthly average differences vary from -0.1 to -0.5 m with most around -0.2 to -0.4 m. The annual RMSD value for both sites is also 0.35 m. Monthly values range from 0.2 to 0.5 m. These statistics are considered acceptable and typical of present hindcast technology. Both the bias and RMSD values could be reduced by calibration of the model. Both input winds and model were applied without prior testing or calibration.

Hindcast values of peak period are higher on the average by about 0.65 sec. Monthly values of bias range from -2.6 to 1.0 sec The average RMSD value for both sites is 3.2 sec with monthly values ranging from 1.9 to 5.0 sec. Thus, there is essentially no bias in the peak period results and the RMSD values are reasonable.

Hindcast values of peak mean direction are biased high with respect to measurements by 14 deg. Monthly values of bias range from 0 to -25 deg. The average annual RMSD value for both sites is about 40 deg. Monthly values range from 26 to 68 deg. The relatively large differences between measured and hindcast values are probably, in part, due to the causes discussed above related to peak spectral parameters.

Overall, WIS results compare favorably with gage measurements. The differences in height are considered within acceptable limits for coastal engineering use, but could be improved with model calibration.

Table 1
Comparison of Hindcast Results to Gage Data at Lake Worth

Month	Hs (m)			Tp (sec)			Dp (deg)		
	Bias	RMSD	#	Bias	RMSD	#	Bias	RMSD	#
Jan	-0.1	0.2	109	0.3	2.5	109	-1	26	109
Feb	-0.3	0.4	160	-1.6	5.0	160	0	27	160
Mar	-0.3	0.4	166	-0.3	2.8	166	-1	32	166
Apr	-0.4	0.4	151	-1.5	3.6	151	-9	27	151
May	-0.3	0.3	120	-0.8	2.4	120	-23	36	120
Jun	-0.2	0.2	139	0.2	3.6	139	-25	40	139
Jul			0			0			0
Aug	-0.2	0.3	169	0.9	4.7	169	-19	62	169
Sep			0			0			0
Oct	-0.3	0.4	184	-1.2	3.9	184	-19	26	184
Nov	-0.2	0.3	164	-0.4	2.9	164	-24	42	117
Dec	-0.4	0.4	183	-2.6	4.5	183	-23	33	183
Year	-0.3	0.3	1535	-0.7	3.6	1535	-14	35	1498

Table 2
Comparison of Hindcast Results to Gage Data at Hallandale

Month	Hs (m)			Tp (sec)			Dp (deg)		
	Bias	RMSD	#	Bias	RMSD	#	Bias	RMSD	#
Jan	-0.2	0.3	150	0.4	1.9	150	-21	32	150
Feb	-0.4	0.5	132	-0.4	2.7	132	-17	55	132
Mar	-0.4	0.5	87	0.7	2.1	87	-6	37	87
Apr	-0.4	0.4	97	-2.2	3.8	97	-6	42	97
May	-0.3	0.4	136	-0.8	2.4	136	-14	42	136
Jun	-0.3	0.4	82	1.0	2.3	82	-2	46	65
Jul			0			0			0
Aug	-0.2	0.2	86	-1.3	3.1	86	-11	68	86
Sep			0			0			0
Oct	-0.4	0.5	163	-0.5	3.1	163	-23	47	125
Nov	-0.4	0.5	132	-0.5	2.7	132	-19	40	132
Dec	-0.5	0.5	133	-1.7	3.8	133	-16	26	119
Year	-0.4	0.4	1198	-0.6	2.8	1198	-14	44	1129

Bias = Gage - Model
RMSD = Root Mean Square Difference
\# = Number of gage values available (6/day possible)

Shoreline Change Modelling

The GENEralized Model for SImulation of Shoreline Changes (GENESIS), (Hanson and Kraus, 1989) is used to predict shoreline changes in Region III of the COFS. This model predicts the location of a depth contour normal to a reference baseline in the along shore direction as a function of time. Since it predicts the location of only one contour (the zero depth or shoreline), it is termed a one-line shoreline change model. The model operates on a length of shoreline sub-divided into a number of cells. The volume transport of sand into and out of each cell is calculated as a function of sources and sinks to the cell and along shore transport through the cell due to wave induced currents.

The information necessary to apply the model is knowledge of; the initial shoreline, a history of offshore wave climate, the cross shore elevation profile, specified volume sand transport at each end of the length of shoreline, and structures on the beach such as jetties and breakwaters.

Assumptions of the model are; that the cross shore profile shape is constant, the shoreward and seaward limits of the profile are constant, sand is transported alongshore by the action of breaking waves, the detailed structure of nearshore circulation can be ignored, and there exists a long-term trend in shoreline evolution. Net change in volume of sand at each alongshore cell results in a change in shoreline position for the cell. Volume change results from a difference in the longshore transport at the lateral sides of each cell. Sand sources or sinks through the shoreward and seaward faces of each cell also result in a change in volume. Longshore transport calculated by the model is a function of wave height and group celerity at breaking, breaking angle with respect to the local shoreline, and the gradient in wave height in the alongshore direction.

Simulation for One Length of Shoreline in Region III

The length of shoreline modeled is bounded by Port Everglades to the north and Baker's Haulover Inlet to the south. This length of shoreline was divided into 137 cells each 500 ft in length. Boundary conditions were imposed at the south jetty at Port Everglades and the north jetty at Bakers Haulover Inlet. The location of seawalls were determined through digitization of aerial photographs taken in October 1989.

Results include cell by cell shoreline positions at intermediate and final time steps, transport volumes, and breaking wave information for selected locations. The breaking wave information reported for each simulation includes average height and angle. Diffractive effects from structures are not included. The seaward and landward most shoreline positions are saved and used in quantifying the envelope of shoreline change predicted for the simulation period. If a history of shoreline positions is available, the model can be calibrated and accuracy quantified.

Predicted Sand Transport Volumes

Transport rates and calculated shoreline positions along the region (as predicted from the measured and hindcast wave data) were compared to determine the potential implications of using the WIS data to predict shoreline change in Region III of the COFS. Simulations were carried out for the 205 days during 1990 in which measured data were available for the Hallandale site. Left and right (north and south) directed transport for each cell, and gross and net values are shown in Figure 3. Trends in the results for left and right transport agree well. Table 3 summarizes the magnitude of various transport estimates.

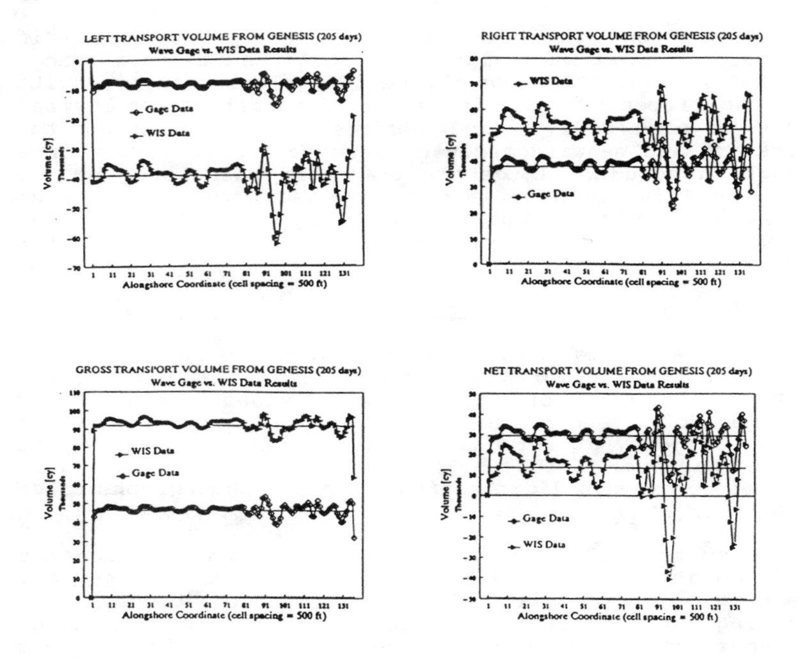

Figure 3 Sand Transport Estimates from GENESIS using WIS and Measured Data

Table 3
Summary of Transport Estimates for Region (CY/205 Days)

	WIS Results	Measured Data
Toward North (Left)	39,200	8,600
Toward South (Right)	52,600	37,900
Gross	91,800	46,500
Net	13,400	29,300

The differences in magnitude may be explained by the positive bias (0.35 m) of WIS wave heights with respect to measurements. Lateral transport through each cell is proportional to breaking wave height to the 5/2 power. Also a threshold is imposed on transport which is proportional to breaking wave height. If the wave model were calibrated to remove this bias, transport values would be lower for these two reasons. The present differences in gross transport will be considered when sediment budgets along the shoreline and at the inlets are developed for

Region III.

Unlike processes which are dominated by gross transport, shoreline change is dependent primarily upon net transport volumes at each cell. Net values from the 205 day simulation were extrapolated to obtain annual estimates of net transport of 52,000 and 24,000 cubic yards from measured and WIS data respectively. These values compare well to previous estimates of 50,000 cubic yards per year as determined in the 1963 U.S. Army Corps of Engineers survey report for Broward County and 35,000 cubic yards from an unpublished report by Mehta.

Predicted Shoreline Change

The initial shoreline location with respect to a baseline, the location of breakwaters along shore with respect to the baseline, and the lengths of the two boundary condition jetties are shown in Figure 4 (top). Also shown in this figure are the predicted shoreline change at each cell using measured data (middle) and WIS results (bottom). Both sets of data indicate the same pattern of shoreline advance and retreat. Differences at any cell range from 0 to about 20 ft, with the average about 4-5 ft. Both indicate a greater potential for critical shoreline recession from cells 83 through 137. Identification of areas of critical recession is the key to the formulation of shore protection alternatives. The results of the present analysis indicate that the use of WIS data in conjunction with GENESIS will facilitate COFS modeling efforts.

Summary

The U.S. Army Corps of Engineers, Jacksonville District, and the Florida Department of Environmental Protection are co-sponsoring studies to evaluate coastal processes in the state. Measured and hindcast wave data were compared at two sites (Lake Worth and Hallandale) in Region III of the Coast of Florida Erosion and Storm Effects Study. It was determined from the analysis that measured wave heights at Lake Worth were approximately 20% higher than at Hallandale. Correlation between measured wave height at Lake Worth and Hallandale was good (r^2=0.85) indicating that the two sites experienced the same wave events. Long period waves occurred more frequently at Lake Worth than at Hallandale. The hindcast wave heights correlated well with measurements (r^2=0.8), but were biased high on average by 0.3-0.4 m, with an average root mean square difference of 0.3-0.4 m.

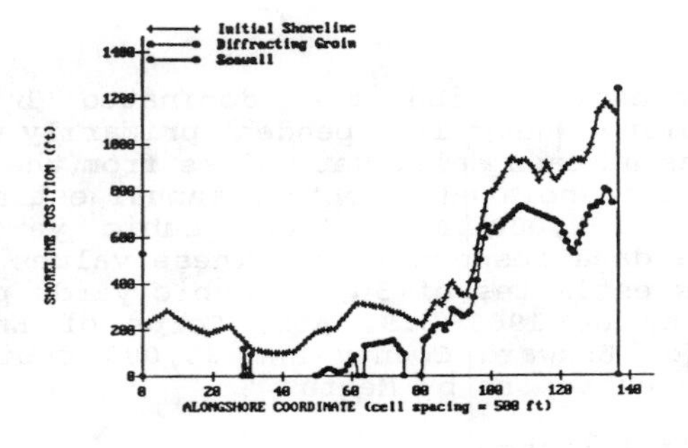

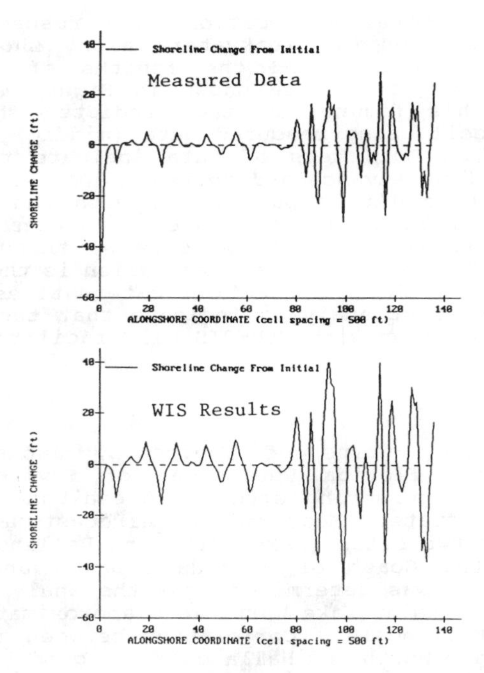

Figure 4 Shoreline Change Results from GENESIS using WIS and Measured Data

The two wave information data bases were used as input for the computer model GENESIS to estimate longshore sediment transport rates and shoreline change. Both sets of wave data (WIS and measured) were used with the GENESIS shoreline change model for the 205 days of available measured data. Results indicated that net longshore transport was directed to the south in both simulations with annual estimated values that agree with accepted estimates for the reach. In conclusion, the revised WIS hindcast can be used with confidence for shoreline change modeling in Region III of the Coast of Florida Erosion and Storm Effects Study.

Acknowledgement

The authors wish to acknowledge the assistance of Ms. Brandon in operating the wave model, and Mr. Gravens for assistance in operating the GENESIS model. The authors acknowledge the Office, Chief of Engineers, U.S. Army Corps of Engineers for permission to publish this paper.

References

Hanson, Hans and Kraus, N. C. 1989 (Dec). "GENESIS: Generalized Model for Simulation Shoreline Change, Report 1" Technical Reference CERC-89-19, US Army Engineer Waterways Experiment Station, Vicksburg, MS.

Hubertz, J. M., Brooks, R. M., Brandon, W. A., and Tracy, B. A. 1993 (Mar). "Hindcast Wave Information for the Atlantic Coast, 1956-1975," WIS Report 30 US Army Engineer Waterways Experiment Station, Vicksburg, MS.

Hubertz, J. M. 1992 (June). "A Users Guide to the WIS Wave Model, Version 2.0," WIS Report 27, US Army Engineer Waterways Experiment Station, Vicksburg, MS.

Hubertz, J. M., Brooks, R. M. 1989 (May). "Gulf of Mexico Hindcast Wave Information," WIS Report 18, US Army Engineer Waterways Experiment Station, Vicksburg, MS.

MEASURED TOTAL WAVE CLIMATE OF A SMALL CRAFT HARBOR

Chuck Mesa[1]

Abstract

Measurements of waves in a small craft harbor were analyzed for the purpose of determining the existing wave climate energy levels causing the excessive agitation of moored small craft and degradation of the harbor infrastructure. Wave measurement instruments were installed at three locations within the harbor and one nearshore location. Frequency domain analysis of the measured pressure time series resulted in one-dimensional energy spectra spanning the range of gravity and infragravity waves. Joint distributions of significant wave height (H_{mo}) and spectral peak frequency, histograms, and cumulative distributions are presented and discussed.

Introduction

It is desirable that current/future design and modification of harbors be based on a more precise understanding of a harbor's wave climate and an analysis of the response of ships, boats, and harbor infrastructure. The design and operation of small craft harbors is critically dependent on an accurate determination of the wave climate both outside and within the harbor. There are few dedicated field wave measurement studies upon which to support current design practices, whereas physical and numerical models are tools often used to attempt to construct the wave energy climate within a small craft harbor. Analytical models can be applied to predict the response of moored

---

[1]Coastal Engineer, U.S. Army Corps of Engineers, Los Angeles District, P.O. Box 2711, Los Angeles, CA 90053.

small craft once the incident harbor wave climate is known. Historically, the design of small craft harbors was based primarily by what worked best in the past; most harbors were not designed/constructed based upon a quantitatively measured climate. Largely that methodology still remains true today. Measured wave climate of existing harbors can be used to develop, refine, and calibrate the physical, numerical, and analytical tools currently in use.

There are currently no generally accepted design criteria for wave climate inside a small craft harbor. Early small craft harbor design guidance [American Society of Civil Engineers, 1969] usually mentioned a "rule-of-thumb" which stated "acceptable" wave heights were on the order of 1 foot.

A study to quantify small craft harbor motion resulted in a provisional quantitative set of wave action criteria [Northwest Hydraulic Consultants Ltd, 1980]. This study included field and hydraulic model measurements of boat motion, mathematical analysis of boat motion, and interviews with small craft operators. Various levels of acceptable wave action were developed which incorporates the frequency of occurrence, wave direction, and wave peak period. To a large extent, the 1 foot rule-of-thumb was maintained.

Shallow draft vessels and harbors have characteristics significantly different than deep draft vessels and harbors; the forces dominating the response of the two classes of vessels/harbors are distinctive. Small craft have natural periods of heave, pitch, and roll which are respondent to waves of very short periods, usually on the order of 2-8 seconds. Small craft harbor infrastructure also is typically responsive to short period wind based waves.

Infragravity waves, or long waves, are generally not critical to small craft harbors unless resonance is excited, or excessive amplitudes and/or horizontal velocities are established. For harbors exposed to the open ocean, the primary driving force of infragravity waves tends to be associated with the grouping of short period wind waves. The most common form of damages is from the larger wave amplitudes or high wave velocities causing excessive stress on boat mooring lines, and the bumping of boat hulls/fenders against piers.

<u>Prototype</u>

Oceanside Harbor is a shallow draft harbor of refuge along the California coast located 80 miles south of Los Angeles and 30 miles north of San Diego (Fig 1). The harbor is used primarily for recreation (approximately 95% recreational, 5% commercial). The harbor protective structures are constructed seaward of the shoreline with the boat basins built inland through the shoreline. The harbor is protected by the 4350 ft long north breakwater and the 1330 ft long south jetty. The small craft harbor encompasses approximately 70 acres of water surface and has depths ranging from 10-20 feet.

The local shoreline orientation is to the southwest. The winter wave climate is dominated by high energy swell and seas from North Pacific extratropical storms arriving from the west and northwest. The summer is characterized by very long period swell from tropical cyclones generated in the equatorial eastern Pacific and long travelling waves from the southern hemisphere. A predominant afternoon onshore breeze generates local seas of very short periods. It is common for the daily nearshore wave climate to be characterized by dual spectral peaks associated with swell and sea.

The entrance channel configuration allows the small craft harbor exposure to direct penetration of waves from the south and southwest. Storm waves have been documented to cause extensive damage to vessels, pilings, floating dock systems, and revetments. Excessive daily wave energy levels tend to fatigue and eventually damage dock structural members, piles, and dock/pile connections. The harbor has experienced excessive maintenance costs above those considered normal in the form of more frequent replacement of piles and dock systems.

The harbor is known to respond to low frequency wave energy. Local observations have long recorded a "surge" problem in various locations throughout the harbor, occurring for a wide range of offshore wave climate conditions. Vertical water motion is generally most severe near the boat basin perimeters; excessive horizontal motion has been noted by the movement of a docked U.S. Coast Guard cutter near the harbor entrance. The reduced harbor infrastructure life may be fatigue related due to cyclical loading from excess torsion and bending between the piling/floating pier system components.

A study to define low frequency energy in Oceanside Harbor resulted in identification of wave frequencies of possible resonance excitation [Mesa, 1992]. This study utilized comparisons between a 3-

dimensional physical model and a numerical harbor oscillation model, and indicated a good correlation between physical model measurements and numerical model predictions.

Field Data Collection and Analysis

The field data collection program was conducted as part of the Oceanside Harbor Storm Damage Reduction and Navigation Improvement project conducted by the U.S. Army Corps of Engineers, Los Angeles District [U.S. Army Corps of Engineers, 1992]. This program focused on the measurement of waves with periods greater than 5 seconds, i.e. those most normally associated with wind based waves, and infragravity waves. This program did not attempt to measure wave periods less than 5 seconds, such as boat wakes, which affect boat agitation. The field data collection program was installed as an element of the Coastal Data Information Program (CDIP) [U.S. Army Corps of Engineers and State of California]. CDIP is a program of oceanographic recording instruments along the coasts of California, Hawaii, Oregon, and Washington cooperatively sponsored by the U.S. Army Corps of Engineers and the State of California, Department of Boating and Waterways. CDIP is operated by the University of California San Diego, Scripps Institution of Oceanography.

The field data collection components (Fig 1) included 1 nearshore S_{xy} slope array and 3 inner harbor single point pressure transducers. The nearshore S_{xy} slope array (CDIP Station 4, hereinafter referred to as the "nearshore") is located approximately 1 mile south of the harbor entrance in 11 m of water and was in place under the currently operating CDIP program. The 3 inner harbor pressure transducers (CDIP Station 68 : Entrance Channel; CDIP Station 69 : South Basin; CDIP Station 70 : North Basin) were placed in 5.9 m, 3.6 m, and 3.3 m of water respectively. The entrance channel location was chosen to best represent the incident conditions to the small craft harbor basins. The basin locations were chosen where wave energy has historically been problematic and potential low frequency wave energy would have the greatest effect.

The field data used in the present study was collected over a 24-month period from August 1991 - July 1993 (with some omissions). Each station utilized a sampling frequency of 1.0 Hz and 8192 samples were collected per record. The recording sensors were cabled to shore and the measured data was polled via telephone at a nominal interval of 3 hours. Spectral parameters were calculated by Fast Fourier Transform techniques, utilizing record lengths of 4096 data points, and a spectral resolution bandwidth of 0.002 Hz. Peak frequency is determined as the frequency associated with the maximum spectral energy density. All recording sensors were synchronized in time resulting in synoptic measurements of the harbor and nearshore oceanographic conditions.

Results

Nearshore

The nearshore S_{xy} slope array recorded 4896 observations during the measurement period. The results of the nearshore slope array are shown in Fig 2. The joint distribution of wave height and peak frequency is shown in Fig 2A. The majority of the measured waves are centered in a relatively narrow range in both the H_{mo} and frequency axes. The higher energy observations, although significantly less in number, tend to be distributed throughout the higher frequency range (> 0.1 Hz) which suggests steeper waves associated with combined swell/sea conditions.

The sample histogram and cumulative distribution of wave heights is shown in Fig 2B-C. The data included a maximum measured wave of 2.61 m; the mean and standard deviation are given as 0.86 m and 0.30 m respectively. The histogram is peaked around the modal value of 0.75 m. The histogram shape is characteristic of nearshore waves. The sample cumulative distribution is plotted on log-normal probability paper and a best fit distribution is shown for illustration.

The histogram of spectral peak periods is shown in Fig 2D. The distribution is relatively peaked around the modal value of 0.067 Hz ($T = 15$ sec), with a very narrow spike of observations at 0.08 Hz. This representation of the data indicates that swell conditions ($f < 0.1$ Hz; $T > 10$ sec) dominate the wave energy climate at this nearshore location. A dual spectral peak wave climate due to swell and sea is known to exist at this location. It was assumed a priori that sea conditions would be of significant severity to dominate some of the measured spectra; a second peak in the histogram was expected. Analysis of individual spectra exhibits the presence of the dual spectral peak condition. A cursory check of individual spectra measured during afternoon times verifies local sea generation.

Entrance

The entrance channel single point pressure sensor recorded 5273 observations during the measurement period. The results of the entrance channel gage are shown in Fig 3. The joint distribution of wave height and peak frequency is shown in Fig 3A. The separation of the distribution into high and low frequency components is immediately apparent. The distribution is triple peaked along the frequency axis. The majority of the observations are centered in a relatively narrow range in the H_{mo} axis. This is expected as a significant portion of the incident wave energy is dissipated by shoaling/breaking and the protection afforded by the entrance breakwater/jetty system.

The sample histogram and cumulative distribution of wave heights is shown in Fig 3B-C. The data included a maximum measured wave of 0.65 m; the mean and standard deviation are given as 0.18 m and 0.08 m respectively. The histogram is skewed around the modal value of 0.125 m. The cumulative distribution is plotted on log-normal probability paper and a best fit distribution is shown for illustration. The sample distribution appears to fit the log-normal probability law well up to approximately the 98% value.

The histogram of spectral peak periods is shown in Fig 3D. The distribution is clearly triple peaked. The two peaks in the wind wave frequency range are noted at 0.059 Hz and 0.081 Hz (T = 17 sec, T = 12.4 sec). It is worth noting that these bracket the modal value of the nearshore location. The most commonly occurring period is observed at frequency 0.009 Hz (T = 111 sec). This very long period is in the range of what is commonly recognized as wave groups or surf beat. This long period was masked by the higher energy, higher frequency waves measured at the nearshore location. As the higher frequency wave energy is dissipated by shoaling/breaking and the entrance protective structures, the lower frequency wave energy becomes significant. Also, long waves, being of second order nature and bound to the wave groups, may become free or unbound at this location manifesting itself as a separate peak in the frequency axis.

South Basin

The south basin single point pressure sensor recorded 4342 observations during the measurement period. The results of the south basin gage are shown in Fig 4. The joint distribution of wave height and peak frequency is shown in Fig 4A. Clearly the majority of measured waves are clustered in a very narrow range in the H_{mo} and frequency axes. The absence of short period waves is notable, since short period waves were measured at the harbor entrance and their presence in the boat basins have been historically noted.

The sample histogram and cumulative distribution of wave heights is shown in Fig 4B-C. The data included a maximum measured wave of 0.29 m; the mean and standard deviation are given as 0.09 m and 0.03 m respectively. The histogram is sharply peaked at the modal value of 0.075 m. Wave heights are expected to be small due to the protected location of the sensor relative to the small craft harbor entrance. Yet, the measured maximum H_{mo} wave height of 0.29 m equals the 1 foot design guidance value most often mentioned. The cumulative distribution is plotted on log-normal probability paper and a best fit distribution shown for illustration. The sample distribution appears to fit the log-normal probability law well up to approximately the 98% value.

The histogram of spectral peak periods is shown in Fig 4D. The distribution is clustered around frequencies in the range of 0.007-0.013 Hz (T = 142-77 sec). The most commonly occurring frequency is 0.007 Hz (T = 142 sec). The long wave dominance is surprising as although the harbor is known to respond to low frequency wave energy, it was not expected to completely control the energy spectra. There exists a very minor distribution of wave heights at frequency 0.049 Hz (T = 20.4 sec). These represent only approximately 7% of the measured observations, but are in the wind wave frequency range.

North Basin

The north basin single point pressure sensor recorded 4915 observations during the measurement period. The results of the north basin gage are shown in Fig 5. The joint distribution of wave height and peak frequency is shown in Fig 5A. It is clear that the majority of the measured waves are clustered in a very narrow range in the H_{mo} and frequency axes; the absence of short period waves is again notable. This basin exhibits a wave climate nearly identical to the south basin.

The sample histogram and cumulative distribution of wave heights is shown in Fig 5B-C. The data included a maximum measured wave of 0.35 m; the mean and standard deviation are given as 0.08 m and 0.03 m respectively. The histogram is sharply peaked at the modal value of 0.075 m, virtually identical to the south basin. The cumulative distribution is plotted on log-normal probability paper and a best fit distribution shown for illustration. The sample distribution appears to fit the log-normal probability law well up to approximately the 98% value.

The histogram of spectral peak periods is shown in Fig 5D. The distribution is clustered around a very narrow band of frequencies in the range of 0.005-0.007 Hz (T = 200-142 sec). The most commonly occurring frequency is 0.005 Hz (T = 200 sec). There exists a very minor occurrence of wave frequencies centered around 0.015 Hz (T = 67 sec) and 0.049 Hz (T = 20.4 sec). These represent approximately 4% and 3% respectively of the total measured observations. The 0.049 Hz measurements match perfectly with those measured in the south basin.

Conclusions

This study conducted synoptic measurements of gravity and infragravity waves inside a small craft harbor and of the nearshore region to quantitatively determine the wave climate causing excessive agitation of moored small craft and degradation of harbor infrastructure. Based on this study:

1. There was no appreciable amount of wind wave energy recorded in the small boat basins during the measurement period. Wind wave energy was dissipated to the extent that low frequency energy tended to control the measured spectra. A synoptic view of measured spectra is shown in Fig 6.

2. The measured wave climates between the separate boat basins were nearly identical in nature.

3. A cursory analysis of the cumulative distributions of wave heights within the harbor suggests the log-normal probability law may be appropriate.

It is hoped that the results of this study will form a quantitative basis for use in the development, refinement, and calibration of the physical, numerical, and analytical tools used in small craft harbor design.

Acknowledgements

The author wishes to acknowledge the District Engineer, U.S. Army Corps of Engineers Los Angeles District, for authorizing publication of this paper. It was prepared as a result of the Oceanside Harbor Storm Damage Reduction and Navigation Improvement study. I would like to thank the following individuals for their assistance and participation in this project: Ms. Julie Thomas and Mr. David Castell at the Scripps Institution of Oceanography, and Mr. Pat McKinney at the Coastal Engineering Research Center. A special 42 thanks is expressed to Ms. Wendy Thompson of the Coastal Engineering Research Center.

References

American Society of Civil Engineers. 1969. Report on Small Craft Harbors. ASCE-Manuals and Reports on Engineering Practice-No. 50.

Mesa, Chuck. March 1992. "A Dual Approach to Low Frequency Energy Definition in a Small Craft Harbor". Proceedings of Coastal Engineering Practice '92, American Society of Civil Engineers, pp 400-411.

Northwest Hydraulic Consultants Ltd. March 1980. Study to Determine Acceptable Wave Climate in Small Craft Harbours. prepared for Canadian Department of Fisheries and Oceans, Small Craft Harbours Branch. Canadian Manuscript Report of Fisheries and Aquatic Sciences No 1581.

U.S. Army Corps of Engineers, Los Angeles District. August 1992. Oceanside Harbor, San Diego County, CA. Storm Damage Reduction and Navigation Improvements Design Memorandum.

U.S. Army Corps of Engineers and State of California. Coastal Data Information Program. Monthly Reports.

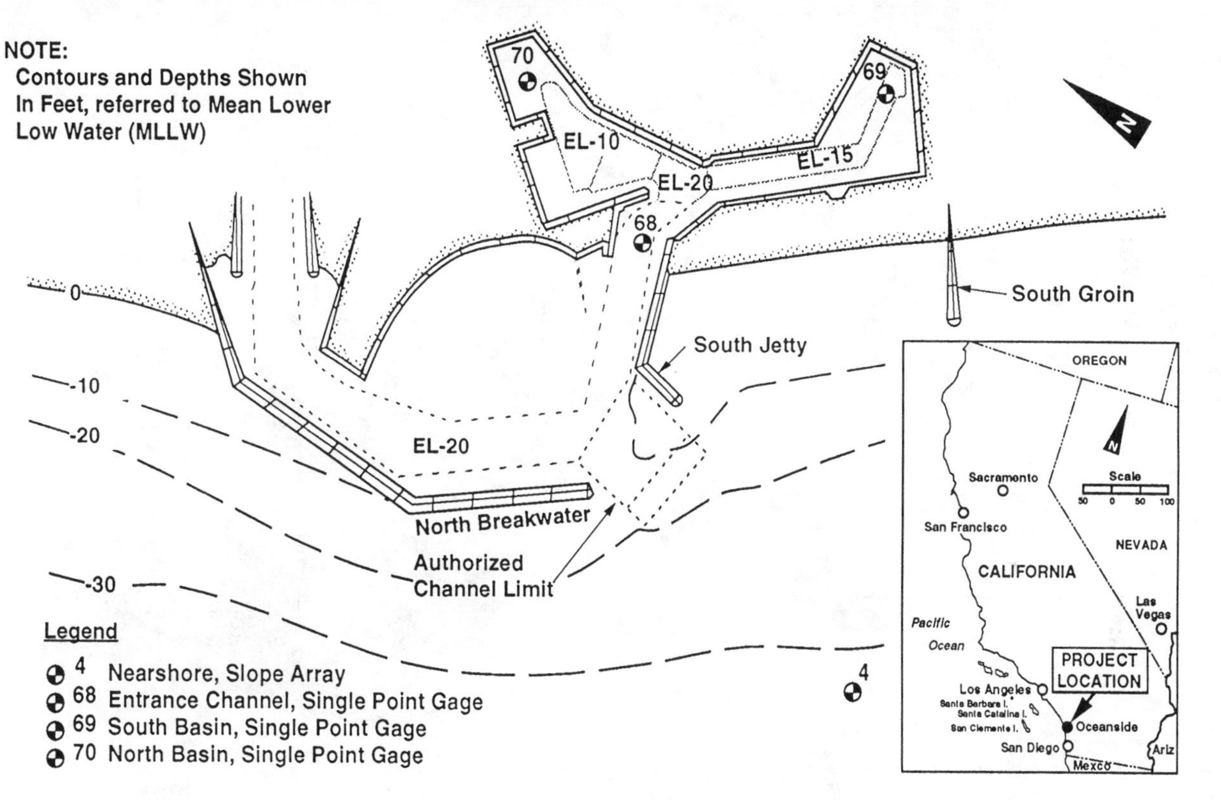

Figure 1 Location and Harbor Gage Layout

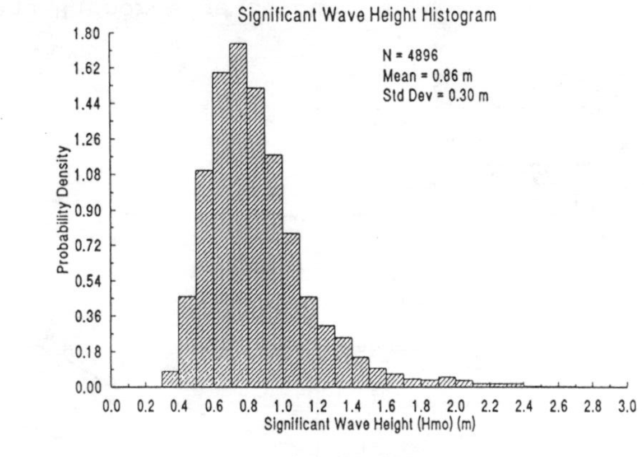

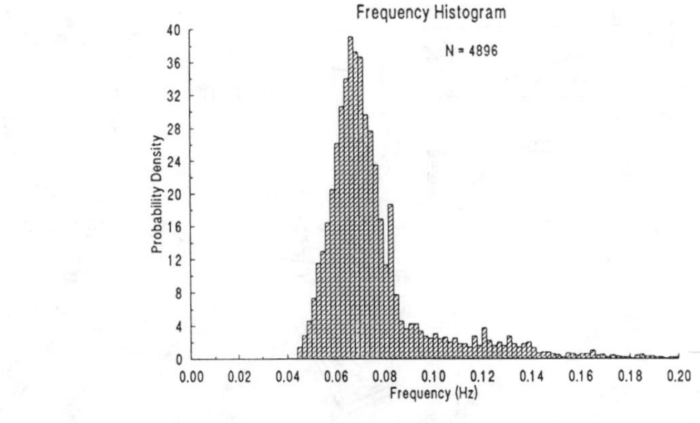

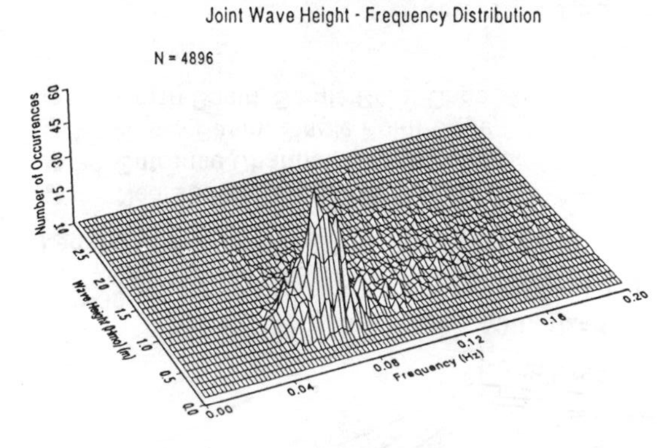

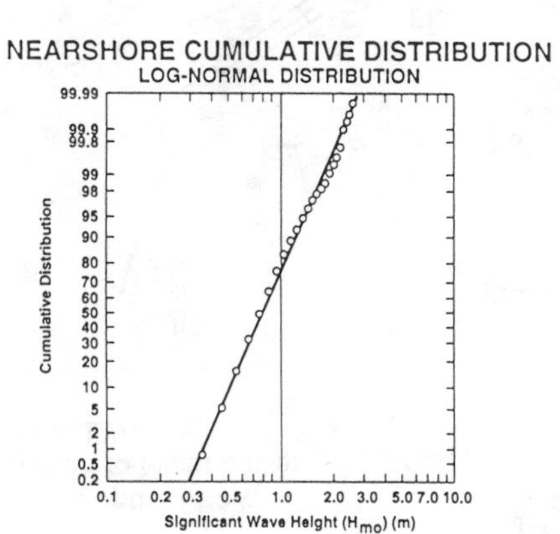

Figure 2 Summary of Nearshore Wave Climate

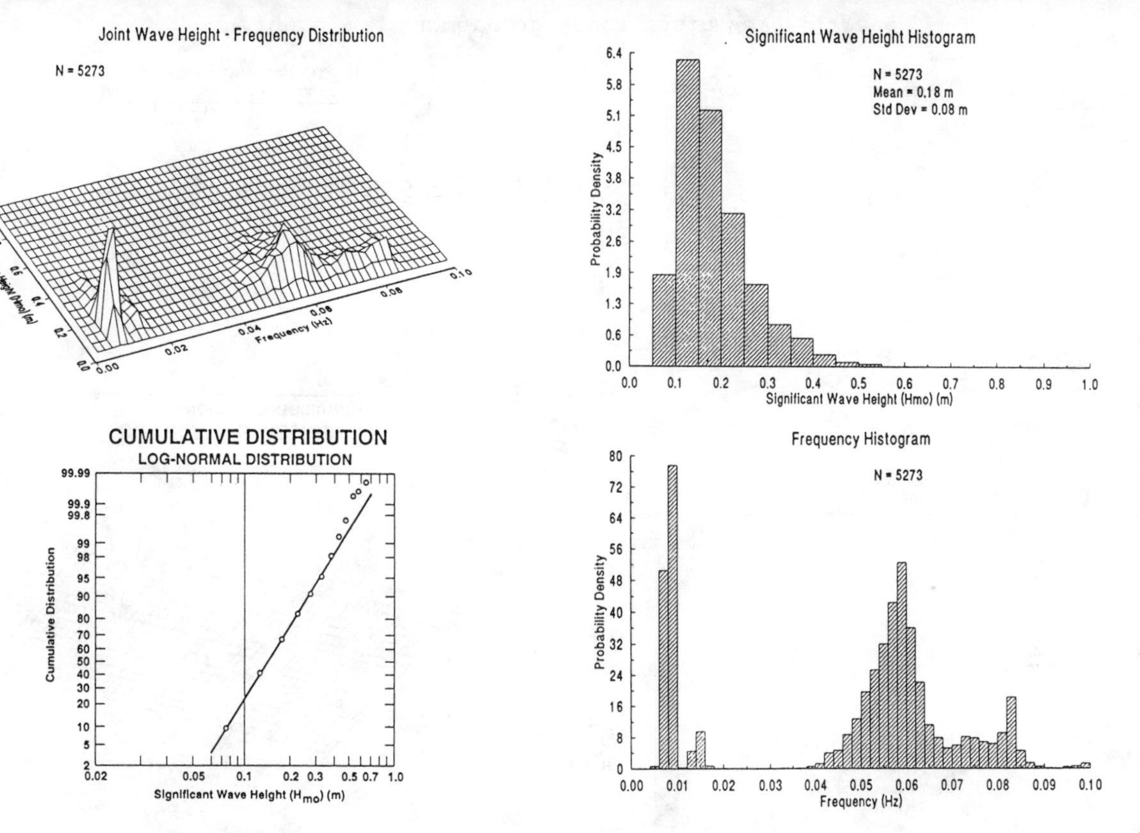

Figure 3 Summary of Entrance Channel Wave Climate

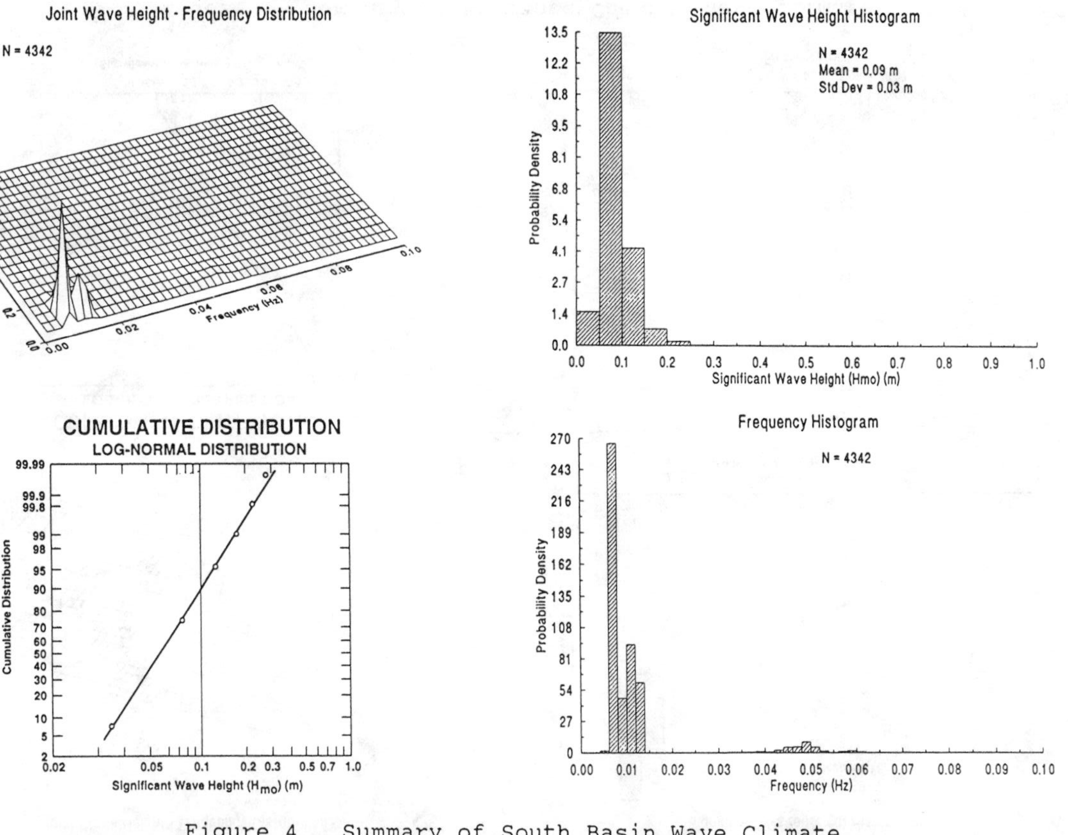

Figure 4 Summary of South Basin Wave Climate

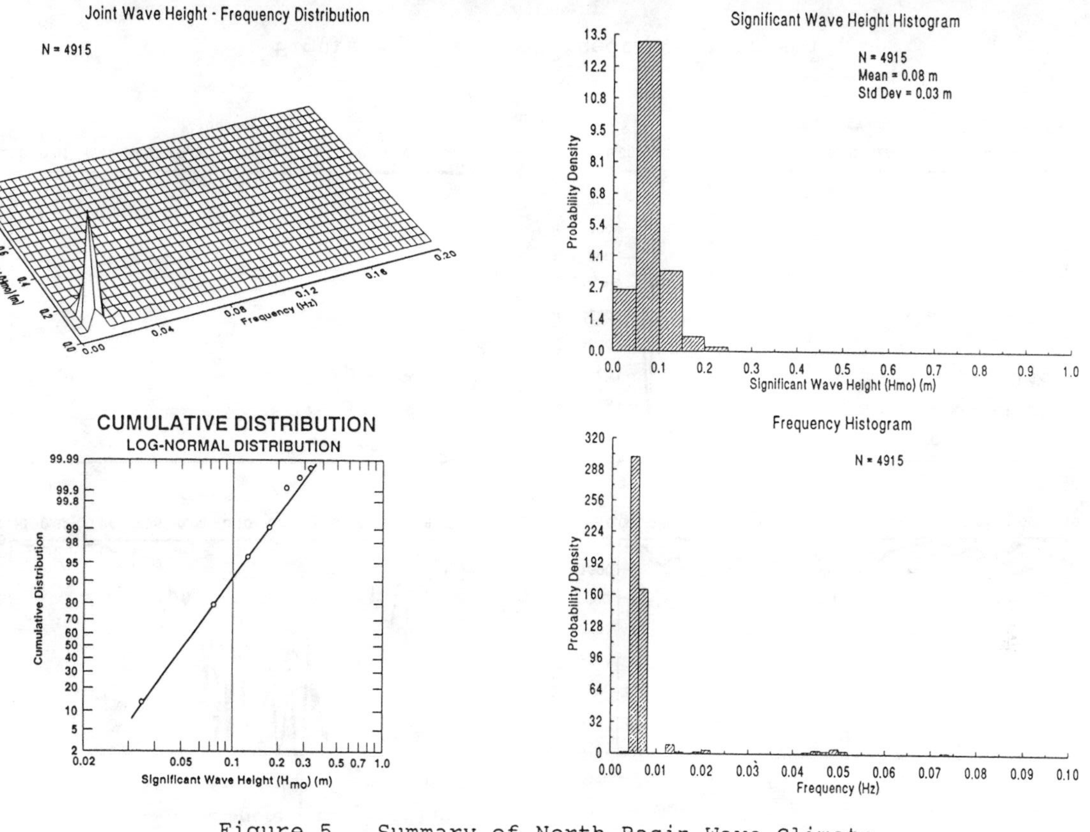

Figure 5 Summary of North Basin Wave Climate

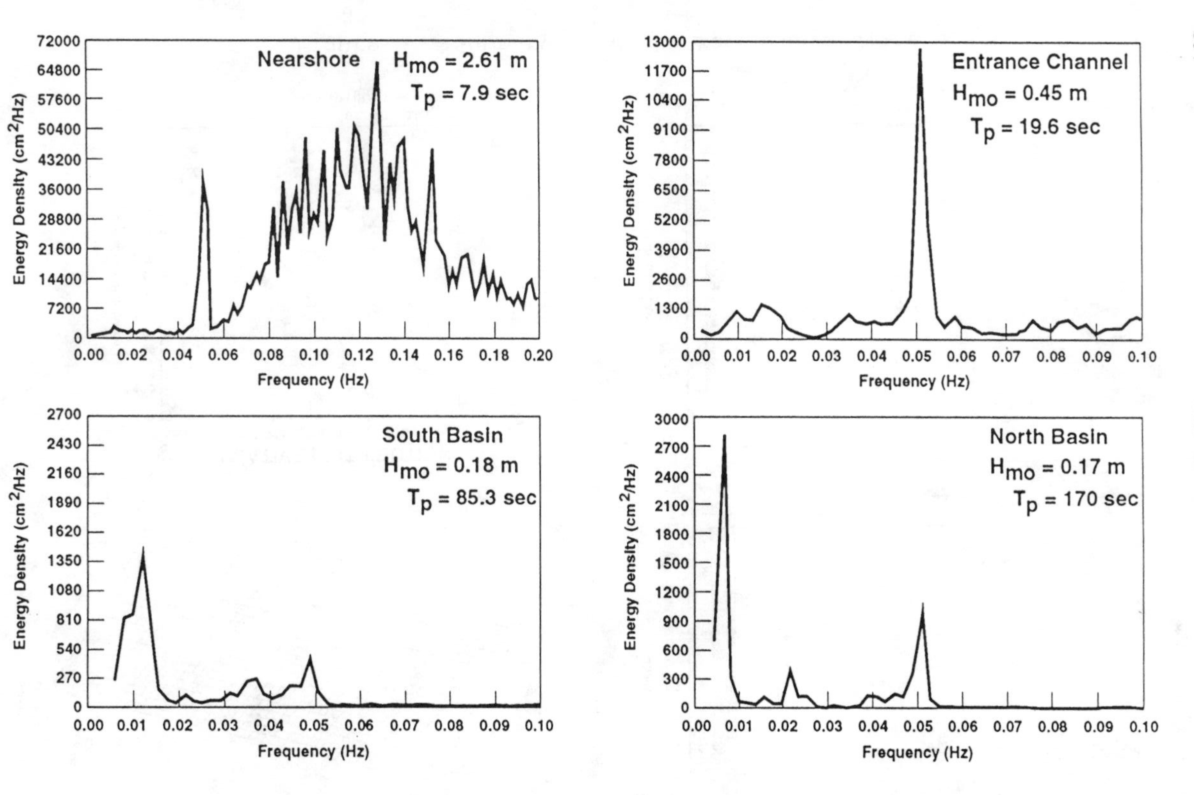

Figure 6 Measured Spectra Comparison

New Technology in Coastal Wave Monitoring

Richard Seymour [1], Member ASCE, David Castel[1],
David McGehee [2], Member ASCE, Juliana Thomas[1], and
William O'Reilly[1]

Abstract

The Coastal Data Information Program (CDIP) is an extensive network for monitoring waves along the Pacific coastlines of the US. The system has evolved substantially since its inception in 1975. The technological innovations in instruments, system control and management, quality control, software, computer hardware, field equipment and installation techniques, data archiving and analysis are described. Some further developments, underway or planned, are also defined.

Introduction

What is now called the Coastal Data Information Program had its beginnings in the mid 1970's with a modest, one node, field wave data collection station. The station was installed at Imperial Beach in San Diego county. This technology development was partially in response to a plea by a keynote speaker in the Waves 1974 conference. In his address, Prof. Robert Wiegel stressed the need for acquiring a nearshore wave climatology for the United States to allow coastal practitioners to make more rational design decisions. Data from the Imperial Beach station, transmitted over normal dial up phone lines, was recorded at a Scripps Institution of Oceanography central facility in La Jolla, CA., on a Nova mini computer. Recorded data were archived for later analysis. The project was supported by the State of California Department of Boating and Waterways and the California Sea Grant program.

Shortly after its inception, the program expanded and added the US Army Corps of Engineers (USACE) as a co-sponsor. As the steward of the nation's coastal infrastructure, the Corps requires reliable, long-term wave measurements for use in planning, designing, and operating coastal projects. Design wave conditions, usually expressed in terms of return intervals, are obtained through extremal analysis of wave histories. The confidence in these projections drops as the desired return interval

[1] Scripps Institution of Oceanography, UCSD, La Jolla, CA 92093.
[2] Coastal Engineering Research Center, USACE, Vicksburg, MS 39180

exceeds about twice the length of the historical record. However, there is seldom time between the inception of a project and the point when design details are finalized to collect sufficiently long wave histories. Wave hindcasts, which are made possible by long-term meteorological observations, are one approach to this requirement. Wave measurements are critical to hindcast validation. Another application for wave data is accurate quantification of conditions during specific events that result in structural damage or operational delays. Finally, laboratory and analytical research into the physics of wave generation, propagation, and transformation requires measurements for calibration and verification. The USACE Field Wave Gaging Program (FWGP) was established with the goal of collecting wave data at sufficient spatial and temporal density to meet and satisfy these requirements for the entire US coastline.

With time, the data collection system grew and evolved into what has become known as the Coastal Data Information Program (CDIP). In its present form, the CDIP has collected data from many field stations, on both the east and west coasts, the Great Lakes and the islands of Hawaii. In 1991, recognizing the value of wave measurements to a broad range of federal, State and local interests, the US Army Corps of Engineers and the California Department of Boating and Waterways entered into a watershed Cooperative Agreement (CA) between the two agencies.

The CA formally recognized that the impact of ocean waves, and the utility of wave information, transcends agency jurisdictions and missions. The data needs of the various State and federal partners overlap, but each agency brings its own set of requirements for the data products. Through the CA, CDIP has evolved technically and procedurally, not only to improve the quality and reduce the costs of wave measurements, but to meet a growing demand for enhancements in capabilities.

Many data user needs can be satisfied by the original thrust of the CDIP - collecting, analyzing, and compiling a database of analyzed wind and wind-wave parameters and statistical permutations of the accumulated histories. Some researchers require special sampling schemes, for example, extended time series to observe long wave phenomena. Harbor oscillation studies can benefit from multiple data sets synchronized in time. Alternately, a separate class of users make operational decisions based on current conditions, and thus require near realtime data access. The early history of the system and a more detailed description of the technology initially utilized is contained in Seymour and Sessions (1976), Seymour (1979), and Seymour et al (1985).

While the original emphasis of the CDIP was wave, current and wind measurements, the system of instrument platforms, telemetry links, and computers can be adapted to obtain other types of data. The benefits of establishing a climatic perspective of the baseline and extremal conditions apply to any coastal environmental parameter. With the collection and processing infrastructure in place, the cost of CDIP monitoring of other physical and chemical processes, such as currents, turbidity, temperature, salinity, water quality, to name a few, is dominated by the constraints of the individual sensor technology.

Historically, this cooperative agreement was based upon a very successful mechanism management of the CDIP in the Scripps Institution of Oceanography, UCSD. As an example, in 1992, the CDIP, using advanced and state of the art technology through a memory-distributed, multi-node network, recorded in excess of one million words of data every day. These environmental data were collected, in realtime, from rugged, high energy nearshore ocean locations, as well as sheltered inland waterways and harbors. Throughout its 18 year history, CDIP reported data from close to 80 separate locations. Most of these are shown in Figure 1.

Parameters measured include wave height, wave period, wave direction, current velocity and wind velocity. In cooperation with the National Weather Service Marine Advisory Program, CDIP provides essential and timely coastal weather information to mariners, fishermen and recreational boaters. Through published and widely distributed monthly and annual reports, the program furnishes coastal engineers and others in the oceanographic community with long-term wind and wave statistics.

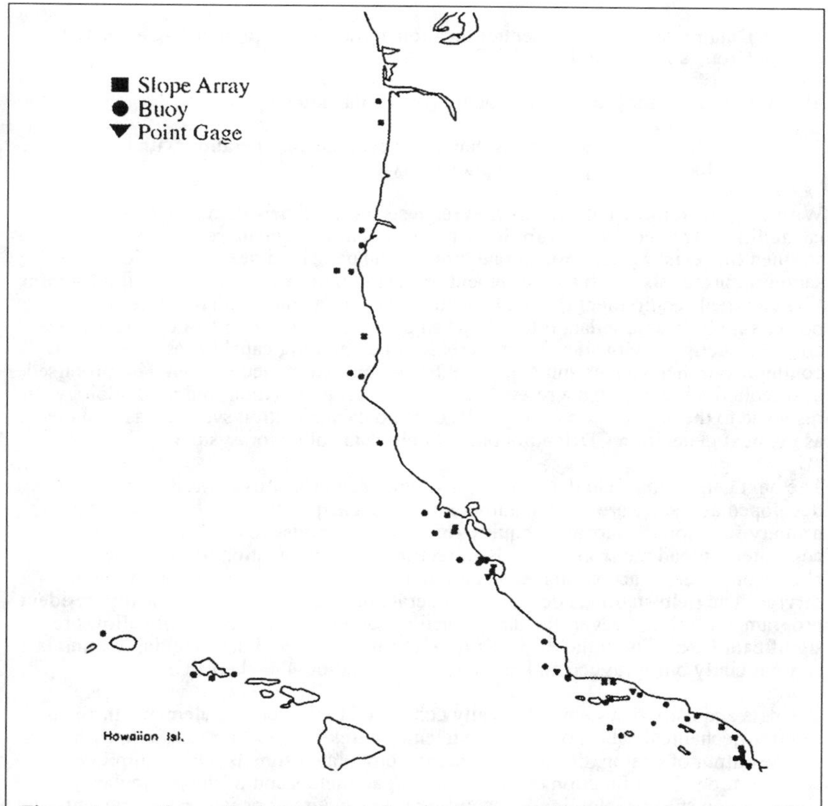

Figure 1. Locations of existing or previously deployed CDIP stations on the Pacific coast.

Data Acquisition And Processing

Data Acquisition

The program's mission and mandate have, for the most part, remained unchanged. The program's primary responsibilities are:

1) Accept input from any sensor type that has an electrical signal as its output,

2) Automatic end-to-end operation of the complete system. Data sampling, transmission, recording, validation, analysis and dissemination, without operator intervention.

3) Data recording in one central location at the Ocean Engineering Research Group's facility in La Jolla, CA.,

4) Quasi realtime availability of analyzed data, and

5) Data links and system costs that allow economical operation with inputs from any location equipped with phone service.

With time, increasing data acquisition requests imposed new demands on system capabilities. The requirement for longer, up-to-continuous data records, severely strained the existing network storage limits. The perceived need for flexible data sampling protocols led to a requirement for a communicative, intelligent field station. The very real requirement for field backup data storage dictated an uninterrupted power supply to assure data integrity when utility services are disrupted. The above concerns, coupled with significant increases in computing capabilities, communication advances and component miniaturization, accompanied by promised cost reductions, suggested a re-evaluation of the programs tools and methodology. In response to the above issues, a new, PC based data acquisition system was designed as the next generation CDIP autonomous field data collection system.

The next generation field data logging system, or Smartstation, was designed and developed as a software driven, autonomous, data acquisition system. The system's primary function is to locally acquire, log, and in response to a call from a host computer, upload the stored data. Bi-directional communication between host and client computers is accomplished via modem connections, through normal phone service. The field station is designed to operate independently, under locally resident program control. However, the dual direction communication capability allows for significant flexibility in the PC execution routines. The field data acquisition unit is a commercially built, rugged and hardened, industrial off-the-shelf PC.

The data acquisition system is typically connected to the sensing elements through electro-mechanical cables or via radio telemetry links. Reporting sensors may have a source output of analog, digital or frequency modulated signals. Data sampling configurations are a function of the observed parameters and of the particular phenomenon studied. Nominally, sampling frequencies for ocean waves, currents and winds measurements are one Hz. Sample length can vary from approximately 34 minutes for ocean directional wave measurements to continuous data for infragravity type wave studies. Typically, a field station is polled eight times daily. At the normal

8192 samples collected for non-directional wave measurements a 43.5 minute data gap is introduced into the three hour cycling period.

The physical functional specifications for the smart station emphasize rugged construction, dependability and reliability. To that end, the system is built to withstand field environmental conditions, insulated from electrical noise, mechanical vibrations and temperature extremes. The single moving component in the system is a hard disk. A removable floppy is an optional feature. Disks are included for long term data storage. However, the system is configured to ignore disk functions in the event of a disk failure. Maintenance is reduced to board level change out functions by unskilled, local field contacts.
Software specifications include a DOS operating system, RS232 or equivalent inputs, minimum 512K ROM disk, a system watch dog timer and up to four megabytes of RAM disk storage. A key component in the data acquisition system is a 16 bit counter-timer board, capable of latching, holding and registering the incoming frequency modulated data signal without altering or interrupting the input signal. Up to 16 signals can be interrogated by one Smartstation. The data lag between the first and last signal is less than 1000th of the sampling interval.

Smartstation capabilities include the ability to alter remotely, through the host computer and phone connections, any number of the field operating specifications. Some of the more relevant features are listed below.

1. Alter the data sampling frequency. Nominally, for most applications, the sampling frequency is 1 Hz. Under certain circumstances, and for trouble shooting purposes, the frequency may be changed from 0.125 Hz. to 10 Hz.

2. Change sample length. Under special conditions, such as during a tsunami alert or tsunami event, long or continuous data may be recorded.

3. Modify the number of data channels returned to the host computer. This feature is particularly useful in the event of a data channel failure. Erroneous data need not be brought back, thus reducing phone connect time.

4. Save all data to hard disk. This is the normal operating mode. In the event very high frequency, multi-channel data is recorded, a savings in system overhead will be achieved by writing data to the RAM disk only.

5. Data compaction option. To reduce data storage volume, subsequent flow over phone lines and reduce operating costs, only differences between successive data points are recorded and transmitted. The number of difference words between bracketing full words is nominally set to ten. Full word spacings may be changed under program control.

6. Chronological information is embedded into each channel's data stream. Standard intervals between successive time tags is five minutes. This interval is optionally changeable.

7. The system takes full advantage of existing, commercially available, modem data transfer integrity and compaction algorithms. Optionally the system can revert to an internal cyclic redundancy check.

8. Packet size transmission is under system control and can be altered from the standard 128 word size packet.

9. Special data runs can be requested from a field station. The request specifies the start and end times for the requested data as well as the data channels desired.

10. The field station can be remotely cold started. Upon restart, the system will boot from ROM, the hard drive or a floppy disk, in that order.

11. A commercial software package allows realtime remote control of the field station. The home keyboard controls the field station. The home screen mimics the field screen. This feature is specially useful during debugging sessions.

The host computer and Smartstation communicate via a header block. A default header is resident on the Smartstation and is returned to the host computer as a prefix to the data stream of every data collection episode. The header, which uniquely describes a station's attributes, contains all the relevant control, acquisition and other critical information about the incoming data set. In the event a change is requested from the field station, the appropriate station header is modified on the host computer and the new instructions are sent out during a normal polling call. The newly sent header becomes the field station default header. The new station configuration will take effect at the following data run. A new header can be sent out as part of a special data request.

Normally, a video display terminal is not part of the field smart station. In certain instances, where local data display is requested, an information screen depicting realtime measurements is displayed. Shown on the screen are the average wave period, the uncorrected (in the case of a pressure sensor) instantaneous wave height and the significant wave height. Data are sensed, according to the header block specifications, by a gauging suite and are telemetered to the shore based Smartstation. At the Smartstation, the data are received through a phase lock loop, electronically conditioned and optionally compacted, according to the header block instructions, and stored locally in RAM and on a hard disk. Storage is based on the "first in first out" principle, with the oldest word overwritten by the latest word as the storage buffer is filled.

In response to a phone query from the Central Station, the Smartstation uploads the latest data buffer. The Central Station data collection computer, a Sun Microsystem's work station, superficially examines the incoming data for obvious defects, such as incomplete transmissions and failed phone connections. A detected fault will trigger a retry call to the Smartstation.

Data Sampling

Stations are polled at a nominal interval of three hours. Event thresholds can be set on a per station basis. Threshold exceedence will cause data from a target station to be collected continuously. Tsunami alerts will cause manual activation of continuous sampling from designated tsunami stations.

Record length is a function of the study objectives. For the old type, memory-distributed system, record length can range from 1024 samples to 16384. For the new, smart stations, maximum record length depends on disk size. For both style stations, instruments intended to define energy or direction of wind-generated waves are sampled at 0.5 or 1.0 Hz. with sample lengths up to 16384. Instruments defining either wind velocity or waves of significantly longer periods sample at 0.125 Hz. and obtain 2048 samples. For ocean waves, this represents an 8 second average sea-level estimate, and covers a sample period of approximately 4.5 hours.

Data Analysis

Each record is objectively and automatically edited before analysis. All incoming raw data is subjected to a battery of objective and rigorous verification and inspection algorithms. Data declared questionable by this inspection process disqualifies the parent data record from further participation in the verification and analysis process. CDIP's policy has been that, with the exception of a very limited number of spikes in the data, no attempt is made to 'repair' the data. Spikes caused by data transmission errors are interpolated by bracketing values. Spikes in excess of one percent of the number of data points cause the run to be rejected. Thresholds are set for maximum expected standard deviation, sequences of equal values, gross mean shifts and sequences without a crossing of the mean. Exceedence of the thresholds causes the run to be rejected. Records from instruments sensing long period waves are detrended to remove the tidal components. Directional arrays are subjected to time domain correlation checks.

Daily exception reports are generated by the system. These reports detail the various data failure modes. Individual data records so identified are visually examined by CDIP personnel. More as an effort to determine sensor status than an attempt to incorporate the data in the analysis stream. At the end of the month, an error summary report is automatically generated by the system software.
All records are analyzed by Fast Fourier Transformation. The Fourier coefficients from pressure sensors are depth corrected by linear theory. Coefficients are combined to produce an energy spectrum which is grouped into the period bands shown in the report.

During the first 14 years, the Coastal Data Information Program analyzed and reported directional data from nearshore slope arrays to produce estimates of Sxy, or the longshore radiation stress. The motivation for this was to enable a reasonably easy estimate of longshore transport (which is related to Sxy in certain models). This assumption is only valid where the contours between the gage and the shore are straight and parallel, a condition that is not met at every location. Further, this analysis depends upon a knowledge of the actual bathymetry near the site, which is subject to change over time. Finally, by providing only the radiation stress data, other valuable properties of the wave field directionality are lost to users of these data. After considerable review, CDIP has decided to alter the directional data analysis and reporting methods.

Beginning January 1992, CDIP adopted a new convention in reporting directional wave data products. This new standard analysis procedure for directional wave gauges produce a two-dimensional energy spectrum from the measured time series. Total energy, significant wave height, peak period and weighted direction are computed and reported.

Data Dissemination

Monthly and annual reports are generated on a routine basis. Project reports are generated upon request. The statistics included in the monthly reports are the following:

1. Energy spectrum grouped into period bands.

2. Vector plots of direction which indicate wave headings and relative energy densities. Angles reported in the directional data are true compass headings towards which the waves are travelling at the location of the measuring instrument.

3. Persistence tables for maximum significant wave heights.

4. Maximum daily significant wave heights.

In 1991 a statistical summary volume reporting all data collected in the 17 year period between 1975 and 1991 was published. 59 stations are characterized in an annual report format.

1. Cumulative Height & Peak Probabilities Table.

2. Height & Peak Period Occurrence Estimates as semilog Functions

3. Seasonal Data. The significant wave heights for each observation at each station were evaluated during each quarter to display seasonal trends.

4. Significant Heights of Sea & Swell. The spectra obtained from each observation were divided arbitrarily at a period of 10 seconds. All energy at periods greater than 10 seconds was labeled as "swell" and all energy less than 10 seconds was labeled "sea".

5. Joint Distribution Table & Plot. Each data run was analyzed for significant wave height and period band containing maximum energy. Figures 2 and 3.illustrate the very large numbers of observations accumulated.

Instrumentation And Deployment Techniques

For nearshore instrumentation the CDIP typically relies on a telemetry cable for both powering the sensor package and recovering the data. While extremely reliable and robust, the requirement for a cable as the data path has presented some unique challenges in instrumentation deployment. For very nearshore applications and where coastal structures such as piers are opportune, gauges can be mounted against available pilings or other permanent structural members. Under these circumstances, installation is a relatively straight forward. Where offshore distance exclude permanent structures, the data cable must travel from the gauge on the ocean bottom, traverse the beach and terminate in a secure beach location. For low to moderate energy beaches, amphibious vehicles (LARC 4) have been used for the difficult task of laying the cable, from the gauge, through the surf zone and on to the beach. The LARC serves as the deployment vehicle in that it carries the sensor package and data cable to the deployment site. Once on site, the slope array is lowered to the bottom and the LARC returns to the beach laying out data cable as it advances. A typical LARC deployment arrangement is shown in Figure 4.

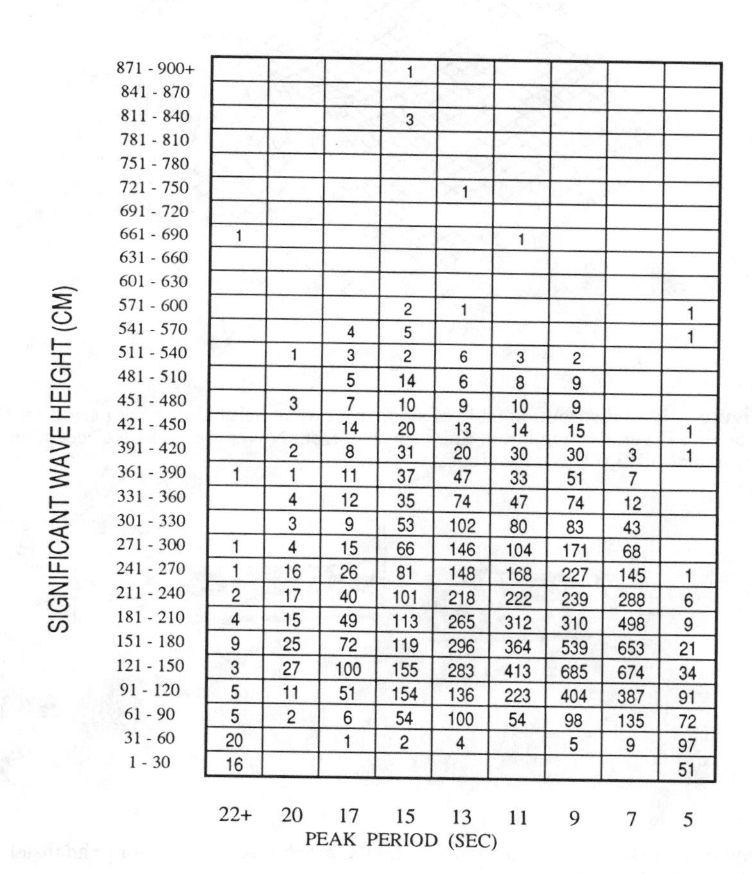

SIGNIFICANT WAVE HEIGHT (CM)	22+	20	17	15	13	11	9	7	5
871 - 900+				1					
841 - 870									
811 - 840				3					
781 - 810									
751 - 780									
721 - 750					1				
691 - 720									
661 - 690	1					1			
631 - 660									
601 - 630									
571 - 600				2	1				1
541 - 570			4	5					1
511 - 540		1	3	2	6	3	2		
481 - 510			5	14	6	8	9		
451 - 480		3	7	10	9	10	9		
421 - 450			14	20	13	14	15		1
391 - 420		2	8	31	20	30	30	3	1
361 - 390	1	1	11	37	47	33	51	7	
331 - 360		4	12	35	74	47	74	12	
301 - 330		3	9	53	102	80	83	43	
271 - 300	1	4	15	66	146	104	171	68	
241 - 270	1	16	26	81	148	168	227	145	1
211 - 240	2	17	40	101	218	222	239	288	6
181 - 210	4	15	49	113	265	312	310	498	9
151 - 180	9	25	72	119	296	364	539	653	21
121 - 150	3	27	100	155	283	413	685	674	34
91 - 120	5	11	51	154	136	223	404	387	91
61 - 90	5	2	6	54	100	54	98	135	72
31 - 60	20		1	2	4		5	9	97
1 - 30	16								51

PEAK PERIOD (SEC)

Figure 2. Joint distribution table of significant wave height and peak period for the Harvest Platform site off Point Conception, CA. The data span from January 1987 to December 1991 for a total of 11,910 observations.

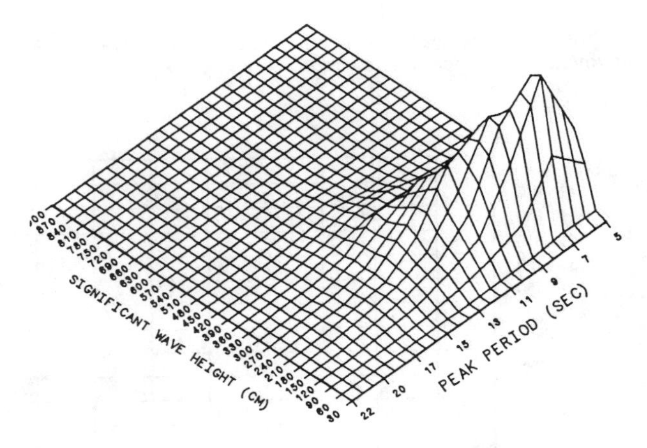

Figure 3. Joint distribution table of significant wave height and peak period for the Coquille River, Oregon buoy. The data span from November 1981 to December 1991 for a total of 13,673 observations.

Figure 4. A slope array mounted on a LARC amphibious truck prior to an installation in Southern California.

For high energy beaches or for wide surf zones a LARC is ineffective. For these applications the CDIP has adopted airborne deployments as the method of choice. Using a heavy lift helicopter, the slope array is carried as an external load while the data cable is carried inside the aircraft, as shown in Figures 5 and 6.

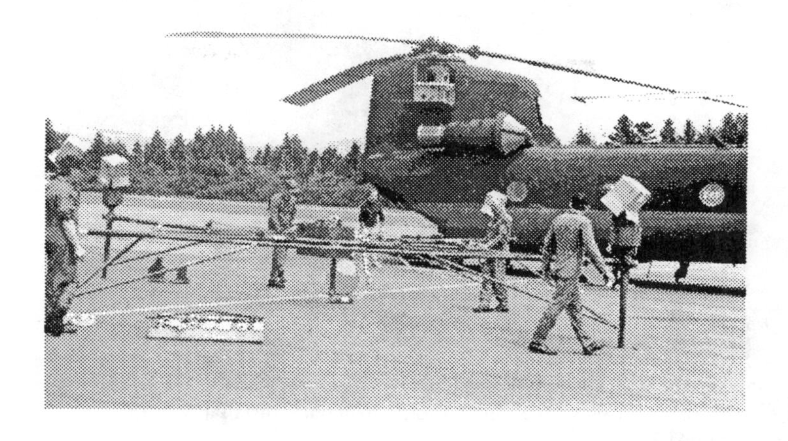

Figure 5. Preparations for installing a slope array by helicopter off the Oregon coast.

Figure 6. Helicopter installation of a slope array.

Over the target location, identified by a marker buoy, the helicopter lowers the slope array to the bottom, disconnects from the slope array and proceeds towards the beach, laying out cable on its way to the shore station. Over the shore station the data cable is severed with a fast acting pneumatic hydraulic cable cutter and is dropped from the helicopter. This method has proven very successful . The CDIP has also used helicopters in the deployment of offshore buoys. This technique is particularly useful during stormy days where bar closures exclude the use of boats as deployment vehicles. Loading of the buoy on the helicopter is shown in Figure 7.

Figure 7. Loading a Waverider buoy prior to helicopter installation off the Washington coast.

In remote, or offshore locations where electricity is unavailable, CDIP has employed solar panels in conjunction with storage batteries to provide power to the sensors. This method works well where cloud cover and vandalism are not an issue. However, battery servicing increase labor costs. Cellular phone service can be used in lieu of regular phone lines. Both the above methods served the network's data collection effort in the Hawaiian islands, as shown in Figure 8.

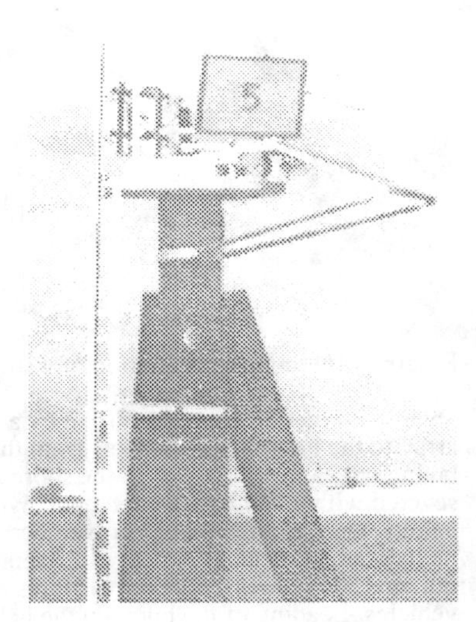

Figure 8. Solar panel powered installation using cellular phone link offshore of Barbers Point harbor in Hawaii.

Deepwater Directional Measurements

The CDIP has measured waves for about ten years from the Harvest Platform - a deep water (approximately 225m depth) oil production platform belonging to Texaco. This site, off Point Conception, has particular value because it is open to direct wave approach from both southern swell and Aleutian storms. Recently, a six-element array was added to the platform to allow high-resolution directional data to be collected. The array configuration and the location of the platform are shown in Figure 9. Almost continuous observations are obtained from this station and relayed to shore through a microwave link.

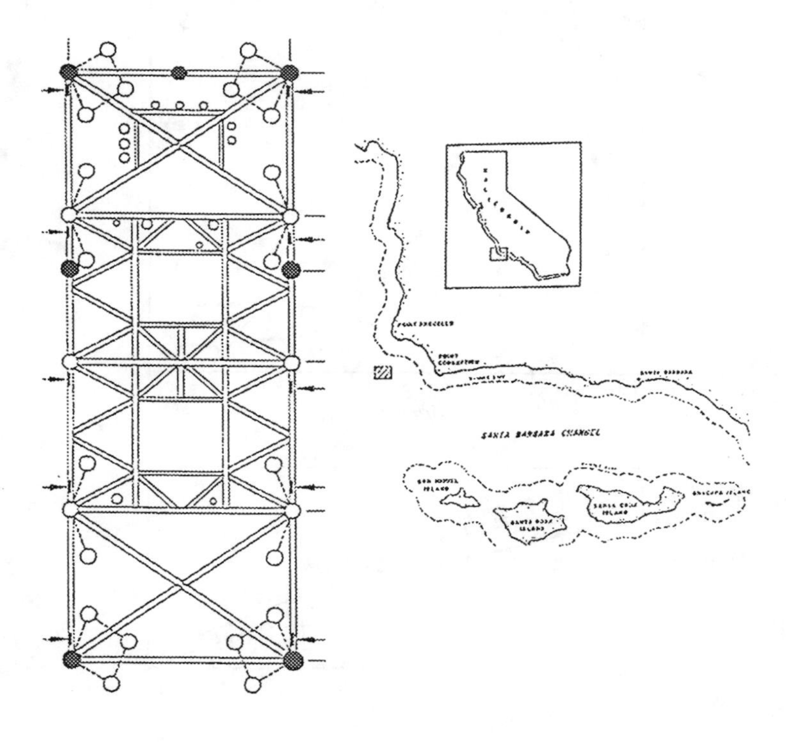

Figure 9. The directional array on the Harvest Platform. The sensors are indicated by the shaded circles in the left diagram, with the maximum distance between sensors of approximately 65 m. The map at the right shows the location of the platform along the California coast.

An example of rather typical spectra plots from the Harvest sensors are shown in Figure 10. The bimodal spectral shape results from a combination of southern swell and a sea-swell train from the northwest. Notice that the sensors track closely in the energetic band and that the noise level at high frequencies is low (the spectra have not been corrected for depth).

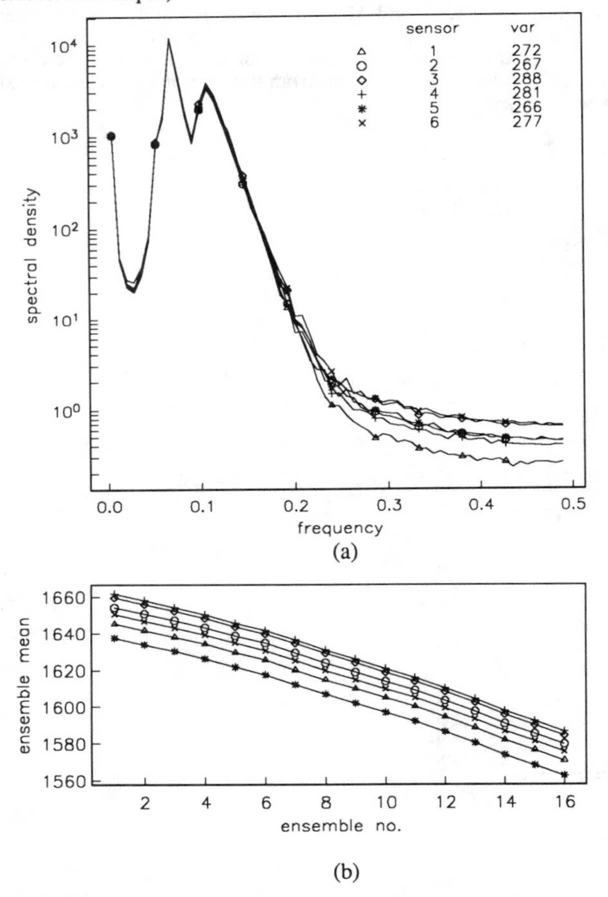

Figure 10. (a) Auto spectra from the Harvest Platform sensors at 2:04 AM, 24 May, 1993. The significant wave height is 1.3m. There are 120 degrees of freedom.
(b) Means of 512 observation ensembles.

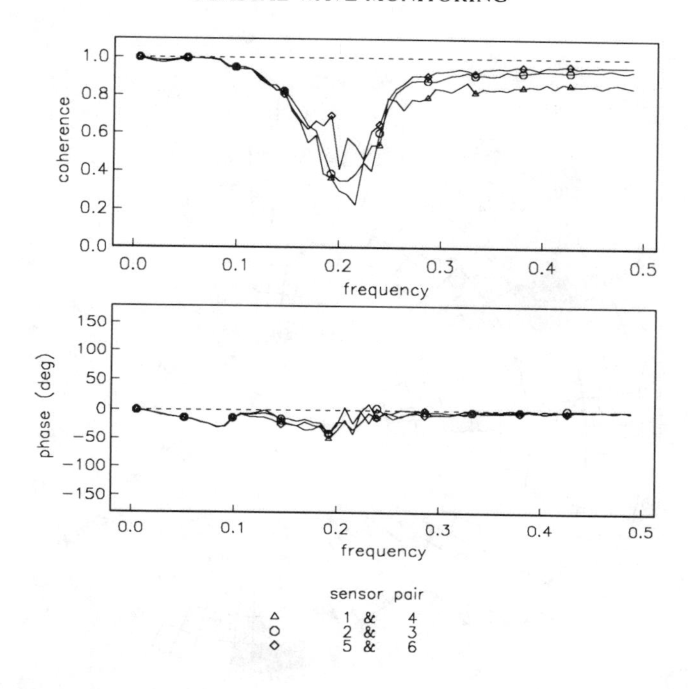

Figure 11. Cross spectra from the data shown in Figure 10.

The phase and coherence relationships among the three redundant short lags in the array are illustrated in Figure 11. As expected, the coherence falls off as the wavelength approaches the sensor spacing. The increasing coherence at very high frequencies (very low energy levels) is due to spectral leakage from the peak of the wave spectrum. This is an artifact of the simple data analysis procedure that was used, and can eliminated by tapering the time series.
To illustrate how complex the typical deep water wave field is off of California, a slope array is formed from four of the platform sensors, and the first two directional moments of the spectrum are calculated following the method of Herbers and Guza (1989). A frequency directional spectrum is then derived from these moments using a maximum entropy estimator (Lygre and Krogstad, 1986). This spectrum is illustrated in Figure 12. The spectrum is truncated at 0.12hz because of the relatively large sensor spacing. Bias errors in the directional moment estimates become significant for wavelengths less than about five times the sensor spacing (Herbers and Guza, 1989). The two wave trains that produce the bimodality in the Figure 10 spectrum are clearly shown in this plot.

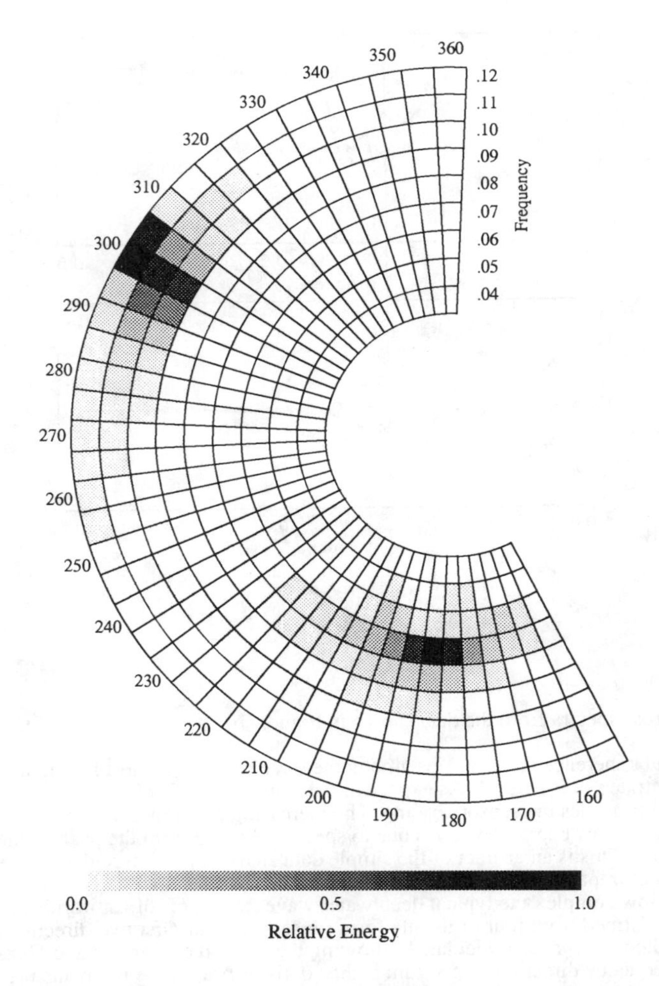

Figure 12. Directional spectrum from the Harvest Platform array spectra of Figure 10.

Future Plans

Nominally, with the exception of buoy information, the CDIP receives data from offshore gauges through telemetry cables. This methodology has been remarkably successful and robust. Unfortunately, cable costs and logistical considerations limit cable lengths to no more than three km. at best. This arbitrary limitation is unnecessarily restrictive. A cable-free telemetry system for data transmission between instrumentation and a CDIP field station could increase the effective data collection range to approximately 50km.

The CDIP is developing a radio telemetry system designed to transmit 10 channels of continuously sampled data. Sampling frequency will be selectable from between 0.5 to 2 hz. The system will be self-contained with enough battery power to last 1 year between service calls. The radio transmitter will reside in a buoy tethered to an instrumentation package via an electro-mechanical mooring cable. Information will flow from the package to the radio transmitter, to be received at a shore side station for storage on board a Smartstation. From there, information will be processed through normal CDIP procedures. This configuration will permit bottom mounted slope arrays, as an example, to be deployed in areas relatively far from shore (e.g. shoals and banks). Alternately, instrumentation can reside along sections of coastline where a shore based field station is impossible or impractical.

Wind Measurements

Offshore winds can be highly variable and unpredictable. This is especially so in storms events off the Pacific coast. Lack of offshore wind observations a a critical, unknown in marine forecasts and reports. Absence of this vital information has, at times, resulted in the damage to boats with accompanying boater injuries and occasional loss of life. This lack of information is of significant concern to the coastal partner states of the CA. CDIP has historically focused on the collection of wave data. However, in regions such as the Southern California Bight, locally generated seas can play a major role on the regional wave climate. A more complete understanding of the spatial variability of winds in the Bight is required for useful now casts and hindcasts of local sea conditions. Thus, the collection of over-water wind conditions is a natural extension of the CDIP database.

To that end, the CDIP is engaged in a development program that will collect, and send to shore, in real time using the concurrently developed telemetry transmission package, data observed by offshore wind buoys.

Self-contained Directional Array

The telemetry system will allow CDIP to expand its real-time wave gaging into areas where the present shore cabling method is not possible. In conjunction with the development of the telemetry package, it is proposed to develop a prototype directional array which uses this module instead of the traditional shore cable. This will permit bottom mounted arrays to be deployed in areas relatively far from shore. The wave data will be transmitted from a surface buoy moored above the slope array. The array will be somewhat larger than the typical 6m square CDIP slope array, and will contain additional sensors in order to test the ability of the telemetry system to transmit large data sets. The prototype will be tested in an appropriate setting within Southern California.

Smartstation Backup Power

In order to economically and efficiently collect wave data in real-time, CDIP uses telephone lines to retrieve data from storage devices at the various wave gage locations. The field storage systems are placed where electricity and phone service is available. This has proven to be a generally reliable method for data acquisition.

However, during periods when utilities fails, wave data can be lost. Not surprisingly, electrical outages are more frequent during storm events, when wave data is often of the greatest interest. Over the last several years, CDIP has developed a Smartstation whose function is to collect and store wave data in the field. CDIP is developing a backup power unit for the Smartstation. In the planning stages is a system allowing data to be acquired and stored during a power outage lasting up to several days. This will require modifying the existing Smartstation so that it can recognize a power outage and operate in a low power mode. The Smartstation will be equipped with a series of trickle-charged automotive batteries as the backup power source.

Central Facility Upgrading And Expansion

Conversion to a UNIX-based SUN system is in progress. A SPARC 1 workstation will replace the PC-based data acquisition system. This workstation will allow greater flexibility and accessibility to the data. Station polling cycle time will be reduced. 1.3 gigabyte 4mm DAT tapes will replace 40 megabyte 9 track tape storage. Networked to the SPARC 1 is a 630 multi-processor server. The FFTs and analysis programs will be performed on the server. Condensed statistics will be archived on disk. At the completion of each data run, the processed data will to be transmitted to the National Weather Service office in San Diego for distribution across their AFOS network. A similar file will also be sent via TCP/IP network to other computers at Scripps and Waterways Experiment Station (WES). These data will be accessible to interested government agencies via modem or TCP/IP network. Optical storage of raw data is under investigation and will be integrated into the system at some future date.

Conclusions

Harsh environmental conditions, the rapid evolution of improved communication and computation technologies, and the special needs of specific coastal engineering projects have driven the development of a variety of new concepts, facilities and techniques for the reliable and economical gathering, archiving and analysis of wave and other coastal data. The experience of the CDIP program, which is approaching the 20 year mark, may provide valuable guidance to other networks with similar objectives.

Acknowledgments

The continuing support of the Coastal Engineering Research Center of the U.S. Army Corps of Engineers and the California Department of Boating and Waterways is gratefully acknowledged. In addition, we wish to thank the U.S. Navy, the National Weather Service, the City and County of San Francisco, Pacific Gas and Electric Company, Texaco USA, and the many others who have provided cooperation, support and services. We would like to offer particular recognition to some of the pioneers of the CDIP concept, including Meredith Sessions, Robert Wiegel, Orville Magoon, George Armstrong, George Domurat, Douglas Pirie and Michael Hemsley.

Appendix. References:

Herbers, T.H.C., and R.T. Guza, 1989. Estimation of wave radiation stresses from slope array data, J. of Geophys. Res., 94(C2), pp 2099-2104.

Lygre, A., and H.E. Krogstad, 1986. Maximum entropy estimation of the directional distribution of ocean wave spectra, JPO, 16, 2052-2060.

Seymour, R. J., 1979. Measuring the nearshore wave climate: California experience. In: Ocean Wave Climate, M. D. Earle and A. Malahoff, eds., Plenum Press, New York, Marine Science, 8: 317-326.

Seymour, R. J. and M. H. Sessions, 1976. Regional network for coastal engineering data. In: Proc. 15th Coastal Engineering Conf.,ASCE, Honolulu, HA, July, Vol. 1, Chap. 5, pp. 60-71.

Seymour, R. J., M. H. Sessions and D. Castel, 1985. Automated remote recording and analysis of coastal data. J. Waterway, Port, Coastal and Ocean Engineering, Proc. ASCE, 111(2): 388-400.

New Program for the Marine Observation
and Prediction Center of the
Central Weather Bureau

Beng-chun Lee[1]

Abstract

With the increase of economic activities on the oceans and seas, the observations of marine phenomena and forecast services become more and more significant. The Central Weather Bureau plans to establish a Marine Observation and Prediction Center in the near future. The proposal has been approved recently from the Executive Yuan. The center will take charge of the integrated planning of wave, tide, and data buoy observing stations; set up and maintain instruments in tide and wave stations; establish a marine phenomena databank and service net; and standardize data formats and develop a marine forecasting system. The forecasts of storm surges, tides , and ocean waves will become a routinely scheduled work to meet the needs of economic activities over the oceans and seas and the development of marine resources.

Introduction

Taiwan, known as "Formosa", is a beautiful island located in the subtropical region. There are more than 20 millions of population residing within the area of 36 thousand square kilometers. Such high density of population cuts short the use of natural resources over the land and signifies the potential

1
Director, Applied Meteorology Division, Central Weather Bureau, 64, Kung-yuan Road, Taipei, Taiwan 100, ROC.

development on oceanic resources.

Marine observation-and-forecast is one of the major tasks of the Central Weather Bureau (CWB) of the Republic of China (ROC). In order to fully support rapid growth of the national economics, The CWB is urgent to set up a complete marine observation network, data bank, and ocean model for the development of marine weather forecasting and marine science/technology. As well, a foundation is set up for future MOPC.

Therefore, CWB desires to expand its understanding towards the development of marine meteorology, including the installation of different types of the tide gauges, wave measurement systems, and in addition, the researches concerning marine meteorology, such as air/sea interaction, wave analysis/forecasting, marine fishery service, and related researches.

The environmental condition of oceanic regions in Taiwan area and its vicinity

Geographically, Taiwan is located at the edge of the continental shelf of the southeast China coast (Fig. 1). However, the environmental conditions of the surrounding oceanic region vary tremendously. For instance, the ocean bottom 50 km away from eastern Taiwan coast rapidly increases to 2000 meters and even deepens to 7000 meters near Ryukyu Trench. A significant and important warm current "Kuroshio" originating from the Philippine Ocean penetrates over the nearby eastern Taiwan ocean and associates with serious drifting sands along the coast. The major branch of the Kuroshio current, which sustantially adjust its speed and intensity by the winter northeasterly monsoon and the summer southwesterly monsoon penetrates through the Taiwan Strait.

An overview of CWB's marine operation function

The current main marine operation functions in the CWB are two folds. One is the marine stations and the other is the marine data requirement and data fallacy. They are briefly described below.

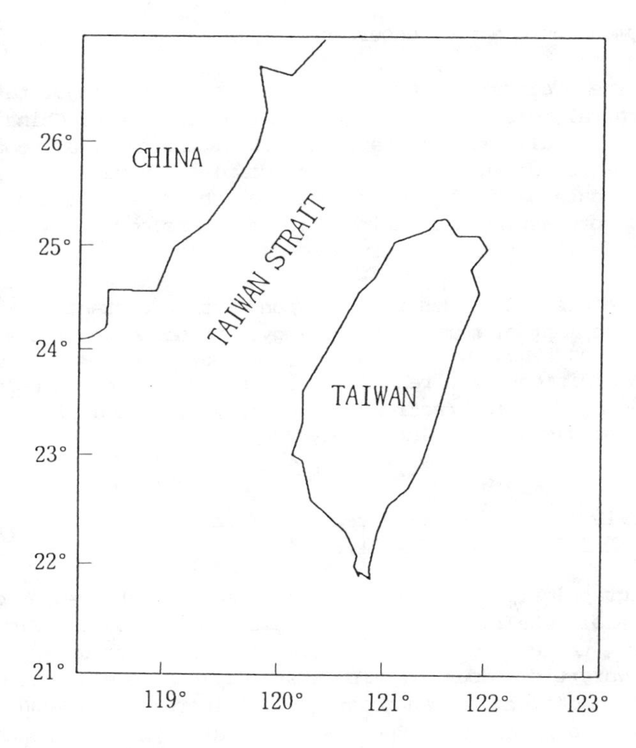

Fig.1 Geographic location of Taiwan

1. Marine stations

The research of marine observations at the CWB of the ROC is still in the developing stage. The main functions of marine observation are as follows:

1) Wave stations: There are four wave stations (Fig. 2), Byitourjeau in the northeastern tip of the island, Chengkung in the southeastern part of the island, Tungchitao over the Taiwan Strait, and Hsiaoliuchu in the southwestern part of the island. Each wave station is equipped with an ultrasonic sensor about one to four km away from the seashore, and the signal is transmitted

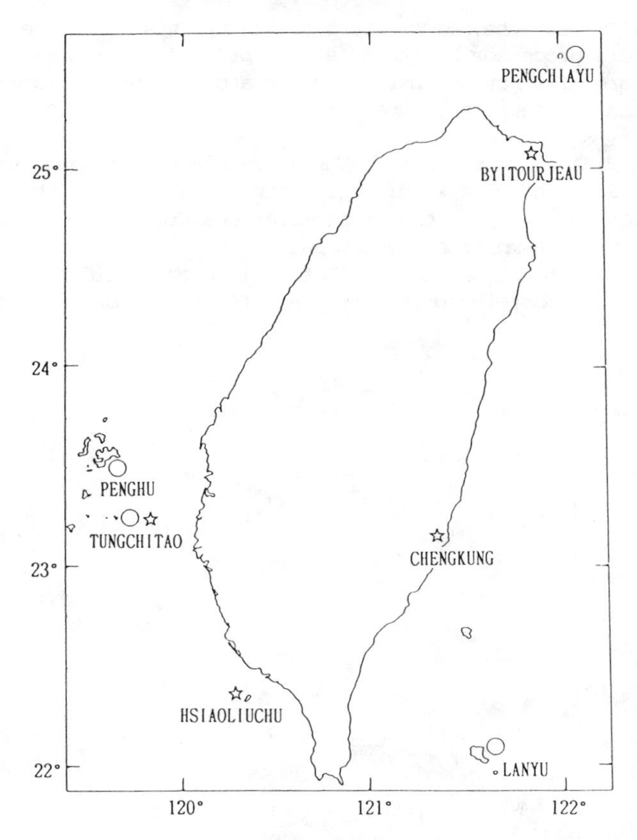

Fig.2 Geographic distribution of wave stations (denoted by ☆) and off-land meteorology stations (denoted by ○)

eventually to the headquarters of the CWB via the cable to the seashore's processor. The data will provide references for the near-shore fishery weather forecasts and for the development of the oceanic resources around Taiwan.

2) Tide gauge stations: There are eight tide gauge stations (Fig. 3) which are directly subordinated to the CWB's jurisdiction. Most of these stations are equipped with an automatic transmmition system and are usually located in the harbor or near the seashore. Data will provide tidal height and timing for port-ins/port-outs and fishings for the general public.

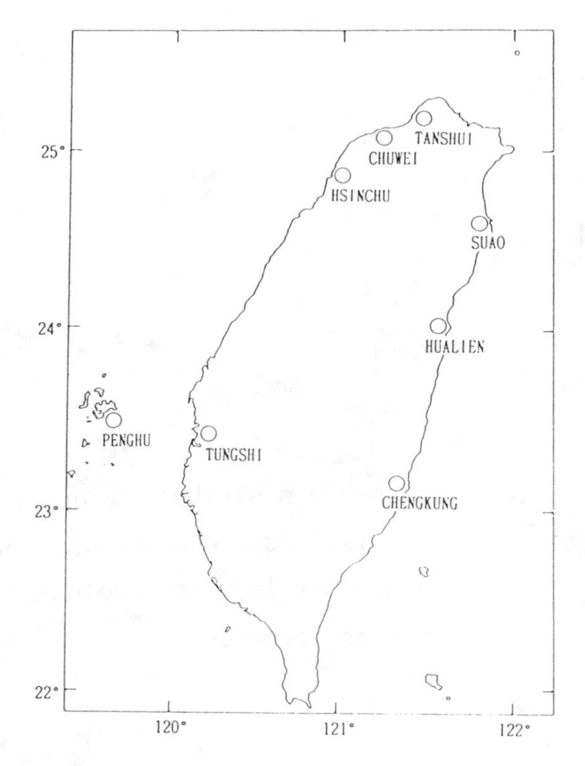

Fig.3 Geographic distribution of tide stations

3) Marine meteorology stations: Penghu, Pengchiyu, Lanyu and Tungchitao (Fig. 2) are four off-island synoptic weather stations to monitor the monsoon weather, typhoon attrack, and abrupt weather change from the open sea.

4) Sea surface temperture (SST) measurements: SST is essential to both the meteorology and the oceanography. CWB can provide the digitalized information under clear sky by using the temperature retrieval method of Japanese GMS data for weather forecasting references.

2. Marine data requirement and data fallacy

1) Data requirement: Marine data requirement for the past five years can be subdivided into two main ranges. Firstly, from the applicational point of view, the percentages of the data are uy 45.2%, 17.2%, 37.2% for scholastic researchers, business programing references, and business execution, respectively (Fig. 4a). Secondly, from the organizational side, the percentages of the data required by related units are 47.2% for the scholastic research, 15% for the civilian enterprises, and 37.8% for the economic establishments (Fig. 4b), respectively.

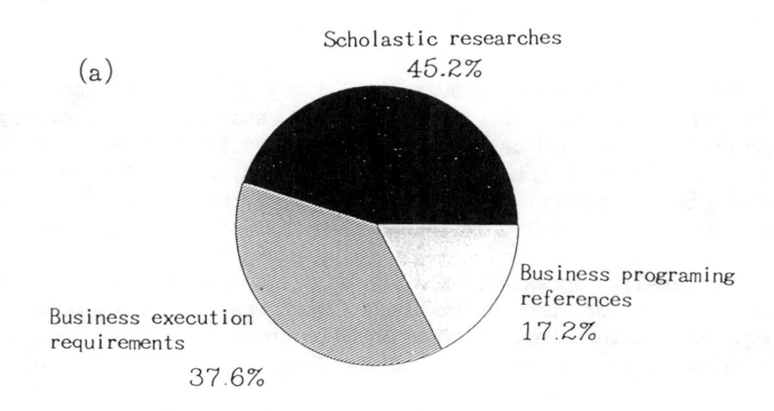

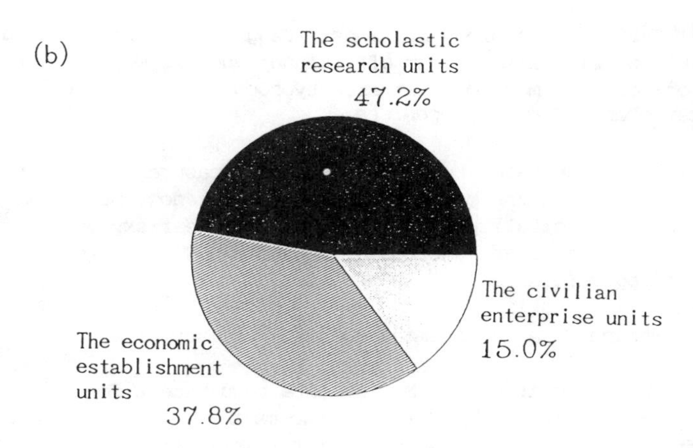

Fig.4 Marine date requrements for the past 5 years

(a): The percentage of the marine data applications

(b): The percentage of the marine data requirements

2) Data fallacy: An averaged fallacy rate for the same time is shown in Fig. 5. For the wave stations, the averaged percentages of data fallacy are 35.78%, 30.36%, 32.00%, 44.60% for the Byitourjeau, Tungchitao island, Hsiaoliuchu, and Chengkung , respectively. The fallacy of the tide stations (not shown), approxmiately, is better than that of the wave stations.

Generally speaking, the causes of lacking in data are: (1) malfunction of the equipments, (2) shortage of man power on equipment maintenence which delays the obstacles evacuation, (3) difficulty in equipment contruction.

The causes of the equipment malfunctions are destroyed by explosions or dumping nets from fishermen in order to harvest fishes. The equipments are also damaged by tourists, strong winds, and surges, etc.

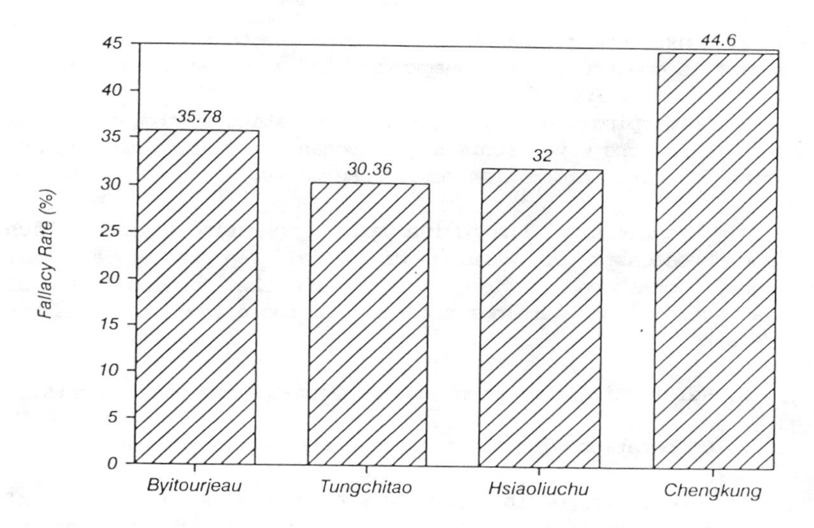

Fig.5 The average fallacy rate for the past 5 years.

Future development on MOPC and expected benefits

As mentioned in the previous section, the origin to introduce this program is due to constant fallacy of marine observation data and this condition can not be met with the national oceanic resources development and the safety of the operational ships and navigation around Taiwan waters. Moreover, there has been a specific responsible organization existing in USA, European, Japan, Australia, and other countries, but not in ROC. Therefore, the direction for the future development on the MOPC in CWB is taken from the following specific aspects:

1. Main points on MOPC and its outlooks

1) Maintaining and establishing the marine observation operating system, such as establishing extra along-shore observation stations, data buoys around Taiwan waters, and completing the automation of the marine observing nets.

2) Constructing data bases of the marine observations, which means file constructions, managements and supplies of the data.

3) Developing the marine forecast operating system, including establishing the wave, storm surge, ocean circulation models, and marine forecast information service networks.

4) Establishing the marine warning operating systems which issue broadcasts of warnings or special reports of abnormal surges, on-shore inflow surges and the proceedings of broadcasting the warnings according to those from the weather forecasts.

5) Reinforcing the technique improvements and developments.

2. Cooperation with the related units within the nation

The CWB is responsible for the wholly integrated planning of the marine observation stations. Therefore, tide stations of Taiwan Provincial Water Conservancy are renewed and annexed by CWB through yearly planned budgets. Thus, it can lead to a data format unification in order to be stored in the MOPC of the CWB and to interchange with the related units. The Taiwan Provincial Harbor Technique Research unit is responsible for the non-real time data processings and research of every harbor wave station. The real-time data are transported to MOPC through CWB weather service nets.

After the above-mentioned procedures, the expected benefits can thus be achieved by

1) establishing the specific responsible marine forecasting units, the collection of the data and forecasts can be unified in order to economically evade the duplicate investments or non-management phenomena.

2) accurately controlling the marine phenomena and broadcasting marine forecasts, the safety of navigation. Besides, the environment of the fishery resources and related information service can also be provided.

3) providing basic data for researches and developments of oceanography to promote the development of marine resources, the planning and execution of the sea-shore construction.

4) elevating the standard of the oceanic technology in order to cope with the current trends of the world.

5) providing the opportunities for oceanic specialists in order to lessen the losses of experts and the cultivation of education from the nation.

And those are the solid foundations for an operational marine observation and prediction center in the world the CWB is going to achieve.

Concluding remarks

The area of the ocean to that of the land is seven to three. Therefore, the future outlook for the development on the ocean is quite optimistic in reality. By following the tendency of marine developments in the world and meeting the requirements for the general public in Taiwan and around the world, the prospect of the MOPC set up in CWB is a timing decision and its future outlook can clearly be perceived in Fig. 6. In this figure shows MOPC obtain many tide stations, wave stations and data buoys. With the help from satellite and oceanic microwave remote sensing techniques on the SST, there will be a complete local marine observation network around Taiwan waters to functionally provide the requirers with all kinds of specific data. Therefore, we can develope the marine forecast operating system by cultivating our own models on waves, storm surges, ocean circulations and provide the general public with a convenient marine forecast information service network. Further more, by actively cooperating internationally into all kinds of researches and timingly upgrading the old observation instruments to meet the newest world trends, Taiwan can fairly contribute what she has in her surrounding waters to the international marine world.

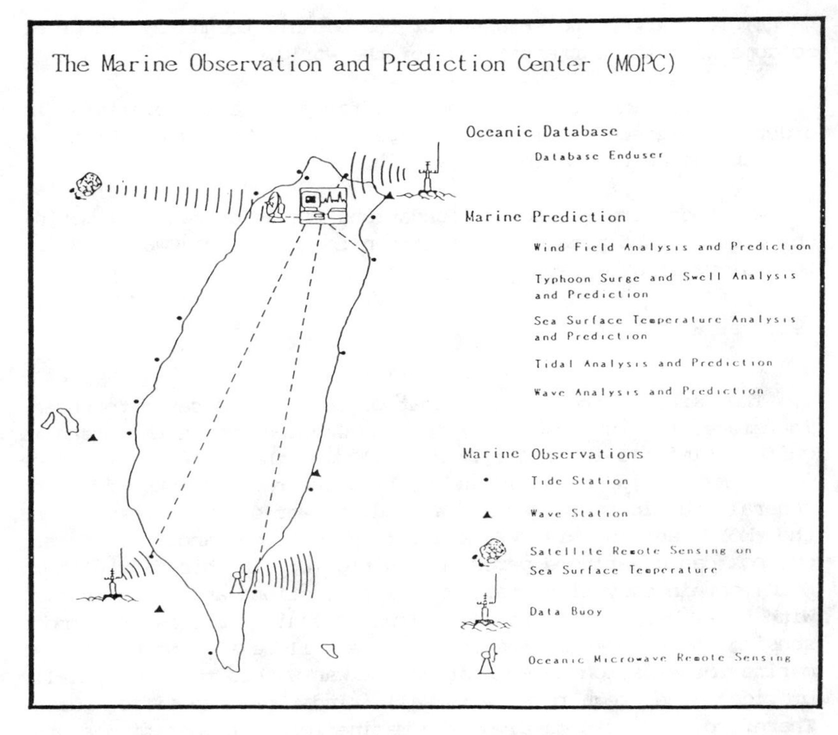

Fig.6 Prospect of the Marine Observation and Prediction Center

CONSERVATION PROPERTIES OF THE MILD SLOPE AND BOUSSINESQ EQUATIONS

Rodney J. Sobey[1]

Abstract

Common water wave evolution equations, such as the mild slope equation, parabolic approximations and Boussinesq equations, are approximate statements of conservation laws. A feasible measure of their adequacy is their ability to predict steady progressive wave propagation over a horizontal bottom, where near-exact kinematics are predicted by steady wave theory. None of the common evolution equations satisfy the conservation laws. The relative errors are surprisingly large and no existing evolution equation is acceptable. A Boussinesq-style equation, with almost acceptable error levels, is suggested. An accurate, robust and viable water wave evolution equation for an uneven bottom is not presently available.

1. Introduction

The mild slope equation, the Boussinesq equation and variations and approximations to these equations are extensively used in the prediction of water wave evolution in the nearshore environment. In principle, these field equations describe wave evolution over an uneven but slowly varying bottom, retaining the capacity to represent shoaling, refraction and diffraction. With suitable measures to represent source/sink terms and boundary conditions, they have been extended to include dissipation and reflection/scattering processes.

The physical scope of such evolution equations is most attractive, but this is not achieved without a very significant computational effort. Both the promise and the computational demands have distracted attention from the fact that these are approximate evolution equations. Efforts to deal with the computational demands, such as parabolic approximations and absorbing boundary conditions, introduce still further approximations.

Frequently, the approximations are so deeply imbedded in the adopted

---

[1]Professor of Civil Engineering, Environmental Engineering Group, University of California, Berkeley, CA 94720, USA

evolution equations that their total impact is unclear. The ultimate theoretical test of any evolution equation or set of equations is how well they satisfy the conservation laws for mass, momentum and energy. For the general problem of an unsteady, spatially-varied wave field over an uneven bottom with a mix of progressive and standing wave modes, a realistic evaluation is not feasible. It is possible only for a steady, spatially-uniform and dissipation-free progressive wave field, where the space and time varying kinematics should match those predicted by nonlinear steady wave theory. Wave fields in the natural environment are complications on this special case.

Because of the approximations in existing evolution equations, it is not expected that any will be exactly satisfied by Fourier wave kinematics. The residual and its distribution throughout the wave cycle provide a revealing measure of the utility of a range of popular evolution equations.

Existing model evolution equations can be grouped into "mild slope" and "Boussinesq" categories. The present paper will briefly review the background and assumptions of each of these categories, together with representative variations or approximations within each category. When reduced to a form consistent with steady, spatially-uniform, dissipation-free, progressive wave propagation over a horizontal bed, their conservation properties are evaluated from Fourier kinematics for moderate wave conditions in deep, transitional and shallow water.

2. Mild Slope Equations

The mild slope genre of model evolution equations is identified principally by the dependence on the linear (or Airy or Stokes I) wave theory to describe aspects of the local kinematics. For time-periodic waves without dissipation or current, the mild slope equation is

$$\frac{\partial}{\partial x}\left(CC_g\frac{\partial F}{\partial x}\right) + \frac{\partial}{\partial y}\left(CC_g\frac{\partial F}{\partial y}\right) + k^2 CC_g F = 0 \qquad (2.1a)$$

in which $F(x,y)$ is the complex amplitude such that the water surface is

$$\eta(x,y,t) = \mathrm{Re}\{F(x,y)\, e^{-i\omega t}\} = \mathrm{Re}\{A(x,y)\, e^{i(kx-\omega t)}\} \qquad (2.2)$$

For waves propagating in or near to the x direction, the complex envelope $A(x,y)$ is an alternative and slowly-varying amplitude measure. In a spatially-uniform wave field propagating in the x direction, the mild slope equation reduces to

$$\frac{\partial^2 F}{\partial x^2} + k^2 F = 0 \qquad \text{or} \qquad \frac{\partial^2 A}{\partial^2 x} + 2ik\frac{\partial A}{\partial x} = 0 \qquad (2.1b,c)$$

The mild slope equation is generally regarded as a linear statement of energy conservation. It can be established from a variational principle statement of energy conservation (Booij 1981, Miles 1991). It can also be established directly from the integral energy conservation equation (Sobey 1993). Earlier derivations by Berkhoff (1972) and Smith and Sprinks (1975) are based on the Laplace equation (a statement of mass conservation) weighted by $f(z) = \cosh k(h+z)$ prior to depth integration. Airy (linear) wave kinematics exactly satisfy Eqs. 2.1b and c.

Nearly exact kinematics for steady progressive waves are provided by the

Fourier approximation wave theory (e.g., Sobey 1989). For the present purposes, three steady waves have been selected, representative of moderate wave conditions in (A) deep, (B) transitional and (C) shallow water. The details are listed in Table 1. In application, all quantities have been normalized by g and $\omega = 2\pi/T$. The complex envelope was determined from the water surface trace by complex demodulation (e.g., Sobey and Liang 1989). In addition, all partial derivatives in the model evolution equations (such as $\partial^2 A/\partial x^2$ in Eq. 2.1c) were evaluated analytically.

Table 1. Steady waves selected for evaluation of evolution equations

Category	h (m)	H (m)	T (s)	$\omega^2 h/g$	$\omega^2 H/g$	H/H_{limit}
(A) Deep	100	10	10	4.0	0.40	0.45
(B) Transitional	10	5	10	0.40	0.20	0.65
(C) Shallow	2	1	10	0.080	0.040	0.60

Current = 0 in all cases.

For deep water Wave A, Fig. 2.1a shows the time histories for the water surface (η), the real parts of terms 1 and 2 in the Mild Slope Eq. 2.1c and the residual balance of terms (Sum), and also the imaginary parts of the same terms. Similar results for transitional depth Wave B and shallow water Wave C are shown in Figs. 2.2a and 2.3a respectively. As all quantities are dimensionless, the residual errors can be measured against the water surface trace. The error varies with the wave phase but a convenient measure of the overall relative error would be $\epsilon = |\text{Peak Sum}|/H$, where both the residual error and the wave height have been normalized by ω and g. If Airy kinematics were appropriate, ϵ would be identically zero. These relative errors are summarized in column (a) of Table 2.

Table 2. Dimensionless relative error $\epsilon = |\text{Peak Sum}|/H$ in evolution equations

Equation	a	b	c	d	e	f	g	h	i
(A) Deep	0.29	0.16	0.18	0	4.3	7.5	4.3	1.0	0.13
(B) Transitional	9.4	3.7	4.0	0	1.4	3.7	1.4	0.95	0.25
(C) Shallow	180	47	47	0	1.0	3.0	1.0	0.63	0.33

Equation labelling corresponds with Figures 2.1 through 2.3

The residual error is smallest in deep water, where Fourier and Airy kinematics differ least, but at 0.29 (29%) it is only marginally tolerable. Energy is not conserved by the Mild Slope equation, and errors are quite severe in transitional ($\epsilon = 9.4$) and shallow ($\epsilon = 180$) water. Unfortunately, the Mild Slope equation is most frequently applied in transitional and shallow water, to problems such as combined refraction and diffraction.

The substantial computational resources demanded by the Mild Slope equation in application to the nearshore environment have encouraged the development of parabolic approximations. For waves propagating at or near to the x direction, existing parabolic approximations include those of Radder (1979)

$$i2kCC_g\frac{\partial A}{\partial x} + \frac{\partial}{\partial y}\left(CC_g\frac{\partial A}{\partial y}\right) + \left[2k(k - k_0)CC_g + i\frac{\partial}{\partial x}(kCC_g)\right]A = 0 \qquad (2.3a)$$

and a weakly-nonlinear extension by Kirby and Dalrymple (1983)

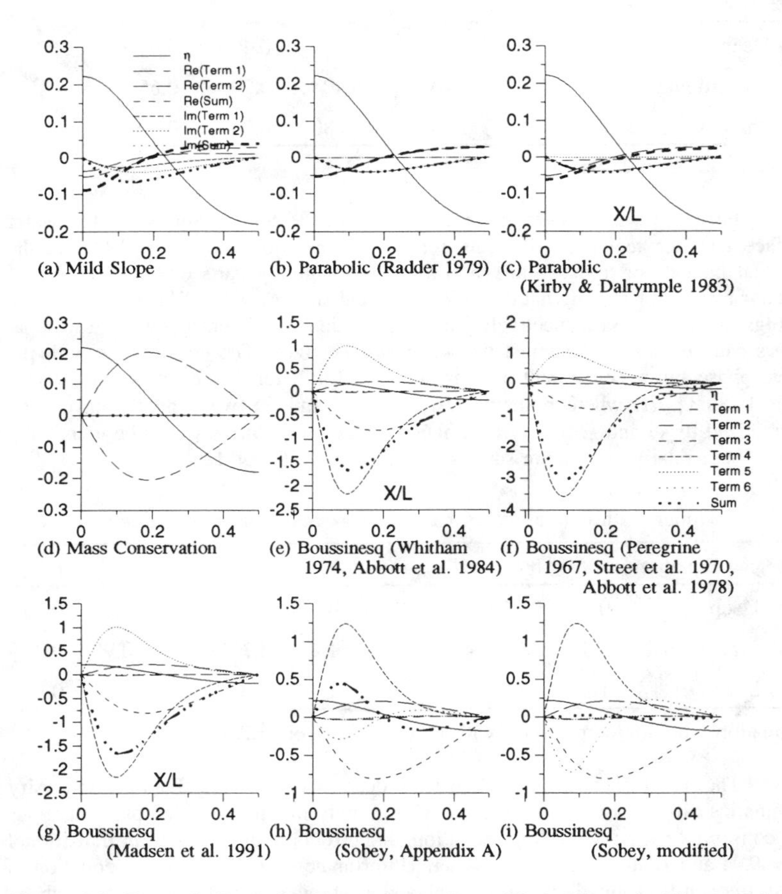

Figure 2.1 Terms in Evolution Equations for Deep Water Wave A. All quantities normalized by ω and g.

$$i2kCC_g\frac{\partial A}{\partial x} + \frac{\partial}{\partial y}\left(CC_g\frac{\partial A}{\partial y}\right) + \left[2k(k-k_0)CC_g + i\frac{\partial}{\partial x}(kCC_g)\right]A - k^4BCC_g|A|^2A = 0 \tag{2.4a}$$

in which $B(kh) = (\cosh 4kh + 8 - 2\tanh^2 kh)/(8\sinh^4 kh)$ and k_0 is a reference wave number. For periodic, spatially uniform waves over a horizontal bed, these become

$$i2k\frac{\partial A}{\partial x} = 0 \qquad \text{and} \qquad i2k\frac{\partial A}{\partial x} - k^4B|A|^2A = 0 \tag{2.3b,2.4b}$$

respectively. Wider angle linear approximations (e.g., Dalrymple and Kirby 1988)

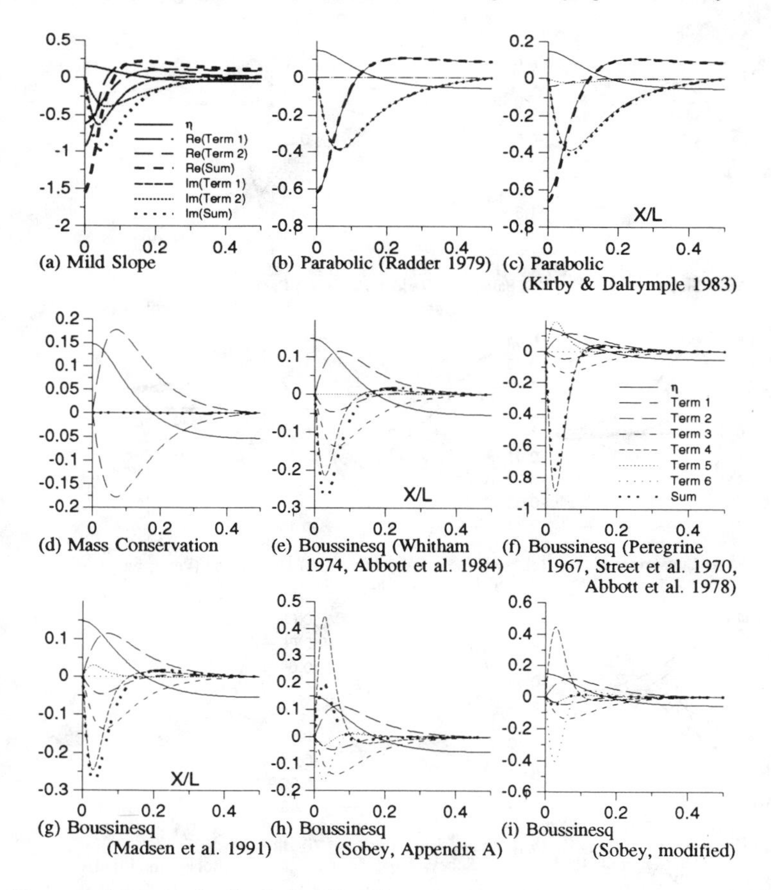

Figure 2.2 Terms in Evolution Equations for Transitional Depth Wave B. All quantities normalized by ω and g.

also reduce to Eq. 2.3b.

Figs. 2.1b through 2.3b show the terms in the Radder parabolic approximation, together with the water surface. The relative errors are listed in Table 2b. The errors are consistently smaller than for the full Mild Slope equation, because of the omission of the $\partial^2 A/\partial x^2$ term. This was expected as the steady progressive wave kinematics adopted for evaluation are consistent with the omission of back-scattered wave modes in the parabolic approximation. Nevertheless, the errors are still only barely tolerable in deep water ($\epsilon = 0.16$) and still unacceptable in transitional ($\epsilon = 3.7$) and shallow ($\epsilon = 47$) water. Error patterns in the parabolic but weakly-nonlinear approximation of Kirby and Dalrymple (Figs. 2.1c through 2.3c, Table 2c)

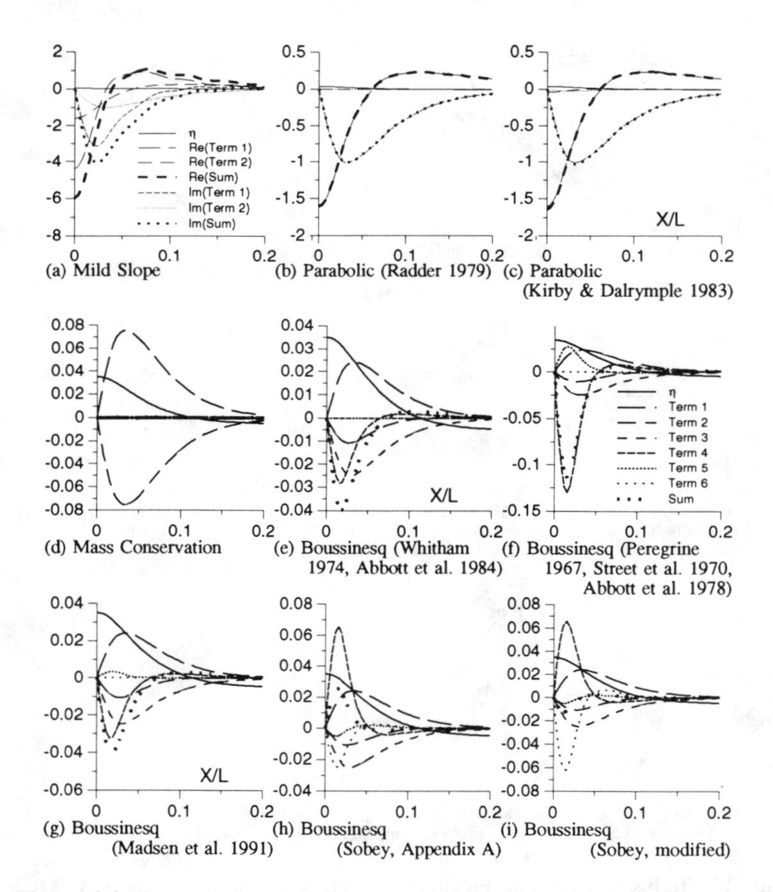

Figure 2.3 Terms in Evolution Equations for Shallow Water Wave C for truncated range $0 \leq X/L \leq 0.2$. All quantities normalized by ω and g.

closely follow the Radder approximation, but with consistently larger ϵ values.

3. Boussinesq Equations

Efforts to establish a set of depth-integrated evolution equations whose validity extends beyond the linear approximation were initiated by Boussinesq in 1876. Attention has mostly been directed to wave evolution in shallow water, but consideration need not be limited to shallow water.

While the water surface is generally described by the local elevation, Boussinesq-style approximations in the literature have used a variety of dependent variables to characterize the local horizontal flow. The depth-integrated flow, the depth-averaged velocity, the velocity at the MWL and the velocity at the bed have all been used. The choice will influence the form of the integral evolution equations and complicate interpretation. There is no established consensus.

For the present discussion, the dependent variables are chosen as the local water surface elevation $\eta(x_\alpha, t)$ [α = 1 and 2] and the local depth-integrated horizontal flow q_α (defined in Eq. A.3). These have the virtue of leading to a relatively simple set of model equations that have much in common with the classical St. Venant equations for long (i.e., shallow water) wave propagation. A wide and potentially confusing range of assumptions has been introduced in attempts to establish a tractable set of "model" evolution equations.

Integral mass conservation is consistently written as

$$\frac{\partial \eta}{\partial t} + \frac{\partial q_\alpha}{\partial x_\alpha} = 0 \tag{3.1}$$

This equation describes mass conservation for a local elemental control volume extending vertically from the bed to the free surface. The initial term is the rate of change of storage within the control volume and the final term is the net outflow from the control volume.

There is no such uniformity with the momentum equation. There are several alternate Boussinesq-style momentum equations. Differences are mostly centered on contributions to the pressure term in the integral momentum equation (Eq. A.2)

$$-\frac{\partial}{\partial x_\alpha}\int_{-h}^{\eta} p\,dz + p|_{-h}\frac{\partial h}{\partial x_\alpha} \tag{3.2}$$

from the temporal component of the vertical acceleration, and the advective components of the vertical acceleration. All rational Boussinesq equations include the gravitational contribution, $-\rho g(h+\eta)\,\partial h/\partial x_\alpha$, which is also a familiar term in the St. Venant or Long Wave Equations. Grouping contributions from the residual pressure terms together as ρD_α, Boussinesq-style momentum equations may be written as

$$\rho\frac{\partial q_\alpha}{\partial t} + \rho\frac{\partial}{\partial x_\beta}\left(\frac{q_\alpha q_\beta}{h+\eta}\right) = -\rho g(h+\eta)\frac{\partial \eta}{\partial x_\alpha} + \rho D_\alpha + \Sigma_{s\alpha} - \Sigma_{b\alpha} \tag{3.3}$$

In addition to the choice of dependent variables, alternative derivations of the Boussinesq momentum equations differ in the details of the dispersion term D.

Without the dispersion term, the mass and momentum equations predict wave propagation at $\sqrt{(g(h+\eta))}$, even in deep water. That the phase speed is not depth but frequency dependent in deep water, where it is also much less than $\sqrt{(g(h+\eta))}$, demonstrates the importance of the dispersion terms in transitional and deep water.

Some common alternatives to the dispersion term are:

(I) Boussinesq, Whitham (1974), Abbott et al. (1984)

$$D_\alpha = -\frac{1}{3}h^2\frac{\partial^3(h+\eta)}{\partial x_\alpha \partial t^2} = \frac{1}{3}h^2\frac{\partial^3 q_\beta}{\partial x_\alpha \partial x_\beta \partial t} \tag{3.4}$$

(II) Peregrine (1967), Street et al. (1970), Abbott et al. (1978)

$$D_\alpha = \frac{1}{2}h(h+\eta)\frac{\partial^3}{\partial x_\alpha \partial x_\beta \partial t}\left(\frac{h+\eta}{h}q_\beta\right) - \frac{1}{6}h(h+\eta)^2\frac{\partial^3}{\partial x_\alpha \partial x_\beta \partial t}\left(\frac{q_\beta}{h}\right) \tag{3.5}$$

(III) Witting (1984; B=1/15), Madsen et al. (1991; B=1/21)

$$D_\alpha = (B+\frac{1}{3})\,h^2\frac{\partial^3 q_\beta}{\partial x_\alpha \partial x_\beta \partial t} + Bgh^3\frac{\partial^3\eta}{\partial x_\alpha \partial x_\beta^2} \tag{3.6}$$

(IV) Sobey (see Appendix A)

$$\begin{aligned} D_\alpha = &+ I_{p1}\frac{\partial}{\partial x_\alpha}\left[(h+\eta)\left(\frac{\partial q_\beta}{\partial x_\beta}\right)^2\right] + I_{p2}\frac{\partial}{\partial x_\alpha}\left[(h+\eta)\frac{\partial}{\partial t}\left((h+\eta)\frac{\partial q_\beta}{\partial x_\beta}\right)\right] \\ &+ I_{p4}\frac{\partial}{\partial x_\alpha}\left[(h+\eta)\frac{\partial}{\partial x_\delta}\left(q_\delta\frac{\partial q_\beta}{\partial x_\beta}\right)\right] \end{aligned} \tag{3.7}$$

Form (I) is essentially the original Boussinesq form. Shallow water wave kinematics are also assumed in the derivation of form (II), and the validity of both forms is accordingly restricted to shallow water. Forms (III) are attempts to extend the validity of Boussinesq-style momentum equations to deeper water, through empirical adjustments to the dispersion term that better approximate the linear (Airy) phase speed in deeper water. Alternative (IV) is a systematic attempt to include all contributions to the vertical pressure structure and to extend the validity beyond shallow water. The shape factors I_{p1}, I_{p2} and I_{p4} (see Fig. A1) change slowly from shallow to deep water; their asymptotic values in shallow water are 1/3.

Evaluation of the conservation properties of each of these equations has again used near-exact kinematics for steady progressive waves. The details are identical with those adopted in the discussion of the Mild Slope equation in the previous section. In application, all quantities have been normalized by g and $\omega = 2\pi/T$ and all partial derivatives in the model evolution equations were evaluated analytically.

The terms in the mass conservation Eq. 3.1 are shown in Figs. 2.1d through 2.3d. As expected from the use of near-exact Fourier kinematics, the residual mass balance is also near-exact throughout. Eq. 3.1 provides an unquestionably adequate description of mass conservation in water of all depths.

The terms in variations I through IV of the Boussinesq/momentum equation are shown in Figs. 2.1e through 2.3e (Whitham 1974, Abbott et al. 1984), Figs. 2.1f through 2.3f (Peregrine 1967, Street et al. 1970, Abbott et al. 1978), Figs. 2.1g through 2.3g (Madsen et al. 1991) and Figs. 2.1h through 2.3h (Sobey, Appendix A) respectively. The terms are numbered from left to right. To simplify the graphical presentation, the sign of Terms 3 through (4 or 5 or) 6 on the right-hand side has been changed so that all terms should sum to zero. Relative errors are summarized in Table 2, columns e through h.

Disturbingly, momentum conservation is consistently violated by all these Boussinesq-style momentum equations. Variations I and III are essentially identical and the empirical correction introduced by Madsen at al. (1991) has no discernible influence. The popular variation II does surprisingly poorly throughout, consistently returning the largest relative error; ϵ = 3.0 (300%) in shallow water where this variation should be most appropriate. Variation IV consistently returns the smallest relative error, from $\epsilon = 0.63$ in shallow water to $\epsilon = 1.0$ in deep water. Nevertheless, these error levels remain unacceptable.

Reduced error levels could be achieved by an empirical modification to variation IV, which increased the I_{p4} shape factor by a factor of 2.5. The results are shown in Figs. 2.1i through 2.3i and Table 2i. Error levels are reduced, now ranging from $\epsilon = 0.33$ in shallow water to $\epsilon = 0.13$ in deep water. On the evidence of the present evaluations, this might be a tolerably appropriate Boussinesq-momentum equation. It has the smallest relative error in shallow water. Its validity also extends to deep water, where, in contrast to all other Boussinesq variations, the relative error is even smaller.

There are a number of examples in the literature of reasonably successful comparisons of experimental measurements with numerical predictions from Boussinesq-style equations. These results would appear to conflict with the present results, though they may not on closer scrutiny. Several observations are appropriate. Firstly, the local error traces reported above are in all cases odd functions about the crest, such that wave-averaged momentum is still conserved. A flow response initiated by momentum balance errors before the crest would be shortly afterwards reversed by equal and opposite errors following the crest. Flow details at the crest may be compromised but the impact may not be apparent elsewhere. Secondly, the spatial scale about the peaks in the local error traces is of order $0.01L$ in shallow water (see, e.g., Figs. 2.3e through 2.3g); the associated temporal scale is of order $0.01T$. To resolve such rapid changes in a numerical model would require spatial and temporal resolutions of similar magnitudes. Space and time steps of order $0.05L$ and $0.05T$, typical of wave models, will numerically filter the response pattern and may significantly mitigate the influence of the local error peaks.

Conclusions

Existing water wave evolution equations, be they "mild slope" or "Boussinesq", do not satisfy the appropriate conservation laws locally for even moderate wave conditions when reduced to steady progressive waves over a horizontal bottom. The mild slope equation and associated parabolic approximations do especially poorly in transitional and shallow water, just where they are commonly applied to model combined refraction and diffraction. Boussinesq-style approximations do better in shallow water but the relative error, of order 100%, cannot be acceptable. A viable water wave evolution equation is not presently available.

Almost acceptable conservation properties are achieved by an empirical modification to a new Boussinesq-style equation. The validity of this equation extends from deep to shallow water.

Appendix A-
Nonlinear Water Wave Evolution Equations without Restriction on Water Depth

This Appendix seeks to establish Boussinesq-style evolution equations for water waves that retain the crucial nonlinearity but do not restrict validity to shallow water. The derivation will be based on the exact integral conservation equations for mass and momentum, respectively

$$\rho\frac{\partial\eta}{\partial t} + \rho\frac{\partial}{\partial x_\alpha}\int_{-h}^{\eta} u_\alpha\, dz = 0 \tag{A.1}$$

and

$$\rho\frac{\partial}{\partial t}\int_{-h}^{\eta} u_\alpha\, dz + \rho\frac{\partial}{\partial x_\beta}\int_{-h}^{\eta} u_\alpha u_\beta\, dz = -\frac{\partial}{\partial x_\alpha}\int_{-h}^{\eta} p\, dz + p|_{-h}\frac{\partial h}{\partial x_\alpha} + \frac{\partial}{\partial x_\beta}\int_{-h}^{\eta}\tau_{\alpha\beta}\, dz + \Sigma_{s\alpha} - \Sigma_{b\alpha} \tag{A.2}$$

in which x_α [α = 1 and 2] is the horizontal position, t is time, $\eta(x_\alpha,t)$ is the water surface elevation above the global SWL (still water level), $u_\alpha(x_\alpha,z,t)$ [α = 1 and 2] is the horizontal velocity, $p(x_\alpha,z,t)$ is the pressure, $h(x_\alpha)$ is the water depth below the global SWL, $\tau_{\alpha\beta}(x_\alpha,z,t)$ is the internal shear, $\Sigma_{s\alpha}$ [α = 1 and 2] is the surface shear and $\Sigma_{b\alpha}$ [α = 1 and 2] is the bed shear. Though literature presentations have rarely adopted this precise approach, the conservation laws and bottom and free surface boundary conditions embodied in these integral equations must remain the foundation of any rational analysis. Further use of these equations must acknowledge that the water surface elevation η will vary with horizontal position and time, and that the velocity field (u_α, w), the pressure p and the shear stress τ will vary with vertical position z and also with horizontal position and time.

Most of the terms in the integral conservation equations involve depth-integration from the local bottom to the local water surface. Truncation of these integrals at the MWL, as in the linear approximation, is not appropriate. Much of the activity in water waves is concentrated near the water surface and the trough-crest region demands careful attention on a process by process basis. Boussinesq-style approximations are frequently associated with enhanced fidelity in the representation of the pressure and advective acceleration terms in the momentum equations. These terms are important, but they are not the only aspects of the integral conservation equations that need careful consideration.

The dependent variables are chosen as the local water surface elevation η and the local depth-integrated horizontal flow

$$q_\alpha = \int_{-h}^{\eta} u_\alpha \, dz \tag{A.3}$$

These same dependent variables are routinely adopted for long (i.e., shallow water) wave propagation, leading in that context to the classical St. Venant equations. The present discussion is nevertheless directed toward short wave propagation, but the common features might profitably be exploited.

The conservation equations are dominated by integrals in z from the local bed to the local water surface. The implicit rationale of most model evolution equations has been the expectation that complete evaluation of these definite integrals will eliminate the z dependence by consistently yielding results that are dependent only on horizontal position x_α and time t.

Separation of the z dependence will guarantee complete evaluation of all the depth integrals in the integral conservation equations and elimination of the z dependence. This suggests a partial separation of variables, typically of the form

$$u_\alpha(x_\alpha, z, t) = U(x_\alpha, t)\, f(z) \tag{A.4}$$

where $U_\alpha(x_\alpha, t)$ and $f(z)$ characterize respectively the horizontal flow and the vertical profile of the horizontal velocity. Though this step is not always highlighted in literature presentations, it is nonetheless pivotal in establishing viable model evolution equations of this genre. Integral methods such as these have been widely successful in pragmatic models of boundary layers, jets and plumes. A partial separation of variables, separating out the dependence on the cross-stream variable, is a necessary assumption of any integral method.

With $\eta(x_\alpha, t)$ and $q(x_\alpha, t)$ as the dependent variables, separating the z coordinate suggests that the horizontal velocity could be represented as

$$u_\alpha(x_\alpha, z, t) = \frac{q_\alpha(x_\alpha, t)}{h(x_\alpha) + \eta(x_\alpha, t)}\, f(z) \tag{A.5a}$$

where the dimensionless function $f(z)$ describes the vertical profile of u_α and

$$\frac{1}{h + \eta} \int_{-h}^{\eta} f(z)\, dz = 1 \tag{A.5b}$$

for consistency with the definition of q (Eq. A.3). This form will permit complete evaluation and simplification of all depth integrals in the conservation equations. No distinction is made, nor can be made, between local wave and local current contributions to the vertical profile of u_α. At some stage of the analysis, explicit descriptions of the vertical structure must be adopted. In the present discussion, this step is deferred to assist in the identification of the eventual influence of the profile assumption.

Though the separation of variables is required if the integral method is to be at all viable, the assumption is quite a reasonable one. It is mplicit in linear wave theory and is also used in the linear mild slope approximation. It is also reasonably consistent with nonlinear steady wave theory, in shallow and even in deep water.

Substituting Eq. A.5a for u_α into Eq. A.1, *integral mass conservation* becomes

$$\frac{\partial \eta}{\partial t} + \frac{\partial q_\alpha}{\partial x_\alpha} = 0 \tag{A.6}$$

Completion of the integrals in the momentum equation requires estimates of the vertical structure of both the vertical velocity and the pressure, vital elements of any Boussinesq approximation. The pressure structure in particular has a decisive influence on the utility of the Boussinesq approximation in describing wave evolution. Integration of the vertical momentum equation from arbitrary level z to the free surface gives the vertical distribution of pressure as

$$p(z, x_\alpha, t) = \rho g(\eta - z) - \rho w^2 + \frac{\partial}{\partial t}\int_z^\eta \rho\, w\, dz' - \frac{\partial}{\partial x_\alpha}\int_z^\eta (\tau_{z\alpha} - \rho\, w u_\alpha)\, dz' - \Sigma_{sz} + \tau_{zz} \tag{A.7}$$

where z' is a dummy variable. Terms one (gravity) and then three (vertical acceleration) on the right side normally provide the major contributions. Similar vertical integration of the mass conservation equation from the bed to arbitrary level z gives the vertical distribution of vertical velocity as

$$\rho\, w(z) = -\frac{\partial}{\partial t}\int_{-h}^{z} \rho\, dz' - \frac{\partial}{\partial x_\alpha}\int_{-h}^{z} \rho\, u_\alpha\, dz' \tag{A.8}$$

The initial term on the right-hand side is zero for a fluid of constant density.

Using the Eq. A.5a separation of variables, the vertical velocity becomes

$$w(z, x_\alpha, t) = -\frac{\partial q_\alpha}{\partial x_\alpha} f_1(z) \quad \text{where } f_1(z) = \frac{1}{h+\eta}\int_{-h}^{z} f(z')\, dz' \tag{A.9}$$

Similarly, the pressure becomes

$$p(z, x_\alpha, t) = \rho g(\eta - z) - \rho\left(\frac{\partial q_\alpha}{\partial x_\alpha}\right)^2 f_1^2(z) - \rho\frac{\partial}{\partial t}\left((h+\eta)\frac{\partial q_\alpha}{\partial x_\alpha}\right) f_2(z) - \frac{\partial}{\partial x_\alpha}\left(q_\alpha \frac{\partial q_\beta}{\partial x_\beta}\right) f_3(z) \tag{A.10a}$$

where

$$f_2(z) = \frac{1}{h+\eta}\int_z^\eta f_1(z')\, dz' \quad \text{and} \quad f_3(z) = \frac{1}{h+\eta}\int_z^\eta f_1(z') f(z')\, dz' \tag{A.10b}$$

Separation of the z dependence remains necessary to guarantee complete evaluation of the more complicated depth integrals in the momentum equation; the f_1 through f_3 profile assumptions follow naturally from Eqs. A.5.

Viscous stresses are completely ignored in steady wave theory and have little influence on wave evolution generally. Omitting terms in the momentum balances involving the viscous shear stresses, $\tau_{z\alpha}$ and $\tau_{\alpha\beta}$, simplifies the following discussion. Both the bed shear Σ_b and the surface shear Σ_s are retained.

Substituting now for w and p in Eq. A.2 gives the horizontal momentum equations in Table 3. The bed slope $\partial h/\partial x_\alpha$ will usually be small (of order 0.01 or less) and the terms on the right-hand side directly multiplied by the bed slope can be omitted. Note however that this does not imply a locally horizontal bed assumption as the kinematic bottom boundary condition for a sloping bed is explicitly accommodated by Eq. A.2. The *horizontal integral momentum conservation* equations then become

$$\rho\frac{\partial q_\alpha}{\partial t} + \rho I_{a1}\frac{\partial}{\partial x_\beta}\left(\frac{q_\alpha q_\beta}{h+\eta}\right) = -\rho g(h+\eta)\frac{\partial \eta}{\partial x_\alpha} + \rho I_{p1}\frac{\partial}{\partial x_\alpha}\left[(h+\eta)\left(\frac{\partial q_\beta}{\partial x_\beta}\right)^2\right] + \rho I_{p2}\frac{\partial}{\partial x_\alpha}\left[(h+\eta)\frac{\partial}{\partial t}\left((h+\eta)\frac{\partial q_\beta}{\partial x_\beta}\right)\right] + \rho I_{p4}\frac{\partial}{\partial x_\alpha}\left[(h+\eta)\frac{\partial}{\partial x_\delta}\left(q_\delta\frac{\partial q_\beta}{\partial x_\beta}\right)\right] + \Sigma_{s\alpha} - \Sigma_{b\alpha} \tag{A.11}$$

All but terms two through four on the right-hand side are familiar from the classical St. Venant equations for long wave evolution. The inertia terms on the left-hand side describe respectively temporal accumulation (storage) of momentum and advection (net outflow) of momentum for the local depth-integrated control volume. The right-hand side sums the horizontal surface forces on the local control volume. Terms 1 through 4 are contributed by the local pressure; in order, they represent contributions from the gravitational acceleration, the temporal component of the vertical acceleration, and the advective components of the vertical acceleration. Term five is an applied stress at the water surface and term six a frictional stress at the bed. In shallow water, the pressure terms are dominated by term one, which is associated with an instantaneous hydrostatic pressure distribution from the local water surface rather than from the MWL.

Table 3. Primitive integral equation for horizontal momentum

$$\rho\frac{\partial q_\alpha}{\partial t} + \rho I_{a1}\frac{\partial}{\partial x_\beta}\left(\frac{q_\alpha q_\beta}{h+\eta}\right) = -\rho g(h+\eta)\frac{\partial \eta}{\partial x_\alpha} + \rho I_{p1}\frac{\partial}{\partial x_\alpha}\left[(h+\eta)\left(\frac{\partial q_\beta}{\partial x_\beta}\right)^2\right] + \rho I_{p2}\frac{\partial}{\partial x_\alpha}\left[(h+\eta)\frac{\partial}{\partial t}\left((h+\eta)\frac{\partial q_\beta}{\partial x_\beta}\right)\right] - \rho I_{p3}\frac{\partial}{\partial t}\left((h+\eta)\frac{\partial q_\beta}{\partial x_\beta}\right)\frac{\partial h}{\partial x_\alpha} + \rho I_{p4}\frac{\partial}{\partial x_\alpha}\left[(h+\eta)\frac{\partial}{\partial x_\delta}\left(q_\delta\frac{\partial q_\beta}{\partial x_\beta}\right)\right] - \rho I_{p5}(h+\eta)\frac{\partial}{\partial x_\delta}\left(q_\delta\frac{\partial q_\beta}{\partial x_\beta}\right)\frac{\partial h}{\partial x_\alpha} + \Sigma_{s\alpha} - \Sigma_{b\alpha}$$

where -

$$I_{a1} = \frac{1}{h+\eta}\int_{-h}^{\eta} f^2(z)\,dz \qquad I_{p1} = \frac{1}{h+\eta}\int_{-h}^{\eta} f_1^2(z)\,dz \qquad I_{p2} = \frac{1}{h+\eta}\int_{-h}^{\eta} f_2(z)\,dz$$

$$I_{p3} = \frac{1}{h+\eta}\int_{-h}^{\eta} f_1(z)\,dz \qquad I_{p4} = \frac{1}{h+\eta}\int_{-h}^{\eta} f_3(z)\,dz \qquad I_{p5} = \frac{1}{h+\eta}\int_{-h}^{\eta} f(z)\,f_1(z)\,dz$$

From the experience of integral methods in steady turbulent shear flows ranging from open channel flow through jets and plumes to pipe flow, there is an expectation that the profile shape factors, the dimensionless definite integrals I_{a1} to I_{p4}, will be almost constant over a wide range of depths, and that these shape factors will not be especially sensitive to the precise details of the $f(z)$ profile. In steady open channel flow, for example, the profile shape factors I_{a1} is the momentum velocity profile correction factor (Henderson 1966).

The relative insensitivity of the profile shape factors to changes in depth can be confirmed with any reasonable choice of profile, say

$$f(z) = \frac{r}{\sinh r} \cosh k(h + z) \tag{A.12}$$

where $r = k(h+\eta)$, as suggested by linear wave theory without current. All shape factors are shown in Fig. A.1 as a function of $\omega^2 h/g$. Note that in shallow and transitional water for $kh < 1$, all of the shape factors are almost constant, I_{a1} at 1, and I_{p1}, I_{p2} and I_{p4} at 1/3. Certainly, local spatial and temporal gradients of these shape factors will be small and it is reasonable to assume that they are locally constant.

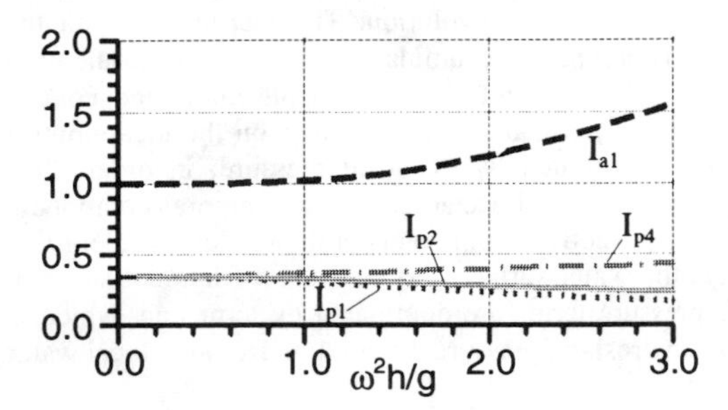

Figure A.1 Shape factors in integral evolution equations

A simple uniform profile, $f(z) = 1$, has proved remarkably successful in pipe and open channel flow and even for buoyant jets (Morton, Taylor and Turner 1956). In the present case, it gives results that are constant and identical to the shallow water asymptote of the cosh profile above. While a uniform profile for horizontal velocity is reasonable for shallow water waves, the cosh profile is credible in deep, transitional and shallow water and does not restrict the Boussinesq equations to shallow water, as has frequently been the case. Locally constant profile shape factors are implicit, where appropriate, in the integral conservation equations for momentum.

There are several alternate presentations of Boussinesq équations in the literature. When transformed to the present choice of dependent variables, the mass equations are identical, as are many terms of the momentum equations. Differences in Boussinesq momentum equations are centered on pressure terms two through four, those contributed by the temporal component of the vertical acceleration, and the advective components of the vertical acceleration. These exclude the gravitational term one, which is also a familiar term in the St. Venant or Long Wave Equations.

References

Abbott, M.B., Petersen, H.M., and Skovgaard, O. (1978). "On the numerical modelling of short waves in shallow water," *Journal of Hydraulic Research*, 18, 173-203.

Abbott, M.B., McCowan, A.D., and Warren, I.R. (1984). "Accuracy of short-wave numerical models," *Journal of Hydraulic Engineering*, 110, 1287-1301.

Berkhoff, J.C.W. (1972). "Computation of combined refraction-diffraction." In *Procs., 13th Conference on Coastal Engineering, Vancouver*. ASCE, New York, 1, 471-490.

Booij, N. (1981). *Gravity waves on water with non-uniform depth and current.* (Communications on Hydraulics, 81-1.) Delft University of Technology, Delft, The Netherlands.

Dalrymple, R.A., and Kirby, J.T. (1988). "Models for very wide-angle water waves and wave diffraction," *Journal of Fluid Mechanics*, 192, 33-50.

Henderson, F.M. (1966). *Open Channel Flow*. MacMillan Publishing, New York.

Kirby, J.T., and Dalrymple, R.A. (1983). "A parabolic equation for the combined refraction-diffraction of Stokes waves by mildly varying topography," *Journal of Fluid Mechanics*, 136, 453-466.

Madsen, P.A., Murray, R., and Sorensen, O.R. (1991). "A new form of the Boussinesq equation with improved linear dispersion characteristics," *Coastal Engineering*, 15, 371-388.

Miles, J.W. (1991). "Variational formulations for gravity waves in water of variable depth," *Journal of Fluid Mechanics*, 232, 681-688.

Morton, B., Taylor, G.I., and Turner, J.S. (1956). "Turbulent gravitational convection from maintained and instantaneous sources," *Procs., Royal Society, London, A*, 234, 1-23.

Peregrine, D.H. (1967). "Long waves on a beach," *Journal of Fluid Mechanics*, 27, 815-827.

Radder, A.C. (1979). "On the parabolic equation method for water wave propagation," *Journal of Fluid Mechanics*, 95, 159-176.

Smith, R., and Sprinks, T. (1975). "Scattering of surface waves by a conical island," *Journal of Fluid Mechanics*, 72, 373-384.

Sobey, R.J. (1989). "Variations on Fourier wave theory," *International Journal for Numerical Methods in Fluids*, 9, 1453-1467.

Sobey, R.J. (1993). "Quantifying coastal and ocean processes (Kevin Stark Memorial Keynote Address)." In *COASTS 1993, Procs., 11th Australasian Conference on Coastal and Ocean Engineering, Townsville, Australia, August.* The Institution of Engineers, Australia, Canberra.

Sobey, R.J., and Liang, H.B. (1989). "The complex envelope of surface gravity waves," *Journal of Waterway, Port, Coastal and Ocean Engineering*, 115, 681-700.

Street, R.L., Chan, R.K.C., and Fromm, J.E. (1970). "Two methods for the computation of the motion of long water waves - A review and applications." In *Procs., 8th Naval Hydrodynamics Symposium, Rome*. 147-187.

Whitham, G.B. (1974). *Linear and Nonlinear Waves*. Wiley, New York.

Witting, J.M. (1984). "A unified model for the evolution of nonlinear water waves," *Journal of Computational Physics*, 56, 203-236.

Nonlinear Decomposition of a 2-D Wave Field

Jun Zhang [1], R.E. Randall , L. Chen, C.A. Spell, J.K. Longridge, & M. Ye

I. Abstract

Based on an understanding of nonlinear wave interaction in a dual component wave, a new numerical scheme allowing for hybrid wave-mode modeling has been developed. In the hybrid wave-mode modeling a conventional wave-mode function is used to describe dominant component waves, and a modulated wave-mode function to describe high-frequency component waves. As a result rapid convergence is obtained for the nonlinear decomposition and free waves distinguishable from bound waves. Using this numerical scheme, predictions of wave kinematics and wave propagation based on a time record of two-dimensional wave elevation are computed and display satisfactory comparison with related laboratory measurements.

II. Introduction

An irregular wave field may be viewed as consisting of many component waves with different wavelengths (or frequencies) and amplitudes. Decomposing a wave field into component waves is essential for studying the dynamics of ocean waves. In this paper, the decomposition of an irregular wave field is investigated for the purpose of predicting its short-distance evolution and kinematics, which is crucial to a variety of offshore and coastal engineering applications. For example, the prediction of wave loads on offshore structures using Morison's equation requires the determination of the kinematics induced by waves in the ambient fluid and the wave evolution over a short distance (e.g. from one leg of a platform to another). The calculation of wave kinematics is important in estimating the migration of dredged material or oil spills. To evaluate sediment transport in the coastal zone and to cap contaminated dredged material on the seabed, the computation of wave kinematics and pressure fields near the sea bottom is useful. Also, to measure wave heights using microwave return from the sea surface, the relationship between the wave-induced particle velocity at the sea surface and the wave elevation must be accurately determined.

[1]Ocean Engineering Program, Department of Civil Engineering, Texas A&M University, College Station, TX 77843-3136. Tele: (409)-845-2168, FAX: (409)-845-6156

Recent studies (Phillips 1981; Longuet-Higgins 1987; Zhang & Melville 1990) on the interaction between short- (wavelength) and long- (wavelength) waves showed that the phases, frequencies, wavelengths and amplitudes of short waves are modulated along the surface of long waves. Thus, the characteristics of component waves, especially those with relatively short wavelengths, are no longer constant but modulated or changed in space and time. The linear decomposition of a wave field into component waves, based on a Fast Fourier Transform (FFT) spectral method, ignores the modulation of component waves. This results in large errors in the prediction of wave kinematics (Donelan et al. 1992). Conventional perturbation (WKB) approaches consider the modulation of component waves as additional "bound" or "forced" waves, and express them in terms of second or higher order solutions. In the case of a narrow-banded wave field, where the frequencies of all significant component waves are close to each other, the solution converges quickly and a truncated solution renders accurate results. However, in the case of a broad-banded wave field, the solution does not converge if truncated at a finite order due to the large modulation in short-wave phases (Zhang et al. 1993). Since ocean waves are often broad-banded and have multiple-peak spectra (Smith & Vincent 1992), the methods based on the assumption of a narrow-banded spectrum may not be reliable when applied to ocean waves. To overcome the convergence difficulty, the modulated wave characteristics (especially phases) of the short component waves are directly modelled in the formulation of leading-order wave components. This is done without invoking the slowly converging solution for those "bound" or "forced" component waves. The direct decomposition of an irregular wave field into modulated component waves is highly nonlinear. A numerical scheme is developed for the nonlinear decomposition. Preliminary numerical results show satisfactory comparison with related laboratory measurements.

A hybrid wave model employing the formulations of both conventional and modulated component waves and the corresponding numerical scheme for decomposing a unidirectional irregular wave field are briefly described in Sections II and III. In Section IV, the numerical results are compared with related laboratory measurements as well as predictions based on other methods currently used by offshore and coastal engineers. Finally, future work and further improvements are discussed.

III. Formulation

The hybrid wave model divides the amplitude and phase spectra of a wave field into several bands as sketched in Figure 1. Each band may have many component waves. The number of component waves within each band is limited by the frequency ratio of longest to shortest component waves. The ratio in our numerical computation is about 0.5 but no less than 0.33. As long as the wave steepness of a long-wave band is much smaller than the square of the frequency ratio of long to short component waves, the decomposition of an irregular wave field is insensitive to changes in each band's width. Component

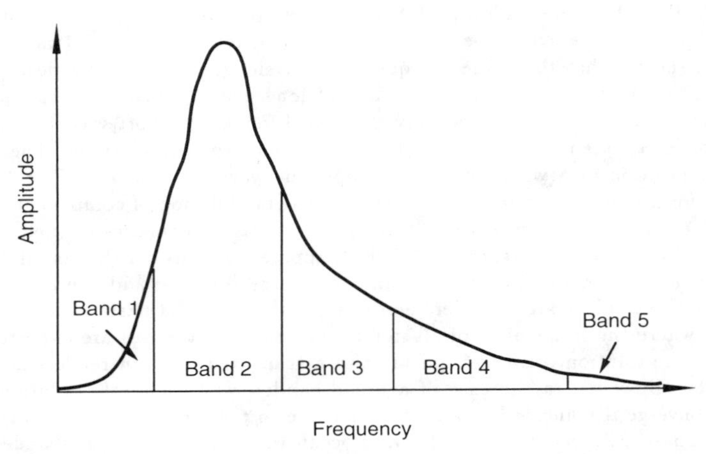

Figure 1. Illustration of wave spectrum band division

waves within each band have close frequencies; therefore, the interaction between any pair of component waves in the same band can still be modeled by " bound" waves, same as in conventional perturbation approaches. In the case of interaction between a pair of component waves from different bands, which are likely to have large frequency differences, the modulation of shorter component waves by longer component waves is significant. Thus, it is directly modeled in the formulation of modulated component waves. The derivation based on our previous studies of the short- and long-wave interaction (Zhang 1990; Zhang et al. 1993) will not be repeated in this paper. Alternatively, we show that the formulation of a hybrid wave model satisfies the governing equations of a unidirectional irregular wave field up to second order. The extension of the solution up to third order is by no means trivial due to the wave instability resulting from the well known quartet wave interaction (Phillips 1960; Hasselmann 1962), and it is now being undertaken.

Assuming that the flow is incompressible and irrotational and that the pressure is constant at the free surface, the governing equations in the rectilinear coordinates (x, z) are:

$$\nabla^2\Phi = 0 \tag{1}$$

$$\frac{\partial^2\Phi}{\partial t^2} + g\frac{\partial\Phi}{\partial z} + 2\nabla\Phi\cdot\frac{\partial}{\partial t}\nabla\Phi + \frac{1}{2}\nabla\Phi\cdot\nabla(\nabla\Phi\cdot\nabla\Phi) = 0, \qquad \text{at} \quad z = \eta \tag{2}$$

$$\eta = -\frac{1}{g}[\frac{\partial\Phi}{\partial t} + \frac{1}{2}(\nabla\Phi\cdot\nabla\Phi)], \qquad \text{at} \quad z = \eta \tag{3}$$

$$\nabla\Phi \to 0, \qquad \text{when } z \to -\infty \tag{4}$$

where the $x-$axis points in the direction of wave propagation and z is positive above calm water level and $z = 0$ at the calm water level. Φ and η are the wave potential and elevation, and g is gravitational acceleration. The formulation has been derived for deep water with respect to shorter component waves and intermediate water with respect to longer component waves. However, to avoid a lengthy formulation, herein we only give the solution for the deep water case with respect to longer component waves. The solution shown below is for a wave field divided into one long-wave band and a short-wave band. However, it should be noted that the solution for multiple short-wave bands can be extended and has been implemented in the numerical scheme used in Section V. Since the wave field is divided into long and short wave parts, the wave potential and surface elevation can be considered as the superposition of the long- and short-wave potentials and elevations,

$$\Phi = \phi_1 + \phi_2 \tag{5}$$

$$\eta = \zeta_1 + \zeta_2 \tag{6}$$

where ϕ_1, ϕ_2 and ζ_1, ζ_2 are the long- and short-wave potential and elevation, respectively.

The long wave potential and elevation up to second-order are given by:

$$\phi_1 = A_1 e^{k_1 z} \sin\theta_1 \tag{7}$$

$$\zeta_1 = a_1 \cos\theta_1 + \frac{1}{2} a_1^2 k_1 \cos 2\theta_1 \tag{8}$$

and the short wave potential and elevation may be written as:

$$\phi_2 = A_2 e^{f_2 k_2} \sin\tilde{\theta}_2 + \mathcal{L} \tag{9}$$

$$\zeta_2 = a_2(1 + a_1 k_1 \cos\theta_1)\cos\tilde{\theta}_2 + \frac{1}{2} a_2^2 k_2 \cos 2\tilde{\theta}_2 + O(\varepsilon_1^2 \varepsilon_3^{-1}) a_2 \tag{10}$$

where

$$f_2 = z(1 + 2a_1 k_1 \cos\theta_1) - a_1 \cos\theta_1 e^{k_1 z} \tag{11}$$

$$\tilde{\theta}_2 = \theta_2 + k_2 a_1 \sin\theta_1 (2 - e^{k_1 z}) \tag{12}$$

$$\mathcal{L} = 2k_2 a_1 A_2 \left\{ \sum_{m=1}^{\infty} \frac{[k_1(z - a_1\cos\theta_1)]^{2m}}{(2m)!} \sin\theta_1 \cos\tilde{\theta}_2 + \sum_{m=1}^{\infty} \frac{[k_1(z - a_1\cos\theta_1)]^{2m+1}}{(2m+1)!} \cos\theta_1 \sin\tilde{\theta}_2 \right\} \tag{13}$$

Using the symbolic manipulation package MACSYMA, we substitute the solutions (5) and (6) into the governing equations (1) through (3) and obtain the following results:

$$\nabla^2\Phi = O(\varepsilon_1^2) A_2 k_2^2 e^{f_2 k_2} \tag{14}$$

$$\frac{\partial^2\Phi}{\partial t^2}+g\frac{\partial\Phi}{\partial z}+2\nabla\Phi\cdot\frac{\partial}{\partial t}\nabla\Phi+\frac{1}{2}\nabla\Phi\cdot\nabla(\nabla\Phi\cdot\nabla\Phi) = O(\varepsilon_1^2\varepsilon_3^{-1})A_2\sigma_2^2, \quad \text{at} \quad z=\eta, \quad (15)$$

$$\eta+\frac{1}{g}[\frac{\partial\Phi}{\partial t}+\frac{1}{2}(\nabla\Phi\cdot\nabla\Phi)] = O(\varepsilon_1^2)a_2, \quad \text{at} \quad z=\eta, \quad (16)$$

The remainders (i.e. errors) of each governing equation are of third order at least and are much smaller than the corresponding remainders produced by conventional perturbation approaches.

IV. Numerical Scheme

The surface elevation of an irregular wave field is equal to the summation of the component waves' elevations. Given the elevation time series at a fixed point, a unidirectional irregular wave field can be decomposed into long wave components and short modulated wave components based on the formulation of the hybrid wave model. Different from FFT methods, the modulated wave characteristics of short component waves depend on those of long component waves. In addition, the characteristics of "bound" waves, which represent the interactions of component waves within the same band, depend on the solution of related component waves. Thus the nonlinear decomposition cannot be easily derived and is accomplished through numerical iteration. The flowchart of our numerical scheme is given in Figure 2, which roughly illustrates the iteration process. Because changes in long component waves are much smaller than those in short component waves and the energy in the long wave band is dominant or relatively strong, the decomposition converges in a few iterations. Once the characteristics of long and short component waves are obtained, each component's surface elevation and kinematics can be predicted using the solutions given in Section II. The resultant elevation and kinematics of an irregular wave field are then calculated as the superposition of all component waves.

In our computation, the wave elevation of an irregular wave field beyond the record duration is assumed to be periodic to the given record. This unrealistic assumption implied, in our formulation, may lead to large errors at the beginning or end of predicted downstream or upstream wave elevation time series, respectively. This is similar to the case of using linear wave theory. The erroneous prediction can be removed by using a shorter time window.

V. Results and Comparison

To examine the hybrid wave model and related numerical schemes, laboratory measurements of irregular wave elevations and kinematics are compared with the numerical predictions. The measurements used in the comparison were conducted at the Norwegian Hydraulic Laboratory (NHL) and the Hydromechanics Laboratory at Texas A&M University in College Station (TAMU). In both experiments, wave elevations are measured by surface piercing wave gages and kinematics are recorded with a laser Doppler anemometery (LDA). A detailed

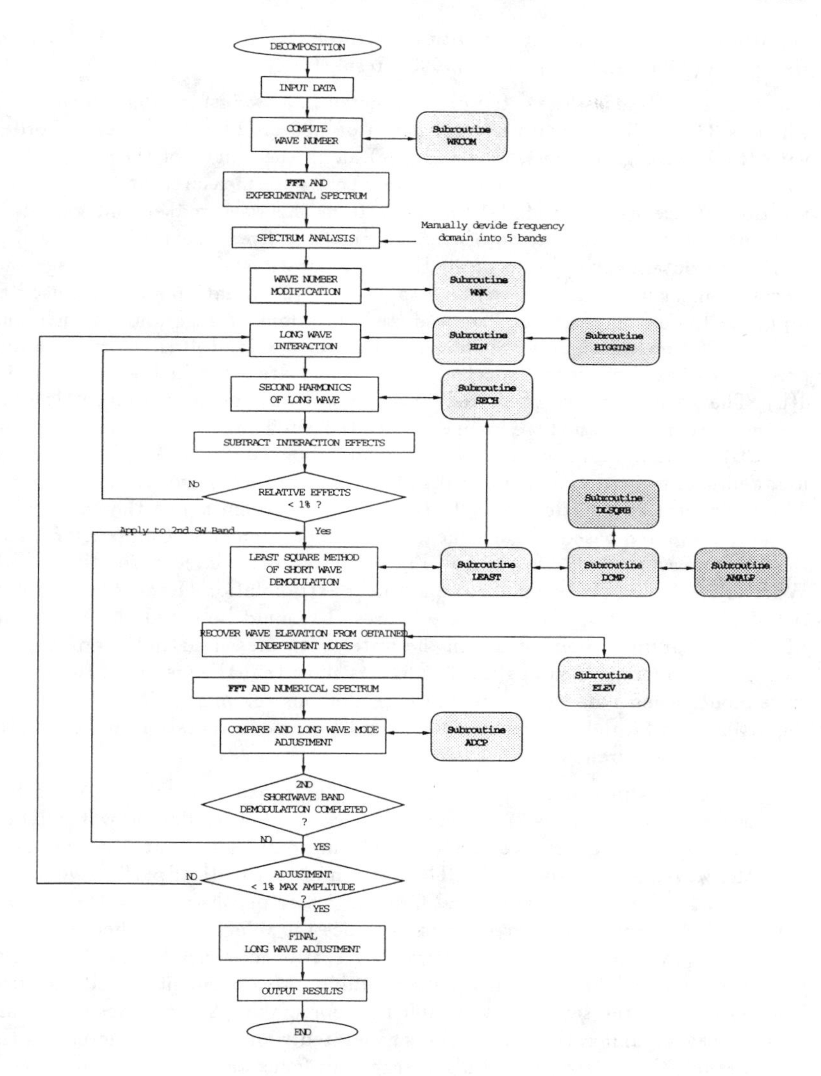

Figure 2. Flow chart of program Decomposition

description of the laboratory facilities and procedures were given by Tzeng & Heideman (1990) and Spell et al. (1993), respectively.

The NHL measurements were conducted in a two-dimensional wave tank, which is 33 m in length with a water depth of 1.30 m. Elevations were recorded with three wave gages (one, four and five) along the length of the tank. Horizontal and vertical particle velocity components were measured at the gage one position. Gage five is around 14 m away from the wave maker and gage four and one are 4 m and 8 m downstream from gage five. Each gage's measurements are obtained at a rate of 40 Hz for a duration of 819.2 sec. Gage one elevation measurements between 30.0 to 81.2 sec (2048 data points) are used as input for the decomposition. The predicted elevations at gage four and five and their related measurements are shown in Figures 3(a) and 4(a). Corresponding predictions based on linear (spectral) wave theory are given in Figures 3(b) and 4(b). The wave elevations predicted by the hybrid wave model are much closer to the measurements than those by linear wave theory. The root mean square errors (RMSE) of the hybrid wave model scheme are about 1.12 and 1.48 cm for gages four and five, while the related RMSEs of linear wave theory are 1.76 and 2.53 cm. The horizontal and vertical velocity time series at 10 cm above the calm water level are predicted based on gage one elevation measurements. Comparisons with the LDA measurements are given in Figures 5 and 6. Related predictions using Wheeler stretching (Wheeler 1970) and linear extrapolation (Rodenbusch & Forristall 1986) are also plotted in these figures. It should be noted that when the LDA's measurement point is not in the water both measured and computed velocities are no longer meaningful. The figures show that the results of the hybrid wave model scheme are the closest to the related measurements. Wheeler stretching gives slightly underestimates results and linear extrapolation overestimates the velocity measurements.

The two-dimensional wave tank used for measurement in TAMU's Hydromechanics laboratory is 37 m long and 0.91 m wide with 0.91 m water depth. Two wave gages (one and two) are exactly 5 m apart and about 9 m and 14 m from the wavemaker, respectively. The horizontal and vertical particle velocities at gage two are measured at several different water depths ranging from 12 cm below to 8 cm above the calm water level. Since the velocities at different depths had to be measured during different wave runs, it is necessary to make sure that wave elevations of different runs are repeatable. Measurement record duration is 40.96 sec and the sample rate is 100 Hz. For making steep waves which are close to wavebreaking, the wave train is generated while gradually increasing the wave period; thus, the wave train's energy may focus and very steep waves can be formed. Using this methodology, steep waves were recorded at gage two and they eventually broke about 1 m behind gage two. Wave elevations measured at gage one are shown in Figure 7, which also shows the gradually increasing wave periods. Using the wave elevation measurements from gage one, the elevation at gage two is predicted using the hybrid wave model scheme and linear wave theory. Predictions are compared with related measurements in Figures 8(a) and

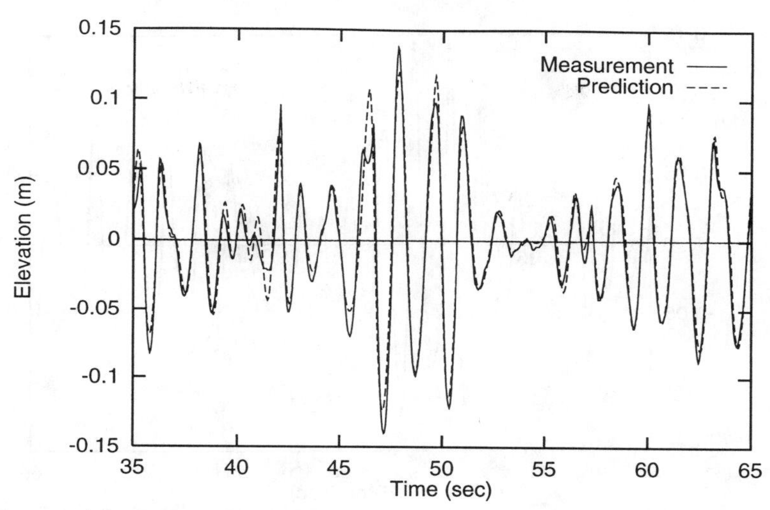

Figure 3(a). NHL Case—Elevation prediction from Gage 1 to Gage 4 based on nonlinear decomposition (Prediction distance: -4.0 m.)

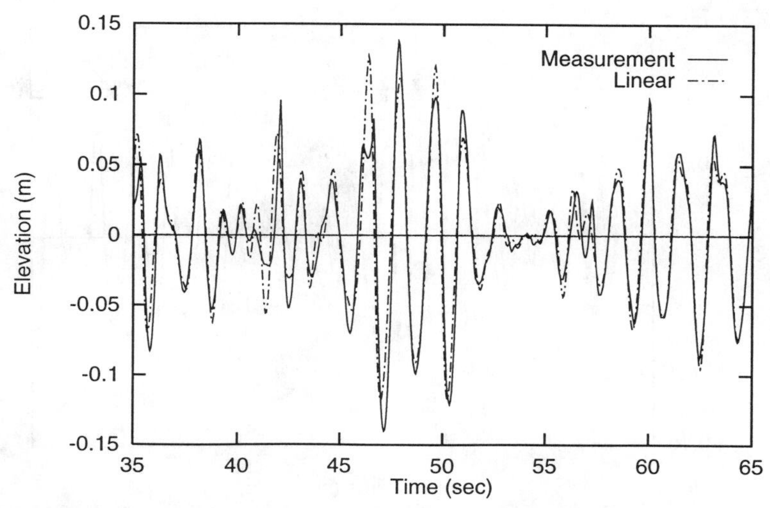

Figure 3(b). NHL Case—Elevation prediction from Gage 1 to Gage 4 based on linear spectral method (Prediction distance: -4.0 m.)

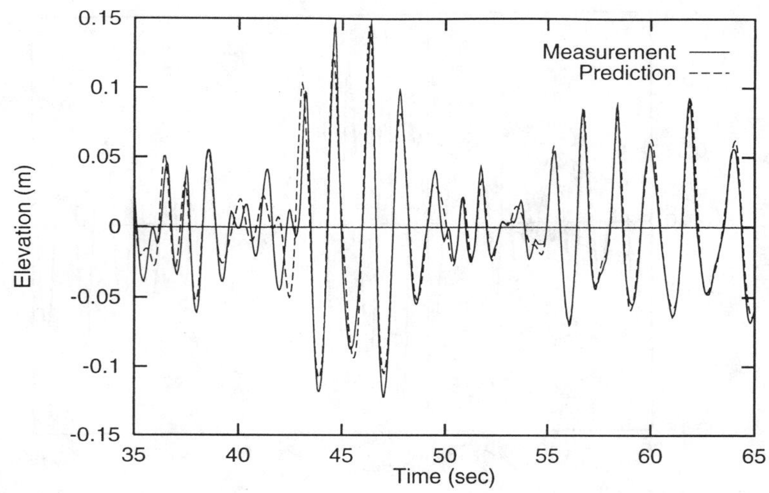

Figure 4(a). NHL Case—Elevation prediction from Gage 1 to Gage 5 based on nonlinear decomposition (Prediction distance: -8.0 m.)

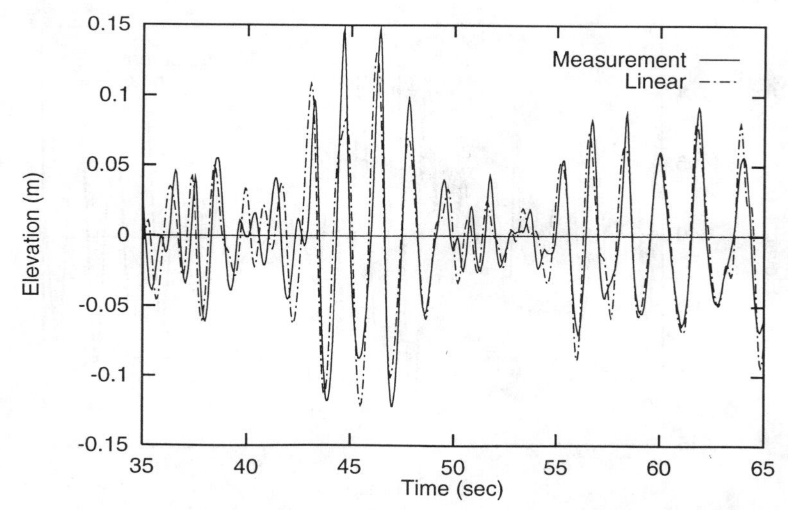

Figure 4(b). NHL Case—Elevation prediction from Gage 1 to Gage 5 based on linear spectral method (Prediction distance: -8.0 m.)

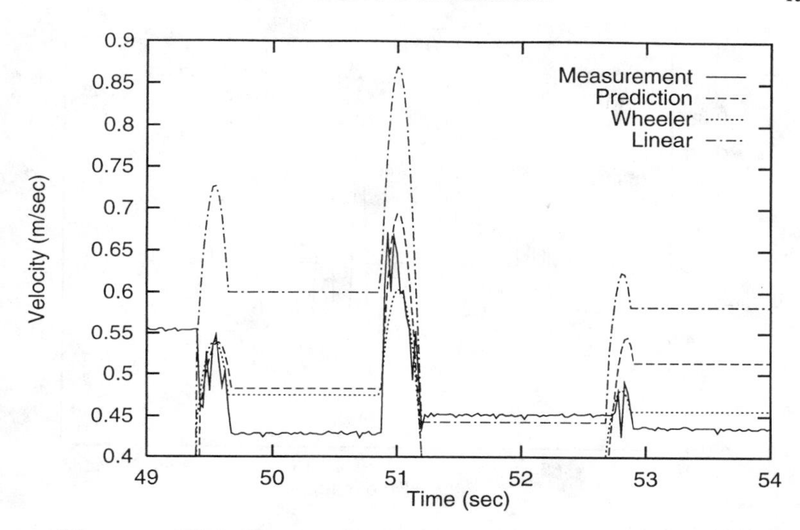

Figure 5. NHL Case—Horizontal velocity comparison (z = 0.1 m.)

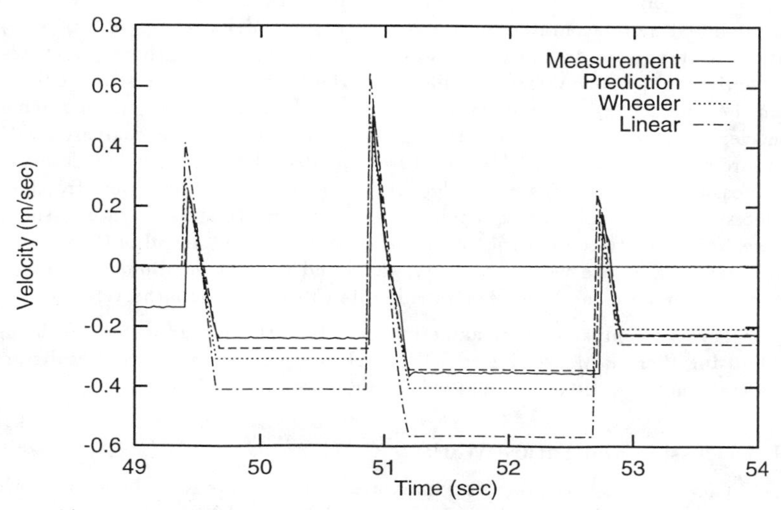

Figure 6. NHL Case—Vertical velocity comparison (z = 0.1 m.)

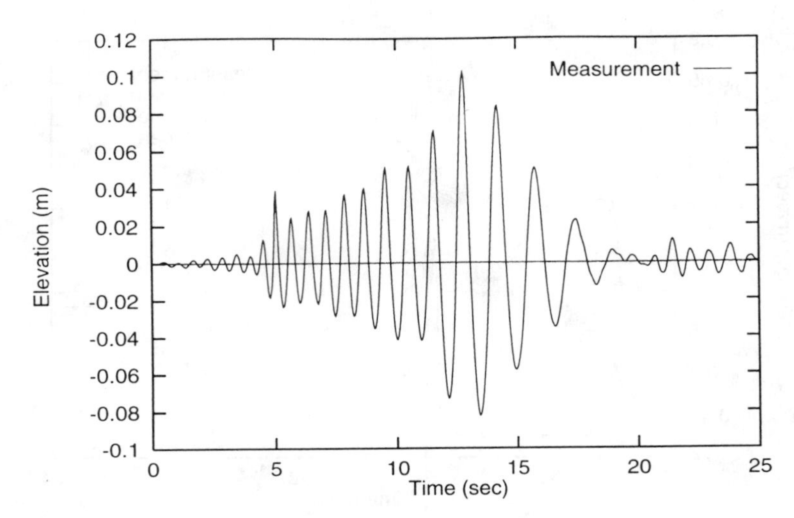

Figure 7. TAMU Case—Wave surface elevation measurement at Gage 1

8(b). The RMSEs of the hybrid wave model scheme and linear wave theory are 1.21 and 1.86 cm, respectively. More importantly, the elevations of steeper waves are predicted more accurately by the hybrid wave model scheme, because steep waves are important to the computation of wave loads on offshore structures. The horizontal particle velocity profile under the steepest wave crest is predicted from the gage two elevation measurements using the hybrid wave model scheme, Wheeler stretching and linear extrapolation. Their results are compared with measurements in Figure 9. The hybrid wave model scheme results are closest to the measurements over the entire depth. The prediction by Wheeler stretching underestimates while linear extrapolation greatly overestimates. Figure 10 shows the vertical velocity comparison occurring at about 0.21 sec ahead of the steepest wave crest, where the vertical velocity was found to be a maximum. A similar trend is observed using Wheeler stretching, but it overpredicts the velocity.

Further comparisons of numerical results with measurements are being made using Pierson Moskowitz or JONSWAP irregular wave trains. Results are similar to those reported here and will be presented in the near future

VI. Discussion and Future Work

Comparisons with laboratory measurements indicate that the hybrid wave model can predict the short-distance wave evolution and kinematics more accurately than linear wave theory and its modified methods such as stretching and extrapolation. It is also shown that the new model can be used for wave fields with

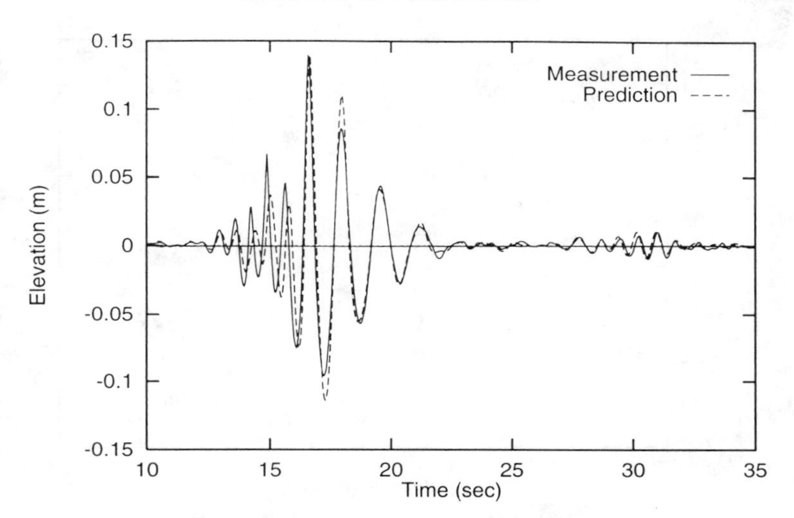

Figure 8(a). TAMU Case—Elevation prediction from Gage 1 to Gage 2 based on nonlinear decomposition (Prediction distance: 5.0 m.)

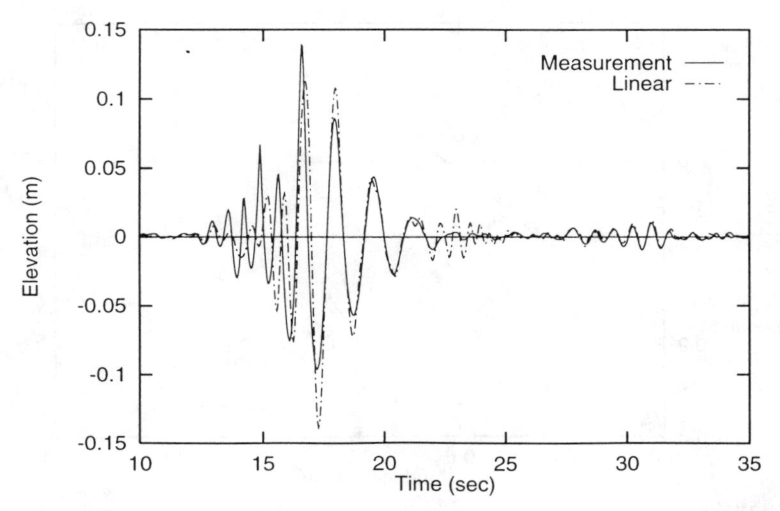

Figure 8(b). TAMU Case—Elevation prediction from Gage 1 to Gage 2 based on linear spectral method (Prediction distance: 5.0 m.)

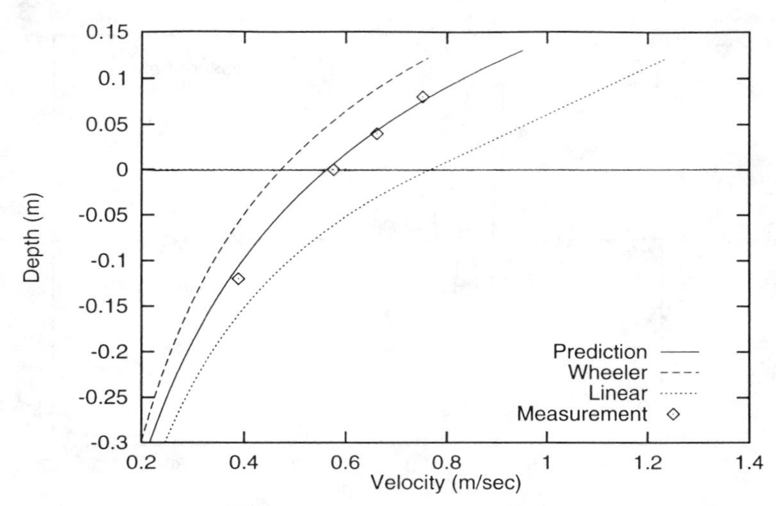

Figure 9. TAMU Case—Horizontal velocity profile at T = 16.57 sec.

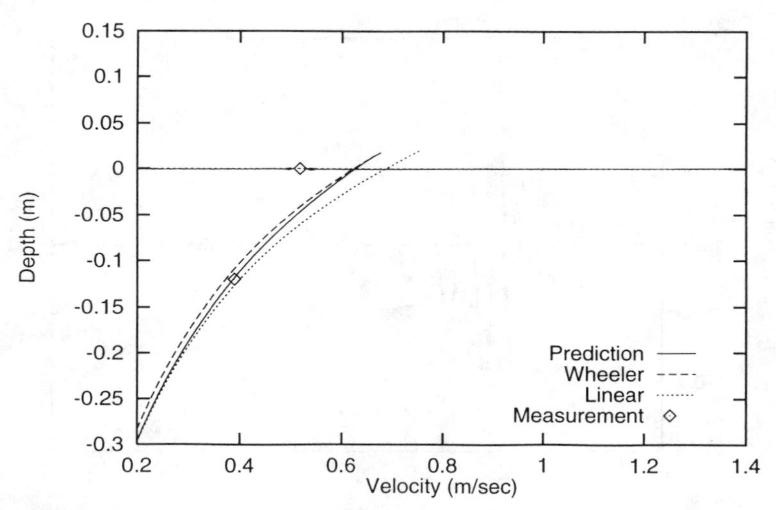

Figure 10. TAMU Case—Vertical velocity profile at T = 16.36 sec.

relatively broad-banded spectra. Since there is no available numerical scheme based on conventional perturbation approaches for computing an irregular wave field, comparisons with its predictions have not yet been made. However, because the hybrid wave model overcomes convergence difficulties present in conventional perturbation approaches, it should be superior to conventional perturbation approaches.

The current hybrid wave model is limited to unidirectional long-crested ocean waves. Ocean waves are often short-crested and wave directionality plays an important role in wave characteristics. The progress made in this study should be viewed as the first step in acquiring the ability to predict the wave evolution, kinematics and pressures associated with ocean waves. The extension of the current model to three dimensions has been proposed. In order to understand wave evolution and interactions in the coastal zone, we plan to extend the current model to allow for changes in water depth and the shallow water case. It is known that the nonlinearity in shallow water is greatly enhanced, due to shoaling effects. The concurrence of multiple wave trains in US coastal zone is about two thirds of time (Thompson 1980). We expect that the extension of the model to the shallow water case will be beneficial to coastal engineering. Finally, we expect that the accuracy of hybrid wave model predictions will be improved when the formulation of the model is accurate to third order. The third order formulation is planned in our future work. In addition, related laboratory experiments are also planned to aid in the understanding wave-wave interaction in ocean waves and the verification of the hybrid wave model.

VII. Acknowledgement

This research is supported by the Offshore Technology Research Center which is sponsored in part by the National Science Foundation Engineering Research Centers Program Grant Number CDR-8721512 and the offshore industry participants, the Texas Advanced Technology Program, and the Joint Industry Program.

VIII. References

1. Donelan, M.A., Anctil, F. and Doering, J.C. (1992) "A simple method for calculating the velocity field beneath irregular waves", *J. Coastal Engr.*, No.16, 399-424

2. Hasselmann, K. (1962) "On the nonlinear energy transfer in a gravity-wave spectrum. Part 1. General theory", *J. Fluid Mech.*, Vol.12, 481-500.

3. Longuet-Higgins, M.S. (1987) "The propagation of short surface waves on longer gravity waves", *J. Fluid Mech.*, Vol.177, 293-306.

4. Phillips, O.M. (1979) "Surface wave physics— A survey", *Flow Research Report No. 145*, Flow Research Company, Division of Flow Industries, Inc,

Kent, Washington.

5. Phillips, O.M. (1981) "The dispersion of short wavelets in presence of a dominant long wave", *J. Fluid Mech.*, Vol.107, 465-485.

6. Phillips, O.M. (1960) "On the dynamics of unsteady gravity waves of finite amplitude", *J. Fluid Mech.*, Vol.9, 193-217

7. Rodenbusch, G. and Forristall, G.Z., (1988) "An empirical for random directional wave kinematics near the free surface", *Offshore Technology Conference*, OTC 5783, 1988.

8. Smith, J.M. and Vincent, C.L. (1992) "Shoaling and decay of two wave trains on beach", *J. Waterways, Ports, Coastal and Ocean Engineering*, ASCE, Vol.118, No.5, 517-533.

9. Steele, K.M., Finn, L.D. and Lambrakos, K.F. (1988) "Compliant tower response prediction procedures", *Proc. 20th Ann. Offshore Technology Conf.*, Houston, Texas, paper 5783.

10. Spell C.A., Zhang, J. and Randall R.E. (1993), submitted to *Phys. Fluids.*

11. Tzeng C.-Y. & Heideman J.C. "Analysis of NHL two-dimensional irregular wave kinematics data", *Private Communication*, 27 February 1990.

12. Thompson, E.F. (1980). "A steady-state wave model for coastal application." *Proc. 21st Coast. Engrg. Conf.*, ASCE, 929-940.

13. Wheeler, J.D., "Method for calculating forces produced by irregular wave", *Journal of Petroleum Technology*, March 1970, 359-367.

14. Zhang, J. and Melville, W.K. (1990) "Evolution of weakly nonlinear short waves on riding long waves", *J. Fluid Mech.*, Vol.214, 321-346.

15. Zhang, J. and Melville, W.K. (1992) "On the stability of weakly nonlinear short waves on finite-amplitude long gravity waves", *J. Fluid Mech.*, Vol.243, 51-72.

16. Zhang, J., Randall, R.E. and Spell, C.A. (1992) "Component wave interactions and irregular wave kinematics", *J. Waterways, Ports, Coastal and Ocean Engineering*, ASCE, Vol.118, No.4, 401-416.

17. Zhang, J., Randall, R.E. Chen, L., Ye, M. and Spell, C.A. (1992) "Hybrid wave-mode modeling and wave kinematics", *Proceedings of 2nd ISOPE Conference*, San Francisco, CA, June 14-19, 1992, Vol.III, 39-44.

18. Zhang, J., Hong, K. & Yue, D. K. P. (1993) "Effects of wavelength ratio on wave modeling", *J. Fluid Mech.*, Vol.248, 107-127.

Second Order Directional Wave Kinematics in Shallow Water

Jagat N. Sharma[1] and R. G. Dean[2]

Abstract

A theory and numerical procedure were developed earlier to represent a real sea by the sum of many linear and second order directional wave components. Earlier publications showed that including short-crested nature of real seas could reduce design loads substantially (up to a factor of 3 for large four pile structures). Comparison of the theory with real sea data has been presented by other investigators showing good agreement. Here, simulated second order directional wave velocities and accelerations for two directional spectra, which could be of interest to coastal and offshore engineers and scientists, are presented and their practical implications discussed.

Simulations were carried out for a moderate sea state in 100 ft water depth. Directional wave kinematics (correct up to second order) at sea surface and seabed are presented and discussed, The maximum nonlinear correction is only 10% of wave height for a moderate seastate simulated here, but would be much higher for a design level seastate, because the nonlinear correction is proportional to wave height squared. The second order corrections increase particle velocities under the wave crest and decrease them under the wave trough. With increasing directionality in wave spectrum, wave particle velocities and accelerations decrease in the inline direction but increase in the transverse direction. The maximum acceleration and velocity in the transverse direction for a cosine-squared directional distribution are nearly equal to the maximum inline values.

The inline spectra of both acceleration and velocity for a cosine-squared directional distribution are reasonably smooth with small bumps corresponding to the sum and difference frequencies. In contrast, the transverse spectra have many peaks, and the energy is distributed over a wider range of frequencies. As expected, both velocity and acceleration spectra near the seabed have lower total energy and appear smoother. Spectra of both inline and transverse accelerations are distributed over a much wider

1. Amoco Production Company, P.O. Box 3385, Tulsa, OK 74102, USA.
2. Coastal Engineering, University of Florida, Gainesville, FL 32611, USA.

range of frequencies than the original wave spectrum. In contrast to velocity spectra, subharmonic frequencies are absent from acceleration spectra.

Several consequences of theoretical and practical significance can be deduced. Second order subharmonic waves associated with a group of waves could be responsible for the negative skewness observed in shallow water random waves, and a lower total wave force below a wave crest. Also, as the velocities and accelerations caused by second order subharmonic waves do not decrease rapidly with depth, they become important for sediment transport problems and forces on pipeline at seabed. Second order super-harmonic wave frequencies are close to the natural frequencies of many offshore structural elements and could be critical in their fatigue analysis.

Introduction

A recent resurgence of interest in directional wave kinematics is evidenced by two workshops (E&P Forum 1989, Torum and Gudmestad 1990) and a few field and lab measurements devoted to the subject. For an up-to--date overview of directional wave kinematics, the reader is referred to a recent article by Gudmestad (1993).

A simple but realistic calculation of wave loading on a single- or multiple-pile group is not available because of nonlinearity, randomness, and directionality of a real sea wave kinematics. Presently, there are two essentially different but complementary methods for computing wave kinematics. One method represents nonlinearities of a single wave composed of a characteristic fundamental period and its higher harmonics. A number of such theories have been developed (Skjelbreia and Hendrickson 1961; Chappelear 1961; Laitone 1960; Dean 1965; and Von Schwind 1972). Dalrymple (1974) extended the stream function approach of Dean to waves on a shear current. Some of these theories adequately account for the nonlinearities; however, they avoid the random and directional characteristics of the sea surface. The second method uses the principle of linear superposition of an infinite number of waves with given frequencies, amplitudes, and directions of propagation but independent phases; the total energy is distributed over a continuum of frequencies and directions. In this manner, a three-dimensional Gaussian sea can be represented fully. However, ignoring the nonlinearities makes the random Gaussian model unrealistic - especially for large waves. The theory of nonlinear directional random waves developed earlier retains both nonlinearity and directionality of random waves. The wave kinematics results presented here and earlier show that the method has successfully represented many important features of a realistic sea.

The theory of nonlinear directional random waves, simulation method, and wave force results were presented earlier, and therefore, are described only briefly here. This paper is devoted mainly to wave kinematics results and their possible applications. The results presented here are limited to simulated velocity and acceleration spectra near sea surface and seabed for both unidirectional and short-crested seas correct up to second order. Finally, the limitations and applications of the method and possible future studies are presented and discussed.

Nonlinear Interaction of Directional Random Waves

Perturbation method was used by Prof Longuet-Higgins (1963) to derive short-crested sea equations in deepwater. Using the same method, we derived equations for any depth. The arbitrary depth equations are more complex and many checks were performed to verify the algebraic derivations.

The nonlinear boundary value problem is solved to the second order by a perturbation method accounting for contributions from nonlinear interactions of linear components with arbitrary frequencies and directions. The second-order directional spectrum is obtained starting with a discrete linear directional spectrum. The second-order wave system is forced by the linear system, i.e., all second-order amplitudes and phases are related to the characteristics of the first-order spectrum. The first- and second-order spectra are added to simulate nonlinear sea surface elevation and water particle kinematics anywhere in the wave field.

Boundary Value Problem Formulation

Irrotational motion of incompressible flows can be described well by a velocity potential or stream function. The velocity components u, v, and w can be defined in terms of the gradients of the velocity potential ϕ, which satisfies the Laplace Equation

$$\frac{\partial^2\phi}{\partial x^2}+\frac{\partial^2\phi}{\partial y^2}+\frac{\partial^2\phi}{\partial z^2} = 0, \text{ where}$$

$$0 \le z \le h+\eta, \text{ and } -\infty \le x, y \le \infty. \quad (1)$$

subject to the following boundary conditions:

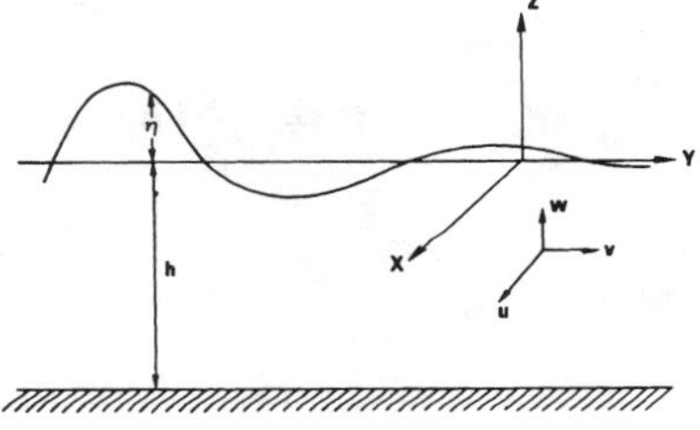

Fig. 1 Definition sketch

Bottom Boundary Condition (BBC).

$$\frac{\partial\phi}{\partial z} = 0, \text{ at } z = -h.$$

Kinematic Free Surface Boundary Condition (KFSBC).

$$\frac{\partial\eta}{\partial t}+u\frac{\partial\eta}{\partial x}+v\frac{\partial\eta}{\partial y} = w, \text{ at } z = \eta(x, y, t).$$

Dynamic Free Surface Boundary Conditions (DFSBC).

$$g\eta + 1/2\,(u^2+v^2+w^2)+\frac{\partial\phi}{\partial t} = -Q(t), \text{ at } z = \eta(x, y, t).$$

Combined Free Surface Boundary Condition (CFSBC).

$$-\frac{\partial^2\phi}{\partial t^2}-g\frac{\partial\phi}{\partial z}-\left(\frac{\partial}{\partial t}+1/2\overline{\nabla}\phi\cdot\overline{\nabla}\right)|\overline{\nabla}\phi|^2+\frac{\partial Q}{\partial t} = 0, \text{ at } z = \eta(x, y, t).$$

Method of Solution

The perturbation method is adopted here for solution of the boundary value problem formulated in the preceding section. This method assumes that all variables can be expanded as a convergent power series of a small parameter, such as water-surface slope, and that the nonlinear CFSBC also can be expanded in a convergent Maclaurin series about the mean water level, z=0, with some small parameter. The random velocity potential ϕ and random sea surface η may be represented by

$$\phi(x, y, z, t) = \phi^{(1)}(x, y, z, t) + \phi^{(2)}(x, y, z, t) + \dots , \quad (2)$$

$$\text{and} \quad \eta(x, y, t) = \eta^{(1)}(x, y, t) + \eta^{(2)}(x, y, t) + \dots , \quad (3)$$

and the boundary conditions can also be represented by similar expansions. On collecting terms of the same order, the original boundary value problem becomes a set of first, second and higher order boundary value problems, which are solved successively to obtain first the familiar linear solution:

$$\phi^{(1)}(x, y, z, t) = \sum_{i=1}^{\infty} b_i \frac{\cosh k_i (h+z)}{\cosh k_i h} \cdot \sin(\bar{k}_i \cdot \bar{x} - \sigma_i t + \varepsilon_i) \quad (4)$$

$$\eta^{(1)}(x, y, t) = \frac{1}{g} \sum_{i=1}^{\infty} b_i \sigma_i \cos(\bar{k}_i \cdot \bar{x} - \sigma_i t + \varepsilon_i) = \sum_{i=1}^{\infty} a_i \cos\psi_i , \quad (5)$$

in which $a_i = \dfrac{b_i \sigma_i}{g}$, $\sigma_i^2 = g|\bar{k}_i| \tanh|\bar{k}_i| h$, and the wave direction θ_i is defined by

$$(k_{i_x}, k_{i_y}) = (|\bar{k}_i| \cos\theta_i, |k_i| \sin\theta_i) . \quad (6)$$

As shown by Birkhoff and Kotik (1951) and Pierson (1955), $\eta^{(1)}$ represents a Gaussian random sea surface if the phases ε_i are independent and uniformly distributed.

Second-Order Solution in Finite Water Depth

Considering the second-order velocity potential

$$\phi^{(2)} = \sum_{i=1}^{\infty} \sum_{i=j}^{\infty} b_{ij} \frac{\cosh|\bar{k}_{ij}| (h+z)}{\cosh|\bar{k}_{ij}| h} \cdot \sin(\bar{k}_{ij} \cdot \bar{x} - \sigma_{ij} t + \varepsilon_{ij}) , \quad (7)$$

which satisfies $\nabla^2 \phi^{(2)} = 0$ and the bottom boundary conditions. Second order wave numbers $\bar{k}_{ij}$ and frequencies σ_{ij} are given by sums and differences of the fundamental vector wave numbers and frequencies. The coefficients b_{ij} are solved in terms of the first-order potential $\phi^{(1)}$ and water elevation $\eta^{(1)}$ for all possible sum and differences of the primary frequencies. The resulting expressions for the second order random velocity potential $\phi^{(2)}$ and random sea surface $\eta^{(2)}$ are available in earlier publications and are reproduced here for completeness.

$$\phi^{(2)} = 1/4 \sum_{i=1}^{\infty} \sum_{j=1}^{\infty} b_i b_j \frac{\cos k_{ij}^{-}(h+z)}{\cosh k_{ij}^{-} h} \cdot \frac{D_{ij}^{-}}{(\sigma_i - \sigma_j)} \sin(\psi_i + \psi_j)$$

$$+ \frac{1}{4} \sum_{i=1}^{\infty} \sum_{i=j}^{\infty} b_i b_j \frac{\cosh k_{ij}^{+}(h+z)}{\cosh k_{ij}^{+} h} \cdot \frac{D_{ij}^{+}}{(\sigma_i + \sigma_j)} \sin(\psi_i + \psi_j) \quad , \quad (8)$$

in which $k_{ij}^{-} = |\bar{k}_i - \bar{k}_j|$, $k_{ij}^{+} = |\bar{k}_i + \bar{k}_j|$, $\psi_i = \bar{k}_i \cdot \bar{x} - \sigma_i t + \varepsilon_i$,

and

$$D_{ij}^{-} = \frac{(\sqrt{R_i} - \sqrt{R_j})\,[\sqrt{R_j}(k_i^2 - R_i^2) - \sqrt{R_i}(k_j^2 - R_j^2)]}{(\sqrt{R_i} - \sqrt{R_j})^2 - k_{ij}^{-} \tanh k_{ij}^{-} h}$$

$$+ \frac{2(\sqrt{R_i} - \sqrt{R_j})^2 (\bar{k}_i \cdot \bar{k}_j + R_i R_j)}{(\sqrt{R_i} - \sqrt{R_j})^2 - k_{ij}^{-} \tanh k_{ij}^{-} h}$$

$$D_{ij}^{+} = \frac{(\sqrt{R_i} + \sqrt{R_j})\,[\sqrt{R_i}(k_j^2 - R_j^2) + \sqrt{R_j}(k_i^2 - R_i^2)\,]}{(\sqrt{R_i} + \sqrt{R_j})^2 - k_{ij}^{+} \tanh k_{ij}^{+}}$$

$$+ \frac{2(\sqrt{R_i} + \sqrt{R_j})^2 (\bar{k}_i \cdot \bar{k}_j - R_i R_j)}{(\sqrt{R_i} + \sqrt{R_j})^2 - k_{ij}^{+} \tanh k_{ij}^{+}}$$

in which $R_i \equiv k_i \tanh k_i h$.

The second-order water surface displacement $\eta^{(2)}$ is

$$\eta^{(2)} = 1/4 \sum_{i=1}^{\infty} \sum_{j=1}^{\infty} a_i a_j \cdot \left(\left[\frac{D_{ij}^{-} - (\bar{k}_i \cdot \bar{k}_j + R_i R_j)}{\sqrt{R_i R_j}} + (R_i + R_j) \right] \right.$$

$$\left. \cos(\psi_i - \psi_j) + \left[\frac{D_{ij}^{+} - (\bar{k}_i \cdot \bar{k}_j - R_i R_j)}{\sqrt{R_i R_j}} + (R_i + R_j) \right] \cos(\psi_i + \psi_j) \right) \tag{9}$$

The specification of the second-order wave field in space and time is now complete and can be used to calculate nonlinear short-crested wave kinematics or force on an offshore structure.It should be noted that the first order waves have independent phases, while the second order phases are equal to the sum and difference of first order phases. In other words, the second order terms are phase-locked with the first order terms.

Simulation Method

The nonlinear short-crested sea simulation begins with the discrete approximation of a 3-dimensional Gaussian random sea surface η:

$$\eta(\bar{x}, t) = \sum_{m=1}^{M} \sum_{n=1}^{N} \sqrt{S(\sigma_m, \theta_n) \Delta\sigma_m \Delta\theta_n} \cdot \cos(\bar{k}_{m,n} \cdot \bar{x} - \sigma_m t + \varepsilon_{m,n}) . \tag{10}$$

Next, the second-order correction terms are calculated using equations presented elsewhere (Sharma). The time history of water elevation η is obtained by calculating the first- and second-order spectra of η, adding amplitudes of the same frequencies, and taking the inverse Fast Fourier Transform. Velocities and accelerations at any point are computed as follows.

a. Velocity and acceleration vectors of each linear and nonlinear component are resolved along the x and y axes.

b. The contributions within each frequency band are added to obtain the Fourier coefficients a_n and b_n.

c. Using the inverse Fast Fourier Transform, the time domain realizations for velocities and accelerations in the x and y directions are obtained.

This simulation method has been implemented in a set of FORTRAN programs. The more details of the computation scheme can be found in Sharma and Dean (1979, 1981).

Difficulties in Simulation

Any stochastic simulation is difficult and short-crested sea simulation is especially more difficult. Even to the first order, the simulated short-crested sea spectrum fluctuates wildly around the target spectrum if arbitrary phases are used (Fig. 2 & 3). This was because of numerical problems associated with

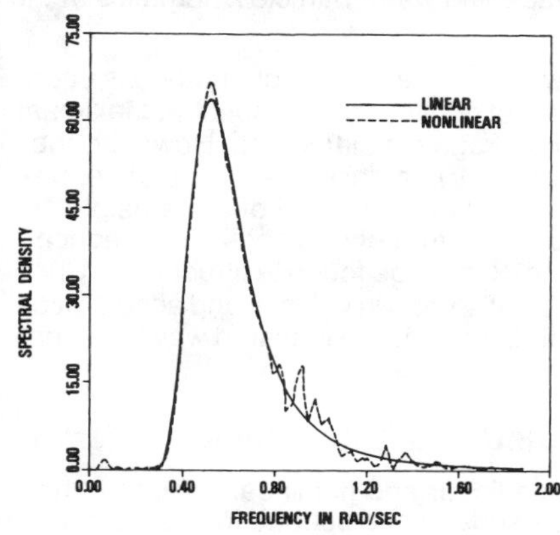

Fig. 2 Target sea surface spectrum.

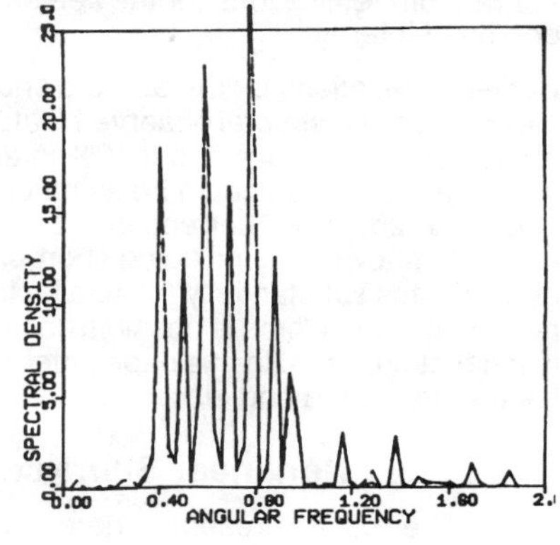

Fig. 3 Simulated spectrum with arbitrary phases.

small number of frequencies and directions. When the total energy of a frequency is replaced by discrete directional components with random phases in simulation, the total energy of that frequency in the simulated sea sample is a function of the finite number of random phases. Earlier (Sharma and Dean 1979, 1981), it was shown that as the number of random phases increases, the simulated total energy approaches the target value. It was also shown that the mean value of a simulated sea tended to zero as the duration of simulation increased. A large number of primary components are required for reasonable accuracy. This in turn requires big CPU memory, long CPU time and hence, a large computing budget.

Single summation method does excellent job of simulating a linear short-crested sea without having to worry about phases. However, the second order components derived from this are unrealistic, because at the second order self-interaction terms are dominant as shown earlier (Sharma and Dean 1979).

Simulation Results

Simulated sea surface and wave force results were presented earlier for four different directional spreads ranging from unidirectionality to a case when the energy is equally distributed in all directions over a half plane. 1,200 cosine waves, with 60 discrete frequencies propagating from 20 discrete directions, constituted the linear wave system. Before assigning phases to this wave system, the set of 1,200 phases was tested for suitability (randomness) in the simulation.

Only 20 linear components having significant energy were used in calculating the nonlinear components. Even then, 160,000 nonlinear components had to be calculated. Adding these to the linear components, we obtain nonlinear random realizations for the sea surface and water particle kinematics at the depths of interest.

Earlier publications by the authors showed that the simulation method is capable of modelling several observed features of nonlinear wave spectra. Maximum nonlinear correction is about 10% in an average simulated sea. However, the nonlinear correction could be as much as 50% in a design level sea state, because the 2nd order correction is proportional to the square of wave height. It was also shown that including short-crested nature of real seas could reduce design loads substantially (up to a factor of 3 for large four pile structures). Such reduction in total force is possible because of spread in velocity and acceleration in a directional sea. In this paper interesting features of simulated wave kinematics are presented and discussed.

Nonlinear Short-crested Sea Kinematics

The results presented here include linear and nonlinear spectra of horizontal velocity and acceleration near sea surface and seabed for both unidirectional and a short crested sea with $\cos^2\theta$ directional distribution. Their significance and possible applications are discussed. Unidirectional spectra are presented for comparison. Unidirectional spectra of horizontal velocity and acceleration near sea surface and seabed have no surprises. The linear spectral density of horizontal velocity has the same shape as the spectral density of water surface elevation η (Fig. 4). In contrast, the nonlinear spectral density

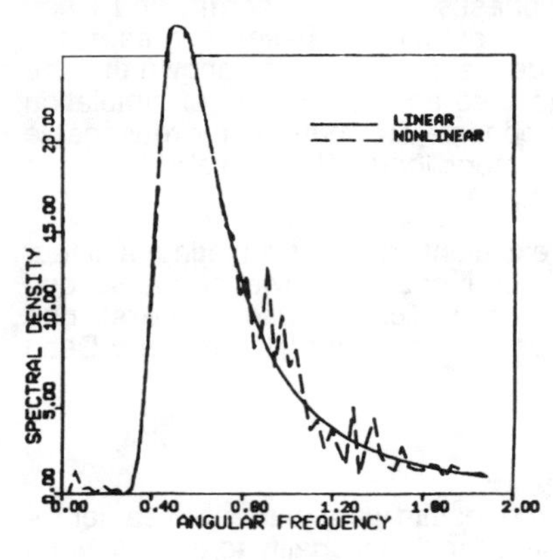

Fig. 4 Unidirectional velocity near seasurface.

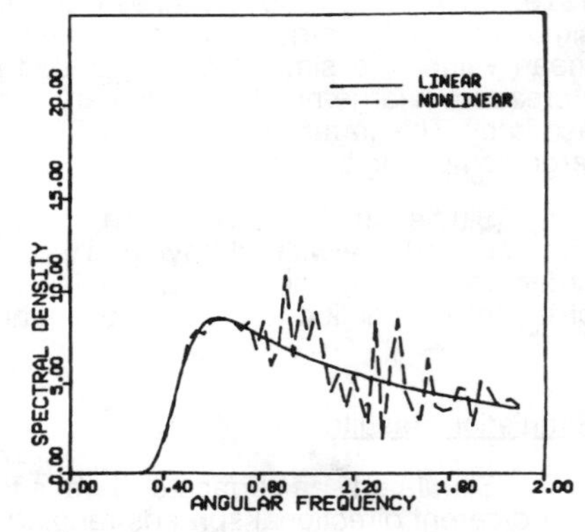

Fig. 5 Unidirectional acceleration near seasurface.

has a small subharmonic peak and super-harmonic fluctuations due to nonlinear interactions. As expected, the acceleration spectral density is wider and contains no subharmonic energy (Fig. 5).

As expected, the spectra near seabed have lower total energy. The subharmonic peak in the velocity spectrum decays very slowly with depth and becomes a larger fraction of the primary peak. However, the super-harmonic frequencies are damped out (Fig. 6). Acceleration spectrum also is very smooth (Fig. 7).

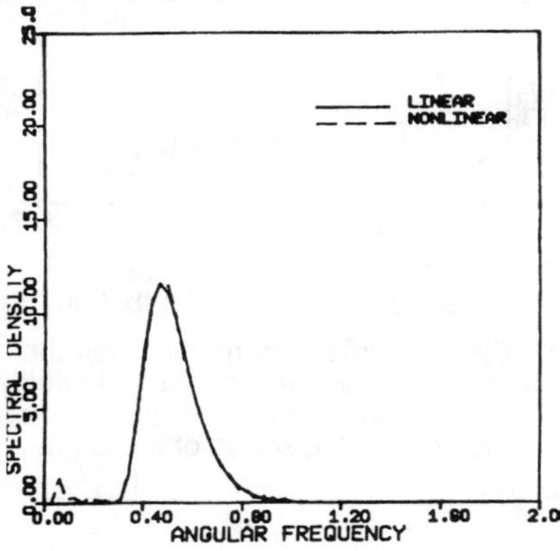

Fig. 6 Unidirectional velocity at seabed.

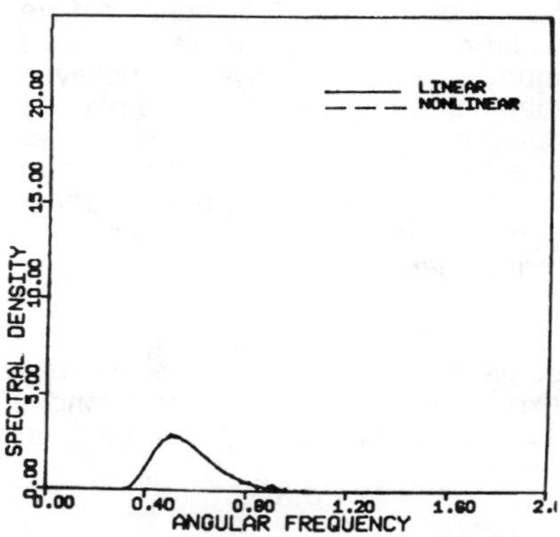

Fig. 7 Unidirectional acceleration at seabed.

Directional Kinematics at Surface

Next, we present and discuss spectra of velocity and acceleration near sea surface in a short-crested sea with a $\cos^2\theta$ distribution of energy. In contrast to a long-crested sea, which has no velocity or acceleration in the transverse direction, the maximum velocity in the transverse direction in a short-crested sea could be comparable to the maximum velocity in the inline direction. Figures 8 and 9 present the spectra of inline velocity and acceleration near sea surface in a short-crested sea with a $\cos^2\theta$ distribution of energy. The inline velocity spectrum is reasonably smooth considering the small number

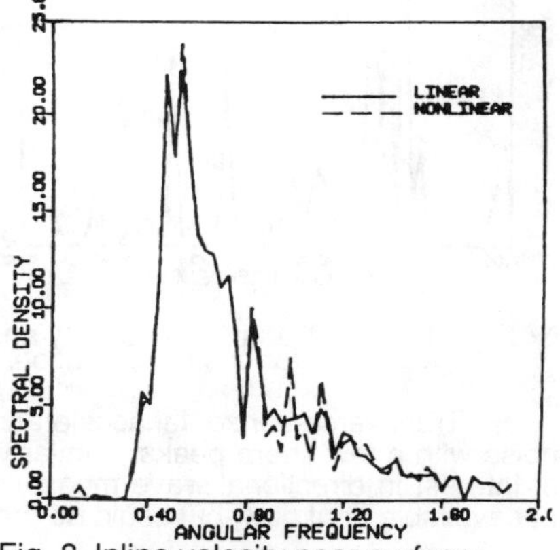

Fig. 8 Inline velocity near surface for $\cos^2\theta$ distribution

of primary waves used in simulation. As expected, the acceleration spectrum is much wider than the velocity spectrum and has many peaks.

Figures 10 and 11 present the spectra of transverse velocity and acceleration near sea surface in a short-crested sea with a $\cos^2\theta$ distribution of energy. Both spectra have wider range of frequencies and have many peaks and valleys. It is believed that increasing the number of primary components would produce smoother spectra. Inline acceleration spectrum has many small peaks, but the transverse acceleration spectrum has many sharp peaks.

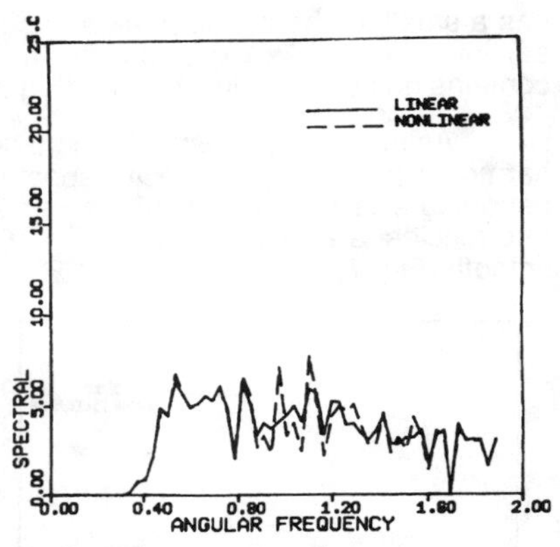

Fig 9. Inline acceleration near surface for $\cos^2\theta$ distribution

Transverse velocity spectrum (Fig. 10) shows a subharmonic peak that contains more energy than a similar peak in the unidirectional spectrum. In field experiments, velocities corresponding to these frequencies have been measured by others. A comparison with field data is beyond the scope of this paper.

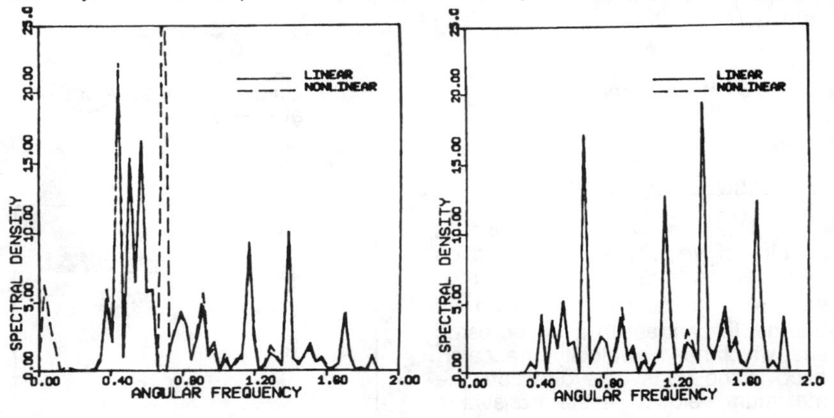

Figs. 10 & 11 Transverse velocity and acceleration spectra near surface for for $\cos^2\theta$ distribution.

Transverse horizontal acceleration spectrum (Fig. 11) approaches white noise with a few sharp peaks. Similar vertical acceleration spectrum would be of interest in directional wave measurement using a wave rider buoy. This is not available right now, but could be produced easily.

Directional Wave kinematics Near Seabed

Simulated velocity and acceleration spectra near seabed are presented next. As expected, all spectra near seabed contain lower total energy and appear smoother. The spectra near seabed contain energy only in frequency bands present in the linear sea spectrum.

Energy due to subharmonic nonlinear terms is present in the inline velocity spectrum, but energy due to superharmonic nonlinear terms is damped out (Fig. 12). The acceleration spectrum does not show any energy in subharmonic frequency bands (Fig. 13) because the subharmonic acceleration terms are very small.

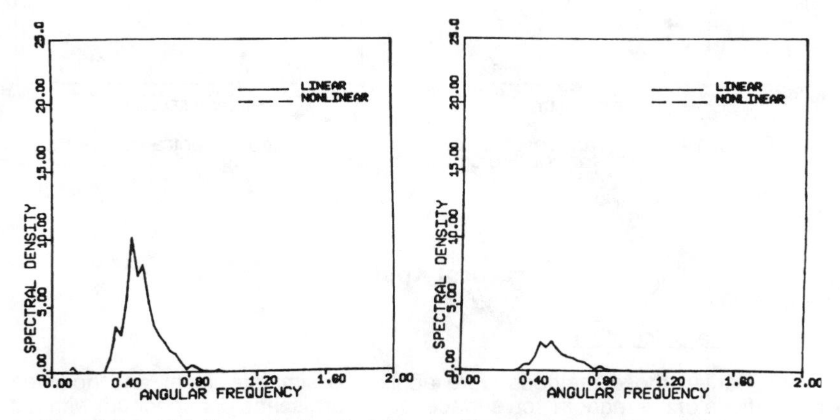

Figs. 12 & 13 Inline velocity and acceleration spectra near seabed for $\cos^2\theta$ distribution.

Transverse spectra near seabed (Fig. 14 & 15) also have energy only in frequency bands present in the linear sea spectrum. This is quite a contrast with the transverse spectra near surface (Fig. 10 & 11) which have very wide bandwidth.

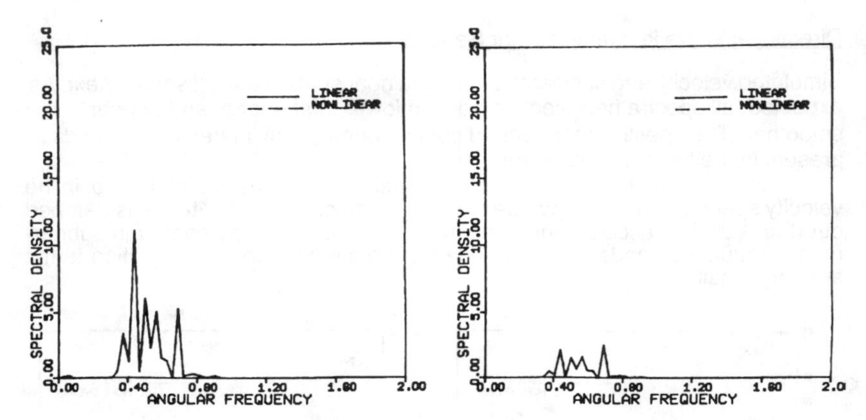

Figss. 14 & 15 Transverse velocity and acceleration spectra near seabed for $\cos^2\theta$ distribution.

Practical Applications

Wave Force Calculation

Earlier, these methods were applied to simulate linear and nonlinear realizations of the random sea surface for the Bretschneider spectrum with four directional spreads, and to compute total wave forces on a single- and four-pile groups (Sharma and Dean 1979 and 1981). The total force on a pile reduced by a factor of 1.0 to 0.61 when the directional spread varied from unidirectional to uniformity over a half plane. For a four-pile group with 60ft (18.2m) separation, the reduction factors were similar to those for the single-pile case. However, for the four-pile group with one pile at each corner of a 300ft (91.4m) square, the reduction factor varied from 0.79 to 0.39 when the directional spread varied from unidirectional to uniformity over a half plane.

The four-pile group with 60ft (18.2m) separation is not very different from the single-pile case because the wavelength associated with the predominant wave energy is nearly 10 times the separation distance. Therefore, the crests of high waves occur almost simultaneously at the four piles. Hence, the maximum total force on the pile group is approximately equal to four times the maximum total force on a single pile.

The separation distance of 300 ft (91.4 m) is more interesting because it is approximately equal to half the wavelength associated with the predominant wave energy. In this case, the crests of highest velocities do not occur simultaneously at the four piles and hence, the total force on the four-pile group is much lower than the two other case considered.

Wave Kinematics for Design

In recent years, considerable effort has been devoted to take advantage of lower design loads by including short-crestedness. Forristall and Rodenbusch (1986) compared the theory with field and laboratory measurements. Further, they used the simulated second order wave kinematics to calibrate an engineering analysis method exploiting the speed of first order short-crested sea simulation. Forristall et. el. used the theory to calibrate a scheme to mimic the full second order directional simulation using only linear 3-D simulation. Their scheme was successfully used to investigate forces on large offshore structures without the difficulties of a full second order simulation. For more details, please refer to the publication cited above.

Possible Future Studies

The original Fortran programs were developed and tested on Buroughs7700 mainframe computer in 1978. Since then, the available computing power and resources have increased beyond anybody's expectations. When the program is adopted to take advantage of these, it is reasonable to expect that many of the limitations mentioned earlier could be overcome easily. A new approach could increase the efficiency and accuracy of programs.

The simulation method has horizontal seabed and no mean current. Adding a sloping bottom could make it more useful in sediment transport studies. Inclusion of a current would make the method more realistic for computing storm wave loads.

Summary and Conclusions

A more realistic sea simulation method has been developed which has many practical applications. In this study, second-order nonlinearities, randomness, and directionality, which are important elements of ocean waves, have been retained. The method was used earlier to calculate wave forces on fixed structures. The total force on a four-pile group with 300ft (91.4m) square was reduced by a factor of 0.79 for a unidirectional to 0.39 for an omnidirectional spectrum. The method has been used by others to realize considerable economy in the design of offshore structures by incorporating the directional effects of a real sea. Here, we presented spectra of simulated second order directional wave velocities and accelerations for $\cos^2\theta$ directional distribution, which could be of interest to coastal and offshore engineers and scientists. Possible practical applications of the method were also discussed. In conclusion, it has been a very useful method which has yet to realize its full potential.

Acknowledgments

Special appreciation is expressed to Amoco Production Co. for coordinating the joint industry support and granting permission to present this work. Partial support for the earlier part of this study was jointly provided by Amoco

Production Co., Continental Oil Co., Earl and Wright Consulting Engineers, Gulf Oil Corp., J. Ray McDermott and Co. Inc., Lloyd's Register of Shipping, Marathon Oil Co., Mobil Research and Development Corp., Phillips Petroleum Co., Shell Oil Co., and Union Oil Co. of California. Their monetary contribution is gratefully acknowledged. We also equally gratefully acknowledge many stimulating and fruitful discussions with colleagues in industry and academia.

References

1. Birkhoff, G. and Kotik, J.: "Fourier Analysis of Wave Trains," Gravity Waves, Natl. Bureau of Standards, Circular No. 521, Washington (1951) 221-234.

2. Chappelear, J. E.: "On the Description of Short-Crested Waves," Beach Erosion Board Technical Memorandum No. 125 (1961).

3. Dalrymple, R. A.: "A Finite Amplitude Wave on a Linear Shear Current," J. Geophys. Res. (Oct. 1974) 79, 4498-4504.

4. Dean, R. G.: "Stream Function Representation of Nonlinear Ocean Waves," J. Geophys. Res. (Sept. 1965) 70, 4561-4572.

5. Dean, R. G. and Dalrymple, R. A.: Water Wave Mechanics for Engineers and Scientists, World Scientific Advanced Series on Ocean Engineering, Vol 2, Singapore, 1991, original 1984 by Prentice Hall Inc., Englewood Cliffs, NJ.

6. E&P Forum: "Wave and Current Kinematics and Loading," E&P Forum Report No.3.12/156, IFP, Paris, 1989.

7. Forristall, G. Z.: "Irregular Wave Kinematics from a Kinematic Boundary Condition Fit (KBCF)," Applied Ocean Research, 7 (1985) 202-12.

8. Forristall, G. Z.: "Kinematics in the Crest of Storm Waves," Proc. of the 20th Coastal Engineering Conference, Taipei, Taiwan, ASCE, New York, Chapter 6, 1986, 208-22.

9. Gallagher, B.: "Generation of Surf Beat by Nonlinear Wave Interactions," J. Fluid Mech. (1971) 49, 1-20.

10. Gudmestad, O. T.: "Measured and Predicted Deep Water Wave Kinematics in Regular and Irregular Seas," Marine Structures, 6 (1993), 1-73.

11. Laitone, E. V.: "The Second Approximation of Cnoidal and Solitary Waves," J. Fluid Mech. (Nov. 1960) 9, 430-444.

12. Longuet-Higgins, M. S.: "The Effect of Nonlinearities on Statistical Distributions in the Theory of Sea Waves," J. Fluid Mech. (1963) 17, 459-480.

13. Pierson, W. J. Jr.: "Wind Generated Gravity Waves," Advances in Geophysics, Academic Press, New York City (1955) 2, 93-178.

14. Rodenbusch, G. & Forristall, G. Z.: "An Empirical Method for Calculation of Random Directional Wave Kinematics near the Free Surface," Proceeding

of the 11th Offshore Technology Conference, Houston, OTC 5097, (1986), 137-46.

15. Sharma, J. N.: "Development and Evaluation of a Procedure for Simulating a Random Directional Second-Order Sea Surface and Associated Wave Force," Ph.D. dissertation, University of Delaware, Newark (1979).
16. Sharma, J. N. and Dean, R. G.: "Development and Evaluation of a Procedure for Simulating a Random Directional Second-Order Sea Surface and Associated Wave Force," Ocean Engineering Report No. 20, University of Delaware, Newark (1979).
17. Sharma, J. N.: "Second Order Directional Seas and Associated Wave Forces," Society of Petroleum Engineers Journal, February 1981, pp129-140.
18. Skjelbreia, L. and Hendrickson, J. A.: "Fifth Order Gravity Wave Theory," Proc., Seventh Conference on Coastal Eng., The Hague (1961) 184-196.
19. Torum, A. & Gudmestad, O. T. (Eds.): "Water Wave Kinematics," NATO ASI Series E, Vol. 178, Kluwer Academic Publishers, Dordrecht, 1990.
20. Von Schwind, J. J. and Reid, R. O.: "Characteristics of Gravity Waves of Permanent Form," J. Geophys. Res. (Jan. 1972) 13, 420-433.

EVALUATION OF DEPTH-LIMITED WAVE BREAKING CRITERIA

George M. Kaminsky[1] and Nicholas C. Kraus[2]

ABSTRACT: Numerous criteria exist for predicting incipient wave breaking. These criteria were verified with a limited number of data sets or under limited beach and wave conditions, and confusion exists as to which of the many criteria to use. In the present study, a large database of wave breaking parameters from 17 laboratory experiments with plane slopes was compiled to yield 416 data points, of which 409 were used in empirical correlations to develop predictive breaking criteria covering the widest possible range available of beach and wave conditions. Selected criteria were evaluated and empirical coefficients adjusted by iterative best-fit procedures involving bin averaging of widely scattered parameter values to reveal functional dependencies. High-predictive capability was obtained using simple criteria expressed in terms of beach slope and wave steepness, and new criteria were developed for three breaking wave ratios or breaker indices: wave height to water depth at incipient breaking; incipient breaking wave height to deep-water wave height; and water depth at incipient breaking to deep-water wave height.

INTRODUCTION

Prediction of incipient depth-limited breaking of individual waves is required in quantitative descriptions of surf zone hydrodynamics and sediment transport. For example, a breaking wave criterion must be applied to calculate the longshore current, longshore sand transport rate, and wave-induced set up. Most engineering applications of breaking wave criteria pertain to individual waves, and such criteria have also been used in mathematical models of random breaking waves by assuming superposition of individual breaking waves (Dally 1990, Larson and Kraus 1991). The present paper focuses on empirically derived criteria for describing incipient breaking of individual (monochromatic) waves.

---

1) Shoreline Engineer, Shorelands and Coastal Zone Management Program, Washington Department of Ecology, P.O. Box 47690, Olympia, WA 98504-7690.
2) Director, Conrad Blucher Institute for Surveying and Science, Texas A&M University – Corpus Christi, 6300 Ocean Drive, Corpus Christi, TX 78412.

In this paper, we consider criteria for predicting the following parameters or "breaker indices" associated with depth-limited incipient breaking:

$$\gamma_b = \frac{H_b}{d_b} \tag{1}$$

$$\Omega_b = \frac{H_b}{H_o} \tag{2}$$

$$\beta_b = \frac{d_b}{H_o} \tag{3}$$

where γ_b is called the *breaker height-to-depth index*; the subscript b denotes the depth-limited breaking condition; H_b is the wave height at breaking; d_b is the water depth at breaking (which may be either the still-water depth or total water depth, depending on the data set); Ω_b is called the *breaker height index*; H_o is the (typically) unrefracted wave height in deep water (the deep-water condition denoted by subscript o); and β_b is called the *breaker depth index*.

In the latter two breaker indices, H_o is considered to be known, so that one is normally solving for γ_b or for H_b or d_b separately. The three indices are algebraically related as $\Omega_b = \gamma_b \beta_b$; however, in forming predictive expressions for each index individually, the resultant predictive expressions may not strictly satisfy this algebraic identity because of reorganization of the data in the bin-averaging procedure used in the best-fit analysis discussed below.

Most existing criteria have been formulated for γ_b and Ω_b, although some predictive expressions are available for calculating H_b and d_b directly. The breaker depth index β_b has been little discussed in the literature (see Goda 1970). The quantity to be predicted is usually expressed in terms of one or more empirical coefficients or functions, the beach slope m, and wave steepness H_o/L_o, where L_o is the wavelength in deep water as given by linear-wave theory. The surf-similarity parameter (Battjes 1974) or Irribarren number $I = m/(H_o/L_o)^{1/2}$ has also been used. More sophisticated criteria contain functions of the unknown on both sides of the equation, requiring iteration for solution. In the present study, we limit discussion to planar beaches. Smith and Kraus (1990, 1991, 1992) investigated wave breaking on planar and barred profiles and find differences in breaker type and indices for beaches with well-developed bars or reefs as compared to breaking on planar beaches.

Many predictive criteria for the breaking indices γ_b and Ω_b can be found in the literature, each evaluated with a certain data set. Simpler criteria are convenient for theoretical developments and for combination with mathematical expressions, whereas more complicated formulas tend to be used in calculation-intensive numerical models. Questions arise on the validity of all the criteria, their range of applicability over the variables expressing them, and their limiting forms if they are used beyond their range of empirical testing. Also, one may ask if the more sophisticated formulas are more accurate.

In the present study a large database was assembled to analyze basic functional dependencies controlling incipient wave breaking. The data set is by far larger than those used in previous works. Selected common predictive expressions for the breaking wave indices are evaluated with the data set, and three best-fit predictive expressions are developed.

PROCEDURE

For this study, we compiled a database of 17 independent data sets giving 416 total points on individual breaking waves in the laboratory. Origins and properties of the data sets are given in Table 1. In the database, beach slope ranges from 1/110 to 1/5, and deep-water wave steepness ranges from 0.0007 to 0.0924.

A censored data set of 409 total points was used to examine general functional dependencies and determine quantitative breaker index criteria. Every data point was quality checked by comparing measured breaker indices with a predicted average index value calculated from common predictive expressions. If a difference greater than 0.30 was obtained, the data point was censored, resulting in the loss of seven points from the complete data set. In visual confirmation, the seven censored points all plotted as outliers.

As pointed out by Smith and Kraus (1991), incipient wave breaking is an instability, and, therefore, inherently subject to wide variation for a small change in physical conditions (for example, slight change in water level and incident wave height). Also, different researchers have used different definitions of incipient breaking, such as where the wave face becomes vertical, or at the point of maximum crest height (Singamsetti and Wind 1980, Smith and Kraus 1990). Some researchers report values of total depth (still-water depth plus deviation in mean-water level), most give still-water depth, and others do not state which depth or definition of breaking waves is used. In a carefully controlled engineering-type laboratory experiment involving periodic waves, Smith and Kraus (1990, 1991) averaged values from 10 consecutive waves and found large scatter both in the data composing the average and among averages. Scatter in values is also expected because subjectivity enters in defining the incipient breaking point, which is often not obvious for waves that are not peaked or do not have a vertical front face, as, for example, if waves begin to spill while still increasing in height.

Table 1. Total compiled database

Data Set	Beach Slope	H_o/L_o	No. Data Points
Munk (1949)	1/110 - 1/14	0.007 - 0.091	16
Iversen (1952)	1/50 - 1/10	0.003 - 0.080	63
Morrison & Crooke (1953)	1/50 - 1/10	0.004 - 0.080	6
Horikawa & Kuo (1966)	1/80 - 1/20	0.006 - 0.073	97
Galvin (1969)	1/50 - 1/5	0.0007 - 0.050	19
Weggel (1972)	1/19.5	0.006 - 0.041	8
Komar & Simmons (c. 1973)	1/28 - 1/9.5	0.003 - 0.066	44
Saeki & Sasaki (1973)	1/50	0.005 - 0.039	2
Iwagaki et al (1974)	1/50 - 1/10	0.005 - 0.073	23
Walker (1974)	1/30	0.001 - 0.038	15
Van Dorn (1976)	1/45 - 1/12	0.001 - 0.031	12
Singamsetti & Wind (1980)	1/40 - 1/5	0.017 - 0.080	95
Mizuguchi (1980)	1/10	0.045	1
Visser (1982)	1/20 - 1/10	0.014 - 0.079	7
Maruyama et al (1983)	1/30	0.091	1
Stive (1985)	1/40	0.031 - 0.032	2
Smith & Kraus (1990)	1/30	0.009 - 0.092	5

Attempts to fit empirical curves to such widely scattered data are difficult and do not inspire confidence in the results. To overcome this problem, in the present work the data were *bin averaged*. In this procedure for developing predictive criteria, bins (widths of 0.05 for γ_b and 0.1 for Ω_b and β_b) were formed as governed by the independent variable (breaker index) under consideration, and average values of the dependent quantity (beach slope, wave steepness) were computed. To display dependencies, beach slope and wave steepness were binned by width of 0.01. The binning procedure was found to be essential for organizing the data and making apparent trends or lack of trends, as described in the next section.

Various predictive expressions appearing in the literature and devised in the present study were fit to the data by iteration to obtain values of empirical multipliers and powers entering the equations. The iteration was done by running "do loops" of fine resolution to determine optimal fitting parameters as defined by the largest coefficient of determination. This procedure is computationally simple and equivalent to nonlinear regression, thereby avoiding distortion in fitting as might be produced in transformations that reduce expressions to forms amenable to linear regression.

RESULTS

Properties of the Data Set

Table 2 summarizes basic statistics of the database. Of particular interest are the average, maximum, and minimum values of the breaker indices. For example, for the complete (uncensored) data set of 416 total points, which includes very steep beach slopes (and possible amplification in wave height due to reflection), the average value of γ_b is 0.89. The maximum and minimum values of γ_b are 1.59 and 0.60, which are considered reasonable estimates of possible extremal asymptotes for this index. The value 1.59 (from Van Dorn 1976) was one of the seven censored data points, and the highest uncensored value is 1.49 (from the data set of Galvin 1969). The only other significant change resulting from the censoring was that the minimum β_b changed from 0.65 in Table 2 to 0.83. For the medium-beach slope data set, the average value of γ_b is 0.79, which is essentially the same as 0.78 derived by McCowan (1891) and used as the standard design value in the United States.

The medium-slope data set summarized in Table 2 comprises 169 data points and is presented only for consideration of reasonable averages and asymptotes for beach slopes typically encountered on sandy beaches. Note that the maximum γ_b value decreased from 1.59 to 1.25 in going from the complete to medium-slope beach data set, whereas maximum values for Ω_b and β_b remain the same.

Table 2. Summary of database statistics

	m	T sec	L_o m	H_o m	H_o/L_o	H_b m	d_b m	γ_b	Ω_b	β_b
Complete Data Set										
Max.	1/5	6.00	56.24	1.37	0.0924	1.50	2.00	1.59	3.78	4.05
Avg.	1/18	1.66	5.15	0.09	0.0301	0.11	0.13	0.89	1.32	1.50
Min.	1/110	0.70	0.76	0.01	0.0007	0.02	0.03	0.60	0.76	0.65
S. D.	--	--	--	--	--	--	--	0.17	0.45	0.42
Medium-Slope Data Set										
Max.	1/30	5.00	39.00	1.21	0.0924	1.50	1.90	1.25	3.78	4.05
Avg.	1/43	1.69	4.97	0.10	0.0301	0.12	0.15	0.79	1.25	1.57
Min.	1/80	0.78	0.95	0.01	0.0007	0.02	0.03	0.60	0.76	0.91
S. D.	--	--	--	--	--	--	--	0.12	0.42	0.41

General properties of the breaker indices are also revealed in their frequency distributions shown in Fig. 1. The index γ_b has a much narrower distribution and range as compared to Ω_b and β_b. Although Ω_b can vary over a relatively wide range, most values lie between 0.9 and 1.6, with a prominent mode at 1.0. The shapes of the distributions are partially an artifact of the capability of laboratory facilities.

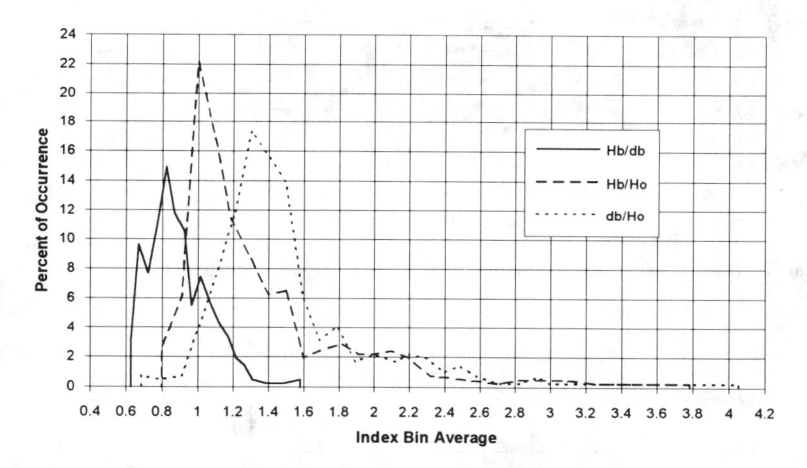

Fig. 1. Frequency distributions of the three breaker indices

An example of the extraordinary scatter in the data is shown in Fig. 2, which plots the height-to-depth index γ_b as a function of deep-water wave steepness for the complete data set. Although a trend is observed of decreasing γ_b with increasing steepness, it is much more apparent after bin averaging, Fig. 3, which plots average γ_b (averaged within each bin) as a function of bin-averaged beach slope and deep-water wave steepness. The γ_b index clearly decreases with increasing wave steepness and increases with increasing beach slope. For clarity, an indicator of scatter among the bin averages is not shown. However, quartile values were determined and typically deviated 0.1 from the bin average (0.2 spread).

Figs. 4 and 5 respectively show plots of average Ω_b and β_b as functions of independently bin-averaged beach slope and deep-water wave steepness. The breaker height index Ω_b decreases with increasing wave steepness but exhibits no dependence on slope. The slope bin averages for Ω_b typically fall within a 0.8-spread of quartile values, whereas the deep-water wave steepness bin averages were contained within a 0.2-quartile spread. The breaker depth index β_b has the trend to decrease with increasing wave steepness and with increasing beach slope, with considerable scatter in the dependence on slope.

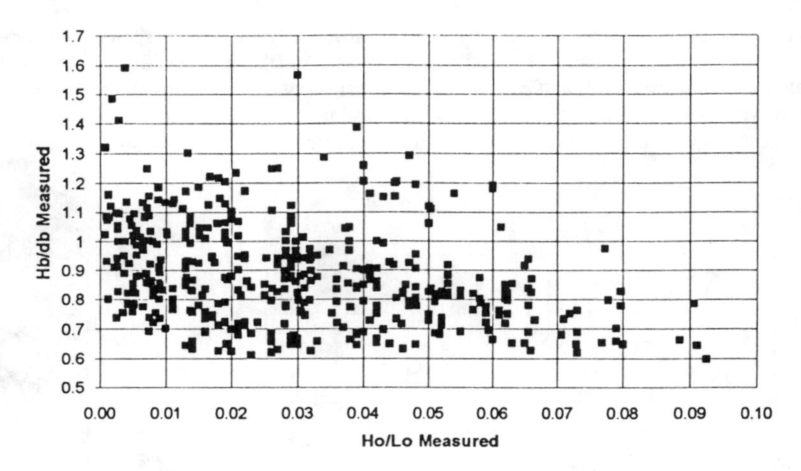

Fig. 2. Breaker height-to-depth index γ_b vs deep-water wave steepness

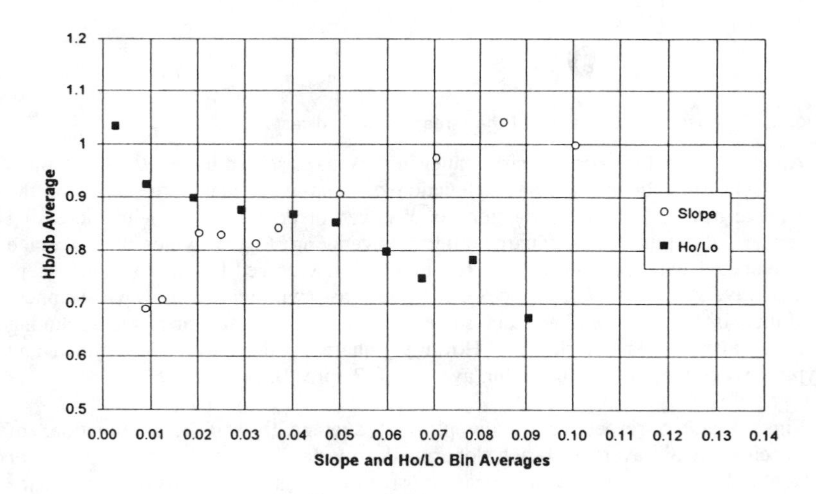

Fig. 3. Average γ_b vs bin-averaged beach slope and deep-water wave steepness

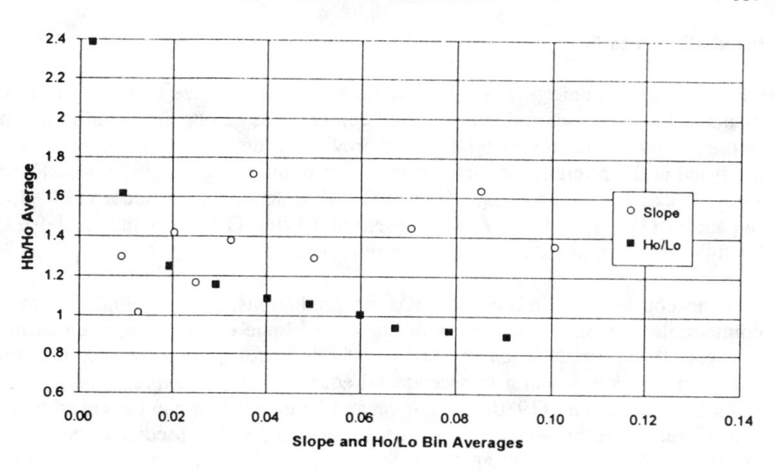

Fig. 4. Breaker height index Ω_b vs bin-averaged beach slope and wave steepness

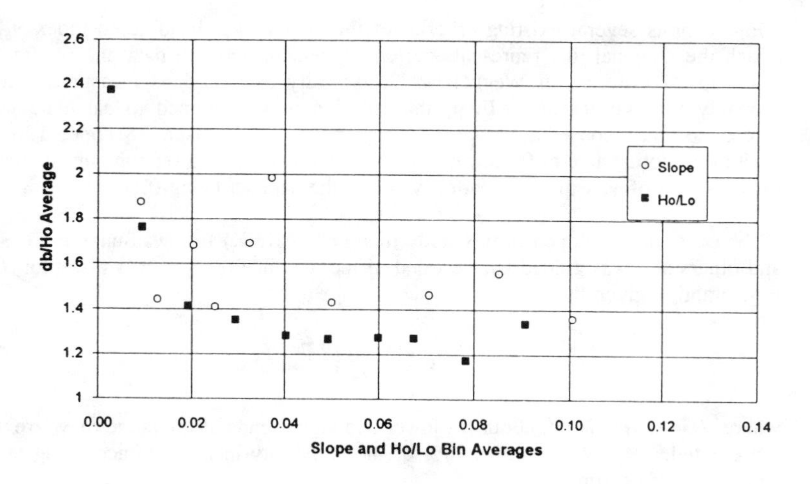

Fig. 5. Breaker depth index β_b vs bin-averaged beach slope and wave steepness

Predictive Criteria

Predictions of numerous existing criteria for γ_b and Ω_b were compared to measurements in the database, and only selected comparisons can be discussed in this paper. In the legends of the next two figures, abbreviations denote authors of the criteria and are listed in the reference section: K&G, Komar and Gaughan (1972); K&K, present study; L&K, Larson and Kraus (1989); O&M, Ostendorf and Madsen (1979); Suna, Sunamura (1983); and S&W, Singamsetti and Wind (1980). Other authors appear by full name and are also listed in the references.

In the course of examining the existing predictions, it was found that most had comparable capability to agree with the data. Some expressions, such as those of Weggel (1972) and Madsen and Ostendorf (1979) for γ_b, involve iteration because the unknown is given in a transcendental equation. Other expressions for γ_b, such as those of Sunamura (1980) and Larson and Kraus (1989), can be solved explicitly. Therefore, in the present study, it was decided to develop predictive equations that are of simple form by fitting to the large data set. Limitations of the developed predictive equations are given in Table 2; that is, caution should be exercised if the criteria are applied beyond their range of empirical determination. In fact, it is recommended that no breaker index be calculated as a value beyond the maxima and minima appearing in the data set and listed in Table 2.

Fig. 6 plots several existing criteria for the breaker height-to-depth index γ_b, in which the diagonal line represents perfect agreement between data and prediction. The popular criterion of Weggel (1972) typically over-predicts measured values, probably for two reasons. First, the criterion was developed to calculate wave forces, so that a conservative approach in estimation was taken. Second, a lower-limit asymptotic value of 0.78 was imposed, whereas the data set contains numerous lower values of γ_b with the smallest value in the data set being 0.60.

The equation developed in this study (denoted as K&K) follows Sunamura (1980) and others and was guided by the trend of the data in Fig. 3. It is also plotted in Fig. 6 and is given by

$$\gamma_b = 1.20\, I^{0.27} \tag{4}$$

where $I = m/(H_o/L_o)^{1/2}$. Both the lower and higher ends of the range of γ_b are well represented. Eq. (4) describes 90% of the variability in the data according to the coefficient of determination.

Fig. 7 similarly plots several existing criteria for the breaker height index Ω_b. The criteria of Goda (1975), Singamsetti and Wind (1980), Sunamura (1980), and the present study describe the data with about equal skill, although all predictions deviate slightly at the lower end of the range, which corresponds to high values of

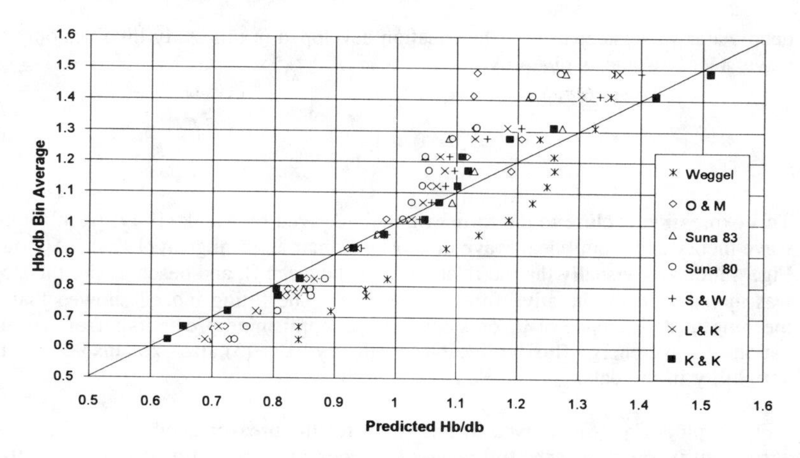

Fig. 6. Measured breaker height-to-depth index γ_b and predictions by several criteria

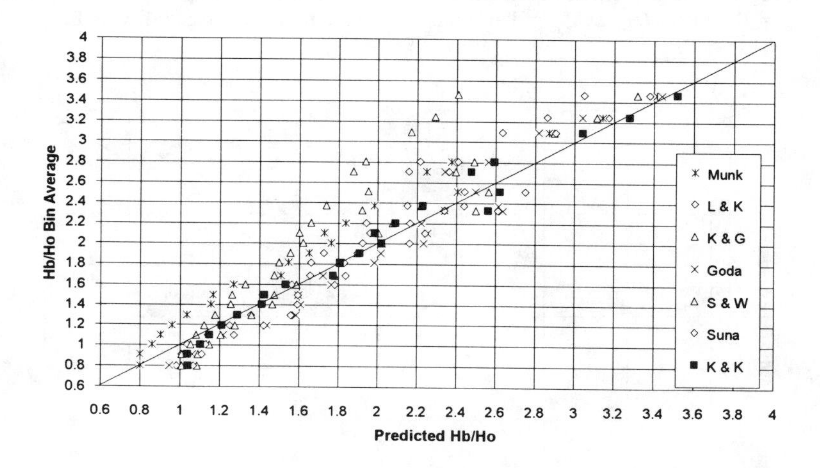

Fig. 7. Measured breaker height index Ω_b and predictions by several criteria

deep-water wave steepness. The equation developed in this study involves only the wave steepness and is given by

$$\Omega_b = 0.46\left(\frac{H_o}{L_o}\right)^{-0.28} \tag{5}$$

This expression is close to the form originally derived by Munk (1949) from solitary wave theory and from linear-wave theory by Komar and Gaughan (1974). Although Fig. 4 indicates visually that no relation exists between Ω_b and beach slope, the slope was included in the iterative fitting procedure. The fitting indeed showed that no meaningful dependence of Ω_b on slope existed, confirming predictions from conservation of wave energy flux by linear-wave theory. Eq. (5) also explains 90% of the variability in the data.

Fig. 8 plots Eq. (6), derived in this study for the breaker depth index β_b. Good agreement is found over the full range of the measured data. Eq. (6) explained 92% of the data variability. For possible use in theoretical work, it may be convenient to have identical powers of beach slope and deep-water wave steepness. The equation $\beta_b = 0.275\ [m\ (H_o/L_o)]^{-0.25}$ has almost the same predictive capability as Eq. (6).

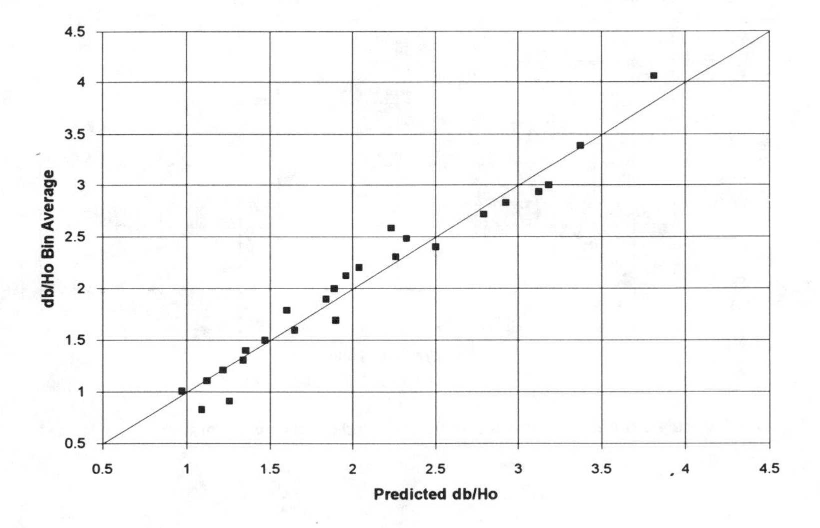

Fig. 8. Comparison of measured breaker depth index β_b and calculation by Eq. (6)

$$\beta_b = 0.30\, m^{-0.25} \left(\frac{H_o}{L_o} \right)^{-0.23} \qquad (6)$$

CONCLUSIONS

In this study, we analyzed a large data set (416 values censored to 409 values) on depth-limited breaking of regular waves incident to plane sloping beaches. The following is a summary of the resultant main conclusions.

1. Bin averaging was a highly useful technique for organizing the widely scattered data, making apparent functional dependencies, and developing predictive equations for wave breaker indices.

Breaker height-to-depth index, γ_b

2. $\gamma_b = 0.78$ is an average or representative value for typical field beach slopes (1/30 to 1/80), confirming the design value in common use.

3. Empirical asymptotes for γ_b are $0.60 < \gamma_b < 1.59$.

4. γ_b was predicted well by the surf-similarity parameter $I = m/(H_o/L_o)^{1/2}$ as $\gamma_b = 1.20\, I^{0.27}$, which explained 90% of the data variability.

5. Iterative (algebraically complicated) expressions did not perform notably better than simple explicit expressions.

Breaker height index, Ω_b

6. $\Omega_b = 1.0$ is a representative value, with Ω_b increasing steeply for deep-water wave steepness less than about 0.02.

7. Empirical asymptotes for Ω_b are $0.76 < \Omega_b < 3.78$.

8. Ω_b showed no dependence on beach slope, and the expression $\Omega_b = 0.46\, (H_o/L_o)^{-0.28}$ explained 90% of the data variability.

Breaker depth index, β_b

9. $\beta_b = 1.25$ is a representative value, but this index has a wide spread with values in the range from 0.9 to 1.8 being most prevalent.

10. Empirical asymptotes for β_b are $0.65 < \beta_b < 4.05$.

11. β_b was predicted well by the expression $\beta_b = 0.30\, m^{-0.25} (H_o / L_o)^{-0.23}$, which explained 92 percent of the data variability.

ACKNOWLEDGEMENTS

We took up work on the quantity β_b at the suggestion of Dr. David Kriebel (U.S. Naval Academy) who had need of a simple predictive expression. This beta's for you, Dave. We would like to thank Ms. Karin Boedecker and Ms. Mary Rogan of the Conrad Blucher Institute for assistance in preparing the camera-ready manuscript.

REFERENCES

Battjes, J. A. 1974. "Surf Similarity," *Proc. 14th Coastal Eng. Conf.*, ASCE, 466-480.

Dally, W. R. 1990. "Random Breaking Waves: A Closed-Form Solution for Planar Beaches," *Coastal Eng.*, 14, 233-263.

Galvin, C. J. 1969. "Breaker Travel and Choice of Design Wave Height," *J. of the Waterways and Harbors Div.*, (95)WW2, 175-200.

Gaughan, M.K., Komar, P.D., and Nath, J.H. 1973. "Breaking Waves: A Review of Theory and Measurements," School of Oceanog., Rep. 73-12, Oregon State U., Corvallis, OR. (contains the Komar and Simmons (c. 1973) data set)

Goda, Y. 1970. "A Synthesis of Breaker Indices," *Trans. Japanese Soc. Civil Engineers*, 2, Part 2, 227-230.

Horikawa, K., and Kuo, C. T. 1966. "A Study of Wave Transformation Inside the Surf Zone," *Proc. 10th Coastal Eng. Conf.*, ASCE, 217-233.

Iversen, H. W. 1952. "Laboratory Study of Breakers," *Gravity Waves*, Circular 521, U.S. Bureau of Standards, 9-32.

Iwagaki, Y., Sakai, T., Tsukioka, K., and Sawai, N. 1974. "Relationship Between Vertical Distribution of Water Particle Velocity and Type of Breakers on Beaches," *Coastal Eng. in Japan*, 17, 51-58.

Komar, P. D., and Gaughan, M. K. 1972. "Airy Wave Theory and Breaker Height Prediction," *Proc. 13th Coastal Eng. Conf.*, ASCE, 405-418.

Komar and Simmons, c. 1973 See Gaughan, et al 1973.

Larson, M., and Kraus, N. C. 1989. "SBEACH: Numerical Model for Simulating Storm-Induced Beach Change, Report 1, Empirical Foundation and Model Development," Tech. Rep. CERC-89-9, U.S. Army Eng. Waterways Expt. Station, Coastal Eng. Res. Center, Vicksburg, MS.

Larson, M., and Kraus, N. C. 1991. "Numerical Model of Longshore Current Over Bar and Trough Beaches," *J. Waterway, Port, Coastal and Ocean Eng.*, 117(4), 326-347.

Mallard, W. W. 1978. "Investigation of the Effect of Beach Slope on the Breaking Height to Depth Ratio," M.S. thesis, Dept. Civil Eng., U. of Del, Newark, DE.

Maruyama, K., Sakakiyama, T., Kajima, R., Saito, S., and Shimizu, T. 1983. "Experimental Study on Wave Height and Water Particle Velocity Near the Surf Zone Using a Large Wave Flume," Civil Eng. Lab. Rep. No. 382034, The Central Research Inst. of Electric Power Industry, Chiba, Japan. (in Japanese)

McCowan, J. 1891. "On the Solitary Wave," *Philos. Mag.*, 5th Series, 32(194), 45-58.
Mizuguchi, M. 1980. "An Heuristic Model of Wave Height Distribution in Surf Zone," *Proc. 17th Coastal Eng. Conf.*, ASCE, 278-289.
Morrison, J.R., and Crooke, R.C. 1953. "The Mechanics of Deep Water, Shallow Water, and Breaking Waves," Beach Erosion Board, Tech. Memo. No. 40, U.S. Army Corps of Engineers. (avail. from Waterways Experiment Station, Vicksburg, MS)
Munk, W. H. 1949. "The Solitary Wave Theory and Its Application to Surf Problems," *Annals of the New York Academy of Sciences*, 51.
Ostendorf, D. W., and Madsen, O. S. 1979. "An Analysis of Longshore Currents and Associated Sediment Transport in the Surf Zone," Sea Grant Rep. 79-13, Mass. Inst. of Tech., Cambridge, MA.
Saeki, H., and Sasaki, M. 1973. "A Study of the Deformation of Waves After Breaking (1)," *Proc. 20th Japanese Conf. on Coastal Eng.*, JSCE, 559-564. (in Japanese)
Singamsetti, S. R., and Wind, H. G. 1980. "Characteristics of Breaking and Shoaling Periodic Waves Normally Incident to Plane Beaches of Constant Slope," Rep. M1371, Delft Hydr. Lab., Delft, The Netherlands.
Smith, E. R., and Kraus, N. C. 1990. "Laboratory Study on Macro-Features of Wave Breaking Over Bars and Artificial Reefs," Tech. Rep. CERC-90-12, U.S. Army Eng. Waterways Experiment Station, Coastal Eng. Res. Center, Vicksburg, MS
Smith, E. R., and Kraus, N. C. 1991. "Laboratory Study of Breaking Waves on Bars and Artificial Reefs," *J. Waterway, Port, Coastal and Ocean Eng.*, 117(4), 307-325.
Smith, E. R., and Kraus, N. C. 1992. "Laboratory Study of Wave Transformation on Barred Beach Profiles," *Proc. 23rd Coastal Eng. Conf.*, ASCE, 630-643.
Stive, M. J. F. 1985. "A Scale Comparison of Wave Breaking on a Beach," *Coastal Eng.*, 9, 151-158.
Sunamura, T. 1980. "A Laboratory Study of Offshore Transport of Sediment and a Model for Eroding Beaches," *Proc. 17th Coastal Eng. Conf.*, ASCE, 1051-1070.
Sunamura, T. 1983. "Determination of Breaker Height and Depth in the Field," *Ann. Report, Inst. of Geoscience*, 8, Univ. of Tsukuba, Tsukuba, Japan.
Van Dorn, W. G. 1976. "Set-Up and Run-Up in Shoaling Breakers," *Proc. 15th Coastal Eng. Conf.*, ASCE, 738-751.
Visser, P. J. 1982. "The Proper Longshore Current in a Wave Basin," Rep. No. 82-1, Lab. of Fluid Mechs., Dept. Civil Eng., Delft U. of Technol., Delft, The Netherlands.
Walker, J. R. 1974. "Wave Transformation Over a Sloping Bottom and Over a Three-Dimensional Shoal," Look Lab.-75-11, U. of Hawaii, Honolulu, HI.
Weggel, J. R. 1972. "Maximum Breaker Height," *J. Waterways, Harbors, and Coastal Eng. Div.*, 98(WW4), 529-548.

WAVES 93
New Orleans, Louisiana
26 July 1993

Opening Remarks by Robert L. Wiegel
Professor Emeritus of Civil Engineering
University of California, Berkeley, CA 94720

Co-chairmen Orville T. Magoon and J. Michael Hemsley
Professor Robert G. Dean
Ladies and Gentlemen

I am very pleased to be here at this conference in New Orleans, on the great Mississippi River, with its vast, fascinating wetlands nearby. I am honored that this conference is dedicated to me, and I want to thank all of you who have worked so hard in its planning, implementation, and preparation of papers - and the many associations and agencies for their sponsorship. I know from personal experience how much work is involved in giving a conference of this scope, and I appreciate it very much. It is a great pleasure to see again so many colleagues that I have worked with over the years, and to meet new members of the community that are advancing our understanding of waves.

The Mississippi River has been in flood again - a very great one, resulting in much suffering and damage. Fortunately the deaths, injury, and damage have been less than they would have been without the extensive flood control system that was in place and operating. Two weeks ago there was a severe earthquake and tsunami in the Sea of Japan, to the west of Hokkaido, which caused a large loss of life and devastation on Okushiri Island. As engineers and scientists, we know that nature always has an episodic event with which to test our defenses for the protection of life and property. We will learn from these terrible events, and prepare better to minimize harmful effects of future ones.

Why did I mention these two natural disasters rather than others? One reason is that they are current. The second reason is that they are waves, one flood and the other impulsively generated. The third reason is that I worked on aspects of these two types of phenomena in my early professional career.

The possibility of a tsunami being generated by the massive underwater explosion in the Bikini tests in 1946 led to my being employed by the University of California at Berkeley (called "Cal" then, not "Berkeley"). It was in June 1946, and I was a 1st Lieutenant on terminal leave from the U.S. Army Ordnance Corps, after four years of service. I went to the university to talk with the adviser I had as an undergraduate student in mechanical engineering, Professor R.G. Folsom, to find out about going to graduate school. He asked if I would be doing anything during the following two weeks, and if not would I like to work for him on an emergency project. I said I was interested, so he told me he had just received a Navy radiogram from Dean Morrough P. O'Brien from the Operation Crossroads Task

Force at Bikini Atoll stating their concern about the possibility of the explosion triggering an underwater landslide down the reef's slope, which might generate a tsunami (called a tidal wave at that time, and in the popular press today). They wanted an experiment made to help them evaluate its likelihood, and possible characteristics. They thought of the possibility of such an event owing to the fact that during preliminary staging in the Hawaiian Islands for the Bikini tests, the earthquake and tsunami of 1 April 1946 occurred near Scotch Cap, Aleutian Islands,Alaska; the tsunami travelled across the North Pacific Ocean, and caused much damage and loss of life in Hawaii. Some of the engineers and scientists were in Hawaii at that time, including John D. Isaacs, and had made a reconnaissance survey of the damage.

I designed, operated, and analyzed the results of a hydraulic model test within the two weeks - fortunately the use of a Froude model was appropriate. I have been at the university ever since, starting as a research engineer and going to graduate school part time - I still go into the office every day, even though I retired six years ago (I recall that I worked for the university even earlier, as a reader in the junior course on fluid mechanics, Spring Semester 1942 - 51 years ago). It was not until many years later, probably in the early 1980's while having lunch with Professor Roger Revelle (who was also at Bikini, as was Professor John D. Isaacs) that I learned that he had worried about the possibility of the generation of a tsunami, and that Mike O'Brien said he would try to have an experiment made.

The group at the university working in the 1940's and early 1950's with O'Brien, Johnson and Isaacs on amphibious operations for the Navy were doing quite a bit of field work, in addition to theoretical and laboratory studies. Owing to this, I worked on all three aspects, on an integrated basis, from which I benefitted greatly. I have always enjoyed and learned much from going on coastal field trips, and still go on them at almost any opportunity, to see the diverse and often unexpected situations that exist in this world. This work expanded into many other types of activities, including the measurement and study of waves, wave-induced forces on coastal and offshore structures and pipelines,tsunamis, mixing processes, surf and beach processes, harbor oscillations, ship motion and mooring line forces, and the functional design of harbors and other coastal facilities.

We did not realize at that time that we were developing two important new branches of civil engineering -"coastal engineering" and "offshore/ocean engineering". They are now firmly established; the 23rd International Conference on Coastal Engineering was held in Venice, Italy in October 1992 under the auspices of the ASCE, and the 25th Offshore Technology Conference was held in Houston, Texas in May 1993, co-sponsored by the ASCE.

The first course I taught was Irrigation 117, at U.C. Davis in the Sacramento Valley; I commuted by train twice a week from Berkeley to Davis. The first course I taught in civil engineering, which was at U.C. Berkeley, was advanced hydraulic structures. So, rivers, floods, dams, and irrigation systems have always interested me professionally. I was very fortunate to have been able to work with Professor Joe W. Johnson during many of my years at the university. Many of you know Joe - you will be pleased to learn that he celebrated his

85th birthday last week in Victoria, BC, Canada, where he now lives to be near his two daughters and his grandchildren.

I would like now to change the subject, and tell you an anecdote about coastal wetlands. The technical problems of trying to save them or develop new ones interests me very much, especially in multiple use situations. In a recent issue of SCIENCE (25 June 1993), there was an article by Leslie Roberts on wetlands, with emphasis on mitigation projects. The author states that one of the experts in the subject claims that plans for wetland restoration or construction often give little attention to the hydrology of the created site, which is the most important factor in a successful project. Other scientists are cited for their statements of a similar nature, about poor specifications for hydrology and vegetation, and that the biggest gap presently is one of insufficient understanding of the interactions of soil, surface water, and ground water on which the ecosystem is dependent.

This is not a new problem, the understanding of coastal wetlands, channels, and navigation in the channels. A graduate student in the History of Science and Technology Department at U.C. Berkeley, Tal Golan, visited me in my office on campus about a month ago to discuss the concepts that an expert witness (John Smeaton, a civil engineer[1]) gave in England on 4 May 1782. There was a law suit at that time (Folkes, Bart. v. Chadd and Others) in regard to the deterioration of a small navigation channel between the North Sea and Wells Harbour, in the county of Norfolk, England, just south of The Wash. One of the questions that had to be resolved by the court was what were the effects on the navigation channel of the embankments constructed to enclose some of the salt marshes, together with the condition and operation of a sluice which had been built and rebuilt to help maintain the small channel. Furthermore, what would be the effects, the benefits (if any), and rate of change with time, of removing the most recently built embankment. The tidal range at that location is quite large, and there is a great amount of sand - it is very different from anything in the USA. As a part of his Ph.D. thesis work, Golan found that this was the first time that the testimony of an expert witness, other than medical doctors, had been admitted by a judge in the history of English jurisprudence (of which ours is derivative); furthermore, the decision of the trial judge was upheld by the higher court (The Court of the King's Bench), in 1782. [Parenthetically, it should be interesting to those who attended the Coastal Zone 93 conference here last week, that this first use of an expert witness in a court case was within the general purview of coastal zone management]. It is very interesting (and difficult) to read the reasoning of John Smeaton, his description in words of a conceptual model of what was occurring from the standpoint of tidal flows, winds, sedimentation, and plants, before there were many tools for the physical/mathematical analysis of such a situation. To close this subject, and tying together the testimony of the expert witness in that early court case, and the recent article on coastal wetlands with its statements of our lack

[1] After I gave the opening remarks, William F. Baird told me that Smeaton was the first person to call himself a civil engineer.

of knowledge of their hydrology, we seem to have made headway very slowly in this important environmental subject owing to its great complexity and the strongly nonlinear interactive processes. We still have much to do in the advancement of our techniques of analysis, including the use of both physical and numerical models; for the graduate students here, don't think we have run out of good thesis topics.

This is a conference on waves. The subject has interested me greatly during the past 47 years. There will many excellent papers presented on waves this week, and I am looking forward to hearing them. I have always been most interested in the nonlinear characteristics of waves, and a number of papers are on some of these aspects. Philosophically, I have no difficulty with nonlinearities. However, I do have a philosophical difficulty with linear superposition - as I have mentioned to my students many times, I always wonder why nature has been so kind to us, to provide us with many situations in which the use of linear superposition permits us to make reasonable predictions, at least in deep water. I hope that one of the papers presented this week will be able convince me why this is so, based on first principles, rather than upon observations. In shallow water, as we all know, linearizations often are not useful, and can lead to gross misinterpretation of the processes and serious errors in predictions.

Thank you again for this honor, and may the conference be a success for all of you.

References

Roberts, Leslie, "Wetlands Trading Is a Loser's Game, Say Ecologists," SCIENCE, Vol. 260, 25 June 1993, pp 1890-1892.

Roscoe, Henry, REPORTS OF CASES ARGUED AND DETERMINED IN THE COURT OF KING'S BENCH, IN THE 22nd, 23rd, 24th, AND 25th YEARS OF THE REIGN OF GEORGE III, from the manuscripts of The Right Hon. Sylvester Douglas, Baron Glenbervie, and also from the manuscripts of Mr. Justice Lawrence, Mr. Justice Le Blanc, Mr. George Wilson, &c, Vol. III, S. Sweet, and Stevens and Sons, Dublin, 1831, pp 157-161 and 340-343.

Skempton, A.W., editor, JOHN SMEATON, FRS, Thomas Telford Ltd., London, England, 1981, 291 pp.

Smeaton, John, REPORTS OF THE LATE JOHN SMEATON, F.R.S., MADE ON VARIOUS OCCASIONS, IN THE COURSE OF HIS EMPLOYMENT AS A CIVIL ENGINEER, Vol. III, Longman, Hurst, Rees, Orme, and Brown, London, 1812, pp 18-37 plus plate.

NDBC Wave Data — Current and Planned

K.E. Steele[1] and Theodore Mettlach[2]

Introduction

The National Data Buoy Center (NDBC) of the National Weather Service (NWS), a part of the U.S. National Oceanic and Atmospheric Administration (NOAA), operates a large number of buoy and fixed-platform environmental data reporting stations in geographical areas of interest to the United States. These areas include the U.S. east and west coasts, Gulf of Mexico, Great Lakes, Gulf of Alaska, Hawaiian Islands, and Guam. Each hour meteorological and oceanographic data are acquired, transmitted to shore via the Geostationary Operational Environmental Satellites (GOES), and distributed to users.

Among the data reported from these stations are ocean wave measurements. Some stations report only nondirectional wave data, but each year more of the buoys are instrumented to report directional data as well. Hourly wave data are quality checked and relayed in real time to NWS offices, and to other users in the U.S. and other countries. In the weeks following acquisition, the data are examined more carefully for quality and then archived at the U.S. National Oceanographic Data Center (NODC) and National Climatic Data Center (NCDC). Since the mid-1970's, large quantities of wave data have been stored at these archival centers.

Real time wave data are used for a wide variety of commercial and U.S. Government purposes, including issuance of warnings of hazardous marine conditions, prediction of favorable conditions for recreational surfing, and ship routing. Scientists and Engineers use the archived data for many purposes, including studies of various aspects of wave dynamics, development of computer models to predict wave conditions from wind fields, verification/ calibration of remote-sensing systems measuring wave data, and design of offshore structures.

Over a period of years, the quality, geographical coverage, and cost effectiveness of wave measurements made by NDBC systems have all improved. New systems superior in design concept to those now in the field are being developed. The purpose of this paper is to

---

[1] Waves Program Manager, National Data Buoy Center, John C. Stennis Space Center, MS 39529-6000 USA

[2] Oceanographer, Applied Technology Division, Computer Sciences Corporation, John C. Stennis Space Center, MS 39529-6000 USA

provide an overview of the NDBC systems now reporting operational data and new systems expected to become operational in the next few years.

Operational Systems

Wave Data Products

NDBC operational nondirectional wave measurement systems report hourly estimates of heave displacement spectral density for contiguous or overlapping frequency bands centered at frequencies ranging from 0.03 Hz up to some maximum frequency. From these estimates, significant wave height, peak period, and average period are derived and reported. In addition to all of the nondirectional data, operational directional systems report estimates of four directional parameters at frequencies up through 0.35 Hz. The four frequency-dependent parameters reported are those in Equation (2) below. The directional spectrum function, in units of m^2 Hz^{-1} $radian^{-1}$, is

$$S(f,\alpha) = C_{11}(f) \cdot D(f,\alpha) \tag{1}$$

in which $C_{11}(f)$ is the nondirectional spectrum function in units of m^2 Hz^{-1}, and

$$D(f,\alpha) \approx \frac{1}{\pi}\left[\frac{1}{2} + r_1 \cos(\alpha - \alpha_1) + r_2 \cos(2(\alpha - \alpha_2))\right] \tag{2}$$

D is the directional spreading function in $radians^{-1}$, its truncated form appearing on the right hand side of Equation (2).

The frequency dependent angles α_1 and α_2 are respectively mean and principal wave direction. When α_1 and α_2 agree, they indicate the main direction of the waves at that frequency, measured clockwise from north around to the direction from which the waves are arriving. The frequency-dependent parameters r_1 and r_2 lie between zero and one. The values of r_1 and r_2 provide information about the spreading of wave directions about the main direction of the waves at a particular frequency.

Types of Measurement Systems

The platforms for most NDBC wave measurement systems are moored buoys. These moored buoys comprise a variety of hull shapes and sizes because the NDBC inventory of hulls has been built up from diverse sources over many years. Each hull type has its own shape and mass distribution, so each has its own hull heave and slope Response Amplitude Operator (RAO) and phase lag functions (NDBC 1986). These functions can also vary somewhat with mooring design, which depends on environmental factors, especially water depth. In shallow water areas of 100 meters or less, all-chain moorings are generally used. In deeper water of 100 meters or more, inverse catenary moorings are generally used (Hamilton 1990).

Hull heave RAO differences, resulting from station-to-station differences in hull type and mooring design, complicate the management of NDBC wave measurement systems. With substantial numbers of buoys to operate, and buoys of different types and moorings being exchanged from time to time at each station, the probability increases that a clerical mistake will lead to a wrong RAO being used to correct measured data.

For nondirectional measurement systems, NDBC uses either laser wavestaffs (Brown & Gustavson 1990) to measure sea surface displacement or linear accelerometers to measure wave heave motion. Laser wavestaffs are used at a few in-water fixed platforms. Linear accelerometers, either fixed in the hull vertical or stabilized parallel to the earth vertical, are used in buoys. Vertical stabilization, when used, is achieved through use of the Hippy 40 sensor built by Datawell in Holland. This bulky, heavy, and expensive—albeit effective and widely used—sensor has built-in mechanical systems for keeping its accelerometer vertical as the buoy and sensor tilt.

Buoy-mounted directional wave measurement systems require, in addition to the measurement of heave acceleration, the measurement of hull azimuth, pitch and roll. NDBC makes use of two different sensor suites for the measurement of these angles. The first suite includes the same Datawell Hippy 40 sensor that measures earth-vertical acceleration; it also measures pitch and roll. The azimuthal angle is provided by magnetic field components measured by two axes of a triaxial magnetometer, corrected for pitch and roll. The two axes are mounted parallel to the bow and starboard axes of the hull. Algorithms in the onboard-buoy microprocessor transform these four sensor outputs to the hull deck east and north slope values needed to calculate directional wave data.

The alternative suite of sensors used for the measurement of azimuth, pitch, and roll consists of just one sensor—the (same) two-axis magnetometer. When the bow and starboard components of magnetic field alone are used to measure azimuth, pitch, and roll angles, an accelerometer with its axis fixed in the hull parallel to the mast is used to measure heave motion. By means of the magnetometer-only (MO) method (Steele & Earle 1991), onboard buoy computer programs extract azimuth, pitch, and roll from only the variations in the magnetic field components seen by the bow and starboard magnetometers.

Buoys producing directional wave data with the MO method have the advantage of being less expensive to assemble and maintain because they do not require the Hippy 40 sensor. However, they do not detect small, low-frequency swells as well as those systems employing the Hippy 40 sensor. As a result, operational use of MO systems is restricted at this time to the U.S. Great Lakes and Gulf of Mexico, where there is little or no swell.

Another circumstance complicating the management of NDBC wave measurement systems is that several different types of onboard buoy electronic systems are presently in service. The various types now in operational use were developed at different times and differ significantly due to improvements resulting from NDBC's field experiences and advances in electronics technologies. The following wave measurement system types are now in operational use:

- General Service Buoy Payload (GSBP) Wave Data Analyzer (WDA)
- Data Acquisition and Control Telemetry (DACT) Wave Analyzer (WA)
- Value Engineered Environmental Payload (VEEP) WA
- DACT Directional Wave Analyzer (DWA)
- DACT DWA, Magnetometer-Only configuration (DWAMO)

The oldest onboard-buoy electronic system is the General Service Buoy Payload (GSBP), developed in the late 1970's. The characteristics of the GSBP are given by Steele *et al.* (1976), NDBO (1978), Steele and Earle (1979), and Magnavox (1979). Although the GSBP WDA produces good quality data, it uses a computationally inefficient covariance method for the estimation of wave spectra, and the hardware is relatively power-hungry. Use of the covariance method for spectral estimates necessitates the use of an unrefined method for the correction of acceleration spectra for noise. Aboard hulls small enough not to filter out the very high frequency heave motions of the water surface, the GSBP can produce estimates of C_{11} at frequencies from 0.03-0.50 Hz in intervals of 0.01 Hz (or, in compact notation, at [0.03(0.01)0.50] Hz). The GSBP has no capability to measure directional spectra, is obsolete, and is being phased out of the NDBC inventory as funds to procure new equipment allow.

The next oldest system comprising wave measurement capability is the Data Acquisition and Control Telemetry (DACT) payload (Magnavox 1986). It was developed in the early 1980's and consumes much less power than the GSBP. The DACT uses a segmented Fast Fourier Transform (FFT) method for the calculation of spectra and has three configurations—nondirectional, Hippy directional, and MO directional. The DACT Wave Analyzer (WA) produces only nondirectional wave data based on estimates of C_{11} at frequencies [0.03(0.01)0.40] Hz. Noise corrections of acceleration spectra are made with an algorithm more refined than is possible using the covariance method for estimating spectra.

The DACT Directional Wave Analyzer (DWA) using the Hippy sensor and two axes of a magnetometer produces, in addition to estimates of C_{11} at WA frequencies, estimates of r_1, r_2, α_1, and α_2 at frequencies [0.03(0.01)0.35] Hz (Steele *et al.* 1985, Tullock & Lau 1986, Steele *et al.* 1992). The DACT DWA Magnetometer Only (DWAMO) firmware is a modified form of the DACT DWA Hippy firmware and produces these same directional wave data using only a fixed accelerometer and two axes of the magnetometer (Steele & Earle 1991).

The most modern of the operational onboard-buoy electronic systems measuring nondirectional wave data is the Value Engineered Environmental Payload (VEEP), developed in the mid- to late 1980's. This payload consumes even less power than the DACT. It has a built-in Wave Analyzer (WA) capability to report estimates of nondirectional spectra produced in a manner virtually identical to that of the DACT WA. Although this payload does not have built-in directional wave measurement capability, it has been designed to allow the relay of directional wave (or any other type of) data produced by a separate module to the VEEP using an RS-232 interface.

NDBC uses a structured system of indices to identify the various hull, sensor, and payload types, so that the computers doing the shore data processing locate the correct RAO's and other site specific data needed to correct wave data coming from a particular station.

Each of the above three payloads normally reports wind speed, wind direction, air temperature, water temperature, and barometric pressure. These meteorological data are very useful in monitoring the correct functioning of on-station wave measurement systems.

Moored Buoy Wave Measurements

Most moored-buoy-mounted wave measurement systems belong to networks of stations that are operated over periods of years as part of the NDBC Moored Buoy Program. From its base budget, NDBC operates one such network of buoys for NWS. The Department of Interior, Minerals Management Service (MMS), provides funds to NDBC for the operation of a separate buoy network located to provide environmental data needed by MMS for the management of offshore oil leases. The National Aeronautics and Space Administration (NASA) provides funds for two buoys operated near Cape Canaveral in support of shuttle launches.

The Coastal Engineering Research Center (CERC), an element of the U.S. Army Corp of Engineers (COE) Waterways Experiment Station, provides funds to NDBC for the operation of a network of approximately 10 buoys measuring directional wave data and for the upgrading of some of the NWS, MMS, and NASA buoys from nondirectional to directional capability. One moored buoy measuring nondirectional wave data is operated for NOAA's undersea habitat project near the Florida Keys.

In addition to the networks of moored buoys providing wave data routinely over periods of years, NDBC deploys moored buoys for other customers, usually for shorter periods of time and for special scientific or engineering purposes. During the winter of 1990-91, NDBC operated three buoys measuring directional wave data in the U.S. Atlantic coastal zone as part of the Surface Wave Dynamics Experiment (SWADE) sponsored by the Office of Naval Research (ONR) and NASA.

During the late summer of 1991, NDBC operated two buoys measuring directional wave data in the Atlantic Ocean for several weeks as part of the ONR High Resolution Remote Sensing Experiment (HRRSE). With ONR support, NDBC reconfigured these two buoys and operated them in a second Atlantic HRRSE during May and June 1993. In these ONR experiments, special equipment, over and above the standard Hippy-based wave measurements, was installed aboard the buoys to acquire time series data for the research project.

NDBC recently deployed and is now operating a 2.4-meter discus DACT DWA buoy in support of the Near Shore Hydrodynamics Project sponsored by the NOAA Great Lakes Environmental Research Laboratory (GLERL).

Fixed Platform Wave Measurements

NDBC operates a number of fixed-platform nondirectional wave measurement systems. Such systems are operated as part of the Coastal-Marine Automated Network (C-MAN). Each of these systems uses a laser wavestaff to measure instantaneous wave displacement. The output from the laser is fed into one of the buoy-type nondirectional wave measurement systems in place of the acceleration analog. Except for the sensor noise and RAO corrections, the operation of these systems is almost identical to that of the buoy-mounted accelerometer systems.

Wave Data Distribution Systems

Wave data from the moored buoys and fixed platforms are both distributed in real time and stored in archives. Data received via the Geostationary Operational Environmental Satellites (GOES) are relayed to the National Weather Service Telecommunications Gateway (NWSTG). There, computer programs correct the nondirectional spectra for spectral noise and hull (for buoy data), electronics, and numerical filtering RAO's. Directional wave data are corrected for frequency-dependent hull and electronic phase lags. Processed data are then checked by computer algorithms for reasonableness and, if they are reasonable, encoded for transmission in real time.

At the NWSTG, directional wave data are encoded and then transmitted around the world using World Meteorological Organization (WMO) standard codes, FM-13 SHIP and FM-65 WAVOB. The WAVEOB message is too long for some purposes, so a shorter, less detailed directional wave message is generated at NWSTG for transmission on the Automation of Field Operations and Services (AFOS) circuit for NWS use. Likewise, messages from buoys producing only nondirectional wave data are sent out on WMO and AFOS circuits. The entire process goes on around the clock, automatically distributing hourly messages in essentially real time.

The wave data received at the NWSTG are also sent to NDBC at Stennis Space Center (SSC). Processing identical to that done at the NWSTG is applied, and for several weeks after receipt of the data, it is examined by automated and manual means at SSC. These data quality assurance procedures are designed to reduce the probability that any bad data resulting from equipment failure or other causes will be placed in the archives. After careful examination, the wave and other data are written on a magnetic tape containing all data that were acquired in a particular month from all NDBC environmental data stations. Approximately one month after the last day of the month in which the data were acquired, copies of this magnetic tape are sent to NODC and NCDC.

To support its internal development efforts, NDBC has the capability to write limited quantities of completely processed directional wave data on 3½- or 5¼-inch floppy disks in an ASCII format. In this form, the data are easy to use with PC programs that are written to investigate some particular aspect of the data.

A number of steps are taken to reduce the probability of bad wave data being archived or reported in real time. Formal procedures are followed in the testing and calibration of measurement systems prior to deployment. Upon deployment of each buoy and during service visits to the buoy between deployment and retrieval, technicians onsite relay the visually observed wave and other environmental conditions by radio to SSC. A data analyst at SSC compares these observations to the data reported by the buoy through the satellite. Normally, the deployment ship stands by until the analyst determines that the buoy-reported wave and other environmental data are in reasonable agreement with visual observations. Typically the ship will stand by for 3 hours unless more pressing business requires it to depart. Should the data appear incorrect, they usually provide clues as to what is wrong, and repairs are attempted before the ship departs, if circumstances allow.

Throughout the period of deployment, every hour of data transmitted via satellite is monitored using a combination of automated and manual techniques. Using empirical relationships based on large quantities of data taken by many buoys, the wave spectral energies reported at wind-spectrum frequencies are automatically examined for reasonableness based on the magnitude of the wind. Wave directions at these higher frequencies are also checked against the wind direction, with which they should generally coincide if the wind is strong and has been steady for a few hours and if no strong currents are present.

Many wave parameters are checked to see that they lie within established normal ranges. Key data from sensor time series records are extracted aboard the buoy and transmitted to shore. These and other housekeeping data are believed to be complete enough to alert NDBC to a failed sensor or electronic system almost immediately. Occasionally, when a sensor may have failed but the numbers it produces are not appreciably different from correct values, the failure may not be recognized immediately. However, users of wave data have contacted NDBC on occasion to say that data from some stations appear low or high. These data are then carefully examined.

Future Wave Measurement Systems

Performance of present NDBC operational wave measurement systems is generally satisfactory, even though some equipment has been in service for over 15 years. On the other hand, the most sophisticated NDBC operational wave measurement systems, the DACT DWA HIPPY and MO, need improvement. The fixed bandwidth of 0.01 Hz is too broad at swell (< 0.10 Hz) frequencies, and the directional upper limit frequency (0.35 Hz) is too low to provide much information about the very earliest stages of wave spectral growth. Record lengths need to be longer for spectral density calculations in the swell range where bandwidths need to be narrow and shorter for spectral density calculations at frequencies where waves can be generated rapidly. Encoding techniques need to be improved to pack more wave information into fewer bits. In addition, an increase in the amount of data sent to shore by the buoy for data quality assurance purposes is needed. For these reasons, and because increases in the inventories of wave measurement equipment are needed to meet an increasing demand for data, NDBC needs a number of units of a new type of wave measurement system. To meet this need, NDBC is now in the last stages of development of a new wave measurement system, described in the following section.

Wave Processing Module (WPM)

Each of the present NDBC operational wave measurement systems is an integral part of an associated payload and cannot operate with some other payload without significant design changes being effected. Thus, refinements of wave measurement techniques have thus far necessitated either costly modifications to existing payloads or the procurement of new payloads. Furthermore, the control functions of the systems software of a payload are complex enough without incorporating the acquisition of wave sensor samples at a fixed rate for 20 minutes. For these reasons, future NDBC wave measurements will be performed by a Wave Processing Module (WPM) dedicated solely to the acquisition and processing of wave data (Chaffin & Bell 1992).

The WPM will be separate from the payload and connect to the payload through an RS-232 interface. The payload can be any device that has a properly designed interface to send a time marker to the WPM and transfer a block of "wave" bits from the WPM back to the payload. The wave sensors will be powered by, and have their outputs monitored by, the WPM. Keying on the time marker provided by the payload, the WPM will sample and process the data from the sensor analogs. The processed data will be placed in a buffer in the WPM, from which the payload will extract the data. Normally, the payload will transmit these data directly to a satellite, but in some near shore applications, the data may be transmitted by radio to another payload on land that will transmit to a satellite.

A majority of NDBC moored-buoy or fixed-platform stations that report wave data report only nondirectional wave data at this time. Thus, in its simplest configuration, the WPM will report nondirectional wave data based on either a laser (or other type of) wavestaff or a linear accelerometer (in the case of a buoy) with its axis mounted parallel to the buoy mast. Either a long or short message will be used to transmit the data depending on the type payload through which the WPM reports and on the application of the data.

To acquire the highest quality directional wave data from buoy stations where there is energy at frequencies below 0.10 Hz, the WPM will use a two-axis magnetometer with the Hippy acceleration, pitch, and roll. For measurements at locations where there is little or no energy at frequencies below 0.10 Hz, a single axis accelerometer fixed in the hull vertical and a two axis magnetometer will be used. These different sensor suites require two different software configurations, which are both an integral part of the WPM design.

Built into the WPM software is the option to choose among several message types. A short message can be placed in the buffer when the WPM is con iected to a payload reporting through a Polar Orbiting Environmental Satellite (POES), which can typically transmit only 256 bits (per message) to shore. A long message can be placed in the buffer for a payload that reports to a geostationary satellite such as a GOES, GMS, or METEOSAT.

The WPM is expected to operate in conjunction with at least two types of payloads that will report through a geostationary satellite. The first is the VEEP, already in the NDBC inventory and used operationally. The second is the Multifunction Acquisition and Reporting System (MARS), a new NDBC payload now being developed.

Although there are no NDBC activities underway to operate the WPM in conjunction with a transmitter that reports through POES using a Service Argos system, the WPM system has been designed with this in mind. For this type of transmitter, the short wave message would be put in the buffer.

For customers needing time series data, NDBC will have available an enhanced version of the VEEP WPM in which two time series recorders will be added (Figure 1). Each time series recorder will record the output of the wave sensors, and the outputs of one of the two wind sensor suites (Table 1). This standard enhanced configuration can be easily modified to record the outputs of sensors specially chosen by customers for their particular research.

Future Wave Measurements Projects

Under discussion now with potential customers is the possible use of NDBC directional wave-measuring buoys in two field research experiments. First, CERC is considering the employment of three NDBC 3-meter buoys in the Sandy Duck experiment scheduled July 1994 through March 1995. These buoys will each include the enhanced configuration of the VEEP WPM measuring time series data described above.

A proposal has been submitted to the Oil Pollution Research Center of the University of Miami for the deployment of one 2.4-meter DACT DWA MO buoy in the Florida Keys in support of the South Florida Oil Spill Research Project. Should this proposal be funded, the buoy will be deployed for one year commencing about September 1993.

References

Brown, H., and Gustavson, R. (1990). "Infrared laser wave height sensor." *Proc. Mar. Instr. '90*, 141-150.

Chaffin, J.N., Bell, W., and Teng, C.C. (1992). "Development of NDBC's Wave Processing Module." *Proc. Mar. Tech. Soc., MTS '92*, 2, 966-970.

Hamilton, G.D. (1990). "Guide to moored buoys and other ocean data acquisition systems." *Report on Marine Science Affairs* No. 16, World Meteorological Organization, Geneva.

Magnavox. (1986). "Technical manual for Data Acquisition and Telemetry (DACT), Amendment 1." Magnavox Electronics Systems Company, Fort Wayne, IN.

Magnavox. (1979). "Technical manual for the General Service Buoy Payload (GSBP), revised." Magnavox Government & Industrial Electronics Company, Fort Wayne, IN.

NDBC. (1986). "A study of hull-mooring data bases for wave measurement systems." Report No. F-344-7. U.S. Dept. of Commerce, National Data Buoy Center, NSTL, MS.

NDBO. (1978). "Summary of NDBO wave measurement development activities." Report F-344-1. U.S. Dept. of Commerce, National Data Buoy Office, NSTL, MS, 61-79.

Steele, K.E, and Earle, M.D. (1991), "Directional ocean wave spectra using buoy azimuth, pitch, and roll derived from magnetic field components." *IEEE J. of Oceanic Engineering*, 16(4), 427-433.

Steele, K.E., and Earle, M.D. (1979). "The status of data produced by NDBO wave data analyzer (WDA) systems." *Proc. IEEE Conf. Oceans '79.*

Steele, K.E., Lau, J.C., and Hsu, Y.H. (1985). "Theory and application of calibration techniques for an NDBC directional wave measurements buoy." *IEEE J. of Oceanic Engineering*, OE-10(4), 382-396.

Steele, K.E., Teng, C.C, Wang, D.W. (1992). "Wave direction measurements using pitch-roll buoys." *Ocean Engng.*, 19(4), 349-375.

Steele, K.E., Wolfgram, P.A., Trampus, A., and Graham, B.S. (1976). "An operational high resolution wave data analyzer system for buoys." *Proc. IEEE Conf. Oceans '76.*

Tullock, K.H., and Lau, J.C. (1986). "Directional wave data acquisition and preprocessing on remote buoy platforms." *Proc. Mar. Data Sys. Intl. Symp., MDS '86.*

Table 1. Parameters Measured by the Time Series Data Recorder (TSDR's)

	TSDR # 1		**TSDR #2**	
	Sensor	**Voltage Range (or Signal)**	**Sensor**	**Voltage Range (or Signal)**
1	Sine Pitch	±10	Sine Pitch	±10
2	Sine Roll	±10	Sine Roll	±10
3	Hippy Acceleration	±10	Hippy Acceleration	±10
4	Mast Acceleration	±10	Mast Acceleration	±10
5	Bow Acceleration	±10	Bow Acceleration	±10
6	Starboard Acceleration	±10	Starboard Acceleration	±10
7	Barometer #1	0-10	Barometer #2	0-10
8	Bow Magnetic Vector	±2.5	Bow Magnetic Vector	±2.5
9	Starboard Magnetic Vector	±2.5	Starboard Magnetic Vector	±2.5
10	Vertical Magnetic Vector	±2.5	Vertical Magnetic Vector	±2.5
11	Air Temperature #1	0-5	Air Temperature #2	0-5
12	Water Temperature #1	0-5	Water Temperature #2	0-5
13*	Wind Speed #1 Wind Speed #1	0-5 Freq.	Wind Speed #2 Wind Speed #2	0-5 Freq.
14	Wind Direction #1	0-5	Wind Direction #2	0-5
15**	Compass #1	Pulse	Compass #2	Pulse

* The TSDR can use either model 5103 or 5203 R.M. Young aerovane. The signal from the model 5103 is a frequency proportional to wind speed. The signal from the model 5203 is a voltage proportional to wind speed.
**The signal from the Digicourse compass is a train of n+1 pulses, where n is the number of degrees azimuth clockwise from north.

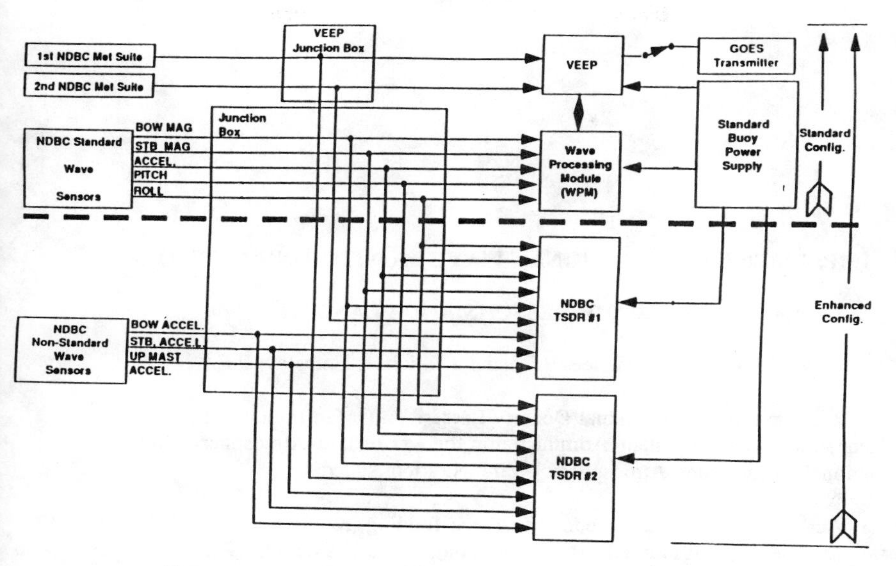

Figure 1. Enhanced VEEP WPM Wave Measurement System

IMPLEMENTING A NATIONAL WAVE MONITORING NETWORK--

SOME LESSONS AND PLANS

David D. McGehee, P.E.[1] and J. Michael Hemsley, P.E.[2]

At the Sixteenth Annual Doherty Lecture on Ocean Policy in May of 1992, Ned A. Ostenso, Assistant Administrator for Ocean and Atmospheric Research, National Oceanic and Atmospheric Administration, stated,

> Academic departments, federal research funding, rewards and even mission agencies are substantially structured along disciplinary lines. Advancing the disciplines of knowledge is necessary. But that alone is no longer sufficient for solving the complex problems we face. This is particularly true in the environmental sciences where everything is connected to everything else and data needs exceed the time and spacial scales common to our research experience.

Five months later, Lieutenant General Arthur E. Williams, US Army Corps of Engineers, said in his Chief's Charge to the 57th meeting of the Coastal Engineering Research Board,

> Most environmental problems involve a chemical and/or biological component and a physical processes component. A holistic approach must consider both components and the way they interact. To assess environmental risk, we must consider potential threat or exposure and resulting impacts. . . . I also want you to recommend other areas where partnerships between environmental and coastal communities can demonstrate our clear commitment to protecting the quality of our environment and move the Corps into the forefront of environmental responsibility.

---

[1]Research Ocean Engineer, US Army Engineer Waterways Experiment Station, 3909 Halls Ferry Road, Vicksburg, MS 39180

[2]Chief, Program Management Division, National Data Buoy Center, Stennis Space Center, MS 39529

This paper will present some ongoing activities in the field of wave data collection that reflect these sentiments.

I. OVERVIEW

While there is a large community of engineers and scientists that have experience in measuring waves, there are only two organizations in the US that have successfully established national-scale, long-term wave observation programs. The US Army Corps of Engineers (CE) funds the US Army Engineer Waterways Experiment Station (WES) Coastal Engineering Research Center (CERC) to manage the Field Wave Gaging Program (FWGP) (McGehee, 1993). The National Oceanic and Atmospheric Administration (NOAA), National Weather Service (NWS) funds the National Data Buoy Center (NDBC) to operate and maintain environmental data buoys in support of weather warnings and forecasts. Table 1 summarizes the charter for each center that authorize implementation of a network of observation stations. Historically, these two agencies have operated their respective networks relatively independently.

CERC	NDBC
• Conceive, plan, and conduct Research and Data Collection in coastal/ocean engineering and near-shore oceanography • Provide Scientific and Engineering Data and Design Criteria • Publish Findings and Data resulting from research and other information concerning coastal phenomena which are useful to the Corps of Engineers and the Public	• To serve as the focal point for Data Buoy and associated automated meteorological monitoring system technology • The Center shall provide support for Data Buoy and Associated Automated Meteorological Systems to meet measurement needs for Research Programs and NOAA's Long-term Operational Needs in Marine and weather areas

Table 1. CERC and NDBC MISSION STATEMENTS

As the steward of the nation's coastal infrastructure, the CE needs reliable, long-term wave measurements for use in planning, designing, and operating coastal projects. Design wave conditions, usually expressed in terms of return intervals, are obtained through extremal analysis of wave histories. The confidence in these projections drops as the desired return interval exceeds about twice the length of the historical record. However, there is seldom time between the inception of a project and finalization of design details to collect sufficiently long wave histories. A commitment is needed to obtain the required long-term measurements for the US coastline, in advance of specific project planning. Wave hindcasts--which are made possible by long-term meteorological observations--such as CERC's successful Wave Information Study, are another approach to this need, but wave measurements are still required for their validation. Another need for wave data is accurate quantification of conditions during specific events that result in damage to structures or delays in operations. Finally, laboratory and analytical research into the physics of wave generation, propagation, and transformation requires measurements for calibration and verification. The FWGP was established with the goal of collecting wave data at sufficient spatial and temporal density to meet these needs for the entire US coastline.

The CE wave gage network began in 1975 with a single station in Southern California operated by the Scripps Institution of Oceanography. This grew into the Coastal Data Information Program (CDIP) with, currently, 23 stations on the west coast and Hawaii. About the same time, the University of Florida began the Coastal Data Network (CDN). At its peak, the CDN operated ten stations in Florida, though it is currently funded for three. In 1993, the Virginia Coastal Information Program (VCIP) was brought on-board with three stations. The CERC also has its own Network for Engineering Monitoring of the Oceans (NEMO), consisting of 14 stations. In addition, the CE maintains mini-networks within several harbors. The FWGP serves as the principal manager and coordinator of these sub-networks, as well as the "clearing house" for CE wave measurement activities to ensure coordination within the CE, and facilitate cooperation with other federal and state partners. The FWGP has initiated Cooperative Agreements with several states for the joint support and management of regional wave networks. These agreements provide an effective mechanism for focusing multiple resources at the state and federal level onto a common need.

WHO	CE	NOAA
WHY	Climatology for Design	Drive/Validate Forecasts
WHAT	Waves/water Levels	Meteorology/Waves
WHERE	Shallow/Intermediate	Deep/Intermediate
HOW	Submarine Pressure Arrays	Surface Buoys
WHEN		
• Collection	Real-Time or Buffered	Real-Time
• Processing	Stored & Post Processed	Near-Real-Time
• Distribution	Monthly/Annual Summary	Near-Real-Time

Table 2. PROFILE OF NETWORK MANAGERS

NDBC has been measuring waves under its moored buoy program since the mid 1970's in support of its basic mission to provide, through NWS, warning and forecast services to the public. It also operates buoys in support of other agencies, principally the CE and the Minerals Management Service. In 1993, 20 of the approximately 60 wave buoys are reporting directional information. Though not a research center, NDBC does conduct development and demonstration of technology directed toward improved sensors, payloads, or platforms.

The distinct missions of these agencies has led to the evolution of two different approaches to the tasks of wave measurement and data management. Table 2 summarizes these differences for the two agencies providing the majority of the US wave data. The CE is principally interested in the wave conditions and water levels at coastal projects, and most of its stations are in shallow (10-20 m) water. Occasionally other parameters, such as currents, water temperature, turbidity, etc., are monitored. The CERC has developed and is still improving gage technology for this extremely hostile environment--typically bottom-mounted pressure arrays. Figure 1 illustrates a CERC-designed pressure array in a trawler-resistant frame mounted on the seafloor.

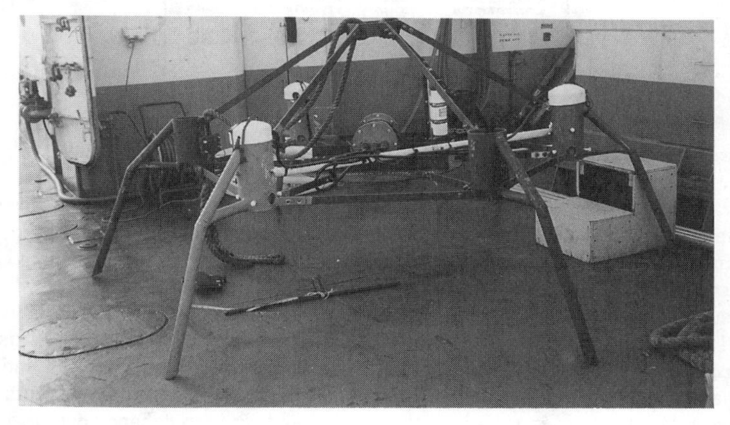

Figure 1. CERC DWG-1

Due to the attenuation of pressure with depth, these instruments are limited to shallow water for measuring wind waves. Occasionally, a CE station includes on-shore meteorological sensors.

NOAA's weather forecasting mission requires that its ocean observations be taken far enough offshore to provide adequate warning of approaching weather systems, and a suite of meteorological sensors must be above the sea surface. Since floating platforms are the only viable choice for these locations, NDBC has developed and continues to refine technology for measuring waves from surface-following buoys. NDBC uses four main hull designs, ranging from 12m to 3m in diameter (Figure 2). Since buoys are not suitable for shallow water deployment, NOAA meets its requirement for coastal *meteorological* data through the similarly-sized network of C-MAN stations. Typically located on shore or shore-connected piers, few C-MAN stations obtain wave measurements.

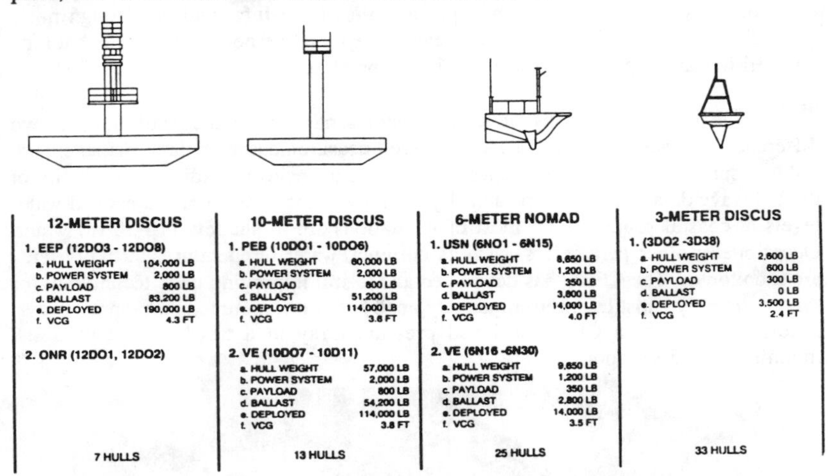

Figure 2. MOORED BUOY HULLS (NDBC, 1992)

Since a forecaster is already dealing with the future, even data distributed in real-time could be considered "old news." NOAA's data management system focuses on timely delivery of its products to users. NDBC utilizes radio telemetry via the Geostationary Operational Environmental Satellite (GOES) to provide hourly measurements to NWS forecasters as quickly as possible. The CE is also interested in "forecasting," but on time scales of decades rather than days. Statistical projections of future events from historic records need not even be started until the data are several years old. However, nearly all the stations of the CE network also transmit data to shore in real-time, though the purpose is to enhance data recovery and quality, not real-time data distribution. The CE's data management system focuses on efficiently assimilating as much data from as many locations as possible, to improve the reliability of the statistics.

Hopefully by now, the reader is concluding, as did the FWGP and NDBC program managers, that there is an obvious synergy awaiting improved coordination in these two networks. Figure 3 shows the network that results when CE and NOAA wave station are combined. The energy in ocean waves marches blithely through this network on its rendezvous with entropy, ignorant that it's crossing political boundaries or agency "territories." The bits comprising a wave record are as content to rest in database limbo as to bounce from sea to space to terminal. The purpose of this paper is to describe results of recent efforts to coordinate these two major operations, and discuss the potential for further evolution toward a functionally integrated national wave measurement network.

II. LESSONS IN COORDINATION

Network Integration: The distinct missions of these two agencies, and the inherent depth limitations of their measurement approaches, resulted in an effective DMZ (Difficult to Measure Zone) at around the 20-m depth contour, with each agency concentrating on its own sector. However, deep to intermediate depth stations are also required by the CE for understanding wave generation, propagation, and transformation. This need had been met with non-directional buoys (Waveriders), but the CE did not have an efficient method of obtaining directional measurements in deeper water. Overlap between the two networks began in 1987 when CERC requested NDBC install one of its new 3-m diameter buoys with directional wave measurement capabilities near Mobile, AL. In 1993, the FWGP will either fully or partially support 18 NDBC buoys with directional wave measurement capability .

Site Selection: Each agency produces a document with the site and schedule of planned station deployments. The FWGP produces a 5-Year plan annually, and the NDBC updates and distributes its Moored Buoy Maintenance Program Schedule quarterly. Coordination takes place at the program manager level through exchange of information on schedules and needs. Often, customers approaching one agency for data services are referred to the other if the data requirements can be better met with that agency's technology.

To date, wave observations are still inadequate to sufficiently monitor our coastline, and any site selected for monitoring is a valuable addition to our knowledge. As the number of stations increases, it is imperative that we efficiently distribute their location to avoid collecting redundant data. SIO has been tasked with investigating the optimization of observation networks using refraction/diffraction modeling and a simulation technique borrowed from metallurgy.

Data Transfer: The data that are distributed in near-real-time by NDBC have undergone extensive automated quality assurance, but have not received the additional quality control measures and inspection provided at NDBC. It is this qualified data from all NDBC stations that is sent to CERC on 9-track tape each month, about 5 weeks after the end of the month collected. The data are reduced and stored using the WES Cray YMP. A similar procedure is followed for the data from CDIP, CDN, VCIP, and NEMO. The FWGP, in turn, produces standard monthly summaries with plots and tables of wave parameters (energy-derived wave height, peak period and peak direction) and directional energy spectra for those stations it is funding. Summaries are mailed monthly to CE districts and state agencies that support specific stations.

The CE mission does not include distribution of data products to the general public; however, the state of California, as co-manager of CDIP, authorizes access to its data

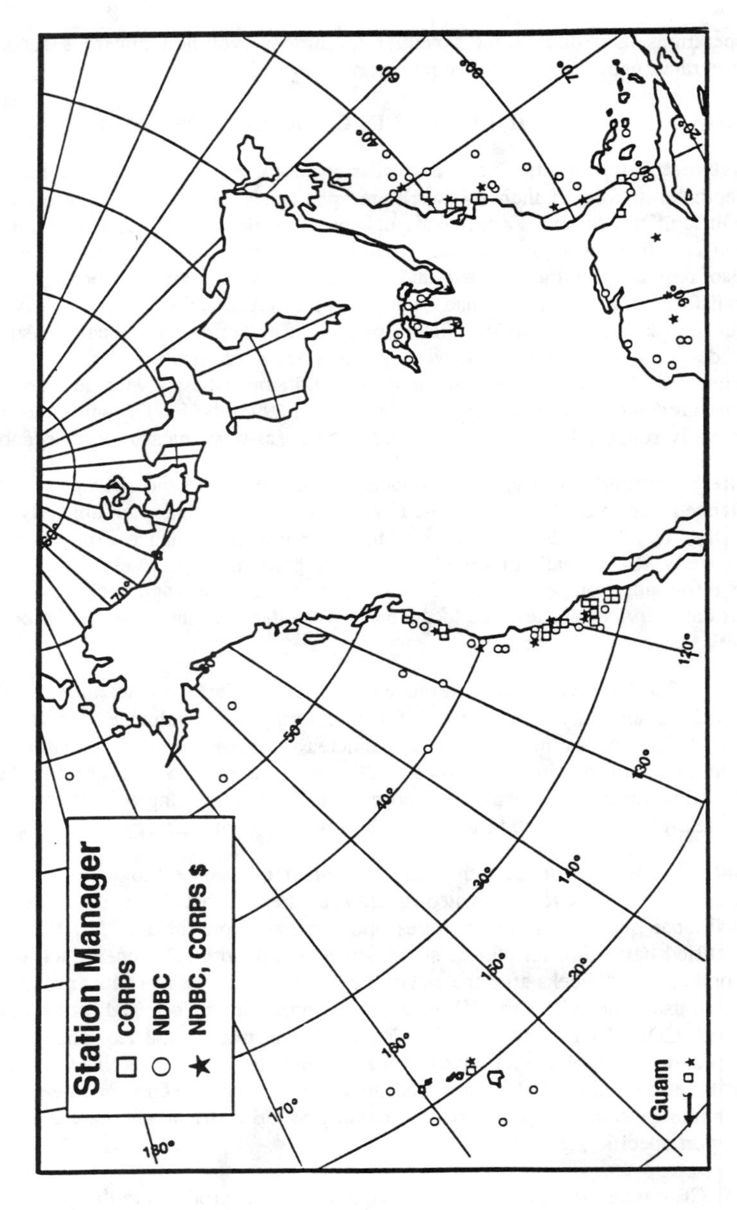

Figure 3. US Network

by the NWS forecast offices in California. The nearshore wave conditions are used by the forecasters to validate the wave conditions predicted from simple wind speed/fetch calculations. In addition, wave conditions from selected CDIP stations are broadcast on NOAA Weather Radio.

Data Products: Since all government agencies are feeling the pressure to reduce costs and improve efficiencies, the demonstrated value of cooperation and integration will ensure that the trend continues. The CE and NOAA already have an effective system for meeting the needs of its in-house clients. If the benefits of this network are going to serve a larger customer base, it would be useful, as a starting point, to better define and quantify present and potential users and their requirements. Table 3 is an attempt to categorize data users in a way that transcends association, mission, or discipline.

The NWS forecaster is the main target of NDBC data, but the term also defines any user who needs a continuous supply of up-to-the-minute data. An operator who is conducting or planning maritime activities needs current and forecast information at his location or his destination. The NWS broadcasts serve the general public and maritime community, but must, by their nature, cover a fairly wide region; they do not, for example, make an airport anemometer redundant. Specialized operators--a dredge operator looking for the precise depth, a ferry captain approaching an exposed waterway, or a loaded trawler entering a rough inlet--could each use data on smaller space and time scales. The engineer-user is exemplified by the CE designer, while the planner may be a land-use regulator.

Products typically used in each function are listed; other products with less frequent or obvious application are followed by question marks. A schedule for delivery of these products is estimated. In spite of the overworked term "high-speed computers," nothing drives up the cost of distributing *qualified* data faster than accelerating its delivery.

III. FUTURE PROSPECTS AND OPPORTUNITIES

This section presents plans and prospects that can contribute to building a workable, efficient, and productive environmental monitoring network for the US coastline. Some are underway and some are in the discussion stage; but are designed to address more comprehensively the needs of a wider base of clients.

Users Survey: To provide a forum for these clients, the FWGP will conduct a national survey in FY94 of wave data (and other parameters) users' needs. The survey will be sent to federal, state and local government agencies and research centers, academic institutions and private industry. The primary goal is to better address the operational and research needs of the CE and its state and federal partners. A secondary goal is to identify those products and services that will also assist the broader coastal engineering community and the general community.

Who	Forecaster	Operator	Engineer	Planner	Scientist
Why	Broadcast & Predict	Economy & Safety	Design & Monitor	Regulate & Predict	Why ask why
What	H,T,D,d E(f,θ)?	H,T,D,d?	H(t),d(t), T(t),D(t)? P(H,T) P(D,d)?	H,d, etc. P(H,d)	S(t) E(f,θ) H,T,D,d
Where	Region	Here	Project	Project	Every-where
How	Network Graphics	Telemetry Voice or Graphics	Media or Network Database	Docu-ment Tables	Media or Network Database
When	Now	Now & Forecast	Variable (soon)	Annual	Later

Table 3. Profile of Wave Data Users; H=Wave Height, T=Wave Period, D=Wave Direction, d=Water Depth, E=Energy, f=Frequency, θ=Direction, S=Sea Surface, t=Time.

Data Management: The time and costs associated with collecting, analyzing, qualifying, distributing, and storing data can be reduced by a well-designed data management system. System design should be guided by the functional requirements resulting from the survey. A distributed relational database architecture is at the heart of a data management system currently under development. It is an approach that insures flexibility and upgradeability. It permits a user to not only access data but to quickly track and cross correlate information *about* the data ranging from the name of the technician that calibrated the sensor that measured a wave record to the address of the latest person to download that record. It eliminates the need for a massive central database because it "knows" where to obtain data from regional databases and "remembers" changes caused by updates in format, extraction protocol, or even location of requested data.

Data Standards: Since a wave record is a representation of a random process, it involves analysis and parameterization that can lead to more diverse results than for other environmental data, such as temperature or salinity. Any serious user of wave data provided by different sources, or by the same source at different times, has come up against the need for data standards. Only a scientific user would be interested in exploiting unique properties of a specific instrument or sampling scheme. To other categories of users, uniqueness is not a virtue. Even for the most basic wave parameter--height--the user must be cognizant of different definitions. When two-dimensional energy spectra are compared between a two stations, or between a station and a model, the process is frequently hampered by different convention in tabulating energy, frequency, or direction bands.

The data management system under development by the FWGP cannot be fully operational until the wave records in the database are standardized. One of the products of the FWGP will be published standards in wave data collection, analysis and reporting of wave records and climatic statistics derived from wave histories. These will be standards in the sense that they will specify procedures to ensure both the *quality* and the *uniformity* of wave data provided by the FWGP. Implementation of these standards will greatly facilitate the use of wave data from various sources, particularly the compilation of climatic statistics. The first product under development is the CERC Wave Data Analysis Standard (WDAS). An intermediate document is a paper describing the current collection and analysis practices of the FWGP-managed networks (Tubman, Earle, McGehee, 1993). The next document will cover program specifications that will address automated quality control and editing functions and specify the output record format.

Though NDBC is providing collaboration and consultation on developing WDAS, the first version will not include procedures applicable to pitch-roll-heave buoys; it will apply to staff arrays, bottom-mounted pressure arrays or pressure plus two-axis current meter gages only. However, standardization is occurring at the data product level, as buoy data is included in standard CERC wave data reports, as described above. Figures 4, 5 and 6 show typical monthly reports from a CERC-operated wave gage at Virginia beach and an NDBC buoy off Norfolk. Uniformity of products generated by accessing these separate databases will be insured through implementation of the CERC WDAS as a contractual specification for all wave data purchased by the FWGP. As funding permits, historical data in the sub-networks' databases will be re-analyzed to conform to the standard.

Standards will also be pursued for statistical treatment of long wave histories. Each of the sub-networks has already produced its own version of a climatic summary. The same consentaneous approach will be directed to the probabilistic analysis to garner the most reliable and useful aspects of each. Development will include testing of various statistical models on subsets of longer datasets to determine the most robust predictors of later conditions.

Instrument technology: Coordination between the agencies has meant additional requirements that have spurred improvements to instruments. Some years ago, the standard NDBC buoy was primarily a meteorological platform that also measured waves. The interest and support for directional waves by the CE and other clients has resulted in improvements in quality and efficiency in providing this product. Several NDBC stations are undergoing evaluation of an innovative, magnetometer-only approach that significantly reduces the sensor cost of the payload. Another

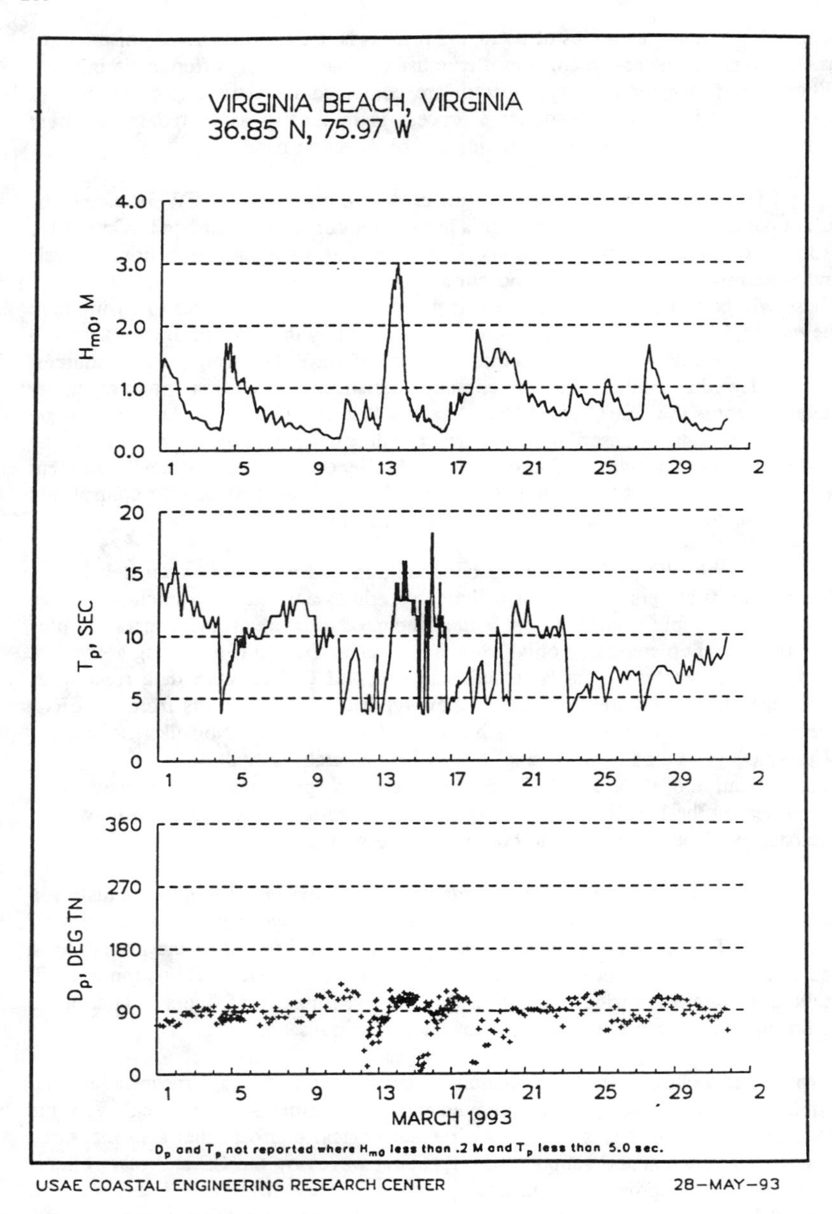

Figure 4. Virginia Beach Monthly Report, March 1993

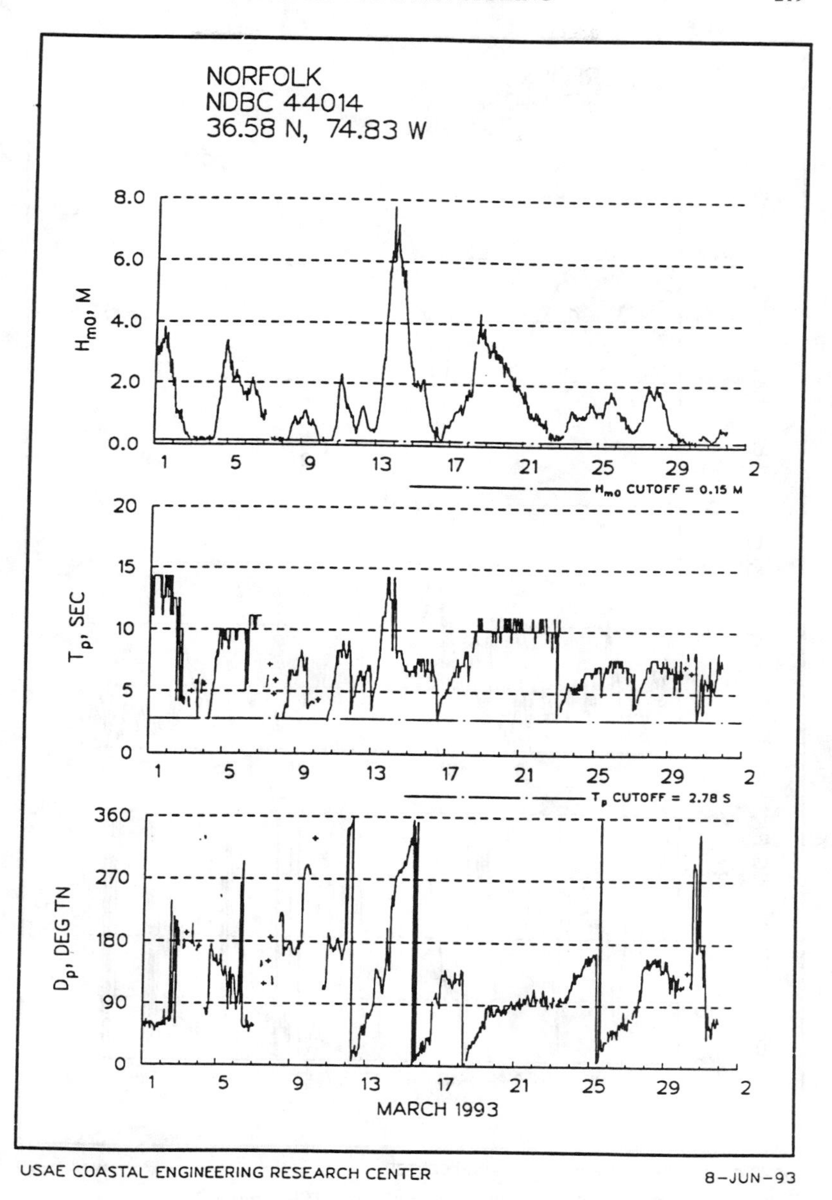

Figure 5. Norfolk, VA Monthly Report, March 1993, page 1

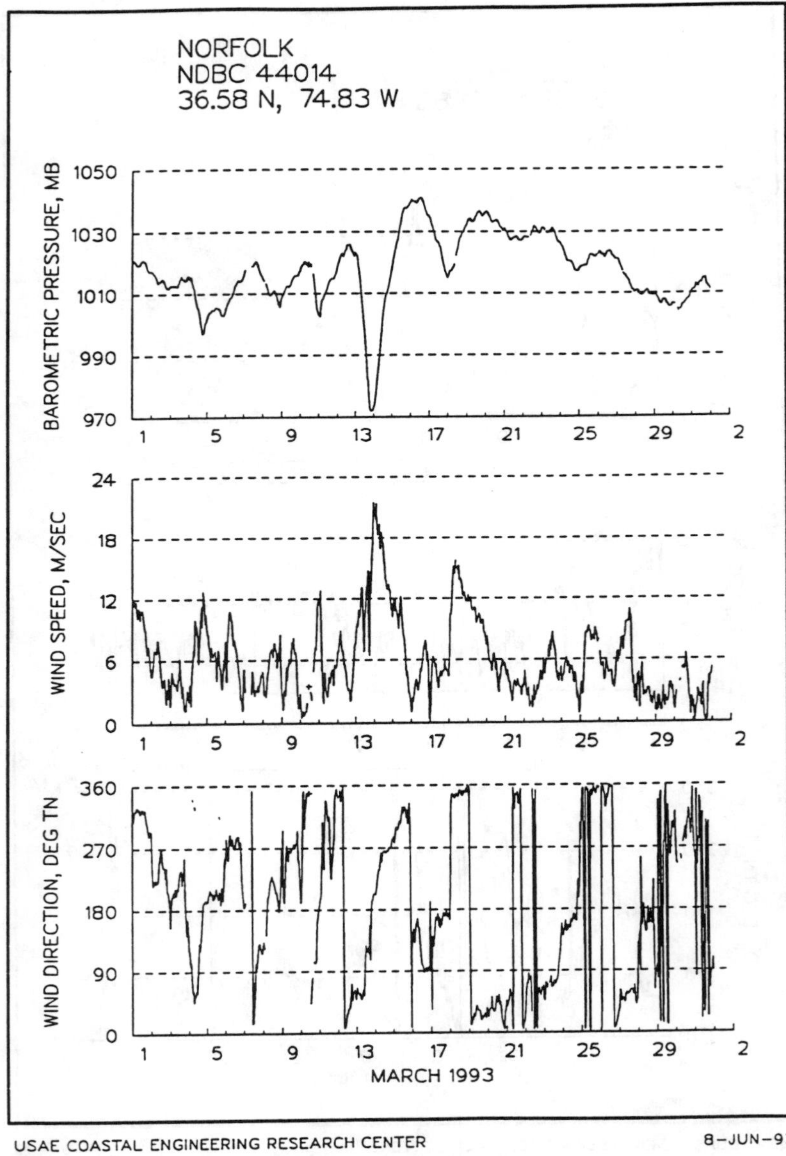

Figure 6. Norfolk, VA Monthly Report, March 1993, page 2

enhancement is the addition of a stabilizing wind vane to better align the buoy with the dominant wind direction.

In 1984, only 25 percent of the operational US wave observation stations obtained directional data. Ten years later, over 75 percent are directional. In fact, the FWGP only installs non-directional stations for special studies, such as long-period harbor oscillation analysis, where the standard definition of wave direction is not meaningful.

In 1992, the CDIP installed a large-baseline, multi-element pressure array in deep water (200 m) on Texaco's Harvest oil platform off Point Conception. This array is designed to provide higher resolution direction to lower frequency energy than has been previously possible. It is also serving as a test site for comparisons between fixed and floating measurement approaches. An NDBC 3-m directional buoy and a directional Waverider by Datawell have been installed nearby and an intercomparison study is underway.

New environmental observations: Recent interest in expanding the role of the existing observation network to include water quality data is driven by technical and economic incentives. Traditional water quality measurements are obtained from periodic "cruises" where samples are taken over a wide area for a short duration. Many environmental and biological scientists are realizing the benefit of continuous, multi-year monitoring from fixed sites in order to establish baselines, and both short and long-term variability of these parameters. Establishing and quantifying the impact on the water quality of a particular activity not possible without an understanding of "normal" levels and fluctuations.

The economics dictate utilization of existing networks, where possible, to obtain these measurements. The majority of the cost in obtaining routine observations is associated with maintaining the infrastructure--the people, facilities, logistic support and data management systems. The sensors and instrumentation represent only a small portion of the total. The largest impediment to this approach at present is the lack of rugged, reliable water quality sensors that can operate for reasonable periods in the ocean environment. NDBC has taken an active role in promoting this concept, and is working with other federal agencies to identify needs and solutions.

IV. CONCLUSIONS

A network of deep and shallow water wave gages, operated by two federal agencies, is in place around the US. Closer coordination in planning, operation and distribution of the products of these two efforts will result in improved utilization and efficiencies for both agencies and the general public. Expansion of the networks beyond their basic meteorological and wave observations could form the basis for a more comprehensive US coastal ocean environmental monitoring system.

Bibliography

McGehee, David D. (1993). "Field Wave Gaging Program: A Conspectus." *The Coastal Ocean Prediction Systems Program: Understanding and Managing Our Coastal Ocean, Volume II, Overview and Invited Papers.* Joint Oceanographic Institutions, Inc. pp. 177-180.

NDBC. 1992. "Moored Buoy Hull Characteristics," National Data Buoy Center 1992 Annual Report, p. 6.

Tubman, M., Earle, M. and McGehee, David D. 1993. See Article titled "The Development of a Wave Data Analysis Standard for a National Wave Measurement Program," this proceedings.

Measurement and Analysis of Wave Data in Los Angeles - Long Beach Harbors

James Rosati III, Paul T. Puckette [1]

1 Introduction

In order to improve the United States' abilities to conduct efficient and profitable international shipping it will be necessary in the coming years to improve and enlarge our present port systems throughout the U.S. In response to this need a feasibility study for the expansion of the Ports of Los Angeles and Long Beach is being conducted by the Los Angeles District of the U.S. Army Corps of Engineers.

The Corps of Engineers has been given the responsibility of providing deeper channels and determining the effects of the construction on the local environment. One major task of the Corps is to investigate how changes in the harbor channel's depth and configuration will affect the harbor's resonance characteristics. Wave oscillations within the harbors may cause serious problems with ship loading/unloading and/or ship-mooring in the harbor.

As part of this investigation wave data from 1984 through the present have been collected and analyzed from several different sites in and around Los Angeles and Long Beach Harbors (Rosati, et. al., 1993). This wave information has been and continues to be used in modifying and improving tools developed to study potential expansions to the harbors (Seabergh, et. al., 1992).

The wave gage sites selected were a compromise between geographical diversity sufficient for physical model calibrations, ease of installation and maintenance, and cost. Some selected sites are not well suited for the measurement of wind waves due to their sheltered locations. Analysis procedures were developed that optimized the calculation of energy spectra for long-period (400-25 seconds) waves in the harbors. A directional wave gage was placed on Platform Edith, which is an oil platform 9 miles offshore of the harbors, to measure the incident wind waves as well as tides and long-period waves.

[1]U S Army Corps of Engineers, Coastal Engineering Research Center, Waterways Experiment Station Vicksburg, Mississippi 39180-6199 USA

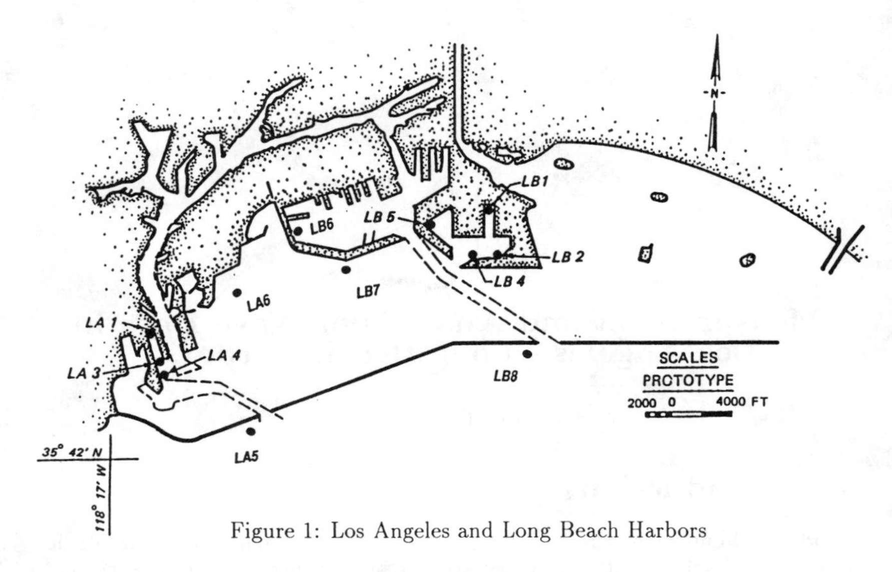

Figure 1: Los Angeles and Long Beach Harbors

2 History of Wave Gaging Systems

Internally Recording Gages In February 1984 wave gaging was initiated for the model enhancement program. SeaData 635-11 (remote) self-recording wave gages were initially installed in the harbors. A highly accurate Paroscientific quartz crystal pressure sensor (Paros, 1976) was strapped to pier pilings at the gage sites. Initially the pressure sensors were cabled to the SeaData instruments which were also strapped to pier pilings but above the water line. Later the remote gages were placed in electrical enclosure boxes dockside which negated the requirement to use a boat to service the wave gages for tape changes (Figure 2). The gages were configured to record 2048 seconds of wave data every 2 hours, sampling once each second. The self recording instrument's internal tape drive reached capacity after one month when recording data at this rate. This rather frequent tape changing and checkout interval was labor intensive and therefore expensive. Gage reliability was also less than desired.

New Technology Gages To reduce the costs and improve the data recovery from the existing wave gages, a wave data collection system utilizing new technology was designed. The new gage system was installed in February of 1988 to work in parallel with the SeaData gage system. In October of 1988 the SeaData tape recording system was removed. At each wave gage location a Remote Transmission Unit (RTU) computer system was installed. The RTUs acquired and stored the data from the existing pressure sensors. A computer system located in the

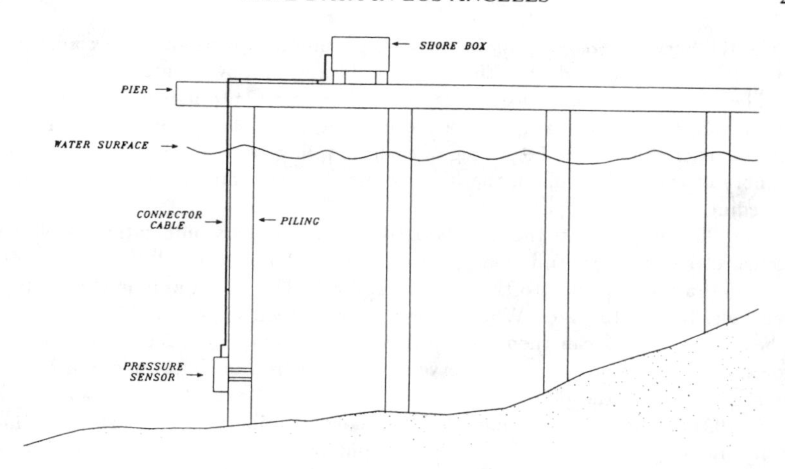

Figure 2: Gage mounted on pier piling with connection to shore box

Port of Long Beach (POLB) administration building interrogates the RTU at each wave gage site via UHF radio link. All system computers use a communications protocol to ensure error-free data transmission. The central computer, named the Real-time Server or RT-Server, analyzes the raw data acquired from each RTU and stores the results. A Digital Equipment Corporation VAX computer in Vicksburg Mississippi at the Corps of Engineers Waterways Experiment Station (WES) in turn calls up the RT-Server and transfers the analyzed data by an error-free network communications protocol over a standard telephone line.

System design details reconciled several competing concerns. The amount of available memory in the RTUs, the storage capability of the RT-Server, the data-link bandwidth limitations, system cost, and the capability to develop and test the required software all contributed to the overall design and capabilities of the system.

Remote Transmission Unit (RTU) At each wave gage location a RTU is used to acquire, store, and send the raw data to the RT-Server, with all these operations possibly occurring simultaneously. Each RTU consists of low-power consumption components which are placed in a sealed electrical enclosure which protects the electronic components. The RTU is based upon a STD-bus, single-board computer utilizing a Z-80 microprocessor. The system also consists of a custom-designed module to interface with the quartz crystal pressure sensor. Additionally the system consists of modules for additional system memory and communications to radio frequency modems which allow transmission of the data to the RT-Server. Batteries allow the RTU to function (without radio communication) for up to three days without loss of data when power outages occur.

The RTU continuously accumulates the frequency output from the pressure sensor and stores each sample in a file existing in solid-state memory. A new data file is created at the beginning of each sampling interval. When the internal memory of the RTU is filled, the RTU erases the data sampled earliest to avoid exceeding its storage capacity.

The RTU implements the KERMIT (daCruz, 1987) communications protocol to ensure error free transmission of raw data to the RT-server. All RTUs listen on the same radio frequency to the RT-server. Each RTU has a unique three letter identifier, LA4, PLF, etc. When a login sequence with the matching identifier is heard, that RTU can then receive further commands in order to transfer its stored data via radio to the RT-Server. After file transfer is complete, a logout message is sent to the RTU.

In 1991 the RTU's were upgraded to a configuration that supports more internal memory and new data acquisition and transmission components. These new components allow for the faster transmission of data and increase system reliability and maintainability. To take advantage of this increased capability, the harbor sampling parameters were changed to .5Hz sample rate, 8192 second record length, and a new record starts recording every four hours.

RT-Server The data storage and analysis system consisted of a Digital Equipment Corporation (DEC) MicroPDP 11/23 computer. It was programmed with the Micropower/Pascal real-time executive. The real-time executive allows for the coordination of several independent tasks within the computer without the overhead associated with an entire general-purpose operating system. This system was subsequently upgraded to a DEC VAXserver 3100 system running the VMS operating system and the INTERBASE relational database. The greater performance of the VAXserver allows the RT-server to perform its tasks for this application without the need for a real-time executive. The use of a relational database to perform tasks whenever possible greatly simplifies programming and adds additional capability to the system. The RT-Server is located in the POLB Administration Building.

The RT-Server has four major functions to perform. First it must communicate with each RTU system. KERMIT server commands are sent to first login, then check the current time (update it if found in error) and request a directory of current files in memory. If the RT-server has not yet acquired any files resident in the RTU, a request is sent for the file contents. After all files are acquired, a logout is performed. This sequence of commands is sent to each RTU. KERMIT implements data checking by packetizing the data as they are sent. If a packet is not received in the correct condition, an error is signaled and the packet is resent up to the retry limit (usually 5 times). If the retry limit is exceeded, then the KERMIT protocol is aborted. The RT-Server stores the time and result of each KERMIT operation in order to compute statistics of communication performance.

The second major function for the RT-Server is to analyze the harbor data which is detailed in the data analysis section.

The third major function is that the RT-Server must be able to respond to data requests from the WES VAX computer at any time. A telephone link is used to establish a connection between the two computers. Both systems run the DECnet software protocol and the RT-Server in Long Beach becomes a node on the WES computer network. Data are transferred to WES over the network link, and by magnetic storage media. In addition to the analyzed data, the system also can send compilations of system performance to the host computer.

The fourth function performed by the RT-Server is to provide a local interface to a single user on a computer terminal locally attached to the RT-Server. The terminal interface performs many of the same functions as the network interface. Additionally, the terminal interface allows a user to type out raw and analyzed data, reconfigure analysis and sensor parameters, and display disk directories. Progress messages indicating the current state of the RT-Server are scrolled at the bottom of the computer terminal which serves as the operator interface station or are saved to a disk file.

Host and development facility WES based computers are used to develop programs for and acquire data from the RT-Server. The RT-Server is polled for data that have not already been acquired. Data obtained by the RT-Server subsequent to the last polling by the WES VAX are transferred via network protocols to the WES VAX computer. Additionally, WES computers are used to create data reports, perform further analysis, and archive the data.

3 Los Angeles and Long Beach Site Selection

The gages in the Los Angeles - Long Beach Harbor are positioned throughout the harbor with each location chosen to monitor a specific aspect of the wave field in the harbor (Figure 1). A primary purpose of the prototype data is to calibrate and verify a physical harbor resonance model of the harbors. Wave gage sites were determined with the needs of the physical model in mind. Also, areas where the Ports were interested in the wave climate from a ship-operations perspective were considered for wave gage sites.

Mounting the pressure sensors on piers allows the structure to protect the pressure sensor and allows for easier maintenance by divers of the underwater components (Figure 2). Since the emphasis of the study is on long-period waves, the interference of the piling with the wave does not present a problem. In 1991 gaging sites were added to the system that required placing pressure sensors in a free-standing configuration on the bottom. A circular pipe welded to the center of a railroad wheel forms the mount, and the pressure sensor is attached to the pipe rising from the center of the railroad wheel. A special mount was designed for a site that is close to a small boat anchoring area and had a muddy bottom. A large circular flat plate was placed on the mudline and anchored from below by a railroad wheel. The pressure sensor is attached to the flat plate which provides a resistant mount in a muddy environment.

Shorebox siting considerations were:

- The availability of 110VAC power
- The ease of running the pressure sensor cable to the shorebox
- The desire to be environmentally sheltered if possible
- The ability to have a line-of-sight view to the POLB administration building for UHF radio communication
- The allowance of easy access by maintenance personnel
- The availability of protection from theft or vandalism
- The desire to minimize interference with Port operations

4 Analysis Methods

Two types of analysis are required, long-period wave analysis to determine the wave characteristics of selected sites within the harbors, and directional wind wave analysis for Platform Edith data to help provide the incident wave conditions to the harbors. Relational database technology is utilized to help identify conditions which influence harbor resonance.

4.1 Harbor Pressure Data

The first and most crucial part of valid wave information is measuring the height of the water at a given point over time. This project uses pressure sensors to accomplish this goal. A time series of pressure measured by the sensor can be directly related to the height of waves passing over the sensor. Through the use of various data processing methods it is possible to convert the pressure time series into statistically valid parameters describing the energy distribution of the waves.

Influences of extremely low-frequency energy, such as tides, are removed from pressure records by linear detrending and a zero frequency rejection filter (Durham, et.al., 1976). This type of filter takes the form:

$$x_i = A(n_i - n_{i-1} + x_{i-1}) \tag{1}$$

where $n_i \equiv$ the ith discrete sample of an input series
$x_i \equiv$ the ith discrete sample of the filtered output series
$A \equiv$ an adjustable coefficient

In order to eliminate phase distortion in the filtered signal it is necessary to pass the series through the filter twice, once in the forward time direction and once in the reverse time direction. The total response function of the filter will therefore be completely described by only the power response function. The z-transform representation , $F(z)$, of the filter is :

$$F(z) = \frac{A(1-z)}{1-Az} \tag{2}$$

where $z = e^{i\omega}$
$\omega = 2\pi f/f_s$ = normalized circular frequency
f_s = sampling frequency
f = discrete Fourier frequency

The response function resulting from passing the series through the filter twice is :

$$|F(z)|^2 = F(z)\tilde{F}(z) = \frac{2A_2(1-\cos\omega)}{1-2A\cos\omega + A_2} \tag{3}$$

where $|\tilde{F}(z)|$ = complex conjugate of $F(z)$.

Since finite length data records are used to simulate an infinite random process, each time series is windowed with a 10 percent cosine bell taper to simulate a stationary periodic time series for the calculation of spectral estimates (Bendat and Piersol, 1971). The problems of decreased resolution, negative energy, and spectral leakage due to sidelobes are minimized by the windowing process.

Windowed pressure time series are transformed from the time domain to the frequency domain by using a Welch method (Welch, 1967) analysis with 50 percent overlapping segments. The pressure spectral estimates, $P(m)$, are transformed into equivalent energy spectral estimates at the surface, $E(m)$, using the linear pressure response function (Dean and Dalrymple, 1984).

4.2 Directional Wave Data Analysis

A PUV directional wave sensor measures three independent variables: pressure(p), **u**-velocity, and **v**-velocity at the gage. Each of these variables is sampled over a 2048 second time period at a sampling rate of 1 Hertz to produce 3 separate time series.

Influences of extremely low frequency energy, such as tides, are removed from **u**-velocity, **v**-velocity,and pressure records by linear detrending and a zero frequency rejection filter as described earlier.

Windowed (as described previously) **u**-velocity, **v**-velocity, and pressure time series are transformed from the time domain to the frequency domain by using the Welch FFT algorithm.

The pressure spectra, $P(m)$, and velocity spectra ($U(m)$ and $V(m)$) are transformed into equivalent energy and velocity spectra at the surface, using the linear pressure K_p and velocity K_u response functions.

The directional wave coefficients are determined by calculating and relating the cross-spectra between the pressure and the velocity time series. Since the pressure and velocity data are in phase the auto- and cross-spectral estimates (Jenkins and Watts, 1968) are expressed by:

$$S_{pp}(m) = P(m)^2 \tag{4}$$

$$S_{uu}(m) = U(m)^2 \tag{5}$$

$$S_{vv}(m) = V(m)^2 \tag{6}$$

$$S_{pu}(m) = P(m)U(m)\cos(\phi_u(m) - \phi_p(m)) \tag{7}$$

where $S_{xx}(m)$ = auto-spectral estimate and $S_{xy}(m)$ = the cross-spectral estimate.

The first five directional Fourier coefficients are determined using a method of analysis similar to that of Longuet-Higgins (Longuet-Higgins et al., 1963). Using the previously defined single-sided auto-spectral estimates, coincident spectral estimates, pressure estimates, pressure and velocity response factors, the five real and imaginary directional coefficients are expressed by:

$$A_0(m) = \frac{G_{pp}(m)}{2\pi K_p^2(m)} \tag{8}$$

$$A_1(m) = \frac{S_{pu}(m)}{\pi K_p(m) K_\mu(m)} \tag{9}$$

$$A_2(m) = \frac{G_{uu}(m) - G_{vv}(m)}{\pi K_\mu^2(m)} \tag{10}$$

$$B_1(m) = \frac{S_{pv}(m)}{\pi K_p(m) K_\mu(m)} \tag{11}$$

$$B_2(m) = \frac{2S_{uv}(m)}{\pi K_\mu^2(m)} \tag{12}$$

where the $A_n(m)$ and $B_n(m)$ are the real and imaginary coefficients respectively and where $G_{xx} = 2S_{xx}$

At the center frequency, m, the frequency spectral estimate (energy density in m^2/Hz) is calculated by:

$$\text{energy density}(m) = \frac{TG_{pp}(m)}{K_p^2(m)} = 2\pi A_o(m) \tag{13}$$

where T is the record length in seconds.

The truncated Fourier series representation (Longuet-Higgins, et al., 1963) of the directional wave spectrum is expressed by:

$$S(m,Q) = A_0(m) + \frac{2}{3}[A_1(m)\cos(Q) + B_1(m)\sin(Q)] + \frac{1}{6}[A_2(m)\cos(2Q) + B_2(m)\sin(2Q)] \quad (14)$$

The values for the wave direction are adjusted to give degrees measured from true North.

The mean wave direction is determined by calculating the mean direction of waves in each band of frequencies and is defined by:

$$Q(m) = \arctan\left(\frac{B_1(m)}{A_1(m)}\right) \quad (15)$$

where $A_1(m)$ and $B_1(m)$ are the directional coefficients which correspond to a particular center frequency. This value is adjusted to give angles in degrees measured clockwise from true North from which the waves are coming. The directional spread is an estimate of the spread of energy about the mean wave direction at a center frequency and is given by:

$$Q(m) = \left|2 - \left[\frac{(A_1^2(m) + B_1^2(m))^{\frac{1}{2}}}{A_0(m)}\right]\right|^{\frac{1}{2}} \quad (16)$$

where $A_0(m)$, $A_1(m)$, and $B_1(m)$ are the directional Fourier coefficients of a particular band (Cartwright, 1963).

4.3 Relational Database

Due to the large amount of wave information currently and historically collected by WES, a relational wave data access and management system has been created with the needs of the Los Angeles - Long Beach data collection effort influencing its design. Relational technology was employed for several reasons.

First, a tool was needed to manage and archive raw sample data and analysis results from several sites for several years as well as the related information needed to transform these data into a usable form.

Data accessibility was a second major consideration in the decision to use a relational database. Related information is maintained within a framework which provides for routine cross-referencing between data objects. For example, if mean spectra for Platform Edith are required for periods when H_{m_o} is between 1 and 2 meters, a query can quickly be formulated to retrieve the appropriate information.

Finally, the utility of relational database design allows for the continued development of the database, the addition of fields and relations without impacting existing software, resulting in reduced programming costs.

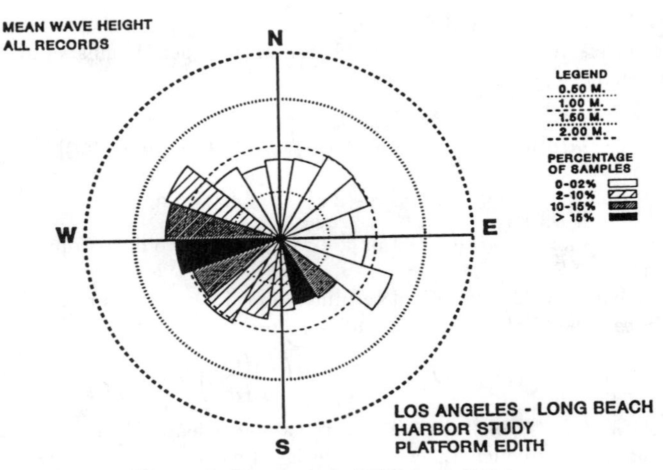

Figure 3: Wave rose of Platform Edith

5 Discussion

Directional Wave Data The directional information gathered from Platform Edith helps to describe the incident wave climate for the Harbors. The rose plot in Figure 3 shows the mean direction at the peak frequency and the significant wave height for each wave record. This plot is one way to summarize the large amount of wave data collected at this site. Other plots not shown describe the temporal variations of the incident wave energy and directions.

The most frequent waves come from two dominant sectors, the S-SE and the W-SW. During the summer months, April thru September, mean significant wave heights are generally one meter or less for all directions. For the winter months, October thru March, mean significant wave heights for all directions are larger than the summer months with the average mean wave height exceeding one meter. For the winter months, waves most frequently come from the West.

Harbor Gages Individual spectra have been computed for each wave record measured in the harbors. In an attempt to describe the wave climate in the harbors, the computed spectra are presented in summary form. (Figure 4) is an example plot of the monthly percent occurrence of the peak frequency (FTp). Peak frequency refers to the frequency which contains the most energy for each computed spectrum. (Figure 5) is an example of the monthly averaged spectra using all of the available computed wave spectra available for a site.

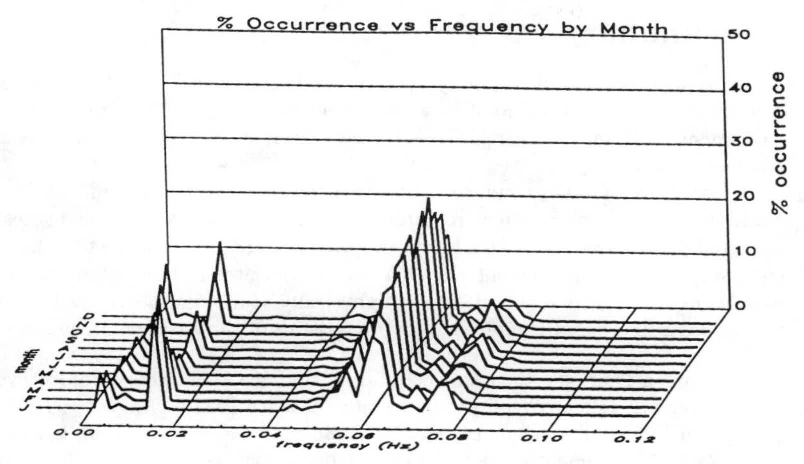

Figure 4: Percent occurrence of peak frequency for gage LB2

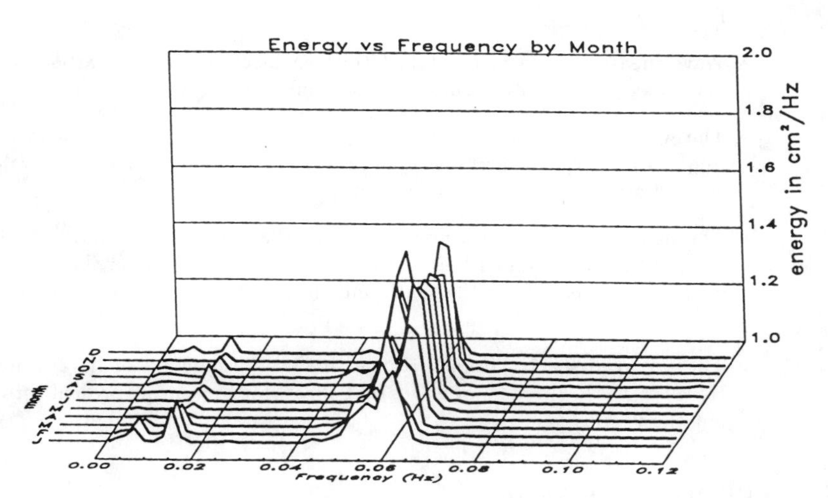

Figure 5: Averaged spectra for gage LB2

6 Summary

Wave data have been successfully acquired and analyzed for LA/LB Harbors. The wave data have been used in various studies of LA/LB. The methods and equipment used for acquiring the data have been documented as well as the analysis methods.

A database of wave conditions has been established which will aid future analysis by reducing the effort required to access the considerable quantities of collected data. Directional data from Platform Edith and two gages at the harbor entrances describe the incident wave climate. Gages sited in the harbors describe harbor response. The acquired data are physically sound and provide a basis for further investigation of the physical characteristics of the harbors. In summary:

- Between February 1984 and February 1988 wave data were sampled every 2 hours at 1Hz for 2048 seconds. From February 1988 to August 1991, data were collected continuously in the harbors at an averaged sample rate of .25Hz. Subsequently, waves in the harbors have been sampled at .5Hz for 8192 seconds. Offshore, data are collected at 1Hz for 2048 seconds every hour but reported every 4 hours.
- Directional wave data were measured from offshore Platform Edith via a PUV gage.
- Wave measurements in a total of thirteen locations in the harbors were performed with highly accurate single point pressure sensors.
- The wave measurement system was optimized to gather and report low-frequency wave data suitable for use by other studies and to establish a wave climatology for LA/LB harbors.
- The directional wave measurements show that during the winter months most of the waves come from the west, but during the summer the majority of the waves come from the south and the average significant wave height is smaller.
- The averaged spectra from the gage sites show some seasonal variations. However, the low-frequency waves that affect ship motion are present throughout the year.

Acknowledgments

The design team for this project included William E. Grogg, William M. Kucharski, and Gary L. Howell. James P. Mckinney performed much of the analysis work. The work described here was performed for studies conducted by U.S. Army Engineer Waterways Experiment Station's Coastal Engineering Research Center for the U.S. Army Engineer District, Los Angeles District, and the Ports of Los Angeles and Long Beach. Permission to publish this paper was granted by the Chief of Engineers, U.S. Army Corps of Engineers.

References

Bendat J.S., and Piersol, A.G. 1971. *Random Data: Analysis and Measurement Procedures*, Wiley, New York, 325.

Cartwright, D.E. 1963. "The Use of Directional Spectra is Studying the Output of a Wave Recorder on a Moving Ship," *Ocean Wave Spectra*, Prentice-Hall, Englewood Cliffs, NJ.

da Cruz, F. 1987. *KERMIT, A File Transfer Protocol*, Digital Press, Bedford, MA.

Dean, R.G., Dalrymple, R.A. 1984. *Wave Wave Mechanics for Engineers and Scientists*, Prentice-Hall, Englewood Cliffs, NJ.

Durham, D.L., et. al. 1976. "Los Angeles and Long Beach Harbors Model Study, Analyses of Wave and Ship Motion Data,", Technical Report H-75-4, U.S. Army Engineer Waterways Experiment Station, Vicksburg, MS.

Jenkins, G.M. and Watts, D.G. 1968. *Spectral Analysis and Its Applications*, Holden-Day, San Francisco, CA.

Longuet-Higgins, M.S., D.E. Cartwright and N.D. Smith , 1963. "Observations of the directional spectrum of sea waves using the motion of a floating buoy," *Ocean Wave Spectra*, Prentice-Hall, Englewood Cliffs, NJ 111-136.

Paros, J.M. 1976."Digital Pressure Transducers," *Measurements and Data*, Issue 56, vol. 10, no. 2 March-April.

Rosati, J., et. al. 1993. "Los Angeles and Long Beach Harbors Model Enhancement Program, Prototype Wave Data Summary", In preparation, U.S. Army Engineer Waterways Experiment Station, Vicksburg, MS.

Seabergh, W.C, et.al. 1992. "Los Angeles-Long Beach Harbors Model Enhancement Program", American Society of Civil Engineers Proceedings, PORTS '92, p. 884-897.

Welch, P.D. 1967. "The Use of Fast Fourier Transform for the Estimation of Power Spectra: A Method Based on Time Averaging Over Short, Modified Periodograms," *IEEE Trans. Audio and Electroacoust.*, vol. AU-15, pp. 70-73.

The TMA Wave-Energy Constraint and Cohesive Shore-Profile Shapes

Charles N Johnson, M ASCE[1]

Abstract. A relation by Andreanov between wave-energy loss and nearshore slope, when combined with the TMA wave-energy connstraint, is shown to closely approximate the classical Bruun/Dean shape. Further, it predicts a bar array and a swash zone, and can be applied to cases of constrained maximum wave-orbital velocity.

Introduction. The study of shore response to wave action has produced an extensive body of literature. Probably the most useful concept to have developed from this body of knowledge is the Bruun rule for equilibrium beach-profile shape and response to water-level fluctuations (2). When dE/dx = constant, the profile is defined by the equation $D = A x^{2/3}$, where D is depth, A is a shape factor commonly taken as a function of beach-particle diameter, and x is distance seaward. With the assumption of no profile-shape change when a storm causes a water-level rise, the Bruun rule has been used to estimate beach recession caused by the storm (11).

This body of knowledge does not apply to shorelines of cohesive materials, such as hard clays, shales, limestones and sandstones. The author's area of concern, the Laurentian Great Lakes of North America, is primarily composed of such materials, with very thin covers of clastic lags. Among the few authors to describe quantitatively such shorelines are Zenkovich (17, 18) and Kamphuis(10). Unfortunately, the descriptions are contradictory. Kamphuis emphasizes the role of suspended sand in abrading the cohesive materrial to a profile similar to that for sand equilibrium. Zenkovich insists that profiles in cohesive materials must be convex-upward, consistent with a constant rate of relative specific-energy dissipation. He cites an empirical equation derived by Andreanov (2) for calculating the relative specific wave energy loss K per wavelength:

$$K \doteq \frac{E1-E2}{E1} \doteq \frac{H1^2L1-H2^2L2}{H1^2L2} = \frac{1.07\left(\frac{dD}{dx}\right)^{0.5}+0.0016}{\sinh\left(\frac{2\pi D_{av}}{L_{av}}\right)} \qquad (1)$$

where: H1 and L1 are wave height and length at starting depth D1; and
H2 and L2 are wave height and length one average wavelength landward, at D2;
Dav and Lav are average depth and wavelength between D1 and D2; and
dD/dx is the average bed slope between D1 and D2.

Andreanov deduced his equation from measurements at Onega and Vyg Lakes in 1931-32. His apparent intent was to calculate wave energy at any specified nearshore location on the basis of deepwater wave conditions. If K is constant and if Dav/Lav is small enough so that sinh (argument) = (argument),then equation (1) may be integrated

[1]Hydraulic Engineer, North Central Division, US Army Corps of Engineers, 111 N Canal St, Chicago, IL 60606

over one wavelength to produce a logarithmic convex-up relationship between D and horizontal coordinate x. Zenkovich clearly believed that this result was important, and discussed at some length (18) the efforts of Longinov (12) to establish a physical basis for such a shape.

In 1970 Saylor and Hands (14) presented an analysis of several years of measurements of sandy barred profiles along the Little Sable Point area of the Lake Michigan east shore. These authors pointed out that the troughs between the bars often were covered with stones and boulders typical of glacial-till-related lag deposits and that the profiles of these deposits were often slightly convex-up. This work was consistent with the results of Andreanov, Longinov, and Zenkovich. Johnson (9) found that the loci of many of these troughs followed a Longinov curve very closely, (Figure 1). This observation motivated the author to explore the possibility of other assumptions for K.

Derivation. Hughes (8) describes the TMA parameterization for estimating depth-limited energy for saturated irregular wave spectra. The mechanism for energy loss is whitecapping due to wave-steepening and wave-wave interaction as the wavelength decreases with decreasing depth. In accordance with Hughes, for shallow water

$$H_{mo}^2 L_p = T_p \sqrt{gD}\, T_p^2 \frac{\alpha g D}{\pi^2}$$

where D is local depth, Lp is local wavelength corresponding to Tp:

$$L_p = T_p \sqrt{gD} \tag{2}$$

g is acceleration of gravity,Tp is period for maximum wave energy density Vw is windspeed, and parameters

$$\alpha = .0078\kappa^{0.49} \approx .0078\kappa^{0.50} \qquad \kappa = \frac{2\pi V_w^2}{gL_p}$$

from which

$$\frac{\left(H_{mo}^2 L_p\right)_2}{\left(H_{mo}^2 L_p\right)_1} = \frac{\alpha_2}{\alpha_1}\left(\frac{D2}{D1}\right)^{1.5} = \left(\frac{D2}{D1}\right)^{-\frac{1}{4}}\left(\frac{D2}{D1}\right)^{\frac{3}{2}}$$

$$K = 1 - \left(\frac{D2}{D1}\right)^{1.25} \tag{3}$$

Hughes (8) points out that in sufficiently shallow depths the waves have entered a universal saturation zone, where solitary-type breaking dominates, such that Hmo=BD, where B is a constant. Then

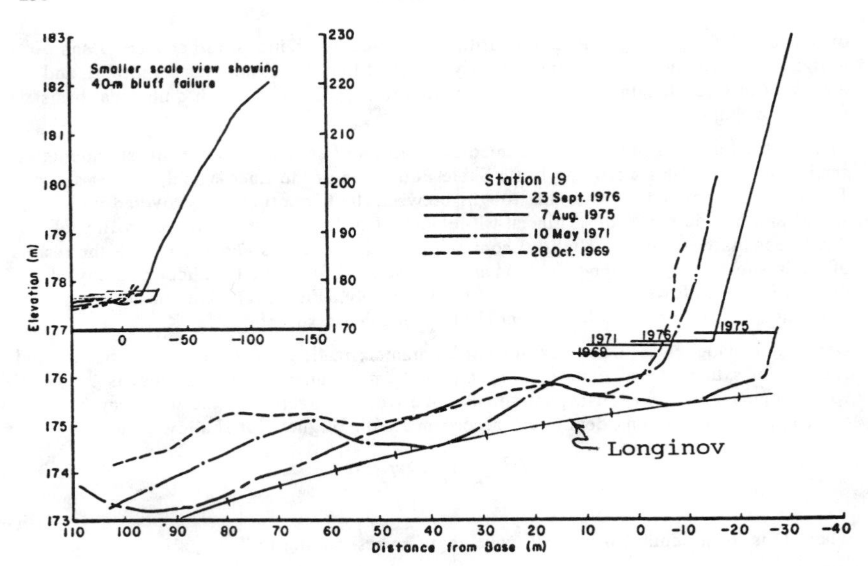

Figure 1. One of Hands' profiles having stony troughs as compared with shape calculated according to Longinov.

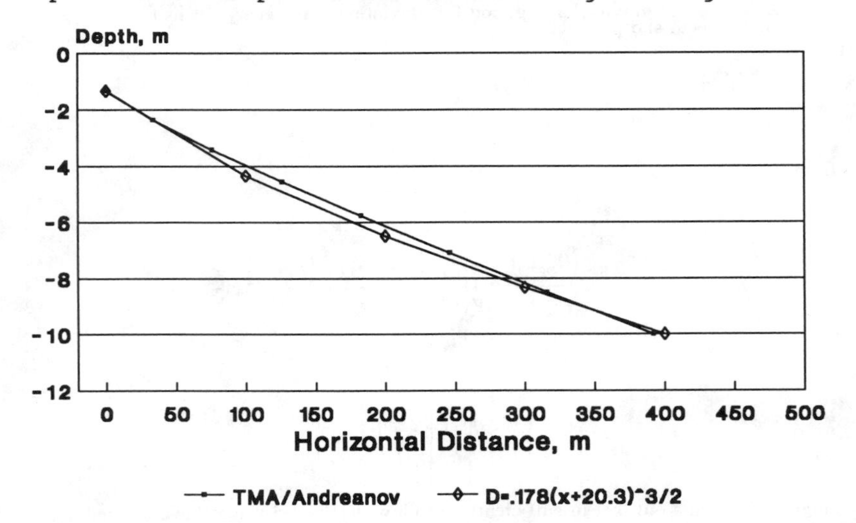

Figure 2. Comparison of profile computed from eq. 6 for Tp=8 sec and j=1.25 with Bruun-Dean shape.

$$K=1-\left(\frac{D2}{D1}\right)^{2.5} \tag{4}$$

The exponent of the (D2/D1) term in K will be allowed to remain constant or increase as depth decreases and will be symbolized by the letter j.

Substituting equation (3) for K in equation (1), taking, for simplicity, the multiplicative constant 1.07 to be 1. and the additive constant 0.0016 to be zero, and confining ourselves to depths for which sinh (argument) approximately equals (argument), we obtain

$$f(D2)=D1-D2-\frac{\pi^2}{\sqrt{g}T_p}(\sqrt{D1}+\sqrt{D2})(D1+D2)\left[1-\left(\frac{D2}{D1}\right)^{1.25}\right]^2 \tag{5}$$

This is an algebraic equation. If we specify T_p and D1, we may iterate numerically to calculate a D2 which satisfies f(D2) = 0. The form of equation (1) precludes negative slopes. The computational delta x is one average wavelength:

$$\Delta x=T_p\frac{\sqrt{g}}{2}(\sqrt{D1}+\sqrt{D2})\doteq L_{av} \tag{6}$$

(Note: Some of the sample computations in this paper are based on an alternative definition of one wavelength:

$$\Delta x=T_p\sqrt{g}\frac{\sqrt{D1+D2}}{2}$$

Use of either definition will give the same qualitative results and almost-identical quantitative results.)

Equation (5) is a discrete dynamical system. The output D2 of each computational step becomes the input D1 for the next step. Equation (5) is valid only where shallow-water small-amplitude wave theory is valid; it cannot take us to the water's edge. To compute a profile we must route several waves from differing D1, say near 12m. At this depth the slope may be assumed linear, so a consistent value of x can be associated with each starting D1. For constant j, the resulting profile shape is concave-up, very similar to that described by Bruun (4) (Figure 2). Thus we have evidence that an irregular-wave spectrum may obey the dE / dx = constant condition that leads to the Bruun-type profile. Also, there appears to be no contradiction between Andreanov's equation and the results of Kamphuis.

For D1 less than about 4.4 m (T_p = 12 sec) a constant value of j = 1.25 can no longer satisfy f(D2)=0. The bed becomes neutrally stable (Figure 3). Further examination of solutions for f(D2) = 0 and j =1.25 show that the solution curve of D1 vs. D2 (Figure 4) is double-valued. The deeper leg will be named the "Bruun-Dean leg" because it corresponds to the shape these investigators defined. The shallow leg will be called the "ridge-and-runnel" leg and will be discussed later. The profile can be carried into shallower D1 only allowing j to increase. This condition is in accordance with the discussion of Hughes. A path through the neutral points for each j gives the gentlest possible breaker law .for each D1. A plot of j versus D1 for neutral stability is a straight line in a log-log plot (Figure 5). The requirement for changing from a constant - j path to a varia ble - j path leads to a rather abrupt change in profile slope. Figure 6, for an 8-sec T_p, shows how this change leads to repeated cusps in the profile. This cusped surface is neutrally stable and not phase-dependent. Because the Andreanov law allows the profile shape to vary with wave-energy-constraint, it is possible that this breaker-driven ridging might mode-lock with the natural resonances in the nearshore water prism (next paragraph). The variable-j zone will be called the "bar zone". The energy gradient dE/dx cannot be constant in the bar zone. If j cannot exceed 2.5, then a deepwater j=2.5 wave condition could not have a bar zone. Thus storm waves would cause a bar system whereas swell waves would suppress it.

Interaction with Surf Beat. Dally (5) derived expressions for positions of the first three nodes and first two antinodes of standing oscillations in a beach prism bounded by a D=A x^ 2/3 curve. These equations are

NODES: $\kappa_e x = |2.84 \quad 11.60 \quad 23.73|$ *ANTINODES*: $|6.52 \quad 17.15|$

$$\kappa_e = \left[\frac{\left(\frac{3}{2} \frac{2\pi}{T_p} \right)^2}{gA} \right]^{\frac{3}{4}} \tag{7}$$

Dally's experiments led him to conclude that the breakpoint- undertow mechanism is more likely than the surf beat mechanism to account for longshore bar formation. This conclusion is consistent with the breaker-law mechanism described above in this paper. However, the neutral stability of the bar-zone surface allows for the possibility of mode-locking. If we consider each cusp in the 12-sec profile to be the intersection of two Bruun-type surfaces with different A, we can plot these A-values versus cusp depth (Figure 7). We can also plot A versus depth for each of Dally's nodes and antinodes. We find that the curve for the third node crosses both cusp lines. This is a very crude analysis, but it indicates that breaker-law-induced bars may be commensurate with surf-beat nodes. If so, mode-locking is very possible. If the coupling between the two effects is sufficiently strong, chaos is also possible (Ecke - 1990).

Swash Zone. When D1 reaches the value for neutral stability with j = 2.5, we assume that the waves are now better described as finite-amplitude bores. We define this D1 - D2 point as the beginning of the "swash zone". Entry into this zone can be from any depth between this D1 and D2. For T_p = 12 sec, for example, D1 = 1.05 m and D2 = 0.66 m. Let us seek to calculate the horizontal and vertical distances to dissipate all of the remaining wave energy. We require K=1. Then the simplified Andreanov equation becomes

$$\frac{\Delta D}{\Delta x} = \left[2\pi \frac{(D+H)}{L} \right]^2 \tag{8}$$

where D + H =D + BD
and B = constant, taken below to be 0.5

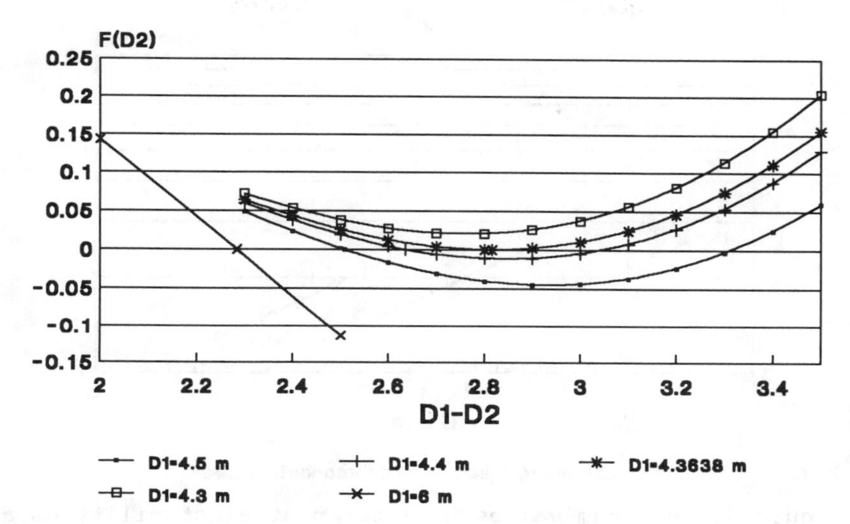

Figure 3. Behavior of eq. 6 for Tp=12 sec and j=1.25 as D1 is decreased.

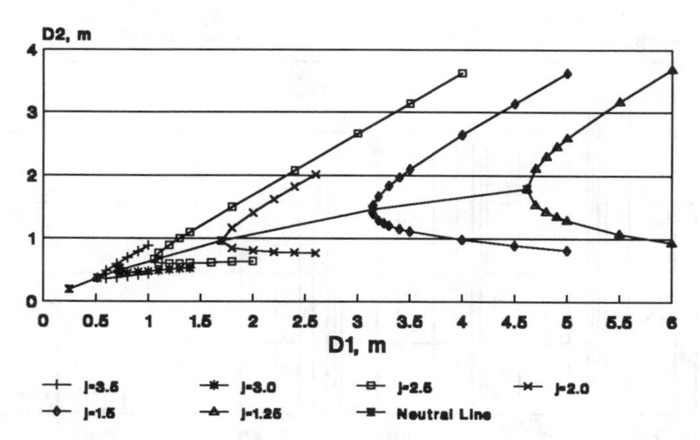

Figure 4. Eq. 6 as a function of D1 and D2 for several values of the exponent j. Also shown is the line connecting the minimum D1 for each j.

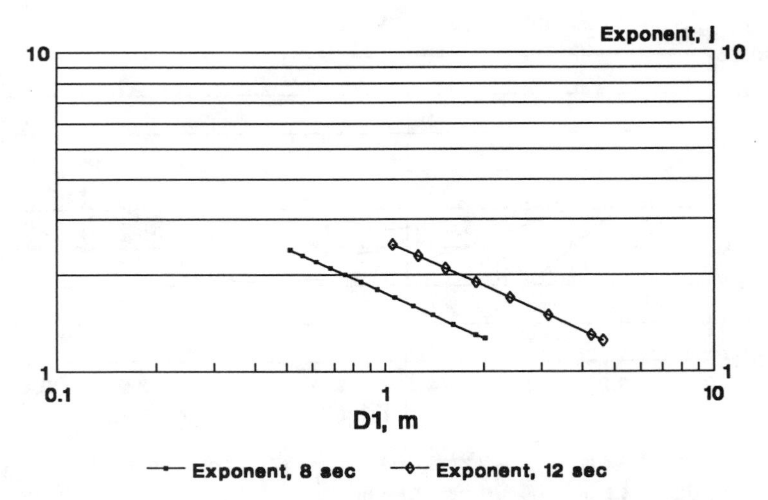

Figure 5. Minimum values of j for neutral stability as a function of D1.

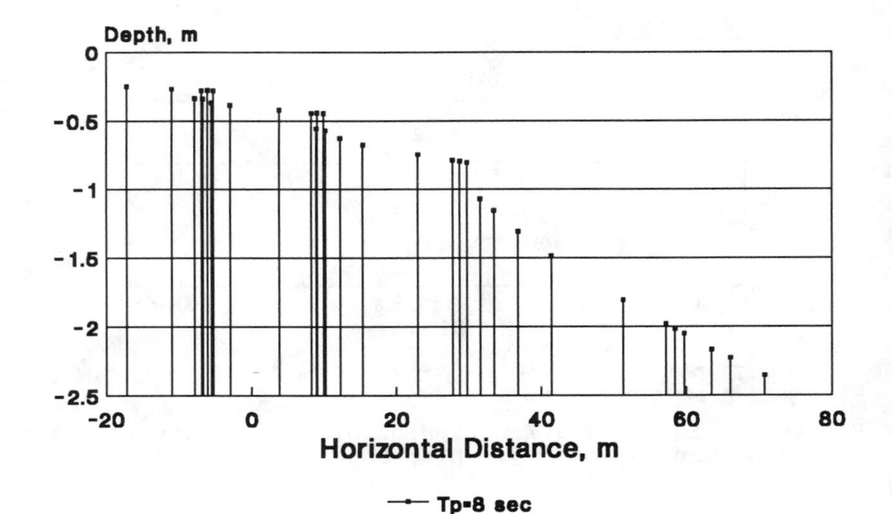

Figure 6. Profile for Tp=8 sec showing effect of requirement for change from constant-j path to variable-j path.

and D1 is swash-entry depth . With

$$\Delta x = L$$

$$\Delta D = \frac{[2\pi D1(1+0.5)]^2}{L} = \frac{[3\pi D1]^2}{L} \equiv D1 + Runup \quad (9)$$

Runup for swash entry from the neutral j=2.5 point is shown in Figure 8. Runup on any beach line is phase-dependent. Highest runup is from the most-landward swash-entry D1. Beach slope is determined by overtopping.

Swash Zone - Effect of Water-Level Drop. On the Figure 4 map a water-level drop displaces a storm beach-profile curve downward and to the left relative to the lines of constant j. If the profile was initially at the neutrally-stable swash-entry point, it would now enter on the shallow, "ridge-and-runnel" leg (Figure 9). If entry is made near the local minimum D2, we find that the runup becomes essentially independent of swash-entry point (Figure 10). A slightly greater drop (Figure 11) reverses the runup dependency, so that the highest runup is always due to the phase from the greatest swash-entry depth.

Swash:Discussion. Equation 9 is as simple as the widely-used Hunt's equation (7,13) but does not indicate a need for correlations with wave heights in deeper water. The changing runup pattern with small water-level changes indicates a natural tendency toward slow oscillations such as surf beat.
If the swash zone in already in equilibrium as in Figure 8 so that the close-in waves are barely not overtopping the beach, any small film or parcel of sand brought in from deeper water would not be carried any farther landward. From this standpoint the swash zone would be stable.
If the water level has gone down to the condition of Figure 10, if the beach crest is barely above overtopping, all waves would run up to that elevation. A small upward waterlevel fluctuation would allow all waves to overtop the beach crest, causing landward movement. From this standpoint the swash zone would be neutrally stable. If the water level has gone down to the condition of Figure 11, and, if the existing beach is barely above overtopping by close-in waves, sand particles would be carried by deeper-entry waves over the ridge. The beach material would tend to be carried landward; thus the "ridge-and-runnel" name to this leg of the Figure 4 map. If overtopping is prevented, the beach slope can become quite steep.

Bar Zone and Bruun-Dean Zone: Effect of Spectral Change at Constant T_p and Water Level. A classical observation of ocean beaches (3) is that they accrete in summer and erode in winter. These changes are attributed to seasonal changes in wave spectra: decayed swell in summer and stormy waves in winter. The map of Figure 12, for 12 sec waves, illustrates the substantial amount of material which must be translated into the nearshore to form the bar zone. This volume is roughly equivalent to the amount mobilized on a storm profile by a water-level rise exceeding 4 meters. This shape-change effect is not easily taken into account by profile-response models incorporating only the Bruun/Dean shape description. As a corollary, whenever the local wave climate is dominated by long-period swell, water's-edge placement of beach nourishment would be a good approximation of natural sand storage.

Bar Zone and Bruun-Dean Zone: Effect of Water Level Change with Constant T_p and Constant Spectral Shape. Figure 13 is a 12-sec j = 1.25 storm-map profile with two different water levels. If the water level drops, the bar zone must migrate seaward, with

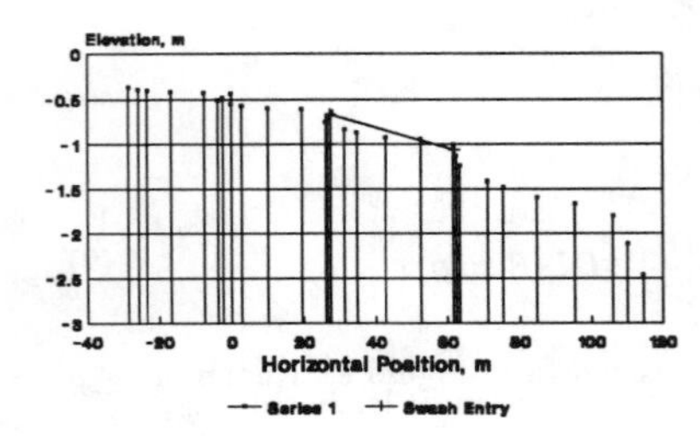

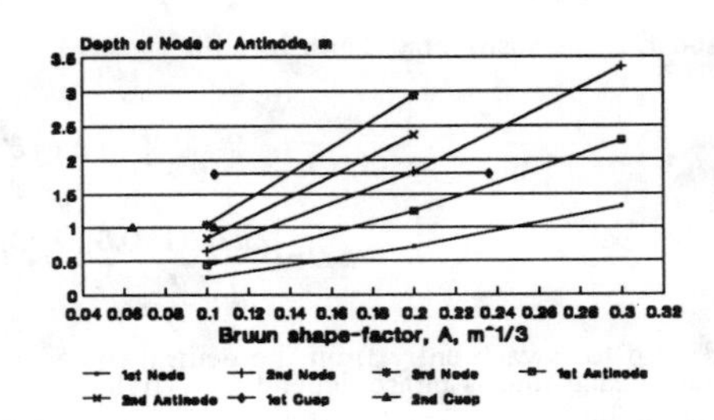

Figure 7: (a) 12-sec profile showing locations of 1st & 2nd cusps and swash-entry zone.

Figure 7: (b) Surf-beat nodes and antinodes by Dally in relation to shape-factor A for cusps in Fig 7(a).

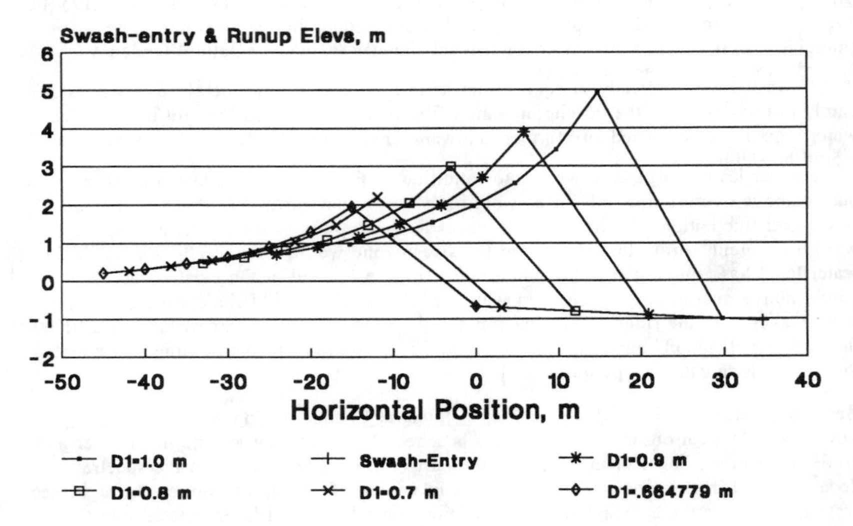

Figure 8. Runup as a function of L for swash- entry from equilibrium beach as calculated from eq 10. Runup from shallowest part of zone is always the greatest.

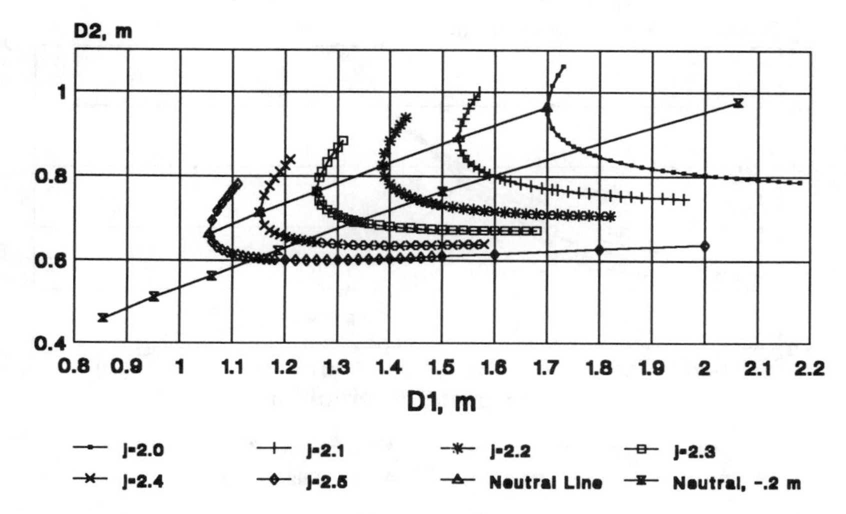

Figure 9. High-resolution plot of D1 vs D2 curves for various j-values between 2.0 and 2.5. Also plotted are loci of minimum-D1 points for each j, and the same path as displaced by a quick 0.2 m water-level drop.

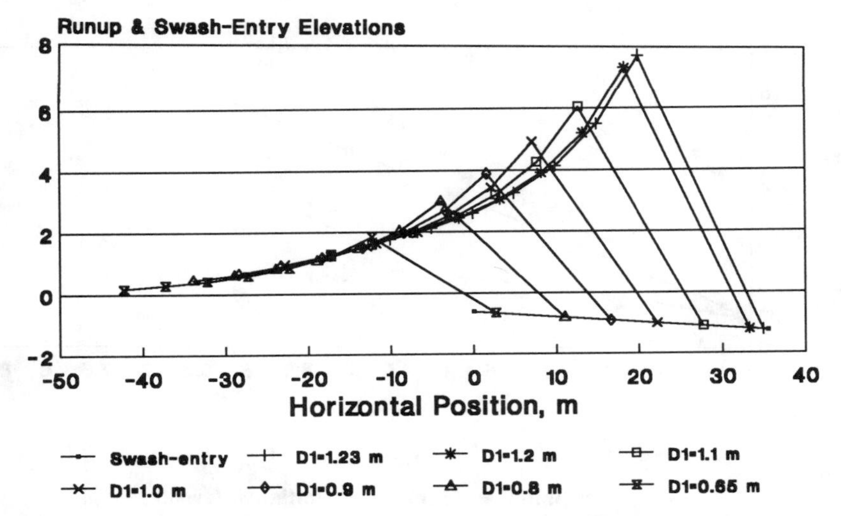

Figure 10. Runup after water-level drop so that loci of minimum D1 enters swash zone at minimum D2 of j=2.5 curve, about D2=0.601 m for Tp=12 sec. Runup is almost independent of where it enters swash zone.

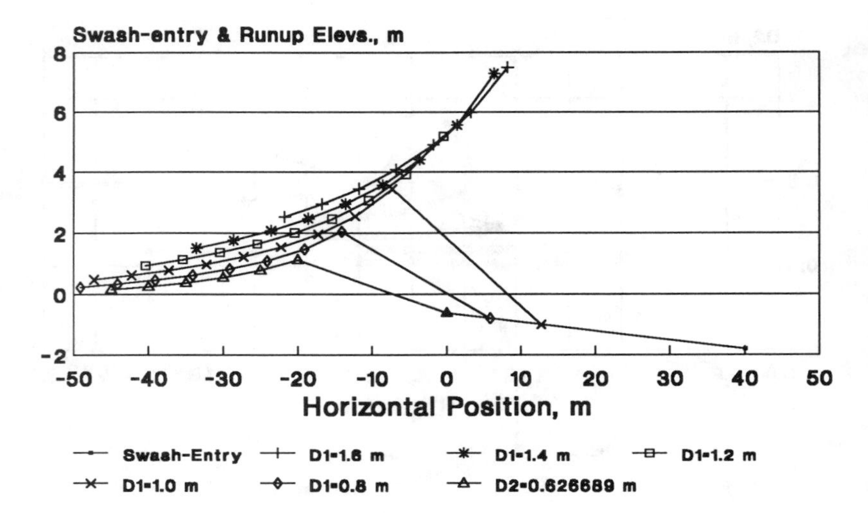

Figure 11. Runup for swash entry after a quick water-level drop of about 0.6 m from equilibrium. Swash-entry is now from seaward of 2nd bar. Runup from deeper parts of swash-entry zone overtops that from shallower parts.

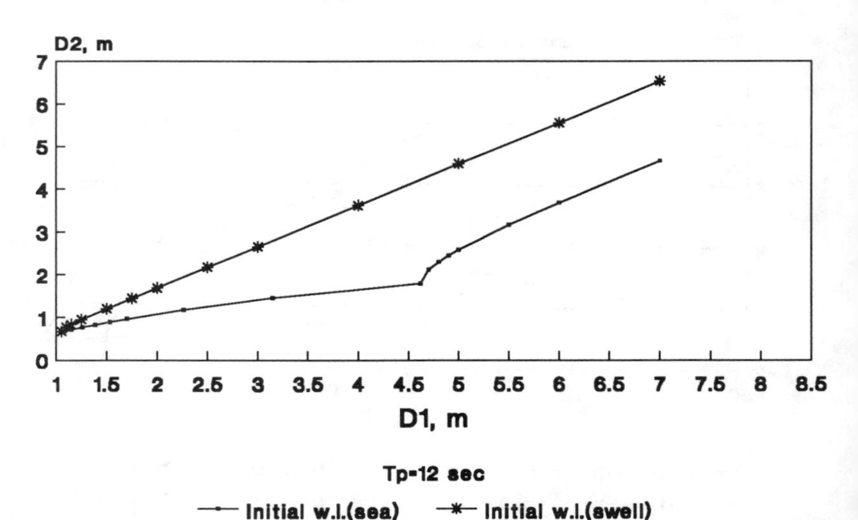

Figure 12. Profile response to differing spectral shapes: sea and swell. Presence of the bar zone makes the sea profile much shallower than the swell profile.

local accretion; and its surface elevation must be lowered. The difference between the response of the bar zone and the Bruun-Dean zone is very pronounced. The bar-zone response is much greater. Figure 14 shows the same process for a j = 2.5 swell-profile map. The cross-sectional changes are much smaller than for the storm profile of Figure 13..

Cohesive Shores. If the profile becomes active or erodible only when maximum wave orbital bed velocity exceeds some value U_{max}, it is cohesive. The classic small amplitude theory equation for this velocity in shallow water is, using equation (2):

$$U_{max}=\frac{HL}{2DT}=\frac{H\sqrt{g}}{2\sqrt{D}} \tag{10}$$

For specified U_{max}, then H is proportional to the square root of the depth. For this rule, the Andreanov K is

$$K=1-\frac{U_{max_2}^{\;2}}{U_{max_1}^{\;2}}\left(\frac{D2}{D1}\right)^{1.5} \tag{11}$$

For transition at the proper depth from an unconstrained TMA profile to a specified Umax profile we have, from the derivations of equations (3) and (11):

$$K=1-\frac{6.507U_{max_2}^{\;2}D2^{1.5}}{T_p^{1.5}D1^{1.25}} \tag{12}$$

Figure (15) illustrates this transition. Once established, a constant Umax profile can also be obeyed by noncohesive material. As we have seen earlier, a constant Umax profile can only be formed if the incident deepwater waves are sea. Umax can be allowed to vary within certain limits. Many combinations of bed slopes, pavement particle-sizes, and depths appear possible. Development of gravel paving can suppress the instability that leads, with noncohesive material, to bar-zone formation. This bar-zone suppression would be consistent with the wave-flume measurements of van der Meer and Pilarczyk (1987) and Ahrens (1993) for gravel and cobble beaches. For transition out,

$$K=1-\frac{T_p^{1.5}D2^{j}}{6.507U_{max_1}^{\;2}D1^{1.5}} \tag{13}$$

This equation has not yet been thoroughly investigated, but it appears to predict a benchlike transition from pavement/cohesive to sand. If sand is not present, equation (13) must be used. Importantly, if Umax2 in equation (13) is too low, no stable profile can be calculated. It is possible to calculate a profile for the minimum stable Umax2. Figure 16 is an example of a computed composite profile for Tp = 12 sec. The profile becomes active at -10 m. From -10 m to -7 m the bed surface is noncohesive in sufficient volume to be piled into equilibrium with TMA waves. From -7 m to -4 m a

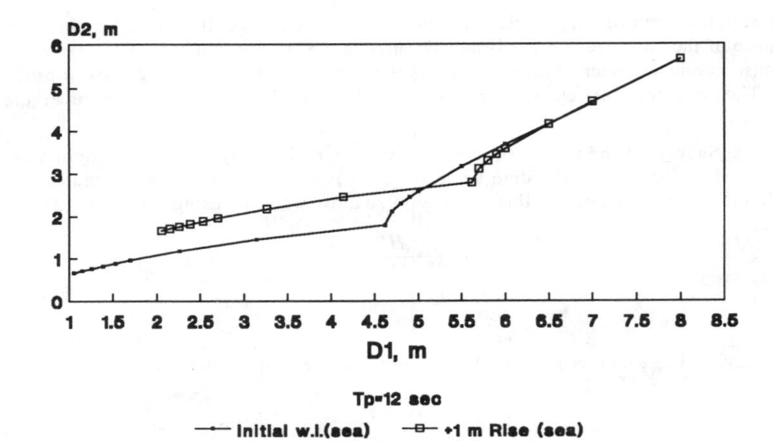

Figure 13. Effect of a 1-m water level change on a storm-beach profile. Tp=12 sec and j=1.25.

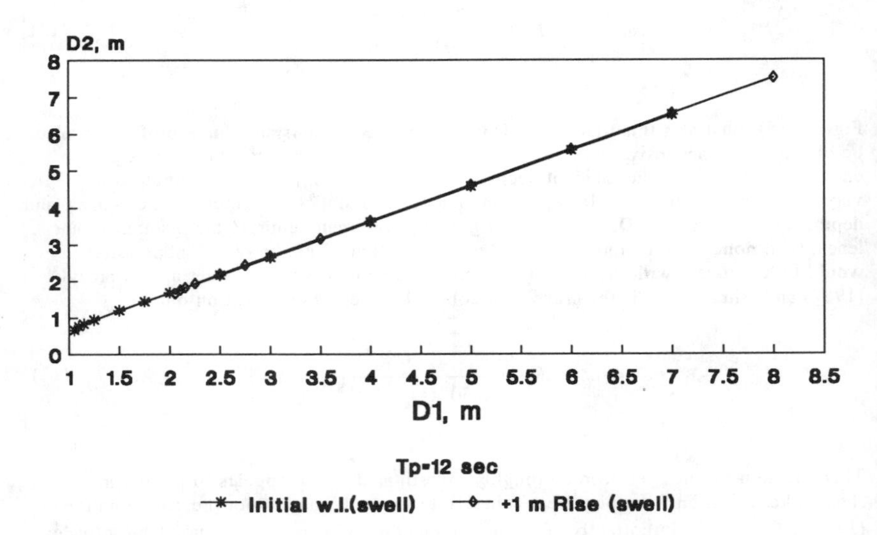

Figure 14. Effect of a 1-m water level change on a swell profile. The calculated effect is much smaller than for a sea profile. Tp=12 sec and j=2.5.

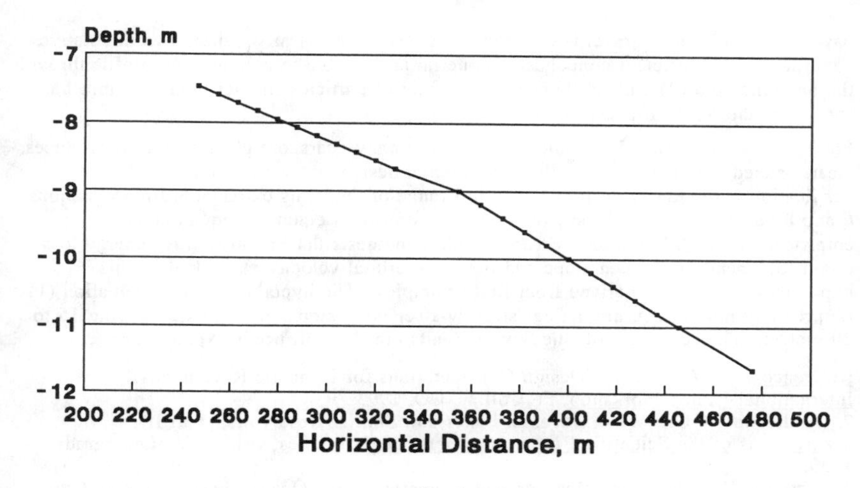

Figure 15. Transition from an unconstrained (noncohesive) bed material below -9 m to a constant maximum wave-orbital velocity Umax=1.915 m/sec surface corresponding to a 14.8 mm gravel pavement.

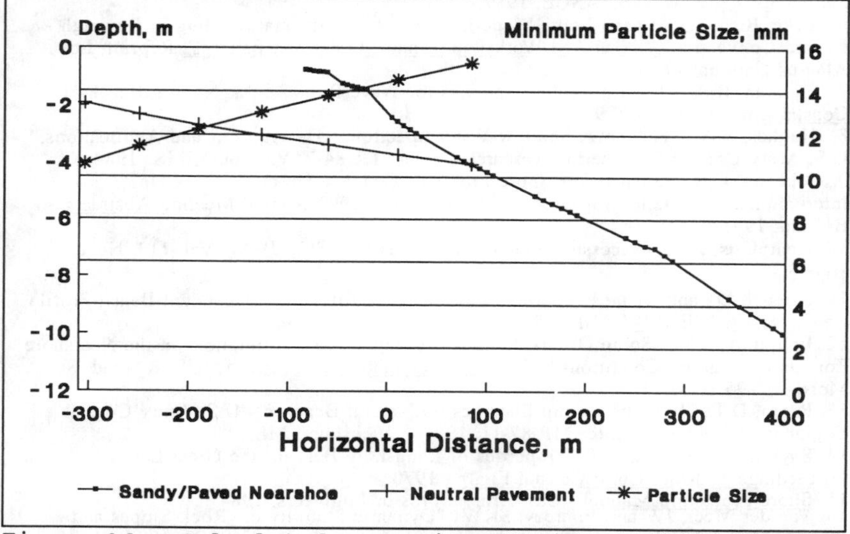

Figure 16. Calculated composite profile: unconstrained TMA -10 m to -7 m; const Umax=1.975 m/sec -7 m to -4 m; then two possible surfaces landward from -4 m: unconstrained (sand); or minimum-size stable pavement.

pavement of 15.3 mm gravel is exposed. Shallower than -4 m we show two alternative conditions: (1) sufficient noncohesive material is available to continue the profile through the bar zone; and (2) a neutrally stable pavement of particles smaller than 15.3 mm has formed as the result of a sand deficit.

Future Work. Equation 5 realistically describes many nearshore phenomena. It produces clearly-stated hypotheses to aid in experimental design and verification. Such verification is obviously desirable. The full range of theorems extractable from equations 6 and 8 has by no means been exhausted. Andreanov's equation (equation 1) is empirical, although based on the very simple hypothesis that relative wave-energy loss ought to depend on the bed slope and the wave orbital velocity at the bed. This hypothesis needs to be derived from first principles. The hyperbolic sine in equation (1) is probably not correct, predicting steadily-steepening slopes in depths exceeding 15 to 20 meters. The convexity of the Saylor-Hands profiles still needs explanation.

References: 1. Ahrens, J., "Design Considerations for Dynamic Revetments," International Riprap Workshop, Ft. Collins, CO, 1993.
2. Andreanov, V.G., "Wind Waves in Lakes and Similar Reservoirs, Part II". Transactions of the Scientific Research Institute of Hydraulics, vol. XXV. Leningrad: 1939.
3. Bascom, W., Waves and Beaches: the Dynamics of the Ocean Surface, Garden City: Anchor Books, 1964.
4. Bruun, P., "The Bruun Rule of Erosion by Sea-Level Rise: A Discussion of Large-Scale Two- and Three-Dimensional Usages" Journal of Coastal Research, Vol. 4, No. 4, 1988.
5. Dally, WR, "Longshore Bar Formation - Surf Beat or Undertow?" Proc. Coastal Sediments '87, New York, ASCE, 1987.
6. Ecke, R.E., "Quasiperiodicity, Mode-Locking, and Universal Scaling in Rayleigh - Benard Convection" NATO ASI Workshop "Chaos, Order and Patterns" Reprint, Los Alamos National Laboratory, 1990.
7. Holman, R.A., "Extreme Value Statistics for Wave Runup on a Natural Beach," Coastal Engineering, Vol. 9.
8. Hughes, S.A., "The TMA Shallow-Water Spectrum: Description and Applications," U.S. Army Coastal Engineering Research Center TR 84-7, Vicksburg, MS 1984.
9. Johnson, C.N. "Great Lakes Shore Profiles: Concave Up or Convex Up?" International Association for Great Lakes Research 1991 Annual Meeting Abstracts. Buffalo, 1991.
10. Kamphuis, J.W., "Recession Rate of Glacial Till," ASCE JWW, Vol. 113, No 1, 1987.
11. Kriebel, D., and Dean, R., "Convolution Method for Time- Dependent Beach-Profile Response," ASCE JWW, Vol. 119, No. 2, 1993.
12. Longinov, V.V., "Some Observations Regarding Wave Deformation in the Nearshore Zone under Natural Conditions." Proc. Inst. Oceanology, Vol. XXI, USSR Acad. Sci, Moscow, 1957.
13. Resio, D.T., "Extreme Runup Statistics on Natural Beaches" U.S. Army Coastal Engineering Research Center MP 87-11, 1987 - Vicksburg, MS.
14. Saylor, J. and Hands, E., "Properties of Longshore Bars in the Great Lakes," Proceedings 12th Int. Conf. Coastal Engrg., 1970
15. Shore Protection Manual, U.S. Army Corps of Engineers, 1984.
16. Van der Meer, JW and Pilarczyk, KW., "Dynamic Stability of Rock Slopes and Gravel Beaches" Delft Hydraulics Laboratory Communication No. 379, 1987.
17. Zenkovich, V.P., Dynamics and Morphology of Sea Shores, Moscow - 1946.
18. Zenkovich V.P., Processes of Coastal Development, New York: Interscience Publishers, 1967.

An Experimental and Analytical Study of the Shoreline Response to Non-Parallel Breakwaters in Oblique Waves

Michael E. McCormick[1]
and
Richard W. Gordon[2]

Abstract

Results of an experimental and analytical study of single breakwaters in oblique waves are presented. Three orientations of the breakwaters are studied, those being parallel to the shoreline, and at plus and minus 45° to the shoreline. The analysis of the experiment is performed using the elliptic shoreline assumption. This analysis is based on data resulting from tests on parallel breakwaters in direct waves. Hence, the best agreement between the experimental and analytical results is for the parallel breakwater orientation.

Introduction

The conceptual designs of single and multiple breakwaters should include a prediction of the response of the shoreline to the placement of the units. There are several analytical tools to use in such a prediction. These analytical techniques are divided into two categories. The first is that which involves numerical methods in the prediction of beach profiles in elevation. Included in this category are the works of Hanson (1989) and Graven, Kraus and Hanson (1991). The former study led to the computer program called "GENESIS". The second category includes the geometric studies of the shoreline in equilibrium. Two of the works included in this category are the works of Hsu and Silvester (1990) and McCormick (1993). The former work is based on the assumption that the equilibrium shoreline under direct wave attack is approximately parabolic, where the shape depends only on the length

----

[1]Professor and Director of Ocean Engineering, U. S. Naval Academy, Annapolis, MD 21402-5042
[2]Ensign, U. S. Navy, U. S. Naval Academy, Annapolis, MD 21402-5042

of the breakwater and its distance from the design shoreline. McCormick (1993) assumes an elliptic equilibrium shoreline, where the major and minor axes depend on the beach slope, deep-water wave steepness and incident wave angle. This analysis led to the computer program called "SHORELYN", which is written in GWBASIC so that it can be used in the field.

The purpose of the present study is to test the McCormick (1993) elliptic shoreline analysis with small-scale experimental data. In the referenced paper the analysis is applied to the eleven-unit segmented breakwater system at Bay Ridge, Maryland, located on the Chesapeake Bay. The scale of the experiment described herein is approximately 1/40 based on the Bay Ridge system.

Experiment

The experimental portion of the present study was performed in the Coastal Engineering Laboratory of the U. S. Naval Academy. The experimental setup is sketched in Figures 1 and 2, where the plan and profile are respectively shown. The inclined bed was constructed of aluminum. A grid of 15 cm x 15 cm was taped on the bed so that the sand movement on the bed could be monitored using photographic methods. The bed had channels on three sides to prevent the sand entering the main basin. There was a 6 cm gap between the tank floor and the lower edge of the bed, as sketched in Figure 2. For the three wave frequencies studied, the measured wave heights in the far field and leeward of the ledge were measured. There was little difference in the two heights, indicating that there was little reflection.

The single breakwater model, sketched in Figures 1 and 2, was constructed of rough stone, varying in "diameter" from 1 cm to 4 cm. The center of the structural bas was located at a distance S from the shoreline on the bare aluminum bed, where $0.85 < S < 0.90\ m$. A sand layer was placed in the lee of the structure, gradually terminating on a line coincident with the toe of the structure on the weather face. The sand layers varied in thickness from approximately 1.25 cm to 2.5 cm for the tests on the parallel breakwater, but was 2.5 cm for the oblique breakwaters. Three wave frequencies (f) were used for each breakwater configuration, those being 1.75 Hz, 2.00 Hz and 2.20 Hz. The respective deep-water wave heights for these frequencies were 3.8 cm (for the parallel breakwater and for $\beta = +45°$, but 3.56 cm for $\beta = 145°$), 5.1 cm and 2.8 cm. For $f = 2.00\ Hz$ *and* $2.20\ Hz$, deep-water conditions existed over the edge.

For all runs, the angle between the wave front and the ledge of the bed was approximately 26°. Furthermore, the far-field waves, where the water depth was 25 cm, were all approximately in deep water. Using Snell's law, the wave angles at the structure were approximately 25° for each frequency.

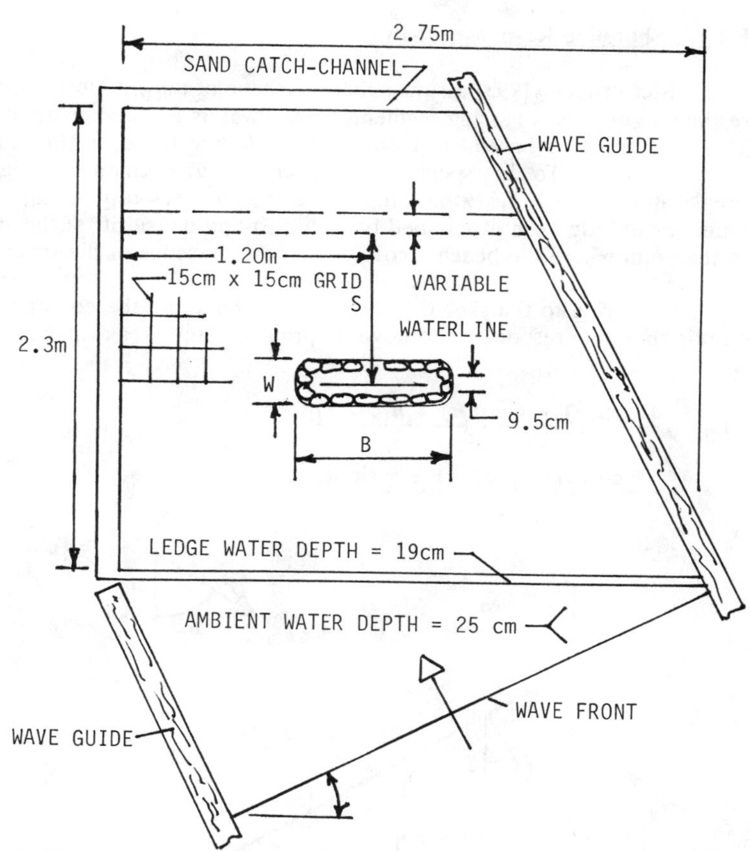

Figure 1. Plan Sketch of the Parallel Breakwater Model and Beach

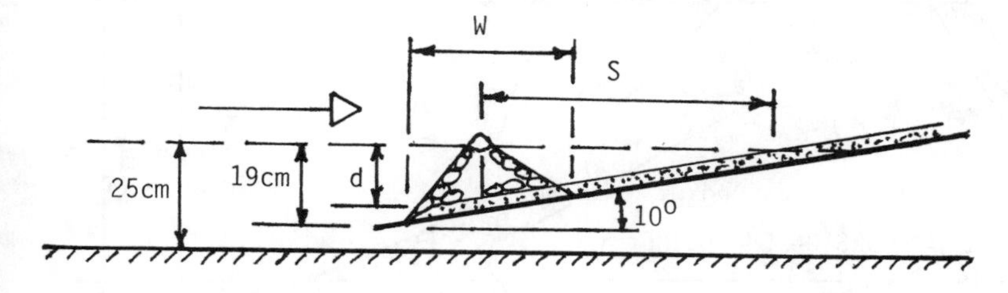

Figure 2. Profile Sketch of the Parallel Breakwater Model and Beach

Elliptic Shoreline Response Analysis

McCormick (1993) assumes that the shape of the equilibrium shoreline responding to waves passing segmented breakwaters is approximately elliptic. This assumption is based on photographs including those of Hardaway and Gunn (1991). To be specific, the pocket bays formed by segmented breakwaters strongly resemble ellipses. For shorelines responding to single units, the elliptic form is assumed both updrift and downdrift of the structure to the point where the beach recovers from the presence of the structure.

Referring to the sketch in Figure 3 for notation, the equation for the elliptic shoreline responding to waves approaching the breakwater at an angle α is

$$1 = \frac{[x\sin(\alpha) + y\cos(\alpha) \mp h\cos(\alpha)]^2}{a^2} + \frac{[x\cos(\alpha) - y\sin(\alpha) \pm h\sin(\alpha)]^2}{b^2} \tag{1}$$

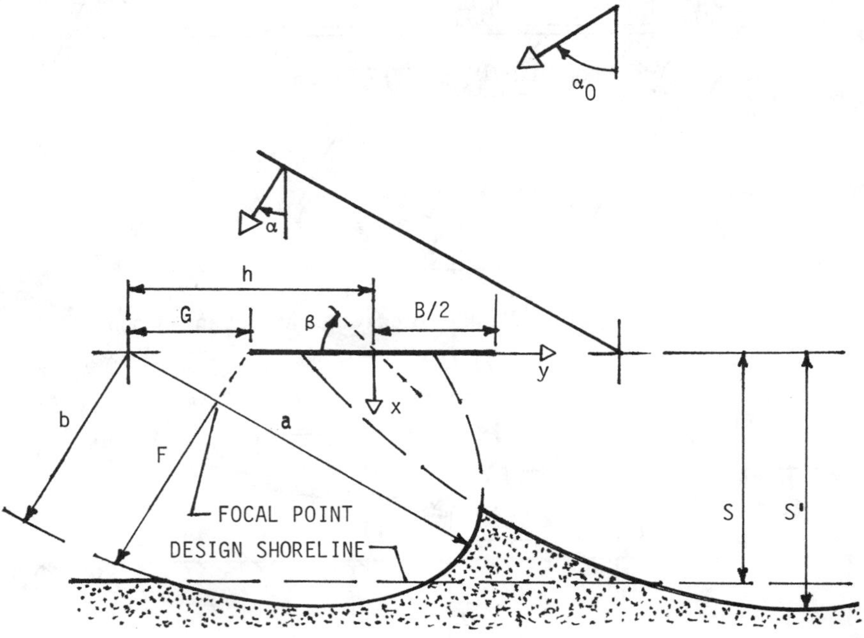

Figure 3. Notation for the Elliptic Shoreline Response Analysis

where a and b are the respective semi-major and semi-minor axes, and h is the distance between the center of the ellipses and the center of the breakwater (the origin of the x-y axes). The upper and lower signs (±) in equation (1) refer to the updrift and downdrift shorelines, respectively. As sketched in Figure 3, the line coincident with the focal length is assumed to pass through the head of the breakwater for any wave angle. The relationship between the focal length and the axes is

$$F = \frac{b^2}{a} \tag{2}$$

while that relating the axes and the "gap", G (the distance between the head and the center of the ellipse), and the axes is

$$a = \sqrt{G^2\cos^2(\alpha) + b^2} \tag{3}$$

The ratio of the semi-minor axis (b) and the design distance (S) between the cap of the structure and the shoreline is

$$\frac{b}{S} = \frac{1}{\cos^2(\alpha)}[\ \cos(\alpha) + 0.2\frac{(H_0/L_0)}{m}\ ]\ \sin(\chi\sqrt{H_0/L_0}) \tag{4}$$

where the design beach slope in the lee of the breakwater is

$$m = \frac{d}{S} \tag{5}$$

and where d is the design water depth beneath the cap of the structure. In equation (4),

$$\chi = -1.92[\frac{S}{B\cos^2(\alpha)}]^2 + 9.92[\frac{S}{B\cos^2(\alpha)}] \tag{6}$$

The minor axis is, then, a function of the deep-water wave steepness (H_0/L_0), the design beach slope (m) and the wave angle (α).

The last relationship in the analysis is that which relates the "gap" (G) and the semi-minor axis (a). From McCormick (1993), that relationship is the following:

$$\frac{G}{b} = \cos(\alpha)\ \{\ \exp[\ \ln(\mu) + \sigma\ln(\frac{H_0/L_0}{m\cos(\alpha)}) - \nu\ (\frac{H_0/L_0}{m\cos(\alpha)})\ ]\ \} \tag{7}$$

where

$$\ln(\mu) = 19.4 \tanh(0.91 \frac{S}{B \cos^2(\alpha)}) \tag{8}$$

$$\nu = 20.0 \tanh(0.99 \frac{S}{B \cos^2(\alpha)}) \tag{9}$$

and

$$\sigma = 17.0 \tanh(0.59 \frac{S}{B \cos^2(\alpha)}) \tag{10}$$

The analytical method described in this section is now applied to the experiment described in the previous section.

Experimental and Analytical Results

Each breakwater orientation-wave property combination was studied for a duration of 6 hours. This duration was chosen to insure near-equilibrium conditions. At the end of each hour of the study (hours 1 through 6), the wave generator was stopped, and a photograph of the beach was taken. It was found that there was little change in the shoreline between hours 5 and 6.

For each breakwater model configuration, the three wave conditions in Table 1 were studied.

TABLE 1
(Wave and Sand Conditions)

	Frequency (Hz)	Height (cm)	β(°)	Sand Thickness (cm)
(a)	1.75	3.80	0	1.25
(b)	1.75	3.80	+45	2.50
(c)	1.75	3.56	-45	2.50
(d)	2.00	5.10	0	1.25
(e)	2.00	5.10	+45,-45	2.50
(f)	2.20	2.80	0,+45,-45	2.50

The values of the distance from the cap to the shoreline (S), the overall breakwater length (B) and the water depth (d) at S for the various orientations are presented in Table 2. Also in that table are the "effective" values of B and S (denoted by the subscripts *up*, for updrift, and *dn*, for downdrift). These effective values are used since the stone used to construct the breakwater model was graded at the angle of repose (approximately 40° to the horizontal), as sketched in Figure 4. Hence, the "updrift" and "downdrift" values of B used in the analysis are the average valued at the mid-depth of the stone on the heads. These values are obtained from

$$B_{up,dn} = B - \frac{d}{\tan(40^{o})} \Big|_{up,dn} \qquad (11)$$

That is, for the purpose of analysis, the breakwaters were considered to have vertical walls at the heads at the mid positions of the slopes. Values obtained from equation (11) for the parallel breakwater are the same for both heads. For the oblique breakwaters, because of the geometry sketched in Figure 4, the effective breakwater lengths are different for the updrift and downdrift heads. The values presented in Table 2 are used in the analysis.

TABLE 2
(Experimental and Analytical Conditions)

Condition	S (cm)	B (cm)	d (cm)	B_{up} (cm)	B_{dn} (cm)	S_{up} (cm)	S_{dn} (cm)
(a)	81	81	14.3	64	64	81	81
(b)	76	89	13.4	80	66	48	100
(c)	71	84	12.5	63	75	95	43
(d)	81	81	14.3	64	64	81	81
(e)	76,71	89,84	13.4,12.5	80,63	66,75	48,95	100,43
(f)	71,76, 71	81,89, 84	12.5,13.4, 12.5	66,80, 63	66,66, 75	71,48, 95	71,100, 43

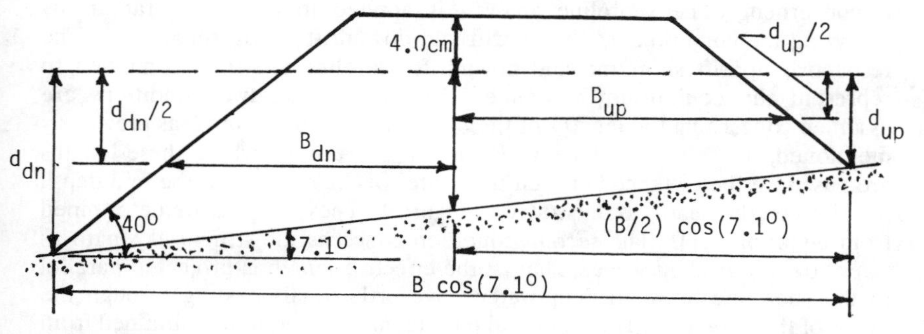

Figure 4. Projected Elevation of the Oblique Breakwater, $\beta = +45°$

For all the model orientations, the cap was approximately 4.0 cm above the still-water level. Hence, from the data in Table 2, the width, W, along the "mudline" was approximately 45 cm, using the 40° angle of repose. As stated earlier in the paper, the sand layer extended to the weather edge of the model.

Experimental and Analytical Data

The experimental data obtained for the conditions described in Tables 1 and 2 are presented in Figures 5 through 7, where the measured equilibrium shorelines are shown for the 1.75 Hz, 2.00 Hz and 2.20 Hz waves, respectively. As previously stated, the data were obtained from photographs taken after 6 hours of running time. In each of these figures, Figures a, b, and c show the data for the respective model angles (β) of 0°, +45° and -45°. For all of the wave and model orientations, salients are shown, with apices either on the line through the passing through the center of the structure or downdrift of that line. Also, it is seen that there are both updrift and downdrift shoreline retreats for all of the experimental conditions. The maximum shoreline retreat is observed for the 2.20 Hz waves, where the deep-water wave steepness is 0.087. The minimum retreat (on average) is observed for the 2.00 Hz waves, where the wave steepness is 0.131. Those waves are near breaking in deep water.

The application of the elliptic shoreline analysis to the parallel breakwaters ($\beta = 0°$) is straight-forward, since this is the condition upon which the equations are derived. As sketched in Figures 5a and 7a, salients are predicted with minor updrift and downdrift retreats. A tombolo is predicted for the 2.00 Hz wave, as sketched in Figure 6a. In that figure, we see both an updrift shoreline advance and a significant downdrift retreat.

The conditions for which $\beta = \pm 45°$ are "off-design" as far as the analysis is concerned. The shoreline analysis is applied to these orientations by analyzing the conditions at the updrift and downdrift heads, separately. The respective updrift shoreline and downdrift shoreline are then combined to represent the equilibrium shoreline. In this process, two conditions are assumed concerning the lengths of the breakwater models. First, as previously mentioned, the "effective" lengths, B_{up} and B_{dn}, are used, where these lengths are based on the distance between the center of the model and the mid-depth positions on the heads, as illustrated in Figure 4. These lengths are determined from equation (11). The second condition concerns the structural length as "seen" by the incident wave. That is, the effective length is projected parallel to the shoreline between deep-water wave orthagonals passing through the heads of the structure. The analytical parallel model lengths are obtained from the following equation:

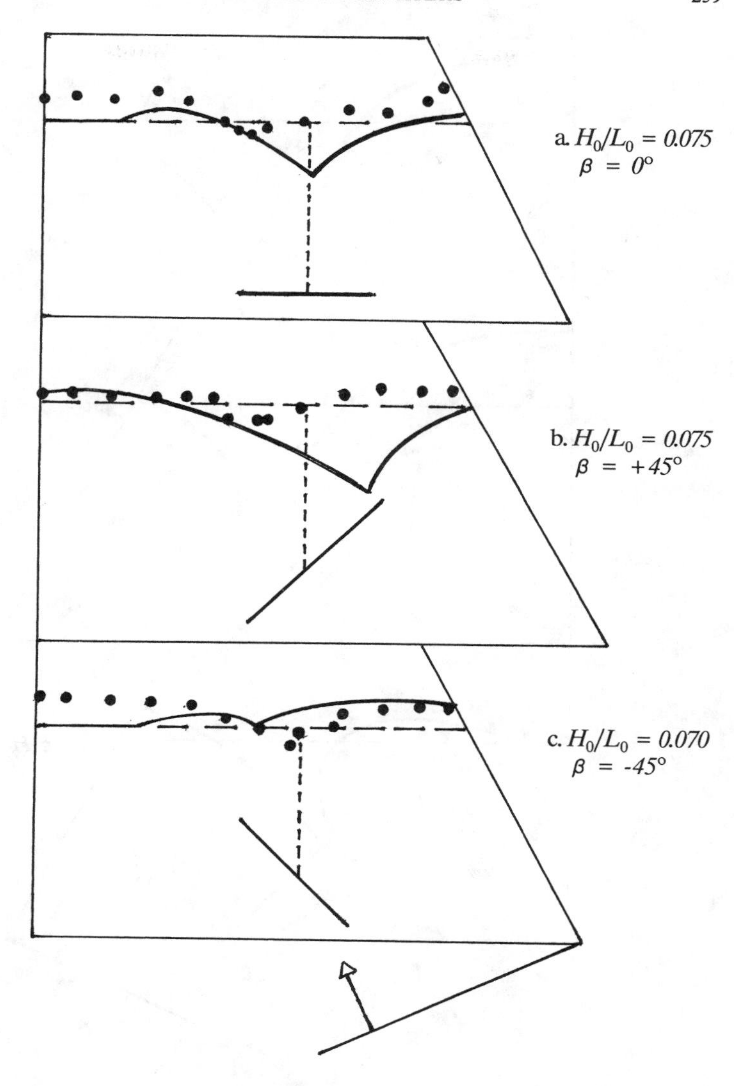

Figure 5. Shoreline Responses for 1.75 Hz Waves

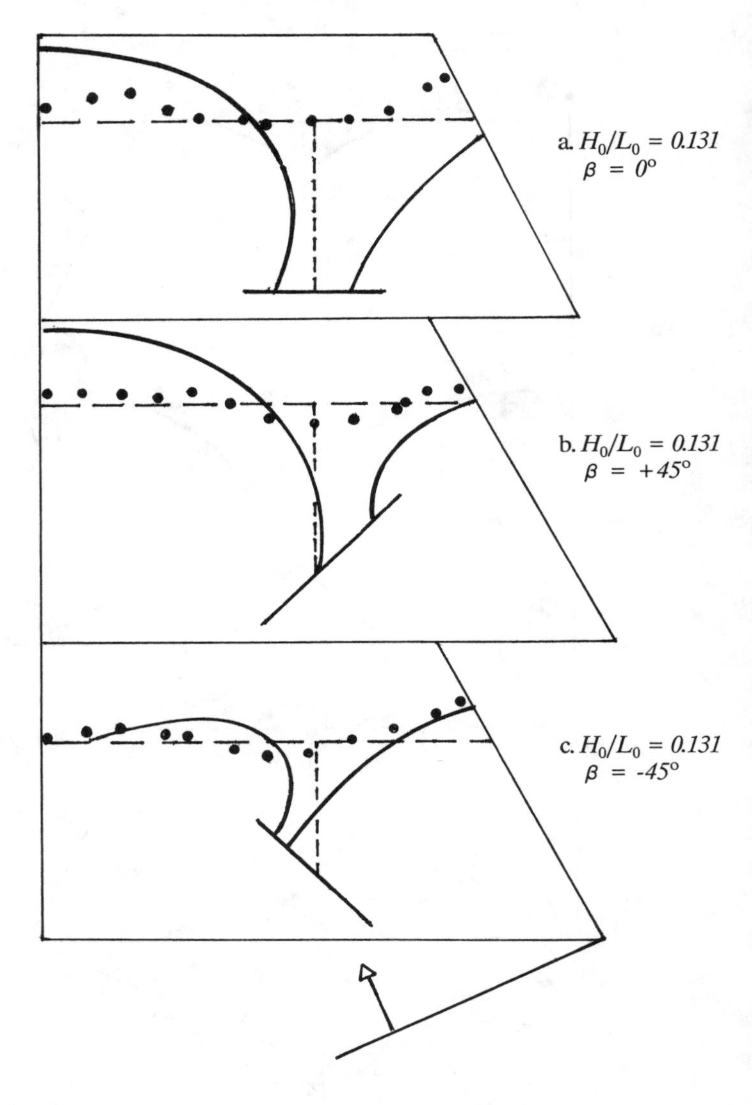

Figure 6. Shoreline Responses for 2.00 Hz Waves

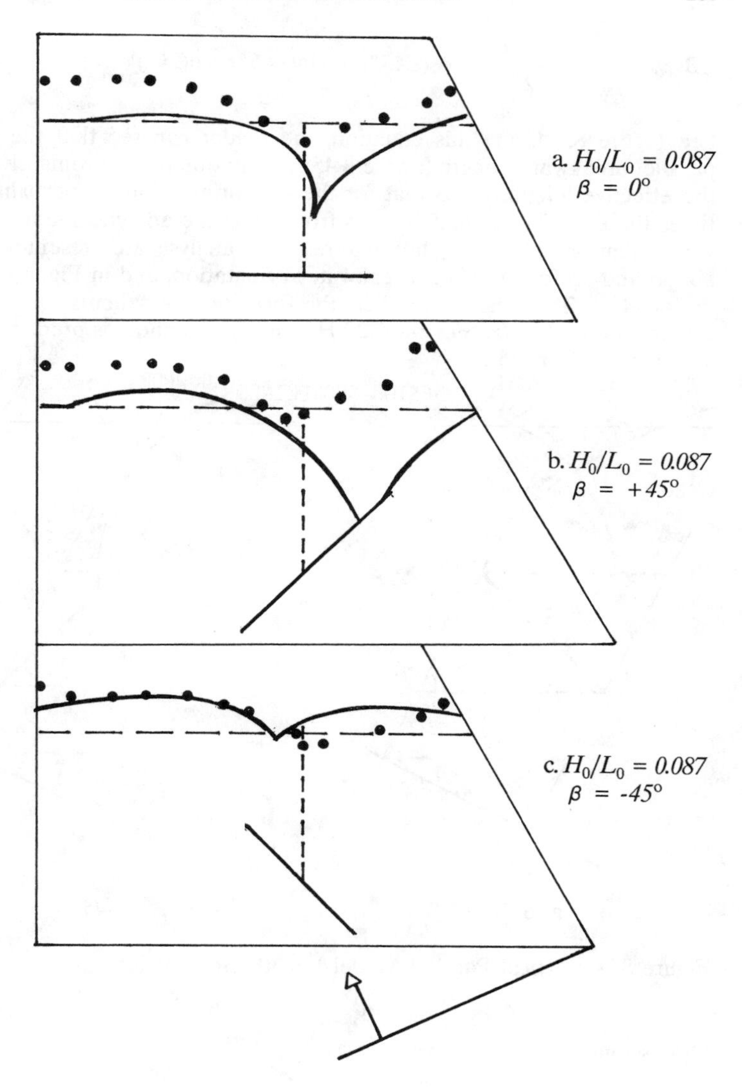

Figure 7. Shoreline Responses for 2.20 Hz Waves

$$B_{\pm 45^\circ} = \frac{(B_{up} + B_{dn})}{2}\{\cos(\pm 45^\circ) + \sin(\pm 45^\circ)\tan(26^\circ)\} \tag{12}$$

See Figure 8. Using this equation, the reader can see that the equivalent parallel breakwater length for the +45° orientation is approximately 1.05 times the effective length, while that for the -45° orientation is approximately 0.36 times the effective length. Results from using the analytical lengths obtained from equation (12) in the shoreline response analysis are presented in Figures 5b, 6b and 7b for the +45° breakwater orientation, and in Figures 5c, 6c and 7c for -45°. The analysis predicts the formation of salients for 1.75 Hz and tombolos for 2.00 Hz. For the 2.20 Hz waves, a tombolo is predicted for +45° and a salient for -45°.

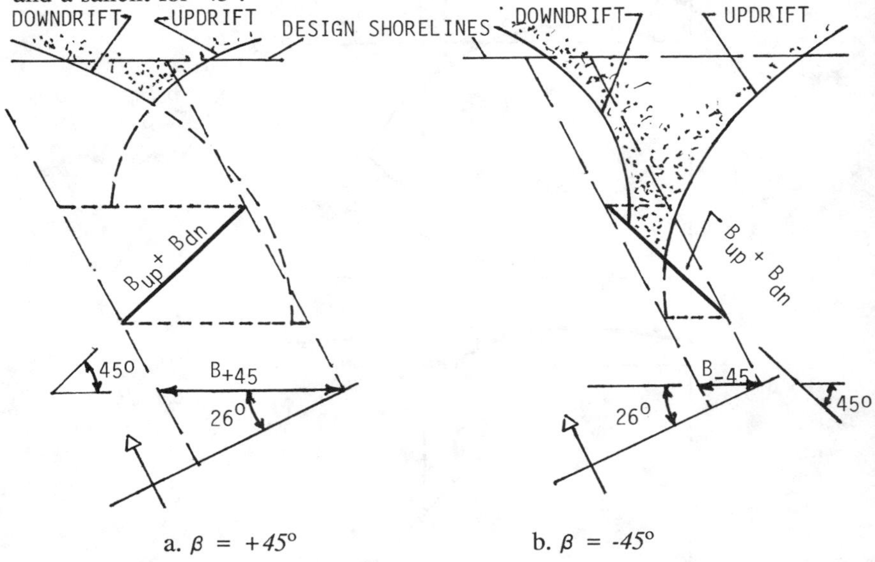

Figure 8. Analytical Parallel Model Length Geometries

Discussion

The elliptic shoreline analysis is based on experimental data obtained using rectangular, impermeable bulkhead-type breakwater models with no overtopping. Since the experiment described herein involves permeable rock models having natural geometries (sides at the angle of repose), the application of the analysis to the present test is very approximate.

Several comments were made to the first author at the symposium concerning the experiment. First, it was pointed out that the sand supply was "finite" in that there was no replenishment of sand in the lee of the models from the updrift surfzone. The finite sand supply, coupled with the significant energy transmission through the structures, probably caused both the formation of the mild salients and the shorline retreat observed for all of the experimental conditions. The second comment was actually a recommendation that the sand grain size be incorporated in the analysis. The inclusion of the grain size in the analysis is, the authors believe, implicitly done. There have been a number of studies, both sponsored and conducted, by the U. S. Army Corps of Engineers that show a definite relationship between mean the grain diameter and the forshore slope, m. Since the slope is included the equations of the analysis, a further inclusion of the grain size not considered to be necessary.

It evident that the analysis must be "fine tuned" by using more experimental and field data to obtain better values of the coefficients in equations (6) through (10). This will be done as more applicable data become available.

Acknowledgements

The authors would like to express their appreciation to Dr. David L. Kriebel of the Naval Academy for his constructive comments and advice.

References

Gravens, M. B., Kraus, N. C. and Hanson, H. (1991). "GENESIS: Generalized Model for Simulating Shoreline Change, Report 2, Workbook and User's Manual", Tech. Rep. CERC-89-19, Report 2 of a series, U. S. Army Engineer Waterways Experiment Station, Vicksburg, MS.

Hansen, H. (1989). "GENESIS - A Generalized Shoreline Change Numerical Model", *J. Coastal Research*, 5(1), pp. 1-27.

Hardaway, C. S. and Gunn, J. R. (1991). "Working Breakwaters", *Civil Engineering*, ASCE, Oct., pp. 64-66.

Hsu, J. R. C. and Silvester, R. (1990). "Accretion Behind Single Offshore Breakwaters", *J. Waterway, Port, Coastal and Ocean Engineering*, ASCE, 116(3), pp. 367-380.

McCormick, M. E. (1993). "Equilibrium Shoreline Response to Breakwaters", *J. Waterway, Port, Coastal and Ocean Engineering*, ASCE, 119(6), in print.

Design of an In-Situ Directional Wave Gage for One Year Deployments

Gary L. Howell [1]

1 Introduction

Wave measurement planning has traditionally forced the engineer to balance the total measurement period against the record interval. This tradeoff results from limitations on data storage or transmission capacity of wave measurement instrumentation. Ideally for design wave statistics, a short measurement interval and long deployment would be possible. Typical performance of internally recording instruments is limited to 3-4 months duration with 4-6 hour record intervals.

This paper describes the design and implementation of a new in-situ, directional wave gage that records hourly directional spectra continuously during a 13-month deployment period. If desired, the gage also can provide real-time data simultaneous with internal analysis and recording. Improved gage reliability and accuracy have also been achieved. Wave height measurements have a resolution of 1-mm and the total system noise is less than 1-cm significant wave height. The principles of operation, data analysis, and field test results of the gage are described in Howell (1992). A description of the hardware and software design is presented here.

2 The case for long deployments.

There are several motivations for longer deployment capability from in-situ wave gages. The requirement arose indirectly as a result of a *top-down* requirements analysis of the wave measurement needs of port and coastal engineers.

Wave data costs Experienced users of wave data want predictable and lower total costs. Novice users often focus on gage acquisition costs in project planning and gage selection. Analysis of costs across a wide range of projects by various

[1]U S Army Corps of Engineers, Coastal Engineering Research Center, Waterways Experiment Station Vicksburg, Mississippi 39180-6199 USA

organizations reveals a consistent pattern. Approximately 70% of the total cost of analyzed data is represented by field activities, and data analysis costs. Often instrument costs represent at little as 15% of total costs. Real cost reductions require improvements in field and analysis costs.

Field costs are proportional to the number of personnel, the length of time on site (including weather delays), and the number of site trips required. Reducing the number of trips by allowing longer periods between gage servicing can result in significant cost reductions. Flexibility in the timing of a field operation to avoid known bad weather periods can minimize the length of field missions.

Data analysis costs are determined by the amount of personnel time required. Fully automated analysis results in the lowest cost. Human error that requires re-analysis of large amounts of data impacts costs and schedules. Traditional instruments place a large burden on the user to manage the recovery, and analysis of data. The user must correctly associate recording media, sites, times, sensor calibrations, etc. with the raw data. As the number of gage sites increase, data handling errors become more frequent. With longer deployments, data management and automated analysis can help lower costs.

Improved wave statistics It is well known that the number of observations strongly affects the confidence of wave parameter *pdf*s. To obtain sufficient observation periods, the record interval is often compromised. In coastal areas, data routinely sampled at hourly intervals has only recently become available. Preliminary indications are that temporal variability is much greater than previously assumed. Statistical estimates of sediment transport and breakwater design waves could be affected by subsampling. This question requires further study.

Extreme event capture For many projects, capturing the extreme event can be more important than detailed coverage of ambient conditions. Inability to access a site to perform routine service has frequently resulted in missing the extreme event because the gage recording media or batteries were exhausted. One reason for the popularity of cable-connected, real-time wave gages derives from the ability to keep them powered and recording continuously. The 13-month deployment capability of the gage described here, permits continuous monitoring even if a scheduled recovery is prevented by weather.

3 Gage design

Long deployment with frequent sampling requires solution of numerous design problems. These include calibration and time keeping stability, biofouling, power consumption, reliability, data storage capacity, and software reliability.

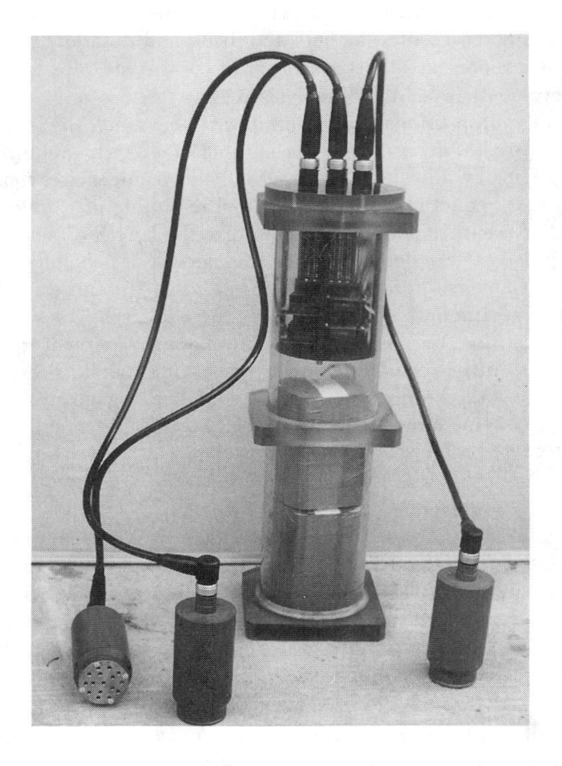

Figure 1: Directional wave gage electronics cylinder and pressure sensors.

3.1 Hardware design.

The gage employs three Paros Digiquartz pressure transducers (Paros, 1976) arranged in a 1.6-m equilateral triangle array. All gage electronics and batteries are contained in a single cylinder approximately 1.3-m long. Figure 1 is a photograph of the electronics cylinder, transducers, and cables. The compact size permits the gage to be deployed in a small, trawler resistant pod. Figure 2 is a photograph of the assembled gage.

Sensor options for compact gages. Small size is the principle reason for the gage's reduced installation costs and long term survivability. Gages based on a single pressure sensor and electromagnetic 2-D current meters, *PUV*s, can be installed in similar sized tripods. However, the current meter is difficult to protect from fishing trawlers, and must be cleaned regularly. Acoustic doppler

Figure 2: Directional wave gage assembled in mounting pod. The outrigger pipes extend into the sediment.

current meters (ADCP) can be installed in small tripods and protected from fishing trawlers. ADCPs provide three components of orbital velocity sufficient to estimate the directional spectrum using techniques similar to the analysis for a *PUV*. However during extreme wave events in shallow water, the entire water column becomes aerated, interfering with the acoustic signal. The design used here is a bottom mounted, short baseline pressure sensor array.

Pressure sensors and wave height accuracy. Wave gages based on pressure transducers have proven to be accurate and reliable in coastal applications. Occasionally, hydrodynamicists voice concerns about pressure transducer measurements because linear wave theory is used to correct the bottom measured pressures to surface displacement. Pressure wave gages are typically employed in intermediate and shallow water depths, and are exposed to waves that clearly violate the assumptions of linear theory. Field wave gages require a theory applicable to directional irregular waves. Common practice for analysis of pressure wave records uses linear theory with the Gaussian random wave model of the sea surface. Estimates of the statistical properties of the sea surface (H_s, T_p, D_p, etc.) are computed by applying linear wave theory to each frequency component of the measured pressure frequency spectrum. This *stochastic linear system theory* approach is different from applying linear theory deterministically to an individual wave.

Many careful field experiments have compared pressure measured wave statistics with buoys, staffs, photo-poles, and other instruments. Results show that when pressure time series data are properly analyzed, and linear theory compensation is applied in the frequency domain, excellent agreement is obtained. Bishop and Donelan (1987) provide a comprehensive review of these experiments, and report on a definitive laboratory study of pressure wave measurements. They confirm that stochastic linear system theory is adequate to compensate pressure measurements and gives reliable estimates of surface wave heights.

Short baseline and wave direction. The second factor contributing to the small size of the gage is the short spacing of the array. Past wave gage arrays have been on the order of wave lengths. Howell (1992) gives details of the short base line with field test results of the effect on directional wave estimates. Using currently available analysis techniques, the short baseline gage gives estimates of mean wave direction comparable to large baseline gages. Mean wave direction can be accurately estimated for each frequency band. As with all three component gages, estimates of directional spreading at a single frequency are not as good as those obtained from arrays with more than three sensors. The implications for typical project measurement scenarios are that the gage accurately resolves multidirectional sea states when waves from different directions also have different periods. This case is typical of the mixed sea and swell predominate at most coastal sites. Examples of cases that may not be accurately resolved would be gage sites in zones of high wave reflection, such as near a breakwater, or high diffraction such as near an offshore island.

Electronics. The instrument architecture exploits advances in microcomputer hardware and software technology. Low power consumption hardware is employed for sensors, data acquisition, internal microcomputer, and data storage systems. Power management software and internal data analysis permit long term deployment with physically small batteries and data storage. Figure 3 is a block diagram of the electronic hardware.

Hardware support for power management is provided by software controllable switches that can power down subsystems, and clocks. The Paros sensors and interface card can be powered off when not required. The EPROM storage subsystem is only powered when data are read or written. The main CPU clock can be stopped by a software command, and automatically restarted by a hardware interrupt. The only hardware that operates continuously is the low power real-time clock.

The Paros Interface Card uses a novel approach to convert the frequency modulated output of the Paros transducer to a digital number. Previous wave gages that use direct frequency counting have been unable to obtain high resolution and high sample rates. The technique employed here is a high resolution digital period measurement which provides 1-mm of resolution of sea water pressure with

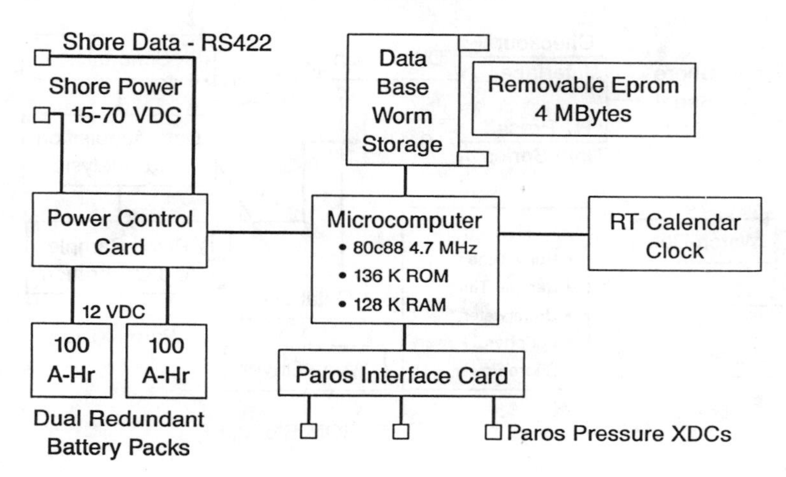

Figure 3: Hardware block diagram.

a maximum sample period of 200-msec. High resolution contributes to the low total system noise of the instrument.

Data storage for the instrument is provided by EPROM memory chips, packaged in removable modules. Once data are programmed into the blank EPROM, it is maintained without power until the modules are erased. This technique provides extremely low bit error rates, and eliminates problems with moving parts or battery backup experienced with previous systems.

3.2 Embedded software

Performance of the gage relies heavily on the embedded software. Modern software engineering techniques were employed in the software design. The design is implemented with the ADA high level programming language. Figure 4 is a simplified block diagram of the code.

Object oriented software design. The code must simultaneously perform real-time data transfer to shore, and internal data analysis and storage. A multitasking design is employed with reusable software *objects*. This design approach minimizes code size, and improves reliability. An important object type is the **Event**. Tasks can wait for events and trigger events. A general purpose **Scheduler** can trigger an event at a preset time, or on a recurring schedule. The **Burst**

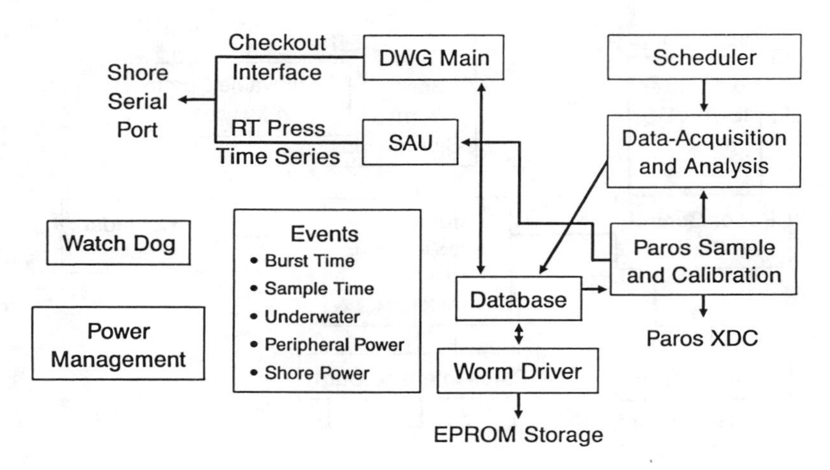

Figure 4: Embedded software block diagram.

Time and **Sample Time** events are handled in this way.

Database. Previous internally recording instruments record data in unformatted blocks of raw data. Although simple to implement, this technique is difficult to adapt to varying sample requirements and sensor suites. Data transfer to analysis programs is time consuming and error prone .

To prepare for gages with deployments up to 13 months, improved data management was made a design requirement. One gage deployed for 12 months will generate 8760 analyzed wave records. Each record consists of autospectra, quadspectra, and water levels. With a design requirement of up to 50 deployments per year, a data management plan is critical.

The approach is to integrate the gage setup, checkout, and recording with the central database used to analyze, report, and store the data. Figure 5 is a cartoon that illustrates the operation. Conceptually, rather than the gage data driving the analysis software, the desired analysis result *pulls* the gage(s) data.

In operation, the site characteristics and measurement requirements are first entered into the database. Independently, calibrations of all pressure transducers are maintained in the database. A gage and set of pressure sensors are then assigned to the site. This data is used to initialize the gage's database on it's EPROM modules. The modules are then installed into the gage.

The gage user interface can then identify itself, its intended site and location,

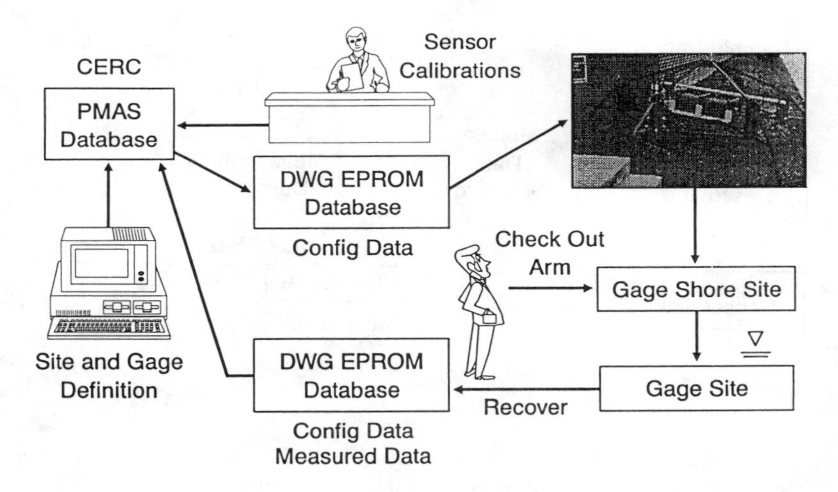

Figure 5: Integrated gage setup, deployment, and analysis system. Parameter definitions in the central database initialize the gage database. When the gage EPROM returns from deployment, data are automatically read, and analyzed.

and the serial numbers and calibration coefficients of its transducers. Built-in-test procedures allow correct operation of the gage to be confirmed before shipping to the project site. On-site, the gage software steps the technician through a checkout procedure that helps assure that the correct gage and sensors are assembled and operational. Part of this procedure requires that the gage correctly sense air pressure from all sensors. This check has been successful in detecting many types of human error. Upon successful completion of all checks, the gage may be **Armed**. The **Armed** state causes the gage to begin sensing for an underwater condition once each hour. Once underwater, data acquisition begins according to the sampling parameters in the database. There are no switches or controls on the gage, not even an on/off switch.

Upon recovery of the gage the EPROM modules are placed in a reader attached to the main database computer. No additional data is input, as all required data has been previously recorded in the database. The computer transfers the data to the main database, associating it with the correct project site and location, and automatically performs the required post analysis and reporting.

Database schema. The gage metadata is a subset of the data definitions used by the Prototype Measurement and Analysis (PMAS) system. PMAS is the

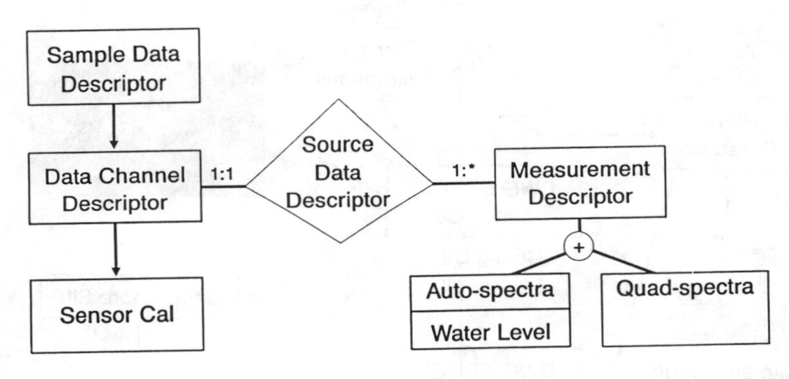

Figure 6: Extended Entity Relationship Set (EER) schema for the gage internal database. Relations are a subset of the analysis computer database schema.

main database for all field measurement and analysis activities at CERC. PMAS uses a relational data base schema based on the Entity Relationship Set (ER) methodology developed by Chen (1976). Extensions for object oriented design by Markowitz and Makowsky (1990) are incorporated. Figure 6 is a ER schema diagram for the gage database. The ER set **Source Data Descriptor** links all time series data required for an analysis data object. Either producer or consumer tasks can enter the schema and obtain information required to create or access objects. For the standard wave gage, analysis objects are auto-spectra, quad-spectra, and water levels. However, the design is completely flexible, allowing additional sample data or analysis objects to be added or removed without impacting existing code. Not shown are housekeeping relations, such as an instrument log that records the time, date, and operator name for important gage events.

Analysis code. Howell (1992) gives theoretical details of the internal data analysis. The implementation in the code is described here. The standard sampling rate for internal analysis is 1-Hz. Data samples are calibrated to pressure in units of depth of seawater with a resolution of 1-mm. A standard water density and atmospheric pressure are assumed. These values may be adjusted in post-processing if required. Welch's method of spectral analysis (Welch, 1967) is used to improve computation time and spectral stability. Other benefits of Welch's method are flexibility in setting record length by increments of the window length (128-sec here), and facilitating a conservative approach to data editing. A burst of data errors can be handled by omitting the erroneous segment rather than choosing between an ambitious data correction procedure or loss of the entire record. A 1792-sec record is divided into 14, 128 second segments. Extensive quality control checking is performed. Wild point editing is permitted for only one successive point and no more than 10 total points per segment. Any segment failing quality control is excluded from further processing. The segments are demeaned and prewhitened with a 10% cosine bell window. Two mean water level estimates are reported. They are computed by the average of the means of the first three and last three segments. This corresponds to a tide measurement with a 6.4-min window average approximately every one half hour. Applying 50% overlapping, 27 segments are used to generate cross-spectral estimates. The Equivalent Degrees of Freedom (EDF) is 39. Co and Quad spectral estimates are subjected to additional quality control checks which make use of the expected similarity of autospectra between individual sensors. The resulting data are saved to the database using data saving assumptions described in Howell (1992).

Implementation requirements to meet power consumption goals require that this code minimize both execution time and size. The majority of the execution time budget is dominated by the Fast Fourier Transform (FFT) routine. Variable range, fixed point arithmetic with 32-bit precision is used for all analysis code computation including the FFT. This provides execution time faster than a floating point coprocessor without the attendant power consumption. The FFT routine uses precomputed sin, cos, and bit reverse tables. Segment computations are carried out as segments become available and without interrupting data acquisition or real-time data transmission.

Real time data access. The wave gage has built in support for optional real time data access. Real time access is automatically enabled by applying external power to the gage. Internal data acquisition, analysis, and storage continues independently of real-time access. Data are output by an RS422 serial interface. Cables up to 2-km long may be driven by the interface, however any external device (eg. telemetry or satellite buoy may be connected). The data format is compatible with a CERC standard known as SAU format. Output data are the water pressure samples from each transducer calibrated in mm of seawater as described above. The allowable range is 0 to 32,767-mm. The real time data is sampled at a fixed rate of 5-Hz. Options in the shore based Remote Transmission Unit (RTU) or PC allow for decimation to lower sample rates.

Power management is the process of controlling the power to devices and stopping the CPU and other devices when not required. It is one of the key techniques which permits long deployments. Because some devices serve multiple functions, and competing or overlapping requirements of different code modules, power management can become complex. Separate power management tasks and events are used to coordinate requirements. The tasks implement a classic multiple access, resource allocation algorithm.

Error checking and failure management. To achieve year long deployments, the embedded code must be highly robust, and have means of dealing with expected and unexpected errors, transient external events, and marginal or failed hardware. Multiple layers of error handling are used. The significant amount of built in checks required by the ADA language definition (ANSI-STD-1815A-1983) are left completely enabled in all code modules. All exceptions are managed at the lowest level possible, and soft recovery achieved if possible. Data acquisition and analysis codes have provisions for dealing with one or more failed sensors. Special hardware prevents a failed cable on one sensor from draining power from the batteries. A hardware watchdog timer is provided which automatically performs a hardware reset if a regular check-in pulse is not triggered by the software watchdog timer task. Forced resets by a software exception handler or the hardware watchdog timer are detected by the initialization code. System initialization and checks are adjusted to preserve operational continuity.

4 Experience and conclusions

System noise and operational depth. It is well known in the analysis of subsurface pressure gages that the high frequency response is limited by water depth and instrument noise. The compensation function increases exponentially with frequency and multiplies the noise into physically unrealistic values for the high frequency portion of the wave spectrum. Common practice is to truncate the spectral estimate at some high frequency value based on water depth. See, for example, Lee and Wang (1984).

Reduced system noise was a design goal of the gage. A complete discussion of the sources of system noise is beyond the scope of this paper. A useful measurement of total system noise of any wave gage can be made by recording a wave record from the instrument while mounted in a still water stand pipe. Data are analyzed by the standard analysis code. An ideal wave gage should report $H_s = 0$. The actual value reported is defined here to be the total system noise. The new gage system noise is typically 1-cm. This is an order of magnitude improvement over previous designs. Due to improved system noise, the gage has successfully operated in 19-m water depths. Stable spectra were obtained with a 0.25-Hz cutoff.

Project use. To date, the gage has been used for four project applications. The last deployment at Redondo Beach, Ca. used four gages in a six month deployment. Analyzed data were available from all four gages within one week after recovery.

User response Efforts to eliminate human error in the setup, deployment, and analysis of gages have been very well received by users. During the design phase, there was a concern that the level of checking may be too extreme. However, based on experience to date, there are plans to increase the amount of internal checking and controls in both the central database and the gage. This technique should be extended to all instruments. Microcomputer and database technology provide the means to reduce the human error associated with field measurements, and create greater confidence in data successfully acquired.

The internal analysis and automated post-analysis has been very successful. The ability to have an analyzed data report of months of data available several days after gage recovery has been very popular.

The long deployment capability has allowed the gage to be used at project sites where wave measurements had been previously impossible. Many areas are exposed to such extreme conditions that safe access may only be possible during a short period each year. The long deployment capability allows gages to be safely deployed and recovered with no loss of data coverage during the extreme weather season.

Conclusion. Modern microcomputer technology and software engineering have been used to develop a directional wave gage which meets the performance and economic requirements of port and coastal engineers. Limitations on deployment duration of in-situ gages have been effectively removed. The technology will make directional wave statistics as available as environmental data for other branches of civil engineering.

Acknowledgments

The design team for this project included William E. Grogg, William M. Kucharski, and Michael D. Carpenter. Jody P. Landreneau implemented hardware improvements in the latest version. The work described here was performed under the Nearshore Directional Wave Measurement Technology work unit, of the U.S. Army Corps of Engineers, Coastal Research Program. Permission to publish this paper was granted by the Chief of Engineers, U.S. Army Corps of Engineers.

References

American National Standards Institute "Reference Manual for the ADA Programming Language,"ANSI/MIL-STD-1815A-1983, Feb. 17, 1983.

Bishop, C.T. and Donelan, M.A. "Measuring Waves with Pressure Transducers," *Coastal Engineering*, 11 (1987), pp 309-328.

Chen, P.P. "The Entity-Relationship Model - Toward a Unified View of Data," *ACM Tran. Database Systems,* Vol. 1, No. 1, Mar, 1976, pp. 9-36.

Howell, G.L. "A New Nearshore Directional Wave Gage," Proc. of the 23rd International Conference on Coastal Engineering, ASCE, 1992,

Lee,D.Y. and Wang,H. "Measurement of surface waves from subsurface gage," Proc. of the 19th Int. Conf. on Coastal Engineering, ASCE, 1984.

Markowitz, V.M. and Makowsky, J.A. "Identifying Extended Entity-Relationship Object Structures in Relational Schemas," *IEEE Trans. on Software Engineering*, vol. 16, no. 8, pp 777-790, August, 1990.

Paros, J.M. "Digital Pressure Transducers," *Measurements and Data*, Issue 56, vol. 10, no. 2 March-April, 1976.

Welch, P.D. "The Use of Fast Fourier Transform for the Estimation of Power Spectra: A Method Based on Time Averaging Over Short, Modified Periodograms," *IEEE Trans. Audio and Electroacoust.*, vol. AU-15, pp. 70-73, June 1967.

Design and Testing of the NDBC Wave Processing Module

Joel N. Chaffin[1], William Bell[1], Kathleen O'Neil[1], and Chung-Chu Teng[2]

Abstract

The U.S. National Data Buoy Center (NDBC) has developed a new wave data acquisition and processing system, called the Wave Processing Module (WPM), using state-of-the-art technologies and the experience accumulated at NDBC over the past 20 years. The WPM is capable of acquiring either directional or nondirectional wave data through various sensors and transmitting the data to various host payloads which, in turn, transmit the data to a shoreside system. The WPM is designed to be compact, inexpensive, expandable, and able to process at high speed. This paper presents the system design of the WPM and describes the WPM prototypes, laboratory tests, and a field test of a prototype WPM deployed on a 3-meter discus buoy. Results from laboratory and field tests show the validity of the WPM design and implementation.

Introduction

Wave data processing techniques onboard data buoys, along with increasing data accuracies, have been continuously evolving in concert with new hardware and software technologies. During the past 20 years, the U.S. National Data Buoy Center (NDBC) has developed a series of payload and wave measurement systems to acquire, process, and report wave data.

Currently, NDBC has three different payload systems in use: the General Service Buoy Payload (GSBP), Data Acquisition Control and Telemetry (DACT) payload, and Value Engineered Environmental Payload (VEEP). These payloads consist of microprocessor-based data acquisition systems that interface with sensors and a transmitter. Each payload has a dedicated wave-measuring subsystem. These subsystems are not interchangeable. The GSBP and VEEP payloads can measure and report nondirectional wave data only, while the DACT can measure and report directional and nondirectional wave data. Also, the payload software is tightly coupled with the wave-measuring software so that upgrading the wave processing software means modifying the payload software, with attendant lengthy software testing. A

[1] Computer Sciences Corporation, Stennis Space Center, MS 39529-6000 USA

[2] National Data Buoy Center, Stennis Space Center, MS 39529-6000 USA

future payload being developed by NDBC will incorporate distributed processing by independent modules and does not require extensive payload software modifications for changes in the modules.

A wave data acquisition and processing system, called the Wave Processing Module (WPM), was developed by NDBC in keeping with the new payload distributed-processing philosophy. The WPM was designed so that wave processing hardware and software can evolve independently of payload hardware and software as requirements and technology change. Conceptually, the WPM consists of a separate hardware and software system with a serial data interface to the host payload. The use of a single wave processing module with existing and future NDBC payloads is intended to standardize wave data acquisition and processing to produce a more uniform and higher quality data product.

The general design requirements for the WPM include:

- capability to measure and process directional and nondirectional wave data using various sensors,
- compatibility with a variety of host payloads,
- in-house design and maintenance using commercially available components,
- compact, inexpensive, expandable, low power consumption, and able to process at high speed.

In this paper, the hardware and software design of the WPM, the prototype WPM, laboratory and on-site testing, and results from a field test are described and presented.

System Design

The WPM consists of four components: processing hardware, interface hardware, processing software, and system software. The block diagram in Figure 1 illustrates these WPM components and their interfaces with each other.

Processing hardware, which is the core of the WPM, consists of a single-board computer platform for data processing, an A/D converter for digitizing the sensor signals, and storage memory. Processing software, loaded in the processing hardware, provides the following functions: accepts time-series sensor data in buffers, processes the acquired data, and encodes and formats the processed data for transmission. Interface hardware connects the processing hardware to sensors and the payload system. It provides the signal conditioning to transform sensor output signals to voltage levels compatible with the A/D converter on the processing hardware. System software provides communication and control between the WPM, host payload, and wave sensors. It also provides an interface between users and the WPM, which allows users to set or modify operating parameters and configurations (i.e., different sensor types, different message types, etc.), through a setup terminal. The WPM system software also provides onboard diagnostic tests and onboard determination of hull magnetic coefficients to simplify the onshore and at-sea testing and maintenance of the system. Additionally, a user friendly interface is designed to simplify the setting and control of configuration/system parameters.

The WPM acquires data from different sensors via an A/D converter, stores the acquired data, processes the raw data, formats the data, and upon interrogation by the host payload, transmits the data to a host payload. The WPM requires a nominal 12-V input (9 to 15 VDC) and converts this input to voltages required by the processor system, signal conditioning, and sensors. The RS-232 serial interface provides communication to the host payload and to a terminal used to set operating modes and parameters.

A Tattletale Model-7, an off-the-shelf single-board computer produced by Onset Computer Corporation, is used as the processing hardware for the WPM. The Tattletale Model-7 features a 32-bit, 16-MHz 68332 processor; 2-Mbyte pseudostatic RAM; 256 Kbyte of flash EEPROM for program storage; tunable system clock; real-time clock; and a 4-channel, 12-bit A/D converter. Since NDBC's buoys must perform unattended for long time periods and report data hourly, power consumption is a vital design criterion for the WPM. The Model-7's power consumption is very low, drawing only 60 mA for intensive computational operations while running at 16 MHz. Its clock frequency can be changed dynamically, allowing data acquisition and control functions to be performed at 320 KHz, which only draws 6 mA. The Model-7 also has a sleep mode in which current drain is less than 1 mA. Table 1 shows the hardware comparison between the WPM and other NDBC payload systems.

The wave processing software installed in the WPM was conceived and funded by NDBC and implemented by MEC Systems Corporation (Earle & Joseph 1989) based on NDBC specifications. The MEC software, designated the Wave Record Analyzer (WRA), was written in C language and designed in a modular format to make future modifications and additions as easy as possible. Since the Model-7 uses a prefix floating point library developed by the manufacturer, the original WRA, designed to use standard infix ANSI, was modified. Currently, the WRA software has six different system configurations representing different wave information (i.e., directional or nondirectional waves), different sensor configurations (e.g., Hippy, magnetometer-only, accelerometer, etc.), and different transmission message formats (i.e., long or short message, 31 or 47 frequency bands). These system configurations can be selected through the system software.

The onboard diagnostic test function, which provides software test functions for the WPM, includes a step-voltage tester and directional wave simulator. The step-voltage tester generates reference voltages to check if the A/D converters work properly. The onboard directional wave simulator generates the digitized sampled data from a heave-pitch-roll sensor (e.g., a Hippy sensor) and a magnetometer to simulate the buoy motion in four sinusoidal wave trains having different wave heights, periods, and directions. These data are then processed by the WPM processing software and transmitted in the normal manner. The onboard directional wave simulator can be used to test WPM software and to verify the correctness of system parameters, formatting, transmission, and system setup.

The onboard determination of the hull magnetic coefficient function that is built into the WPM determines the magnetic effects of the buoy hull and electronic systems and develops coefficients to correct the magnetometer data for these effects (Remond & Teng 1990). Usually, this software function is performed prior to deployment of the buoy, then these coefficients are loaded into the WPM for correct onboard processing.

Prototype WPM Systems

Prototype WPM systems were built using the Tattletale Model-7 as the processing hardware and the WRA as the onboard processing software. An 8"x10"x4" aluminum box was used as the system box to house the processing hardware, interface hardware, and two sensors (a magnetometer and an accelerometer). Figure 2 shows a sketch of the WPM system box and hardware setup.

The Tattletale Model-7 includes four analog input channels for data acquisition. The Magnetometer-Only (MO) directional wave measurement system, which requires only three input channels (i.e., heave and two magnetometer signals), was selected to be the wave measurement configuration for the first WPM prototype system (abbreviated as the WPM/MO system). More details about the MO technique can be found in Steele and Earle (1991) and Teng *et al.* (1991). Since 5 input channels are needed for a traditional directional wave measurement system using a pitch-roll sensor, the number of A/D channels was increased to 11 for the second prototype version by multiplexing one A/D channel. The second prototype WPM system was the WPM/Hippy system which includes a Datawell Hippy heave-pitch-roll sensor and a magnetometer as the wave measurement sensors. The prototype WPM systems were tested originally with the VEEP payload system. Since the RS-232 communication interface is used, eventually the WPM can be interfaced with other payloads.

Originally, the WRA onboard processing software was designed to handle wave data with 31 variable frequency bandwidths (Earle & Joseph 1989). Due to the demand for high precision and the capability of the powerful WPM hardware, NDBC has expanded the WPM wave data format. Currently, the WPM collects 4,096 samples from the connected sensors for 40 minutes at a sampling rate of 1.7066 Hz, converts this data into engineering units in floating point format, and stores this data in arrays. From this 40-minute record, the WPM performs three Fast Fourier Transforms (FFT), generating 47 frequency bands with different bandwidths (see Table 3). The 40-minute record of 4,096 samples of sensor data is used to provide high resolution spectra for swells covering the 0.03 Hz to 0.09 Hz band in 0.005 Hz bandwidths. The second FFT is performed on the most recent 20-minute record and provides data on wave frequencies from 0.1 to 0.35 Hz in 0.01 Hz bandwidths. The third FFT is performed on the latest 10-minute record and provides wave data on the high frequency portion, which is strongly affected by local winds. In addition to these data, the WPM examines the array of time series sensor data, determines the mean, minimum, maximum, and standard deviation of the sensor data, and transmits the data back to shore.

Laboratory and Onsite Testing

One of the WPM design requirements is high speed processing. To test the processing speed of the WPM, the WRA wave processing software was run on both the WPM and two PC's using simulated time-series wave data with 4,096 data points. Table 2 provides the processing times for the prototype WPM and two different desktop PC systems (i.e., 80386-based and 8088-based). This table shows that the WPM has high-speed processing power which will allow new data processing techniques to be included in future WPM system designs without significantly affecting the overall time budget for NDBC payloads.

The prototype WPM systems were fully tested by using two NDBC Directional Wave Simulators: one is the Regular Directional Wave Simulator which can simulate two sinusoidal wave trains with different wave heights, periods, and directions, and the other is the Random Directional Wave Simulator. Both simulators feature a PC with an 8-channel D/A plug-in board and can generate heave, pitch, roll, and the two horizontal components of a magnetometer. The following four phases for the simulator tests were conducted:

(1) Wave Simulator -> WPM/MO -> PC
(2) Wave Simulator -> WPM/MO -> VEEP -> PC
(3) Wave Simulator -> WPM/MO -> VEEP -> GOES -> PC
(4) Wave Simulator -> WPM/MO -> VEEP -> GOES -> DG

in which PC represents shoreside processing using a desktop personal computer, DG represents NDBC routine shoreside processing using a Data General super mini-computer, and GOES (Geostationary Orbiting Environmental Satellite) represents data being transmitted through the GOES system.

In the first test phase, test data generated by the wave simulator were input to the WPM. Processed and formatted data were stored and processed using the shoreside processing software on a PC. In the second phase, the processed and formatted data were transmitted through a host VEEP payload before shoreside processing. In the third phase, processed and formatted data were transmitted to the GOES satellite through the VEEP. After the data were received by the ground station, shoreside processing was conducted using a PC. In the fourth phase, data were transmitted through the VEEP to the GOES satellite to the shoreside system for processing on a super mini-computer, which is the standard NDBC operational data processing.

In addition to the wave simulator tests, the WPM was also tested using the NDBC Ocean Wave Instrumentation Facility (OWIF), which is a Ferris-wheel-type test device. Due to the strong local magnetic field around the OWIF, outputs from the magnetometer were ignored, and only wave heights from the WPM were examined.

Buoy spin tests were performed in the canal at Stennis Space Center after the WPM was integrated into a 3-meter buoy hull. The buoy spin tests are designed to obtain buoy hull magnetic constants, which are used to compensate magnetometer data for buoy hull and electronics effects (Remond & Teng 1990). These tests are performed by slowly spinning the buoy in water while collecting data from the magnetometer and a gyrocompass. After the hull magnetic constants were obtained and loaded into the processing software, direction verification tests were conducted by comparing the buoy azimuths reported from the WPM with those measured by a gyrocompass (with the buoy at fixed headings). The results from these tests described above showed that the overall WPM design and its implementation are correct and acceptable.

Field Test

One prototype WPM/MO was interfaced with a VEEP payload and installed in a standard NDBC 3-meter buoy for its first field test. An NDBC DACT payload with a DWA/Hippy directional wave subsystem was also installed in the buoy, as a reference for evaluating the performance of the WPM/MO system. The test buoy was deployed off the California coast (station 46014) on June 26, 1992. Since that date, both wave systems have acquired and transmitted wave data through the GOES satellite each hour.

Time series plots of the significant wave heights (H_s) and average wave periods (T_a) for both wave systems between 0000 UTC, July 1, 1992, and 2300 UTC, July 8, 1992 (see Figure 3), show good correlation between the two systems. Figure 4 shows the time series plots of the mean wave directions (α_1) and wave energies (C_{11}) at a frequency of 0.20 Hz for the same time frame. Note that the DWA/Hippy system measured pitch and roll angles from a Hippy sensor, while the WPM/MO estimated the pitch and roll angles from the outputs of a magnetometer. As shown in Figure 4, the mean wave directions from the WPM/MO system track those from the DWA/Hippy system quite well. The large difference of mean wave direction that occurred at approximately 1800 UTC, on July 4, 1992, can be attributed to the low wave energy value (C_{11}). For an MO system, the estimated mean wave direction may have a larger error when the corresponding wave energy is low, as presented by Teng *et al.* (1991). For wave frequencies other than 0.20 Hz, similar results as shown in Figure 4 were observed for frequencies greater than 0.12 Hz. For lower frequencies, mean wave directions did not match with those reported by the DWA/Hippy wave system. This is because the MO technique works better for high frequency waves and may be non-representative for low frequency waves (Teng *et al.* 1991). The results from the field test show that wave data from the prototype WPM/MO system are comparable to those from the proven DWA/Hippy wave measurement system, especially for high frequency waves. Based on the nature of the MO technique, it can be concluded that the prototype WPM/MO system functions properly. Since the main purpose of the first field test for the WPM/MO was to test the WPM design concept and its implementation, no detailed data analysis was made.

To further test the WPM's performance and reliability, additional field testing for the prototype WPM/Hippy system is scheduled for October and November 1993, in the Gulf of Mexico and/or somewhere off the California coast.

Summary

In summary, the WPM is a powerful and effective wave data acquisition and processing system. Its flexibility and expandability will provide NDBC with an advanced wave system for many years to come. Other advantages of the WPM include:

- quick development time for future software modifications.
- field serviceability with limited risk to other measurement systems onboard the buoy.
- integration compatibility with NDBC's future payloads as a "smart sensor" module.
- processing hardware and software upgrade capability provided, thus eliminating the need to modify payload hardware and software.

Based on results from both the laboratory and field tests, it can be concluded that the WPM system works, and the validity and implementation of the WPM design are proven.

Acknowledgements

The authors would like to thank Mr. Jim Sharp and Mr. Ralph Dagnall of NDBC for their engineering management and support.

References

Earle, M.D., and Joseph, D.E., Jr. (1989). "Wave record analyzer (WRA) software," Report prepared by MEC Systems Corporation for the National Data Buoy Center under Subcontract from Computer Sciences Corporation.

Remond, F.X., and Teng, C.C. (1990). "Automatic determination of buoy hull magnetic constants." *Proceedings of Marine Instrumentation '90*, San Diego, California, 151-157.

Steele, K.E., and Earle, M.D. (1991). "Directional ocean wave spectra using buoy azimuth, pitch, and roll derived from magnetic field components." *IEEE J. of Oceanic Engineering*, 16(4), 427-433.

Teng, C.C., Dagnall, R.J., and Remond, F.X. (1991). "Field evaluation of the magnetometer-only directional wave system from buoys." *Proceedings of Oceans '91*, New Orleans, Louisiana, 1216-1224.

Table 1. NDBC Hardware System Corporation

	DACT (1982)	VEEP (1987)	WPM (1992)
Processor	NSC800	V30 70116	68332
Processor Register Size	8-bit	16-bit	32-bit
Processor Clock Frequency	0.768 MHz	1.8 MHz	16 MHz to 0.16 MHz dynamically programmable[1]
Programmable Memory	64 KB EPROM	128 KB EPROM	512 KB FLASH EEPROM[2]
Data Storage RAM	1 KB	128 KB	2512 KB
Non-Volatile RAM	256 bytes	512 bytes	512 bytes
A/D Converter	12-bit (binary)	4½ digit decimal	12-bit (binary) autocalibrating
Real Time Clock	Yes	No	Yes
Software Language	PLM Z80 assembly	C 8086 assembly	C 68020 assembly

(1) System clock programmable from 0.16 MHz to 16 MHz in 0.16 MHz increments.
(2) FLASH EEPROM provides the same function as ultraviolet EPROM; however, the device does not have to be removed from the circuit to accomplish reprogramming.

Table 2. Wave Processing Execution Time

WPM	16 MHz	68332	87 seconds
IBM PS/2	16 MHz	80386	91 seconds
IBM XT	5 MHz	8088	977 seconds

Table 3. Operational Bands for WPM/WRA Directional and Nondirectional Wave Measurement Systems

BAND NUMBER	Frequency Band (Hz) Low Limit (Hz)	High Limit (Hz)	Center (Hz)	Center Period (SEC)	Band Width (Hz)	Record Lengths (MIN)	Degrees of Freedom
Noise Band	.010	.030	.02	50	.02	40	96
1	.030	.035	.0325	30.77	.005	40	24
2	.035	.040	.0375	26.67	.005	40	24
3	.040	.045	.0425	23.53	.005	40	24
4	.045	.050	.0475	21.05	.005	40	24
5	.050	.055	.0525	19.05	.005	40	24
6	.055	.060	.0575	17.39	.005	40	24
7	.060	.065	.0625	16.00	.005	40	24
8	.065	.070	.0675	14.81	.005	40	24
9	.070	.075	.0725	13.79	.005	40	24
10	.075	.080	.0775	12.90	.005	40	24
11	.080	.085	.0825	12.12	.005	40	24
12	.085	.090	.0875	11.43	.005	40	24
13	.090	.095	.0925	10.81	.005	40	24
14	.095	.105	.10	10	.01	20	24
15	.105	.115	.11	9.09	.01	20	24
16	.115	.125	.12	8.33	.01	20	24
17	.125	.135	.13	7.69	.01	20	24
18	.135	.145	.14	7.14	.01	20	24
19	.145	.155	.15	6.67	.01	20	24
20	.155	.165	.16	6.25	.01	20	24
21	.165	.175	.17	5.88	.01	20	24
22	.175	.185	.18	5.56	.01	20	24
23	.185	.195	.19	5.26	.01	20	24
24	.195	.205	.20	5.00	.01	20	24
25	.205	.215	.21	4.76	.01	20	24
26	.215	.225	.22	4.54	.01	20	24
27	.225	.235	.23	4.35	.01	20	24
28	.235	.245	.24	4.17	.01	20	24
29	.245	.255	.25	4.00	.01	20	24
30	.255	.265	.26	3.85	.01	20	24
31	.265	.275	.27	3.70	.01	20	24
32	.275	.285	.28	3.57	.01	20	24
33	.285	.295	.29	3.45	.01	20	24
34	.295	.305	.30	3.33	.01	20	24
35	.305	.315	.31	3.22	.01	20	24
36	.315	.325	.32	3.13	.01	20	24
37	.325	.335	.33	3.03	.01	20	24
38	.335	.345	.34	2.94	.01	20	24
39	.345	.355	.35	2.86	.01	20	24
40	.355	.375	.365	2.74	.02	10	24
41	.375	.395	.385	2.60	.02	10	24
42	.395	.415	.405	2.47	.02	10	24
43	.415	.435	.425	2.35	.02	10	24
44	.435	.455	.445	2.25	.02	10	24
45	.455	.475	.465	2.15	.02	10	24
46	.475	.495	.485	2.06	.02	10	24

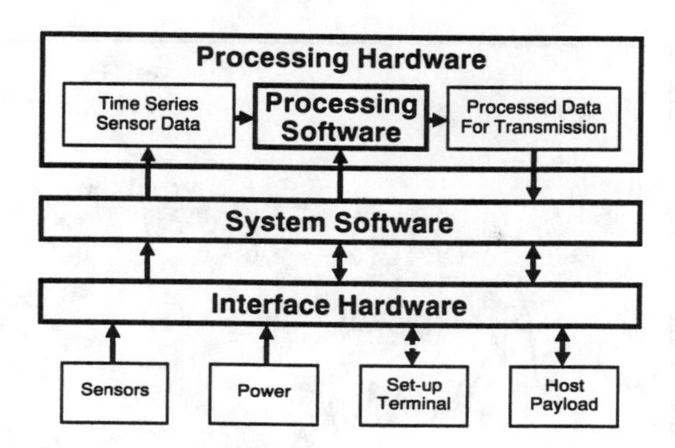

Figure 1. Block Diagram of Wave Processing Model Elements and Interfaces.

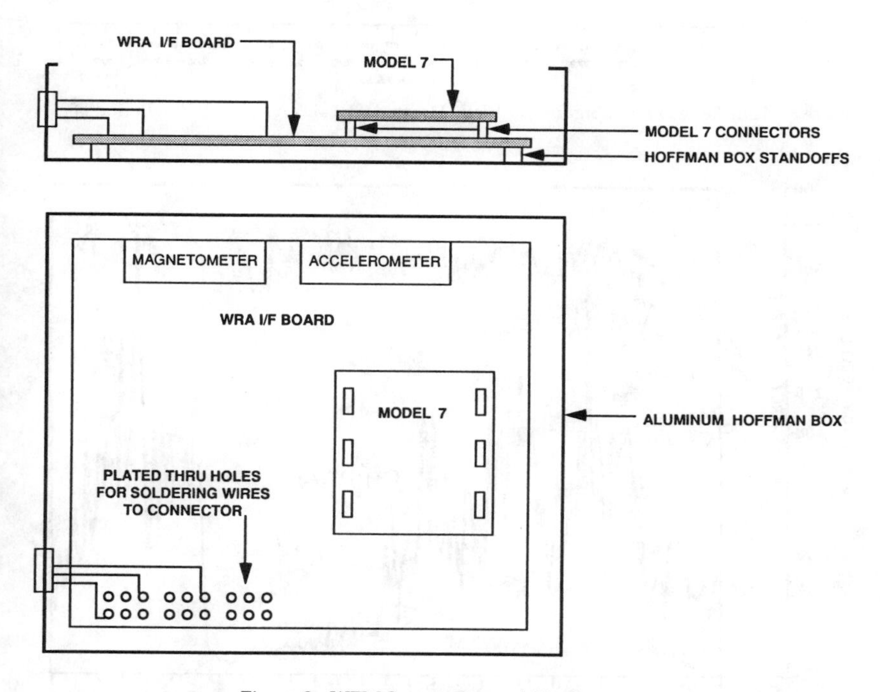

Figure 2. WPM System Box and Hardware.

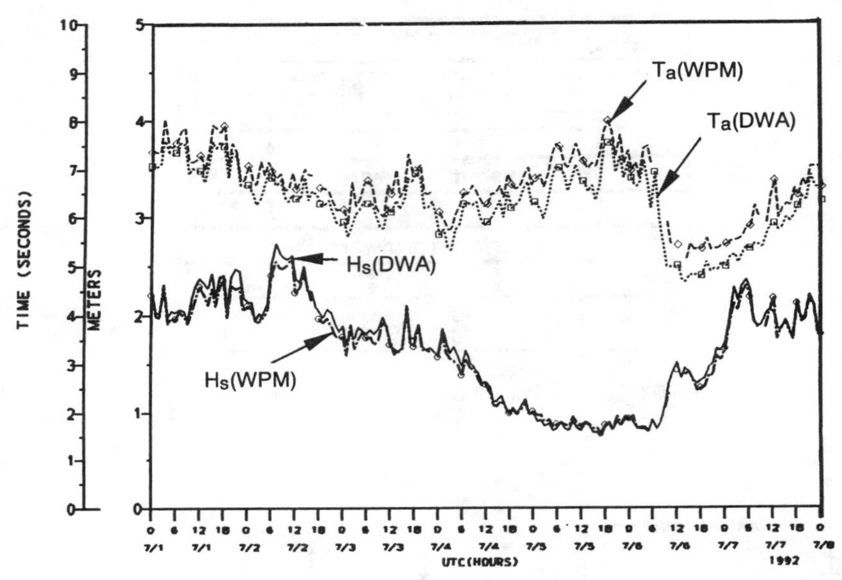

Figure 3. Time Series Plot of Significant Wave Heights (H_s) and Average Wave Period (T_a).

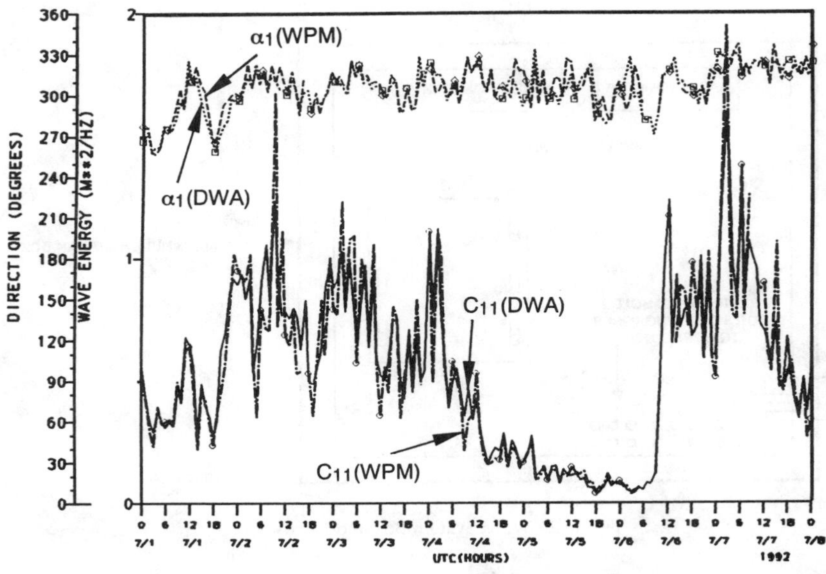

Figure 4. Time Series Plot of Mean Wave Directions (α_1) and Wave Energies (C_{11}) at 0.20 Hz.

A Comparison of Three Wave-Measuring Buoys

Anand Gnanadesikan[1] and Eugene A. Terray[2]

Abstract

We compare the performance of three wave-measuring buoys: a Seatex Wavescan™, an Endeco/YSI Wavetrack™, and a taut-moored 3m discus deployed in the Gulf of Alaska during November of 1991. The response of the Seatex buoy is well understood and we use it as a standard of comparison. Nondirectional parameters such as the significant wave height and peak frequency compare extremely well. We compare the directional response between the Seatex and Endeco buoys, showing that they agree well with each other and with the wind direction at high frequencies (0.25-0.33Hz), but are sometimes different at the spectral peak. The tilt response of the three buoys is studied in some detail. A simple model of the discus buoy is described, showing that the buoy transfer function is strongly affected by mooring tension. The Endeco buoy is reasonably described as a tilt follower for periods between 3 and 7 seconds but deviates strongly from this response outside this band.

Introduction

The basic idea of measuring ocean surface gravity waves from the pitch and roll of a buoy was first proposed by Longuet-Higgens, Cartwright, and Smith in a seminal paper in 1963. The basic idea was to use the spectra and cross-spectra of the heave and tilt to infer the directional spectrum. If the directional spectrum at one frequency is broken down into Fourier coefficients, the first coefficient (corresponding to the overall level of the

---

[1]Graduate Research Assistant, MIT/WHOI Joint Program in Physical Oceanography, Clark 214B, Woods Hole Oceanographic Institution, Woods Hole, MA 02543.

[2] Research Specialist, Department of Applied Ocean Science and Engineering, Woods Hole Oceanographic Institution, Woods Hole, MA 02543.

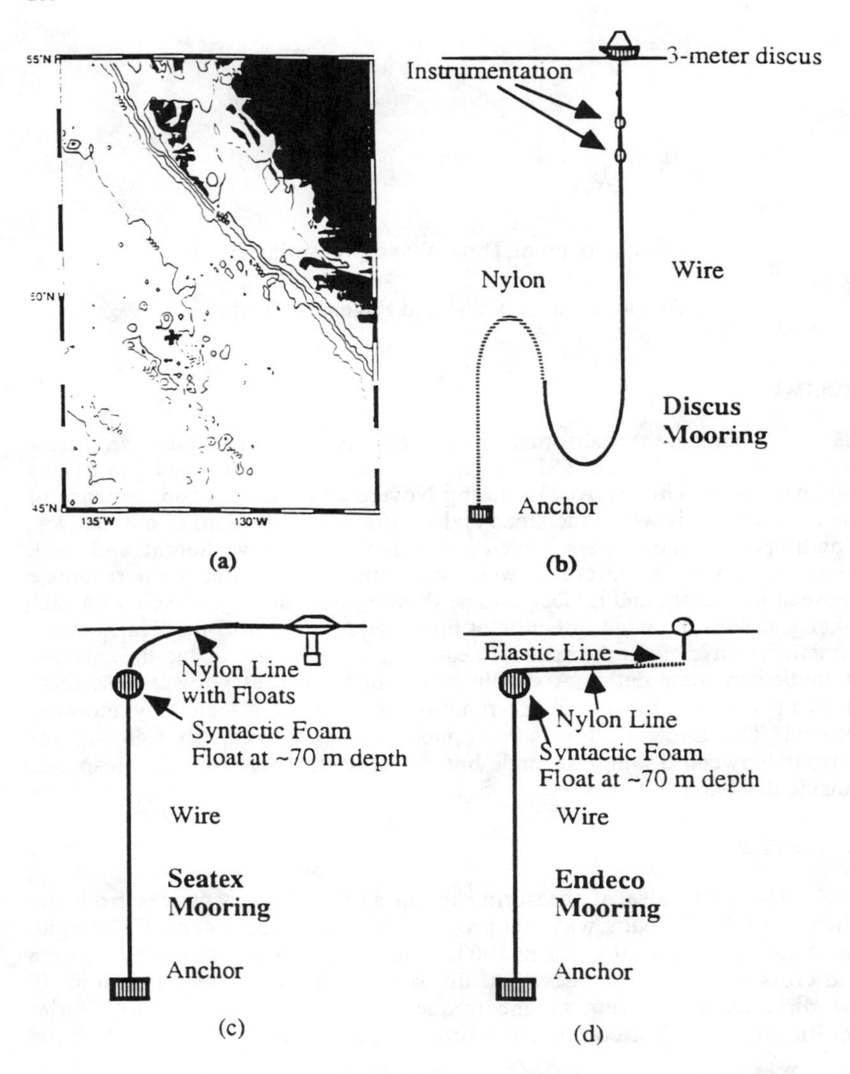

Figure 1: Location of the Experiment and schematics of the moorings. (a) Location of the mooring array off Vancouver Island. (b) Discus Mooring (c) Seatex Mooring (d) Endeco Mooring

spectrum) is inferred from the heave spectrum, the second (corresponding to the mean propagation direction) is obtained from the cross-spectra of heave and tilt and the third (corresponding to a measure of directional spread) is inferred from the tilt spectra and cross-spectra. The measured spectra and cross-spectra are dependent on the buoy response. Consequently, in recent years, much effort has gone into characterizing the response of pitch-roll buoys to waves. (Barstow and Krogstad, 1984; Steele et al.,1985; Steele et al.,1992; Wang and Leonard,1992).

This paper will seek to understand why the response of certain buoys deviates from following the tilt exactly. We will focus on two mechanisms, resonance and mooring tension, and examine how they work in three buoys. We will show that both these mechanisms may cause major changes in amplitude and phase of the tilt response.

Experimental Description

During the late fall of 1991, a group from the Woods Hole Oceanographic Institution deployed three buoys, a Wavescan™, manufactured by Seatex A/S, a Wavetrack ™, manufactured by Endeco/YSI, and a standard 3m NDBC discus hull in the southern Gulf of Alaska in support of the 1991 Acoustic Surface Reverberation Experiment (ASREX 91).

Figure 1 shows the experiment site and schematics of the three moorings which were deployed in 3200m of water. The discus buoy was taut-moored using an inverse catenary mooring, in which buoyant line is deployed at the base of the mooring with non-buoyant wire above it. The result is to increase the ability of the mooring to respond to changes in the current without producing large changes in tension. The discus buoy carried two meteorological packages and the mooring line was instrumented with current meters and temperature sensors. The tension at the buoy was about 2x 10^4 N. The moorings for the Seatex and Endeco buoys both consisted of subsurface floats at a nominal depth of 70 meters, with a tether running to the surface. In the case of the Seatex buoy, the tether was buoyant and had a surface expression of about 50 meters. This resulted in an extremely compliant system which to first order decoupled the buoy motion from the subsurface mooring. The Endeco buoy was tethered to the subsurface float using a chain connected to an extensible cord about 10 meters in length, which in turn led to a long nylon tether. As with the Seatex buoy, the intent was to provide some measure of decoupling between the surface float and the mooring.

Figure 2 shows simple cartoons of the three buoys, with important features highlighted. The discus buoy, shown at left, has a standard NDBC 3-meter hull. The buoy is outfitted with a tower top with meterological sensors and a vane which tends to make it turn into the wind. The attachment point is at the base of a bridle roughly 1.5 meters below the water surface. The design is extremely similar to that used by the National Data Buoy Center. Heave and tilts were measured using a Datawell Hippy 40.

The Seatex buoy is a standard pitch-roll buoy. It consists of a two-part hull molded from polyethelene. The two foam-filled halves separate along the vertical plane and are clamped around a central instrument well. The mooring was attached laterally to the side of the buoy, stability being provided by a 2m keel weighing approximately 300 kg. The instrumentation package consisted of a Datawell Hippy 120 sensor to measure heave and pitch and a two-axis fluxgate compass. The buoy is described in more detail in Barstow et al., (1991).

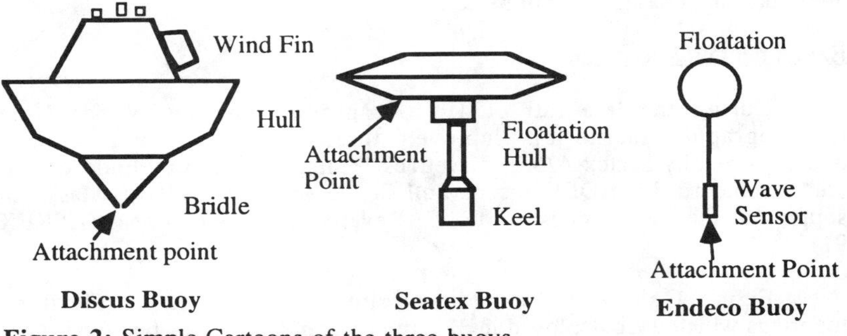

Figure 2: Simple Cartoons of the three buoys.

The Endeco buoy is the smallest of the three buoys. It consists of a floatation hull roughly 1m in diameter containing electronics and batteries. The sensor package, which weighs about 12 kg, is on the end of a 2m stalk and consists of a gimballed vertical accelerometer, a fluxgate compass, and two electrolytic tilt sensors. The mooring line was attached to the base of the sensor package.

This paper will discuss some of the issues the authors found to be of importance in understanding the response of these wave-measuring systems. We will focus on two ways in which the response of the buoy may be changed. 1.) Resonances. 2) Mooring tension.

Comparing the Three Buoys

We begin by considering some of the nondirectional data from the buoys. In all cases the data presented was post-processed using our own software. Figure 3 shows time series of significant wave height and period

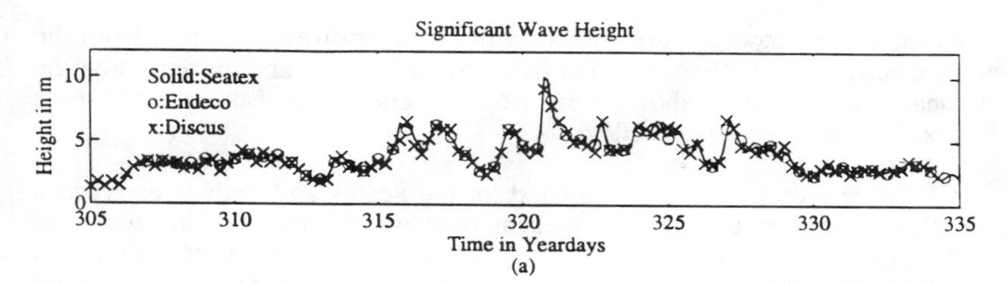

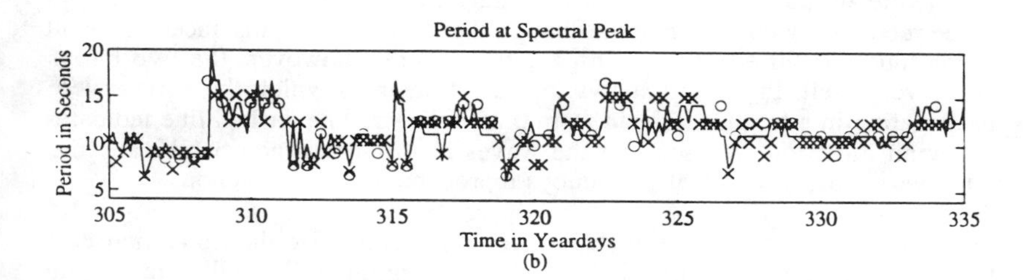

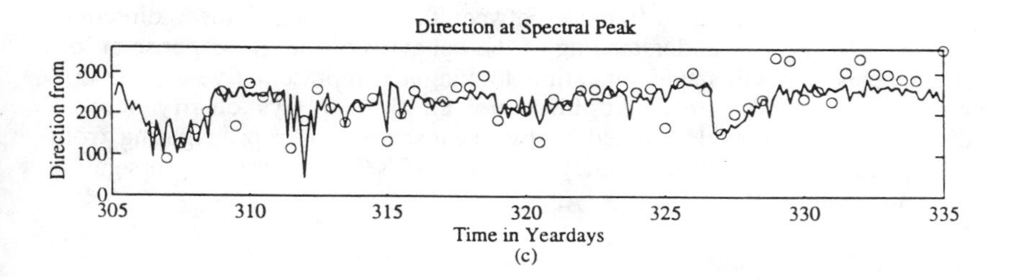

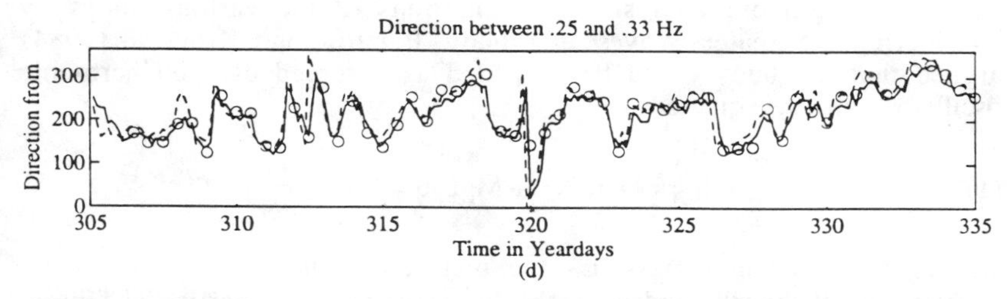

Figure 3: Comparisons between the three buoys.Solid is Seatex, o, Endeco, x, Discus (a) Significant wave height (b) Period of Spectral Peak (c) Direction from which waves at the spectral peak are propagating. (d) Direction from which waves in a frequency band from 0.25-0.33 Hz are propagating. Dashed line indicates the wind direction.

from the three buoys. Figure 3a shows the significant wave height from the Seatex buoy asa solid line, and that from the Endeco (o) and discus (x) on the vertical axis. Figure 3b shows the period at the spectral peak from the three buoys. The agreement is excellent.

Wave directions were computed for the Seatex and Endeco buoys as a function of frequency. At the present time we do not feel that we have sufficiently validated the discus transfer function to present directional estimates from this buoy. The direction from which the waves at the spectral peak were propagating is shown in Figure 3c. We see that the two buoys agree relatively well in general, although there are isolated instances of ment by as much as 90 degrees. At higher frequencies, however, the two buoys agree very well. Figure 3d shows the direction from which the waves were propagating in a frequency band from 0.25-0.33 Hz. The dashed line indicates the wind direction. We see that the waves in this frequency band track the wind very closely, and that both buoys reproduce this phenomenon.

In the remainder of this paper, we will characterize the tilt response of these buoys and present models for understanding them. We will demonstrate that the presence of a taut mooring substantially affects the dynamics of the discus buoy, and that the difference between the Seatex and Endeco directions is due to our lack of understanding of the behavior of the tilt response at low frequencies. We will spend some time lookingat a representative case, 0000Z on November 5th, 1991. At this time, all three buoys clearly show a directional spectrum dominated by two peaks, a windsea propagating from the southeast and a low-frequency swell (12-15 seconds period) propagating from the southwest.

Behavior of the Seatex Buoy

We begin our analysis of the motions of the various buoys by considering the Seatex Wavescan™ buoy. Barstow and Krogstad (1984) argued that the buoy could be modelled as a forced damped harmonic oscillator using the equation

$$I_b \frac{\partial^2 \theta}{\partial t^2} + K \frac{\partial \theta}{\partial t} + MgL_c \theta = MgL_c \frac{\partial \eta}{\partial x} \quad (1)$$

where M is the buoy mass, L_C the metacentric height, K a damping coefficient, θ the tilt and I_b is the buoy moment of inertia. The basic assumption involved in this equation is that the horizontal motion of the buoy is nearly equal to that of surface water particles. There is a resonance at

$$\omega_{RT} = \sqrt{MgL_c/I_b} \quad (2)$$

We have omitted terms from equation 1 having to do with added mass since they are typically small. If the frequency of forcing is much less than ω_{RT}, the buoy tilt will be almost exactly equal to to the surface slope and hence 90 degrees out of phase with the heave. If the frequency of forcing is equal to ω_{RT}, the buoy will overrespond and the tilt will be in phase with the heave. If the frequency of forcing is higher than ω_{RT}, the buoy will underrespond and the measured tilt will be 180 degrees out of phase with the surface tilt.

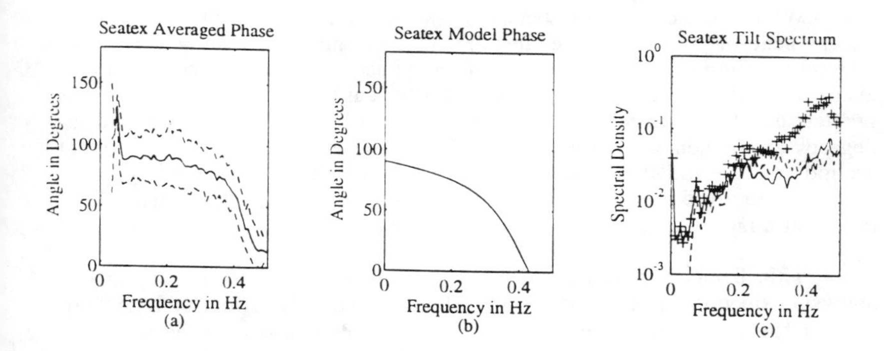

Figure 4: Modelling the behavior of the Seatex buoy. (a) Mean heave-pitch (solid) phase for all cases where the coherence is more than).6. ±1 standard deviation is shown by the dashed lines. (b) Predicted phase from best-fit simple harmonic oscillator. (c) Tilt spectra, November 5th,1991 at 00Z. Dashed line is predicted from heave spectrum, + is the measured spectrum, solid is the measured spectrum corrected using harmonic oscillator model.

Figure 4a shows the average of the absolute magnitude of the relative phase between heave and pitch averaged over the course of the experiment for cases when the coherence between heave and pitch was higher than 0.6. The solid lines respresent the mean phase and the dashed lines are plus and minus one standard deviation. We see that the relative phase is generally close to 90 degrees from frequencies of about 0.08 to about 0.35 Hz, when it begins to dip towards zero. At low frequencies, there are fewer instances of strong pitch/tilt coherence, so that the mean deviates somewhat more from 90 degrees. Figure 4b shows the predicted phase from the simple harmonic oscillator model in equation (1) where the parameters ω_{RT}=0.43Hz and K/I_b=0.2 are taken from Krogstad (1992), based on analysis of a subset of the data presented in this paper. Figure 4c shows the spectrum of the tilt measured by the buoy (+), predicted from the heave spectrum (dashed) and the measured spectrum corrected by the simple harmonic oscillator model (solid). Clearly treating the buoy as a simple harmonic oscillator accounts for most of the difference between the measured tilt and that predicted from linear theory. There are some differences between the predicted and measured phases. The phase change occurs quite a bit more quickly than would simply be predicted from the linear harmonic oscillator. It is our feeling that the problem is most likely the incorrect specification of the drag.

Behavior of the Discus Buoy

While the Seatex buoy behaves essentially as a good pitch-roll buoy except near the tilt resonance, the behavior of the discus buoy was quite different. Figure 5a shows the relative phase between heave and pitch averaged over the course of the experiment for cases when the coherence was greater than 0.6. The relative phase between heave and pitch is around 20 degrees at frequencies between 0.1-0.2Hz. It rises to about 50 degrees at around 0.45Hz and then falls off rapidly. This behavior is in many ways similar to that reported by Steele et al. (1992) except that the sharp falloff occurs at a higher frequency than in their paper.

This behaviour was extremely puzzling to us when we began our analysis– since if one considers a free-floating body, the effect of both drag and tilt forces is to force the tilt 90 degrees out of phase with the heave. It was only after including the effects of mooring tension that we were able to derive a qualitatively accurate model of the buoy motion.

The presence of a taut mooring changes the mooring dynamics substantially. In the first place, it means that the buoy is no longer as free to move laterally. This in turn means that there will be large relative velocitieswith a corresponding increase in drag forces. Additionally, the tension on the line will act to increase the righting moment. We let T be the tension on the mooring line, M_a the added mass of the buoy, L_d the distance between the attachment point and the center of drag, and L_1 the distance between the center of buoyancy and the mooring attachment point. If the buoy now pivots about the attachment point, equation (1) is transformed to

$$(3)\quad \left(I_b+(M+M_a)L_1^2\right)\frac{\partial^2\theta}{\partial t^2} - K\frac{\partial\theta}{\partial t} - (Mg+T_1)L_1\theta = g\,(M+M_a)L_1\frac{\partial\eta}{\partial x} + Drag{*}L_d$$

For a taut-moored discus, gM_a is small compared with T. Additionally, in the absence of mooring tension, this buoy is unstable and will tend to capsize. Since a relatively moderate amount of tension (of order 1000-3000N) is often sufficient to prevent this from happening we know that $(M+M_a)L_1^2$ is large compared with I_b. The presence of a taut mooring will thus have the following effects

•The frequency at which the resonance occurs will shift to higher frequencies.

•The buoy will may underrespond to the actual tilt, if tilt forces are larger than drag forces.

•The presence of drag forces due to the buoy's lack of freedom to surge with the waves may produce motion in phase with the horizontal velocities, and so in phase with the heave.

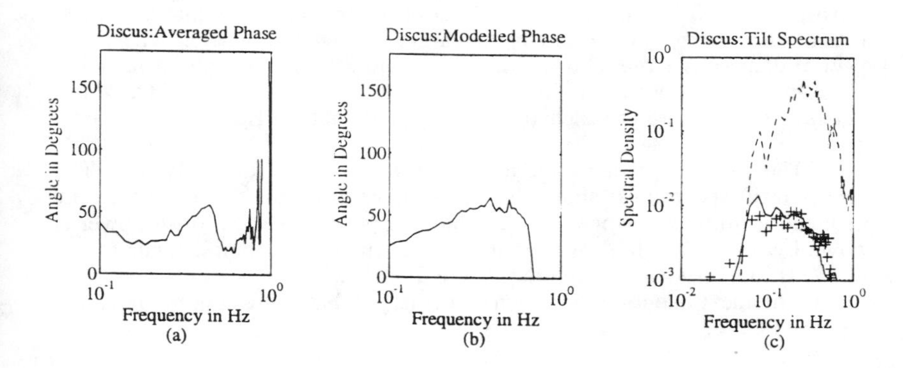

Figure 5: Modelling the behavior of the Discus buoy. (a) Mean heave-pitch (solid) phase for all cases where the coherence is more than .6. (b) Predicted phase from model on November 5th,1991 at 00Z. (c) Tilt spectra, November 5th,1991 at 00Z. Dashed line is predicted from heave spectrum, + is the measured spectrum, solid is the prediction from model.

Are drag forces really important? A quick order of magnitude calculation shows that they can be. We let the tilt force be of order $Mgka=Ma\omega^2$, where a is the amplitude, ω the frequency and k the horizontal wavenumber of the waves. The drag force is of order $\rho Aa^2\omega^2$, is the cross-sectional area of the submerged portion of the buoy. If V is the volume of the submerged part of the buoy, then the ratio of these two terms is

$$\frac{\text{Drag Force}}{\text{Tilt Force}} \sim \frac{Aa}{V} \tag{4}$$

Since V/A is of order a meter, the implication is that in high sea states near the peak of the spectrum drag forces will become more important than tilt forces. This in turn implies that the relative phase between heave and pitch can be quite far from 90 degrees.

A model identical to that given in equation 3 was used to fit the tilt spectrum and phase difference. It was found that the model gave qualitatively good results for values of the drag, moment of inertia, and damping coefficient which were in order of magnitude agreement with a priori estimates based on the physical parameters of the buoy. The results of the model are shown in Figures 5b and 5c for our test case of 00Z November 5th,1991. We see that the phase rises slowly from about 25 degrees at 0.1 Hz to around 55 degrees at 045 Hz and then falls off rapidly. The rise is much slower than that seen in Figure 5a (which closely approximates the observed heave-pitch phase for this case). Analysis of the model shows that the phase shift is due to the drag forces being larger at low frequencies where the spectral density is large, and the tilt forces being larger at high frequencies where the spectral density is lower. There is a resonance at 0.63 Hz, which accounts for the sharp phase change starting at about 0.45 Hz.

The model reproduces the measured tilt spectrum relatively well. The measured tilt spectrum is shown by the +-marks in Figure 5c. The spectrum predicted from linear theory assuming the buoy to be a perfect tilt-follower is shown by the dashed line. We see that the discus buoy underresponds to the tilt. This low response is captured by the model, where it occurs because the righting moment due to mooring tension is larger than the tilting moment due to surface slope.

Behavior of the Endeco Buoy

In contrast with the discus buoys discussed in the preceding sections, the dynamics of the Endeco/YSI Wavetrack buoy has been the least well-studied, and as a consequence we began with fewer preconceptions about how it would respond, preferring instead to be guided by the data. As discussed earlier, the Wavetrack consists of an approximately 1m diameter sphere connected to mooring cable via a freely-flooding pipe. This "sting" has an outside diameter of 5 cm and a length of roughly 2 m.

The phase and coherence between the tilts (in both east and north directions) and the heave are shown in Figures 6a and 6b for the case study data (5 November, 1991, 00Z). All estimates shown have 32 degrees-of-freedom. The computations have been carried out independently of Endeco's processing software, and only the published amplitude and phase corrections associated with the electronics have been applied to the raw heave and tilt signals (Endeco/YSI, 1992).

In general the tilts are seen to be in quadrature with the heave over a band of frequencies between 0.15-0.5 Hz, with coherence of 0.5 or larger. Although not shown here, the coherence above 0.5 Hz is small, and the phase estimates vary rapidly. Below 0.15 Hz the phase drifts away from 90 degrees. This qualitative behavior of both coherence and phase is generic to both the present data set and was also observed in data obtained from a freely-drifting Wavetrack.

As discussed earlier, the waves on 5 November were a bi-directional system consisting of swell (12-15s) from the southwest together with a wind sea from the southeast. Hence, even in the absence of phase errors, we expect that the phase of the east tilt will shift by 180 degrees between high and low frequencies, whereas the phase of the north tilt should remain at 90 degrees. This may explain why the drift in the east phase begins at somewhat higher frequency than that of the north.

The fact that the tilts are mostly in quadrature with the surface elevation suggests that the Wavetrack is acting as a slope-follower over the range of frequencies from roughly 0.15 to 0.5 Hz. Intuitively, this behavior might be expected if the mooring inhibited the lateral motion of the bottom of the sting, so that the buoy acted as an "inverted pendulum", pivoting about the mooring attachment point (we imagine, however, that the pivot is free to move vertically). Then, since the pressure force on the sphere is proportional to the wave slope (i.e. the sphere essentially follows the local fluid motion), the response of the system will be 90 degrees out of phase with the heave. However, the persistent phase deviation at low frequencies indicates a substantial departure from this simple picture. The coincidence of this frequency band with the typical frequencies of oceanic swell make it of interest to inquire more closely into the actual dynamical balance of forces on the buoy.

The behavior of spherical buoys having stings has been addressed recently by Wang and Leonard (1992). These authors presented the results of a direct time-domain integration of the equations of motion, including quadratic drag terms. In general their solution showed that the presence of the sting dampened the heave and tilt resonances, and that the tilt approximately followed the wave slope. However, they did not give a detailed discussion of the phase.

We have formulated a frequency-domain model for the Wavetrack response function based on the treatment of Wang and Leonard (1992) but using a linearized drag. We summarize below a few of the preliminary results from this model. The magnitude of the non-directional tilt response (i.e. for the sum of the north and east tilt spectra) is shown in Figure 6c. The vertical stripes on this figure summarize the tilt "response amplitude operator" (estimated as the square root of the non-directional tilt to heave spectral ratio) over the entire data set of 135 runs. The dashed curve shows

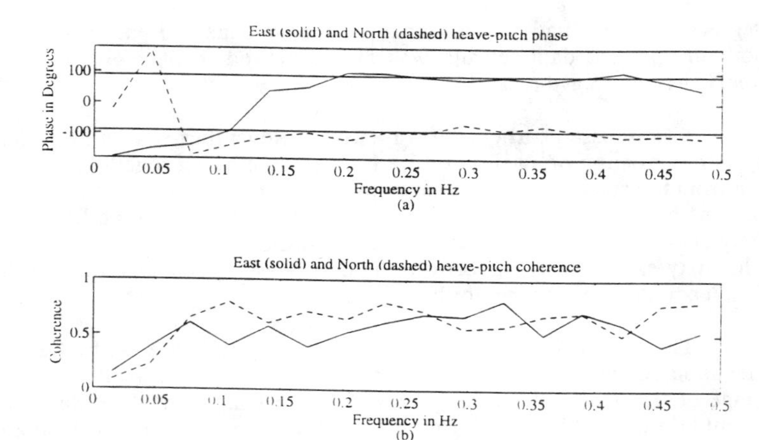

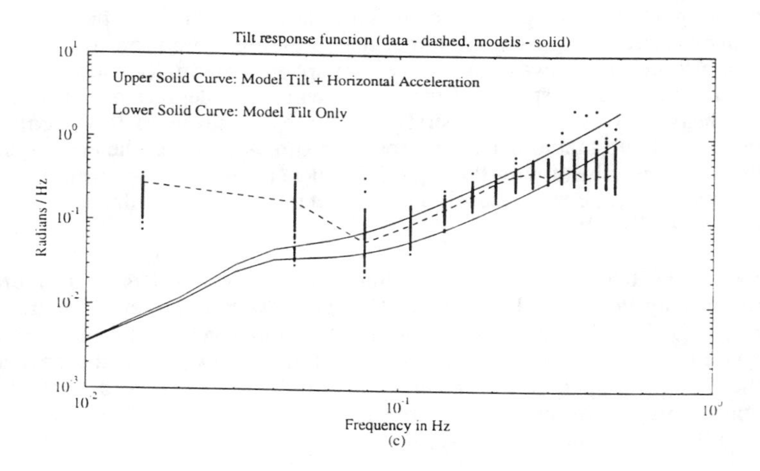

Figure 6: Characterizing and modelling the Endeco response. (a) Phase between East (solid) and North (dashed) tilts and heave at 00Z on November 5th,1991. (b) Coherence between East (solid) and North (dashed) tilts and heave at 00Z on November 5th,1991. (c) Tilt response function. Dots show response function over the entire deployment, lower solid line model response function including tilt only, upper solid line model response function including contamination of the tilt sensor by horizontal acceleration. The dashed line shows the response at 00Z on November 5th,1991.

the data for 5 November, 00Z. The lower solid curve is the model prediction for the tilt response. The model is seen to systematically underpredict the observed response, except at the highest frequencies where it gives an over-response.

We believe the error at high frequencies is due to our use of the long-wavelength limit of the hydrodynamic added mass and damping coefficients. This approximation breaks down at frequencies corresponding to kd > 1, where k is the wavenumber of the wave and d is the diameter of the sphere. For example, the damping coefficients in surge and heave, which vanish at zero frequency, rise to roughly 0.2 and 0.7, respectively, at kd = 1 (corresponding in our case to 0.5 Hz). For kd = 2 the surge damping increases to 0.75 (Garrison, 1974). We propose that the discrepancy between the measured and predicted tilt amplitude response in the mid-frequency range from roughly 0.07 - 0.3 Hz is due to the contamination of the tilt measurement by acceleration. The Wavetrack buoy measures tilt using an orthogonal pair of electrolytic inclinometers that are fixed with respect to the sensor can at the bottom of the stalk. Hence these devices falsely report lateral accelerations as tilt. An examination of the calculated surge response (not shown) indicates that the lateral motions of the buoy are not negligible. We have added this false contribution to the tilt response to get the upper solid curve in Figure 6c. It is interesting to note that this addition substantially improves the agreement with the data over the mid-frequency range.

Although we do not show it here, the computed tilt/heave phase is 90 degrees above 0.1 Hz, decreasing rapidly toward toward zero at lower frequencies. This behavior is in qualitative, although not quantitative, agreement with the data shown in Figure 6a. The reason for this is not clear, although we suspect that errors in phase are responsible for the scatter seen in the Wavetrack directional estimates at low frequencies (Figure 3c). At sufficiently low frequencies the phase of the response should be dominated by the drag and tension. Although the currents at the experiment site varied from a few cm/s to a peak of 0.8 m/s, we have failed to see any clear correlation between the error in mean direction (relative to the Wavescan estimate) and the variations in current. The fact that the heave-tilt phase has a similar form for the case in which the Wavetrack is freely-drifting leads us to suspect that variations in drag and tension are not the cause of the error.

We consider that the results of the spectral model warrant its continued development. Issues that need to be addressed are a more realistic treatment of the mooring line, including the chain, the incorporation of frequency-dependent hydrodynamic coefficients, and the extension of the model to three dimensions in order to include cases where the mean current and wave directions are not coincident. We also plan to investigate whether a more realistic drag model is necessary by comparing our model output to the results of a time-domain approach..

Conclusions

Tilts measured from a buoy moored from the side with a compliant mooring. will tend to follow the surface slope except near and above the tilt resonance. The technique of Barstow and Krogstad (1984) of modelling the buoy as a simple harmonic oscillator, calculating a transfer function and applying it to all measurements works well. It is necessary to keep track of where resonances can occur in order to properly estimate the actual spectra and cross-spectra of heave and tilt.

The presence of mooring tension changes the righting moment of the buoy, causing the frequency of the tilt resonance to change. It also limits the ability of the buoy to follow water particles. As a result drag forces will cause the heave and pitch to be in phase. Since the effect of drag forces will depend on sea-state and current and the tilt resonance depends on the moooring tension it is be important to recompute the transfer function frequently, as done by Steele et al., 1985,1992.

While the models presented here for the discus and Endeco buoys are still under development, they reproduce a number of the features of the observed response. It is our hope that by further refining the physics of these models to include more realistic drag and cable dynamics, we will be able to move from diagnostic models which give rough physical understanding to prognostic models which can be used to correct the buoy response.

Acknowledgments

The authors wish to thank Capt. Clampitt and the crew of the *Thomas Thompson* for assistance during the deployment and recovery of the moorings. George Tupper performed the mooring design and John Kemp supervised the deployment and recovery of the mooring array. Nan Galbraith helped with the data analysis. We also wish to thank both Seatex A/S and Endeco/YSI for their generous assistance. This work was funded by the Office of Naval Research under contracts N0014-91-J-1891 (Acoustics Surface Reverberation Program) and N0014-90-J-1495 (Surface Waves Processes Program).

Bibliography

Barstow, S.E. and H. Krogstad, (1984), General Analysis of Directional Ocean Wave Data from Heave/Pitch/Roll Buoys, *Modeling Ident. and Contr.*, 5:47-70.

Barstow, S.E., G. Ueland, H.E. Krogstad, and B.A. Fossum, (1991) The WAVESCAN Second Generation Wave Buoy, *IEEE J. Ocean. Eng.*, 16:254-266.

Berteaux, H.O., (1991), *Coastal and Oceanic Buoy Engineering*,, Available from H.O. Berteaux, P.O. Box 182, Woods Hole, MA.

Garrison, C.J., (1974), Hydrodynamics of Large Objects in the Sea: Part I–Hydrodynamic Analysis, *J. Hydronautics*, 8:5-12.

Krogstad, H.E., (1992) *Calibration of the WAVESCAN Directional Wave Buoy*, SINTEF Report STF-10-F92005.Report available from Seatex A/S.

Longuet-Higgens, M.S., D.E. Cartwright, and N.D. Smith (1963), Observations of the Directional Spectrum of Sea Waves Using the Motions of a Floating Buoy, In *Ocean Wave Spectra*, pp. 111-136, Prentice-Hall, Englewood Cliffs, NJ.

Steele, K.E., J.C.-K. Lau and Y.-H. Hsu, (1985) Theory and Application of Calibration Techniques for an NDBC Directional Wave Measuring Buoy, *IEEE J. Ocean. Eng.*, 382-396.

Steele, K.E., C.-C. Teng and D.W.C. Wang, (1992) Wave direction measurements using pitch-roll buoys, *Ocean Eng.*, 19:349-375.

Wang, H., and J.W. Leonard, (1992) Time Domain Solutions for the Motion of Directional Wave Buoys, in *Proc. 11th Intl. Conf. on Offshore Mechanics and Arctic Engineering*, Vol.1 pp.127-134.

A Sonobuoy-Sized Expendable Air-Deployable Directional Wave Sensor

Marshall D. Earle[1], Ralph H. Orton[2], Harry D. Selsor[3], Kenneth E. Steele[4]

Abstract

Obtaining wave information from small expendable air-deployable buoys is important for numerous military operations, civilian applications, and research purposes. Such buoys may be deployed when and where wave data are needed. Under U.S. Navy sponsorship, a sonobuoy-sized expendable air-deployable directional wave sensor is being developed. Prototype buoys have been designed and three prototypes have been built for initial wave tank and field testing. Before deployment, the buoy is approximately 90 cm in length by 12 cm in diameter and weighs less than 30 lbs. A nitrogen-inflated floatation collar provides buoyancy after deployment. Small accelerometers, tilt sensors that respond to local acceleration including wave acceleration, and a fluxgate compass are used as sensors for the prototypes. Directional wave information will be calculated within the buoy and transmitted to shore via ARGOS satellite using data compression techniques. Directional wave spectra at high frequencies will be used to estimate wind speed and direction to avoid use of an anemometer.

---

[1]Senior Oceanographer and Vice President, Neptune Sciences Inc., 1181 Robert Blvd., Slidell, LA 70458

[2]Senior Design Engineer, METOCEAN Data Systems Ltd., P.O. Box 2427 D.E.P.S., 40 Fielding Av., Dartmouth, Nova Scotia, Canada B2W 4A5

[3]Deputy Director, Tactical Oceanographic Warfare Support Program Office, Code 7410, Naval Research Laboratory, Stennis Space Center, MS 39529

[4]Waves Program Manager, National Data Buoy Center, Stennis Space Center, MS 39529

Introduction and Background

In 1991, U.S. Navy requirements specified the need for a series of A-sized (approximately 90 cm in length by 12 cm in diameter) sonobuoy style Mini-Drifting Data Buoys (MDDB's) to provide real-time environmental data for Naval operations. A meteorological-oceanographic version (air temperature, sea surface temperature, and atmospheric pressure measurements), an omni-directional ambient noise version (equipped with a hydrophone suspended at 100 m depth), and a subsurface temperature versus depth version (equipped with a 300 m thermistor string) have been developed (Selsor, 1993). Further development of these buoys, including possible addition of wind speed and direction measurements, is continuing (Selsor, 1993). The meteorological-oceanographic MDDB has been air-certified for deployment from P-3 and S-3 aircraft. AN/WSQ-6 is the Navy series designation for MDDB's.

Particularly in shallow water, waves pose the greatest environmental threat to Naval operations such as amphibious landings, mine warfare, mine countermeasures, ship operations, and special warfare. Capability to make wave measurements during particular meteorological situations (e.g. severe extratropical storms and hurricanes) is needed for improving and testing numerical wave forecasting computer models. Capability to make wave measurements at locations and in situations of interest has additional scientific applications for coastal processes, air-sea interaction, and remote sensing verification studies. In late 1992, development of a wave measurement MDDB to meet these needs and applications was started.

Design Goals and Constraints

Table 1 lists design goals and constraints of the wave measurement MDDB that is being developed.

Table 1
Design Goals and Constraints

Design Goals and Constraints
A-sized sonobuoy size (pre-deployment size, 90 cm length by 12 cm diameter)
Air-deployment water impact shock considerations
Relatively low cost (target cost \$5,000 - \$6,000, desirable cost < \$4,000)
Operating lifetime (few days minimum, few weeks desirable)
Maximum utilization of existing MDDB features and hardware
Data transmission via ARGOS satellite with on-buoy data processing
Non-directional and directional wave measurements

Unusual buoy motion associated with an MDDB's small size and configuration will likely cause the most important problems to overcome. An inflatable nitrogen-filled floatation collar provides buoyancy and causes the buoy to behave somewhat as a pitch/roll buoy. Unlike larger wave measurement buoys whose hulls partially attenuate very high frequency waves, the buoy responds well to these waves. Theory is inadequate to model buoy motion considering its geometry with its inflated floatation collar, non-uniform mass distribution, and viscous damping.

The requirement that the buoy be inexpensive enough to be expendable also poses challenges to obtaining accurate wave measurements. Large and expensive pitch/roll sensors (e.g. HIPPY sensors made by Datawell and used in larger buoys) cannot be used. Multiple angular rate sensors and accelerometers could be accommodated within the buoy, but presently available angular rate sensors are too costly. Steele and Earle (1991) describe a magnetometer approach for obtaining buoy angular motions, but this approach could have insufficient accuracy caused by buoy azimuthal motion at frequencies with non-negligible wave variance.

Prototype wave measurement MDDB's that have been built and that are being used for initial wave tank and field testing provide for storage and downloading (without satellite transmission) of digital time series data from all wave-related sensors so that buoy wave responses can be quantified. On-buoy data processing and satellite transmission of results will be added. Appropriate data processing algorithms will correct for buoy motion effects.

While the buoys are called MDDB's, a shallow water moored version is planned after capability to measure waves with such small buoys is demonstrated. There is room within an outer case that is jettisoned after water entry for a mooring line and the case or a weight within the case may serve as an anchor.

Scientific applications of directional wave information usually require high quality directional wave spectra, but most Naval applications for which MDDB's would be employed could be satisfied by the following basic wave parameters: significant wave height, dominant wave period corresponding to the frequency with maximum spectral density, and a primary wave direction (e.g. mean wave direction corresponding to the frequency with maximum spectral density). Although there is little uncertainty that these parameters can be provided with suitable accuracy, design goals have been set higher to provide scientifically better and more complete data without substantial buoy cost increases.

Measurement and Analysis Approaches

Work with theory, computer simulations, and measured data to develop and test wave measurement and analysis procedures used by the National Data Buoy Center (NDBC) indicates that non-directional wave information can be obtained from a wave measurement MDDB instrumented with a buoy-fixed accelerometer. Obtaining accurate directional wave information is thus a main development objective. Several directional wave measurement and analysis approaches could be used.

Directional wave measurement instrumentation used with pitch/roll buoys often consists of a vertically-stabilized accelerometer and pitch/roll sensors. Versions of the Datawell HIPPY sensor that are used in several commercially available wave measurement buoys and in some NDBC buoys are the best examples of this type of instrumentation. A compass or magnetometer provides buoy azimuth for transforming buoy pitch and roll to buoy tilts in two mutually perpendicular earth-fixed horizontal directions. Data processing involves standard calculations of cross-spectra between buoy tilts in the earth-fixed horizontal directions and the vertical component of buoy acceleration (or buoy heave obtained from double integration of the vertical acceleration component) as well as consideration of buoy responses. This approach has been described often (e.g. Longuet-Higgins et al., 1963; Steele et al., 1985; Steele et al., 1992). Large and expensive vertically-stabilized sensors are not suitable for a wave measurement MDDB. However, MDDB microprocessor capability accommodates software that performs these analyses so that developed electronics could be used with larger buoys and more expensive sensors.

Measurement of buoy angular rotation rates about and acceleration components along three mutually perpendicular buoy-fixed axes, as well as measurement of buoy azimuth provides sufficient information to obtain buoy vertical acceleration and tilts in earth-fixed coordinates (e.g. Earle and Longman, 1986). Multiple solid-state angular rate sensors and accelerometers can be accommodated within an MDDB, but presently available solid-state angular rate sensors are too costly. Developed electronics provides for digitization of eight wave measurement channels so that three angular rates, three acceleration components, and two magnetic field components could be utilized when lower cost angular rate sensors are available.

A magnetometer approach for obtaining buoy angular motions (pitch, roll, and azimuth) without use of a sensor specifically to measure buoy pitch and roll has been developed (Steele and Earle, 1991). A buoy-fixed accelerometer would measure the approximate vertical acceleration component. Using NDBC wave data, Wang et al. (1993) show that this approach generally provides good directional wave results. If used with a wave measurement MDDB, this approach could have insufficient accuracy caused by buoy azimuthal motion at frequencies with non-negligible wave variance. MDDB microprocessor capability will permit use of this approach if buoy wave tank and field tests show that it can be used.

Most Naval applications, but not scientific applications, could be satisfied by reasonably accurate values of significant wave height, dominant wave period, and a primary wave direction. These basic wave parameters could likely be determined by a simplified sensor system consisting of a single buoy-fixed accelerometer, two mutually-perpendicular fluid tilt sensors, and a compass or magnetometer. The tilt sensors would not measure true buoy tilt, but would respond to the local acceleration which is nearly perpendicular to the sea surface (e.g. Tucker, 1959; Earle and Longman, 1983). Wave height and period would be obtained by standard non-directional spectral analysis of acceleration data with corrections for fixing of the accelerometer similar to those used by NDBC. If a wave measurement MDDB has a wave response between that of a pitch/roll buoy and a buoy that responds mainly to wave orbital velocity drag forces, a directional probability distribution could provide primary wave direction with use of acceleration data to remove directional ambiguities (e.g. tilts in the direction of wave advance near wave crests and opposite tilts near wave troughs). While this approach could be extended to estimate primary wave directions as a function of frequency using frequency domain calculations, it would not provide standard directional wave spectra.

Prototype wave measurement MDDB's that have been built for testing have been designed to permit evaluation of the best approach to use including a multiple accelerometer approach that has apparently not been used with other buoys. Two mutually-perpendicular fluid tilt sensors provide estimates of buoy tilt deviations from wave slopes assuming linear wave theory. Three mutually-perpendicular acceleration components are measured by accelerometers and buoy azimuth is provided by a gimballed fluxgate compass (possibly later by a magnetometer). Assuming linear wave theory, numerical solutions of equations involving azimuth, acceleration components, and tilt deviations provide estimates of three earth-fixed acceleration components. Non-directional and directional wave information can then be obtained using procedures that are analogous to those used with pitch/roll buoy data.

Wave measurement MDDB's will not transmit complete directional wave spectra due to ARGOS message length limitations. However, directional spectra could be calculated within the buoy using cross-spectral analysis of measured time series. Cross-spectral analysis results can be used with several directional spectra calculation techniques such as the directional Fourier coefficient method (e.g. Longuet-Higgins et al., 1963), maximum likelihood methods (e.g. Oltman-Shay and Guza, 1984), maximum entropy methods (e.g. Lygre and Krogstad, 1986), and other methods (e.g. Benoit, 1993). Non-iterative methods could be applied within an MDDB, but iterative methods have not been examined for MDDB use.

Wind speed and direction will be estimated from non-directional spectral density levels and mean wave directions at high frequencies using extensions of NDBC procedures for wind and wave data quality control (e.g. Palao and Gilhousen, 1993). Field tests will determine whether estimates are accurate enough for Naval use.

Prototype Design

Prototype wave measurement MDDB data acquisition specifications are given in Table 2.

Table 2
Prototype Data Acquisition Specifications

Sampling rate per channel	5.12 Hz
Maximum number of wave channels	8
Analog to digital resolution	12 bits
Data acquisition frequency	Every hour
Date points per acquisition cycle	4096
Record length	800 s (13.33 minutes)
Sensors	3 accelerometers 2 fluid tilt sensors fluxgate compass strain gage atmospheric pressure sensor air temperature thermistor sea temperature thermistor

The fluxgate compass is used for expediency. It may be replaced by a solid-state magnetometer. Two input channels are used by the compass, and would be used by a magnetometer, so that there is a spare wave channel. All wave channels have matched electronic anti-aliasing filters.

Figure 1 illustrates a wave measurement MDDB as packaged before deployment, a pre-deployment cross-section, and a post-deployment cross-section. Prototype buoys for initial testing do not include a parachute nor the clamshells (that are jettisoned) around the floatation collar. Wave tank and field tests are conducted with the mast in its deployed position and an inflated floatation collar. Detachable anti-rotation wings (i.e. fins) and vertical dampers of various sizes are being tested. Tests are being made in which the outer pre-deployment case serves as a subsurface sea anchor for the buoy. There is room between the buoy itself and the outer case for storage of spring-loaded wings and dampers as well as line for deployed suspension of the outer case as a sea anchor.

Three prototype wave measurement MDDB's have been built for initial wave tank and field testing. Figure 2 shows two of the prototypes.

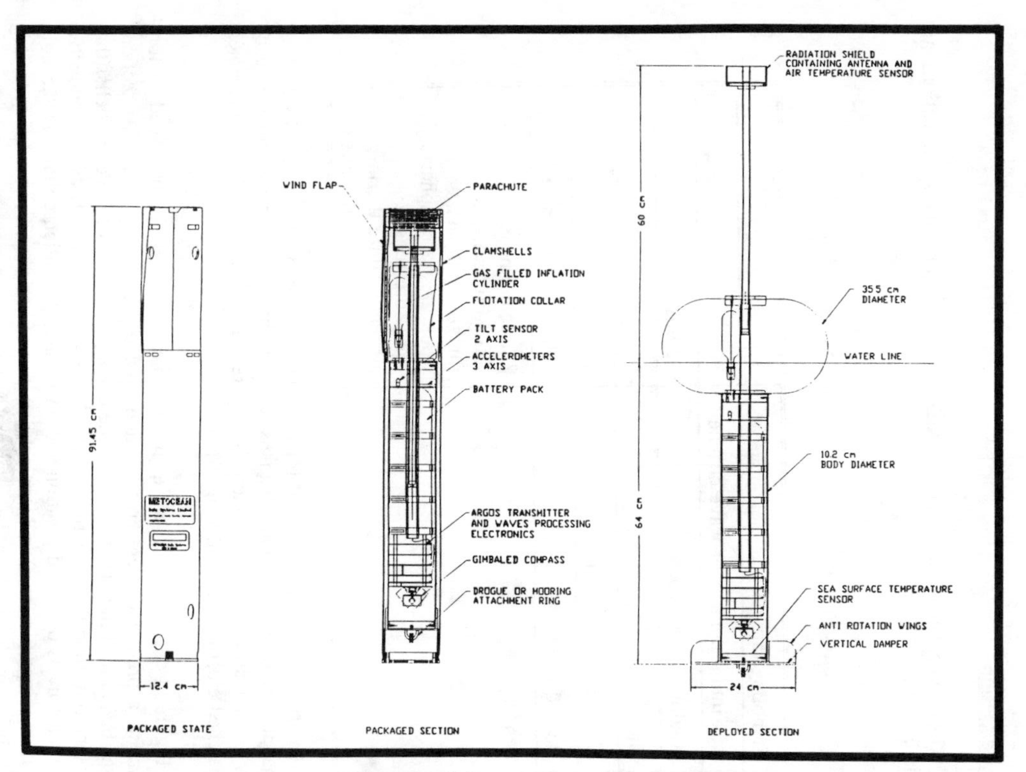

Figure 1. Wave Measurement MDDB. The outer package before deployment, a pre-deployment cross-section, and a post-deployment cross-section are illustrated.

Figure 2. Two Wave Measurement MDDB Prototypes.
The buoy on the left has anti-rotation wings.

Understanding MDDB motion in response to waves is a goal. Several wave measurement buoy descriptions document the need for wave-induced buoy motion information (e.g. Stewart, 1977; Steele et al., 1985; Steele et al., 1992). A high sampling rate is used initially to facilitate analysis of digital time series data from each sensor. Prototype buoys internally store each sensor's digital time series data rather than processing these data. A watertight connector at the top of the buoy mast permits connecting a cable to a notebook computer's serial port. Using the notebook computer, data acquisition procedures are programmed and data are downloaded to the notebook computer's hard disk. Programs used during initial wave tank and field tests plot the data on the computer's screen and perform basic statistical and spectral calculations. Stored data can be analyzed later to study buoy motion and to test described analysis approaches. During initial wave tank and field tests, data will be downloaded after each data acquisition cycle. Initial field tests will not be conducted in high sea states, but later field tests are planned during higher sea states using on-buoy data processing and satellite transmission.

Wave-related electronics boards consist of: a signal conditioning board that filters out high frequency signal components; a data acquisition board that performs analog to digital conversions, stores data in random access memory (RAM), and communicates with a PC via a serial interface; and a digital signal processing (microprocessor) board. The digital signal processing board uses a Motorola CMOS 68000 microprocessor with up to 512 kbytes of programmable read only memory (PROM) and up to 512 kbytes of RAM. The board has been designed to be double-sided, rather than multi-layer, to reduce later production costs. It has not been placed in the three constructed buoys since it is not needed for acquisition and downloading of digital time series data.

Two electronics boards are used from previous MDDB versions. One board contains an Intel 8031 microcontroller which controls overall buoy activities. This board also acquires and processes atmospheric pressure, air temperature, and sea surface temperature data. Another board transmits processed non-wave and wave information to ARGOS satellites.

On-buoy software is being written in C language. Desktop computer simulations of data processing that could occur, including tests with versions of NDBC's on-buoy data processing algorithms, show that processing can be completed within 20 to 120 s depending on microprocessor clock speed.

Wave measurement MDDB's have been designed so that the battery pack, electronics boards, and sensors other than the fluxgate compass can be moved within the case to improve wave response and/or to meet air deployment mass distribution constraints. An expendable weight below the buoy itself can also be used to adjust mass distribution to meet air certification requirements before the parachute opens.

Only 256 bits of information can be transmitted in an ARGOS message. Directional wave spectra or the cross-spectral parameters from which directional wave spectra are calculated cannot be fit into one or a few messages. Three single message formats were developed and a version of a paired (two) message format that was previously developed for NDBC was considered. For Naval applications, transmitted information should fit into one message. Developed single message formats include: an even period message, an even frequency message with particular ordering of the frequencies and transmission of information only for frequencies with non-negligible wave variance, and a simple format consisting of wave parameters. The even period format that is shown in Table 3 will be used initially. The logarithmic encoding schemes are used by NDBC for compression of spectral analysis results for transmission by GOES satellite. Message formats could be changed by changing the on-buoy software.

Mean wave directions rather than directional spectra will be transmitted. It is expected that buoy responses corrections can be made using versions of NDBC's procedures (e.g. Steele et al., 1992). However, for an axially-symmetric buoy, mean wave directions are independent of buoy heave and pitch/roll amplitude and phase responses if the net phase response difference between heave and pitch/roll is between -90° and $+90^{\circ}$ (e.g. Kuik et al., 1988; Tucker, 1989). Accurate mean wave directions should be obtained even if response corrections have some errors.

Table 3
Even Period ARGOS Message Format

Information	Encoding (if used)	Bits
Date quality assurance:		
Message checksum		12
Error indicators		
Horizontal acceleration		1
Vertical acceleration		1
Tilts		2
Compass		1
Mean azimuth, 6° resolution	linear	6
Bit subtotal		23
Wave/wind information:		
Maximum spectral density band number		4
Maximum spectral density	logarithmic	10
Other 15 spectral densities normalized to maximum, 6 bits each	logarithmic	90
16 mean wave directions, 6° resolution, 6 bits each	linear	96
Crossing wave indicator at maximum spectral density		1
Wind speed, 1 m/s resolution	linear	6
Wind direction, 6° resolution	linear	6
Bit subtotal		213
Other information:		
Atmospheric pressure, 1 mb resolution	linear	8
Air temperature, 1° resolution	linear	6
Sea temperature, 1° resolution	linear	6
Bit subtotal		20
Total bits		256
Spectrum periods (s): 3, 4, 5, 6, 7, 8, 9, 10, 11, 12, 14, 16, 18, 20, 22, 24		

Power requirements were determined assuming hourly data acquisition with ARGOS transmissions disabled during data acquisition to avoid radio frequency interference. Prototype buoy power consumption is 46 (mA-hr)/hr. Alkaline C cell batteries provide a capacity of 28 A-hr so that buoy lifetime is approximately 25 days. This lifetime is suitable for Naval applications and most scientific applications for which a small expendable wave measurement buoy would be used.

The total cost of all components (hull, sensors, electronics, etc.) for a prototype buoy is approximately $4,000. Wave measurement MDDB's can likely be built for roughly $4,000 in production quantities.

Further Development

The planned schedule is for air-deployable wave measurement MDDB's to be available for Navy fleet use in exercises and demonstrations by October-December, 1994. These buoys would still be considered prototypes by the Navy pending transition to production and operational status after passing military specification tests and air deployment certification procedures. Table 4 summarizes planned development activities.

Table 4
Development Plan Summary

Activity	Estimated Completion Date(s)
Initial wave tank tests (U.S. Naval Academy)	August, 1993
Initial field tests (CERC Field Research Facility)	September, 1993
Test evaluations, design improvements, on-buoy software coding, addition of ARGOS and air-deployment capability	January, 1994
Possible directional wave basin tests and tests near an NDBC buoy, air-deployment field test	June, 1994
Participation in DUCK94 using non-air-deployed buoys outside surf zone to provide data for surf modeling	August and October, 1994
Delivery for fleet use	October - December, 1994

Initial wave tank tests are being made in the U.S. Naval Academy Hydromechanics Laboratory 128 m long high performance towing/wave tank. These tests are being conducted using sinusoidal waves and waves representing realistic spectra with heights up to approximately 1 m.

The U.S. Army Engineer Waterways Experiment Station Coastal Engineering Research Center (CERC) Field Research Facility (FRF) at Duck, NC, is being used for initial field tests. For these tests, prototype wave measurement MDDB's are compliantly moored near wave measurement instrumentation maintained by the FRF.

Wave measurement MDDB technology could be used with larger standard drifting data buoys as well as with non-air-deployable small buoys that have better wave following characteristics than MDDB's. For example, inexpensive discus-shaped hulls could be utilized. If somewhat larger buoys are not considered expendable, more expensive and better sensors (e.g. accelerometers, angular rate sensors, and magnetometers) could be utilized.

Summary and Conclusions

A sonobuoy-sized expendable air-deployable directional wave sensor is being developed as part of a program to make Mini-Drifting Data Buoys (MDDB's) available for operational Naval use. Prototype buoys have been designed and three prototypes have been built for initial wave tank and field testing. Unusual buoy motion related to the buoy's small size and configuration will likely cause the most important problems to overcome. Several measurement and analysis approaches have been examined. Constructed prototypes provide for storage and downloading of digital time series data from all wave-related sensors so that buoy wave responses can be quantified and the most appropriate measurement and analysis approaches can be selected. Initial wave tank and field tests of three constructed prototypes are now being conducted.

The technology that is under development has additional military, civilian, and scientific applications if incorporated into moored buoys, standard drifting data buoys, non-air-deployable buoys, and non-expendable buoys. Wave measurement MDDB development has considered these applications so that developed technology can be put to wider use.

Acknowledgements

Development of non-wave measurement MDDB's is co-sponsored by the Chief of Naval Operations, Oceanographer of the Navy, through the Naval Research Laboratory's Tactical Oceanographic Warfare Support Program Office (TOWS) and the Defense Electronics Division, Industry Science and Technology, Canada. METOCEAN Data Systems Ltd. is the contractor for this development. The Chief of Naval Operations, Oceanographer of the Navy, through TOWS is sponsoring development of wave measurement MDDB's using Neptune Sciences Inc. as the prime contractor and METOCEAN Data Systems Ltd. as a subcontractor. MDDB testing to meet air certification and military specification requirements is provided by the Naval Air Warfare Center, Aircraft Division, Indianapolis. Colin Frame and Andrew Keast of METOCEAN Data Systems are performing much of the electronic and mechanical development. Joseph Eckard, Eileen Kennelly, and David Zwack of Neptune Sciences are providing software and testing support. William Popovich of the Naval Research Laboratory, Stennis Space Center, is assisting in testing and satellite data relay aspects.

References

Benoit, M., "Extensive Comparison of Directional Wave Analysis Methods from Gauge Array Data," *Proceedings WAVES 93 Symposium*, New Orleans, LA, 1993.

Earle, M.D., and Longman, R.W., "An Analysis of Short Pendulum Heave Sensors," *Proceedings Symposium on Buoy Technology*, Marine Technology Society and NOAA Data Buoy Center, New Orleans, LA, 1983.

Earle, M.D., and Longman, R.W., "An Alternate Hull Pitch/Roll System for Use with NDBC DACT/DWA Systems," MEC Systems Corporation (now Neptune Sciences, Inc.) report for the National Data Buoy Center, Stennis Space Center, MS, 1986.

Kuik, A.J., Van Vledder, G.P., and Holthuijsen, L.H., "A Method for the Routine Analysis of Pitch-and-Roll Buoy Wave Data," *Journal of Physical Oceanography*, Vol. 18, pp. 1020-1034, 1988.

Longuet-Higgins, M.S., Cartwright, D.E., and Smith, N.D., "Observations of the Directional Spectrum of Sea Waves Using the Motions of a Floating Buoy," *Ocean Wave Spectra*, Prentice-Hall, pp. 111-136, 1963.

Lygre, A., and Krogstad, H.E., "Maximum Entropy Estimation of the Directional Distribution in Ocean Wave Spectra," *Journal of Physical Oceanography*, Vol. 16, pp. 2052-2060, 1986.

Oltman-Shay, J., and Guza, R.T., "A Data-Adaptive Ocean Wave Directional-Spectrum Estimator for Pitch and Roll Type Measurements," *Journal of Physical Oceanography*, Vol. 14, pp. 1800-1810, 1984.

Palao, I.M., and Gilhousen, D.B.,"The Re-Derivation of the NDBC Wind-Wave Algorithm" *Proceedings WAVES 93 Symposium*, New Orleans, LA, 1993.

Selsor, H., "An Overview of the US Navy's AN/WSQ-6 (Series) Drifting Buoy Program," *Proceedings International Maritime Defence Conference*, 1993.

Steele, K.E., Lau, J.C., and Hsu, L., "Theory and Application of Calibration Techniques for an NDBC Directional Wave Measurements Buoy," *IEEE Journal of Oceanic Engineering*, Vol. OE-10, pp. 382-396, 1985.

Steele, K.E., and Earle, M.D., "Directional Ocean Wave Spectra Using Buoy Azimuth, Pitch, and Roll Derived from Magnetic Field Components," *IEEE Journal of Oceanic Engineering*, Vol. 16, pp. 427-433, 1991.

Steele, K.E., Teng, C.C., and Wang, D.W.C., "Wave Direction Measurements Using Pitch-Roll Buoys," *Ocean Engineering*, Vol. 19, pp. 349-375, 1992.

Stewart, R.H., "A Discus-Hulled Wave Measuring Buoy," *Ocean Engineering*, Vol. 4, pp. 101-107, 1977.

Tucker, M.J., "The Accuracy of Wave Measurements Made with Vertical Accelerometers," *Deep-Sea Research*, Vol. 5, pp. 185-192, 1959.

Tucker, M.J., "Interpreting Directional Data From Large Pitch-Roll-Heave Buoys," *Ocean Engineering*, Vol. 16, pp. 173-192, 1989.

Wang, D.C., Teng, C.C., and Ladner, R., "Buoy Directional Wave Measurements Using Magnetic Field Components," *Proceedings WAVES 93 Symposium*, New Orleans, LA, 1993.

Buoy Directional Wave Measurements Using Magnetic Field Components

David W. Wang[1], C.C. Teng[2], and R. Ladner[1]

Abstract

Directional wave data can be estimated from using the heave, pitch, and roll motion of a discus buoy, which are conventionally measured by a pitch-roll sensor. To reduce the size and cost, this study presents a method to determine the buoy's azimuth, pitch, and roll angles from the two magnetic field components along two orthogonal axes parallel to the deck of a buoy measured by a fixed magnetometer. An evaluation of three months of directional wave measurements from a 3-meter discus buoy is presented. The directional wave data from using the magnetic components agree well with those derived from measurements by a pitch-roll sensor.

Introduction

Use of a pitch-roll buoy is a reliable and widely accepted approach for *in situ* directional wave measurements in deep waters. The buoy's heave, pitch, and roll motions induced by ocean surface waves provide the needed information for estimation of wave direction (Longuet-Higgins et al. 1963). The pitch and roll motions are usually measured by an expensive and bulky pitch-roll sensor. The National Data Buoy Center (NDBC) has been developing a method to estimate the pitch and roll angles from measurements of the magnetic field components (Burdette 1976; Hsu 1986a; Burdette 1987; Steele 1990; Steele and Earle 1991). By using this method, referred to as the MO (Magnetometer-Only) method, the directional wave measurement system only needs a small and relatively inexpensive magnetometer to obtain the pitch and roll information. As a result, the acquisition and operational costs of directional wave measurements are greatly reduced and the NDBC buoy network can provide directional wave measurements at a significantly lower cost. As an example, the configurations and costs of a typical magnetometer and the Datawell Hippy 40 II are listed in Table 1.

Numerical simulations have been used to evaluate the MO method (Hsu 1986b; Earle 1992). In addition, along the coasts of the United States, several field deployments equipped with wave measurement system using the MO method have been carried out (Earle 1990; Lipa

---

[1] Computer Sciences Corporation, Stennis Space Center, MS 39529-6000

[2] National Data Buoy Center, Stennis Space Center, MS 39529-6000

1990; Mettlach et al. 1991; Teng et al. 1991; Teng et al. 1992; Wang 1993). The MO method shows good performance based on evaluations of field data, which were collected under limited marine environment conditions. This study presents the results of an evaluation of the MO method from 3 months of directional wave measurements at an NDBC buoy station located off Monterey Bay, California (36.8° N., 122.4° W.), where the marine environment conditions are complicated due to the coexistence of long-period swells and local wind-generated waves during this three months (Wang 1992).

The directional wave data analyzed in this study were collected by two directional measurement systems, which were collocated on an NDBC 3-meter discus buoy deployed at station 46042 off Monterey Bay, California, during July 1991. The first system is a standard NDBC directional wave measurement system, which was equipped with a Datawell Hippy 40 II sensor for the measurement of heave, pitch, and roll. This system is referred to as the Hippy system (or DWA/Hippy). The other measurement system was equipped with a strapped-down accelerometer measuring buoy heave acceleration and a General Oceanics triaxial magnetometer measuring the magnetic vector for the estimation of pitch and roll by the MO method. This system is referred to as the MO system (or DWA/MO). Details of these two systems can be found in Steele (1990) and Steele et al. (1990). The data used in the analysis were collected in October, November, and December 1991. A total of 2140 hours of wave data was used.

Theory

Steele and Lau (1986) showed that the magnetic field vector at the mounting point inside the buoy hull provided by a magnetometer can be approximated by

$$\begin{aligned} B_i = b_{io} + B_{ez}\,[\,b_{i1}\sin(P) - b_{i2}\cos(P)\sin(R)\,] \\ + B_{ey}\,\{\,-b_{i2}\cos(R)\sin(A) \\ + [b_{i1}\cos(P) + b_{i2}\sin(P)\sin(R)]\cos(A)\,\} \end{aligned} \tag{1}$$

where B_i are the vector components along buoy bow (i=1) and starboard axis (i=2), A, P, and R are buoy azimuth, pitch, and roll angle, respectively; B_{ey} and B_{ez} are the horizontal and vertical components of the local earth's magnetic field; b_{10} and b_{20} are the constants for the fixed hull magnetic field correction; and b_{11}, b_{22}, b_{12}, and b_{22} are the constants for the induced hull magnetic field correction. Details of obtaining the correction constants can be found in Remond and Teng (1990).

It is assumed that the magnetic field vector, B_i, consists of high- and low-frequency part, which is expressed as

$$B_i = \overline{B}_i + B_i' \tag{2}$$

where $\overline{B}_i$ is the low-frequency part of B_i due to the buoy's azimuth motion (yaw) and B_i' is the high-frequency part of B_i due to the buoy's tilting motions (pitch and roll).

The MO method determines the buoy pitch and roll from the high-frequency part of B_i and the buoy azimuth from the low-frequency part of B_i. In other words, the buoy yaw, pitch, and roll motion are not determined from measurements of a pitch-roll sensor. Instead, they can be determined from measurements of a magnetometer.

It is assumed that the time average pitch and roll angle are zero and the pitch and roll are small. Therefore,

$$\sin(A)=\frac{-b_{11}(\overline{B_2}-b_{20})+b_{21}(\overline{B_1}-b_{10})}{\Delta\ B_{ey}} \tag{3}$$

and

$$\cos(A)=\frac{b_{22}(\overline{B_1}-b_{10})-b_{12}(\overline{B_2}-b_{20})}{\Delta\ B_{ey}} \tag{4}$$

where

$$\Delta=b_{11}b_{22}-b_{12}b_{21} \quad . \tag{5}$$

The azimuth angle, A, is then determined by

$$A=\tan^{-1}[\sin(A),\cos(A)] \quad . \tag{6}$$

The pitch, P, and roll, R, can be determined by

$$\sin(P)=\frac{b_{22}B_1^{'}-b_{12}B_2^{'}}{\Delta\ B_{ez}} \tag{7}$$

and

$$\sin(R)=\frac{-b_{11}B_2^{'}+b_{21}B_1^{'}}{\Delta\ B_{ez}} \quad . \tag{8}$$

Details of the derivation of the MO method can be found in Steele (1990) and Steele and Earle (1991).

To show the application of above equations, measurements of pitch and roll from a Datawell Hippy 40 II and buoy hull magnetic components from a magnetometer aboard an NDBC 3-meter discus buoy were used. Figures 1 and 2 show the time series of B_1 and B_2. As can be seen, the B_1 and B_2 show a time variation consisting of a low- and a high-frequency part. A low pass filter with a cutoff frequency of 0.03 Hz was used to separate B_1 and B_2 into a low- and a high-frequency part. Figures 3 and 4 show the comparison of the pitch and roll derived from the high-frequency part of B_1 and B_2 by Equations (7) and (8) and those measured by the Datawell Hippy 40 II sensor. Figure 5 shows the buoy azimuth derived from the low-frequency part of B_1 and B_2 by Equation (6) and the measured buoy azimuth. Very good agreement is shown between the estimates from the MO method and those measured data. Steele and Earle (1991) showed the standard deviation of the difference of pitch and roll between the MO method and the Hippy measurement is about 2 degrees.

Data Analysis

The directional wave spectrum, $S(f,\alpha)$, can be expressed as the product of the nondirectional wave spectrum, $S(f)$, and a directional wave distribution function, $D(f,\alpha)$, which is

$$S(f,\alpha) = S(f)D(f,\alpha) \tag{9}$$

The $D(f,\alpha)$ can be expressed as

$$D(f,\alpha)=\frac{1}{\pi}[\frac{1}{2}+R_1 \cos(\alpha-\alpha_1)+R_2 \cos 2(\alpha-\alpha_2)] \tag{10}$$

where α_1 and α_2 are the frequency-dependent wave directions measured clockwise from true north around to the direction from which the wave is coming. The R_1 and R_2 are the two frequency-dependent parameters related to the directional spreading width. The α_1 is usually referred to as the mean wave direction. To more explicitly display the directional spreading, a directional spreading width parameter related to R_1 is used, which is

$$\sigma=\sqrt{2-2R_1} \tag{11}$$

For a narrow directional distribution, σ approximates to the root-mean-square of direction distribution about the mean wave direction (Ewing 1986). For general engineering application, α_1 and σ are the two most important parameters. The nondirectional wave spectral density, $S(f)$, is computed from 0.03 Hz to 0.40 Hz. The α_1 and σ are computed from 0.03 Hz to 0.35 Hz.

Results and Discussion

To show the performance of the MO method under various sea states, selected directional wave data under the conditions of severe sea, wind wave coexisting with swell, and turning winds are presented here. In late November 1991, strong winds from the northwest generated severe seas with a significant wave height, H_s, reaching 6.43 m at 1600 UTC, November 29, 1991, which was the highest sea state during the evaluation. Figures 6a-6c show directional wave data from the Hippy system and the MO system. The wind speed, U, was 18.2 m/s with a direction, *dir*, of 315 degrees. The nondirectional wave spectrum, $S(f)$, mean wave direction, α_1, and directional spreading width, σ, from the MO system agree well with those from the Hippy system. Figures 7a-7c show directional wave data collected when the local wind wave coexisted with long-period swell. The $S(f)$ displays three peaks at 0.07, 0.11, and 0.19 Hz. The wave directions at the high-frequency end of $S(f)$ aligned into the local wind of 142 degrees, wave directions around 0.11 Hz were from about 220 degrees, and wave directions around 0.07 Hz were from about 290 degrees. The wave directions from the MO system agree well with those from the Hippy system. The directional spreading width from the MO system also show reasonable agreement with those from the Hippy system, as shown in Figure 7c.

Figures 8 and 9 show the time series of wind speed and wind direction from November 16 to November 19, 1991. The wind speeds varied between 3 to 14 m/s while wind direction gradually shifted from about 75 to 350 degrees. Figures 10 and 11 show the comparison of wave direction at 0.30 and 0.35 Hz between the MO system and the Hippy system. As the wind direction continuously shifted, the wave directions from the MO system agreed well with those from the Hippy system and closely followed the local wind direction.

The scatterplots of α_1 of the Hippy system versus MO system from this 2,140-hour data set at frequencies of 0.08, 0.12, 0.20, and 0.30 Hz are shown in Figures 12a-12d. As can be seen, there is a very good agreement at 0.12, 0.20, and 0.30 Hz. At 0.08 Hz, the α_1 of the MO system shows less agreement with the Hippy system. Figures 13a-13d show the scatterplots of σ of the Hippy system versus the MO system at 0.08, 0.12, 0.20, and 0.30 Hz. As can be seen, σ from the MO system agrees well with σ from the Hippy system at 0.12, 0.20, and 0.30 Hz. A larger data scatter is seen at 0.08 Hz.

Statistical summaries of the difference of α_1 and σ between the two systems from 0.03 to 0.35 Hz are shown in Figures 14a and 14b. The mean difference of α_1 between the two systems is less than 1 degree at frequencies higher than 0.05 Hz. The standard deviation of the difference generally remains around 10 degrees at frequencies higher than 0.09 Hz. Figures 15a and 15b show statistics of the difference of directional spreading width, σ, between the two systems normalized by the σ of the Hippy system. At frequencies higher than 0.17 Hz, the mean difference remains constant and small (less than 1 percent). Even though the difference gradually increases as the frequency decreases from 0.16 to 0.1 Hz, the mean difference is still less than 10 percent with a standard deviation less than 20 percent. In general, the wave direction and directional spreading width at frequencies higher than 0.1 Hz from the MO system agree very well with those from the Hippy system.

The comparison of directional wave data between the Hippy system and the MO system shows less agreement at frequencies lower than 0.1 Hz. This is largely due to weak wave energy at lower frequencies. Figure 16 shows the time variation of spectral density due to arrivals of long-period swell at 0.08 Hz from the Hippy system from October 10 to 15. Figure 17 shows the wave direction at 0.08 Hz from both systems. As can be seen, the wave directions from the MO system agree well with those from the Hippy system when the wave energy was high on October 12 and 13. Figure 18 shows the difference of wave direction between the Hippy system and the MO system versus the wave energy density at 0.08 Hz. As indicated in the figure, larger differences (greater than 10 degrees) of wave direction between the two systems occurred when wave energy density was less than 1 m^2/Hz, which is equivalent to a 0.08-Hz monochromatic wave with a height of 0.28 m.

This indicates that the MO system can provide very good estimates of directional wave information at a lower frequency as long as the associated wave energy exceeds a certain threshold, which can be empirically determined from data comparisons.

Concluding Remarks

This evaluation, based on 2,140 hours of directional wave data obtained under various marine environmental conditions, demonstrates that the MO method works very well, especially at frequencies higher than 0.1 Hz. At lower frequencies, the MO method also works well when the energy exceeds a certain threshold.

Using the MO method to estimate wave directional information from the magnetometer outputs requires that the magnetic vector variation inside a buoy hull induced by buoy azimuth motion (yaw) is harmonically different from that induced by buoy pitch and roll motions. In other words, the buoy yaw motion occurs at a rate much slower than that of the pitch and roll motion. Hsu (1986b) showed, based on numerical simulation, that a slower buoy yaw motion provides a better estimate of directional wave parameters with the MO method. Hence, the buoy dynamics induced by the marine environmental forces are important factors in the success of applying the MO method. A better buoy hull design, which is more responsive to low-frequency waves and has a slower yaw motion, can definitely improve the directional wave estimation from the MO method, especially for lower frequencies.

Acknowledgements

The authors would like to thank Mr. Ken Steele for supporting this study and Mr. Jim Sharp and Mr. Ralph Dagnall for their engineering management and support of the field test. The valuable comments from the CSC Review Committee are also appreciated.

References

Burdette, E.L. (1976). "A study of the feasibility of using a triaxial magnetometer as an attitude sensor in a buoy mounted system for the measurement of ocean wave directional spectra." NDBC Technical Report.

Earle, M.D. (1990). "Evaluation and application of time series data from the NDBC prototype SWADE buoy." NDBC Technical Report.

Earle, M.D., and Herchenroder, B.E. (1991). "Investigation of filtering techniques for magnetometer-only determination of buoy heading." NDBC Internal Report.

Ewing, J.A. (1986). "Presentation and interpretation of directional wave data." *Underwater Technology*, 12(3).

Hsu, Y.H.L. (1986a). "Development of algorithms for use of a magnetometer measuring buoy pitch and roll motions." NDBC Technical Report DB-3505-3.

Hsu, Y.H.L. (1986b). "A technique for the detection of inaccurate data being produced by the No-Hippy DWA system." NDBC Technical Report DB-3510.

Lipa, B. (1990). "Comparison of Hippy 40 and magnetometer-only systems on a 12-meter discus buoy." NDBC Technical Report.

Longuet-Higgins, M.S., Cartwright, D.E., and Smith, D.E. (1963). "Observations of the directional spectrum of sea waves using the motions of a floating buoy," in *Ocean Wave Spectra*, Prentice-Hall, Englewood Cliffs, NJ.

Mettlach, T., Gilhousen, D., and Wang, D. (1991). "Evaluation of the National Data Buoy Center Coastal Buoy," in *Proceedings of MTS '91*, 1208-1215.

Remond, F.X., and Teng, C. (1990). "Automatic determination of buoy hull magnetic constants," in *Proceedings of Marine Instrumentation '90*, 151-157.

Steele, K.E. (1990). "Measurement of buoy azimuth, pitch, and roll angles using magnetic field components." NDBC Technical Report 1804-91.03, 16 pp.

Steele, K.E., and Earle, M.D. (1991). "Estimation of directional ocean wave spectra using azimuth, pitch, and roll angles derived from magnetic field components only." *IEEE J. of Oceanic Engineering*, 16(4), 427-433.

Steele, K.E., and Lau, J. (1986). "Buoy azimuth measurements-correction for residual and induced magnetism," in *Proceedings of Marine Data Systems International Symposium*, 271-276.

Steele, K.E., Wang, D.W., Teng, C., and Lang, N. (1990). "Directional wave measurements with NDBC 3-meter discus buoys." NDBC Technical Report 1804-01.05.

Teng, C.C., Dagnall, R., and Remond, F. (1991). "Field evaluation of the magnetometer-only directional wave system from buoys," in *Proceedings of MTS '91*, 1216-1224.

Teng, C.C., Timpe, G.L., and Dagnall, R.J. (1992). "Field test and evaluation of NDBC's Coastal Buoy," in *Proceedings of MTS '92*, 943-948.

Wang, D.W. (1992). "Long-term wave statistics from an NDBC buoy station," in *Proceedings of MTS '92*, 477-485.

Wang, D.W. (1993). "Field evaluation of the wave processing module magnetometer-only version (WPM/MO)." NDBC Technical Report 1804-01-06.

Table 1. Comparison of the two systems.

	Datawell Hippy 40 Mark II	General Oceanics Magnetometer
SIZE	0.553 m (height) 0.378 m (diameter) = 0.24 m^3	10.3 cm * 5cm * 3.8 cm = 196 cm^3
WEIGHT	36 Kg	0.23 Kg
PRICE (approximate)	$ 20,000	$ 1,500

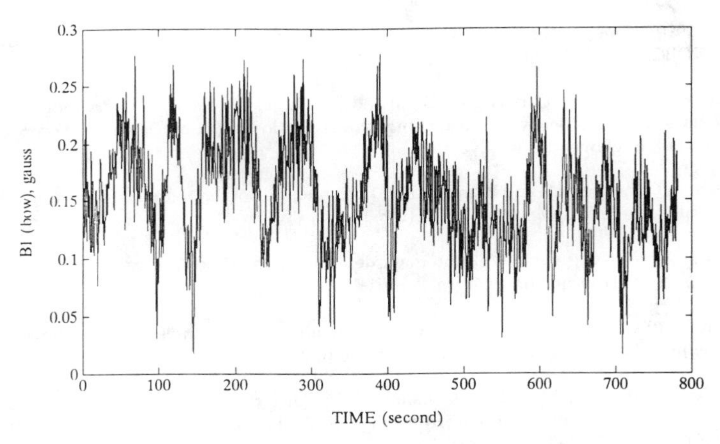

Figure 1. Time series of magnetic vector B_1.

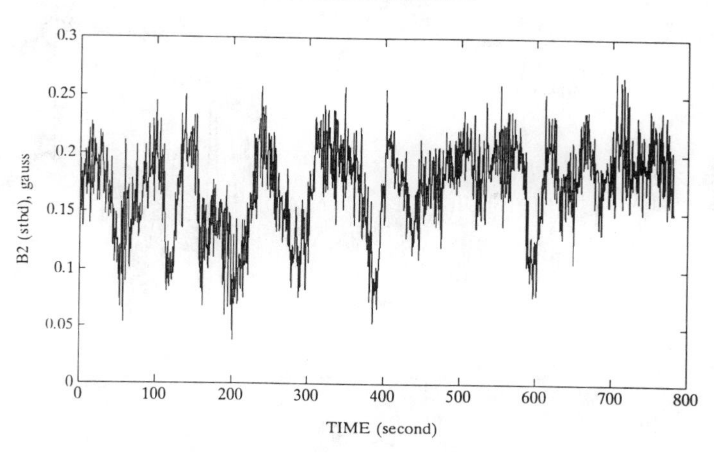

Figure 2. Time series of magnetic vector B_2.

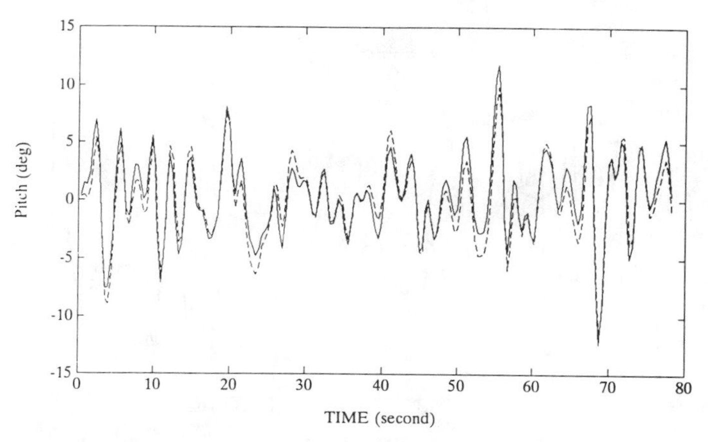

Figure 3. Time series of buoy pitch angle measured by Hippy 40 (solid line) and estimated by MO method (dashed line).

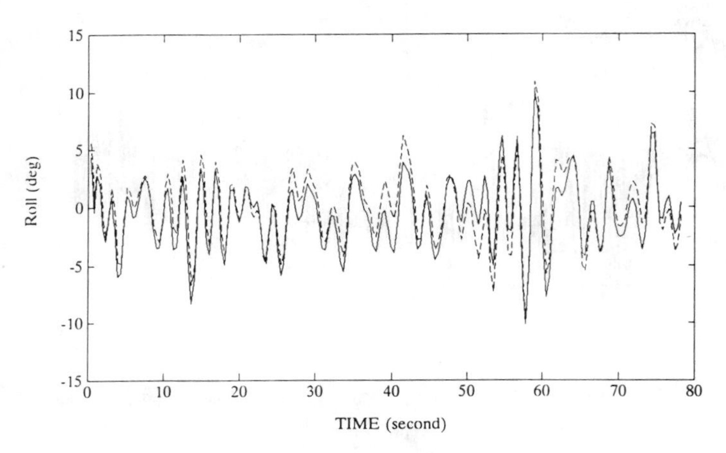

Figure 4. Time series of buoy roll angle measured by Hippy 40 (solid line) and estimated by MO method (dashed line).

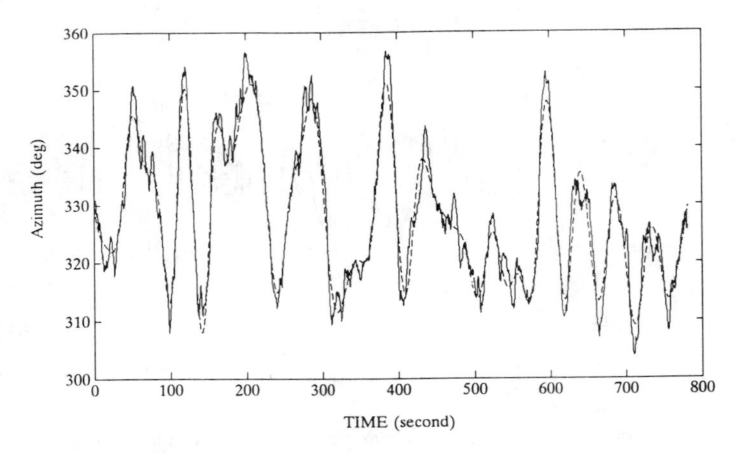

Figure 5. Time series of buoy azimuthal angle computed using the pitch and roll from Hippy 40 (solid line) and using the pitch and roll estimated by MO method (dashed line).

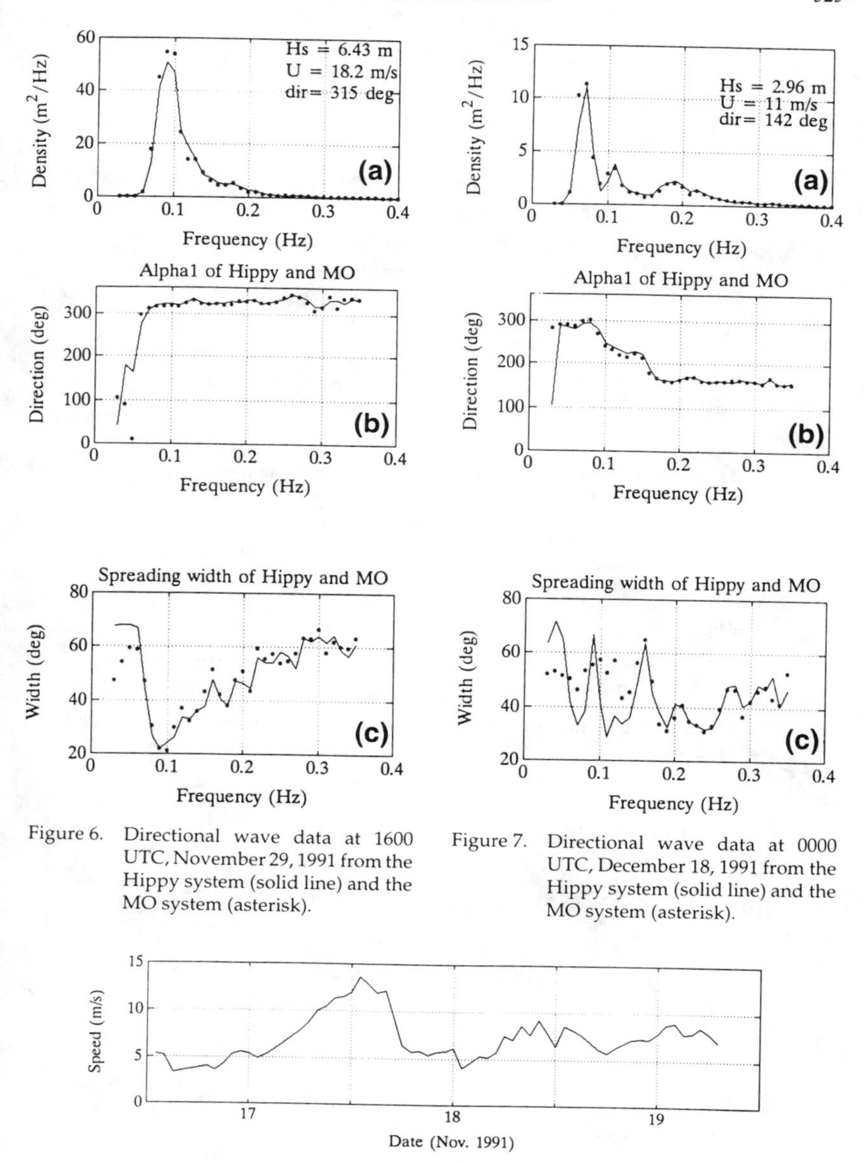

Figure 6. Directional wave data at 1600 UTC, November 29, 1991 from the Hippy system (solid line) and the MO system (asterisk).

Figure 7. Directional wave data at 0000 UTC, December 18, 1991 from the Hippy system (solid line) and the MO system (asterisk).

Figure 8. Time series of wind speed at 5 meter height.

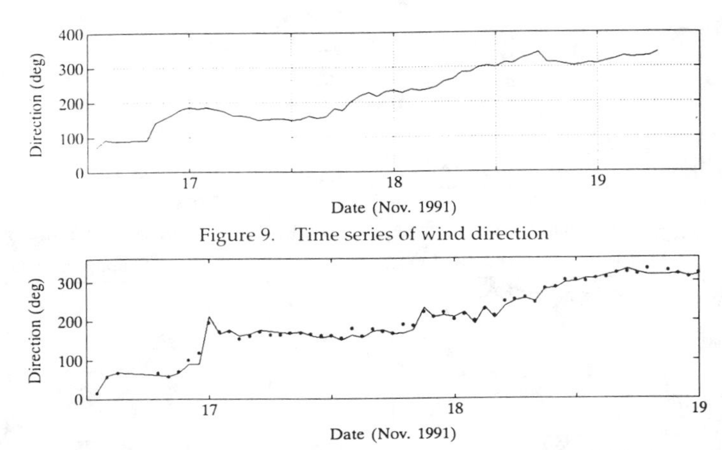

Figure 9. Time series of wind direction

Figure 10. Mean wave direction at 0.30 Hz from the Hippy system (solid line) and the MO system (asterisk).

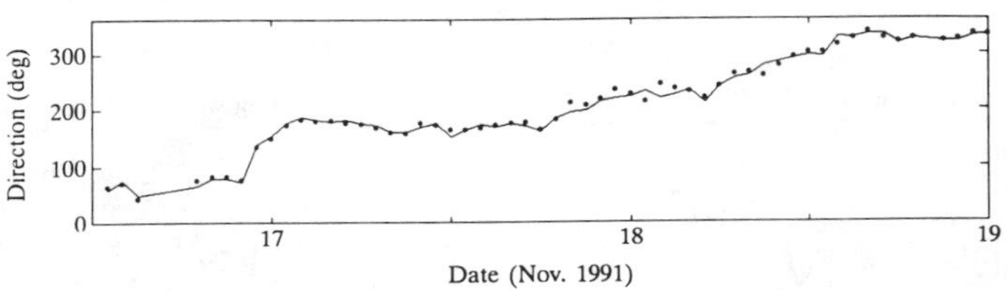

Figure 11. Mean wave direction, α_1, at 0.30 Hz from the Hippy system (solid line) and the MO system (asterisk).

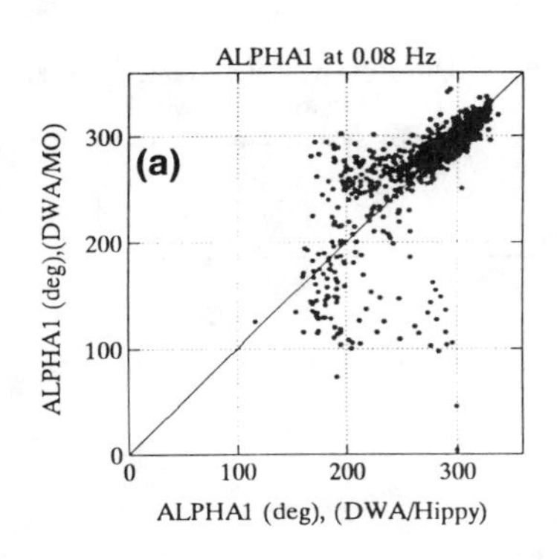

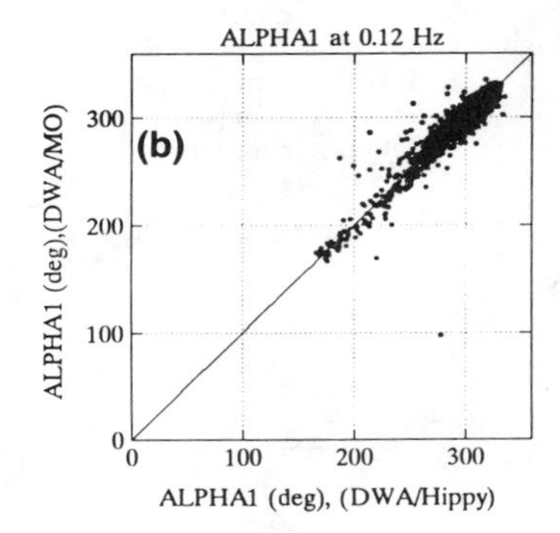

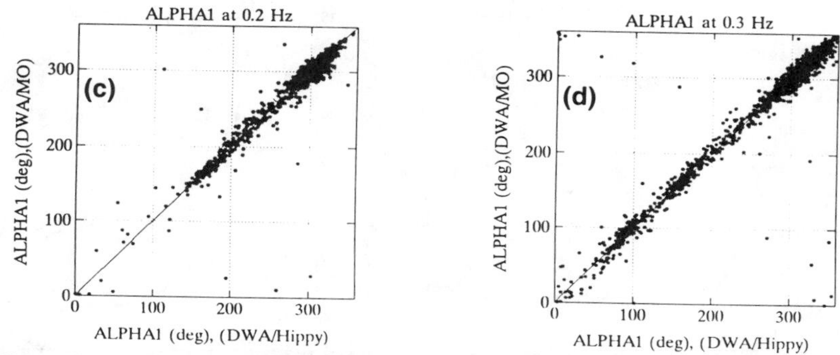

Figure 12. Mean wave direction, α_1, of the Hippy system versus the MO system at (a) 0.08 Hz, (b) 0.12 Hz, (c) 0.20 Hz, and (d) 0.30 Hz.

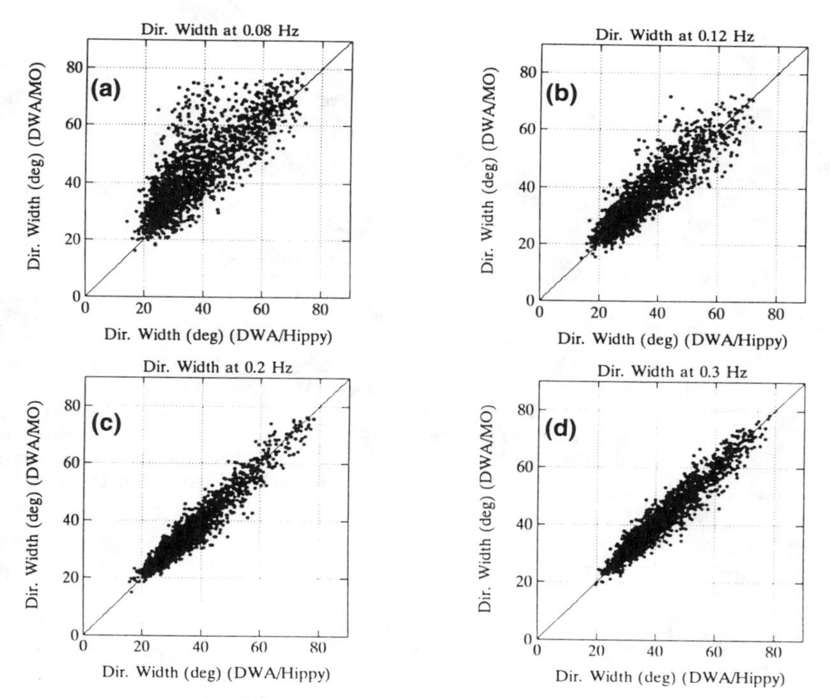

Figure 13. Directional spreading width, σ, the Hippy system versus the MO system at (a) 0.08 Hz, (b) 0.12 Hz, (c) 0.20 Hz, and (d) 0.30 Hz.

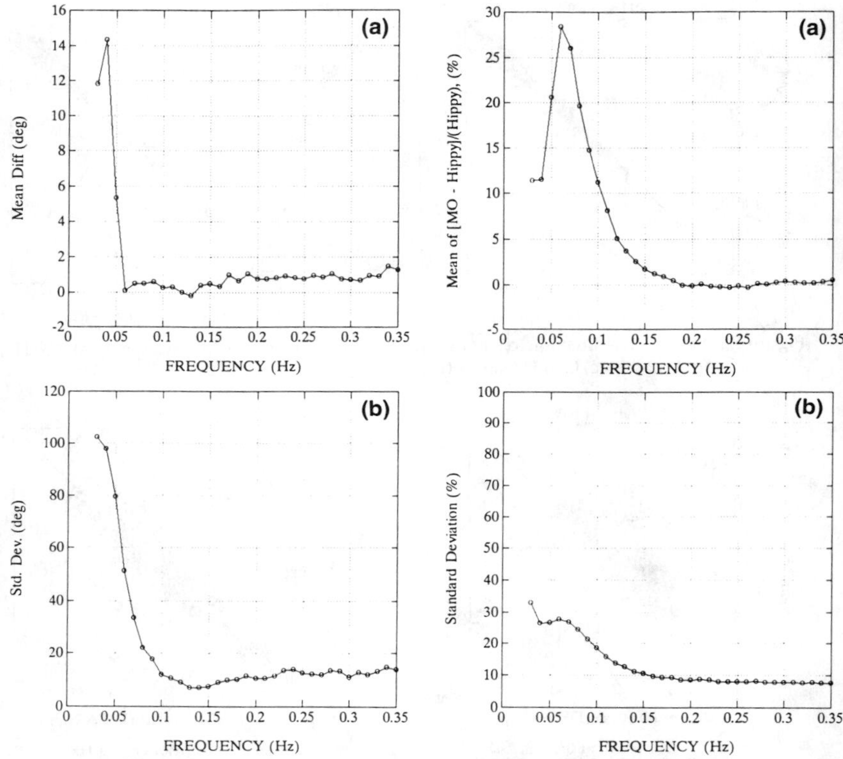

Figure 14. Statistical summary of difference of α_1 between the Hippy system and the MO system (a) mean and (b) standard deviation of the difference.

Figure 15. Statistical summary of difference of σ between the Hippy system and the MO system (a) mean and (b) standard deviation of the ratio of difference to σ of the Hippy system.

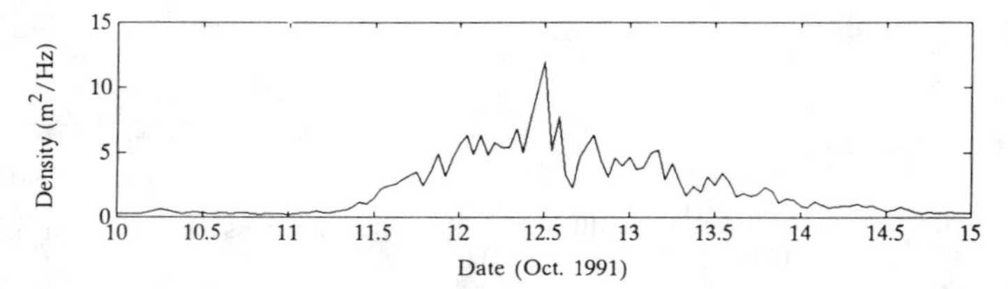

Figure 16. Time series of spectral density at 0.08 Hz of the Hippy system.

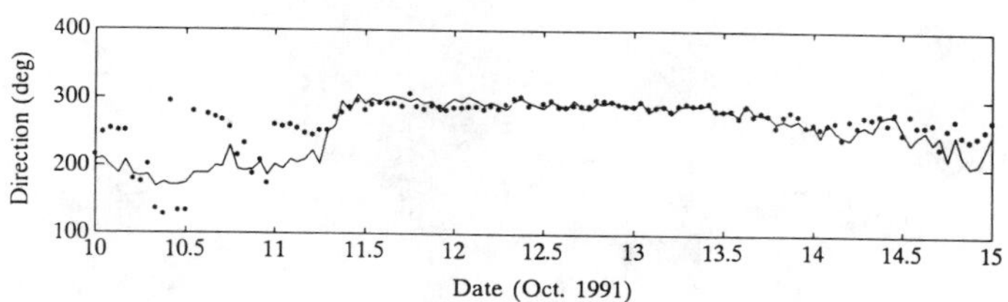

Figure 17. Time series of α_1 at 0.08 Hz from the Hippy system (solid line) and the MO system (asterisk).

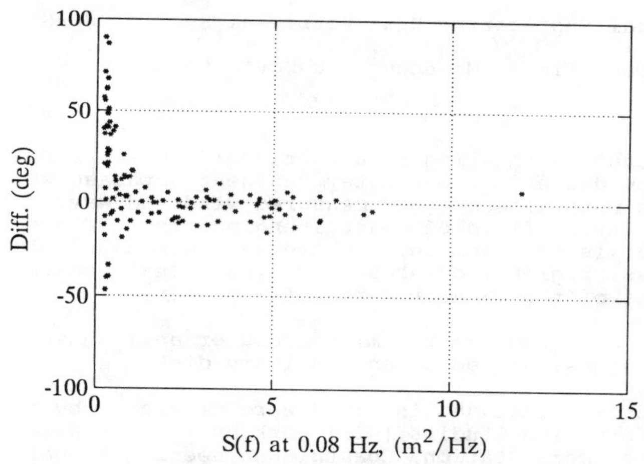

Figure 18. Difference of α_1 at 0.08 Hz between the Hippy system and the MO system versus the spectral density at 0.08 Hz of the Hippy system

Coastal Engineering Data Retrieval System(CEDRS)

Danielle S. McAneny[1] and Doyle L. Jones[2]

Abstract

The Coastal Engineering Data Retrieval System (CEDRS) is a menu driven microcomputer resident database which provides both hindcast and measured wind and wave data for use in the field of coastal engineering. The general goal is to assemble, archive and make available via CEDRS, regional databases for individual coastal districts of the US Army Corps of Engineers.

CEDRS is designed for the microcomputer environment with databases stored on an auxiliary disk.

Hindcast data included in CEDRS were developed by the Wave Information Studies (WIS) work unit, USAE Waterways Experiment Station, Coastal Engineering Research Center (CERC). The databases also contain measured data from the CERC Field Wave Gage Program, the National Oceanic and Atmospheric Administration, and visual data from the Littoral Environment Observation System (LEO).

Introduction

Coastal databases are recognized as important support tools for design and maintenance of coastal projects.

----

[1]Computer Specialist, Coastal Engineering Research Center, US Army Engineer Waterways Experiment Station, 3909 Halls Ferry Road, Vicksburg, Mississippi 39180-6199.

[2]Physical Scientist, Coastal Engineering Research Center, US Army Engineer Waterways Experiment Station, 3909 Halls Ferry Road, Vicksburg, Mississippi 39180-6199

The Coastal Engineering Data Retrieval System (CEDRS) has been developed to provide convenient microcomputer access to a voluminous database of coastal data to meet these needs for Corps of Engineers offices. CEDRS consists of a series of individual regional systems containing data applicable to each Corps of Engineers coastal district. This regional approach provides a comprehensive, long-term database of both hindcast and measured data from a number of sources, specifically for the district's area of responsibility.

Each system resides on a microcomputer at the district office and consists of a basic interactive user interface module, unique site-specific information for each region, and a regional database of measured and hindcast wind and wave data. An interface program provides menu-driven data access and complete on-line documentation. A printed User's Guide is also provided for all regional databases.

Pilot Implementation

A pilot version of CEDRS containing data applicable to the west coast of Florida (Gulf of Mexico) was developed for the Jacksonville District using FORTRAN and sequential ASCII files. This system and all of its supporting files were installed on an auxiliary optical disk drive attached to a 286 microcomputer.

Current Implementation

Subsequent research and testing have resulted in the conversion of the pilot system database to a format consistent with relational database management system (RDBMS) concepts.

Storage medium for the converted system has been changed to an auxiliary hard disk to take advantage of technological advances which now provide for very large, very fast hard disks at prices comparable to or lower than the cost of optical disks.

Implementation of RDBMS concepts has been accomplished by linking the original CEDRS FORTRAN-based user interface module to a new database created using ORACLE RDBMS software. This implementation has greatly reduced the time required for data access.

Hardware

CEDRS is designed to reside on a standard microcomputer

now available at many Corps of Engineers offices. CEDRS requires an AT with at least 3 megabytes of memory, a math co-processor, EGA or VGA monitor, one empty slot on the motherboard and MS-DOS/PC-DOS version 4.0 or higher. If more than one external disk drive is being installed, DOS version 5.0 or higher is required. A printer is not required for execution of CEDRS, but is desirable for practical use.

Software

The interactive CEDRS driver module is written in FORTRAN, with databases constructed using the ORACLE RDBMS. Linkage to the databases by CEDRS is coded as Structured Query Language (SQL) instructions. The ORACLE ProFORTRAN precompiler then converts the SQL instructions to FORTRAN statements necessary to access the database. While CEDRS is in operation the code languages of FORTRAN and SQL are not apparent to the user and require no special knowledge of languages or special keystrokes.

ACES Compatibility

One of the primary goals of CEDRS is to furnish database support for CERC's Automated Coastal Engineering System (ACES). ACES is a coastal engineering design and analysis system which provides a full range of computer tools for increasing the accuracy, reliability, and cost-effectiveness of Corps coastal engineering endeavors. Initial development of the ACES library of computer-based tools and its continued expansion and evolvement have been guided by a committee of Corps coastal experts. ACES is in use Corps-wide as well as in the private sector. CEDRS development efforts have closely followed ACES design principles both in software and hardware elements; and continued coordination is ensured for the future.

Hindcast Data

The CEDRS databases contain both measured data from several sources and computer model generated hindcast data. At the present time CEDRS contains only wind and wave data.

Each CEDRS database contains appropriate regional data from the Wave Information Study (WIS) from more than 300 locations along the US coastlines. These data are a time series produced by WIS wave model, WISWAVE 2.0. For the Atlantic Ocean, Pacific Ocean and the Gulf of

Mexico the time series is a hindcast for the 20-year period 1956-1975. For the Great Lakes a 32-year period 1956-1987 is used for the hindcast. Parameters stored in CEDRS for WIS data are consistent with WIS Standard Hindcast Data Format. See WIS references for a complete description of WIS parameters and formats.

The WIS hindcasts contain data for every three hours for either a 20-year or 32-year period (58,440 or 93,504 records per station). Data storage requirements for individual CEDRS regions range from 600 to 2000 megabytes.

In addition to the WIS time series of wave and wind parameters, there are also available tables of basic statistics calculated from the total time series. Hurricane statistics from separate WIS hindcasts (Abel et al, 1989) were made for the Gulf of Mexico and Atlantic coasts.

Measured Data

In addition to hindcast values, the databases contain measurements as described below.

a. CERC Field Wave Gage Program (FWGP). The FWGP includes the California Coastal Data Information Program (Scripps), the Florida Coastal Data Network (University of Florida), and other CERC site-specific measured data sets. Measurements are usually recorded at six hour intervals, using various devices such as buoys, pressure devices and current meters.

b. National Oceanic and Atmospheric Administration (NOAA). The NOAA data set includes data from a network of moored buoys operated by the National Data Buoy Center (NDBC) since the 1970's and from stations of the Coastal-Marine Automated Network (C-MAN) which became operational in March 1983. For CEDRS only winds and waves (height/period/direction) from this voluminous meteorological data set have been extracted. Early records were made every three hours but some of the more recent data is recorded hourly.

Visual Data

The Littoral Environment Observation System (LEO) was established in 1968, and observations have been made at more than 200 sites along the coasts of the United States. The LEO system uses trained observers to obtain daily visual observations of such coastal variables as breaker height, wave period, direction of wave approach, wind speed and direction, longshore current velocity and beach slope. Observers have obtained this data using simple, inexpensive equipment and for some data, e.g., wave height and direction, observers are asked to simply record a visual estimate. Observations are made either one or two times per day. These data should be used with caution.

Operational Systems

The following table provides the dates of installation and the Corps of Engineers offices in which CEDRS has been placed.

Jacksonville	Sep 1991	Charleston	Jan 1993
New York	Oct 1991	Savannah	Jan 1993
Mobile	Jan 1992	Wilmington	Mar 1993
New Orleans	Jan 1992	Norfolk	Mar 1993
Galveston	Mar 1992	Baltimore	Apr 1993
Chicago	May 1992	Philadelphia	Apr 1993
Detroit	Jul 1992	New England Div	May 1993
Buffalo	Aug 1992	Los Angeles	Aug 1993

Other district offices that will have CEDRS systems installed before the end of 1993 are San Francisco, Portland and Seattle. The district office at Anchorage, Alaska, and the Pacific Ocean Division, Fort Shafter, Hawaii, are scheduled to receive CEDRS systems early in 1994.

User Options

The user has seven options during the operation of CEDRS:

1) Draw a map of the region
2) Display the exact location of the stations
3) Extract all or part of a time series
4) Display an x-y plot of data
5) Display precomputed statistics
6) Display hurricane data
7) Exit to DOS

Option 1 allows the user to display one or more maps of the region of interest showing the approximate locations of the hindcast or measured stations.

Option 2 displays the exact latitude and longitude of the stations within a region of interest, along with the water depth of the station and the period of data availability.

Option 3 allows the user to extract data from CEDRS and write it to an external source. The user has control over the type of data extracted, the number of stations, the time interval of interest and the data destination. Data may be extracted for intervals of one month, six months, one year or several years. The size of the extracted file may be very large, depending upon the period selected. Measured data may not be continuous; CEDRS indicates the gaps that occur in the measured data. This data may be used as input for a numerical model, imported into a spreadsheet, used as input for an ACES module, or as input to some other user-designed process. The data may be extracted as SI (metric) or common English units.

Option 4 allows the user to produce x-y plots of various parameters. Plotting of data requires user input very similar to that of option 3.

Option 5 provides the user with precomputed basic statistics derived from the WIS hindcast. Options include percent occurrence of wave height and period by direction, mean wave height by month and year, maximum wave height by month and year, 20-year summary statistics and return period tables.

Option 6 provides hurricane data for the Gulf of Mexico and the Atlantic coast from separate WIS hindcasts. Options include maximum wave height, wave period, wave direction; return period by station, and track maps of hurricanes that occurred during the hindcast period.

Option 7 allows the user to exit the program.

Future

CEDRS will continue to provide update and maintenance support to keep databases current. Additional measured data and new and/or additional hindcast time series data will be added to the regional databases as it becomes available. Future plans include addition of other data (currents, bathymetry, water levels, etc.).

Investigations are currently under way to determine the feasibility of creating a central database at the Coastal Engineering Research Center in Vicksburg, Mississippi, and allowing user access through currently installed networks.

Availability

CEDRS is not available for public distribution at this time. Data requests should be directed to appropriate Corps of Engineers Offices with current installations.

Acknowledgements

This work was performed at the Coastal Engineering Research Center of the U. S. Army Engineer Waterways Experiment Station, and was funded by the Coastal Field Data Collection Program, Office of the Chief of Engineers. Permission was granted by the Chief of Engineers to publish this paper. Citation of trade names does not constitute an official endorsement or approval of the use of such commercial products by the United States Government.

References

Abel, C. E., Tracy, B. A., Vincent, C. L. and Jensen, R. E. 1989(Apr). "Hurricane Hindcast Methodology and Wave Statistics for Atlantic and Gulf Hurricanes from 1956-1975," WIS Report 19, US Army Engineer Waterways Experiment Station, Vicksburg, MS.

Brooks, R. M. and Corson, W. D. 1984(Sep). "Summary of Archived Atlantic Coast Wave Information Study Pressure, Wind, Wave, and Water Level Data, "WIS Report 13, US Army Engineer Waterways Experiment Station, Vicksburg, MS.

"Coastal Data Information Program, Monthly Reports," (January 1976 to date.) Monthly Summary Report Nos. 1-167..., University of California San Diego, Scripps Institute of Oceanography, La Jolla, CA.

Driver, D. B., Reinhard, R. D., and Hubertz, J. M. 1991(Oct). "Hindcast Wave Information for the Great Lakes: Lake Erie," WIS Report 22, US Army Engineer Waterways Experiment Station, Vicksburg, MS.

Driver, D. B., Reinhard, R. D., and Hubertz, J. M. 1992(Jan). "Hindcast Wave Information for the Great Lakes: Lake Superior," WIS Report 23, US Army Engineer Waterways Experiment Station, Vicksburg, MS.

Hemsley, J. Michael and Brooks, Rebecca M. 1989(Fall). "Waves for Coastal Design in the United States," Journal of Coastal Research, Vol. 5, No. 4., The Coastal Education and Research Foundation (CERF), Fort Lauderdale, FL.

Hubertz, Jon M. and Brooks, Rebecca M. 1989(May). "Gulf of Mexico Hindcast Wave Information," WIS Report 18, US Army Engineer Waterways Experiment Station, Vicksburg, MS.

Hubertz, J. M. and Brandon, W. A. 1992(June). "A Comparison of Measured and Hindcast Wave Conditions at Lake Worth and Hallandale, Florida During 1990. Final Report, Coast of Florida Study, Region III. Prepared for US Army Engineer District, Jacksonville. US Army Engineer Waterways Experiment Station, Vicksburg, MS.

Hubertz, J. M., Brooks, R. M., Brandon, W. A., and Tracy, B. A. 1993(Mar). "Hindcast Wave Information for the Atlantic Coast (1956-1975," WIS Report 30, US Army Engineer Waterways Experiment Station, Vicksburg, MS.

Hubertz, J. M., Driver, D. B., and Reinhard, R. D. 1991(Oct). "Hindcast Wave Information for the Great Lakes: Lake Michigan," WIS Report 24, US Army Engineer Waterways Experiment Station, Vicksburg, MS.

Leenknecht, David A. and Szuwalski, Andre. 1992(Sep). "Automated Coastal Engineering System: Technical Reference," Version 1.07, US Army Engineer Waterways Experiment Station, Vicksburg, MS.

Leenknecht, David A., Szuwalski, Andre and Sherlock, Ann R. 1992(Sep). "Automated Coastal Engineering System: User's Guide," Version 1.07, US Army Engineer Waterways Experiment Station, Vicksburg, MS.

McAneny, Danielle S. 1986(Nov). "Sea-State Engineering Analysis System (SEAS), Revised Edition 1," WIS Report 10, US Army Engineer Waterways Experiment Station, Vicksburg, MS.

"NDBC Data Availability Summary," 1989(Mar). Report 1801-24-02, Rev. C, National Data Buoy Center, National Oceanic and Atmospheric Administration, US Department of Commerce, Stennis Space Center, MS.

Reinhard, R. D, Driver, D. B., and Hubertz, J. M. 1991(Dec). "Hindcast Wave Information for the Great Lakes: Lake Ontario," WIS Report 25, US Army Engineer Waterways Experiment Station, Vicksburg, MS.

Reinhard, R. D, Driver, D. B., and Hubertz, J. M. 1991(Dec). "Hindcast Wave Information for the Great Lakes: Lake Huron," WIS Report 26, US Army Engineer Waterways Experiment Station, Vicksburg, MS.

Sherlock, Ann R. and Szuwalski, Andre. 1987(Apr). "A User's Guide to the Littoral Environment Observation Retrieval System," Instruction Report CERC-87-3, US Army Engineer Waterways Experiment Station, Vicksburg, MS.

"University of Florida Coastal Data Network, Wave Data Report." 1981 to date, Coastal and Oceanographic Engineering Department, University of Florida, Gainesville, FL.

Wave Attenuation at the Mobile Stable Berm

Gregory L. Williams, P.E. Associate Member, ASCE[1]
Cheryl Burke Pollock, Associate Member ASCE[1]

ABSTRACT

This paper describes work examining the ability of submerged berms to "filter" certain erosive wave energy while allowing accretionary type waves to pass unhindered. Analysis of wave data from the National Nearshore Berm Demonstration Project's Mobile Stable Berm shows that overall attenuation is greater for waves of increased steepness, H/L. Application of erosion/accretion criteria developed by Kraus, Larson and Kriebel (1991) and Zwamborn et al. (1970) quantifies the attenuation of these erosive, higher energy waves.

INTRODUCTION

The Mobile Stable Berm (or Mobile Outer Mound) was constructed as part of the National Nearshore Berm Demonstration Project (established in 1986) to investigate the benefits of placing dredged material in nearshore waters to serve as a wave energy dissipating device (Figure 1). An underwater structure such as a berm is typically used to induce propagating waves to shoal, increase in steepness (wave heights increase relative to wave lengths) and possibly cause breaking. Wave energy is dissipated through the processes of wave shoaling, breaking and reforming thus leaving a remaining transmitted wave with a reduced height containing less energy. This principle can be observed on coasts with nearshore linear bars where wave breaking is observed offshore. These transmitted waves are smaller in height and energy and may therefore be less likely to cause

[1] Coastal Engineering Research Center, U.S. Army Engineer Waterways Experiment Station, 3909 Halls Ferry Road, Vicksburg, MS 39180-6199

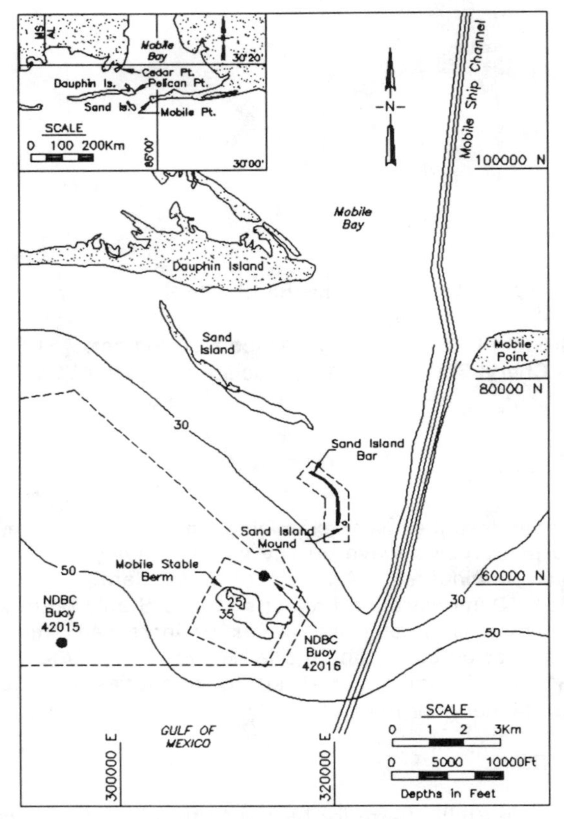

Figure 1. Location map

erosion on the adjacent beach. The lower energy waves may actually carry sediment shoreward where it can be deposited as part of a beach building processes. Attenuation of wave energy resulting from wave propagation across submerged aquatic features, namely, nearshore berms, has been documented in Burke and Williams (1992), Burke, McLellan and Clausner (1991), McLellan, Pope and Burke (1990), Zwamborn, Fromme and Fitzpatrick (1970) and others.

Construction of the Mobile Stable Berm was completed in 1989 with material dredged from the Mobile Ship Channel. Final dimensions were 300 m x 2750 m at the crest and 1.6 km x 4.0 km along the base with approximately 6 m of relief (Burke, McLellan and Clausner, 1991). Post construction monitoring included hydrographic surveys, wave data collection, side-scan sonar, sediment sampling and fisheries monitoring. The work described in this paper only involves wave data analysis.

Wave data were collected by two National Data Buoy Center (NDBC) 3-meter directional wave buoys. An offshore buoy (No. 42015) was deployed in April 1987 approximately 6.5 km southwest of the berm in 16.0 m of water and measured incident wave energy. An inshore buoy (No. 42016) was deployed in April 1988 in the lee and near the midpoint of the berm in 13.0 m of water. This buoy measured transmitted wave energy. Both buoys collected hourly data until September 1990. The buoy locations and their relation to the berm and shore are shown in Figure 1. McGehee, McKinney and Grogg (in prep) details overall instrumentation for the National Nearshore Berm Demonstration Project including instruments utilized, accuracy and downtime.

DATA ANALYSIS

Initial data analysis consisted of several steps of data manipulation to obtain a final working data set. These steps included: (1) removing wave data not propagating from an offshore direction; (2) removing all data collected prior to completion of berm construction; (3) removing waves that were determined to be too small to be affected by the berm; and (4) aligning the offshore and inshore data with respect to time for direct offshore-inshore comparisons.

Using the final working data set, Burke and Williams (1992) present comparison plots of H_{m0}(Offshore) versus H_{m0}(Inshore). H_{m0} is an energy-based significant wave height defined as four times the square root of the area under the energy spectrum. The initial plot, Figure 2, shows the linear scatter associated with this comparison plot. A linear best fit line is also shown along with a line of 45° [ie. H_{m0}(Offshore) $=H_{m0}$(Inshore)]. This plot indicates that the transmitted wave energy is less than the incident wave energy (ie. H_{m0} Offshore > H_{m0} Inshore). However, to fully examine the berm's ability to attenuate wave energy, Burke and Williams (1992) estimated the inshore wave heights that would have occurred due to natural shoaling had the berm not been in place. Figure 3 shows how the actual measured data (with a berm) compares to two shoaling approximations (without a berm). Linear best fit lines are plotted for (1) the actual measured data, (2) a linear shoaling approximation and (3) a TMA (Texel, Marsen and ARSLOE) shoaling approximation by Hughes and Miller (1987). These theoretical shoaling approximations utilized the wave data from the offshore buoy to calculate a wave height for the location of the inshore buoy as if no berm existed. This figure demonstrates that the

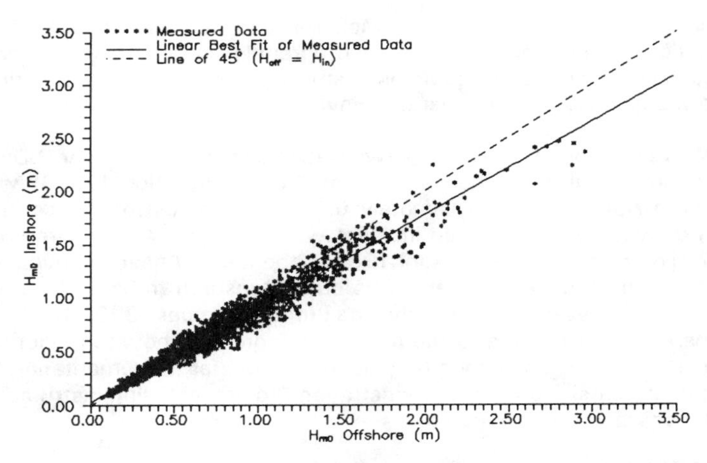

Figure 2. H_{m0} Offshore versus H_{m0} Inshore plotted with a linear best fit line (from Burke and Williams, 1992)

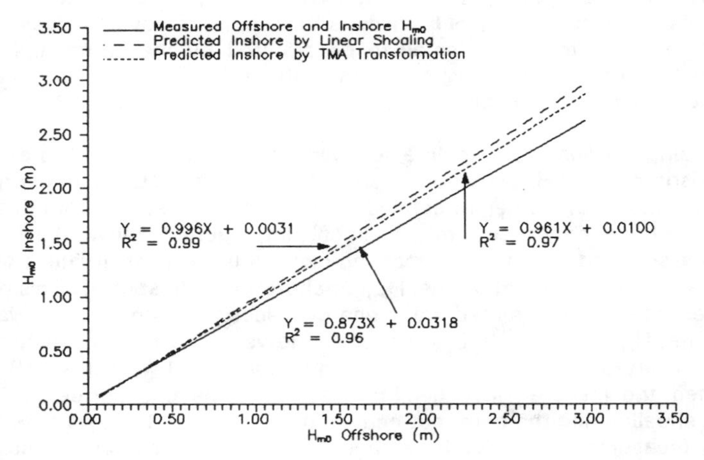

Figure 3. Linear best fit lines for measured H_{m0} Offshore vs. H_{m0} Inshore and two shoaling methods (modified from Burke and Williams, 1992)

actual transmitted inshore wave heights (measured by the inshore buoy) were in fact smaller than those predicted by the shoaling approximations for no berm conditions. Therefore, wave attenuation by the Mobile Stable Berm is supported.

To investigate the potential filtering effect of the berm, each data pair (offshore/inshore height, period, direction) was categorized into period classes based on the offshore wave period. Period ranges were: 3.1--3.9, 4.0--4.9, 5.0--5.9, 6.0--6.9, 7.0--7.9, 8.0--8.9, 9.0--9.9, and greater than 10.0 seconds. Figure 4 shows linear best fit plots for actual H_{m0}(Offshore) versus H_{m0}(Inshore) and the TMA, no-berm approximation for each of these period classes. If wave period were a suitable indicator of erosive/ accretionary conditions, one would expect a greater difference between the "actual data" line and the TMA, no-berm approximated line.

Because, period alone does not appear to be an appropriate indicator of erosive or accretionary waves, the data pairs were then grouped into steepness (H/L) classes based on the steepness of the offshore wave condition. Figure 5 shows similar linear best fit plots for actual H_{m0}(Offshore) versus H_{m0}(Inshore) and the TMA, no-berm approximation for each of the steepness classes. Whereas the period classes showed no apparent difference of attenuation, the steepness classes appear to show an increase in attenuation with an increase in steepness. That is, as steepness increases, the difference between the "actual data" line and the TMA, no-berm approximated line increases, thus indicating wave attenuation. Wave steepness alone is not completely adequate to differentiate between erosive and accretionary waves, but, in general, steeper waves are indicative of more energy which may be related to greater potential for erosion.

DISCUSSION

To examine the filtering effect of the berm, the known wave conditions must be distinguished between erosive and accretionary. Two sets of known wave conditions exist for the Mobile Stable Berm, namely the offshore data set and the inshore data set. By comparing the number of erosive waves offshore with the number of erosive waves inshore (or, analogously the number of accretionary waves offshore with the number of accretionary waves inshore), an assessment can be made on the concept of wave filtering.

Criteria to differentiate between erosive and accretionary wave conditions requires an onshore location where information related to sediment grain size and beach slope can be obtained. Since the waves in this area ultimately impacted Dauphin Island, AL, a report by Douglas

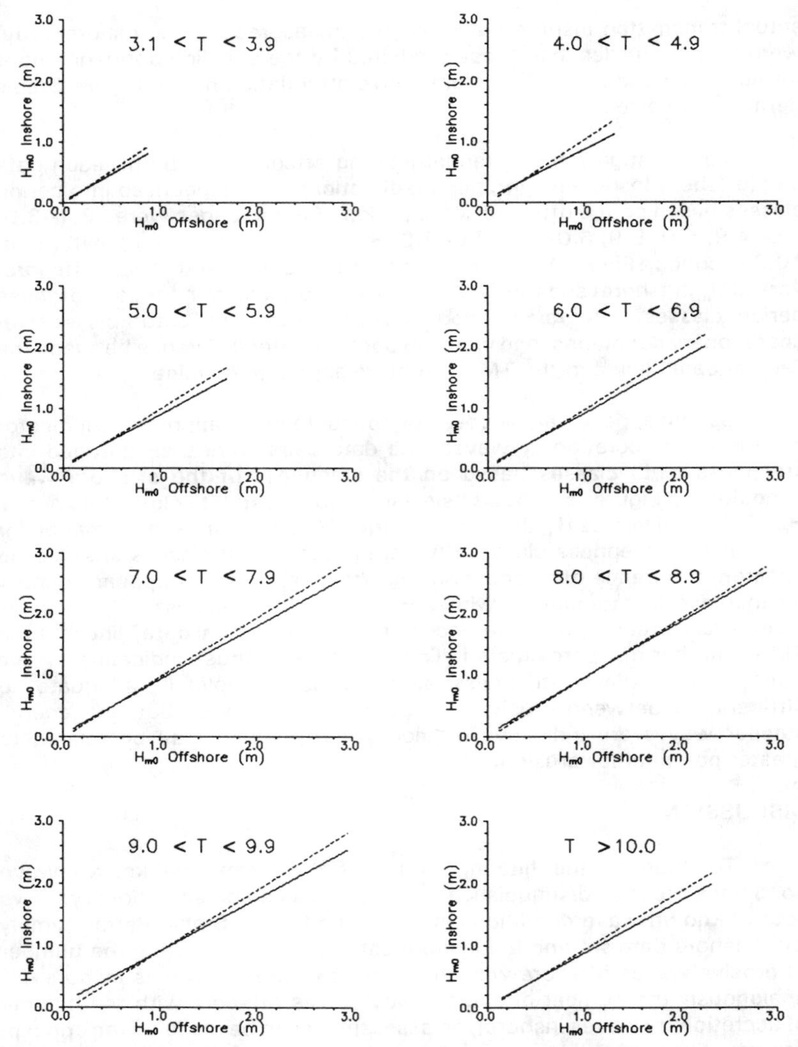

Figure 4. H_{m0}Offshore vs H_{m0}Inshore (actual ——; TMA ----) for period classes

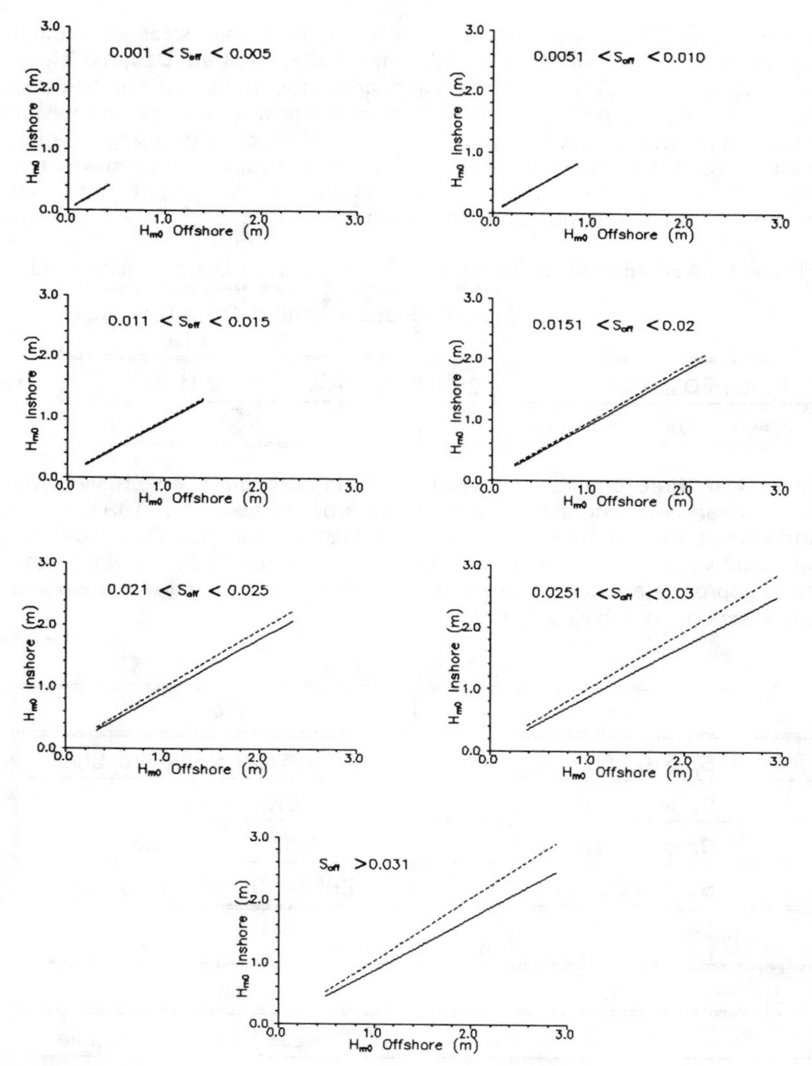

Figure 5 H_{m0}Offshore vs H_{m0}Inshore (actual ——; TMA ----) for steepness classes (S_{off} = H_{off}/L_{off})

and Haubner (1992) was used to estimate typical grain sizes and beach slopes for this area. Douglas and Haubner (1992) separate Dauphin Island into two reaches: (1) the eastern reach, protected by Sand Island and the nearby shoals and having a relatively smaller beach grain size and milder beach slope and (2) the western "unprotected" reach with a larger beach grain size and steeper beach slope. Average values for eastern and western Dauphin Island respectively were estimated for input into an erosion/accretion criterion and are shown in Table 1.

Table 1. Average Beach Slopes and Grain Sizes for Dauphin Island, AL

	Beach Slope	Median Grain Size (d_{50}) mm (ϕ)
Eastern Dauphin Island	2.5 (1:40)	0.29 (1.8)
Western Dauphin Island	3.8 (1:26)	0.37 (1.43)

The three criteria employed to distinguish between erosive and accretionary wave conditions were taken from Kraus et al. (1991) (two) and Zwamborn et al. (1970). Kraus et al. (1991) describes several criteria based on various parameters. Though no specific guidance is given on the most appropriate criteria, the two used here may be the most known and are described in Tables 2 and 3.

Table 2. S_o - N_o Criteria

If:	Then:
$S_o > 0.00014\ N_o^3$	ACCRETION highly probable
$S_o > 0.00027\ N_o^3$	ACCRETION probable
$S_o \leq 0.00027\ N_o^3$	EROSION probable
$S_o < 0.00054\ N_o^3$	EROSION highly probable

Table 3. N_o Criteria

If:	Then:
$N_o < 2.4$	ACCRETION highly probable
$N_o < 3.2$	ACCRETION probable
$N_o \geq 3.2$	EROSION probable
$N_o > 4.0$	EROSION highly probable

The Zwamborn et al. (1970) criteria (Figure 6) was developed using data collected from Durban, South Africa coupled with Wiegel's (1964) relationship between grain sizes and beach slopes.

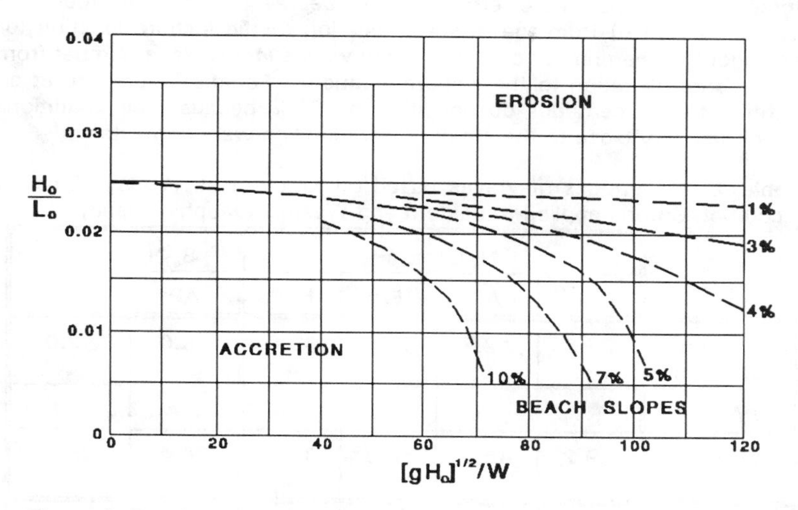

Figure 6 Erosion/accretion Criteria from Zwamborn et al. (1970)

The Kraus et al. (1991) criteria were based on the deepwater wave steepness[2], H_o/L_o, and fall speed parameter or "Dean Number," $H_o/(wT)$. Kraus et al. (1991) refer to the deepwater steepness and the Dean Number as S_o and N_o, respectively. The Zwamborn et al. (1970) criteria is based on beach slope, deep water wave steepness, and the parameter$(gH_o)^{1/2}/w$.

Both of the Kraus et al. (1991) criteria and the Zwamborn et al. (1970) criteria necessitate the use of a deepwater wave height and length. Therefore, each measured wave condition from both the offshore and inshore wave buoys was transformed (reversed shoaled) to deepwater using the TMA transformation. These deepwater wave conditions were

---

2

H_o--deepwater wave height
L_o--deepwater wave length
w --sediment fall velocity
g --acceleration due to gravity

S_o--H_o/L_o
N_o--$H_o/(wT)$
T --wave period

applied to the erosion/accretion criteria to determine how many erosive and accretionary waves existed offshore and inshore of the berm, respectively.

Tables 4 and 5 show the results of applying the measured data to the three criteria for for both eastern and western Dauphin Island. In all three cases the number of erosive waves *decreases* in number (between 1.7% and 6.4%) from the offshore location to the inshore location (or analogously, the number of accretionary waves *increases* in number from the offshore location to the inshore location). For the Zwamborn et al. criteria, the percentages do not sum to 100% because the conditions which fell very close to the erosion/accretion line were neglected.

Table 4. Percentages of waves based on Kraus et al. (1991) Erosion/ Accretion Criteria applied to Eastern and Western Dauphin Island, AL.

	N_o Criteria				S_o-N_o Criteria	
	AHP*	AP	EP	EHP	AP	EP
Eastern Offshore	27.5	42.7	57.3	39.6	58.0	42.0
Δ %	1.7	3.0	3.0	5.4	6.4	6.4
Eastern Inshore	29.2	45.7	54.3	34.2	64.4	35.6
Western Offshore	40.4	63.0	37.0	17.9	86.7	13.3
Δ %	2.9	6.3	6.3	5.6	3.4	3.4
Western Inshore	43.3	69.3	30.7	12.3	90.1	9.9

*AHP--Accretion highly probable
AP --Accretion probable
EP --Erosion probable
EHP--Erosion highly probable

These results support the concept of the filtering effect of a berm for wave energy attenuation. Figures 2 and 3, from Burke and Williams (1992), show that overall the berm is contributing to reduced wave energy. Additionally, by employing erosion/ accretion criteria applicable to Dauphin Island, the increased attenuation (in numbers) of erosive waves over non-erosive waves is confirmed. The small number of erosive wave reductions indicated from Tables 4 and 5 is mostly due to the fact that the Mobile

Table 5. Percentages of waves based on Zwamborn et al. (1970) Erosion/Accretion Criteria applied to Eastern and Western Dauphin Island, AL.

	Accretion	Erosion
Eastern Offshore	76.8	10.1
Δ %	5.2	3.9
Eastern Inshore	82.0	6.2
Western Offshore	73.6	10.4
Δ %	5.6	4.0
Western Inshore	79.2	6.4

Stable Berm was constructed in approximately 14 to 15 meters of water with total relief of roughly 7 meters. This remaining water depth (7 meters) is greater than would typically be expected for a nearshore berm design.

CONCLUSION

Previous work by Burke and Williams (1992), Burke, McLellan and Clausner (1991), McLellan, Pope and Burke (1990) and Zwamborn, Fromme and Fitzpatrick (1970) has supported the intuition of wave energy attenuation from submerged berms. The work presented here, however, follows Nakamura, Shiraishi and Sasaki (1966) by examining to what extent erosive energy is attenuated more than accretionary energy. Tables 4 and 5 indicate that, in fact, the numbers of erosive waves is reduced and the numbers of accretionary waves increased from offshore to inshore, indicating that through the shoaling process the relief of the berm is converting some of the erosive waves to accretionary waves. Therefore, proper design and construction of berms could protect against the stormier conditions where erosive conditions persist, yet allow the lower energy accretionary type waves to pass unhindered.

The results from this study show that a small percentage of difference (less than 7%) was observed between the offshore and inshore numbers of erosive/accretionary waves. This may be due primarily to the location and relative water depth of the Mobile Stable Berm. Nakamura, Shiraishi and Sasaki (1966) found that the damping effects of a submerged dike were significantly larger for breaking waves than for non-breaking waves. For non-breaking waves, wave transmission (ie. energy dissipation) was primarily reduced by reflection and friction. For breaking to occur at the Mobile Stable Berm, wave heights would have had to be roughly 78%

of the water column depth (7 meters) or 5.46 meters. By inspection of the data used in this analysis there were no measured H_{m0}s greater than 3.0 meters. Therefore, this small percentage difference between the offshore and inshore erosive/accretionary waves is probably due to the lack of wave breaking at the site. For berms with greater relief relative to water depth or those placed in waters shallow enough to induce breaking, a larger difference in the number of converted waves could be expected.

ACKNOWLEDGEMENTS

The authors wish to acknowledge the USAE District, Mobile who provided funding for this project. Thanks to Mr. Thomas Richardson, Ms. Joan Pope, Dr. Yen-hsi Chu and Mr. James Clausner who provided review and technical support for this paper. Thanks also to Ms. Melissa Moore who helped prepare figures for this paper. Permission was granted by the Chief of Engineers to publish this paper.

REFERENCES

Burke, C.E., McLellan, T.N. and Clausner, J.E. 1991. "Nearshore Berms - Update of the United States Experience," Proceedings of CEDA-PIANC Accessible Harbours, Dredging Days Conference, Amsterdam, The Netherlands.

Burke, C.E. and Williams, G.L. 1992. "Nearshore Berms - Wave Breaking and Beach Building," Proceedings of ASCE Ports '92 Conference, Seattle, Washington.

Douglas, S.L. and Haubner, D.R. 1992. "Coastal Processes of Dauphin Island, Alabama," College of Engineering Report Number 92-1, Department of Civil Engineering, University of South Alabama, Mobile, Alabama

Hughes, S.A. and Miller, H.C. 1987. "Transformation of Significant Wave Heights," Journal of Waterway, Port, Coastal and Ocean Engineering, ASCE, Vol. 113, No. 6, pp 588-605.

Kraus, N.C., Larson, M. and Kriebel, D.L. 1991. "Evaluation of Beach Erosion and Accretion Predictors," Proceedings of Coastal Sediments '91, ASCE, pp 572-587.

McGehee, D., McKinney, J.P. and Grogg, W.E. (in prep). "Monitoring of Waves and Currents at the Dredged Material Mounds Offshore of Mobile, Alabama," DRP Report, USAE Waterways Experiment Station, Coastal

Engineering Research Center, Vicksburg, Mississippi.

McLellan, T.N., Pope, M.K. and Burke, C.E. 1990. "Benefits of Nearshore Placement," Proceedings of Third Annual national Conference on Beach Preservation Technology, St. Petersburg, Florida.

Nakamura, M., Shiraishi, H. and Sasaki, Y. 1966. "Wave Damping Effect of Submerged Dike," Proceedings of Tenth Conference on Coastal Engineering, ASCE, Vol. 1, pp 254-267.

Wiegel, R.L. 1964. Oceanographical Engineering, Prentice-Hall, Inc., Englewood Cliffs, NJ.

Zwamborn, J.A., Fromme, G.A.W. and Fitzpatrick, J.B. 1970. "Underwater Mound for the Protection of Durban's Beaches," Proceedings of the Twelfth Coastal Engineering Conference, ASCE, pp 975-994.

LOW-FREQUENCY FLUCTUATIONS OF SUSPENDED SAND AND WAVE GROUPS IN THE SURF ZONE

R.D.Kos'yan[1], S.Yu.Kuznetsov[2] and N.V.Pykhov[2]

Abstract

Field observations of the sand suspension near the bottom in surf zone are presented and discussed. Calculated spectra of suspended sand concentration and wave envelope was found to be similar and show that sand was more intensively suspended at the wave groups frequencies (≈0.013 Hz for our conditions) and at the frequencies lower than 0.004 Hz. Peculiarities of fluctuations of suspended sand median diameter are discussed.

Introduction

Modeling of sediment transport processes in the coastal zone for different time scales during storm is of great importance for the solving both fundamental and applied problems. Today this problem is solved satisfactorily only at a level of regularities which are based on the averaged parameters of wind waves and swell. The results of recent field measurements (Hanes, Huntley, 1986; Beach, Sternberg, 1988; Hanes, 1991) have showed that the most intensive sand suspension from the bottom is observed not at the wind and swell wave frequencies, but at the frequencies of wave groups or infragravity waves. Field and laboratory data testify to the importance of low-frequency waves in the cross-shore sediment transport (Shi, Larson, 1984; Jaffe, Sallenger, 1992; Osborne, Greenwood, 1992 a,b; Sato, 1992; Shibayama, Okayasu, Kashiwagi, 1992). The problem of sediment movement under the influence of low-frequency waves is not practically studied till now, and is very actual today. The question about the contribution of various section of the surface wave spectrum and near bottom velocities spectrum to the formation of suspended sediment concentration field keeps to be a blank space in this problem. We believe that its solution will help to understand physical aspects of sediment transport under storm wave and to approach to the more founded modeling than earlier.

Some results of our research of low-frequency changes of suspended sand concentration in the Black Sea, in the region of 50 km southward from Varna city (Bulgaria), are considered in the present report.

---

[1]The Southern Branch of the P.P.Shirshov Institute of Oceanology, Russian Academy of Sciences. Gelendzhik-7, Krasnodar region, 353470 Russia.

[2]The P.P.Shirshov Institute of Oceanology, Russian Academy of Sciences. 23 Krasikova str., Moscow, 117218 Russia.

Technique and methods

Underwater slope profile along the pier where the measurements during storm were carried out is shown at Fig.1. Synchronous records of suspended sand concentration, water particle velocities and surface waves were made in surf zone of swell and wind waves at the depth (h) from 1 to 3 m (Kos'yan,Pykhov,1991). Concentration and velocities were measured at the distance from the bottom z=0.1-0.3 m .

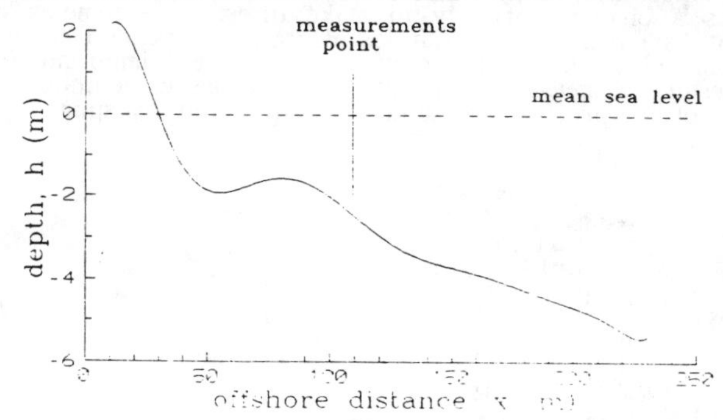

Figure 1. Beach profile and gauges location.

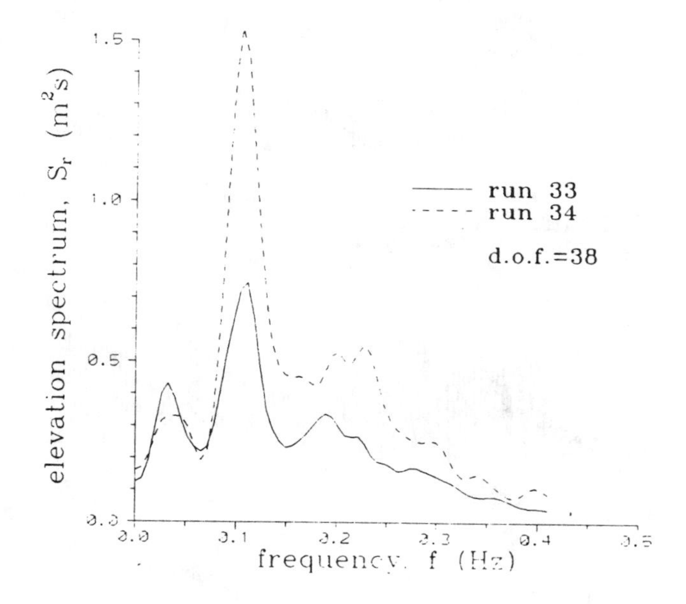

Figure 2. Typical free surface elevation spectra in surf zone, swell.

Suspended sand concentration was measured by continuous pumping of water with suspended sediments. The sampling time interval was changed from 12 s to 120 s, depending on the the duration of measurements. A minimum sampling time interval of 12 s was found in response to the necessity to get sand sample enough for sieve analysis. Such long sampling interval is not enough for the determination of the concentration changes at the surface wave frequency. This point is a shortcoming of considered measuring method in comparison with optical and acoustic ones. But on the other hand, this method of suspended sand sampling gives a chance not only to make direct measurements of concentration but to observe temporal variations of granulometric and mineral composition of suspended sand. The latter information is very important for interpretation the results of measuring by optical and acoustic methods and for correct modeling of the processes of sand suspending and transport.

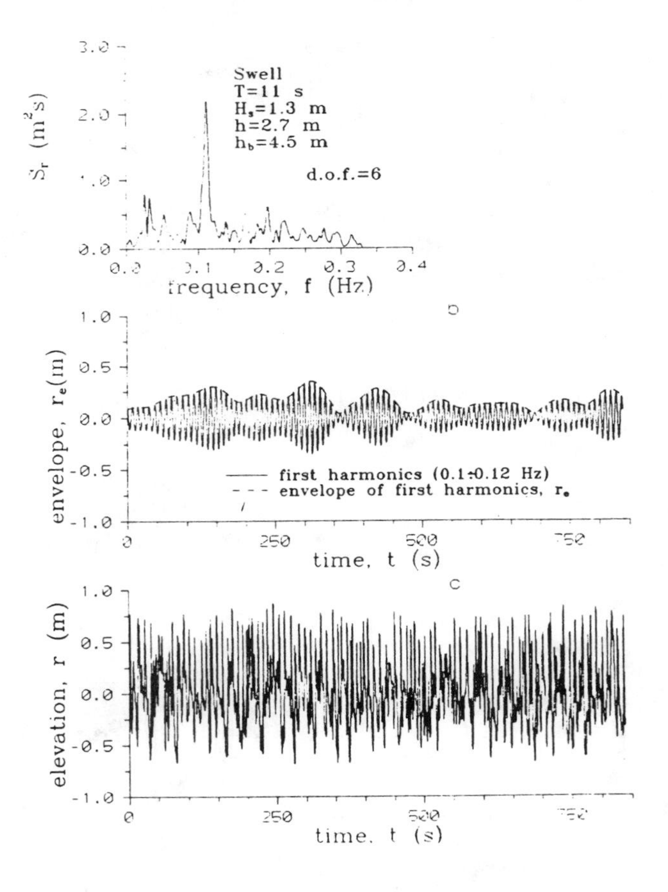

Figure 3. Free surface elevation (a), its spectrum (b) and its first harmonics with envelope (c). Run 33, bore (d.o.f.-degrees of freedom).

Measurements of wave velocities in the near bottom area was conducted with the help of dynamic pressure gauges with the perceptive element in the form of disk, 2 cm in diameter. Free surface elevations were measured by resistance type wire gauges.

The separation of wave groups was done by plotting of the first harmonic envelopes of surface elevations and wave velocities in the following way.

Typical for our measurements free surface elevations spectra (Fig.2) show that the most significant wave energy is concentrated in a frequency band of the first harmonics (0.11 Hz) but not at the lower frequencies or at the second and higher harmonics, and therefore it is looking probable that the groups of the first harmonics must determine the fluctuations of the concentration.

A spectrum of free surface elevations for the bore zone is given at the right side of Fig.3. It is plotted with greater frequency resolution than in the previous case (Fig.2). The spectrum has a very narrow frequency range of the first harmonics, and band filtering in the frequency range of 0.1 - 0.12 Hz has been used for its separation and plotting of its envelope curve. Boundary frequencies for other runs were in the range of 0.08 - 0.13 Hz. Envelope curves ($r_e(t)$) of the first harmonics ($r_1(t)$) were calculated by a formula $r_e(t)=[r_1^2(t)+\tilde{r}^2(t)]^{1/2}$, were ~ denotes the Hilbert transform. In this figure (Fig.3) an example of wave chronogram ($r(t)$) and envelope of the first harmonics for run 33 are given.

Results

In Fig.4 chronograms of suspended sand concentration and the first harmonics envelope for run 33 measured in the bore zone of swell are given. Significant wave height and period was: H =1.3 m, T=11 s. Sampling time interval for concentration measurements was 15 s. A median diameter of suspended sand was nearly constant during the whole period of measuring and was equaled to 0.34 mm, the root - mean - square deviation was 0.04 mm. Chronogram shows that fluctuations of concentration and the first harmonics envelope are changing practically synchronously.

The similar results for the breaking zone of swell are shown in Fig.5. The wave parameters in the measurements point was H =1.7 m, T=10.5 sec.

The spectra of suspended sand concentration, free surface elevations, near bottom water particle velocity and the spectra of the first harmonics envelopes for the bore zone (run 33) are given in Fig.6. On the concentration spectra one can point out two characteristic sections: extremely low-frequency fluctuations with the frequency f<0.004 Hz and relatively small peak at f=0.011 Hz. The same regularity may be observed on the spectra of the first harmonic envelope of free surface elevations and near bottom velocity. The frequency of 0.011 Hz corresponds to the frequency of wave group and more lower one <0.004 Hz is probably determined also by infragravity waves.

The similar spectra for the run 34 are shown in Fig.7. The only difference from the previous case is a relatively small energy of fluctuations at the lowest frequencies in the envelope spectra. Strictly speaking a statistical reliability of these exposed wave (but not the concentration) spectra is not large owing to the limits of duration of records. But coincidence visible period of concentration fluctuations and wave groups period (100-150 s) are confirmed by many chronograms of suspended sand concentration. One of such examples is shown in Fig.8.

Fluctuations of concentration with the period of 10-15 minutes (f=0.001-0.0015 Hz) can be observed on the concentration chronograms too (Fig.9), the discreteness of measuring is 2 minutes.

In these measurements we have not received an exact answer for the question whether the mean diameter of suspended sand changes synchronously with concentration, or not.

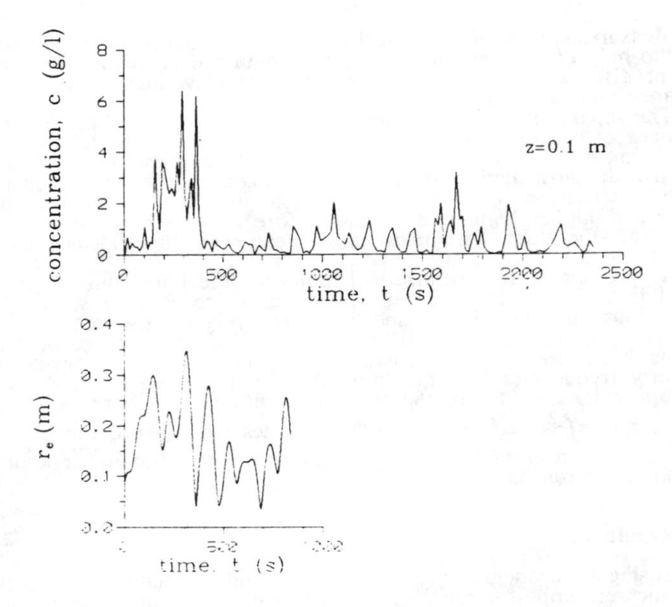

Figure 4. Suspended sand concentration and envelope of first wave harmonics. Run 33, bore.

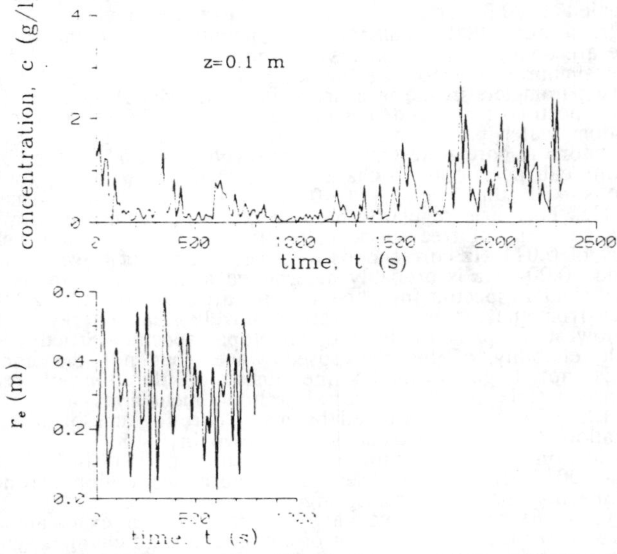

Figure 5. Suspended sand concentration and envelope of first wave harmonics. Run 34, breaking zone.

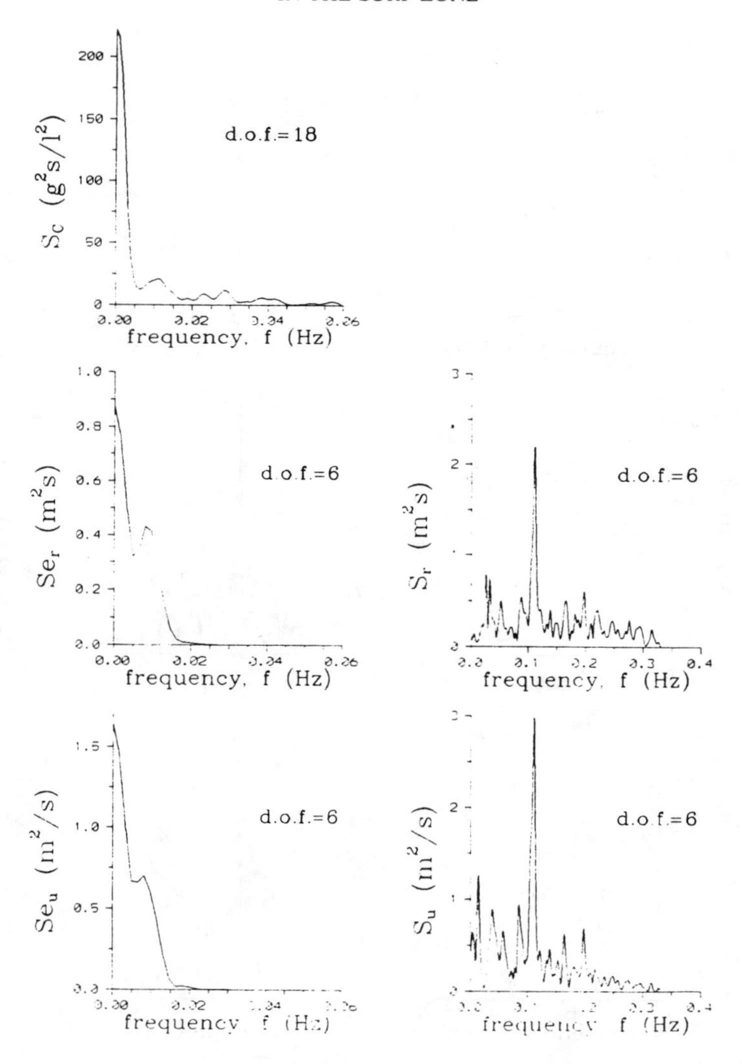

Figure 6. Concentration spectrum and spectra of free surface elevation, near-bottom water particle velocity and of envelopes of their first harmonics. Run 33, bore.

Fig.10 shows the chronogram of concentration and a mean diameter for the breaking zone of swell wave. A satisfactory correlation between concentration fluctuation and a mean size is observed.

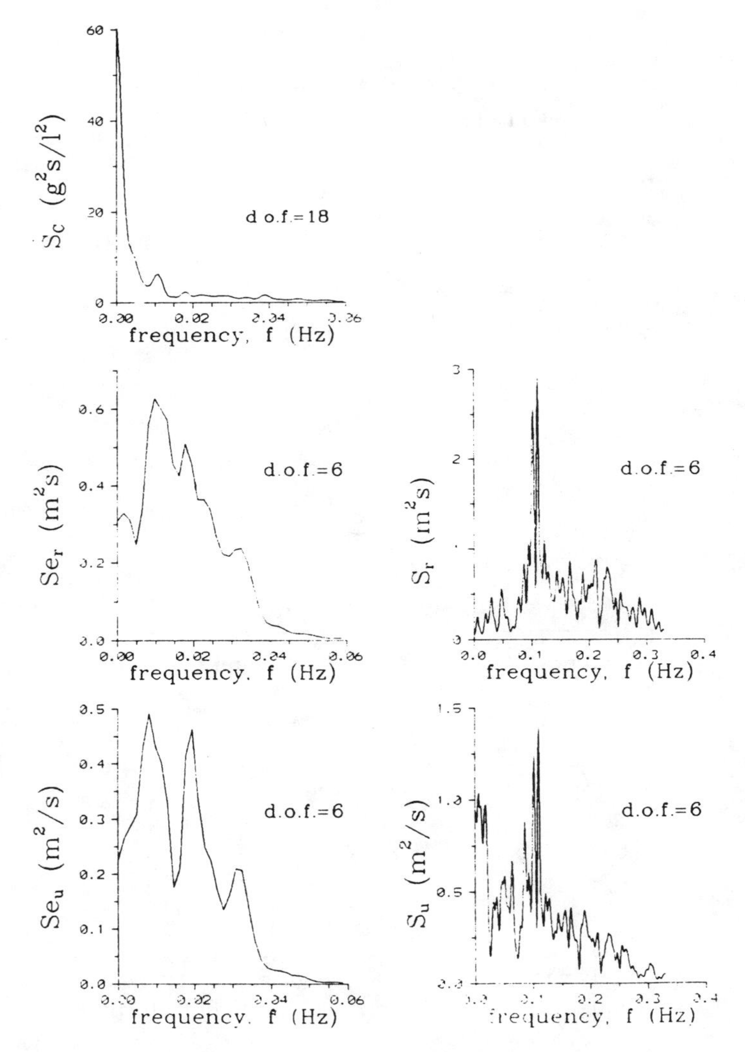

Figure 7. Concentration spectrum and spectra of free surface elevation, near-bottom water particle velocity and of envelopes of their first harmonics. Run 34, breaking zone.

Typical distribution of suspended particles on fractions in the maximum and minimum of concentration is shown in Fig. 11. Coarse-grained sand fractions are present in the peaks of concentration, but they are absent in the concentration minimum. This testifies to the fact that suspension composition is probably determined by local sediment lifting above the bottom when wave groups with the greatest height are passing.

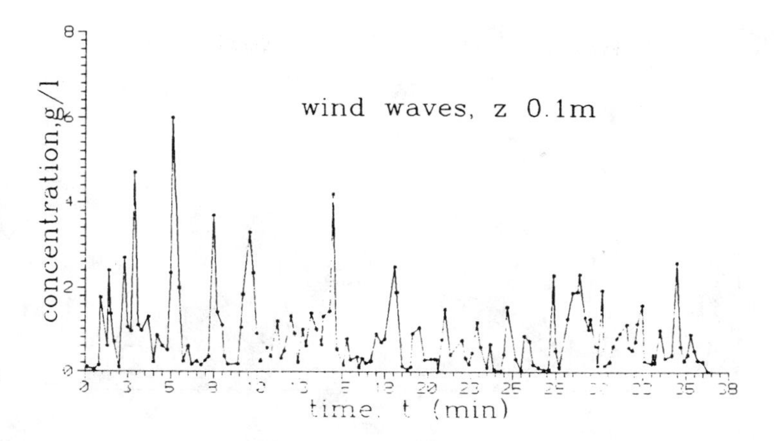

Figure 8. Temporal variations of suspended sand concentration in surf zone, wind waves.

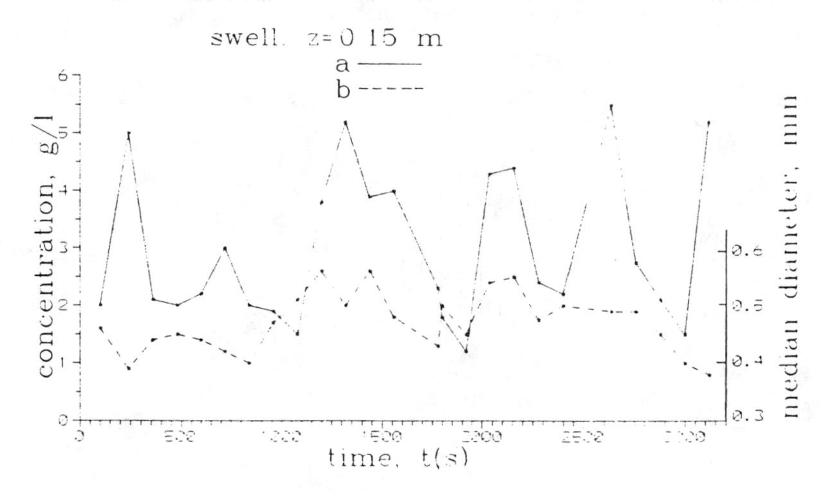

Figure 9. Temporal variations of concentration (a) and median diameter of suspended sand (b) in surf zone, swell.

Similar chronograms for the breaking zone of wind wave are given in Fig.12.

In contrast to previous case there is no apparent synchronism in fluctuation of concentration and a mean size of suspended sand.

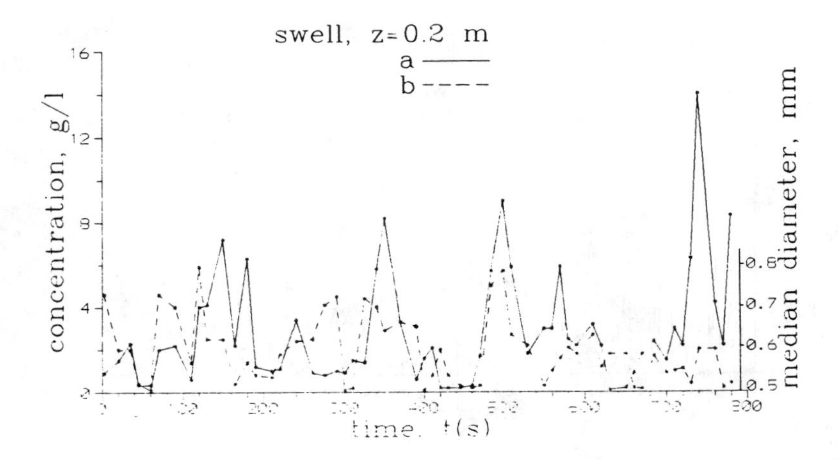

Figure 10. Temporal variations of concentration (a) and median diameter of suspended sand (b) in surf zone, swell.

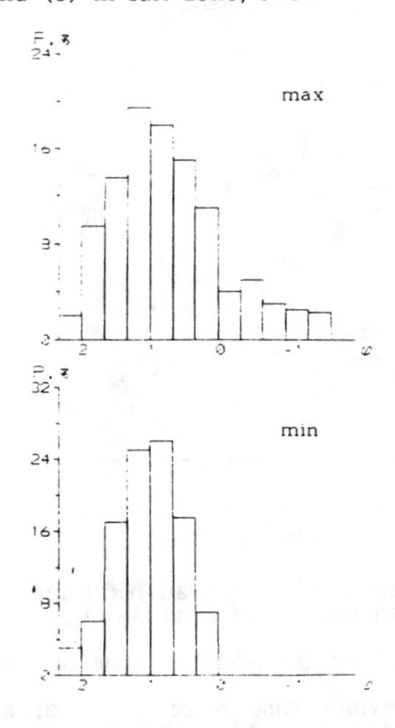

Figure 11. Distribution of suspended sand by fractions for case nearly synchronous variations of concentration and median diameter.

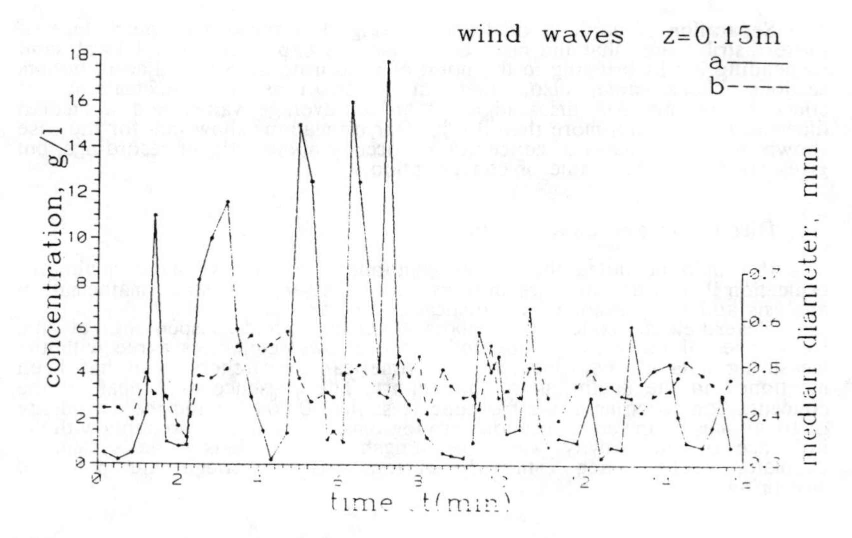

Figure 12. Temporal variations of concentration (a) and median diameter of suspended sand (b) in surf zone, wind waves.

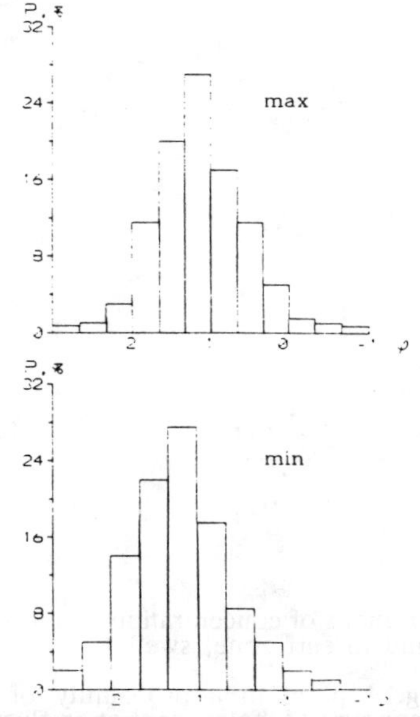

Figure 13. Distribution of suspended sand by fractions for case uncorrelated variations of concentration and median diameter.

Separation of particles on fractions (Fig. 13) shows the coincidence of those distributions, that the most likely may be explained by not local sand suspending but its bringing to the point of measuring from the adjacent bottom sections. Let's note also, that in these runs the fluctuations of concentration are 3-5 times higher than an average value, and a median diameter changes not more than 30 %. Our estimations show that for the case shown in Fig.12 peaks of concentration occupy about 30% of recordings, but gives 60 % of mean value of concentration.

Discussion and conclusions

It should be noted that above mentioned results have more qualitative estimation because longer recordings are necessary for strict mathematical analysis and for guarantee of statistical reliability.

Nevertheless, conclusions about intensive sand suspending at the frequencies of the wave groups and at more low frequencies agree with the measuring data of the Canadian and American researchers, who has been mentioned in the beginning of this report. The presence of a peak in the concentration spectrum at the frequency less than 0.004 Hz and the periodicity of 10 minutes on the concentration chronograms are connected probably with the influence of infragravity waves or longshore migrations the system of circulatory cells, when the rip current passes through the point of measuring.

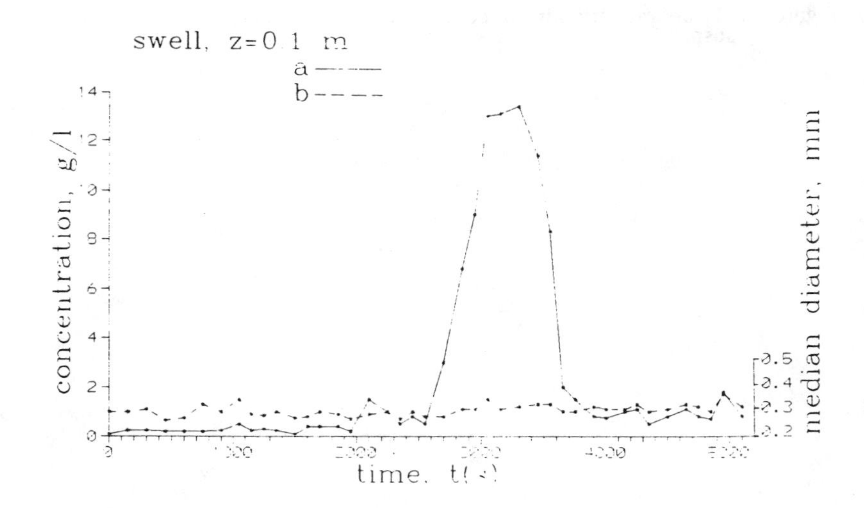

Figure 14. Temporal variations of concentration and median diameter of suspended sand in surf zone, swell.

Data given in Fig.14 point to a probability of the latter assumption. Fig. 14 shows the chronograms of the concentration fluctuation and of a median diameter of suspended sand during 85 minutes in the swell break point at the depth of 1.2 m. A median diameter of bottom sand in the point of measuring was 0.35 mm. Sampling has been made from the horizon of 10 cm above the bottom. The sampling time interval was 2 minutes. Obtained data indicate that when a median diameter of suspended sand is almost invariable one a distinct peak 20 minutes long is observed on the concentration chronograms. In its maximum the sand concentration is higher for one order than before and after it. Visual observations showed that this peak coincided in time with the action of rip current, which was traced in the place of measuring by more muddy water. Longitudinal shift of rip current was caused probably by the reformation of coastal cusps when the storm was dumping and wind waves were changed by swell waves.

The results of our research and published data show that synchronous measuring of concentration fields, velocity and surface waves are necessary during storm for the reception of trustworthy, statistically reliable regularities of sand suspending and transport at the different frequencies. Such measurements must be some hours long for different types of submerged slope and for different coarseness of the material.

The solution of this problem, to our mind, depends mainly on the breadth of such measurements just in the field conditions because the modeling of these processes at the low frequency in laboratory is impossible now. On that basis one can reveal the whole spectrum of spatial-temporal scales of sediment suspending and transport under real marine conditions which are necessary for modeling of such processes.

Acknowledgement

This work was supported by Fund of Fundamental Research of the Russian Academy of Sciences, Grant N. 93-05-8133.

References

Beach R.A. and R.W.Sternberg, 1988. Suspended sediment transport in the surf zone: response to cross - shore infragravity motion. Marine Geology, v.80, pp.61-79.

Hanes D.M. and D.A.Huntley, 1986. Continuous measurements of suspended sand concentration in the wave dominated nearshore environment. Continental Shelf Research, v.6, N.4, pp.585-596.

Hanes D.M., 1991. Suspension of sand due to wave groups. Journal of Geophysical Research, v.96, N.c5, pp.8911-8915.

Jaffe B. and A.Sallenger, 1993. The contribution of suspension events to sediment transport in the surf zone. Proc. 23th Coastal Eng. Conf., v.2,pp.2680-2693.

Kos'yan R.D. and N.V.Pykhov, 1991. Hydrogenous sediment shift in the nearshore zone. M.:Science, 280p.(in Russian).

Osborne P.D. and B.Greenwood, 1992a. Frequency dependent cross-shore suspended sediment transport. 1. A non - barred shoreface. Marine Geology, v.106, pp.1-24.

Osborne P.D. and B.Greenwood, 1992b. Frequency dependent cross-shore suspended sediment transport. 2. A barred shoreface. Marine Geology, v.106, pp.25-51.

Sato Sh., 1993. Sand transport under grouping waves. Proc. 23th Coastal Eng. Conf., v.2, pp.2680-2693.

Shibayama T., A.Okayasu and M.Kashiwagi, 1993. Long period wave and sand transport in the surf zone. Proc. 23th Coastal Eng. Conf., v.2, pp.2438-2449.

Anomalous Dispersion Paradox in Shallow Water Gravity Waves Shoaling Over Slopping Bottom

S.Yu.Kuznetsov[1] and N.S.Speransky[2]

Abstract

Peculiarities of celerity spectra of shoaling gravity waves are considered. The results of flume and field experiments focused on the measurements of celerity spectra of shoaling gravity waves reliable established the existence of anomalous dispersion phenomenon when the celerity grows with the frequency. Anomalous dispersion was detected when the wave are transforming from the symmetrical about the vertical direction form to the sow-tooth ones due to shoaling process and can be explained by not the faster movements of high frequency harmonics but the birth new ones with the phase $\pi/2$ relative the primary wave harmonic.

Introduction

Process of waves shoaling has linear and nonlinear aspects. The second aspect includes problem of high harmonics generation and change of their phase (f_j) relatively to the primary harmonic (j is a number of harmonic, j =1 corresponds to the primary one). As it was showed by laboratory experiment of Flick, Guza, Inman (F-G-I,1981) that a phase difference between harmonics does appear in waves shoaling over a slopping bottom. In

----

[1]Research physicist of the P.P.Shirshov Oceanology Institute, Russian Academy of Sciences, 23 Krasikova str., Moscow, 117589 Russia

[2]Senior research physicist of the P.P.Shirshov Oceanology Institute, Russian Academy of Sciences, 23 Krasikova str., Moscow, 117589 Russia

relatively deep water f_j =0, then f_j grow as the waves shoal and in case of shallow water f_j tends to be equaled to $\pi(j-1)/2$. Simple analysis shows that this effect may be caused by either 1)very strong normal dispersion or 2)week anomalous(!) dispersion. The first assumption contradicts to ideas of shallow water waves dynamics that requires waves to be weakly normal dispersive (or nondispersive in limited condition). The second assumption corresponds practically to existence of a new effect.

Meanwhile the field measurement of Thornton and Guza (T-G, 1982) demonstrates the constancy of celerity spectra in surf zone that contradicts to F-G-I results since change of f_j value means motion of harmonics with different phase speed. This contradiction stimulated our laboratory and field measurements. The purpose was to find answer on a question whether anomalous dispersion phenomenon does exist and if it exist to find an explanation for it.

Experiments

Laboratory measurements were conducted at Hydrodynamics Laboratory of Qingdao Ocean University, Qingdao, China. Wave channel had 40 m length, 1.2 m width and 0.70 m of maximal depth over horizontal bottom (Fig.1). A piston type wave maker produced regular waves

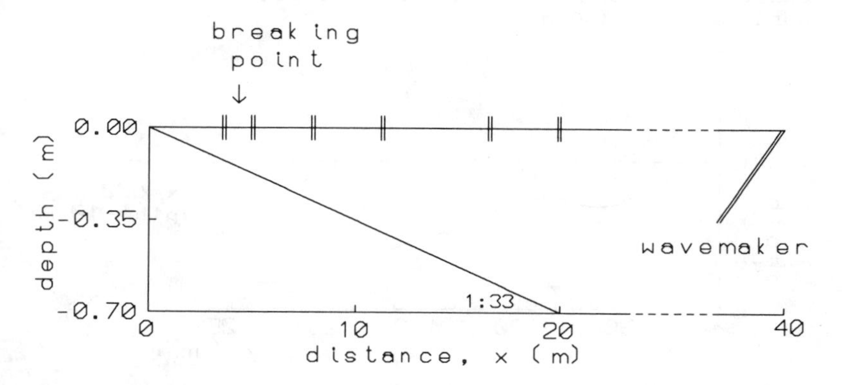

Figure 1. Laboratory experiment sketch. Wave gauges are shown by shot double vertical lines.

with high stability. 20 m from the wave maker, the horizontal bottom was conjugated with the slopping section, which value of sloping was 1:30. Wave gauges of capacity type were positioned at six points on the sloping section at a distance from one another of 1.5 m to 5.1 m (Fig.1).

Regular waves with period 2, 3 and 4 seconds were generated in duration of different runs. Wave height varied between 0.12 m and 0.14 m. Surface oscillations were registered simultaneously by all gauges with sampling frequency 10 Hz. The duration of a record was equaled to the time of the propagation of 50 waves.

Field measurements were conducted at Schcorpilovci, the Black sea coast near Varna, Bulgaria. The research pier of Institute of Oceanology of the Bulgarian Academy of Sciences is located on the straight 12-km long coast with near parallel contours. The pier and consequently resistance type wave gauges were located on the line perpendicular to the general shore line propagation. The offshore location, depth, and distance between instruments used are shown in Fig.2. One of the typical sequences of measurements are considered in here. The direction in which wind waves ran was perpendicular to the shore. Data was recorded digitally in duration of 900 s of measurements, with sampling frequency 3.3 Hz. Errors of synchronization were less than 0.005 s. Average values of wave characteristics, measured at special tower located at 10 m depth (600 m offshore) are: significant wave height - 1.77 m, frequency of primary peak of energy density - 0.13 Hz. Waves was spilling at point with coordinate x = 150 m.

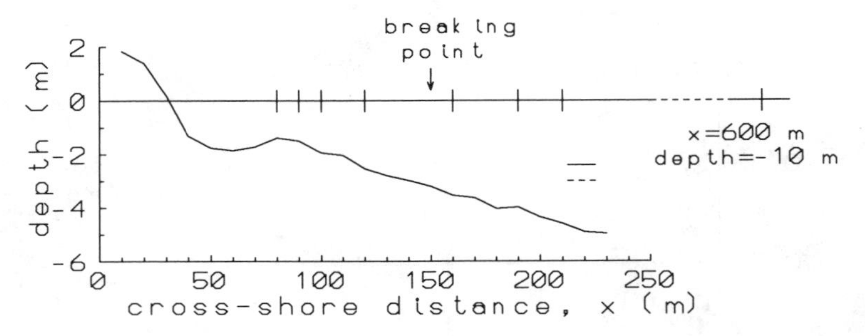

Figure 2. Beach profile and wave gauges location (vertical lines) along research pier on the Black sea coast.

Data processing methods

The processing of the laboratory and field data was the same procedure. Energy density spectra ($S(f)$) of free surface elevation, coherence functions $\gamma^2(f)$ between surface elevations in neighboring points of measurements, phase spectra $\phi(f)$ were calculated by standard methods (Bendat, Piersol,1986). Spectral window width equaled 0.05 Hz, decrees of freedom was about 90 and 30-40 for field and laboratory data respectively. Celerity spectra were calculated by formulae

$$c(f)= \frac{\Delta x}{\tau_0+\phi(f)/2\pi f} , \tag{1}$$

where Δx is a base length, τ_0 is the previous relative shift of data series to improve results of phase analysis. The value of τ_0 was chosen to be equal to argument of the absolute maximum of cross-correlation function between surface elevations measured in two neighboring points. The base length Δx was variant 1.50 m to 5.10 m in flume measurements and equaled 20 m or 40 m in field measurements. Error of the method of celerity spectrum definition not exceed 4% for field condition and not exceed 2% for laboratory ones. Measured values were compared with linear theory calculated values.

Results

In the flume experiment wave form was changed in duration of propagation over the horizontal section. Surface elevation spectrum had several lines on multiple frequencies at the beginning of the slopping section (x=20 m) and is shown at Fig.3.

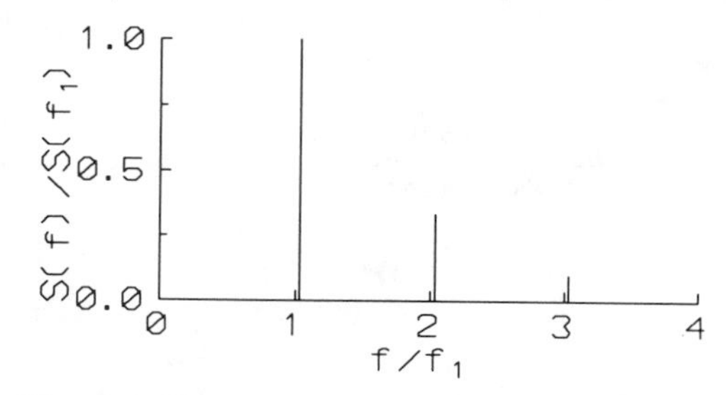

Figure 3. Normalized spectrum of surfase elevation measured at point x=20 m (laboratory experiment). The main spectrum peak frequency is denote as f_1.

Fig.4 demonstrate variations in celerity of the first (c_1) and second (c_2) harmonics with the distance x from the shore. At the lower part of the section ($x>x_0$, x_0=14-15 m) inequality $c_2(x)<c_1(x)$ takes place and demonstrates normal dispersion at this area. Higher up on the slope the difference between c_1 and c_2 values decreases and near the point $x=x_0$ they become equal to each other (nondispersive motion). Onshore direction ($x<x_0$) the second harmonic celerity becomes more than the first one and the difference $\Delta c=c_2- c_1$ increases further on the slope up to x=10 m, where Δc has maximal value 0.35 m/s. The corresponding value of the relative deviation ($\Delta c/c$) equals 22%, and is far greater than the error of measurement. The coherence function magnitude was more than 0.90 in all laboratory runs. This means that the existence of anomalous dispersion effect ("AD-effect") was quite reliable fixed. Upper on the slope this effect became weaker, and Δc-values becomes about zero in a point of wave breaking (x=5 m).

Waves with other tested periods demonstrated the similar peculiarities of celerity variations.

Celerity spectra measured in field conditions are shown in Fig.5 together with energy density spectra and

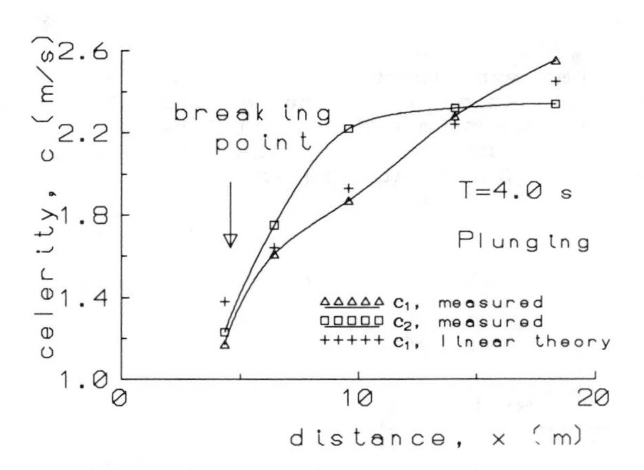

Figure 4. Flume variance of celerities of harmonics over plane slopping bottom.

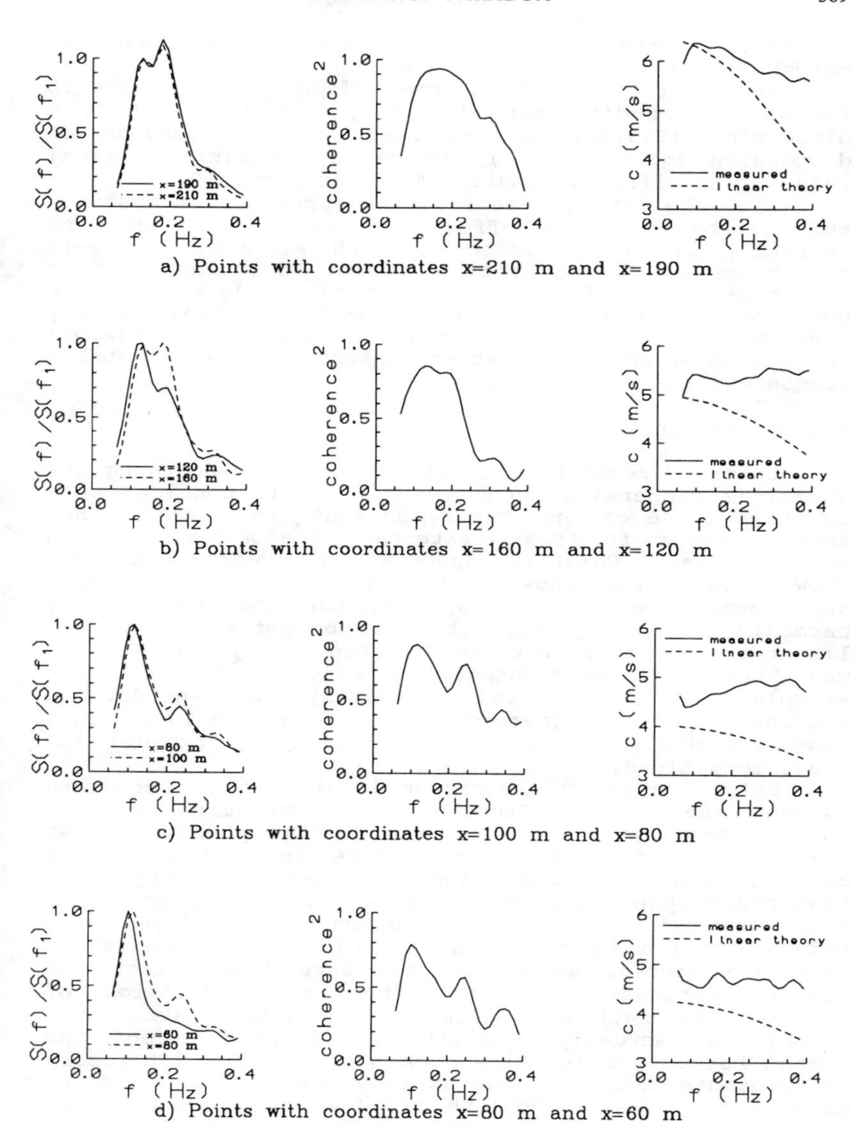

Figure 5. Sea surface elevation spectra, coherence functions and celerity spectra for wind wave condition at the different distance from the shore.

coherence functions of sea surface elevations at neighbors points of measurements.

Transitions from the normal dispersion (Fig.5a) to the absence of dispersion (Fig.5b), then to the anomaly dispersion (Fig.5c) and at least to the absence of dispersion again (Fig.5d) can be seen clearly. Maximal value of AD-effect intensity ($\Delta c/c$) is about 10%.

Thus the data of the field measurements demonstrate the existence of AD-effect as well, although the intensity of the AD-effect is not so high in field conditions as in regular flume waves. It is important to note that the AD-effect occurs simultaneously with strong nonlinearity of wave motion. Actually spectral density functions that were measured in points of maximal appearance of AD-effect, showed intensive peaks on double frequency.

Discussion

Before discussing the strange result concerning the anomalous dispersion phenomenon, we shall consider data on phase speed of the primary harmonic to study whether anomaly exist for some "external" reason, e.g. media peculiarities. Celerity spectra that were measured, showed high coincidence of the primary harmonic celerity and linear theory (Fig. 4,5). It means that the primary harmonic has a dispersion that can be determined by the linear properties of wave motion. Previously this feature was fixed with measurements by many researchers (for example, ref. Thornton and Guza (1982) and Kuznetsov and Speransky (1990)). Therefore it may be concluded that anomalous dispersion was found under quite typical and usual conditions.

Returning to the AD-effect itself, it is necessary to note that the AD-effect was detected quite reliably and the central point is to explain the phenomenon. At first glance the AD-effect appears to contradict the basic statement of the classical gravity wave dynamics that phase speed must not increase when frequency grows. This conclusion however, is concerned with free waves only and not with the high harmonic of shoaling waves that are coupled waves. At the same time there is another contradiction since the ideal condition of nonlinear interaction is equality of coupling waves phase velocities. Actually equality of the primary and the second harmonic celerities took place in our laboratory measurements (Fig. 4), but at only one "point". Onshore direction from this "point" $c(f_2)$-value becomes bigger than $c(f_1)$. It means that the conditions of nonlinear interaction between these harmonics are getting worse.

The strange experimental result contradicts conventional idea of nonlinearity growth onshore direction and appears to present a paradox - a paradox of anomalous dispersion of shoaling gravity waves.

Let us consider the theory of nonlinear waves on shallow water to find an explanation of the AD-effect. According to Brehovskih and Goncharov (1989), the equation (2) would be based on one-dimension analysis

$$\frac{\partial u}{\partial t} + \mathbf{L}u = -\varepsilon u \frac{\partial u}{\partial r} \quad , \tag{2}$$

where u is horizontal component of Eulerian velocity, t is a time, r is a coordinate which increases in direction of waves running, ε is a small parameter ($\varepsilon \ll 1$), and $\mathbf{L}$ is a linear operator which corresponds to definite type of dispersion. For example, one can easily obtain Kortweg-de Vries equation, Burgers equation, etc. from (2) putting in it corresponding operator $\mathbf{L}$.

The solution of the equation (2) is found following Brehovskih and Goncharov by the method of expansion on a small parameter in the form of $u=\varepsilon u_1+\varepsilon^2 u_2+\varepsilon^3 u_3+\ldots$ under the supposition that the initial form of the wave ($t=0$) is as follows

$$u_1(r,0)=a \exp(ikr) + \text{c.c.} \tag{3}$$

Here $\omega(k)$ is a dispersion relation for linear waves, k is a wave number, $i=\sqrt{-1}$ and c.c. is a complex conjugate term. When dispersion is not negligible ($\Delta\omega=2\omega(k)-\omega(2k)\neq 0$), the second term of the expansion is equal to

$$u_2(r,t)=-2ika^2 \frac{\sin(\Delta\omega t/2)}{\Delta\omega} \exp\{i[2kr - \frac{\omega(2k)+2\omega(k)}{2}]\} + \\ + \text{c.c.} \tag{4}$$

The comparison of expressions (3) and (4) shows that the second harmonic is generated with the phase $\phi(f_2)=\pi/2$ when the primary harmonic has a phase $\phi(f_1)=0$. Beginning from this point we are able to develop a qualified model of the AD-effect. Let us suppose, for simplicity, that gravity waves with $\phi(f_2)=0$ are generated artificially in shallow water by a programmed wave maker. Accordingly to the theoretical results mentioned above, the second harmonic with phase $\phi(f_2)=\pi/2$ (with respect to the phase of the primary harmonic) should appear. As a result two different second harmonics will take place: the "old" second harmonic ($\phi_{old}(f_2)=0$) and the "new" second harmonic ($\phi_{new}(f_2)=0$) that appeared as a result of nonlinear interaction of the primary harmonic with itself. The superposition of these components shows the "summary" second harmonic, and phase of it depends on amplitude relation of "new" and "old" components. It is not difficult to show that the relative increase of the "new" component results the "summary" harmonic phase tending to be $\pi/2$ under any phase of the "old" second harmonic.

Returning to reality we can say that whole zone of sea surface, located offshore then anomalous dispersion zone may be considered as a kind of wave marker.

The new property of waves shoaling over slopping bottom may be stated as follows: the change of phase in the duration of the "summary" second harmonic propagation from one point to another is determined not by the time of running only, but by the intensity of the "new" second harmonic growth. Certainly, the formulae (1) cannot distinguish "new" and "old" harmonics and operates with the "summary" second harmonic wholly. Because of more fast increase of it's phase (in comparison with "ordinary" case of the only second harmonic) due to "new" harmonic appearance and growth, the formulae (1) gives anomaly high celerity of the second harmonic. Phase velocity of "old" and "new" second harmonics themselves may be less then the celerity of the primary harmonic. As the "new" component propagates onshore direction and its intensity grows, the significance of the "old" component becomes less and less and therefore the phase of "summary" second harmonic tends to $\pi/2$. This conclusion is in agreement with result of Flick et al. (1981) and Elgar, Guza (1985). Due to the relative attenuation of the "old" harmonic value, anomalous dispersion of the "summary" second harmonic decreases since the main role now belongs to the "new" harmonic alone. The additional mechanism of phase shift vanishes. Thus this conclusion of the model agrees qualitatively with the result of our flume and field measurement (Fig.4,5).

From the point of view of the model the role of initial wave spectrum width may be understood as well. Actually the case of a narrow spectrum was considered already. In the case of a wide spectrum there are three types of wave component with a frequency of the second harmonic: 1) free component with phase velocity less than celerity of the first harmonic, 2) coupled "old" second harmonic and 3) coupled "new" second harmonic. The last two give "summary" coupled harmonic with celerity greater than the celerity of the first harmonic as it was shown before. Mixing of the free and coupled harmonics may attenuate or cancel out AD-effect depending on relation of harmonic amplitudes, if the frequency of one of the free components in wide spectrum is about the frequency of the coupled second harmonic.

This is very likely the reason that the AD-effect was not found in the field measurements of Thornton and Guza (1982).

Conclusions

Thus our qualified model of waves shoaling over a sloping bottom, which was developed to explain the AD-paradox shows that the standard method of phase spectrum determination based on cross- spectral analysis of two surface elevation series gives a paradoxical result because it is not adequate to the complex nonlinear phenomenon of gravity waves running over a shallow water.

References

Bendat, J.S. and Piersol, A.G., 1986. Random data. Analysis and measurement procedures. Ed. by John Willey&Sons, N.Y., 537 pp.

Brehovskih, L.M. and Goncharov, V.V., 1982. Introduction in continuum mechanics. Nauka, Moscow, 335 pp.

Elgar, S. and Guza, R.T., 1985. Observation of bispectra of shoaling surface gravity waves. J. Fluid Mech., 181: pp. 425-448.

Flick, R.E., Guza, R.T. and Inman, D.L., 1981. Evolution and velocity measurement of laboratory shoaling waves. J. Geophys. Res., 86(C5): pp.4149-4160.

Kuznetsov, S.Yu. and Speransky, N.S., 1990. Phase speed of free and coupled waves on shallow water. In N.A.Aibulatov (Editor), "The modern shelf sedimentary processes". Nauka, Moscow, pp. 180-186.

Thornton, E.B. and Guza, R.T., 1982. Energy saturation and phase speed measured on a natural beach. J. Geophys. Res., 87(C12): pp. 9499-9508.

ON HURRICANE-GENERATED SEAS

Michel K. Ochi*

Abstract

Two different situations in the relationship between wind speed and sea severity during hurricanes (and tropical cyclones) are clarified from analysis of measured data obtained by NOAA buoys. From analysis of 400 wave spectra obtained from measured data during hurricanes, a wave spectral formulation specifically applicable for hurricane-generated seas is derived in the form of the JONSWAP spectral formulation and as a function of significant wave height and modal frequency. A family of wave spectra which can be used for design of marine systems during hurricanes is also developed.

Introduction

Wind-generated seas during hurricanes are significantly different from those observed in ordinary storms because the input source of energy generating the waves is advancing at a speed of 5 to 12 knots. The rate of change of wind speed measured at a location in the moving path of a hurricane is much faster than that observed during an ordinary storm. The time duration, therefore, of a given wind speed during hurricane is extremely short in contrast to seas generated by ordinary winds blowing continuously for several hours with constant speed.

Many studies have been carried out on hurricane-generated seas, primarily through hindcasting and forecasting techniques. These include Cardone, Pierson and Ward (1976), Bretschneider and Tamaye (1976), Young and Sobey (1981), Ross and Cardone (1978) and Young (1988), among others. Although these studies provide valuable information for individual hurricanes, it is not possible

* Professor, Coastal and Oceanographic Engineering Department, University of Florida

to draw general conclusions regarding the severity of the sea and the shape of wave spectra during hurricanes.

The purpose of this study is to find the relationship between wind speed and sea severity during hurricanes in order to provide information on simultaneous loadings associated with winds and waves for the design and safe operation of ships and offshore structures. For this, analysis is carried out on measured data obtained by NOAA buoys deployed in the Gulf of Mexico and the Atlantic Ocean. It is also the purpose of this study to derive a wave spectral formulation specifically applicable for hurricane-generated seas. This is accomplished by analysis of 400 wave spectra obtained during hurricanes.

Sea Severity During Hurricanes

It is generally known that sea severity generated by hurricanes depends on several factors of the hurricane such as maximum sustained wind speed, radius of maximum winds, central pressure, forward speed, etc. The results of analysis of wind speed and sea severity during hurricanes, however, indicate that there exists a relatively simple relationship between the sea severity and mean wind speed. In fact, there are two different situations; one is the growing stage of hurricane-generated seas in which the wind speed is increasing at an extremely high rate but the sea severity is comparatively moderate, the other is the sea condition resulting from continuous winds of mild severity blowing for one week or longer then followed by a storm, usually a tropical cyclone.

We may first consider the sea condition at a given location as the hurricane approaches. We define this situation here as the growing stage of the hurricane. As an example, let us examine the relationship between wind and sea severity observed in Hurricane ELOISE. Figure 1 shows the track of Hurricane ELOISE taken from a report published by NOAA Data Buoy Office (NOAA 1975). The hurricane crossed the Gulf of Mexico and its center passed within 10 miles (16 kilometers) of Buoy EB 10 as shown in the figure. The relationship between the mean wind speed at a 10 meter-height and the significant wave height obtained by the buoy at one hour time intervals is shown in Figure 2.

The white circle in Figure 2 indicates the wind speed and significant wave height in the growing stage of the hurricane. It can be seen in the figure that the sea severity increases almost linearly with increase in wind speed during the growing stage, and becomes the severest (significant wave height 8.8 m) when the wind speed becomes maximum (35.2 m/sec, 68.6 knots) and is then followed by a transition stage. That is, the wind speed reduces in magnitude to a great extent (from 35.2 m/sec to 8 m/sec) within two hours when the hurricane eye passes near the buoy. The sea state follows the change in energy source by reducing in severity from a

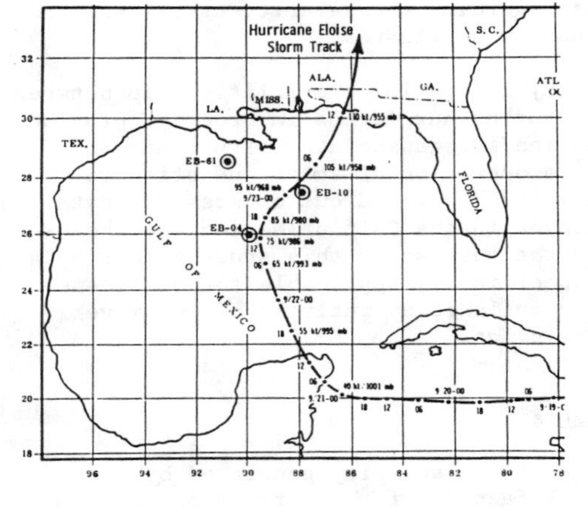

Figure 1:

Track of Hurricane ELOISE (From NOAA Data Report 1975)

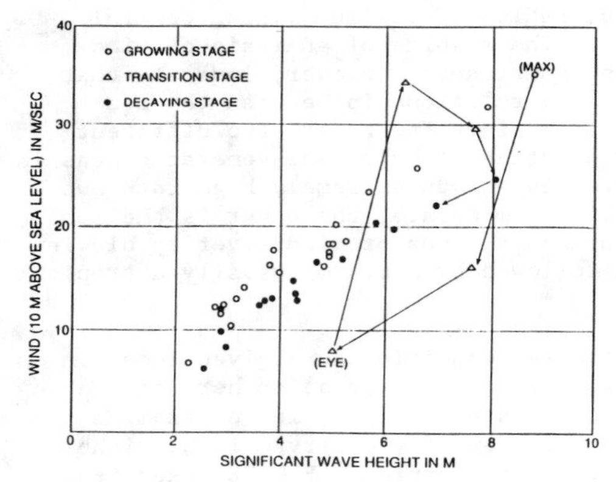

Figure 2:

Relationship between mean wind speed and significant wave height, Hurricane ELOISE

significant wave height of 8.8 m to 5.0 m. Then, during the following two hours, the wind speed as well as the sea state come back to nearly the same level as they were in the growing stage. After that, the sea severity reduces almost linearly with reduction in wind speed. Thus, it can be seen that the severity of sea state is subject to wind speed during the hurricane.

The wind speed - sea severity relationship during the growing stage of hurricanes is obtained for an additional five hurricanes and the results are summarized in Figure 3. The data shown in the figure are all measured by NOAA buoys in deep water. Some wind speeds measured at a 5-meter height above the buoy are converted to that for a 10 m height. Some wind speeds included in the figure were below hurricane level (33.5 m/sec, 65 knots) when they passed over the buoys, but the wind severity reached hurricane level later.

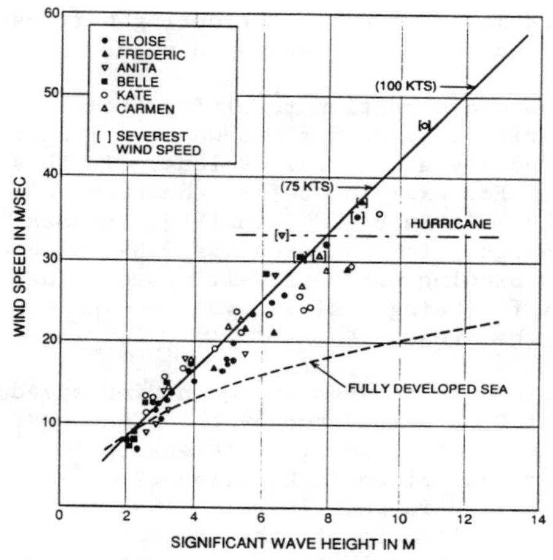

Figure 3: Relationship between mean wind speed and significant wave height obtained in various hurricanes

As can be seen in Figure 3, the sea severity increases almost linearly with increase in wind speed during the growing stage of hurricanes. Since no appreciable scatter can be observed in the data, a probability analysis of data may not be required. By drawing the average line in the figure, the significant wave height during the growing stage of the hurricane can be simply obtained as a function of the mean wind speed at a 10 meter level as,

$$H_s = 0.235\,\overline{U}_{10}. \tag{1}$$

Included also in Figure 3 is the functional relationship between wind speed and significant wave height for fully-developed seas obtained from the Pierson-Moskowitz spectrum (1964). That is, the original Pierson-Moskowitz's spectral formulation for fully developed seas is given as a function of the mean wind speed referred to a 19.6 meter height above sea level. Under the assumption of a narrow-band spectrum, however, the formula can be presented as a function of significant wave height. Hence, we can drive the relationship between the mean wind speed and significant wave height for fully developed seas. By converting the wind speed to that at a 10 meter height, we have the following relationship:

$$H_s = 0.237\,(\overline{U}_{10}^2/g). \tag{2}$$

As expected, Figure 3 shows that the sea severity during the growing stage of a hurricane for a given wind speed is much less than that for fully developed seas. This is because the time duration of a given wind speed is extremely short during hurricanes in comparison with that required for fully developed seas.

Next, let us consider the sea condition resulting from continuous winds of mild severity blowing for one week or longer and then followed by storm, usually a tropical cyclone. In this case, the sea becomes severe. For example, before the tropical cyclone GLORIA passed near the NOAA Buoy 41002 in 1985, the sea condition (significant wave height) in that area was 2.5-4.0 meters for 10 days with consistently blowing winds of 7-11 m/sec. That is, the sea had the potential for being easily augmented in severity when GLORIA came to the area.

Figure 4 shows the relationship between the mean wind speed and significant wave height of GLORIA measured by the NOAA buoy. Included also in the figure is the functional relationship applicable for fully-developed seas given in Equation (2). As can be seen, the wind velocity-sea relationship is very close to that applicable for a fully-developed situation in this example. This does not imply, however, that the shape of the wave spectrum is very close to that of a fully-developed sea.

It is given by Pierson-Moskowitz that a sufficiently long time duration is prerequisite for generating the fully-developed wave spectrum. For example, at least 42 hours is required at a mean wind speed, U_{10} of 20 m/sec for a sea to become fully-developed and to have the spectral shape given by Pierson, Neumann and James (1955). It is not realistic to expect such a long sustained wind speed during a tropical cyclone. Therefore, even though the wind-wave relationship is close to that given in Eq.(2), the shape of the spectrum is different from that for a fully-developed sea as will be shown in the next section.

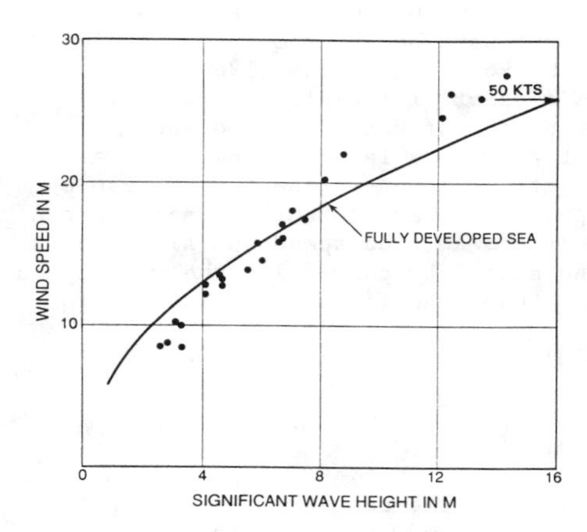

Figure 4:

Relationship between mean wind speed and significant wave height, Tropical cyclone GLORIA

Wave Spectra During Hurricanes

It is of considerable interest to examine the shapes of wave spectra during hurricanes, since the rate of change of wind speed is much shorter during hurricanes than that observed during ordinary storms. As an example, Figure 5 shows a comparison between a wave spectrum obtained during Hurricane ELOISE and various spectra obtained during ordinary storms in the North Atlantic. All spectra are obtained in deep water and have the same severity; significant wave height of 8.8 meters. As can be seen, the wave energy is concentrated primarily in the neighborhood of the peak frequency during the hurricane as contrasted with the energy being spread over a wide frequency range, including double peaks, for wave spectra obtained during ordinary storms.

One may expect the shape of wave spectra during hurricanes to be extremely random; however, the results of analysis of measured data in the open ocean have shown that there is a consistent trend in their shape, particularly in severe sea conditions. For example, Figure 6 shows a number of observations of modal frequency

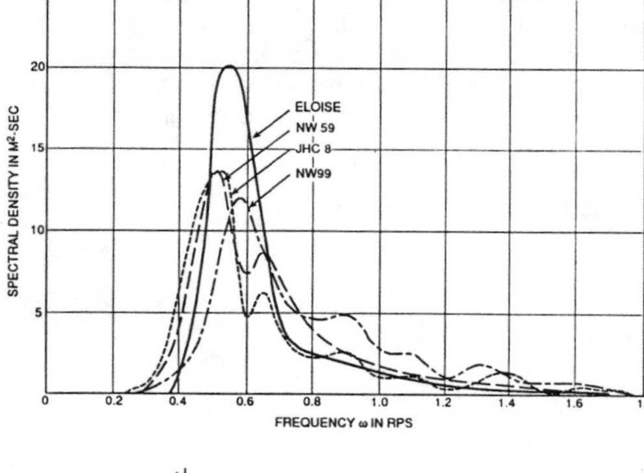

Figure 5:

Comparison between Hurricane ELOISE wave spectrum and wave spectra in ordinary storms having the same significant wave height, 8.8 m

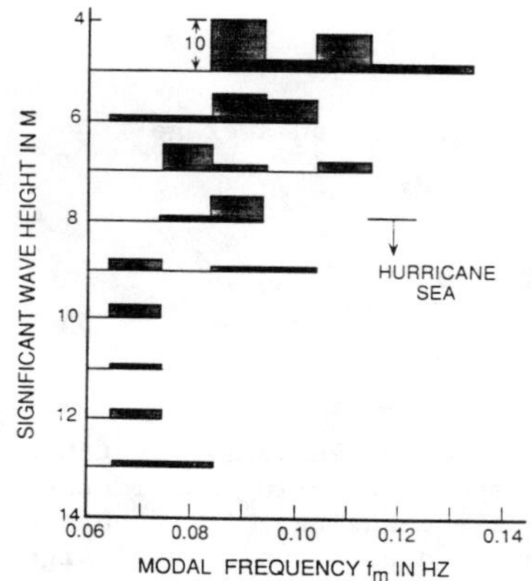

Figure 6:

Number of modal frequencies for specified significant wave height (From Foster 1982)

of wave spectra for one-meter intervals of significant wave height. It can be seen that the location of modal frequencies during hurricanes is concentrated in the neighborhood of 0.07 Hz in severe seas.

Antani (1981) analyzed the shape of approximately 400 wave spectra obtained during hurricanes. In his analysis, the spectral shape is compared to various mathematical formulations such as the one-parameter (Pierson and Moskowitz 1964), the two-parameter, the three-parameter, the six-parameter (Ochi and Hubble 1976) and the JONSWAP spectra (Hasselmann et al. 1973). The parameters involved in these spectral formulations are determined so that the sum of the squared differences between the mathematical formulation and the observed spectra are minimal. The results of his analyses indicate that the six-parameter and the JONSWAP formulations well represent the wide variety of spectral shapes associated with the stages of growth, eye passage and decay of hurricane-generated seas.

Furthermore, Foster (1982) systematically analyzed the values of parameters involved in the JONSWAP spectral formulation and derived several significant conclusions. Based on the results of studies carried out by Antani and Foster, it is concluded that the shape of wave spectra during hurricanes can be well represented by the JONSWAP spectral formulations; however, the values of parameters involved in the formulation are different from those originally by Hasselmann as follows.

The JONSWAP spectral formulation is given as,

$$S(f) = \alpha \frac{g^2}{(2\pi)^4} \exp\left\{-1.25(f_m / f)^4\right\} \times \gamma^{\exp\left\{-(f - f_m)^2 / 2(\sigma f_m)^2\right\}}, \tag{3}$$

where

γ = peak-shape parameter, 3.30 as an average
α = $0.076\ (\bar{x})^{-0.22}$
σ = 0.07 for $f \leq f_m$, and 0.09 for $f > f_m$
f_m = $3.5(g/\bar{U})(\bar{x})^{-0.33}$
$\bar{x}$ = dimensionless fetch = $gx/\bar{U}^2$
x = fetch length, and
$\bar{U}$ = mean wind speed.

The parameter α is a function of dimensionless fetch length which is extremely difficult to evaluate during hurricanes since the mean wind speed is continuously changing. It is obtained, however, that α can be well presented as a function of significant wave height, H_s, and modal frequency, f_m, as follows (Foster 1982):

$$\alpha = 4.5\, H_s^2\, f_m^4, \tag{4}$$

where the constant carries the units of $(\mathrm{sec}^2/\mathrm{meter})^2$.

It is also found that the peak-shape parameter γ substantially deviates from the mean value 3.3 given in Eq. (3). Figure 7 shows the histogram of the peak-shape parameter evaluated by fitting the measured spectra by the JONSWAP formulation. As can be seen in the figure, the values of γ for hurricane-generated seas range from 0.60 to 4.0 which is significantly smaller than that obtained in the JONSWAP Project. It is noted that values of γ greater than 3.0 shown in the figure are primarily from the decay stage of a specific hurricane —— Hurricane BELLE.

Figure 8 shows the peak energy density, $S(f_m)$, obtained from measured data in hurricanes as a function of significant wave height. As can be seen, there is a strong correlation between the peak energy and significant wave height, and it can be presented by

$$S(f_m) = 0.75\, H_s^{2.34}. \tag{5}$$

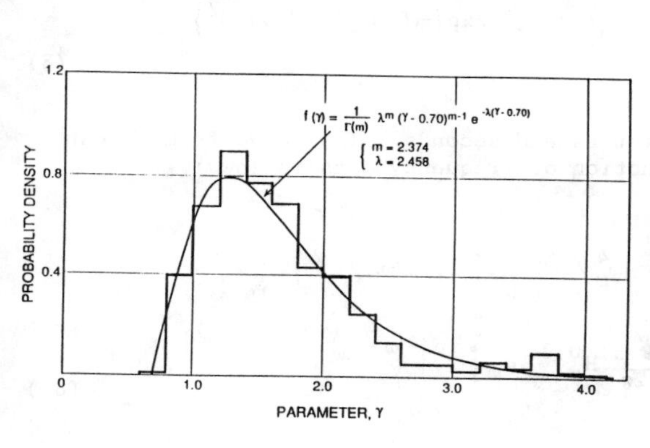

Figure 7:

Comparison of probability density function of Parameter of the JONSWAP spectrum (Data from Foster 1982)

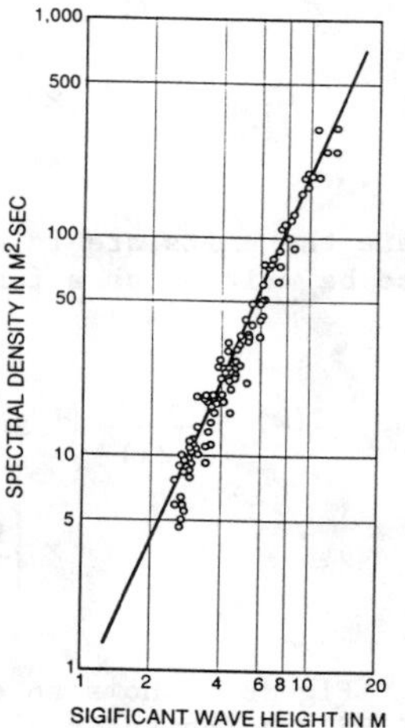

Figure 8:

Wave energy densities at modal frequency versus significant wave height (From Foster 1982)

On the other hand, the peak energy density of the JONSWAP spectrum can be obtained from Eq. (3) as

$$S(f_m) = \alpha \frac{g^2}{(2\pi)^4 f_m^5} e^{-1.25} \cdot \gamma . \tag{6}$$

Thus, from Eqs. (4) through (6), the peak-shape parameter can be expressed as a function of significant wave height and modal frequency as follows:

$$\gamma = 9.5 H_s^{0.34} f_m . \tag{7}$$

By using these relationships, we may derive a wave spectral formulation specifically applicable for hurricane-generated seas in the form of the JONSWAP formulation and as a function of significant wave height and modal frequency as follows:

$$S(f) = \frac{4.5}{(2\pi)^4} (H_s g)^2 (f_m^4 / f^5) \exp\left\{-1.25 (f_m / f)^4\right\} \times (9.5 H_s^{0.34} f_m)^{\exp\left\{-(f - f_m)^2 / 2(\sigma f_m)^2\right\}} , \tag{8}$$

where the units are in meters and seconds. The above formula can also be written as a function of frequency ω as follows:

$$S(\omega) = \frac{4.5}{(2\pi)^4} (H_s g)^2 (\omega_m^4 / \omega^5) \exp\left\{-1.25 (\omega_m / \omega)^4\right\} \times \left(\frac{9.5}{2\pi} H_s^{0.34} \omega_m\right)^{\exp\left\{-(\omega - \omega_m)^2 / 2(\sigma \omega_m)^2\right\}} . \tag{8'}$$

Figure 9 shows an example of comparison between Hurricane ELOISE and the spectral formulation given in Eq.(8'). Another two comparisons between the spectral formulation and spectra obtained during Hurricanes KATE and GLORIA are shown in Figures 10 and 11, respectively. Figure 10 is the case for a strong wind speed, 42.0 m/sec (81.7 knots), while Figure 11 is the case for a severe sea state, significant wave height of 14.3 meters. A good agreement between measure spectra and the formulation given in Eq.(8') can be seen in all these cases.

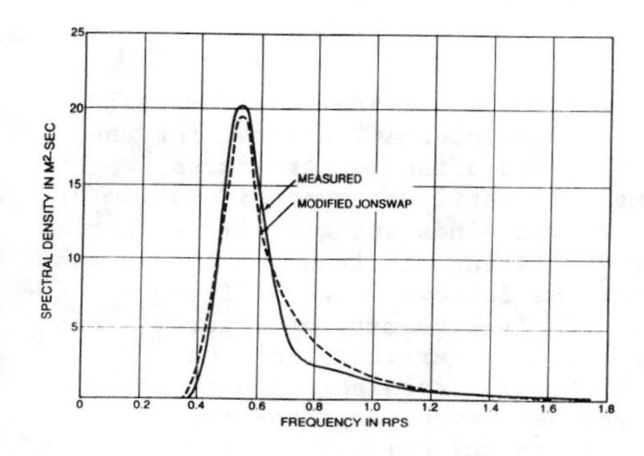

Figure 9:

Comparison between measured and modified JONSWAP spectra, Hurricane ELOISE (Significant wave height 8.8 m)

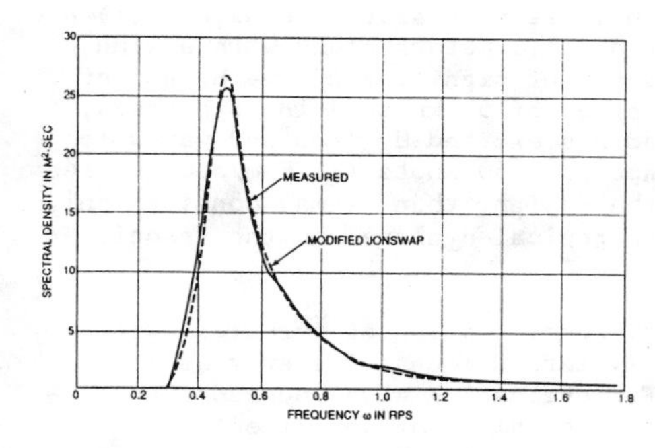

Figure 10:

Comparison between measured and modified JONSWAP spectra, Hurricane KATE (Significant wave height 10.0 m)

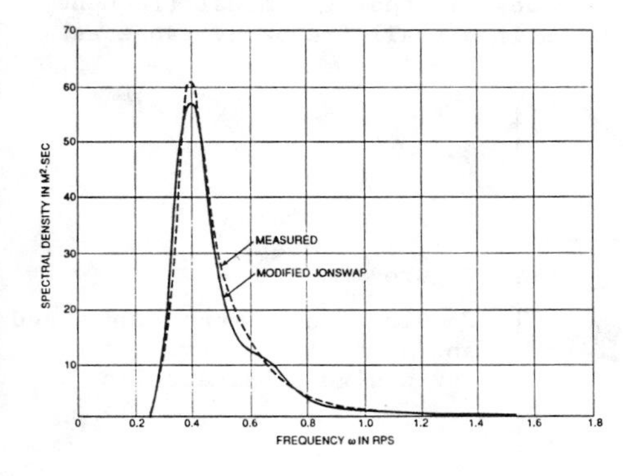

Figure 11:

Comparison between measured and modified JONSWAP spectra, Hurricane GLORIA (Significant wave height 14.3 m)

Design Application

There is no doubt that estimation of severe environmental conditions during hurricanes provides information vital for the design and safe operation of ships and offshore structures, hereafter called marine systems. In particular, consideration of simultaneous loadings associated with winds and waves is essential for marine system design. For evaluating simultaneous loadings of winds and waves, we may consider the following two different situations shown in Figure 12 which is a summary of Figures 3 and 4. One is the severest wind speed in a hurricane and the associated sea severity given by Eq.(1) (straight line in the figure), the other is the severest sea expected in a tropical cyclone and the associated wind speed which is given by Eq.(2).

As a practical application, suppose a hurricane wind speed of 100 knots (51.4 m/sec) is chosen as the design wind speed for a marine system in a seaway, then it is necessary to consider (i) a sea of significant wave height of 12.1 meters along with a wind speed of 100 knots, and (ii) a sea of significant wave height of 16.0 meters along with a wind speed of 50 knots. In both cases, evaluation of simultaneous loadings excited by wind and waves is highly desirable. If a wind speed of 50 knots (25.7 m/sec) or less is considered sufficient for the design, then we may consider only the severest sea expected in a tropical cyclone and the associated wind speed given by Eq.(2).

For evaluating wind loads, consideration of turbulent wind spectra carrying a significantly large amount of energy at low frequencies is recommended. For evaluating wave-induced loads during hurricanes, the concept of a family of wave spectra developed based on the results obtained in the previous section may be applied. That is, one of the features of wave spectra observed during hurricane-generated severe seas is that the modal frequency where the spectrum peaks occurs is in a small frequency range as

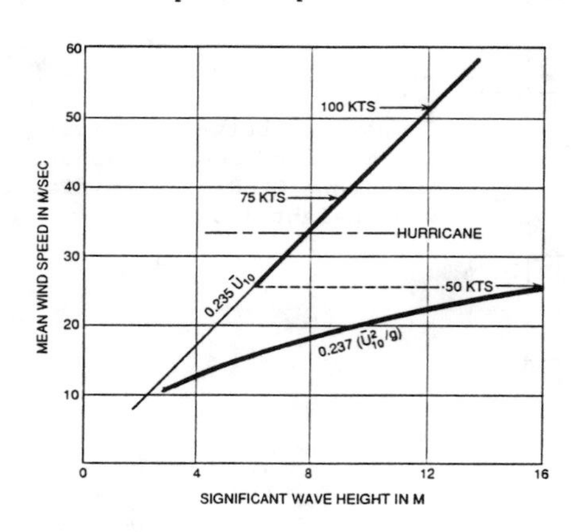

Figure 12:

Combination of mean wind speed and significant wave height for design consideration shown in Figure 6. In particular, the frequency range is extremely limited for severe seas of significant wave heights greater than 10 meters. Since the location of the modal frequency in severe seas appears to be indepndent of significant wave height, we may choose the three modal frequencies 0.063, 0.073 and 0.083 Hz (0.40, 0.46 and 0.52 rps, respectively) for design consideration. The results of analysis indicate that the probabilities of occurrence of these modal frequencies are 0.75, 0.17 and 0.08, respectively. By using these values of modal frequencies, a family of wave spectra consisting of three members can be generated by Eqs.(8) or (8′) for a specified significant wave height.

As an example, Figure 13 shows two sets of wave spectra which can be considered for a design wind speed of 100 knots. One is a family having a significant wave height of 12.1 meters for which a simultaneous loading associated with a wind speed of 100 knots (51.4 m/sec) should be considered; the other is a family having a significant wave height of 16.0 meters for which a wind-induced loading associated with wind speed of 50 knots (25.7 m/sec) should be concurrently considered.

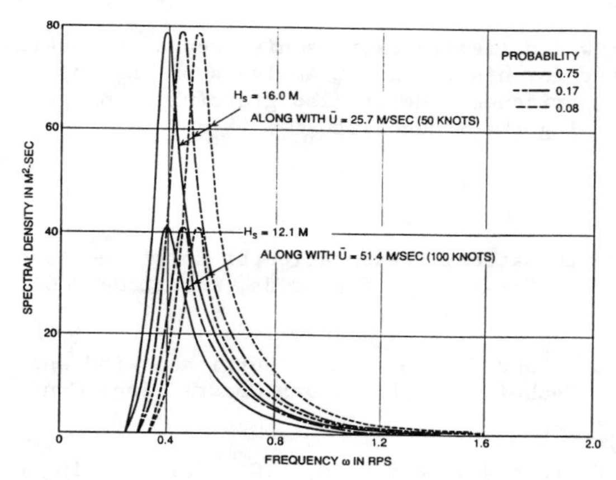

Figure 13: Family of wave spectra for design mean wind speed of 100 knots

Conclusions

This paper addresses the relationship between hurricane (or tropical cyclone) wind speed and sea seveity as well as the shape of wave spectra during hurricanes. It was found from analysis of measured data obtained by NOAA buoys that there are two different situations in the relationship between sea severity and mean wind speed. One is the growing stage of hurricanes in which the wind speed is increasing at an extremely high rate but the sea severity is comparatively mild; the functional relationship between wind speed and sea severity is given by Eq.(1). The other is the sea condition resulting from continuous winds of mild severity blowing for one week or longer then followed by a tropical storm in which the sea condition is very severe but the magnitude of wind speed is much less than that of a hurricane; the functional relationship between wind speed and sea severity is given by Eq.(2).

For the design and safe operation of marine systems, therefore, it is highly desirable to consider simultaneous loadings associated with winds and waves under these two circumstances.

From analysis of 400 wave spectra obtained from measured data during hurricanes, a wave spectral formulation specifically applicable for hurricane-generated seas is derived in the form of the JONSWAP formulation but as a function of significant wave height and modal frequency. Comparisons between wave spectra obtained in Hurricanes ELOISE, KATE and GLORIA and computed spectra by the formulation show good agreement.

Acknowledgements

The author would like to express his deep appreciation to Mr. Emmett Foster for his excellent analysis on the shape of wave spectra during hurricanes. He is also grateful to Ms. Laura Dickinson for typing the manuscript.

References

Antani, J.K., "Mathematical Presentation of Hurricane Associated Wave Spectra", University of Florida, *Report UFL/COEL*-81/007, 1981.

Bretschneider, C.L. and Tamaye, E.E., "Hurricane Wind and Wave Forecasting Techniques", *Proc. 5th Coastal Eng. Conf.*, 1976, pp. 202-237.

Cardone, V.J., Pierson W.J. and Ward, E.G., "Hindcasting the Directional Spectra of Hurricane-Generated Seas", *J. Petroleum Tech*. Vol. 28, 1976, pp. 385-395.

Foster, M.F., "JONSWAP Spectral Formulation Applied to Hurricane-Generated Seas", University of Florida, *Report UFL/COEL*-82/004, 1982.

Hasselmann, K. et al., *Measurements of Wind-Wave Growth and Swell Decay During the Joint North Sea Wave Project (JONSWAP)*, Deutsches Hydrograph. Inst., 1973.

NOAA Data Report, "Buoy Observations during Hurricane ELOISE", National Oceanographic Data Center, 1975.

Ochi, M.K. and Hubble, N.E., "On Six-Parameter Wave Spectra", *Proc. 15th International Coastal Eng. Conference*, Vol. 1, 1976, pp. 301-328.

Pierson, W.J., Neumann, G. and James, R.W., *Practical Methods for Observing and Forecasting Ocean Waves*. Hydrographic Office, U.S. Navy, Publication #603, 1955.

Pierson, W.J. and Moskowitz, L., "A Proposed Spectral Form for Fully Developed Wind Seas Based on the Similarity Theory of S.A. Kitaigorodskii", *Journal Geoph. Res.*, Vol. 69, 1964, pp. 5181-5190.

Ross, D. and Cardone, V., "A Comparison of Parametric and Spectral Hurricane Wave Prediction Products", *Turbulent Fluxes through Sea Surface, Wave Dynamics and Prediction*, 1978.

Young, I.R., "Parametric Hurricane Wave Prediction Model", *J. Waterway, Port, Coastal & Ocean Eng.*, Vol. 114, No. 5, 1988, pp. 637-652.

Young, I.R. and Sobey, R.J., "The Numerical Prediction of Tropical Cyclone Wind-Waves", *Res. Bulletin No. CS20*, University of North Queensland, 1981.

A Method for Locating Spikes in a Measured Time Series

James P. McKinney[1]

Abstract

Raw directional wave sample time series are frequently contaminated with spikes from various sources. The standard procedure in correctly editing these time series includes locating and correcting spikes using some interpolation method. This paper presents a procedure for locating spikes based on an assumption that wave data are well distributed about the time series mean.

Introduction

The Prototype Measurement and Analysis Branch (PMAB) of the U.S. Army Engineer Coastal Engineering Research Center (CERC) has been collecting remotely measured time series of pressure and horizontal water particle velocities since its inception in 1983. These data have been collected using a variety of instruments including self-contained tape recording gages, self-contained solid-state memory gages, and near-real-time remote transmission units.

Plotted data points which fall outside the general envelope of the time series and are caused by phenomena other than water pressure are points which can be thought of as spikes. Data spikes have been found to contaminate time series from each type of collection device. The causes of these spikes include electronic noise from various sources, bad memory on memory boards, and erroneous transmission of data. The particular problem which provided motivation for this spike location method resulted from an error in a transmission protocol program.

Method

This spike location method is based on the assumption that good sample data are well-distributed about the time series mean for a sufficiently long sample period. A good pressure time series collected from Ocean City, Maryland in January, 1992 is displayed in Figure 1. Figure 2 shows a ranked distribution of the sample data values from the time series mean. Note the smooth shape of the distribution. Generally, points lie within 3 standard deviations from the mean.

Figure 3 also shows a plot of pressure data collected from Ocean City, Maryland, in January, 1992. The spikes in this plot result from misinterpretation of a particular raw data word value by the transmission

---

[1] Mathematician, Coastal Engineering Research Center, USAE Waterways Experiment Station, 3909 Halls Ferry Road, Vicksburg, MS, 39801

protocol and are not necessarily outside the range of valid values. The plot of the ranked distribution of this time series, Figure 4, shows an unusual jump in the distribution with a flat spot at the end. The index point numbers of the data which caused these flat spots can be determined by utilizing a secondary array of point index numbers of the original time series and making the same element swaps required to rank the distribution time series.

The spikes in this distribution can be determined by examining the pointwise difference between successive points in the ranked distribution. Downward spikes are the first to be located. We find the index number (min) in the distributed time series where the maximum difference occurs for negative differences from the mean, see Figure 5. All points prior to min in the distributed time series are tagged as spikes. For upward spikes, we find the index number (max) in the distributed time series where the maximum difference occurs for positive differences from the mean, see Figure 6. For elements following max in the distributed time series, the corresponding points in the original time series are flagged for data editing. Figure 7 shows the edited data.

Variations of cut off criteria can be utilized to determine which points to edit. Instead of locating the maximum differences in the distributed time series as we did in the above example, one may determine the first and last pointwise differences exceeding 0.2 - 0.5 standard deviations or some other hardwired criteria.

Steps

The following steps should be included in code utilizing this method:

- Calculate time series mean
- Load rows of sorting array with point-wise difference from the mean and original point indices
- Sort rows according to ascending difference from the mean
- Determine cutoff criteria
- Determine first good point
- Determine last good point
- Identify spikes
- Take appropriate action

Other Tests Required

Other test for data quality may be required in addition to this form of spike checking. Figures 8, 9, and 10 show raw data collected at Sarasota, Florida, the distribution time series and the edited time series. In spite of obvious problems with the raw time series, the distribution series data are well-distributed about the mean however, a mean-shift test or standard deviation test for data segments would have eliminated the record prior to the spike checking routine, which failed to catch the bad data.

Acknowledgement

Permission to publish this paper was granted by the Chief of Engineers, United States Army Corps of Engineers.

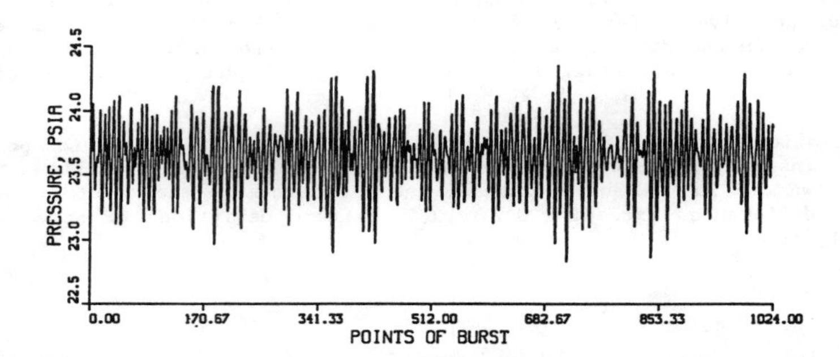

Figure 1: Time series plot of quality pressure data.

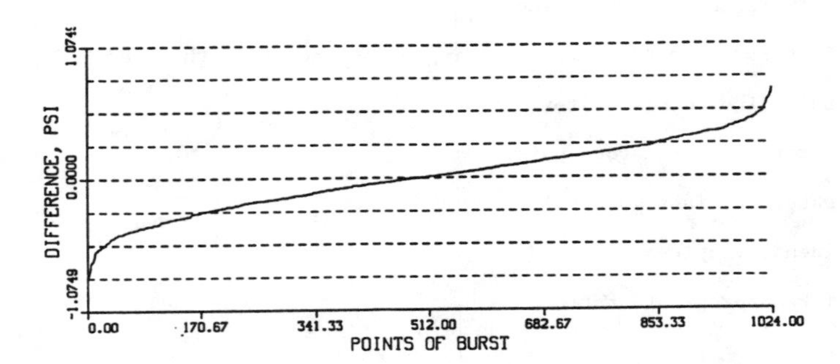

Figure 2: Ranked distribution time series plot of quality pressure data.

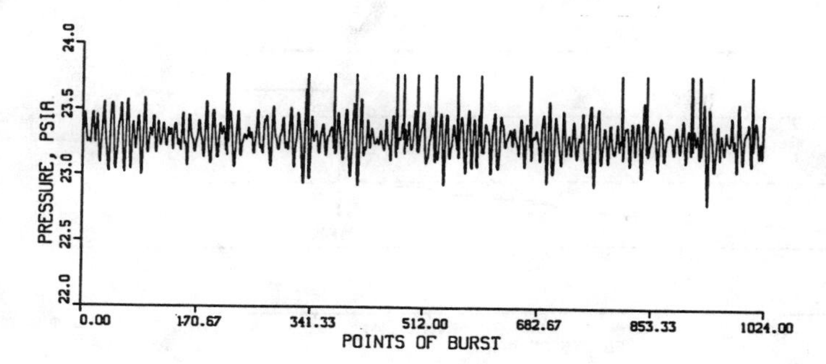

Figure 3: Plot of pressure time series contaminated with "spikes" resulting from erroneous transmission protocol procedure.

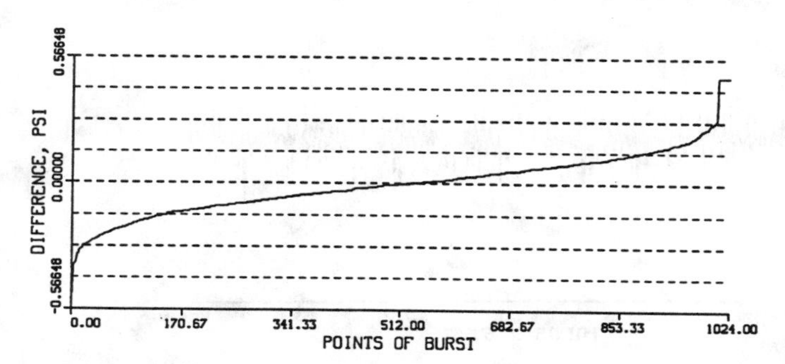

Figure 4: Ranked distribution time series of spikey data shown in Figure 3.

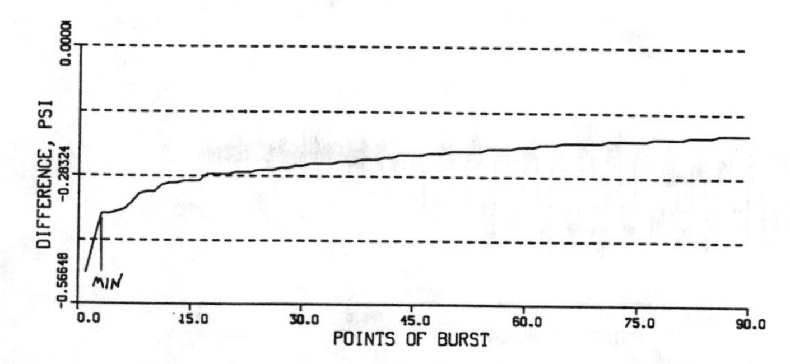

Figure 5: Plot showing location of downward spikes. MIN is the index number first of quality point in the ranked distribution.

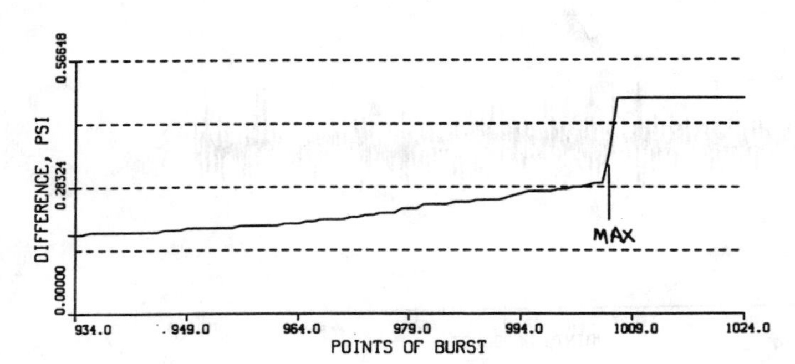

Figure 6: Plot showing location of upward spikes. MAX is the index number of last quality point in the ranked distribution.

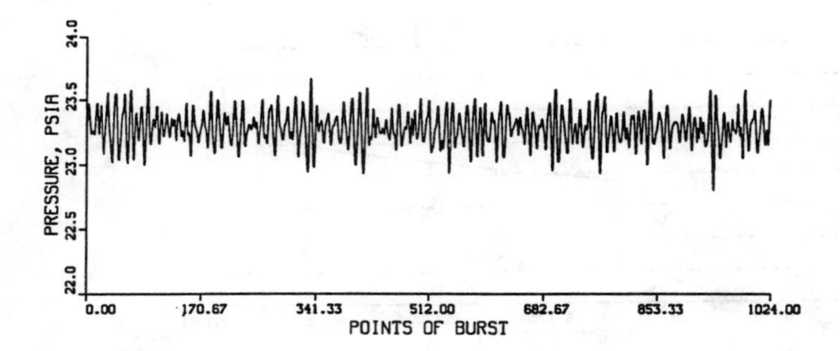

Figure 7: Plot showing edited pressure time series shown in Figure 3.

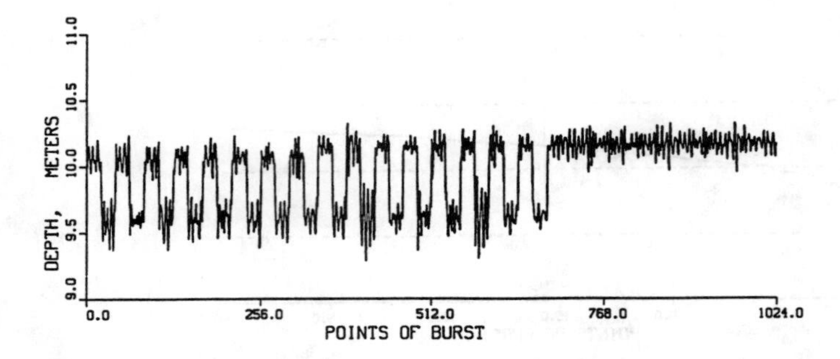

Figure 8: Plot of pressure time series stored on a bad memory board.

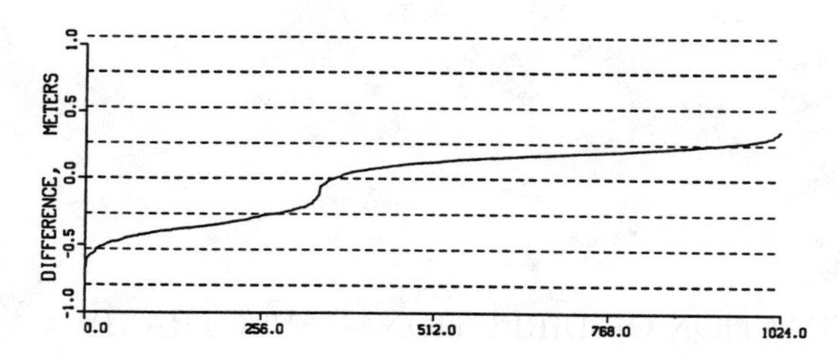

Figure 9: Ranked distribution plot of time series displayed in Figure 8.

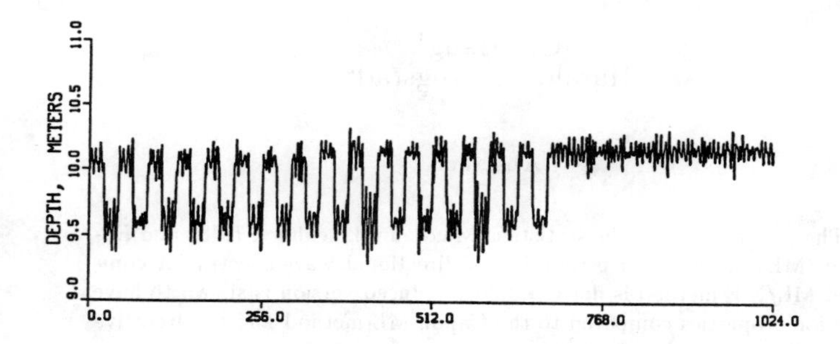

Figure 10: Plot of edited time series shown in Figure 8.

ESTIMATION OF DIRECTIONAL SPECTRA BY ML/ME-METHODS

Ola Haug[1]
Harald E. Krogstad[2]

Abstract

The paper discusses the statistical Maximum Likelihood (ML) and Entropy (ME) methods for estimation of directional wave spectra. A combined ML/ME method is derived, and a reduced version is shown to have superior properties compared to the Capon ML-method and the Iterative ML-method.

Introduction

The Capon Maximum Likelihood method for estimation of directional wave spectra from a spatial array of wave recorders is well known [9, 12, 14, 15, 18]. However, the meaning of *maximum likelihood* (ML) in this case is only remotely connected with its usage in statistics. On the other hand, it is also well known that the discrete Fourier transforms obtained from multivariate time series of wave recordings are asymptotically complex Gaussian variables with covariance matrices essentially equal to the cross spectra of the time series. Viewing the Fourier coefficients as multivariate Gaussian observations, thus suggests using methods from statistical inference for the estimation of the cross spectra. ML estimation of covariance matrices is treated in all textbooks of multivariate statistics (*e.g.* [2]), and, of particular interest in the present case, by Burg *et al.* [8]. The idea of using statistical ML to estimate cross spectra has been around for some time [3, 4, 14].

[1]Project Engineer, OCEANOR A/S, Pirsenteret, N-7005 Trondheim, Norway
[2]Research Manager, SINTEF Industrial Mathematics, N-7034 Trondheim, Norway

It turns out that the ML-principle alone is not enough to ensure a unique solution for the directional wave spectrum. It must therefore be combined with a second optimizing principle in order to be really useful. The *Maximum Entropy* (ME) principle provides such a criterion, and it has turned out that the *combined ML/ME-method* reduces to the common way of processing wave data from single point triplets, *e.g.* heave/pitch/roll buoys [10].

The present paper will discuss aspects of the above principles for the estimation of directional wave spectra from spatial arrays of wave recorders. Most real valued results from [8] generalize to the present complex valued context. An additional benefit of ML estimation is the invariance principle which ensures that parameters derived from ML-estimators are ML-estimators as well.

The full ML/ME-optimization problem, which turns out to be rather tough, has not been solved explicitly in the general case. However, a sub-optimal method which coincides with the full method for single point triplets is shown to give better performance than the Iterative Maximum Likelihood Method [15] when applied to spatial arrays.

Wave measurements

We consider a stochastic deep water ocean wave field obeying Gaussian linear wave theory. Such a field is characterized by a *wavenumber spectrum* $\Psi(\boldsymbol{k})$ and a dispersion relation $\omega^2 = gk$ where ω is the angular frequency, $\boldsymbol{k}$ the wavenumber and g the acceleration of gravity. All wave-related properties are connected to the surface elevation by linear filters characterized by known *transfer functions* [5]. The wave spectrum may also be written in terms of ω and the direction, θ, of the wavenumber as $E(\omega,\theta) = S(\omega)D(\theta,\omega)$ where S is the *frequency spectrum* and D the *directional distribution*,

$$D(\theta,\omega) = \frac{1}{2\pi}[1 + 2\sum_{n=1}^{\infty}\{a_n(\omega)\cos(n\theta) + b_n(\omega)\sin(n\theta)\}] = \frac{1}{2\pi}\sum_{n=-\infty}^{\infty} c_n(\omega)e^{in\theta}. \quad (1)$$

In situ measurements typically consist of simultaneously measured time series $\boldsymbol{W}(t) = \{W_m(t)\}_{m=1}^{M}$ of surface elevation, slope, velocity *etc.*, taken at fixed, possibly different locations. The series are finite, sampled realizations of a corresponding set of stochastic processes with cross spectra that can be expressed in terms of the directional spectrum and the transfer functions. For appropriate data recording procedures with sampling frequency ω_s and N data points, the discrete Fourier coefficients of $\boldsymbol{W}$, $\widehat{\boldsymbol{W}}(k), k = 0, \cdots, N-1$ are zero mean, multivariate complex Gaussian variables with covariance matrices essentially equal to the cross spectra:

$$\boldsymbol{\Sigma}(\omega_k) = \int_{\theta=0}^{2\pi} \boldsymbol{h}(\theta,\omega_k)\boldsymbol{h}^H(\theta,\omega_k)S(\omega_k)D(\theta,\omega_k)d\theta \quad (2)$$

where the vector $\boldsymbol{h}$ consists of the transfer functions and location factors of the form $\exp(i\boldsymbol{k}\cdot\boldsymbol{x}_m)$, $m = 1, \cdots, M$, and $\omega_k = k\omega_s/N$. Moreover, $\widehat{\boldsymbol{W}}(\omega_k)$ is virtually independent of $\widehat{\boldsymbol{W}}(\omega_{k'})$ if $k \neq \pm k'$ [7]. The basic objective of wave measurements is to infer S and D from estimates of $\boldsymbol{\Sigma}$.

The *standard estimate* for the cross spectrum is [7]

$$\widehat{\boldsymbol{\Sigma}}(\omega) = \frac{1}{\nu}\sum_{\beta=1}^{\nu} \widehat{\boldsymbol{W}}(\omega_\beta)\widehat{\boldsymbol{W}}^H(\omega_\beta) \tag{3}$$

where ω_β runs over ω_k–s surrounding ω. In the following we frequently omit the ω-dependence for clarity. Algorithms that directly utilize the standard estimate are the Capon ML-method in the generalization of Isobe [12], and the Iterative ML-method [15, 20]. The Capon estimate for the directional distribution is

$$\widehat{D}_{MLM}(\theta) = \kappa/\boldsymbol{h}^H(\theta)\widehat{\boldsymbol{\Sigma}}^g\boldsymbol{h}(\theta) \tag{4}$$

where κ is a normalization factor and ^{g} signifies a generalized inverse. The Capon ML-method is inconsistent with the data in the sense that $\widehat{D}_{MLM}$ is not in accordance with the cross spectrum computed from the same distribution. The Iterative ML-method is intended to make up for this deficiency. Introducing the nonlinear operator $M\ :\ D \to D_{MLM}$ by

$$D \longrightarrow \boldsymbol{\Sigma}_D = \int_0^{2\pi} \boldsymbol{h}(\theta)\boldsymbol{h}^H(\theta)D(\theta)\,d\theta \longrightarrow D_{MLM} = \kappa/\boldsymbol{h}^H(\theta)\boldsymbol{\Sigma}_D^g\boldsymbol{h}(\theta), \tag{5}$$

the Iterative ML-estimate is obtained by "solving" $M(D) = \widehat{D}_{MLM}$ by means of the iteration

$$\begin{aligned} D_{n+1} &= D_n + \omega_R[\widehat{D}_{MLM} - M(D_n)] \quad , n \geq 0, \\ D_0 &= \widehat{D}_{MLM} \end{aligned} \tag{6}$$

where ω_R is a relaxation parameter around 1 [15].

The maximum likelihood and entropy principles

As mentioned in the introduction, the cross spectrum, considered as the covariance matrix of the discrete Fourier coefficients of $\boldsymbol{W}$, suggests to use methods from statistical inference. The maximum likelihood estimate results from maximizing the likelihood function, which, after taking the logarithm and leaving out the irrelevant terms and the dependence on ω, takes the form [8, 15]

$$L(\boldsymbol{\Sigma}, \widehat{\boldsymbol{\Sigma}}) = -\log|\boldsymbol{\Sigma}| - \mathrm{tr}(\boldsymbol{\Sigma}^{-1}\widehat{\boldsymbol{\Sigma}}). \tag{7}$$

From the theory of real covariance matrix estimation with multivariate Gaussian variables it is known that the standard estimate *is* the ML-estimate if $\boldsymbol{\Sigma}$ is allowed to vary over all positive definite matrices [2]. A similar result is also true for complex Gaussian variables. Note that the Fourier coefficients only occur through the standard estimate. However, in our case, the solution must conform to the special structure in Eqn. 2. In a much cited paper Burg *et al.* [8] present several important results for maximum likelihood estimation of structured covariance matrices. Related work, apparently unknown to Burg *et al.*, includes a series of papers by T.W. Anderson, see *e.g.* [1]. Most results in Burg's paper generalize to the present situation. Thus, the ML-estimates for S and D are solutions of

$$\max(L(\boldsymbol{\Sigma}, \widehat{\boldsymbol{\Sigma}})) \tag{8}$$

subject to

$$S \geq 0, \quad D \geq 0, \quad \text{and} \quad \int_0^{2\pi} D(\theta)\, d\theta = 1, \tag{9}$$

where the structure of $\boldsymbol{\Sigma}$ satisfies Eqn. 2. If we write $\boldsymbol{\Sigma} = S\boldsymbol{\Sigma}_D$ where $\boldsymbol{\Sigma}_D = \int_0^{2\pi} \boldsymbol{h}(\theta)\boldsymbol{h}^H(\theta)D(\theta)\, d\theta$, it follows that the estimate for $\boldsymbol{\Sigma}_D$ is the solution, $\boldsymbol{\Sigma}_{DML}$, of

$$\max \left(\frac{1}{|\boldsymbol{\Sigma}_D| \mathrm{tr}(\boldsymbol{\Sigma}_D^{-1}\widehat{\boldsymbol{\Sigma}})^M} \right), \tag{10}$$

and the estimate for S is

$$\mathrm{tr}(\boldsymbol{\Sigma}_{DML}^{-1}\widehat{\boldsymbol{\Sigma}})/M. \tag{11}$$

In many cases there is often a direct and simple way of estimating S such that the actual ML estimation is only used for D. However, as shown in [10], the above estimate for S may be preferable with respect to sampling variability.

No simple general algorithm for the above problem for $\boldsymbol{\Sigma}_D$ seems to exist although certain transfer functions admit explicit solutions. In general, the functional will have at least one maximum, but uniqueness has not been proved. Even a unique matrix solution to the ML problem does not guarantee a unique solution for D since several directional distributions may result in the same cross spectral matrix (Examples are easily constructed in the case of single point triplets, see below). In order to arrive at a unique, and in some way optimal, directional distribution among possible candidates, another selection criterion has to be invoked. The Maximum Entropy principle is one such choice.

In [17] the following entropy functional was used

$$\mathcal{H}(D) = \int_0^{2\pi} \log D(\theta)\, d\theta \tag{12}$$

and the ME problem now consists of maximizing $\mathcal{H}(D)$ subject to

$$\int_0^{2\pi} \boldsymbol{h}(\theta)\boldsymbol{h}^H(\theta)D(\theta)\, d\theta = \boldsymbol{\Sigma}_{DML}. \tag{13}$$

The functional is bounded above, and if there is at least one directional distribution D_{ML} such that $\mathcal{H}(D_{ML}) > -\infty$, the ME problem will have a solution. Since $\boldsymbol{\Sigma}_{DML}$ is Hermitian, there are $N(N+1)/2$ linear constraints and by introducing a Hermitian matrix of Lagrange multipliers, $\boldsymbol{\Lambda}$, variation of $\mathcal{H}$ with respect to D yields

$$D_{ME}(\theta) = \frac{1}{\boldsymbol{h}^H(\theta)\boldsymbol{\Lambda}\boldsymbol{h}(\theta)}. \tag{14}$$

Apart from special cases, no explicit solution for $\boldsymbol{\Lambda}$ fulfilling Eqn. 13 is known. It is also possible to introduce Eqn. 14 into the ML functional and maximize with respect to all possible matrices $\boldsymbol{\Lambda}$, but the advantage is not obvious. A simplified method based on this idea is presented in the next section.

The entropy functional in Eqn. 12 is not the only one that could be used. Good results have also been experienced [13] with the functional

$$\mathcal{H}(D) = -\int_0^{2\pi} D(\theta)\log D(\theta)\,d\theta \tag{15}$$

leading to directional distributions of the form

$$D_{ME}(\theta) = e^{\boldsymbol{h}^H(\theta)\boldsymbol{\Lambda}\boldsymbol{h}(\theta)}. \tag{16}$$

An ME-method for spatial arrays of elevation recorders which used this form of the entropy, although not in connection with an ML estimation of the cross spectrum, is given in [19].

In the special case that $\boldsymbol{h}(\theta) = \{e^{in\theta}\}_{n=0}^{K}$, the constraints take the form

$$\int_0^{2\pi} e^{-in\theta}\,D(\theta)\,d\theta = c_n\,, \quad n = 0,\ldots,K, \tag{17}$$

and the ME solution resulting from the definition in Eqn. 12 is

$$D_{ME}(\theta) = \frac{1}{2\pi}\frac{\sigma_e^2}{|1-\phi_1 e^{-i\theta} - \cdots - \phi_K e^{-iK\theta}|^2}, \tag{18}$$

where the parameters $\phi_1,\ldots,\phi_K$ and σ_e^2 are obtained from the Yule-Walker equations [6]. Single point triplets may be reduced to this situation, and with $K = 2$, Eqn. 18 is the exact ME estimate in this case. The modified definition of entropy does not give an explicit solution although it is easily seen to be of the form

$$D_{MEmod}(\theta) = e^{p(\theta)} \tag{19}$$

where $p(\theta)$ is a real trigonometric polynomial of degree K. See [18] for an extensive treatment of the Capon ML-method and various formulations of the ME principle.

The combined ML/ME method outlined above has several appealing properties. First of all, both ML and ME are well-proven principles in statistical inference. Moreover, often basic directional parameters are obtained directly from the ML-matrix, and since one-to-one functions of ML-estimators are ML-estimators [2], we immediately know that the estimators for the derived parameters are also optimal in the ML sense.

It may be proved that if measurements are subject to linear time-invariant filters, this does not affect the solution for the directional spectrum as long as the transformation is non-singular. That is, if the matrix of transfer functions for the filters, $\boldsymbol{R}(\omega)$, is non-singular for a particular frequency, then the new ML solution is simply $\boldsymbol{R}^H \boldsymbol{\Sigma}_{ML} \boldsymbol{R}$ and the final ME directional estimate is unchanged [11]. It may turn out advantageous to utilize this principle in practical wave data analysis in order to improve the numerics and the stability of the algorithms, see [19].

Heave/pitch/roll data, or more generally data from single point triplets, allow explicit solutions to the above ML and ME problems as long as the wavenumber is treated as an independent parameter and not restricted by the dispersion relation [10]. In particular, it has turned out that the routinely used expressions for the Fourier coefficients introduced by R.B. Long [16] in this case *are* ML-estimators. From the above comments, also derived parameters such as the mean direction and directional spread will be ML-estimators. In this case, the ML-problem has a unique and simple solution. The problem gets much more difficult and has to be solved numerically if the wavenumber is restricted by the dispersion relation [10].

The constrained ML method

The numerical solutions to the ML and ME problems are rather elaborate and time-consuming. A sub-optimal method may be constructed by restricting the directional distributions to a finite parameter set and maximizing the ML functional over this particular set. Any family of positive functions could in principle be used, in particular two-parameter families like the *cos-2s*-distributions. More general families consist of functions of the forms in Eqns. 18 and 19.

So far, only the form given in Eqn. 18 has been checked out. This method thus consists of choosing K and then maximizing the ML-functional over all such functions. The optimization problem defined in terms of the K Fourier coefficients is numerically tricky because of the constraint $D(\theta) \geq 0$. However, Fourier coefficients of positive functions can be re-cast in terms of so-called Schur coefficients (or reflection coefficients) $\{s_1, \cdots, s_K\}$ which may be chosen independently from the complex unit disc. By finally introducing a new set of variables

$q_k = s_k/(1- \mid s_k \mid)$, $k \leq K$, we obtain an unconstrained optimization problem in $2K$ independent real variables (Incidentally, the family of functions following from the modified definition of entropy (Eqn. 19) seems to be even simpler to apply). In the numerical experiments reported below, the numerical solution was carried out using the POWELL subroutine from [21]. This routine is based on function values only, so that no expression for the gradient is needed. The algorithm follows a scheme of successive line minimizations.

As the constrained ML-method with $K = 2$ reduces to the common method for single point triplets, we already have extensive experience for real data. However, it is obvious that K has to be chosen reasonably depending on the array configuration. Using $K = 3$ in the single point triplet case gives a flat maximum of the ML functional and a non-unique solution. On the contrary, including too few Fourier coefficients will reduce the ability to resolve complicated directional distributions.

Numerical Experiments

A series of numerical experiments comparing the constrained ML-method (CMLM) to the Capon ML-method (MLM) and the iterative ML-method (IMLM) have been carried out and some selected results are presented below. For a spatial array, there are several effects that should be considered:

- Array orientation effects
- Resolution of complicated distributions
- Sampling variability
- Array dimension effects
- Tendency to peak splitting

The methodology for the experiments is to generate synthetic cross spectral matrices $\boldsymbol{\Sigma}$ using known directional distributions. These matrices are either used directly in the likelihood function, or one may simulate sample cross spectral matrices by factoring $\boldsymbol{\Sigma} = \boldsymbol{L}\boldsymbol{L}^H$ and adding products of complex "Fourier coefficients" of the form $\boldsymbol{Le}$ where $\boldsymbol{e}$ is a simulated vector of independent standard complex Gaussian variables.

The arrays for the experiments reported below are given in Fig. 1. The array dimensions are later expressed in terms of the dimensionless parameter Δ defined as the array diameter times the wavenumber.

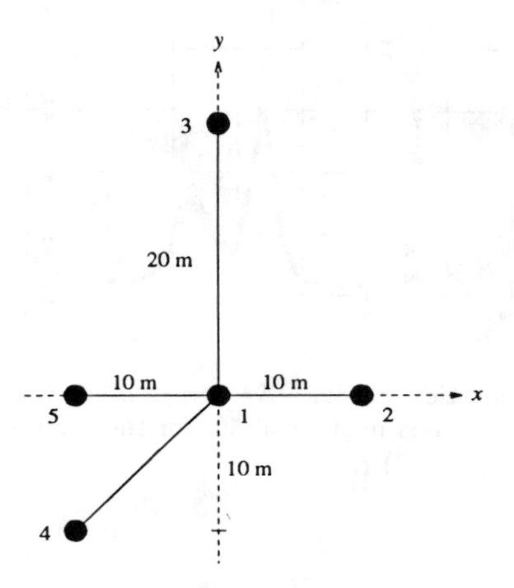

Figure 1: Arrays used in the numerical experiments. A4: locations 1-4 ; A5: locations 1-5.

Array orientation effects occur because arrays are not symmetric. Such effects are illustrated in Fig. 2 and we note that the CMLM may be considerably less influenced than the MLM or even the IMLM.

The ability to resolve complicated directional distributions is an important property of a method, and the three methods are compared for a rather complicated three peaked distribution in Fig. 3. The results are based on the exact cross spectral matrix for $\Delta = 1$, and $K = 4$ is used for the CMLM (Obviously, a smaller K would show poorer resolution). In this particular case, the CMLM greatly outperforms the other methods.

The resolving power and stability will be strongly dependent on Δ. A series of numerical experiments were carried out using simulated sample cross spectral matrices. It turned out that both the IMLM and CMLM worked similarly for $\Delta = \mathcal{O}(1)$, but the range of possible Δ−values was considerably larger for the CMLM. It is difficult to claim that this will be the case in general, however. Whereas the IMLM estimates broke down due to noise as Δ was increased, the CMLM could still provide reasonable solutions, see Fig. 4.

However, the experiments revealed that for very large Δ-values, the algorithm converged in a few cases to directional distributions far from the true distribution. Investigating the functional around these spurious solutions indicated that the likelihood function could have several local maxima. Consequently, the final

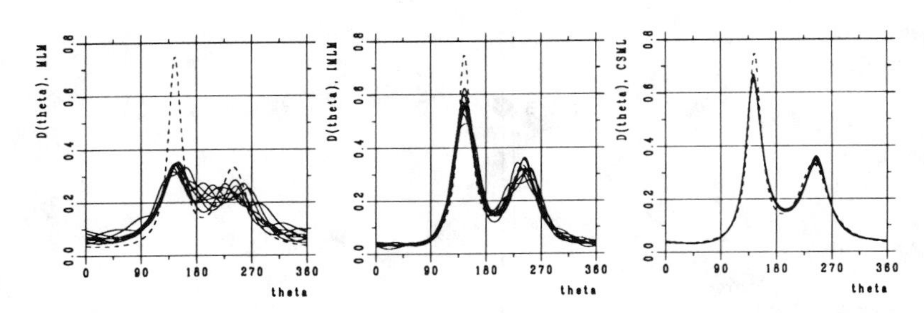

Figure 2: Array orientation effects. A4 array, $\Delta = 1$. Exact cross spectral matrices. The array is turned in steps of 50° for the same wave field. MLM left; IMLM middle; CMLM($K = 3$) right.

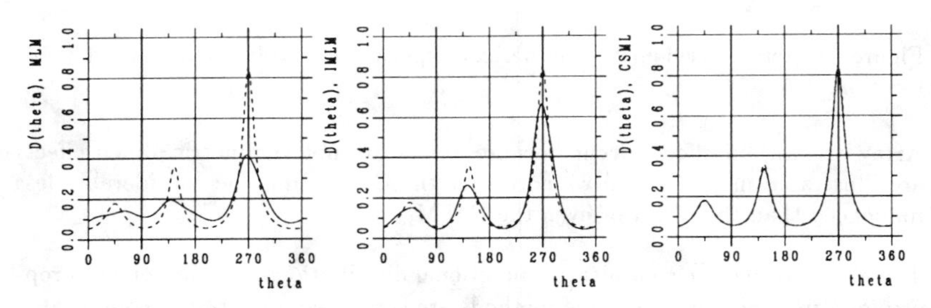

Figure 3: Complicated directional distribution. A5 array, $\Delta = 1$. Exact cross spectral matrix. MLM left; IMLM middle; CMLM($K = 4$) right.

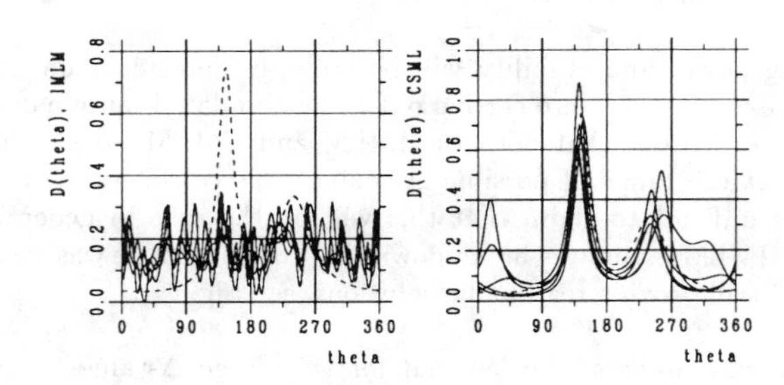

Figure 4: Breakdown effects. A4 array, $\Delta = 10$, 100 DOFs in the sample cross spectral matrices. IMLM left; CMLM($K = 3$) right.

estimate is in these cases more sensitive to the starting point of the optimization.

The sampling variability for highly non-linear methods like the Capon MLM, the IMLM and the CMLM is difficult to compute analytically. As a general principle, the sampling variance for $D(\theta)$ goes as the inverse of the degrees of freedom (DOF) in the cross spectral matrices. The numerical experiments revealed that the sampling variability was about the same for the IMLM and CMLM methods when both worked properly.

The peak splitting for the ME-method when applied to the Fourier coefficients of a box-car or a *cos-2s* distribution is well known. The CMLM, applied to single point triplet data will obviously have the same behaviour. An example of this using input from a *cos-2s* distribution for the array A4 is shown in Fig. 5. A tendency to a splitting into three peaks is clear in the CMLM estimates.

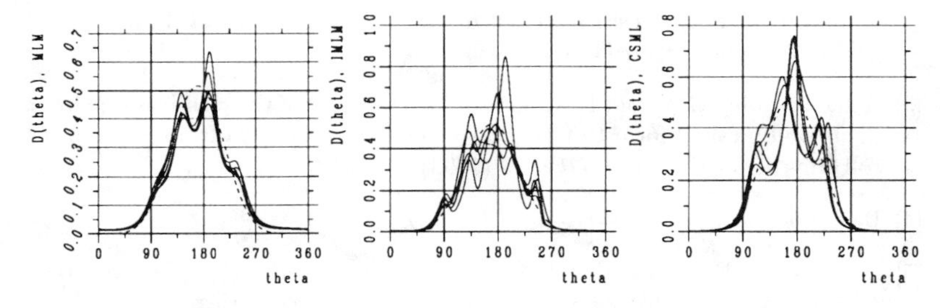

Figure 5: Peak splitting. *cos-2s* directional distribution. A4 array, $\Delta = 1$, 100 DOFs in the sample cross spectral matrices. MLM left; IMLM middle; CMLM($K = 3$) right.

Conclusions

The paper has introduced a general ML/ME-methodology for estimation of directional spectra. Both the ML and ME principles are well-proven in statistical inference, and the methodology may be fruitful in other situations as well. Two convenient properties are that parameters derived from ML-estimators are also ML-estimators, and that the results are essentially invariant to linear filtering of the data series.

Neither an explicit solution nor a simple numerical algorithm has so far been derived for the general ML/ME problem. Since over the last years it has been realized that ML-functionals may be rather ill-behaved numerically, this may presently be a drawback for routine use.

The suboptimal CMLM provides a manageable technique that reduces to the standard method of analysing single point triplets. For an optimal number of Fourier coefficients, the CMLM seems to outperform the MLM and the IMLM methods when applied to synthetic data. The numerical experiments indicate that choosing the number of complex Fourier coefficients to be one less than the number of sensors in the array is a fair compromise between stability and resolving power. So far, however, the experience on real data from spatial arrays is limited.

References

[1] Anderson, T.W., "Asymptotically efficient estimation of covariance matrices with linear structure", *Ann. of Statistics*, Vol. 1, no. 1 (1973) pp. 135-141

[2] Anderson, T.W., *"An introduction to Multivariate Statistical Analysis"*, 2nd ed., J. Wiley and Sons (1984)

[3] Athanassoulis, G.A. and E.K. Skarsoulis, "The use of the classical maximum likelihood method for estimating directional wave spectra from heave-pitch-roll time series", *Proc. ISOPE'93* (1993)

[4] Baggeroer, A., *pers. comm.*

[5] Borgman, L.E., "Directional spectra from wave sensors", in *Ocean Wave Climate*, Vol. 88 (1979) pp. 269-300, Marine Science Series, Plenum Press, New York.

[6] Box, G.E.P. and G.M. Jenkins, *Time Series Analysis, Forecasting and Control*, Holden-Day Inc., San Francisco (1976)

[7] Brillinger, D.R., *Time Series. Data Analysis and Theory*, Holt, Rinehart and Wilson Inc., New York (1975)

[8] Burg, J.P. and D.G. Luenberger and D.L. Wenger, "Estimation of Structured Covariance Matrices", *Proc. IEEE*, Vol. 70, no. 9 (1982) pp. 963-974

[9] Capon, J., "Maximum-likelihood spectral estimation", in *Nonlinear Methods in Spectral Analysis*, Topics in Applied Physics, Vol. 34, ed. S. Heykin, Springer, Berlin, Heidelberg, New York (1979) pp. 155-179

[10] Glad, I.K. and H.E. Krogstad, "The Maximum-Likelihood Property of Estimators of Wave Parameters from Heave, Pitch and Roll Buoys", *Journ. Atmos. Ocean. Techn.*, Vol. 9, no. 2 (1992) pp. 169-173

[11] Haug, O., *Directional Wave Spectra from Arrays of Wave Recorders*, Diploma Thesis, Division of Mathematical Sciences, The Norwegian Institute of Technology (1993)

[12] Isobe, M., K. Kondo and K. Horikawa, "Extension of the MLM for estimating directional wave spectra", in *Proc. Symp. on Description and Modelling of Directional Seas*, Paper no. A6, Lyngby, Denmark (1984)

[13] Konube K. and N. Hashimoto, "Estimation of directional spectra from maximum entropy principle", *Proc. 5th OMAE Symposium*, Tokyo (1986) pp. 80-85

[14] Krogstad, H.E., "Maximum likelihood estimation of ocean wave spectra from general arrays of wave gauges", *Mod. Ident. Control*, Vol. 9 (1988) pp. 81-97

[15] Krogstad, H.E., R.L. Gordon and M.C. Miller, "High-resolution directional wave spectra from horizontally mounted acoustic doppler current meters" *J. Atmos. Ocean. Techn.*, Vol. 5, no. 4 (1988) pp. 340-352

[16] Long, R.B., "The statistical evaluation of directional estimates derived from pitch/roll buoy data", *J. Phys. Ocean.*, Vol. 9 (1980) pp. 373-381

[17] Lygre, A. and H.E. Krogstad, "Maximum entropy estimation of the directional distribution in ocean wave spectra", *J. Phys. Ocean.*, Vol. 16 (1986) pp. 2052-2060

[18] McDonough, R.N., "Application of the maximum likelihood method and the maximum entropy method to array processing", in *Nonlinear Methods in Spectral Analysis*, Topics in Applied Physics Vol. 34, ed. S. Haykin, Springer, Berlin, Heidelberg, New York, (1979) pp. 181-244

[19] Nwogu, O., "Maximum entropy estimation of directional wave spectra from an array of wave probes", *Appl. Ocean Res.*, Vol. 11, no. 4 (1986) pp. 176-182

[20] Pawka, S.S., "Island shadows in wave directional spectra", *J. Geophys. Res.*, Vol. 88 (1983) pp. 2579-2591

[21] Press, W.H. *et al.*, *Numerical recipes: The Art of Scientific Computing*, Cambridge University Press (1986)

Implementation and Validation of a Global Third-Generation Wave Model at Fleet Numerical Oceanography Center

Paul A. Wittmann and R. Michael Clancy[1]

Abstract

The third-generation Wave Model (WAM) is currently being implemented at the U.S. Navy's Fleet Numerical Oceanography Center (FLENUMOCEANCEN) to replace the operational Global Spectral Ocean Wave Model (GSOWM). For the purpose of comparison, the WAM was run in parallel with the GSOWM for two test periods, February-March 1992 and August-September 1992. For these tests, the WAM was implemented on a one degree global grid and forced by wind stress from the Navy Operational Global Atmospheric Prediction System (NOGAPS) model. The model was run in near real time, producing wave forecasts to 48 hours twice per day, with output saved every 6 hours. The significant wave heights from WAM and GSOWM were compared with data from 60 moored buoys located in various regions of the North Atlantic and North Pacific. Based on these comparisons, the WAM consistently produced more accurate wave predictions than the GSOWM.

Introduction

The GSOWM has been providing the Navy with routine operational wave forecasts since 1985 (Clancy *et al.* 1986, Clancy and Sadler 1992). The GSOWM is a "first-generation" model based on the work of W.J. Pierson and others, and employs both theoretical and empirical formulations of wave growth and dissipation (Pierson 1982). Since the development of the GSOWM, a more complete understanding of the spectral energy balance has emerged, which includes the theory of non-linear wave interactions (Komen *et al.* 1984). First generation

----

[1]Ocean Models Division, Fleet Numerical Oceanography Center, 7 Grace Hopper Avenue, Monterey, CA 93943-5501.

models, such as the GSOWM, neglect the nonlinear source function, while tuning the input source function to achieve realistic results. Second generation models use a simplified parameterization of the nonlinear transfer source term. The Sea Wave Modeling Project (SWAMP) study revealed limitations of both first and second generation models (SWAMP Group 1985), and led to the development of the WAM, which uses a parameterization of the exact nonlinear transfer source function (WAMDI Group 1988).

The WAM has been run at over 35 institutions around the world, 10 of which have implemented the model for operational use. It has been verified for many different regions by Zambresky 1989 and Gunther *et al*. 1992. The WAM was used to predict wave characteristics during the Surface Wave Dynamics Experiment (SWADE), using surface wind stress from NOGAPS (Jensen *et al*. 1991). Predictions from WAM and GSOWM, along with those from seven other wave models, were compared with directional wave energy spectra observed during the Labrador Extreme Waves Experiment (LEWEX; Beal 1991). In LEWEX, all wave models were forced by a common set of surface winds in an attempt to reveal model differences. Although both WAM and GSOWM predictions agreed well with observed wave heights, WAM was more accurate in reproducing certain aspects of the measured wave energy spectra (Wittmann and Clancy 1991a; Zambresky 1991).

A regional implementation of the WAM for the Mediterranean Sea became operational on the Cyber 205 computer at FLENUMOCEANCEN in 1990 (Clancy and Wittmann 1990, Wittmann and Clancy 1991b). However, due to the near saturation of the Cyber 205 with operational workload, a global implementation of the WAM was not viable on this machine. In October 1992, a Cray Y-MP C90 supercomputer was installed at FLENUMOCEANCEN, providing the opportunity to run a global version of the WAM model operationally.

Implementation

The WAM was run on a global 1.0° spherical grid, using a 20 minute propagation and source term time step. Surface wind stress fields were generated by the NOGAPS planetary boundary layer (NPBL) contained within NOGAPS 3.2 (Hogan et al. 1991). The vertical flux parameterization of NPBL is based on the work of Louis *et al*. (1982). The wind stress fields were input to WAM every three hours during the forecasts. The model was run twice a day, updated by the 12 hour forecast from the previous run. Forecasts were produced to 48 hours and output fields were saved every 6 hours. The model was run on the Cray Y-MP 8 computer at the Stennis Space Center in Mississippi. This configuration of the WAM used 32 CPU minutes per forecast day, running at 175 MFLOPS on one processor of the Cray Y-MP 8.

The GSOWM is run operationally every 6 hours, alternating between forecast mode and analysis mode (Clancy *et al*. 1986). The forecast runs produce wave predictions out to 72 hours in forecast time. The analysis runs update the wave history using the latest 6 hour wind analysis. The GSOWM is forced by surface winds produced by the Global Surface Contact Layer Interface (GSCLI) boundary layer model, which is based on the work of Cardone (1969). The GSCLI winds are derived from variables produced by NOGAPS. The GSOWM runs on a global 2.5° spherical grid with a three hour time step on the FLENUMOCEANCEN Cyber 205 computer.

Figure 1 shows the NOGAPS surface pressure field for 10 March 1992 for the Gulf of Alaska region. Figures 2 and 3 are the GSOWM and WAM wave height and direction fields for the same time, and illustrate the different resolutions of the two wave models. As expected, more spatial structure can be seen in the wave fields produced by WAM, due to the model's finer grid. The area encompassed by the 3 m wave height contour is much larger in GSOWM than WAM. Also, islands, such as the Aleutian and Hawaiian chains, are represented on the WAM grid but not on the GSOWM grid.

Verification

The National Data Buoy Center (NDBC) and others maintain approximately 60 moored buoys in the North Atlantic and North Pacific which routinely report wave height and wind speed. These observations are compared to the GSCLI/GSOWM and NPBL/WAM wind/wave models at forecast times of 0, 24, and 48 hours. The accuracy of this buoy data is approximately 0.5 m s^{-1} or 5% for wind speed and 0.5 m for wave height (Hamilton 1980). Wind speed observations and model predictions are adjusted to a reference height of 10 m assuming a logarithmic wind profile.

The buoys are grouped into seven different regions. The mean error (i.e., bias), the root mean square (RMS) error, and the scatter index, which is the standard deviation of the error divided by the mean of the corresponding observed values (Janssen *et al*. 1984), are computed for the wave heights and wind speeds in each region.

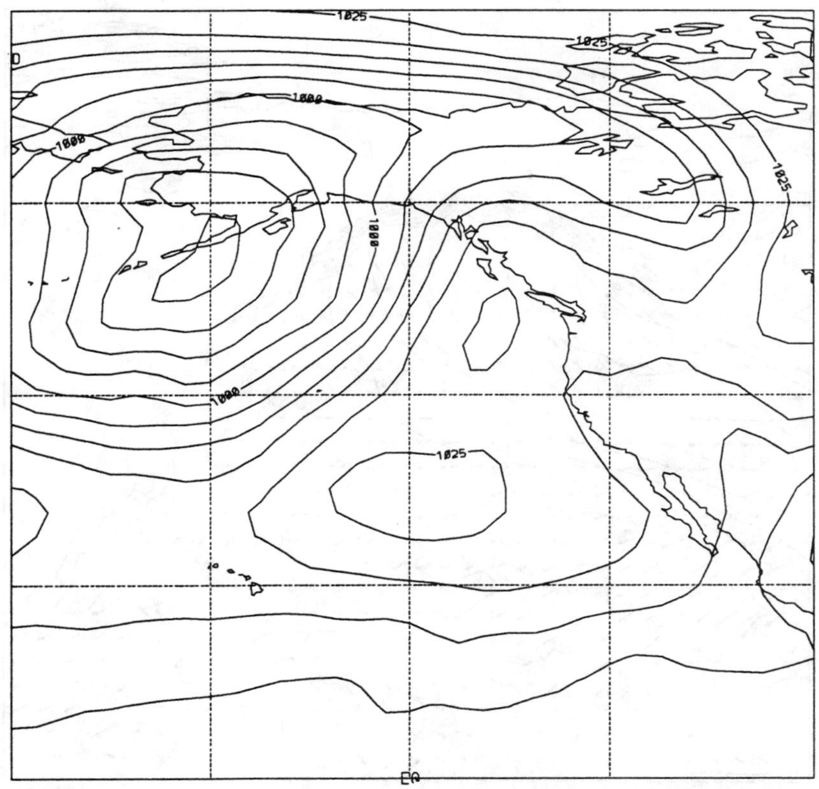

Figure 1. NOGAPS surface pressure for the northeast Pacific at 0000 GMT 10 March 1992. The contour interval is 25 mb.

Table 1 summarizes these statistics for the winter test period, for the analysis (forecast time zero hours). The GSOWM has a positive bias in all areas except Hawaii. The largest GSOWM biases are in swell-dominated regions, such as the US West Coast and the Gulf of Alaska. This result is consistent with the tendency of the GSOWM to develop too much energy at low frequencies, resulting in over prediction of wave heights. The overall bias for the WAM model is near zero (0.09 m) compared to 0.44 m for the GSOWM. RMS errors for WAM are significantly less than GSOWM in all regions, except the NE Atlantic and Gulf of Mexico where they are nearly the same. The largest differences are again in the two swell dominated regions, US West Coast and Gulf of Alaska, where the

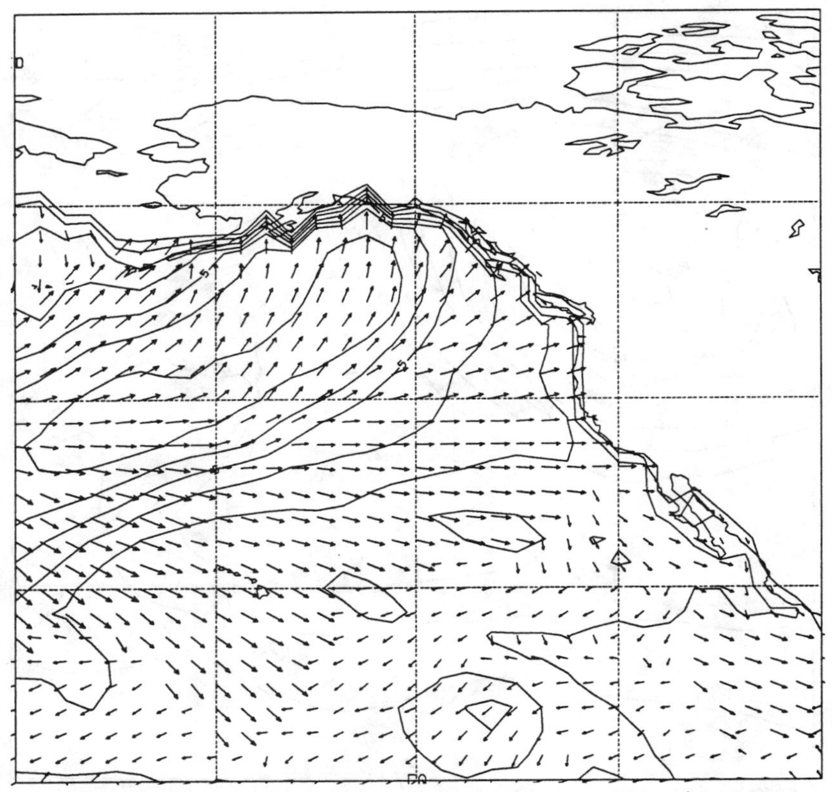

Figure 2. GSOWM significant wave height and primary wave direction for the northeast Pacific at 0000 GMT 10 March 1992. The contour interval is 1 m.

large bias of the GSOWM contributes to the RMS errors. However, the scatter index, which is independent of the bias and often used as a measure of model skill, is also smaller for the WAM model in most areas. In general, the statistics indicate that WAM performs better than GSOWM.

The wind speed statistics for the winter period are shown in Table 2. The biases of the GSCLI and NPBL vary widely with region. Overall, however, the GSCLI has a negative bias and NPBL has a positive bias. The RMS errors for NPBL are generally somewhat less than those of GSCLI, with the exception of the Gulf of Alaska and the NE Atlantic regions. The scatter index shows a slight improvement of the NPBL winds over the GSCLI winds for

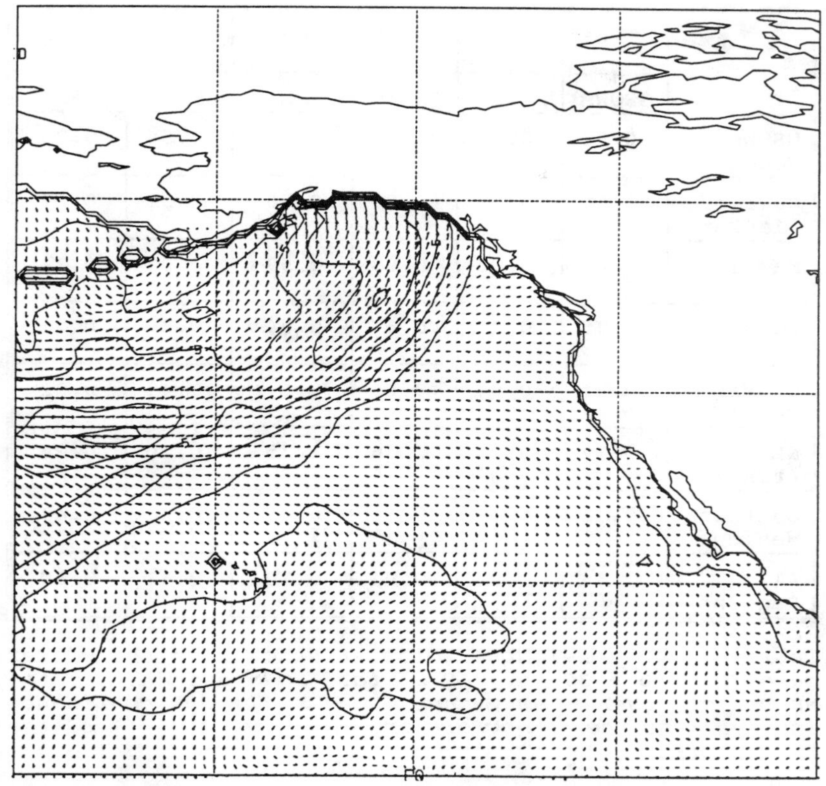

Figure 3. WAM significant wave height and primary wave direction for the northeast Pacific at 0000 GMT 10 March 1992. The contour interval is 1 m.

most areas.

The wave height statistics for the summer period are shown in Table 3. In general, the mean and RMS errors are smaller than those of the winter period due to the less energetic wind and wave activity in the northern hemisphere. The exception is the West Pacific, which was impacted by six typhoons passing through the region during the summer test period. The same trends in the statistics of both models relative to each other, as seen in the winter period, are evident in the summer period. The WAM has significantly smaller errors than GSOWM in most regions.

	Mean Error (m)		RMS Error (m)		Scatter Index	
	GSOWM	WAM	GSOWM	WAM	GSOWM	WAM
US West Coast	0.92	0.26	1.15	0.62	0.56	0.29
Gulf of Alaska	0.56	0.17	1.03	0.76	0.36	0.26
Hawaii	-0.09	-0.03	0.63	0.59	0.23	0.21
West Pacific	0.25	0.17	0.67	0.54	0.39	0.31
US East Coast	0.31	-0.04	0.86	0.67	0.41	0.32
NE Atlantic	0.17	-0.03	1.36	1.35	0.26	0.26
Gulf of Mexico	0.12	0.13	0.46	0.49	0.41	0.43
All Regions	0.44	0.09	0.94	0.70	0.41	0.30

Table 1. Wave height error statistics from the winter case (20 Feb - 20 Mar, 1992) for GSOWM and WAM. Total number of observations equals 3350. The 95% confidence interval for the RMS errors are approximately ±5% of the RMS errors.

Wind speed statistics for the summer case are given in Table 4. The RMS wind speed errors are smaller in the summer than in the winter, with the exception of the West Pacific. Like the winter case, the NPBL winds are slightly better than the GSCLI winds. The GSCLI wind speeds have a negative bias in all but two regions (Table 4) while the GSOWM wave heights have a positive bias in all regions (Table 3).

Note that, in contrast to the results presented here, Clancy and Sadler (1992) reported a negative GSOWM wave height bias for the period January 1987 through August 1990. This earlier study was based on comparison of GSOWM 24-hour forecast wave heights with buoy observations, and most likely reflected the presence of a negative bias in the GSCLI wind fields used to drive GSOWM. As no changes have been made in GSOWM since its original operational implementation in 1985, we conclude that the shift from a negative to a positive bias in the

	Mean Error (m s^{-1})		RMS Error (m s^{-1})		Scatter Index	
	GSCLI	NPBL	GSCLI	NPBL	GSCLI	NPBL
US West Coast	-1.01	0.24	3.03	2.92	0.55	0.52
Gulf of Alaska	0.29	0.84	2.74	3.06	0.37	0.41
Hawaii	-1.29	-0.02	1.66	1.53	0.23	0.21
West Pacific	-0.40	0.24	2.56	2.45	0.37	0.34
US East Coast	-0.10	0.94	2.95	2.76	0.37	0.35
NE Atlantic	2.45	2.19	3.51	3.66	0.34	0.36
Gulf of Mexico	-1.26	0.57	2.60	2.40	0.39	0.36
All Regions	-0.36	0.63	2.83	2.76	0.39	0.38

Table 2. Wind speed error statistics from the winter case (20 Feb - 20 Mar, 1992) for winds used to drive GSOWM and WAM. Total number of observations equals 3114. The 95% confidence interval for the RMS errors are approximately ±5% of the RMS errors.

wave fields reflects changes in the NOGAPS, and hence GSCLI, wind fields.

Figure 4 gives the RMS wave height errors and wind speed errors as a function of forecast time, calculated over all regions, for both winter and summer. These overall results, based on over 18000 buoy reports, clearly show that the wave heights predicted by WAM are more accurate than those predicted by GSOWM. Note that the WAM 48-hour forecast wave heights are more accurate than the GSOWM 0-hour forecast (i.e., nowcast) wave heights.

Figure 4 also shows that, in a relative sense, the wave height errors do not grow with forecast time as fast as the wind speed errors. This may seem surprising at first since the height of the local wind sea generally varies with the square of the local wind speed, and should thus be sensitive to errors in the local winds.

	Mean Error (m)		RMS Error (m)		Scatter Index	
	GSOWM	WAM	GSOWM	WAM	GSOWM	WAM
US East Coast	0.45	0.12	0.63	0.46	0.43	0.31
Gulf of Alaska	0.46	0.09	0.73	0.51	0.39	0.28
Hawaii	0.09	-0.14	0.54	0.47	0.26	0.23
West Pacific	0.68	0.06	1.27	0.71	0.71	0.40
US East Coast	0.27	-0.06	0.58	0.41	0.49	0.34
NE Atlantic	0.64	-0.29	1.08	0.93	0.37	0.32
Gulf of Mexico	0.22	0.08	0.45	0.41	0.48	0.51
All Regions	0.36	0.02	0.70	0.51	0.45	0.33

Table 3. Wave height error statistics from the summer case (1 Aug - 16 Sep, 1992) for GSOWM and WAM. Total number of observations equals 4516. The 95% confidence interval for the RMS errors are approximately ±5% of the RMS errors.

In fact, it reflects the nonlocal process of swell propagation in the model. Following its generation by the analyzed winds used to carry the model's history fields forward in time (and which are generally the most accurate winds available), swell does not interact strongly with the wind. Since the local wave height forecast in the model is a effectively a combination of sea and swell forecasts, while only the sea height forecast is sensitive to the local winds, the local wave height errors do not grow as rapidly with forecast time as the local wind speed errors. Thus, the useful forecast time intervals for wave height predictions may generally extend beyond those of the weather prediction models which drive them. This has obvious implications for support of long-range Optimum Track Ship Routing (OTSR).

	Mean Error (m s^{-1})		RMS Error (m s^{-1})		Scatter Index	
	GSCLI	NPBL	GSCLI	NPBL	GSCLI	NPBL
US West Coast	0.06	-0.58	2.86	2.48	0.51	0.46
Gulf of Alaska	-0.14	0.06	2.07	2.10	0.29	0.30
Hawaii	-1.14	-0.54	2.28	2.01	0.30	0.26
West Pacific	-0.24	-0.39	2.94	2.53	0.45	0.40
US East Coast	-0.69	0.12	2.70	2.03	0.47	0.36
NE Atlantic	1.53	1.04	2.50	2.42	0.31	0.32
Gulf of Mexico	-1.06	0.59	2.27	2.26	0.47	0.46
All Regions	-0.33	0.04	2.54	2.21	0.41	0.36

Table 4. Wind speed error statistics from the summer case (1 Aug - 16 Sep, 1992) for GSOWM and WAM. Total number of observations equals 4333. The 95% confidence interval for the RMS errors are approximately ±5% of the RMS errors.

Conclusion

Based on comparison with data from moored buoys, the wave heights predicted by WAM consistently have smaller mean errors, RMS errors and scatter index values than those predicted by GSOWM. Unlike the GSOWM, which exhibits a positive bias as a systematic model tendency, WAM shows no systematic bias in wave height. The error statistics arrived at here are consistent with those derived by Gunther *et al*. (1991) in a recent verification study of the WAM at ECMWF.

The WAM is currently being run in a pre-operational test mode on the FLENUMOCEANCEN Cray Y-MP C90. The code has been modified to run concurrently on six processors using a solid state disk for storage of large work arrays. In this configuration, the global implementation of the model on a 1° spherical grid with 15° angular resolution for the directional spectra uses 14.6 minutes

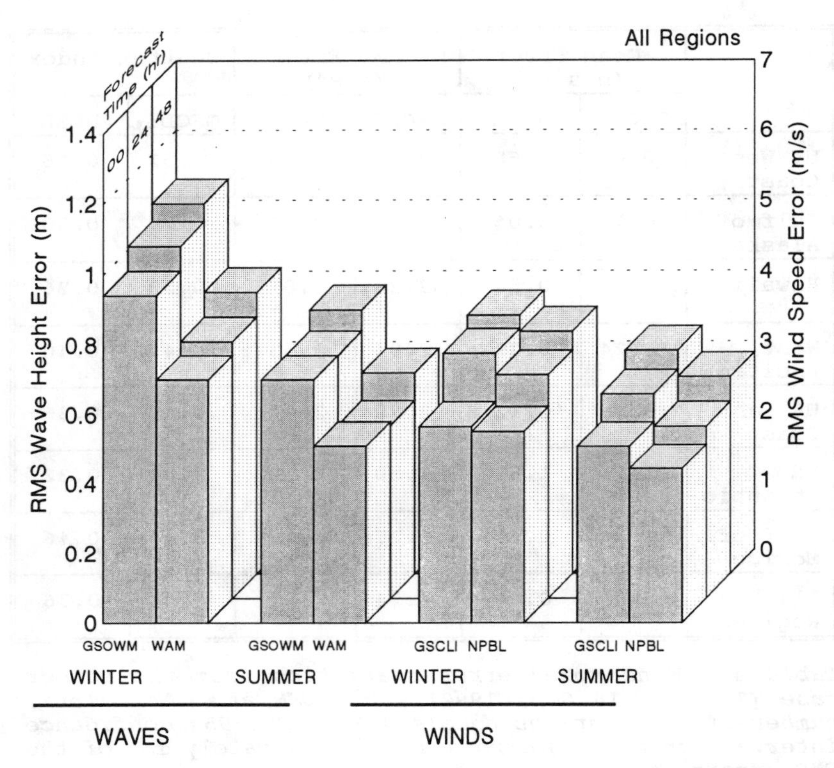

Figure 4. RMS wave-height and wind-speed errors for GSOWM/WAM and GSCLI/NPBL for the winter (20 Feb - 20 Mar, 1992) and summer (1 Aug - 16 Sep, 1992) test periods. Results are based on comparison against over 18000 reports from approximately 60 moored buoys.

of CPU time and 6.4 minutes of wallclock time for a 24 hour forecast. The WAM is also being interfaced with the FLENUMOCEANCEN Integrated Stored Information System (ISIS) relational data base management system to provide random access to forecast directional wave energy spectra and related parameters for support of OTSR and other applications.

Future plans include a significant increase in the spatial resolution of NOGAPS (and hence the winds used to drive WAM), high-resolution nested regional implementations of WAM forced by a high-resolution regional atmospheric model in support of coastal wave and surf prediction, assimilation of wave data into WAM, and coupling of WAM with ocean circulation models to handle wave-current interactions. Ultimately, we expect WAM to be integrated into NOGAPS to account for two-way interactions between winds and waves and to support coupled air-sea data assimilation (see Clancy and Plante 1993).

Appendix. - References

1. Beal R.C., "LEWEX: Motivation, Objectives and Results," In, *Directional Ocean Wave Spectra*, R.C. Beal (ed.), Johns Hopkins Univ. Press, Baltimore, MD, 1991, 218 pp.

2. Cardone, V.J., "Specification of the Wind Distribution in the Marine Boundary Layer for Wave Forecasting," Technical Report 69-1, New York University, School of Engineering and Science, New York, N.Y., 1969, 131 pp.

3. Clancy, R.M., Kaitala, J.E., and Zambresky, L.F., "The Fleet Numerical Oceanography Center Global Spectral Ocean Wave Model," *Bulletin of the American Meteorological Society, 67,* 1986, pp. 498-512.

4. Clancy, R.M., and P.W. Wittmann, "FNOC Implements Advanced Wave Model," *Naval Oceanography Command News, 10(11),* Commander Naval Oceanography Command, Stennis Space Center, MS 39529-5000, 1990, pp. 1-4.

5. Clancy, R.M., and W.D. Sadler, "The Fleet Numerical Oceanography Center Suite of Oceanographic Models and Products," *Weather and Forecasting, 7,* 1992, pp. 307-327.

6. Clancy, R.M., and R.J. Plante, "Evolution of Coupled Air-Sea Models at Fleet Numerical Oceanography Center," Conference Proceedings, MTS '93 (available from Marine Technology Society, 1828 L Street, N.W., Suite 906 Washington, DC 20036), 1993.

7. Gunther H., P. Lionello, P.A.E.M. Janssen, L. Bertotti, C. Bruning, J.C. Carretero, L. Cavaleri, A. Guillaume, B. Hansen, S. Hasselmann, K. Hasselmann, M. de las Heras, A. Hollingsworth, M. Holt, J.M. Lefevre, R.Portz, "Implementation of a Third Generation Ocean Wave Model at the European Centre for Medium-Range Weather Forecasts. ECMWF

Tech. Report No. 68, 1992.

8. Hamilton, G.D., "NOAA Data Buoy Office Programs," *Bulletin of the American Meteorological Society, 61,* 1980, pp. 1012-1017.

9. Hogan, T.F., and T.E. Rosmond, "The Description of the Navy's Operational Global Atmospheric Prediction System's Spectral Forecast Model," *Monthly Weather Review, 119,* 1991, pp. 1786-1815.

10. Janssen, P.A., G.J. Komen, and W.J. De Voogt, "An Operational Coupled Hybrid Wave Prediction Model," *Journal of Geophysical Research, 89,* 1984, pp. 3635-3654.

11. Jensen R.E., C.L Vincent, and M.J. Caruso, "A Wind-Wave Forecast for the Surface Wave Dynamics Experiment," Conference Proceedings, MTS '91 (available from Marine Technology Society, 1828 L Street, N.W., Suite 906 Washington, DC 20036), 1991.

12. Komen, G.J., S. Hasselmann and K. Hasselmann, "On the Existence of a Fully Developed Windsea Spectrum," *Journal of Physical Oceanography, 14,* 1984, pp. 1271-1285.

13. Louis, J.F., M. Tiedtke and J.F. Geleyn, "A Short History of the Operational PBL Parameterization at ECMWF," Workshop on Planetary Boundary Parameterization, ECMWF, Reading, England, 1982, pp. 59-79.

14. Pierson, W.J., "The Spectral Ocean Wave Model (SOWM), a Northern Hemisphere Computer Model for Specifying and Forecasting Ocean Wave Spectra," Technical Report DTNSRDC-82/011, David W. Taylor Naval Ship Research and Development Center, Bethesda, MD 20084, 1982, 186 pp.

15. The SWAMP Group, "Sea Wave Modelling Project (SWAMP). An Intercomparison Study of Wind Wave Prediction Models, Part 1: Principal Results and Conclusions," *Ocean Wave Modelling*, Plenum Press, 1985, 256 pp.

16. The WAMDI Group, "The WAM Model - A Third Generation Ocean Wave Prediction Model," *Journal of Physical Oceanography, 18,* 1988, pp. 1775-1810.

17. Wittmann, P.A. and R.M Clancy, "Predictions from

the GSOWM during LEWEX," In, *Directional Ocean Wave Spectra*, R.C. Beal (ed.), Johns Hopkins Univ. Press, Baltimore, MD, 1991a, 218 pp.

18. Wittmann, P.A. and R.M Clancy, "Implementation and Validation of a Third-Generation Wave Model for the Mediterranean Sea," Conference Proceedings, MTS-91 (available from Marine Technology Society, 1828 L Street, N.W., Suite 906 Washington, DC 20036), 1991b.

19. Zambresky L.F., "A Verification Study of the Global WAM Model, December 1987 - November 1988," ECMWF Tech. Report. No. 63, 1989.

20. Zambresky L.F. 1991: An Evaluation of Two WAM Hindcasts for LEWEX," In, *Directional Ocean Wave Spectra*, R.C. Beal (ed.), Johns Hopkins Univ. Press, Baltimore, MD, 1991, 218 pp.

Predicting Wave Spectra With A Third Generation Spectral Wave Model

Charles L. Vincent, PhD, M. ASCE[1]
Robert E. Jensen, PhD
Hans C. Graber, PhD

Abstract

The ability of the third generation spectral wave model, 3GWAM, to simulate frequency and directional spectra during a northeast storm of October 1990 along the U.S. mid-Atlantic coast is investigated. 3GWAM had been observed to predict wave heights very well during the storm which provided an opportunity to evaluate the quality of the spectral predictions. The storm conditions are represented by high speeds (25 m/s wind speeds) with a distinct wind direction shift. In general, the model performed well with under-predictions in low frequency wave components most likely due to small under-estimates in wave height. The model simulated the shape of the observed directional spectra within the assumed uncertainty of the observations.

---

[1]Vincent, Charles L., PhD, Senior Research Scientist, Jensen, Robert A., PhD, Research Hydraulic Engineer, U.S. Army Engineer Waterways Experiment Station, Coastal Engineering Research Center, 3909 Halls Ferry Road, Vicksburg, MS 39180-6199; Graber, Hans C., PhD, Associate Professor, Division of Applied Marine Physics, Rosenstiel School of Marine and Atmospheric Science, 4600 Rickenbacker Causeway, Miami, FL 53149-1098

Introduction

Numerical models for predicting ocean wave conditions are often evaluated in terms of the error characteristics of significant wave height prediction, and occasionally significant mean or peak period (or equivalent frequency). Only rarely are the model directional spectral predictions evaluated. During the Surface Wave Dynamics Experiment (SWADE) (Weller et al., 1991), data were obtained in a severe northeast storm off the North Carolina - Virginia coast, 25-28 October 1990. Initial simulations with the 3GWAM model (WAMDI, 1988) using winds from the U.S. Navy Fleet Numerical Oceanographic Center (FNOC) and European Center for Medium Range Weather Forecasts (ECMWF) indicated that the model performed well. Further refinement of the forecast winds by Cardone et al., in preparation, improved the model outputs in terms of heights and periods.

Our interest in the model's ability to predict spectra is motivated by two factors. First, the spectra in themselves are of value for a variety of engineering problems including offshore structure design. Second, and perhaps more importantly, the wave model actually predicts the spectral components individually based on theoretical forms of spectral source functions; hence, the quality of the prediction is inherently informative about the validity of the model physics.

Evaluation of the model's ability to predict spectral characteristics is difficult since the area under the spectrum is related to the square of wave height. Errors of 15 percent in wave height begin to distort the spectra so much that little information can be gleaned from comparisons. The strategy employed in this paper is to compare the spectra for a case where there is small error in wave heights. Any consistent differences would then shed light on uncertainties in the formulation of the spectral source terms. It is emphasized that there is no reason to expect that model and predicted spectral shapes will agree simply because the wave height is correctly estimated.

Model Characteristics

The model used was 3GWAM (WAMDI, 1988). The model was run on a one-fourth deg grid over the region shown in Figure 1, which was selected to be large enough to contain the storm. The model was run on a time step of 240 seconds. Wind input from AES Canada was provided on a 0.5 deg grid every hour for the period 20-31 October 1990 (Cardone, et al., in preparation).

In evaluating the model performance, it must be noted that this storm was originally forecast using fields from FNOC and ECMWF, both of which produced reasonable results. The wind field was further refined by Dr. Cardone, and hence these adjustments are not a "blind" experiment. In no instance were any of the parameters internal to 3GWAM tuned to improve its performance.

Wave Measurements

For SWADE, the existing array of National Data Buoy Center (NDBC) buoys along the Atlantic Coast was supplemented with additional buoys in the Mid-Atlantic Bight region. The distribution of wave measuring stations available to provide data for evaluation of 3GWAM is also shown in Figure 1. In this paper, overall model performance was evaluated at buoys 41001, 41006, 44001, 44008, 44011, 44014, and 44015. Evaluation of the directional spectral characteristics were based on the buoys 44001, 44014, and 44015 which were added as part of the SWADE experiment.

For all but the directional buoys listed above, wave height, period and frequency spectrum calculations used the standard NDBC analyses. The approach of Drennan, Graber, and Donelan (in preparation) was used to provide estimates of the directional spectrum. For a first pass, a Maximum Entropy Method (MEM) is used. If that fails, a Maximum Likelihood estimator is applied. For all of the spectra presented in this paper, the MEM approach worked. To make the paper length manageable, discussion of the directional spectrum results is limited to buoy 44014 because the results are consistent with the results at the other two directional buoys.

Wave Height and Period Comparisons

Figure 2 provides a history of significant wave height (H_s), peak period (T_p), and mean period (T_m) at 44014 for the storm along with hindcast values. Overall, the hindcast curve predicts the measurements with only a few cases being more than 10 percent different. Statistics (Table 1) indicate a slight ($^-$0.27 m) bias in that hindcast wave heights are lower than the observed and a correlation of 0.95. The largest errors in T_p occur when there is a dual peak in the frequency spectrum, in which case, insignificant differences in spectral densities lead to large jumps in the position of the selected peak frequency and its inverse period. The T_m is a more stable parameter, and the intercomparisons are very good with the bias in T_m showing that overall the hindcast T_m is slightly lower ($^-$0.37 s) with a correlation of 0.88. The wind statistics (Table 1) indicate that overall the hindcast winds were biased, slightly high ($^+$0.44 m/s). In summary, the hindcast waves are slightly lower and shorter than the observed although the modeled winds are slightly higher than the observed.

Frequency Spectra

Comparison of frequency spectra consists of overplotting a pair of spectra. There is no accepted process for comparing pairs of spectra and then aggregating the results over several hundred spectra to arrive at a meaningful set of statistics. Here, we will simply subtract the observed spectrum from the hindcast, and then perform a simulated three-dimensional plot of the difference versus frequency versus time for this storm. The results for buoy 44011 shown in Figure 3 are similar to those found for the buoys not shown. The overall impression left from the plot is that the observed density is larger than that hindcast. Furthermore, the error is larger for lower frequencies. This pattern expresses in frequency space the equivalent statistics in height and period. In order for the modeled wave heights to be biased low, the overall spectral densities must also be low. In order for the modeled periods to be biased somewhat low, the error in spectral space would be distributed more towards the lower frequencies.

Field studies (JONSWAP, 1973; Toba,1972) show an intrinsic relationship between height and period for an actively growing wind sea. The error in period can result from at least two possible causes: the source terms in 3GWAM are somewhat off balance (for a given wave height they produce too low a period) or the error simply results from an under-prediction of wave height (i.e., the model relationship between height and period is correct, but since the height was too low the period is low also). The dimensionless parameters of height and period were computed and compared to the Toba and JONSWAP curve. For the active wind sea, the observed and model data scatter between the trends of the two references and follow the same tendencies. It is concluded that the model and observed height-period relation is the same, and the error seen both in the spectra and period most likely results from the underestimate of the wave height.

Directional Spectral Comparisons

Comparisons of the predicted and observed directional spectra at buoy 44014 for three times in the storm are shown in Figures 4-6. Each figure consists of four panes. The top pair are plots of the directional spectrum and the frequency spectrum for the buoy; the bottom pair are the corresponding spectra predicted by 3GWAM. The directional spectral plots consist of contours of spectral density and a vector showing the direction from which the wind is blowing. The contours are chosen automatically to define shape. The frequency spectral densities, which are the integration through angle space of the directional spectrum, are plotted in physical units. The convention for plotting the directional spectra places the density with the frequency radially scaled from 0 Hz at the center to 0.4 Hz on the box edge. The compass radials represent the direction to which the waves are traveling. The legend beneath each directional spectrum provides in order: wind speed in m/s, wind direction, mean direction of the waves, and peak direction. The legend above each frequency spectrum gives Hs (in m) and T_p (in s) rounded to an integer. Thus, for the top pair in Figure 4, the wind is blowing from 19 deg at 14 m/s and the waves are 3 m high, at 7 seconds, with a mean direction of 215 deg, that is waves propagating out of NNE towards SSW. In this case, the peak direction is also 215 deg.

Figure 4 is from the early growth phase of the storm. Both the predicted and observed directional spectrum are roughly symmetric about the wind direction. The differences in mean direction between observed and predicted spectra is 9 deg. The predicted spectrum does not appear as broad (120 vs 180 deg) as the observed. The frequency spectra are both single peaked and have similar shape.

Figure 5 is some ten hours later. Wave height has doubled. At this point, the difference in mean direction is 17 deg, and the predicted directional spread is still less than that seen at the buoy. The observed spectrum is more asymmetric to the wind direction than the model spectrum. However, examination of the sequence of wind vectors leading up to this point indicates that in the model case, the wind direction began its shift several hours before the observed; hence, the model sea state is being forced to turn earlier. Thus the difference seen at this stage is related to a difference in wind input, not necessarily model response.

Figure 6 is nine hours later. The primary difference is that the observed spectrum is somewhat broader in both frequency and direction space. In the observed frequency spectrum, a secondary peak appears to be evolving at 0.15 Hz that is not seen in the predicted spectrum.

Judging the quality of these comparisons is inherently subjective because there is no widely accepted method for quantifying the comparisons statistically. Recognizing that the observations are from an accelerometer buoy that has been passed though a MEM analysis, it is felt that the predictions are in reasonable agreement. It is not possible at this point to say definitively that the model under-predicts the spread. The appearance of the secondary peak is seen at the other buoys, so the inability of the model to replicate this should be investigated further. However, it should be noted that the principal features of the frequency and directional spectra are reproduced and that the mean directions are within 10-20 deg. Given all the uncertainties in wind input and in measuring waves, this is state-of-the-art quality.

The objective of the paper was to evaluate the ability of the model to predict the spectra with the intent of uncovering any deficiency in the spectral source term physics. For this case, which involved high wind speeds and a significant wind shift, the predictions were very good. The need to make a major change seems unwarranted.

Summary

The ability of 3GWAM in simulating frequency and directional spectra of waves during the October 1990 Northeast storm off the U.S.mid-Atlantic coast has been evaluated against observations. The errors in the frequency and directional spectra, in general, are within the range that can be assigned confident measurement. Consistent errors in under-prediction of the spectral densities of low frequency components may be explained by the model's small bias to underestimate wave height.

Since the observed wave height data were used in part of the analysis for refining the wind fields, comparisons of modeled and measured heights are not meaningful. The slight tendency for the simulations of height to be biased low when the winds are slightly biased high is noted, and further evaluation with other data sets is warranted.

Acknowledgements

The spectral wave modeling work was conducted at the U.S. Army Engineer Waterways Experiment Station's Coastal Engineering Research Center under the Coastal Research and Development Program, Upgrading of Discrete Spectral Hindcasting Models work unit. The data and model runs presented here could not have been completed without the significant achievements and contributions of the many SWADE investigators, scientists and technicians who collaborated in this experiment and afterward. Permission was granted by the Office of the Chief of Engineers, U.S. Army Corps of Engineers to publish this information. Special thanks to Ms. Claudette Doiron for preparing the manuscript.

Appendix. References

Cardone, V., Graber, H.C., Jensen, R.A., Hasselmann, S., and Caruso, M. (In preparation). "In Search of the True Surface Wind Field: Ocean Wave Modeling Perspective."

Drennan, W.M., Graber, H.C., and Donelan, M.A. (In preparation). "On the Measurement of Directional Wave Spectra from Pitch-Roll and Acceleration Buoys."

Hasselmann, K., et al. (1973). "Measurements of Wind-Wave Growth and Swell Decay During the Joint North Sea Wave Project (JONSWAP)," Deutschic Hydrographisches Institut, Hamburg, 95 pp.

Toba, Y. (1972). "Local Balance in the Air-Sea Boundary Processes, I. On the Growth Process of Wind Waves," *Journal of Oceanographic Society of Japan*, 28, 109-120.

WAMDI Group (1988). "The WAM Model - A Third Generation Ocean Wave Prediction Model," *Journal of Physical Oceanography*," 18, 1775-1810.

Weller, R.A., et al., 1991, Riding the Crest: A Tale of Two Experiments, *Bulletin of American Meteorological Society*, 72, 163-183.

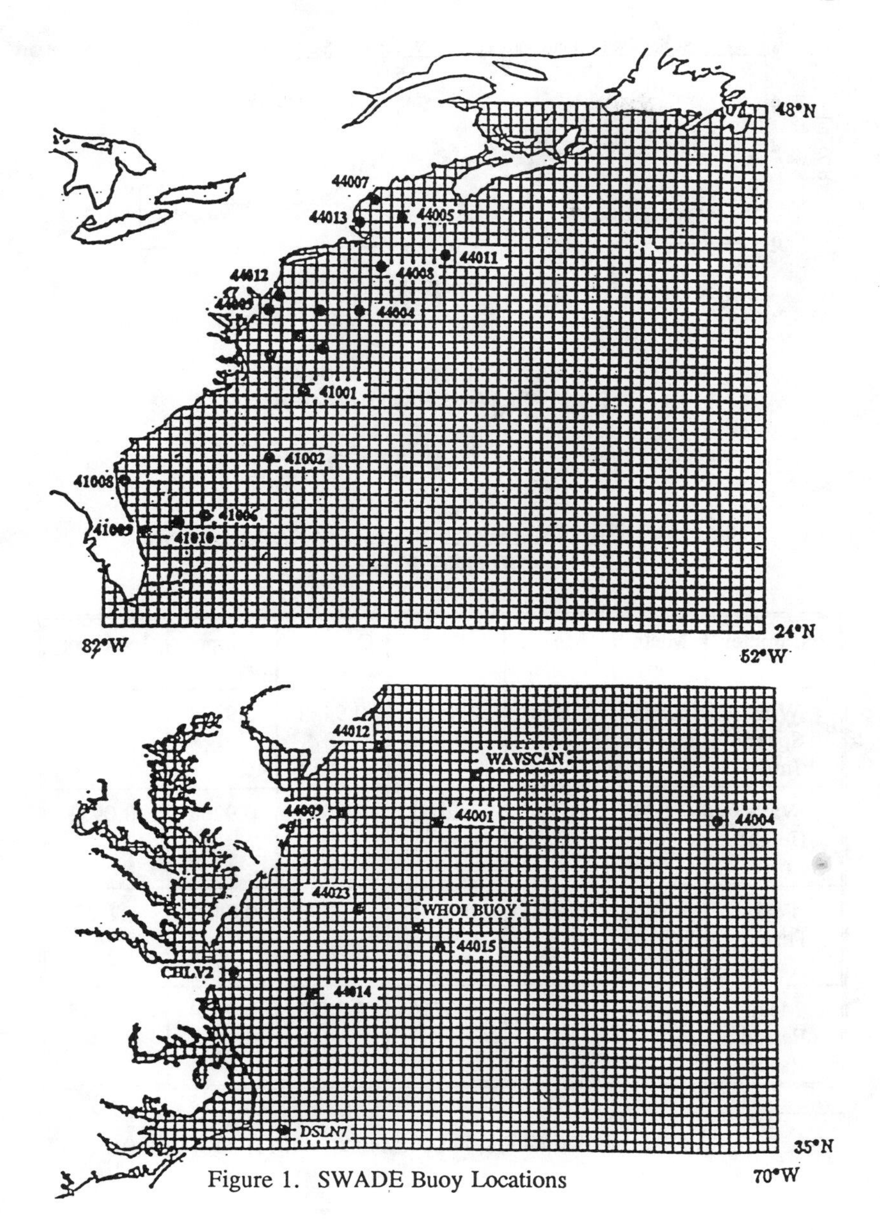

Figure 1. SWADE Buoy Locations

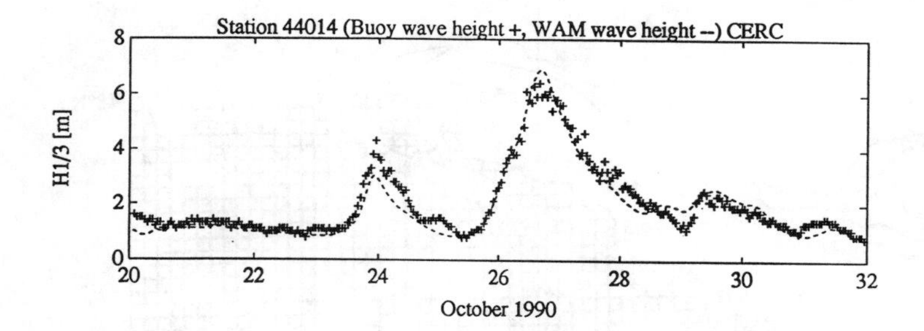

Figure 2. Wave Height at 44014

Table 1
STATISTICAL SUMMARY FOR *ALL*[2] NDBC BUOYS DURING IOP-1

Parameter	Mean Buoy	Mean Model	Bias	Corr Coef	Slope	Intercept
Wind Speed (m/s)	7.874	8.317	0.4425	0.9442	0.9724	0.6602
Wave Height (m)	2.334	2.066	-0.2681	0.9535	0.9268	-0.0973
Peak Period (sec)	8.940	8.421	-0.5196	0.7796	0.7491	1.7237
Mean Period (sec)	7.499	7.128	-0.3716	0.8799	0.9584	-0.0598

[2] Buoys: 41001, 41006 44001, 44004, 44008, 44011, 44014, 44015.

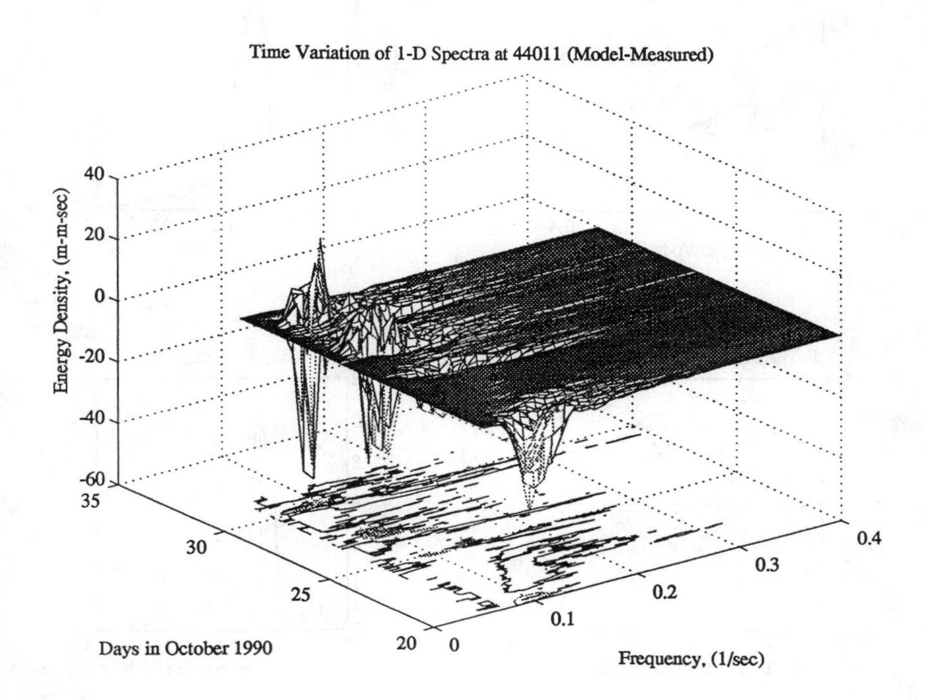

Figure 3. Model versus Observed Spectra

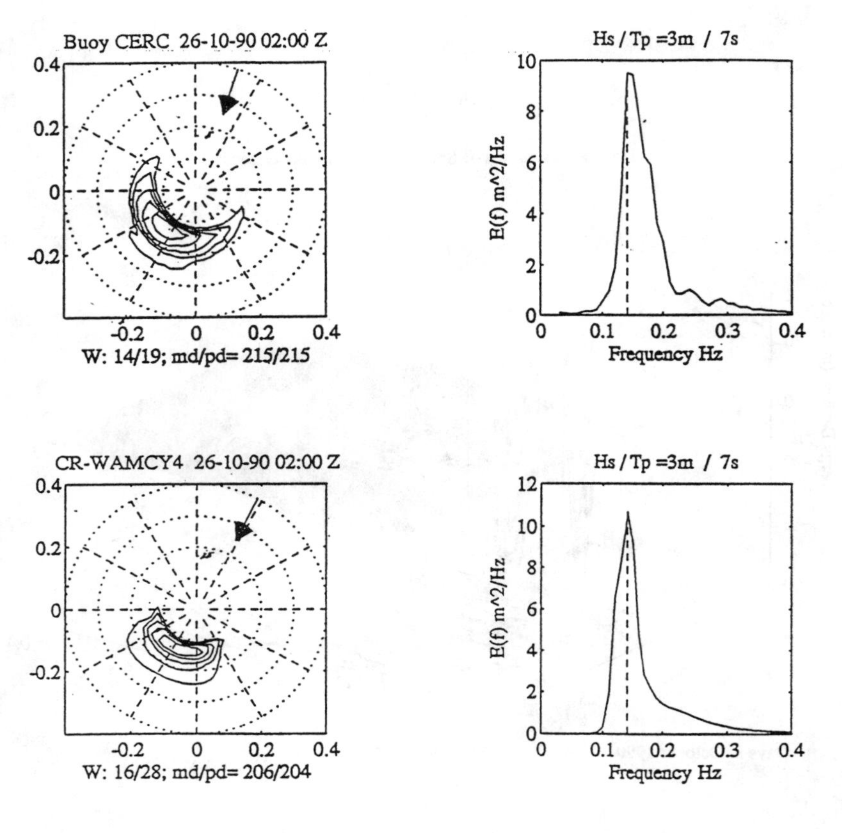

Figure 4. Comparison of Modeled and Measured Spectra

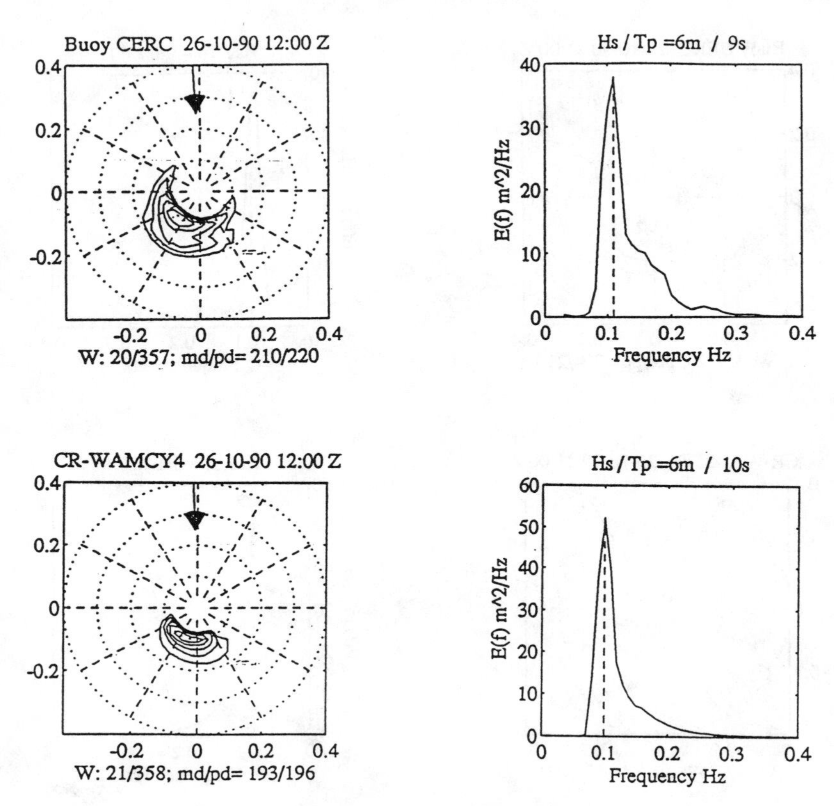

Figure 5. Comparison of Modeled and Measured Spectra

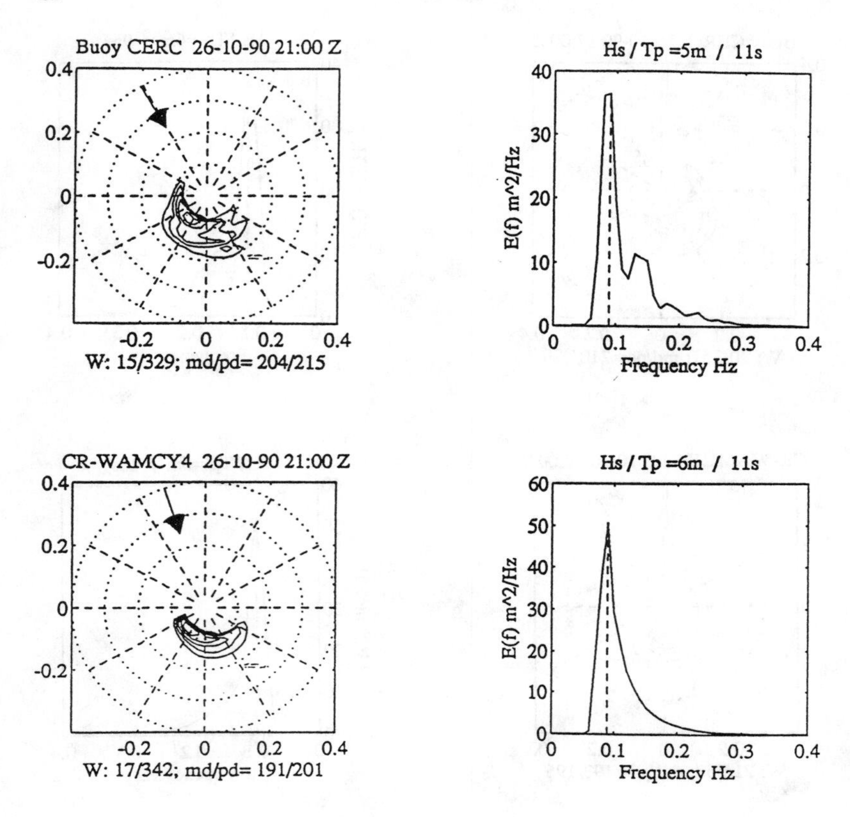

Figure 6. Comparison of Modeled and Measured Spectra

Evaluation of a Third-Generation Wave Model for the U.S. Atlantic Coast

R.E. Jensen[1], S.H. Houston[2], C.L. Vincent[3], M.D. Powell[2]

Abstract

Accurate description of surface wind fields in ocean spectral wave modelling is of critical importance to estimating storm-induced wave conditions along a coastal reach. A new approach for the construction of a surface wind field is investigated by the Hurricane Research Division (HRD) of NOAA using the Spectral Application of Finite-Element Representation (Powell et al. 1991, Houston and Powell 1993). Because this technique requires input of wind measurements, comparison of the method's results to observations used in the method would not be realistic. Thus, the HRD winds are evaluated from simulations of the October 1991 storm using 3GWAM (WAMDIG 1988), and comparing the results to in-situ buoy wave measurements.

Introduction

During the final week of October 1991, a powerful extra-tropical storm system developed in the Atlantic Ocean. This storm began when a low-pressure area formed on a cold front that pushed east of Canada into the Atlantic Ocean. This occurred at the same time that Hurricane Grace was reaching its maximum intensity a few hundred kilometers south of the system (Pasch and Avila 1992). The system

---

[1] Research Hydraulic Engineer, USAE Waterways Experiment Station, Coastal Engineering Research Center, Vicksburg, MS 39180-6199.

[2] Meteorologist, NOAA/AOML/Hurricane Research Division, Miami, FL.

[3] Senior Scientist, USAE Waterways Experiment Station, Coastal Engineering Research Center, Vicksburg, MS 39180-6199.

appeared to deepen rapidly when it encountered moisture provided by Hurricane Grace's circulation. This storm, which will be referred to as the Halloween Storm of 1991 (HS91), proved to be one of the most powerful storm systems along the Atlantic seaboard. Recorded maximum wind gusts were in excess of 30 m/s, and energy based wave heights (H_{mo}) exceeded 15 m, near the generation region. Along the US coastline, massive shoreline destruction took place, and, as far south as the Florida coast, long period swell was evident. The meteorological complexities of this storm were so great, that most of the forecast centers had a difficult time estimating the central pressure of the storm as well as the storm's position.

This storm has been recently studied (Wang and Mettlach 1993), and modeled in a hindcast mode by Vincent et al. (1992), and Cardone et al. (1993). In both works, 3GWAM was used in the wave simulations, and provided very satisfactory results, although the wave model results tended to be overestimated in both. The wind fields used in both studies ranged from Fleet Numerical Oceanography Center (FNOC) forecast global wind stress fields, to kinematically analyzed surface wind fields, respectively. This is the first real test of the HRD winds method for *end-product* use in an extra-tropical storm. Therefore, rather than force an inter-wind field comparison between the HRD method, and the two aforementioned wind fields, the aim of this paper is to determine:

a. The relative accuracy in HRD surface wind fields for an extra-tropical storm.
b. The plausiblity of determining wind field accuracy based on wave model results.
c. The accuracy of 3GWAM estimates of measured wave parameters based on HRD surface wind fields.

Wind Field Specification

The wind fields produced for the HS91 were derived from the Spectral Application of Finite-Element Representation (SAFER) method developed by Ooyama (1987), and described by DeMaria et al (1992). The SAFER method represents the dependent variables by truncated series of cubic beta splines. The application of this technique to surface wind analyses is described in Powell et al. (1991) and Houston and Powell (1993). The surface wind analyses produce a low-pass filtered meso-scale wind field with high-frequency, small-scale features removed. For wave modelling applications, 10-min mean wind measurements were assumed to represent the time scale of the wave-producing wind field and were associated with time scales that approximate the meso-scale spectral gap (Pierson 1983, Champagne-Philippe 1989). An empirical 10-min gust factor relationship was developed to estimate the maximum 10-min wind speed (V_{M10}) that would occur over some longer time period based upon the relationship of consecutive 10-min mean wind speeds to the hourly mean for all National Data Buoy Center (NDBC) Coastal-Marine Automated Network (C-MAN) station measurements collected in tropical cyclones since 1985, and the works of Durst (1960) and Krayer and Marshall (1992). A time scale was computed for all meso-scale analysis wind speed periods given by the ratio of twice the analysis filter wavelength to the low-pass filtered analysis wind speed. For time scales over 10 min, V_{M10} was then computed based on the gust factor relationship. This field can be visualized by considering continuous wind measurements available from a dense network in a stationary storm; if the highest 10-min winds for each station over an hour were contoured, the resulting wind field would be similar to that described here.

The National Weather Service's (NWS) Aviation Model (AVN) surface wind analyses (Petersen and Stackpole 1989) were available on a 1° grid for the HS91 at 0000 and 1200 UTC (Coordinated Universal Time) each day. However, the AVN wind fields were not considered to be reliable because the analyzed center of the extra-tropical storm was often displaced > 200 km from the observed center. In addition, the analysis did not resolve the surface winds for Hurricane Grace and the Unnamed Hurricane of 1991 (Pasch and Avila 1992) shown with HS91 in Figure 1. Hence, the AVN winds were only used as the background field for surface wind analyses. Input wind data for the surface wind analysis were obtained from moored NDBC and Canadian Atmospheric Environment Service (CAES) buoys, C-MAN land based stations, and ship reports. Additional surface wind observations which were provided from Air Force reconnaissance flight-level wind measurements of Hurricane Grace and the Unnamed Hurricane were adjusted to the surface. In some cases, wind speed data from the Special Sensor Microwave/Image (SSM/I) were available and were also included.

All surface wind observations were adjusted to the 10 m level with an appropriate roughness length, z_0, using a logarithmetic wind profile relationship for neutral conditions. For anemometers with over-water flow, z_0 was set at 0.001m, while z_0 was set at 0.03m for flow over land. The aircraft winds for Hurricane Grace and the Unnamed Hurricane were adjusted from flight level (~1.5km) to the surface with a planetary boundary layer model (PBL) described by Powell (1980). This PBL model estimates surface speeds over water as a function of atmospheric stability conditions and surface stress. Storm inflow at the surface was accounted for by backing the flight-level wind directions by 15°, based on comparisons of wind direction from reconnaissance aircraft flight-level winds and surface observations.

Analyses were run every 6 hr in a fixed domain, 25 - 60° N, 51 - 82° W, which is slightly larger than the area of available AVN fields. This was needed because of the two inner grid meshes required for the SAFER method. The first mesh was 333 km^2 and the second was 667 km^2, centered on HS91 as it moved within the boundaries of the analysis domain (Figure 2, an example shown for 30 October 1200 UTC). The filter wavelengths, λ, used for each mesh were: 1° for mesh 1,

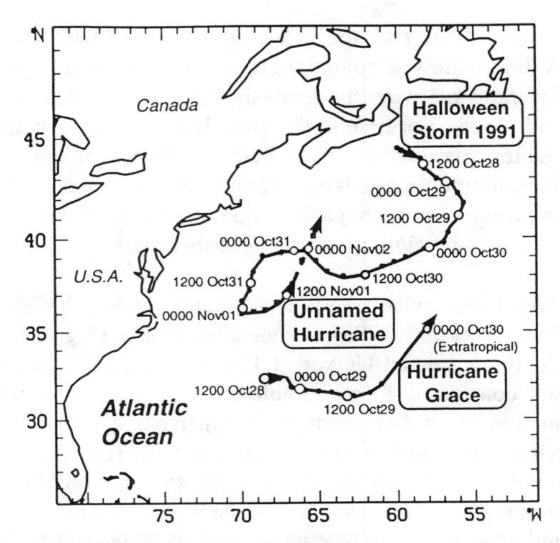

Figure 1. Tracks for HS91, Hurricane Grace, and the Unnamed Hurricane during 28 October-2 November (from Pasch and Avila 1992). Times shown are 12-hr intervals in UTC.

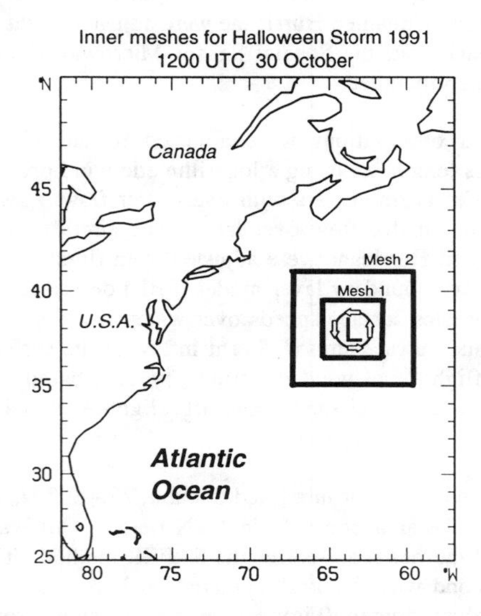

Figure 2. Example of the two inner meshes (mesh 1 = 333km^2 and mesh 2 = 667km^2) centered on HS91, 1200 UTC on 30 October.

2° for mesh 2, and 4° for the remaining area in the domain. All of the surface data (buoys, C-MAN, and ship observations) were included in a ± 3 hr window centered on the analysis time, and the data were positioned in locations according to a storm-relative coordinate system based on the HS91 storm track (Figure 1). The surface wind fields for Hurricane Grace and the Unnamed Hurricane, which were primarily based on the Air Force reconnaissance flight-level winds adjusted to the surface, were analyzed separately for each time. These analyzed fields were included in the larger domain used for HS91 as gridded data covering 444 km^2 areas centered on each of these hurricanes when applicable. The AVN winds were used as the background field with a relative weight of 0.020 for these analyses. Because the center of the Halloween storm in AVN was > 200-km from the actual center in many instances, the AVN grid data within a 250-km radius of the storm center were removed. Subjectively determined data based on all available surface wind observations were used in a Barnes interpolation scheme to fill in the gaps due to the removal of these AVN grid points within 250 km^2 of the HS91 center.

There were instances where SSM/I surface wind speed data were available. Because wind directions were not available with these data, a surface wind analysis was first produced for each time, and the wind direction was interpolated from the analyzed wind field for each SSM/I data point. Then the analysis was run again with all of the SSM/I wind speeds and interpolated wind directions included with the original data and the AVN wind field.

The final step of the process was to map the wind fields from a moving coordinate system to a fixed grid with spatial resolution of 0.25° in both longitude and latitude, with a time-step resolution of 6 hr. An example of the final analyzed V_{M10} field for 30 October 1200 UTC is shown in Figure 3.

Wave Model Specification

The wind fields generated from the HRD method use in situ measurements. A laborious validation technique could be used that selectively withholds information from site to site, and then compare the resulting HRD wind field estimates to the measurements. Uncertainties arise using this evaluation study, because of the complexities in the meteorology found in the HS91, the inter-dependency between measurements, and the accurate representation of the wind field.

An alternate test in the accuracy level of the HRD wind fields would be from the analysis in end-products, such as wave parameters. Knowing scaling relationships between wind speeds and wave parameters, one can, in general terms infer error bounds in the wind estimates. This does not suggest that the wave model is a perfect tool, because no matter how sophisticated a wave model is, there are deficiencies. It is the relative magnitude of those deficiencies in relation to a bad estimate in a wind field that must be identified.

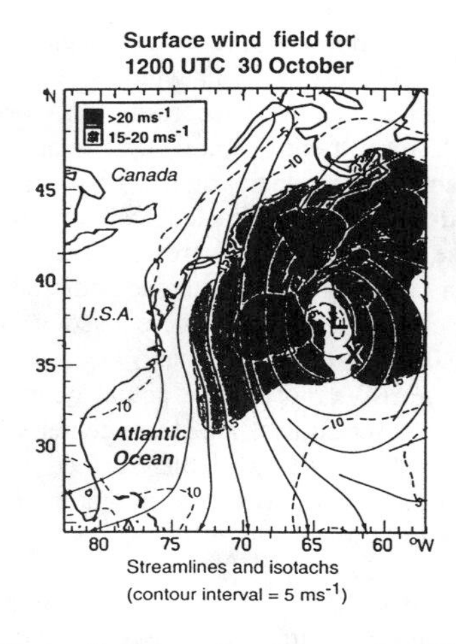

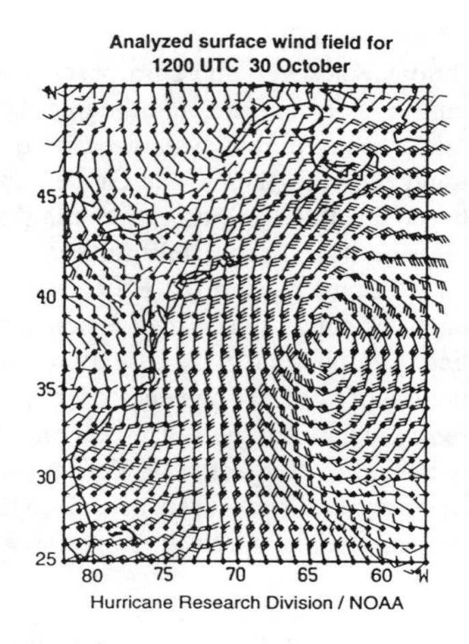

Figure 3. Example of the analyzed V_{M10} field for 1200 UTC on 30 October. The **L** in the analysis is the observed center of HS91, while **X** represents the center of the AVN wind field.

The wave model selected for this work is 3GWAM (WAMDIG 1988), developed principally by a consortium of European scientists. The model is presently being used at various forecasting centers on a daily basis. 3GWAM was also extensively used in the Surface Wave Dynamics Experiment (SWADE), in a forecast and hindcast mode (Graber et al. 1992). The model solves the Radiative Transfer Equations, for the time rate of change in a two-dimensional spectrum over a fixed gridded system.

The 3GWAM was run in three phases. The first phase was run on an Atlantic Ocean Basin grid with spatial resolution of 1°, using FNOC Global wind stress fields ($\Delta x = 1°$, and $\Delta t = 3$hr) as input, minimizing uncertainties in residual energy entering the HRD region. These fields were used from 22 October to the start date of the HRD simulation interval reducing *spin-up* required in the wave model. Both wind fields (FNOC and HRD) were interpolated to a 1-hr time-step in 3GWAM. All model simulations assumed deep water, and principal comparisons are to those locations in a deepwater environment.

Twelve NDBC buoy stations (Figure 4) were used as a basis for the wave model comparison. Three of the buoy locations were directional, and all but one location, based on average peak frequency measurements and water depths, were representative of wave conditions in finite depths. Mean wave parameters (H_{mo}, the peak, T_p, and mean T_m wave periods) are used for initial comparisons between

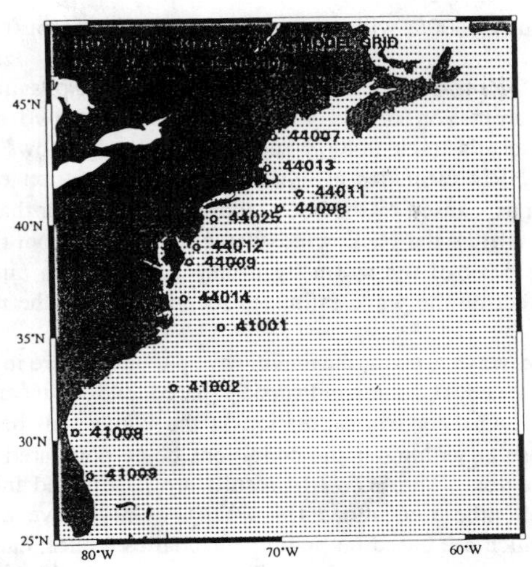

Figure 4. NDBC buoy sites, wind and wave model grid.

model and measurements. Time plots, scatter plots, statistical tests, and non-dimensional plots are used in the analysis. Some of the results from the various statistical tests (e.g. bias, root-mean-square error (RMS Error), scatter index (SI), correlation coefficient (Corr Coef), the slope and intercept of a least squares fit of the paired model) and measurement results are shown in Tables 1 - 3. Three individual buoy locations (44011, 41001 and 41009) are selected for further analysis.

Wind and Wave Model Comparison

Although the measurements from each NDBC buoy were used in the HRD method, it is instructive to compare the model results to the actual measurements. This is done in Table 1, and results are also displayed in scatter plots (Figure 5) for all buoy locations. There is a spread about the 45° (exact correlation) line is approximately 2.5 m/s variation. There is a large underestimate in wind speeds at 44008, which will have a severe impact on the wave estimates discussed later. In general, the wind direction estimates from the HRD winds compare favorably to the buoy measurements. It is also interesting to note the tight distribution about 200° from all measurement/model estimates. In general, from the statistics (Table 1), the HRD method shows near symmetry in wind speeds of about ± 2 m/s. The RMS error ranges from 1 to a high of 3.78m/s located at 44008. In general, the scatter index falls well within an acceptable range of 8-25 percent. Again, the linear regression of the HRD wind estimates to measurements shows good correlation, with the exception of buoy 44008, with a slope of 0.5, and intercept Of 6.6m/s.

These types of comparisons have a tendency to mask phasing problems in a model result. Examination of the wind speed traces at two stations in close proximity to the storm's track (44008 and 44011) clearly shows (Figure 6) good agreement of the HRD estimates at 44011 with the exception of one point just before the peak winds, caused by the buoy recording no data for that particular hour. At 44008 (SW of 44011), the HRD winds depict the trends from the buoy, and yet underestimates the peak of the storm by nearly 17m/s. This underestimate will, have a rippling effect on the wave estimates found at the southern buoys

For active wind-wave growth, the mean wave parameters are in simplistic terms, scaled to the wind speed, H_{mo} the wind speed squared, and T_p linearly with the wind speed. Thus, any under- or overestimation in the winds will be reflected in the wave parameter results, as long as the conditions being compared are in the active wave generation region. The 3GWAM results are partitioned into three regions, North, Mid, and Southern locations (Figure 4). Only wave conditions in the Northern region which are based on nondimensional estimates, can be classified as in the active wave generation region. The more southerly located buoys are responding to swell energy propagating out of the HS91. For the Northern buoys, the authors find a disparity in the previous presumption. Small biases in the wind speeds, either positive or negative, may in fact lead to an opposite effect, thus indicating that either the wind field outside the domain of measurement influence is incorrectly simulated, or 3GWAM cannot resolve wind-wave growth properly. Trends in the HRD records from 44008 point to the earlier hypothesis.

The statistical tests (Tables 2 and 3) and comparisons of 3GWAM to measurements in scatter plots (Figure 7) reveal quite reasonable results. Biases in the H_{mo} results range from -0.6 to 0.8 m, with accompanying RMS errors of 0.4 to 0.9 m. The scatter indexes once again are quite good, with the exception to 41008, a distinct southern, shallow water buoy (18m water depth). Plotting the paired 3GWAM wave parameter estimates to the measurements (Figure 7) for the Northern (upper panel), Central (middle panel), and Southern (lower panel) locations, shows distinct trends. The H_{mo} model results are well correlated to the measurements up to approximately 8 m, where results depart from the *line of perfect correlation*. That departure originates from buoy 44011. The three outlying points indicating an over-estimate in wave model results, are from buoy 44008. This trend follows a similar pattern, progressing through the Central buoys and finally in the Southern buoy results. Although time is removed from these scatter plots, one does see a strong phasing problem in model to measurement comparisons. The T_p results show similar trends, with a stronger scatter, supported by larger the RMS errors and lower correlation coefficients found in Table 3. Most of the scatter in Figure 7 for the T_p results are caused principally by phasing. A further examination of this is can be found in comparison of time plots of H_{mo} results for buoys 44011, 41002, and 41009.

Table 1. Wind Speed Statistical Summary

BUOY	Buoy Mean	Bias	RMS Error	S I	Corr Coef	Slope	Intercept
44007	13.22	0.20	1.73	13	0.79	0.84	2.28
44013	15.15	-1.18	1.14	8	0.92	0.95	-0.45
44011	13.20	2.24	1.94	15	0.96	0.88	3.87
44008	15.50	-1.18	3.78	24	0.81	0.50	6.61
44025	11.89	0.04	1.26	11	0.92	1.01	-0.09
44012	11.99	-1.66	1.40	12	0.89	0.95	-1.03
44009	11.81	-0.88	1.45	12	0.88	0.97	-0.57
44014	10.92	0.40	1.21	11	0.91	1.05	-0.11
41001	12.84	2.02	1.12	9	0.88	0.90	3.34
41002	10.03	1.56	0.89	9	0.95	1.16	-0.10
41008	6.82	-1.03	1.63	24	0.82	0.74	0.73
41009	6.06	-0.59	1.39	23	0.89	0.88	0.13

Table 2. Energy-Based Wave Height Statistical Summary

BUOY	Buoy Mean	Bias	RMS Error	S I	Corr Coef	Slope	Intercept
44007	2.93	0.70	0.42	14	0.96	0.98	0.76
44013	4.05	-0.61	0.56	14	0.94	0.80	0.20
44011	6.41	-0.17	0.80	13	0.96	0.98	-0.04
44008	5.21	0.10	0.91	17	0.88	0.79	1.17
44025	3.16	-0.26	0.51	16	0.87	1.32	-1.26
44012	2.84	-0.23	0.48	17	0.76	0.84	0.23
44009	3.24	0.08	0.50	15	0.85	1.13	-0.34
44014	4.52	-0.43	0.64	14	0.88	0.70	0.94
41001	5.65	0.26	0.81	14	0.83	1.04	0.04
41002	5.27	-0.11	0.96	18	0.67	0.61	1.96
41008	1.61	0.07	0.57	35	0.51	0.47	0.92
41009	3.17	0.08	0.74	23	0.63	0.91	0.37

Table 3. Peak Spectral Wave Period Statistical Summary

BUOY	Buoy Mean	Bias	RMS Error	S I	Corr Coef	Slope	Intercept
44007	11.60	-1.50	3.11	27	0.64	0.43	5.13
44013	11.52	0.68	2.69	23	0.65	0.88	2.06
44011	12.77	-0.61	1.48	12	0.81	0.75	2.52
44008	11.50	0.71	1.76	15	0.66	1.02	1.44
44025	9.22	1.98	4.26	46	0.17	0.49	6.67
44012	11.75	1.14	3.58	30	0.58	0.54	6.56
44009	10.81	2.19	2.67	25	0.64	0.95	2.70
44014	14.07	-0.48	1.74	12	0.81	0.85	1.67
41001	13.15	0.18	2.74	21	0.64	0.47	7.10
41002	14.85	-0.53	2.26	15	0.71	0.59	5.57
41008	12.78	1.90	3.75	29	0.57	0.34	10.27
41009	15.07	-0.63	2.05	14	0.77	0.67	4.40

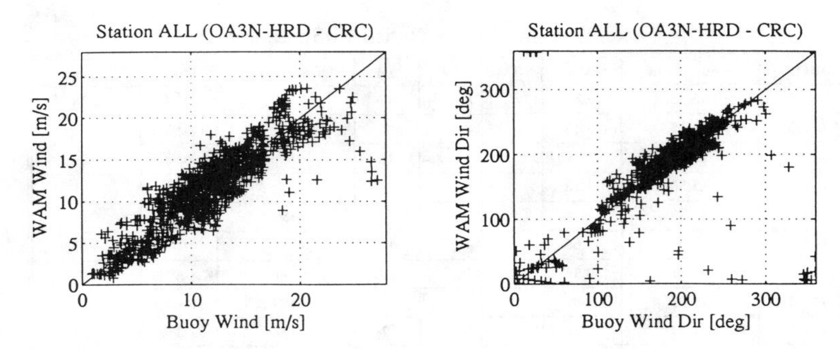

Figure 5. Wind speed and wind direction comparison at all buoy sites.

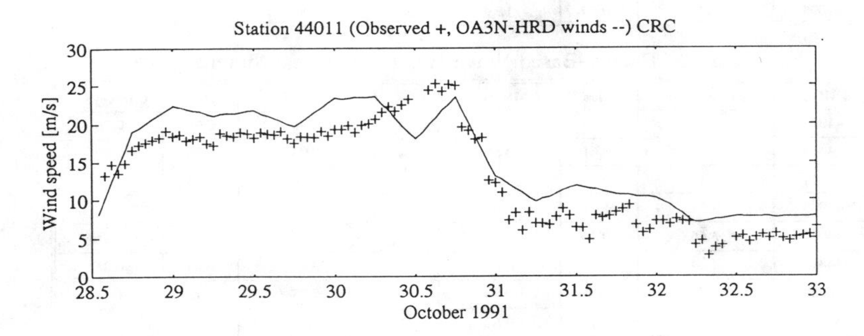

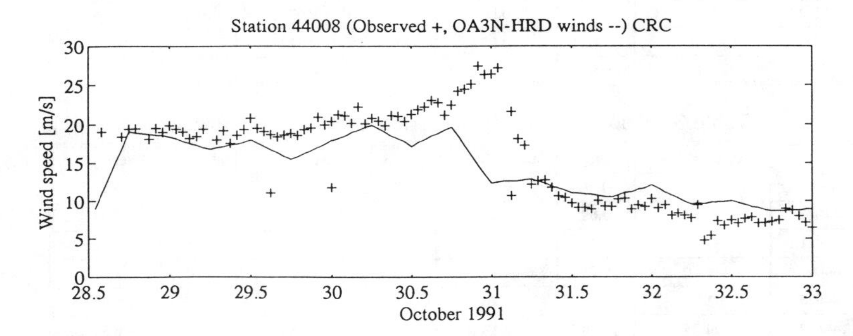

Figure 6. Time plot of HRD wind speed estimates at 44011 (upper panel), and 44008 (lower panel).

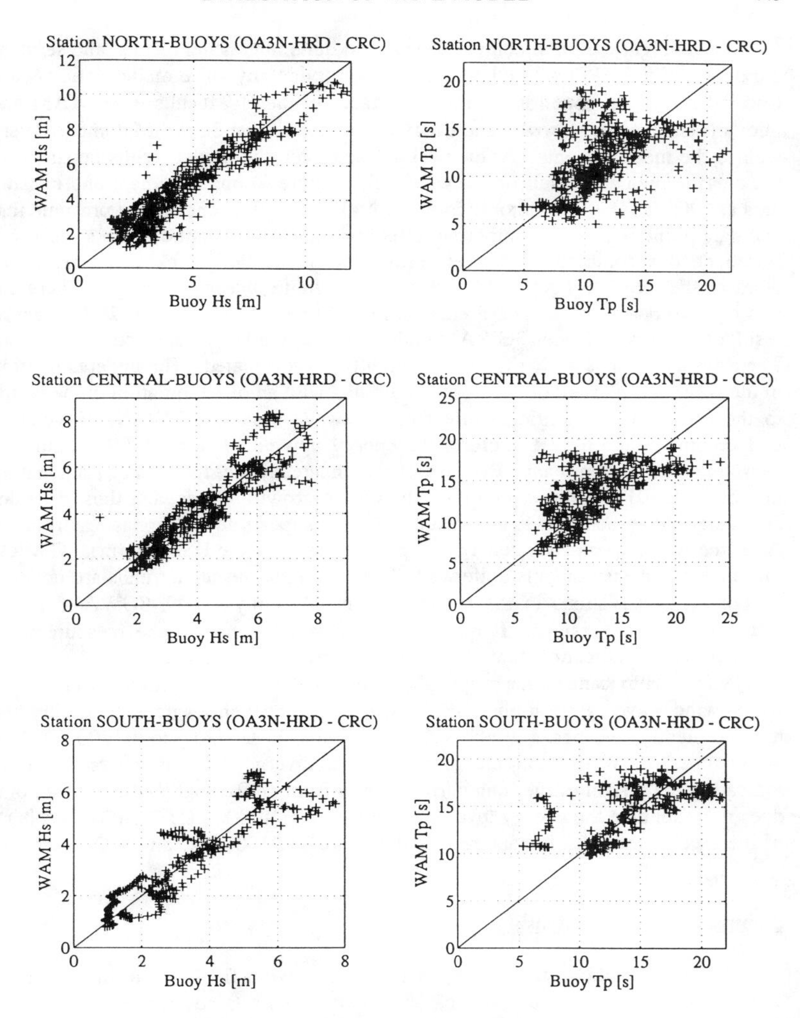

Figure 7. Energy based wave height (H_{mo}) and peak spectral period (T_p) scatter plots for Northern (top panel), Central (middle panel), and Southern (lower panel) buoys.

Wave height temporal comparisons reveal quite interesting results, and serve as a true test of the HRD method, as well as identifying any wave model deficiencies. Located in the top panel of Figure 8 are traces of the H_{mo} results in 3GWAM and buoy measurements. Growth and decay of the storm from 3GWAM compares very well to the measurements. At the peak of the storm, the model results depart from the measurements, a result of the wind field (Figure 6) underestimate and missing data at 1200 UTC on 30 October for buoy 44011. Clearly, one cannot presume that one data point has such an important effect on the entire Atlantic storm simulation. Based on this comparison, it does seem rather plausible. Looking further down wind of the HS91 at buoy 41001, there is a definite phasing problem. There are two relative peaks in the measurements, the first near 1200 UTC on 29 October, a result of Hurricane Grace. 3GWAM underestimates the H_{mo} by approximately 1.5m. The estimated winds at that time were slightly overestimated. The underestimation in the waves at the peak of the storm, combined with an overestimation in the winds to the south has a significant effect on the wave model. 3GWAM, in general, underestimates the T_p. Therefore, the energy upwind of buoy 41001 should be passive, or swell dominant. Because of the underestimate, the energy propagating into buoy 41001 remains within the active wave-growth regime, and thus, does not attenuate, but rather grows, and results in what appears to be a phasing problem. After the second peak (from energy propagating in from the HS91 storm), 3GWAM estimates the measurements quite well. Further south, the same trends are noticed, supporting observations in the central region. At buoy 41009, 3GWAM underestimates the H_{mo} results. It does however, depict trends in the measurements adequately, an indication that the HRD winds are well represented. The wave energy from Hurricane Grace propagates into 41009, where it combines with modest local wind-wave generation (HRD winds in excellent agreement with the measurements), and peaks to about 4 m for nearly 24 hr. At about 1200 UTC on 30 October, 3GWAM and the measurements diverge. Local winds decrease, indicating the arrival of the long period energy from HS91. 3GWAM, at that point does not attenuate the energy level (to the minimum at 0000 UTC on 31 October), or increase in intensity to capture the final peak on 2100 UTC on 31 October.

Summary and Conclusions

The HRD method was developed to estimate wind fields for tropical storms. The method for these meteorological situations uses to its advantage aircraft wind measurements. Hence, application of the HRD method to the HS91 is very different, both in terms of the meteorological conditions, and the fact that it is a large, synoptic-scale storm combined with measurements that are stationary. It is remarkable that the method resolves such a complex situation as well as it does in a first-time attempt. Improvements in the HRD-generated wind field specification can be derived from having access to higher temporal resolution initial wind fields. The AVN wind time-step was 6 hr, too coarse for this meteorological event.

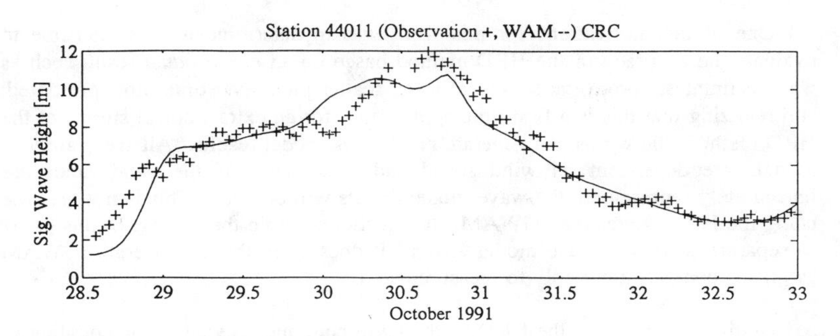

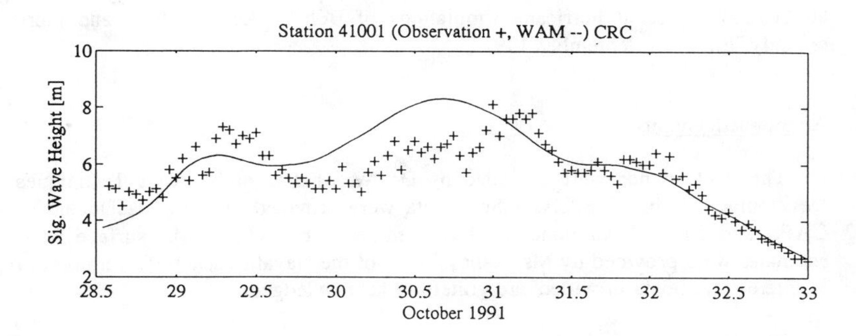

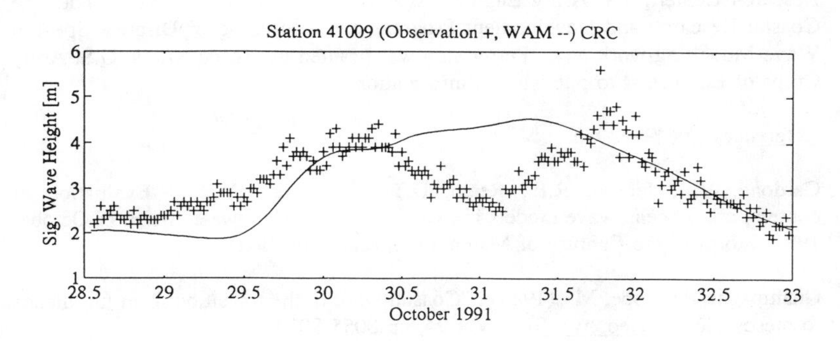

Figure 8. Time plot of H_{mo} at buoy site 44011 (upper panel), 41001 (middle panel), and 41009 (lower panel).

One of the motivations for this work was to determine if it is plausible to evaluate the accuracy of the HRD method based on an *end-product* result such as wave estimates. From the statistical tests, scatter plots, and time plots presented, and realizing that this is a first-time application (for an extra-tropical storm) of the HRD method, the winds can generate good wave model results. All spectral wave models are dependent on wind speed and directions. If the wind fields are inaccurately represented, the wave model results will be poor. This is not the case using the HRD winds and 3GWAM. It is difficult within the context of this paper to separate wind and wave model errors. It does seem that the overall 3GWAM estimates compare favorably to measurements.

Additional testing of the HRD method will continue, as well as the coupling of this method to 3GWAM. It is envisioned that this particular study will grow, with application to recent hurricane simulations of Bob in August 1991, and more recently, Emily in September 1993.

Acknowledgements

The AVN winds were provided by Dr. W. Shaffer of NOAA's Techniques Development Lab. The CAES buoy data were provided by Ken MacDonald of CAES and Dr. V.J. Cardone of Oceanweather, Inc. The SSM/I surface wind estimates were provided by Ms. Nancy Baker of the Naval Research Laboratory, in Monterey. All data providers are gratefully acknowledged.

The spectral wave modelling work was conducted at the Coastal Engineering Research Center, U.S. Army Engineer Waterways Experiment Station, under the Coastal Research and Development Program, the Upgrading of Discrete Spectral Wave Modelling work unit. Permission was granted by Headquarters, U.S. Army Corps of Engineers, to publish this information.

References

Cardone, V.J., Jensen, R.E., Resio, D.T., and Swail, V.K. "Evaluation of contemporary ocean wave models in extreme events: Halloween Storm of October 1991; Storm of the Century of March 1993, (in publication).

Champagne-Phillippe, M. (1989). "Coastal wind in the transition form turbulence to mesoscale," J. Geophys. Res., Vol 94, pp 8055-8074.

DeMaria, M., Aberson, S.M., Ooyama, K.V., and Lord, S.J. (1992). "A nested spectral model for hurricane track forecasting," Mon. Wea. Rev. Vol 120, pp 1628-1643.

Durst, C.S. (1960). "Wind speeds over short periods of time," Meteor. Mag.

Vol 89, pp 181-186.

Graber, H.C., Caruso M.J. and Jensen, R.E. (1992). "Surface wave simulations during the October storm in SWADE," Proceedings of MTS'91, Marine Technology Society, pp 147-153.

Houston, S.H., and Powell, M.D. (1993). "Observed and modeled wind and water level responses from Tropical Storm Marco (1990)," submitted to Wea. Forecast.

Krayer, W.R. and Marshall, R.D. (1992). "Gust factors applied to hurricane winds, "Bull. of AMS, Vol 73, pp 613-617.

Ooyama, K.V. (1987). "Scale-controlled objective analysis," Mon. Wea. Rev.,Vol 115, pp 2479-2506.

Pasch, R.J., and Avila, L.A. (1992). "Atlantic hurricane season of 1991," Mon. Wea. Rev.,Vol 120, pp 2671-2687.

Peterson, R.E. and Stackpole, J.D. (1989). "Overview of the NMC Production Suite," Wea. Forecasting, Vol 4, pp 313-322.

Pierson, W.J. (1983). "The measurement of synoptic scale wind over the ocean," J. Geophys. Res., Vol 108, pp 757-766.

Powell, M.D. (1980). "Evaluations of diagnostic marine boundary layer models applied to hurricanes," Mon. Wea. Rev., Vol 108, pp 757-766.

Powell, M.D., Dodge, P.P., and Black, M.L. (1991). "The landfall of Hurricane Hugo in the Carolinas," Wea. Forecasting, Vol 6, pp 379-399.

Vincent, C.L., Jensen, R.E., and Wittmann, P.A. (1992). "A wind wave hindcast for the Halloween Northeaster in 1991 in the Atlantic ocean coastal waters," Supplement 3rd International Workshop on Wave Hindcasting and Forecasting, Environment Canada, Atmospheric Environment Service, pp 89-100.

Wang, W.C., and Mettlach, T. (1992). "The Halloween Storm: data observations from NDBC Stations," Supplement 3rd International Workshop on Wave Hindcasting and Forecasting, Environment Canada, Atmospheric Environment Service, pp 71-88.

WAMDIG (The WAM Development and Implementation Group) (1988). "The WAM Model - a third generation ocean wave prediction model," J. Phys. Oceanog., Vol 18, pp 1775-1810.

Wave Monitoring in The Southern California Bight

W. C. O'Reilly [1], Member ASCE, R. J. Seymour [1], Member ASCE, R. T. Guza [1], D. Castel [1]

Abstract

An overview of recent research by the Coastal Data Information Program (CDIP), aimed at effectively monitoring wave conditions in the Southern California Bight, is presented. The topographic and bathymetric complexity of the Bight presently limits the utility of *in situ* measurements to the immediate vicinity of the observations. A long-term goal of the CDIP research program is to provide useful estimates of the wave and wind field throughout the southern California region from a limited number of *in situ* observations. Ongoing and planned CDIP research projects, involving numerical wave models, deep ocean directional wave measurements, and regional wind data are described.

Introduction

Numerous wave measurements have been made within the Southern California Bight over the last decade by NOAA and the Coastal Data Information Program (CDIP). However, shoreward propagating surface gravity wave energy is focussed and diffused by offshore banks and blocked by islands in this bathymetrically complicated coastal region. Wave energy can therefore be very different at sheltered sites separated by only a few kilometers, and a prohibitively large number of *in situ* instruments would be required to monitor all locations of interest for engineering, boater safety and other purposes.

Determining the wave climate within the Bight is further complicated by locally generated waves. Although the incident deep ocean wave field is usually the

[1] Scripps Institution of Oceanography, UCSD 0214, La Jolla, CA 92093.

dominant source of wave energy within the Bight, there are important exceptions. Intense local storms with strong and spatially variable winds can generate energetic seas on large pre-existing swell (Seymour, 1989). Many wind measurements throughout the Bight will be required to study local wave generation in these cases, and present wind-wave generation models may be inaccurate.

Wave conditions within the Bight can be conceptually divided into swell waves arriving from outside the islands and local seas generated inside the Bight. One approach to estimating swell conditions in coastal regions is to initialize a numerical wave propagation model with the offshore (unsheltered) frequency-directional spectrum, $S_o(f,\theta)$. However, for complex bathymetry, the model simulations can be highly sensitive to the details of $S_o(f,\theta)$. This may degrade model predictions which are based on input conditions specified with conventional (fundamentally low resolution) measurement systems, such as the ubiquitous pitch-and-roll buoy.

The Southern California Wave Experiment (Aug. 91 - Feb. 92), jointly sponsored by The California Department of Boating and Waterways, Sea Grant, and The Army Corps of Engineers, was designed in part to test methodologies for predicting swell conditions within the Bight from directional buoy measurements of S_o. Preliminary results from this experiment are encouraging; however, the expected sensitivity of wave conditions in the Bight to small errors in the peak direction of S_o is clearly seen.

Geographical Setting

The Southern California Bight extends from approximately 32°N to Point Conception (34.5°N) on the west coast of the U.S. (Figure 1). The topography and bathymetry of the Bight are characterized by numerous offshore islands and shallow banks, a narrow continental shelf, and coastal submarine canyons.

The complex bathymetry results in an equally complicated regional wave field. The islands block a significant amount of the incident deep water wave energy and the wave field is influenced by offshore banks, the irregular shaped continental shelf, and numerous coastal submarine canyons. The spatial variability of the resulting nearshore wave field is often dramatic, with large changes in wave energy taking place over only a few kilometers.

The wind field in the Bight is profoundly affected by the rugged topography of both the offshore islands and the coastal mainland. However, a very limited number of wind measurements are presently available to define these spatially varying winds. Therefore, properly initializing wave generation models in the Bight remains a difficult and poorly understood problem.

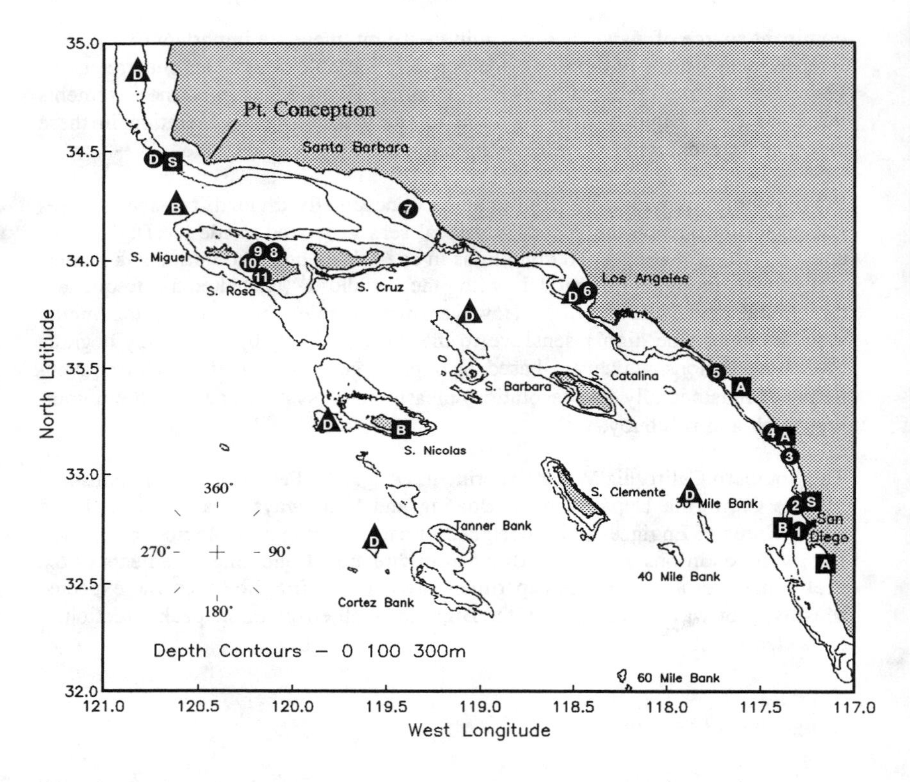

Figure 1. Locations of measurement stations in the Southern California Bight during the winter 1991-1992. "B" represents nondirectional buoys, "D" directional buoys, "S" single-point pressure gages and "A", directional arrays. The triangles are gages maintained by NOAA or the Navy, squares, the Coastal Data Information Program, and the numbered circles are for single-point gages deployed as part of the Southern California Wave Experiment.

Utilizing Historical Wave Measurements

Since the mid-1970's, routine wave measurements have been made in the southern California region. Most of this data has been collected by the CDIP, NOAA, and the U.S. Army Corps of Engineers (see the paper by Seymour et. al. in these conference proceedings). The purpose of these measurements is to provide quasi real-time and historical wave information to engineers, coastal planners, and mariners. Over the past decade, the number of wave gaging stations has grown considerably, and the Bight is arguably the most heavily monitored wave climate in the U.S..

A common problem facing coastal engineers is translating historical wave data collected at one more sites in a coastal region to a study site where there are few or no direct measurements. While a wave measurement made along an open coastline may be relevant for a large section of coast, measurements made in areas with complex bathymetry are often quite site specific. Under these conditions, simple interpolation methods will not accurately predict waves at uninstrumented sites.

In an attempt to effectively utilize wave data collected along complex coastlines, data interpretation schemes based on inverse mathematical methods are presently being studied by CDIP. Inverse methods encompass a wide range of techniques from linear programming (solving simple sets of linear equations) to more sophisticated forms of inverse theory which have found important uses in such fields as geophysics and medical tomography. The inverse techniques being applied in southern California borrow heavily from a branch of oceanography known as acoustic tomography, where acoustic sources and receivers are strategically placed in the ocean for the purpose of mapping, for example, the mesoscale sound speed field (Munk and Wunsch, 1978).

The goal of a southern California inverse model is to estimate the entire regional wave field using a network of wave gages. In addition, by applying analogous procedures to those developed for designing acoustic tomography experiments (Barth and Wunch, 1990), the best future gaging locations can be selected in an objective manner.

The inverse approach is being used to study swell waves from distant sources. Locally generated seas are also important to the overall southern California wave climate and can be incorporated into the inverse methodology. However, for the purpose of evaluating the viability of the inverse approach, we have limited our present scope to wave propagation and avoided the additional uncertainties of wind-wave generation models.

Linear Wave Propagation Models

The inaccuracy of simple interpolation schemes between measurement stations in southern California means that some other "model" must be used to relate the measurements to each other and to wave conditions throughout the region.

Two well known linear wave propagation methods, spectral refraction (Longuet-Higgins, 1957) and spectral refraction-diffraction (Izumiya and Horikawa, 1987), have been adapted for use in southern California. These models assume S_o is homogeneous in deep water outside the islands and no local generation of wave energy. A comparison of these two models (O'Reilly and Guza, 1993) in southern California, using simulated incident directional spectra, showed that they predict surprisingly similar results across most of the Bight. The agreement generally is best between the models when S_o is broad in direction and frequency. Simulated southern hemisphere swell, which are narrow in both frequency and directional upon their arrival in California, produced the largest differences between the model predictions.

The Southern California Wave Experiment

In the fall and winter of 1991-92, an extensive field experiment was undertaken to test the accuracy of the spectral wave propagation models and assess the viability of using inverse methods to improve estimates of the Bight-wide wave field. The existing wave monitoring stations were supplemented with 11 additional single-point pressure gages (i.e. nondirectional), deployed throughout the Bight as part of the Southern California Wave Experiment (Figure 1). In addition, a Datawell directional waverider buoy was deployed by the U.S. Navy in deep water west of Pt. Conception.

Preliminary results indicate that a linear model which includes island blocking and spectral refraction yields useful predictions of wave energy across most of the region. The exception was the east end of the Santa Barbara Channel (Site 7, Figure 1) where both the spectral refraction and refraction-diffraction wave models underpredicted significant wave heights of swell from the northwest by 50%. The reasons for this are still being studied and CDIP hopes to redeploy instrumentation at this site in the future.

The month of January, 1992 was suitable for a field verification of models because the deep water wave conditions were often dominated by swell from a single, distant source, which are optimal conditions for estimating directional spectra with low-resolution directional buoy data. Figure 2 shows the hourly measurement of wave energy in the 10-20*s* period band (swell wave energy) at Batiquitos Lagoon (Site 3 , Figure 1), for January, 1992. Also shown is the predicted wave energy based on the spectral refraction model (Figure 2, lower panel). This model was initialized using hourly, iterative maximum likelihood estimates (IMLE, Oltman-Shay and Guza, 1984) of S_o from an unsheltered, deep ocean directional buoy. The total wave energy is significantly overpredicted by the wave model in this case. However, if the hourly estimates of S_o are rotated 8 degrees to the North (e.g. deep water wave energy which was originally

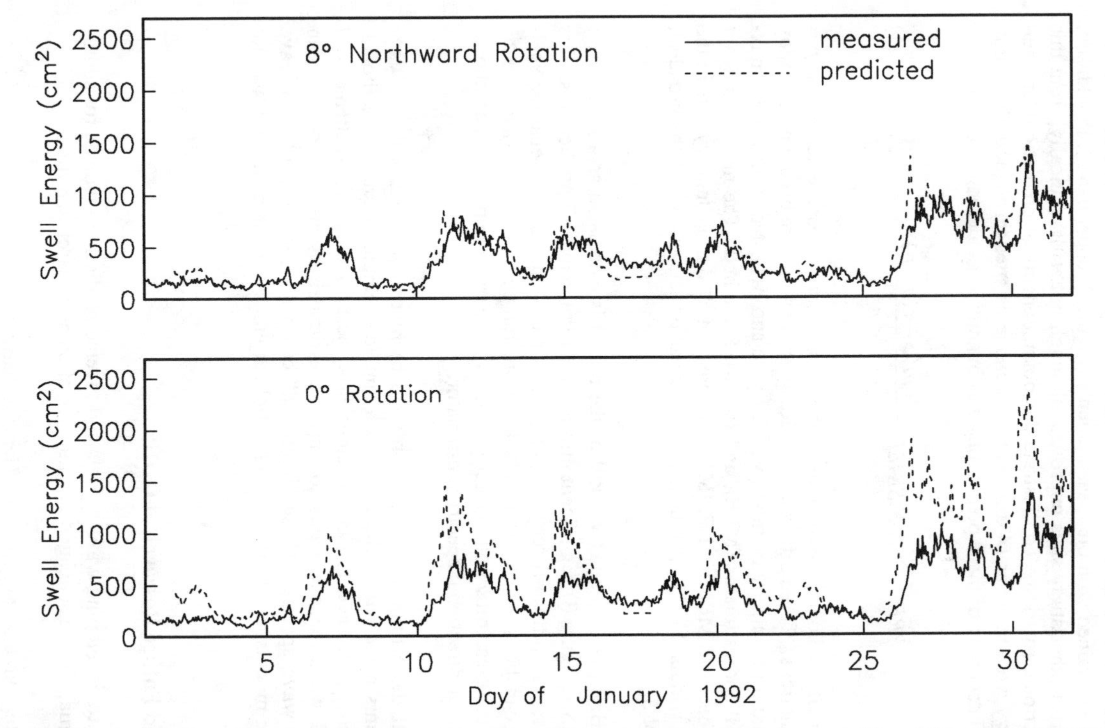

Figure 2. Measured (solid line) and predicted (dashed) swell energy (total energy between 10 and 25s periods) in 30m depth near Batiquitos Lagoon (Site 3, Figure 1). Predictions are based on spectral estimates from a deep water directional buoy. The upper panel is for predictions based on the same deep water spectra that have been rotated to the north by 8 deg.. (From O'Reilly, 1993)

estimated to arrive from 270° now arrives from 278° etc.), then much better agreement is obtained with the measurements (Figure 3, upper panel). In addition, the fit to most other study sites improved, although not as dramatically, from this slight directional rotation. To illustrate this point, concurrent predictions of swell wave energy were made at 14 different measurements sites throughout the Bight. The overall accuracy of these predictions was quantified as a predictive skill,

$$PREDICTIVE\ SKILL = \frac{<m^2> - <(m-p)^2>}{<m^2>} \qquad (1)$$

were <> = all the hourly predictions at the 14 sites, m = measurements, and p = model predictions. Three different directional estimators were used with the buoy data; A maximum entropy method (MEM, Lygre and Krogstad, 1986) and a maximum likelihood estimator (MLE), and the IMLE method. The most significant difference between MEM, MLE, IMLE estimates of S_o during January was in the directional widths. MEM generally makes the narrowest, and MEM the broadest, estimates of S_o.

When the Bight-wide predictive skill of the model is calculated as a function of the rotated S_o (Figure 3), significant improvement in the overall predictive skill is seen for a small northward rotation ($\approx 5°$). This result shows how sensitive sheltered sites can be to errors in S_o. In addition, for unidirectional wave events, measuring the mean direction correctly appears to be more important than choosing a particular directional spectrum estimator.

Rotating all the estimates of S_o, in order to be more consistent with the wave measurements made *within* the Bight, is a simple example of an inverse model. O'Reilly and McGehee (1993) consider how best to combine sheltered and unsheltered wave measurements to improve estimates of the true deep ocean directional wave spectra. These "improved" estimates can be used with wave propagation models to estimate the wave field at uninstrumented sites throughout the Bight.

Present and Future Research at CDIP

In addition to developing wave models and inverse prediction methods for southern California, CDIP is engaged in a number of other research activities.

- Deep Water Directional Data from Harvest Platform

In November, 1992, Harvest Platform, an oil drilling platform in 200m water depth west of Pt. Conception, was equipped with an array of 6 pressure transducers. CDIP hopes to use this data to initialize the linear wave models and make real-time predictions of Bight-wide swell conditions during the winter months (when swell approach California from the west-northwest). In addition, the

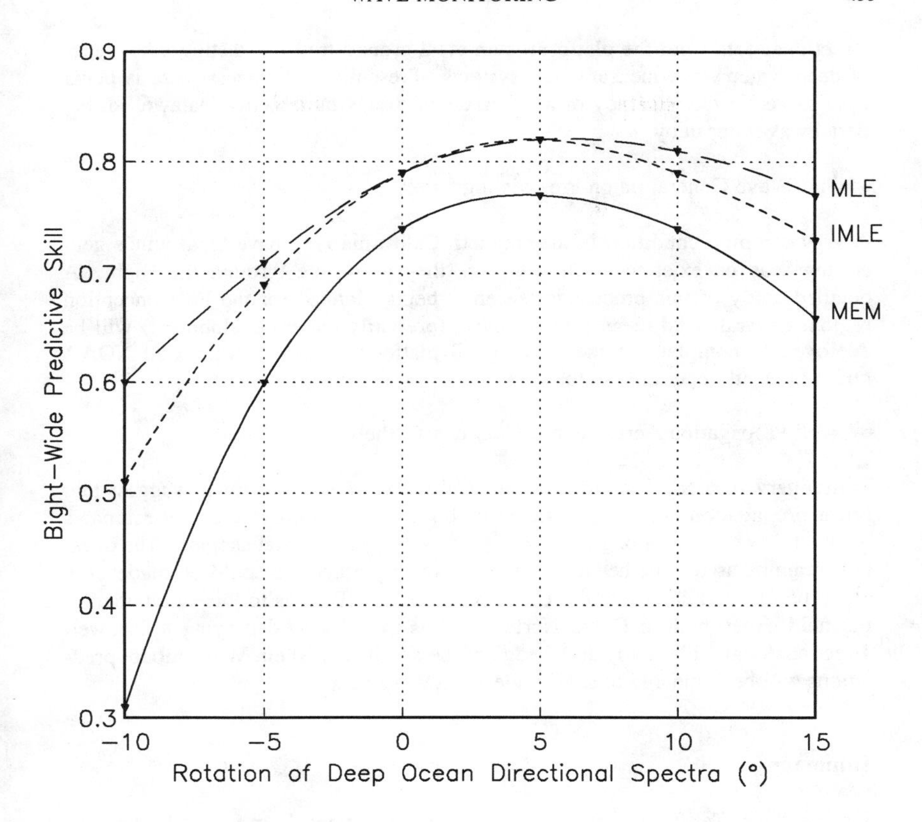

Figure 3. Variation in overall skill of spectral refraction model predictions when the deep water directional spectra are rotated to the south (-) and north (+). The skill value is averaged over 744 hourly predictions and 14 different measurement sites during the month of January, 1992.

directional data from the platform is an ideal ground-truth for testing other types of deep water wave measurement systems. Presently, the Harvest data is being used to verify the accuracy of a NOAA 3-m discus buoy and a Datawell directional waverider buoy.

- Wind-Wave Generation on Pre-existing Swell

A very common condition in the southern California is to have local winds generating seas on existing swell waves as they propagate through the Bight. A detailed study of this process is presently being planned for the Pt. Conception region. Several wind measurement buoys (presently under development) will be deployed to complement the existing oil platform instrumentation and NOAA buoys presently operating in the area.

- Swell Propagation Across Broad Continental Shelves

Preliminary results from the Southern California wave experiment suggest that linear propagation models, initialized by buoy data, can provide useful estimates of swell wave energy along coastlines with narrow continental shelves. The question remains as to whether or not similar swell predictions could be made on a broad continental shelf like the U.S. east coast. CDIP plans to join in an upcoming field experiment at Duck, North Carolina in 1994 by deploying a Datawell directional waverider buoy at the edge of the continental shelf. Wave model predictions will be compared to shelf-wide swell wave data.

Summary

The Southern California Bight is one of the most intensely monitored wave climates in the world. However, the topography and bathymetry of the region produce spatially complex wind and wave fields. Therefore, a combination of numerical wave models and inverse mathematical methods are being developed to fully utilize the historical wave data. Preliminary results from a recent field experiment indicate that linear wave propagation models are a viable method for predicting swell wave conditions across narrow continental shelves. The field results also demonstrate that wave predictions at highly sheltered sites are sensitive to small directional errors (≈5°) in estimates of S_o. Ongoing CDIP research includes the field verification of deep water wave measurement instrumentation, the development of inexpensive wind monitoring buoys, the study of wind-wave generation on existing swell, and the verification of linear wave propagation models on broad continental shelves.

Acknowledgements

The continuing support of the Coastal Engineering Research Center of the U.S. Army Corps of Engineers and the California Department of Boating and

Waterways (CDBW) is gratefully acknowledged. The wave models and inverse methodologies have been developed as part of CDBW's continuing program to promote boating safety and access and to conduct shoreline erosion studies. The Southern California Wave Experiment was jointly sponsored by CDBW and Sea Grant.

This work is the result of research sponsored in part by the National Oceanic and Atmospheric Administration, National Sea Grant College Program, Department of Commerce, under grant number NA89AA-D-SG138, project number R/CZ-90, through the California Sea Grant College, and in part the California State Resources Agency. The U.S. Government is authorized to reproduce and distribute this report for governmental purposes.

References

Barth N., and C. Wunsch, 1990. Oceanographic experiment design by simulated annealing, JPO, 20, 1249-1263.

Izumiya T., and K. Horikawa, 1987. On the transformation of directional waves under combined refraction and diffraction, Coastal Eng. in Japan., 30(1), 49-65.

Longuet-Higgins, M. S., 1957. On the transformation of a continuous spectrum by refraction, Proc. Camb. Phil. Soc., 53(1), 226-229.

Lygre A., and H. E. Krogstad, 1986. Maximum entropy estimation of the directional distribution in ocean wave spectra, JPO, 16, 2052-2060.

Munk W., and C. Wunsch, 1978. Ocean acoustic tomography; a scheme for large scale modeling, Deep-Sea Res., 23, 123-161.

Oltman-Shay J., and R. T. Guza, 1984. A data-adaptive ocean wave directional-spectrum estimator for pitch and roll type measurements, JPO, 14(11), 1800-1810.

O'Reilly W. C., and R. T. Guza, 1983. A comparison of two spectral wave models in the Southern California Bight, Coastal Engineering, 19, 263-282.

O'Reilly W. C., 1993, The southern California wave climate: effects of islands and bathymetry, Shore and Beach, 61(3), 14-19.

O'Reilly W.C., and D. D. McGehee, 1993. The design of regional wave monitoring networks: a case study for the Southern California Bight, CERC Tech. Rep., U.S. Army Corps of Engineers, In Press.

Seymour, R. J., 1989, Wave observations in the storm of 17-18 January 1988, Shore and Beach, 57(4), 10-13.

A Comparison of Wave Statistics in Uni- and Multidirectional Shoaling Seas

Andrew M. Cornett and Okey G. Nwogu
NRC Coastal Engineering Program, Building M32
Montreal Road, Ottawa, Ontario, K1A 0R6, Canada

Abstract

The transformation of six unidirectional and multidirectional seas propagating over a 1:25 planar beach are examined in a series of laboratory experiments. In all cases, the mean wave direction is normal to the bathymetric contours. Changes to frequency spectra, directional distributions, wave asymmetries, and the distribution of wave heights and crest elevations, are examined through the shoaling and surf zones. Wave multidirectionality greatly reduces the growth of long-waves, and slightly reduces the growth of wave energy at higher harmonics of the fundamental peak frequency. In all cases, unidirectional waves exhibited greater increases in wave heights and crest elevations through the shoaling zone. The effect of multidirectionality on wave asymmetries, and the distribution of wave heights and crest elevations, is small compared to the overall changes throughout the nearshore.

Introduction

As irregular waves propagate into shallow water, they undergo transformations due to shoaling, refraction, breaking, frictional dissipation and wave-wave interactions, that result in changes to the frequency spectrum, directional distribution, and probability distributions of wave heights and crest amplitudes. As well, nonlinear effects are more pronounced in shallow water, and act to distort wave profiles, introducing both vertical and horizontal asymmetries. The effect of wave directionality on these transformation processes, and on the resulting shallow water wave characteristics, has not yet been completely discerned. This paper attempts to contribute to this understanding through an examination of laboratory measurements of unidirectional and multidirectional waves shoaling on a planar beach.

Freilich et al. (1990) investigate the contributions of nonlinearities to the transformation of directional spectra on a nearly planar bathymetry using measurements obtained in the field at Torrey Pines, California. They conclude that linear refraction theory inaccurately predicts both the shapes of directional spectra in shallow water, and the variances in some frequency bands. They suggest that near-resonant triad interactions in the shoaling multidirectional wave field are responsible for the prominent nonlinear features in the directional spectra observed in shallow water. Elgar et al. (1992) describe a laboratory simulation of the Torrey Pines field experiment, and conclude that the evolution of multidirectional seas in shallow water can be well modelled in the laboratory.

Briggs and Smith (1990) report a laboratory study on the transformation of fifty-four directional spectra in shallow water over a planar beach. They conclude that directional spreading results in a reduction of the nonlinear phase coupling between modes, and a pronounced reduction in the formation of subharmonics.

The depth-limited, breaking wave height of short-crested waves has been studied by Halliwell and Machen (1981), who found that the height of short-crested breaking waves can be significantly higher than that of long-crested breaking waves on the same planar beach slope. They also observe that large short-crested breaking waves are widely distributed in two dimensional space, and therefore not likely to be satisfactorily recorded by floating buoys.

Laboratory measurements of unidirectional (2D) and multidirectional (3D) waves propagating on a planar bathymetry are used to study the effects of wave directionality on nearshore wave transformations. The analysis focuses on changes to the wave spectrum, the spreading function, the skewness of the water surface elevation, and the distribution of wave heights and crest elevations, throughout the shoaling and breaking zones.

Experiments

Experiments were conducted in the multidirectional wave basin of the NRC Coastal Engineering Laboratory (formerly Hydraulics Laboratory) at Ottawa, to investigate the characteristics of unidirectional (2D) and multidirectional (3D) shoaling waves. The rectangular basin is 30 m wide, 20 m long, and 3 m deep, and is equipped with a sixty-segment directional wave generator, described in detail by Miles et al. (1986). Progressive wave absorbers, consisting of numerous, vertical, porous, metal screens, are installed along the perimeter of the basin not occupied by the wave generator. Jamieson and Mansard (1987) demonstrate that these efficient wave absorbers are characterized by reflection coefficients of 5% or less for most wave conditions. In some circumstances, impermeable side walls, installed at either end of the wave generator, are used to improve the quality of wave simulations within the test area.

The test set-up is sketched in Figure 1. A nearly planar bathymetry, with an impermeable concrete skin, was constructed parallel to the wave generator at a slope of 1:25. A 4.6 m length of horizontal bottom was provided between the wave boards and the toe of the planar slope. A constant water depth h = 0.56 m was maintained at the wave generator for all tests. Water surface elevations η were measured at twenty-three stations along a linear array, normal to the beach slope, at the basin centreline. Three bi-axial, electromagnetic, velocity meters were installed along this array (coincident with probes 1, 6 and 13) for directional wave measurement. Water depths

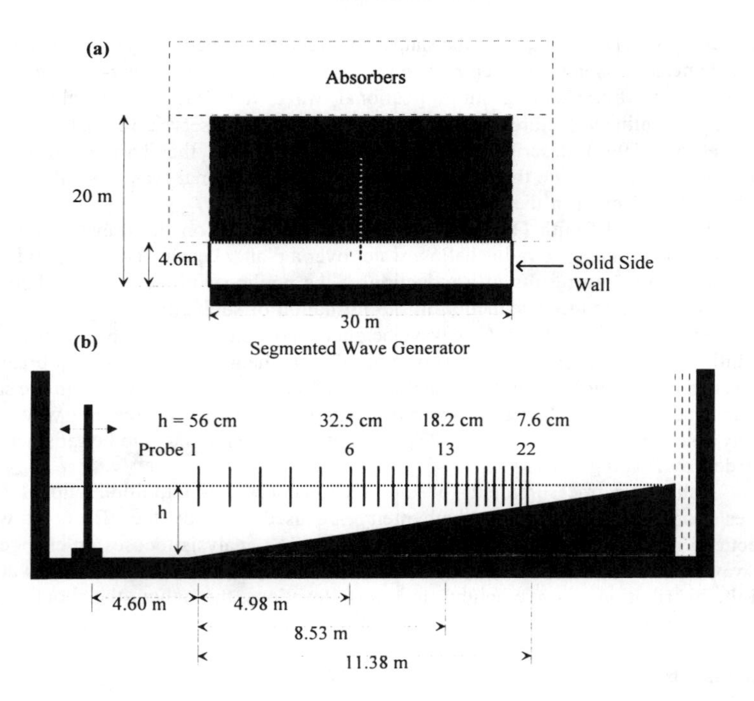

Figure 1. Sketch of the experiment.

at each probe were obtained from a survey of the sloping beach.

Wave Synthesis

Matched pairs of unidirectional and multidirectional wave records, with identical frequency spectra and time-domain characteristics ($\eta_{2D}(t) = \eta_{3D}(t)$), were synthesized using GEDAP wave generation software (Miles, 1989). The numerical synthesis and generation of matched 2D and 3D seas is described by Cornett (1992). Matched simulations enable effects caused by the three-dimensional structure of the wave field to be isolated from those caused by differences in the time-domain character of the waves. Matched simulations ensure that $\eta_{2D}(t) \approx \eta_{3D}(t)$ at one location, in this case, at the toe of the slope along the basin centreline (at probe 1).

Multidirectional waves were synthesized from a parametric directional wave spectrum model using a random phase implementation of the single summation method. The target directional wave spectra $S(f,\theta)$ were specified in terms of a JONSWAP frequency spectrum $S(f)$ paired with a cosine-power spreading function $D(f,\theta)$ according to

$$S(f,\theta) = S(f) \cdot D(f,\theta) \tag{1}$$

with

$$D(f,\theta) = \frac{\Gamma(s(f)+1)}{\sqrt{\pi}\,\Gamma(s(f)+0.5)} \cos^{2s(f)}(\theta - \bar{\theta}(f)) \quad \text{for} \quad |\theta - \bar{\theta}(f)| < \frac{\pi}{2} \tag{2}$$

$$S(f) = \frac{\alpha g^2}{(2\pi)^4 f^5} \exp\left[-\frac{5}{4}\left(\frac{f}{f_p}\right)^{-4}\right] \gamma^a \tag{3}$$

in which

$$a = \exp\left[-\frac{(f-f_p)^2}{2\sigma^2 f_p^{\,2}}\right] \tag{4}$$

$$\sigma = \begin{cases} \sigma_a = 0.07 & \text{for } f \le f_p \\ \sigma_b = 0.09 & \text{for } f > f_p \end{cases} \tag{5}$$

In the above, f = frequency, f_p = peak frequency, g = gravitational acceleration, Γ = the gamma function, θ = direction of wave propagation, $\bar{\theta}$ = mean wave angle, s = spreading index, and α, γ are parameters of the JONSWAP spectrum. The spectrum can be specified using γ, f_p, and an estimate of the significant wave height H_{m0}, defined as

$$H_{m0} = 4\left[\int_0^{\infty} S(f)\,df\right]^{\frac{1}{2}} \tag{6}$$

in which case α is scaled to satisfy equations 3-6. The spreading function (2) is symmetrical about the mean wave angle $\bar{\theta}$, and its width is controlled by the spreading index s, with $s \sim 1$ for a broad spread of wave energy and $s \rightarrow \infty$ for unidirectional waves.

Water surface elevation time series $\eta(t)$ of 819.2 s duration were synthesized using a linear summation of N sinusoidal components, each defined by an amplitude A_n, circular frequency $\omega_n = 2\pi f_n$, propagation direction θ_n, and phase ε_n, according to the single summation model

$$\eta(x,y,t) = \sum_{n=1}^{N} A_n \cos\left\{\omega_n t - k_n\left[(x-x_r)\cos\theta_n + (y-y_r)\sin\theta_n\right] + \varepsilon_n\right\} \tag{7}$$

where the wave number k_n is related to ω_n by the linear dispersion relation $g k_n \tanh(k_n h) = \omega_n^{\,2}$, h = water depth, (x, y) are spatial coordinates, and (x_r, y_r) are coordinates of a reference location in the basin. Amplitudes were chosen explicitly from the frequency spectrum, directions were distributed systematically to approximate the spreading function (or set to $\bar{\theta}$ for unidirectional waves), and phases were selected at random between 0 and 2π.

Table 1. Specifications for 2D and 3D seas.

Name	f_p (Hz)	H_{m0} (cm)	γ	s	$\bar{\theta}$ (deg)
M1	0.80	7.5	3.3	1	0
M2	0.67	10	3.3	1	0
M3	0.40	10	3.3	1	0
I1	0.80	7.5	3.3	∞	0
I2	0.67	10	3.3	∞	0
I3	0.40	10	3.3	∞	0

Three pairs of matched multidirectional and unidirectional wave fields have been selected for the present analysis. The parameters used to synthesize these seas are summarized in Table 1. Note that for these seas, s and $\bar{\theta}$ are constant over all frequencies. Wave field I3 is a unidirectional version of M3, containing the same sinusoidal components, in which all propagation directions have been constrained to be $\bar{\theta}$. I1-M1 and I2-M2 are similarly matched pairs.

Spectral Transformations

As unidirectional irregular waves propagate from deep to shallow water, they are transformed by shoaling, refraction, breaking, energy dissipation, and the growth of lower and higher harmonics. The effects of some of these transformations can be observed in terms of changes to the frequency spectrum in different water depths. Spectra for I3 waves at probes 1, 13 and 22 (h = 56.0, 18.2 and 7.6 cm) are shown in Figure 2. Probe 1 is located at the toe of the slope, probe 13 corresponds approximately to the transition between zones of shoaling and breaking, while probe 22 is well inside the surf zone. The spectrum at probe 1 has a strong fundamental peak near 0.40 Hz, and low secondary peaks near 0.08 Hz and 0.80 Hz, indicating small amplitude, second-order, low and high frequency harmonics. The dramatic growth of non-linear low and high frequency harmonics in shallow water dominate the changes to the spectrum observed at probe 13, where substantial energy at harmonics of the fundamental up to the third harmonic at 1.60 Hz are observed. Wave breaking and energy dissipation are the dominant processes contributing to the transformed spectrum observed at probe 22. The fundamental peak is dramatically reduced, however, more wave energy remains below 0.28 Hz and above 0.95 Hz, than in the original deep-water spectrum. Similar low frequency energy exists at probes 13 and 22, but the long waves are distributed over a broader range of frequencies in the post-breaking domain.

Some differences can be observed in the transformation of shallow water frequency spectra for multidirectional and unidirectional seas, particularly for longer-period seas, for which non-linearities are more important. Figure 3 shows the ratio of the spectra for M3 and I3 waves $S_{M3}(f)/S_{I3}(f)$, observed at probes 1, 13 and 22. At the toe of the slope (probe 1), the 2D and 3D waves have very similar energy content

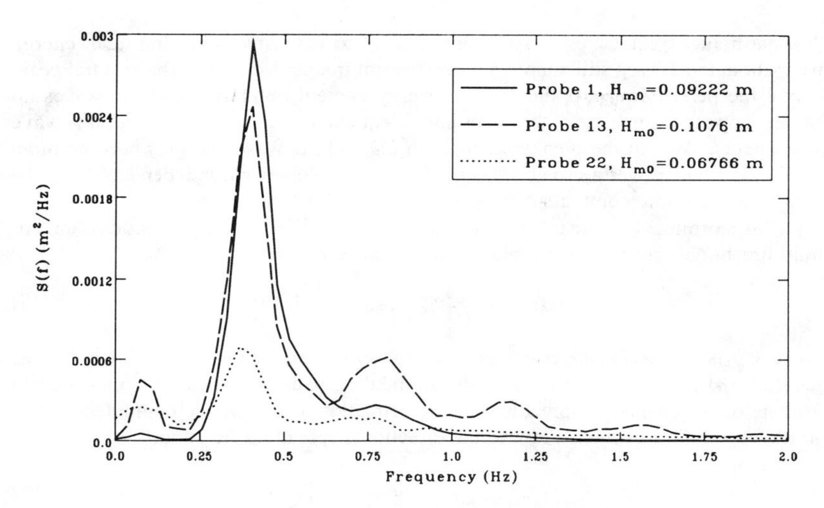

Figure 2. Spectra of I3 waves at probes 1, 13 and 22.

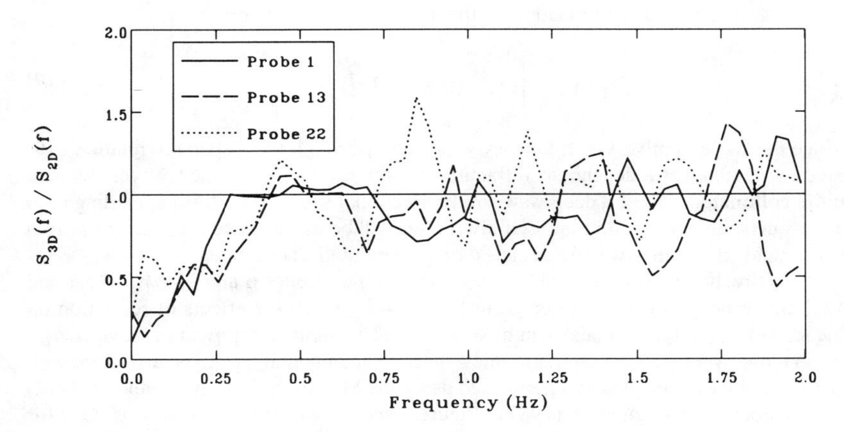

Figure 3. $S_{M3}(f)/S_{I3}(f)$ observed at probes 1, 13 and 22.

over the fundamental frequency band between 0.30 and 0.65 Hz. Before shoaling, the 3D waves support less long wave energy, and less energy near the first (0.80 Hz) and second (1.20) Hz) harmonics of the fundamental peak. At probe 13, near the transition between the shoaling and breaking zones, the reduced long-wave activity in 3D waves continues, and has penetrated upwards in frequency, very near to the spectral peak at 0.40 Hz. The 3D sea continues to support less energy at harmonics of the fundamental peak, moreover, this phenomenon has extended up to the third (1.6 Hz) and fourth (2.0

Hz) harmonic frequencies. At probe 22, after extensive breaking and energy dissipation, the 3D sea still supports less energy at frequencies below the spectral peak. Above the peak frequency, the relative energy content of the 3D and 2D waves no longer appears correlated to harmonic frequencies of the fundamental wave components. Within the breaker zone, both 2D and 3D wave energies become more evenly distributed over a broad range of frequencies, somewhat independent from the growth of harmonic components in evidence after shoaling.

According to linear finite-depth theory (LFDT), the transformation of multidirectional spectra over a planar slope can be estimated by

$$S(f,\theta) = \frac{k\, C_{g0}}{k_0\, C_g} S_0\left\{f, \sin^{-1}\left[\frac{k}{k_0}\sin\theta\right]\right\} \tag{8}$$

where C_g is the group celerity, and the subscript zero denotes initial conditions. Equation (8) includes the effects of linear shoaling and refraction. An approximate estimate of the change in spreading, due to refraction alone, can be computed from a simple application of Snell's law to the directional spreading function:

$$\sigma_\theta = \sin^{-1}\left(\frac{k_0}{k}\sin\sigma_{\theta_0}\right) \tag{9}$$

where σ_θ is the standard deviation of the spreading function, given by

$$\sigma_\theta(f) = \left(\int_{\bar{\theta}-\pi/2}^{\bar{\theta}+\pi/2} D(f,\theta)\,(\theta-\bar{\theta})^2\,d\theta\right)^{1/2} \tag{10}$$

Since the wave number k is frequency dependent (through the dispersion relation), (9) predicts a frequency dependent refraction in which lower frequency waves become more collimated. Thus, a deep water multidirectional sea with uniform spreading over all frequencies, propagating into shallow water, will be transformed into a more narrow wave field in which lower frequencies are preferentially collimated.

Directional spectra of M3 waves, measured at probes 6 and 13 (h = 32.5 and 18.2 cm respectively), are presented in Figure 4, in which the effects of refraction on the spread of the spectral peak can be observed. The nonlinear growth of wave energy in shallow water, at higher harmonics of the fundamental peak, is also apparent. Directional wave analysis was performed using the Maximum Entropy Method (MEM) as described by Nwogu et al. (1987). Figure 5 shows the MEM estimates of $\sigma_\theta(f)$ for the M3 spectra of Figure 4. Between probes 6 and 13, σ_θ near the spectral peak at 0.40 Hz is reduced by approximately 5.4°, from 23.4° to 18°. The reduction in spreading is reasonably stable between 0.30 and 1.25 Hz. Outside this band, directional wave analysis becomes unreliable due to low signal levels. Inspection of $\sigma_\theta(f)$ for M1, M2 and M3 seas, indicates that dips in $\sigma_\theta(f)$ are somewhat correlated to harmonic frequencies of the fundamental spectral peak. This suggests that the high frequency wave components that grow in shallow water are more collimated along the mean wave direction than the background sea state. This is consistent with the triad interaction mechanism for the growth of nonlinear wave components, discussed by Freilich et al. (1990).

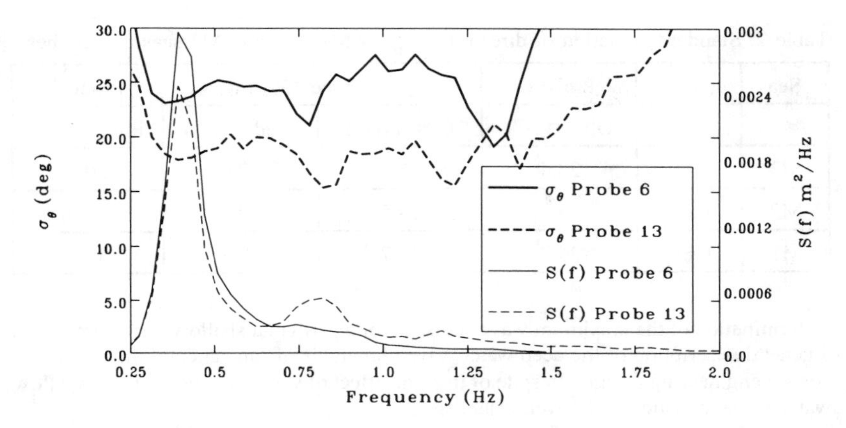

Figure 4. σ_θ(f) observed for M3 waves at probes 6 and 13.

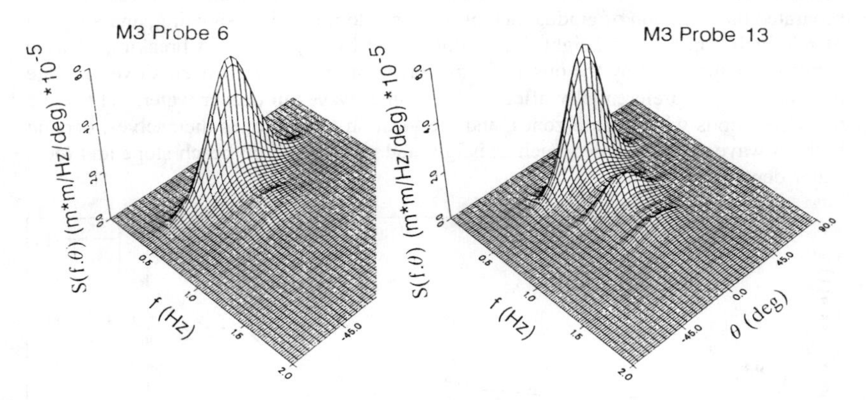

Table 2 compares observed changes in σ_θ at the spectral peak between probes 6 and 13, to predictions using (9). Estimates of σ_θ at probe 1, based on observations at probe 6, are also provided. Unfortunately, directional wave analysis at probe 1 is not possible due to instrument failure, however, consideration of the target spreading function used in wave synthesis (s = 1) provides an estimate of $\sigma_\theta = 32.3°$ at probe 1 for all frequencies. The observed narrowing of the shoaling directional spectra compare favourably to predictions based on Snell's law.

Shoaling Wave Statistics

The performance of coastal structures, such as rubble-mound breakwaters, are often sensitive only to the largest wave heights and crest elevations in a sea state.

Table 2. Standard deviation of direction at f_p for M1, M2 and M3 seas at 3 probes.

Sea	f_p (Hz)	Probe 6	Probe 13		Probe 1
		Obs. σ_θ (°)	Obs. σ_θ (°)	Pred. σ_θ (°)	Pred. σ_θ (°)
M3	0.40	23.4	18.0	17.6	30.5
M2	0.67	25.9	19.5	20.0	32.0
M1	0.80	25.7	20.7	20.3	30.3

Determination of the maximum wave statistics for a particular shallow water site, given a general description of the deep water wave climate, is of fundamental importance to coastal engineering design. In spite of this, the effect of wave directionality on shallow water wave statistics is not well established.

Waves propagating from deep to shallow water over a planar bathymetry experience transformations which first tend to increase wave heights, until breaking commences, at which point heights decay rapidly. The nearshore zone can thus be separated into a region of gradual height increase, dominated by shoaling, and a region of relatively rapid wave height decay, dominated by depth limited breaking. These zones are demarcated by the onset of wave breaking, which itself occurs over a range of water depths, preferentially affecting the higher waves in deeper water. The wave transformations within these zones, and the location of the zones themselves, depend both on wave characteristics, such as height and period, and on beach slope and local water depth.

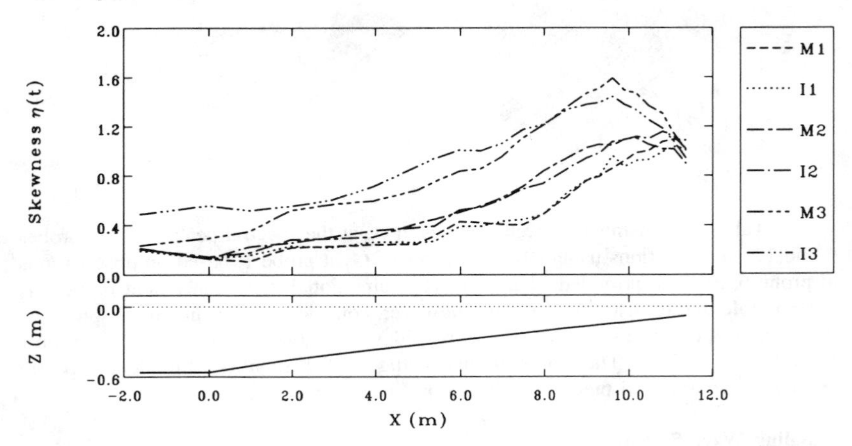

Figure 6. Skewness of $\eta(t)$ over a 1:25 slope for 6 seas.

Nonlinear effects in shallow water distort wave profiles, creating higher, more peaked crests and flatter, broader troughs. This leads to a change in the probability distribution of the surface elevation, from a Gaussian distribution in deep water, to a skewed, non-Gaussian distribution in shallow water. The distortion and asymmetry of waves in shallow water is an important driving mechanism for many coastal processes, such as sediment suspension and transport. A useful measure of the nonlinearity of a wave field, indicating the overall horizontal wave asymmetry, is the skewness of the water surface λ_η, defined by

$$\lambda_\eta = \frac{1}{\sigma_\eta^3}\int_{-\infty}^{\infty}\eta^3 p(\eta)\,d\eta = \sum_{i=1}^{N-1}(\eta_i-\bar{\eta})^3 \Big/ \left[\sum_{i=1}^{N-1}(\eta_i-\bar{\eta})^2\right]^{3/2} \tag{11}$$

where σ_η is the standard deviation of $\eta(t)$, $p(\eta)$ is the probability density function for $\eta(t)$ and $\bar{\eta}$ is the mean value of $\eta(t)$. The skewness of the water surface time derivative $\lambda_{d\eta/dt}$ provides an equally useful measure of the overall vertical asymmetry of a wave field. $\lambda_{d\eta/dt}$ describes the forward tilting of shallow water waves, and has been named 'atiltness' by Goda (1986). Figures 6 and 7 show the evolution of these skewness parameters over the planar bathymetry for the six seas of Table 1. The skewness of $\eta(t)$ increases with decreasing water depth for all seas, rising to a maximum well within the surf zone, indicating the growth of nonlinear wave components and horizontal wave asymmetry. The growth of water surface skewness is strongest for the lower frequency M3 and I3 waves, but does not appear to depend significantly on the directionality of the wave field.

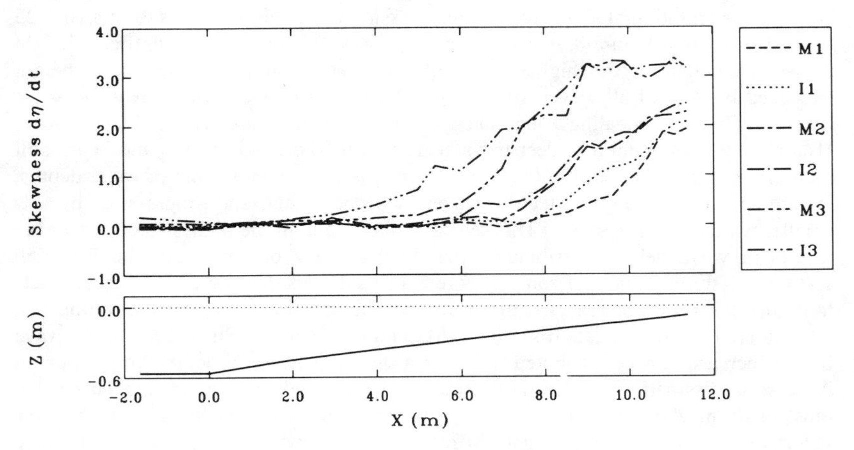

Figure 7. Skewness of $d\eta/dt$ over a 1:25 slope for 6 seas.

Figure 7 indicates that the water surface time derivative remains unskewed until just before the breaker zone, however, during breaking, atiltness increases, reflecting the forward lean of wave crests prior to and during breaking. There is some indication that the growth of vertical asymmetry is delayed to slightly shallower water depths for multidirectional waves. This observation is consistent with the idea that 3D waves are more stable and can propagate into shallower water before leaning forward and breaking. Well within the surf zone, wave directionality does not seem to influence the overall horizontal or vertical wave asymmetry.

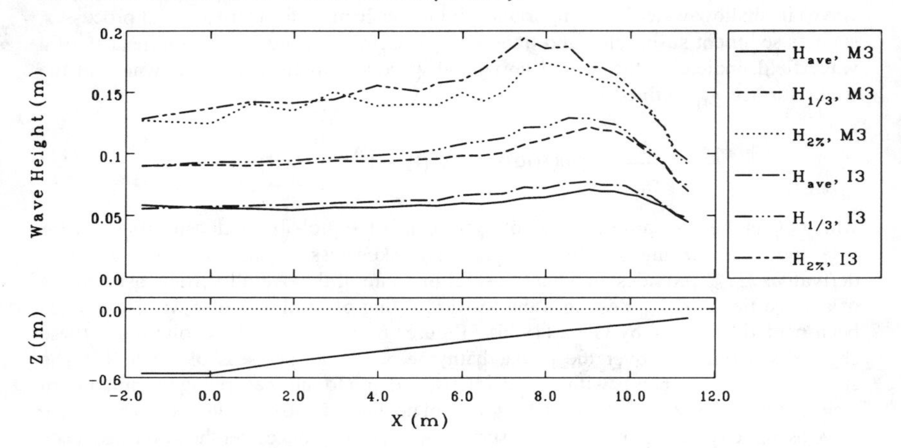

Figure 8. Wave height statistics over the 1:25 slope for I3 and M3 seas.

The evolution of I3 and M3 zero-upcrossing wave heights across the planar 1:25 beach is shown in Figure 8, where $\bar{H}$ = average wave height, $H_{1/3}$ = significant height = average height of the highest third of all waves, and $H_{2\%}$ = threshold height exceeded by 2 % of all waves. This figure demonstrates the usual increase in wave heights within the shoaling zone, and the subsequent decrease within the surf zone. $H_{2\%}$ rises to a maximum in deeper water, further offshore, while $H_{1/3}$ and $\bar{H}$ are still increasing. Thus, the onset of breaking occurs gradually over a band of water depths, beginning with the largest, steepest waves in deeper water, and progressing towards smaller waves closer to shore. This selective dissipation of the highest, steepest waves results in wave height distributions that progressively diverge from the Rayleigh distribution through the surf zone. Figure 8 also suggests that wave directionality acts to reduce the height increases realized through the shoaling zone, an observation that is reinforced by similar data from the I2-M2 and I1-M1 pairs. Such reduction in wave height increase can be attributed to either a decrease in wave height growth, or an increase in dissipation. Refraction of the obliquely incident wave components is the most likely mechanism acting to reduce the growth of multidirectional wave heights. Alternatively, it is possible that diffraction effects within the wave basin have contributed to the reduction in wave height increase observed for 3D waves.

Crest elevations η_c for the M3 and I3 seas are shown in Figure 9. The growth and decay of crest elevations over the planar bathymetry parallels the evolution of wave heights, except that η_c experiences larger changes.

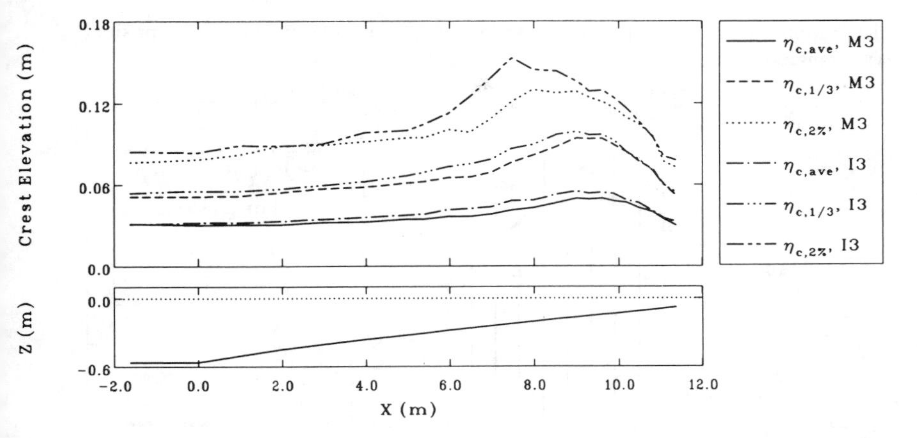

Figure 9. Crest elevation statistics over the 1:25 slope for I3 and M3 seas.

Distribution of Wave Heights and Crest Elevations

The distribution of wave heights and crest elevations of deep water waves are known to closely follow the Rayleigh probability distribution, for which the probability density of heights, p(H), can be expressed in terms of the average zero-crossing height $\bar{H}$, as

$$p(H) = \frac{\pi H}{2\bar{H}^2} \exp\left[-\frac{\pi}{4}\left(\frac{H}{\bar{H}}\right)^2\right] \tag{12}$$

In shallow water, heights and crest elevations are known to deviate from the Rayleigh distribution, particularly after the onset of breaking, when the Rayleigh distribution over-predicts the presence of high waves and crest elevations. Although many authors have proposed alternative distributions for the surf zone (e.g. Dally and Dean (1988), Hameed and Baba (1985) and Hughes and Borgman (1987)) there does not appear to be consensus on a single model that can be easily applied to all conditions. Moreover, Thornton and Guza (1983) conclude, based on analysis of wave data collected at Torrey Pines Beach, that the Rayleigh distribution can be used to give a reasonable description of shallow water waves, even within the surf zone. In the following, observed wave heights and crest elevations will be compared to the Rayleigh distribution to highlight changes through the nearshore zone, and effects due to wave directionality.

At the toe of the slope, wave heights for all six seas are well described by the Rayleigh distribution. For the shorter period I1-M1 and I2-M2 seas, deviations from the Rayleigh distribution become pronounced only well within the surf zone. For the case of longer period I3 and M3 seas, significant deviations from the Rayleigh distribution, particularly at the high wave height tail, were observed over a surprisingly wide range of water depths.

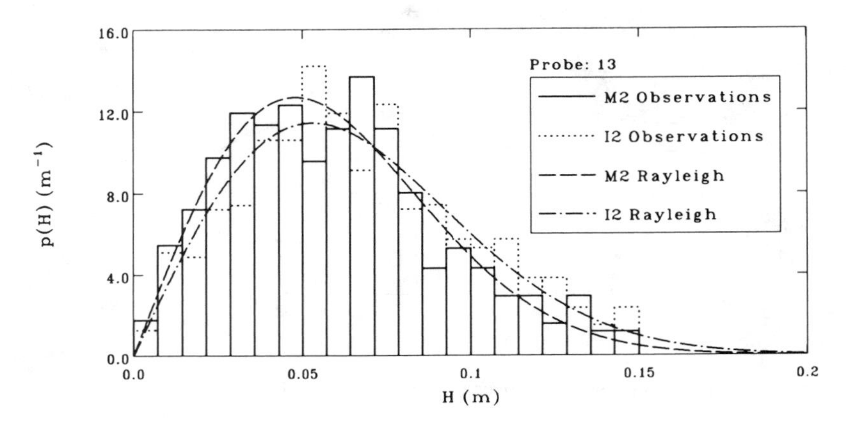

Figure 10. Distributions of wave height observed at probe 13 in I2 and M2 seas.

Wave heights observed in M2 and I2 seas, at probe 13, located near to the transition between shoaling and breaking, are presented in Figure 10, compared to Rayleigh distributions fitted using (11). Both 2D and 3D wave heights agree reasonably well to their respective Rayleigh estimates, however, as discussed previously, the 2D sea supports larger wave heights.

Estimates of wave height statistics based on the Rayleigh distribution, such as the 2 % exceedence threshold $H_{2\%,ray}$, can be expressed as $H_{n,ray} = \alpha_n \bar{H}$, where n and α_n are summarized in Table 3. The accuracy of the Rayleigh estimate of a wave height statistic can be quantified in terms of a relative wave height $H_{n,rel}$, given by

$$H_{n,rel} = \frac{H_n}{H_{n,ray}} = \frac{H_n}{\alpha_n \bar{H}} \tag{13}$$

Table 3. Rayleigh statistics.

n	$\alpha_n = H_{n,ray}/\bar{H}$
1/3	1.598
10 %	1.712
5 %	1.953
2 %	2.232
1 %	2.421

The evolution of three relative wave height statistics in both M2 and I2 waves are plotted in Figure 11 as a function of location on the slope. This Figure shows that the Rayleigh distribution can be used to accurately estimate extreme wave heights outside the breaker zone for these seas, but within the breaker zone, Rayleigh estimates over-predict $H_{1\%}$ by up to 30 %. Preferential decay of the largest waves at the beginning of the breaker zone (near X = 8.5 m) is also exhibited. Relative wave height statistics for the lower frequency M3 and I3 seas (not shown) indicate that Rayleigh estimates of extreme wave

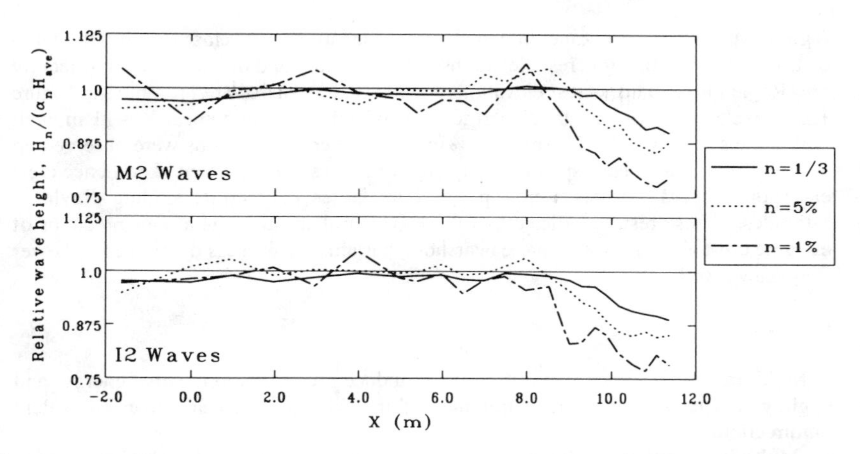

Figure 11. Relative wave height statistics over the 1:25 slope for M2 and I2 waves.

heights are fairly accurate up to the $H_{5\%}$ level, both within the shoaling and surf zones. Rayleigh estimates at the $H_{2\%}$ and $H_{1\%}$ level are much less precise, leading to under-prediction of observed heights in the shoaling zone, and over-prediction in the surf zone. In general, the effects of wave directionality are small in comparison to the overall changes in wave height distribution throughout the nearshore zone.

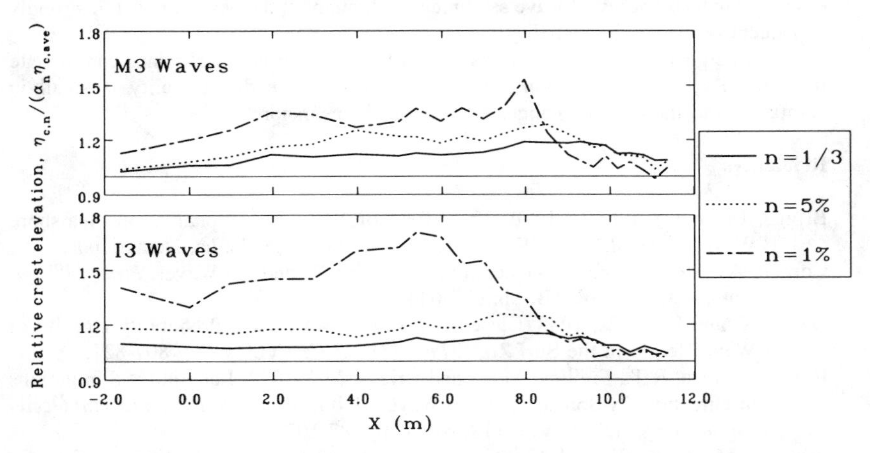

Figure 12. Relative crest elevation statistics over the 1:25 slope for M3 and I3 seas.

Rayleigh estimates of large crest elevations under-predict observed values within the shoaling zone for all six seas, and often over-predict observed values well into the breaker zone. The data suggests that for each wave condition, there is a transition region within the surf zone in which crest elevations are close to the Rayleigh distribution, even though crest elevations both nearshore and offshore are significantly non-Rayleigh. Relative crest elevations for M3 and I3 waves are presented in Figure 12. These crest elevations are closest to the Rayleigh distribution near X = 11 m, well within the surf zone. For I3 waves, 1 % exceedence crest elevations were observed up to 70 % greater than corresponding Rayleigh estimates. One percent exceedence crest elevations for M3 waves were up to 50 % larger than corresponding Rayleigh estimates. These results indicate that the Rayleigh distribution is a poor predictor of extreme crest elevations within the nearshore, particularly for seas dominated by lower frequency swell.

Conclusions

- Multidirectional waves in shoaling water induce much less long-wave energy, and slightly less energy at higher harmonics of the fundamental peak, than equivalent unidirectional waves.
- Multidirectional wave heights and crest elevations exhibit slightly less growth through the shoaling zone than equivalent unidirectional wave heights and crest elevations.
- The onset of wave height decay due to breaking is slightly delayed for multidirectional waves.
- Snell's law can be used to estimate the change in directional spread of shoaling multidirectional waves.
- The growth and decay of wave asymmetries throughout the nearshore is not strongly dependent on wave directionality.
- Wave heights and crest elevations within the shoaling and surf zones can deviate from the Rayleigh distribution. The influence of wave directionality is small in comparison to the overall changes throughout the nearshore.

References

Briggs, M. and Smith. J., 1990. The Effect of Wave Directionality on Nearshore Waves. *Proc. 22nd ICCE*. Vol. 1, pp. 267-280. The Hague, Netherlands.

Cornett, A. 1992. Laboratory Generation of Similar 2D and 3D Waves. *Proc. 11th Int. Conf. OMAE*. Vol. 1B, pp. 637-644.

Dally, W. and Dean, R., 1988. Closed-Form Solutions for the Probability Density of Wave Height in the Surf Zone. *Proc. 21st ICCE*. Vol. 1, pp. 807-821.

Elgar, S., Guza, R.T., Freilich, M.H., and Briggs, M.J. 1992. Laboratory Simulations of Directionally Spread Shoaling Waves. *J. Waterway, Port, Coastal and Ocean Engineering, ASCE*. Vol. 118, No. 1, pp. 87-103.

Freilich, M., Guza, R. and Elgar, S., 1990. Observations of Nonlinear Effects in Directional Spectra of Shoaling Gravity Waves. *J. Geophysical Research*. Vol. 95, No. C6, pp. 9645-9656.

Goda, Y. 1986. Effect of Wave Tilting on Zero-Crossing Wave Heights and Periods. *Coastal Engineering in Japan*. Vol. 29, pp. 79-90.

Halliwell, A. and Machen, P., 1981. Short-Crested Waves in Water of Limited Depth. *Proc. Instn. Civ. Engrs*. Part 2, No. 71, pp. 663-674.

Hameed, T. and Baba, M., 1985. Wave Height Distribution in Shallow Water. *Ocean Engineering*. Vol. 12, No. 4, pp. 309-319.

Hughes, S. and Borgman, L., 1987. Beta-Rayleigh Distribution for Shallow Water Wave Heights. *Proc. ASCE Spec. Conf. on Coastal Hydrodynamics*. pp. 17-31.

Jamieson, W. and Mansard, E., 1987. An Efficient Upright Wave Absorber. *Proc. ASCE Spec. Conf. on Coastal Hydrodynamics*. pp. 124-139.

Miles, M., Laurich, P. and Funke, E., 1986. A Multi-Mode Segmented Wave Generator for the NRC Hydraulics Laboratory. *Proc. 21st American Towing Tank Conference*. Washington, U.S.A.

Miles, M. 1989. User Guide for GEDAP Version 2.0 Wave Generation Software. *NRC Technical Report LM-HY-034*. Ottawa, Canada.

Nwogu, O. 1993. Second-Order Directional Wave Interactions in Shallow Water. submitted to *Applied Ocean Research*.

Nwogu, O., Mansard, E., Miles, M. and Isaacson, M. 1987. Estimation of Directional Wave Spectra by the Maximum Entropy Method. *Proc. IAHR Seminar on Wave Analysis and Generation in Laboratory Basins*. pp. 209-235.

Thornton, E., and Guza, R., 1983. Transformation of Wave Height Distribution. *J. Geophysical Research*. Vol. 88, No. C10, pp. 5925-5938.

Experimental Study of Monochromatic Wave-Ebb Current Interaction

Michael J. Briggs[1], Member and Philip L.-F. Liu[2], Fellow

Abstract

Laboratory experiments were conducted in a 26-m by 36-m basin with a 1:30 plane beach to study the interaction of monochromatic waves and an ebb current in and near a shallow entrance channel. Two entrance channel configurations were tested: one with a 1:1 side slope and the other with vertical sides. Data were collected for three current-only, four wave-only, and twelve wave-current cases for each entrance channel configuration. Good agreement was obtained between these laboratory data and numerical model predictions. The effect of the ebb current on waves is to increase the wave height as much as 60 percent and to increase higher harmonics in the spectra.

Introduction

The coastal zone involves interactions between winds, waves, currents, structures and sediment. To develop a sound coastal management plan for shoreline stabilization and protection near inlets, it is essential to have a better understanding of the complicated physics which occur between waves and currents in coastal waters. In the vicinity of tidal inlets and river mouths, currents can significantly modify wave amplitudes, form, and directions. Wave steepness decreases when the current flows in the same direction as the waves and

[1]Research Hydraulic Engineer, USAE Waterways Experiment Station, 3909 Halls Ferry Rd., Vicksburg, MS 39180-6199, 601/634-2005

[2]Professor, School of Civil and Environmental Engineering, Cornell University, Ithaca, NY 14853, 607/255-5090

increases when the current is opposed to the wave travel. If the waves are large relative to the currents, they can change the characteristics of the current field.

Numerical models in the form of mild-slope equations for monochromatic waves propagating over a slowly varying depth and current field have been developed using a parabolic approximation. Recent advances have included Boussinesq equations for long waves interacting with a spatially-varying current field over a varying depth. These numerical methods incorporate nonlinearities, bottom friction, and the effect of waves on currents. However, the study of wave-current interaction is far from complete as laboratory or field study of three-dimensional wave-current interaction to verify these theories is very rare.

Sakai and Saeki (1984) measured the effect of opposing currents on wave height transformation over a 1:30 sloping beach for a range of wave periods and steepness. They found an increase in wave height and decay rate in the presence of the opposing current. Willis (1988) conducted monochromatic wave and ebb current tests in a 1-m wide rectangular entrance channel cut in a 1 on 30 sloping beach. Current measurements were averaged over three minutes to obtain a quantitative picture of the mean currents. They experienced major problems with the stability of the current field due to large scale meandering motions.

Lai et al. (1989) conducted flume tests of kinematics of wave-current interactions for strong interactions with waves propagating with and against the current. They found the influence of the waves on the mean current profiles was small, although opposing waves would give a slightly lower current. They observed a drastic change in the spectral shape, especially higher harmonics, following wave breaking in the presence of opposing currents. Their experiments confirmed that blockage of waves by a current when the ratio of depth-averaged current velocity to wave celerity without currents approaches -0.25.

The directional spectral wave basin at the U.S. Army Engineer Waterways Experiment Station's Coastal Engineering Research Center (CERC) was used to conduct wave-current interaction tests of monochromatic waves in shallow water. Steady ebb currents were generated through an entrance channel cut in a 1 on 30 sloping beach. Two entrance channel configurations were tested: one with a 1 on 1 slope and the other with vertical sides. Small amplitude waves with different frequencies and angles of incidence were generated. Measurements of current velocity and wave amplitude were made, with and without wave-current interaction, using capacitance wave staffs and triaxial, ultrasonic current meters at 42 locations. Dye visualization and video recordings were used. Data were collected for three current-only, four wave-only, and twelve wave-current cases for each entrance channel configuration. Prior to collecting data, a calibration phase was conducted to verify the wave and current conditions.

These laboratory investigations provide a high quality data set which will be used to gain better insight into the complicated nonlinear dynamics of wave-current interactions, provide guidance, and verify theories and numerical models. In this paper, the experimental set-up including basin layout, wave and current conditions, and instrumentation is described.

Next, a brief description of the numerical model is given. Finally, laboratory and numerical results for wave height transformation and spectral evolution are presented and discussed.

Experimental Design

Model set-up

A three-dimensional, physical model of a tidal inlet was constructed in CERC's 26 m by 36 m directional spectral wave basin (Fig. 1). The fixed-bed model included a 12.4-m-long flat section, a 1 on 30 beach (i.e. $\theta = 1.9°$) with plane parallel contours, and an unprotected entrance channel. Water depth was 36 cm in the deep, flat portion of the model, decreasing to 10 cm in the 1 m wide, 5.5 m long entrance channel. All depths are relative to the still water level (SWL). Two entrance channel sidewall configurations were tested: one with a 1 on 1 slope and the other with vertical sides. Both were designed to have equal cross-sectional areas at the waterline.

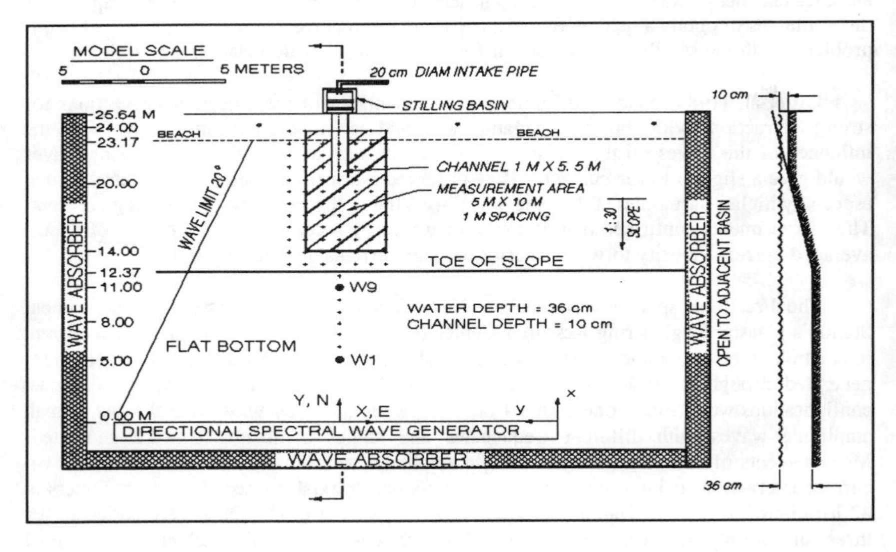

Fig. 1 Basin layout

The X-axis (X) or East direction (E) of the right-hand, global coordinate system is parallel to the DSWG and the Y-axis (Y) or North direction (N) is coincident with the entrance channel centerline. A local coordinate system (x-axis and y-axis) has its origin at the eastern end of the DSWG (see Fig. 1). It is rotated 270° clockwise from the global system.

The rear of the basin (i.e. behind the DSWG) was lined with wave absorber backed by a concrete wall. The basin sides were lined with absorber and one side was open to the adjacent basin. Thus, wave energy was able to propagate away from the test area into the adjacent larger basin with minimal reflections from distant vertical walls and associated basin cross-seiching.

Wavemaker

Waves were generated with the directional spectral wave generator (DSWG), an electronically-controlled, electromechanical system, described in detail by Outlaw (1984) and Briggs and Hampton (1987). It is 27.43 m long, and consists of 60 paddles in for portable modules, each 46 cm wide and 76 cm tall. The paddles are individually driven at each of the 61 joints in translational motion (i.e. piston mode) by electric motors using the "snake" principle. Flexible plastic seals slide in guides between each paddle to provide a smooth, continuous wave front.

Current-generating system

A 22 KW pump, with a capability of 0.16 m^3/s, was used to generate ebb currents. Water was pumped from an intake manifold in an adjacent basin, approximately 55 m away, to the stilling basin upstream of the inlet (see Fig. 1). This intake manifold was approximately 30 m long and aligned parallel to the entrance channel. It was constructed from 20 cm diameter PVC piping, which terminated in multiport diffusers with 2.5 cm diameter holes spaced uniformly at 30 to 61 cm intervals. The flow was controlled by an orifice plate and a manometer.

The stilling basin had two baffled compartments to minimize turbulence and straighten the flow. The flow came in over the top through the intake pipe into the bottom of the first compartment. It then flowed under the first baffle into the second compartment before exiting out the top into the inlet. The water level was designed to be higher in the larger, first compartment.

Instrumentation

Wave gages. Twenty-three capacitance wave gages were used to measure surface elevations in a 5 m by 6 m measurement area adjacent to and seaward of the inlet. This measurement area was comprised of six cross-shore transects, labeled 2 through 7 (X=-2 to 3), and seven rows oriented longshore, labeled 1 through 7 (Y=14 to 20, h=30.6 to 10.6 cm). Transect 4 was aligned with the entrance channel centerline (X=0). A uniform spacing of 1 m was selected between transects and rows. A wave gage was located at each of the transect-row intersections in this grid for a total of forty-two locations. Two gage layouts (A and B for entrance channel 1 and H and I for entrance channel 2) were required to cover the entire measurement area because of the number of available wave gages.

Wave gages 1 and 9 were located along the entrance channel centerline at Y=5 m and Y=11 m, respectively, to record incident wave conditions for the four wave only cases. The measured wave heights from these two gages were averaged from both entrance channel configurations.

Current meters. Two triaxial, ultrasonic current meters were used to calibrate the ebb currents and to quantify the effect of the waves on the currents. These current meters are manufactured by Sensordata A/S, Bergen, Norway, as the Minilab SD-12 system. The system consists of a display unit, an instrument cable, a probe housing, and the 3-axis probe.

Wave and current conditions

Four small amplitude wave trains were generated using the DSWG. These waves had wave periods of 0.75 and 1.50 sec, directions of 0° and 20° (relative to normal to the DSWG), and wave heights of 5.0 cm. The wave height was selected based on the maximum wave which, theoretically, would not break at the shallowest depth of 10.6 cm. Steady ebb currents of 10, 20 and 30 cm/sec were generated. Table 1 lists the wave and current parameters for the nineteen combinations of wave-only, current-only, and wave-current interaction which were tested for each entrance channel configuration.

Table 1
Test Program Summary

Case Name	Period sec	Height cm	Dir deg	Cur. Flow cm/sec
Wave Only				
10	0.75	5	0	0
20	0.75	5	20	0
30	1.50	5	0	0
40	1.50	5	20	0
Current Only				
01	-	-	-	10
02	-	-	-	20
03	-	-	-	30
Wave and Current				
11	0.75	5	0	10
21	0.75	5	20	10
31	1.50	5	0	10
41	1.50	5	20	10
12	0.75	5	0	20
22	0.75	5	20	20
32	1.50	5	0	20
42	1.50	5	20	20
13	0.75	5	0	30
23	0.75	5	20	30
33	1.50	5	0	30
43	1.50	5	20	30

Wave calibration. In addition to the two wave gages 1 and 9, a linear array of seven wave gages (gages 2 to 8) was used to calibrate the four wave conditions. This linear array was patterned after the Field Research facility linear array (Oltman-Shay 1987) and had lag spacings of 2-3-1-7-5-1/2. The unit lag was equal to 30 cm. It was aligned parallel to and 8 m from the DSWG in the flat section of the model.

Control signal durations of 750 sec were created for each of the wave conditions. Data were collected for 300 sec at a sampling rate of 10 Hz after a waiting time of 300 sec. Table 2 compares measured (zero-downcrossing) to target wave parameters for each wave case. The percent deviation between target and measured values is also given. Agreement is very good for all wave parameters, with wave period matching exactly and wave direction within 1.5 percent of the target values. Wave heights can vary spatially as much as 25 percent for basin measurements (Sand 1979). The maximum variation for these waves varied between 4 and 10 percent, with the larger value for case 40.

Table 2
Calibrated Wave Parameters

Test Case	Wave Period, sec			Wave Height, cm			Wave Direction, deg		
	Target	Measured	% Dev	Target	Measured	% Dev	Target	Measured	% Dev
10	0.75	0.75	0.00	5.00	5.03	0.60	0.00	0.00	0.00
20	0.75	0.75	0.00	5.00	5.06	1.20	20.00	19.82	0.90
30	1.50	1.50	0.00	5.00	5.24	4.80	0.00	0.00	0.00
40	1.50	1.50	0.00	5.00	5.09	1.80	20.00	19.73	1.35

Waves were also checked to insure their stability with time. Comparisons of wave period and wave height for the first 75 sec, first 150 sec, and total duration of 300 sec indicated very little deviation in these parameters.

Current calibration. Currents were calibrated at two locations in the entrance channel, corresponding to Y=22 and 24 m and the centerline. Flow passing the meter at Y=24 m was contained by the channel side walls. Beyond the intersection of the 1 on 30 slope with the 10-cm water level, the flow was no longer contained by the entrance channel. Therefore, calibration of the flow was based on the meter closest to the stilling basin at Y=24 m. The top of the "U-shaped" support frame of the current meters was positioned 5 cm below the surface. To minimize potential interference from this support frame, the closed end of the "U" was positioned downstream of the current flow toward the DSWG.

Current data were also collected for 300 sec at a sampling rate of 10 Hz after a waiting time of 300 sec. The current meter u-, v-, and w-velocity channels were converted to the global coordinate system. Current magnitude was calculated from all three channels. Current direction was calculated from the u- and v-velocity channels, measured clockwise from north. Current flows from the direction indicated. Table 3 lists 5-minute resultant current magnitudes and directions for the three current cases for the entrance channel 1 configuration.

Table 3
Calibrated Current Parameters

Test Case	Speed, cm/s		Dir, deg	
	Target	Measured	Target	Measured
01	10	11	0	-1
02	20	20	0	0
03	30	28	0	0

Numerical Model

The mild-slope equation for a monochromatic wave propagating over a slowly varying topography and current field is given as

$$\nabla \cdot (C\,C_g\,\nabla\phi) - \vec{U}\cdot\nabla(\vec{U}\cdot\nabla\phi) + [2i\omega - \nabla\cdot\vec{U}]\,\vec{U}\cdot\nabla\phi + (\omega^2 - \sigma^2 + k^2 CC_g + i\omega\nabla\cdot\vec{U})\,\phi = 0 \qquad (1)$$

where

$$\Phi_w = -\frac{ig\cosh k(z+h)}{\cosh kh}\,\phi(x,y)\,e^{-i\omega t}$$

is the wave velocity potential. In the above equation, ∇ represents the two-dimensional gradient, i.e. $\nabla = (\partial/\partial x, \partial/\partial y)$. The wave number k satisfies the following dispersion relation.

$$\sigma^2 = gk\tanh kh, \qquad \sigma = \omega - \vec{k}\cdot\vec{U} \qquad ($$

and

$$CC_g = \frac{g}{2k}\tanh kh\left(1 + \frac{2kh}{\sinh 2kh}\right) \qquad ($$

We consider the current field only having an on-offshore component i.e. $\bar{U} = (U,0)$. Then eq. (1) can be rewritten as

$$(CC_g - U^2)\frac{\partial^2\phi}{\partial x^2} + \left[\frac{\partial(CC_g)}{\partial x} - 2U\frac{\partial U}{\partial x} + 2i\omega U\right]\frac{\partial\phi}{\partial x} + CC_g\frac{\partial^2\phi}{\partial y^2} + (\omega^2 - \sigma^2 + k^2 CC_g + i\omega\frac{\partial U}{\partial x})\,\phi + \frac{\partial(CC_g)}{\partial y}\frac{\partial\phi}{\partial y} = 0 \qquad ($$

If we assume that the waves also propagate primarily in on-offshore direction (i.e. x-direction), ϕ can be rewritten as

$$\phi = R(x,y)\,e^{i\int\bar{k}dx} \qquad ($$

where $\bar{k}(x)$ is the average value of k(x,y) in the y-direction. R(x,y) varies slowly in both x- and y-directions because of the assumption of slowly varying topography and current. To apply the parabolic approximation, we assume

$$\frac{\partial R}{\partial x} \sim O(|\nabla h|), \quad \frac{\partial^2 R}{\partial x^2} \sim O(|\nabla h|^2), \quad \frac{\partial^2 R}{\partial y^2} \sim O(|\nabla h|) \tag{6}$$

Substituting eq. (5) into eq. (4) and using the above assumption, we obtain

$$A\frac{\partial R}{\partial x} + BR + CC_g\frac{\partial^2 R}{\partial y^2} + \frac{\partial (CC_g)}{\partial y}\frac{\partial R}{\partial y} = 0 \tag{7}$$

here

$$A = 2i\,[\bar{k}(CC_g - U^2) + \omega U]$$

$$B = i\frac{\partial}{\partial x}[\bar{k}(CC_g - U^2)] + (CC_g - U^2)(k^2 - \bar{k}^2) + 2\omega U(k - \bar{k}) + i\omega\frac{\partial U}{\partial x}$$

q. (7) can then be rewritten as

$$\frac{\partial R}{\partial x} = -\frac{BR + CC_g\dfrac{\partial^2 R}{\partial y^2} + \dfrac{\partial (CC_g)}{\partial y}\dfrac{\partial R}{\partial y}}{A} \tag{8}$$

hen U = 0, eq. (8) is reduced to the small-angle parabolic approximation model for onochromatic waves propagating on a slowly varying bottom.

To solve eq. (8), either a Crank-Nicolson scheme or Chebyshev pseudo-spectral method ill work well. The relation between the wave amplitude and ϕ is given as

$$\eta = \omega\phi + iU\frac{\partial\phi}{\partial x} = \left[(\omega - \bar{k}U)R + iU\frac{\partial R}{\partial x}\right]e^{i\int \bar{k}dx} \tag{9}$$

the numerical computation, we replace k_x by k and the dispersion relation can be proximated as

$$\omega^2 = (\sigma + \bar{U}\cdot\bar{k})^2 = \sigma + Uk_x = \sigma + Uk \tag{10}$$

here

$$\sigma^2 = (\omega - Uk)^2 = gk\tanh kh$$

Results and Analysis

In this section, wave height transformation and spectral evolution for numerical and physical model data are presented and compared. Because of space limitations in this paper, a representative case is used as an example. Wave and current parameters for this case are T=1.50 s, H=5 cm, θ=0°, and U=10 cm/s. Assuming a model scale of 1 to 25, this would correspond to a prototype wave and current of T=7.5 s, H=1.25 m, and U=1 kt.

Wave height transformation

Current profile. The 10 cm/s cross-shore velocity profile assumed in the numerical model is shown in Fig. 2. Longshore sections at four water depths ranging from 12.7 to 32.7 cm seaward of the channel entrance are given. The centerline of the entrance channel is parallel to X=0 and the sides are at X=$\pm$0.5. The velocity profiles are more peaked closer to the mouth of the channel and broaden further offshore. The profile corresponding to the physical model measurements inside the channel in 10 cm water depth would be steeper with a peak at 10 cm/s and a width of $\pm$50 cm. The cross-shore velocity profiles for the 20 cm/s and 30 cm/s ebb currents (not shown) have similar shapes but are more peaked with larger magnitudes.

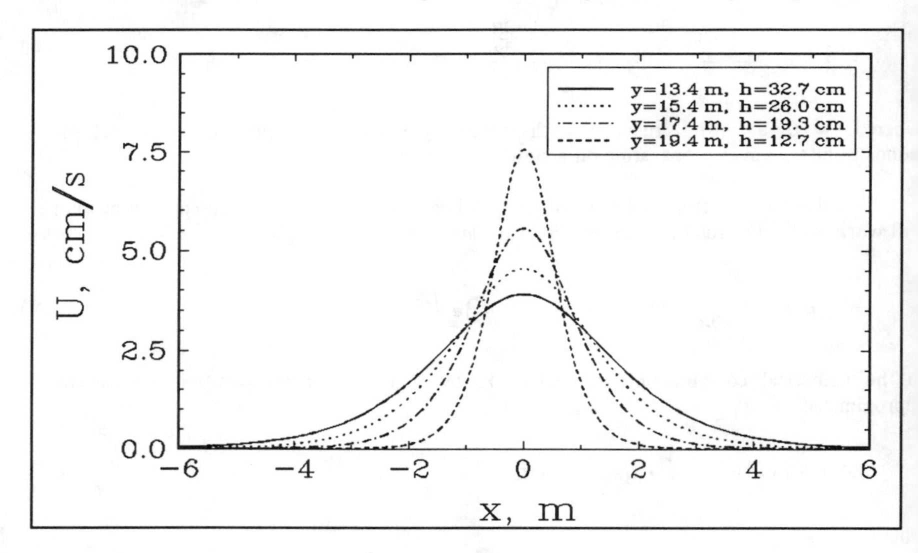

Fig. 2 Numerical model cross-shore ebb current profiles

Wave amplitude comparisons. Fig. 3 compares normalized wave amplitude A/A_0 for wave only and wave plus 10 cm/s ebb current conditions for the numerical and physical models. For the physical model, normalized amplitude was obtained by dividing the measured wave heights $\bar{H}_d$ along the channel centerline (X=0) by incident wave height H_0 from gages 1 and 9 for each case.

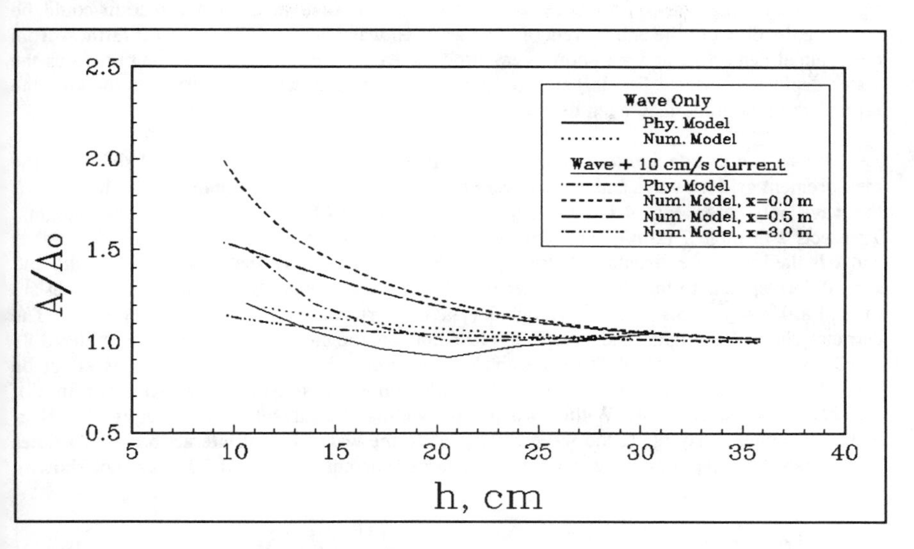

Fig. 3 Numerical and physical model normalized wave amplitude

The effects of wave refraction and shoaling on wave transformation are seen by examining the two wave-only curves. Agreement is very good between numerical and physical models with wave height increases of 15 percent from the point farthest offshore (h=30.6 cm) to the point nearest to where the channel daylights (h=10.6 cm). The wave height increase is slightly less for entrance channel 2 (not shown).

The effect of the ebb current on the wave height is shown in the three curves for the numerical model at x=0, 0.5, and 3 m and the physical model at x=0. For the physical model, wave height increases by 42 percent between the same two locations. For ebb currents of 30 cm/s (not shown), wave height increased by 71 percent to a value of A/A_0=1.9. Measured wave heights were slightly larger for entrance channel 2. These values are consistent with earlier laboratory measurements by Briggs, et al. (1993) for a 4 cm/s ebb current in a 27-cm-deep, 4.1-m-wide, sloped channel.

The numerical model predictions were somewhat higher than the physical model measurements. Along the centerline, the normalized amplitude approaches a value of 1.9, 25 percent greater than the measured 1.5 value. However, the predicted amplitude for the cross-shore transect at X=0.5, corresponding to the projection of the channel sidewalls, is very close to the measured value. Thus, a small variation in the positioning of the laboratory gages off of the centerline could have had a significant effect on the measured wave heights. Another possible explanation for this discrepancy is that the physical model indicated that the currents meandered in direction after leaving the entrance channel. Since the numerical model only considers currents with on-offshore components, this change in current direction could be reducing the effect of the ebb current on the measured wave heights. The transect farthest from the channel centerline at x=3 shows very little increase in normalized wave amplitude as the water depth decreases. For larger ebb currents and smaller wave periods (not shown), the normalized amplitude may even decrease.

Wave amplitude contour plot. For the physical model data, wave heights in the measurement grid for wave and current were normalized by the corresponding wave heights for the wave only case. Fig. 4 is a three-dimensional surface of these normalized wave heights. Transects 2 through 7 (x-axis) are represented by k_0x, where k_0 is the deepwater wavenumber and x is the longshore distance. Increments of $k_0x=2.5$ are equivalent to a channel width and $k_0x=0$ corresponds to the channel centerline. The projected sides of the entrance channel are located at $k_0x=\pm1.25$. Rows 1 through 7 (y-axis) were converted to relative depth kh. The entrance channel daylights at kh=0.45 (h=10.6 cm) and the measurement grid ends at kh=0.81 (h=30.6 cm). As the strength of the ebb current diminishes flowing seaward, its effect on wave height is reduced. Wave height amplification on either side of the projected confines of the channel is also reduced. Within two channel widths, the current effect is negligible. Near where the channel daylights, the wave height due to the wave-current interaction is 1.24 times larger than the wave only condition. For the larger current velocity of 30 cm/s (not shown), this ratio is as large as 1.50.

Spectral evolution

As waves shoal and break, higher harmonics are formed as the wave becomes more nonlinear. Fig. 5 shows semi-log plots of the measured frequency spectra along the centerline for the wave only condition at water depths of h=30.6 and 10.6 cm. As the wave shoals, first (f=1.33 Hz), second (f=2.00 Hz), and third (f=2.67 Hz) harmonics grow as energy is transferred from the fundamental mode at $f_p=0.67$ Hz due to nonlinear coupling between frequencies. The number above each harmonic peak represents the ratio of the spectral ordinate in shallow water relative to the deeper gage location. Although there is little change in the fundamental mode (i.e. 1), the higher harmonics show appreciable increases. The first harmonic increases 19 times and the second and third harmonics increase 1273 and 928 times, respectively. Although initially there is significantly less energy in these modes than in the fundamental, the increase in energy is considerable after shoaling and contributes to the nonlinear growth of the wave. Also, energy appears to have increased uniformly in all subharmonics as well. This result is consistent with wave transformation results on a plane beach (Briggs and Smith 1990) which indicated that directional spreading tends to reduce the growth of subharmonics as waves shoal. Since these waves are characterized by very narrow directional spreading, the subharmonics grow as the wave shoals.

Fig. 6 shows the effect of the 10 cm/s ebb current on this nonlinear wave transformation for the shallowest water depth of h=10.6 cm. The wave only curve is repeated for convenience. The effect of the ebb current on spectral evolution is most noticeable in the second and third harmonics, although not as pronounced as that due to shoaling alone. The ebb

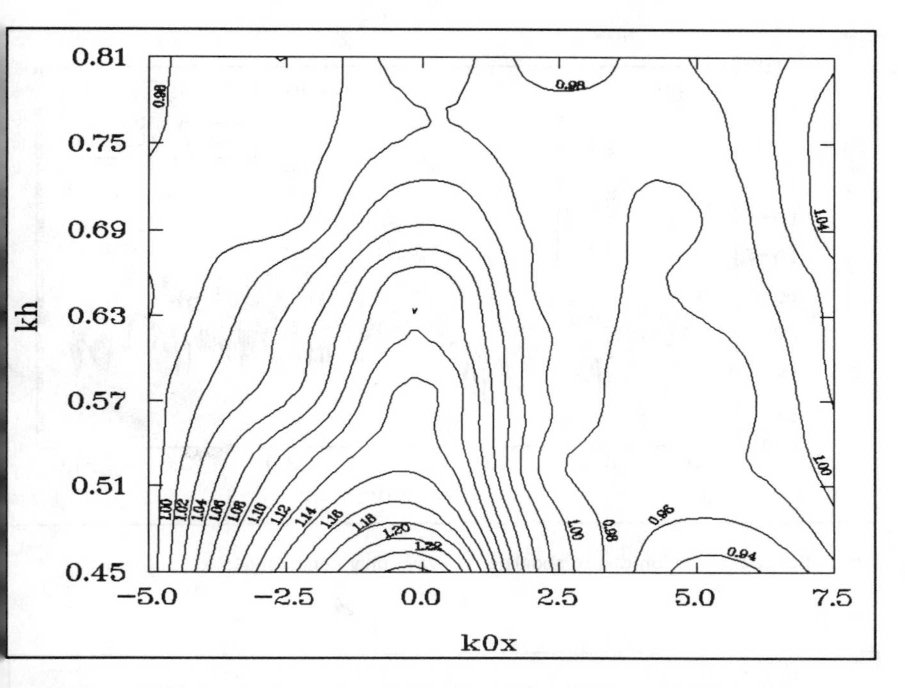

Fig. 4 Three-dimensional surface of normalized wave heights for 10 cm/s ebb current

current of U=30 cm/s (not shown) produces spectral peak amplifications of 2 to 8 in the fundamental and higher harmonics. Also, for some cases there were increases of two orders of magnitude in the spectral floors between peaks due to the ebb current. Thus, the presence of an ebb current promotes the nonlinear growth of the fundamental, higher harmonics, and subharmonics as the wave shoals.

Summary and Conclusions

In the vicinity of tidal inlets and river mouths, currents can significantly modify wave amplitude, form, and direction. Laboratory experiments were conducted in a 26-m by 36-m basin to study the interaction of ebb currents flowing through an idealized entrance channel with monochromatic waves in the nearshore region where the channel daylights on a 1:30 plane beach. Two entrance channel configurations were tested: one with a 1:1 side slope and the other with vertical sides. Data were collected for three current-only, four wave-only, and twelve wave-current cases for each entrance channel configuration. Reasonable agreement was

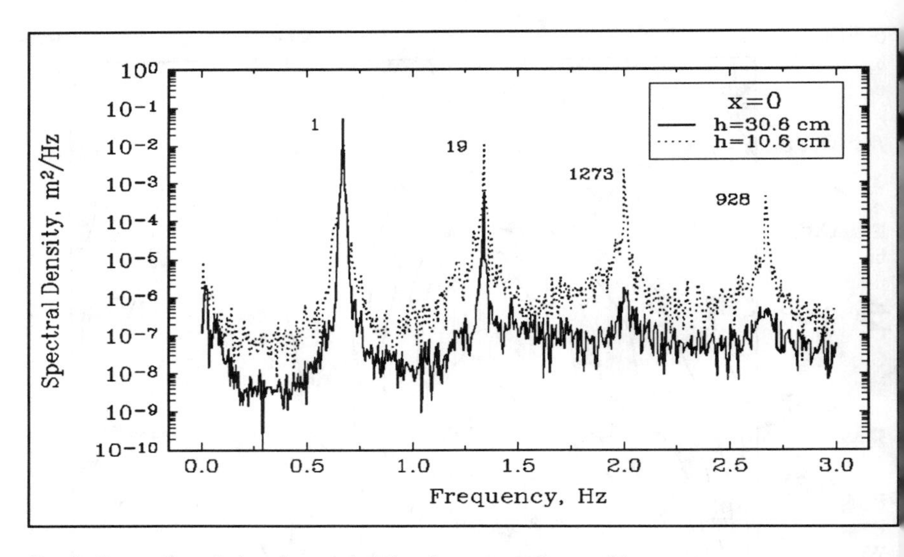

Fig. 5 Spectral evolution due to shoaling for wave only conditions

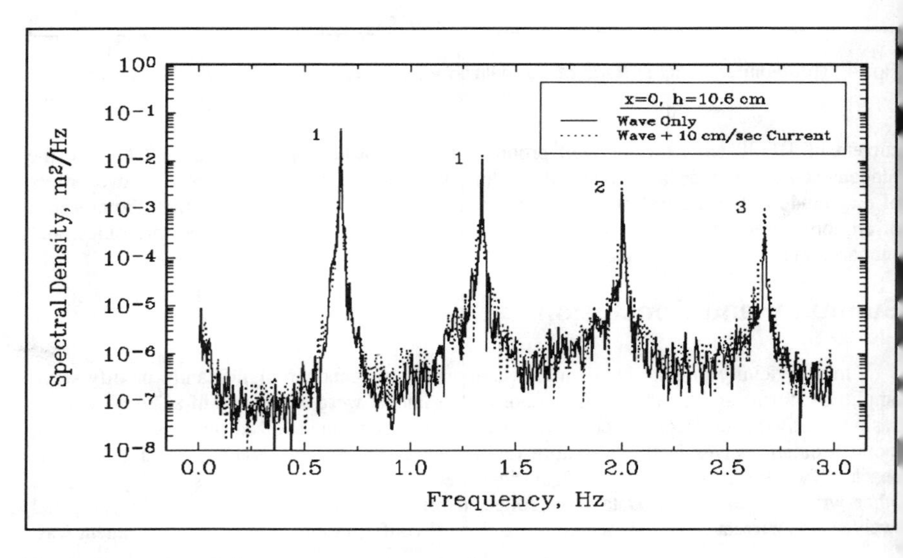

Fig. 6 Spectral evolution due to wave-current interaction

obtained between the laboratory data and predictions from a numerical model. The ebb current had little effect on wave period, but significantly increased the wave height and the nonlinearity of the wave as evidenced by the spectral evolution.

Acknowledgements

The authors wish to acknowledge the Office, Chief of Engineers, U.S. Army Corps of Engineers, for authorizing publication of this paper. It was prepared as part of the "Experimental and Theoretical Studies of Wave-Current Systems," study funded by Cornell University through the U.S. Army Engineer Waterways Experiment Station and the Laboratory Simulation of Nearshore Waves work unit in the Harbor Entrances and Coastal Channels Program of the Civil Works Research and Development Program. We would like to thank Ms. Debra R. Green, Mr. Glen Mize, Mr. Dave Daily, and Mr. Gordie Harkins for their help during this project.

References

Briggs, M. J. and Hampton, Mary L. 1987. "Directional Spectral Wave Generator Basin Response to Monochromatic Waves," Technical Report CERC-87-6, US Army Engineer Waterways Experiment Station, Vicksburg, MS., pp 1-90.

Briggs, M. J. and Smith, J.M. 1990. "The Effect of Wave Directionality on Nearshore Waves," Presented at 22nd ICCE, Delft, The Netherlands, 2-6 July.

Briggs, M. J., Thompson, E.F., Green, D.R., and Lillycrop, L. 1993. "Laboratory Description of Harbor Idealized Tests, Volume I: Main Text and Appendices A Through C" Technical Report CERC-93-1, US Army Engineer Waterways Experiment Station, Vicksburg, MS., pp 1-64.

IAHR, International Association of Hydraulic Research Working Group on Wave Generation and Analysis. 1986 (Jan). "List of Sea State Parameters," Supplement to Bulletin No. 52, pp 1-24.

Lai, R.J., Long, S.R., and Huang, N.E. 1989. "Laboratory Studies of Wave-current Interaction: Kinematics of the Strong Interaction," J. Geophysical Research, Vol. 94, No. C11, November, pp 16,201-16,214.

Long, C. E. 1986. "Laboratory Wave Generation and Analysis: An Instructional Report for Unidirectional Wave Generation and Analysis," (Unpublished Manuscript), US Army Engineer Waterways Experiment Station, Vicksburg, MS, pp 1-32.

Oltman-Shay, J. M. 1987. "Linear Arrays: Wind Wave Directional Measurement Systems," Seminar at US Army Engineer Waterways Experiment Station, Vicksburg, MS, pp 1-28.

Outlaw, D.G. 1984. "A Portable Directional Irregular Wave Generator for Wave Basins." Symposium on Description and Modeling of Directional Seas, Paper No. B-3, Technical University of Denmark.

Outlaw, D. G., and Briggs, M. J. 1986. "Directional Irregular Wave Generator Design for Shallow Wave Basins," 21st American Towing Tank Conference, Washington, DC, Aug.

Sakai, S., and Saeki, H. 1984. "Effects of Opposing Current on Wave Transformation on a Sloping Bed," Proc. 19th Conf. on Coastal Engineering, ASCE, Vol. 1, pp. 1132-1148.

Sand, S. E. 1979. "Three-Dimensional Deterministic Structure of Ocean Waves," Series Paper No. 24, Institute of Hydrodynamics and Hydraulic Engineering, Technical University of Denmark, pp 1-177.

Willis, D.H. 1988. "Experimental and Theoretical Studies of Wave-Current Systems," Technical Report, Hydraulics Laboratory, National Research Council of Canada, Ottawa, Ontario, Canada.

Wave Runup on and Wave Reflection from Coastal Structures

John P. Ahrens[1], William N. Seelig[2], Donald L. Ward[3], and William Allsop[4]

Abstract

Wave runup and wave reflection are two of the most important, easily quantifiable variables that influence the design of coastal structures. Unfortunately there are very few comprehensive studies of these variables which give definitive and consistent guidance on prediction.

This paper uses laboratory data from an extensive study of wave runup at the Coastal Engineering Research Center (CERC) to develop predictive equations for runup and reflection. The CERC study used irregular waves on plane, smooth, impermeable and non-overtopped slopes. These slopes ranged from 1 on 1 (vertical : horizontal) to 1 on 4.

Predictive equations for wave runup and reflection were developed from the CERC data. The criteria used to develop the equations were that the equations fit the data well, be relatively simple, approach logical limiting values, and be consistent with the physics of the phenomenon as it is currently understood. Generally the criteria have been satisfied. In addition, the equations have been compared to other data sets. Most comparisons showed that the prediction equations gave usefully accurate estimates of values and trends. However, the comparisons also suggest that one of the equations could be improved. The paper shows how this improvement was implemented.

---

[1]Specialist, Coastal Processes, NOAA, Sea Grant, 1315 East-West Highway, Silver Spring, MD 20910
[2]Civil Engineer, Naval Facilities Engineering Command, Washington Navy Yard, Washington, DC 20374-2121
[3]Civil Engineer, CEWES-CW-R, Coastal Eng. Research Center, 3909 Halls Ferry Road, Vicksburg, MS 39180-6199
[4]Leader, Coastal Structure Section, Hydraulic Research Wallingford, OX108BA

I Introduction and Background

Even though wave runup and reflection are two of the most functional quantitative variables that can be used to design coastal structures or evaluate their performance, it is difficult to find comprehensive and consistent methods of prediction. Data from an extensive study of irregular wave runup at CERC is used to develop prediction equations for wave runup and wave reflection for plane, smooth, impermeable and non-overtopped slopes. The criteria adopted for the equations to predict runup and reflection were: 1. The equations should provide a good fit to the data (but not at the expense of criteria 3 or 4). 2. The equations should be relatively simple and not include variables with questionable or marginal influence. 3. The equations should approach logical limiting values which will provide some protection against strange predictions when they are used beyond the range of data employed in their development. 4. The equations should be consistent with the physics of the phenomenon as it is currently understood, eg. Battjes (1974), Madsen and White (1976).

Only the simplest type of structure is considered in this paper. However, this elementary approach seems like a logical first step toward a more fundamental understanding of wave interaction with more complex structures. For example, the equations developed in this paper can be regarded as the limiting case of plane, rough, permeable structures for the limits where the roughness, permeability and overtopping approach zero.

II Test Setup, Conditions, and Procedures

Wave runup data were collected in a 61 cm wide test channel within a 4.47m wide by 42.68m long by 1.22m high wave tank at the Coastal Engineering Research Center (CERC). A considerable amount of absorber material was used in the wave tank to suppress wave reflection and long wave oscillation. Runup and reflection data were collected on plane, smooth, impermeable and non-overtopped slopes of 1 on 1, 1 on 1.5, 1 on 2, 1 on 2.5, 1 on 3, and 1 on 4.

Three wave gages were used in the test channel to resolve the incident and reflected wave spectrum using the method of Goda and Suzuki (1976).

Table 1, range of basic variables

Variable	Range
Cotθ	1.0 - 4.0
d_s	45 - 75 cm
H_{mo}	4.1 - 19.1 cm
T_p	1.1 - 4.5 sec
Q_p	1.5 - 6.6

In Table 1 Cotθ is the cotangent of the angle theta, θ, that the slope makes with the horizontal plane, d_s is the water depth at the toe of the slope, H_{mo} is the zeroth moment incident wave height, T_p is the period of peak energy density of the incident wave spectrum, and Q_p is Goda's spectral peakedness parameter, Goda (1985).

The runup board was made of smooth, reinforced polyvinylchloride (PVC) plastic with copper runup sensors 1.0 cm apart along the slope. Sensors were mounted flush with the surface of the board so as to not interfere with the runup and rundown process. For each test runup data was collected for only 256 seconds, which restricts the usefulness of the more extreme values of the runup record, eg., the two percent runup.

Further details related to test setup, conditions, and procedures are given in Ahrens and Titus (1981) and Ahrens (1983).

III Results from Analysis

Overview

Theoretical considerations indicate that it is illogical to lump together for analysis, data that includes both breaking and non-breaking wave conditions, Battjes (1974). Although some data sets follow reasonably consistent trends there is a reversal of influence that makes mixing breaker types an undesirable practice, e.g. relative runup increases with decreasing slope for non-breaking waves and decreases with decreasing slope for breaking wave conditions. The surf parameter is used to partition the data for analysis. The surf parameter is defined in this study as,

$$\xi = \tan\theta/\sqrt{(H_{mo}/L_o)}$$

where tanθ is the tangent of the slope angle θ and L_o is the deep water wave length defined by,

$L_o = gT_p^2/2\Pi$, where g is the acceleration of gravity.

After some trial and error experimentation it was found that a good partition of data was obtained using surf parameters 2.5 or smaller for breaking wave conditions and using surf parameters 4.0 or greater for non-breaking wave conditions. In between breaking and non-breaking is an intermediate wave condition referred to as transitional.

Figure 1 gives a conceptual overview of the partitioning of the data and the resulting analysis. Rundown efficiency, the ratio of the reflected zeroth moment wave height to significant runup, was investigated as part of this study but is not discussed in this paper.

Runup Formula for Breaking Waves, $\xi \leq 2.5$

Relative runup, R_s/H_{mo} can be formulated entirely in terms of the surf parameter,

$$R_s/H_{mo} = 3.86\xi/(1+1.12\xi) \qquad \text{Eq.1}$$

Where R_s, the significant runup, is the average of the highest one-third of the runups based on up-crosses of the still water level on the runup slope. Equation 1 was calculated from 74 tests, i.e. N = 74, and explains about 49 percent of the variance in the relative runup. When Equation 1 is solved for R_s the correlation between predicted and observed runup is about 0.98.

Reflection Formula for Breaking Waves, $\xi \leq 2.5$

The reflection coefficient, K_r is a function of the surf parameter and the slope and is predicted by,

$$K_r = 1/(1+ 1.57\xi^{-1.18}\exp(0.329\text{Cot}\theta)) \qquad \text{Eq.2}$$

Equation 2 was calculated from 74 tests and explains about 94 percent of the variance in the reflection coefficient.

Runup Formula for Non-Breaking Waves, $\xi \geq 4.0$

The equation for predicting relative runup is given by,

$$R_s/H_{mo} = \exp(2.48X_p+0.446(\text{Cos}\theta)^{3.5}+0.194\Pi\ ) \qquad \text{Eq.3}$$

where $X_p = d_s\text{Cot}\theta/L_p - (d_s\text{Cot}\theta/L_p)^2$,

WAVE CONDITIONS

VARIABLE	Breaking $\xi \leq 2.5$	Transitional $2.5 < \xi < 4.0$	Non-Breaking $\xi \geq 4.0$
Wave Reflection K_r	$f(\xi, Cot\theta)$ Eq. 2	Eq. 5	$f(d_s Cot\theta / L_p, \Pi)$ Eq. 4
Relative Runup R_s/H_{mo}	$f(\xi)$ Eq. 6	Eq. 5	$f(d_s Cot\theta / L_p, Cos\theta, \Pi)$ Eq. 3
Relative 2% Runup R_2/H_s R_2/H_{mo}	$f(\xi)$ Eq. 7	Eq. 5	$f(d_s Cot\theta / L_p, Cos\theta, \Pi)$ Eq. 8

Figure 1 Overview of the partitioning of data and results of analysis.

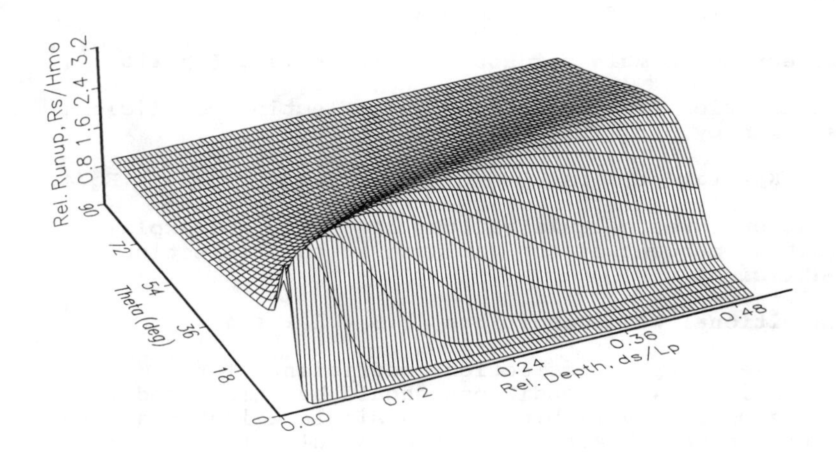

Figure 2 Surface plot of relative runup, R_s/H_{mo}, as a function of slope angle, θ, and relative depth, d_s/L_p, for non-breaking wave conditions using Equation 3.

$\Pi = (H_{mo}/L_p)/(\tanh(2\pi d_s/L_p))^3$, Goda's (1985) wave non-linear

parameter, L_p is the wave length computed from Airy theory using the period of peak energy density and the depth at the toe of the structure, and Cosθ is the cosine of the angle theta. Equation 3 was calculated from 144 tests and explains 74 percent of the variance in the relative runup. When Equation 3 is solved for the significant runup the correlation between predicted and observed is about 0.96.

Figure 2 shows a surface plot of Equation 3. Goda's wave non-liner parameter is not included in Figure 2 so that the plot can include the two variables, the slope angle and relative depth, which have the most influence on relative runup. For the most non-linear wave conditions tested, $\Pi \approx 0.5$, the non-linear term would increase the relative runup about ten percent. In Figure 2 it can be seen that the relative runup, without the non-linear term, approaches the logical limits of 1.0 for vertical walls (ninety degrees) and 0.0 for slopes of zero degrees.

Reflection Formula for Non-Breaking Waves, $\xi \geq 4.0$

The equation for predicting the reflection coefficient is given by,

$$K_r = \exp((d_s \mathrm{Cot}\theta/L_p)(-0.686-3.37\Pi)) \qquad \text{Eq.4}$$

Equation 4 was calculated from 144 tests and explains about 62 percent of the variance in the reflection coefficient.

Transitional Wave Conditions, $2.5 < \xi < 4.0$

Relative runup or reflection coefficients for transitional wave conditions should be calculated for the respective variable using a simple weighted average of the values obtained from the formulae for breaking and non-breaking conditions, i.e.,

$$\text{value} = ((4.0-\xi)/1.5)(\text{breaking wave value}) + ((\xi-2.5)/1.5)(\text{non-breaking wave value}) \qquad \text{Eq. 5.}$$

IV Comparison with Other Data Sets and Discussion

Kamphuis and Mohamed (1978)

Kamphuis and Mohamed did not tabulate the significant wave runup, but did tabulate the two percent and mean runup, where the two percent runup is the elevation exceeded by two percent of the runups. In Figure 3 the mean runup and two percent runup from the study of Kamphuis and Mohamed for non-breaking wave conditions, i.e. $\xi > 4.0$, are compared to the predicted significant runup by Equation 3. Since the predicted trend falls between the mean and two percent runup values of Kamphuis and Mohamed, the comparison indicates that Equation 3 predicts reasonable values.

Not all of the data from the study of Kamphuis and Mohamed were in the non-breaking range. Figure 4 compares the relative runup for $\xi < 4.0$ from their study to the predicted values obtained using Equations 1 and 3, with Equation 5. Again the predicted trend falls between the data of Kamphuis and Mohamed indicating that the predicted values are reasonable.

Mase (1989)

Mase's data was collected on relatively flat slopes where the dominant variable is the surf parameter. Figure 5 shows the relative significant runup from Mase's study with Equation 1 versus the surf parameter. Equation 1 predicts relative runup somewhat higher than Mase's data around surf parameters of about 1.0. The higher relative runup shown in this paper is probably due to the definition of runup used in the data of Ahrens and Titus (1981). There was only one runup counted between two upcrosses of the still water line on the runup slope. For small values of the surf parameter, where wave setup is significant, there may be three to five incident waves for each runup. Therefore, the definition used in this study will result in high values for the significant runup. This tendency was also pointed out by Klein-Breteler (1988) who investigated the data from this study as tabulated in Ahrens and Titus. It is concluded that some modification to Equation 1 is necessary for predictions to conform to the more commonly used definition where each prominent maximum in the runup registration is counted as a runup. This modification will be shown in the comparison to runup data from two large wave tanks.

Large Wave Tank Runup Studies

One study was conducted in the large wave tank in Hannover, Germany, Führböter, et. al (1989) and the other was conducted in the Deltaflume in the Netherlands, with data discussed by van der Meer and Stam (1992). Both studies used irregular waves, with

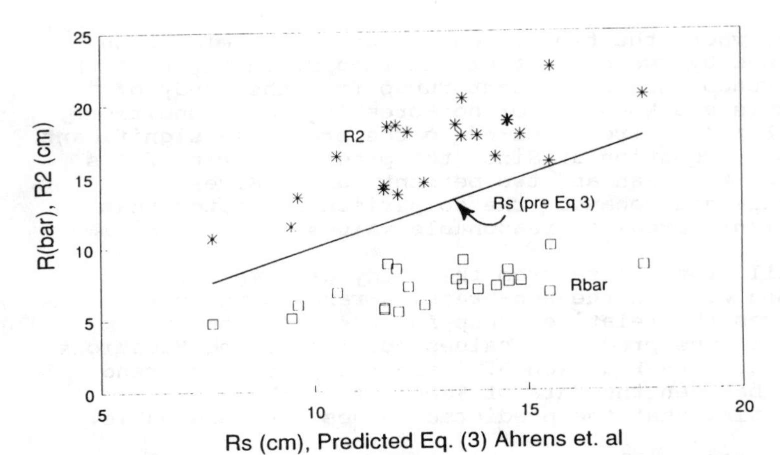

Figure 3 Comparison of predicted R_s using Equation 3 with data from Kamphuis and Mohamed (1978), for non-breaking wave conditions.

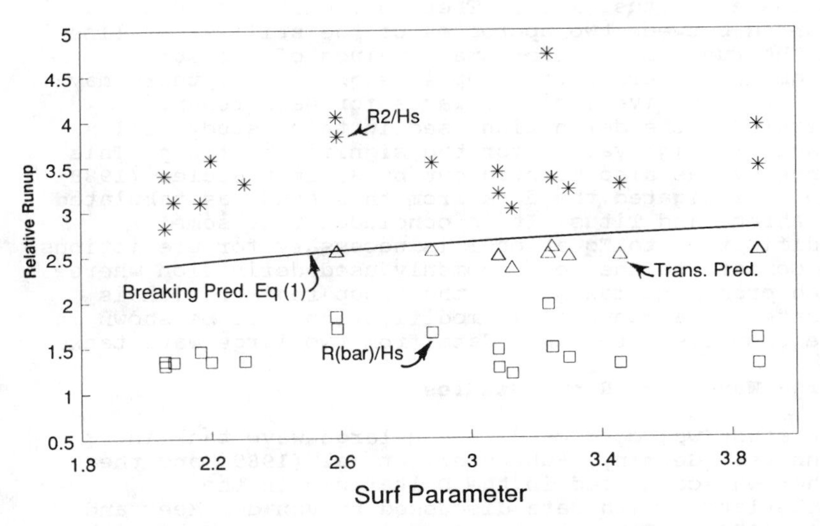

Figure 4 Comparison of predicted relative runup, R_s/H_{mo}, with data from Kamphuis and Mohamed (1978), for breaking and transitional wave conditions. Equation 1 used to predict breaking wave runup.

long wave suppression, on smooth 1 on 6 slopes. Figure 6 shows the relative runup versus the surf parameter for these studies. The runup data includes the two percent runup, the 13% runup, and the 50% runup which are identified by symbol. The data trends shown for each runup percentile are of the same form as Equation 1; Figure 6 indicates that this form is a very satisfactory way to fit the trends. The trend curve shown for the 13% runup, Equation 6, follows close to the lower edge of Mase's (1989) data shown in Figure 5, is consistent with the large wave tank data shown in Figure 6, and predicts values similar to Equation 1 for surf parameters around 2.5 where Equation 1 was confirmed by the data of Kamphuis and Mohamed (1978).

It is recommended that Equation 1 be replaced for predicting runup for breaking wave conditions by Equation 6 which is given by,

$$R_s/H_{mo} = 1.45\xi/(1+0.212\xi) \qquad \text{Eq. 6.}$$

Figure 7 is similar to Figure 4 except that Equation 6 is used instead of Equation 1 to predict breaking wave runup. It can be seen in Figure 7 that Equation 6 does a satisfactory job of predicting relative runup for breaking wave conditions and contributes to a satisfactory prediction for transitional wave conditions when compared to the data of Kamphuis and Mohamed (1978).

Allsop and Channell (1989)

Allsop and Channell measured reflection coefficients in 27 tests with plane, smooth, impermeable slopes. The slopes used were 1 on 1.5, 1 on 2.0, and 1 on 2.5. The surf parameter ranged from 1.75 to 10.20 for the study, which produced five tests with breaking wave conditions, eleven tests with transitional wave conditions, and eleven tests with non-breaking wave conditions. Figure 8 shows predicted reflection coefficients versus observed reflection coefficients. Predicted reflection coefficients were computed using Equations 2 and 4 and the weighted average for transitional wave conditions, i.e. Equation 5. It can be seen in Figure 8 that the procedure produces remarkably good estimates of the reflection coefficient over a wide range of data.

V Two Percent Runup

The elevation exceeded by two percent of the runup extrema is called the two percent runup and is denoted R_2. The two percent runup was not the focus of this

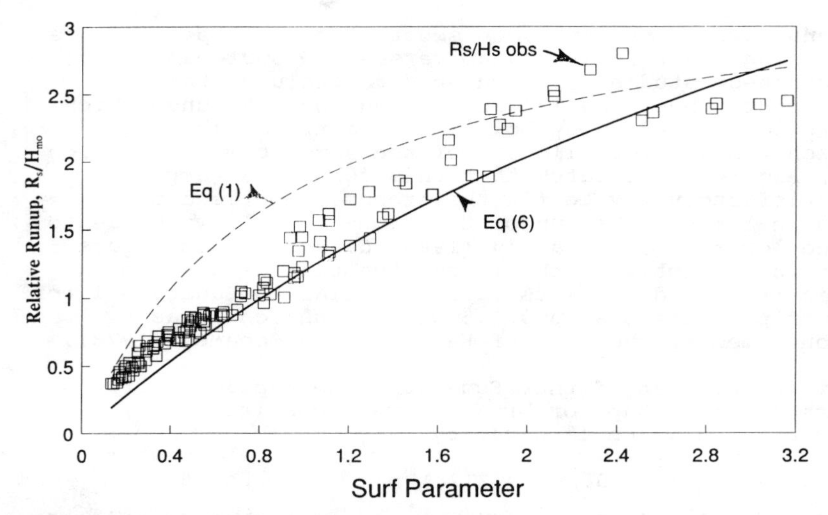

Figure 5 Comparison of predicted relative runup for Equations 1 and 5 with Mase's (1989) data.

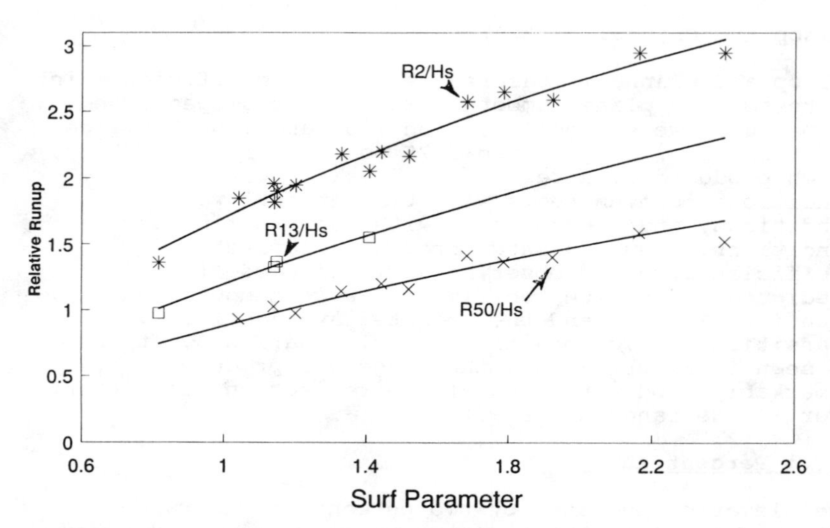

Figure 6 Comparison of predicted relative runup for Equations 6 and 7 with large wave tank data for breaking wave conditions.

study and the short duration of data collection for each test does not allow a very stable estimate of this variable. However, since there is considerable interest in extreme values of wave runup the following estimates are provided: For breaking wave conditions the curve shown in Figure 6 has the equation,

$$R_2/H_s = 2.26\xi/(1+0.324\xi) \qquad \text{Eq. 7}$$

For non-breaking wave conditions,

$$R_2/R_s(\text{pred}) = 1.61 \pm 0.24 \qquad \text{Eq. 8}$$

where R_s(pred) is the significant runup predicted using Equation 3. The coefficient is 1.61 with a standard deviation of 0.24, based on a sample size of 144 tests.

Equation 7 is fit to the large wave tank data of Führböter, et al.(1989) and van der Meer and Stam (1992) and is not subject to the limitations imposed by the short duration of runup data collection of the CERC study. It should be noted that Equation 7 is normalized by the significant wave height and not the zeroth moment wave height. The two wave heights are probably almost equal in the relatively deep water conditions of this study, generally $d_s/H_{mo} > 3$, but can differ considerably in shallow water.

Equation 8 gives, for all practical purposes, an upper bound to the data of Kamphuis and Mohamed (1978). It is not clear why Equation 8, which gives a good fit to the data of Ahrens and Titus (1981), should yield high estimates of the two percent runup measured by Kamphuis and Mohamed.

VI Summary, Conclusions, and Recommendations

This paper develops simple procedures and formulae to predict wave runup and wave reflection from plane, smooth, impermeable and non-overtopped slopes. The predictions are compared to other published data sets and are found to generally give usefully accurate values and trends. For one situation where runup estimates were not satisfactory the source of the problem was identified and a modified equation was developed based largely on the data of other researchers.

This study showed that existing data could be used to develop empirical equations for runup and reflection that fit the data well, are simple, approach logical

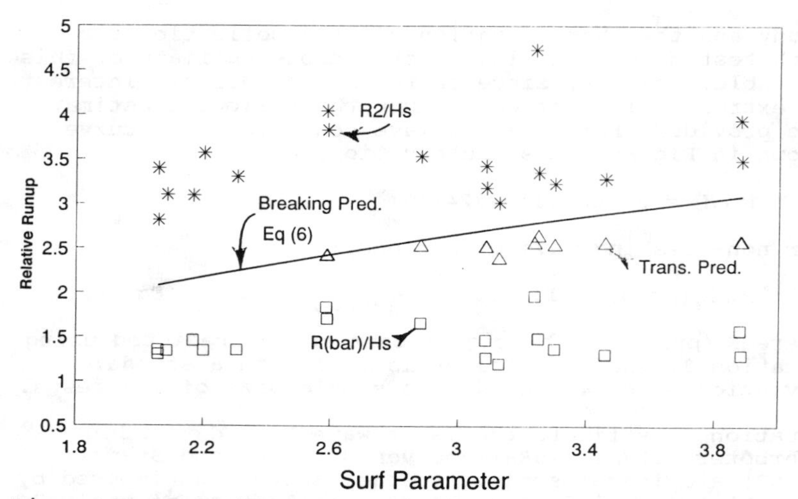

Figure 7 Comparison of predicted relative runup, R_s/H_{mo}, with data from Kamphuis and Mohamed (1978), for breaking and transitional wave conditions. Equation 6 used to predict breaking wave runup.

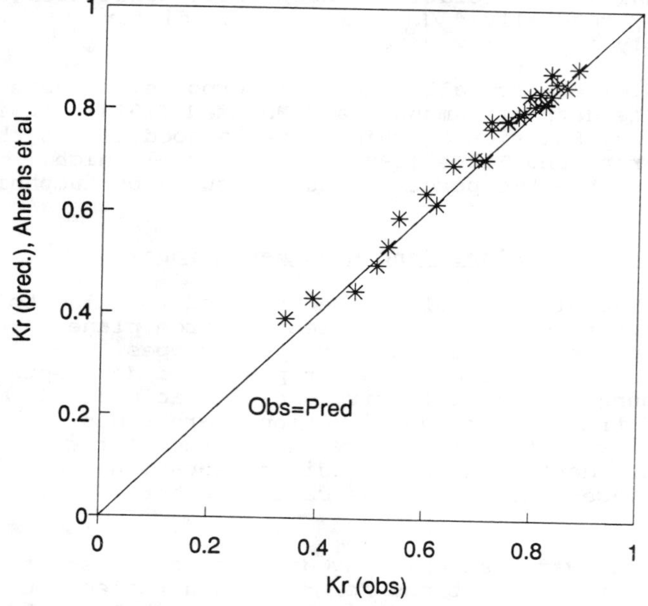

Figure 8 Comparison of predicted reflection coefficients with data of Allsop and Channell (1989).

limits, and are consistent with current understanding of the physical phenomenon. Because of this approach the predictions should be good for a wide range of wave conditions and structure slopes. The equations are for plane smooth slopes, but it is anticipated that they can be used as a starting point for further study of runup and reflection from impermeable slopes with rough porous armor layers.

The equations indicate the variables which are important for wave runup and reflection on simple structures. By omission the equations suggest that spectral type/shape and wave grouping are not particularly important. However, the data used and the method of analysis were not particularly sensitive to the influence of these variables and further laboratory tests are required to remedy this short coming. A limitation of all the data sets used was that there were no really severe wave conditions relative to the water depth. Data is badly needed for, $0.33 \leq H_{mo}/d_s \leq 0.60$, which includes the design conditions for many structures built in the United States.

VI Acknowledgements

The authors acknowledge the help of Geri Taylor of the NOAA National Sea Grant Office in typing and putting this paper together. The senior author acknowledges the support for this research from the NOAA National Sea Grant Office. Ole Madsen provided information and data on monochromatic wave runup and reflection.

VII References

Ahrens, J.P.(1983), "Wave Runup on Idealized Structures", Proceedings of Coastal Structures '83, Arlington, VA, pp. 923 - 938.

Ahrens, J.P., and Titus, M.F.(1981), "Lab Data Report: Irregular Wave Runup on Plane, Smooth Slopes", unpublished CERC laboratory report.

Allsop, N.W.H., and Channell, A.R. (1989),"Wave Reflection in Harbours", Hydraulics Research Wallingford, Report OD 102.

Battjes, J.A.(1974),"Wave Runup and Overtopping", Technical Advisory Committee on Inundation, Rijkswaterstaat, the Hague, Netherlands.

Führbôter, A., Sparboom, U., and Witte, H.H.(1989), "Großer Wellenkanal Hannover: Versuchsergebnisse über den Wellenauflauf auf glatten und rauhen Deichböschungen der Neigung 1:6", Die Küste, Sonderdruck aus Heft 50/1989, Westholsteinische Verlagsanstalt Boyens & Co., Heide

Goda, Y.(1985), "Random Seas and the Design of Maritime Structures", University of Tokyo Press.

Goda, Y., and Suzuki, Y.(1976), "Estimation of Incident and Reflected Waves in Random Wave Experiments", Proceedings 15th Coastal Engineering Conference, Honolulu, HI, pp. 828 - 845.

Kamphuis, J.W., and Mohamed, N.(1978), "Runup of Irregular Waves on Plane, Smooth Slope", ASCE Journal of Waterway, Port, Coastal, and Ocean Div., Vol. 104, No. WW2, pp 135 - 146.

Klein-Breteler, M. (1988) personal communication

Madsen, O.S., and White, S.M.(1976), "Reflection and Transmission Characteristics of Porous Rubble-Mound Breakwaters", CERC MR 76-5.

Mase, H.(1989), "Random Wave Runup Height on Gentle Slope", ASCE Journal of Waterway, Port, Coastal, and Ocean Engineering, Vol. 115, No. 5, pp 649 - 661.

van der Meer, J.W., and Stam, C.M.(1992), "Wave Runup on Smooth and Rock Slopes of Coastal Structures", ASCE Journal of Waterway, Port, Coastal, and Ocean Engineering, Vol 118, No. 5, pp 534- 550.

DECAY OF RANDOM WAVES ON NON-MONOTONIC PROFILES

By Magnus Larson[1]

ABSTRACT: A model is presented to calculate the decay of random waves in the surf zone with special focus on non-monotonic profiles. A wave-by-wave approach is employed to derive a model that requires the transformation of only one representative wave height, the root-mean-square wave height, and no assumptions are needed regarding the shape of the probability density function. Data from large wave tank tests (SUPERTANK) and field measurements off the coast of North Carolina (DELILAH) are employed to validate the model. Comparisons between measurements and calculations are satisfactory with a slight overprediction for cases where the zone of wave breaking is narrow over a non-monotonic feature such as a bar or a mound.

INTRODUCTION

Models describing wave decay in the surf zone differ mainly in the formulation of the energy dissipation due to breaking and whether they were developed for monochromatic or random waves (Battjes and Janssen 1978, Thornton and Guza 1983, Svendsen 1984, Dally et al. 1985). One type of models that has been successfully employed to simulate transformation of random waves in the surf zone is based on a wave-by-wave approach (Dally 1990, 1992, Larson and Kraus 1991). This method has been verified with high-quality field data (Dally 1992) and has the advantage of not requiring an assumption about the shape of the probability density function (pdf) in the surf zone. Also, the wave-by-wave approach may be generalized to beach profiles where the depth is not increasing monotonically with distance offshore, whereas it is less straight-forward to apply wave decay models that require assumptions about the pdf to non-monotonic profiles.

---

[1] Associate Professor, Department of Water Resources Engineering, Lund Institute of Technology, University of Lund, Box 118, S-221 00 Lund, SWEDEN; Tel. +46 46 108729, Fax. +46 46 104435.

The objective of this paper is to present a model for the decay of random waves in the surf zone with special focus on non-monotonic beach profiles. A wave-by-wave approach is used as a starting point to derive a model that involves the transformation of only one representative wave, the root-mean-square wave height, and this model produces almost identical results to a complete Monte-Carlo simulation approach with many individual waves (Larson 1993a, 1993b). The model is validated with large wave tank data from SUPERTANK and field data from DELILAH.

THEORETICAL CONSIDERATIONS

In the offshore, where wave breaking is negligible, a Rayleigh pdf is assumed to characterize the variation in wave height H (Longuet-Higgins 1952). Thus, the random sea is considered to be narrow-banded in frequency and direction, specified through the peak spectral wave period (T) and the mean incident wave angle (θ), respectively. The shape of the Rayleigh pdf is uniquely determined by the root-mean-square (rms) wave height (H_{rms}), and if the pdf is represented by a large number of individual waves N, the empirical estimate of the rms wave height is,

$$H_{rms}^2 = \frac{1}{N}\sum_{i=1}^{N} H_i^2 \tag{1}$$

where H_i is the height of the individual wave i.

As the waves approach shore, however, higher waves will experience depth-limited breaking and loose energy, making the Rayleigh pdf less suitable for describing variation in wave height (Hughes and Borgman 1987). In the surf zone, an estimate of H_{rms} from Equation 1 will encompass both broken and nonbroken waves. The following terminology is employed here regarding wave breaking: unbroken refers to a wave that has never been broken, whereas nonbroken waves include both unbroken and reformed waves. If the number of nonbroken waves at a specific location x is $n(x)$ and the number of broken waves is $q(x)$, where $N = n(x) + q(x)$ = constant, since the number of waves is not changing across shore, Equation 1 may be developed to yield:

$$H_{rms}^2 = \frac{1}{N}\sum_{i=1}^{n(x)} H_i^2 + \frac{1}{N}\sum_{i=1}^{q(x)} H_i^2 \tag{2}$$

The ratio of broken waves is defined as $\alpha = q/N$ and the ratio of nonbroken waves is $n/N = (N - q)/N = 1 - \alpha$. The following definitions are introduced,

$$H_n^2 = \frac{1}{n}\sum_{i=1}^{n(x)} H_i^2 \tag{3}$$

$$H_q^2 = \frac{1}{q}\sum_{i=1}^{q(x)} H_i^2 \qquad (4)$$

where H_n and H_q is the rms wave height for nonbroken and broken waves, respectively. Using Equations 3 and 4, Equation 2 may be expressed:

$$H_{rms}^2 = (1-\alpha)H_n^2 + \alpha H_q^2 \qquad (5)$$

In the general case, the nonbroken waves consist both of unbroken waves and waves that once were broken but reformed at some seaward point. If the number of unbroken waves at a specific location x is denoted $m(x)$ and the number of reformed waves $r(x)$, then the number of nonbroken waves is $n(x) = m(x) + r(x)$, and $N = m(x) + r(x) + q(x) =$ constant. Equation 2 is modified to yield:

$$H_{rms}^2 = \frac{1}{N}\sum_{i=1}^{m(x)} H_i^2 + \frac{1}{N}\sum_{i=1}^{r(x)} H_i^2 + \frac{1}{N}\sum_{i=1}^{q(x)} H_i^2 \qquad (6)$$

Defining the ratio of unbroken, reformed, and broken waves as $\beta = m/N$, $\mu = r/N$, and $\alpha = q/N$, respectively, where $\beta + \mu + \alpha = 1$, Equation 6 may be written,

$$H_{rms}^2 = \beta H_m^2 + \mu H_r^2 + \alpha H_q^2 \qquad (7)$$

where H_m and H_r is the rms wave height for the unbroken and reformed waves, respectively.

Representing the Rayleigh distribution in the offshore with a collection of N waves, each individual wave should be transformed according to,

$$\frac{d}{dx}(F_i \cos\theta) = 0 \qquad (8)$$

if the wave is nonbroken, and adopting the breaker decay model by Dally et al. (1985),

$$\frac{d}{dx}(F_i \cos\theta) = \frac{\kappa}{d}(F_i - F_s) \qquad (9)$$

if the wave is broken, where F_i is the energy flux for the individual wave i, θ the mean incident wave angle, d the total water depth including wave setup, F_s the stable wave energy flux, and κ an empirical parameter (equal to about 0.15). The wave energy flux may be expressed using linear wave theory,

$$F_i = \frac{1}{8}\rho g H_i^2 C_g \tag{10}$$

in which ρ is the density of water, g the acceleration of gravity, and C_g the group speed. The stable wave energy flux F_s may be written,

$$F_s = \frac{1}{8}\rho g(\Gamma d)^2 C_g \tag{11}$$

assuming that the stable wave height is $H_s = \Gamma d$, where Γ is a coefficient typically set to 0.4.

Simplifying Equations 8 and 9 by dividing with 1/8 ρg and adding all broken and nonbroken waves separately at a specific location x yields:

$$\sum_{i=1}^{n(x)} \frac{d}{dx}\left(H_i^2 C_g \cos\theta\right) = 0 \tag{12}$$

$$\sum_{i=1}^{q(x)} \frac{d}{dx}\left(H_i^2 C_g \cos\theta\right) = \sum_{i=1}^{q(x)} \frac{\kappa}{d}\left(H_i^2 C_g - \Gamma^2 d^2 C_g\right) \tag{13}$$

Furthermore, Equations 12 and 13 may be added together,

$$\sum_{i=1}^{N} \frac{d}{dx}\left(H_i^2 C_g \cos\theta\right) = \sum_{i=1}^{q(x)} \frac{\kappa}{d}\left(H_i^2 C_g - \Gamma^2 d^2 C_g\right) \tag{14}$$

where $N = n + q$ was used in the first summation. Dividing Equation 14 with N, moving the derivative out from the summation, and using Equation 5 gives,

$$\frac{d}{dx}\left(H_{rms}^2 C_g \cos\theta\right) = \frac{\kappa}{d}\left[H_{rms}^2 C_g - (1-\alpha)H_n^2 C_g - \alpha\Gamma^2 d^2 C_g\right] \tag{15}$$

Introducing the rms energy flux and a definition of the stable energy flux for random waves, Equation 15 could be written in a form equivalent to Equation 9, which is valid for a single wave,

$$\frac{d}{dx}\left(F_{rms}\cos\theta\right) = \frac{\kappa}{d}\left(F_{rms} - F_{stab}\right) \tag{16}$$

where:

$$F_{rms} = \frac{1}{8}\rho g H_{rms}^2 C_g \tag{17}$$

$$F_{stab} = \frac{1}{8}\rho g\left[(1-\alpha)H_n^2 + \alpha\Gamma^2 d^2\right]C_g \tag{18}$$

If all waves are broken, i.e., $\alpha = 1$, and $H_n = 0$, Equation 16 reduces to the original breaker decay model proposed by Dally et al. (1985).

In general, H_{rms} is solved for by employing numerical methods. The root-mean-square wave height of the nonbroken waves, H_n, has to be specified at each point across shore together with the ratio of broken waves α. This specification is straightforward for a beach profile with a depth that increases monotonically with distance offshore; however, for a non-monotonic beach profile, such as a barred profile, an empirical closure relationship must be added to model wave reformation.

A monotonic profile will never allow wave reformation, and the ratio of broken waves will increase monotonically as the waves propagate onshore. The Rayleigh pdf is transformed from the offshore point and truncated at a wave height corresponding to the height of incipient breaking waves at the point of interest, producing a ratio of broken waves,

$$\alpha = e^{-\left(\frac{\gamma_b d}{H_x}\right)^2} \tag{19}$$

where γ_b is the ratio between wave height and water depth at incipient breaking for a single wave and H_x is the root-mean-square wave height neglecting breaking given by,

$$H_x^2 = \frac{C_{gd}}{C_g}\frac{\cos\theta_d}{\cos\theta}H_d^2 \tag{20}$$

where H is the rms wave height ignoring wave breaking, and the subscript d refers to an offshore location where the Rayleigh pdf is valid. The root-mean-square wave height for the unbroken waves determined from the truncated Rayleigh pdf is:

$$H_n^2 = \frac{1}{1-\alpha}\left[H_x^2 - \alpha\left(H_x^2 + \gamma_b^2 d^2\right)\right] \tag{21}$$

Using Equation 21 to replace H_n in Equation 15 yields:

$$\frac{d}{dx}\left(H_{rms}^2 C_g \cos\theta\right) = \frac{\kappa}{d}\left[H_{rms}^2 C_g - (1-\alpha)H_x^2 C_g + \alpha d^2\left(\gamma_b^2 - \Gamma^2\right)C_g\right] \tag{22}$$

The rms height for broken and nonbroken waves, H_{rms}, is the only unknown in Equation 22, whereas α and H_x are calculated explicitly at each point across shore from Equations 19 and 20, respectively.

If the profile depth is not increasing monotonically with distance offshore, such as along a barred beach profile, H_n and α in Equation 15 may not be calculated directly from the Rayleigh pdf as in the case of a monotonic profile. A predictive equation is needed to determine the amount of waves that reform along negatively sloping sections of the beach. The following equation was chosen to model the effect of wave reformation,

$$\mu = 1 - \beta_k - \alpha_r \left(\frac{F_{rms} - F_{stab}}{F_{rms,r} - F_{stab,r}} \right)^{\lambda} \tag{23}$$

valid for $F_{rms} > F_{stab}$, where λ is an empirical coefficient and subscript r refers to the closest seaward point x_r where wave reformation starts. Equation 23 implies that all waves that are broken at x_r reform as F_{rms} approaches F_{stab}, which is one requirement for a predictive equation. Once F_{rms} falls below F_{stab} energy dissipation ceases and no broken waves should exist at that point.

As before, Equations 16-18 are solved to yield H_{rms}; however, since H_{rms} results from summing up both unbroken, reformed, and broken waves, writing F_{stab} in a slightly different form is more convenient. Using Equation 7 instead of Equation 5 in the derivation of Equation 16, F_{stab} may be expressed as:

$$F_{stab} = \frac{1}{8} \rho g \left[\beta H_m^2 + \mu H_r^2 + (1 - \beta - \mu) \Gamma^2 d^2 \right] C_g \tag{24}$$

Thus, to determine H_{rms} the ratio of unbroken (β) and reformed waves (μ), as well as the respective rms wave heights (H_m and H_r), must be predicted before H_{rms} is calculated.

When a negative slope is encountered, wave reformation occurs and μ increases simultaneously as α decreases. The ratio of unbroken waves remains fixed, and is determined from Equation 19 by using the smallest depth seaward of the studied point. Equation 20 is used to calculate H_n, also employing the smallest seaward depth. The rms height of the reformed waves H_r is estimated by transforming the complete pdf for reformed waves as the waves propagate onshore, and computing H_r at each point from the empirical pdf. As the reformed waves experience shoaling again, they will eventually break, and this is described in the model by truncating the pdf for the reformed waves at $\gamma_b d$ (waves with heights above this threshold become broken waves again). The empirical pdf for the reformed waves is constructed by calculating the ratio of waves reforming between two grid points using Equation 23, simultaneously as the reformed wave height is determined from Γd.

VALIDATION OF RANDOM WAVE MODEL

Two data sets were employed to validate the random wave model for non-monotonic beach profiles encompassing laboratory measurements from SUPERTANK (Kraus et al. 1992) and field measurements from DELILAH (Smith et al. 1993). The laboratory tests were performed in a large wave tank under normally incident waves and for varying profile shapes, whereas the field tests were carried out on an open, barred coast where the waves approached the beach at an angle.

Standard values were used in all calculations with the random wave model without any calibration to evaluate how the model would perform in a truly predictive situation. The coefficients in the breaker decay model by Dally were set to $\kappa=0.15$ and $\Gamma=0.4$ according to Dally et al. (1985), Ebersole (1987), and Larson and Kraus (1991), and the coefficient controlling wave reformation was set to $\lambda=0.5$ based on Larson (1993a, 1993b). The ratio between wave height and water depth at incipient breaking for an individual wave was assumed to be $\gamma_b=0.78$.

SUPERTANK Data Set

Four cases from the SUPERTANK Data Collection Project (Kraus et al. 1992) were selected to validate the random wave model for non-monotonic beach profiles. SUPERTANK was conducted to investigate cross-shore hydrodynamic and sediment transport processes using the large wave tank at Oregon State University, Corvallis, Oregon. A sandy beach was constructed in one end of the tank, which is 104 m long, 3.7 m wide, and 4.6 m deep, and waves were generated at the other end. Both monochromatic and random waves were employed, and the tank was densely instrumented to record the cross-shore hydrodynamics including wave and current properties. A large number of beach profile surveys were taken during the twenty major data collection runs made, where each run typically consisted of several minor cases.

The large wave tank was instrumented with 16 resistance wave gages during all experimental runs of SUPERTANK, and 10 capacitance gages were employed in the very nearshore during most of the runs. The capacitance gages were mounted around the still-water shoreline, and these gages covered the swash zone to determine runup properties and fluctuations in the sand surface elevation. Comparisons between model predictions and measured wave heights in this paper focus on the rms wave height for nonbroken and broken waves, and only wave heights measured by the resistance gages are used in the comparison. The cases selected for comparison involved the larger generated rms wave heights to obtain marked wave breaking at the largest possible number of resistance gages. For the cases with smaller generated rms wave heights a majority of the individual waves would break shoreward of the resistance gages, and the measured data would be less suitable to test the random wave decay model.

Table 1 summarizes the cases used from SUPERTANK for the model comparison in terms of run and case number, target rms wave height and peak spectral wave period. Random wave decay was studied across two basic types of non-monotonic profile shapes, namely a nearshore bar and an offshore mound. During Run ST10 a nearshore bar was present, whereas Runs STJ0 and STK0 involved a narrow-crested and broad-crested offshore mound, respectively. The measured rms wave heights were averages for time periods between 20 and 70 min depending on the studied case. The surveyed profile at the end of a case was employed in the wave calculations; changes in the profile shape were minor during the time of wave action for the cases studied.

Table 1. Wave properties for studied SUPERTANK cases.

Run No.	Case No.	Rms Wave Height (m)	Peak Period (s)
ST10	A0517A	0.57	3.0
ST10	A0914A	0.64	4.5
STJ0	S0913A	0.49	3.0
STK0	S1208B	0.49	3.0

In the following, the calculated and measured rms wave height (H_{rms}) for each case are presented together with the surveyed beach profile. All cross-shore distances presented in the figures refer to the origin of the wave tank (the end where the beach was constructed), and the elevations are given with respect to the still-water shoreline. Some brief comments are provided for each of the studied cases.

Case A0517A. A pronounced bar was present with the crest located at a water depth of about 0.6 m, and Figure 1 summarizes the results of the calculations. Only the two most shoreward gages seem to have been influenced by wave breaking, but the wave model reproduces the measured wave height decay faithfully. The calculated wave energy dissipation (not shown) has two distinct peaks with a minimum in between corresponding to the marked trough shoreward of the bar crest. The ratio of broken waves increases seaward of the bar but as the trough is encountered, wave reformation occurs and more than 1/3 of the waves are calculated to reform in the trough.

Case A0914A. This case belongs to the same run as the previous case, but the bar was located further offshore implying that a larger number of the gages were exposed to breaking waves. Also, a longer wave period was employed (T=4.5 s) and H_{rms} was slightly larger. The decay in H_{rms} (Figure 2) through the surf zone is well reproduced, as in the previously discussed case; however, it seems like nonlinear

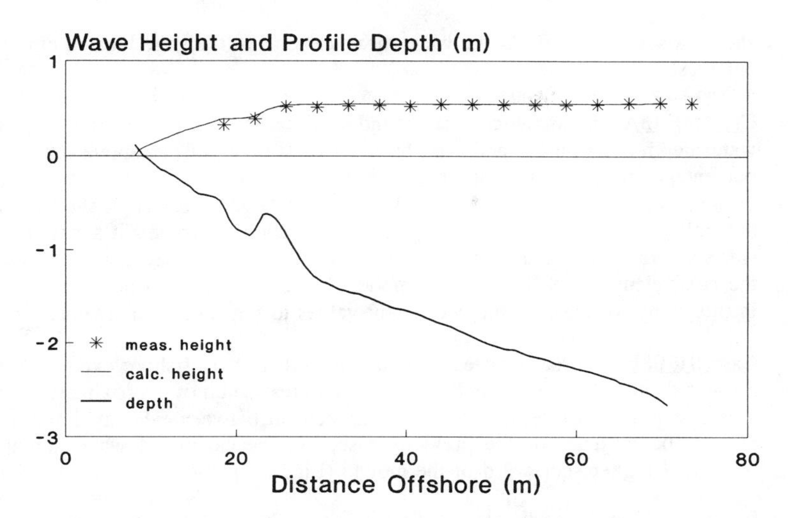

Figure 1. Calculated and measured rms wave height for SUPERTANK Case A0517A together with the beach profile.

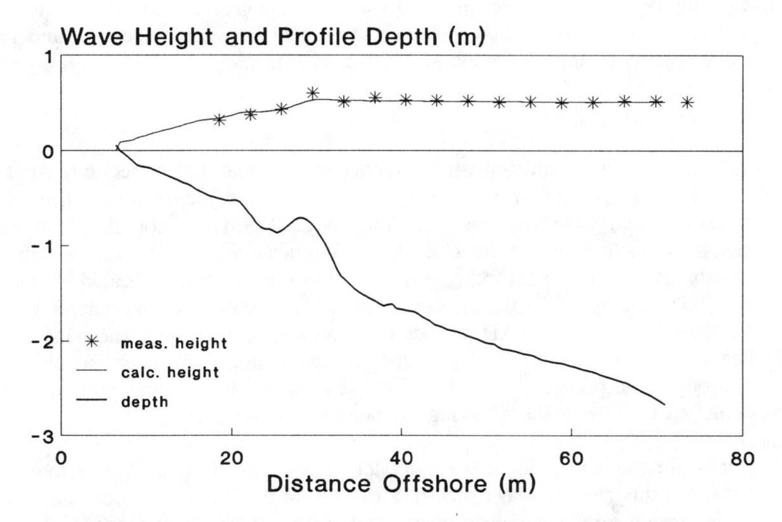

Figure 2. Calculated and measured rms wave height for SUPERTANK Case A0914A together with the beach profile.

shoaling somewhat affects the wave measurements at the gage located closest to the bar crest.

Case S0913A. A narrow-crested mound was constructed in the offshore between a water depth of about 1.5 and 2 m, having a crest depth of just above 0.5 m. A large percentage of the waves broke on the mound and a marked reduction in H_{rms} occurred across the mound (Figure 3). The calculated wave height showed the same general trend as the measured height, but the reduction in height across the mound was smaller in the calculations. This discrepancy may be corrected for by modifying the coefficients in the Dally wave model (Larson 1993b), but no attempt was done in this study to fine-tune the coefficient values to improve the agreement.

Case S1208B. A broad-crested mound was built in the offshore extending between a water depth of about 1.2 and 2.2 m, with a crest width of approximately 12 m and a crest depth of just below 0.5 m. The agreement between calculated and measured H_{rms} is better than for the previous case, but the calculated wave height is still somewhat higher shoreward of the mound (Figure 4).

In summary, the random wave model satisfactorily reproduced the measured cross-shore variation in the rms wave height for the non-monotonic profiles studied, including profiles with both a nearshore bar and an offshore mound. For the mound cases, the calculated H_{rms} was consistently somewhat larger than the measured H_{rms}, where the wave model predicts more rapid reformation than what actually occurs. This discrepancy was more marked for a narrow-crested mound than for a broad-crested mound, which showed better agreement between calculated and measured wave heights (compare Larson and Kraus 1991).

DELILAH Data Set

In October 1990, a multi-institutional cooperative field data collection project named DELILAH (Duck Experiment on Low-frequency and Incident-band Longshore and Across-shore Hydrodynamics; see Smith et al. 1993) was conducted on the barred nearshore bathymetry at the U.S. Army Engineer Waterways Experiment Station, Coastal Engineering Research Center, Field Research Facility located in Duck, North Carolina, facing the Atlantic Ocean on a long sandy barrier island beach. The objectives of the DELILAH project were to measure the wave- and wind-forced 3-D nearshore dynamics and to monitor the bathymetric response to the operating hydrodynamic processes. A dense array of current meters and pressure gages was deployed in the nearshore during the period October 1-19, 1990.

For evaluating the random wave model, measurements taken on October 14 were used. On this day the wave spectra measured at the 8-m depth were narrow banded in frequency with symmetric directional distributions about a mean oblique wave direction. The unimodal, swell wave trains are favorable for modeling wave transformation with linear wave propagation. On October 14, the significant wave

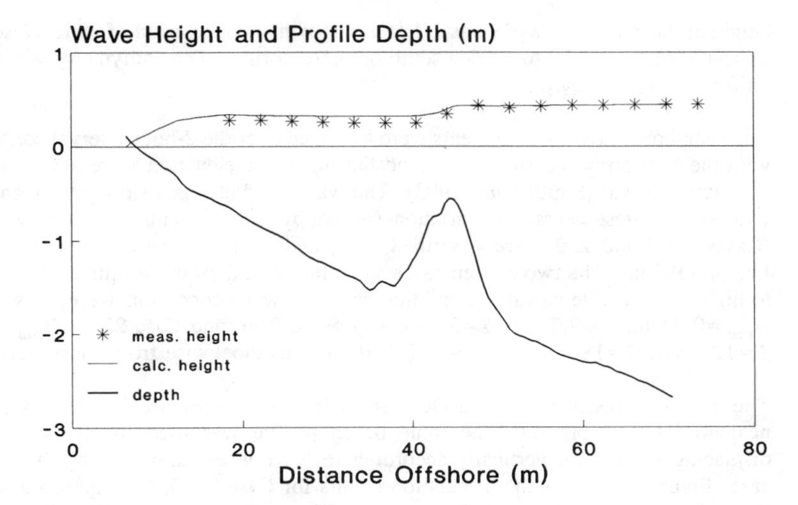

Figure 3. Calculated and measured rms wave height for SUPERTANK Case S0913A together with the beach profile.

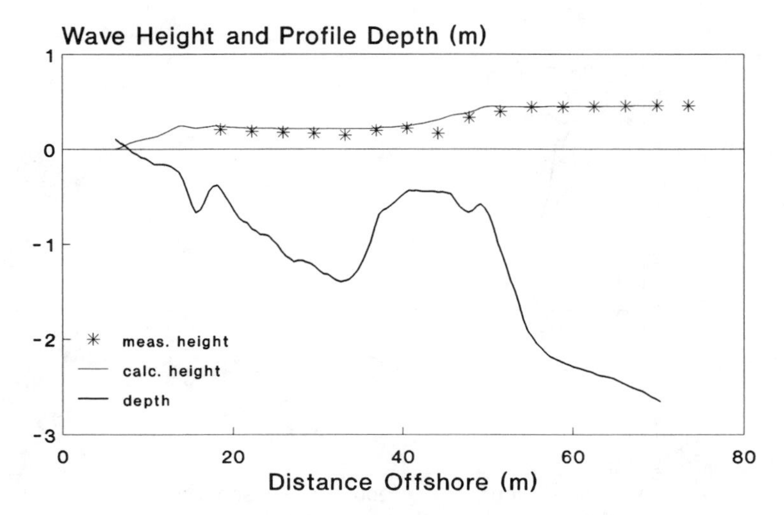

Figure 4. Calculated and measured rms wave height for SUPERTANK Case S1208B together with the beach profile.

height at the 8-m gage depth was 1.1 to 1.5 m with a peak period of 10 to 12 sec and a mean direction of 10 to 30 deg south of shore-normal. The bathymetry was nearly uniform alongshore.

The nearshore wave measurements were averaged over the 2-hour interval concurrent with the 8-m array measurements, and the input tide elevation were averaged over the same interval (Smith et al. 1993). The wave field changed slowly through time, with the greatest nearshore variation caused by tide. Calculations for two cases (Cases 0100 and 2200) are described here, and additional cases are presented in Larson (1993a). The two chosen cases show the typical model results corresponding to high and low tide elevation, and the measured wave conditions were; Case 0100: H_{rms}=0.94 m, T=9.7 sec, θ=32.0 deg, *tide*=0.2 m, and Case 2200: H_{rms}=0.79, T=12.0 sec, θ=18.0 deg, *tide*=-0.5 (θ defined as clockwise from shore normal).

The measured beach profile on October 14 is shown in Figure 5, where a distinct nearshore bar is present. The same beach profile was used in all calculations, displacing the profile vertically according to the tide elevation during the specific case. Figure 6 displays the calculation results for Case 0100; the calculated H_{rms} is shown together with the measured rms wave height. The calculated H_{rms} differs from the measured heights inshore of the bar, indicating that the energy dissipation is underestimated in the model by Dally et al. (1985) in this region. As for the SUPERTANK cases with the offshore mound, improved agreement could be obtained if the coefficient values κ and Γ were calibrated instead of using the standard values.

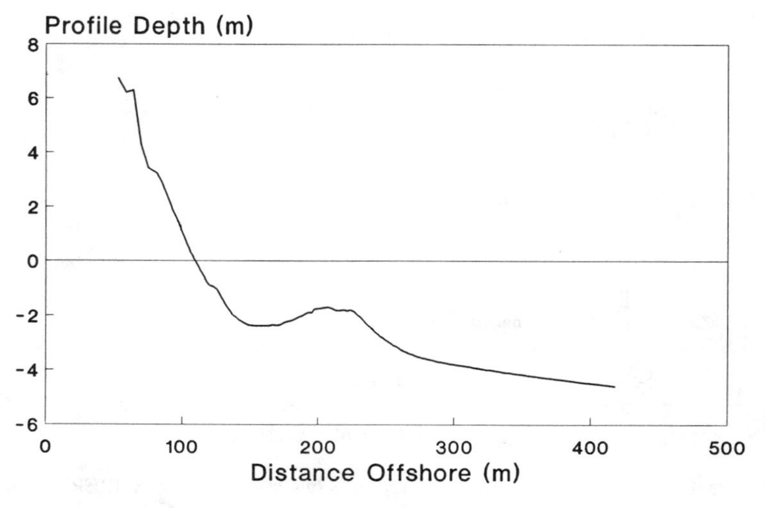

Figure 5. Beach profile measured during DELILAH (October 14, 1990).

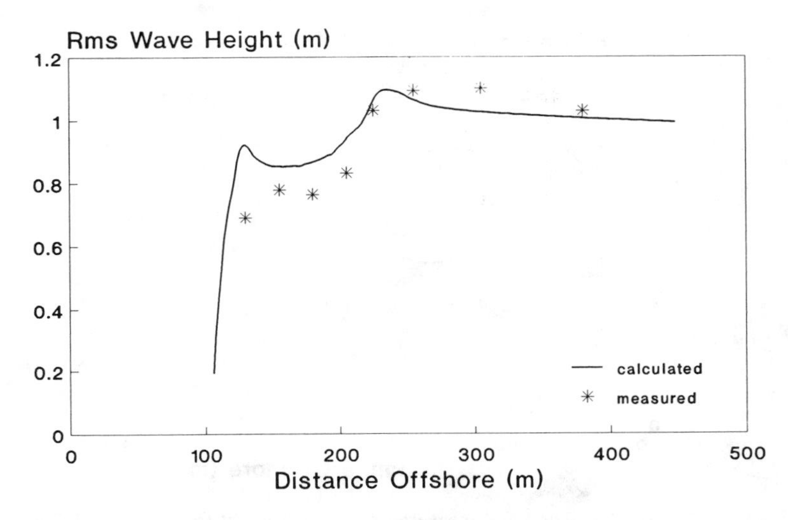

Figure 6. Calculated and measured rms wave height for DELILAH Case 0100.

For the case with low tide elevation (Case 2200), the agreement between the calculated and the measured wave height is considerably better (Figure 7). In Case 2200 around 50% of the waves are predicted to break seaward of the bar, whereas only about 20% of the waves were calculated to break in this region for Case 0100.

CONCLUDING REMARKS

The random wave model reproduced the measured rms wave heights across shore satisfactorily for the studied laboratory and field data, which included a nearshore bar and an offshore mound for the laboratory tests and a nearshore bar for the field tests. For non-monotonic profile features, such as bars and mounds, the largest differences between measured and calculated rms wave height occurred when a feature produced a narrow zone of breaking. In these cases, the model does not predict a large enough energy dissipation using standard values on model coefficients to reduce the calculated wave height to agree with the measurements. This deficiency in the model behavior may in part be corrected through modifications of the coefficient values. Furthermore, the description of the wave transformation during the initial phase after incipient breaking could be improved in the model. The model is expected to produce best results after the initial rapid transformation of the breaking wave when a bore-like wave has developed.

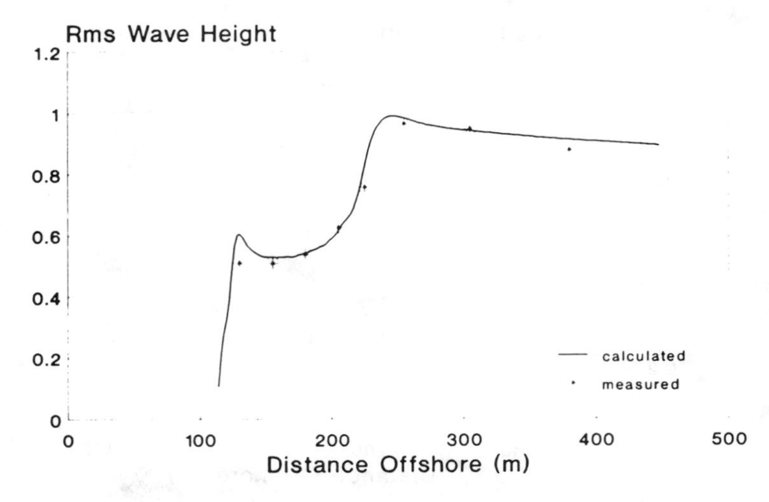

Figure 7. Calculated and measured rms wave height for DELILAH Case 2200.

ACKNOWLEDGEMENTS

The SUPERTANK laboratory data collection project was conducted at the O.H. Hinsdale Wave Research Laboratory, Oregon State University, under the direction of Dr. Nicholas C. Kraus. The collection of the high-quality data set was accomplished through the expert planning and execution of Nicholas C. Kraus, William R. Dally, David L. Kriebel, William G. McDougal, Jane M. Smith, and Charles K. Sollitt. The DELILAH data is courtesy of Jane M. Smith. The research presented in this paper was conducted under the Calculation of Cross-Shore Sediment Transport and Beach Profile Change Processes Work Unit 32530 and the Shoreline and Beach Topography Response Modeling Work Unit 32592, Shore Protection and Restoration Program, Coastal Engineering Research Center, U.S. Army Engineer Waterways Experiment Station. Contract coordination was provided by the European Research Office of the U.S. Army in London under Contracts DAJA45-92-C-0017 and DAJA45-93-C-0013.

REFERENCES

Battjes, J.A. and Janssen, J.P.F.M. (1978). "Energy Loss and Setup Due to Breaking of Random Waves." *Proceedings of the 16th Coastal Engineering Conference*, ASCE, 569-587.

Dally, W.R. (1990). "Random Breaking Waves: A Closed-Form Solution for Planar Beaches." *Coastal Engineering*, 14, 233-263.

Dally, W.R. (1992). "Random Breaking Waves: Field Verification of a Wave-by-Wave Algorithm for Engineering Application." *Coastal Engineering*, 16, 369-397.

Dally, W.R., Dean, R.G., and Dalrymple, R.A. (1985). "Wave Height Variation Across Beaches of Arbitrary Profile." *Journal of Geophysical Research*, 90 (C6), 11917-11927.

Ebersole, B.A. (1987). "Measurement and Prediction of Wave Height Decay in the Surf Zone." *Proceedings of Coastal Hydrodynamics '87*, ASCE, 1-16.

Hughes, S.A. and Borgman, L.E. (1987). "Beta-Rayleigh Distribution for Shallow Water Waves." *Proceedings of Coastal Hydrodynamics '87*, ASCE, 17-31.

Kraus, N.C, Smith, J.M., and Sollitt, C.K. (1992). "SUPERTANK Laboratory Data Collection Project," *Proceedings of the 23rd Coastal Engineering Conference*, American Society of Civil Engineers, pp 2191-2204.

Larson, M. and Kraus, N.C. (1991). "Numerical Model of Longshore Current for Bar and Trough Beaches." *Journal of Waterways, Port, Coastal, and Ocean Engineering*, 117 (4), 326-347.

Larson, M. (1993a). "Model for Decay of Irregular Waves in the Surf Zone," Report No. 3167, Department of Water Resources Engineering, Lund Institute of Technology, Lund University, Lund, Sweden.

Larson, M. (1993b). "Model for Decay of Random Waves in the Surf Zone," *Journal of Waterways, Port, Coastal, and Ocean Engineering*, (submitted).

Longuet-Higgins, M.S. (1952). "On the Statistical Distribution of the Heights of Sea Waves." *Journal of Marine Research*, 11, 245-266.

Smith, J.M., Larson, M., and Kraus N.C. (1993). "Longshore Current on a Barred Beach: Field Measurements and Calculation." *Journal of Geophysical Research*, (in press).

Svendsen, I.A. (1984). "Wave Heights and Set-Up in a Surf Zone." *Coastal Engineering*, 8, 303-329.

Thornton, E.B. and Guza, R.T. (1983). "Transformation of Wave Height Distribution." *Journal of Geophysical Research*, 88 (C10), 5925-5938.

PROPAGATION OF LINEAR GRAVITY WAVES OVER SLOWLY VARYING DEPTH AND CURRENTS

Francisco J. Rivero and Agustín S.-Arcilla [1]

ABSTRACT

A detailed derivation of some kinematic and dynamic properties of linear surface gravity waves is presented for the case of combined refraction-diffraction and wave-current interaction. The obtained expressions, which are exact whithin the framework of linear wave theory in the mild-slope approximation, differ from those normally given in the literature (e.g. Dingemans (1985), Liu (1990) and Chae & Jeong (1990), among others). The observed differences appear to be particularly relevant when horizontal gradients of wave amplitude and ambient current velocity field are not negligible. A new small parameter θ , termed 'phase lag', is defined as a measure of horizontal inhomogeneities (due to currents and depth variations) and their effects. A numerical model for linear waves based on the re-derived kinematic and dynamic equations has been implemented to evaluate the significance of the newly obtained correction terms.

1. INTRODUCTION

Numerical wave propagation models are nowadays common engineering

[1] *Associate Professor and Professor, Laboratori d'Enginyeria Marítima (LIM), Universitat Politècnica de Catalunya, Campus Nord, Mod. D-1, 08034 Barcelona, Spain*

practise to predict wave behaviour in coastal regions. These models simulate the transformation of wave characteristics from the location where wave data are collected to the site of concern. In the last decades a large number of numerical codes have been developed with a varying degree of approximation in the description of several physical processes, such as shoaling, refraction, diffraction, wave-current interaction, etc... Among these models, those based on linear wave theory have enjoyed great success in engineering applications due to their low cost/performance ratio.

Much of the early work in this area was based on linear wave ray theory, which gives a general understanding of the wave propagation phenomenon. Ray models compute wave transformation due to bottom refraction by tracing wave rays from deep water to the shore, and assuming that the energy flux is conserved between two adjacent rays. This theory, however, ceases to be valid when ray crossings (caustics) occur.

The physical process of diffraction was explained by Battjes (1968) as a redistribution of the wavenumber vector $\mathbf{K}$ -and henceforth, a reorientation of wave rays- due to horizontal variations of wave amplitude, the energy flux still being conserved between adjacent rays. Battjes also gave expressions to correct the wavenumber modulus $|\mathbf{K}|$ and the group velocity $\mathbf{C_g}$ in zones with local convergence or divergence of wave rays. These expressions were obtained from the Helmholtz equation, which corresponds to waves propagating in a homogeneous medium.

A more adequate equation to describe combined refraction and diffraction effects from deep to shallow water was derived by Berkhoff (1972) using the socalled mild-slope approximation. The mild-slope equation confirmed Battjes findings, and gave more correct $|\mathbf{K}|$ and $\mathbf{C_g}$ expressions for the case of waves propagating in an inhomogeneous medium.

A fundamental step towards the understanding of wave-current interacting processes was given by Bretherton and Garrett (1968) using a Lagrangian approach for nondiffracted progressive waves. The most significant result was the identification of the wave action and its conservation along wave rays.

Booij (1981) extended the mild-slope equation to include the effect of varying currents. Kirby (1984) showed that Booij had used an improper form of the dynamic free surface boundary condition, and derived an improved

equation which has been shown to be the correct governing equation for linear gravity waves propagating over slowly varying depth and currents.

While considerable research has gone into the study of refraction-diffraction and wave-current interaction separately, the combined physical process is not yet fully understood. This becomes clear through inspection of different papers dealing with that subject: (Kirby, 1984), (Dingemans, 1985), (Kostense et al., 1988), (Liu, 1990), (Li and Anastasiou, 1992), among others.

The aim of this paper is, therefore, to highlight some inconsistencies that have appeared in the literature, and to advance in the understanding of the physical processes that linear gravity waves undergo in their propagation over slowly varying depth and currents.

2. THE LINEAR WAVE PROBLEM

The governing equations and boundary conditions for small amplitude water waves propagating over a region with gradually varying depth and currents will be summarized. Cartesian coordinates $(\mathbf{x}, z) = (x, y, z)$ referred to the mean free surface $z = 0$, where z is positive upwards, will be employed. The undisturbed water depth is described by $z = -h(\mathbf{x})$ and the free surface displacement is denoted by $z = \eta(\mathbf{x}, t)$. Assuming that the fluid is inviscid and incompressible and that wave and current motions are irrotational, a velocity potential Φ may be introduced, from which wave velocities are defined. The instantaneous horizontal $\mathbf{u}$ and vertical w fluid velocities are written as follows :

$$\mathbf{u} = \mathbf{U} + \nabla\Phi \tag{1.a}$$

$$w = W + \frac{\partial \Phi}{\partial z} \tag{1.b}$$

where $\mathbf{U}$ and W are, respectively, the horizontal and vertical components of the current velocity vector, and ∇ denotes the horizontal gradient operator.

Some scale parameters, representative of physical processes involved in wave propagation, are defined in what follows:

- **$\varepsilon = k\,A$ is the Stokes steepness parameter, where k is the separation factor -equal to the wavenumber modulus in absence of** diffraction-, and A is the wave amplitude. ε is a measure of wave nonlinearity.

- $\delta = \frac{|\nabla h|}{kh}$ is related to the rate of change of depth over the space of a wavelength. This parameter characterizes the mildness of bottom slope.

- $\nu' = \frac{|\mathbf{U}|}{c}$, where $c = \frac{\omega}{k}$ is the wave phase celerity and ω is the absolute wave angular frequency, acts as an indicator of wave-current interacting processes.

- $\nu = \delta\nu'$ is a more complete measure of the influence of horizontal inhomogeneities on waves, since it accounts for the combined effects of currents and bottom variations in the wave propagation space.

The following dimensionless independent variables are adopted herein using ω^{-1} as the time scale, k_o^{-1} as the vertical length scale, and $(\delta k_o)^{-1}$ as the horizontal length scale :

$$(\mathbf{x}, z) = (\delta \mathbf{x}^*, z^*)k_o \qquad t = \omega t^*$$

where all variables with asterisks represent physical quantities and $k_o = \frac{\omega^2}{g}$ is the deep-water wavenumber value; g is the gravity acceleration.

Reference values for the wave amplitude, A_o, and current velocity, $|\mathbf{U_o}|$, are employed to normalize all variables associated with wave and current motion, viz.:

$$\eta^* = \eta A_o = \varepsilon \frac{\eta}{k_o}$$

$$\Phi^* = A_o \frac{\omega}{k_o} \Phi = \varepsilon \frac{\omega}{k_o^2} \Phi$$

$$\mathbf{U}^* = |\mathbf{U_o}| \mathbf{U}$$

$$W^* = \delta |\mathbf{U_o}| W$$

Assuming ε to be small $(\varepsilon \ll 1)$, and the current field to be nearly-horizontal, the linearized governing equations for the wave velocity potential Φ are: (Longuet-Higgins and Stewart, 1961), (Liu, 1990)

$$\delta^2\nabla^2\Phi + \frac{\partial^2\Phi}{\partial z^2} = 0 \quad \text{for} \quad -h < z < 0 \tag{2}$$

$$\frac{\partial\eta}{\partial t} + \nu(\mathbf{U}\cdot\nabla\eta) + \nu(\nabla\cdot\mathbf{U})\eta - \frac{\partial\Phi}{\partial z} = 0 \quad \text{at} \quad z = 0 \tag{3}$$

$$\frac{\partial\Phi}{\partial t} + \nu(\mathbf{U}\cdot\nabla\Phi) + \eta = 0 \quad \text{at} \quad z = 0 \tag{4}$$

$$\frac{\partial\Phi}{\partial z} + \delta^2(\nabla h\cdot\nabla\Phi) = 0 \quad \text{at} \quad z = -h \tag{5}$$

The above set of equations may be converted into a second-order differential equation for Φ :

$$\delta^2\nabla^2\Phi + \frac{\partial^2\Phi}{\partial z^2} = 0 \quad \text{for} \quad -h < z < 0 \tag{6}$$

with the following boundary conditions :

$$\frac{D^2\Phi}{Dt^2} + \nu(\nabla\cdot\mathbf{U})\frac{D\Phi}{Dt} + \frac{\partial\Phi}{\partial z} = 0 \quad \text{at} \quad z = 0 \tag{7}$$

$$\frac{\partial\Phi}{\partial z} + \delta^2(\nabla h\cdot\nabla\Phi) = 0 \quad \text{at} \quad z = -h \tag{8}$$

and an explicit expression for the free surface elevation :

$$\eta = \frac{D\Phi}{Dt} \tag{9}$$

where $\frac{D}{Dt} = \frac{\partial}{\partial t} + \nu(\mathbf{U}\cdot\nabla\ )$ is the total time-derivative.

3. SIMPLE HARMONIC WAVES

Applying the separation-of-variables technique to obtain approximate solutions for Eqs. (6), (7), (8)

$$\Phi(\mathbf{x}, z, t) = \phi(\mathbf{x}, t) \cdot f(z) \tag{10}$$

leads to consider the Sturm-Liouville problem for the profile function $f(z)$ in the region $-h < z < 0$ (Smith & Sprinks, 1975). The solution associated with the propagating wave mode is given by

$$f(z) = \frac{\cosh[k(z+h)]}{\cosh(kh)} \tag{11}$$

where k is the separation factor that satisfies the local dispersion relation given below (16).

Assuming nearly-plane waves, which are harmonic in time, the wave velocity potencial Φ at any level z may be defined in terms of its amplitude $\Phi_o\ f(z)$ and a phase function S_ϕ :

$$\Phi(\mathbf{x}, z, t) = \phi(\mathbf{x}, t) f(z) = \phi_0(\mathbf{x}, t)\ e^{iS_\phi(\mathbf{x}, t)} f(z) \tag{12}$$

where, unless otherwise specified, the real part is considered.

The wavenumber vector $\mathbf{K}_\phi$ associated with Φ , and the absolute angular frequency ω are defined in terms of the S_ϕ derivatives:

$$\mathbf{K}_\phi = \nabla S_\phi \tag{13}$$

$$\omega = -\frac{\partial S_\phi}{\partial t} \tag{14}$$

The relative (intrinsic) angular frequency σ_ϕ is found from Doppler's relation :

$$\sigma_\phi = -\frac{DS_\phi}{Dt} = \omega - \mathbf{U} \cdot \nabla S_\phi = \omega - \mathbf{U} \cdot \mathbf{K}_\phi \tag{15}$$

The dispersion relation is obtained from the real part of the free surface boundary condition (7), after substitution of the Φ expression (12):

$$\sigma_\phi^2 = \sigma^2 + \frac{1}{\phi_o}\left(\frac{D^2\phi_o}{Dt^2} + \nu(\nabla \cdot \mathbf{U})\frac{D\phi_o}{Dt}\right) \tag{16}$$

where $\sigma^2 = k\tanh(kh)$ is the well-known (dimensionless) dispersion relation in linear gravity wave theory.

The fact that σ_ϕ may differ from the classical value σ for small-amplitude waves over slowly varying depth and currents, was first pointed out by Liu (1990), although he did not provide any explicit relationship between σ and σ_ϕ (the "effective" intrinsic frequency, Σ, in his notation).

Assuming now a stationary wave field, it can be seen that all correction terms in (16) are of order $O(\nu^2)$:

$$\sigma_\phi^2 = \sigma^2 + \nu^2 \frac{1}{\phi_o} \nabla \cdot \Big((\mathbf{U} \cdot \nabla \phi_o) \mathbf{U} \Big) \tag{17}$$

The expression for the free surface elevation η is found from the dynamic free surface boundary condition (9) :

$$\eta = -\frac{\partial \phi}{\partial t} + \nu (\mathbf{U} \cdot \nabla \phi) \tag{18}$$

$$\eta = i \sigma_\phi \phi_o (1 + i\nu\theta) e^{iS_\phi} \tag{19}$$

$$\theta = \frac{\mathbf{U} \cdot \nabla \phi_o}{\sigma_\phi \phi_o}$$

Writing η in terms of an amplitude A and a phase function S_η :

$$\eta(\mathbf{x}, t) = A(\mathbf{x}, t) e^{iS_\eta(\mathbf{x}, t)} \tag{20}$$

it is found that

$$A = i\sigma_\phi \phi_o (1 + \nu^2 \theta^2)^{\frac{1}{2}} = i\sigma_\phi \phi_o (1 + \frac{1}{2}\nu^2\theta^2) + O(\nu^4) \tag{21}$$

$$S_\eta = S_\phi + \tan^{-1}(\nu\theta) = S_\phi + \nu\theta + O(\nu^3) \tag{22}$$

from which, within the framework of linear wave theory in the mild-slope approximation, the following initial conclusions may be derived:

(i) The common relationship $A = i\sigma\phi_o$ (or $\phi_o = -ig\frac{A}{\sigma}$, in dimensional form), which has been rarely disputed, is strictly valid up to order $O(\nu)$, and approximate only for higher orders of approximation in terms of ν . The use of the standard (A, ϕ_o) relationship appears, thus, to be inconsistent with the distinction between σ_ϕ and σ (Liu, 1990), whose differences are of order $O(\nu^2)$.

(ii) Whenever ambient currents are present, there exists in general a phase lag between wave velocities (in terms of Φ) and the free surface elevation (η). To the best of the authors' knowledge, this lag has never been mentioned before in the literature within this context. Up to order $O(\nu)$, the phase lag can be estimated -in dimensional form- by $\theta = \frac{\mathbf{U}\cdot\nabla\phi_o}{\sigma_\phi\ \phi_o}$, which accounts for the combined effects of refraction-diffraction and wave-current interaction.

According to the aforementioned findings, the wavenumber vector $\mathbf{K}_\eta$ and wave frequencies (absolute ω and relative σ_η), defined from the free surface elevation phase function, may be expressed as:

$$\nabla S_\eta = \mathbf{K}_\eta = \mathbf{K}_\phi + \nu\nabla\theta + O(\nu^3) \tag{24}$$

$$-\frac{\partial S_\eta}{\partial t} = \omega = -\frac{\partial S_\phi}{\partial t} \tag{24}$$

$$-\frac{DS_\eta}{Dt} = \sigma_\eta = \sigma_\phi - \nu^2(\mathbf{U}\cdot\nabla\theta) + O(\nu^4) \tag{25}$$

4. THE MILD-SLOPE EQUATION

Kirby (1984) derived a hyperbolic equation for wave propagation through varying depth and currents within the mild-slope approximation. The socalled time-dependent mild-slope equation (hereafter referred to as MSE, for shortness) reads in non-dimensional form

$$\frac{D^2\phi}{Dt^2} + \nu(\nabla\cdot\mathbf{U})\frac{D\phi}{Dt} - \delta^2\nabla\cdot(cc_g(\nabla\phi)) + k^2(c^2 - cc_g)\phi \ = \ 0 \tag{26}$$

where $c = \frac{\sigma}{k}$ and $c_g = \frac{\partial\sigma}{\partial k}$ are, respectively, the phase and group velocity in linear wave ray theory (i.e. k= $|\mathbf{K}|$). All remaining variables have been previously defined.

Kirby also enhanced the validity of equation (26) by showing that it correctly leads to the wave action conservation equation (Bretherton & Garrett, 1968) without further approximations. It should be mentioned, however, that his derivation was accurate up to order $O(\nu)$, in our notation.

A more correct, though incomplete, demonstration is given in Liu (1990). An exact derivation -within this context- of the wave action conservation principle and the corresponding eikonal-equivalent equation, obtained from the MSE, is given below.

After substituting the wave potential expression at $z = 0$, ϕ -as given by equation (12)-, into the MSE (26), and separating real and imaginary parts, it follows :

(i) from the real part,

$$|\nabla S_\phi|^2 = \mathbf{K}_\phi{}^2 \; = k^2 + \delta^2 \frac{\nabla \cdot (cc_g \nabla \phi_o)}{cc_g \theta_o} +$$

$$+ \frac{1}{cc_g} \left(\sigma_\phi^2 - \sigma^2 - \frac{1}{\phi_o} \Big(\frac{D^2 \phi_o}{Dt^2} + \nu \frac{D \phi_o}{Dt} (\nabla \cdot \mathbf{U}) \Big) \right) \tag{27}$$

Eq. (27) is equivalent to the eikonal equation in geometrical wave ray theory. It recovers the relation between the vertical structure of the flow - in terms of the separation factor k - and the wave propagation space -in terms of the 'effective' wavenumber vector $\mathbf{K}_\phi$ - which was split off due to the separation-of-variables approach (10) applied to Eq. (6). An expression similar to (27) was given in [Liu (1990), Eq. (35)] using the standard (A , ϕ_o) relationship ($\phi_o = -ig\frac{A}{\sigma}$). However, Liu did not notice that, according to the new dispersion relation (16), the term inside brackets in Eq. (27) drops out. The eikonal-equivalent equation should thus read:

$$\mathbf{K}_\phi{}^2 \; = \; k^2 + \delta^2 \frac{\nabla \cdot (cc_g \nabla \phi_o)}{cc_g \phi_o} \tag{28}$$

or in dimensional form

$$\mathbf{K}_\phi{}^2 \; = \; k^2 + \frac{\nabla \cdot (cc_g \nabla \phi_o)}{cc_g \phi_o} \tag{29}$$

(ii) from the imaginary part,

$$\frac{\partial \mathcal{A}}{\partial t} + \nabla \cdot \Big((\mathbf{C_g} + \mathbf{U}) \mathcal{A} \Big) = 0 \tag{30}$$

which is the appropiate form of the wave action conservation principle (WACP), and where

$$\mathbf{C_g} = c_g \, \frac{\sigma}{\sigma_\phi} \, \frac{\mathbf{K}_\phi}{k} \tag{31}$$

is the 'effective' group velocity, first obtained by Liu (1990), and

$$\mathcal{A} = \sigma_\phi \, \phi_o^2 \tag{32}$$

is the correct form of the wave action density. Liu's derivation is only approximate according to the above findings -see Eq. (21)- since he uses

$$\mathcal{A} = \frac{A^2}{\sigma}\Big(\frac{\sigma_\phi}{\sigma}\Big) \tag{33}$$

instead of

$$\mathcal{A} = \frac{A^2}{\sigma_\phi} \, \frac{1}{1+\nu^2\theta^2} = \frac{A^2}{\sigma_\phi}\Big((1-\nu^2\theta^2+O(\nu^4)\Big) \tag{34}$$

which follows directly from (21) and (32) without further approximations.

5. PRELIMINARY NUMERICAL RESULTS

A numerical model based on these re-derived equations has been implemented to evaluate the newly obtained correction terms. Details of the numerical procedure will not be given here. Suffice it to say that it is based on the following two equations :

(i) Irrotationality of the 'effective' wavenumber vector $\mathbf{K}_\phi$:

$$\nabla \times \mathbf{K}_\phi = 0 \tag{35}$$

(ii) Conservation of the wave action density $\mathcal{A}$, in which a dissipation -sink- term $\mathcal{D}$ has been included and modelled according to Battjes & Janssen (1978) :

$$\frac{\partial \mathcal{A}}{\partial t} + \nabla \cdot \Big((\mathbf{C_g}+\mathbf{U})\mathcal{A}\Big) + \frac{\mathcal{D}}{\sigma_\phi} = 0 \tag{36}$$

Numerical results correspond to the case of normally incident waves on a plane beach with a rip current system and a slope of $\frac{1}{50}$. This problem was originally studied in Arthur (1950). The current velocity field is given by

$$U_x = -1.097\left(2-\Big(\frac{y}{76.2}\Big)^2\right) F\Big(\frac{y}{76.2}\Big) \int_o^{\frac{x}{7.62}} F(\sigma)\, d\sigma \qquad (m/s)$$

$$U_y = 0.144\ y\ F\left(\frac{y}{76.2}\right)\ F\left(\frac{x}{7.62}\right) \qquad (m/s)$$

with

$$F(\sigma) = \frac{1}{\sqrt{2\pi}}\ e^{\frac{-\sigma^2}{2}}$$

as proposed in the original paper (see Fig. 1). This figure also shows Arthur's results, which were obtained with linear wave ray theory for a wave train period with 8 seconds. Since this ray method only considers refraction and shoaling effects, unlimited wave heights are predicted where the rays cross each other.

Using the developed numerical model computations were performed for normally incident waves with 8s period and 0.2m height at the incoming wave boundary, as originally proposed by Arthur. Energy dissipation effects due to wave breaking were taken into account to obtain realistic results in very shallow water. Fig. 2 shows the calculated wave height distribution in which the strong convergence of wave energy flux due to current-refraction can be observed. Diffraction effects -due basically to horizontal gradients of the amplitude- are responsible for limiting wave heights by redistributing wave action fluxes (rather than wave energy). To investigate the importance of the newly obtained correction terms, a contour plot of the 'phase lag' θ is shown in Fig. 3. The maximum value obtained for θ was 0.07, which corresponds to a phase lag of about 4 degrees.

6. CONCLUSIONS

New expressions for some kinematic and dynamic properties of linear surface gravity waves over slowly varying depth and currents have been obtained, which are exact within the mild-slope approximation. These expressions have been derived from a set of linear wave equations including current effects (Longuet-Higgins and Stewart, 1961) and based on the time dependent mild-slope equation (Kirby, 1984). A new small parameter ν, which takes into account the effects of horizontal inhomogeneities (due to depth and current variations) on waves, has been defined. This parameter allows to assess the order of approximation of similar expressions proposed in the literature (e.g. Liu, 1990) with respect to the exact equations.

Some of the obtained results are summarized in the following five conclusions :

1. The fact that the intrinsic wave frequency, σ_ϕ , differs from that given in classical linear theory $\sigma = \sqrt{gk\tanh(kh)}$ (Liu, 1990) has been confirmed, and an explicit relaionship between σ_ϕ and σ has been found. Differences appear at order $O(\nu^2)$.

2. In the presence of ambient currents, the free surface elevation η and wave orbital velocities have, in general, a small phase lag, due to the combined effects of currents and horizontal amplitude variations. Up to order $O(\nu)$, this phase lag is given by $\theta = \frac{\mathbf{U}\cdot\nabla\phi_o}{\sigma_\phi\,\phi_o}$.

3. The relationship between the free-surface elevation amplitude A and the wave velocity potential amplitude at MWL, ϕ_o , is found to differ from the common expression $\phi_o = -ig\frac{A}{\sigma}$. An exact relationship -in this context- has been obtained.

4. The eikonal-equivalent equation for combined refraction-diffraction and wave-current interaction has been derived from the MSE and found to differ from other expressions given in the literature.

5. The wave action conservation equation has also been derived from the MSE, and more correct forms of the group velocity and wave action density have been obtained.

6. Numerical results based on the aforementioned expressions, have also been presented to study the effects of a rip current on the propagation of waves normally incident on a plane beach (Arthur, 1950). The phase lag was calculated throughout the domain giving maximum values of about 4 degrees.

ACKNOWLEDGEMENTS

This work was undertaken as part of the MAST G-6 Coastal Morphodynamics research programme. It was funded jointly by the Programa de Clima Maritimo (PCM) of the Spanish Ministry of Public Works (MOPT) and by the Commission of the European Communities (D.G. XII) under contract no. MAST-0035-C.

FIGURES

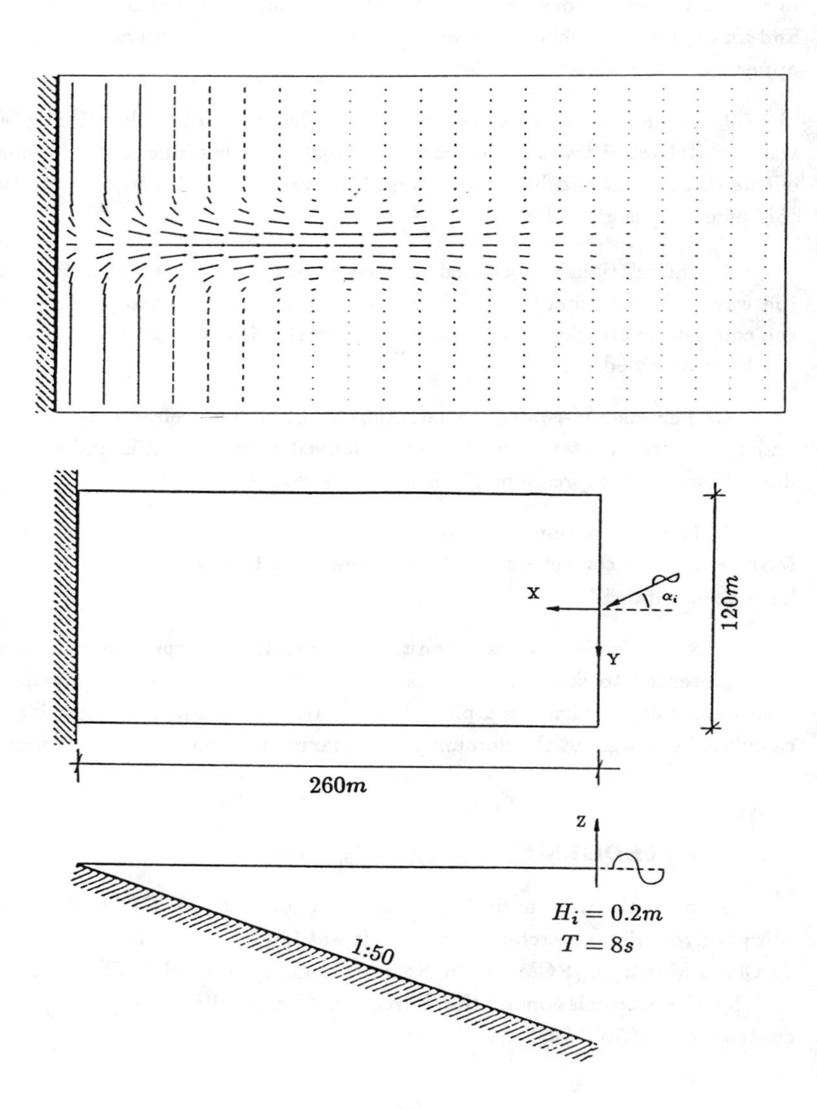

Figure 1. Definition sketch of test case. Arthur (1950)

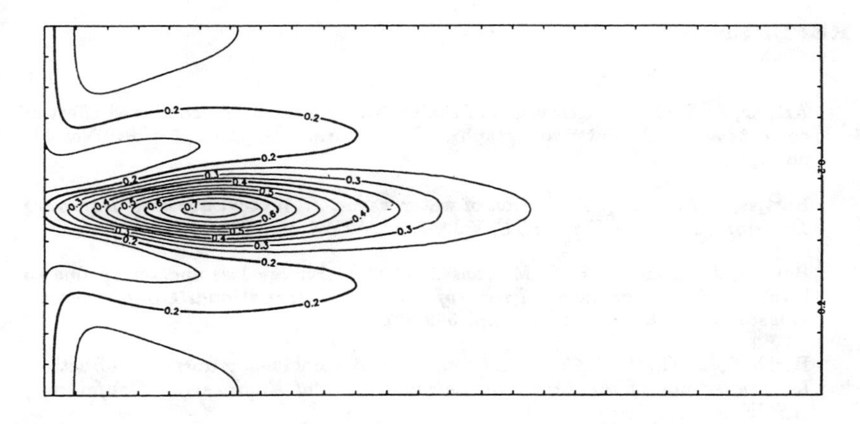

Figure 2. Contour plot of wave height distribution (in m).
Normally incident waves ($\alpha_o = 0$)

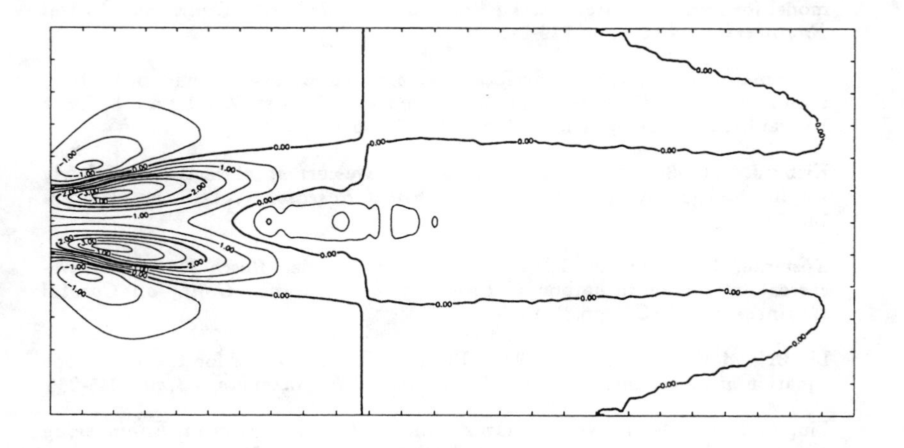

Figure 3. Contour plot of θ parameter -phase lag- (in deg)
Normally incident waves ($\alpha_o = 0$)

REFERENCES

Arthur, R. S. (1950). Refraction of shallow water waves: The combined effect of currents and underwater topography. *Trans. Am. Geophys. Union.*, Vol.**31**, no. 4, pp. 549-552.

Battjes, J. A. (1968). Refraction of water waves, *J. Waterways and Harbours Division. Proc. of the ASCE*, Vol.**94**, no. WW4, pp. 437-451

Battjes, J. A. and J. P. F. M. Janssen (1978). Energy loss and set-up due to breaking of random waves. *Proc. of the 16th International Conference on Coastal Engineering, ASCE*, pp. 569-587

Berkhoff, J. C. W. (1972). Computation of combined refraction diffraction. In *Proceedings of the 13th International Coastal Engineering Conference, ASCE*, pp. 471-490

Booij, N. (1981). Gravity waves on water with non-uniform depth and current. Report No. 81-1, Delft University of Technology, The Netherlands, 130 pp.

Bretherton, F. P. and C. J. R. Garret (1968). Wave trains in inhomogeneous moving media. *Proc. Roy. Soc. London*, Series A, **302**, pp. 529-554

Chae, J.W. and S.T. Jeong (1992). Current-depth refraction and diffraction model for irregular water waves. *Proc. of the 23rd Int. Conf. on Coastal Engineering, ASCE*, pp.129-141

Dingemans, M. W. (1985). Surface wave propagation over an uneven bottom; evolution of two-dimensional horizontal models. Report W301 part 5, Delft Hydraulics Laboratory, The Netherlands, 117 pp.

Kirby, J. T. (1984). A note on linear surface wave-current interaction over slowly varying topography. *Journal of Geophysical Research*, Vol.**89**, no. C1, pp. 745-747

Kostense, J. K., M. W. Dingemans, and P. van den Bosch (1988). Wave-current interaction in harbours. *Proc. of the 21th Int. Conf. on Coastal Engineering, ASCE*, pp.32-46

Li, B. and K. Anastasiou (1992). Efficient elliptic solvers for the mild-slope equation using the multigrid technique. *Coastal Engineering*, **16**, pp. 245-266

Liu, P. L.-F. (1990). Wave transformation. *The Sea: Ocean Engineering Science*. John Wiley & Sons, Inc. Vol.**19**, Ch. 2, pp. 27-63

Longuet-Higgins, M.S. and R.W. Stewart (1961). The changes in amplitude of short gravity waves on steady non-uniform currents. *J. of Fluid Mechanics*, Vol.**10**, pp. 529-549

Smith, R. and T. Sprinks (1975). Scattering of surface waves by a conical island. *J. of Fluid Mechanics*, Vol.**72**, part 2, pp. 373-384

The Swept Sine Wave Testing Technique in Hydrodynamic Applications

Nicholas Haritos[1]
Okey Nwogu[2] & Andrew M. Cornett[2]

Abstract

This paper provides a detailed description of a simple yet powerful testing technique, referred to as Swept Sine Wave (SSW) testing, highlighting its major attributes that make it attractive when compared with more "standard" wave testing methods. The method is particularly suited to the experimental identification of the dynamic response characteristics of simple structural systems such as the transfer function characteristics of a wave maker and the "large-amplitude" non-linear dynamic response characteristics of simple models of compliant offshore structural systems, example of which are presented in the paper.

Introduction

Swept Sine Waves (SSWs) have been commonly used by electrical and mechanical engineers to study the behaviour of oscillatory phenomena, (White, 1971; Kandianis, 1971). For example, modern signal generators, which may form part of a dynamic physical testing system, often include a SSW or Chirp signal option for the control of the input excitation in such a system. In recent times, the technique has also been found to be a particularly convenient and expedient method for use in the study of hydrodynamic related phenomena that exhibit dynamic or frequency-dependent characteristics, (Haritos, 1988; Haritos, 1989; Haritos et al, 1989).

This paper provides a detailed description of this simple yet powerful testing technique highlighting its major attributes that make it attractive when compared with more "standard" testing approaches (such as regular wave testing, irregular wave testing and the use of "snapshot" wavetrains or Gaussian wave packets) and draws upon some recent applications of the method when applied to the hydrodynamic context by way of illustration. These applications include the experimental identification of the wave making transfer function characteristics of a Salter duck style wave generator (as used in the Michell laboratory of the Univesity of Melbourne), and the "large-amplitude" non-linear dynamic response characteristics of

[1] Reader in Civil Engineering, University of Melbourne, Grattan St., Parkville 3052, AUSTRALIA
[2] Research Engineers, Hydraulics Laboratory, National Research Council of Canada, Montreal Rd. Ottawa, CANADA K1A OR6

simple compliant offshore structural systems. In particular, SSWs are used to study the response of model vertical bottom-pivoted surface-piercing cylinders tested both at the Hydraulics Laboratory of the National Research Council of Canada in Ottawa and at the Michell laboratory of the University of Melbourne. In addition, reference is also made to numerical modelling using the SSW concept that has been conducted using easy to generate special purpose code that can execute on more modest computing platforms, such as the popular IBM compatible PCs, by way of illustrating the versatility of the method as a numerical technique.

Description of Swept Sine Waves

The following description presents the basic principal features of the SSW technique and its implementation in the generation of waves with properties suited to the study of hydrodynamic phenomena. Further detail may be found in Haritos, (1988).

Consider a wave surface trace $\eta(x,t)$ in depth of water h and conforming to the generalised description:

$$\eta(x,t) = A(t) \cos (\kappa(t).x - \Omega(t).t) \tag{1}$$

in which $A(t)$ is the time varying amplitude, $\kappa(t)$ is the time varying wave number and $\Omega(t)$ is related to circular frequency $\omega(t)$ at time t via the condition

$$\omega(t) = \frac{\partial(\Omega(t).t)}{\partial t} \tag{2}$$

$\omega(t)$ also satisfies the dispersion relationship for small amplitude linear waves given by

$$\omega^2(t) = g.\kappa(t) \tanh\left(\kappa(t).h\right) \tag{3}$$

A special case for the description above would be that of an Airy wave in which $\omega(t)=\omega_o$, $\kappa(t)=\kappa_o$ and $A(t)=a_o$, so that

$$\eta(x,t) = a_o \cos\left(\kappa_o x - \omega_o t\right) \tag{4}$$

or

$$\eta(t) = a_o \cos\left(\omega_o t\right) \tag{5}$$

for $\eta(t)$ defined at the origin, $x = 0$.

While it is possible to select an $\Omega(t)$ variation that may conform to a more complex description, a rather simple concept for producing an SSW is one in which $\omega(t)$ is chosen to vary linearly with respect to time over prescribed frequency limits in a time length of record of T. If $\omega_1 = 2\pi f_1$ and $\omega_2 = 2\pi f_2$ represent the start and end values of circular frequency for such an SSW, then

$$\omega(t) = \omega_1 + \left(\frac{\omega_2 - \omega_1}{T}\right).t \tag{6}$$

which, from eqn. (2), requires

$$\Omega(t) = \omega_1 + \left(\frac{\omega_2 - \omega_1}{2T}\right).t \qquad (7)$$

The frequency at any time t within the record of this form of SSW is therefore given by

$$f(t) = f_1 + \left(\frac{f_2 - f_1}{T}\right).t \qquad (8)$$

Equation (8) clearly demonstrates the one-to-one correspondence between time and frequency in the definition of this *linear-frequency* form of SSW which is a property important for developments that are to follow.

The time (or frequency) dependent wave amplitude, A(t) (or A(f)), in the case of a linear frequency SSW may be chosen to conform to any desired characteristic, eg constant wave amplitude (for "general purpose" testing) or constant inertia force amplitude (for testing the dynamic response characteristics of compliant bottom-pivoted test cylinders).

Implementation of the Linear Frequency SSW in the Wave Tank

The control signal that generates a linear-frequency SSW in a wave tank is normally applied so that sweeping occurs from its high frequency to its low frequency end in order to reduce the effects of wave reflections from the beach (or wave absorbers at the end of the tank). Reflections from low frequency waves tend to be more pronounced than those from higher frequency waves for most beach types, hence the preference for having the low frequency end of an SSW generated last in the wavetrain produced. Figure 1 depicts a schematic of such a constant amplitude linear-frequency SSW.

If wave tank reflections are to be avoided altogether, the data acquisition time would have to be limited to the time required by the incident waves to be reflected back to the location of the experiment being conducted in the wave tank. Although the use of efficient wave absorbers can significantly reduce the effects of wave reflections, thereby allowing extended recording periods, they do not entirely eliminate these reflections. In practice, waves at a test site are a combination of incident and reflected components. A technique proposed by the principal author which capitalises on the linear frequency characteristics of the general purpose SSW, (Haritos & Hunt, 1993), allows identification and removal of reflected wave influences from single point measurements of the waves (and of other quantities that have been digitally recorded during the conduct of an SSW experiment), thus improving the "quality" of the data being processed: an added benefit of SSW testing.

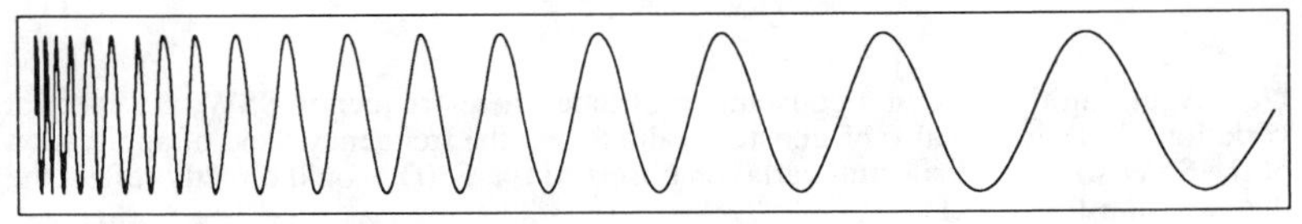

Figure 1. Schematic of a Linear-Frequency SSW with Constant Amplitude

Spectral Variation of a SSW

The amplitude at any time t in a linear-frequency SSW can be selected to directly conform to any desired spectral variation $S_\eta(f)$ because of the one-to-one correspondence between time and frequency associated with this form of SSW generation as given by eqn. (8). Now if $f_1 < f < f_2$ represents the active frequency band for the SSW under consideration and $\Delta f = f_2 - f_1$, the required selection for the frequency-dependent amplitude becomes

$$A(f) = \sqrt{2 S_\eta(f) \Delta f} \qquad (9)$$

From inspection of eqn. (9), a *constant amplitude* SSW would be associated with a *constant amplitude* surface elevation spectrum over the frequency bandwidth of application of this SSW.

The Determination of Transfer Functions using the SSW Testing Technique

Suppose quantity q(t) is linearly related to surface elevation at x = 0, which takes the form of a linear-frequency SSW, then the trace q(t) becomes:

$$q(t) = H_{q\eta}(f)\, A(f) \cos(\Omega(t).t - \phi_{q\eta}(f)) \qquad (10)$$

where the amplitude/phase characteristics of the linear process governing the relationship between $\eta(t)$ and q(t) are given by $(H_{q\eta}(f), \phi_{q\eta}(f))$.

It is obvious that these characteristics are imbedded directly in the time trace q(t) as an amplitude amplification $H_{q\eta}(f)$ of the SSW at a point in time t corresponding to frequency f (as determined by eqn. (8) for the linear-frequency SSW concerned) and a phase shift $\phi_{q\eta}(f)$ relative to this SSW. Consequently these characteristics can be obtained directly by inspection of the traces for $\eta(t)$ and q(t).

In the case of an SSW with a constant amplitude, say $A(f) = A_o$, eqn. (10) suggests that the temporal variation for q(t) would *directly* reflect the variation in $H_{q\eta}(f)$ over the active frequency band of this SSW. It is this particular feature of the constant amplitude linear-frequency form of SSW that has been found to be very useful to the investigation of the frequency response characteristics of physical models being tested in the wave tank as well as several other applications. (Some of these applications will be described a little later in this paper).

As an alternative, the transfer function for the process governing the relationship between $\eta(t)$ and q(t) can be obtained from the more traditional approach of a spectral description such as given by:

$$H^2_{q\eta}(f) = \frac{S_q(f)}{S_\eta(f)} \qquad (11)$$

Again, in the case of a constant amplitude linear-frequency SSW, the spectral variation $S_\eta(f)$ would also be constant valued over the frequency band of application of the SSW so that the spectral variation observed for $S_q(f)$ would directly reflect the the variation for $H^2_{q\eta}(f)$.

Applications of the SSW Testing Technique

The SSW testing technique has many attributes which make it well suited to hydrodynamic applications. One such attribute is that frequency and time occur in parallel therefore the transition from the time domain to the frequency domain can be made effortlessly especially in the case of linear-frequency SSWs. In addition because a band of frequencies are involved in generating SSWs, application of the method is more-or-less equivalent to conducting a large number of successive regular wave tests. These attributes have been capitalised on in the conduct of several experimental programs using SSWs, an overview of which is provided below.

Studies of the Response of Compliant Structural Systems using SSWs

A comprehensive experimental study of the dynamic response characteristics of a range of compliant bottom-pivoted cylinders of the type depicted in Figure 2 has recently been completed at the University of Melbourne. This has involved experimental programs that have been performed both at the hydraulics laboratory of National Research Council (NRC) of Canada in Ottawa and in the Michell laboratory of the University of Melbourne. The study has principally been aimed at identification of fluid/structure interaction effects and their influence on the damping characteristics of the test cylinders involved. Some of the results obtained from these two experimental programs detailing the role played by the SSW testing is provided below.

NRC Test Cylinder

Figure 3 presents a schematic of the NRC segmented test cylinder depicting key features of the geometry. A particular feature of this test cylinder is that it contains some nine independent segments instrumented for the in-line and transverse force acting on each, seven of which were completely submerged during testing. In addition, the four support links restraining the top of the cylinder were paired on either side of the cylinder to form an orthogonally disposed cruciform the axes of which were oriented to be either parallel or orthogonal to the sides of the wave tank. The far ends of each link were then connected to individual force transducers themselves mounted on an independent rigid frame. The links consisted of either rigid rods to provide rigid restraint, or springs that permitted controlled levels of compliancy to be introduced to the cylinder.

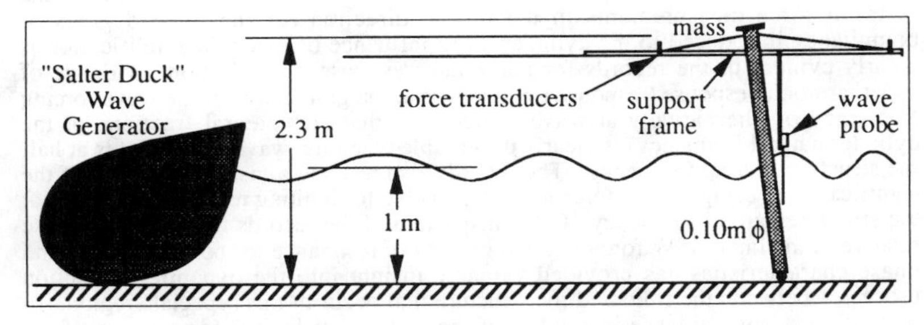

Figure 2. Schematic of a Compliant Bottom-pivoted Cylinder in an SSW Test

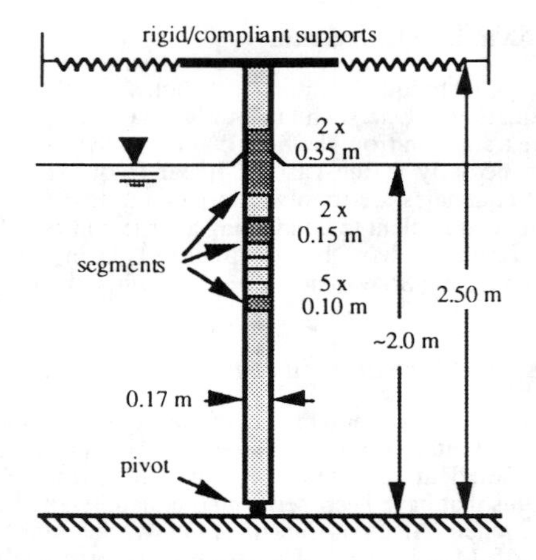

Figure 3. Schematic of NRC Segmented Test Cylinder

A "constant overall cylinder inertia force" linear-frequency version of SSW testing was conducted on this cylinder as well as a series of uni-directional and multi-directional Pierson-Moskowitz (P-M) waves for both rigid support conditions and for three separate degrees of compliancy (natural frequencies f_o of 0.61, 0.66 and 0.76 Hz, respectively). The choice of this style of SSW was made on the basis that any departures from a constant-value of in-line restraint force would be associated with non-linearities in the character of the forcing say from drag or non-linear features of the waves themselves.

The waves were generated over a 819.2 second duration and were linearly ramped in frequency from 1.25 down to 0.25 Hz, (Haritos et al., 1989). The first and last $1/8^{th}$ of the record was scaled by a "half-cosine bell" of the appropriate shape in order to reduce "leakage" effects when conducting a spectral analysis of the recorded time series of measurements from these experiments. Data series were captured at a 10 Hz sampling rate to yield 8192 data points.

Figure 4 provides sample traces of the input uni-directional SSW and of the restraint force measurements in the in-line direction for the three degrees of compliancy introduced to the cylinder. The influence of dynamic amplification is clearly evident in the records for compliant response. In addition, evidence of superharmonic response (response frequency is at integral factors of the wave forcing frequency or alternatively at wave frequencies that are integral fractions of the cylinder natural frequency) is clearly observable when the wave frequency is at half the structure natural frequency. The records also provide a visual indication of the significance of damping, the force term responsible for limiting response amplitude at the structure natural frequency. Close inspection of the records for in-line response relative to the input SSW forcing in the vicinity of resonance for both amplitude and phase characteristics has provided valuable insight into the dynamic interaction processes between the fluid and the structure that occur in this "special" region. In particular, strong evidence in support of the "relative velocity" formulation of the

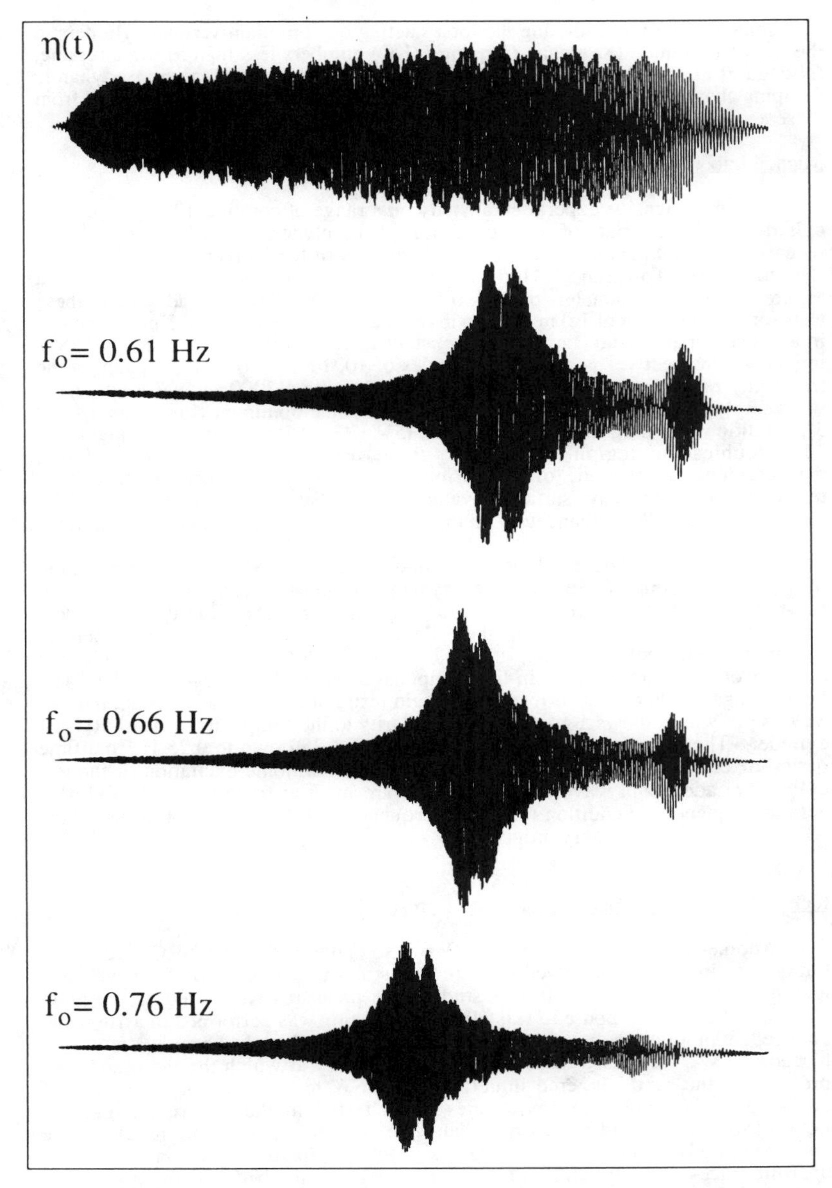

Figure 4. Traces for "Constant Inertia Force" SSW and In-line Restraint Spring Force

Morison equation for predicting the forces acting on compliant vertical cylinders in the inertia dominant (Keulegan Carpenter (KC) numbers less than 5) force regime, (Nwogu et al., 1992), and in support of a simplified model of the hydrodynamic damping characteristics of cylinders of this type, (Haritos, 1991[b]) have emerged from these tests.

Michell Laboratory Test Program

A comprehensive experimental study on a range of compliant bottom-pivoted cylinders under a variety of wave conditions that included sets of SSW signals has recently been completed in the Michell Laboratory of the University of Melbourne (Haritos, 1991[a]; Townsend & Haritos, 1993). The experimental setup is depicted in Figure 2. Cylinder diameters of 0.15, 0.10, 0.05 and 0.025 m were adopted in these tests for a water depth of 1.0 m with up to three cylinders being tested simultaneously in any one experimental run. Three "constant-inertia force linear frequency" SSW inputs with respective "nominal" amplitudes of 10, 14 and 20 mm all spanning the frequency range from 1.75 Hz down to 0.25 Hz over a 2000 second long period were employed in this study. As well as inertia force dominant conditions, (KC < 5), conditions of "drag force significance", (5 < KC < 15), and otherwise known as "the troublesome region", applied to these tests. In all cases considered, measurements were made of the spring restraint force on each support force transducer and of the wave surface elevation profile adjacent to each cylinder with up to 3 cylinders tested simultaneously in the wave tank in any single experimental run.

Again, strong evidence of the applicability of the "relative velocity" formulation of the Morison equation for the forced dynamic response of compliant cylinders of this type and for models of the hydrodynamic damping observed in these cylinders has been facilitated by the use of the SSW testing technique as a complement to irregular wave testing, (Yang, 1990; Townsend, 1993). Figure 5 provides a comparison of the trace for the in-line compliant response of a "drag-significant" test cylinder against that of its corresponding rigid restraint force for a "constant inertia" SSW input which shows a great deal of similarity to the sample traces for NRC test cylinder. The extended linear frequency range of 1.75 down to 0.25 Hz in a time period of 2000 seconds provides evidence of superharmonic excitation of the test cylinder at additional wave frequencies that are integral fractions of the cylinder natural frequency (in addition to a wave frequency of $f_o/2$ that was observed in the NRC test cylinder laboratory program).

Identification of Sand Bed Liquefication using SSW Testing

Another recent application of SSW's (also conducted at the NRC Hydraulics Laboratory in Ottawa) involved the effective stress response of sands under wave loading, (Davies, 1989). In this study, pore pressures were used as the main indicator of seabed response to wave loading. Testing was performed in a 1m wide, 2m deep, 30m long channel on very fine, uniform sand which had been hydraulically placed. SSW's were used to determine the frequency at which the increased pore pressure in the sand triggered liquefaction. SSW testing had the advantage of allowing a wide range of frequencies to be tested in the one run, triggering liquefaction of the sand more often than time consuming "straight" regular wave testing methods. Figure 6, from Davies (1989), depicts sample wave and pore pressure traces from the SSW tests used to trigger sand bed liquefication and to assist with the identification of the transfer function for pore pressure response.

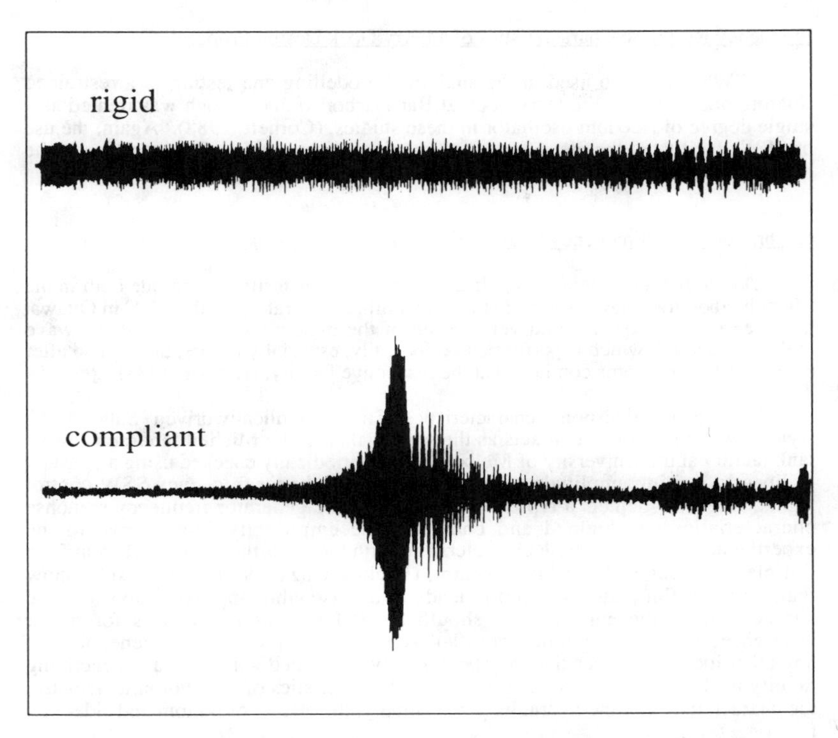

Figure 5. Comparison of the In-line Rigid and Compliant Restraint Force Traces from "Constant Inertia Force" SSW (Michell Lab.)

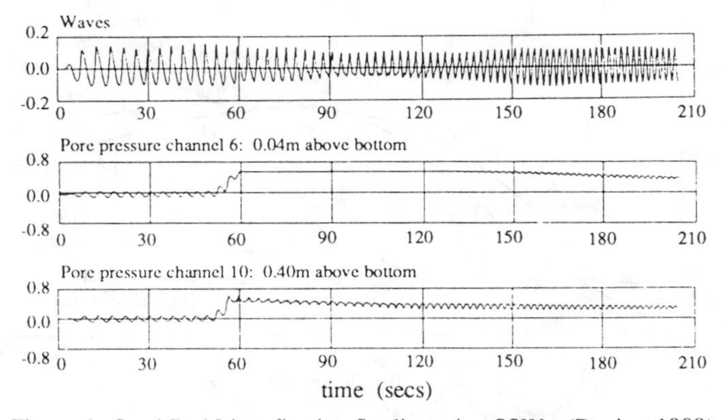

Figure 6. Sand Bed Liquefication Studies using SSWs (Davies, 1989)

Frequency Response Characteristics of a Ferry Dock Design Proposal

SSW's were also used in the analytical modelling and testing of a restrained floating body (that of the ferry dock at Bar Harbor Maine) which was treated as a single degree of freedom oscillator in these studies, (Cornett, 1988). Again, the use of SSWs facilitated the identification of the dynamic response characteristics of the model dock under test.

Calibration of the Frequency Characteristics of a "Salter Duck" Wave Generator

One of the most frequent applications of the SSW testing technique both in the Michell laboratory wave tank and at the Hydraulics Laboratory of the NRC in Ottawa, is the expedient experimental verification of the dynamic calibration of the wave maker concerned, which is performed periodically, especially before, during and after extensive test programs conducted at the respective facility, (Haritos, 1988).

The frequency response characteristics of the hydraulically driven "Salter duck" style of wave generator that acts as the wavemaker in the Michell laboratory wave tank facility at the University of Melbourne are periodically checked using a constant or linearly varying amplitude of stroke version of a linear frequency SSW control signal. Figure 7 depicts a typical calibration for the generator frequency response characteristics so obtained and compares the empirically fitted curve to the experimental results for these characteristics with those for the theoretical form for a flat-plate generator of similar geometry, (Flick & Guza, 1980). The results show that, whilst a flat-plate assumption leads to a reasonable approximation to these characteristics, the empirical fit should be preferred as it accounts for minor differences associated with the "non-flat" active surface profile of the generator and any other local influences that may be uniquely associated with the wave generating facility itself, (eg the servo-valve response characteristics of the hydraulic actuator, the characteristics of the hydraulic power supply, the effects of bottom and side-wall friction, etc).

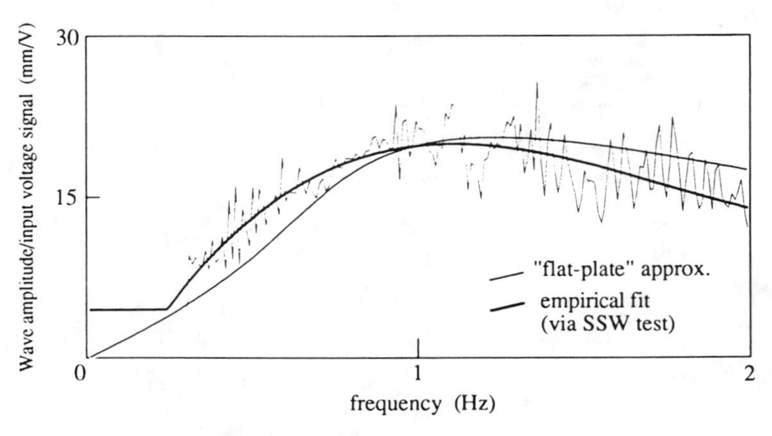

Figure 7. Frequency Response Characteristics of a "Salter Duck" Wave Generator

An SSW signal has also been applied to determine the dynamic calibration characteristics (linear transfer function describing the magnitude and phase response) of the servo control system of the wave generators at the Hydraulics Laboratory of the NRC in Ottawa. (The detailed transfer function characteristics are used in the calculation of drive signals necessary to produce specified wave trains). The amplitude of the SSW signal used for the calibration is chosen to vary gradually in time so as to produce a resultant wave height about 50% of the maximum wave height capability of the wave machine, (see Figure 8). The maximum wave height is limited by the stroke of the wave machine at lower frequencies and the breaking wave height at higher frequencies. By measuring the actuator displacement for a given SSW, a transfer function that includes the response of the digital to analogue converter and analogue filters as well as the servo dynamics of the wave machine can be determined. Figure 9 shows an example of the measured magnitude and phase angle

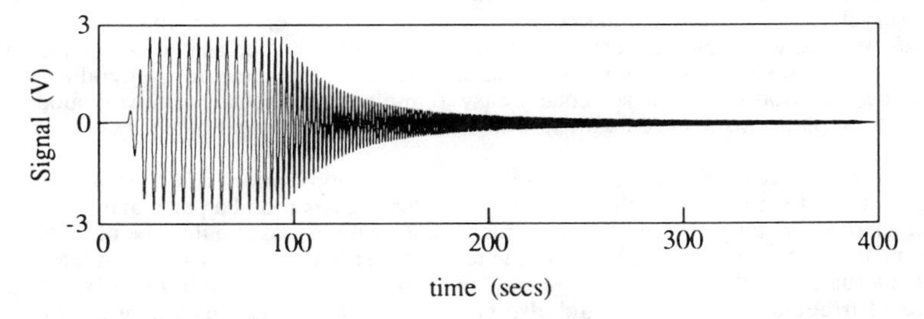

Figure 8. SSW Signal used for Determination of NRC Wave Maker Servo Dynamics

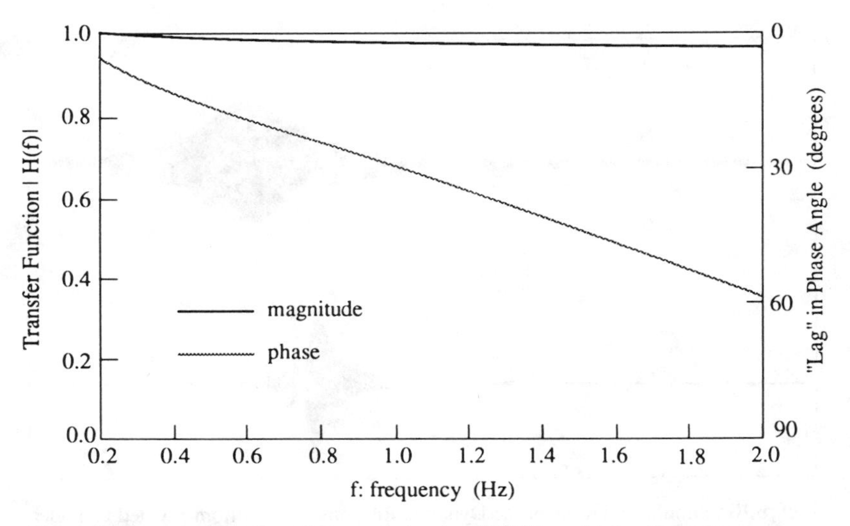

Figure 9. Transfer Function Characteristics for NRC Wave Maker Servo Dynamics

of this hybrid servo dynamics transfer function. It is observed that the magnitude decreases by about 3% over the observed frequency range while the phase angle changes by about 50° in this same frequency range.

Numerical Simulation Studies of Hydrodynamic Forcing Involving the Use of SSWs

Numerical simulations of the response of compliant bottom-pivoted surface-piercing cylinders that account for the non-linear features in the fluid-structure interaction based upon the "relative velocity" formulation of the Morison equation of hydrodynamic forcing have also been conducted, (Haritos, 1990; Haritos, 1991[b]).

These simulations have largely substantiated the characteristics observed in the response of cylinders of this type, and tested in the laboratory programs described above. See Figure 10 for a typical trace of the simulated response of a modelled cylinder depicting similar characteristics to those observed from experimental observations and lending some confidence to application of the resultant computer code to more wide-ranging geometrical conditions than would be able to be conducted in the laboratory test programs. It should also be noted that the computer coding for these simulation studies is particularly easy to implement requiring relatively modest hardware platforms to execute the resultant code, (eg IBM compatible PCs).

From this brief overview of some of the recent applications of the SSW testing technique, it would appear that this method of testing has found a particular niche in the laboratory study of dynamic phenomena, but by no means should it be deemed to be restricted to such studies. Indeed, the technique should be seriously considered as an alternative to regular wave testing and as a useful adjunct to studies involving the use of irregular waves. It should also be considered in situations where the time-history of the disturbance to the system under investigation is important to the process being studied (as in the case of geotechnical instability).

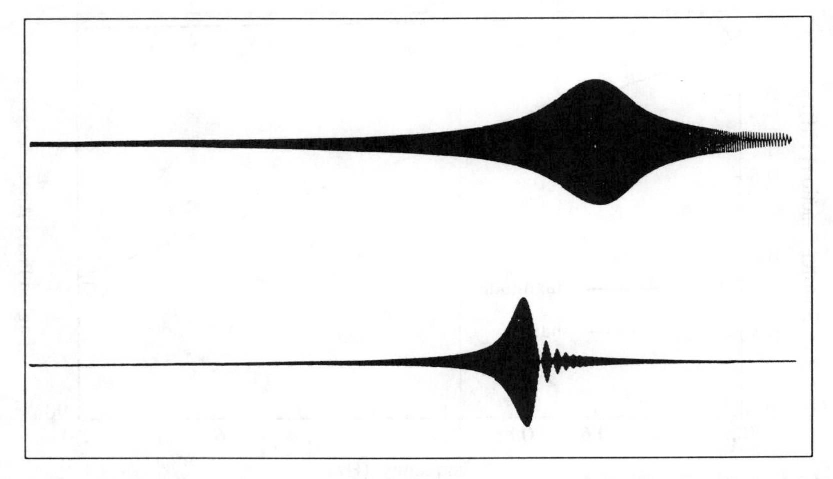

Figure 10. Simulated Traces of the Dynamic Response of a Bottom-pivoted Cylinder under Excitation by SSWs

Conclusions

The SSW testing technique has recently emerged as a powerful easily implement tool in hydrodynamic applications that is ideally suited to the study of phenomena exhibiting frequency dependent characteristics. The applications to which this testing technique has been recently applied include such diverse topics of interest as: the evaluation of the detailed dynamic response characteristics of compliant bottom-pivoted cylinders (that have included studies directed towards a better understanding of models for hydrodynamic damping and for the basic fluid structure interaction in the models for forcing of such cylinders), and the identification of hydrodynamic conditions leading to the instability condition of sand liquefication in sea beds.

Primarily because of their utility as a versatile testing technique, use of suitably configured SSWs should be seriously considered as a viable option in *any* hydrodynamic testing application.

References

Cornett, A.M. (1988). Analytical Modelling of a Restrained Floating Body as a Single Degree of Freedom Oscillator, Technical Report, TR-HY-022: Division of Mech. Eng., NRC, Ottawa, Canada 1989.

Davies, M. (1989). Effective Stress Response of Sands under Wave Loading - Physical Testing, Proc. XXIII Congress Intl. Association of Hydraulic Research, Ottawa, August, pp C185-C192.

Flick, R.E. & Guza, R.T. (1980). Paddle Generated Waves in Laboratory Channels, Proc. ASCE Jl. Waterways, Port, Coastal & Ocean Div, Vol WW1, pp 79-97.

Haritos, N. (1988). The "Swept Sine Wave" Technique in Hydrodynamic Applications, Research Report, TR/Struct/01/88: Department of Civil & Agric. Eng., University of Melbourne, Australia.

Haritos, N. (1989). Using the Swept Sine Wave Concept to Determine Basic System Behaviour, Proc. 9th Australasian. Conf. on Coastal & Ocean Engineering, Adelaide, pp 241-245.

Haritos, N. (1990). Post-Accident Structural Response to Dynamic Loading, Australasian Conf. on the Mechs. of Structures & Materials, Brisbane, September, pp 109-114.

Haritos, N. (1991[a]). Swept Sine Wave Testing of Compliant Bottom-Pivoted Cylinders, Proc. 1st Intl. Offshore & Polar Engineering Conference, Edinburgh, August, Vol III, pp 378-383.

Haritos, N. (1991[b]). Non-Linear Response Characteristics of Compliant Offshore Structural Systems via the Swept Sine Wave Technique, Proc. 10th Offshore Mechanics and Arctic Engineering Conference 1991, Stavanger, Norway, June, Vol I, pp 301-307.

Haritos, N., Cornett, A.M. & Nwogu, O. (1989). The Swept Sine Wave - An Invaluable Tool for Hydrodynamic Research Proc. XXIII Congress Intl. Association of Hydraulic Research, Ottawa, August, pp C485-C492.

Haritos, N. & Hunt, B. (1993). The SSW Technique for Determining the Influence of Wave Reflections, Proc. 13th Australasian Conf. on the Mechs. of Structures & Materials, Wollongong, July, pp 357-364.

Kandianis, F. (1971). Frequency Response of Structures and the Effects of Noise on the Estimates from Transient Testing, Jl. of Sound & Vibration, Vol 15, No 2, pp 203-216.

Nwogu, O., Cornett, A. & Haritos, N. (1992). Response of a Compliant Cylinder in Irregular Waves, Proc. Offshore Mechanics & Arctic Engineering, Calgary, June, Vol I, part A, pp 189-196.

Townsend, M. (1993). The Effect of Structural Damping on the Hydrodynamic Properties of Dynamically Responding Vertical Cylinders, M.Eng. Sci. Thesis, Dept. of Civil & Environmental Engineering, The University of Melbourne.

Townsend, M. & Haritos, N. (1993). The Influence of Structural Damping on the Response of Compliant Cylinders in Waves, Proc. 11th Australasian Conf. on Coastal & Ocean Engineering, Townsville, August, pp 289-294..

White, R.G. (1971). Evaluation of the Dynamic Characteristics of Structures by Transient Testing, Jl. of Sound & Vibration, Vol 15, No 2, pp 147-161.

Yang, H. (1990). The Study of Hydrodynamic Damping of Bottom-Pivoted Cylinders, M.Eng. Sci. Thesis, Dept. of Civil & Environmental Engineering, The University of Melbourne.

The Development of a Wave Data Analysis Standard for a National Wave Measurement Program

Michael Tubman[1], Marshall Earle[2] and David McGehee[1]

Abstract

The Field Wave Gaging Program (FWGP) at the U.S. Army Corps of Engineers' Coastal Engineering Research Center (CERC) collects and analyzes wave data from around the U.S. coastline to contribute to a National wave data base. The FWGP relies not only on its in-house gaging effort to provide data, but also uses several regional gaging networks to extend its geographical coverage. Questions regarding the quality of statistical treatments or model comparisons arise when data are used from sources using different analysis techniques. To maximize the usefulness of the FWGP's wave data to the coastal engineering community, improve the integration of efforts between the various regional networks and maintain the quality of the data, the FWGP is developing a standard for the analysis and reporting of the data. This paper describes the development of the standard for the FWGP.

Introduction

The U.S. Army Corps of Engineers uses wave data in planning, designing, and operating coastal projects. However, there is seldom enough time between the inception of a project and the need to use the data, to collect sufficiently long records to establish wave conditions for engineering purposes. To address this problem, the

----

[1] USAE Waterways Experiment Station, 3909 Halls Ferry Rd., Vicksburg, MS 39180

[2] Neptune Sciences Inc., 1181 Robert Blvd., Slidell, LA 70458

the Corps established the FWGP to provide data to meet anticipated requirements.

The FWGP is presently composed of three gaging networks. The largest network began in 1975 with a single station in Southern California operated by the Scripps Institution of Oceanography. This grew into the Coastal Data Information Program (CDIP) with, currently, 23 stations on the West Coast and Hawaii. About the same time, the University of Florida began the Florida Coastal Data Network (FCDN). At its peak, the FCDN operated 10 stations in Florida, though it is currently funded for 3. The CERC also has its own network called the Network for Engineering Monitoring of the Oceans (NEMO), consisting of 14 stations. In addition, the Corps maintains small networks within several harbors.

To effectively use wave data statistics in solving engineering problems, it is desirable that they represent wave conditions in a manner consistent with those used to develop the theories and techniques applied to the problems. Data in the FWGP data base are from different sources, locations, and times and the methods used to make the measurements, analyze the data and report the results affect the consistency of the ways that the data represent wave conditions. The FWGP is working to address these sources of differences, so as to ensure the uniformity, quality and usefulness of the wave data and the data products, regardless of source. In particular, a focused effort is underway to develop a standard for analyzing and reporting directional wave data.

Method

The wave data analysis standard is being developed by:

1. Preparing a detailed mathematical description of generally accepted procedures for analyzing and reporting directional wave data, termed a tutorial.
2. Holding a workshop during which representatives from each of the FWGP networks use the detailed description as a framework and reference and describe how they analyze directional wave data.
3. Evaluating the significance of differences in analysis methodology with test data.
4. Preparing a final detailed mathematical description of a standard wave data analysis procedure from the original document with changes made as necessary to utilize the best procedures as demonstrated by the tests.

The tutorial, which is a FWGP working document rather than a final report, is a detailed mathematical description of directional wave data analysis, presented according to the steps in Table 1. Nondirectional wave data analysis is considered as a subset of directional wave data analysis. The workshop was held in February, 1993. Participating in the workshop were representatives from the University of Florida (representing FCDN), Scripps Institution of Oceanography (representing CDIP) and the CERC (representing NEMO). The National Data Buoy Center also participated in the workshop, since the Corps uses NDBC data and the FWGP supports several of their buoys. The analysis practices for pitch-roll-heave buoys share many steps with fixed sensors, however, the procedures for hull response corrections are sufficiently different to complicate comparisons. As a result, the workshop concentrated on analysis procedures for bottom mounted pressure arrays and single pressure transducers integrated with two axis current meters (PUV gages). The analysis procedures documented during the workshop were confined by an assumption that the measured time series had been appropriately corrected for instrument specific effects.

- Data collection
- Data quality assurance
- Mean removal
- Trend removal
- Segmenting
- Windowing
- Fast Fourier Transform (FFT)
- Correction for window use
- Cross-spectral calculations
- Averaging over segments
- Directional spectra calculations
- Parameter transformations
- Final parameter calculations
- Reporting

Table 1. - Analysis Steps

Results

The data collection parameters for the networks in the FWGP are summarized in Table 2. All three networks use PUV gages. The FCDN presently uses them exclusively where as CDIP and NEMO are also using multiple pressure gage slope arrays. There are some minor differences with regard to record length and storm mode threshold which are a result of regional differences in wave conditions. The West Coast typically has longer and larger waves than the East Coast and Gulf of Mexico, therefore they take longer records and use site specific storm wave criteria. The networks produce records at different intervals; they are 6, 4 and 3 hours, as shown in the table.

Presentations at the workshop showed that the networks' followed very similar procedures in analyzing the collected data. All three used data quality assurance procedures, as a first step, which checked for out-of-bounds values, flat spots, spikes and jumps. It was noted however that there were probably significant differences in the details of how each network did these checks and how they treated data with problems. The FCDN transformed their coordinate system from instrument coordinates to wave compass coordinates as a second step, whereas, the other two did this after the directional spectra calculations. There was only one other exception to the three networks following the same sequence of steps as shown in Table 1; i.e., CDIP does not use windowing.

Parameters	FCDN	NEMO	CDIP
Data points per record	1024	1024	variable 1024 - 8192 2048 typical
Sampling rate	1 Hz	1 Hz	variable 0.125 - 2 Hz 1 Hz typical
Record interval normal mode	1 hour, collected 6 hours, retained	4 hours	3 hours
Record interval storm mode	1 hour	1 hour	continuous data for time period wanted
Storm mode threshold	variable $H_{mo} > 2$ m typical	$H_{mo} > 1$ m, Great Lakes $H_{mo} > 1.5$ m, U.S. East Coast	variable
Record time	after even hours, EST	after even hours, UTC, and after even hours, local time, at some stations	variable dependent on time to poll all stations

Table 2. Data Collection Parameters

Table 3. provides additional details about how the analysis steps are executed. The NEMO demeans each segment, whereas the other two demean the entire record. The CDIP does not use trend removal for records shorter than 34 minutes, whereas the other two use a linear trend removal. The FCDN detrends the entire record, then does a two minute moving average of each segment, whereas NEMO detrends each

	FCDN	NEMO	CDIP
Segmenting	15 50% overlapping segments, 128 points each	15 50% overlapping segments, 128 points each	50% overlapping segments, number of segments and points depends on record length
Mean removal method	demean entire record	demean each segment	demean entire record after following mean test based on H_{mo}
Record rejection based on mean	reject if mean falls outside of normal tide range by more than 2 m	pressure gages and slope arrays - none, puv gages - reject if p <14.7 or >31 psia	reject if a mean difference for segments > 10% of H_{mo}
Trend removal method	linear trend removal for entire record, then 2 min. moving average demeaning of each segment	linear for each segment	none for records < 34 min., linear trend removal for entire record for records > 34 min.
Windows	10% cosine taper for each segment	10% cosine taper for each segment	none
Correction for window use	variance correction in frequency domain	variance correction in frequency domain	not needed
FFT method	IEEE (1979)	IEEE (1979)	IEEE (1979)
Frequency bands	64 bands, df = 1/128 Hz	64 bands, df = 1/128 Hz	8 bands in 3 s period intervals, 4 - 22 s and 22+ s
Pressure gage cutoff frequency	max. linear wave correction factors = 125 or 1000 depending on gage	none	max. linear wave correction factor = 10
Directional analysis method	Longuet-Higgins (1963)	Longuet-Higgins (1963)	Longuet Higgins (1963)

Table 3. Data Analysis Methods

segment. The CDIP uses 8 frequency bands, whereas the other two use 64.

From a users point of view, some of the most significant differences are in how the three networks report data; these are summarized in Table 4. Table 4 shows differences in the way the networks report representative wave directions and in the wave direction conventions. The CDIP reports the spectral density weighted average as representative wave direction, whereas the other two report mean wave direction at the peak period. The wave direction convention for NEMO is the direction from which the waves come relative to degrees true, whereas FCDN reports the direction waves go relative to magnetic north, and CDIP reports the direction toward which waves go relative to degrees true.

	FCDN	NEMO	CDIP
Data storage	time series, five Longuet-Higgins analysis coefficients as functions of frequency	time series, cross-spectra between individual sensors, nondirectional spectra, H_{mo}, peak period, representative direction	time series, online data bank of climatological statistics
Significant wave height	$H_{mo} = 4 [m_o]^{1/2}$	$H_{mo} = 4 [m_o]^{1/2}$	$H_{mo} = 4 [m_o]^{1/2}$
Peak period	corresponds to center frequency of spectral band with maximum spectral density	corresponds to center frequency of spectral band with maximum spectral density	corresponds to center frequency of spectral band with maximum spectral density
Representative wave direction	mean wave direction at peak period	mean wave direction at peak period	spectral density weighted average of orthogonal mean wave direction components converted to direction
Wave direction convention	direction toward which waves go relative to magnetic north	direction from which waves come relative to degrees true	direction toward which waves go relative to degrees true

Table 4. Data Reporting Products

Conclusions

There are some differences in the ways the networks analyze and report directional wave data, but the general feeling of the participants in the workshop was that these differences can be resolved. The FWGP intends to do this by providing standard test data records to the networks for them to analyze using their analysis procedures to evaluate the significance of the existing differences. The workshop revealed a need to determine what differences exist in the details of the data quality assurance procedures to resolve whether these significantly affect the analysis results and application of these results. The FWGP wave data analysis standard will document the procedures used by the participating networks, help maintain the consistency and quality of the data and improve the effective use of the FWGP's data products.

Acknowledgements

The work presented herein was conducted under the Coastal Field Data Collection Program, United States Army Corps of Engineers, Coastal Engineering Research Center, U.S.A.E. Waterways Experinment Station. Permission was granted by the Office of Chief Engineers, U.S. Army Corps of Engineers, to publish this information.

References

IEEE, 1979. Digital Signal Processing Committee(ed.), *Programs for Digital Signal Processing* (Ch. 1), IEEE Acoustics, Speech and Signal Processing Society, IEE Press, New York, NY.

Longuet-Higgens, M.S., D.E. Cartwright, and N.D.Smith, 1963. "Observations of the Directional Spectrum of Sea Waves Using the MOtions of a Floating Buoy,"*Ocean Wave Spectra*, Prentice-Hall, Englewood Cliffs, NJ.

COMPARISONS OF DIRECTIONAL WAVE ANALYSIS METHODS

Taerim Kim[1], Lihwa Lin[2], and Hsiang Wang[3]

Abstract

Four methods analyzing directional wave spectrum were tested and compared through numerical simulation. The four methods are (i) the truncated Fourier series (TFS), (ii) the Longuet-Higgins parametric model method (LHM), (iii) the maximum entropy method developed based on the autoregressive process of directional waves (MEM I), and (iv) the maximum entropy method developed based on a probability density distribution of directional waves (MEM II). These methods were all tested using the first five frequency-dependent Fourier coefficients from the Fourier series expansion of a directional spectrum. The result showed that the MEM II appears to have the better overall performance. However, the MEM II also often has convergence problem when dealing with real data. For application, this convergence problem can be overcome by using a maximum entropy approximation method suggested here.

Introduction

There are several different measuring systems utilized today in finding the directional information of the sea. For instance, a heave/pitch/roll buoy has been used in the open ocean while wave gage array or pressure transducer and bi-axial current meter are often deployed in coastal water to collect directional wave data. However, all of these measuring systems are capable of collecting only a limited directional information. From the measured data, a finite Fourier series of directional spectrum can then be resolved whereas an infinite long series is required to represent the true spectrum.

Based upon the measured Fourier series coefficients, several methods analyzing directional spectrum were developed. Among them, the simplest was to express the directional spectrum in terms of the truncated Fourier series (TFS). However, the method was seen often to result in negative values in the estimated spectrum. A dif-

---

[1]Graduate Assistant, [2]Research Scientist, [3]Professor, Coastal and Oceanographic Engineering Department, University of Florida, Gainesville, FL 32611, USA

ferent approach leading to non-negative spectral values was from the Longuet-Higgins parametric model which employed a cosine-square-power function to represent the directional distribution of a wave spectrum. This Longuet-Higgins method (LHM), on the other hand, was only suitable for a smoothed, symmetric distribution of directional spectrum in each computed frequency band. Another approach following the suggestion of Borgman (1982) was to use the maximum entropy estimation technique. In this approach, the maximum entropy method is considered to be superior to other methods since the method can solve for any asymmetric, multi-mode directional wave system.

As a matter of fact, a number of spectral estimators have been developed based on the maximum entropy approach method. Two of them were evaluated here, one from Lygre and Krogstad (1986) who applied the maximum entropy method under the assumption of a complex Gaussian, stationary process of directional waves and the other was from Kobune and Hashimoto (1986) regarding the directional distribution of wave spectrum as a probability density function. The two maximum entropy estimators derived by Lygre and Krogstad, and by Kobune and Hashimoto were designated here as the MEM I and MEM II, respectively.

In this paper, the methods of TFS, LHM, MEM I and MEM II for estimating directional wave spectrum were tested and compared against each other for a variety of simulated cases. An approximation to the MEM II estimator was also introduced when the original method encounters divergence problems in real situation.

Methods Estimating Directional Spectrum

Mathematically, a directional spectrum $E(\sigma, \phi)$, with σ and ϕ denoting the frequency and direction, respectively, can be expressed in terms of an infinite Fourier series as

$$E(\sigma, \phi) = \frac{A_0(\sigma)}{2} + \sum_{n=1}^{\infty}[A_n(\sigma)\cos(n\phi) + B_n(\sigma)\sin(n\phi)], \tag{1}$$

where $A_0(\sigma)$, $A_n(\sigma)$ and $B_n(\sigma)$ are the frequency dependent Fourier coefficients. However, based on the wave data measured at sea, the estimation of directional spectrum is limited to the use of a truncated Fourier series when a few Fourier coefficients can only be determined.

Four methods analyzing the directional spectrum are to be compared and evaluated through computer simulation. The four methods are the truncated Fourier series (TFS), the Longuet-Higgens parametric model (LHM), and the two maximum entropy approach methods (MEM I and II). In order to be consistent, the four methods are to be compared based upon the first five Fourier coefficients, $A_0(\sigma)$, $A_1(\sigma)$, $B_1(\sigma)$, $A_2(\sigma)$, and $B_2(\sigma)$.

(1)TFS (truncated Fourier series)

Based on the first five Fourier coefficients determined from the measured data, a truncated five-term Fourier series can be utilized to approximate the directional spectrum.

$$E(\sigma, \phi) = \frac{A_0(\sigma)}{2} + \sum_{n=1}^{2}[A_n(\sigma)\cos(n\phi) + B_n(\sigma)\sin(n\phi)].$$

This estimator is able to show an asymmetric directional spectrum. However, the use of the estimator may result in obtaining negative values for the non-negative spectrum (Longuet-Higgins, et al., 1963).

(2)LHM (Longuet-Higgins parametric model method)
The parametric model proposed by Longuet-Higgins (1963) is

$$E(\sigma,\phi) = E(\sigma)\frac{2^{2s-1}}{\pi}\frac{\Gamma^2(s+1)}{\Gamma(2s+1)}\cos^{2s}\frac{\phi-\phi_o}{2},$$

where Γ indicates the Gamma function, $s(\sigma)$ and $\phi_o(\sigma)$ are the directional spreading parameter and the symmetric center direction, respectively. As a first order approximation, the parameters s and ϕ_o can be solved from

$$s = \frac{C_1}{1-C_1}, \quad \text{and} \quad \phi_o = \tan^{-1}\frac{B_1}{A_1},$$

with

$$C_1 = \frac{s}{s+1} = \frac{\sqrt{A_1^2+B_1^2}}{A_0},$$

(3)MEM I (Maximum Entropy Approach, Method I)
By defining the entropy of a sea system based upon an autoregressive process of directional waves, Lygre and Krogstad (1986) showed that

$$M(\text{entropy}) = -\int_0^{2\pi} \ln H(\sigma,\phi)\mathrm{d}\phi.$$

and

$$E(\sigma,\phi) = E(\sigma)H(\sigma,\phi), \qquad H(\sigma,\phi) = \frac{1-d_1c_1^*-d_2c_2^*}{2\pi|1-d_1e^{-i\phi}-d_2e^{-2i\phi}|^2},$$

with

$$c_1 = \frac{A_1}{A_0}+i\frac{B_1}{A_0}, \quad c_2 = \frac{A_2}{A_0}+i\frac{B_2}{A_0}, \quad d_1 = \frac{(c_1-c_2c_1^*)}{(1-|c_1|^2)}, \quad d_2 = c_2 - c_1d_1,$$

where the asterisk indicates a complex conjugate.

(4)MEM II (Maximum Entropy Approach, Method II)
By defining the entropy with a probability density function of a directional wave system, Kobune and Hashimoto (1986) showed that

$$M(\text{entropy}) = -\int_0^{2\pi} H(\sigma,\phi)\ln H(\sigma,\phi)\mathrm{d}\phi,$$

and the directional distribution function H is determined by maximizing the defined entropy. Using the Lagrange's multipliers, λ_j, it can be shown that

$$H(\sigma,\phi) = \exp[-1-\sum_{j=0}^{4}\lambda_j(\sigma)\alpha_j(\phi)] \tag{2}$$

where $\alpha_0(\phi) = 1$, $\alpha_1(\phi) = \cos\phi$, $\alpha_2(\phi) = \sin\phi$, $\alpha_3(\phi) = \cos 2\phi$, $\alpha_4(\phi) = \sin 2\phi$, and λ_j need to be determined by solving a set of nonlinear equations:

$$\int_0^{2\pi} \exp[-1 - \sum_{j=0}^{4} \lambda_j(\sigma)\alpha_j(\phi)]\alpha_i(\phi)\mathrm{d}\phi = \beta_i(\sigma), \quad i = 0,1,2,3,4. \tag{3}$$

with all the β_j defined as $\beta_0(\sigma) = 1$, $\beta_1(\sigma) = A_1(\sigma)/A_0(\sigma)$, $\beta_2(\sigma) = B_1(\sigma)/A_0(\sigma)$, $\beta_3(\sigma) = A_2(\sigma)/A_0(\sigma)$, $\beta_2(\sigma) = B_1(\sigma)/A_0(\sigma)$. The solution of λ from the above equations can be obtained only by iteration method.

It is noted here that ,based upon the five measured Fourier coefficients, the directional spreading function H determined from both MEM I and II method can have at most two modal peaks, which can be understood by taking $\partial H/\partial\phi = 0$. Using the maximum entropy technique to estimate the directional spectrum is also attractive in that the wave spectrum does not have to be symmetrical in direction. This certainly indicates a better approach than the conventional LHM model of which a directional distribution function is bounded by a symmetrical function. Therefore, the maximum entropy method is particularly attractive in shallow water applications where wave spectra are usually very asymmetric in nature.

Comparison of Methods by Numerical Simulation

The aforementioned four methods estimating the directional spectrum were tested and compared with each other through computer simulation. The target spectra utilized for the simulation were from the following three types:

Type A: Uniform distribution

$$E(\phi) = \begin{cases} a, & \text{if } |\phi| \le \phi_a; \\ 0, & \text{otherwise.} \end{cases}$$

Type B: Cosine-square-power function

$$E(\phi) = \sum_n D_n \cos^{2s_n}(\frac{\phi - \phi_n}{2}), \quad D_n = \text{constant}.$$

Type C: Fractionary function

$$E(\phi) = \sum_n \frac{(1 - b_n^2)}{2\pi(1 - 2b_n\cos(\phi - \phi_n) + b_n^2)}, \quad b_n = \text{constant}.$$

Figure 1 shows the result of comparisons of all four tested methods for the unimodal type target spectra. The ordinate was normalized as dividing the computed spectrum by the largest value of the target spectrum. In the case of the uniform distribution (Type A), both MEM I and MEM II showed two peaks while TFS and LHM yielded single peak. For the cosine-square function (Type B), the MEM II and LHM showed good estimation whereas the MEM I still resulted a double peak distribution. The TFS showed satisfactory result only for wide target spectrum. The result from TFS became poor and showed negative spectral values when applying to narrow target spectrum. On the other hand, the MEM I estimator, which was seen to result very biased picture to the target spectra of both Type A and B, showed a nearly perfect estimation to the target spectrum of the fractionary function (Type C).

Figures 2 and 3 show the comparisons of tested methods for the bimodal type target spectra with two peaks having the same and different strength, respectively. Here, the MEM II spectra generally gave the closest estimation to the target spectra of the Type A and B except the Type C. Both TFS and LHM were not giving satisfactory result in any circumstances. The MEM I estimator was seen to overestimate the peaks for target spectra of Type A and B, but was best to estimate the spectra of Type C.

Figure 4 shows the simulation result for the triple peak system of Type B distribution. When the three peaks were located in close distance and, together, looked like an unimodal distribution, the MEM II estimator still indicated the best result while the MEM I showed a false two peak distribution. When three peaks were all dominant and distinguishable from each other, none of the four methods was good for estimation of the target spectrum.

MEM II Approximation

Based on computer simulation, the MEM II approach was seen to be the most satisfactory among the four tested methods to estimate the directional sea from the measured data. The only shortcoming of MEM II for the general application was the necessary of solving a set of nonlinear equations by numerical iterations, which often encounters divergence problem in real situation. The problem can be largely overcome by using an approximation scheme for solving the Lagrange's multipliers which is the primary algorithm in the iteration process. It can be shown that by expanding the exponential term appearing in Eq.(3) to the second order as

$$\int_0^{2\pi} [\beta_i(\sigma) - \alpha_i(\phi)] \cdot \{1 - \sum_{j=1}^{4} \lambda_j(\sigma)\alpha_j(\phi) + \frac{[\sum_{j=1}^{4} \lambda_j(\sigma)\alpha_j(\phi)]^2}{2}\} = 0,$$

an approximation to compute the Lagrange's multipliers, or λ_i, is from the following equations:

$$\begin{aligned}
\lambda_1 &= 2\beta_1\beta_3 + 2\beta_2\beta_4 - 2\beta_1(1 + \sum_{j=1}^{4} \beta_j^2), \\
\lambda_2 &= 2\beta_1\beta_4 - 2\beta_2\beta_3 - 2\beta_2(1 + \sum_{j=1}^{4} \beta_j^2), \\
\lambda_3 &= \beta_1^2 - \beta_2^2 - 2\beta_3(1 + \sum_{j=1}^{4} \beta_j^2), \\
\lambda_4 &= 2\beta_1\beta_2 - 2\beta_4(1 + \sum_{j=1}^{4} \beta_j^2),
\end{aligned}$$

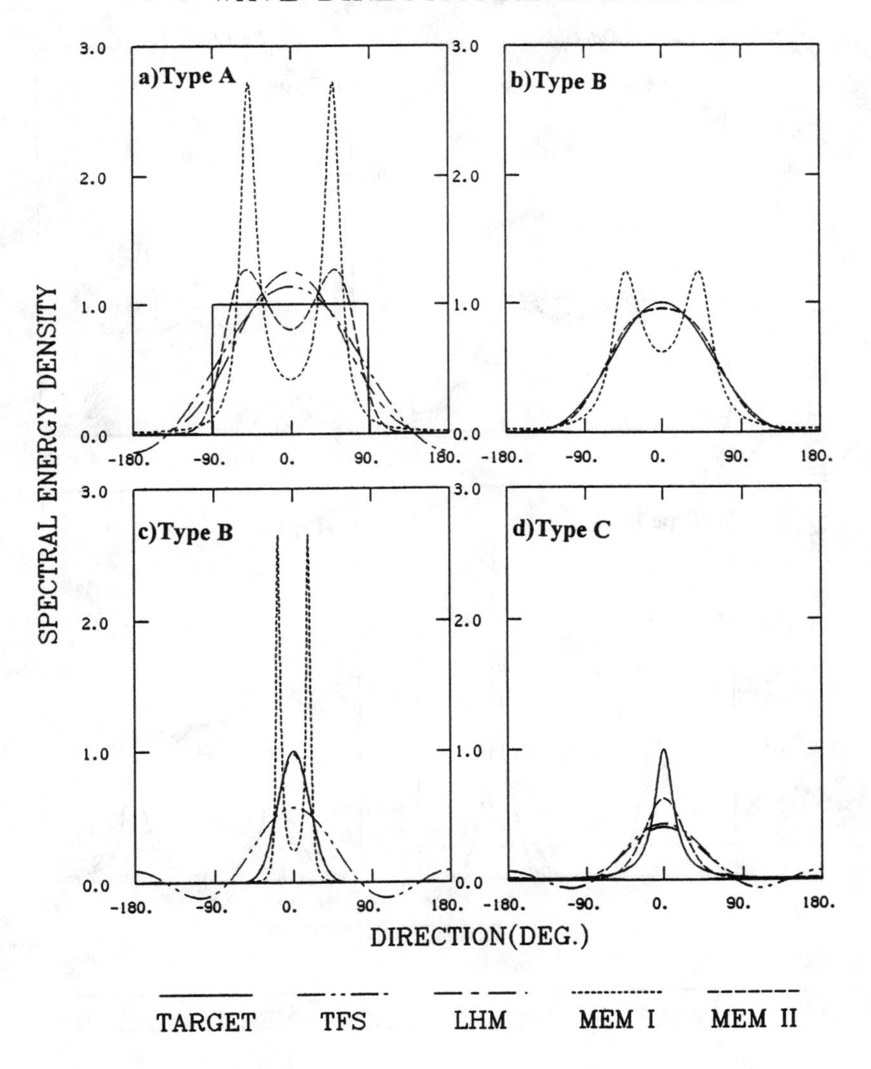

Figure 1: Comparison of simulation results for unimodal directional spectra.

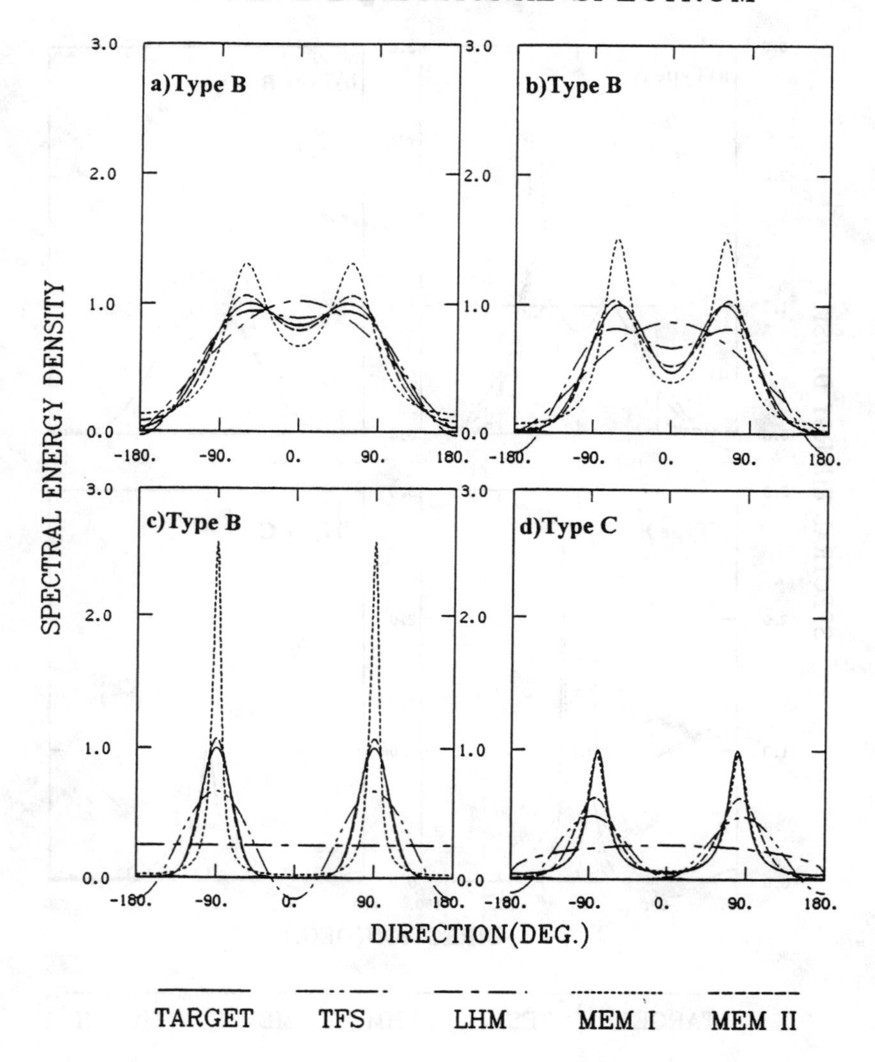

Figure 2: Comparison of simulation results for two identical strength peak spectra.

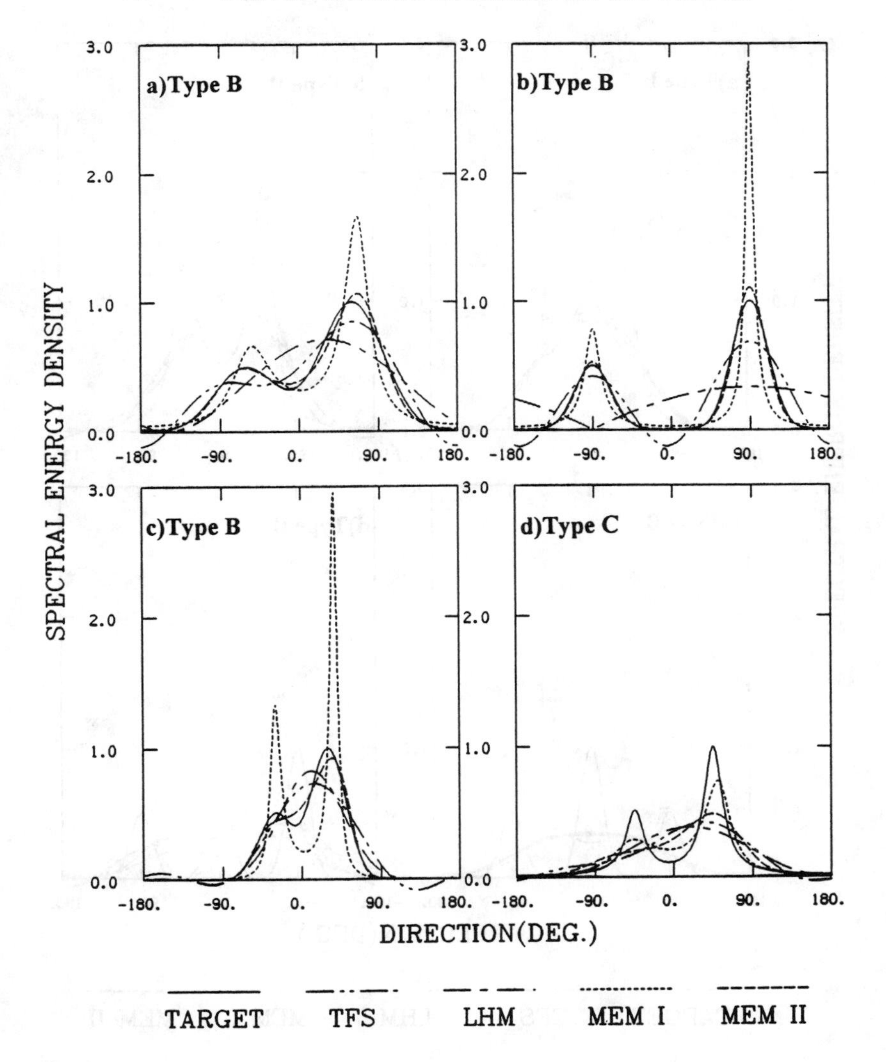

Figure 3: Comparison of simulation results for two different strength peak spectra.

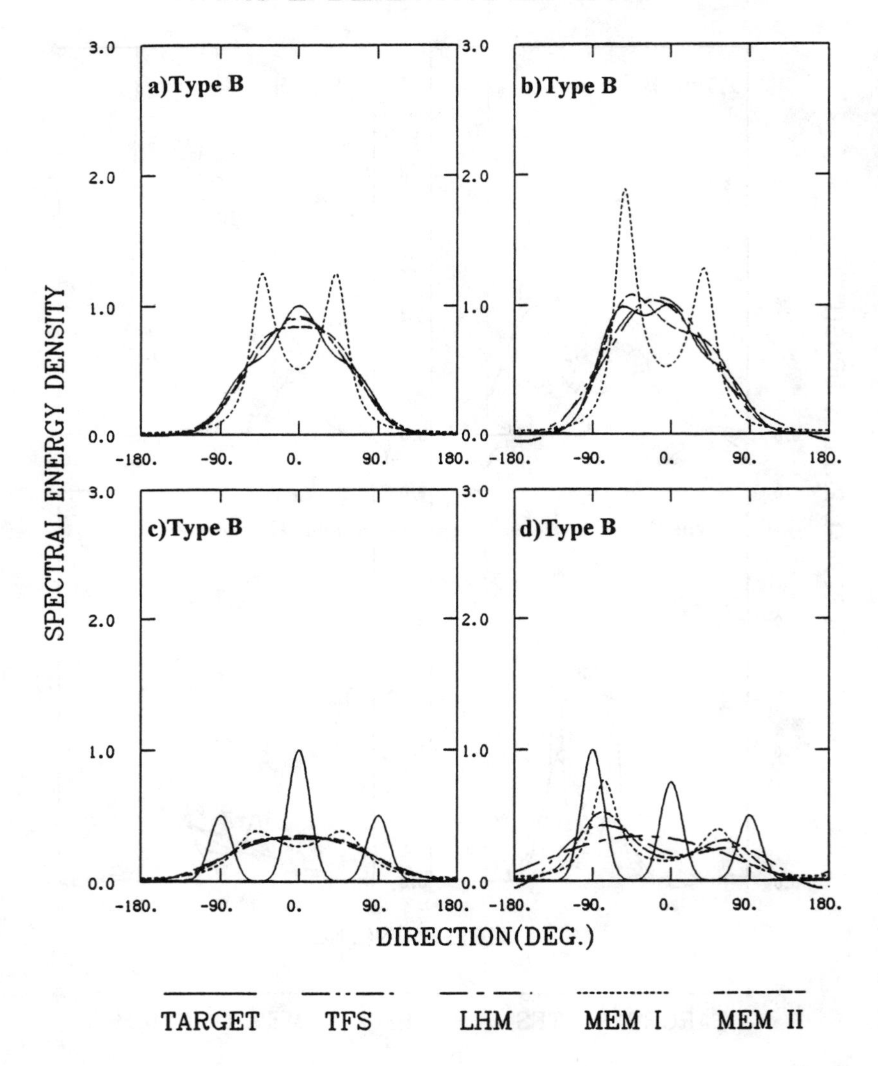

Figure 4: Comparison of simulation results for triple peak spectra.

and

$$\lambda_0 = \ln\{\int_{-\pi}^{\pi} \exp[\sum_{j=1}^{4} \lambda_j(\sigma)\alpha_j(\phi)]d\phi\} - 1.$$

Figure 5 shows the comparison of the MEM II spectra and the MEM II approximation spectra along with the target spectra of the cosine-square-power type (Type B). Although the approximation method was not identical to the MEM II, it generally gave good results to the unimodal and bimodal target spectra. For application , this approximation method could be used in the condition when the MEM II failed to converge.

An example using the combination of the MEM II and its approximation method to simulate a target spectrum for the entire frequency and directional components is shown in Figure 6. The target spectrum contains two different wave systems. One wave system has a dominant peak near the low frequency end and a main direction towards north. The other wave system has a peak at higher frequency heading to the west direction. The result of this simulation also shows a good agreement between the computed and the target spectrum. The comparisons are also shown at four different frequencies in Figure 7 together with the estimates from the TFS, LHM, and MEM I methods. The MEM II method is clearly superior in all cases.

Another example showing the estimate of the directional spectrum based on the real measured sea data is given in Figure 8. The data were from offshore Perdido Key near Pensacola, Florida. The wave direction was seen to have changed between the midnight and the 6am on Feb. 5th 1992. The switch of wave direction from NW to N in the higher frequency region was found to agree with a rapid change of wind direction during this period.

Conclusions

Four methods analyzing directional wave spectrum were compared through numerical simulation. The four methods included the truncated Fourier series (TFS), the Longuet-Higgins parametric model (LHM), and the two maximum entropy methods developed based on the assumption of autoregressive process for directional waves and the assumption of using a probability density function for directional distribution (namely, the MEM I and MEM II, respectively, in the paper). All of the four methods were tested based on the first five frequency-dependent Fourier coefficients from the Fourier series expansion of a directional spectrum. The findings from the simulation study were summarized below:

(1) For the four methods tested, the LHM is known to be suitable for symmetric, single peak distribution. The TFS often has the shortcoming of producing negative spectral values. The MEM I has a tendency of overestimating the spectral peak and of creating false peaks. The MEM II shows an overall better performance than the other three methods. However, it has a nonconvergence problem in real data handling.

(2) Both MEM I and II are able to describe asymmetric, multiple peaked spectra but the MEM I is only suitable for sharp-peaked spectra.

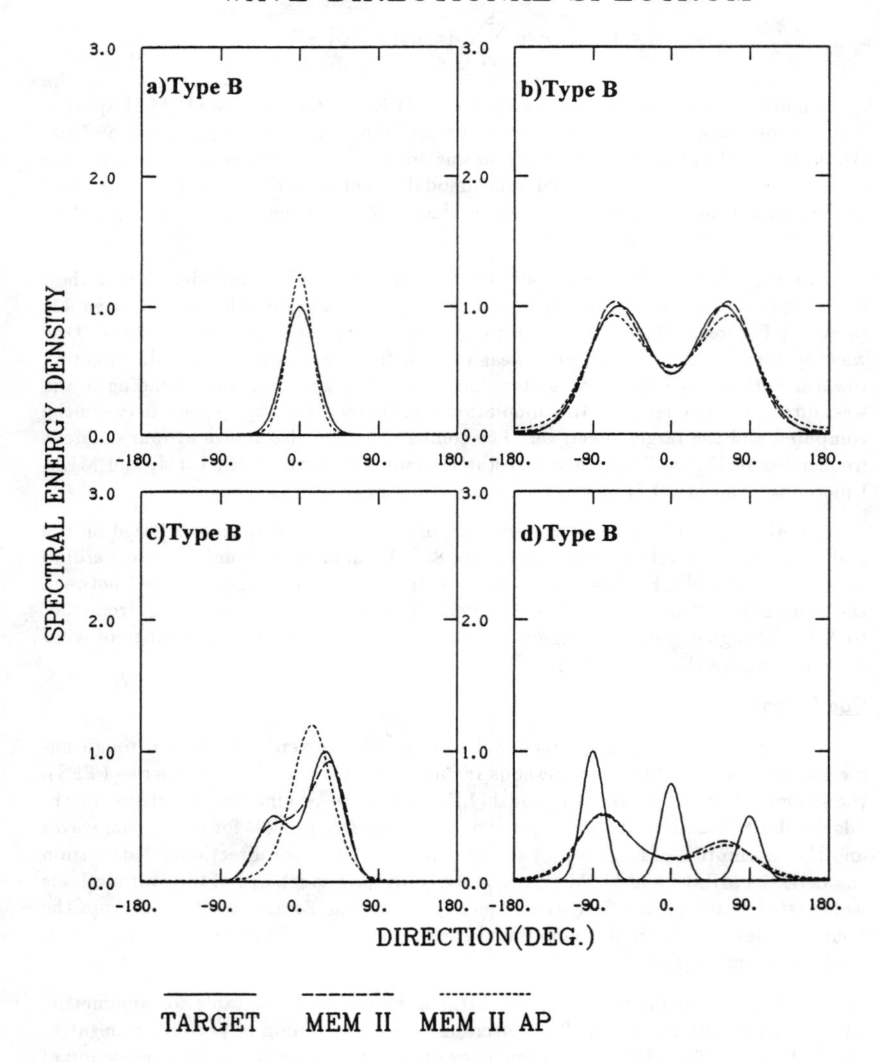

Figure 5: Comparison of the MEM II and its approximation method.

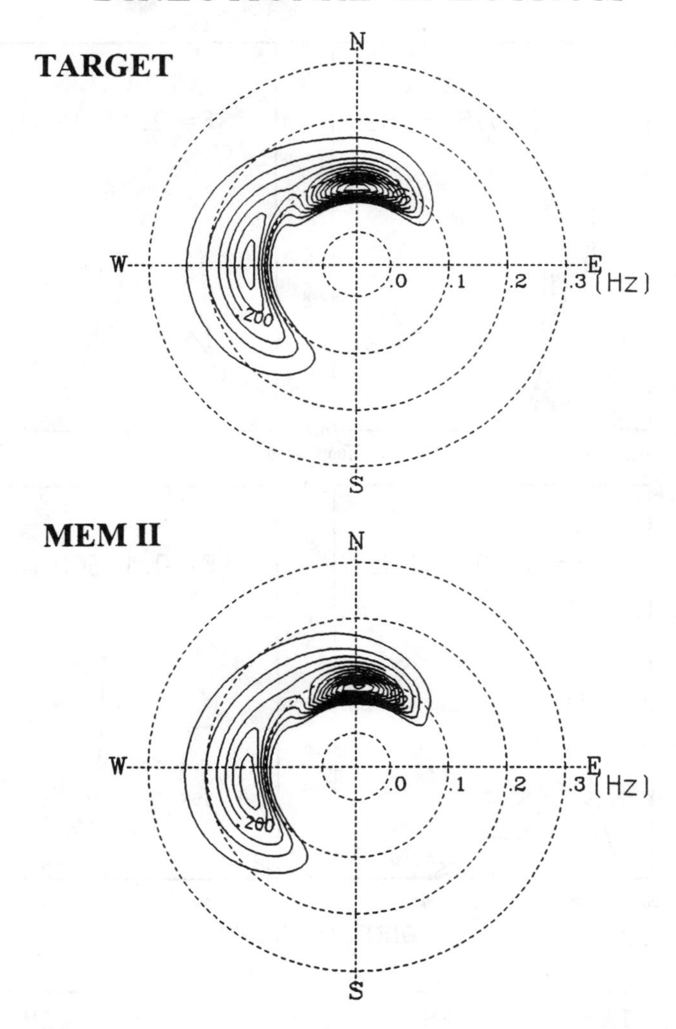

Figure 6: Comparison of MEM II and target spectra with two different directional wave systems.

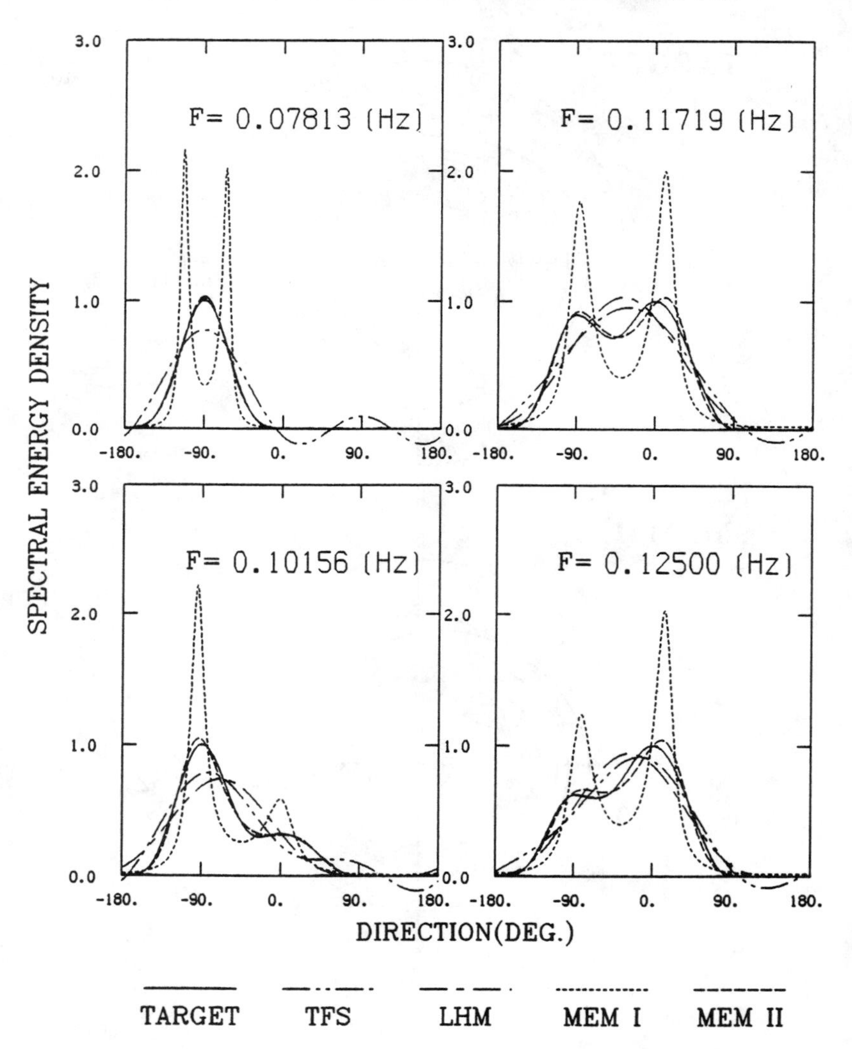

Figure 7: Comparison of simulated and target spectra having two directional wave systems for four different frequencies.

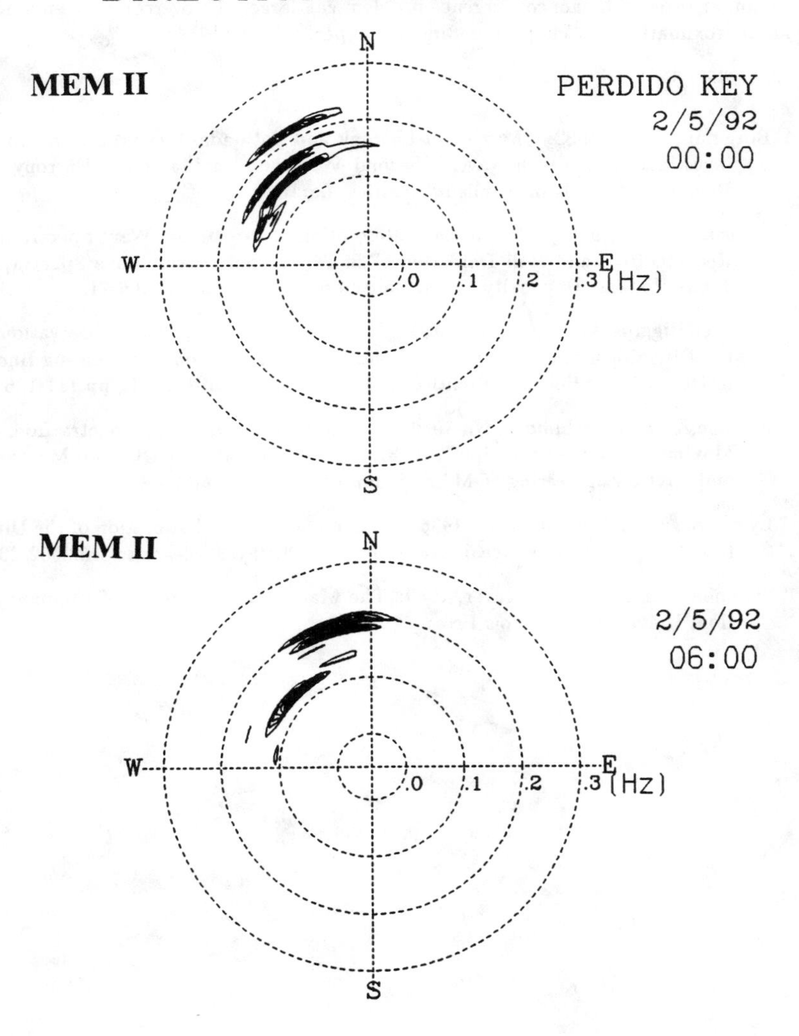

Figure 8: Application of the MEM II to the measured sea data.

(3) For application to measured sea data the MEM II is judged to be superior than the other three. The nonconvergence problem can largely be overcome by employing an approximation scheme proposed in this paper.

References

Borgman, L. E., 1982: "Maximum-Entropy and data-adaptive procedures in the investigation of ocean waves." Second Workshop on Maximum Entropy and Bayesian Methods in Applied Statistics, Laramie.

Krogstad, H.E., 1989:"Reliability and Resolution of Directional Wave Spectra from Heave, Pitch, and Roll Data Buoys",in Directional Ocean Wave Spectra,The Johns Hopkins University Press, Baltimore and London, PP.66-71.

Longuet-Higgins, M. S., Cartwright, D. E., and Smith, N. D., 1963:"Observations of the Directional Spectrum of Sea Waves Using The Motion of a Floating Buoy", in Ocean Wave Spectra, Prentice Hall, Englewood Cliffs, N. J., pp.111-136.

Kobune,K., and Hashimoto, N., 1986: "Estimation of directional spectra from the Maximum Entropy principle", in Proc. 5th International Offshore Mechanics and Arctic Engineering (OMAE) Symposium, Tokyo,pp.80-85.

Lygre, A., and Krogstad, H. E., 1986: "Maximum Entropy Estimation of the Directional Distribution in Ocean Wave Spectra", J.Phys. Oceanogr. 16,2052-2060.

Shannon, C.E., and W. Weaver, 1949: The Mathematical Theory of Communication, University of Illinois Press.

The Re-derivation of the NDBC Wind-Wave Algorithm

Ian M. Palao[1] and David B. Gilhousen[2]

Abstract

The National Data Buoy Center (NDBC) currently manages a network of 57 moored buoys and approximately 7 Coastal-Marine Automated Network stations on offshore platforms which transmit wind speed, wind direction, and spectral wave energy data. As with all geophysical measurements, quality control of the data is necessary. In 1986, NDBC derived a wind-wave algorithm, based upon the relationship between the high-frequency wave energy and the wind speed, that identifies erroneous wave data. This algorithm flags wave data when the data fall beyond preset upper and lower limits for the corresponding wind speed. The algorithm performs well for most sea and atmospheric conditions, but it has limitations, especially during light winds and at stations with severe fetch limitations. Furthermore, an algorithm was needed that was more representative of the nonlinear wind / wave relationship.

Therefore, NDBC initiated a study to develop a new method of quality control. Whereas the previous method compared an hourly wave energy value to linearly derived maximum and minimum values for all buoys, the new method computes a station-specific, long-term probability density distribution of acceptable energy values related to a given wind speed. This distribution is computed using at least 6 years of hourly, buoy-measured wind speed and high-frequency wave energy observations. Current hourly pairs of wind speed and wave energy data are compared to the climatological "energy envelope" represented by the 1-percent contour line of the normalized, joint probability density distribution. Wind speed and wave energy observations falling outside this contour are identified as questionable. The data analyst then uses various meteorological products, such as weather maps and satellite images, to determine whether the data are in error.

The result of this study clearly shows that the new method successfully models the nonlinearity of the wind / wave relationship which the previous algorithm could not. Because of its sensitivity to small changes in wave energy, the new procedure can also detect when incorrect empirical constants have been applied to the wave energy. This new approach will enable NDBC to expeditiously detect anemometer failures, wave system failures, or wave processing problems.

---

[1] Computer Sciences Corporation, John C. Stennis Space Center, MS 39529-6000

[2] National Data Buoy Center, John C. Stennis Space Center, MS 39529-6000

Introduction

Instruments onboard NDBC buoys and at Coastal-Marine Automated Network (C-MAN) sites measure and transmit more than 4 million individual data observations per year. These data are monitored within a day of their receipt to remove any degraded data from real-time release and subsequent archival. Because of the large amount of data collected, a totally manual review of all measurements is not feasible. A completely automated validation procedure is not feasible either because air-sea interaction is influenced by many factors which are difficult to model in an automated system, such as coastal topography and bathymetry. Therefore, a man-machine mix is used where automated validation procedures "flag" suspect data and produce graphics for manual review and possible action (Gilhousen 1988).

Among the most difficult measurements to validate are wave measurements and wind measurements made when the station has only a single functioning anemometer. Early in NDBC's history, it was recognized that a cross validation of wind speed and high-frequency wave energy would help this monitoring process (Steele and Marks 1979). The primary function of the new quality control procedure is to check for significant data discrepancies. Not only can such a technique detect significant anemometer and wave accelerometer malfunctions, it can detect errors associated with all digital processing applied to the wave energy, such as unrefined noise corrections and hull-transfer functions.

Background

In 1986, NDBC developed the first operational "wind-wave algorithm" (Lang 1987). This algorithm compared wind-wave energy observations to the square of the 4-hour average wind speed; i.e., the average of the present hour and the three previous hours. The sum of the spectral density within the frequency range of 0.20-0.27 Hz was used to represent the energy of wind-generated waves. These frequencies were thought to exclude all energy originating from swell. Data used for the calculation of the wave energy limits consisted of 22 months of wave energy and wind speed observations from various NDBC buoy sites.

The 1986 algorithm was constructed in several steps. Three linear least squares lines through three segments of the data cloud were determined. These segments were chosen to best linearize the nonlinear data cloud. Next, the upper limit was calculated so that not more than 3 percent of the data fell beyond it. The energy differences between the upper limit and the linear least squares lines were then used to determine the lower limit. As a result of the original grouping of data, the upper and lower limits are "universal"; i.e., they do not vary according to buoy hull type or station location (Figure 1).

This algorithm performs well but has limitations, especially at stations with severe fetch restrictions. During these cases, wave energy will be less than if the station had unlimited fetch. Therefore, a range of wave energy values can correspond to any given wind speed. The three linear least squares lines cannot account for this range of wave energy. Because of this limitation, many observations are incorrectly flagged as erroneous. Due to the magnitude of the false flags, a new algorithm was suggested that incorporates some of the characteristics of the initial algorithm, while reducing the severity of its limitation.

The New Wind-Wave Validation Procedure

The basic structure of the new wind-wave procedure is the same as that of the initial algorithm: to compare an observed wave energy value to upper and lower limits based on the square of the corresponding 4-hour average wind speed. The initial limits are calculated using at least 6 years of quality controlled, archived hourly wind speed and wave energy data for a minimum of eight stations. Since the energy of low-frequency waves is depth limited, a higher frequency range (0.30-0.35 Hz) was used with this technique so that NDBC could validate data transmitted from buoys located in shallow water. When the sum of the wave energy within the given frequency range is plotted against the square of the 4-hour average wind speed, the wind-speed/wave-energy relationship is visualized. The plot is then divided into sectors of wave energy and wind speed. Then, the conditional probability density distribution is calculated for each sector. This density distribution decomposes the entire data cloud into probabilities of occurrences per wind speed/wave energy sector. The upper and lower limits of wave energy corresponding to a given wind speed are then delineated by the 1-percent contour line (Figure 2).

The upper and lower limits of wave energy calculated by this procedure vary with buoy hull type or geographic characteristics of the deployment site. Through many analyses, it was determined that buoys of like hull type (3-meter discus, 6-meter NOMAD, or 10-meter discus) exhibit similar "wave response" characteristics, and that stations in the Great Lakes require unique quality control limits because there is no low-frequency swell. Therefore, the above categories of NDBC buoys merit unique wind-wave energy limits. NDBC realizes that, for their large discus buoys, the hull transfer function at high frequencies are not as refined as they could be. This does not imply that NDBC's significant wave height data are biased according to hull type or station location. Recall that the sum total energy of the frequency range used in this quality control procedure is only a small percentage of the total energy used to compute significant wave height.

Unlike the previous algorithm, which is run on a daily basis to flag individual observations, the primary purpose of this procedure is to check for gross sensor errors. On a monthly basis, the new wind-wave procedure will be used as a final check for sensor errors before the data are sent to the national archive centers. Not only are anemometer and wave accelerometer malfunctions detected, errors associated with all digital processing applied to the wave energy, such as noise correction factors and hull transfer functions can also be discovered.

With regular use of the new wind-wave procedure, the data analysts will develop familiarity with each station's wind-wave relationship, as visualized by the algorithm, allowing them to recognize idiosyncracies in the graphical output. As an example, in testing the new technique, it was noticed that wave energy for a given site in the Great Lakes dramatically decreased after one 6-meter NOMAD hull was replaced with another 6-meter NOMAD hull (Figure 3). Further investigation revealed that the accelerometer (the wave sensor) was placed significantly off-center in the latter buoy. NDBC quickly solved this problem by correctly placing the accelerometer at the center of buoyancy. In another example, an incorrect noise correction function applied to wave data at NDBC station 44014 was detected (Figure 4). As a result, the correct noise correction function was quickly applied. This procedure is also a useful tool for the timely answering of users' questions and comments regarding whether or not wave energy from a particular buoy is too high or low. Human intervention is also a necessary factor in specialized situations, such as when the energy observation falls above the upper limit but the surrounding circumstances account for the differences, as with severe storms.

Conclusions

Quality control of wind-wave energy data will never be a completely automated procedure at NDBC, especially at coastal locations. Air-sea interaction is influenced by many factors, such as bottom topography of the basin, the surrounding topography of the land, the atmospheric conditions, etc. Therefore, human intervention in the quality control procedure is necessary to integrate the many influencing factors before determining the validity of the data.

The new wind-wave procedure is a great supplement to NDBC's linear least squares algorithm. It is used as a monthly check while the algorithm is used for daily quality control.

It's many attributes provide NDBC with a tool that is very sensitive to slight changes in wave energy. First, the contour lines are not universal. The Great Lakes buoys and the different hull types have unique wave response characteristics which can not be accurately compensated for with empirical constants. Second, the various contour lines are nonlinear. A linear quality control technique can not and will not accurately control the quality of a nonlinear process, such as wind and wave interaction. Next, the procedure is conceptually simple and its graphical nature allows for the intervention of human reasoning during special cases when the wave energy data may exceed the climatological upper limit. Finally, a large mainframe computer is not necessary. This procedure can be run on a standard 486-type PC using commercially available graphics software.

References

Gilhousen, D.B. (1988). Quality control of meteorological data from automated marine systems. *Preprints of the Fourth Int. Conf. on Interactive Information and Processing Systems for Meteorology, Oceanography, and Hydrology*, AMS, Anaheim, CA, 248-253.

Lang, N.C. (1987). An algorithm for the quality checking of wind speeds measured at sea against measured wave spectral energy. *IEEE J. of Ocean. Eng.*, OE-12, No. 4, 560-567.

Steele, K.E. and Marks,G.E. (1979). Detection of NDBC wave measurement system malfunctions. *Proceedings of Oceans '79*, MTS and IEEE, 226-236.

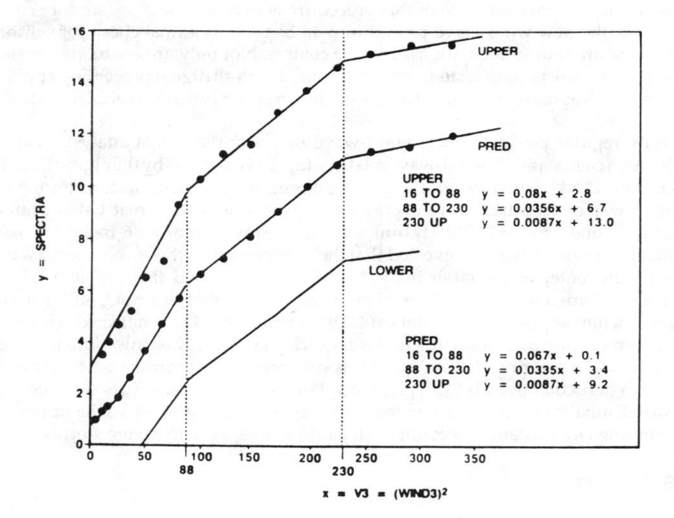

Figure 1. Graph of upper and lower bounds of the 1986 algorithm. "V3" corresponds to the square of the 4-hour average wind speed (Lang 1987).

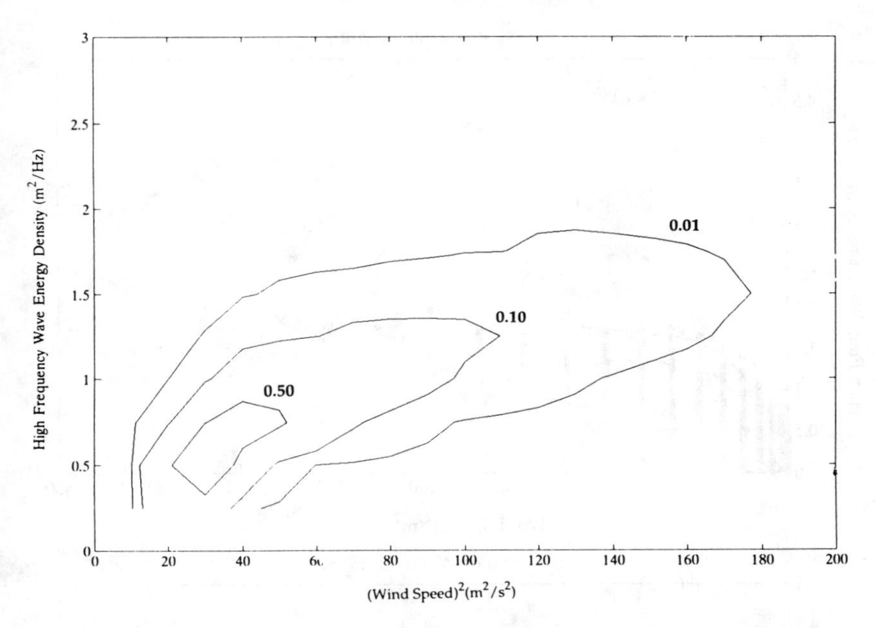

Figure 2. Contour lines for the 3-meter discus buoy used in the 1992 wind-wave procedure.

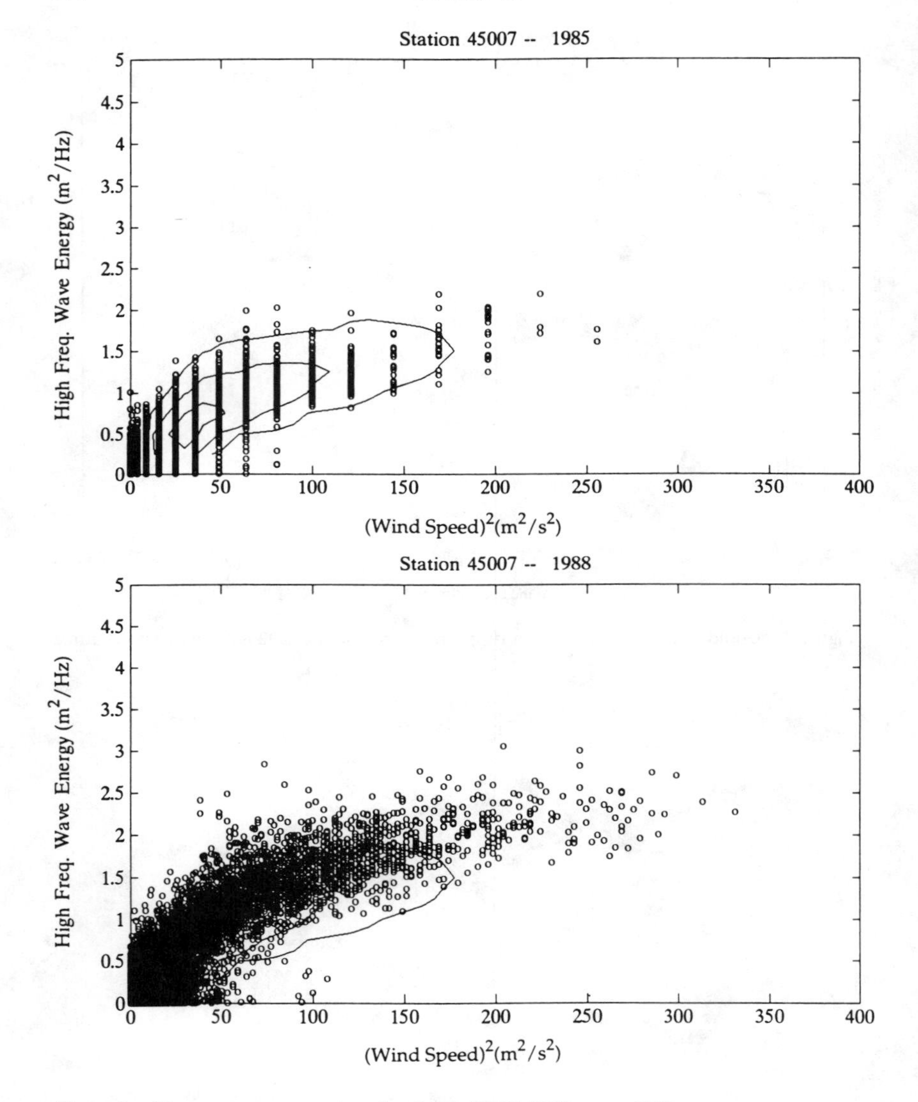

Figure 3. Wave energy comparison for station 45007, 1985 versus 1988.

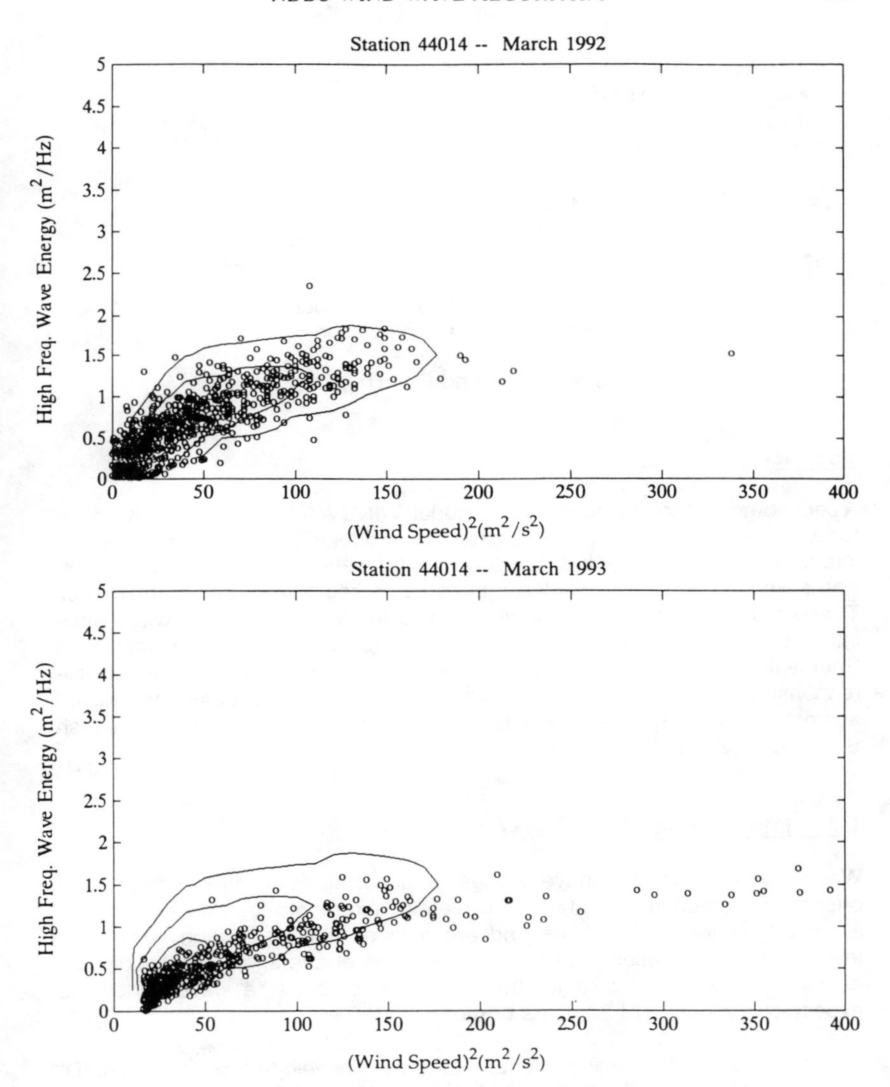

Figure 4. Wave energy comparison for station 44014, March 1992 versus March 1993.

A second-generation wave model for coastal wave prediction

J A Ewing[1] and R C Hague[1]

Abstract

A second-generation numerical wave model, WINDWAVE, has been developed for wave prediction in coastal regions. The model is based on the numerical integration of the energy balance equation for waves in water of finite depth. Three Source terms are included in the physical representations of the model. These terms describe wave growth due to the wind, nonlinear wave-wave interactions and dissipation due to bottom friction in shallow water. The Source term for wave-wave interactions is based on a simple, but directionally-responsive, formulation given by Young (1988). The model has been tested against the SWAMP (1985) results and for real winds for storms in the Irish Sea and elsewhere.

1 Introduction

Wave prediction models have a wide range of applications in coastal and offshore engineering, as data from wave measurements are only available at a limited number of locations and are also of limited duration. Hindcast techniques can be used both for the estimation of average wave conditions, or wave climate, and also for the estimation of extreme wave conditions expected in a period of 20 years or more.

The most advanced wave model has been developed by the WAMDI Group (1988). In this third-generation wave model, the Source term for wave-wave interactions, S_{nl}, is computed using an improved, but still rapid, method based on the discrete interaction approximation of Hasselmann et al. (1985).

[1] HR Wallingford Ltd, Wallingford, Oxfordshire OX10 8BA, UK

The dissipation Source term was chosen to close the energy balance equation without the need to specify a limiting spectral form. The computing power necessary to run the third generation model, WAM, is not available for extensive engineering applications. Thus at the present time engineers must make use of existing second-generation wave models. WINDWAVE uses a good approximation to S_{nl} which is responsive to changes in the spectral shape in frequency and direction, based on the scheme given in Young (1988).

The Source term for wave growth from the wind is taken from the work of Snyder et al. (1981), but a limiting wave spectrum is used in place of the dissipation function for wave breaking. The rapid evaluation of the Source terms used in this second-generation wave model, WINDWAVE, allows its application to engineering problems. Earlier models employed at HR Wallingford do not include the detailed physics of wind-wave growth and interaction in wind fields which vary in time and space. Furthermore, previous wave models were designed to provide information at a specific site, while the present model gives results at all grid points. In addition to the Source terms previously considered, a bottom friction term is included in shallow water. A full description of WINDWAVE is given in Ewing and Hague (1992).

There are a number of different applications which should benefit from the use of a second-generation wave model. In many coastal engineering studies wave information is required on a fine mesh over an area in shallow water where both local wind-wave growth and bottom dissipation are present. The model is also able to include water level changes due to tides but not the influence of tidal currents.

The model produces estimates of the wave spectrum which may be needed for the design of coastal and offshore structures where the wave period and direction are important parameters. The second-generation model can also be used in studies of the joint probability of occurrence of waves with other phenomena such as winds, tides and currents.

2 Model formulation

A second-generation wave model was chosen in view of the need to limit the calculations of the Source terms to those compatible with practical applications where hindcast studies need to be made for long time periods.

The model, is based on the solution of the energy balance equation for water waves:

$$\frac{DE}{Dt} = S \qquad (1)$$

Here $E(\underline{x},t,f,\theta)$ is the directional wave spectrum, where $\underline{x}$ is the position in space, t is the time and (f,θ) are the frequency and direction of the waves. S is the Source function which includes the physical processes necessary to describe the growth, interaction and decay of waves.

The solution of equation (1) is obtained by an explicit finite difference method with two stages. In the first stage, the homogeneous equation $DE/Dt \equiv 0$, is solved for the propagation of each component of the discretised directional spectrum at each time step. In the second stage, the Source function, S, is evaluated after each propagation step.

2.1 Propagation scheme

The left hand side of equation (1) can be expanded to :

$$\frac{DE}{Dt} = \frac{\partial E}{\partial t} + \nabla.(\underline{Cg}E) - \frac{\partial}{\partial \theta}[(\underline{Cg}.\nabla\theta)\, E\,]$$

where $\underline{Cg}$ is the group velocity in shallow water.

The first term on the right hand side is the local rate of change of the spectrum. The second term, $\nabla.(\underline{Cg}E)$, is the propagation term modified to take into account variations in group velocity. These terms are integrated using a modified Lax-Wendroff scheme which is stable for Courant Numbers less than 1.

The last term, $\partial(\underline{Cg}.\nabla\theta)E]/\partial\theta$, is the refraction term and is integrated following the method adopted by Golding (1983). The change $\Delta\theta$ is not allowed to exceed 30° in a single time step.

2.2 The Source terms

Following the work in JONSWAP (Hasselmann et al., 1973), the Source Function can be represented by the sum of four terms :

$$S = S_{in} + S_{nl} + S_{ds} + S_{bot} \tag{2}$$

where S_{in} describes the wind input to the waves, S_{nl} represents nonlinear wave-wave interactions, S_{ds} represents the dissipation due to wave breaking and S_{bot} describes the dissipation in shallow water due to bottom processes. The influence of tidal currents is not considered in equation (2) although the effect of varying tidal levels is included in some of the hindcasts (Ewing and Hague, 1992).

2.2.1 The atmospheric input

For S_{in} we take the expression given by Snyder et al (1981) and also used in the third-generation wave model, WAM (WAMDI Group 1988) :

$$S_{in} = \beta E, \text{ where } \beta = 0.25[28u_{*}\cos(\theta-\theta_w)/C-1]\omega. \qquad (3)$$

Here we use the friction velocity, u_*, rather than U_{10}. θ is the wave direction and θ_w is the wind direction. $|\theta-\theta_w| \leq \pi/2$. ω is the radian frequency, $2\pi f$, where f is in Hz.

The friction velocity is obtained from : $u_* = C_D^{1/2}.U_{10}$, where the drag coefficient can be taken as :

$$10^3.C_D = \begin{cases} 1.2875 & U_{10} < 7.5 \text{ m/s} \\ (0.8+0.065U_{10}) & U_{10} \geq 7.5 \text{ m/s} \end{cases}$$

2.2.2 Nonlinear wave-wave interactions

As the rigorous evaluation of this term is not possible except for a few idealised cases, we must use an approximation to S_{nl} which is computationally efficient.

The method proposed by Young (1988) has been followed. In this method the forward lobe of S_{nl} is represented by a simple triangular distribution, as shown in Figure 1. Then S_{nl} can be approximated by the linear relations :

$f_1 = 0.74$
$f_2 = \max(0.94, \ 1.17-0.13\gamma)$
$f_3 = \max(1.01, \ 1.457-0.167\gamma)$

where γ is the JONSWAP "overshoot" parameter. The scaling of the overall value of S_{nl} then depends on the JONSWAP parameters α and f_m in a well-known way, for any spectrum of JONSWAP form. Finally, the directional dependence of S_{nl} is easily obtained by linear interpolation from Table 1 of Young (1988).

This parameterisation of S_{n1} assumes the JONSWAP parameters of the wind-sea are available at each time step of the model calculations. This is achieved by finding the wind-sea spectral peak, f_m, in the high frequency region above $0.8f_{PM}$, where f_{PM} is the Pierson-Moskowitz (PM) frequency given by:

$$f_{PM} = 0.13g/U_{10}.$$

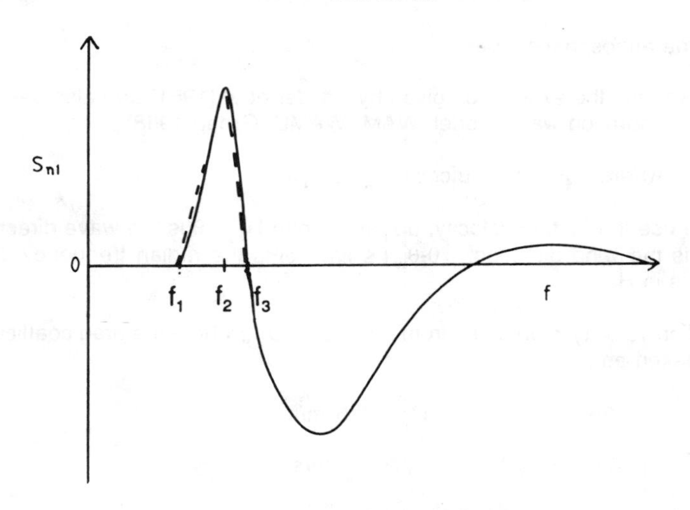

Figure 1 Source function for wave-wave interactions (Young, 1988)

α and γ can then be found from the relations : where $\upsilon = U_{10}f_m/g$.

$\alpha = 0.033\upsilon^{0.666}$ [Hasselmann et al. (1976)]
$\gamma = 4.42\upsilon^{0.429}$ [Mitsuyasu et al. (1980)]

(U_{10} is the mean wind speed at an elevation of 10m .g is the acceleration due to gravity). However, the above relation given for γ does not approach unity near full development. It was therefore found necessary to include an extra condition, namely that $\gamma = 1$ when $\upsilon \leq 0.16$. This condition ensures a transition to a fully-developed state.

2.2.3 Limiting spectrum

The dissipation due to wave breaking, S_{ds}, is represented in a simple manner by a limiting wave spectrum at high frequencies beyond which the waves cannot grow. The saturation spectrum of Phillips (1958) is given by :

$$E(f) = \alpha g^2(2\pi)^{-4}f^{-5} \tag{4}$$

A value of α = 0.01 was chosen appropriate to wind-seas (Hasselmann et al., 1973).

In shallow water the limiting spectrum is taken as equation (4) multiplied by the shallow water function, Φ, of Kitaigorodskii et al. (1975). These authors showed that the concept of a saturation range could be extended to water of finite depth, h, by means of the function, Φ, given by:

$$\Phi = \frac{\tanh^2 kh}{(1 + 2kh / \sinh 2kh)}$$

where k satisfies the linear dispersion relation $\omega^2 = gk \tanh kh$.

The one-dimensional spectrum is multiplied by a directional spreading function $(8/3\pi)\cos^4\theta$ with $|\theta| \leq \pi/2$. This form of limiting spectrum is more general and less restrictive than the more commonly imposed Pierson-Moskowitz spectrum, which is wind speed dependent.

2.2.4 Bottom friction

Bottom friction in shallow water was modelled using the Source term, S_{bf}, derived from JONSWAP, namely,

$$S_{bf} = -\Gamma \left(\frac{\omega}{g \sinh kh}\right)^2 E$$

where k is the wave number in water of depth h, and ω is the radian frequency.

Two values of Γ are currently used in second-generation wave models. In the JONSWAP study a value of $\Gamma = 0.038\ m^2/s^3$ was obtained. Subsequent work by Bouws and Komen (1983) proposed that $\Gamma = 0.067\ m^2/s^3$ was more relevant in storm seas. In this report we use a value of $\Gamma = 0.067\ m^2/s^3$.

2.3 Model computations and initialisation procedure

The choice of model frequencies used in this paper depends on the application. In general however, the lowest wave frequency is 0.05 Hz and the bandwidth Δf=0.01 Hz. The lowest wave frequency determines the time step for the finite difference scheme since the Courant Number has to satisfy the stability criterion. Twelve wave directions were selected: that is, a directional resolution of 30°.

The wave growth term, S_{in}, gives rise to an exponential growth assuming the existence of a sea state. All calculations have therefore to be initialised with a 'background' sea state from which the waves develop. This sea state was taken as a uniform spectral density of 0.02 m^2/Hz per direction band apart from a local peak of 0.04 m^2/Hz per direction at the third highest frequency (this local peak is necessary to enable the Source Term for S_{nl} to be activated). The model calculations were found to be insensitive to the choice of initial spectrum, the influence of which disappears after a few time steps.

In order to reduce repetitive calculations, look-up tables for information on various parameters were established. These tables were divided up by depth and involved the pre-calculation of group velocity, phase velocity, wave number, and refraction coefficients at each of the grid depths with a step of 1m up to the maximum depth in the grid.

3 Model validation for idealised wind fields

WINDWAVE was first tested under idealised wind fields for both fetch-limited and duration-limited conditions in deep and shallow water. In the case of fetch-limited conditions, the results from JONSWAP are often used for checking the "growth curve" of wave models in deep water. For duration-limited situations and also for shallow water cases, a data base of field measurements is not available for validation: in this case we make comparisons with other second-generation models and with the third-generation model, WAM.

3.1 Fetch-limited conditions

3.1.1 Deep water

The calculations were made for constant wind speeds of 10 m/s and 20 m/s. For comparison with SWAMP, we use the following dimensionless variables, based on friction velocity, u_*

nondimensional total wave energy, $E^* = g^2E/u_*^4$
nondimensional peak frequency, $f_p^* = u_*f_p/g$
nondimensional fetch, $x^* = gx/u_*^2$

where E is the total wave energy, f_p is the peak frequency and x is the fetch. The model grid was chosen to be a square ocean with 20 x 20 mesh points.

Figure 2 shows the results for the nondimensional energy in comparison with the SWAMP tests. The calculations are made for different choices of wind speed and mesh interval. The calculations for duration-limited conditions are given in Ewing and Hague (1992).

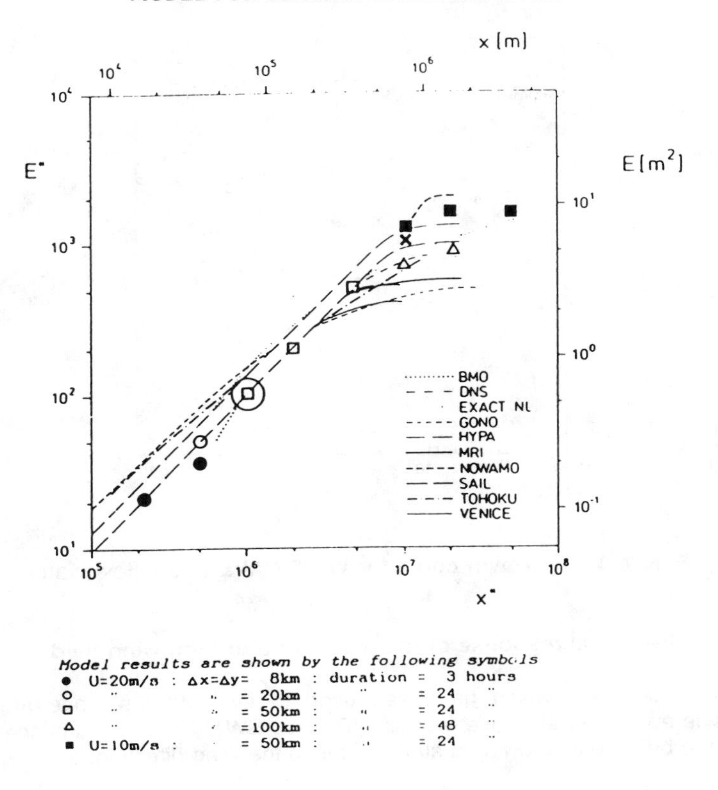

Figure 2 Growth curve for WINDWAVE compared with SWAMP (1985)

3.1.2 Shallow water

The calculations for shallow water were made with a bottom friction value of $\Gamma = 0.067\ m^2/s^3$ appropriate to storm seas (Bouws and Komen, 1983).

The computations were carried out for nondimensional depths $D^* = Dg/u_*^2$ = 200, 400, 800 and 1600, as given in Figure 7 of the WAM paper. For a wind speed of 20 m/s, the calculations over a range of mesh intervals allow coverage of the nondimensional variable x^*. (To compare the results with WAM f_p is converted to mean frequency $\bar{f}$ using the relation $\bar{f} = 1.295 f_p$ which is appropriate for a JONSWAP spectrum). The results are shown in Figure 3. There is reasonable agreement with the results of $\bar{f}^*$ from WAM: however, for the wave energy, E^* the results differ from WAM at the shallowest depth, $D^* = 200$.

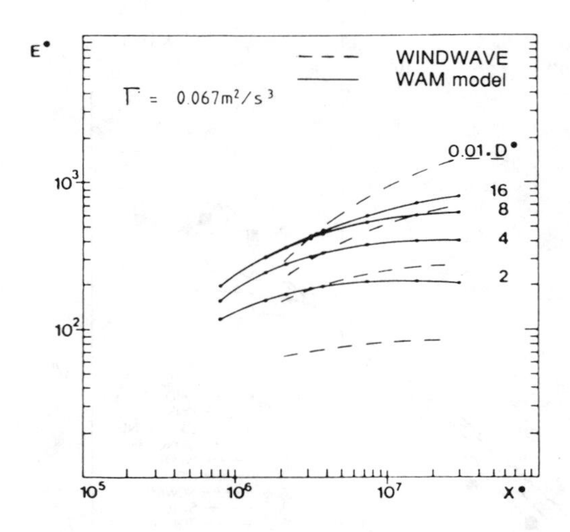

Figure 3 Growth curve for WINDWAVE in shallow water

3.2 Directional response of the model to a turning wind field

An important test which has been used to evaluate the wave models considered in the Sea Wave Modelling Project (SWAMP, 1985) is concerned with the behaviour following a sudden shift in the wind direction.

The particular case, referred to as Case VII, is discussed in SWAMP. A uniform wind, U_{10}=20m/s, blows in a steady direction over an infinite ocean until the wind-sea spectral peak reaches $2f_{PM}$, where f_{PM} = 0.13g/U is the Pierson-Moskowitz frequency. At this time, the wind direction is changed by 90° and the integrations continue for another 10 hours. When the wind direction turns, the previous wind-sea becomes effectively swell and a new wind-sea begins to grow. The purpose of the test is to study the relaxation of the wave field towards a steady state which approaches a fully-developed wind sea in the new wind direction.

The model was run for two different cases. In the first case the Source term, S_{nl}, acted in the wind direction. In the second case S_{nl} acted in the mean wave direction. Comparisons with the model EXACT-NL (given in Fig. 10.19 from SWAMP (1985)) are given in Ewing and Hague (1992). It was found that the closest agreement to EXACT-NL was obtained for the second-generation model when S_{nl} acts in the wind direction. This scheme is used in the hindcast calculations for situations where rapidly turning wind fields are important.

4 Model validation for real wind fields

WINDWAVE has been validated against measured wave data for real wind fields. Three sets of wind data were considered and the full set of results are given in Ewing and Hague (1992). The "SWIM" (Bouws et al., 1985) storm of 20-27 November 1981 is considered.

In the SWIM study, an intercomparison was made between three wave models and wave measurements in the North Sea. The three wave models were those of the UK Meteorological Office (BMO), the KNMI (GONO) and the model of the Max-Planck-Institute, Hamburg (HYPAS). All these wave models can be classified as second-generation models.

This paper compares the results at stations Fulmar and Gorm with model calculations from GONO and WINDWAVE.

4.1 Wind data and model grid

The wind data for the storm was supplied by the UK Meteorological Office, at 3-hourly intervals from the then operational numerical weather prediction model. Surface winds were estimated using the relations given by Findlater et al. (1966). The data was defined on a polar stereographic projection with a grid spacing of approximately 50 km. The model grid was chosen to be the same as that used for the wind fields. (The bathymetry was obtained from data supplied by Tidal Group of HR Wallingford).

Fourteen wave frequencies from 0.05 Hz (with an interval of 0.02 Hz) up to 0.31 Hz were chosen to represent the spectrum. A time step of 9 minutes was chosen. A constant bottom friction factor of $\Gamma = 0.067\ m^2/s^3$ was used in the calculations which included wave refraction.

4.2 Hindcast results

Figure 4 shows the comparison of results of significant wave height from WINDWAVE and GONO against measured values. Station Fulmar is in deep water in the northern North Sea. Gorm is in the southern North Sea where the depth of water is 35m. At Fulmar, WINDWAVE estimates are below those of the measurements and of the model, GONO. At Gorm, however, the results from WINDWAVE are in close agreement with both the wave measurements and GONO, especially if an alternative, neighbouring grid point is considered. The rms error and scatter index at Fulmar were found to be 1.64m and 27% respectively. For Gorm the corresponding values were 0.93m and 23%.

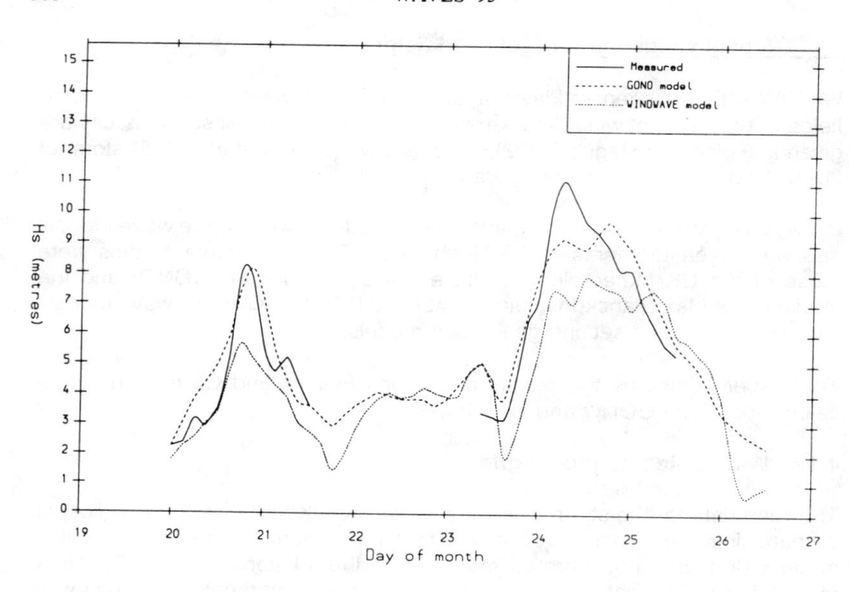

Figure 4a Comparison of wave heights at Fulmar

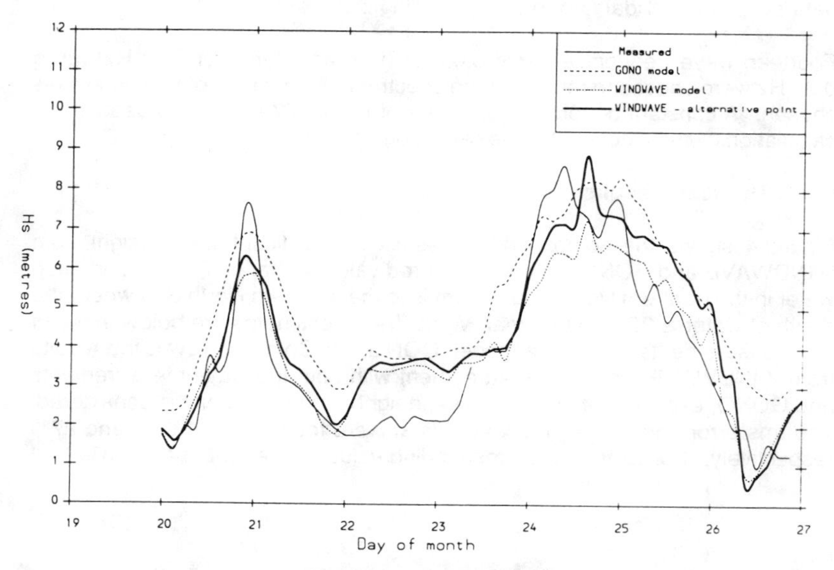

Figure 4b Comparison of wave heights at Gorm

5 Conclusions

5.1 General

A second-generation numerical wave model, WINDWAVE, has been developed for wave prediction in coastal regions.

The model is based on the numerical integration of the energy balance equation for waves in water of finite depth. The propagation of waves is obtained using a second-order Lax-Wendroff scheme with wave refraction included as an additional term in shallow water.

Three Source terms are included in the physical representations of the model. These terms describe wave growth due to the wind, nonlinear wave-wave interactions and dissipation due to bottom friction in shallow water. The Source term for wave-wave interactions is based on a simple, but directionally-responsive, formulation given by Young (1988). A limiting wave spectrum, based on Phillips (1958) saturation spectrum, is used to parameterise the effect of wave breaking.

WINDWAVE was tested in uniform winds for fetch- and duration-limited conditions. The results showed good agreement with other wave models given in SWAMP (1985) and JONSWAP for wave energy and peak frequency. Satisfactory comparisons were obtained in shallow water with results from the third-generation (WAM) model.

WINDWAVE was further tested against real winds for the SWIM storm. In general the model results showed good agreement in the southern North Sea where the water depth is less than 40 m. In the northern North Sea the model values of significant wave height were generally less than the measured values. The Scatter Index was about 35% for locations in the northern North sea and 20% in the southern North sea.

5.2 Uses of the model

It is envisaged that the model could be used in the following situations:

(i) for real time forecasting in coastal areas for which the UK Meteorological Office European Fine Mesh Wave Model has too coarse a grid. The model could be driven at the open sea boundaries by the Meteorological Office Model.

(ii) for studies where wave prediction is needed at a large number of locations.

(iii) for studies where swell is important.

(iv) for studies where there are offshore banks or where refraction is important over a considerable part of the wave generation area.

5.3 Future development of the model

An additional Source term to represent wave breaking over offshore banks will be included in WINDWAVE. A study is in progress with comparisons of wave breaking over the Shipwash Bank in the southern North Sea.

6 Acknowledgements

Mr E Bouws, KMNI, kindly supplied additional wave data used for the SWIM comparisons. The Meteorological Office, Bracknell supplied wind data for the SWIM storm.

We thank Dr Steve Foreman and Dr Martin Holt of the Meteorological Office for their advice on the analysis of wind fields. We are grateful to Dr Alan Brampton, Dr Peter Hawkes and Mr G Gilbert for many helpful discussions.

This work was funded by the Ministry of Agriculture, Fisheries and Food as part of HR Wallingford's Flood Defence Research Commission (FD0702) during the period 1990-1992.

7 References

Bouws E, Ephraums J J, Ewing J A, Francis P E, Gunther H, Janssen P A E M, Komen G J, Rosenthal W and de Voogt W J P. 'A shallow water inter-comparison of three numerical wave prediction models (SWIM)'. Quart. J Roy Meteor : Soc, 111, 1087-1112, 1985.

Bouws E and Komen, G J. 'On the balance between growth and dissipation in an extreme depth-limited wind-sea in the southern North Sea'. J Phys. Oceanogr, 13, 1653-1658, 1983.

Ewing J A and Hague R C. 'The development of an improved wave model for coastal wave prediction'. HR Wallingford Report SR 295, May 1992.

Findlater J et al. 'Surface and 900mb wind relations'. Met Office, Bracknell, Scientific Paper No 23, 1966.

Golding B W. 'A wave prediction scheme for real time sea state forecasting.' Quart. J Roy. Meteor Soc, 109, 393-416, 1983.

Hasselmann K et XV al. 'Measurements of wind-wave growth and swell decay during the Joint North Sea Wave Project (JONSWAP)'. Dt Hydrogr Z, A8 (12), 1973.

Hasselmann K et al. 'A parametric wave prediction model'. J Phys Oceanogr, 6, 200-228, 1976.

Hasselmann K and Hasselmann S. 'Computations and parameterizations of the nonlinear energy transfer in a gravity wave spectrum. Part I. A new method for efficient computations of the exact nonlinear transfer'. J Phys Oceanogr, 15, 1369-1377, 1985.

Kitaigorodskii S A, Krasitskii V P and Zaslavskii M M. 'On Phillips' theory of equilibrium range in the spectra of wind-generated waves'. J Phys Oceanogr, 5, 410-420, 1975.

Mitsuyasu H et al. 'Observations of the power spectrum of ocean waves using a cloverleaf buoy'. J Phys Oceanogr, 10, 286-296, 1980.

Phillips O M. 'The equilibrium range in the spectrum of wind-generated waves'. J Fluid Mech, 2, 417-445, 1958.

Snyder R L, Dobson F W, Elliot J A and Long R B. 'Array measurements of atmospheric pressure fluctuations above surface gravity waves'. J Fluid Mech, 102, 1-59, 1981.

SWAMP Group. 'The Sea Wave Modelling Project (SWAMP) : Principal results and conclusions'. Plenum Press, pp 256, 1985.

WAMDI Group. 'The WAM model - A third generation ocean wave prediction model'. J Phys Oceanogr, 18, 1775-1810, 1988.

Young I R. 'A shallow water spectral wave model'. J Geophys Res, 93, 5113-5129, 1988.

Numerical Modeling of Waves in Harbors

Edward F. Thompson[1], M. ASCE, H. S. Chen[2], M. ASCE, and Lori L. Hadley[1]

Abstract

The numerical wave model HARBD has been used extensively for estimating waves in harbors. The model is used to simulate laboratory tests of long waves in a rectangular harbor with variable entrance width. The sensitivity of the model estimates to various input parameters is documented. New gridding and visualization software for use with the model is discussed.

Introduction

Numerical models are a powerful tool for estimating the wave response of harbors. The effects of both long waves (periods on the order of 25 s to 5 min) and short waves (periods from about 5 s to 25 s) can be considered. The U.S. Army Corps of Engineers (CE) has used the model HARBD (Chen 1986) extensively for both long and short wave harbor response.

The performance of HARBD was originally tested by comparing with available, but limited, physical model data. More recently, Briggs, et al. (1992) evaluated both HARBD and physical model results against prototype measurements in Barbers Point Harbor, HI. Laboratory results published by Lepelletier (1980) afford another valuable opportunity for evaluating HARBD with long waves. The laboratory tests include a steady state evaluation of a simple rectangular harbor with two symmetric breakwaters

---

[1] Hyd. Engr., U.S. Army Engineer Waterways Expt. Station, Vicksburg, MS 39180.

[2] Physical Scientist, NOAA/NWS/NMC21, Room 206, 5200 Auth Road, Camp Springs, MD 20746.

at the mouth. Only resonant long waves were considered. Lepelletier (1980) and Lepelletier and Raichlen (1987) also presented a time-dependent numerical model which satisfactorily reproduced their laboratory tests. The model includes an explicit energy loss term (negative source term) at the harbor entrance.

The main objectives of this paper are to evaluate HARBD sensitivity to key parameters, bottom friction and boundary reflection coefficient, and to compare HARBD predictions with the steady state laboratory results of Lepelletier (1980). New software systems the CE is using for finite element grid preparation and visualization of model results are also discussed.

Numerical Model HARBD

Chen and Mei (1974) developed a hybrid element numerical model for estimating linear, steady state, regular, long waves in a shallow offshore harbor. The model solves the Helmholtz equation as a boundary value problem. Houston (1981) adapted the model for use with short waves by solving Berkhoff's (1972) mild slope equation. Chen (1986) further modified the model to include bottom friction and variable boundary reflectivity. As a result of these contributions, the model has become a flexible, powerful tool for studying harbor response to both long and short waves. The model was documented for users by Chen and Houston (1987). The form of the model which incorporates the general depth dispersion relation is named HARBD. The model has been used in a variety of harbor oscillation, tsunami, and short wave applications (e.g. Briggs, et al. 1992; Houston, 1978; Lillycrop and Boc, 1992).

The HARBD model domain consists of a harbor region (Region A) and a far region (Region B) (Figure 1). The harbor region is covered by a triangular finite element grid. The boundary between the harbor and far regions is a semicircle which encompasses the harbor entrance and other nearshore features of interest. The far region extends from a semi-infinite straight coastline at either end of the semicircle out to infinity.

Input to HARBD includes specification of a finite element grid, wave conditions to be tested, and desired output. Some particularly important input parameters are summarized in Table 1. Reflection coefficient, K_r, along the solid boundaries and the coastline of the infinite half-plane beyond the grid domain is generally set

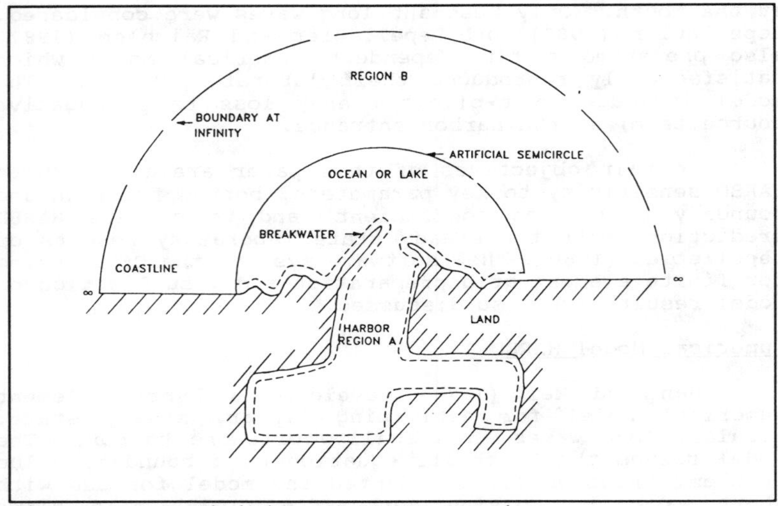

Figure 1. Definition of HARBD Domain

according to the physical characteristics of the boundary. Guidance for setting the bottom friction parameter, β, is less well established. It is typically set to a single constant value over all elements. For long waves, β may range from zero to 10.

The model HARBD has a number of limitations which reduce its accuracy or make it unsuitable for some applications. Primary concerns related to long wave modeling include lack of wave transmission through structures, lack of explicit energy loss terms at constricted entrances, and uncertainty in choosing values for parameters in Table 1.

Comparison with Lee's (1969) Laboratory Tests

Lee (1969) studied the response of a simple, flat-bottomed, rectangular harbor to incident long waves. Laboratory data were collected at the center of the harbor back wall (Figure 2). The test harbor was fully open, corresponding to a/b=1.0, where a = entrance width and b = harbor width. Similar tests were conducted by Ippen and Goda (1963).

Table 1. Critical HARBD Input Parameters

Parameter	Where Specified
Bottom friction, β	Every element
Boundary reflection, K_r	Every element on solid boundary
Coastline reflection	Single value
Depth in infinite region	Single value

Chen (1986) compared the harbor response predicted by his formulation and by frictionless, fully reflective potential theory to the laboratory data (Figure 3). The x- and y-axes in the figure are given in terms of $k\ell$ and A_{amp}, where

k = wavenumber;
= $2\pi/L$;
L = wavelength;
ℓ = harbor length;
A_{amp} = amplification factor;
= ratio of local wave amplitude to incident wave amplitude.

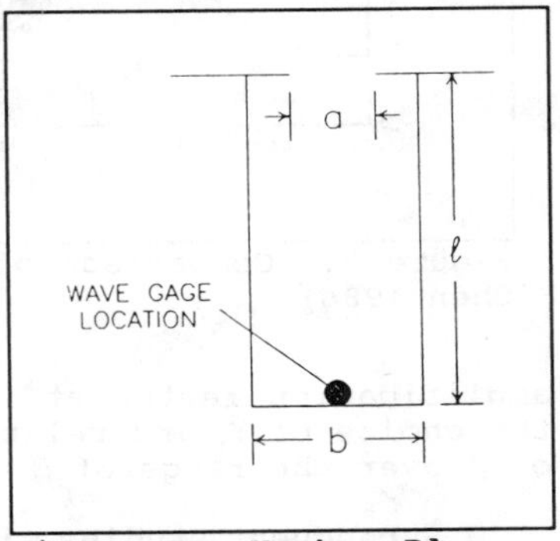

WAVE GAGE LOCATION, a, b, ℓ

Figure 2. Harbor Plan for Laboratory Tests

The value of A_{amp} is divided by 2 to follow the historical convention of referencing harbor wave amplification to wave amplitude which would be present along a straight coastline (which is twice the incident amplitude for a fully reflective coast).

The range of wave conditions considered includes the first and second resonant modes of harbor oscillation. It was demonstrated that, with proper choice of K_r and β, Chen's (1986) formulation provides a good fit to the laboratory data. Chen (1986) also demonstrated that the

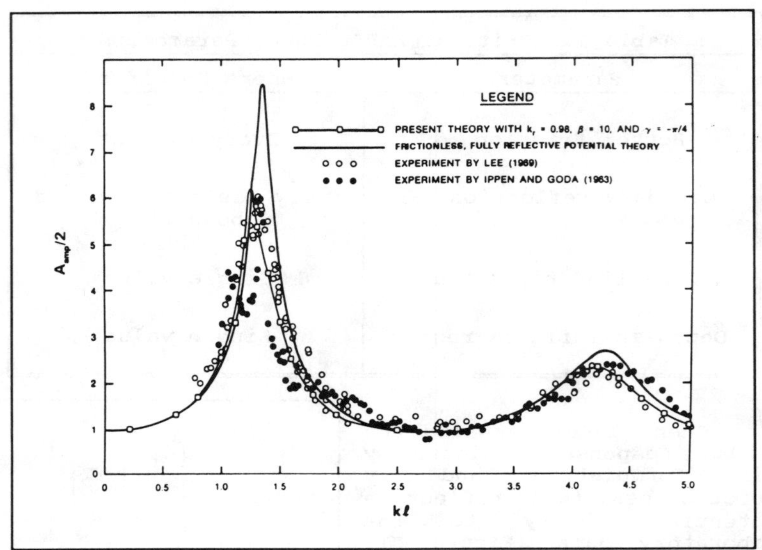

Figure 3. Comparison of Amplification Factors (after Chen 1986)

amplification factor at resonance is quite sensitive to the choice of K_r and relatively insensitive to the choice of β over the range of β values from 0 to 10.

Subsequent studies with HARBD have indicated that the value of β=10 used by Chen (1986) is unrealistically high and a value of less than 1 would be more appropriate. Lee's (1969) tests were modeled in this study using HARBD and smaller values of β. Selected results are given in Table 2. The sensitivity of $A_{amp}/2$ to K_r is evident. The value β=0.05 was taken as a reasonable estimate of bottom friction. The corresponding K_r value of 0.990 for a good approximation to Lee's (1969) tests also seems realistic.

<u>Comparison with Lepelletier's (1980) Laboratory Tests</u>

More recent laboratory studies of several harbor shapes were conducted by Lepelletier (1980). The tests included a simple rectangular harbor with varying degrees of entrance constriction. The width of the entrance gap ranged from 0.2 to 1.0 times the harbor width, the latter case being a fully open harbor with no breakwaters. The

Table 2. HARBD Amplification Factors for Peak of Fundamental Resonant Mode, Lee's (1969) Rectangular Harbor		
β	K_r	$A_{amp}/2$
0.05	0.985	5.5
0.05	0.990	6.1
1.00	0.990	6.0
10.0	0.990	5.7
0.05	0.995	6.9
Lee (1969)		6.0

harbor bottom was flat.

A finite element grid suitable for numerical modeling of various entrance gaps was developed (Figure 4). Dimensions of the harbor are the same as Lepelletier's (1980) laboratory model:

$b = 7.6$ cm
$\ell = 38.0$ cm
$h = 10.0$ cm

where h = water depth. A wave period of 2 s was used, corresponding to the fundamental resonant mode of the harbor.

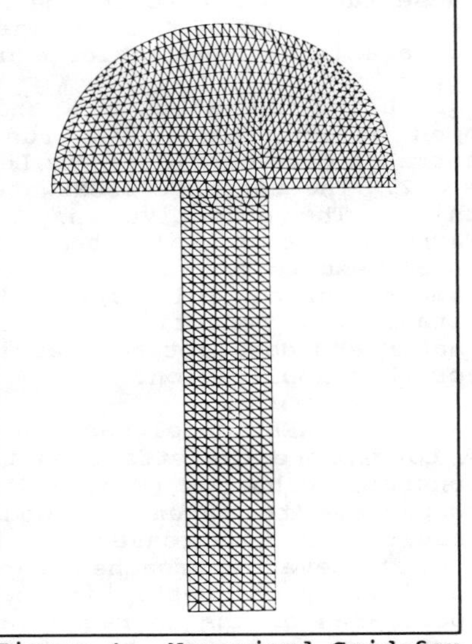

Figure 4. Numerical Grid for Lepelletier's (1980) Harbor

The amplification factor reported by Lepelletier (1980) is half as big as Lee's (1969). Since both tests were performed in the same basin, the same bottom friction should apply, β=0.05. However there is good justification for a difference in boundary reflection coefficient between the two tests. Lepelletier (1980)

Table 3. Summary of Amplification Factors for Peak of Fundamental Resonant Mode in Fully Open Rectangular Harbor

Laboratory		Numerical Model		
Test	Amp. Factor	β	K_r	$A_{amp}/2$
Lee (1969)	6.0	0.05	0.990	6.1
Lepelletier (1980)	3.0	0.05	0.955	3.1

constructed the model from 9.5-mm (3/8-in) lucite while Lee (1969) used 19-mm (3/4-in) plywood. Contrary to Lee's model, Lepelletier's model was not sealed to the bottom and possibly some other joints were not sealed either. These factors act to reduce reflections in Lepelletier's tests. A value of K_r=0.955 was found to give HARBD results comparable to Lepelletier's data (Table 3).

Using the values of K_r and β established for the fully open harbor, HARBD was run for the most constricted entrance case considered by Lepelletier (1980), namely a/b = 0.2. The entrance breakwaters in HARBD were infinitely thin. The objective was to assess the importance of HARBD's lack of explicit entrance energy loss terms. The HARBD results for the constricted entrance were nearly the same as for the fully open entrance whereas Lepelletier's data show a reduction by a factor of 2.5. It is clear that HARBD does not adequately represent entrance losses for this application.

The HARBD model was run for several other values of K_r to explore the effect on the relative response of the constricted harbor (Figure 5). For K_r=1, the results are comparable to Ippen and Goda's (1963) numerical model results. As K_r decreases, the relative response decreases, but it never approaches the laboratory results. The possibility of greatly increasing bottom friction in the local area of the entrance was also explored. The β was increased from 0.05 to 100 for 17 elements in the flow path at the entrance. The phase difference parameter, γ, associated with bottom friction (Chen 1986) was set equal to $-\pi/4$, as in other model runs. The increased bottom friction failed to remove sufficient energy. Some other alternatives for adapting HARBD to approximate the

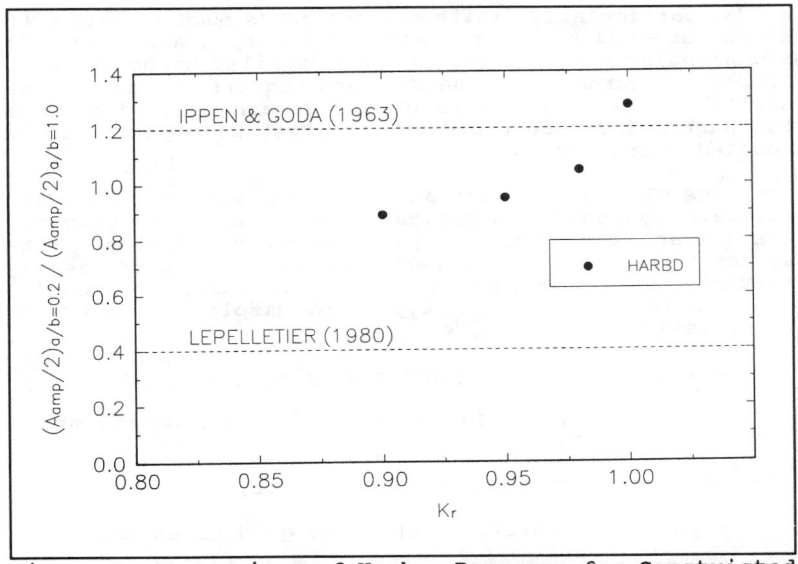

Figure 5. Comparison of Harbor Response for Constricted and Fully Open Entrance

constricted entrance laboratory tests are presently being considered.

While HARBD under represented entrance losses in the laboratory tests, these losses may be exaggerated relative to losses in actual harbor entrances. The laboratory harbor entrance, with very thin, solid, vertical breakwaters and a flat bottom can be expected to exaggerate the entrance loss phenomenon relative to a typical entrance defined by porous, partially reflecting structures and a bottom with a channel and sloping sides. Good comparisons of HARBD long wave results with exceptionally detailed field and laboratory measurements at Barbers Point Harbor, HI, were recently reported by Briggs, et al. (1993). Parameter values used in HARBD were K_r=1.0 and $\beta \leq 0.03$.

Gridding and Visualization Software

Valuable new tools for developing numerical model grids and displaying model input and output fields are becoming available. The tools were badly needed because grids, particularly finite element grids such as used with HARBD, are tedious to prepare and modify. Also there is a continuing need for larger more detailed harbor models. Grid size increases as the size and required detail in the harbor increase, the size of the domain modeled outside the harbor increases, and the shortest wave period to be modeled decreases.

The CE is developing an automated, workstation-based software system for numerical model grid generation and visualization of model input and output fields. The system is interactive and menu-driven. It is sufficiently flexible to accommodate a variety of numerical models, grid manipulations, and types of display. Principal benefits include:

- speed in creating and modifying grids;
- spatial visualization of grids, bathymetry, and results;
- flexibility in grid element size;
- ease in identifying specific grid nodes and elements.

The software is being developed under contract with the Center for Coastal and Land-Margin Research, Oregon Graduate Institute, Portland, OR.

Some capabilities of the visualization software are illustrated using a flat bottom, square harbor with an asymmetric, constricted entrance. Amplification factor contours over the harbor in response to a selected incident wave frequency are shown in Figure 6. Large amplification is evident just inside the entrance and in the southeast corner. The corresponding phase plot shows a clear standing wave pattern with a node running approximately between the northeast and southwest corners of the harbor (Figure 7). Because of the large amplification factors relative to those at nearby frequencies (not shown) and the presence and location of a single node near the middle of the harbor, it is evident that these results represent resonance at the second mode of oscillation. The harbor response can also be displayed as an animated time series in which surface elevations are color-coded. Optionally, results from a number of frequencies can be linearly combined to give a composite harbor response to a spectrum of incident wave conditions.

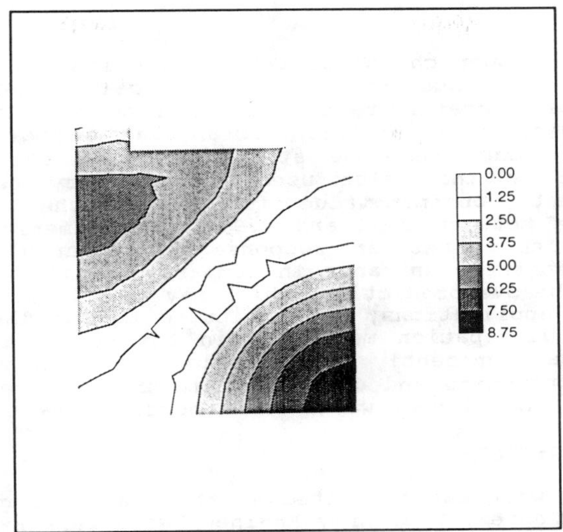

Figure 6. Amplification Factor Contours

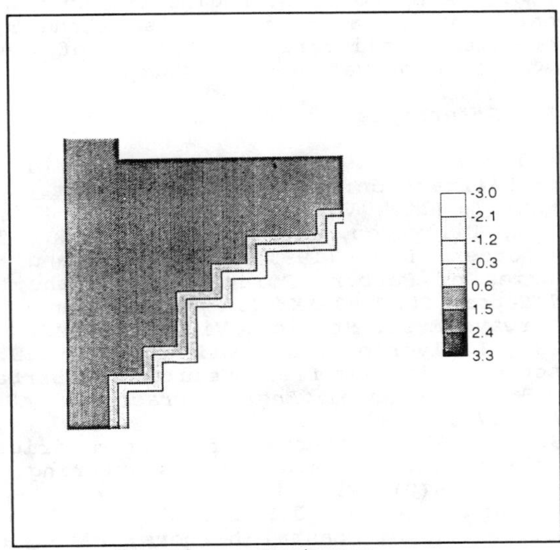

Figure 7. Phase Contours

Summary

With proper choice of bottom friction and boundary reflection parameters, the numerical model HARBD reproduces laboratory tests of resonant wave conditions in fully open, flat bottom, rectangular harbors. The numerical model response at resonance is particularly sensitive to the value used for boundary reflection coefficient. Optimum values for matching the laboratory tests are between 0.95 and 0.99. The numerical model overpredicts the resonant response of a rectangular harbor with constricted entrance in comparison with laboratory tests. The overprediction, which may be a lesser concern in field applications, is attributed to the lack of an entrance dissipation mechanism in the numerical model. New software presently under development greatly aids in grid construction and display of information related to numerical modeling of waves in harbors.

Acknowledgements

This work was conducted at the Coastal Engineering Research Center, U.S. Army Engineer Waterways Experiment Station, under the Numerical Model Maintenance Program. Permission was granted by the Office of the Chief of Engineers, U.S. Army Corps of Engineers, to publish this information. Special appreciation is extended to Prof. Fredric Raichlen, California Institute of Technology, Pasadena, CA, for motivating this study.

Appendix I. References

Berkhoff, J.C.W. (1972). "Computation of Combined Refraction-Diffraction," Proc., 13th Int. Conf. on Coastal Engrg., ASCE, Vol. 1, 471-490.

Briggs, M.J., Lillycrop, L.S., Harkins, G.R., Thompson, E.F., and Green, D.R. (1993). "Physical and numerical model studies of Barbers Point Harbor, Oahu, Hawaii." *Technical Report CERC-93-XXX* (in publication), U.S. Army Engr. Waterways Expt. Station, Vicksburg, MS.

Briggs, M.J., Lillycrop, L.S., and McGehee, D.D. (1992). "Comparison of model and field results for Barbers Point Harbor." *Proc., Coastal Engrg. Practice*, ASCE, Long Beach, CA, 387-99.

Chen, H.S. (1986). "Effects of bottom friction and boundary absorption on water wave scattering." *Applied Ocean Research*, 8(2), 99-104.

Chen, H.S., and Houston, J.R. (1987). "Calculation of water oscillation in coastal harbors; HARBS and HARBD user's manual." *Instruction Report CERC-87-2*, U.S. Army

Engr. Waterways Expt. Station, Vicksburg, MS.

Chen, H.S., and Mei, C.C. (1974). "Oscillations and wave forces in an offshore harbor." *Report 190*, R.M. Parsons Lab. Water Resources and Hydraulics, Mass. Inst. Technology, Cambridge, Mass.

Houston, J.R. (1978). "Interaction of tsunamis with the Hawaiian islands calculated by a finite-element numerical model." *J. Phys. Oceanography* 8(1), 93-101.

Houston, J.R. (1981). "Combined refraction and diffraction of short waves using the finite elements method." *Appl. Ocean Res.*, 3(4), 163-170.

Ippen, A.T., and Goda, Y. (1963). "Wave induced oscillations in harbors: the solution for a rectangular harbor connected to the open-sea." *Report 59*, Hydrodynamics Lab., Mass. Inst. Technology, Cambridge, Mass.

Lee, J.J. (1969). "Wave Induced Oscillations in Harbors of Arbitrary Shape." *Report No. KH-R-20*, W.M. Keck Lab. of Hydraulics and Water Resources, Calif. Inst. of Technology, Pasadena, CA.

Lepelletier, T.G. (1980). "Tsunamis - harbor oscillations induced by nonlinear transient long waves." *Report No. KH-R-41*, W.M. Keck Lab. of Hydraulics and Water Resources, Calif. Inst. of Technology, Pasadena, CA.

Lepelletier, T.G., and Raichlen, F. (1987). "Harbor oscillations induced by nonlinear transient long waves." *J. Wtrway., Port, Coast., and Oc. Engrg.*, ASCE, 113(4), 381-400.

Lillycrop, L.S., and Boc, S.J. (1992). "Numerical modeling of proposed Kawaihae Harbor, HI." *Proc., Coastal Engrg. Practice*, ASCE, Long Beach, CA, 412-24.

Experimental Study of Undertow and Turbulence Intensity Under Irregular Waves in the Surf Zone

Nels J. Sultan[1] and Francis C.K. Ting[2]

ABSTRACT

Experiments are performed in a two-dimensional wave tank with a sloping beach. Irregular waves are generated and the water particle velocity is measured in the surf zone with a laser Doppler velocimeter. Using ensemble averaging, the turbulent kinetic energy, k, time series is then calculated. Results are compared with previously published experiments using regular waves. There is a significant temporal variation of k throughout the surf zone. However, k, is more uniform at points closer to the shoreline. Further from shore k is more intermittent and discrete "bursts" of k caused by individual breaking waves seem to decay in an exponential manner. The use of ensemble averaging to study turbulence under irregular waves is found to be feasible although time consuming.

INTRODUCTION

This paper deals with the study of turbulent flows created by random wave breaking on a plane slope. The primary objective is to investigate the turbulence dynamics in the surf zone and the interaction of the turbulence with the waves and the currents. The results presented herein relate to the initial portion of the study dealing with experimental procedures for measuring the turbulence and undertow in the surf zone. In particular, the turbulent kinetic energy, k, (also

[1]Graduate Student, Ocean Engineering Program, Civil Engineering Dept., Texas A&M University, College Station, Texas 77843.

[2]Assistant Professor, Ocean Engineering Program, Civil Engineering Dept., Texas A&M University, College Station, Texas 77843.

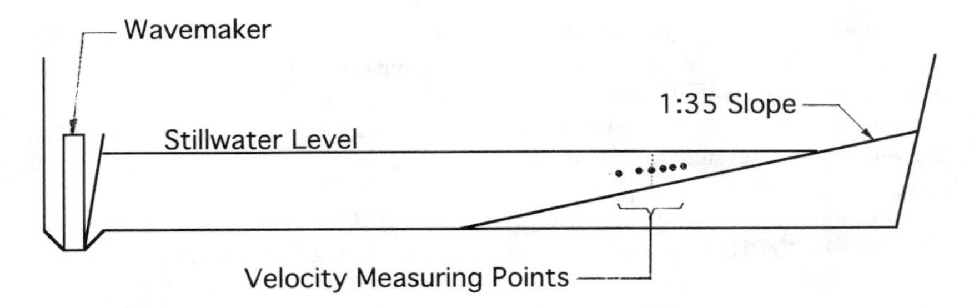

Figure 1: Wave Tank - Schematic

called the turbulence intensity) is calculated from the measurements. k varies in time and space and is defined by:

$$k = \frac{1}{2}\left(\overline{u'^2} + \overline{v'^2} + \overline{w'^2}\right) \tag{1}$$

where u', v', and w' are the turbulent fluctuations in the velocity time series calculated from an ensemble averaging process. (u is horizontal and w is vertical. The positive direction is upwards and towards the shoreline.) This is the same k that is often applied in sediment transport studies and is the same k used in numerical models of turbulence, such as $k - \epsilon$ models.

Previous studies have measured the turbulent kinetic energy under *regular* waves in the surf zone. Svendsen (1987) compares the results of a number of investigators. This study is different in that it measures the turbulence under irregular waves.

EXPERIMENT PROCEDURE

Laboratory experiments were performed in a two-dimensional wave tank at Texas A&M University (Figure 1). The overall dimensions of the tank are 91 cm wide and 36 m long. The depth of water was 46 cm for all experiments. Approximately half the tank length was occupied by a removable false bottom with a painted plywood surface sloping at 1:35.

Water particle velocities were measured using a two-component Argon-Ion laser Doppler velocimeter equipped with tracker type electronics. Six horizontal locations were chosen in the surf zone. At these locations measurements were made midway between the bottom and stillwater level. At one horizontal location data was taken at eight points in a vertical line. The exact locations of all thirteen measuring points are given in figures 8 and 9 and are defined by specifying ξ/h for elevation in the water column and h/H_{mo} for horizontal location in the surf zone. (ξ is the distance above the bottom, h is the local stillwater depth, and H_{mo} is the spectral significant wave height outside the surf zone.) The measuring point was 10 cm from the inside face of the glass sidewall. Additionally, a capacitance type wave gage measured the water surface elevation (synoptic with the velocity measurements) at each location.

Data was acquired and analyzed using a 486 based personal computer. A sampling rate of 100 Hz was used for each channel. During experiments, entrained air bubbles due to wave breaking usually cause the velocity signal to "drop-out" when the bubbles pass through the laser beam. Drop-outs also occur when the laser beams are above the water surface. Dropouts are usually detected by the electronic lock detector. The velocity signal drop-outs were filled in by linear interpolation for the times when the laser beams were below the water surface. Typically, the lock detector indicated a below surface drop-out 4% of the time and the laser beams were never above the water surface. For the measuring point which was highest in the water column ($\xi/h = 0.875$) the drop-out rate was 19% and the laser beams were in addition above the water surface 28% of the time.

The exact same incident wave time series was used for all experiments. In other words, the exact same voltage time series was sent to the wavemaker for every experiment run. Figure 2 is the measured wave spectrum (outside the surf zone in the flat part of the wave tank) that was used. This has a TMA spectral shape (Bouws, et al. 1985) with a spectral significant wave height, H_{mo}, of 9.34 cm, a peak period, T_p, of 3.0 s and a peak enhancement factor, γ, of 7.0. ($\gamma = 7.0$ provides a somewhat narrower spectral peak than the typical JONSWAP spectrum.) The average height of the highest one-third waves, $H_{1/3}$, was 9.22 cm. The exact same incident wave time series was used for all experiments.

It was desired to have $H_{mo} = 14.5$ cm in order to match the wave height and period used by Stive and Wind (1982) in their experiments done using regular waves. However, the wave generator was only capable of getting to $H_{mo} = 9.34$ cm.

The velocity time series, after ensemble averaging, is divided into three components. For the horizontal direction this gives:

$$u = \bar{u} + \tilde{u} + u' \tag{2}$$

where $\bar{u}$ is the "undertow" or the mean velocity averaged over the entire time series. $\bar{u} + \tilde{u}$ is the deterministic or repeatable component, and u' is the random, turbulent fluctuation. The data acquisition and ensemble averaging process which results in equation 1 is as follows:

1. Run the wavemaker for 300 seconds and record the velocity and water surface elevations. Start acquiring data shortly before the first wave reaches the measuring point.

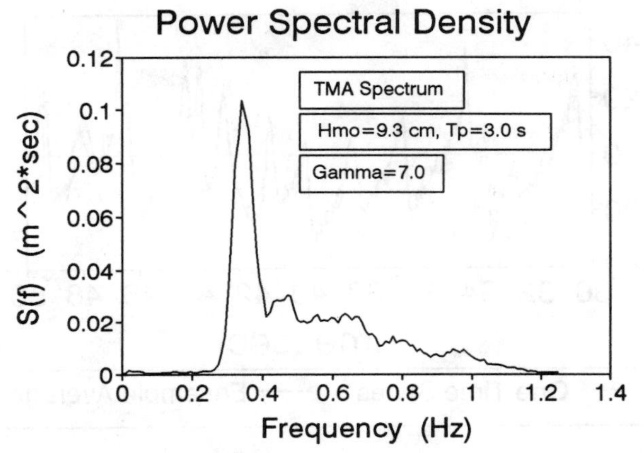

Figure 2: Measured Spectrum - Outside Surf Zone

2. Wait 12 minutes for the water in the tank to settle down then run the exact same signal again and record data. Repeat this 15 times.

3. Average all 15 time series to get the ensemble average or mean flow time series, $(\bar{u} + \tilde{u})$.

4. Subtract this mean flow time series from each of the raw time series to get 15 different turbulent fluctuation time series, (u').

5. Square the values in each of these time series and take the ensemble average to get one time series of $\overline{u'^2}$.

6. Because the v component is not measured, estimate k with the equation $k = 1.33 \cdot \frac{1}{2}\left(\overline{u'^2} + \overline{w'^2}\right)$. after Svendsen (1987) and Stive and Wind (1982).

This process gives a time series of k at one location. There are a total of 13 measuring points in this study.

Figure 3 shows an ensemble average time series and one of the individual time series that is part of the ensemble average. The turbulent fluctuations are evident. A similar graph for a point *outside* the surf zone (not shown) has no turbulence, the two lines overlap almost perfectly. The wavemaker is thus highly repeatable. Note that the horizontal location in the surf zone is indicated by the ratio of the local stillwater depth, h, to the measured spectral significant wave height, H_{mo}, outside the surf zone.

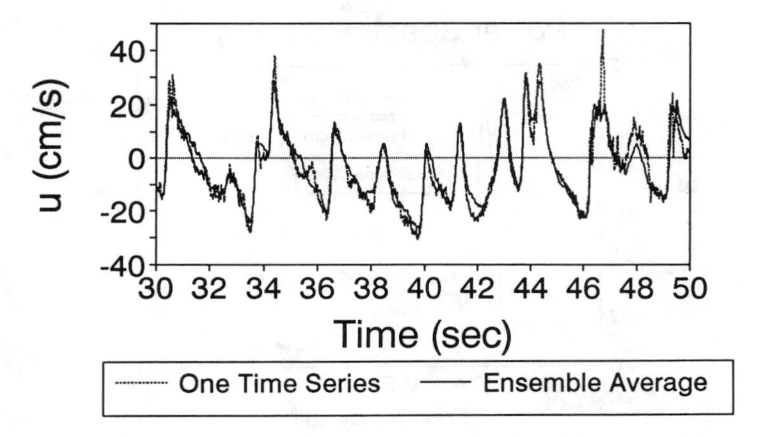

Figure 3: Ensemble Average Horizontal Velocity - Inside Surf Zone at $h/H_{mo} = 0.75$, $\xi/h = 0.50$

RESULTS

Figure 4 plots part of the time series of k, $\bar{u} + \tilde{u}$, and water surface elevation, η for one location. As expected the horizontal velocity and water surface are highly coherent. The turbulent kinetic energy is more random and exhibits strong temporal variation. Through comparison with plots at other locations, k also shows strong spatial variation. For locations closer to shore, a plot of k looks more like that part of figure 4 between 30 and 40 seconds. That is, the level of k is more uniform. For locations further away from shore where wave breaking is less frequent, a plot of k is more like that part of figure 4 at 47 and 49 seconds. Away from shore the breaking events are more discrete. A single breaker causes k to increase rapidly from near zero to a certain level and then decay in an exponential manner over about 8 seconds back to near zero.

In performing the experiments, one must select a length of time series as well as the number of time series to be used in the ensemble average. Figure 5 shows how the undertow, the set-up, and the time mean of k vary with the length of time series analyzed. The value of $\bar{k}$ reaches a nearly constant value after only 100 seconds. The set-up and undertow are small relative to the typical velocities and water surface elevations and show higher variability with length of time series analyzed. A length of 300 seconds seems reasonable and is long enough to allow development of the low frequency waves which occur in the surf zone (discussed further below).

In selecting the number of time series to include in the ensemble average, one

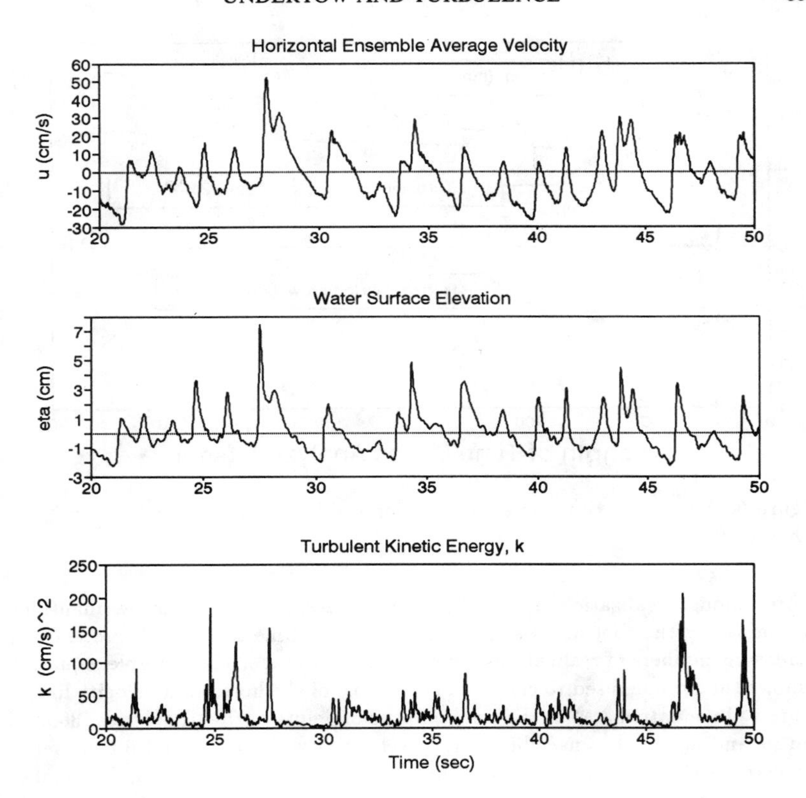

Figure 4: Relative Phase of η, $\tilde{u}$ and k at $h/H_{mo} = 0.75$, $\xi/h = 0.50$

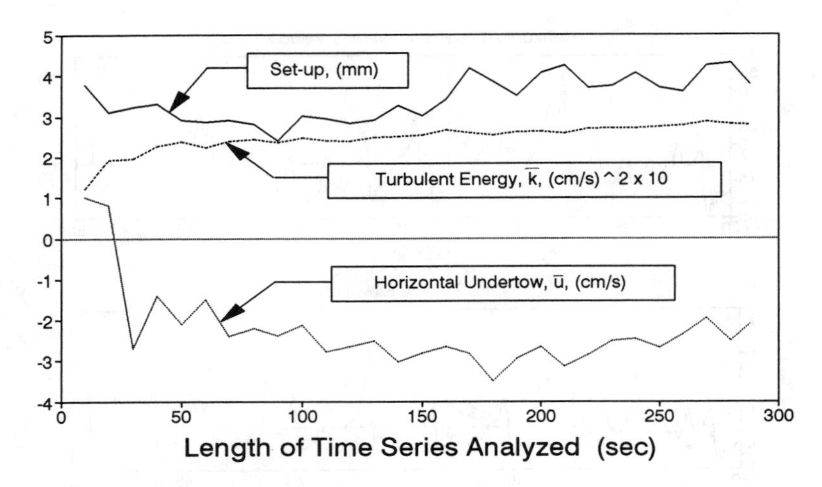

Figure 5: Length of time series needed for stable statistics at $h/H_{mo} = 0.75$, $\xi/h = 0.50$

desires enough realizations to provide a stable average. This can be quantified by comparing the root-mean-square error between time series with successively increasing numbers of realizations included in the average. For example, equation 3 shows the formula used to calculate point No.$\tilde{6}$ of the horizontal velocity line in figure 6. In equation 3, N is the total number of points in the time series (30,000) and $\tilde{u}_5$ and $\tilde{u}_6$ are the ensemble average velocities with only 5 and 6 time series respectively. The lines in figure 6 would approach zero in theory as the number of realizations approaches infinity. The selection of 15 time series for this study seems reasonable.

$$\frac{\text{rms difference}}{\text{rms average}} = \frac{\frac{1}{N}\sum_{i=1}^{N}\left(\tilde{u}_6 - \tilde{u}_5\right)^2}{\frac{1}{N}\sum_{i=1}^{N}\tilde{u}_5^2} \tag{3}$$

Figure 7 plots the water surface elevation time series after it has been low-pass filtered at 0.05 Hz. Clearly evident are low frequency oscillations which are of the same order of magnitude as the still water depth. These low frequency waves exist in nature and affect the location of individual breaking waves by altering the local water depth. The important thing to note is that these oscillations are repeatable. They recur at the same time and with the same amplitude in successive runs of the wavemaker. Therefore, the oscillations are not picked up as part of the turbulence.

A comparison with previous turbulence studies for *regular* waves is made in figure 8. The value of k averaged over the time series is non-dimensionalized

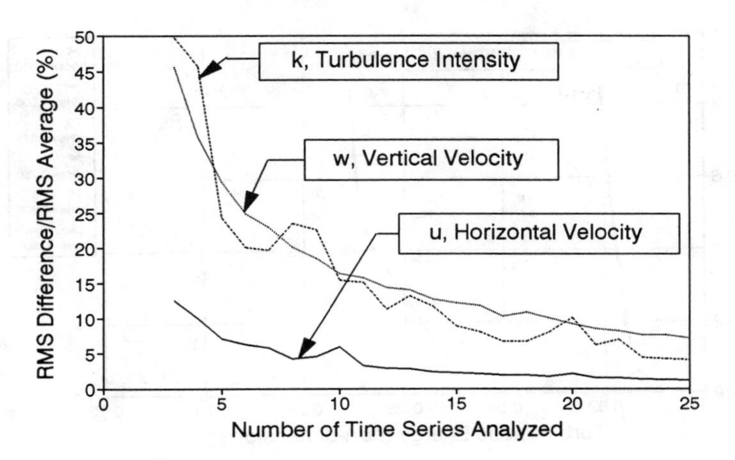

Figure 6: Number of realizations needed for stable ensemble average at $h/H_{mo} = 0.75$, $\xi/h = 0.50$

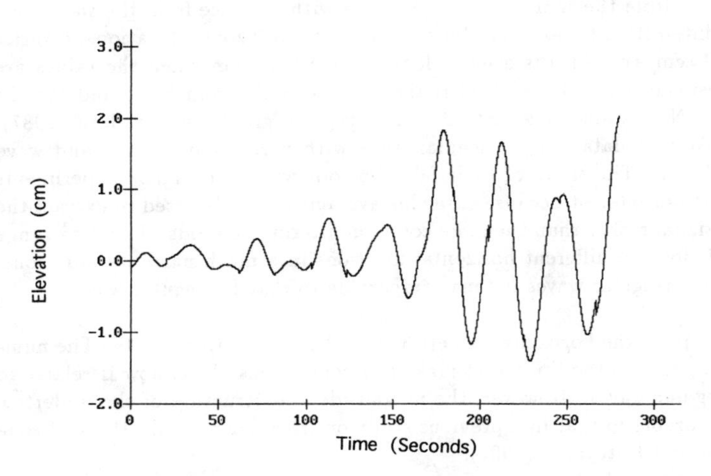

Figure 7: Water surface elevation inside surf zone, Low-pass filtered (0.05 Hz), at $h/H_{mo} = 0.50$, $\xi/h = 0.50$

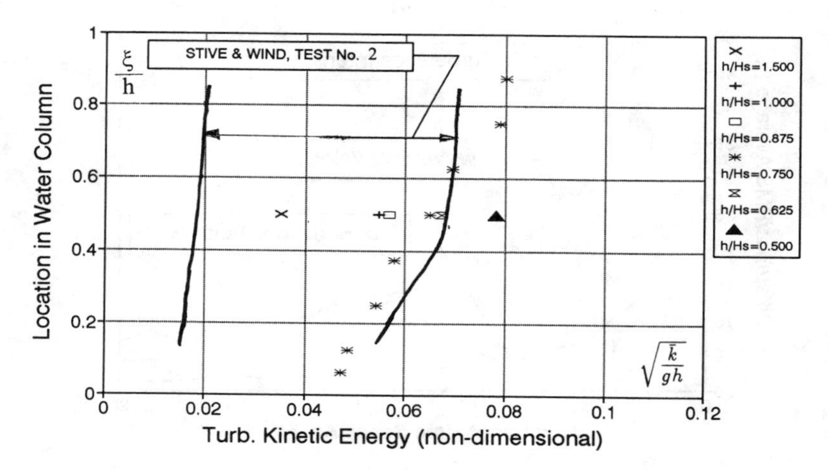

Figure 8: Comparison of turbulence intensity for regular and irregular waves.

through division by the local wave celerity for shallow water, $\sqrt{gh}$. All thirteen measuring points used in this study are plotted. The level of $\sqrt{\bar{k}/gh}$ decreases with distance from the shoreline and decreases with distance from the water surface. It is interesting to note that for the 6 different horizontal locations, $\bar{k}$ varies from 26-29 $(\mathrm{cm/s})^2$ which is a much lower variability than when the values are non-dimensionalized. Also plotted is the range of data from Stive and Wind's (1982) Test No.$\tilde{2}$ which was plotted in this type of graph by Svendsen (1987). Stive and Wind's data is for a regular wave with period 3.0 seconds and wave height 14.5 cm. The slope of their false bottom was 1:40. Their experiments are of particular interest because ensemble averaging was also used to extract the turbulent signal rather than the more common filtering methods. The data range indicated is for five different horizontal locations in a much narrower surf zone. The data for irregular waves is thus comparable to that for regular waves.

Figure 9 plots the horizontal undertow for all 13 measuring points. The number of data points is too limited to make any conclusions about how it relates to data for regular waves. However, the magnitude and structure of the undertow seems comparable to that in a previous study on irregular waves in the surf zone by Okayasu and Katayama (1992).

An interesting question is the correlation between the turbulence and the ensemble average velocity. Considering the common conception of turbulent energy stirring up sediment and keeping it in suspension for transport by the undertow, the relationship between the mean flow and k is important. Figure 10 plots all 30,000 data points in the time series of k and $\bar{u} + \tilde{u}$ for a location relatively

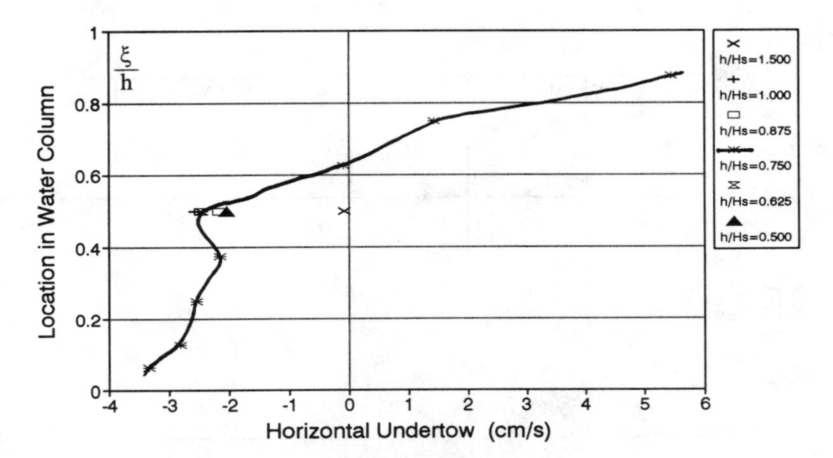

Figure 9: Horizontal undertow in surf zone under irregular waves.

close to the water surface. In this plot the negative velocities correspond with generally lower, more uniform turbulence intensities while the positive velocities correspond with higher, more varied turbulence intensities. For locations closer to the bottom, there is typically less scatter of the points with positive $\bar{u}+\tilde{u}$ and the graph looks more symmetrical about a vertical line through $\bar{u}$.

A statistical correlation coefficient, ρ, can be calculated for both the positive and negative velocities to quantify the strength of the linear relationship between $\bar{u}+\tilde{u}$ and k. $\rho = 1$ indicates a perfect positive correlation and $\rho = 0$ indicates no linear correlation. A slightly stronger correlation is observed for positive rather than negative velocities. However correlations are relatively weak. ρ typically varies from -0.15 to 0.25. Higher in the water column, positive velocity correlations are typically higher. $\rho \approx 0.5$ for positive velocities at $\xi/h = 0.875$.

CONCLUSIONS

The study of turbulence under irregular waves using the ensemble averaging technique is feasible although time consuming. There is significant temporal and spatial variation of the turbulent kinetic energy, k. Individual breaking events seem to have k decay in an exponential manner. Nearer to shore, the level of k is more uniform throughout the time series.

ACKNOWLEDGMENTS

This authors acknowledge the support of the Texas Higher Education Coordinating Board through Grant No. 999903-261.

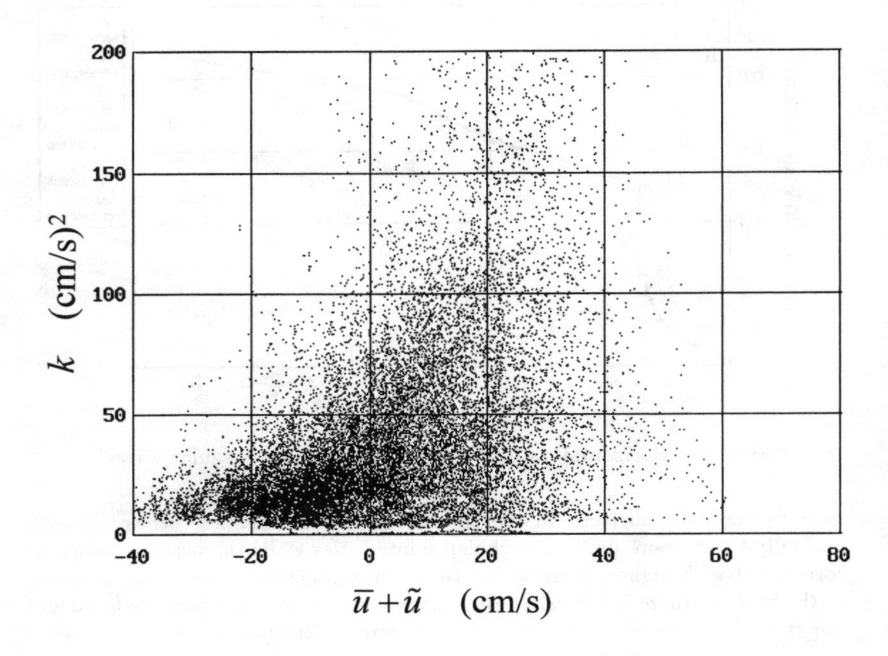

Figure 10: Correlation - Turbulent intensity and horizontal velocity at $h/H_{mo} = 0.50$, $\xi/h = 0.875$

REFERENCES

Bouws, E., Gunther, H., Rosenthal, W., and Vincent, C.L. (1985). "Similarity of the Wind Wave Spectrum in Finite Depth Water. Part I-Spectral Form." *Journal of Geophysical Research*, 90(C1), 975-986.

Okayasu, A. and Katayama, H. (1992). "Distribution of Undertow and Long-Wave Component Velocity Due to Random Waves." *Proceedings of the 23rd International Coastal Engineering Conference*, ASCE, New York, N.Y., 883-893.

Stive, M.J.F. and Wind, H.J. (1982). "A Study of Radiation Stress and Set-up in the Surf Zone." *Coastal Engineering*, 6, 1-25.

Svendsen, I.A. (1987). "Analysis of Surf Zone Turbulence." *Journal of Geophysical Research*, 92(C5), 5115-5124.

Physical Modelling of Multidirectional Waves

A. M. Cornett, M. D. Miles, O. G. Nwogu
NRC Coastal Engineering Program, Building M32,
Montreal Road, Ottawa, Ontario, Canada, K1A 0R6

Abstract

A sixty-segment wave generator has been used since 1985 at the Coastal Engineering Laboratory of the National Research Council (NRC) of Canada to generate multidirectional waves, and study their effects on coastal and ocean structures. The multidirectional wave-maker is installed on one side of a rectangular wave basin that is 30 m wide, 20 m long, and 3 m deep. This paper presents a comprehensive review of the synthesis, generation, measurement and analysis techniques that have been developed at NRC to improve the physical modelling of multidirectional seas.

Introduction

It is becoming more common to account for the directional structure of real ocean waves in coastal and ocean engineering design. Physical model tests of the performance of coastal and ocean structures in multidirectional seas play an important role in this development. Funke and Mansard (1992) review the contribution of physical model tests in multidirectional seas to improved design. Accurate physical modelling of realistic multidirectional seas is crucial if reliable design guidelines are to be achieved.

Accurate modelling of waves in laboratory basins relies on both numerical and physical skills. Numerical skills include the specification of sea states, the synthesis of waves, the calculation of wave machine command signals, the analysis of waves, and modelling of the wave basin. Physical skills include the generation and measurement of waves. Figure 1 presents a summary of these skills, and also identifies four distinct directional wave spectra involved in the modelling effort: specified; target; synthesized; and measured. The specified spectrum describes the sea state in nature that is being modelled. The target spectrum describes the sea state that is being simulated in the laboratory; this will differ from the specified spectrum due to the model scale, and perhaps due to constraints imposed by the geometry of the wave basin and capabilities of the wave generator. The synthesized spectrum results from directional wave analysis of numerically synthesized wave time series, while the measured spectrum results from analysis of signals measured in the basin. These four

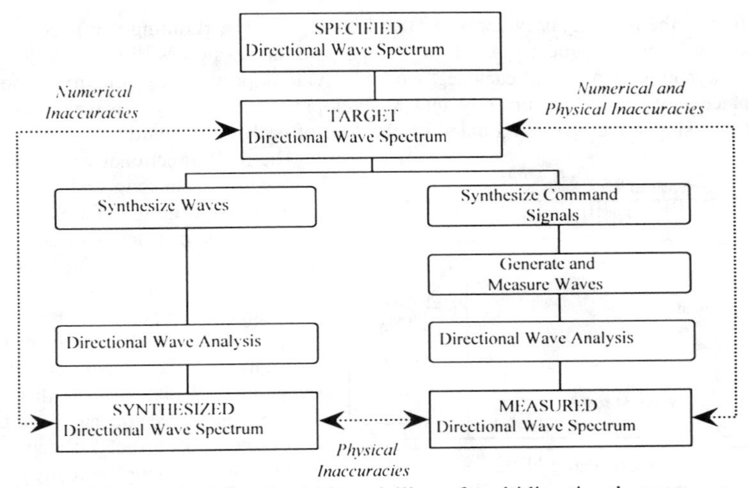

Figure 1. Skills required for physical modelling of multidirectional waves.

spectra will differ due to inaccuracies of the numerical and physical activities involved.

NRC Multidirectional Wave Basin

A system to generate multidirectional waves has been operating since 1985 at the Coastal Engineering (formerly Hydraulics) Laboratory of the National Research Council of Canada. This wave generator, described by Miles et al. (1986), has sixty independent wave boards which are driven by hydraulic actuators. Each wave board is 0.5 m wide and 2.2 m high and the wave machine can be adjusted vertically to accommodate any water depth from 0.3 m to 2.8 m. Three different wave board articulation modes (piston, hinged flapper and combination) are available to handle a wide range of wavelengths and water depths. The relatively large stroke and power of this machine allow the generation of regular waves with heights up to 0.75 m and irregular waves with significant wave heights up to 0.5 m.

A distributed microprocessor-based control system provides direct digital control of each segment of the wave machine. Wave synthesis is performed on a host workstation and the resulting wave board position drive signals are then downloaded over Ethernet to the various microprocessors that perform the real-time servo control of the individual segments. The digital control system has several features to simplify maintenance, including automatic calibration of all feedback sensors and functions to monitor system performance.

A major benefit of the digital control system has been the ability to easily implement various improvements over the years by software upgrades. A recent example has been the addition of an automatic dynamic calibration feature that measures the frequency response of each segment by comparing the actual wave board motion to the motion commanded by the drive signal. The resulting complex transfer functions are subsequently used to compensate drive signals for the individual servo-dynamic response of each segment to provide improved wave quality. Another enhancement, which is currently under development, is an active wave absorption feature using wave probes mounted on the face of each wave board.

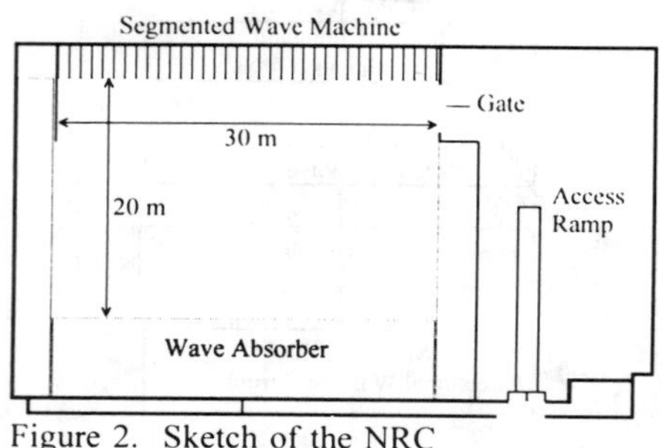

Figure 2. Sketch of the NRC Multidirectional Wave Basin

The multidirectional wave basin is 30 m wide, 20 m long and 3 m deep. As shown in Figure 2, the segmented wave machine is installed on one 30 m side and passive wave absorbers are installed on the remaining three sides. It should be emphasized that good quality wave absorbers are just as important as the wave generation system in a facility such as this. The absorbers used here consist of multiple vertical perforated metal sheets with progressively varying porosity. This type of absorber is particularly well suited to variable depth wave basins. Jamieson and Mansard (1987) investigated the performance of this wave absorber over a wide range of regular and irregular wave conditions and found that reflection coefficients of 5% or less were generally achieved for all but the longest and highest waves.

Multidirectional Wave Synthesis

Multidirectional seas can be specified in a number of forms, for example, as a directional wave spectrum or as a time series of water surface elevation with specific directional spreading characteristics. It is desirable to be able to accommodate directional wave data in different forms from different sources. These include: parametric models; hindcast data; and field measurements from either a heave-pitch-roll buoy, a wave gauge array, or a hybrid array such as a wave gauge in combination with an orthogonal axis velocity meter. Custom software is sometimes required to convert unusual specification formats into formats that are compatible with existing synthesis programs.

Multidirectional seas are commonly specified using a frequency spectrum $S(f)$ and a directional spreading function $D(f,\theta)$ to construct a directional wave spectrum $S(f,\theta)$ according to

$$S(f,\theta) = S(f)\cdot D(f,\theta) \quad (1)$$

Both $S(f)$ and $D(f,\theta)$ can be specified by parametric and non-parametric methods. Popular parametric spectral models include the JONSWAP spectrum for deep water applications and the TMA spectrum for shallow water depths. Parametric models of directional spreading include the cosine-power function, defined by

$$D(f,\theta) = \frac{\Gamma(s(f)+1)}{\sqrt{\pi}\,\Gamma(s(f)+0.5)}\cos^{2s(f)}(\theta-\theta_o(f)) \quad \text{for} \quad |\theta-\theta_o(f)| < \frac{\pi}{2} \tag{2}$$

where Γ is the gamma function, $\theta_0(f)$ is the frequency-dependent mean wave direction, and s(f) is the frequency-dependent spreading index. Small values of s(f) specify a broad spread of wave energy with direction that is always symmetrical about $\theta_0(f)$. Figure 3 shows a parametric spreading function for which s(f) decreases from 4.5 to 0.2, and $\theta_0(f)$ increases from 0° to 47°, between 0.07 Hz and 0.20 Hz. Otherwise, s(f) and $\theta_0(f)$ remain constant.

The standard deviation of wave direction, $\sigma_\theta(f)$, defined by

$$\sigma_\theta(f) = \left(\int_{\theta_0-\pi/2}^{\theta_0+\pi/2} D(f,\theta)\,[\theta(f)-\theta_0(f)]^2\, d\theta \right)^{1/2} \tag{3}$$

is a general measure of the directional spread of a wave field, that can be applied to both parametric and non-parametric spreading functions. $\sigma_\theta(f)$ is inversely proportional to the spreading index, such that s = 1 corresponds to σ_θ = 32.3°, while s = 5 is equivalent to σ_θ = 17.2°.

A variety of single and double summation models can be applied to synthesize wave conditions for random seas from a directional wave spectrum. In all models, the elevation of the water surface $\eta(x,y,t)$ is specified by a linear summation of sinusoidal components, each defined by a specific amplitude A, circular frequency ω, phase ε, and direction of propagation θ. The discretized form of the standard double summation model can be written as

$$\eta(x,y,t) = \sum_{i=1}^{N}\sum_{j=1}^{M} A_{ij}\cos\left\{\omega_i t - k_i[(x-x_r)\cos\theta_j + (y-y_r)\sin\theta_j] + \varepsilon_{ij}\right\} \tag{4}$$

where the wave number k is related to the circular frequency ω by the linear dispersion relation k tanh(kh) = ω^2/g, h = water depth, and (x_r, y_r) are co-ordinates of a reference point in the basin. According to this model, waves at each of N frequencies are distributed over M directions. Different double summation synthesis methods can be obtained by varying the way in which the amplitudes A_{ij}, directions θ_j, and phases ε_{ij} are selected. In all cases, they are chosen to approximate the target directional spectrum $S(f,\theta)$. As shown by Jefferys (1987), the wave field produced by this model is neither ergodic nor spatially homogeneous for finite values of M and N. Miles and Funke (1987) propose modified double summation models that reduce the artificial phase-locking in any given realization, and provide more realistic simulations. In general, double summation models produce wave fields that approach spatial homogeneity only at very long record lengths. It should be noted that the spatial inhomogeneity is primarily due to finite record length and is not an inherent limitation of the double summation method. Short records of natural short-crested seas will exhibit similar spatial variability.

Single summation methods can be used to generate multidirectional waves that are spatially homogeneous at much shorter record lengths. The discretized form of the standard single summation model can be written

$$\eta(x,y,t) = \sum_{n=1}^{N} A_n \cos\left\{\omega_n t - k_n\left[(x-x_r)\cos\theta_n + (y-y_r)\sin\theta_n\right] + \varepsilon_n\right\} \quad (5)$$

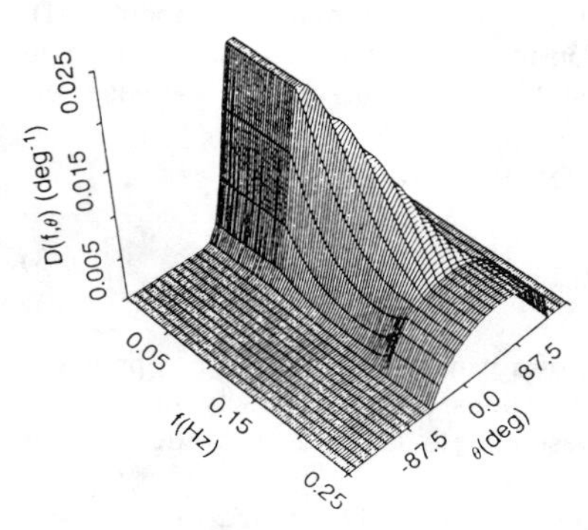

Figure 3. Parametric spreading function $D(f,\theta)$.

The principal feature of single summation models is that each wave frequency is constrained to propagate in a single direction. Single summation models are most often implemented through the random phase (RP) method of wave synthesis in order to match the target frequency spectrum. Miles (1989) investigates four different RP single summation synthesis methods. According to the RP method, the amplitude of each frequency component is chosen to satisfy the target frequency spectrum, while the phases are selected at random between 0 and 2π. Single summation models are less commonly applied with other 2D synthesis methods such as the "random complex spectrum" or "filtered white noise" methods.

The present wave synthesis software at NRC allows the desired multidirectional wave field to be specified in three different ways:

- frequency spectrum S(f) and spreading function D(f,theta) [stochastic];
- wave elevation record $\eta(t)$ and spreading function D(f,theta) [partially deterministic];
- wave elevation record $\eta(t)$ and orthogonal horizontal velocity components u(t) and v(t) [deterministic].

In the case of a spectral specification, either single or double summation models can be used with either uniformly-spaced wave angles or random wave angles. Regardless of the method selected to synthesize the wave field at the test site, the same basic procedure is used to calculate the corresponding drive signals for each segment of the wave generator. This is based on the "Snake Principle" method for segmented wave machines as originally described by Biesel (1954).

Drive signal synthesis is performed mainly in the frequency domain using FFT techniques. Each sinusoidal component of the multidirectional wave field is first propagated from the test site back to the centre of each segment using the linear dispersion relation. A complex hydrodynamic transfer function is then applied to obtain the corresponding wave board displacement. This transfer function is based on linear wave theory and depends on the frequency, the water depth, the wave board articulation mode, the wave machine elevation and the wave propagation angle. The wave board motions for each wave component are then added together to obtain the total wave board motion. Another transfer function is then applied to compensate for the servo-dynamic response of each segment, based on the dynamic calibration feature mentioned previously. This procedure is then repeated for each segment to obtain the complete set of drive signals required to reproduce the desired multidirectional wave field at the test site.

Practical Synthesis Techniques

In many practical situations, it is often necessary to modify the desired sea state to account for physical limitations of the wave generator and basin. These modifications are consistent with the philosophy that it is better to suppress the generation of wave components at frequencies and directions that cannot be properly controlled than to have these components diffract and reflect within the basin, corrupting the multidirectional simulation. Three constraints that are commonly applied are: maximum and minimum wave angle; maximum and minimum wave frequency; and the Biesel limit. At NRC, wave angles are typically constrained by $\theta_{min} < \theta < \theta_{max}$, where θ_{max} and θ_{min} ~±65°, and wave frequencies are limited to approximately $0.25 < f < 2$ Hz. The Biesel limit defines the minimum wavelength L_{min} that can be generated by a segmented wave machine without producing spurious secondary waves, as

$$L_{min} = b\left(\sqrt{2} + \left|\sin(\theta)\right|\right) \tag{6}$$

where b is the wave board width.

The energy that is clipped due to each of these constraints is generally redistributed over the remaining frequencies and directions to preserve the overall variance of the specified sea state. The resulting truncated directional spectrum is known as the target spectrum. A primary goal of multidirectional wave synthesis is to minimize the amount of wave energy that must be redirected through these various truncations. Where circumstances permit, truncations can sometimes be minimized by judicious rotation of the target sea state within the basin, so that the mean wave direction is no longer normal to the wave generator. Figures 4 and 5 show the target directional spreading function and target directional spectrum synthesized from 3D wave hindcasts of hurricane Betsy (Gulf of Mexico, 1965), rotated to minimize truncation so that the mean wave direction is at -22.1° relative to wave board normal. The narrow, low frequency peak of this complex sea state propagates at ~-45° relative to wave board normal, while the higher frequency waves are broadly spread over a wide range of directions.

Directional Wave Measurement and Analysis

The estimation of the directional characteristics of a wave field requires the simultaneous measurement of several wave properties such as the surface elevation, slope, curvature or orbital velocities. In laboratory basins, it is usually most convenient to use either an array of wave gauges or a combination of a wave gauge and a two-axis current meter. The amplitude ratio or phase difference between different measured

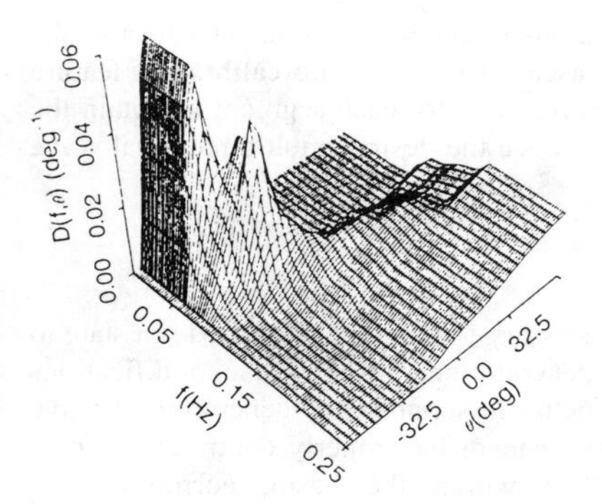

Figure 4. Target directional spreading function D(f,θ) for Betsy seas.

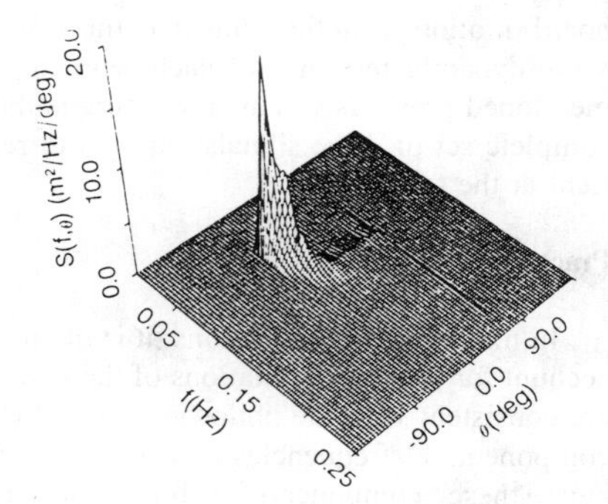

Figure 5. Target directional spectrum S(f,θ) for Betsy seas.

quantities can be used to determine the direction of wave propagation. In an irregular sea state with different frequency components, the amplitude and phase information for any given pair of sensors is contained in the cross-spectrum. For measurements of the surface elevation (η) and two orthogonal components of the horizontal velocity (u, v), the cross-spectra is related to the directional distribution through the circular moments of the distribution (e.g. Nwogu et al., 1987):

$$\int_{-\pi}^{\pi} D(\omega,\theta)\, d\theta = 1 \tag{7}$$

$$\int_{-\pi}^{\pi} D(\omega,\theta) \cos\theta\, d\theta = \Re[S_{\eta u}(\omega)] / h(\omega) S_{\eta\eta}(\omega) \tag{8}$$

$$\int_{-\pi}^{\pi} D(\omega,\theta) \sin\theta\, d\theta = \Re[S_{\eta v}(\omega)] / h(\omega) S_{\eta\eta}(\omega) \tag{9}$$

$$\int_{-\pi}^{\pi} D(\omega,\theta) \cos 2\theta\, d\theta = S_{uu}(\omega) - S_{vv}(\omega) / h^2(\omega) S_{\eta\eta}(\omega) \tag{10}$$

$$\int_{-\pi}^{\pi} D(\omega,\theta) \sin 2\theta\, d\theta = 2\Re[S_{uv}(\omega)] / h^2(\omega) S_{\eta\eta}(\omega) \tag{11}$$

where $h(\omega) = [S_{uu}(\omega) + S_{vv}(\omega) / S_{\eta\eta}(\omega)]^{1/2}$ is a transfer function that relates the orbital velocity to the surface elevation. We need to determine the directional distribution over the range $[-\pi, \pi]$ from five known moments. There are an infinite number of distributions that will satisfy the moment equations. Most of the analysis methods assume a distribution *a priori*, and then calculate the parameters of the distribution from the moments. The form of the distribution can be derived from variational principles (e.g. least-squares or maximum entropy); other theoretical principles (e.g. maximum likelihood); or arbitrarily assumed (e.g. cosine power model). The Fourier series distribution can be derived from a least-squares principle, and is given by:

$$D(\theta) = \mu_0 + \mu_1 \cos\theta + \mu_2 \sin\theta + \mu_3 \cos 2\theta + \mu_4 \sin 2\theta \qquad (12)$$

The Maximum Entropy Method (MEM) uses the following exponential distribution:

$$D(\theta) = \exp\{-(\mu_0 + \mu_1 \cos\theta + \mu_2 \sin\theta + \mu_3 \cos 2\theta + \mu_4 \sin 2\theta)\} \qquad (13)$$

The distribution derived from the Maximum Likelihood Method (MLM) can be written as:

$$D(\theta) = \frac{1}{\mu_0 + \mu_1 \cos\theta + \mu_2 \sin\theta + \mu_3 \cos^2\theta + \mu_4 \sin 2\theta + \mu_5 \sin^2\theta} \qquad (14)$$

The parameters of the MLM distribution are determined from the inverse of the cross-spectral density matrix and do not satisfy the moment equations. An iterative technique can be applied to the MLM ensure that the moment equations are satisfied (e.g. Oltman-Shay and Guza, 1984).

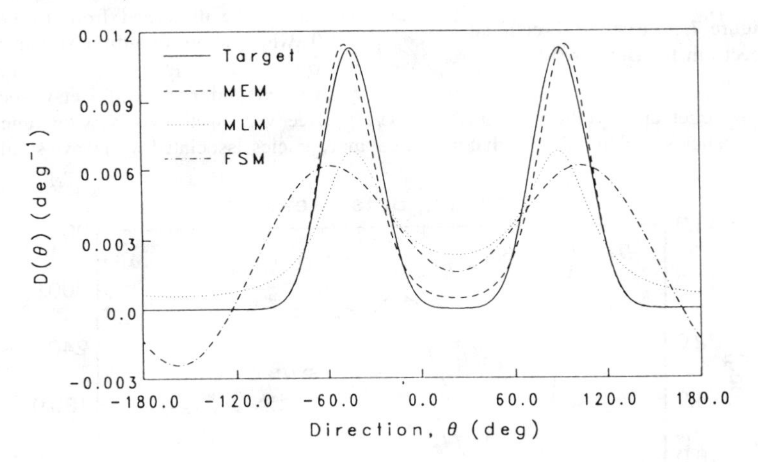

Figure 6. Fourier, MLM and MEM estimates of a bimodal spreading function.

The different analysis methods can be evaluated on their ability to resolve numerically simulated data. Figure 6 shows a comparison of a target bimodal distribution with the Fourier series, MLM and MEM estimated distributions. The MEM gives the closest match to the target distribution. Other numerical tests (Kobune and Hashimoto (1986), Nwogu et al. (1987), van Tonder and Pos (1991), Benoit (1992)) have also shown that the MEM has the highest directional resolution of all the existing methods for a three component array measurement. The MEM has also been applied to wave gauge array measurements (Nwogu, 1989). Irrespective of the resolving power of the different methods, Ochoa and Delgado-Gonzalez (1990) have noted that the more fundamental question of whether any of the assumed distributions is consistent with the physical processes occurring in nature is yet to be resolved.

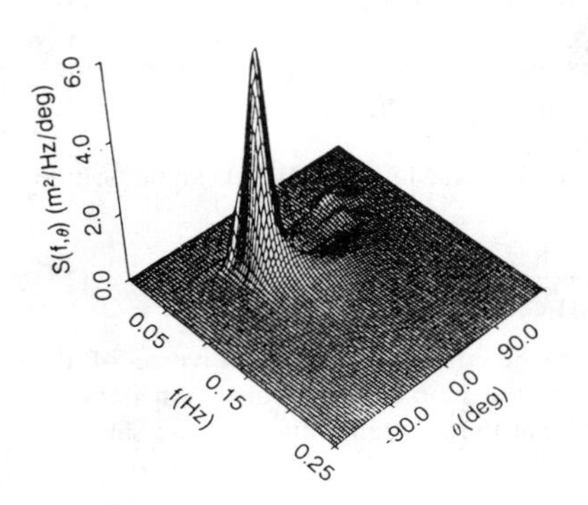

Figure 7. Measured directional spectrum for Betsy seas.

Figure 7 shows the measured directional spectrum for hurricane Betsy seas, computed using the MEM technique applied to coincident $\eta(t)$, u(t) and v(t) signals, that can be compared to the target spectrum of Figure 4. The overall shape of the target spectrum is preserved, however, the height of the measured spectral peak is reduced, and fine details of spectral shape are blurred. The measured and target spectra differ due to inaccuracies associated with the synthesis, generation, measurement and analysis of the sea.

More quantitative comparison of the spectra can be discerned from Figures 8 and 9, which show two-dimensional plots of $\theta_0(f)$ and $\sigma_\theta(f)$ for the target, synthesized and measured Betsy spectra. The target and synthesized $\theta_0(f)$ and $\sigma_\theta(f)$ agree very well over a wide range of frequencies, which indicates that numerical inaccuracies associated with wave synthesis

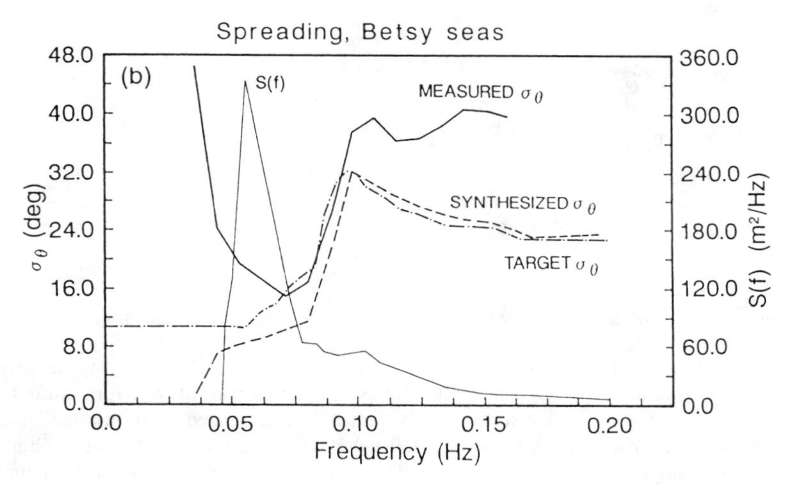

Figure 8. Target, synthesized, and measured spreading $\sigma_\theta(f)$ for Betsy seas.

and analysis are small. Greater discrepancies between the target and measured $\theta_0(f)$ and $\sigma_\theta(f)$ indicate the contribution of physical factors associated with wave generation and measurement, to the inaccuracy of the wave simulation. Considering the challenging nature of the Betsy sea, generally good agreement between measured and target values is achieved at frequencies with significant energy content.

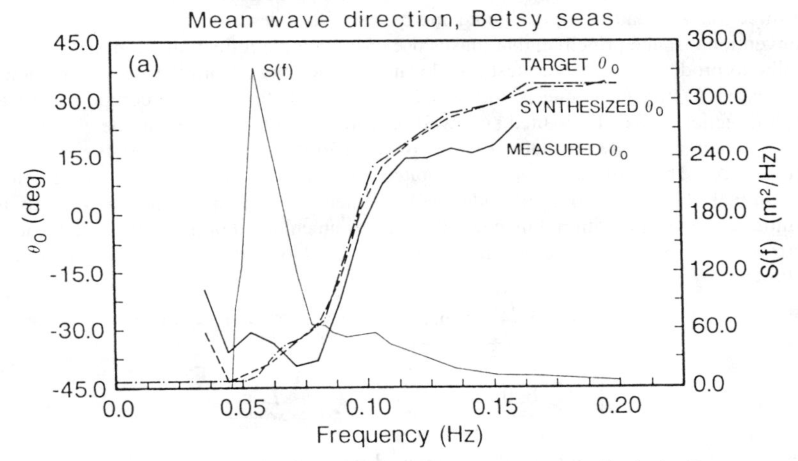

Figure 9. Target, synthesized and measured mean wave angle $\theta_0(f)$ for Betsy seas.

Numerical Modelling of Multidirectional Wave Basins

The two main factors that limit the accuracy of multidirectional wave generation are diffraction and reflection. In contrast to the case of unidirectional 2D wave generation, wave diffraction can cause significant spatial variability in multidirectional wave fields because the side walls of the basin can no longer act as wave guides. Numerical modelling is a valuable tool for investigating these effects.

Isaacson (1989) has developed an efficient linear diffraction method that can be used to model wave basins of constant depth with segmented wave machines on one or more sides. This numerical model has been verified by an extensive experimental investigation carried out at the NRC basin by Hiraishi et al. (1992). The model is particularly useful for investigating various techniques that can be used to reduce wave diffraction effects at the test site. A new version has recently been developed by Isaacson and Mathai (1993) that can model regular wave generation in basins with partially reflecting passive wave absorbers. This model is currently being extended at NRC to handle the case of irregular multidirectional waves. Numerical modelling has proven to be a very effective tool for designing optimum test configurations in our existing basin and also for designing new multidirectional wave basins.

Special Wave Generation Techniques - Side Wall Reflection

Multidirectional waves have generally been produced using board motions computed by the snake principle. This technique has several limitations, including an inability to account for wave reflections and diffraction, and a rather small optimal testing area located very near to the wave generator, particularly if the maximum angle of directional spread is large. Funke and Miles (1987) describe a modification to the conventional snake principle, that makes use of intentional reflections from partial side walls, to produce an enlarged test area located further away from the wave generator. The basis for the technique is illustrated in Figure 10 for a single component of a multidirectional sea. Unwanted side wall reflections are suppressed at one end of the wave generator, while an indirect wave, incident to the side wall, is produced at the other end. The indirect wave is generated to line up with the direct wave after intentional reflection from the side wall. When applied to all components in a multidirectional simulation, this method produces an enlarged optimal test area located further away from the wave generator. The radius R of the optimal test area can be estimated by

$$R = \begin{cases} \dfrac{\cos\theta\left(\frac{L_B}{2} + l\tan\theta\right)}{1 + \sin\theta} & \text{for } 0 \le l < \dfrac{L_B\cos\theta}{2} \\ \dfrac{L_B\cos\theta}{2} & \text{for } l \ge \dfrac{L_B\cos\theta}{2} \end{cases} \tag{15}$$

where L_B is the total length of the segmented wave generator, l is the length of side walls, and θ is the maximum wave angle in the simulation. Typical values in the multidirectional wave basin at NRC are L_B = 30 m, l = 6 m, and θ = 65°, for which R = 6.2 m. This is a substantial improvement over the test area radius, R = 3.8 m, obtained without side walls.

An experimental investigation by Hiraishi et al. (1992) has verified that this technique does provide a larger useful testing area than the snake principle. The spatial variability of wave height near the centre of the test area is not significantly reduced, however, since both methods are subject to similar wave diffraction errors.

Special Wave Generation Techniques - Dalrymple's Method

Dalrymple (1989) describes a more sophisticated side wall reflection technique that can generate an oblique planar wave train across the full width of the basin at a pre-selected distance from a segmented wave generator. This method is based on the mild-slope equation, and can be used in basins with a sloping bottom topography that is normal to the wave boards. Oblique wave quality is greatly improved with this technique, because the effects of linear wave diffraction, refraction and shoaling are included. Mansard et al. (1992, 1993) use numerical diffraction modelling of the NRC basin in conjunction with physical wave simulations to validate Dalrymple's method for the generation of oblique, long-crested, regular waves. Wave height contours for an oblique regular wave train, generated by the snake principle and the Dalrymple method,

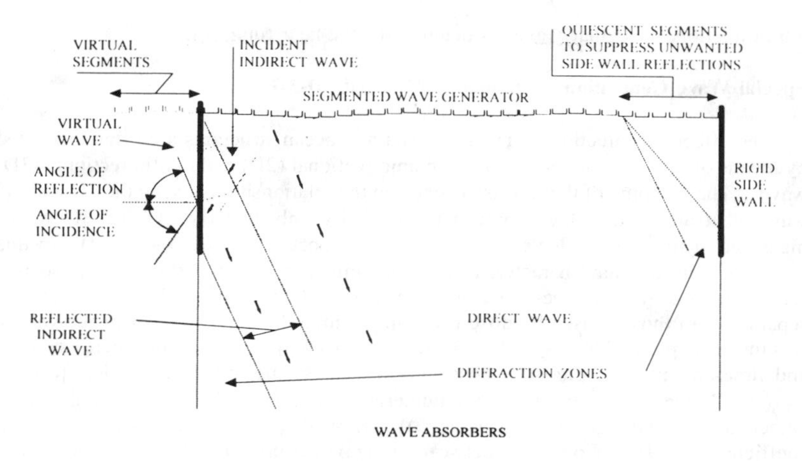

Figure 10. Sketch of a side-wall reflection technique.

are shown in Figures 11 and 12, which clearly show the much larger homogeneous area that can be obtained with the latter technique.

Although it is somewhat computationally intensive, the Dalrymple method could also be used for multidirectional irregular wave generation by applying it to each individual component in the linear superposition. It is expected that larger optimal test areas and more accurate multidirectional wave simulations, particularly in shoaling water depths, will be realized through such an implementation. The Dalrymple method is currently in use for oblique unidirectional waves and development of a version for

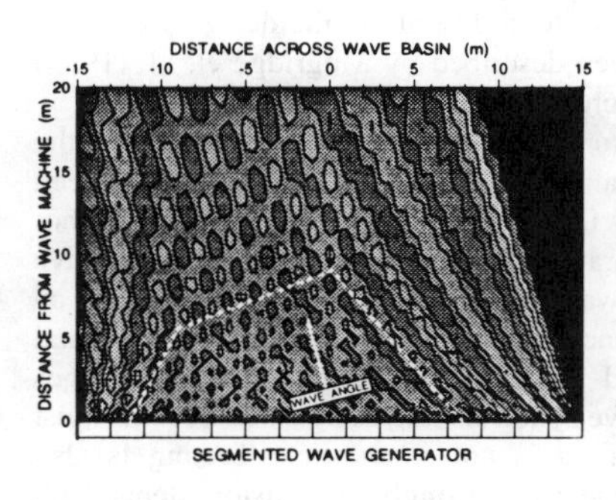

Figure 11. Numerical simulation of normalized wave height using the snake principle.

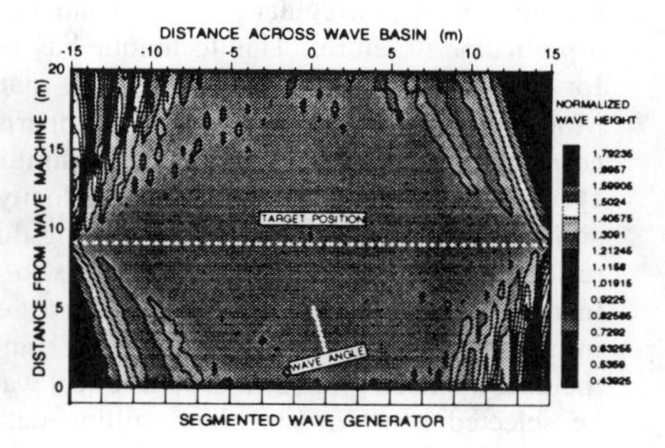

Figure 12. Numerical simulation of normalized wave height using Dalrymple's method.

multidirectional irregular waves is planned for the near future.

Special Wave Generation Techniques - Matched 2D-3D Waves

The effects of directional seas on coastal and ocean structures are often evaluated by means of physical model tests in both unidirectional (2D) and multidirectional (3D) waves. The outcome of these tests depends on the relationship between the 2D and 3D waves that are used. One approach is to specify only that the 2D and 3D waves measured at the test site have similar frequency spectra, ie. $S_{2D}(f) \approx S_{3D}(f)$. In this case, the time domain characteristics of the simulations may differ greatly, so that effects caused by the three-dimensional structure of the wave field are difficult to separate from those caused by differences in the time-domain character of the waves. It is therefore preferable to specify 2D and 3D realizations for which the energy content and time-domain characteristics measured at the test site are identical, ie. $\eta_{2D}(t) = \eta_{3D}(t)$. Cornett (1992) examines the numerical synthesis and physical simulation of 'matched' 2D and 3D seas for which $\eta_{2D}(t) \approx \eta_{3D}(t)$. An average correlation coefficient of 0.91 was observed between the $\eta(t)$ measured in 2D and 3D versions of fourteen different 680 s long matched simulations. While perfectly matched simulations can be synthesized numerically, the match of waves generated in a basin is degraded by various distortions associated with wave generation, wave propagation, diffraction, wave breaking, wave reflections, nonlinear wave interactions, measurement errors, and constraints imposed by the basin geometry. Furthermore, the 2D and 3D wave fields are subject to different distortions which causes a deterioration in their match.

Special Wave Generation Techniques - Wave Snapshots

The objective of a snapshot simulation is to generate just the portion of a longer wave record containing a special wave or wave group of particular interest; for example, a large irregular wave that satisfies criteria describing the design waves for a particular structure. This technique has been described by Mogridge et. al. (1989) for 2D waves. Snapshot simulations of matched 2D and 3D waves are described by Mogridge and Cornett (1989). Comparisons between snapshot waves and the corresponding portion of long duration simulations show excellent agreement between $\eta(t)$ and orthogonal components of velocity u(t), v(t). Advantages of the snapshot simulation technique over traditional long-duration simulations include: reduced wave reflections and associated improvements to wave quality; reduced sampling time; an increase in the number of large wave encounters; enhanced repeatability; and improvement to the match between 2D and 3D wave realizations. Disadvantages include: the need for additional numerical wave synthesis; the design wave events must be selected *a priori*; and the resulting data set is not stochastic. Testing is also restricted to fixed structures, for which response is not strongly time-history dependent.

The first step in the method is the selection of a short portion of wave activity containing the event of interest from a long record measured in the field or wave basin, or from a record generated through numerical synthesis. This insures that the selected event is a valid realization drawn from the family of events making up the population for the particular design sea state. Special wave board control signals for the snapshot are synthesized by considering the propagation of each constituent wave component from the wave boards to the test site at its linear group celerity. The propagation time required for the highest frequency component to travel at the maximum wave angle to the furthest board dictates the duration of wave board activity required prior to reproduction of the snapshot at the test site. For example, the duration required for a 2 Hz deep water wave to travel 20 m is 51 s.

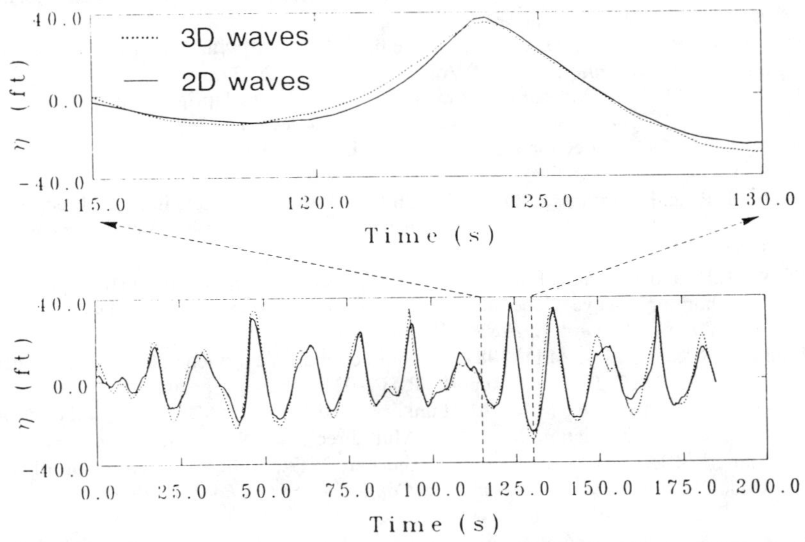

Figure 13. Water surface elevations $\eta(t)$ measured for a matched wave snapshot.

Figure 13 shows water surface elevations measured during 185 s (prototype scale at 1:28) snapshots of matched 2D and 3D waves. The design wave of interest is plotted at an expanded time scale that shows the excellent agreement between the measured uni- and multidirectional versions of $\eta(t)$. Structures tested in these two snapshots may respond differently, due to differences in the directional structure of kinematics, but not the time-domain water surface elevation profile.

References

Benoit, M. 1992. Practical Comparative Performance Survey of Methods Used for Estimating Directional Wave Spectra from Heave-Pitch-Roll Data. *Proc. 23rd Int. Conference on Coastal Engineering*, Venice, Italy, pp. 62-75.

Biesel, F. 1954. Wave Machines. *Proc. First Conference on Ships and Waves*. Hoboken, NJ, U.S.A.

Cornett, A.M. 1992. Laboratory Generation of Similar 2D and 3D Waves. *Proc. 11th International Conference on Offshore Mechanics and Arctic Engineering - 1992.* Vol. 1B, pp. 637-644.

Cornett, A.M. 1988. Dynamic Response of a Compliant Tower Test Structure in Two- and Three-Dimensional Seas. *Proc. 1988 Ocean Structural Dynamics Symposium.* Oregon. pp. 245-260.

Cornett, A.M. and Miles, M.D. 1991. Simulation of Hurricane Seas in a Multidirectional Wave Basin. *Journal of Offshore Mechanics and Arctic Engineering.* Vol. 113, pp. 219-227.

Dalrymple, R.A. 1989. Directional Wavemaker Theory with Sidewall Reflection. *Journal of Hydraulic Research.* Vol. 27, No. 1, pp. 23-34.

Jamieson, W.W. and Mansard, E.P.D. 1987. An Efficient Upright Wave Absorber. *Proc. ASCE Conference on Coastal Hydrodynamics.* pp. 124-139.

Jefferys, E.R. 1987. Directional Seas Should be Ergodic. *Applied Ocean Research.* Vol. 9, No. 4.

Funke, E.R. and Mansard, E.P.D. 1992. On the Testing of Models in Multidirectional Seas. *Proc. 23rd International Conference on Coastal Engineering.* Venice. pp. 3455-3467.

Funke, E.R. and Mansard, E.P.D. 1987. A Rationale for the Deterministic Approach to Laboratory Wave Generation. *Proc. IAHR Seminar on Wave Analysis and Generation in Laboratory Basins.* Lausanne. pp. 153-195.

Funke, E.R. and Miles, M.D. 1987. Multidirectional Wave Generation with Corner Reflectors. NRC Technical Report TR-HY-021, Ottawa, Canada.

Hiraishi, T., Mansard, E., Miles, M., Funke, E. and Isaacson, M. 1992. Validation of a Numerical Diffraction Model for Multidirectional Wave Generation. Part 1: Numerical Evaluation of the Model, and Part 2: Experimental Verification of the Model Results. *Proc. 2nd International Conference of Offshore and Polar Engineers.* San Francisco, Vol. 3, pp. 154-169.

Isaacson, M. 1989. Prediction of Directional Waves Due to a Segmented Wave Generator. *Proc. 23rd IAHR Congress*, Ottawa. Vol. C, pp. 435-442.

Isaacson, M. and Mathai, T. 1993. Wave Field in a Laboratory Wave Basin with Partially Reflecting Boundaries. *Proc. 3rd International Conference of Offshore and Polar Engineers.* Singapore. Vol. 3, pp. 669-676.

Kobune, K., and Hashimoto, N. 1986. Estimation of Directional Spectra from the Maximum Entropy Principle. *Proc. 5th Int. Offshore Mechanics and Arctic Engineering Symposium*, Tokyo, Japan, Vol. I, pp. 80-85.

Mansard, E., Miles, M. and Dalrymple, R. 1992. Numerical Validation of Directional Wavemaker Theory with Sidewall Reflections. *Proc. 23rd International Conference on Coastal Engineering.* Venice. pp. 3468-3481.

Mansard, E.P.D. and Miles, M.D. 1993. Experimental Validation of Directional Wavemaker Theory with Sidewall Reflections. *Proc. 3rd International Conference of Offshore and Polar Engineers.* Singapore. Vol. 3, pp. 686-693.

Miles, M.D. 1989. A Note on Directional Random Wave Synthesis by the Single-Summation Method. *Proc. 23rd IAHR Congress.* Ottawa. Vol. C, pp. 243-250.

Miles, M.D. 1990. Numerical Models for Synthesis of Directional Seas. NRC

Hydraulics Laboratory Report No. 31610.

Miles, M.D., Laurich, P.H. and Funke E.R. 1986. A Multi Mode Segmented Wave Generator for the NRC Hydraulics Laboratory, *Proc. 21st American towing Tank Conference*. Washington, U.S.A.

Miles, M.D. and Funke, E.R. 1987. A Comparison of Methods for Synthesis of Directional Seas. *Proc. 6th Int. Conf. on Offshore Mechanics and Arctic Engineering*. Houston. Vol. II, pp. 247-255.

Mogridge, G.R., Funke, E.R. and Bryce, P.W. 1989. Wave Simulation for Run-up and Deck Clearance Tests on a Gravity Base Structure. *Proc. 8th Int. Conf. on Offshore Mechanics and Arctic Engineering*. The Hague. Vol. II, pp. 563-571.

Mogridge, G.R. and Cornett, A.M. 1989. Multidirectional Wave Loading on the Jacket and Deck of a Space-Frame Offshore Platform. NRC Technical Report CTR-HY-035, Ottawa, Canada.

Nwogu, O., Mansard, E.P.D., Miles, M.D. and Isaacson, M. 1987. Estimation of Directional Wave Spectra by the Maximum Entropy Method. *Proc. IAHR Seminar on Wave Analysis and Generation in Laboratory Basins*. Lausanne, Switzerland. pp. 363-376.

Nwogu, O. 1989. Maximum Entropy Estimation of Directional Wave Spectra from an Array of Wave Probes. *Applied Ocean Research*, Vol. 11, No. 4, pp. 176-182.

Ochoa, J., and Delgado-Gonzalez, O.E. 1990. Pitfalls in the Estimation of Wind Wave Directional Spectra by Variational Principles. *Applied Ocean Research*, Vol. 12, No. 4, pp. 180-187.

Oltman-Shay, J., and Guza, R.T. 1984. A Data-Adaptive Ocean Wave Directional Spectrum Estimator for Pitch-Roll Type Measurements. *Journal of Physical Oceanography*, Vol. 14, pp. 1800-1810.

Sand, S.E. and Mynett, A.E. 1987. Directional Wave Generation and Analysis. *Proc. IAHR Seminar on Wave Analysis and Generation in Laboratory Basins*. Lausanne. pp. 209-235.

van Tonder, A., and Pos, J. 1991. A Cost-Effective Nearshore Directional Wave and Orbital Sensor. *Proc. 3rd Int. Conference on Coastal and Port Engineering in Developing Countries*, Mombassa, Kenya, Vol. II, pp. 1258-1272.

A SPECTRAL WAVE MODEL FOR THE COASTAL ZONE

L.H. Holthuijsen, N. Booij and R.C. Ris

Delft University of Technology
Stevinweg 1, 2628 CN, Delft, the Netherlands

Abstract

Spectral wave models that represent the evolution of the waves on a grid are superior in several respects to conventional wave ray models. Spectral models on a grid have been developed for applications in the deep ocean and for shelf seas. However, they are not economically feasible in coastal waters due to numerical limitations. We present the first step in the implementation of a version that does not have these limitations. We remove time as an independent variable (reducing the computations to stationary or quasi-stationary computations, which is proper considering the residence time of the waves in the area) and we use an unconditionally stable propagation scheme. The propagation scheme is successfully tested in academic cases, including a case with complete reversal of wave direction. As a preliminary test of propagation in an observed field case, computations are carried out for waves travelling across and around an extended (5 km) shoal. With a limited representation of the bottom induced processes (bottom friction and surf dissipation), realistic results are obtained for the significant wave height. This test also shows the relevance of the planned implementation of wave-wave interactions (in particular triad interactions) and wind generation. The model is planned to be optionally second- or third-generation (with or without predefined spectral constraints).

Introduction

In traditional coastal wave models wave components are traced from deep water into shallow water with wave rays. This often results in chaotic wave ray patterns which are difficult to interpret. Using a wave model that is formulated on a regular grid avoids these problems because of the inherent spatial smoothing. Such a model is also attractive because it allows an efficient representation of the random, short-crested character of the waves with their generation and dissipation. Such models are fairly common for wave computations in oceans and shelf seas. Excellent examples are the third-generation wave models WAM and WAVEWATCH (WAMDI group, 1988; Tolman, 1991). This type of model is based on the discrete spectral Eulerian energy (or action) balance equation of the waves with sources and sinks representing the dissipation and generation of the waves. The qualification "third-generation" refers to the fact that in these models the wave spectrum is allowed to develop without a priori constraints. Effects of shallow water and currents such as refraction and bottom friction have been included in these models. They can therefore be used for coastal applications. However, the numerical finite difference schemes for computing the wave propagation in these models is subject to the Courant stability criterion. For oceanic

applications this implies for a typical resolution of 100 km and waves with periods of 15-25 s, a time step of about 30 min. For coastal applications with a typical resolution of 100 m (sand bars, surf zone, estuaries etc.) the time step would be about 10 s for the same waves in 10 m water depth. These computations would therefore require considerable computational effort. The situation is aggravated by the fact that coastal grids regularly contain as many as 20.000 grid points whereas oceanic grids contain typically only 2000 grid pints. The application of models such as WAM or WAVEWATCH requires therefore several orders of magnitude more computer effort in coastal applications than in oceanic applications. This renders this type of model impractical for many coastal applications, in particular in a consulting engineering environment.
For this reason we developed a few years ago a model that is computationally feasible for such applications while retaining as much of the basic characteristics of oceanic models as possible (the HISWA model; Holthuijsen et al., 1989). This was achieved by removing time as an independent variable and by limiting the degrees of freedom of the wave spectrum. The first measure reduces the basic action balance equation (see below) to a stationary equation. This is acceptable for any coastal wave model since the residence time of the waves in coastal areas is usually far less than the time scale of variations of the wave boundary condition, the ambient wind or the tide. To reduce the number of degrees of freedom we parameterized the wave spectrum in each spectral direction and limited the directional sector of the spectrum. The result was a fairly successful model (it is used by some 25 institutions) but the penalties were that (a) propagation is limited to a directional sector of about 120°, (b) multimodal wave fields cannot be accommodated (e.g. wind sea and swell) and (c) the computational grid has to be oriented in the mean wave direction which is operationally inconvenient. These measures were forced by the available computer capacity at that time and by the available knowledge of the spectral development of waves in very shallow water (in particular in the surf zone). Developments on both points have now led us to start the development of a fully spectral model (called SWAN for Simulating Waves in the Near shore).

The SWAN wave model

The SWAN wave model is conceived to be a coastal wave model that is fully discrete spectral (over the total range of wave frequencies and directions (360°)) and computationally feasible in a consulting environment within the next 3 - 5 years (return time of a typical computation should be less than 15 - 30 min on a desk-top computer). The wave propagation in the model is based on linear wave theory, including the effects of currents (hence a formulation in terms of action density rather than energy density). The formulation of generation and dissipation of the waves may still be limited by the available computer capacity. The model will therefore be optionally second-generation (predefined spectral constraints) or, for larger computer capacities, third-generation (no predefined spectral constraints).

The SWAN model is based on the discrete spectral action balance equation which is a generalization of the energy balance equation in the presence of currents (Whitham, 1974; Phillips, 1977; Mei, 1983; Longuet-Higgins and Stewart, 1961, 1962; Bretherton and Garrett, 1968; Hasselmann et al., 1973). This formulation takes into account the interaction between waves and current through the radiation stresses (which are implicit in the formulation). In its general form the action balance equation is:

$$\frac{\partial N(\sigma,\theta)}{\partial t} + \frac{\partial C_x N(\sigma,\theta)}{\partial x} + \frac{\partial C_y N(\sigma,\theta)}{\partial y} + \frac{\partial C_\sigma N(\sigma,\theta)}{\partial \sigma} + \frac{\partial C_\theta N(\sigma,\theta)}{\partial \theta} = \frac{S(\sigma,\theta)}{\sigma} \tag{1}$$

in which $N(\sigma, \theta)$ is the action density of the waves, defined as the energy density $E(\sigma, \theta)$ divided by the relative frequency σ, as function of this relative frequency and direction θ (normal to the wave crest). The left-hand-side represents the local rate of change of the action density, rectilinear propagation in geographic space (terms with C_x and C_y), shifting of the relative frequency due to variations in depth and currents (term with C_σ) and refraction (term with C_θ). The expressions are taken from linear theory (e.g. Christoffersen, 1982; Mei, 1983). For rectilinear propagation the expression is:

$$\underline{C} = \underline{C}_g + \underline{U} \tag{2}$$

where $\underline{C}_g$ is the group velocity in the absence of currents, $\underline{U}$ is the ambient current speed and $\underline{C} = (C_x, C_y)$. For the frequency shift the expression is

$$C_\sigma = \frac{\partial \sigma}{\partial d}\left[\frac{\partial d}{\partial t} + U_x \frac{\partial d}{\partial x} + U_y \frac{\partial d}{\partial y}\right] - C_g \underline{k}.\frac{\partial \underline{U}}{\partial s} \tag{3}$$

where $\underline{k}$ is the wave number, d is depth and s is a coordinate normal to the wave crest. For refraction the expression is:

$$C_\theta = -\frac{1}{k}\left[\frac{\partial \sigma}{\partial d}\frac{\partial d}{\partial m} + \underline{k}.\frac{\partial \underline{U}}{\partial m}\right] \tag{4}$$

where m is a coordinate along the wave crest. The right-hand-side of Eq. 1 (the net production of wave action) represents all effects of generation and dissipation of the waves. The processes which need to be included are: wave generation by wind, wave-wave interactions (quadruplet and triad) and dissipation (breaking, bottom- and current-induced). These can all be taken directly from existing third-generation models such as WAM or WAVEWATCH with the exception of depth-induced breaking and triad interactions. The modelling of these processes will be based on the work of Battjes and others (e.g., Beji and Battjes, 1993; Battjes et al. 1993; Madsen et al., 1991; Abreu et al., 1992).

For the reasons given above, we consider the situation to be stationary (nonstationary situations can usually be treated as quasi-stationary with repeated runs of the model). We accordingly remove time as an independent variable. The basic equations that are implemented in SWAN are therefore (in addition to Eq. 4):

$$\frac{\partial C_x N(\sigma,\theta)}{\partial x} + \frac{\partial C_y N(\sigma,\theta)}{\partial y} + \frac{\partial C_\sigma N(\sigma,\theta)}{\partial \sigma} + \frac{\partial C_\theta N(\sigma,\theta)}{\partial \theta} = \frac{S(\sigma,\theta)}{\sigma} \tag{5}$$

$$C_\sigma = \frac{\partial \sigma}{\partial d}\left[U_x \frac{\partial d}{\partial x} + U_y \frac{\partial d}{\partial y}\right] - C_g \underline{k}.\frac{\partial \underline{U}}{\partial s} \tag{6}$$

In its present state the model computes the wave propagation (without currents) and the effects of bottom friction and bottom-induced breaking (Hasselmann et al., 1973 and Battjes and Janssen, 1978).

The numerical propagation scheme in SWAN

In the SWAN wave model the wave action propagates through the geographic space (rectilinear with C_x and C_y) and through the spectral space (refraction with C_θ). The numerical technique for the propagation through the geographic space is somewhat similar to the techniques that have been used extensively in parabolic refraction-diffraction models (e.g. Radder, 1979; Kirby, 1984) and in the HISWA model. In these models a forward marching method is used in which the computations progress line-by-line in the x -direction which must be chosen roughly parallel to the mean wave direction. In HISWA the state in one geographic grid point (x_i, y_j) is determined from three up-wave points at (x_i, y_{j-1}), $(x_{i-1/2}, y_{j-1/2})$ and $(x_{i+1/2}, y_{j-1/2})$ (see Fig. 1; forward stepping one-sweep technique).

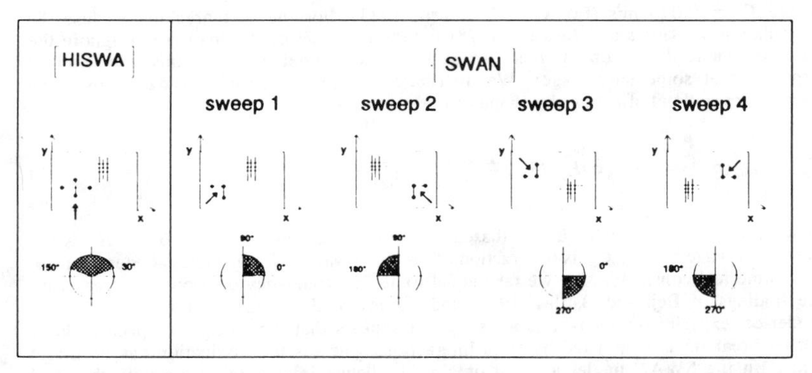

Fig. 1 Numerical schemes for wave propagation in geographic space in HISWA (left panel) and SWAN (right panel).

For stability reasons the propagation is therefore limited to a directional sector of $\Delta\alpha = \arctan(\Delta y\ /\ \Delta x)$ to either side of the direction of the x -axis (about 60° on either side in practise). This excludes the propagation of waves from widely differing directions such as swell at large angles with a wind sea or reflected waves.

To overcome this problem (which is also operationally inconvenient as the computational grid has to be aligned with the mean wave direction), we use in the SWAN model a somewhat similar scheme in a sequence of four 90° intervals. In this scheme the state in the grid point (x_i, y_j) is determined from the state in two up-wave grid points (x_{i-1}, y_j) and (x_i, y_{j-1}). The computation is now stable for wave directions in the 90° directional interval between the lines $x = x_i$ and $y = y_j$ (see Fig. 1). Waves from these directions (i.e. in one directional quadrant of the two-dimensional wave spectrum) are propagated with this scheme over the entire geographic region. By rotating the scheme over an interval of 90°, the next directional quadrant in the spectrum is propagated. This is repeated twice more to cover all four quadrants (four-sweep technique). Waves in each directional quadrant are thus propagated over the entire geographic domain. In the absence of refraction this would be sufficient to propagate the waves. In the presence of refraction, wave action can shift from one quadrant to another. This is taken into account in SWAN by repeating the computations with converging results (iterative four-sweep technique). Typically we choose a change of less than 1% in significant wave height and mean wave period in all geographic grid points to terminate the iteration. Refraction, which is modelled in the above action balance equation as propagation in θ -space is implemented with a central implicit scheme for each directional quadrant.

The source terms are implemented in the model as subroutines that can be readily activated to represent either second-generation formulations or third-generation formulations. In the present version we have implemented bottom friction and bottom-induced breaking (surf breaking) which are adequate as second- and third-generation formulations. For the bottom friction we use the JONSWAP formulation (Hasselmann et al., 1973):

$$S_{bottom\,friction}(\sigma,\theta) = \Gamma \frac{\sigma^2 E(\sigma,\theta)}{g^2 \sinh^2(kd)} \tag{7}$$

where $\Gamma = 0.076\ m^2s^{-3}$(Bouws and Komen, 1983). For the surf breaking we use the formulation of Battjes and Janssen (1978) for the total energy dissipation (we ignore the steepness-induced fraction of their expression as we intend to model this with another expression at some later stage). We distribute this depth-induced dissipation over the frequencies without affecting the shape of the spectrum:

$$S_{surf\,breaking}(\sigma,\theta) = -\alpha\, Q_b \bar{\sigma} H_m^2 / (8\pi)\, \frac{E(\sigma,\theta)}{E_{total}} \tag{8}$$

in which $\bar{\sigma}$ is the mean frequency (based on the first moment of $E(\sigma, \theta)$), H_m is the maximum wave height, Q_b is the fraction of breaking waves, E_{tot} is the total energy and α is a numerical constant which we take equal to 1. The frequency distribution agrees with the findings of Beji and Battjes (1993) and Battjes et al. (1993). In their physical and numerical experiments (at laboratory scale), it appears that the change in spectral shape during breaking is dominated by triad interactions whereas the dissipation has no direct effect. In the SWAN model we will obtain the change in spectral shape with the triad interactions at some later stage of the development of the SWAN model. This implies that at present the SWAN model does not change the mean frequency during surf breaking (in contrast to the HISWA model where a simple model for frequency change is used).

Results

We consider three situations to demonstrate the performance of the propagation scheme in SWAN. First we consider a harmonic, long-crested wave approaching a plane beach (i.e. all wave energy concentrated in one frequency bin with a $\cos^{64}(\theta)$ - directional distribution; the significant wave height $H_s = 1$ m and the wave period $\bar{T} = 5$ s). Bottom friction and surf breaking are not activated in this test so that for normal incidence shoaling dominates. This permits a direct comparison with a conventional analytical solution of the linear theory for the total energy (shown as the significant wave height $H_s = 4\sqrt{E_{tot}}$). When these waves approach the beach from another direction, refraction occurs and again a comparison with an analytical solution of the linear theory is possible. The results are shown in Figs. 2 and 3 with an excellent agreement with the linear wave theory. These tests were carried out with the computational grid oriented in the direction of the incoming wave.

To show that the results are independent of this orientation, we repeated the computations for orientations that are 90°, 180° and 270° different from the original. We plotted the results in Figs. 2 and 3 but the differences with the original computations are not visible at the scale of these Figs. To show that the interaction between the directional quadrants is properly implemented (shift of action from one quadrant to another due to refraction), we

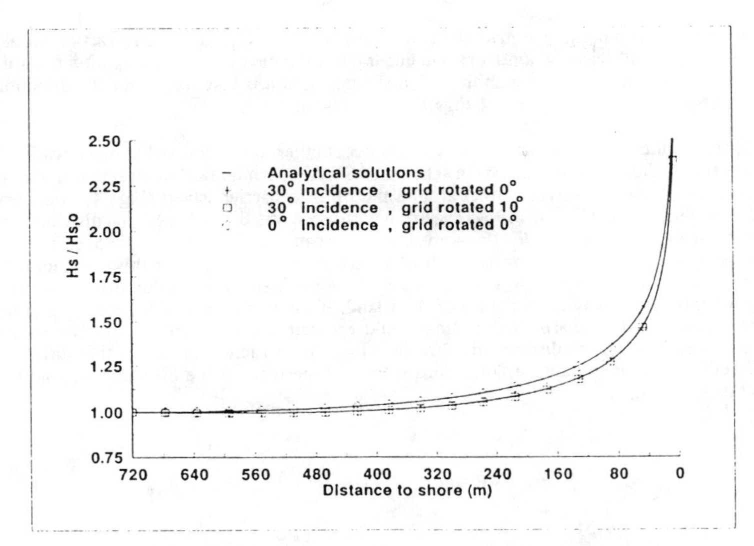

Fig. 2 The significant wave height from the results of SWAN for a plane beach compared with analytical solutions (normalized with the incident significant wave height).

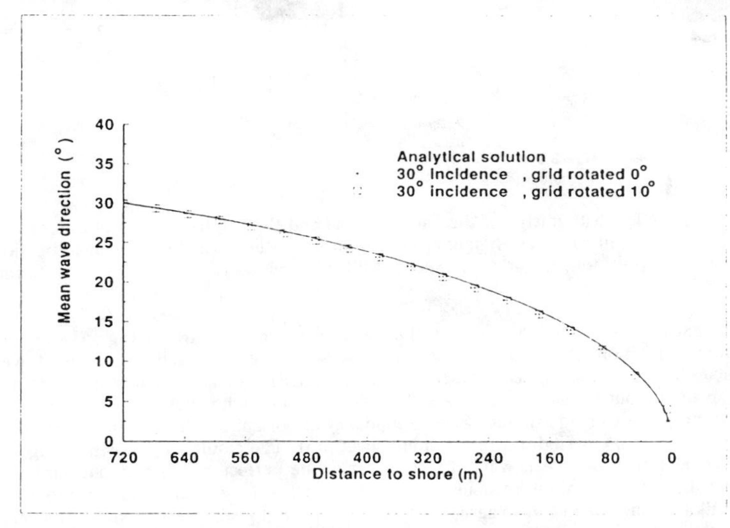

Fig. 3 The mean wave direction from the results of SWAN for a plane beach compared with analytical solutions.

also rotated the computational grid 10° clockwise. In this particular case refraction would carry practically all wave action from one quadrant to the next during propagation towards the beach. A slight difference with the original computation is just visible for the direction in the last grid point at the scale of Figs. 2 and 3 (about 1.5°).

In addition to the above academic case we consider another that is somewhat more realistic. It is chosen to show that the refraction scheme of SWAN permits fairly extreme refraction. Random, short-crested waves refract around the tip of a barrier island (Fig. 4; boundary condition: Pierson-Moskowitz (1964) spectrum with a $\cos^2(\theta)$ - directional distribution, the significant wave height H_s = 4 m and the mean wave period $\bar{T}$ = 5 s, normal incidence). Both bottom friction and depth-induced breaking are active for this computation. The results are shown in Fig. 4 where two aspects are evident. First is the diffuse "energy shadow" spreading away from the tip of the island. If the waves were long-crested and no refraction occurred, a sharp shadow line would emanate from the island tip. The short-crestedness of the waves diffuse this shadow while the refraction enhances this diffusive character. Second is the almost complete reversal of the mean wave direction beyond the island due to refraction.

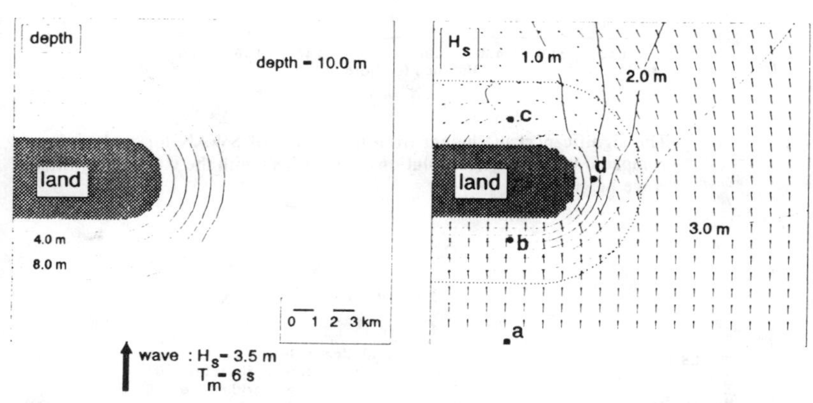

Fig. 4 The bathymetry of the barrier island and the results of SWAN in terms of significant wave height (1 m contour line interval) and mean wave direction (vectors). Locations of spectral information are indicated with A, B, C and D.

At four characteristic locations we give the two-dimensional spectra in Fig. 5, showing continuous refraction effects in the spectrum with stronger refraction for the lower frequencies than for the higher frequencies (as expected). An analytical solution is not readily available but instead we compare the SWAN results with those of a conventional wave ray program (for a harmonic wave component at normal incidence with a period of 7 s which is the peak period of the incoming spectrum). The results are shown in Fig. 6. The agreement with the mean wave direction cannot be perfect as other frequencies and directions than the spectral peak frequency and direction affect the mean wave directions but the overall agreement is fair. The agreement with the spectral peak direction should be good as here the refraction of the spectral peak can be identified (although dissipation may somewhat affect the peak frequency). In fact, the agreement is excellent as shown in Fig. 5.

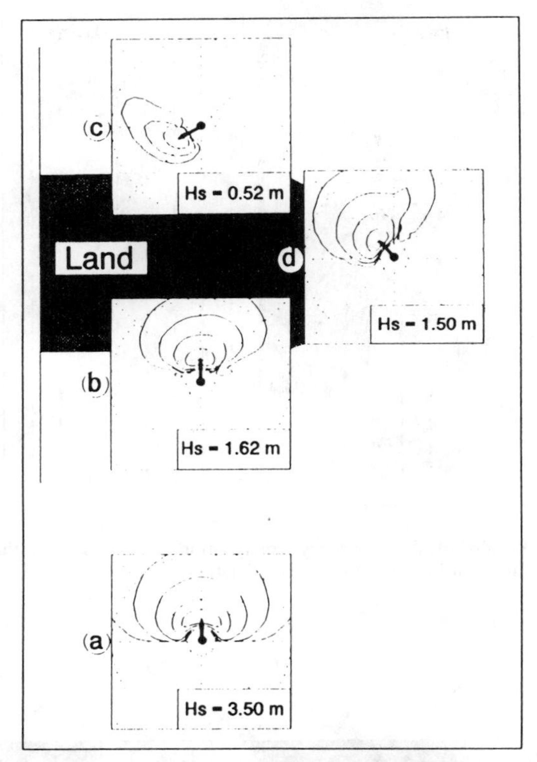

Fig. 5 Spectra from the SWAN computations at locations A, B, C and D (polar plots with maximum frequency at 0.5 Hz; logarithmic contour line interval; geographically positioned). The wave directions from the wave ray computation are indicated with a vector originating in the origin of the spectral plane.

The last case we consider is essentially a preliminary test using an observed situation in the Haringvliet (one of the branches of the Rhine delta, see Fig. 7). This test has also been published for the HISWA model in Holthuijsen et al. (1989). Waves that have been generated by a nearby storm approach the estuary and travel across and around an extended shoal (3 km scale; practically no currents during the observations).

The incoming significant wave height H_s = 3.38 m and the mean wave period $\overline{T}$ = 7 s from NW direction. The observed width of the directional energy distribution (pitch-and-roll buoy) is 31° (frequency integrated). We correspondingly use a $\cos^2(\theta)$ -directional distribution. As the observations show that a considerable fraction of the wave energy is dissipated over the shoal, we activate the bottom friction and the surf breaking for the SWAN computation. The results are given in Figs. 7 and 8. The computed significant wave

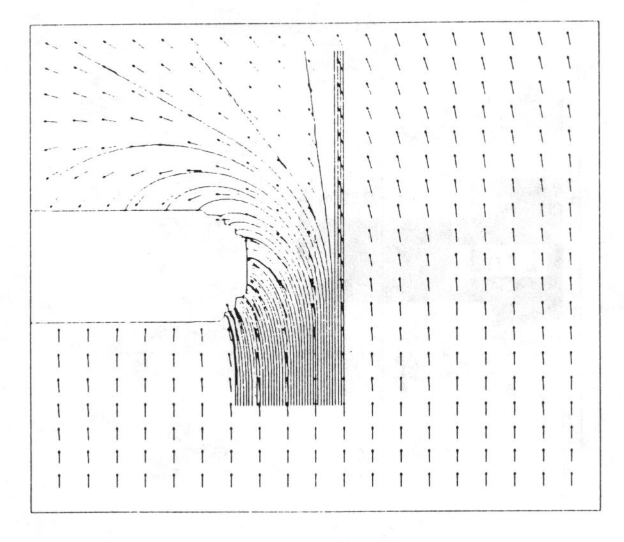

Fig. 6 Results of the wave ray computation compared with the mean wave direction from the SWAN computation.

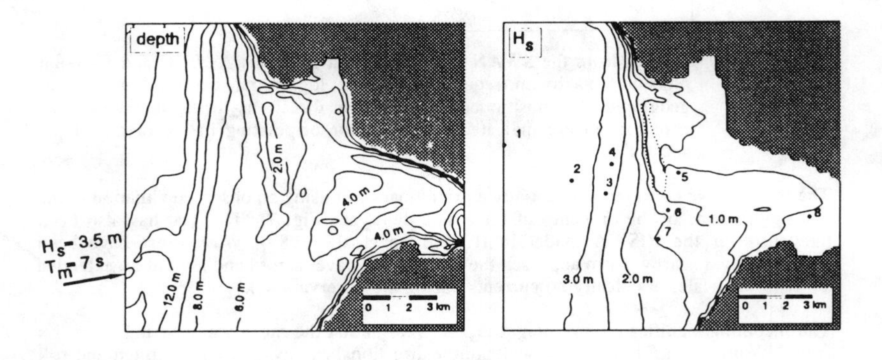

Fig. 7 The bathymetry of the Haringvliet area (contour line interval 2 m) and results of the SWAN computation in terms of significant wave height.

height agrees fairly well with the observations (except at station 8 where HISWA computations have shown that wind generation is required). It is obvious that the computed mean wave period does not compare well with the observations. As indicated above, this is almost certainly due to the absence of triad interactions in the SWAN model (this effect is parameterized in the HISWA model which gives therefore better results in this case). This in turn affects refraction but, considering the results, not to the extent that the significant wave height is poorly predicted.

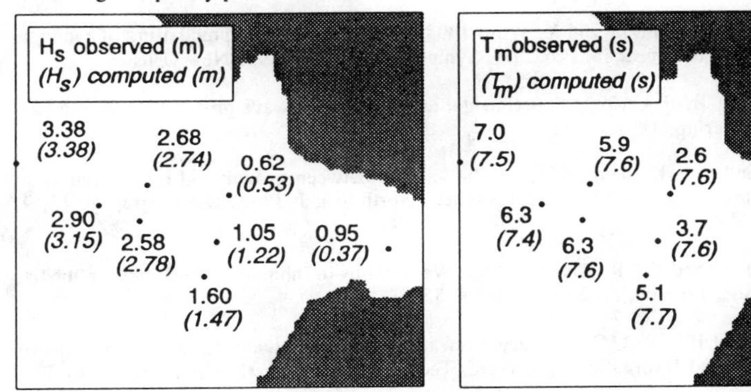

Fig. 8 The results of the SWAN computation compared with observations for significant wave height and mean wave period (observations in brackets).

Conclusions and future work

We have described the first steps in the development of the fully spectral wave model SWAN for coastal applications in a consulting environment. The model will eventually be optionally second- and third-generation with an unconditionally stable propagation scheme for stationary situations (nonstationary situations can be simulated as quasi-stationary with repeated model runs).

At present the propagation scheme has been implemented and tested with good results in two academic cases including one where the wave reverse direction due to refraction. The agreement with linear theory is excellent. A preliminary test has been carried out for an observed situation in the field in which waves propagate over and around an extended shoal (1 - 3 m depth over 3 km) with considerable dissipation. A comparison with the observations show that the dissipation is well modelled but that the planned implementation of the triad interactions is essential for the computation of the frequency evolution of the waves.

The next step in the development of the SWAN model is to implement second-generation formulations for the generation, dissipation and wave-wave interactions. The expressions can to a large extent be taken from other wave models except the effects of triad interactions which will be parameterized on the basis of the work of Battjes and others. When these sources and sinks have been tested, the effects of currents will be added.

References

Abreu, M., A. Larraza and E. Thornton, Nonlinear transformation of directional spectra in shallow water, J. Geophysical Res., 97 (C10), pp. 15,579 - 15,589

Battjes, J.A. and J.P.F.M. Janssen, 1978, Energy loss and set-up due to breaking of random waves, Proc. 16th Int. Conf. Coastal Engineering, Hamburg, pp. 569 - 587

Battjes, J.A., Y. Eldeberky and Y. Won, 1993, Spectral Boussinesq modelling of random, breaking waves, to appear in: Proc. Int. Symposium WAVES '93, New Orleans

Beji, S and J.A. Battjes, 1993, Experimental investigation of wave propagation over a bar, Coastal Engineering, 19, pp. 151 - 162

Bouws, E. and G.J. Komen, 1983, On the balance between growth and dissipation in an extreme depth-limited wind-sea in the southern North Sea, J. Phys. Oceanography, 13, 9, pp. 1653 - 1658

Bretherton, F.P. and C.J.R. Garrett, 1968, Wave trains in inhomogeneous moving media, Proc. Roy. Soc. London, A, 302, pp. 529 - 554

Christoffersen, J.B., 1982, Current depth refraction of dissipative water waves, Institute of Hydrodynamic and Hydraulic Engineering, Techn. Univ. Denmark, Series paper no. 30.

Hasselmann, K., T.P. Barnett, E. Bouws, H. Karlson, D.E. Cartwright, K. Enke, J.A. Ewing, H. Gienapp, D.E. Hasselmann, P. Kruseman, A. Meerburg, P. Muller, D.J. Olbers, K.Richter, W. Sell and H. Walden, 1973, Measurements of wind wave growth and swell decay during the Joint North Sea Wave Project (JONSWAP), Erganzungsheft zur Deutschen Hydrographischen Zeitschrift, 12

Holthuijsen, L.H., N. Booij and T.H.C. Herbers, 1989, A prediction model for stationary, short-crested waves in shallow water with ambient currents, Coastal Engineering, 13, pp. 23 - 54

Kirby, J.T., 1986, Higher-order approximation in the parabolic equation method for water waves, J. Geophysical Res., 91(C1), pp. 933 - 952

Longuet-Higgins, M.S. and R.W. Stewart, 1961, The changes in amplitude of short gravity waves on steady non-uniform currents, J. Fluid Mech., 10, pp. 529 - 549

Longuet-Higgins, M.S. and R.W. Stewart, 1962, Radiation stresses and mass transport in gravity waves, with applications to "surf beats", J. Fluid Mech., 13, pp. 481 - 504

Madsen, P.A., R. Murray and O.R. Sorensen, 1991, A new form of the Boussinesq equations with improved linear dispersion characteristics, Coastal Engineering, 15, 4, pp. 371 - 388

Mei, C.C., 1983, The applied dynamics of ocean surface waves, Wiley, New York

Phillips, O.M., 1977, The dynamics of the upper ocean, Cambridge University Press

Pierson and Moskowitz, 1964, A proposed spectral form for fully developed wind seas based on the similarity theory of S.A. Kitaigorodskii, J. Geophysical Res., 69 (24), pp. 5181 - 5190

Radder, A.C., 1979, On the parabolic equation method for water-wave propagation, J. Fluid Mech., 95, pp. 159 - 176

Tolman, H.L., 1991, A third-generation model for wind waves on slowly varying, unsteady, and inhomogeneous depths and currents, J. Phys. Oceanography, 21 (6), pp. 782 - 797

WAMDI group (Hasselmann et al.), 1988, The WAM model - a third generation ocean wave prediction model, J. Physical Oceanography, 18, pp. 1775 - 1810

Whitham, G.B., 1974, Linear and nonlinear waves, Wiley, New York

SHOALING OF WAVE SPECTRA IN FRONT OF REFLECTIVE STRUCTURES

Ernest R. Smith[1] *and Steven A. Hughes*[1]

ABSTRACT: Breakwater and jetty design depend in part on the incident significant wave height expected to reach the structure. This height is not always known and must be estimated from a deepwater condition obtained from either gage measurements or wave hindcasts. Numerical models have been developed to transform the wave condition from deep water to the shallow water depth. These models typically are calibrated and verified using laboratory and/or field studies with plane slopes or irregular profiles from which wave reflection is small. However, it is not known how well these wave transformation techniques predict the incident wave spectrum in the presence of significant wave reflection.

This paper presents results of wave transformation laboratory experiments conducted on a plane slope fronting a reflective structure. Incident and reflected significant wave heights were measured at various depths seaward of the structure using the co-located velocity method described by Hughes (1993). Transformation of the measured incident wave height was compared to estimates made using linear transformation (Airy 1845) and random wave transformation (Goda 1975). For the random wave conditions tested, both transformation methods provided estimates that compared well with estimates based on measurements made at each location. The inverse shoaling of the reflected wave was also compared to predictions of the linear theory and Goda random wave transformation with less favorable results.

INTRODUCTION

Several numerical models have been developed to predict random wave transformation from deepwater to breaking and from breaking through the surf zone

(1) Research Hydraulic Engineer, Coastal Engineering Research Center, U.S. Army Engineer Waterways Experiment Station, 3909 Halls Ferry Road, Vicksburg, MS 39180-6199; Tel (601) 634-4030 FAX (601) 634-2055

(Svendsen 1984, Dally, Dean, and Dalrymple 1985). These models were typically calibrated with and verified by laboratory and field studies utilizing plane slopes or irregular profiles. However, the bottom slopes over which the waves transformed were gentle and reflection in the incident wave frequency bands was minor. Waves propagating toward coastal structures, such as breakwaters, encounter a significant reflected wave that often causes breaking to occur when the waves impinge upon the steep breakwater slope, rather than over the fronting beach slope.

Design of breakwaters and jetties and the size and placement of armor stone on these structures is dependent on the incident significant wave height expected at the structure. In the initial design, this wave height is determined by transforming deepwater wave heights available either from gage measurements or from wave hindcasts, and any effects due to wave reflection are neglected. This paper examines whether these transformation techniques can successfully predict shoaling of random wave incident significant wave height in front of a highly reflective revetment.

EXPERIMENTS

Incident and reflected wave spectra were estimated using a Laser Doppler Velocimeter (LDV) and the co-located velocity method developed by Hughes (1993 this volume). Horizontal and vertical velocities were measured with the LDV and used to calculate incident and reflected zero-moment wave heights, H_{mo}. The co-located velocity method is similar to the multiple wave gage method of Goda and Suzuki (1976). The waves are assumed to be non-breaking normally incident to the reflective structure, and the method requires that sufficient energy is present in both the horizontal and vertical components. The procedure is not valid for long waves or measurements near bottom because the vertical component is too small, and during analysis, estimates were manually restricted to the frequency range containing appreciable energy. The LDV probe is a single source and can be moved to estimate reflections in various portions of the tank with ease. Perhaps the greatest advantage of the LDV over the wave gage method is that the LDV can be applied for reflection measurements on a mild slope, whereas the Goda and Suzuki method is restricted to horizontal bottoms because of spatial wavelength variations over a sloping bottom.

The LDV used in the present study measured the horizontal and vertical velocities necessary for analysis of incident and reflected spectra. A single laser beam is generated and separated into three different wavelengths (for three-dimensional capability), and the frequency of each wavelength is shifted. The shifted and unshifted beams are transmitted through optical fibers to a probe, which focuses the beams on a small sample volume in which water particle motion is detected. The probe can be positioned with ease to any location along the tank and any water depth in the water column.

Tests were conducted in a 45.7-m long, 0.46-m wide, 0.91-m deep glass-walled wave tank. The tank included a 1-on-30 concrete-capped slope, separated from the generator by a 21-m horizontal section. An electro-mechanical hydraulic wave generator with a piston-type wave board was used to generate waves. Displacement of the board and production of waves was controlled by electronic signals generated and transmitted by a micro-computer. Six random wave conditions with peak periods, T_p, ranging from 1 to 2 sec were generated in a constant depth of 40 cm in the horizontal section.

A smooth, solid board was placed at a 1-on-2 slope with the toe located in a depth of 16.7 cm. The solid barrier was selected to represent a reflective structure with no porosity and represents the upper limit of reflection for the given slope. Wave velocity data were collected at a 40-Hz rate per channel for 204 sec, but only the first 102 sec were analyzed. Measurements were obtained at mid-depth for the locations shown in Figure 1. Mid-depth measurements were taken to assure sufficient energy was present in the vertical component and to avoid increased nonlinearities in the upper portion of the water column. The procedure for data collection for each test was to generate waves, collect and analyze data at one location, reposition the probe to another location, and repeat the procedure.

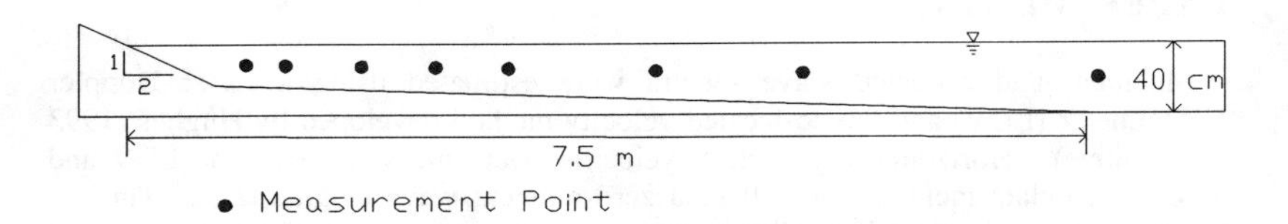

Figure 1. LDV measurement locations

RESULTS

Estimated incident and reflected zero-moment wave heights, H_{mo}, for the six wave conditions are shown in Table 1 with the depths at which the data were collected. Shallow-water measurements were limited to depths offshore of the breaking region. Cases W1 through W4 were conducted with relatively high waves for which a significant decrease in H_{mo} was observed as the waves shoaled. This indicates energy saturation on the onset of large-scale breaking on the 1 on 30 slope. Lower waves were used for Cases W5 and W6, and the waves either broke on the structure or shoreward of the shallowest measurement location.

Wave Spectra

Typical incident and reflected spectra are shown in Figures 2a to c for depths of 40, 28, and 20 cm, respectively for Case W3 (T_p = 1.5 sec). The incident spectra for this condition vary little as the waves progress from deeper to shallow water. However, the reflected spectrum is much higher at d = 20 cm than at the 40 and 28 cm depths.

Incident wave spectra at all depths for Case W4 (T_p = 1.25 sec) are shown in Figure 3, which shows the energy at the peak frequency varies little with depth. The

Table 1
Measured Wave Heights

Case	W1		W2		W3		W4		W5		W6	
T_p (sec)	1.0		2.0		1.5		1.25		1.75		2.0	
d (cm)	H_i (cm)	H_r (cm)	H_i (cm)	H_r (cm)	H_i (cm)	H_r (cm)	H_i (cm)	H_r (cm)	H_i (cm)	H_r (cm)	H_i (cm)	H_r (cm)
40	6.25	3.31	8.47	4.73	7.12	4.55	6.85	4.40	6.53	4.76	7.03	4.69
32	6.05	3.40	8.80	5.26	7.38	4.99	6.68	4.48	6.66	4.85	7.30	5.18
28	6.04	3.44	8.69	5.26	7.14	4.93	6.73	4.62	6.53	4.93	7.33	5.24
24	5.92	3.46	8.58	5.88	7.02	4.72	6.68	4.77	6.83	5.31	7.44	5.80
22	-	-	-	-	-	-	-	-	-	-	7.63	6.40
20	5.67	3.55	7.98	6.71	7.02	5.77	6.20	4.36	7.24	6.37	7.66	6.86
19	-	-	7.69	7.06	-	-	-	-	-	-	-	-
18	-	-	-	-	6.67	5.80	-	-	6.85	6.43	-	-
17	5.53	3.58	-	-	-	-	-	-	-	-	-	-

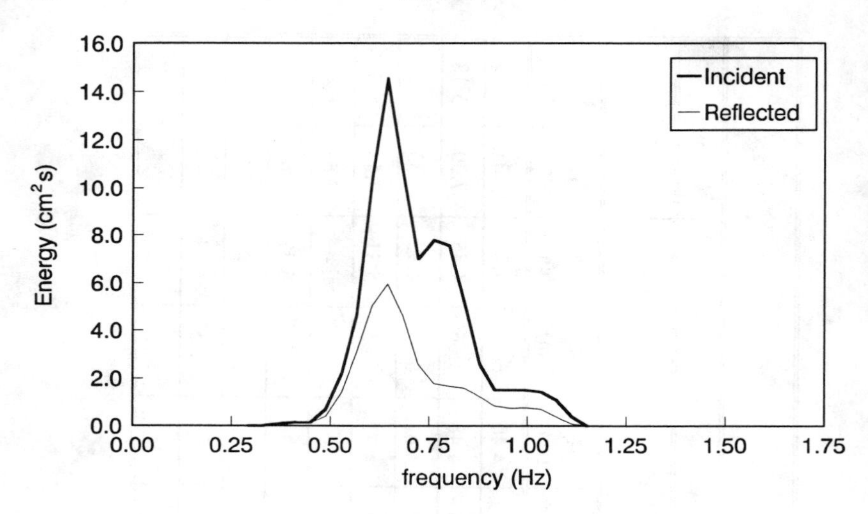

Figure 2a. Incident and reflected wave spectra, T_p = 1.5 sec, d = 40 cm

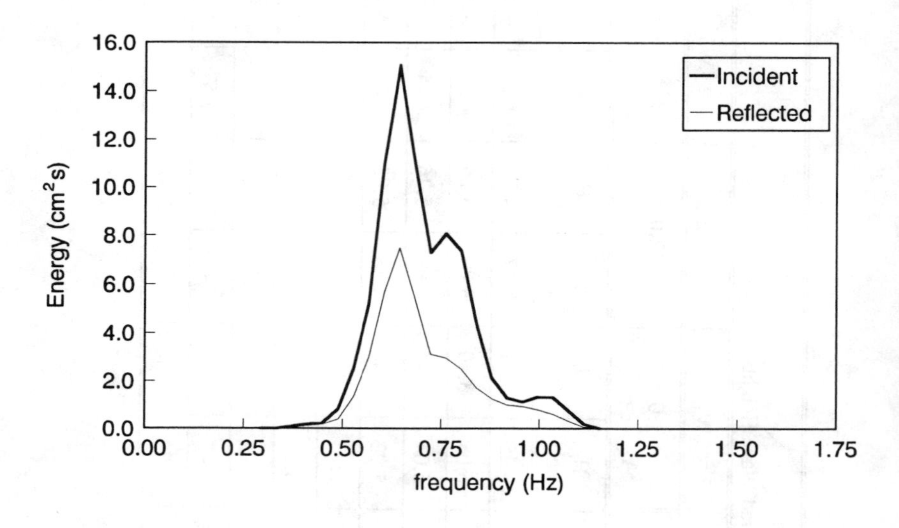

Figure 2b. Incident and reflected wave spectra, T_p = 1.5 sec, d = 28 cm

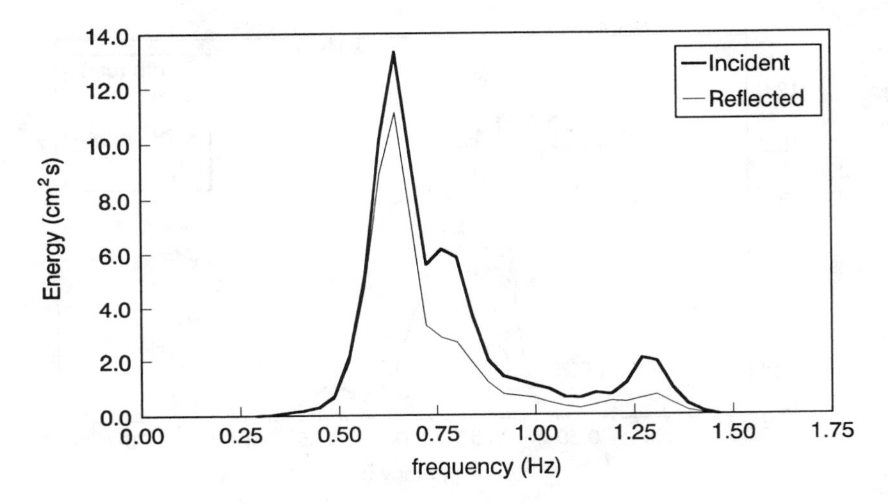

Figure 2c. Incident and reflected wave spectra, T_p = 1.5 sec, d = 20 cm

period is relatively short, and significant shoaling is not expected. Figure 4 shows incident wave spectra at all depths for a longer period, T_p = 1.75 sec (Case W5). Shoaling at the peak frequency is more apparent for the wave condition, with the greatest energy at the peak in the shallow-water depth.

Reflected wave spectra for Case W1 (T_p = 1 sec) are shown in Figure 5 for all water depths. Although shoaling of incident spectra for shorter waves was minimal, the figure shows that differences in energy at the peak is more explicit for reflected waves. The difference is greater for a longer period of 1.75 sec, shown in Figure 6. Energy at the peak is highest at the 20-cm depth, decreases significantly at d = 24 cm, and decreases further and remains near constant in deeper water. For both short- and long-period wave conditions, inverse shoaling was observed in which wave energy at the peak frequency decreased as the reflected wave progressed into deeper water.

Wave Height Transformation

Incident zero-moment wave heights were compared to linear transformation (Airy 1845) and to the Goda (1975) random wave transformation technique. The two transformation models were selected because both are well known and widely used. Both models are available in the Automated Coastal Engineering System (ACES) computer program (Leenknecht, Szuwalski, and Sherlock, 1992). The models were developed for plane sloping beaches for which reflection is minimal. Additionally, breaker criteria is included in both models (the ACES program incorporates the breaker criteria of Singamsetti and Wind (1980) and Weggel (1972) in the linear

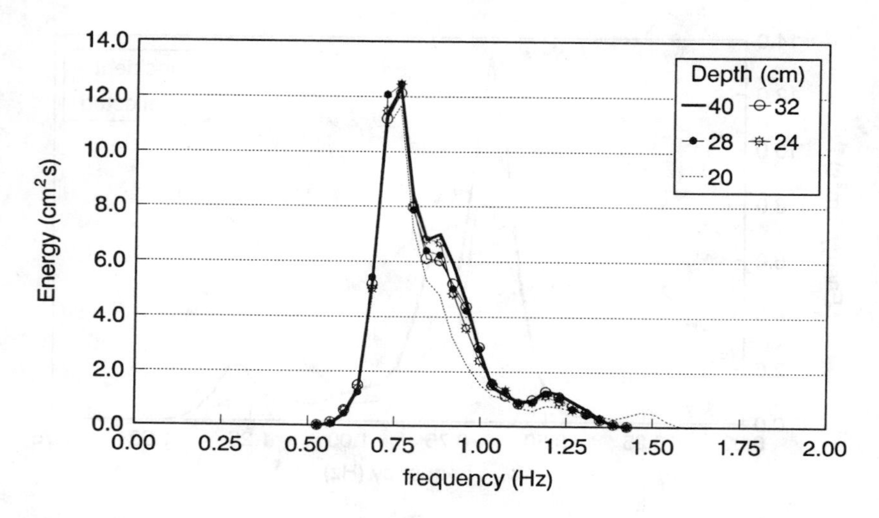

Figure 3. Incident wave spectra at all depths, T_p = 1.25 sec

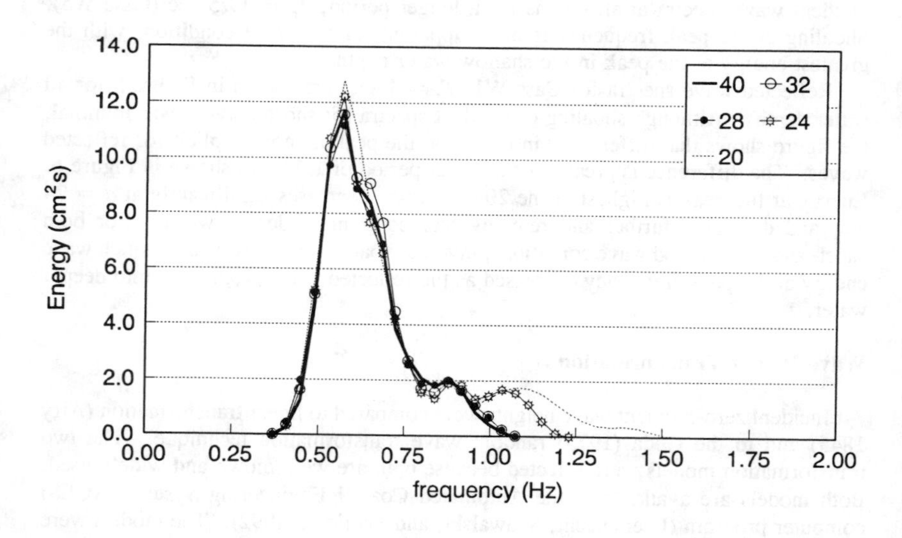

Figure 4. Incident wave spectra at all depths, T_p = 1.75 sec

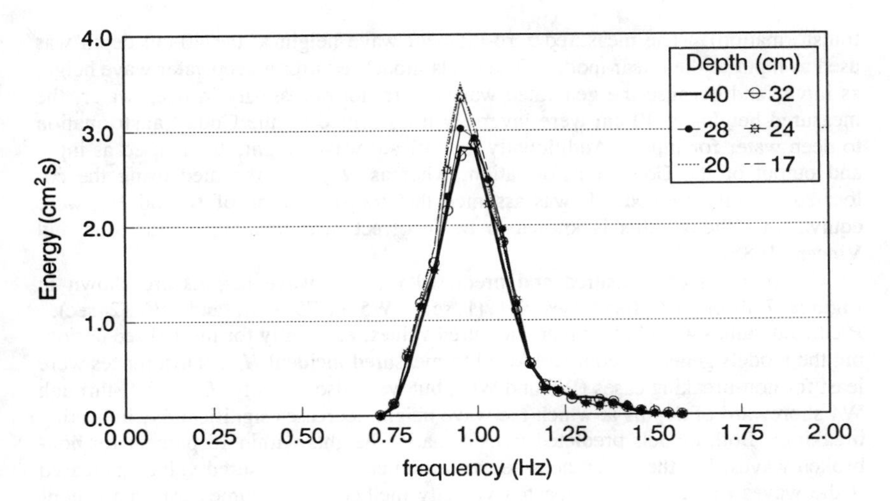

Figure 5. Reflected wave spectra at all depths, $T_p = 1.0$ sec

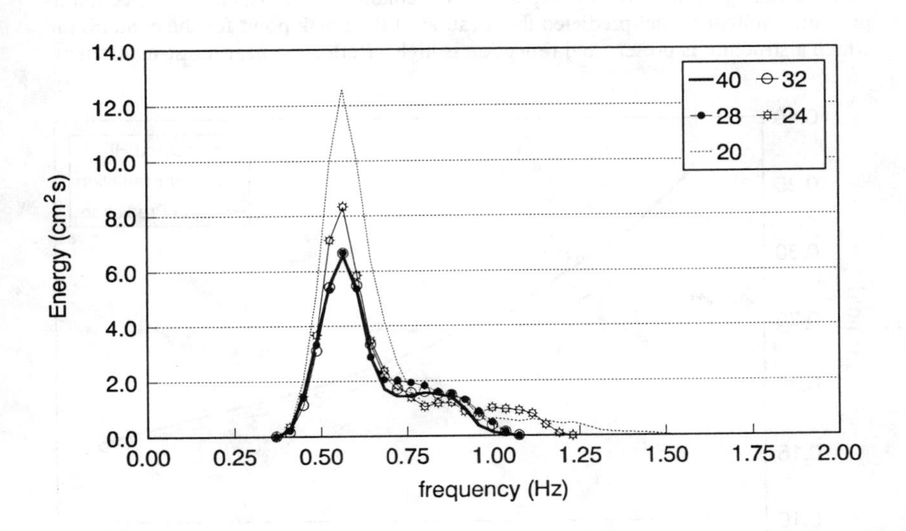

Figure 6. Reflected wave spectra at all depths, $T_p = 1.75$ sec

transformation). The measured zero-moment wave height at the 40-cm depth was used as input to the linear model. The Goda model requires a deepwater wave height as input, and because the generated waves were not necessarily in deep water, the measured heights at 40 cm were inversely transformed by the Goda transformation to deep water for input. Additionally, significant wave height, H_s, is used as input and output of the Goda transformation, whereas H_{mo} is calculated using the co-located velocity method. It was assumed that transformation of H_s and H_{mo} were equivalent although this is known to be incorrect near breaking (Thompson and Vincent 1985).

Comparisons of measured and predicted incident wave heights are shown in Figures 7 through 9 for Cases W1 (1 sec), W5 (1.75 sec), and W6 (2 sec). Predicted values were higher than measured values, especially for the 1.0 sec period, but the models generally compared well to measured incident H_{mo}. Differences were least for non-breaking cases (W5 and W6), but were also small for Cases W1 through W4 shoreward of depths at which the wave height decreases significantly, indicating breaking. Both models predicted transformation heights within 10 percent for non-broken waves, but the differences between predicted and measured values increased if the waves broke. The co-located velocity method also assumes waves are non-breaking, and this may have contributed to this observed difference.

The dissimilarity between modeled and measured values for broken waves can be attributed to the presence of the structure. Both transformation models account for wave breaking, but only for depth limited conditions in which no reflection is present. Neither model predicted the location of the break point for the condition in which a structure is present and reflection is high. Reflection near the structure may

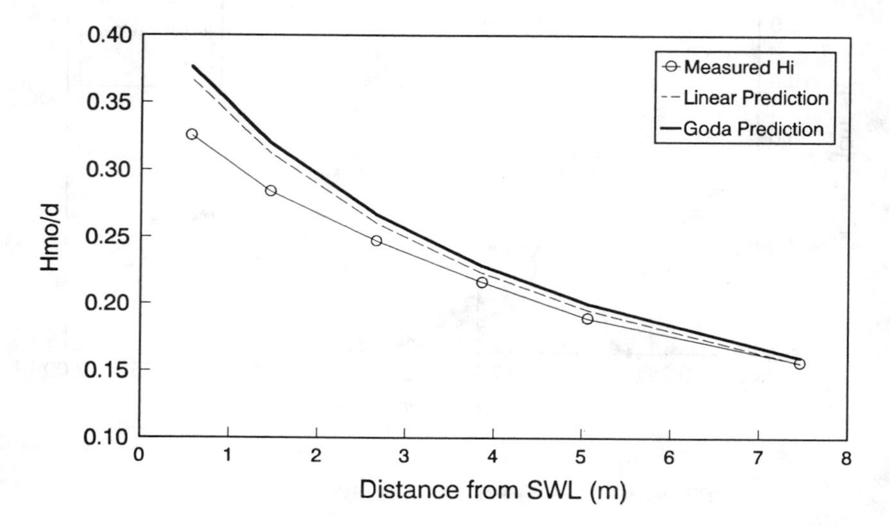

Figure 7. Comparison of measured and predicted incident waves, T_p = 1.0 sec

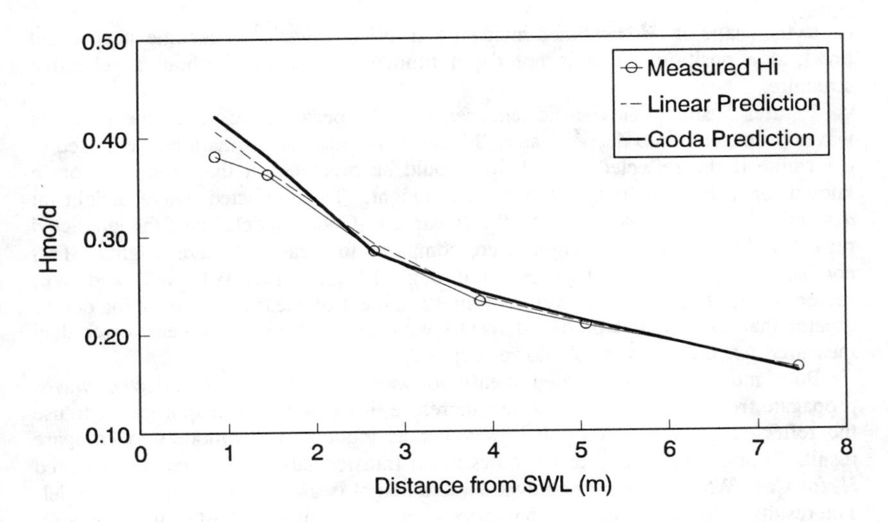

Figure 8. Comparison of measured and predicted incident waves, $T_p = 1.75$ sec

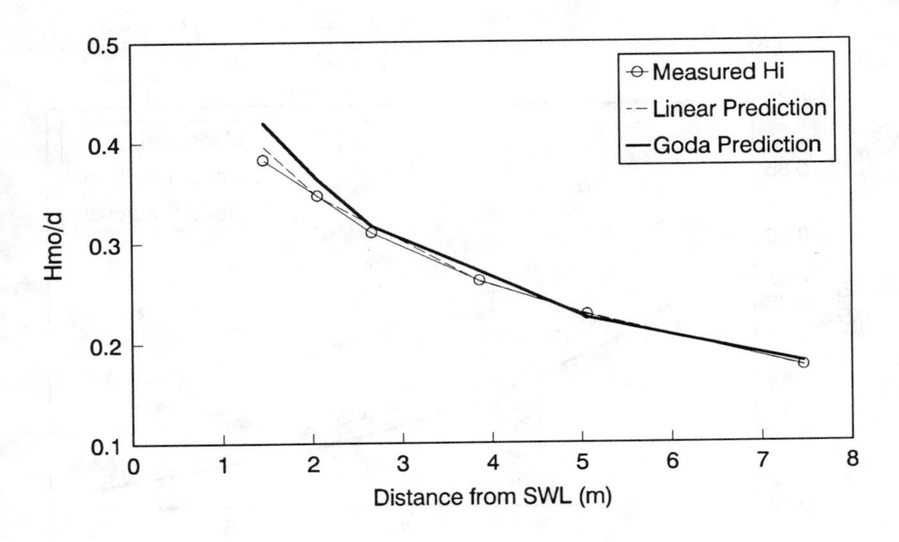

Figure 9. Comparison of measured and predicted incident waves, $T_p = 2.0$ sec

cause the shape of the combined incident and reflected waves to become steeper and break at a position which is not depth limited for beaches without a reflective structure.

Figures 5 and 6 show reflected energy at the peak decreased as the reflected waves propagated into deeper water. Linear and Goda transformations were used to determine if the reflected wave height could be predicted at the structure from a known or estimated offshore reflected height. The reflected wave height at d = 40 cm was used as input to the linear and Goda models, and the predicted reflected significant wave heights were compared to measured wave heights. This comparison is shown in Figures 10 through 12 for Cases W1, W5, and W6, respectively. Model results were within 12 percent of measured results for depths greater than 24 cm, but predicted results were as much as 25 percent lower than measured wave heights for shallower depths.

Both models underestimated measured wave reflection, but reflected waves propagate from shallow to deep water; therefore, it would be more appropriate to use the reflected wave height in shallow water as input into the models to compare results. For example, Figure 13 shows linear transformation overpredicts measured H_r for Case W5 if the most shoreward wave height is used as input into the model. The results of this comparison show reflected wave height is highest near a structure, such as a breakwater or jetty, and the reflected wave height decreases significantly as it moves into deeper water. Neither model was appropriate to estimate inverse shoaling of the reflected wave.

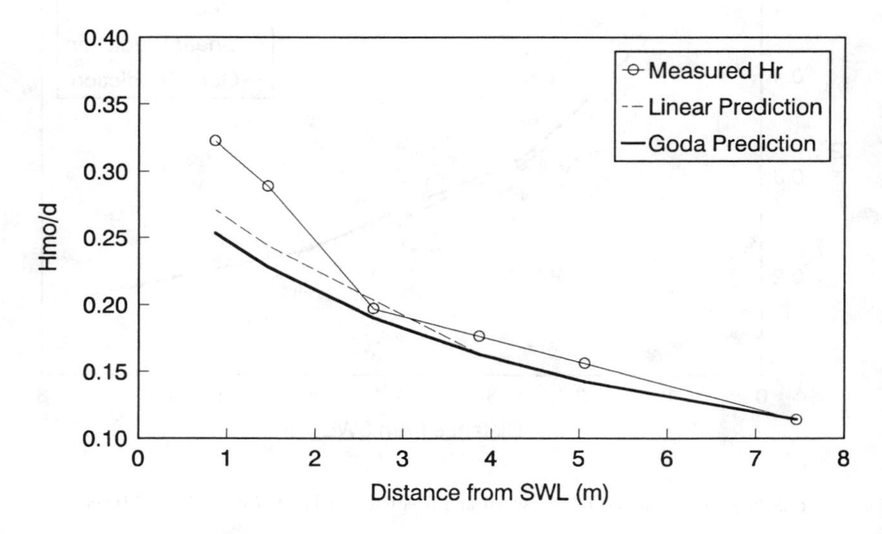

Figure 10. Comparison of measured and predicted reflected waves, T_p = 1.0 sec

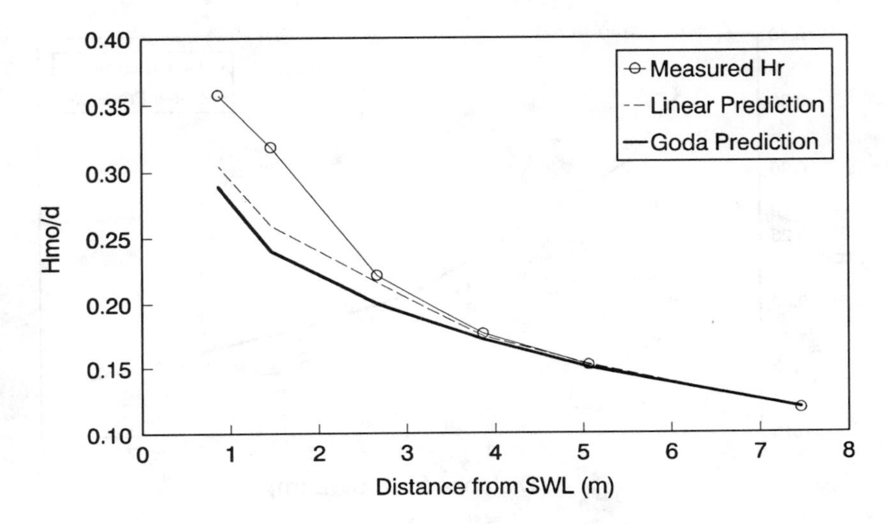

Figure 11. Comparison of measured and predicted reflected waves, $T_p = 1.75$ sec

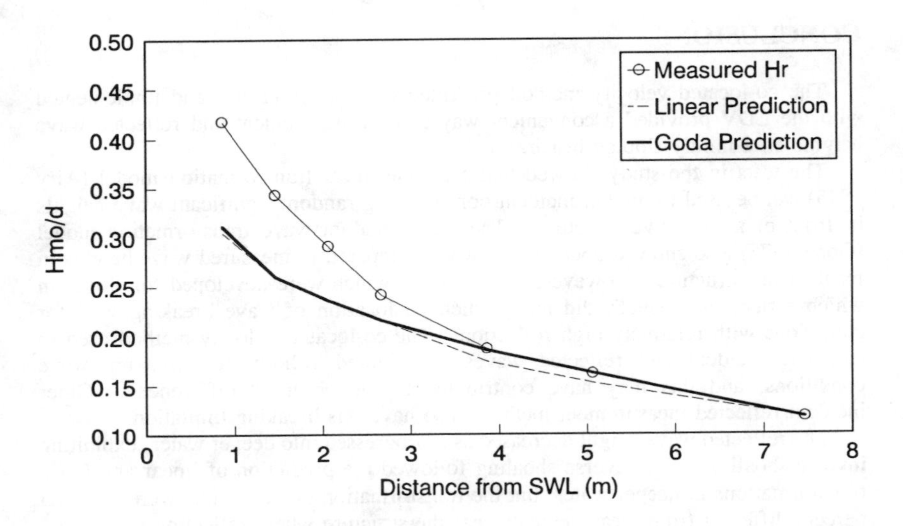

Figure 12. Comparsion of measured and predicted reflected waves, $T_p = 2.0$ sec

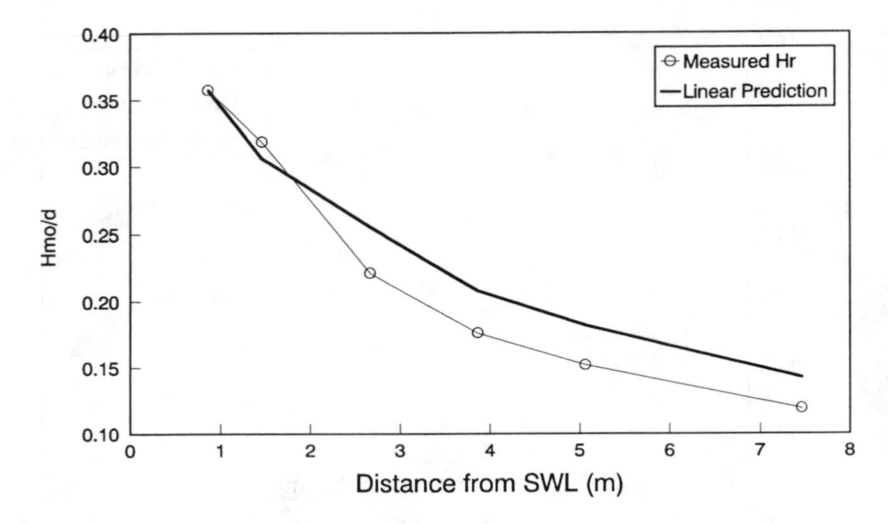

Figure 13. Comparison of measured and predicted reflected waves using shoreward depth as input, T_p = 1.75 sec

CONCLUSIONS

The co-located velocity method presented by Hughes (1993) and implemented with the LDV provided a convenient way to measure incident and reflected wave heights on a mildly sloping beach.

The tests in the study showed that a simple linear transformation model (Airy 1845) can be used to predict incident non-breaking random significant wave heights in front of a reflective structure. The Goda random wave transformation model (Goda 1975) also showed good results when compared to measured wave heights in front of the structure. However, both models, which were developed for slopes in which reflection is small, did not predict the location of wave breaking well for conditions with relatively high reflection. The co-located velocity method used to calculate incident and reflected waves has limited validity for breaking wave conditions, and this may have contributed to the observed differences. Other incident/reflected measurement methods also have this breaking limitation.

The reflected wave height decreased as it progressed into deeper water, exhibiting inverse shoaling. The inverse shoaling followed the prediction of linear and Goda transformations in deeper water, but the transformation model results were 15 to 25 percent different from measurements near the structure where reflection was high.

Results of the comparison indicated transformation of incident waves on a plane-sloping beach were approximately the same as incident non-breaking waves shoaling in front of a reflective structure. The reflection produced by the structure had little effect on the non-broken incident wave heights, however, the reflected wave transformation could not be predicted using plane-beach models. The interaction of the incident and reflected waves may account for differences, such as breaker location, between predicted transformation results of plane and irregular profiles and measured heights at a structure. The models did not predict breaker location if front of the structure, and without this knowledge, it becomes more difficult to specify depth-limited waves critical for structural design. Further research is required on breakpoint location in front of structures before transformation of waves can be adequately modeled.

ACKNOWLEDGEMENTS

The authors thank Mr. John Evans, Engineering Technician, Coastal Engineering Research Center, for his assistance and advice in the laboratory. Permission was granted by the Chief of Engineers to publish this information.

REFERENCES

Airy, G. B. 1845. "Tides and Waves," *Encyclopedia Metropolitana*, Vol. 192, pp. 241-396.

Dally, W. R., Dean, R. G., and Dalrymple, R. A. 1985. "Wave Height Variation Across Beaches of Arbitrary Profile," *Journal of Geophysical Research*, Vol. 90, No. C6, pp. 11917-11927.

Goda, Y. 1975. "Irregular Wave Deformation in the Surf Zone," *Coastal Engineering in Japan*, Vol. 18, pp 13-26.

Goda Y., and Suzuki, Y. 1976. "Estimation of Incident and Reflected Waves in Random Wave Experiments," *Proceedings of the 15th Coastal Engineering Conference*, ASCE, pp. 828-845.

Hughes, S. A. 1993. "Estimating Laboratory Wave Reflection Using Laser Doppler," *Proceedings of Waves '93*, ASCE.

Leenknecht, D. A., Szuwalski, A., and Sherlock, A. R. 1992. "Automated Coastal Engineering System, "User's Guide, Version 1.06, US Army Engineer Waterways Experiment Station, Corps of Engineers, 3909 Halls Ferry Road, Vicksburg, MS.

Singamsetti, S. R., and Wind, H. G. 1980. "Characteristics of Shoaling and Breaking Periodic Waves Normally Incident to Plane Beaches of Constant Slope," Breaking Waves Publication No. M1371, Waterstaat, The Netherlands, pp. 23-27

Svendsen, I. A. 1984. "Wave Heights and Set-Up in a Surf Zone," *Coastal Engineering*, Vol 8, pp. 303-329.

Thompson, E. F., and Vincent, C. L. 1985. "Significant Wave Height for Shallow Water Design," *Journal of Waterways, Port, Coastal, and Ocean, Engineering*, ASCE, Vol 111, No. 5, pp. 828-842.

Weggel, J. R. 1972. Maximum Breaker Height, "*Journal of Waterways, Harbors and Coastal Engineering Division*," ASCE, Vol. 98, No. WW4, pp 529-548.

Waves Propagating on an Adverse Jet

Fredric Raichlen[1]

Introduction

A laboratory investigation was conducted of the propagation of waves on an adverse three-dimensional jet. The problem which was simulated was the ebb-tide flow from a tidal inlet into a constant depth ocean and the effect of such a flow on an incoming waves. The interaction of waves and currents near the entrance to inlets is important with respect to its effect on wave shoaling and thus both small and large boat navigation and on sediment transport within and in the nearshore region of a coastal inlet.

Various aspects of this problem have been studied in the past, and a reasonable summary of the earlier literature is presented by Bruun (1978). Bruun (1978) discusses at some length the problem of sediment transport in the vicinity of inlets along with inlet-fluid mechanics and the interaction of waves with currents. The review paper by Peregrine (1976) and the annotated bibliography presented by Peregrine, Jonsson, and Galvin (1983) discuss the interaction of waves and currents although not specifically the case of waves propagating on a three-dimensional jet where the current is in a direction opposite to the direction of wave propagation.

A large number of investigators have studied jets and plumes discharging into an ambient fluid. These have ranged from studies of jets which are lighter, neutrally buoyant, and heavier than the fluid into which it discharges; only two will be mentioned here. Albertson et al (1950) discuss the problem of a three-dimensional turbulent jet discharging into an infinite fluid without boundary friction. In a fundamental way, this is similar to the expanding ebb-tide flow from an inlet. The work of Mehta and Ozsoy (1978) discuss the hydrodynamics of a two-dimensional turbulent jet including the effects of bottom friction. The effects of bottom friction appear important and increase the rate of spreading of the jet, since the areal extent of the adverse velocity field and the corresponding velocity distribution has a significant effect on the incident wave characteristics.

The objectives of the limited experiments reported herein were essentially twofold. First it was of interest to design and test a wave machine which could reasonably pass a simulated ebb-tide flow through the wave generator while producing acceptable small amplitude waves. The study of wave-current interaction

---

[1]Professor of Civil Engineering, W.M.Keck Laboratory of Hydraulics and Water Resources, California Institute of Technology, Pasadena, CA 91125

in the laboratory is usually complicated by the fact that either a reasonable wave system propagates on a developing current or *vice versa*. An effort was made in this study to develop a wave machine which transmits the current though the generator while waves are being generated. Only in this way would the simulated ebb-tide jet spread unhindered and without the influence of the recirculating pump system. The second, and main objective, was to explore several aspects of the interaction of monochromatic small amplitude waves with an adverse three-dimensional jet representing the ebb-tide flow from a coastal inlet. The results reported herein are of a more exploratory nature pointing to some of the salient features of such an interaction and suggesting areas of future study.

Experimental Equipment and Procedures

Experiments have been conducted in a tilting flume with a working section which is approximately 17.2 m long, 85 cm wide, and 30.5 cm deep. A wave generator is located at one end of the flume and mounted to it; thus, as the flume tilts the wave machine tilts with it. In that way shoaling waves can be generated easily, and with the proper type of generator waves can be caused to break at a specified location within a wave tank. However, for the experiments reported herein the flume was kept horizontal. The flume is re-circulating using a variable speed pump, a gate valve for flow control, and an orifice meter to monitor the flow rate.

A channel 10.07 cm wide and 3.7 m long has been installed at the upstream end of this flume with one flume wall forming a sidewall of the channel. A streamlined entrance guides the flow into this narrow 10 cm wide channel. A vertical wall perpendicular to the flume sidewalls closes the remaining 75 cm width of the flume at its upstream end. A beach composed of rock and constructed at a slope of 1:10 on the flume-side of this 75 cm long wall dissipates energy from waves generated at the other end of the flume which do not propagate into the narrow channel. A photograph of this arrangement taken from the inlet end of the flume looking toward the wave generator is presented in Figure 1.

A unique wave machine is located at the downstream end of the flume producing waves which propagate in a direction opposite to the current. It is a vertical bulkhead type wave machine connected to a variable speed electric motor by a Scotch-yoke which provides the means of changing the stroke of the generator. Thus, the wave generator, at the present time, is non-programmable and can be moved only in a nearly sinusoidal manner. The bulkhead is constructed of two sets of eleven 5 cm wide and 6.4 mm thick moveable overlapping aluminum slats similar to a "venetian blind" mounted to a frame which travels on circular rails fixed to the flume walls. The slats are mounted in a frame whose width was divided into two sections with eleven slats approximately 40 cm long in each section. (The wave generator has some similarities to that proposed by Sarpkaya (1957), but there are several important differences.) The angle of the slats can be adjusted individually from fully closed to fully open and locked in place for a given experiment. When the slats are fully closed the generator operates as a standard bulkhead generator. If desired, the slats near the bottom can be opened while those near the free surface are kept closed and/or one side of the generator can have slats adjusted differently from the other. During these experiments all slats were partially opened the same amount with an opening between slats of about 6.4 mm. This allowed the current from the simulated inlet-jet to travel through the wave machine over the full depth of flow and full width of the flume and thence into the pump inlet. In this way a uniform flow across the channel was induced well downstream from the channel exit. The generator can be seen at the top of Figure 1, and a close-up view is shown in the photograph in Figure 2.

Resistance wave gages were used to measure water-surface time histories. These gages were mounted to instrument carriages which could be moved along the flume on circular rails affixed to the sidewalls of the flume; the rails had been adjusted to be parallel to the water surface to within ± 0.3 mm when the flume was horizontal. The instrument holder on one carriage could be moved laterally across the flume.

Velocities were measured using a Nixon miniature current flow meter with a propeller cage approximately 15 mm in diameter. The propeller meter has a measurable range of about 2.5 cm/sec to 150 cm/sec. All velocity measurements were made at a distance of 0.37 times the depth measured from the bottom. If the velocity profile is logarithmic, the velocity at that depth would be about equal to the depthwise average velocity.

Presentation and Discussion of Results

Several aspects of the operation of the wave generator were investigated. Water-surface time histories obtained with a resistance wave gage located about 52 depths from the wave generator without an imposed current are presented in Figure 3. The ratio of the depth to the wave length, h/L, varied from 0.086 to 0.496 for these measurements, and the wave heights were about 0.5 cm in a depth of 13.5 cm. Although it is somewhat difficult to assess the wave form from the simple presentation shown in Figure 3, it is apparent that the waves generated are essentially monochromatic for the range of relative depths shown. This was at first glance somewhat surprising considering the "leaky" nature of the wave generator. The wave machine appeared to generate waves which were more monochromatic for larger depth-to-wave length ratios. This probably should be expected, since a piston type wave generator moving sinusoidally generally produces waves with higher harmonics at small ratios of depth to wave length, but the waves away from the generation region are generally better formed as deepwater conditions are approached.

The wave maker efficiency was investigated for $0.56 < kh < 1.23$, where k is the wave number, $2\pi/L$. The variation of the ratio of the wave height, H, to the stroke of the wave machine, 2S, with kh is presented in Figure 4. Experimental data are presented for a slat opening of 6.4 mm. The "leaky" characteristics of the wave machine can be expressed in terms of the ratio of the side clearance between the wave generator and the side-walls of the flume to the depth, Δ_b/h, and the ratio of total clearance between adjacent submerged slats and between the wave generator and the bottom, Δ_s/b, where b is the width of the flume. For the experiments conducted the water depth was approximately $h \approx 13.5$ cm and the width of the flume was $b = 85$ cm, and the leakage parameters were: $\Delta_b/h = 0.047$ and $\Delta_s/b = 0.037$. Three curves are presented in Figure 4 to which the data can be compared. The solid curve is labeled the piston theory and corresponds to the small amplitude wave theory for a solid vertical bulkhead wave generator (see, e.g., Dean and Dalrymple (1984)). It is noted, as would be expected, that the wave machine is significantly less efficient than the theory predicts for all kh investigated. The two dashed curves correspond to a theory presented by Madsen (1970) for a solid bulkhead wave generator with clearance around the sides and the bottom of the values mentioned above. The curves correspond to half-strokes of 1 cm and 2 cm which are the range of motion of the generator in the experiments conducted. The data appear to agree reasonably well with the theory of Madsen (1970) even though, for the application of the theory, all of the leakage between the slats of the generator throughout the depth was assumed to be concentrated at the bottom as if all leakage had actually occurred there.

The variation of the velocity both downstream and laterally for the three-dimensional jet issuing from the channel has been measured using the miniature propeller meter described earlier for the condition without waves. Due to the nature of such a device it was not considered appropriate to use this to evaluate the velocities under the influence of both waves and currents. The variation of the normalized velocity in the downstream direction (approximately along the centerline of the channel with width b_o) with relative downstream distance $x/2b_o$ is shown in Figure 5. (Twice the channel width, i.e., $2b_o$, is used to normalize distance in an attempt to transform the data obtained experimentally to the case of a jet with width $2b_o$ centered in a channel.) The velocity at $y/2b_o = 0.215$ is denoted as U_{cl} and is normalized by the average channel exit velocity, U_o. The velocity ratio remains approximately constant and equal to unity until about $x/2b_o \approx 2.5$ (or five channel widths downstream). This region probably constitutes the potential core of the two-dimensional jet. A curve is presented in Figure 5 corresponding to the square root variation of the jet velocity as proposed by Albertson et al (1950) corresponding to a linear growth of the jet width. It is seen that the potential core as described by the data is shorter than that obtained by Albertson et al (1950). The data depart significantly from this curve for $x/2b_o > 10$, and for $x/2b_o = 20$ the centerline velocity is less than one-half that reported by Albertson et al (1950). As discussed by Mehta and Ozsoy (1978) bottom friction is important, and it probably contributes to the observed increased spreading compared to that described by Albertson et al (1950). Aside from this behavior of the jet, it is interesting to note that at $x/2b_o \approx 13$ the centerline velocity increases and then begins to rapidly decrease in a downstream direction. The physical reason for this behavior is not obvious.

The variation of the velocity in a direction perpendicular to the direction of wave propagation for several downstream locations is presented in Figure 6. In the upper part of Figure 6 the variation of the velocity with lateral distance is shown for the downstream positions of: $x/2b_o = 1.36$, 6.32, and 11.29. The spreading of the jet is generally as expected although perhaps one would have expected a greater decrease in the velocities for $x/2b_o = 11.29$ compared to 6.32 than was observed. The centerline measurements presented in Figure 6 confirm those shown in Figure 5. At the station nearest the channel exit the velocity is a maximum on the centerline of the channel ($y/b_o = 0.5$) decreasing in both directions. This shows the difference between the case investigated and that of a symmetric channel of width $2b_o$ where for the latter a maximum velocity at $y/b_o = 0$ would be expected. The velocities normalized by the maximum are presented in the lower portion of Figure 6 as a function of lateral distance; Gaussian curves are fitted to these data. (All curves have been fitted to a maximum at $y/b_o = 0.5$; however, this choice of the location of the maximum may not be appropriate at the downstream locations.)

The wave height was measured at two downstream locations , $x/2b_o = 4.06$ and 30.82, near the centerline of the inlet channel, ($y/2b_o = 0.19$), for various adverse jet currents. The variation of the wave height is shown in the upper part of Figure 7 for the two downstream locations as a function of the average velocity, U_o, at the channel exit, $x/2b_o = 0$, normalized by the wave phase speed, C, for a depth of h = 14.92 cm and a wave period of T = 1 sec or: h/L = 0.137. In the lower portion of Figure 7 the observed wave height is normalized by the wave height without the current; the current velocity used is that which is predicted by the half-power law shown in Figure 5 for the location of interest. It is seen that there is a tendency to collapse the data for the two locations. It is interesting to note that the wave height is increased by nearly a factor of 2.5 near the channel exit by a current only 7% of the phase speed. This corresponds to a ratio of current velocity at the channel exit to the wave group velocity of: $U_o/C_g = 0.09$.

The variation of wave height with distance downstream and upstream from the channel exit is presented in Figure 8 for a constant channel discharge velocity, U_o = 22.84 cm/sec, and for a wave phase speed of 103.8 cm/sec in a depth of h = 13.27 cm.

In the upper portion of the figure the ordinate is the wave height and the abscissa is the relative distance measured from the channel exit. The ratio of channel discharge velocity to wave group velocity for this experiment was: $U_o/C_g = 0.26$. For waves without a current, it is seen that the wave height generally decreases in a monotonic fashion from a height of about 1.25 cm nearly 40 channel widths offshore ($x/2b_o \approx 20$) to about unity nearly 30 channel widths upstream from the channel exit ($x/2b_o \approx -15$). In the lower portion of Figure 8 the ordinate is the relative wave height, i.e., the ratio of the wave height at a given location with the current to that without the current, and the abscissa is the relative distance from the channel exit. The relative wave height increases shoreward from unity at $x/2b_o = 18$ to about 2.25 at $x/2b_o = 6$, then decreases before increasing to a maximum near the channel exit. An interesting feature of this variation is the monotonic decrease in relative wave height as one proceeds up the channel. This does not appear to be due to friction, since the incident wave height without the current exhibits very little change as it propagates up the channel. The velocity distribution within the channel, both laterally, depthwise, and along the channel has not been determined to establish in a definitive way the sensitivity of the waves to small velocity differences. It appears that the increase in wave height for $0 < x/2b_o < 6.0$, i.e., outside of the channel, is in a region of monotonically decreasing velocity as seen in Figure 5. Thus, this maximum must be related more to the refractive effect of this adverse current on the incident waves than to variations in the current distribution. This type of behavior was discussed to some extent by Yoon and Liu (1986) in their analytical study of the interaction of waves with a rip current.

The lateral distribution of wave height is presented in Figure 9 with and without a current. In the upper portion of the figure the actual wave height is presented as a function of the normalized distance from the sidewall of the flume. As with the variation of the velocities in a lateral direction presented in Figure 6, the distance is normalized by the channel width, b_o, rather than twice this value. Thus, $y/b_o = 1$ corresponds to the sidewall of the channel. The lateral wave height distribution is presented at three distances downstream from the exit: $x/2b_o = 4.0$, 12.06, and 18.0. For the condition without the current, as shown in the upper portion of Figure 9, the wave height is reasonably constant across the width of the flume at the three downstream locations. The lower portion of Figure 9 shows the variation of the ratio of wave heights with and without the current as a function of normalized lateral distance. For $x/2b_o = 4.0$ and 12.06 the wave height is greater than unity for $y/b_o < 3$ and less than unity for $y/b_o > 3$. At the furthest station downstream ($x/2b_o = 18.0$) there is still a lateral variation of relative wave height but, as one would expect, the variation is not as large as it is closer to the exit. This variation is due both to the effects of refraction and the circulation induced by the discharging jet. As the jet expands into the flume it entrains fluid from the region for $y/b_o > 1.0$. This in turn induces a weak but favorable current in that region which could contribute to the reduced wave height observed in that region.

Conclusions

The following major conclusions may be drawn from this study:

1. A slat type wave machine which permits flow through the wave generator plate appears to work well for such a study of wave-current interactions. The performance of the generator is described reasonably well by an available theory.

2. The jet discharge from a channel located asymmetrically in the flume appears to agree reasonably well with results for a two-dimensional symmetric turbulent jet.

3. The experimental arrangement used controls the jet discharge by the use of the training wall, and in this manner allows for the isolation of certain aspects of the problem.

4. Even for small ratios of the average velocity at the channel exit to the wave celerity, i.e., less than 10 %, the incident wave height with the adverse current is more than twice what it is without the current.

5. In the region seaward of the channel exit a wave maximum occurs caused primarily by the effects of refraction.

6. The wave height decreases significantly as waves propagate up the channel for the case with an adverse current. This effect apparently is not due to friction.

7. There is a significant lateral variation of wave height across the flume caused by the jet which may be due to both wave refraction and the shoreward velocities induced by the effects of the entrainment of the ambient fluid by the jet which discharges from the channel exit.

References

Albertson, M.L, Dai, Y.B., Jensen, R.A. and Rouse, H, "Diffusion of Submerged Jets", The Transactions of the ASCE, Vol. 115, 1950.

Bruun, P., "Stability of Tidal Inlets", Elsevier Scientific Publishing Co., New York, 1978.

Dean, R.G. and Dalrymple, R.A., "Water Wave Mechanics for Engineers and Scientists", Prentice-Hall, Inc., 1984.

Madsen, O.S., "Waves Generated by a Piston-Type Wavemaker", Proceedings of the Twelfth Coastal Engineering Conference, Washington, D.C., 1970.

Mehta, A.J. and Ozsoy, E., "Inlet Hydraulics", Ch. 3 of Bruun, P., "Stability of Tidal Inlets", Elsevier Scientific Publishing Co., New York, 1978.

Peregrine, D.H., "Interaction of Water Waves and Currents", Advances in Applied Mechanics, Vol. 16, 1976.

Peregrine, D.H., Jonsson, I.G., and Galvin, C.J., "Annotated Bibliography on Wave-Current Interaction", Misc. Report No. 83-7, U.S. Army Corps of Engineers, Coastal Engineering Research Center, March 1983.

Sarpkaya, T., "Oscillatory Gravity Waves in Flowing Water", Transactions of the ASCE, Vol. 122, 1957

Yoon, S.B. and Liu, P.L.-F., Wave and Current Interactions in Shallow Water", Proceedings of the Twentieth Coastal Engineering Conference, Washington, D.C., 1986.

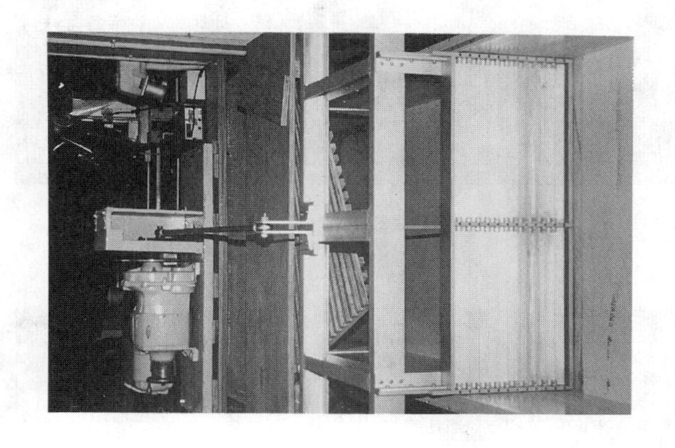

Figure 2 Photograph of "leaky" wave generator

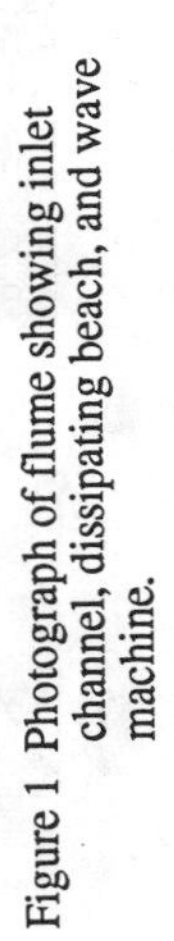

Figure 1 Photograph of flume showing inlet channel, dissipating beach, and wave machine.

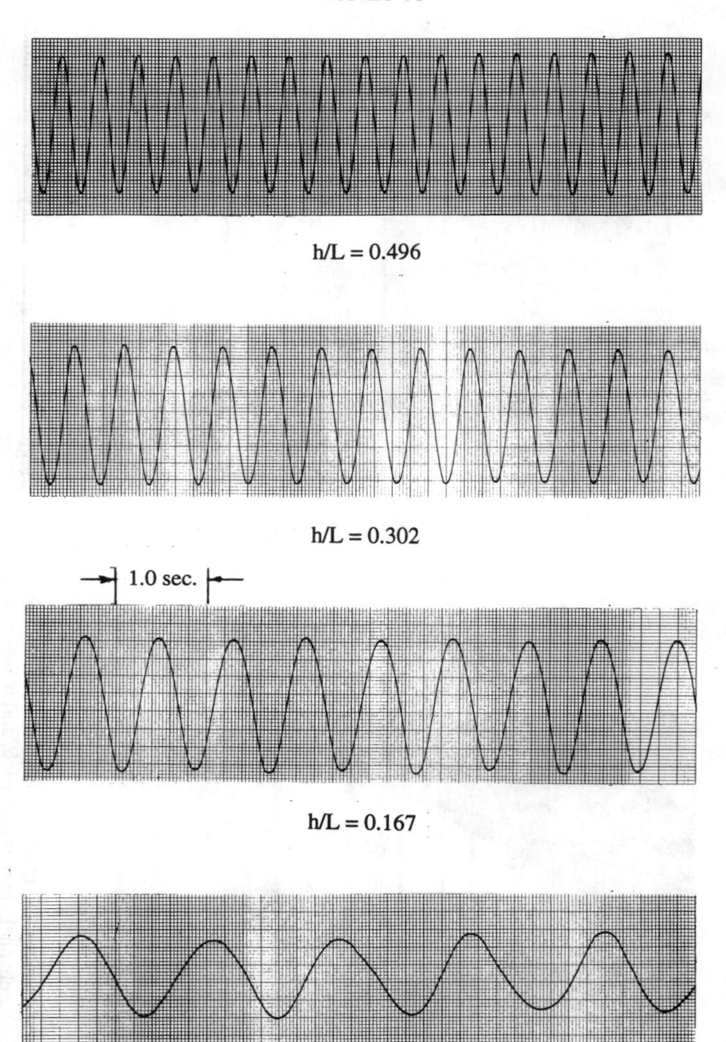

Figure 3 Water surface time histories for a range of periodic small amplitude waves

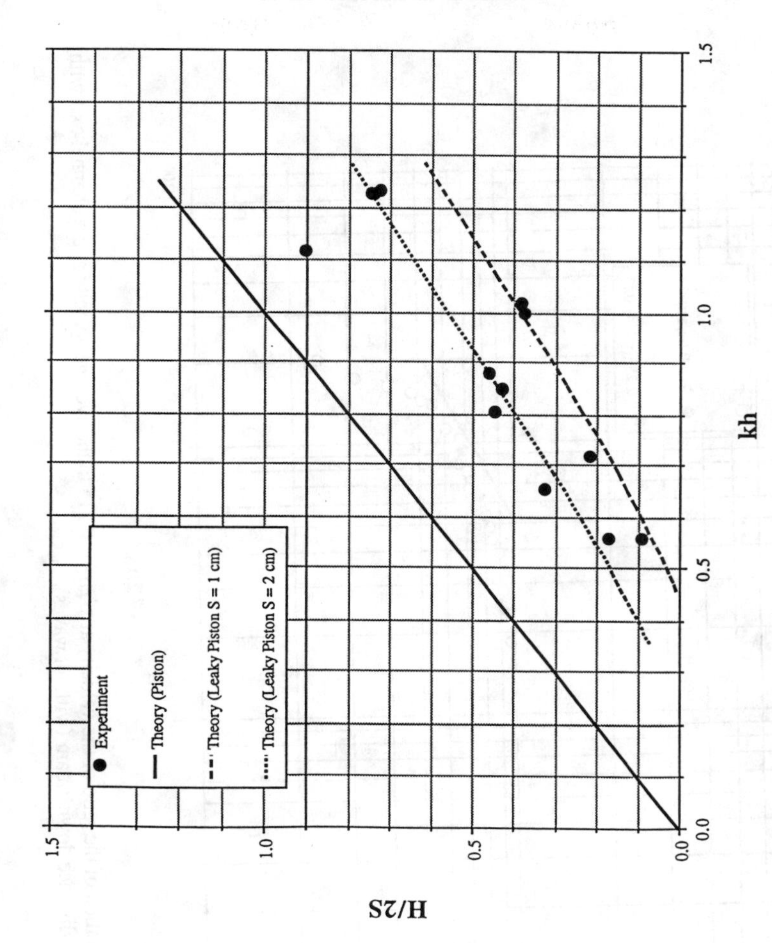

Figure 4 The variation of the ratio of wave height to wave machine stroke with kh

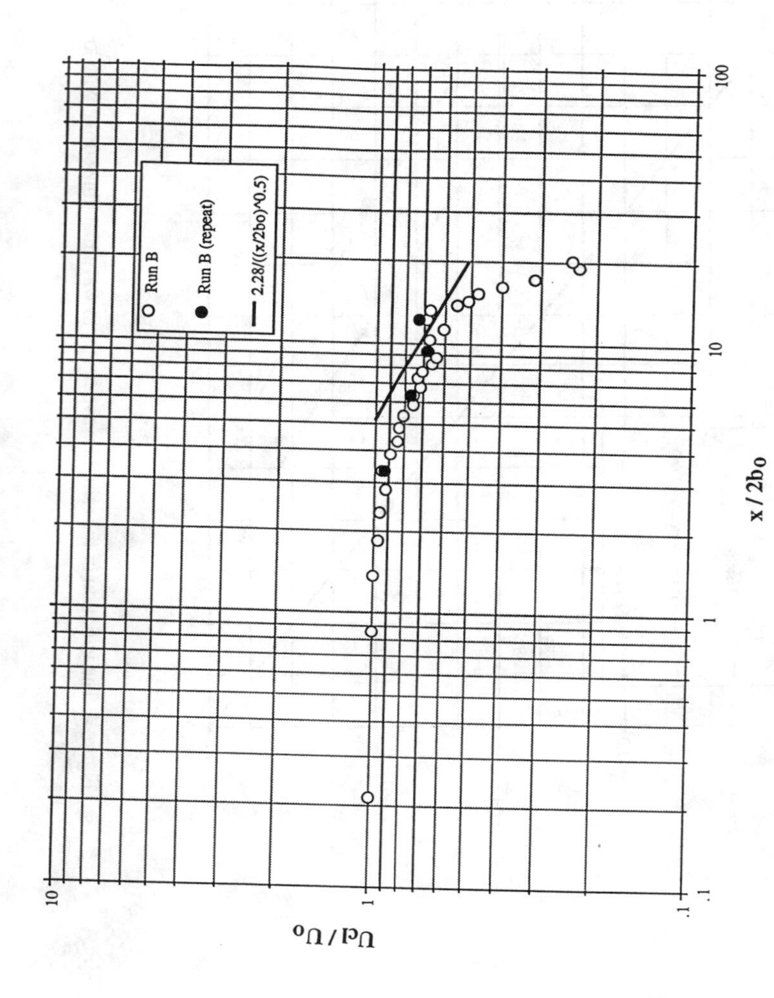

Figure 5 The variation of the ratio of the velocity at $y/2b_o = 0.215$ to the average velocity at the channel exit with relative distance downstream (without waves)

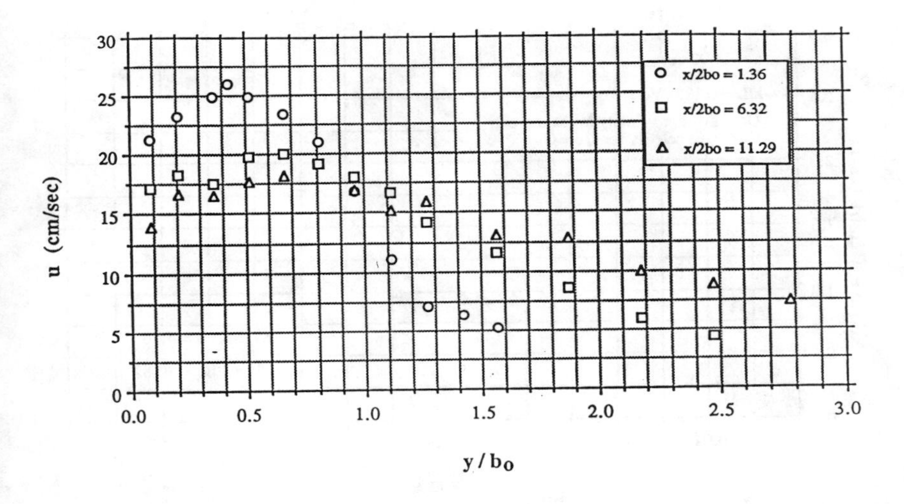

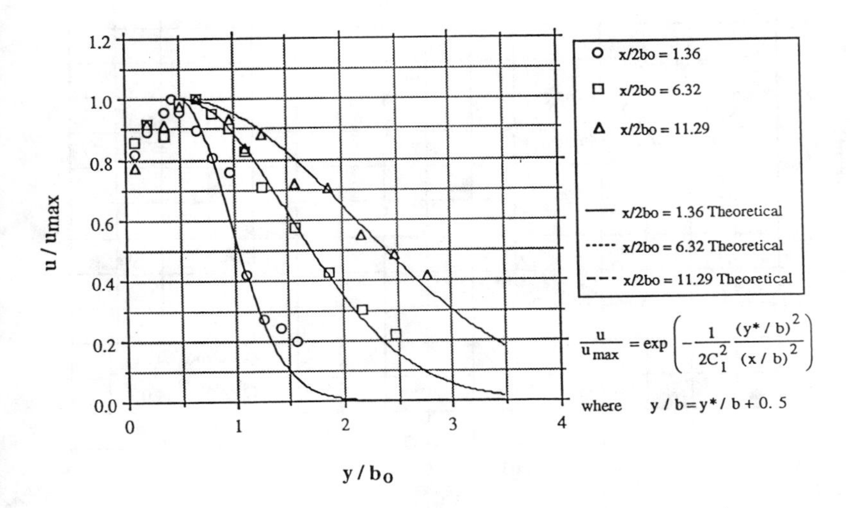

$$\frac{u}{u_{max}} = \exp\left(-\frac{1}{2C_1^2}\frac{(y^*/b)^2}{(x/b)^2}\right)$$

where $y/b = y^*/b + 0.5$

Figure 6 The variation of the velocity and of the normalized velocity with relative lateral distance at several downstream locations (without waves)

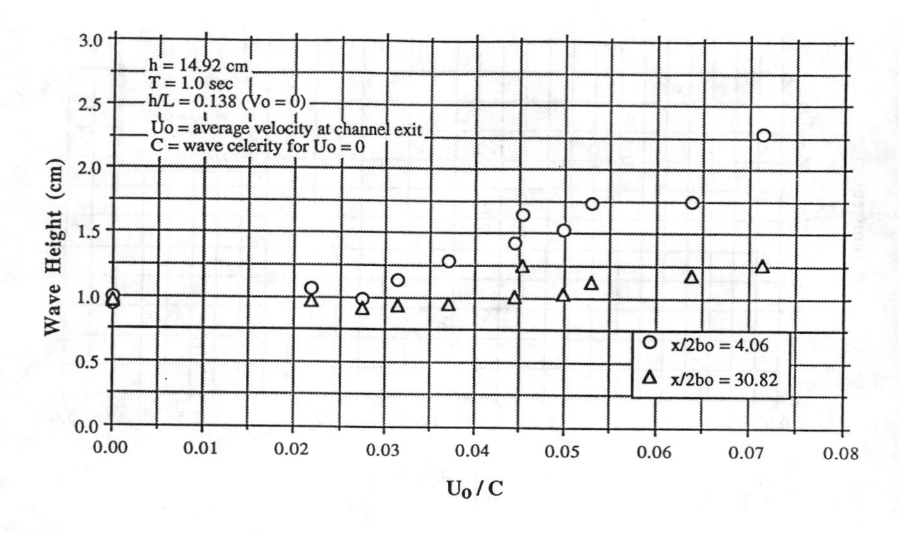

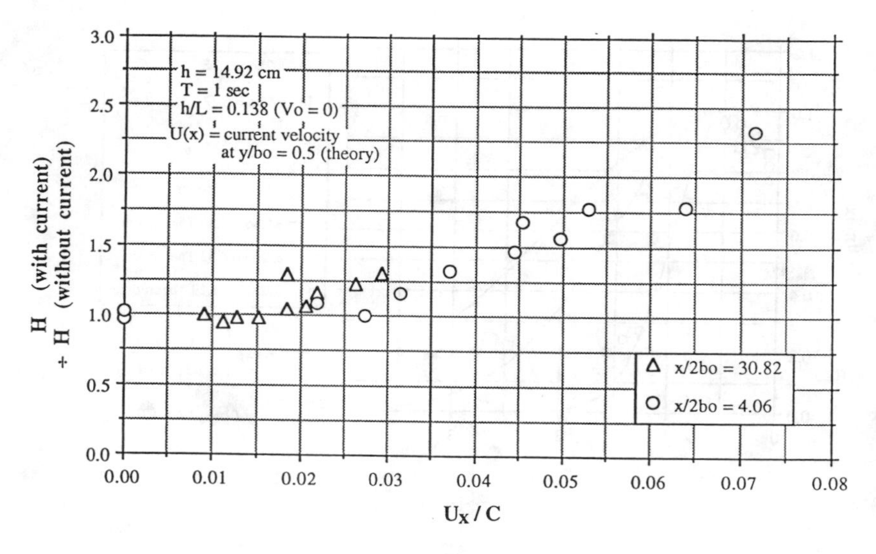

Figure 7 The variation of the wave height with the ratio of the average velocity at the channel exit to the wave celerity, and the variation of the relative wave height with the ratio of the velocity predicted at the given location to the wave celerity.

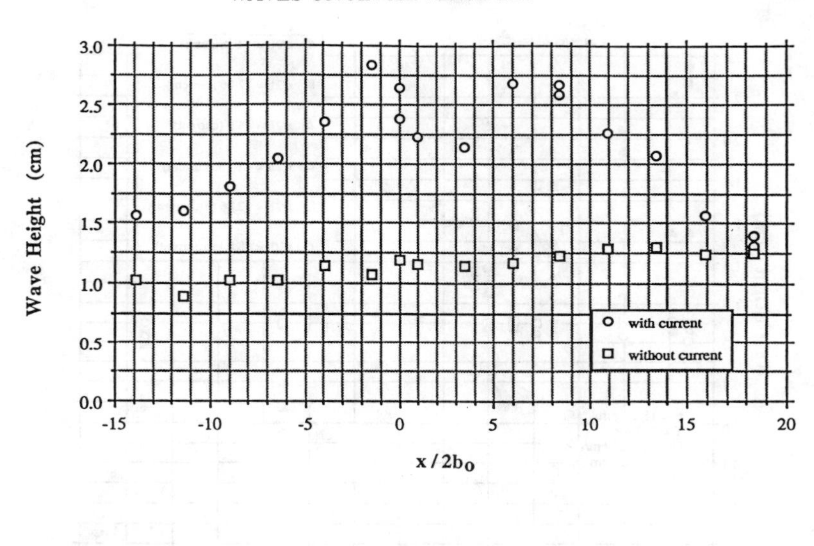

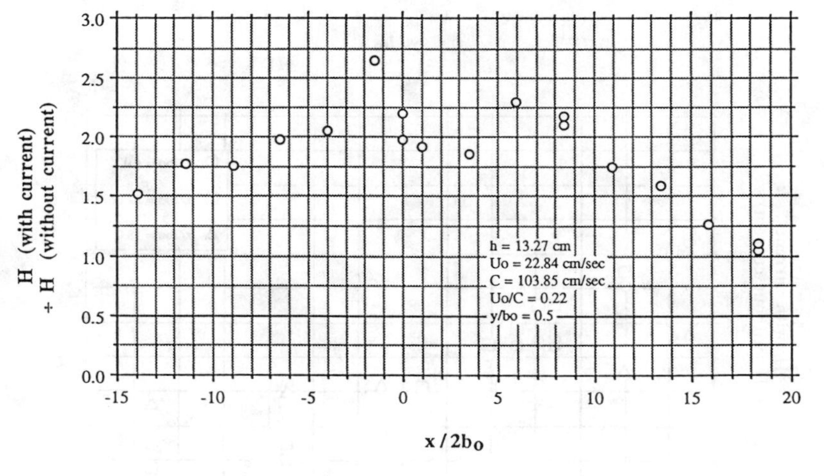

Figure 8 The variation of the wave height and of the relative wave height with relative distance from the channel exit

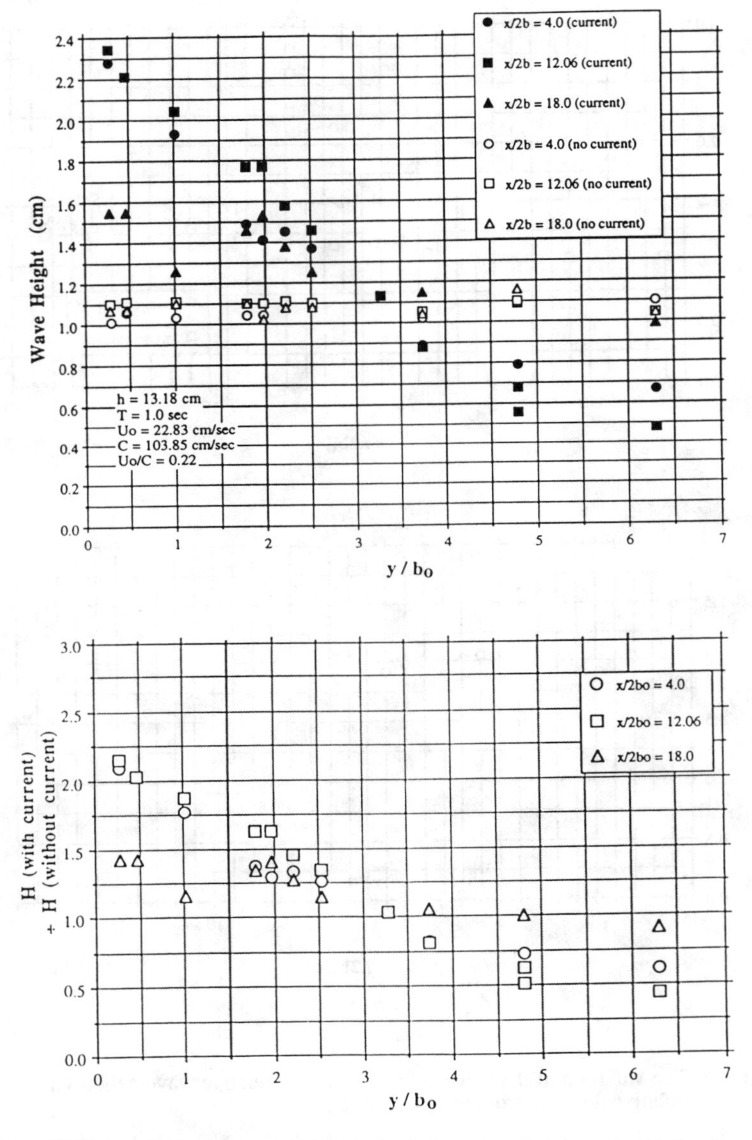

Figure 9 The variation of wave height and relative wave height with relative later distance across the flume at several downstream locations

The Time Dependent Ray Method for Calculation of Wave Transformation on Water of Varying Depth and Current [1]

Yung Y. Chao [2]

Abstract

This paper focuses on an aspect of the numerical calculation of wave refraction using the ray tracing or the semi-Lagrangian approach. The conventional ray method for steady state depth and current fields is extended to the unsteady situation. It is shown that a factor which characterizes the unsteadiness of the medium can be included explicitly in the calculation of the wave energy amplification. The contribution associated with this factor on wave energy is ${c_o}^2/(\mathbf{C_g} + \mathbf{U})^2$, where $\mathbf{C_g}$ and $\mathbf{U}$ are the wave group velocity and current velocity, respectively, and c_o is the phase velocity of the wave-like fluctuation of currents and depth (e.g., caused by tides) in the direction of the ray path, i.e., in the direction of wave energy propagation.

Introduction

Efforts to include the effects of wave refraction due to spatially and/or temporally varied current and water depth in a numerical wave prediction model have been made by a number of researchers (see, e.g., Collins, 1972; Cavaleri and Rizzoli, 1981; Chen and Wang, 1983; Tolman, 1989,1991; Hubbert and Wolf, 1991). Numerical schemes for wave propagation and refraction calculations employed by these researchers can be conveniently classified into two types of approach. The Eulerian approach applies a finite difference scheme to obtain solutions at each grid point over the area of interest simultaneously. The Lagrangian approach uses the ray tracing technique to derive solutions at each specified point independently.

---

[1] OPC Contribution No.81

[2] National Meteorological Center, NWS/NOAA, 5200 Auth Road, Camp Springs, MD 20746

The advantage of the Eulerian approach is that it provides a synoptic view of the wave pattern involving all frequency and directional components over the entire area and allows addition of the various source and sink terms including nonlinear wave-wave interactions. However, finite difference schemes for the convective transport equation have a common problem - the solution always involves unavoidable numerical errors (numerical dispersion and diffusion). As a result, the physical significance of their solutions is obscured.

In contrast, the Lagrangian approach (or what comes to the same thing: the method of characteristics or the ray method) follows a wave packet along the wave ray, thus avoiding the troublesome convective term and providing a strong physical realization of wave propagation. It provides a true path and time for any choice of wave parameters arriving at a specific point. Consequently, the loss of spatial generality is compensated by an increase in temporal accuracy.

There are limitations to this approach. It does not provide the synoptic view of the wave field over the entire area of interest at each computational time step. It is incapable of modeling the effects of source terms which involve all spectral components at each time step, such as nonlinear wave-wave interactions and white capping. It is also difficult to specify quantitatively the energy amplification factor in the vicinity of ray crossings and caustics which may occur if waves propagate over a complicated bottom topography or current field.

An early effort to combine the strengths and eliminate the shortcomings of both the finite difference and ray tracing techniques has been made by Barnett et al. (1969, also see, Allender et al.,1985). They developed a scheme which constructed a set of parallel rays over the ocean for each specified wave direction. Along each ray, the distance between ray points is determined according to the group velocity of the given frequency and time interval. The wave energy, considered as particles, hop from ray point to ray point at appropriate multiples of the time step. The wind field is provided on a net of grid points defined over the area of interest. At each given time step, the wave spectrum at any grid point is accumulated from the nearest ray point for each frequency and direction. After the spectrum is constructed, the processes of wave growth, dissipation and wave-wave energy transfer are executed. The spectral components are then re-distributed back to the appropriate ray points. Their approach can be applied only in deep water without currents and under steady state conditions since it assumes the constancy of wave direction along the rays.

It is of interest to note that the so-called semi-Lagrangian technique, which has gained widespread recognition as an efficient way to integrate the primitive equations of numerical weather prediction, also has been applied to the numerical prediction of ocean waves in recent years. Basically, it is an extension of the classical treatment of the refraction problem at a given site using the backward ray tracing technique (Dorrestein, 1960). With this approach, each grid point of a grid mesh covered the entire area of concern is considered to be a target point to receive the wave energy of all spectral components in a specified time step from various source locations. The source location of each component (the departure point) is determined by the method of ray back-tracing for the given time step. The localized physical processes such as wave growth by wind, dissipation due to whitecapping and wave-wave energy transfer are calculated at each grid point. The calculated spectral energy densities at all grid points serve to provide information for deriving the spectral energy density at each source location. Since it is impossible to have the departure point always coincide with a certain grid point, an appropriate interpolation procedure is performed to obtain the energy content at the departure point based on the values at neighboring grid points. A feasibility study of this approach along with different methods for spatial interpolation as a part of this system, has been studied by Ryabinin (1991).

The scheme is potentially attractive because of its high accuracy and efficiency in computation by allowing large time intervals and low spectral resolutions. In shallow water, the energy dissipation due to bottom friction can be computed along the rays. The effect of sub-grid irregularity in bottom topography and/or surface currents on the wave energy amplification can be calculated explicitly, and the problem of crossing rays and caustics can be avoided as the wave propagation distance is relatively short in a given time step. In addition, the underlying grid mesh can be arranged irregularly depending on the accuracy of the input data and the requirements for wave information in the area of interest.

A problem in applying this semi-Lagrangian approach, however, remains to be solved. In connection with the study of wave-current interaction problem in the southern North Sea, Tolman (1990) pointed out that for the large scale continental shelf unsteadiness of current and water depth induced by tides must be considered if the time scale of variation in current and depth (typically 12 hours) is not large compared to the travel time of the waves through the area of interest. The wave energy amplification factor derived from the conventional ray method is derived based on the assumption of steady state water depth and flow conditions. Therefore the effect of unsteadiness in depth and currents cannot be evaluated. In this paper, the time dependent ray method is derived to express explicitly factors involved in calculating the change of wave spectrum due to the existence of unsteady and irrelegular depth and current fields.

Basic Equations

The change of wave field due to the presence of varying currents and bathymetry can be specified based on the dynamic conservation of wave action and the kinematic conservation of wave number or wave crests along characteristic curves or rays (Bretherton and Garrett, 1969; Phillips, 1977). The path of a ray is determined by simultaneous solution of the following set of equations:

$$\frac{dx_j}{dt} = \frac{\partial \omega}{\partial k_j} = c_{g_j} + u_j, \tag{1}$$

$$\frac{dk_i}{dt} = -\frac{\partial \omega}{\partial \lambda}\frac{\partial \lambda}{\partial x_i} = -\frac{\partial \sigma}{\partial h}\frac{\partial h}{\partial x_i} - k_j\frac{\partial u_j}{\partial x_i}, \tag{2}$$

and

$$\frac{d\omega}{dt} = \frac{\partial \omega}{\partial \lambda}\frac{\partial \lambda}{\partial t} = \frac{\partial \sigma}{\partial h}\frac{\partial h}{\partial t} + k_i\frac{\partial u_i}{\partial t}, \tag{3}$$

where

$$\frac{d}{dt} \equiv \frac{\partial}{\partial t} + \frac{\partial \omega}{\partial k_j}\frac{\partial}{\partial x_j}. \tag{4}$$

Here the wave-number vector $\mathbf{k} = (k_1, k_2)$, group velocity $\mathbf{C_g} = (c_{g_1}, c_{g_2})$, flow velocity $\mathbf{U} = (u_1, u_2)$, and the horizontal cartesian coordinates $\mathbf{x} = (x_1, x_2)$. The apparent frequency, ω, is given by $\omega(\mathbf{k}, \lambda) = \sigma + \mathbf{k} \cdot \mathbf{U}$, where σ, the intrinsic frequency of waves in a frame of reference moving with flow velocity $\mathbf{U}(\mathbf{x}, t)$, obeys the dispersion relationship, $\sigma = (gk \tanh kh)^{1/2}$, where g is the gravitaional acceleration, $k = \mid \mathbf{k} \mid$, and $h(\mathbf{x}, t)$ is water depth. $\lambda(\mathbf{x}, t)$ represents local properties of the medium, i.e., h and $\mathbf{U}$. Equation 3 indicates that if water depth and current velocity do not vary with time, ω remains constant along the rays.

Conservation of wave action for a slowly varying wavetrain of small amplitude can be expressed in terms of rays as

$$\frac{d}{dt}\left(\frac{E}{\sigma}\right) + \left(\frac{E}{\sigma}\right) \nabla \cdot (\mathbf{C_g} + \mathbf{U}) = 0. \tag{5}$$

E is the local wave energy per unit area (proportional to the square of the wave amplitude). The wave action is defined as E/σ. For a continuous spectrum, the energy density of a group of waves whose wave-numbers lie in the element of area δA of the wave-number plane, specified by the vectors $\mathbf{k}$, $\mathbf{k} + \delta\mathbf{k}'$, and $\mathbf{k} + \delta\mathbf{k}''$ is given by

$$\delta E(\mathbf{k}) = \rho g F(\mathbf{k})\, \delta A, \tag{6}$$

in which

$$\delta A = |\delta\mathbf{k}' \times \delta\mathbf{k}''|. \tag{7}$$

$F(\mathbf{k})$ is the spectral density and ρ the water density. By applying the kinematic conservation principle, Phillips (1977) has shown that

$$\frac{d}{dt}\delta A + \delta A\, \nabla \cdot (\mathbf{C_g} + \mathbf{U}) = 0. \tag{8}$$

Therefore by substituting eqs.(6) and (8) into eq.(5) we have

$$\frac{d}{dt}\left(\frac{F(\mathbf{k})}{\sigma}\right) = 0. \tag{9}$$

Equation 9 expresses the conservation of spectral wave action density along the ray. In the absence of a current, $F(\mathbf{k})$ remains constant along the ray. This result was first demonstrated by Longuet-Higgins (1957).

Time Dependent Ray Method

Equation 8 cannot be directly integrated along a ray because knowledge of neighboring solutions is required to determine the divergence of velocity. Further complications arise if variations of currents and/or water depth with time cannot be ignored. Under these conditions, ω and $\mathbf{C_g} + \mathbf{U}$ are not independent of time. In order to solve this problem in a general manner without making a usual assumption that ω is independent of time, the approach of Shen and Keller (1975) is employed. We begin by defining

$$\frac{dt}{d\gamma_1} = \mu. \tag{10}$$

Here $\mu(t, x, y)$ is an arbitrary but non-zero proportionality function. The choice of μ determines the nature of the parameter γ_1 as we shall see. Equations 1, 2, 3 and 8 consequently can be expressed, respectively, as

$$\frac{dx_j}{d\gamma_1} = \mu(c_{g_j} + u_j), \tag{11}$$

$$\frac{dk_i}{d\gamma_1} = -\mu\left(\frac{\partial\sigma}{\partial h}\frac{\partial h}{\partial x_i} - k_j\frac{\partial u_j}{\partial x_i}\right), \tag{12}$$

$$\frac{d\omega}{d\gamma_1} = \mu\left(\frac{\partial\sigma}{\partial h}\frac{\partial h}{\partial t} + k_i\frac{\partial u_i}{\partial t}\right), \tag{13}$$

and

$$\mu^{-1}\frac{d}{d\gamma_1}\delta A + \delta A\,\nabla\cdot(\mathbf{C_g} + \mathbf{U}) = 0, \tag{14}$$

where

$$\frac{d}{d\gamma_1} \equiv \mu\left\{\frac{\partial}{\partial t} + \frac{\partial\omega}{\partial k_j}\frac{\partial}{\partial x_j}\right\}. \tag{15}$$

We now introduce the Jacobian

$$J(\gamma_1, \gamma_2, \gamma_3) = \frac{\partial(x_1, x_2, x_3)}{\partial(\gamma_1, \gamma_2, \gamma_3)} = det\left(\frac{\partial x_i}{\partial\gamma_j}\right), \qquad i, j = 1, 2, 3. \tag{16}$$

Here, for convenience, we have set $\mathbf{r} = (\mathbf{x}, c_o t) = (x_1, x_2, x_3)$, where c_o is a constant reference speed associated with the ambient medium, and $(\gamma_1, \gamma_2, \gamma_3)$, is a set of parameters describing a point on a particular ray. γ_1 is the running parameter which varies along the ray and γ_2 and γ_3 are labeling parameters specifying a particular ray and are constants on each ray of the family of rays. In terms of these parameters, a point on a ray of a family can be respresented by $\mathbf{r} = \mathbf{r}(\gamma_1, \gamma_2, \gamma_3)$. For fixed values of γ_2 and γ_3, this represents the equation of a ray.

We observe that the determinant can be expanded in terms of the cofactors such that

$$\sum_{j=1}^{3}\frac{\partial x_m}{\partial\gamma_j}cof\frac{\partial x_i}{\partial\gamma_j} = J\delta_{im}. \tag{17}$$

Here δ_{im} is the Kronecker symbol and cof denotes the cofactor. If $i = m$, $\delta = 1$ and equation 17 follows from the rule for the expansion of a determinant by cofactors. If $i \neq m$, it follows from the fact that a determinant with two identical rows vanishes. We now differentiate equation 16 with respect to γ_1, observing that the derivative of a determinant is the sum of the derivatives of all its elements, each multiplied by its cofactor. Then we write $\partial/\partial\gamma_j = \sum_{m=1}^{3}(\partial x_m/\partial\gamma_j)\partial/\partial x_m$ and obtain

$$\begin{aligned}\frac{\partial J}{\partial\gamma_1} &= \sum_{i,j=1}^{3}\frac{\partial^2 x_i}{\partial\gamma_j\partial\gamma_1} cof \frac{\partial x_i}{\partial\gamma_j} = \sum_{i,j=1}^{3}\frac{\partial}{\partial\gamma_j}\left(\frac{\partial x_i}{\partial\gamma_1}\right) cof \frac{\partial x_i}{\partial\gamma_j} \\ &= \sum_{i,j,m=1}^{3}\frac{\partial x_m}{\partial\gamma_j}\frac{\partial}{\partial x_m}\left(\frac{\partial x_i}{\partial\gamma_1}\right) cof \frac{\partial x_i}{\partial\gamma_j} = \sum_{i,m=1}^{3} J\delta_{im}\frac{\partial}{\partial x_m}\left(\frac{\partial x_i}{\partial\gamma_1}\right) \\ &= J\sum_{i=1}^{3}\frac{\partial}{\partial x_i}\left(\frac{\partial x_i}{\partial\gamma_1}\right) \end{aligned} \tag{18}$$

Here we have used equation 17 in equation 18 and evaluate the sum over j and then the sum over m. Finally we use equations 10 and 11 and find

$$\begin{aligned}\frac{\partial J}{\partial\gamma_1} &= J\left\{\frac{\partial}{\partial t}\left(\frac{dt}{d\gamma}\right) + \frac{\partial}{\partial x}\left(\frac{dx}{d\gamma}\right) + \frac{\partial}{\partial y}\left(\frac{dy}{d\gamma}\right)\right\} \\ &= J\left\{\frac{\partial\mu}{\partial t} + \nabla\cdot\mu(\mathbf{C_g} + \mathbf{U})\right\} \\ &= J\left\{\frac{d\mu}{dt} + \mu\nabla\cdot(\mathbf{C_g} + \mathbf{U})\right\}. \end{aligned} \tag{19}$$

By using $d\mu/dt = (d\mu/d\gamma)(d\gamma/dt) = (d\mu/d\gamma)\mu^{-1}$ in equation 19 and rearranging the equation, we have

$$\nabla\cdot(\mathbf{C_g} + \mathbf{U}) = \frac{1}{\mu}\left\{\frac{1}{J}\frac{dJ}{d\gamma} - \frac{1}{\mu}\frac{d\mu}{d\gamma}\right\}. \tag{20}$$

Therefore equation 14 becomes

$$\frac{d}{d\gamma_1}\ln\left(\frac{J\delta A}{\mu}\right) = 0, \tag{21}$$

which states that the quantity $J\delta A/\mu$ is conserved along the space-time ray. Since equation 5 has the same form as equation 8 we can conclude that

$$\frac{JF\delta A}{\mu\sigma} = constant \tag{22}$$

along the space-time ray.

Interpretation of $J(\gamma)$

The Jacobian J in equation 22 can be given a geometric interpretation. We envisage a curved surface $f(x, y, t)$ formed by a rectifiable wave action front as it moves forward during time δt. A point on this surface $\mathbf{x}$ can be identified by the running parameter γ_1 and the labeling parameters γ_2 and γ_3. The parameter γ_1 gives the location of the point on a ray whose initial position on a wave front is given by parameter γ_2. This action front is specified by γ_3 related to some initial time. This wave action front resembles the so-called initial manifold in the context of solving some initial value problems.

We note that the Jacobian defined by equation 16 can also be expressed as a triple scalar product

$$J = \frac{d\mathbf{r}}{d\gamma_1} \cdot \left(\frac{\partial \mathbf{r}}{\partial \gamma_2} \times \frac{\partial \mathbf{r}}{\partial \gamma_3} \right) = \frac{\partial \mathbf{r}}{\partial \gamma_3} \cdot \left(\frac{d\mathbf{r}}{d\gamma_1} \times \frac{\partial \mathbf{r}}{\partial \gamma_2} \right). \tag{23}$$

Since

$$\left| \frac{d\mathbf{r}}{d\gamma_1} \right|^2 = \left(c_o \frac{dt}{d\gamma_1} \right)^2 + \left(\frac{dx}{d\gamma_1} \right)^2 + \left(\frac{dy}{d\gamma_1} \right)^2 = \mu^2 [c_o^2 + (\mathbf{C_g} + \mathbf{U})^2], \tag{24}$$

if we choose

$$\mu = [c_o^2 + (\mathbf{C_g} + \mathbf{U})^2]^{-1/2}, \tag{25}$$

then it follows from equation 24 that $(d\mathbf{r}/d\gamma)^2 = 1$. Therefore, $d\mathbf{r}/d\gamma_1 = \mathbf{s}$ is the unit tangent vector to a ray, and the parameter γ_1 is the arclength of the space-time ray. Since we are concerned with the change of wave characteristics associated with this initial manifold as time passes by, we may choose the parameter γ_3 to coincide with the time coordinate such that $\partial \mathbf{r}/\partial \gamma_3 = (0, 0, 1) = \mathbf{n}$, i.e., the unit vector in the direction of the time coordinate. Furthermore, the time coordinate can be considered, without loss of generality, to be perpendicular to the horizontal plane formed by the x and y coordinate, i.e., $c_o t$ represents the z-axis. In this way, $\mathbf{n}$ is the unit vector normal to the horizontal plane. The term $\partial \mathbf{x}/\partial \gamma_2$ represents a vector tangent to the action front from which a family of rays is spreading out. Thus the width between two adjacent rays, $d\nu$, can be expressed as [3]

$$d\nu = \left| \mathbf{n} \cdot (\mathbf{s} \times \frac{\partial \mathbf{x}}{\partial \gamma_2}) d\gamma_2 \right| = \left| J d\gamma_2 \right|. \tag{26}$$

Here $d\gamma_2$ is the infinitesimal interval of length identifying two adjacent rays with parameter values γ_2 and $\gamma_2 + d\gamma_2$ along the initial manifold. Since $d\gamma_2$ is a constant, it follows from Eqs. (22) and (6) that

$$\frac{\delta E}{\sigma} d\nu \mid \mathbf{C_g} + \mathbf{U} \mid \left[1 + \frac{c_o^2}{|\mathbf{C_g} + \mathbf{U}|^2} \right]^{1/2} = constant. \tag{27}$$

[3] A triple scalar product $\mathbf{n} \cdot (\mathbf{A} \times \mathbf{B})$ represents an area of the orthogonal projection of the parallelogram determined by vectors $\mathbf{A}$ and $\mathbf{B}$ onto a plane whose unit normal is $\mathbf{n}$.

The Unsteadiness Factor

The factor $c_o/(\mathbf{C_g}+\mathbf{U}) \equiv \Psi$ in (27) can be considered as a contribution to the wave height amplification due to the unsteadiness of the ambient medium. This factor is equivalent to the one suggested by Tolman(1990) as a measure of the unsteadiness of the depth or the current field. The major cause of unsteadiness in water depth and current in the coastal region is tides and tidal currents. The value c_o in this situation represents the phase speed of the tide in the direction of wave propagation, i.e., $c_o \equiv C_t \cos\alpha \sim \sqrt{gh}\cos\alpha$, where α is the angle between the tide and the direction of wave propagation . Since $\sqrt{gh} > \mid \mathbf{C_g}+\mathbf{U} \mid$, Ψ is in the order of one or larger.

To provide a rough idea about the magnitude of Ψ, we consider a situation where waves propagate in water of constant depth, say h = 25,m following the direction of tide. We take the amplitude of tides to be $A' = 0.5\ m$ in the open ocean where the depth is typically $h' = 4000\ m$, then $c_o' = 198\ m/s$ and $U' = c_o'A'/h' = 2.5\ cm/s$ (Bowden, 1983). The corresponding values in water of $h = 25\ m$ are $A = A'(h/h')^{1/4} = 1.8\ m$, $c_o = c_o'(h/h')^{1/2} = 15.7\ m/s$, $U = U'(h'/h)^{3/4} = 1.1\ m/s$. For waves of period 15 seconds propagated in the water of 25 meter depths, the group velocity is $C_g = 12.4\ m/s$, we have $\Psi = 1.2$. For waves of period 10 seconds, $C_g = 9.4\ m/s$, $\Psi = 1.5$ and for 5 second waves, $C_g = 3.9\ m/s$, $\Psi = 3.1$. Thus, the effect of unsteadiness in the ambient medium on short waves can be substantial.

References

Allender, J.H., T.P. Barnett, and M.Lybanon: The DNS model: An improved spectral model for ocean wave prediction. In *Ocean Wave Modeling*, SWAMP group, Plenum Press. 235-248.

Barnett, T.P., C.H. Holland, and P. Yager(1969): A general technique for wind wave prediction to the South China Sea. Final report, contract N62306-68-C0285, U.S. Naval Oceanographic Office, Washington, D.C.

Bowden, K. F. (1983): *Physical Oceanography of Coastal Waters.* John Wiley and Sons, New York, 302 pp.

Bretherton, F. P. and C. J. R. Garrett (1969): Wavetrains in inhomogeneous moving media. Proc. Roy. Soc. A.302, 529-554.

Cavaleri, L. and P.M. Rizzoli(1981): Wind wave prediction in shallow water: Theory and applications. J. Geophys. Res. 86, 10961-10973.

Chen, Y.H. and H.Wang(1983): Numerical model for nonstationary shallow water wave spectral transformation. J. Geophys. Res., 889(C14), 9851-9863.

Collins, J.I. (1972): Prediction of shallow-water spectra. J. Geophys. Res. 77, 2693-2707.

Dorrestein, R.(1960): Simplified method of determining refraction coefficients for sea waves. J. Geophys. Res. 65(2), 637-642.

Hubbert, K.P. and J. Wolf (1991) : Numerical investigation of depth and current refraction waves. J. Geophys. Res. 96, 2737- 2748.

Longuet-Higgins, M. S. (1957): On the transformation of a continuous spectrum by refraction. Proc. Camb. Phil. Soc. 53(1), 226-229.

Phillips, O. M. (1977): *The Dynamics of the Upper Ocean.* Cambridge University Press, 336 pp.

Ryabinin, V.E. (1991) : Semi-Lagrangian algorithms for discrete spectral sea wave models. Soviet Meteorology and Hydrology, 8, 56-64.

Shen, M.C. and J.B. Keller(1975): Uniform ray theory of surface, internal and acoustic wave propagation in a rotating ocean or atmosphere. SIAM J. Appl. Math. 28(4), 857-875.

Tolman H.L. (1989): The numerical model WAVEWATTCH: a third generation model for the hindcasting of wind waves on tides in shelf seas. Communications on Hydraulic and Geotechnical Engineering. Delft University of Technology, Rep. no. 89-2, 72 pp.

Tolman H.L. (1990): The influence of unsteady depths and currents of tides on wind-wave propagation in shelf seas. J. Phys. Oceano., 20, 1166-1174.

Tolman H.L. (1991): The third-generation model for wind waves on slowly varying, unsteady, and inhomogeneous depths and currents. J. Phys. Oceano., 21(6), 782-797.

WAVE RUN-UP AND REFLECTION ON A PERMEABLE SEA WALL

by

Mitsuo Takezawa[1], Susumu Kubota[2], and Shintaro Hotta[1]

ABSTRACT

The swash oscillation on a sloping seawall made of step-shaped concrete blocks, waves and water particle velocity in front of the seawall were measured. It was inferred that incident waves form two-dimensional standing waves with the anti-node in the swash slope. Results of wave separation indicated the individual wave reflection predominantly occurred. Wave reflection ratio from the seawall was similar to the value of natural beach which had a steep beach face. This may be caused by an effect of permeability of the seawall.

INTRODUCTION

The construction of sloping seawalls consisting of precast concrete blocks, where permeability or roughness is increased to dissipate wave energy, have recently become popular in Japan. The main purposes of this type of construction are to reduce the wave run-up height and to prevent local scour occurring at the foot of the seawall. Furthermore, wave-induced sea bottom topographical change in front of the seawall is reduced by the decreased rate of reflection. However, the properties of swash oscillation and reflection on the permeable slope of a seawall are not

[1]Dr.Eng., Professor, Dept. of Civil Engrg., College of Science and Technology, Nihon University.

[2]Dr.Eng., Assistant Professor, Dept. of Civil Engrg., College of Science and Technology, Nihon University.
Kanda-surugadai 1-8-14, Chiyoda-Ku, Tokyo 101 Japan.

fully understood, although limited measurements from laboratory experiments and field observations have been reported. In order to obtain a data record long enough in time to determine the statistical quantities of waves on seawalls and to predict the rate of reflection from the seawall, field measurements were carried out on a sloping seawall, constructed by precast concrete blocks, using electrical and photographic measurement techniques. The purpose of this paper is to describe the measurement system, characteristics of swash oscillation on the slope, and reflection rate from the seawall.

FIELD OBSERVATIONS

The field observation was carried out on May 6, 1989, at a beach facing the Pacific Ocean, located at the northern end of the Kujukuri Coast, at a distance of about 80 Km from Tokyo (Fig. 1). Hereafter this field observation will be abbreviated as KU89. At this beach, a sloping seawall with a gradient of 1/3 near the mean sea water level was covered by step-shaped concrete blocks called "Cross Akmon Flat Type" shown in Photo. 1 and Fig. 2.

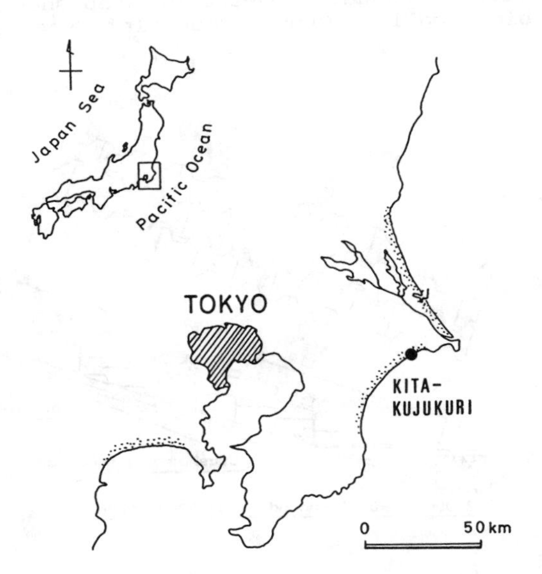

Fig. 1 Location map of field site.

Photo. 1 Sloping seawall covered by step-shaped concrete block called "Cross Akmon Flat Type".

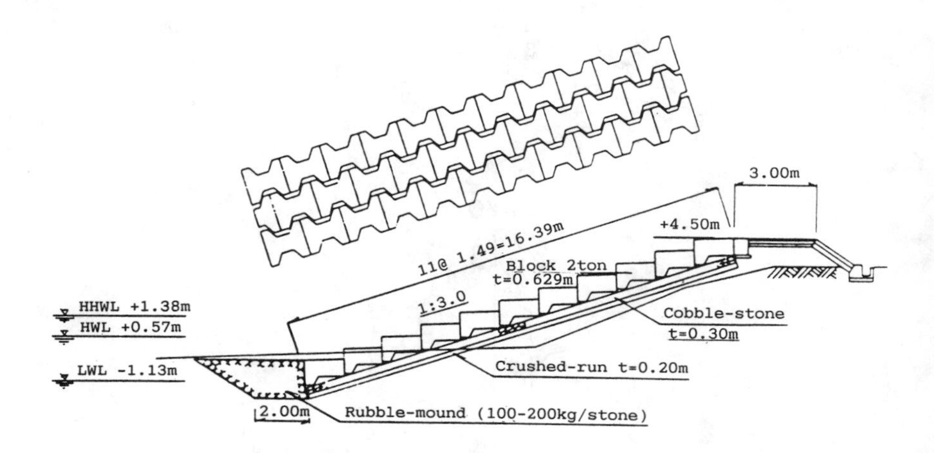

Fig. 2 Typical section of the sloping seawall.

Figure 3 shows the sea bottom topography, and the instrumentation system at the field observation site. A target marker stick array for photographing the wave run-up was prepared on the slope normal to the seawall (see Fig. 4). A capacitance-type swash gage (CSG) was also stretched along the stick array. Along the seaward extension of the stick array, six poles (St.1 to St.6) were installed to be used for wave measurements (Photo. 2). Poles at St.3 and St.4 were equipped with an electromagnetic current meter (EMCM) and a capacitance-type wave gage (CWG)(Photo.3). A set consisting of a couple of 16 mm memo-motion cameras, which can be used alternatively for long time period photographing, was used to shot waves on the slope from the side. Two sets of cameras were employed in photographing the poles in the sea (Photo. 4).

Data collection was started on 16:50:00 and ended 17:40:40, giving a about 50 min experimental duration. The tide was full in this time zone. The wave condition was swell and the angle of wave direction from normal to the shoreline was small. Average breaker line was located near the St.3 on a bar. The type of breaker was plunging.

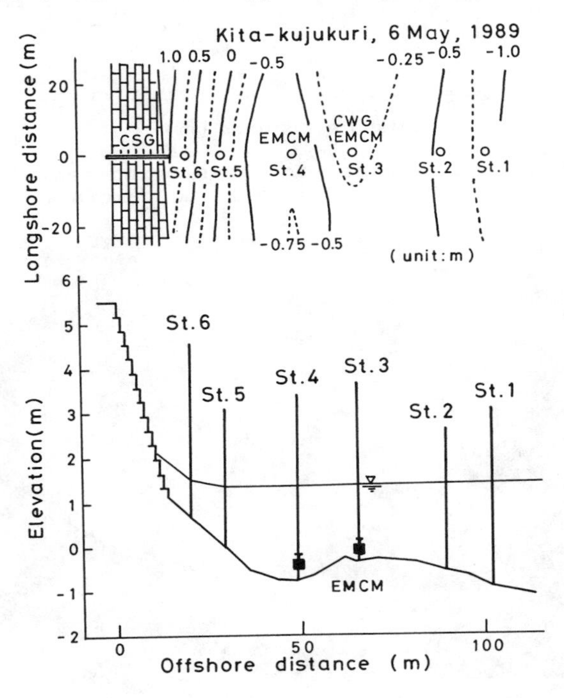

Fig. 3 Beach profile, plan, and arrangement of instruments.

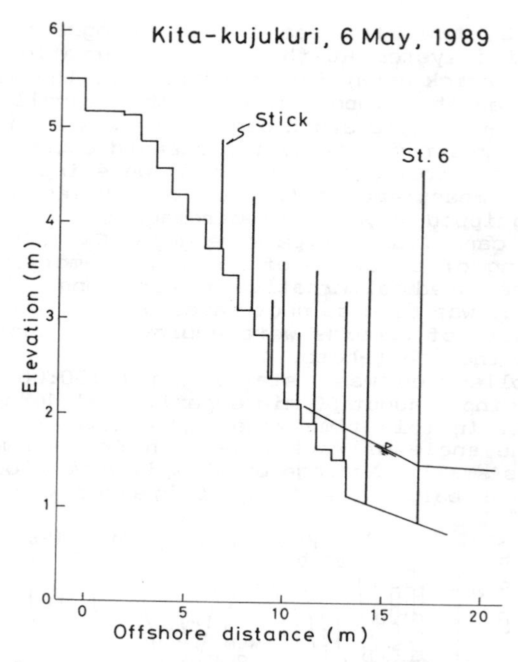

Fig. 4 Stick array on the sloping seawall for photographing by 16 mm cameras.

Photo. 2 Target poles and wave condition before measurement.

Photo. 3 Capacitance type wave gage.
(electromagnetic current meter is installed at the bottom near the wave gage.)

Photo. 4 Two couple of 16 mm memo-motion cameras for photographing the sea surface variations at the target poles.

Output of the electrical instrument was recorded on an open-reel digital data recorder. The data sampling interval was 0.2 s.

Sand in front of the seawall was collected. Sieve analysis indicated the sand was well-sorted with a median diameter of 0.103 mm.

RESULTS AND DISCUSSION

1)Raw data

Figure 5 shows a portion of sea water surface variations (η) from St.2 through St.4 and the swash oscillation. Water particle velocities (u and v) at St.3 and St.4 are also shown in Fig. 5. u and v are on-offshore and longshore component of water particle velocity. The sea water surface variation at St.3 were measured by CWG, and other water surface records were from 16 mm films. It is seen that η and u at the same point are in phase. And values of v are smaller than that of u. It suggests on-offshore two-dimensional phenomenon was predominant. Swash oscillations were of the vertical distance converted from measured horizontal distance using averaged slope. Swash profile has a step-type character corresponding to the slope of the seawall.

Figure 6 shows the on-offshore distribution of significant wave height and period. In this figure, the values of the incident and reflected wave separated by the method based on quasi-nonlinear long-wave theory are also plotted. Propagating from St.2 to St.6, waves did not change in height and period for significant value. These value at the swash zone increased significantly.

Figure 7 shows power spectral density functions. The main power at St.2 lies in the range from 0.09 Hz to 0.15 Hz. This power decreases considerably from St.2 to St.3. This change is due to the wave breaking. Power in the range lower than about 0.9 Hz increases greatly in the surf zone and swash zone. At St.3 and St.4 located in the surf zone, peaks and deep depression in the power density function alternately appear. These frequencies are 0.05 Hz and 0.07 Hz. From these figures, existence of two-dimensional standing waves with the anti-node in the swash slope is suggested. That is, the two-dimensional standing wave with the frequency of 0.07 Hz has anti-node at St.3 and node at St.4. Another standing wave with the frequency of 0.05 Hz has anti-node at St.4 and node at St.3.

Figure 8 shows the cross spectral density function between the sea surface elevation and the on-offshore component of water particle velocity at St.3. An abrupt fall in coherence and a sudden change in phase function at corresponding frequencies are also characteristic features of two-dimensional standing waves (Hotta et al., 1981).

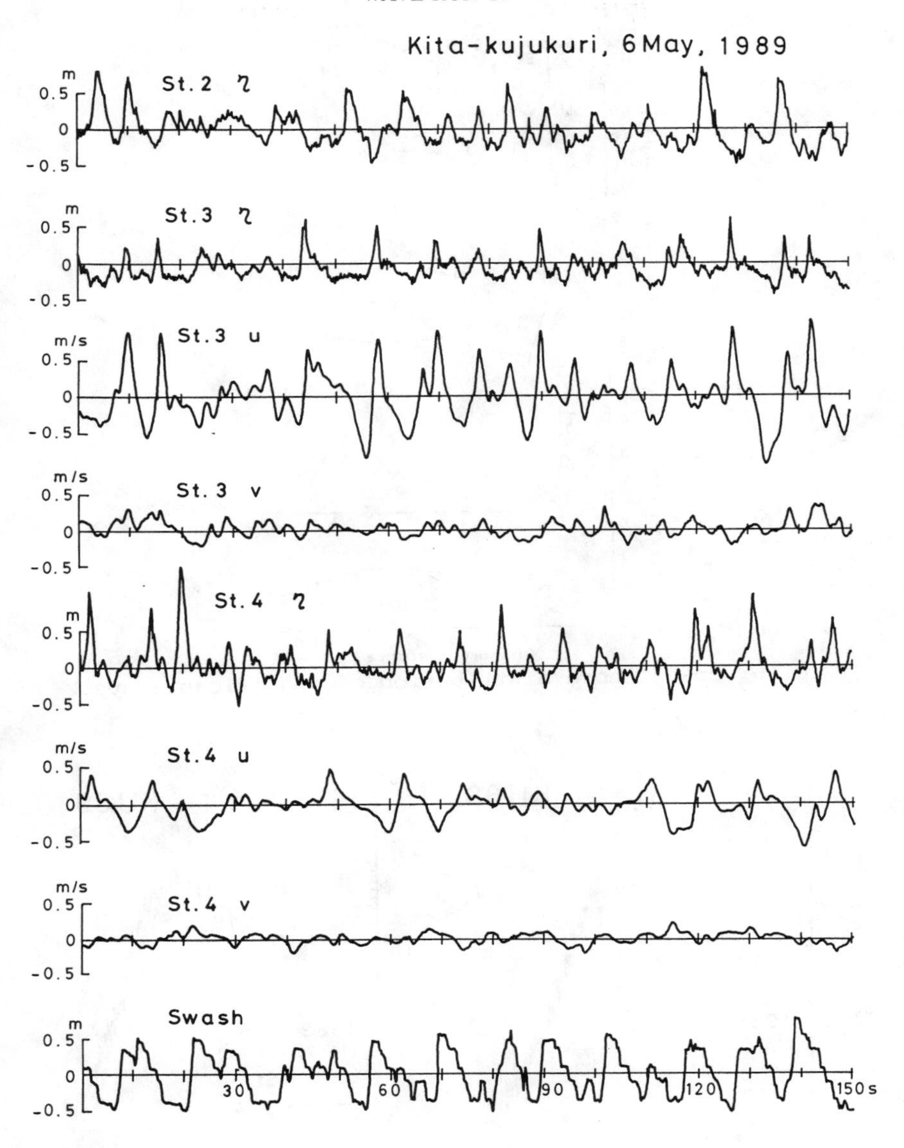

Fig. 5 Example of raw data.

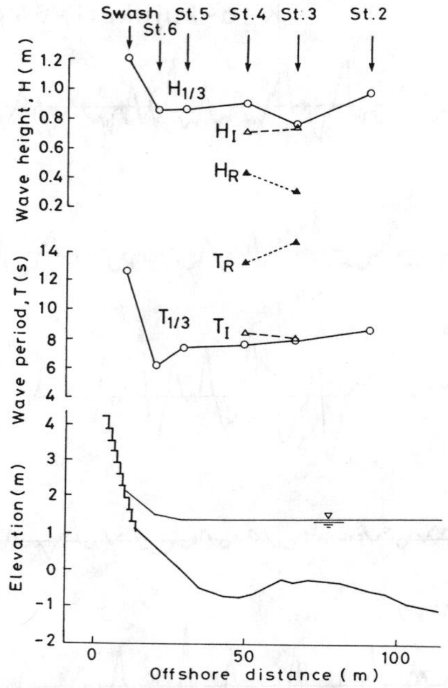

Fig. 6 On-offshore distribution of significant wave.

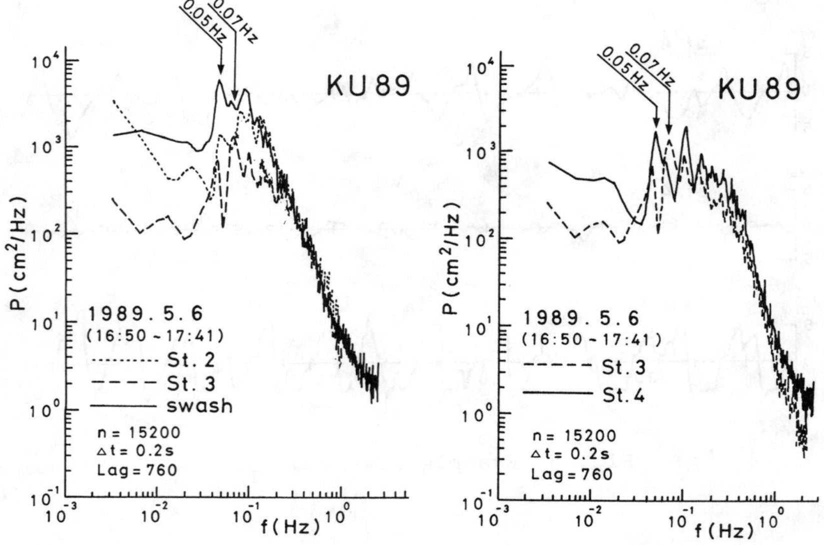

Fig. 7 Power spectral density functions.

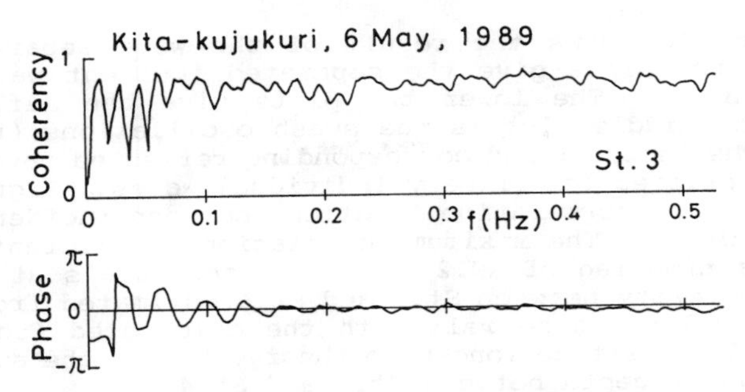

Fig. 8 Cross spectra between surface elevation and on-offshore component of water particle velocity at St.4.

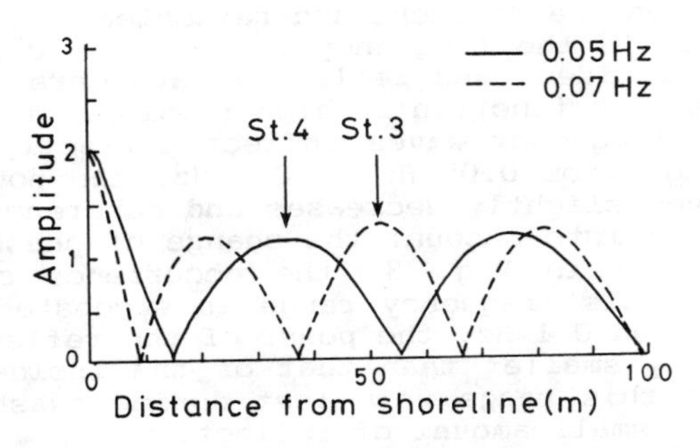

Fig. 9 On-offshore distribution of standing wave amplitude.

Figure 9 shows the on-offshore distribution of standing wave amplitude with frequencies of 0.07 Hz and 0.05 Hz, calculated by the method based on linear long two-dimensional standing wave theory (Mizuguchi, 1984). Positions of the anti-node and the node agree well with the observed data.

2) Reflected wave

In above section, indications that waves in the surf zone form two-dimensional standing waves, during this experiment were given. Separation of the incident and reflected waves was carried out using the method based on the quasi-nonlinear long-wave theory (Kubota et al., 1990).

Figure 10 shows the result of the wave separation. The upper two plots give the separated incident waves at St.3 and St.4. The lower two plots give the reflected waves. The middle plot is the swash oscillations (run-up waves). The incident and corresponding reflected waves are easily identified for primary individual waves. Figure 11 shows the cross-correlation functions between incident and reflected waves. The maximum correlation coefficients are found at a time lag of 20.2 s at St.4 and 26.6 s at St.3. The wave celerity between St.3 and St.4 estimated from the above two values agree well with the calculated from the linear small amplitude long-wave theory, taking the average value of water depth between St.3 and St.4.

Figure 12 shows the power spectral density function of incident and reflected waves at St.3 and St.4, and the swash oscillation. Figure 13 shows the cross spectral function between the incident and reflected waves. At the both stations, in the frequency range below 0.04 Hz, the power of the incident and reflected waves are almost the same and coherence function is higher value. This implies that the low frequency waves reflect perfectly. In the frequency range from 0.05 Hz to 0.1 Hz, the power of the reflected waves slightly decreases and coherence function is 1.0. Taking into account the change of phase function between η and u in Fig. 8, the occurrence of partial reflection in this frequency range is suggested. In the range higher than 0.1 Hz, the power of the reflected waves is considerably smaller than that of the incident waves. The energy in this range were lost during swash process, resulting in a small amount of reflection.

Figure 14 shows the joint distribution of wave height and period for the incident and reflected waves and the swash oscillations by the zero-up and zero-down crossing methods. Tendency that wave distribution of the surf zone has two maxima (Hotta et al., 1982) is not clearly seen. The wave height and period at the swash zone are larger than those of the incident wave at St.4. However, the reflected wave heights at St.4 and St.3 are smaller than that of swash oscillation though the reflected wave period is the same as swash period. The number of defined waves somewhat decrease in the swash oscillation fields compared to the incident wave field. The number of reflected waves is similar to that of swash oscillations.

The ratio of the square root of the mean sea surface variation of the reflected and incident waves, which implies an average reflection coefficient, were 0.48 at St.3 and 0.59 at St.4. These values approximately agrees with those obtained on a natural beach in which incident waves were swell and the beach face slope is steep (Takezawa et al., 1988).

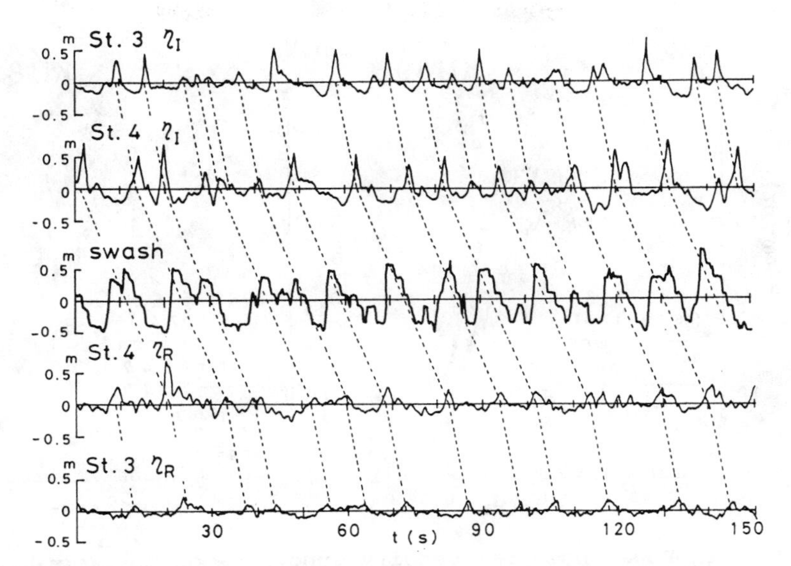

Fig.10 Example of the resolved incident and reflected waves.

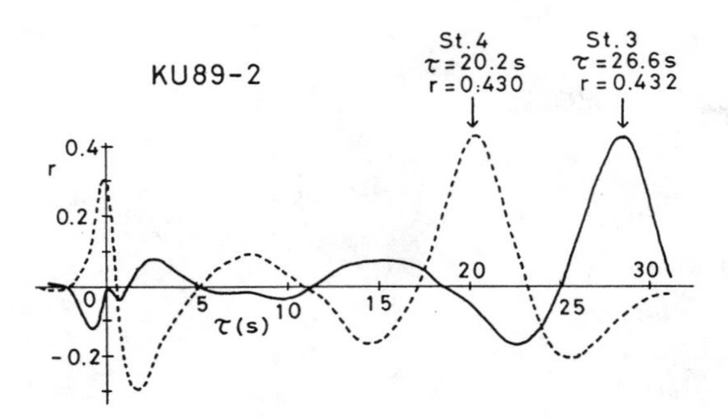

Fig.11 Cross-spectral function between the incident and reflected waves.

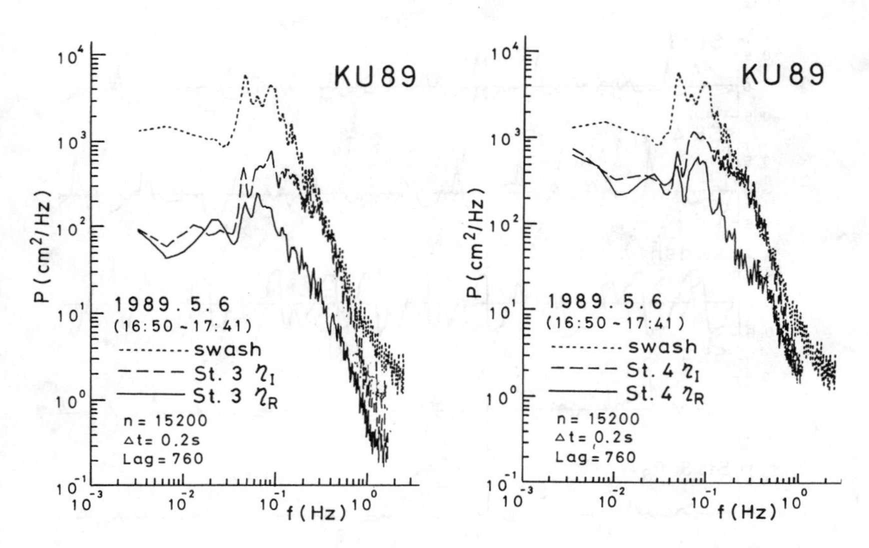

Fig.12 Power spectral density functions of the incident and reflected waves and swash oscillations.

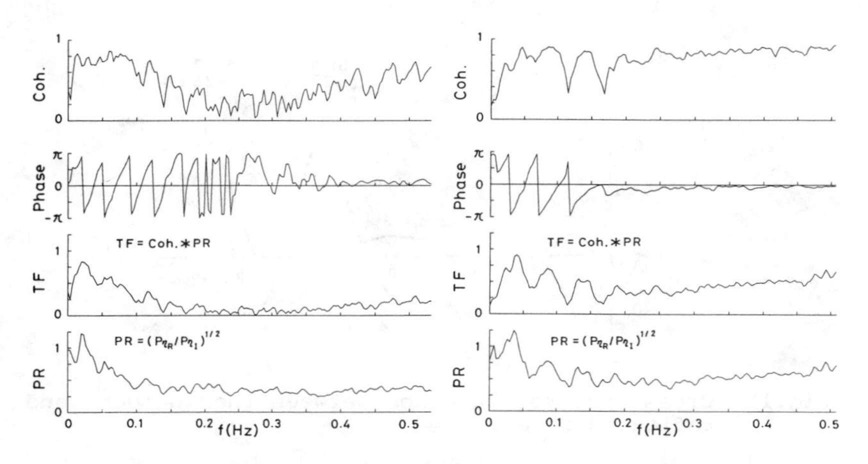

Fig.13 Cross-correlation function between incident and reflected waves.

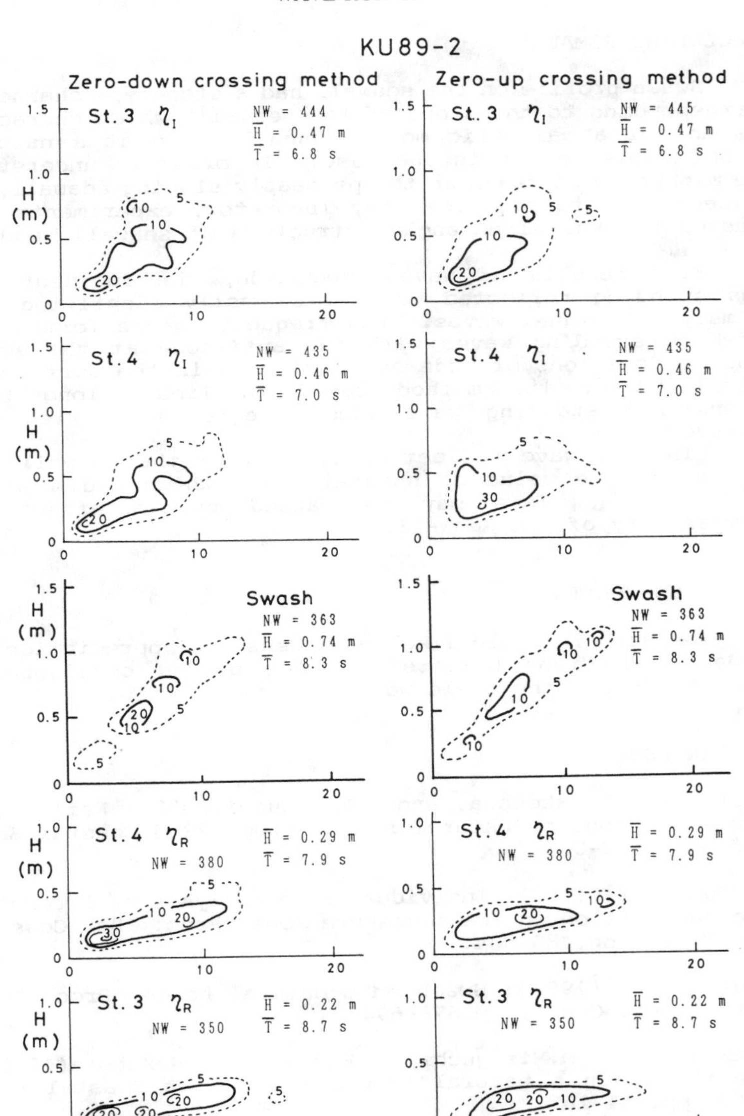

Fig.14 Joint distribution of wave height and period for the incident and reflected waves and swash oscillations.

CONCLUDING REMARKS

Swash profile on the seawall had a step-type character corresponding to the slope of the seawall, which character was not of a parabolic motion usually seen at a natural beach. This result is one case. In order to understand the mechanism of swash on the permeably sloping seawall, it is necessary to practice many laboratory experiments for kind of blocks, slope angle, structure of seawall, and so on.

From results of wave separation, the incident and corresponding reflected waves are easily identified for primary individual waves. Low frequency waves formed on-offshore standing waves with the anti-node at the swash slope. Positions of node and anti-node in the surf zone, calculated by the method based on linear long two-dimensional standing wave theory, agree well with the observed data.

Finally, wave reflection ratio from the seawall was similar to the value of natural beach which had a steep beach face. This may be caused by an effect of permeability of the seawall.

ACKNOWLEDGEMENTS

The authors would like to express our appreciation to students of Nihon University who provided considerable supports during the field work.

REFERENCES

Kubota, S., M.Takezawa, and M.Mizuguchi(1990): Reflection from swash zone on natural beach, Proc. 22nd Coastal Eng. Conf., pp.570-583.

Mizuguchi, M.(1982): Individual wave analysis of irregular wave deformation in the nearshore zone, Proc. 18th Coastal Eng. Conf., pp.485-504.

Mizuguchi, M.(1984): Swash on a natural beach, Proc. 19th Coastal Eng. Conf., pp.678-694.

Takezawa, M., M.Mizuguchi, S.Hotta, and S.Kubota(1988): Wave run-up on a natural beach, Proc. 21st Coastal Eng. Conf., pp.151-165.

An Offshore Island Wave Sheltering Model

Andrew L. Kadib [1], Member ASCE

Abstract

The wave climate along Southern California Coastal area is greatly modified by the existing systems of natural islands shown in **figure 1.** As deep water waves approach the nearshore area, they generally experience changes in height and direction resulting from refraction and shoaling. Wave heights are further modified by the sheltering effect of these islands.

This paper presents the development, verification and application of a model which quantifies the wave height reduction as caused by the offshore islands. An Island Wave Sheltering Coefficient (K_{is}) was developed. The Coefficient (K_{is}) was found to be a function of a dimensionless number describing the wave period, island geometry and deep water wave direction.

---

(1) Chief, North Coast Section, Coastal Resources Branch, Planning Division, U. S. Army Corps of Engineers, Los Angeles District, P.O. Box 2711, Los Angeles, California 90053-2325

Introduction

When waves arrive at an offshore island, they undergo some degrees of reflection, dissipation and diffraction. Generally speaking, waves may be considered blocked by the islands geometric shadow which extends from the wave source to the shoreline. In real situations, as waves approach an isolated offshore island, they will diffract around both ends of the island, and their height will change from full incident magnitude, just outside the geometric shadow line, to small proportions next to the leeside of the island. As the diffracted waves proceed along their orthogonals (which are radial to the tip of the island), the wave height will experience additional changes due to energy spreading. These orthogonals, infinite in number, are straight when no refraction is taking place.

The subject of island sheltering has attracted the interest of many coastal engineers and scientists (Hsao et al. 1980, Pawka and Guza, 1983, Goda, 1985 and O'Reiley, 1988). Of particular interest, is the method used by Goda (1985), in which a spreading parameter as a function of wave steepness and period was developed. The effect of island geometry and dimension were not considered in his analysis. Another interesting study has been conducted since 1988 by Guza and O'Reilly to develop a numerical model, using traditional wave refraction methods and available wave data from hindcasts and measurements to determine island sheltering effects. A major objective of the ongoing study by Guza and O'Reilly is to develop such a model and verify its accuracy through deployment of additional instrumentation.

The need for a practical model capable of reasonably quantifying the wave field in the leeside of Southern California offshore islands is vital to the understanding of the existing complex nearshore wave climate. This paper has recognized this fact and the need to develop an engineering solution to this problem.

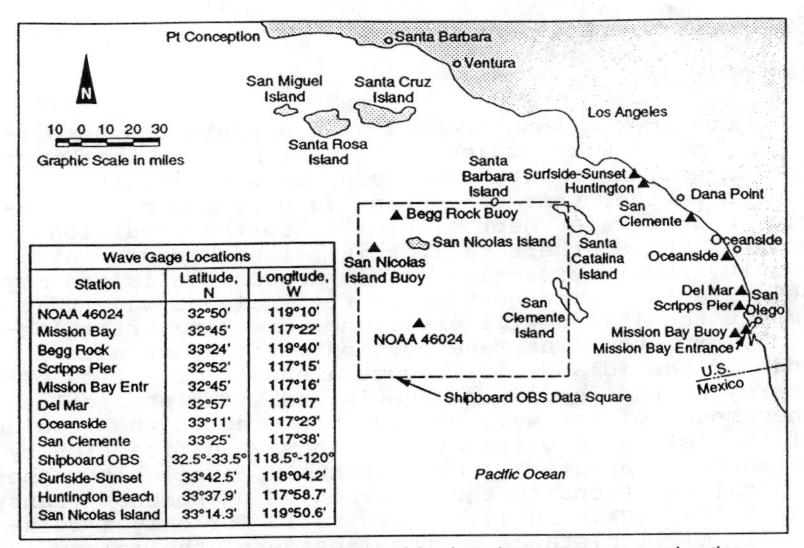

Wave Gage Locations

Station	Latitude, N	Longitude, W
NOAA 46024	32°50'	119°10'
Mission Bay	32°45'	117°22'
Begg Rock	33°24'	119°40'
Scripps Pier	32°52'	117°15'
Mission Bay Entr	32°45'	117°16'
Del Mar	32°57'	117°17'
Oceanside	33°11'	117°23'
San Clemente	33°25'	117°38'
Shipboard OBS	32.5°-33.5°	118.5°-120°
Surfside-Sunset	33°42.5'	118°04.2'
Huntington Beach	33°37.9'	117°58.7'
San Nicolas Island	33°14.3'	119°50.6'

Figure 1. Southern California offshore islands and wave measurement locations.

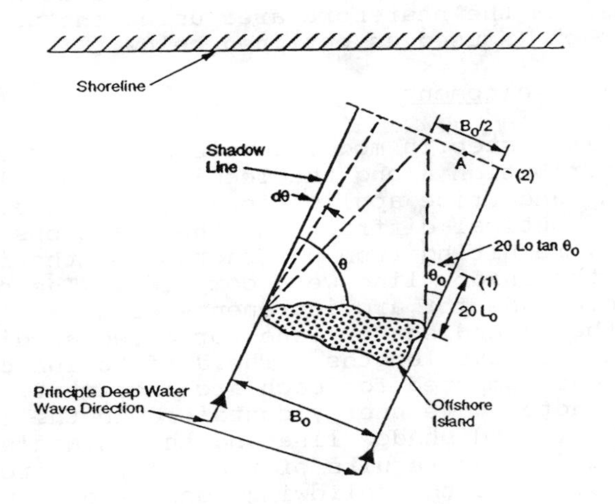

Figure 2. Wave spreading along the leeside of an offshore island.

Method Development

(a) Concept

Consider a long-crested wave approaching an offshore island with projected width B_o, perpendicular to the direction of wave advance as show in **figure 2**. Assume further that the island is surrounded by relatively deep water depths, similar to the conditions at most of the Southern California island system. As a result, local refraction effect around the island may be neglected. The portion of the incident wave that impact the island will experience breaking or reflection, or both. The wave portion moving past either ends of the island will become a source of a flow of energy to the leeside of the island. The physical phenomenon of the wave energy spreading in the leeside of the island is quite similar to wave diffraction by offshore breakwaters. Unfortunately, available mathematical solutions to the diffraction phenomenon (Penny and Price, 1944 and 1952), are valid for only a relatively small distance to the leeside of the island (approximately 10 to 20 times the wave length). Most of the Southern California islands are located at distances in excess of 20 miles from the shoreline. A method of transferring wave energy from valid limits of the diffraction effect near the island to some locations close to the nearshore area using the wave energy flux approach is used as outlined below.

(b) Model Development

The first step in model development considered the energy distribution along the leeside of an island. Using Penny and Price applications of the Somerfeld solution of optical diffraction, the fractions of the wave energy resulting from diffraction of the incident wave from the shadow line were computed. The computations were made for arc increments of $d0 = 5°$, starting from the island shadow line for a redial distance, R, equal to 20 wave lengths. The diffraction coefficient, K, was computed for each arc increment. The total diffracted wave energy contained in the region between the island shadow line and the island's leeside was estimated. The results of the analysis are shown in **figure 3**, where the following conclusions and assumptions are made:

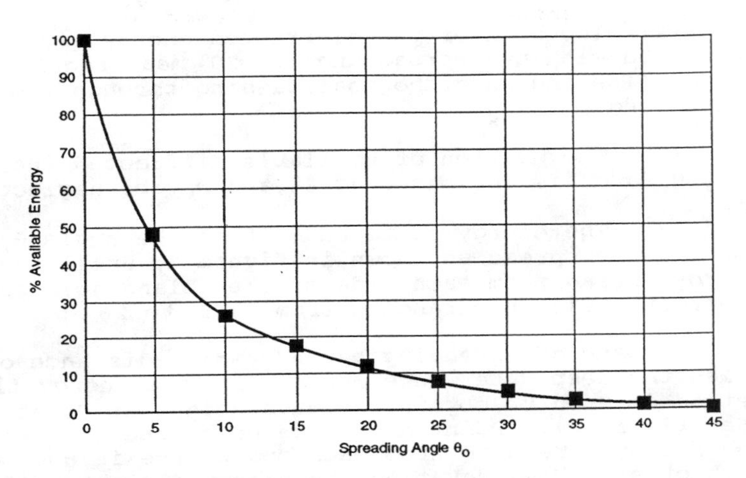

Figure 3. Percent energy available for spreading vs arc angle from island shadow line.

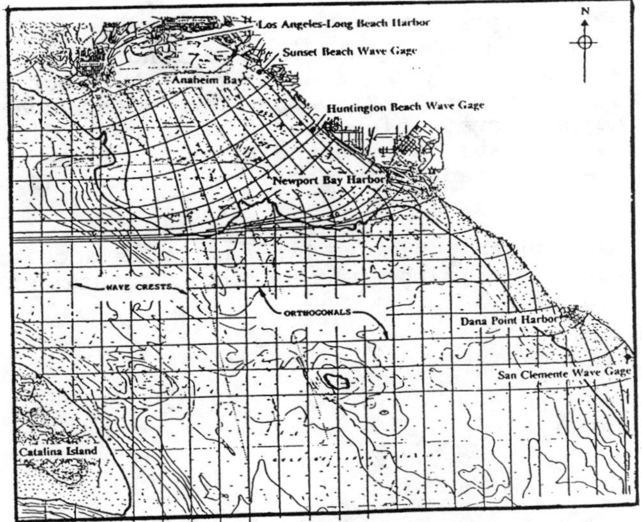

Figure 4 Refraction Diagram with direction of 270 degrees @ T = 15 seconds

(1) The bulk of the diffracted energy (90%) is contained within a spread angle θ=20° measured from the shadow line and an orthogonal passing through the island end.

(2) The portion of available diffracted energy beyond θ_o=25° is estimated at 8% and may be neglected.

(3) The energy contained within the angle θ_o will continue to spread as shown in **figure 2**, until the energy spread from each side of the island join each other at a distance ranging from $1.5B_o$ to $2B_o$.

For energy spreading beyond radial distance of 20 times the deep water wave length, L_o, the energy flux method is used to obtain the wave height at Point A (See **figure 2**). Point A is located at the zone where the energy spreading from each end of the island meet each other. Now, equating the energy flux at sections (1) and (2) of **figure 2**, the following equation is obtained:

$$20L_o \tan\theta_o K^2 H_o^2 = \frac{B_o}{2} H_A^2 \qquad (1)$$

where; L_o is the deep water wave Length, θ_o is the angle of wave spreading, K is an average diffraction coefficient at section 1, H_o is the deep water wave height, B_o is the projected island length along the direction of deep water wave advance and H_A is the wave height at point A, unaffected by local refraction and shoaling.

Now substituting L_o by $\frac{gT^2}{2\pi}$ (g is acceleration of gravity and T is the wave period), equation (1) can be written as:

$$\frac{H_A}{H_o} = C\left[\frac{gT^2}{B_o}\right]^{1/2} = K_{is} \qquad (2)$$

where C is a factor depending on the wave spreading angle θ_o and the wave diffraction coefficient at section 1, K_{is} is defined as an island sheltering, coefficient and gT^2/B_o is named Island Wave Number (IWN).

Table 1 gives the computed values of C using K = 0.60.

Table 1. Variation of C and θ_o

Spreading Angle 0_o (degrees)	C
15	0.814
20	0.940
22.5	1.00

The C value is considered in this study to be a site specific parameter ranging from 0.814-1.00, and was determined to have a value of 1.00 for the Santa Catalina and San Clemente Islands. This corresponds to a spreading angle, θ_o, of 22.5° and K = 0.60 as obtained from the model verification process outlined in the following section.

Model Verification

The validity of equation is examined in this section. Proper verification of equation (2) utilized actual field data, where simultaneous deepwater and nearshore wave measurements are available. The Southern California Coastal Area shown in **figure 1**, presents an ideal case for such data. Wave gages installed by the U. S. Army Corps of Engineers and the State of California Department of Boating and Waterways, and maintained, and operated by the Scripps Institute of Oceanography (SIO), include Begg Rock Buoy, San Nicholas Island Buoy, and San Clemente, Huntington Beach and Surfside Sunset wave gages.

Table 2. Model Verification Using Sun Set & San Clemente Gages

date	gage	measured offshore wav H(mt)	dir.	T(sec)	Hg(mt) actual	Kis actual	IWN gT^2/B_0	Kis(model) $(\theta)_0=15$	$(\theta)_0=20$	$(\theta)_0=22.5$
F13,83	ss	5.15	225	17	1.58	0.30	0.08	0.25	0.29	0.31
J17,88	ss	10.12	270	15	2.84	0.30	0.11	0.27	0.31	0.33
J17,88	sc	10.12	270	15	3.5	0.37	0.11	0.27	0.31	0.33
J27,83	ss	7.258	285	20	3.4	0.47	0.18	0.36	0.42	0.45
M13,92	sc	2.69	271	15	0.7	0.29	0.10	0.26	0.3	0.32
M18,92	sc	2.77	273	13	0.72	0.26	0.08	0.25	0.29	0.31
M28,92	sc	2.39	280	15	0.64	0.26	0.11	0.27	0.31	0.33

ss Surfside gage data
sc Sanclemente gage data
Offshore wave data from Begg Rock & San Nicholis gages

The location of these gages is shown in **figure 1**. These data are collected and summarized each month in reports available from SIO. This wave measuring network is also referred to as the Coastal Data Information Program (CDIP). Field wave data were utilized to verify the accuracy of developed model using the following steps:

1. A coastal reach extending from Dana Point Harbor to the San Gabriel River (South of Los Angeles - Long Beach Harbors) was selected for the analysis. **Figure 4** shows the three nearshore wave gages which are located along the nearshore area of Orange County, California. These gages were deployed in water depths of approximately 10 meters (30 feet). **Figure 4** also presents an example of wave refraction along the nearshore area of Orange County.

2. The wave height, H_g, at any of these location can be expressed by the equation:

$$H_{gage}=H_o K_s K_r K_{is} \tag{3}$$

where K_s and K_r are the shoaling and refraction coefficient respectively, at the gage location, and the other terms as previously defined.

3. Both the refraction and shoaling coefficient were completed for a combination of deepwater direction and periods.

4. The CDIP wave data were searched to select the recorded major storms and deep water waves in excess of 2 meters (6 feet) in height.

5. These data were summarized as shown in Table 2, and the corresponding actual K_{is} values were calculated using equation (3) i.e,

$$K_{is}=\frac{H_{gage}}{H_o K_s K_r} \tag{4}$$

where H_{gage} is the recorded nearshore gage wave height and H_o is the corresponding and measured deep water wave height (Begg Rock or San Nicholis).

6. The corresponding values of Island Wave Number, were computed for both islands using the applicable wave period, wave direction and island projected length. These values are also shown in Table 2.

7. The model K_{is} values were computed using equation 2 for θ_o = 15°, 20°, and 22.5°. These correspond to C values of 0.814, 0.94, and 1.0, respectively.

8. By examining the data presented in Table 2, it is concluded that the model predicts the Island Sheltering Coefficient K_{is} with a high degree of accuracy for 0_o = 22.5° and C = 1.00.

Model Application

Figure 5-a presents the main model results which are readly applicable to an offshore island with projected width, B_o, ranging from 4 to 20 kilometers and for wave periods ranging from 8 to 20 seconds. The curves shown in **figure 5-a** were computed using equation (2) with C value = 1.0.

Figure 5-b gives a schematic presentation for B_o and shows the limits of the coastal reach affected by the island sheltering. Application of the results shown in **figure 5-a** to estimate the K_{is} value may follow the following steps:

1. Find the B_o value corresponding to the desired deepwater wave direction.

2. The method is applicable along a Coastal area bounded by the shoreline, the island shadow Line and a line located at least 1.5 B_o shoreward of the island (the hatched area of **Figure 5-b**).

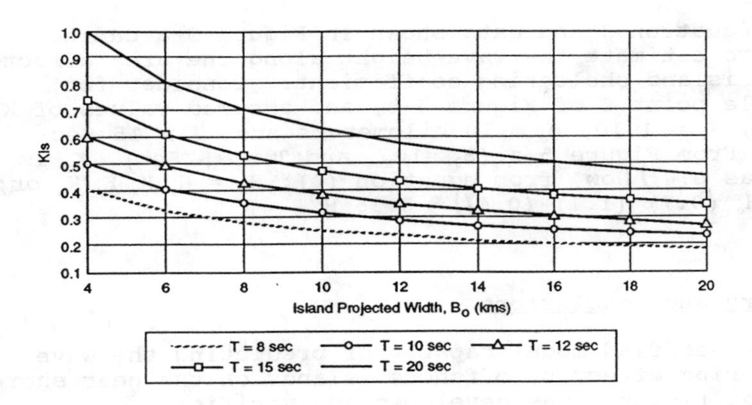

Figure 5-a. The island sheltering coefficient, Kis.

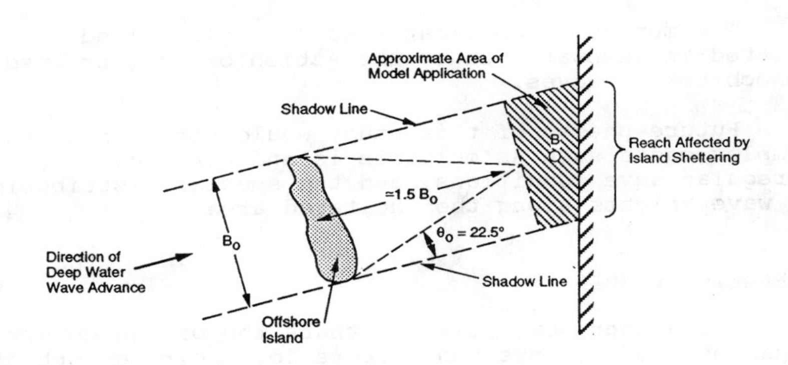

Figure 5-b. Schematic example of model application.

3. Equation 3 and data shown in Figure 5-a can be used to estimate the wave height along the area affected by island sheltering coefficient. Consider for example point B of Figure 5-b, and assumed values of K_r = 0.7, K_s = 1.10, B_o = 10 kilometers and, T = 15 Seconds. From Figure 5-a (B_0 =10_{Kms} and T = 15 Sec), K_{is} is read as 0.47) Now, from equation (3); $H_B = H_o K_s K_r K_{is}$ or; $H_B = H_o$ (0.7) (1.1) (0.47) = 0.36 H_o.

Summary and Conclusions

1. A verified model capable of predicting the wave sheltering effect of offshore islands on the near shore coastal regions, was developed and verified.

2. The Island effect on wave height reduction was found to be a fuction of gT^2/B_o which is a dimensionless number (Island Wave Number) describing the wave period T, and the projected Island length along the direction of deep water wave advances.

3. The method is applicable to a single island located in deepwater under the action of long crested monochromatic waves.

4. Future phases of this study would consider more complex cases such as multiple island arrangements, irregular wave conditions; and the special distribution of wave heights along the sheltered area.

Acknowledgements

The author would like to thank the Orange County Department of Beaches and harbors for their support of the study. Mr. Robert Atkins, Ms. Elenaor Encinas and Ms. Domitila Sanchez of the United States Army Corps of Engineers, Los Angeles, District assisted in the preparation of the refraction diagrams and typing the manuscript.

References

Arthur, R.S., 1951, "The Effect of Islands on Surface Waves," Bulletin of the Scripps Institution of Oceanography, University of California, La Jolla, CA, Vol 6, No. 1, pp 1-26.

Goda, Y., 1985, "Random Seas and Design of Martime Structures" University of Tokyo Press.

Guza, R.T. and O'Reiley, W. C., 1988, "Southern California Waves: Model Verification and Utilization," Proposal to California Sea Grant Program.

Hsiao, et al, 1980 "An Investigation of Wave Sheltering by Islands" proceedings of the 17th Coastal Engineering Conference, ASCE, pp 840-849.

Penny, W. G. and A. T. Price, "Diffraction of Sea Waves by Breakwaters", Directorate of Miscellanous Weapons Development, Technical History No. 26, Artificial Harbors Section 3D, 1944.

Penny, W. G. and A. T. Price, "The Diffraction Theory of Sea Waves by Breakwaters and Shelter afforded by Breakwaters", Phil Trans., Roy Soc. (London), Ser. A, 244 (March 1952), 236-53.

EVOLUTIONARY FOURIER ANALYSIS OF WAVE DATA

Leon E. Borgman[1], Mihail Petrakos[2], and Chao Li[3]

ABSTRACT: A new method is presented for analyzing nonstationary water level height time series. The usual assumption of pseudo-stationary intervals is avoided and the wave spectra can be changing very rapidly. The time series, $\eta(t)$, is represented in a finite Fourier series, summed on m, for which the Fourier coefficients, $A(m,t)$, are themselves functions of time. The technique combines methods from finite element theory with Fourier analysis by representing the $A(m,t)$ as linear combinations of basis function over each of a sequence of finite intervals (or elements). The multipliers of the basis functions depend on m only, and not t. The method is applied here primarily to the estimation of frequency spectra, although some discussion is given for the use of the procedure in the estimation of directional wave spectra.

INTRODUCTION

Several techniques have been proposed for estimating spectra in nonstationary time series. These include evolutionary spectral estimation (Priestley, 1981, 1988), complex demodulation (Bloomfield, 1976) and wavelet analysis (Daubechies, 1992). These are all based on techniques closely related to a sequence of narrow-band filters, each centered at a selected frequency. A quite different procedure was developed by Brown (1967) based on a multiplier of a stationary spectra which is a linear function of time, but has unknown

1) Statistics Department and Department of Geology and Geophysics, University of Wyoming, Laramie, WY
2) Statistics Department, University of Wyoming, Laramie, WY
3) Mathematics Department, Teikyo Loretto Heights University, Denver, CO

coefficients. The coefficients are computed from an estimate of the time-varying variance.

The present study was directed at the ultimate goal of being able to estimate the directional wave spectral density over space, time, and frequency as it changes so fast that it is not possible to work with conventional 18 minute pseudo-stationary time intervals. It was desired that it be possible to follow the time-varying spectra second-by-second, or minute-by-minute. This should allow the study of a hurricane moving along the coast or the rapid build up of waves as a squall passes. It should also allow an investigation of the detailed behavior of wave spectra in a wave tank as the generators begin creation of waves, the waves reach a stationary interval of behavior, and finally reflections develop to the point where the wave train degenerates too far from the target spectra.

Most methods for estimating the directional wave spectra take the Fourier coefficients in time as the initial data from which to proceed. Therefore, this study ultimately was concerned with the estimation of time-varying Fourier coefficients, rather than the wave spectra, itself. The spectra can be computed from the coefficients, of course, but the coefficients provide much more information than does the spectra by itself. A detailed study was made of evolutionary spectral analysis and complex demodulation. However, although the methods are most elegant for their purpose, they did not seem to be suitable for the goals of this study. Consequently, a completely different approach was taken.

A technique much more in the spirit of that developed by Brown was discovered. However, contrary to the methods of Brown, this new technique, called here evolutionary Fourier analysis, allows different time variation at each frequency and polynomial time variation of any order. The technique combines procedures from finite element methods with finite Fourier transform relations. The real and imaginary parts of the complex fourier coefficient are represented separately piecewise with polynomials in time.

THEORY

Let the water level elevations, $\eta(t)$, be represented in terms of the time-varying complex-valued Fourier coefficients, $A(m,t)$, by the equation

$$\eta(t) = \sum_{m=0}^{N-1} A(m,t)\, e^{i2\pi mt/T} \qquad (1)$$

The square root of -1 is denoted here by i. The parameter, T, is the period for the first harmonic in the Fourier series. It will be assumed that A(N-m, t) is the complex-conjugate of A(m,t) for all t. This guarantees that $\eta(t)$ will be real valued. For the statistical properties of $\eta(t)$, it will be a model assumption that the A(m,t) for different m, all less than or equal to N/2, are independent of each other, and the real and imaginary part of A(m,t) are independent. This is consistent with the linear statistical theory of waves for the stationalry case (Borgman, 1990, Appendix) and should hold in the limit in the present formulation as the dependence on time of A(m,t) goes to zero.

The total interval of data to be analyzed will be divided up into S segments of equal time length. Each of the S segments will be further divided into J intervals, each of length, T. Thus, the segments will have length JT and the total data interval will have length JST. The basic analysis procedure will represent the time variation of A(m,t) over a segment as a linear combination of basis functions, where the multipliers of the basis functions depend on m only, and not t. That is, on a given segment

$$A(m,t) = \sum_{k=0}^{K} B_k(m)\, f_k(t) \tag{2}$$

Typically, K=3 (cubic) or K=2 (quadratic). In the studies reported here, cubic splines are used with K=3, and Lagrangian polynomials are used with K=2. However, the method is general, and does not require these choices.

Let the total interval of data be digitized with time increments Δt so that $T=N\Delta t$. Let the segments be numbered s=0, 1, 2, ... , S-1, and the intervals in each segment be indexed as j=0, 1, 2,... , J-1. Within each of the J intervals, let the time increments of size Δt be numbered n=0, 1, 2, ... , N-1. Then the water level at each time step can be indexed on (s,j,n) as

$$\eta_{s,j,n} = \eta\left([sJN + jN + n]\,\Delta t\right) \tag{3}$$

The same basis functions are used in each segment. Thus, $f_k(t)$ may be written $f_k([jN+n]\Delta t)$, since the basis function will have the same value at the same position within each segment and the expression does not really depend on s.

The exponential in eq. (1) becomes with $T=N\Delta t$,

$$e^{i2\pi m[sJN + jN + n]\Delta t/N\Delta t} = e^{i2\pi mn/N} \tag{4}$$

Combining all this, one gets for eq. (1) for the specified segment

$$\eta_{s,j,n} = \sum_{m=0}^{N-1} \sum_{k=0}^{K} B_k(m)\ f_k([jN+n]\Delta t)\ e^{i2\pi mn/N} \tag{5}$$

The terms in eq (5) can be rearranged to obtain

$$\eta_{s,j,n} = \sum_{k=0}^{K} f_k([jN+n]\Delta t)\ b_k(n) \tag{6}$$

with

$$b_k(n) = \sum_{m=0}^{N-1} B_k(m)\ e^{i2\pi mn/N}$$
$$B_k(m) = \frac{1}{N}\sum_{n=0}^{N-1} b_k(n)\ e^{-i2\pi mn/N} \tag{7}$$

It will be shown that a system of linear equations can be used to solve eq. (6) for $b_k(n)$, given the $\eta_{s,j,n}$, and then eq. (7) can be computed to get $B_k(m)$. When this is substituted into eq.(2), the time-varying Fourier coefficients are obtained. The time-varying (2-sided) frequency spectral density is the modulus-squared of the FFT coefficients, multiplied by $N\Delta t$. This will make two times the frequency integral of the spectra from zero to infinity, at a given time, equal the variance. of $\eta(t)$.

THE QUADRATIC BASIS FUNCTIONS

The quadratic case will be discussed first. Let segment s extend over the time interval (t_s , t_{s+1}) and define $h = t_{s+1} - t_s$. The quantity, h, is the same for each segment by construction. Define $\tau = (t - t_s)/h$. Then τ ranges from 0 to 1.0 as t varies across the segment.

Each segment will involve 3 $B_k(m)$ sequences. For convenience, the subscript k on B(m) will be replaced with 2s+k. A similar change in subscripts will be made on the corresponding b(n). The B(m) frequency sequences relevant to segment s will be selected as

$$
\begin{aligned}
B_{2s}(m) &= A(m, t_s) \\
B_{2s+1}(m) &= A\left(m, \frac{t_s + t_{s+1}}{2}\right) \\
B_{2s+2}(m) &= A(m, t_{s+1})
\end{aligned} \tag{8}
$$

There is a B(m) sequence at the beginning, middle, and end of the segment. The beginning and end sequences are common to the adjoining segments. The quadratic basis function can be constructed as Lagrangian interpolants (Prenter, 1992, p.30) to give

$$
\begin{aligned}
f_0(\tau) &= 2\tau^2 - 3\tau + 1 \\
f_1(\tau) &= 4\tau - 4\tau^2 \\
f_2(\tau) &= 2\tau^2 - \tau
\end{aligned} \tag{9}
$$

With these polynomials, the estimate of A(m,t) is represented across segment s with

$$
\begin{aligned}
\hat{A}(m,t) &= B_{2s}(m)\,[2\tau^2 - 3\tau + 1] \\
&\quad + B_{2s+1}(m)\,[4\tau - 4\tau^2] \\
&\quad + B_{2s+2}(m)\,[2\tau^2 - \tau]
\end{aligned} \tag{10}
$$

The representation is clearly quadratic, and at $\tau = 0$, 0.5, and 1.0, the required B(m) sequences at those times are given. In the application of these quadratic basis functions, it seems reasonable to require continuity in the first derivative at the boundary between two segments. Let $t = t_s$. The derivative of the representation at this time in the previous segment is, from eq. (10) with s replaced by s-1,

$$
[B_{2s-2}(m) - 4B_{2s-1}(m) + 3B_{2s}(m)]/h \tag{11}
$$

The corresponding derivative in the segment to the right of $t = t_s$ at its beginning

$$[-3B_{2s}(m) + 4B_{2s+1}(m) - B_{2s+2}(m)]/h \qquad \textbf{(12)}$$

If these are equated to each other, one get the constraint equation on the B(m) sequences in order to require first derivative continuity at t_s. This gives

$$\begin{aligned} &B_{2s-2}(m) - 4B_{2s-1}(m) \\ &+ 6B_{2s}(m) - 4B_{2s+1}(m) + B_{2s+2}(m) = 0 \end{aligned} \qquad \textbf{(13)}$$

THE SYSTEM OF EQUATIONS (QUADRATIC CASE)

The examples were done with 8 segments (S=8), and 8 intervals in each segment (J=8). N was set to 128 and Δt was picked as 0.5. The system of equations to solve for $b_{2s+k}(n)$ will be illustrated for this case. Equation (6) provides the primary starting point. With s and n fixed, consider the equation as j varies from 0 to J-1=7, in this case. There will be 8 equations, one for each j. Within that segment, the n-th value is being selected from each interval. For fixed n, this gives

$$\begin{bmatrix} \eta_{s,0,n} \\ \eta_{s,1,n} \\ \eta_{s,2,n} \\ \eta_{s,3,n} \\ \eta_{s,4,n} \\ \eta_{s,5,n} \\ \eta_{s,6,n} \\ \eta_{s,7,n} \end{bmatrix} = \begin{bmatrix} f_0(\tau_0) & f_1(\tau_0) & f_2(\tau_0) \\ f_0(\tau_1) & f_1(\tau_1) & f_2(\tau_1) \\ f_0(\tau_2) & f_1(\tau_2) & f_2(\tau_2) \\ f_0(\tau_3) & f_1(\tau_3) & f_2(\tau_3) \\ f_0(\tau_4) & f_1(\tau_4) & f_2(\tau_4) \\ f_0(\tau_5) & f_1(\tau_5) & f_2(\tau_5) \\ f_0(\tau_6) & f_1(\tau_6) & f_2(\tau_6) \\ f_0(\tau_7) & f_1(\tau_7) & f_2(\tau_7) \end{bmatrix} \begin{bmatrix} b_{2s}(n) \\ b_{2s+1}(n) \\ b_{2s+2}(n) \end{bmatrix} \qquad \textbf{(14)}$$

where $\tau_j = (jN+n)/(JN)$, for each j. The significant point here is that the f_k matrix is the same for each segment, s=0, 1, 2, ... , S-1.

The simultaneous equations for all $b_{2s+k}(n)$ at that n can now be written. Let η_s denote the vector left of the equal sign in eq. (14), and, in the f_k matrix, let F specify the first column, G denote the second column, and H represent the third column . Finally, let β stand for the vector of all the $b_{2s+k}(n)$, ordered from subscript value 0 to 2S+1 in ascending sequence. The total system of equations can be written

$$
\begin{bmatrix} \eta_0 \\ \eta_1 \\ \eta_2 \\ \eta_3 \\ \eta_4 \\ \eta_5 \\ \eta_0 \\ \eta_6 \\ \eta_7 \end{bmatrix}
=
\begin{bmatrix}
F & G & H & O & O & O & O & O & O & O & O & O & O & O & O & O & O \\
O & O & F & G & H & O & O & O & O & O & O & O & O & O & O & O & O \\
O & O & O & O & F & G & H & O & O & O & O & O & O & O & O & O & O \\
O & O & O & O & O & O & F & G & H & O & O & O & O & O & O & O & O \\
O & O & O & O & O & O & O & O & F & G & H & O & O & O & O & O & O \\
O & O & O & O & O & O & O & O & O & O & F & G & H & O & O & O & O \\
O & O & O & O & O & O & O & O & O & O & O & O & F & G & H & O & O \\
O & O & O & O & O & O & O & O & O & O & O & O & O & O & F & G & H
\end{bmatrix}
\boldsymbol{\beta} \qquad (15)
$$

The vector **Y** will be used to represent the vector left of the equal sign in eq. (15) and X will stand for the matrix to the right of the equal sign. Then, the equation can be expressed as $\mathbf{Y} = \mathrm{X}\ \beta$. This will be JS equations in 2S+1 unknowns. With the example values cited previously, their will be 64 equations in 17 unknown $b_{2s+k}(n)$ values at that selected n. There are more equations than unknowns, so it is natural to turn to a least-square solution. However, it is desirable to enforce the continuity in first derivative to produce a "smooth" curve at segment boundaries as the estimate of A(m,t).

The constraints in eq. (13) can be expressed for all the vector β as

$$
\begin{bmatrix}
1 & -4 & 6 & -4 & 1 & 0 & 0 & 0 & 0 & 0 & 0 & 0 & 0 & 0 & 0 & 0 & 0 \\
0 & 0 & 1 & -4 & 6 & -4 & 1 & 0 & 0 & 0 & 0 & 0 & 0 & 0 & 0 & 0 & 0 \\
0 & 0 & 0 & 0 & 1 & -4 & 6 & -4 & 1 & 0 & 0 & 0 & 0 & 0 & 0 & 0 & 0 \\
0 & 0 & 0 & 0 & 0 & 0 & 1 & -4 & 6 & -4 & 1 & 0 & 0 & 0 & 0 & 0 & 0 \\
0 & 0 & 0 & 0 & 0 & 0 & 0 & 0 & 1 & -4 & 6 & -4 & 1 & 0 & 0 & 0 & 0 \\
0 & 0 & 0 & 0 & 0 & 0 & 0 & 0 & 0 & 0 & 1 & -4 & 6 & -4 & 1 & 0 & 0 \\
0 & 0 & 0 & 0 & 0 & 0 & 0 & 0 & 0 & 0 & 0 & 0 & 1 & -4 & 6 & -4 & 1
\end{bmatrix}
\boldsymbol{\beta} = \mathbf{0} \qquad (16)
$$

This is S-1 equations in the 2S+1 unknowns (for the example, 7 equations in 17 unknowns). Let Z denote the matrix in eq. (16), so that the constraint equation can be expressed as $Z\beta = \mathbf{0}$.

The basic estimation procedure will be to produce a least-square solution to

$Y=X\beta$, while requiring the solution to satisfy $Z\beta = \mathbf{0}$. This is achieved by the method of Lagrangian multipliers by minimizing

$$\begin{aligned} Q^* &= |Y - X\boldsymbol{\beta}|^2 + 2\,[Z\boldsymbol{\beta}]^T \boldsymbol{\lambda} \\ &= Y^T Y - 2\boldsymbol{\beta}^T X^T Y + \boldsymbol{\beta}^T X^T X \boldsymbol{\beta} + 2\,[Z\boldsymbol{\beta}]^T \boldsymbol{\lambda} \end{aligned} \tag{17}$$

and then also requiring that $Z\beta = \mathbf{0}$ be true. Setting the derivative of eq. (17) with respect to β equal to zero, and attaching the constraint, gives

$$\begin{aligned} X^T X \boldsymbol{\beta} + Z^T \boldsymbol{\lambda} &= X^T Y \\ Z\boldsymbol{\beta} &= \mathbf{0} \end{aligned} \tag{18}$$

In these equations, the superscript T means transpose. All vectors are column vectors, and must be transposed to get row vectors. The equation (18) can be expressed as a single matrix equation (with O denoting a matrix of zeros) as

$$\begin{bmatrix} X^T X & Z^T \\ Z & O \end{bmatrix} \begin{bmatrix} \boldsymbol{\beta} \\ \boldsymbol{\lambda} \end{bmatrix} = \begin{bmatrix} X^T Y \\ \mathbf{0} \end{bmatrix} \tag{19}$$

The determination of estimates of $b_{2s+k}(n)$, for all s and k at a fixed n integer, is obtained from the solution of eq. (19).

The b_{2s+k} (n) are solved for n=0, then for n=1, and n=2, and so forth in sequence until the final n=N-1 is completed. The N value should have been selected so that $\Delta f = 1/(N\Delta t)$ is the frequency resolution desired in the final A(m,t) estimates. Alternately, N could be taken larger than this, and smoothing over m used to bring the A(m,t) to the appropriate resolution and noise suppression (degrees of freedom).

The $b_{2s+k}(n)$ are resequenced or rearranged into sequences on n, ant then Fourier inverted by eq. (7) to get $B_{2s+k}(m)$. The estimate of A(m,t) on segment s is then provided by

$$\hat{A}(m,t) = \sum_{k=0}^{2} B_{2s+k}(m)\; f_k([t - t_s]/h) \tag{20}$$

for this quadratic case. The time variation of the estimates is ordinarily exhibited as a long sequence at increment Δt, or some decimation of it. In the examples, it was convenient to display every fourth value of the total data input sequence of 8192 values. Since the time increment on the input was 0.5 seconds, this allowed graphing the estimate of A(m,t) every 2 seconds.

The estimate of A(m,t) will attempt to follow the noise in the input data. If that has too much fluctuation, it may be necessary to make further smoothing both on frequency, (or m), and on time. This is a common task in spectral estimation in the pseudo-stationary case, so it is no different here.

CUBIC SPLINE BASIS FUNCTIONS

The steps in the procedure for the cubic are very similar to those for the quadratic. Therefore, the various formulas will be stated with very little discussion. It takes four constants to define a cubic over a specified segment (or fine element). It is convenient here to use the value of A(m,t) and the second time derivative of A(m,t) at each segment boundary. specifically

$$\begin{aligned} B_{2s}(m) &= A(m, t_s) \\ B_{2s+1}(m) &= \frac{d^2}{dt^2}[A(m, t_s)] \\ B_{2s+2}(m) &= A(m, t_{s+1}) \\ B_{2s+3}(m) &= \frac{d^2}{dt^2}[A(m, t_{s+1})] \end{aligned} \tag{21}$$

The corresponding cubic polynomials are

$$\begin{aligned} f_0(\tau) &= 1 - \tau \\ f_1(\tau) &= -(2\tau - 3\tau^2 + \tau^3)h^2/6 \\ f_2(\tau) &= \tau \\ f_3(\tau) &= -(\tau - \tau^3)h^2/6 \end{aligned} \tag{22}$$

One could parameterize on the first derivative rather than the second, if that were desirable, or use the values of A(m,t) at the beginning, one-third, two-thirds, and end of the segment, if that were preferred. However, the above choice is consistent with the cubic spline formulation often used (Carnahan and Wilkes, 1973, p.307-310).

The formulation is continuous in function value and second derivative on t by the structure of the basis function representation. Traditional cubic splines are obtained if continuity in first derivative is also enforced by constraint equations. In the segment to the left of t_s the first derivative at $t=t_s$ is

$$-\frac{B_{2s-2}(m)}{h} + \frac{hB_{2s-1}(m)}{6} + \frac{B_{2s}(m)}{h} + \frac{hB_{2s+1}(m)}{3} \quad \textbf{(23)}$$

The corresponding expression in the segment to the right of t_s, evaluated at $t=t_s$, is

$$-\frac{B_{2s}(m)}{h} - \frac{hB_{2s+1}(m)}{3} + \frac{B_{2s+2}(m)}{h} - \frac{hB_{2s+3}(m)}{6} \quad \textbf{(24)}$$

If these are equated, one gets the equation forcing continuity in first derivative between the two segments meeting at t_s .

$$\begin{aligned} &-B_{2s-2}(m) + 2B_{2s}(m) - B_{2s+2}(m) + \\ &\frac{h^2}{6}\left[B_{2s-1}(m) + 4B_{2s+1}(m) + B_{2s+3}(m)\right] = 0 \end{aligned} \quad \textbf{(25)}$$

THE SYSTEM OF EQUATIONS (CUBIC CASE)

As before, let β stand for the vector of all the $b_{2s+k}(n)$, ordered from subscript value of 0 to 2S+2 in ascending sequence. The X matrix and the Z matrix for system and constraints may be exhibited as follows. Let F and G be defined as

$$F = \begin{bmatrix} f_0(\tau_0) & f_1(\tau_0) \\ f_0(\tau_1) & f_1(\tau_1) \\ f_0(\tau_2) & f_1(\tau_2) \\ f_0(\tau_3) & f_1(\tau_3) \\ f_0(\tau_4) & f_1(\tau_4) \\ f_0(\tau_5) & f_1(\tau_5) \\ f_0(\tau_6) & f_1(\tau_6) \\ f_0(\tau_7) & f_1(\tau_7) \end{bmatrix} \quad \textbf{(26)}$$

$$G = \begin{bmatrix} f_2(\tau_0) & f_3(\tau_0) \\ f_2(\tau_1) & f_3(\tau_1) \\ f_2(\tau_2) & f_3(\tau_2) \\ f_2(\tau_3) & f_3(\tau_3) \\ f_2(\tau_4) & f_3(\tau_4) \\ f_2(\tau_5) & f_3(\tau_5) \\ f_2(\tau_6) & f_3(\tau_6) \\ f_2(\tau_7) & f_3(\tau_7) \end{bmatrix} \tag{27}$$

The X matrix may be then written

$$X = \begin{bmatrix} F & G & O & O & O & O & O & O & O \\ O & F & G & O & O & O & O & O & O \\ O & O & F & G & O & O & O & O & O \\ O & O & O & F & G & O & O & O & O \\ O & O & O & O & F & G & O & O & O \\ O & O & O & O & O & F & G & O & O \\ O & O & O & O & O & O & F & G & O \\ O & O & O & O & O & O & O & F & G \end{bmatrix} \tag{28}$$

For the Z matrix, let H = (-1 , h^2 / 6) and L = (2 , $2h^2$ / 3) be row vectors. Then Z can be written

$$Z = \begin{bmatrix} H & L & H & O & O & O & O & O & O \\ O & H & L & H & O & O & O & O & O \\ O & O & H & L & H & O & O & O & O \\ O & O & O & H & L & H & O & O & O \\ O & O & O & O & H & L & H & O & O \\ O & O & O & O & O & H & L & H & O \\ O & O & O & O & O & O & H & L & H \end{bmatrix} \tag{29}$$

The procedure from here on is identical with that used for the quadratic case.

Equation (19) is solved for β, the b's are rearranged in sequences on n and Fourier inverted to get the B's. Finally the estimate of A(m,t) on segment s is obtained as

$$\hat{A}(m,t) = \sum_{k=0}^{3} B_{2s+k}(m)\; f_k([t - t_s]/h) \qquad \textbf{(30)}$$

EXAMPLES

First a deterministic case, which is exactly quadratic,is studied. In the absence of noise the method should give the Fourier coefficients used to generate the data. The results are presented in Fig. 2 while the input data is shown in Fig.1

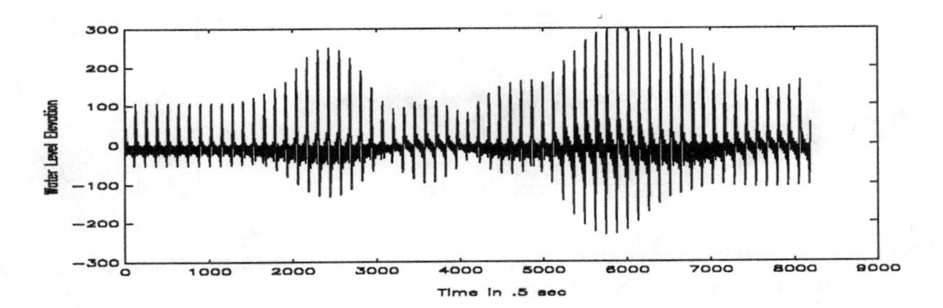

Figure 1. Input data for the deterministic example.

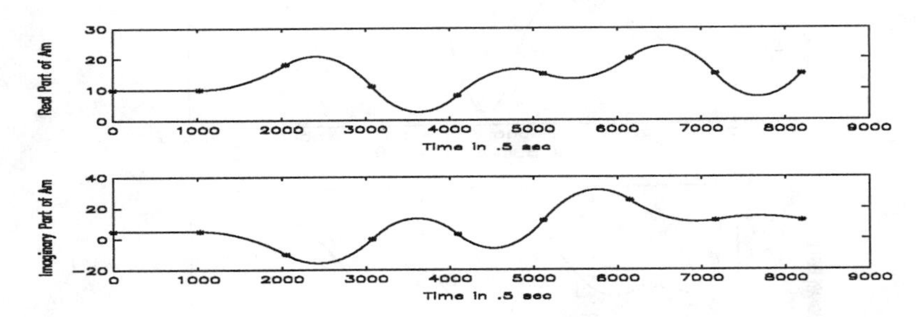

Figure 2 The input and output Fourier coefficients. The asterisks show segment boundaries. The curves lie exactly on top of each other.

Next random data is analyzed. For that a stationary time series was generated using a Pierson-Moskowitz spectra with modal frequency .09 Hz and variance 6.0 . An envelop function in time is used to provide the nonstationary character of the data. The variance as a function of time along with the time series is provided in Fig. 3 while the results are given in Fig. 4-6.

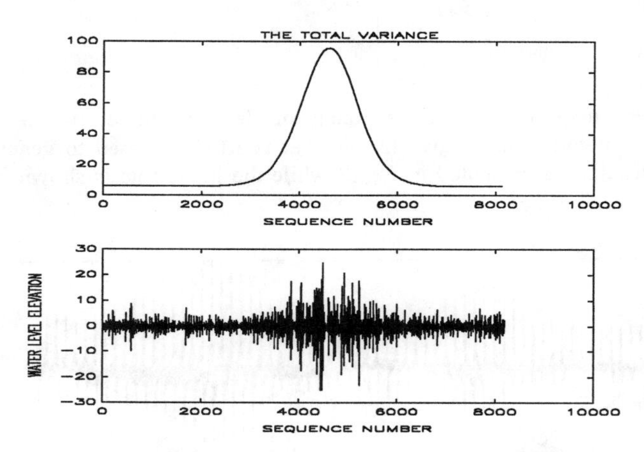

Figure 3. The variance envelope (upper figure) and a time series realization (lower figure) for Example No. 1.

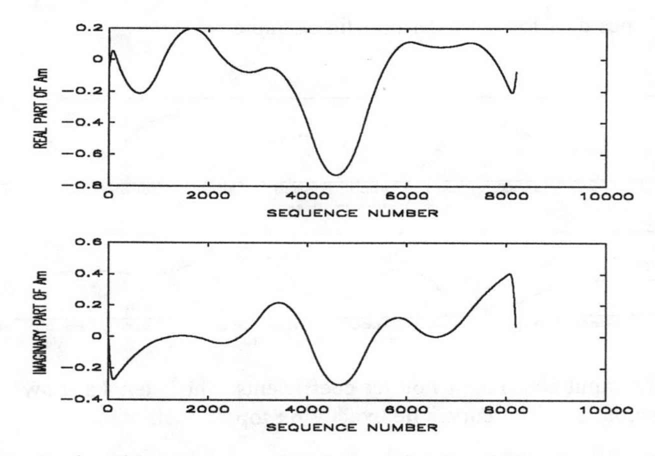

Figure 4. Real and imaginary parts of the estimate of A(m,t) at modal frequency as a function of time, for Example No. 1.

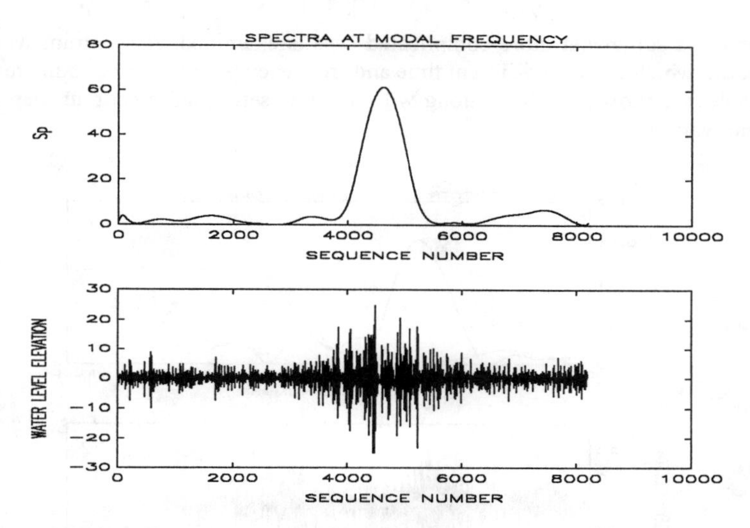

Figure 5. Example No. 1 graph of spectral time variation at the mode. For comparison, the realization analyzed is plotted again below.

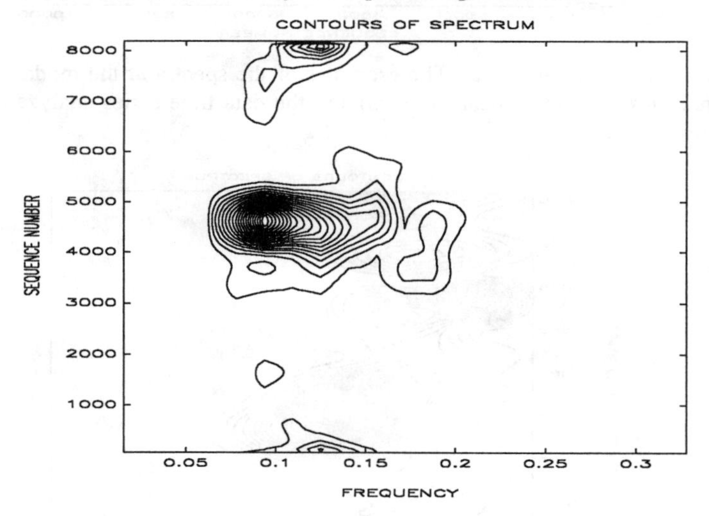

Figure 6. Contour map for the estimate of A(m,t) in Example No.1.

Finally a somewhat more complicated case is examined. A spectrum with two modes which occur in different time and frequency points is to be estimated. The results are shown in Fig. 7 along with the time series and a contour map of the time varying spectra.

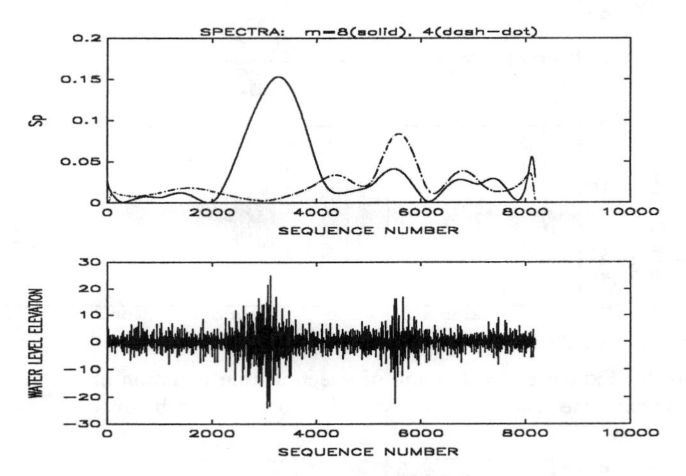

Figure 7. Example No. 2. The estimates of the spectra at the modal frequencies of the two peaks (above) and the data time series analyzed (below).

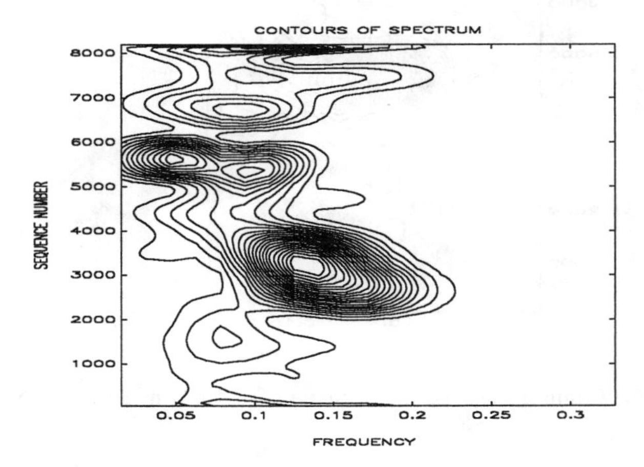

Figure 8. Example No. 2. Contour map of the estimate of A(m,t) for the two-peaked example.

CONCLUDING COMMENTS

(1) This study presents the first results in a continuing investigation of evolutionary Fourier analysis of nonstationary data. Much remains to be done. Even in the estimation of the frequency spectral density given here, there are questions of sampling error, confidence intervals, optimum smoothing procedures, estimate distortion due to leakage as it affects amplitude and phase of the spectra, and similar topics.

(2) If measurements from space arrays of wave gages form the input time series, in principle, each time series can be analyzed by the evolutionary Fourier procedures to get the A(m,t) for each gage. The set of these coefficients at a given frequency and time can be used in standard procedures for the estimation of directional spectra (Borgman, 1979, 1982) to characterize the time-varying directional wave train. However, there are many practical details of the procedure that are sure to arise when the method is applied to actual field data. Questions of phase distortion in the Fourier coefficients are particularly relevant in this context. It will also be important to cooperate with other researchers investigating wave generation in meteorological conditions that are changing rapidly in time to determine the usefulness of the technique in assisting such studies.

(3) A further extension of the procedures to nonstationarity in space and time is planned. In this extension, the Fourier coefficient may be written A(m,x,y,t). The assumption will be that the data is measured on equal increments in time, but irregularly over the 2-D (x,y)-space. The investigation of this formulation is just now starting in connection with environmental studies. However, considerable work has been done in the Fourier structure of such an (x,y,t) space over the last eight years in another study (now completed) which developed large scale conditional simulations of ocean currents. That study (Borgman and Faucette, 1993; Borgman, Miller, Signorini, and Faucette, in press) assumed stationarity, but much of the theoretical structure and notation from that investigation can be carried over to the evolutionary nonstationary case.

(4) The evolutionary Fourier analysis procedures have great flexibility and can be crafted to fit the particular application. Different basis functions can be selected and, in particular, other polynomial orders than the quadratic or cubic can be chosen if they are more appropriate. Additional constraints can be imposed. One useful modification investigated in the study, but not reported in

the paper, is to impose the constraint that the first three or more B(m) sequences are identical to each other. This has the effect of making the first, or several, segments have a stationary spectra and FFT coefficients. It is easy to add additional lines to the Z matrix to impose these selected constraints.

(5) The computer code for the analysis was written in MATLAB (The Mathworks, INC., Natick, Mass. 01760). Any investigators desiring a copy of the program are invited to contact the Senior Author. The code is non-proprietary.

ACKNOWLEDGEMENTS

This research was supported in part by the Offshore Technology Research Center, NSF Engineering Research Centers Program Grant No. CDR-8721512. Some aspects of the study were also supported by the EPSCoR Groundwater studies at the University of Wyoming, NSF Grant No. EHR-9108774, in anticipation that the evolutionary Fourier Analysis technique will find useful applications in environmental studies.

The division of labor in the study was as follows. The senior author developed the theory. The second author prepared the examples and assisted in checking derivations. The third author made a very thorough review of Priestley's work to determine if it could be applied to the problem a hand. The second and third authors together examined closely complex demodulation as a possible tool.

REFERENCES

Bloomfield, P. (1976) Fourier Analysis of Time Series: An Introduction, John Wiley & Sons, New York.

Borgman, Leon E. (1979) Directional Spectra from Wave Sensors. Ocean Wave Climate. Vol. 8, pp. 269-300, Marine Sci. Series, Plenum Press, New York.

Borgman, Leon E. (1982) Maximum-Entropy and data-Adaptive Procedures in the Investigation of Ocean Waves*. Proc. Second Workshop on Maximum Entropy and Bayesian Methods in Applied Statistics, August 1982, Physics Department, University of Wyoming, Laramie, Wyoming.

Borgman, Leon E. (1990) Irregular ocean waves: kinematics and forces. The Seas: v. 9, Ocean Engineering Science, 90 pages, John Wiley and Sons, Pub.

Borgman, L. and Faucette, C (1993) Basic Mathematics and Statistical Theory for Finite Fourier Coefficients of Gaussian Vector Functions, chapter 2, Computational Stochastic Mechanics (Cheng, H. D. and Yang, C. Y., editors), Elsevier Applied Science, New York.

Borgman, L. and Faucette, C (1993) Multidimensional Simulation of Gaussian Vector Random Functions in Frequency Domain, chapter 3, Computational Stochastic Mechanics (Cheng, H. D. and Yang, C. Y., editors), Elsevier Applied Science, New York.

Borgman, L. E.; Miller, C. D.; Signorini, S. R.; and Faucette, (in press) Stochastic Interpolation As a Means to Estimate Oceanic Fields*, Canadian Jour.of Oceanography.

Brown, LLoyd J. (1967) Methods for the Analysis of Non-Stationary Time Series With Applications to Oceanography, Ph.D. Thesis, University of California, Davis; Also published as Hydraulic Engineering Laboratory Report HEL 16-3, College of Engineering, University of California, Berkeley. 135pp.

Carnahan, Brice, and Wilkes, James O. (1973) Digital Computing and Numerical Methods, Johm Wiley & Sons, New York, 477p.

Daubechies, Ingrid (1992) Ten Lectures on Wavelets, CBMS-NSF Regional Conference Series in Applied Mathematics, Society for Industrial and Applied Mathematics, Philadelphia, Pennsylvania. 367pp.

Prenter, P. M. (1975) Splines and Variational Methods, Wiley Interscience Pub., John Wiley & Sons, New York, 323pp.

Priestley, M. B. (1981) Spectral Analysis and Time Series, Academic Press, New York, 653pp.

Priestley, M. B. (1988) Non-Linear and Non-Stationary Time Series Analysis, Academic Press, New York, 237 pp.

A NEW CORRECTION PROCEDURE FOR SHIPBORNE WAVE RECORDER DATA

E G Pitt[1]

Abstract

A new correction procedure for Shipborne Wave Recorder (SBWR) wave height and spectral data is described. The method is based on a reassessment of the available frequency response measurements using a Froude-type frequency scaling. The paper discusses the direct application of the correction function to spectral data as well as describing a method of applying the correction to archive SBWR data which consists of Hs and Tz only. It is found that the new correction gives a considerable improvement in accuracy. This work may be seen as a sequel to the paper by van Aken and Bouws [1] presented at Waves 74 which was based on a more limited data set.

Introduction

The Shipborne Wave Recorder (SBWR) was devised in the early 1950's, Tucker [10], and for the first time enabled wave measurements to be made on a routine basis in offshore areas. Since then it has been used extensively for the measurement of waves in sea areas around the British Isles and in the North-East Atlantic, as well as in other parts of the world. Where a suitable station-keeping ship is available the SBWR remains a uniquely cost-effective method of measuring the waves, and because the instrument is protected within the hull the continuity of the measured data series has often been very high. Although the number of available ships has

---

[1]Applied Wave Research, West Sussex RH20 2AD, United Kingdom

declined in recent years some measurement programmes continue, notably one at Ocean Station Lima in the north-east Atlantic. Thus the performance of the SBWR continues to be of interest both because of the measurement programmes which are presently underway and because of the extensive historical archive.

Principle of operation and frequency response

The SBWR consists of two instrument packages which are located within the hull below the waterline on either side of the ship at an alongships station which is close to the pitch axis. Each package consists of an accelerometer mounted in gimbals so as to measure the vertical acceleration and a pressure sensor which communicates with the ocean via a hole in the ship's side. These holes are positioned at the smallest distance below the mean water line which ensures that they remain submerged at all times. In practice this results in a depth of between one and three metres depending on the ship.

The four signals (two from each side) are led to a central computer unit where the accelerations are doubly integrated to give the displacement at each instrument package. These signals, now two displacement and two pressure (scaled as pressure head in metres), are added and it is this composite signal which gives a measure of the wave height. Design and calibration aspects of the system (which uses analogue electronic techniques throughout) are discussed in Haine [5] and Crisp [3]; the use of vertical accelerometers mounted on short-period pendula is discussed by Tucker [12].

The use of an accelerometer and a pressure sensor on either side cancels out roll signals (both in acceleration and pressure) as well as providing some compensation for pressure fluctuations due to waves reflected from one side of the ship.

This a complex system and in order to gain some insight into its frequency response we model the system by a single pressure sensor and a single accelerometer mounted on a spar buoy, Figure 1. It is assumed in what follows that the wave heights and the heave of the ship are very small with respect to the wavelength.

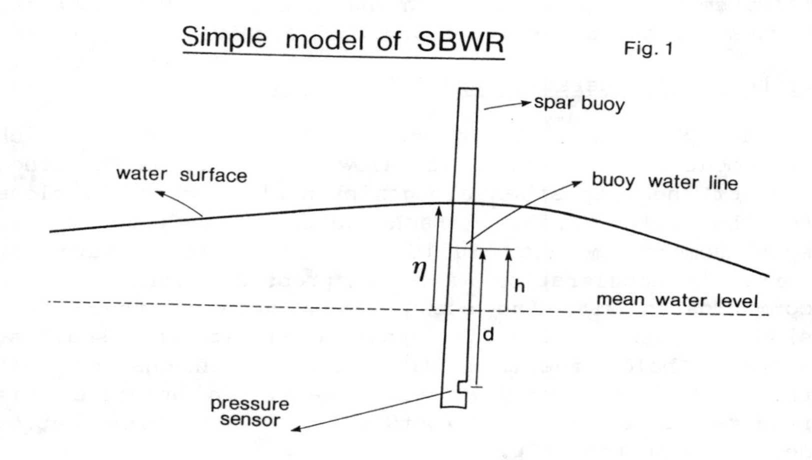

If the sea surface elevation above the mean level is given by $\eta(t)$ and the heave of the buoy is $h(t)$, then the pressure at the pressure sensor (in metres of sea water) is given by:

$$p(t) = d - h(t) + \eta(t) * r_H(t) \tag{1}$$

where d is the mean depth of immersion of the pressure sensor and r_H is the impulse response function of the pressure attenuation with depth. * indicates convolution.

The response of the pressure sensor can be written:

$$p'(t) = p(t) * r_E(t) \tag{2}$$

ie $$p'(t) = \{d - h(t) + \eta(t) * r_H(t)\} * r_E(t) \tag{3}$$

Likewise, the response of the heave sensor can be written

$$h'(t) = h(t) * r_E(t)$$

where r_E is the impulse response of the accelerometer/double integrator. In practice this is just the response of the double integrator with respect to a perfect double integrator [3]. The equivalent filter must be included in the pressure signal path and this is reflected in equation 2.

The sum signal is given by:

$$s'(t) = h'(t) + p'(t) = \eta(t) * r_H(t) * r_E(t)$$

where we have ignored the constant d.

Taking Fourier transforms we get

$$S(f) = W(f).R_H(f).R_E(f)$$

where W is the Fourier transform of η. Forming the spectra we get:

$$S_W(f) = S_S(f) \times \frac{1}{|R_H|^2} \times \frac{1}{|R_E|^2} \qquad (4)$$

Equation 4 is used to correct the measurements of the SBWR. Since R_E can be calculated, the problem reduces to determining R_H, the transfer function of the measured pressure fluctuations to the surface waves.

The simplest assumption that can be made is that the pressure beneath the waves is the same as it would be in the absence of the ship. Then according to the linearised theory of water waves and assuming the ship is operating in deep water, the amplitude response, R_H, is given by:

$$R_H(k) = \exp(-kd)$$

ie,

$$R_H(f) = \exp \frac{(-(2\pi f)^2 d)}{g} \qquad (5)$$

where k is the wave number
f is the frequency of the waves
d is the mean depth of immersion of the pressure sensors
g is the acceleration due to gravity

Some early measurements of the response suggested that it fell more quickly with increasing frequency than is indicated by equation 5 and so the depth was multiplied by a factor α (usually called k). The routine analysis of SBWR data at the Institute of Oceanographic Sciences used this 'modified' form of the classical response with α = 2.5 to give

$$R_H(f) = \exp \frac{(-2.5(2\pi f)^2 d)}{g} \qquad (6)$$

Empirical determinations of the frequency response

Table 1 sets out details of the ships on which the frequency response measurements which are considered in this work were made. The most recent measurements were those undertaken on the Channel lightvessel and reported by Crisp [3]. The measurements made onboard Ocean Weather Ship (OWS) Cumulus were reported by van Aken and Bouws [1] at Waves 74. These, together with measurements made on OWS Weather Reporter by Canham et al [2], were discussed by Crisp [3]. In addition to these, Pitt [8] and [9] considered measurements made on SS Cairndhu and RV Ernest Holt.

Table 1: Main dimensions of the 5 ships (in metres)

Ship	Pressure Depth, d	Length	Beam	Draught
Channel	2.0	35.0	8.7	3.5
Cumulus	1.5	62.0	12.6	4.5
Weather Reporter	2.2	72.0	10.9	4.3
Ernest Holt	2.0	53.3	9.1	4.4
Cairndhu	3.7	128.0	18.3	5.8

Note: All the ships were operating in deep water (1000m or more) except for the Channel lightvessel which was in 60m.

If, following Crisp [3], we define a scaled frequency variable:

$$\xi_2 = \left(\frac{2\pi d}{g}\right)^{1/2} f \tag{7}$$

equation 5 gives

$$R^2(\xi_2) = \exp(-4\pi\xi_2^2) \tag{8}$$

and equation 6 gives

$$R^2(\xi_2) = \exp(-10\pi\xi_2^2) \tag{9}$$

In Figure 2 are plotted the experimental estimates against ξ_2 as well as equations 8 and 9. It will be seen that the measured responses do not agree well with either of the exponential forms.

FIGURE 2: Whole data set plotted against ξ_2

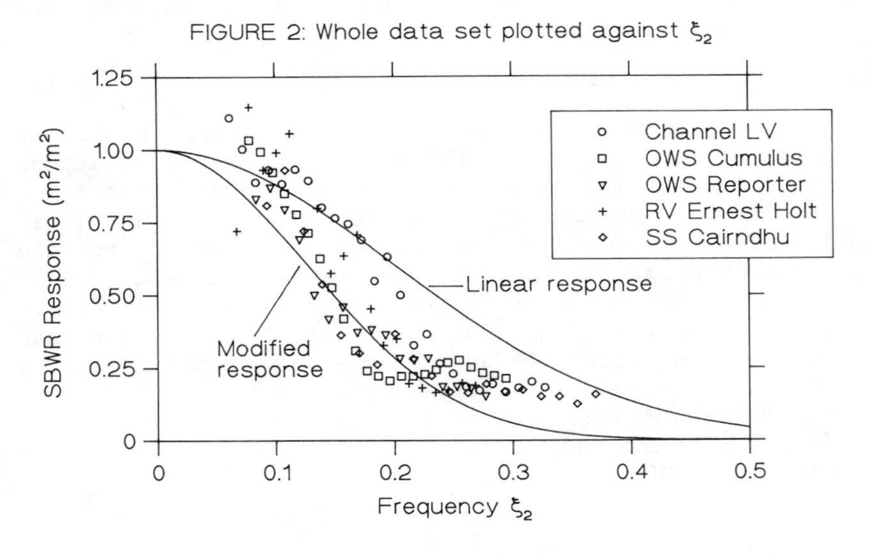

Equation 7 represents a Froude number scaling of the frequency with d as the length scale; in ship motion work the ship's length L is more often used and a scaling based on this was tried with rather satisfactory results.

Further combinations of dimensions were explored and finally a scaling based on the use of the harmonic mean of L and d was selected:

$$\xi_4 = \left(\frac{2\pi}{g}\right)^{1/2} (Ld)^{1/4} f \qquad (10)$$

Figure 3 shows $|R_H|^2$ plotted against ξ_4 (the fitted line is equation 11 below).

Fitting formulae to the measured response data

Since neither the 'classical' hydrodynamic attenuation formula nor its 'modified' form fit the experimental data well it was decided to use a completely empirical approach. A function was sought which tended to unity at very low frequencies (this was expected on physical grounds) and which tended to a constant at high frequencies. After some experimentation the following form was adopted:

$$R_H^2 = 1 - A_0\{1 - \exp[-A_1\xi_4 - A_2\xi_4^2 - A_3\xi_4^3]\} \qquad (11)$$

The experimental data can be expected to be progressively less reliable as the frequency decreases below 0.1 Hz, and so in order to force the fitted curve to adopt a physically reasonable shape at the lower frequencies a group of five points was added to the data for each ship. These were specified at frequencies 0.0, 0.005, 0.010, 0.015, 0.020 Hz, the response being set to unity. Table 2 gives the values of the four constants and the rms error. N is the number of points including the five manufactured ones. Equation 11 was fitted to the data from each of the five ships and to the whole data set. The fit obtained for the whole data set is shown in Figure 3.

Table 2. Fitting constants for equation 11 for the five ships

Ship	A_0	A_1	A_2	A_3	ε	N
Channel	0.8468	0.4876	-6.4058	26.691	0.0443	30
Cumulus	0.7734	1.0832	-19.048	64.048	0.0260	28
W. Reporter	0.8258	-1.0047	8.9790	3.7864	0.0440	22
E. Holt	0.8373	2.213	-21.234	54.182	0.0877	23
Cairndhu	0.8211	0.2709	-7.4233	40.096	0.0531	24
All ships	0.8103	0.5072	-8.2000	35.790	0.0690	127

FIGURE 3: Whole data set plotted against ξ_4 with Eq 11 fitted

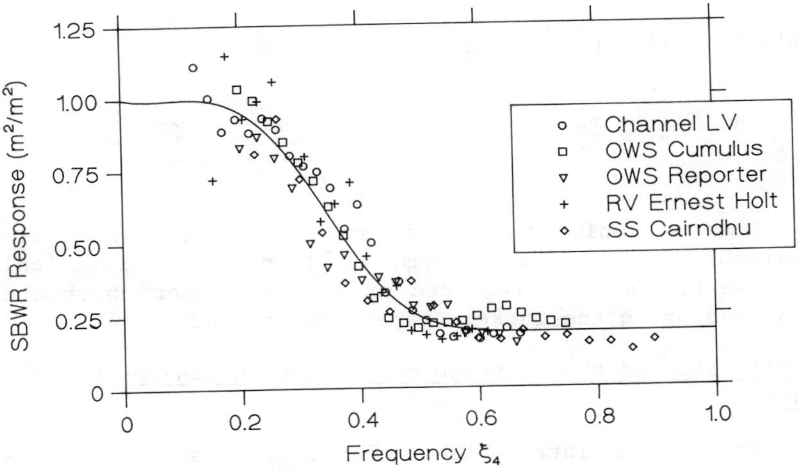

Application of the response correction to spectral data

For SBWR data which have been subjected to spectral analysis the application of the empirical frequency response correction is straightforward. If an experimental determination of the frequency response is available for a particular installation equation 11 is fitted to those data and used to correct the spectra. If not, the fitting constants for the whole data set may be used.

Then

$$S_i = S_i' \times \frac{1}{R_H^2(\xi_{4i})} \times \frac{1}{|R_E(f_i)|^2}$$

Where R_H^2 is given by equation 11 and S' is the uncorrected spectrum, and then

$$m_n = \Delta f \sum S_i f_i^n$$

$$Hs(\text{Spectral}) = 4\sqrt{m_o}$$

$$Tz(\text{Spectral}) = \left(\frac{m_o}{m_2}\right)^{1/2}$$

$$T_1 = \frac{m_o}{m_1}$$

However only the more recent measurements are available in spectral form, most of the historical archive having been recorded on pen and paper chart and analysed using the Tucker-Draper (TD) method.

Application of the response correction to non-spectral data

Before the introduction of microprocessors at SBWR installations the wave data were recorded as 12-minute samples using pen and paper charts. The charts were analysed 'by eye' using the TD method which gives estimates of Hs, the significant wave height and Tz', the apparent (low-pass filtered) mean zero-crossing period, see Tucker [12], Draper [4]. The heights were corrected using Equation 6 evaluated at a characteristic frequency f_c, where $f_c = 1/Tz'$.

$$Hs(TD) = Hs'(TD) \times \frac{1}{R_H(f_c)} \times \frac{1}{|R_E(f_c)|}$$

Where $f_c = 1/Tz'$

R_H is given by equation 6 and Hs' is the uncorrected Hs. We refer to this as a 'scalar' correction, in contrast to the 'full' correction where the spectrum is corrected frequency by frequency.

After the introduction of spectral analysis at the SBWR installations chart roll recording was maintained as a back-up and to allow comparisons between the two methods. Using these data it has been possible to investigate the operation of the scalar correction method and to show that there is very little difference between the scalar results and the spectral method if an appropriate characteristic frequency is defined and a small residual bias in the scalar method is allowed for. Data from the SBWR installations on the Channel LV, the Seven Stones LV, the Dowsing LV and from OWS Lima were used and the details of these investigations are reported in [8] and [9].

On the basis of this work the (re)correction equation for archive Hs(TD) data can be formulated as follows.

$$Hs_R = \frac{Hs'(TD)}{\{\sqrt{S_{SF}} \times R_H(\xi_4(f_c)) \times |R_E(f_c)|\}} \qquad (13)$$

where Hs_R is the corrected value of Hs and Hs'(TD) is the uncorrected TD estimate of Hs.
$\sqrt{S_{SF}}$ is an empirical coefficient which corrects for the bias in the scalar correction method [8].

$$f_c = \frac{1}{Tz(TD) \times ST_1'Tz}$$

where $ST_1'Tz$ is the empirical coefficient relating T_1 from the uncorrected spectrum to Tz(TD).

For the four stations used in the development of the method, S_{SF} had a mean value of 0.8925 with standard deviation 0.014. ST_1'Tz varied between 1.008 for the Dowsing LV to 1.123 at the Channel LV.

Accuracy of the new correction procedure

This section discusses the accuracy of the recorrected values of Hs(TD). Figure 4 is a plot of Hs(TD) against Hs from simultaneous spectral data for Seven Stones LV. There is a large discrepency between the two whose origin lies mainly in errors in the old frequency response. The pressure sensors were comparatively deep in the Seven Stones LV, d=2.6 m, and the over-correction using the old scheme is large. For vessels with shallow pressure sensors, eg the Dowsing LV, the over-correction was much smaller.

Figure 5 shows the recorrected values of Hs(TD) plotted against the spectral results for Seven Stones. There is a considerable improvement. Table 3 shows the results of regression analyses for the four sites.

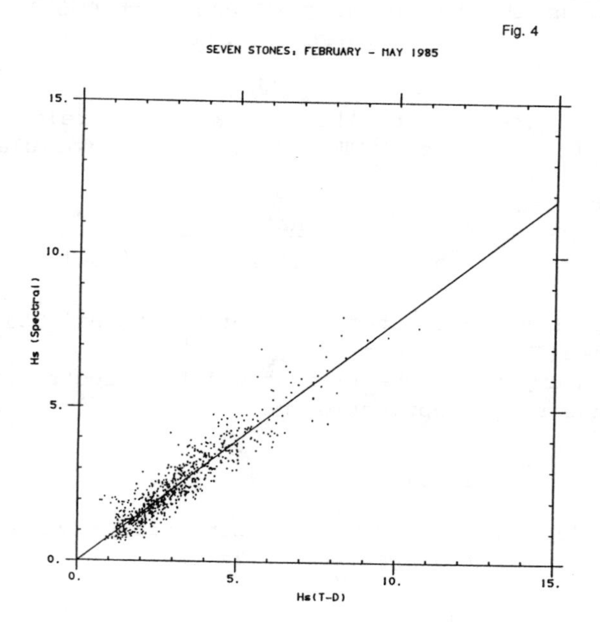

Table 3: Comparison between recorrected Hs(TD) and spectral Hs

Station	Slope (Hs_R:Hs(spectral))	Correlation Coeff
OWS Lima	1.0482	0.9611
Seven Stones LV	1.0026	0.9347
Dowsing LV	0.9903	0.9751
Channel LV	0.9505	0.9690

Fig. 5

SEVEN STONES: FEBUARY - APRIL 1985

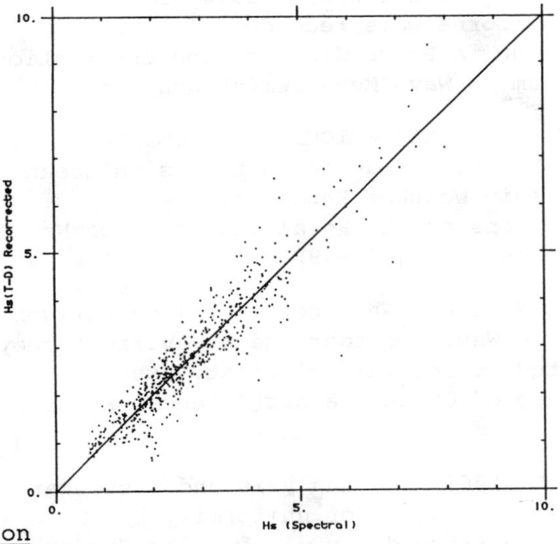

Conclusion

By using a non-dimesional frequency scaling it has been possible to reconcile the available experimental determinations of the frequency response of the SBWR. The resulting response does not match the exponential form which has been used in the past and so an empirical formula was developed to fit the experimental data. A correction scheme based on this empirical formulation has been developed for use with historical Hs(TD) data and leads to much better agreement with simultaneously recorded spectral results. Comparisons with other data sources including satellite altimeter data are described in [9].

Acknowledgements

The research which forms the basis of this paper was undertaken at the Institute of Oceanographic Sciences and was supported by the UK Department of Energy. Thanks are due to Elsevier Science Publishers Ltd for permission to reproduce material from [9].

References

[1] van AKEN H M and BOUWS E 1974 Frequency response of a shipborne wave recorder.
Pp 281-300 in Proceedings of the International Symposium on Wave Measurement and Analysis.

[2] CANHAM H J S, CARTWRIGHT D E, GOODRICH G J, and HOGBEN N 1962 Seakeeping trials on ocean weathership Weather Reporter.
Transactions of the Royal Institute of Naval Architects, 104, 447-492.

[3] CRISP G N 1987 An experimental comparison of a Shipborne Wave Recorder and a Waverider buoy conducted at the Channel Light Vessel.
Institute of Oceanographic Sciences Report, No. 235, 181pp

[4] DRAPER L 1967 The analysis and presentation of wave data - a plea for uniformity.
Pp 1-11 in Proceedings of the 10th Conference on Coastal Engineering, Tokyo, September 1966.
Volume 1. New York: American Society of Civil Engineers.

[5] HAINE R A 1980 Second Generation Shipborne Wave Recorder
Transducer Technology, 2, 25-28.

[6] NPL 1965 Report on analysis of ship motion trials data for 'Ernest Holt'.
National Physical Laboratory Coordinating Committee for Research into the Seagoing Qualities of Ships, Report No 7, 25pp.
(Unpublished manuscript)

[7] NPL 1965 Report on analysis of ship motion trials data for 'Cairndhu'.
National Physical Laboratory Coordinating Committee for Research into the Seagoing Qualities of Ships, Report No 6, 21pp.
(Unpublished manuscript)

[8] PITT E G 1988 The application of empirically determined frequency response functions to SBWR data.
Institute of Oceanographic Sciences Report, No. 259, 82pp.

[9] PITT E G 1991 A new empirically-based correction procedure for shipborne wave recorder data.
Applied Ocean Research, 13, No 4, 162-174.

[10] TUCKER M J 1956 A shipborne wave recorder.
Transactions of the Royal Institute of Naval Architects, 98, 236-250.

[11] TUCKER M J 1959 The accuracy of wave measurements made with vertical accelerometers.
Deep Sea Research, 5, 185-192.

[12] TUCKER M J 1963 Analysis of records of sea waves.
Proceedings of the Institute of Civil Engineers, 26 305-316.

EXTENSIVE COMPARISON OF DIRECTIONAL WAVE ANALYSIS METHODS FROM GAUGE ARRAY DATA

Michel BENOIT [1]

Abstract

In order to settle an efficient operational wave gauge array for estimating directional wave spectrum, several directional analysis methods are implemented and compared. These methods are based on different modelling approaches including simple and sophisticated ones : Fourier Series decomposition, Fit to unimodal or bimodal parametric models, Variational Fitting Technique, Maximum Likelihood Methods, Maximum Entropy Method and Bayesian Approach. The comparison is performed following a three steps procedure : first on exact cross-spectral simulations, then on numerical sea-surface elevation time series and finally on laboratory measurements. Various directional shapes are reproduced at each step. The influence of several parameters for measurement and analysis is examined. Practical recommendations for operational processing are underlined.

1. Introduction

Among the systems dedicated to the measure of directional wave spectrum the wave gauge array may be interesting in that it offers the possibility to record a greater number of information than classical single-point systems (heave-pitch-roll buoy or 2D-currentmeter associated with a wave probe). The wave gauge array however requires a spatial treatment based on an assumption of spatial homogeneity whereas single point measurement systems do not. Wave gauge arrays may be used in laboratory (Nwogu, 1989) or in situ, with pressure transducers lying on the bottom for instance (Long and Hasselmann, 1979 ; Hashimoto *et al.*, 1987 ; Howell, 1992).

Several approaches are proposed to extract an estimate of directional spectrum from the recorded time series of wave elevation, each one being claimed by their authors to be the best one. The aim of this study is to compare a large number of these methods on the same data both for numerical and laboratory sea-state simulations, to examine the influence of several practical parameters and to estimate the ability of the methods to estimate various target directional spectra.

This work follows a previous study of the same type on directional wave analysis methods applied to heave-pitch-roll buoy measurements (Benoit, 1992).

[1] Research Engineer
EDF - Laboratoire National d'Hydraulique, 6, quai Watier 78400 CHATOU, FRANCE

2. Problem formulation

The directional wave spectrum $S(f,\theta)$ is a function of wave frequency f and direction of propagation θ. The following classical decomposition is used :

$$S(f,\theta) = E(f).D(f,\theta) \tag{1}$$

in which E(f) is the one-sided frequency spectrum that may be estimated by a single record of sea-surface elevation. It is related to the directional spectrum by :

$$E(f) = \int_0^{2\pi} S(f,\theta)\, d\theta \tag{2}$$

$D(f,\theta)$ is the Directional Spreading Function (DSF) satisfying :

$$D(f,\theta) \geq 0 \ \text{ over } [\,0\,,2\pi\,] \quad \text{and} \quad \int_0^{2\pi} D(f,\theta)\, d\theta = 1 \tag{3}$$

The following pseudo-integral relation may be written between the directional spectrum and the sea-surface elevation η :

$$\eta(x,y,t) = \int_0^{\infty}\int_0^{2\pi} \sqrt{2.S(f,\theta).df.d\theta}\ .\cos\left[2\pi ft - k.(x.\cos\theta + y.\sin\theta) + \varphi\right] \tag{4}$$

The wave probe array is able to record at the same time the sea-surface elevation at N different locations (see Definition Sketch on Figure 1).

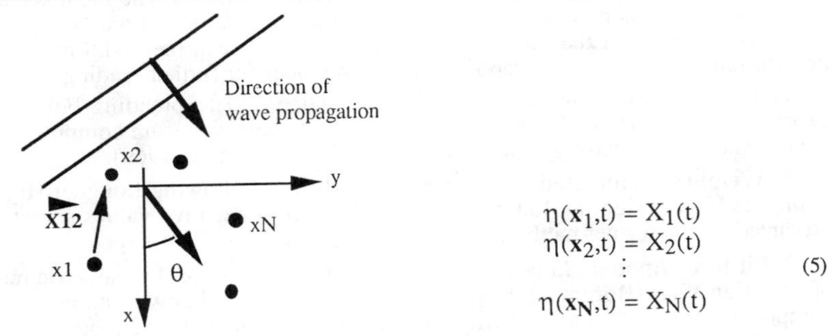

$$\begin{aligned} \eta(\mathbf{x_1},t) &= X_1(t) \\ \eta(\mathbf{x_2},t) &= X_2(t) \\ &\vdots \\ \eta(\mathbf{x_N},t) &= X_N(t) \end{aligned} \tag{5}$$

Figure 1 : Definition Sketch

The calculation of the cross-spectra $\widehat{G}_{ij}(f)$ between each couple (X_i, X_j) is performed by spectral analysis procedures. The following expression may be extracted between the "measured" complex cross-spectra $\widehat{G}_{ij}(f)$ and the unknown directional spectrum $S(f,\theta)$:

$$\widehat{G}_{ij}(f) = \widehat{C}_{ij}(f) - \widehat{Q}_{ij}(f) = \int_0^{2\pi} S(f,\theta).\exp(-\vec{\mathbf{k}}.\overrightarrow{\mathbf{x_{ij}}})\, d\theta \quad i = 1,....,N \text{ and } j \geq i \tag{6}$$

After cross-spectral analysis of N elevation time series one obtains :

* **N auto-spectra** $\widehat{C}_{ii}(f)$ i=1,........, N, which are real quantities ($\widehat{Q}_{ii}(f)=0$) giving estimations of the non-directional frequency spectrum E(f).
* **M = N.(N-1)/2 complex cross-spectra** whose real parts $\widehat{C}_{ij}(f)$ (j>i) are called "coïncident spectral density functions" or "co-spectra" and imaginary parts $\widehat{Q}_{ij}(f)$ are called "quadrature spectral density functions" or "quad-spectra".

The above cross-spectra are normalized in the following way to get 2.M relations with the Directional Spreading Function D(f,θ) :

$$C_{ij}(f) = \frac{\widehat{C}_{ij}(f)}{\sqrt{\widehat{C}_{ii}(f).\widehat{C}_{jj}(f)}} = \int_0^{2\pi} D(f,\theta).\cos(\vec{\mathbf{k}}.\overrightarrow{\mathbf{x}_{ij}})\, d\theta \quad i = 1,\ldots,N-1 \text{ and } j > i$$

$$Q_{ij}(f) = \frac{\widehat{Q}_{ij}(f)}{\sqrt{\widehat{C}_{ii}(f).\widehat{C}_{jj}(f)}} = \int_0^{2\pi} D(f,\theta).\sin(\vec{\mathbf{k}}.\overrightarrow{\mathbf{x}_{ij}})\, d\theta \quad i = 1,\ldots,N-1 \text{ and } j > i \tag{7}$$

From the cross-spectral analysis step it is possible to get for each frequency 2.M real coefficients related to the unknown Spreading Function at this frequency through (7). Estimating a continuous function over [0 , 2π] satisfying (3) with only 2.M integral properties from (7) lies at the core of directional wave analysis.

3. Directional Analysis Methods used in this study

As it is not possible to find a unique solution to the difficult problem exposed above, several techniques based on different theoretical backgrounds and modelling approaches have been proposed by various authors. The methods taken into account in this comparative study cover most of the operational or research methods. As it is not the scope of this paper to detail each method, only a short description of its principle is reported, with references for further reading.

1. Truncated Fourier Series (TFS) : the Directional Spreading Function is expressed as a truncated Fourier series whose first coefficients are computed from the co- and quad-spectra by a least squares method (Borgman, 1969).

2. Weighted truncated Fourier Series (WFS) : Following Longuet-Higgins *et al.*. (1963) a weighting function is applied to avoid negative values sometimes produced by the former method.

3. Fit to Unimodal Gaussian Model (1MFG) : A unimodal parametric model of gaussian type (Borgman, 1969) is used whose two unknown parameters are computed from the first two Fourier coefficients of the Spreading Function.

4. Fit to Bimodal Gaussian Model (2MF2) : A bimodal parametric model obtained from linear combination of two unimodal Gaussian-type models is used. Its five unknown parameters are calculated from the spectral cross-correlation coefficients by a least-squares method (Benoit, 1991).

5. Maximum Likelihood Method (MLM) : By this method the spreading function is expressed as a linear combination of the cross-spectra. The weighting coefficients are calculated with the condition of unity gain of the estimator in the absence of noise (Krogstad, 1988).

6. Iterative Maximum Likelihood Method (IMLM2) : The estimator produced by the former method is iteratively modified to let its cross-correlation coefficients become closer to the ones obtained from the data (Krogstad, 1988).

7. Maximum Entropy Method (MEM) : The directional estimate is found by minimizing an entropy function under the constraints given by the cross-correlation coefficients (Nwogu, 1989).

8. Bayesian Directional Method (BDM) : With this statistical technique used for regression analysis (Hashimoto *et al.*, 1987), no *a priori* assumption is made about the spreading function which is considered as a piecewise-constant function over $[0, 2\pi]$. The unknown values of $D(f,\theta)$ on each of the K segments dividing $[0, 2\pi]$ are obtained by considering the constraints of the spectral cross-correlation coefficients and an additional condition on the smoothness of $D(f,\theta)$.

9. Variational Fitting Technique - Long-Hasselmann Method (LHM) : Long and Hasselmann (1979) developed this method by which an initial simple estimate is iteratively modified to minimize a "nastiness" function that takes into account the various conditions on the Directional Spreading Function.

4. Description of the validation procedure

This comparative study is based on a three steps validation procedure. In the processing chain of figure 2, we first enter at point 1 by computing directly with the software COQUAD auto- and cross-spectra from a given analytical model of directional spectrum. This allows us to appreciate the directional analysis methods in the case of a "perfect" spectral analysis.

In a second step we include spectral analysis computations through the software SPECAN. The chain is entered at point 2 by computing numerical sea-state through the software DIRSEA. At this step we introduce records of finite duration, signal sampling and spectral analysis.

In the third step we finally enter the chain at point 3 by performing laboratory measurements in the LNH Directional Wave Basin. The practical problems and limitations in signal generation and measurement are added, but we still have a target spectrum to compare to the results of directional analysis.

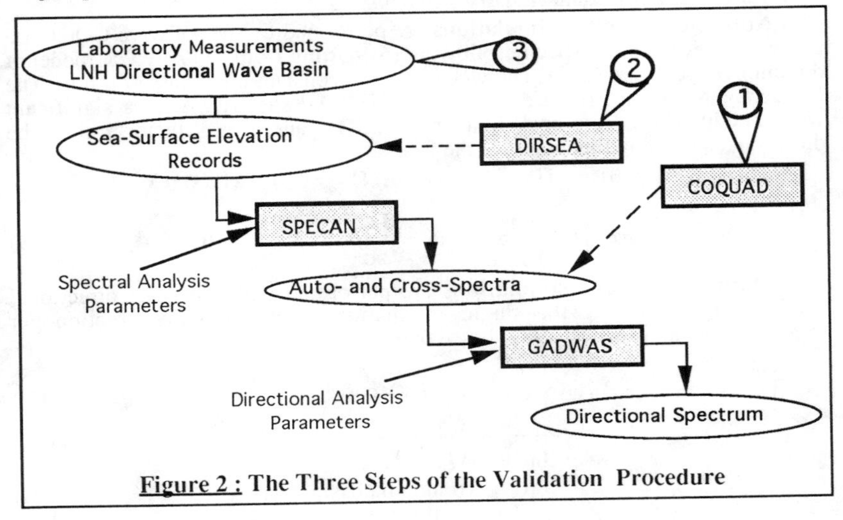

Figure 2 : The Three Steps of the Validation Procedure

Several parameters that may have an influence on the quality of directional analysis are taken into account. They are summarized in Table 1. We also report in this table the various steps at which the influence of these parameters is examined.

Parameters of interest	STEP 1 Num. Spectral Simulations	STEP 2 Num. Wave Simulations	STEP 3 Laboratory Measurements
TYPE OF METHOD	X	X	X
DIRECTIONAL SHAPE	X	X	X
NUMBER OF GAUGES	X		
ARRAY DIMENSION	X		
ARRAY ORIENTATION	X		
SPECTRAL ESTIMATOR		X	X
NOISE LEVEL		X	
ERROR vs CPU TIME	X	X	X

Table 1 : List of parameters whose influence is examined

The volume of tests and results of this study is very important and due to the limited size of this paper only the most significant results will be presented and highlightened here.

5. Description of the tests

5.1 Numerical Spectral and Wave Simulations :

a) **Numerical spectral simulations** are performed by the software COQUAD that computes the cross-spectra following (6) starting from an analytical model of directional spectrum $S(f,\theta)$. The water depth is constant at a value of 0.6 m. The frequency spectrum $E(f)$ is a classical JONSWAP spectrum with a significant wave height of 0.1 m, a peak frequency of 1 Hz and a peak-factor $\gamma=3.3$. The directional spreading function is of the form :

$$\Pi_{\sigma1,\theta1,\sigma2,\theta2,\lambda}(\theta) = \lambda.\Pi_{\sigma1,\theta1}(\theta) + (1-\lambda).\Pi_{\sigma2,\theta2}(\theta) \quad \text{with } 0 \leq \lambda \leq 1 \tag{8}$$

$$\text{with : } \quad \Pi_{\sigma,\alpha}(\theta) = \frac{1}{\sqrt{2\pi}\,\sigma} \exp\left(-\frac{(\theta-\alpha)^2}{2.\sigma^2}\right) \quad \text{if } \theta \in [\alpha-\pi ; \alpha+\pi] \tag{9}$$

b) **Numerical wave simulations** are performed with the same directional spectrum by following the single summation method (single direction per frequency method) (Miles, 1989) based on a discretization of (4) :

$$\eta(x,y,t) = \sum_{n=1}^{N} A_n \cos(2\pi.f_n.t - k_n(x.\cos\theta_n + y.\sin\theta_n) + \varphi_n) \tag{10}$$

with :

$$A_n = \sqrt{2\,S(f_n,\theta_n)\,\Delta f_n\,\Delta\theta_n}$$

$$\varphi_n = 2\pi\,U[0,1] \quad \text{(random phase)} \tag{11}$$

θ_n are of the form $k.\Delta\theta$, but randomly distributed over $[0 , 2\pi]$

A Gaussian Real White Noise with constant spectral density from zero to the Nyquist frequency may be added to the simulated series. The noise level is expressed in terms of a percentage of the significant wave height.

The noise series added to the numerical simulated signals are uncorrelated.

The simulated time series have a time step of 0.05 s and a duration of 819.2 s (16384 points per signal). This represents a record of around 820 waves with 20 points per wave (at peak frequency).

Both for spectral and wave simulations three test-cases based on three different shapes of the spreading function are examined. The first one rather represents a wind-sea with a large spread around a main direction of propagation. In the second case the spread around the main direction is much lower indicating a rather "old" swell. In the third case we have a crossed-sea with two main directions. The parameters of the DSF model (8) for these test-cases are reported in Table 2.

NUMERICAL TEST-CASES	$\sigma 1$	$\theta 1$	$\sigma 2$	$\theta 2$	λ
Num. Test-Case 1 Unimodal Broad DSF	15.°	0.°			1.
Num. Test-Case 2 Unimodal Thin DSF	50.°	0.°			1.
Num. Test-Case 3 Bimodal DSF	12.°	-70.°	45.°	50.°	0.5

Table 2 : Parameters of the numerical simulated DSF

5.2 Laboratory Simulations :

Laboratory tests are performed at LNH in the 30x50 m 3D Wave Facility. The segmented wavemaker is set up along the largest side of the basin and is composed of 56 piston-type paddles of width 0.4 m. Reflective walls are installed at each side of the wavemaker to increase the work area through the use of a corner reflection method.

The water depth is 0.6 m as for numerical simulations. The frequency spectrum is the same JONSWAP spectrum. The Directional Spreading Function is however somewhat different. Instead of the Gaussian model (9), we use a $\cos^{2s}$-type model:

$$\Lambda_{s,\alpha}(\theta) = \frac{1}{\Delta} \cos^{2s}\left(\frac{\pi}{2}\frac{\theta-\alpha}{\theta_{max}}\right) \quad \text{if } \theta \in [\alpha-\pi \, ; \, \alpha+\pi] \tag{12}$$

Four test-cases are processed with the parameters indicated in Table 3.

LABORATORY TEST-CASES	s1	$\theta 1$	s2	$\theta 2$	λ
Lab. Test-Case 1 Unimodal Broad DSF	1	0.°			1.
Lab. Test-Case 2 Unimodal Medium DSF	5	0.°			1.
Lab. Test-Case 3 Unimodal Thin DSF	20	0.°			1
Lab. Test-Case 4 Bimodal DSF	30	-45.°	5	20.°	0.5

Table 3 : Parameters of the laboratory simulated DSF

5.3 Error and CPU Time Measurement :

The calculations are performed on a UNIX-type workstation. The TFS method is choosen as reference for CPU time comparison. For each comparison of methods we use the relative CPU Time compared to the TFS method.

To get a significant indicator of estimation error we use the Weighted Average Percent Error (WAPE) as proposed by Oltman-Shay and Guza (1984) :

$$\text{WAPE} = \frac{\sum_{\theta} \left| \widehat{D}(\theta) - D(\theta) \right|}{\sum_{\theta} D(\theta)} \times 100 \tag{13}$$

in which $D(\theta)$ is the target DSF and $\widehat{D}(\theta)$ is the estimated one.

6. Effect of directional shape (numerical spectral simulations)

The three numerical test-cases presented above (Table 2) are used to examine the effect of directional shape on analysis methods. This comparison is presented here on the first step of the study (anlysis of numerical spectral simulations), but similar results are obtained on numerical wave simulations and then on laboratory measurements. We use here the 5-gauges array described on Figure 4 with R/L=20% (L is the wavelength at peak frequency). The results (WAPE vs relative CPU Time) for the nine methods are plotted on Figure 3.

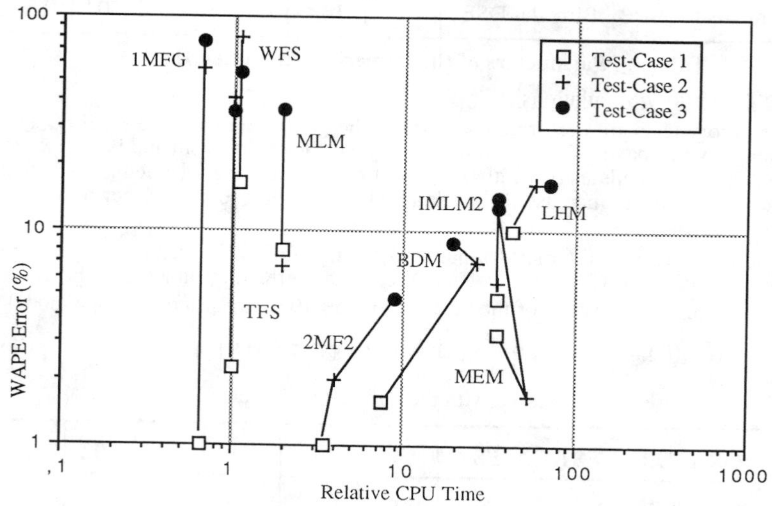

Figure 3 : Influence of directional shape on numerical spectral simulations - 5 Gauges Array .. R/L=20%

Test-case 1 is satisfied by almost all the methods (less than 10% WAPE except for WFS). Test-case 2 (Unimodal Thin DSF) seems the most difficult to overcome for most of the methods. On the bimodal case five methods (MLM, 2MF2, BDM, IMLM2, MEM) give good estimates with less than 10% WAPE. Methods 1MFG, TFS, WFS and MLM are rather sensitive to DSF shape whereas the other ones, and in particular 2MF2, BDM, IMLM2 and MEM, seem more stable and reliable.

7. Effect of number of wave gauges (numerical spectral simulations)

Three different arrays are used with 4, 5 and 6 gauges (See Figure 4).

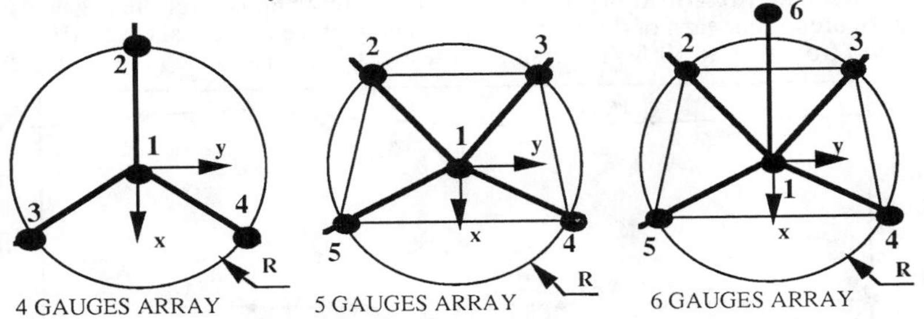

Figure 4 : the three gauge arrays

Figure 5 is an example of comparative results (WAPE error vs CPU Time) for numerical case 2 (Unimodal Thin DSF) with a ratio R/L=20% for the three arrays (L is the wavelength at peak frequency fp).

The general trend is a decrease of WAPE and an increase of CPU Time as the number of probes increases (WFS, TFS, LHM, IMLM2). There is however little effect of the gauge number on the CPU Time for the 2MF2, BDM and MEM methods. The 5-gauges array clearly improves the quality of estimation. In the meantime, except for the TFS, WFS and IMLM2, the 6-gauges array does not exhibit definite improvements. As conclusion of this part, the 5-probes array is considered as a good compromise and is kept for the following tests.

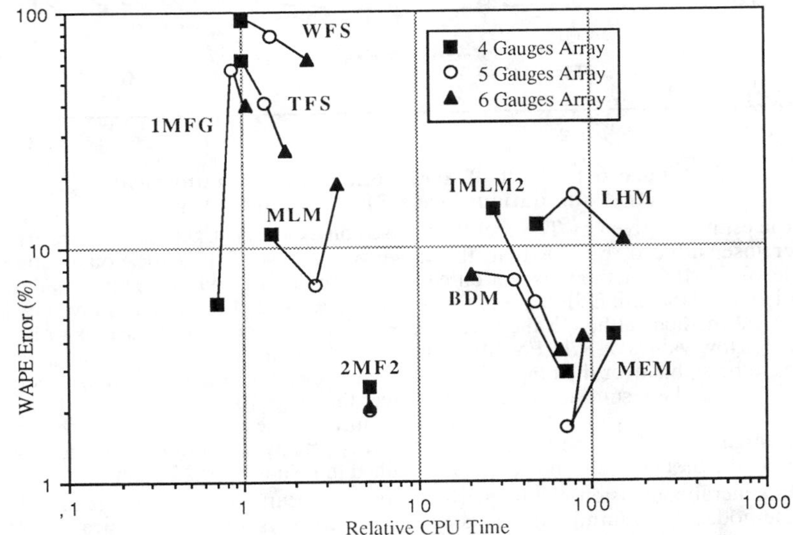

Figure 5 : Influence of wave probe array on numerical spectral simulation (Case 2)

8. Influence of array dimension (numerical spectral simulations)

In order to test the effect of array dimension (represented by the ratio R/L) several tests are performed on numerical spectral simulations with the 5-gauges array (figure 4). For each of the nine methods and the three numerical cases, R/L is increased from 10% to 50% . The results for test-case 3 are plotted on Figure 6.

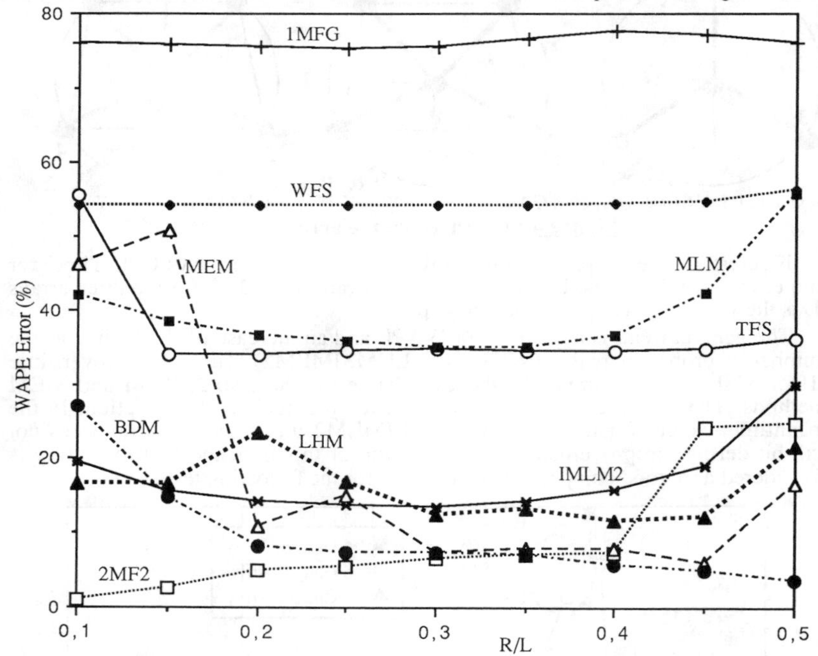

Figure 6 : Effect of array dimension on numerical spectral simulations (case 3) - 5 Gauges Array

The estimates given by TFS and WFS based on exact spectral cross-spectra are rather insensitive of ratio R/L in the range of interest. This same conclusion applies for 1MFG method, except on case 2 where the method overestimates the peak height of the thin DSF close of the limits of the range. The estimates given by the 2MF2 method (although they are in every case of good quality) seems to be better for low values of R/L. For MLM and IMLM2 methods, the estimation error seems to be stable over the range [0.20 ; 0.40] and increases out of this range. For MEM method the results are also constant over the range [0.20 ; 0.45]. The BDM resolution increases with ratio R/L, but the estimate does not vary so much for values higher than 0.2. The behaviour of LHM is somewhat more complicated to describe, the best estimates however are obtained in the range [0.25 ; 0.40].

As general conclusion, the range [0.20 ; 0.40] appears to be acceptable for all the methods. As minimizing the size of the array is often a constraint of operational use, the lower limit of this range R/L=20% is taken as the basic value for the following tests of this sutdy. This aspect however should receive the same attention on laboratory analysis where the input data are not that pure.

9. Influence of spectral estimator (numerical wave simulations)

When starting validation steps 2 and 3 (analysis of numerical wave simulations and laboratory measurements) spectral analysis of time series must be performed. Spectral estimators based on the technique of the periodogramm are used and compared in this study. The whole simulated sequence is quite long (819.2 s or 16384 points) and allows to perform precise computations. Special techniques as signal segmenting, frequency band averaging, data windowing and segment overlapping are used to improve the quality of estimation.

In particular the series may be partitioned in segments to average the raw spectra obtained by simple FFT on each segment. Several segment length are considered with the known rule : when the number of segments increases the estimated spectra are smoother, but the length of segment decreases and the frequency bandwidth is larger. Under the same degrees of freedom it appears that band averaging gives better estimates of spectra than segment averaging does.

Figure 7 shows results (auto- and cross-spectra) of two estimators. The first one is based on a averaging of 32 segments, has a bandwidth of 0.039 Hz and 64 degrees of freedom. It appears very satisfactory and sufficient for the estimation of auto-spectrum, but still shows scattering in the estimation of cross-spectra. Although its spectral resolution is twice the one of the former, the spectral estimator of figure 7-b clearly improves the estimation of cross-spectra. This estimator combines segment overlapping and averaging, frequency band averaging and data windowing and has a number of degrees of freedom greater than 150.

As major conclusion of this section the need for a good spectral analysis is emphasized. The record duration should be longer than for conventional auto-spectral analysis because it appears that spectral estimation of cross-spectra requires more data on input to produce reliable estimate. It is advised to perform few directional analysis on spectral estimates with minimum variance than a lot of directional analysis on spectra with strong variance.

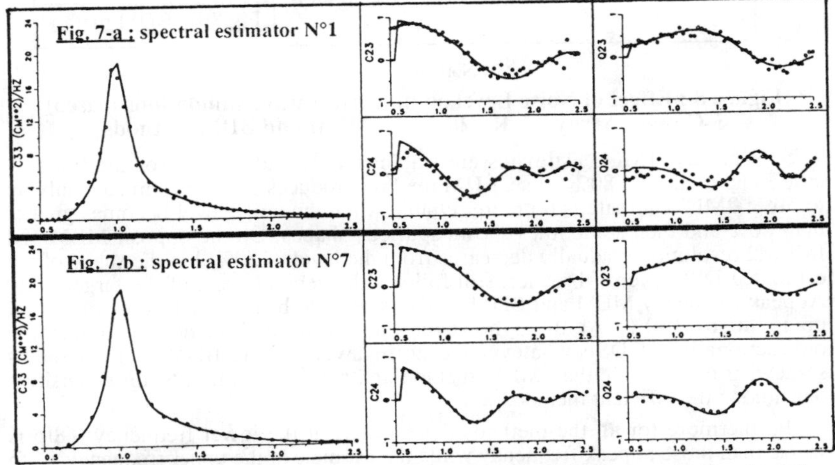

Figure 7 : Examples of auto- and cross-spectral estimates obtained on the same numerical wave simulations

10. Influence of noise level (numerical wave simulations)

The effect of noise level on directional estimation is examined during the second step of validation (numerical wave simulations). The 5-gauges array with R/L=20% is used and Gaussian Real White Noise is added to the wave records simulated by the software DIRSEA (see § 5.1 b).

The spectral estimator defined in section 9 is used for the spectral analysis step by SPECAN. The directional analysis is presented at four frequencies : both frequencies surrounding the target peak frequency fp (fp- and fp+), a lower frequency at 0.8fp and an upper one at 1.3fp. The WAPE as a function of noise level on case 3 is plotted on Figure 8 for MLM (8-a) and BDM (8-b) methods.

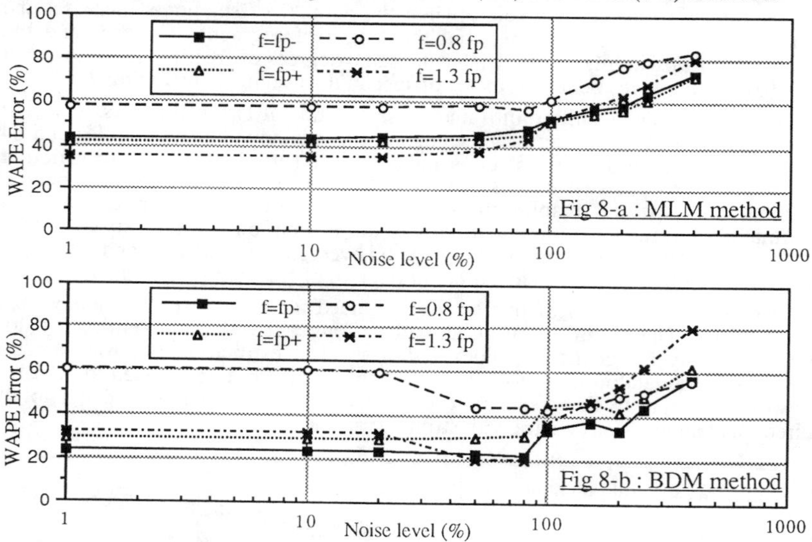

Figure 8 : Effect of Noise Level on numerical wave simulations (case3)
5-Gauges Array - R/=20% - MLM and BDM methods

TFS, WFS and 1MFG estimates are stable (but of bad quality for the bimodal case 3) up to around 80% noise. LHM method produces reliable estimates only up to 50%. 2MF2 estimate is no more reliable when the noise level becomes greater than 80% and then sometimes shows spurious shapes. On the opposit MLM and IMLM2 resolution gradually decreases from the level of 50%, but the shape of the estimated DSF, even if broader, still follows the general trend of the target DSF. At peak frequency MEM and BDM methods seem to be able to handle noise level up to 80 or 100%. At 0.8fp however the MEM method do not detect the bidirectionality of DSF whatever the noise level is. The BDM method seems somewhat more stable than MEM method and appears as the less noise sensitive one among the efficient methods.

Furthermore for all the methods the estimate at the lower frequency 0.8fp is worse than those at peak frequency while the estimate at the upper frequency 1.3fp is of the same quality and sometimes better than those at peak frequency. This partly confirms the choice of section 8 for an array with rather little R/L.

11. Application to laboratory measurements

As described in §5.2 four laboratory tests are performed with various DSF shapes. The three first ones are unimodal with different directional spreads and the fourth one is a severe bimodal case with 65° between the main directions of propagation. We use the same 5-gauges array as for previous steps, with R/L=20%. The peak frequency of the JONSWAP spectra is 1 Hz. Wave time series are recorded on the same basis as for numerical wave simulations : time step : 0.05 s, record length : 819.2 s or 16384 points..

The following observations are extracted from the directional analysis :

- unimodal cases : On the first case (broad DSF), all the methods produce acceptable (TFS, WFS, MLM, LHM), good (IMLM2, MEM) or very good (1MFG, 2MF2, BDM) results. On the second case, the sharper peak of DSF is underestimated by TFS, WFS and LHM methods while the other methods still give estimations of good quality. Figure 10 shows examples of 2D-directional wave spectra for this second case. On the third unimodal case (very sharp DSF), TFS, WFS, 1MFG, LHM and MLM methods are not able to estimate the sharp nature of the DSF. Only 2MF2, IMLM2, MEM and MEM give reliable estimates.

- bimodal case : although the fourth case is rather severe, some methods (IMLM2, 2MF2, MEM and BDM) produce acceptable estimations. The DSF estimated by these methods at both frequencies surrounding the peak frequency are plotted on Figure 9. The other methods hardly detect the bidirectionality (TFS, MLM, LHM) or give broad unimodal DSF (WFS, 1MFG).

These preliminary laboratory tests show the capability of the 5-gauges array as designed on numerical simulations to measure directional waves with different angular spreadings. Some advanced directional analysis methods (IMLM2, 2MF2, MEM and BDM) also reveal their ability to produce reliable estimates on these cases. However some more laboratory tests are necessary (in particular on bimodal seas) to fully appreciate these methods and confirm or alter these first conclusions.

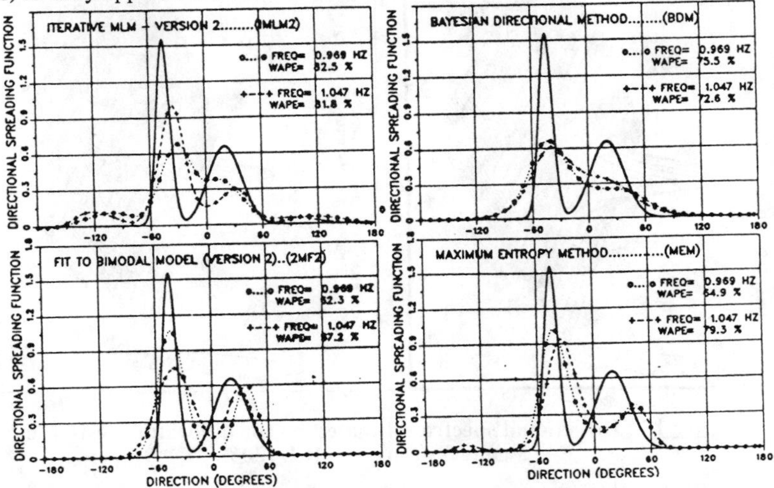

Figure 9 : DSF estimated at peak frequency on the laboratory test-case 4 5-gauges array - R/L=20% - IMLM2, 2MF2, MEM and BDM methods

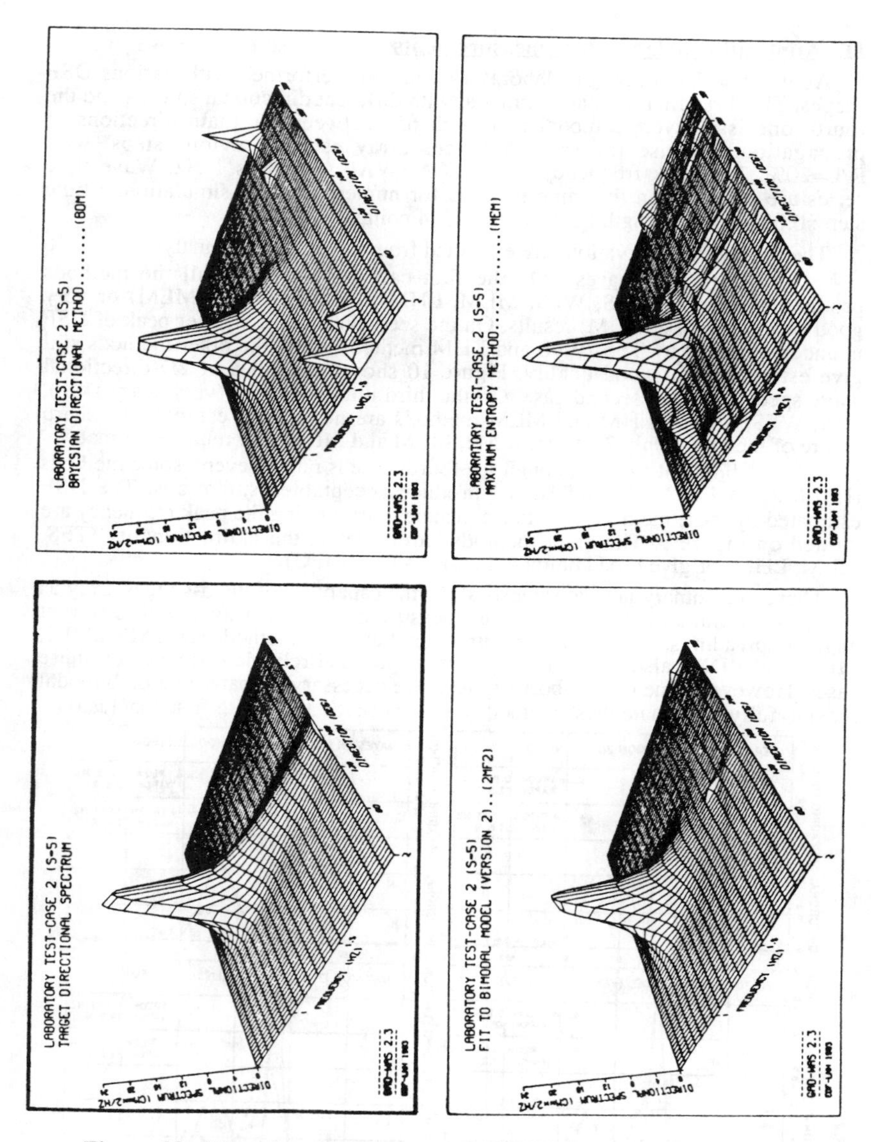

Figure 10 : Directional spectra estimated on the laboratory test-case 2
5-gauges array - R/L=20%

12. Conclusions

As conclusion of this study the following points are underlined :

ARRAY DESIGN : Several recommendations are derived from the tests :

- number of wave gauges : the range [4 ; 6] is reasonable for operational use. An array of 5 gauges as used in this study may be advised.
- array arrangement : this point was not studied, but the same 5-gauges array as Nwogu (1989) was used with success in this work.
- array dimension : a ratio R/L of 0.2 to 0.3 seems effective for operational use.

SPECTRAL ANALYSIS : this step appears to be very important. There is a need for precise and rather long records to perform spectral analysis with a high degrees of freedom and minimum variance. This part should receive great care.

DIRECTIONAL ANALYSIS METHODS : their comparison over a good panel of cases shows that they all have different characteristics and behaviours. The sensitivity on some parameters has been partly studied in this work. Some of the conclusions collected at each of the three steps of the validation procedure are put together in Table 4. Various criteria of appreciation are considered and five of the most efficient methods (2MF2, IMLM2, MEM, BDM and LHM) are placed in the various columns depending on their performances. Although the separation is somewhat arbitrary, this table allows an overview of their characteristics.

Characteristic of method	LEVEL OF SENSITIVITY LOW				HIGH
DSF shape sensitivity		IMLM2 BDM	2MF2 MEM	LHM	
Noise sensitivity	BDM	IMLM2	2MF2 MEM	LHM	
Computation time		2MF2	BDM	IMLM2 MEM	LHM
Implementation difficulty			IMLM2	2MF2 BDM	LHM MEM
Need for tuning		IMLM2 MEM	BDM	2MF2 LHM	
Need for more validation		MEM BDM	IMLM2	LHM	2MF2

Table 4 : Overview of the characteristics of some advanced methods

Concerning the various methods the following comments are expressed :

Fourier Series Methods : Although they are fast methods, the TFS and WFS are not advised for operational use. The WFS method often gives broad peaks and is unable to estimate thin DSF. The TFS usually exhibits the trends of the DSF, but sometimes produces negative values.

Fit to Parametric Model : The 1MFG method (Fit to Unimodal Gaussian Model) is the fastest one, but only produces good results when the target spreading function is unimodal. The 2MF2 method (Fit to Bimodal Gaussian Model) usually gives proper estimates for a short computing time. This promising method however needs more validation cases to improve its robustness in all cases.

Maximum Likelihood Methods : The MLM method usually exhibits acceptable estimates with rather short CPU Time. Its iterative refinement (IMLM2) actually improves the estimation if well tuned. This latter method may be advised for practical use although it requests more computing time.

Maximum Entropy Method : The MEM method often achieves good estimations of the target DSF, but may require some tuning in some cases. On the other hand the CPU time is high and its implementation is not straightforward. This method however is recommended for operational use.

Bayesian Directional Method : Although rather difficult to implement, the BDM method appears to be the most powerful above the methods considered in this study. It may produce good estimates with rather short CPU time if well tuned. This is the prefered method among the ones tested in this work.

Variational Fitting Technique of Long and Hasselmann : the use of this method requires care in the implementation, use and interpretation of results. The estimates often contain spurious peaks. This method does not show definite advantages compared to its rather long and difficult implementation.

13. Acknowledgements

This study is a joint research program between EDF-LNH and the French "Secrétariat d'Etat chargé de la Mer (STNMTE)".

14. References

BENOIT M. (1991) : Comparison of methods for estimating directional wave-spectrum. Part 1 : Ability of the methods to build theoretical spectra with heave-pitch-roll data by assuming a perfect spectral analysis. *Report. EDF-LNH, HE42/91.32. (in french)*

BENOIT M. (1992) : Practical comparative performance survey of methods used for estimating directional wave spectra from heave-pitch-roll data. *Proc. 23rd Int. Conf. on Coastal Eng., Vol 1, pp 62-75, Venice (Italy)*

BORGMAN L.E. (1969) : Directional spectra models for design use. *Offshore Technology Conference, Houston, Texas*

HASHIMOTO N., KOBUNE K., KAMEYAMA Y. (1987) : Estimation of directional spectrum using the Bayesian approach and its application to field data analysis. *Report of the Port and Harbour Research Institute, vol 26, pp 57-100.*

HOWELL G.L. (1992) : A new nearshore directional wave gage. *Proc. 23rd Int. Conf. on Coastal Eng., Vol 1, pp 295-307, Venice (Italy)*

KROGSTAD H.E. (1988): Maximum likelihood estimation of ocean wave spectra from general arrays of wave gauges. *Modelling, Identification and Control, vol 9, pp 81-97*

LONG R.B., HASSELMANN K. (1979) : A variational technique for extracting directional spectra from multicomponent wave data. *J. Phys. Oceanog., vol 9, pp 373-381*

LONGUET-HIGGINS M.S., CARTWRIGHT D.E., SMITH N.D. (1963) : Observations of the directional spectrum of sea waves using the motions of a floating buoy. *Ocean Wave Spectra, Prentice-Hall, pp 111-136.*

MILES M.D. (1989) : A note on directional random wave synthesis by the Single Summation Method. *Proc. 23rd IAHR Congress, Ottawa, Canada , Vol C, pp243-250*

NWOGU O.U. (1989) : Maximum entropy estimation of directional wave spectra from an array of wave probes. *Applied Ocean Research, vol 11, N°4, pp 176-182*

OLTMAN-SHAY J., GUZA R.T. (1984) : A data-adaptative ocean wave directional spectrum estimator for pitch-roll type measurements. *J. Phys. Oceanogr., vol 14, pp 1800-1810.*

NONLINEAR EFFECTS ON WAVE ENVELOPE AND PHASE

By M. Aziz Tayfun,[1] Member, ASCE

Abstract

The representation of second-order random waves is examined in terms of the effect of second-order nonlinearities on the wave envelope and phase. Theoretical expressions describing the joint, marginal, and conditional distributions of wave envelope and phase are derived systematically, correct to first order in the rms surface slope. Some immediate implications of these results are discussed in detail. In particular, it is found that the wave envelope is Rayleigh-distributed, as in the case of linear waves, whereas the wave-phase distribution is nonuniform over the interval $(0,2\pi)$. As the rms surface slope and thereby the surface skewness increase, the phase distribution deviates from uniformity progressively, indicating an excess of values near the mean phase π and corresponding symmetrical deficiencies away from the mean toward 0 and 2π. Comparisons with four sets of wave data gathered in the Gulf of Mexico during a hurricane provide a favorable confirmation of these theoretical results, and thus reinforce the validity of the second-order random wave model.

INTRODUCTION

In the linear Gaussian wave theory, it is known that the wave envelope and phase are statistically independent, the envelope heights are Rayleigh-distributed, and the phases are uniformly random over an interval of 2π. In the nonlinear case, the surface elevations and associated statistics exhibit non-Gaussian characteristics. One would then also expect the envelope and phase statistics to be modified accordingly. Following this line of thought, Tayfun and Lo (1990) have shown that nonlinearities do affect the phase statistics noticeably, e.g. leading to an excess of values near the mean phase. But, the effect of nonlinearities on the envelope statistics, if any, appeared relatively less significant. Though comparisons with observational data seemed to support these conclusions, they were based on an approximate form of the second-order theory, appropriate to unidirectional waves with narrow-banded characteristics. Further, some results on the statistical cumulants associated with the surface elevations contained high-order terms that cannot possibly be justified on the basis of the second-order theory alone. Such drawbacks of the Tayfun-Lo model may thus warrant a more general systematic analysis of the effect of nonlinearities on the wave envelope and phase.

The envelope-phase problem is reconsidered here in a manner consistent with the second-order theory and without any unrealistic

---

[1] Prof., Dept. of Civil Engrg., Coll. of Engrg. Petroleum, Kuwait Univ., P.O. Box 5969, 13060 Safat, Kuwait.

limitations on wave directionality or spectral bandwidth. However, only deep-water waves are considered since the associated physics and algebra are considerably simpler. This also helps avoid additional complications that would otherwise arise from the presence of a finite or shallow water depth, in particular, regarding the validity of the second- or higher-order random wave theories in such cases. First, a brief description of the second-order theory, some key definitions, and parameters relevant to the analysis of wave envelope and phase are given. Then, the derivation of theoretical expressions for the joint, marginal, and conditional statistics of the wave envelope and phase follows in a straightforward manner via the characteristic function expansion technique. This procedure is detailed in Tayfun and Lo (1990), and corresponds to a simpler version of the more general two-variable expansion scheme devised by Longuet-Higgins (1963). The nature of these theoretical results are then discussed and, subsequently, compared with wave data gathered during hurricane Camille in the Gulf of Mexico in 1969.

SECOND-ORDER RANDOM WAVE THEORY

Surface Elevation in Time

The surface elevation from the mean level is given by

$$\eta = \eta(\underline{\mathbf{x}}, t) = \eta_1 + \eta_2 \tag{1}$$

where $\underline{\mathbf{x}}=(x,y)=$ horizontal coordinates fixed on the mean surface level; t=time; η_1 and η_2= the first- and second-order solutions, respectively. If η is statistically homogeneous and observed at some fixed location $\underline{\mathbf{x}}_0$ as a function of time only, then there is no loss of generality in shifting the origin of $\underline{\mathbf{x}}$ so that $\underline{\mathbf{x}}_0=(0,0)$. On this basis, the first-order linear (Gaussian) solution can be expressed as

$$\eta_1 = \lim_{N\to\infty} \sum_{n=1}^{N} c_n \cos(\omega_n t + \epsilon_n) \tag{2}$$

where c_n= amplitude of the nth wave component; ω_n= frequency in rad/s; and, ϵ_n= independent random phases uniformly distributed over the interval $(0,2\pi)$.

Any two components of η_1 with frequencies ω_m and ω_n generate two nonlinear nonresonant components of frequencies $\omega_m+\omega_n$ and $\omega_m-\omega_n$. The second-order solution comprises such components and has the form

$$\eta_2 = \eta_2^+ + \eta_2^- \tag{3a}$$

$$\eta_2^\pm = \lim_{N\to\infty} \frac{1}{4} \sum_{m=1}^{N}\sum_{n=1}^{N} c_m c_n K_{m,n}^\pm \cos[(\omega_m \pm \omega_n)t + (\epsilon_m \pm \epsilon_n)] \tag{3b}$$

$$K_{m,n}^\pm = K^\pm(\underline{k}_m, \underline{k}_n) = K^\pm[k_m, k_n, \cos(\theta_m - \theta_n)] \tag{3c}$$

where $\underline{k}_m$, $\underline{k}_n$ = horizontal wave-number vectors with moduli k_m and k_n; θ_m, θ_n= directions of $\underline{k}_m$ and $\underline{k}_n$, measured positive counterclockwise from the x-axis; and, $K^{\pm}$ represent the interaction coefficients or kernels whose explicit forms are well known and thus not repeated here for economy of space (see e.g. Longuet-Higgins 1963, Tayfun 1990).

In deep water,

$$\omega_n^2 = gk_n \tag{4}$$

where g= gravitational acceleration. Also, the amplitude c_n is related to various forms of the surface elevation spectrum (density). In particular, if t is fixed while η_1 is viewed as a function of **x**, then as $N \to \infty$

$$E_1(\underline{k})\ d\underline{k} = \frac{1}{2} c_n^2 \tag{5}$$

where E_1= two-dimensional wave-number spectrum; $\underline{k}=(k_x, k_y)$ with k_x and k_y designating the components of $\underline{k}$ in the x and y directions; and, $d\underline{k}=dk_x dk_y$. If, on the other hand, η_1 is observed at fixed **x** as a function of t only (as in the present case), then as $N \to \infty$

$$F_1(\omega)\ d\omega = \frac{1}{2} c_n^2 \tag{6}$$

where F_1 = one-dimensional frequency spectrum. By using a polar description for the wave-number plane, the two spectra so defined can be related to one another in the form

$$F_1(\omega)\ d\omega = k\, dk \int E_1(\underline{k})\ d\theta \tag{7}$$

where $\theta=\tan^{-1}(k_y/k_x)$; $\underline{k}=(k\cos\theta,\ k\sin\theta)$; and, $\omega^2=gk$ as the continuous limit of (4) when $N \to \infty$.

Wave Envelope and Phase

The Hilbert transform or conjugate of η in time is defined by

$$\hat{\eta} = \frac{1}{\pi} \int_{-\infty}^{\infty} \frac{\eta(\tau)}{t-\tau}\, d\tau \tag{8}$$

where the Cauchy principal value of the integral is implied. In the present case, it can be shown that

$$\hat{\eta} = \hat{\eta}_1 + \hat{\eta}_2 = \hat{\eta}_1 + \hat{\eta}_2^{+} + \hat{\eta}_2^{-} \tag{9a}$$

$$\hat{\eta}_1 = \lim_{N\to\infty} \sum_{n=1}^{N} c_n \sin(\omega_n t+\epsilon_n) \tag{9b}$$

$$\hat{\eta}_2^{\pm} = \lim_{N\to\infty} \frac{1}{4} \sum_{m=1}^{N} \sum_{n=1}^{N} c_m c_n K_{m,n}^{\pm} \sin[|\omega_m \pm \omega_n| t + (\epsilon_m \pm \epsilon_n)] \qquad (9c)$$

Evidently, $<\eta>=<\hat{\eta}>=0$, where the angled brackets stand for the expectation operator. Further,

$$\langle\eta\hat{\eta}\rangle = 0 \qquad (10a)$$

$$\eta_{rms}^2 = \langle\eta^2\rangle = \langle\hat{\eta}^2\rangle \qquad (10b)$$

Consider now the complex process

$$W = \eta + i\hat{\eta} = A \exp(i\phi) \qquad (11)$$

where $i^2 = -1$, and

$$A(t) = (\eta^2 + \hat{\eta}^2)^{1/2} \quad , \quad (A \geq 0) \qquad (12a)$$

$$\phi(t) = \tan^{-1}\left(\frac{\hat{\eta}}{\eta}\right) \quad , \quad (0 \leq \phi < 2\pi) \qquad (12b)$$

simply correspond to the modulus and argument of W, respectively. Thus,

$$\eta(t) = A(t) \cos\phi(t) \qquad (13)$$

from which it follows that A and ϕ are interpreted as the wave envelope (or amplitude) and phase. Obviously, from (12a),

$$\langle A^2\rangle = A_{rms}^2 = 2\,\eta_{rms}^2 \qquad (14)$$

Spectral Moments and Bounds

Let the ordinary moments of F_l and E_l be defined respectively by m_j and M_j (j=0,1,...) so that $m_{2j}=g^j M_j$ by virtue of the dispersion relation. The probability structure of η_l and its spatial/temporal properties depend on these moments. For example,

$$M_0 = m_0 = \langle\eta_1^2\rangle \qquad (15a)$$

$$M_2 = \langle|\underline{\nabla}\eta_1|^2\rangle = \mu^2 \qquad (15b)$$

$$m_4 = \langle\ddot{\eta}_1^2\rangle = g^2\mu^2 \qquad (15c)$$

where $\underline{\nabla}=(\partial/\partial x, \partial/\partial y)$= horizontal gradient operator; $\ddot{\eta}_l = \partial^2\eta_l/\partial t^2$; and, μ= rms surface slope (gradient).

Consider now the mean-square of η: Because η_1 and η_2 are orthogonal,

$$\langle\eta^2\rangle = \langle\eta_1{}^2\rangle + \langle\eta_2{}^2\rangle \tag{16}$$

The first term on the right is m_0 obviously, and the second term representing the nonlinear correction follows after some algebra from (3) and (5) as

$$\langle\eta_2{}^2\rangle = \frac{1}{4}\int_{\underline{k}}\int_{\underline{k}'} G(\underline{k},\underline{k}')\, E_1(\underline{k})\, E_1(\underline{k}')\, d\underline{k}d\underline{k}' \tag{17a}$$

where

$$G(\underline{k},\underline{k}') = [K^+(\underline{k},\underline{k}')]^2 + [K^-(\underline{k},\underline{k}')]^2 \tag{17b}$$

The maximum $2(k^2+k'^2)$ of G occurs for the co-linear components of the first-order field, corresponding to $\theta-\theta'= 0, \pm\pi$ (Tayfun 1990). Thus,

$$0 < G(\underline{k},\underline{k}') \leq 2(k^2+k'^2) \tag{18}$$

which is substituted in (17a), and the resulting expression is simplified further via (15) to obtain

$$0 < \langle\eta_2{}^2\rangle \leq m_0\,\mu^2 \tag{19}$$

In general then,

$$m_0 < \langle\eta^2\rangle \leq m_0\,(1+\mu^2) \tag{20}$$

The upper bound holds as a strict equality for the theoretical limit case of a unidirectional wave field. Note further that Phillips' (1985) equilibrium arguments coupled with the trend of oceanic data at wind speeds greater than 10 m/s indicate that $\mu^2 \leq 0.06$ as an upper-limit asymptote. This implies first that the contribution of second-order nonlinearities to $\langle\eta^2\rangle$ should be less than 6% under oceanic conditions of local statistical equilibrium. Further, μ is a key nonlinearity measure and can be used as a parameter for an order-of-magnitude analysis of η_1 and η_2, and also of their conjugates, to which all the results of this section apply identically. Formally, a zero-mean random variable, say, X with the rms μ^j ($j\geq 0$) is $O(\mu^j)$ in the sense that $|X|\leq\beta\mu^j$ nearly always for a sufficiently large constant $\beta>0$. For instance, if X is Gaussian, $\text{Prob}\{|X|>\beta\mu^j\}\approx 0$ for $\beta>4\sim5$. In the more general case where X is non-Gaussian, one can obviously resort to Tchebycheff's inequality as a conservative basis for the same argument. For the present model, η_1 and $\hat{\eta}_1$ are zero-mean Gaussian with the rms $m_0^{1/2}$, while η_2 and $\hat{\eta}_2$ are zero-mean non-Gaussian with the upper-limit rms $m_0^{1/2}\mu$. Thus, $\eta_1/m_0^{1/2}$ and $\hat{\eta}_1/m_0^{1/2}$ are O(1), whereas $\eta_2/m_0^{1/2}$, $\hat{\eta}_2/m_0^{1/2}$, and so $\eta_2^{\pm}/m_0^{1/2}$ and $\hat{\eta}_2^{\pm}/m_0^{1/2}$ are all $O(\mu)$ at most.

DISTRIBUTIONS OF ENVELOPE AND PHASE

Joint Moments

The joint moments of η and $\hat{\eta}$ are defined by

$$\kappa_{rs} = \langle \eta^r \hat{\eta}^s \rangle \qquad (r,s=0,1,\ldots) \tag{21}$$

In the present model,

$$\kappa_{10} = \kappa_{01} = \kappa_{11} = 0 \tag{22a}$$

$$\kappa_{20} = \kappa_{02} = \eta^2_{rms} \tag{22b}$$

For r+s=3, an expansion of κ_{rs} in terms of the components η_1, $\eta_2^{\pm}$, $\hat{\eta}_1$, and $\hat{\eta}_2^{\pm}$ gives

$$\kappa_{03} = \kappa_{21} = 0 \tag{23a}$$

and, correct to $O(\mu)$,

$$\kappa_{30} = 3\,\kappa_{12} \tag{23b}$$

where

$$\kappa_{12} = \int_{\underline{k}} \int_{\underline{k}'} K(\underline{k},\underline{k}')\, E_1(\underline{k})\, E_1(\underline{k}')\, d\underline{k} d\underline{k}' \tag{24a}$$

with

$$K(\underline{k},\underline{k}') = \frac{1}{2}\,[\,K^+(\underline{k},\underline{k}') + K^-(\underline{k},\underline{k}')\,] \tag{24b}$$

For r+s ≥ 4 , all the components that contribute to the joint moments κ_{rs} are found to be $O(\mu^2)$ at most. Thus, they need not be considered any further since η is correct to $O(\mu)$ only. Bearing this in mind, it is convenient to normalize κ_{rs} in the form

$$\lambda_{rs} = \frac{\kappa_{rs}}{\eta_{rms}^{\;r+s}} \tag{25}$$

Clearly then, $\lambda_{rs}=0$ for r+s=1,2, and 3 except for $\lambda_{20}=\lambda_{02}=1$, and $\lambda_{30}=3\lambda_{12}$, which is recognized as the coefficient of skewness.

Densities of η and $\hat{\eta}$

To $O(\mu)$, the joint probability density of $\xi=\eta/\eta_{rms}$ and $\hat{\xi}=\hat{\eta}/\eta_{rms}$ is of the form (Tayfun and Lo 1990)

$$p_{\xi\hat{\xi}} = \frac{e^{-\frac{1}{2}(\xi^2+\hat{\xi}^2)}}{2\pi} \left[1+\frac{1}{3!}\sum_{j=0}^{3} \frac{3!}{(3-j)!\,j!}\lambda_{(3-j)j}H_{(3-j)}(\xi)\,H_j(\hat{\xi})\right] \tag{26}$$

where H_j= jth-order Hermite polynomial. Because $\lambda_{30}=3\lambda_{12}$ and $\lambda_{(3-j)j}=0$ for j=1 and 3, (26) is simplified to

$$p_{\xi\hat{\xi}} = \frac{e^{-\frac{1}{2}(\xi^2+\hat{\xi}^2)}}{2\pi} \left[1+\frac{\lambda_3}{6}\xi(\xi^2+\hat{\xi}^2-4)\right] \tag{27}$$

where $\lambda_3=\lambda_{30}$ for notational ease, and $-\infty<\xi,\hat{\xi}<\infty$. The marginal densities of ξ and $\hat{\xi}$ follow easily by integration from (27) as

$$p_{\xi} = \frac{e^{-\frac{1}{2}\xi^2}}{\sqrt{2\pi}} \left[1+\frac{\lambda_3}{6}\xi(\xi^2-3)\right] \tag{28a}$$

$$p_{\hat{\xi}} = \frac{e^{-\frac{1}{2}\hat{\xi}^2}}{\sqrt{2\pi}} \tag{28b}$$

The first of these, which is well known and has been articulated by Longuet-Higgins (1963) and others, quite simply shows that ξ is non-Gaussian with a positive skewness, since $\lambda_3 > 0$ in general. The second indicates that although $\hat{\xi}$ is nonlinear, it is nevertheless Gaussian to $O(\mu)$. Departures of $p_{\hat{\xi}}$ from the Gaussian shape (e.g. in terms of an excess at the mode) can be shown to be $O(\mu^2)$.

Densities of Wave Envelope and Phase

Let $\zeta=A/A_{rms}$, where $A_{rms}=\sqrt{2}\eta_{rms}$. So, from (13)

$$\xi = \sqrt{2}\,\zeta\cos\phi \tag{29a}$$

$$\hat{\xi} = \sqrt{2}\,\zeta\sin\phi \tag{29b}$$

A standard change of variables from $(\xi,\hat{\xi})$ to (ζ,ϕ) with the Jacobian 2ζ yields the joint probability density of ζ and ϕ in the form

$$p_{\zeta\phi} = \frac{\zeta}{\pi}e^{-\zeta^2}\left[1+\frac{\sqrt{2}}{3}\lambda_3\zeta(\zeta^2-2)\cos\phi\right] \tag{30}$$

where $\zeta\geq 0$, and $0\leq\phi<2\pi$. The leading term of (30) is the same as the joint density in the <u>linear</u> case, for which the corresponding marginal densities are

$$p_\zeta = 2\zeta e^{-\zeta^2} \tag{31}$$

$$p_\phi = \frac{1}{2\pi} \tag{32}$$

Thus, $p_{\zeta\phi}=p_\zeta p_\phi$, and ζ and ϕ are (statistically) independent. In the <u>nonlinear</u> case, the marginal density of ζ has the same form as (31), whereas

$$p_\phi = \frac{1}{2\pi}\left(1 - \frac{1}{6}\sqrt{\frac{\pi}{2}}\,\lambda_3 \cos\phi \right) \tag{33}$$

Generally then, $p_{\zeta\phi}$ cannot be expressed as a product of p_ζ and p_ϕ . Thus, ζ and ϕ are not independent, though they are uncorrelated since

$$\mathrm{cov}(\zeta,\phi) = 0 \tag{34}$$

To $O(\mu)$, ζ is Rayleigh-distributed in either case. For the linear field, p_ϕ is uniform with $\langle\phi\rangle=\pi$ and

$$\mathrm{var}(\phi) = \frac{\pi^2}{3} \tag{35}$$

In the nonlinear case, it is still true that $\langle\phi\rangle=\pi$, but

$$\mathrm{var}(\phi) = \frac{\pi^2}{3} - \frac{1}{3}\sqrt{\frac{\pi}{2}}\,\lambda_3 \tag{36}$$

and p_ϕ is deformed symmetrically with respect to $\langle\phi\rangle=\pi$, showing a maximum at $\phi=\pi$ and minima at $\phi=0$ and 2π. Accordingly, var(ϕ) is reduced, as is clearly seen from a comparison of (35) and (36). Specific examples illustrating the explicit forms of p_ζ and p_ϕ will be given later in the next section.

The conditional density $p_{\zeta|\phi}$ of ζ, given ϕ, has the form

$$p_{\zeta|\phi} = p_\zeta \frac{1 + \frac{\sqrt{2}}{3}\lambda_3 \zeta(\zeta^2-2)\cos\phi}{1 - \frac{1}{6}\sqrt{\frac{\pi}{2}}\,\lambda_3\cos\phi} \tag{37}$$

As $\lambda_3 \to 0$, $p_{\zeta|\phi} \to p_\zeta$, which is also true for $\phi=\pi/2$ and $3\pi/2$ corresponding to zero crossings of η .

The conditional density of ϕ, given ζ, is

$$p_{\phi|\zeta} = \frac{1}{2\pi} \left[1 + \frac{\sqrt{2}}{3} \lambda_3 \zeta (\zeta^2 - 2) \cos\phi \right] \tag{38}$$

Similarly, as $\lambda_3 \to 0$ and at zero crossings of η , $p_{\phi|\zeta} \to 1/2\pi$. Note further that for $\zeta > \sqrt{2}$, $p_{\phi|\zeta}$ has maxima at $\phi=0,2\pi$, and a minimum at $\phi=\pi$, and shows an excess of values for $0 < \phi < \pi/2$ <u>and</u> $3\pi/2 < \phi < 2\pi$, and a corresponding deficiency for $\pi/2 < \phi < 3\pi/2$. These intervals coincide with the crest ($\eta>0$) and trough ($\eta<0$) segments of the surface profile. For $\zeta < \sqrt{2}$, the preceding arguments are simply reversed. Incidentally, the particular value $\zeta=\sqrt{2}$ is very nearly the same as the mean $\zeta_{1/3} = 1.001\sqrt{2}$ of the 1/3 highest ζ values. Thus, it appears that over any time interval where $\zeta \geq \zeta_{1/3}$, η tends to stay longer above the mean level $\langle\eta\rangle=0$ than below it. Clearly, the opposite is true when $\zeta<\zeta_{1/3}$.

Now, let p_ζ^- denote the conditional density of ζ, given that $\eta<0$, which occurs if and only if $\pi/2 < \phi < 3\pi/2$. Similarly, define p_ζ^+ as the conditional density of ζ, given $\eta>0$, which occurs if and only if $0 < \phi < \pi/2$ <u>and</u> $3\pi/2 < \phi < 2\pi$. As the spectrum bandwidth narrows and approaches zero, these densities tend to describe the distributions of crest and trough amplitudes. In the same narrow-band limit of the linear theory, the latter quantities are both described by the Rayleigh law p_ζ. Now, note that $p_{\zeta\phi}$ and p_ϕ are symmetrical with respect to $\phi=\pi$. Thus, it follows by definition that

$$p_\zeta^+ \int_0^{\pi/2} p_\phi \, d\phi = \int_0^{\pi/2} p_{\zeta\phi} \, d\phi \tag{39a}$$

$$p_\zeta^- \int_{\pi/2}^{\pi} p_\phi \, d\phi = \int_{\pi/2}^{\pi} p_{\zeta\phi} \, d\phi \tag{39b}$$

In view of (30) and (33), these will lead to

$$\frac{p_\zeta^\pm}{p_\zeta} = \Omega^\pm = \frac{1 \pm \frac{2\sqrt{2}}{3\pi} \lambda_3 \zeta (\zeta^2 - 2)}{1 \mp \frac{\lambda_3}{3\sqrt{2\pi}}} \tag{40}$$

It is seen that if $\lambda_3 \to 0$, then $\Omega^\pm \to 1$ so that $p_\zeta^\pm \to p_\zeta$ as in the linear theory. In the more general case, $\Omega^+<1$ ($\Omega^->1$) if $0.228<\zeta<1.287$, and $\Omega^+>1$ ($\Omega^-<1$) otherwise. Thus, the probability mass under p_ζ^+ flattens out slightly, and shows an excess of values for $0<\zeta<0.228$ <u>and</u> $\zeta>1.287$, and a deficiency for $0.228<\zeta<1.287$. As for p_ζ^-, it quite simply exhibits just the opposite behavior. In either case, the physical picture implied appears to be in accord with the vertically skew nature of the surface profile in terms higher more pointed crests and shallower more rounded troughs. A comparison of p_ζ^+, p_ζ^-, and p_ζ for $\lambda_3=0.3$ is shown in Fig. 1. If the mean and the rms values associated with $p_\zeta^\pm$ are denoted by $\langle\zeta\rangle^\pm$ and $\zeta_{rms}^\pm$, then

$$\frac{\langle\zeta\rangle^{\pm}}{\langle\zeta\rangle} = \left(1 \mp \frac{\lambda_3}{3\sqrt{2\pi}} \right)^{-1} \tag{41a}$$

$$\frac{\zeta_{rms}{}^{\pm}}{\zeta_{rms}} = \left(1 \pm \frac{\lambda_3}{2\sqrt{2\pi}} \right)^{\frac{1}{2}} \left(1 \mp \frac{\lambda_3}{3\sqrt{2\pi}} \right)^{-1} \tag{41b}$$

where $\langle\zeta\rangle=\sqrt{\pi}/2$ and $\zeta_{rms}=1$. Thus, for the case shown in Fig. 1, the mean of ζ, given $\eta>0$ ($\eta<0$), is about 4% more (less) than $\langle\zeta\rangle$; and, the rms of ζ, given $\eta>0$ ($\eta<0$), is 5% more (less) than ζ_{rms} .

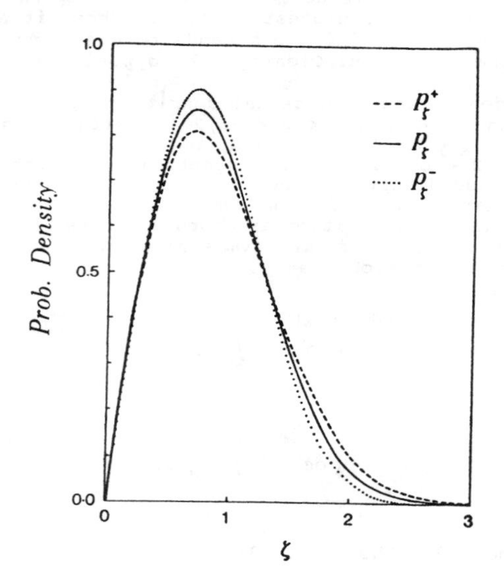

Fig 1. A Comparison of $p_\zeta^{\pm}$ with p_ζ for $\lambda_3=0.3$.

COMPARISONS WITH DATA

The wave data utilized in the comparisons to follow were obtained with the Ocean Data Gathering Program during hurricane Camille in the Gulf of Mexico in 1969 (for details, see e.g. Hamilton and Ward 1974). The original record was arranged in hourly files designated as camwave.nn, nn being the file number. The present analysis is based on four slightly overlapping segments, each comprising 2^{12}=4,096 data points sampled at 1 Hz and starting at 1-hr intervals coincident with the beginning of files nn=12, 13, 14, and 15. These are hereafter referred to as wave data 1, 2, 3, and 4, respectively. Wave data 3 and 4 are the same as the data considered earlier by Tayfun and Lo (1990). The previous analysis of these included the estimates of spectra and histograms of surface displacements, which are not repeated here. The analysis here will first deal with the observed values of various parameters of interest such as spectral moments, skewness coefficient, and surface slope. Attention is then focused on a comparison of the observed histograms of wave envelope and phase with their theoretical predictions.

The observed values of the spectral moments, skewness coefficient, and mean-square surface slope are summarized in Table 1. The trend of m_0 values in column (2) suggests that the wave field around the measurement site was in a nonstationary state, building up and reaching an extreme as the hurricane approached the site. The values in column (5) indicate that max μ^2 < 0.05, including the extreme conditions implied by data 4 with waves as large as 21~22 m in height.

Table 1. Observed Parameters of Wave Data 1, 2, 3, and 4.

Wave Data (1)	m_0 (m^2) (2)	m_2 (m^2/s^2) (3)	m_4 (m^2/s^4) (4)	m_4/g^2 (5)	λ_3 (6)
1	6.110	2.622	2.751	0.028	0.154
2	7.283	3.064	3.201	0.033	0.126
3	7.634	3.602	3.948	0.041	0.254
4	10.128	4.509	4.766	0.050	0.262

The construction of wave-envelope/phase time series from a given time series of the surface displacement follows standard procedures described in detail elsewhere (Tayfun 1986). The present theory suggests that envelope heights scaled with the rms envelope height is Rayleigh-distributed, as in (31). In view of Table 1, the segmental rms

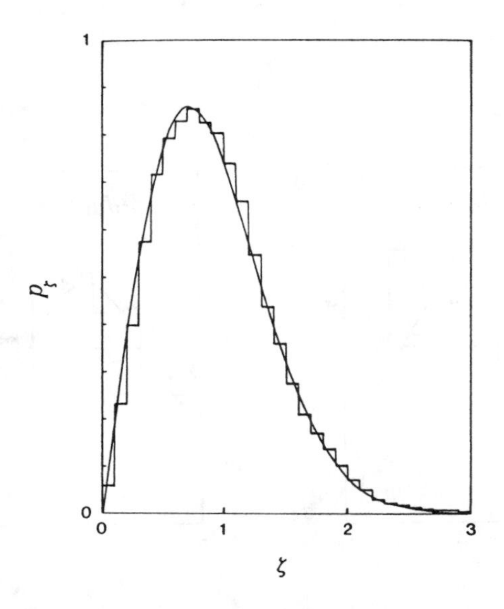

Fig 2. Rayleigh Density p_ζ versus Histogram of Scaled Envelopes from Composite of Wave Data 1, 2, 3 & 4.

envelope height, viz. $(2m_0)^{1/2}$ varies from one data segment to the next. Nevertheless, if the theory is correct, then a composite of the envelopes generated from each of the four data segments separately and scaled with the appropriate segmental rms must have the same distribution. The envelope histogram of the composite data so generated is compared with the Rayleigh density in Fig. 2. The comparison is seen to be very favorable, with the observed $\langle\zeta\rangle=0.885$ and $var(\zeta)=0.215$ agreeing closely with the theoretical $\langle\zeta\rangle=0.886$ and $var(\zeta)=0.214$.

The theoretical wave-phase density depends on the skewness coefficient, which also varies from one data segment to the next. Thus, the histogram of each data must be computed separately for a proper comparison with the corresponding theoretical density. These are shown in Fig. 3. In all cases, the observed and theoretical means differ by

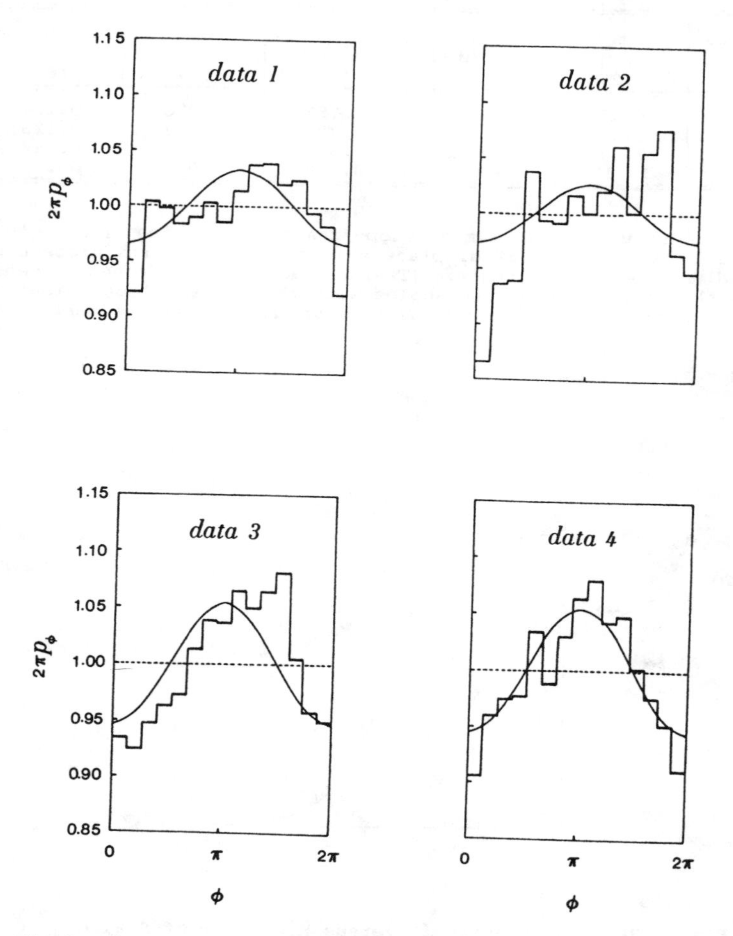

Fig 3. Theoretical p_ϕ versus Observed Wave-Phase Histograms.

less than 1.2% , and the variances by at most 1%. It is seen that the overall trend of the histograms is predicted reasonably well by the theoretical densities, particularly for wave data 3 and 4 with the larger skewness coefficients. For wave data 1 and 2 characterized with the relatively small skewness coefficients, the apparent instability of histogram ordinates makes it difficult to establish a clear quantitative correspondence with the theoretical predictions. This is also true to some extent for wave data 3. In all cases of Fig. 3, the (relative) standard error of histogram ordinates is about 6% of the corresponding theoretical values. Unfortunately, this is either much larger than or nearly comparable to the magnitudes by which the theoretical phase density differs from uniformity in all four cases. In theory, the stability of wave-phase histograms can be improved improved significantly by increasing both the record length and the sampling rate substantially. Of course, this would be feasible if one did not have to compromise between the conflicting requirements of statistical stability and stationarity. That, clearly, is not the case in the present applications in view of the nonstationary nature of the hurricane-generated wave field. The record length (1.14 hr) employed is already much larger than those commonly used in most wave analyses. Ideally, an ensemble of shorter wave records measured simultaneously at different sites within the wave field would have been more suitable for the present purposes.

SUMMARY AND CONCLUSIONS

The second-order random wave model was examined systematically in terms of the effect of weak nonlinearities on various statistics related to the wave-surface displacement, its envelope, and phase. Then, the theoretical expressions describing the joint, marginal, and conditional probability structure of the wave envelope and phase were derived, correct to first-order in the rms surface slope. The basic nature and some immediate implications of these results were discussed. In particular, it was shown that the wave envelope is Rayleigh-ditributed as in the case of linear waves, whereas the phase distribution is nonuniform and deviates from uniformity in a progressive manner as the rms surface slope and thereby the surface skewness increase. The present comparisons with four sets of field data two of which are typical of extreme seas provide a favorable confirmation of these theoretical predictions, and thus of the second-order random wave model. This appears so despite the fact that the theoretical model is representative of waves traveling freely of any surface wind forces or viscous stresses. The validity of other theoretical results on the conditional distributions of the wave envelope and phase remains to be checked with observational data, perhaps, in a future study. Some of these, for example those related to the distributions of the envelope heights/amplitudes associated with the crest segments of the surface profile should be of practical interest to engineers designing offshore structures.

ACKNOWLEDGMENTS

The Camille data was furnished by G. Z. Forristall.

REFERENCES

Hamilton, R.C., and Ward,E.G. (1974). " Ocean Data Gathering Program: Quality and reduction of data." Proc., Fourth Annual Offshore Tech. Conf., Houston, Tex., 749-769.

Longuet-Higgins, M. S. (1963). "The effect of non-linearities on statistical distributions in the theory of sea waves." J. Fluid Mech., 17(3), 459-480.

Phillips, O. M. (1985). " Spectral and statistical properties of the equilibrium range in wind-generated gravity waves." J.Fluid Mech., 156, 505=531.

Tayfun, M. A. (1986). " On narrow-band representation of ocean waves." J. Geophys. Res., 91(C6), 7743-7759.

Tayfun, M. A., and Lo, J-M. (1990). " Nonlinear effects on wave envelope and phase." J. Wtrwy., Port, Coast., and Oc. Engrg., ASCE, 116(1), 79-100.

Tayfun, M. A. (1990). " High-wave-number/frequency attenuation of wind-wave spectra." J. Wtrwy., Port, Coast., and Oc. Engrg., ASCE, 116(3), 381-398.

MODELING BOTTOM FRICTION IN WIND-WAVE MODELS [1]

Hendrik L. Tolman [2]

ABSTRACT

Effects of bottom friction in wind-wave models are investigated, with an emphasis on wave-induced bottom roughnesses (moveable-bed effects). A state-of-the-art bottom friction model is defined, based on literature. An analysis of this model indicates that, initial ripple-formation is important for swell propagation, but that moveable-bed effects are less important for depth-limited wind-seas. The small spatial decay scales associated with swell call for a sub-grid approach in (large-scale) numerical models. A sub-grid model is developed and applied successfully to swell and wind-sea cases, removing (unrealistically) large effects of sediment parameters in the later cases. Finally, implications for wave observations and sediment transport are discussed briefly.

1 Introduction

Wind-waves in oceans and shelf seas are generally described with their surface elevation ("energy") spectra, the development of which is described using a spectral balance equation. In shallow water wave-bottom interactions become a potentially important source term in the wave energy balance. An early review of such source terms is given by Shemdin et al. (1978), who consider percolation, bottom motion, bottom-friction and scattering of wave energy. For sandy bottoms, as found in many shelf seas, Shemdin et al. (1978) expect

---

[1] OPC Contribution No.76.

[2] UCAR visiting scientist, Marine Prediction Branch, Development Division, NOAA/NMC21, 5200 Auth Road Room 206, Camp Springs, MD 20746, USA.

bottom-friction to be dominant, in particular when the near-bottom wave motion is sufficiently strong to generate sediment transport and corresponding bed-forms (ripple-formation). In fact, only ripple-formation can explain the large range of friction factors observed for swell in nature [Shemdin et al. (1978)], and the large friction factors for laboratory experiments with irregular waves [Madsen and Rosengaus (1988), Madsen et al. (1990)]. However, in modeling bottom friction in numerical wind-wave models, the attention is usually focussed on hydrodynamic aspects of the source term, assuming that the physical bottom roughness is known [e.g., Cavaleri and Lionello (1990), Weber (1991a,b)]. To the knowledge of the present author, efforts to explicitly model moveable-bed bottom roughnesses are presented by Graber and Madsen (1988) and Tolman (1989) only.

The present study seeks to investigate bottom friction in wind-wave models with an emphasis on moveable-bed effects. To this end, a state-of the art model is defined in section 2. In section 3, this model is analyzed with respect to occurrence of roughness regimes and space scales of decay for bottom friction. It is show that typical swell can be associated to both smooth beds and wave-induced sand ripples, and that depth-limited wind-seas are generally associated with washed-out ripples and sheet-flow roughness. It is shown, that initial ripple-formation might result in preferred wave heights for swell propagation in shelf seas away from the coast, when bottom slopes are small. The corresponding decay scales [O(10 km)], call for a sub-grid approach in numerical models. A sub-grid model is briefly described in section 4, and applied successfully to swell propagation and depth-limited wind-seas in section 5. The wind-sea cases furthermore indicate, that a sub-grid approach is essential to avoid unrealistically strong dependencies of depth-limited wave heights on sediment parameters. Finally, the present results and implications for wave observations and sediment transport studies are discussed briefly in section 6. Note that the presentation and discussion of results has to be cursory due to space limitations. The results of this study will be presented in full elsewhere.

2 A local bottom friction model

In the present study, the hydrodynamic bottom friction source term of Madsen et al. (1988) is used. This model is selected because (i) it is a simple model, yet it explicitly depends on the Nikuradse equivalent sand grain roughness k_N and (ii) for consistency with the roughness model below. This model relates the source term S_l to the surface elevation spectrum F using a drag-law approach (the subscript l denoting "local" for later comparison with a sub-grid model)

$$S_1 = -f_w u_r \frac{\omega^2}{2g\sinh^2 kd} F , \tag{1}$$

$$f_w = \frac{0.08}{\mathrm{Ker}^2\left(2\sqrt{\zeta_0}\right) + \mathrm{Kei}^2\left(2\sqrt{\zeta_0}\right)} , \tag{2}$$

$$\zeta_0 = \frac{1}{21.2\kappa\sqrt{f_w}} \frac{k_N}{a_r} , \tag{3}$$

$$u_r = \left\{\int \frac{2\,\omega^2}{\sinh^2 kd} F\right\}^{\frac{1}{2}} \quad , \quad a_r = \left\{\int \frac{2}{\sinh^2 kd} F\right\}^{\frac{1}{2}} \quad , \tag{4}$$

where $\omega = 2\pi f$ is the radian frequency, d is the depth, f_w is the wave friction factor, κ is the Von Kàrmàn constant, Ker and Kei are Kelvin functions of the zeroth order and u_r and a_r are the representative near-bottom orbital velocity and amplitude, obtained by integration over F. Note that f_w is a function of k_N/a_r only, and that f_w is constant for $k_N/a_r > 1$ ($f_w = 0.236$ in the present model). Note furthermore, that this model shows a relation between the roughness k_N and Weber's "dissipation coefficient" $C \equiv f_w u_r$ similar to that of the most advanced eddy viscosity models of Weber (1991a), the main differences being a moderate intra-spectral variation of C which is neglected here and a systematic difference between friction factors for identical roughnesses k_N, which could be interpreted as a different definition of the bottom roughness (figures not presented here).

Grant and Madsen (1982, henceforth denoted as GM) developed a semi-empirical moveable-bed roughness model based on observations for monochromatic waves. This model relates k_N to the Shields number ψ, which is defined here as [Cf. Madsen et al. (1990)]

$$\psi = \frac{f_w' u_r^2}{2(s-1)gD} , \tag{5}$$

where s is the relative density of the sediment compared to water (2.65 for quartz sands), D is a representative grain diameter and the prime indicates that the friction factor is based on skin friction, i.e., using $k_N = D$ in Eq. (3). The

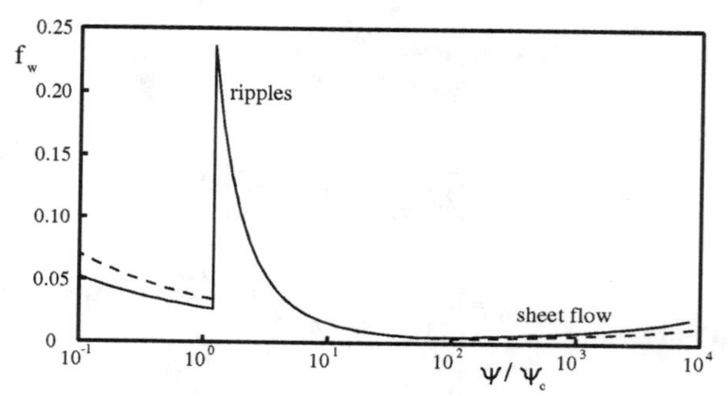

Fig. 1 Friction factors f_w as a function of the normalized Shields number ψ/ψ_c for grain diameters D = 0.1 mm (dashed line) and D = 0.4 mm (solid line) for T = 10 s, ψ_c = 0.05 (clean sand) and $k_{N,0}$ = 0.01 m.

critical Shields number for initial sediment motion ψ_c is estimated as $\psi_c \approx 0.04 \sim 0.06$ for clean, well-sorted sands [e.g., Madsen and Grant (1976), Glenn and Grant (1987)], but can become larger than 0.2 for bioturbated or multimodal sands [e.g., Drake and Cacchione (1986), Cacchione et al. (1987), Gross et al., (1992)]. If no sediment motion occurs ($\psi < \psi_c$), the bottom is assumed to be smooth [Graber and Madsen (1988) assume $k_N = D$], otherwise the roughness is comprised of ripple-roughness and sheet-flow roughness (equations not reproduced here). However, this implementation of the GM model does not seem realistic for practical conditions, because: (i) Ripples are generally much smoother for irregular waves than for monochromatic waves [e.g., Dingler and Inman (1976), Nielsen (1981), Madsen et al. (1990), Ribberink and Al-Salem (1990)]. (ii) The sheet-flow roughness term appears to over-estimate more recent data by an order of magnitude [e.g., Wiberg and Rubin (1989)]. (iii) Roughnesses for conditions without wave-induced sediment motion are typically much larger than the representative grain diameter due to bioturbation, current-induced ripples and relict bed forms [e.g., Amos et al. (1988)]. The present study therefore uses a roughness formulation similar to the GM model, but based on more recent studies, with for $\psi > 1.2\psi_c$

$$\frac{k_N}{a_r} = 1.5\left(\frac{\psi}{\psi_c}\right)^{-2.5} + 0.0655\left(\frac{u_r^2}{(s-1)ga_r}\right)^{1.4} . \tag{6}$$

The first term represents ripple-roughness [Madsen et al. (1990)] and the second term represents sheet-flow roughness [based on Wilson (1989), derivation not presented here]. For $\psi < 1.2\psi_c$ a constant "base roughness" $k_{N,0} >> D$ is assumed, which is typically 0.01 m. The behavior of the present ripple-roughness model is illustrated in Fig. 1. If near-bottom wave-motion is too weak to generate sediment transport ($\psi/\psi_c < 1.2$), the base-roughness $k_{N,0}$ generally results in friction factors O(0.02). At conditions of initial ripple-formation ($\psi/\psi_c \approx 1.2$) steep, well developed ripples are formed, resulting in a large relative roughness ($k_N/a_r \approx 1$) and corresponding large friction factors $f_w \approx 0.2$. For more severe near-bottom wave-motion, ripples are washed-out rapidly, so that large ripple-induced friction factors occur in a narrow range of normalized Shields numbers only. At even higher Shields number sheet-flow becomes important, and the friction factors become fairly insensitive to ψ/ψ_c.

3 Analysis of the local model

To analyze the effects of bottom friction in general and moveable beds in particular, the occurrence of roughness regimes is investigated. To promote insight, wave conditions are expressed in terms of mean wave parameters such as the significant wave height H_s ($= 4\sqrt{E}$, $E = \int F$), and the wavenumber and frequency corresponding to the spectral peak (k_p and f_p). In terms of these parameters, Eqs. (4) and (5) become

$$a_r = \frac{\alpha_a H_s}{2^{3/2}} \frac{1}{\sinh k_p d} \quad , \quad u_r = \frac{\alpha_u H_s}{2} \left(\frac{g}{d} \frac{k_p d}{\sinh 2k_p d} \right)^{1/2} , \tag{7}$$

$$\psi = \frac{\alpha_u^2}{8(s-1)} \frac{H_s^2}{dD} f_w \frac{k_p d}{\sinh 2k_p d} , \tag{8}$$

where α_a and α_u are shape factors. For near-monochromatic swells $\alpha_a \equiv \alpha_u \equiv 1$ and for typical TMA spectra [Bouws et al (1985)] $0.7 < \alpha_a , \alpha_u < 1$ (figures not presented here). Using (8), it is easily shown that typical swell conditions result in both smooth beds without sediment motion, and in rough beds related to initial ripple formation. Swells can result in significant sheet-flow roughnesses in the surf zone only. Similarly, wind-seas (described using a TMA spectrum with given steepness $k_p H_s$) can be accompanied by roughnesses ranging from smooth beds, to conditions with significant sheet-flow roughnesses (figures not presented here).

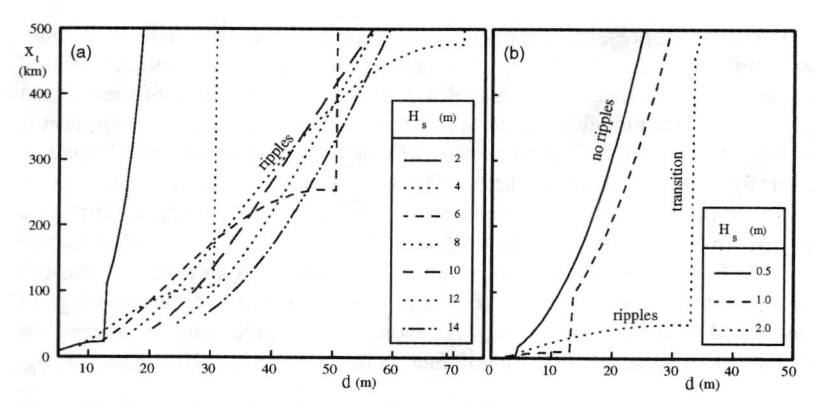

Fig. 2 Decay scales x_t for wind-seas (panel a, TMA spectra with $\gamma = 3.3$, $E = \frac{1}{4}\alpha k_p^{-2}$ and $\alpha = 0.015$) and swell (panel b, $f_p^{-1} = 12$ s). Wave heights as shown, $\psi_c = 0.05$, $D = 0.3$ mm, $k_{N,0} = 0.01$ m.

Although the occurrence of roughness regimes is interesting, it does not identify the importance of bottom friction (or ripple generation) in the overall energy balance of the wave field. This importance can be assessed without a full analysis of all source terms by analyzing decay scales related to bottom friction. An overall time scale for decay t_d can be estimated from Eq. (1) as $t_d \equiv E/\int S_1$. The corresponding spatial decay scale x_d is defined using the group velocity for the spectral peak frequency $c_{g,p}$ as $x_d \equiv c_{g,p} t_d$. After some straightforward algebraic manipulations, this decay scale becomes

$$x_d = \frac{2\sqrt{2}}{\alpha_u^3} \frac{d^2}{f_w H_s} G(k_p d) \quad , \quad G(z) = \frac{\sinh z}{z}\left[1 + \frac{\sinh 2z}{2z}\right] \quad . \tag{9}$$

In Fig. 2 decay scales x_d are presented for several wind-sea cases [panel a, TMA spectra, Bouws et al. (1985)] and several swell cases (panel b, semi-monochromatic, $f_p^{-1} = 12$ s). The sediment parameters ($k_{N,0} = 0.01$ m, $D = 0.3$ mm and $\psi_c = 0.05$) represent fairly fine, clean sand.

Fig. 2a indicates, that decay scales for wind-seas related to smooth beds are typically O(10^3 km), except for low wave heights in extremely shallow water. Such scales are generally much larger than bathymetric scales for the corresponding depths, making smooth-bed bottom friction irrelevant in the overall energy balance for wind-seas. Within the ripple regime, decay scales are

O(10^2 km), which is generally relevant in shelf seas. Severely depth-limited wind-seas are expected to generate conditions with normalized Shields numbers O(10) (figures not presented here). In such conditions, the friction factor is not very sensitive to the Shields number (see Fig. 1). Hence the friction factors and depth-limited wave heights for wind-seas are not expected to be sensitive to sediment parameters through moveable bed effects.

Fig. 2b indicates, that both smooth and rippled bottoms result in relevantly small decay scales for swell [O(100 km) and O(10 km), respectively]. As both roughness regimes are potentially important for swell propagation, the discontinuous behavior of the roughness model needs to be discussed. The roughness model (6) implies that the bottom roughness adjusts instantaneously to the wave conditions, i.e., that the time scales of ripple-adjustment are smaller that the time scales of evolution of the wave field. Within the ripple-regime, this appears to be reasonable, in conditions of initial ripple-formation (i.e., near the discontinuity of the model), it is not. In such conditions, slowly intensifying wave conditions will eventually result in initial sediment motion. This results in a rapid increase of roughness and hence in a moderation of wave conditions. Ripple build-up will stop as soon as conditions of initial sediment motion are no longer met, *regardless of the actual ripple-roughness*. Hence, ripple-roughness is partially determined by the spatial energy balance, and not solely by the local wave conditions. The corresponding wave height is the critical wave height for initial sediment motion H_c, which follows from Eq. (8) by substituting $\psi = 1.2\psi_c$

$$\frac{H_c^2}{dD} = 8(s-1)\frac{1.2\,\psi_c}{\alpha_u^2 f_w}\frac{\sinh 2k_p d}{k_p d} \tag{10}$$

If the bathymetric scales are larger than the decay scales corresponding to full ripple development, , this critical wave height becomes a practical maximum swell height. This is illustrated in Fig. 3, which shows swell height calculated with the following simple one-dimensional propagation model for the swell energy E.

$$\frac{\partial c_g E}{\partial x} = \int_{\text{spectrum}} S_1 \, , \tag{11}$$

where the right hand side is checked and corrected for the above mechanism of roughness generation in the regime of initial ripple-formation (details not presented here).

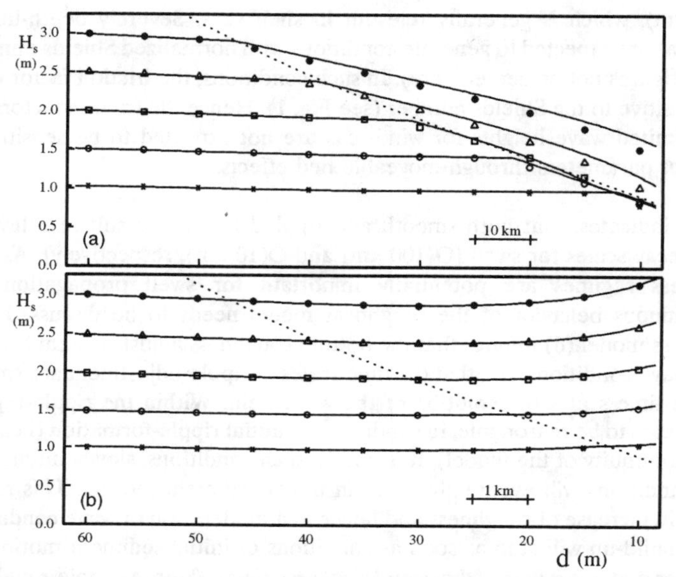

Fig. 3 One-dimensional swell propagation over a bottom with a constant slope for a "small" slope (5 10^{-4}, panel a) and a "large" slope (5 10^{-3}, panel b). Exact solution (solid lines), sub-grid numerical model (symbols) and the critical wave height for initial sediment motion H_c (dotted line). Sediment as in Fig. 2.

The solid lines in Fig. 3 represent results of this model (swell propagation form deep to shallow water) and the dotted line represents the critical wave height H_c of Eq. (10). Sediment conditions and the bottom slope are assumed constant. Note that H_c decreases monotonically with d, but not as a function of $d^{1/2}$, because f_w' implicitly varies with both H and $k_p d$.

For bottoms with a small slope (Fig. 3a) and lower input wave heights ($H_{s,d=60\text{ m}}$ < 2 m), the swell height closely follows the critical wave height, once this wave height is reached. For larger input wave heights, the reaching of the critical wave height results in a noticeable increase in wave energy dissipation, but the wave heights remain larger than H_c. This is explained as decay scales corresponding to well-developed ripples increase with increasing wave height (Fig. 2b). Apparently, the decay scale at the location where the wave height reaches H_c exceeds bathymetic scales for $H_{s,d=60\text{ m}}$ > 2 m. For much smaller bathymetric scales (i.e., larger slopes), swell heights are not noticeably influenced by H_c (Fig. 3b).

Considering the above, moveable-bed effects are potentially important, in particular for swell propagation away from the coast. Application of moveable-bed bottom friction in large-scale numerical wave models would require a sub-grid approach, because (i) the decay scale related to fully developed ripples [O(10 km)] can be significantly smaller than the grid resolution (typically 25 km or larger) and because (ii) conditions near the discontinuity in the model might not result in a roughness representative for an entire grid-box with variable sediments, depths and wave conditions. A sub-grid version of the present bottom friction source term is outlined below.

4 A sub-grid model

To account for sub-grid variations of bottom friction, an average source term representative for a grid-box has been derived, based on the local (instantaneous) application of the discontinuous roughness model (6). The derivation of a full model and the application of subsequent simplifications will be presented elsewhere. Here, only the background of the model is discussed, and the suggested source term is presented.

The roughness model (6) is a discontinuous function of the normalized Shields number $\psi_n \equiv \psi/\psi_c$. A representative (continuous) source term for a grid box is obtained by locally applying (6), using statistical properties of ψ_n and Bayes' theorem. Statistical properties of ψ_n are governed by statistical properties of d, D, ψ_c and the spectrum F [Eqs. (5) and (1) through (4)]. However, due to the integral nature of the Shields number, the two-dimensional spectrum F can be replaced by integral wave parameters H_s and f_p, assuming that the shape factors α_a and α_u are constant for the grid box. Given the law of large numbers, the pdf of ψ_n closely follows the normal distribution (as is easily confirmed using Monte Carlo simulations). Its mean value is estimated from the mean values of d, D, ψ_c, H_s and f_p and its spread σ_ψ is estimated from the corresponding spreads by linearizing (8). Using the pdf of ψ_n and Bayes' theorem, a general representative source term is obtained. After some straight-forward simplifications, the resulting model consists of the hydrodynamic model of Eqs. (1) through (4), combined with the a representative roughness $k_{N,r}$

$$\frac{k_{N,r}}{a_r} = P_I \frac{k_{N,0}}{a_r} + P_{II} \left[1.5\, \psi_{n,r}^{-2.5} + 0.0655 \left(\frac{u_r^2}{(s-1)g a_r} \right)^{1.4} \right], \tag{12}$$

$$\psi_{n,r} = \psi_n + p\left(\frac{1.2-\psi_n}{\sigma_\psi}\right)\frac{\sigma_\psi}{P_{II}} , \tag{13}$$

where P_I (P_{II}) represents the probability that $\psi_n < 1.2$ ($\psi_n \geq 1.2$), $\psi_{n,r}$ is the representative normalized Shields number for the ripple regime and $p(..)$ represent the standard normal pdf. P_I and P_{II} are calculated assuming a normal distribution of ψ_n. Other parameters in these equations follow directly from the (mean) depth, sediment parameters and the spectrum at the grid point. To evaluate Eqs. (12), (13) and (13), an expression for the spread σ_ψ is required. Formally, this expression depends on spreads of all five input parameters to the Shields number, as well as their correlations. For practical purposes, the following (semi-empirical) expression is suggested.

$$\frac{\sigma_\psi}{\psi_n} \approx \left[\sigma_{r,0}^2 + \left(\frac{k_p d}{\tanh k_p d} \frac{\sigma_{d,g} + \sigma_{d,s}}{d} \right)^2 \right]^{1/2} , \tag{14}$$

where $\sigma_{r,0}$ is a representative (normalized) spread, describing the combined variabilities of D, ψ_c and H_s, $\sigma_{d,g}$ represents the variability of the depth at the grid scale (which can be estimated from depths at surrounding points) and $\sigma_{d,s}$ represents an additional sub-grid variability of the depth. This particular formulation is suggested because (i) σ_ψ / ψ_n is directly related to the spread of D, ψ_c and H_s, whereas its relation to σ_d is a strong function of $k_p d$, because (ii) the latter results in systematically different behavior for swell and wind-seas and because (iii) $\sigma_{d,g}$ dominates σ_d and is easily obtained from the model grid.

5 Applications

To test the sub-grid bottom-friction source term presented in the previous section, it has been implemented in the third-generation wave model WAVEWATCH [Tolman (1989), (1991)]. For cases without currents this model solves an energy balance equation for the spectrum $F(f,\theta)$

$$\frac{\partial F(f,\theta)}{\partial t} + \nabla_x\left[c_g F(f,\theta)\right] = S(f,\theta) , \tag{15}$$

where S represents the net source term. The present version of this model uses source terms identical to those of cycle 4 of the WAM model [WAMDI group (1988)], as described in detail by Mastenbroek et al. (1993). It furthermore includes improved numerical schemes for propagation and source term integration [Tolman (1992)]. With this model, several idealized swell and wind-sea cases have been assessed.

5.1 Swell propagation

First, the numerical model has been used to simulate the swell propagation cases presented in Fig. 3. For the small-slope case (Fig. 3a), a grid increment $\Delta x = 10$ km and a time step $\Delta t = 6$ min have been used. For the large-slope case, Δx and Δt were reduced by a factor 10. σ_ψ / ψ_n was obtained from Eq. (14), using $\sigma_{r,0} = \sigma_{d,s} = 0$, and with $\sigma_{d,g}$ calculated from the actual bottom profile. Steady results where obtained by defining a constant deep-water boundary condition, and running the model for a sufficiently long time interval.

The numerical results (symbols in Fig. 3) follow the "exact" solution (solid lines) closely, clearly identifying the different effects of initial ripple formation at the different scales. The numerical results diverge somewhat from the exact solution for the highest wave heights in the small-slope case only (solid circles and triangles in Fig. 3a). Such divergence is implicit to the moveable-bed bottom friction model for wave heights slightly larger than H_c: an overestimation of dissipation will draw the wave height closer to conditions of initial ripple formation, resulting in a (significantly) increased dissipation, drawing the wave height even closer to H_c.

5.2 Depth-limited wind-seas

Depth-limited wave heights are assessed by considering steady wave spectra for constant wind speeds, water depths and sediment parameters assuming quasi-homogeneous conditions [i.e., neglecting the second term in Eq. (15)]. Sub-grid variability of depth and sediment parameters, however, is assumed to exists. The spectrum is discretized using 24 directions ($\Delta\theta = 15°$) and 25 frequencies, ranging from 0.04 Hz to 0.45 Hz with an increment $\Delta f = 0.1 f$ (Cf. the WAM model). The model independently determines the integration time step ($\Delta t \leq$ 15 min). Computations start from an arbitrary (small) JONSWAP spectrum, and are performed until a steady solution is reached. As an illustration, depth-limited wave heights H_d are presented as a function of the wind speed at 10 m height U_{10} in Fig. 4 for a case with $d = 20$ m, $D = 0.2$ mm, $\psi_c = 0.05$ (clean sand) and $k_{N,0} = 0.01$ m. Results are presented for the discontinuous model (dashed line), the sub-grid model with $\sigma_{r,0} = 0.2$ and $(\sigma_{d,g} + \sigma_{d,s})/d = 0.2$ (solid

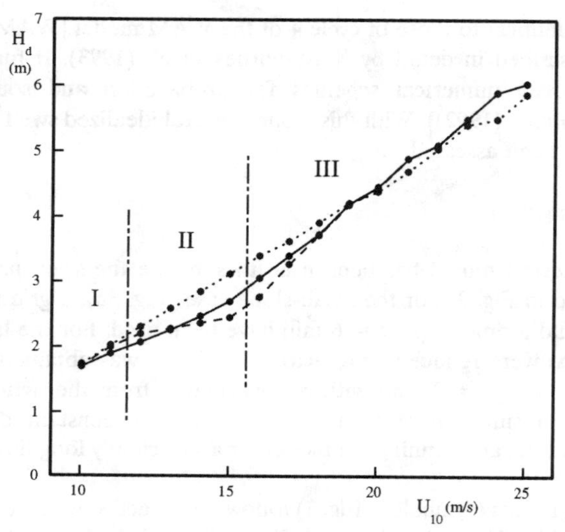

Fig. 4 Numerical simulation of depth-limited wave heights H_d for homogeneous conditions as a function of the wind speed U_{10}. The present sub-grid model (solid line, $\sigma_{r,0}$ = 0.2 and $(\sigma_{d,g}+\sigma_{d,s})/d$ = 0.2), the discontinuous model on which it is based (dashed line) and a constant roughness model ($k_N \equiv k_{N,0}$, dotted line). d = 20 m, D = 0.2 mm, ψ_c = 0.05 and $k_{N,0}$ = 0.01 m. The symbols represent the actual model results.

line) and a (conventional) model with a constant roughness length k_N = 0.01 m (dotted line).

The results for the discontinuous model in Fig. 4 (dashed line) show clearly discontinuous behavior. For low wind speeds (range marked as I) near-bottom wave motion is insufficient to move sediment, and H_d increases with U_{10}. For a fairly broad range of intermediate wind speeds (range II in Fig. 4), conditions of initial ripple formation occur, where the actual bottom roughness is governed by the source term balance. In this range, the dependency of H_d on U_{10} is practically negligible. Note that the results in range II appear to be independent of initial conditions (figures not presented here), so that the discontinuous behaviour does not seem to generate "chaotic" behavior (bifurcations). For the higher wind speeds of region III, the wind is sufficiently strong to "break through" the discontinuity and reach the ripple regime, where H_d again increases with U_{10}. As conditions of initial ripple-formation are sensitive to sediment parameters, H_d of the discontinuous model is potentially sensitive to sediment parameters. This is an artifact of the application of the discontinuous model to the entire grid-box, as follows from a comparison with the sub-grid model. Although the latter model (solid line) still shows some effects of the moveable bed roughness when compared to the constant roughness model (dotted lines), such effects are mild and do not show any discontinuities. Note that much smaller spreads of the Shields number than applied here are sufficient to remove the discontinuities of H_d related to the discontinuous nature of the local model.

6 Discussion and conclusions

An analysis of the decay scales related to bottom friction for a state-of-the-art moveable-bed bottom-friction model indicates that initial ripple-formation is potentially important for swell propagation over mildly sloping bottoms. In fact, the discontinuous transition between flat beds and sand ripples results in a non-local mechanism of roughness generation and energy decay. It is unlikely, that such a mechanism is described accurately by previous models, which do not explicitly consider moveable-bed effects. For depth-limited wind-seas moveable-bed effects are less important, because such wave conditions are generally accompanied by washed-out ripples, where the roughness is fairly insensitive to sediment or wave parameters.

Initial ripple-formation can results in length scales of energy decay, which are significantly smaller than the resolution of typical wind-wave models. Therefore, a sub-grid version of the above moveable-bed bottom-friction model has been developed. This model is shown to reproduce energy decay for swell propagation in conditions of initial ripple-formation, in spite of the somewhat unstable model characteristics in such conditions. Model results furthermore indicate, that a sub-grid model is essential to avoid an unrealistically strong dependency of wind-sea wave-heights on sediment parameters for mildly depth-limited conditions.

Moveable-bed effects are also expected to be important in analyzing wave observations. Roughness generation by swell in conditions of initial ripple-formation represents a clearly different dissipation mechanism than the conventional constant roughness concept. Analyzing results as presented in Fig. 3a in a conventional way by fitting a single friction factor gives results without a physical meaning. Clearly, sediment data is imperative in interpreting swell decay data. Observations of depth-limited wind-seas generally consider extremely shallow water and high wind speeds, typically corresponding to range

III in Fig. 4. In such conditions, constant roughness and moveable-bed models (dotted and solid lines, respectively) result in similar wave heights H_d, indicating that sediment parameters are not expected to have a large impact on such observations. In more mildly depth-limited conditions (range II in Fig. 4), this is not necessarily the case, making such conditions interesting for further research.

Finally, the mechanism for roughness generation in conditions of initial ripple formation has an interesting implication for sediment transport. In the Sediment transport literature, large wave-generated roughnesses are usually implied from large suspended sediment concentrations. If, however, roughnesses are related to the spatial energy balance or the source term balance, sediment transport is by definition small, but roughnesses can reach their maximum.

Acknowledgements. The present study was initiated while the author held an NRC resident research associateship at NASA / Goddard Space Flight Center, and was finished at NOAA / NMC.

REFERENCES

Amos, C.L., A.J. Bowen, D.A. Huntley and C.F.M. Lewis, 1988: Ripple generation under combined influence of waves and currents on the Canadian continental shelf. *Cont. Shelf Res.*, **8**, 1129-1153.

Bouws, E., H. Günter, W. Rosenthal and C.L. Vincent, 1985: Similarity of the wind wave spectrum in finite depth water. 1: Spectral form. *J. Geophys. Res.*, **90**, 975-986.

Cacchione, D.A. , W.D. Grant, D.E. Drake and S.M. Glenn, 1987: Storm-dominated bottom boundary layer dynamics on the northern California continental shelf: Measurements and predictions. *J. Geophys. Res.*, **92**, 1817-1827.

Cavaleri, L. and P. Lionello, 1990: Linear and non-linear approach to bottom friction in wave motion: Critical intercomparison. to *Estuarine, Coastal and Shelf Science*, **30**, 355-367.

Drake, D.E. and D.A. Cacchione, 1986: Field observations of bed shear stress and sediment resuspension on shelves. *Cont. Shelf Res.*, **6**, 415-429.

Dingler, J.R. and D.L. Inman, 1976: Wave-formed ripples in near-shore sands. *Proc. 15th Int. Conf. Coastal Eng.*, ASCE, Honolulu, 2109-2126.

Glenn, S.M. and W.D. Grant, 1987: A suspended sediment stratification correction for combined wave and current flows. *J. Geophys. Res.*, **92**, 8244-8264.

Graber, H.C., and O.S. Madsen, 1988: A finite-depth wind-wave model. Part I: Model description. *J. Phys. Oceanogr.*, **18**, 1465-1483.

Grant, W.D., and O.S. Madsen, 1982: Movable bed roughness in unsteady oscillatory flow. *J. Geophys. Res.*, **87**, 469-481.

Gross, T.F., A.E. Isley and C.R. Sherwood, 1992: Estimation of stress and bed roughness during storms on the northern California shelf. *Cont. Shelf Res.*, **12**, 389-413.

Madsen, O.S., and W.D. Grant, 1976: Quantitative description of sediment transport by waves. *Proc. 15th Int. Conf. Coastal Eng.*, ASCE, 1093-1112.

Madsen, O.S., and M.M. Rosengaus, 1988: Spectral wave attenuation by bottom friction: experiments. *Proc. 21st Int. Conf. Coastal Eng.*, ASCE, Malaga, 849-857.

——, Y.-K. Poon and H.C. Graber, 1988: Spectral wave attenuation by bottom friction: theory. *Proc. 21st Int. Conf. Coastal Eng.*, ASCE, Malaga, 492-504.

——, P.P. Mathiesen and M.M. Rosengaus, 1990: Movable bed friction factors for spectral waves. *Proc. 22st Int. Conf. Coastal Eng.*, ASCE, Delft, 420-429.

Mastenbroek, C., G. Burgers and P.A.E.M. Janssen, 1993: The dynamic coupling of a wave model and a storm surge model through the atmospheric boundary layer. *J. Phys. Oceanogr.*, in press.

Nielsen, P., 1981: Dynamics and geometry of wave-generated ripples. *J. Geophys. Res.*, **86**, 6467-6472.

Ribberink, J.S. and A. Al-Salem, 1990: Bedforms, sediment concentrations and sediment transport in simulated wave conditions. *Proc. 22nd Int. Conf. Coastal Eng.*, ASCE, Delft, 2318-2331.

Shemdin, O., K. Hasselmann, S.V. Hsiao and K. Heterich, 1978: Nonlinear and linear bottom interaction effects in shallow water, in : Turbulent fluxes through the sea surface, wave dynamics and prediction. NATO Conf. Ser. V, Vol 1, 347-365.

Tolman, H.L., 1989: The numerical model WAVEWATCH: a third generation model for the hindcasting of wind waves on tides in shelf seas. *Communications on Hydraulic and Geotechnical Engineering*, Delft Univ. of Techn., ISSN 0169-6548, Rep. No. **89-2**, 72 pp.

——, 1991: A third-generation model for wind waves on slowly varying, unsteady and inhomogeneous depths and currents. *J. Phys. Oceanogr.*, **21**, 782-797.

——, 1992: Effects of numerics on the physics in a third-generation wind-wave model. *J. Phys. Oceanogr.*, **22**, 1095-1111.

WAMDI group, 1988: The WAM model - a third generation ocean wave prediction model. *J. Phys. Oceanogr.*, **18**, 1775-1810.

Weber, S.L., 1991a: Eddy-viscosity and drag-law models for random ocean wave dissipation. *J. Fluid Mech.*, **232**, 73-98.

——, 1991b: Bottom friction for wind sea and swell in extreme depth-limited situations. *J. Phys. Oceanogr.*, **21**, 149-172.

Wiberg, P.L. and D.M. Rubin, 1989: Bed roughness produced by saltating sediment. *J. Geophys. Res.*, **94**, 5011-5016.

Wilson, K.C., 1989: Friction on wave induced sheet flow. *Coastal engineering*, **13**, 371-379.

Second Order Irregular–Wave Generation in Flumes - Computation of Transfer Functions by an asymptotic summation method

Hemming A. Schäffer †

ABSTRACT

Recently the theory for second order irregular–wave generation (including superharmonics as well as subharmonics) has been rederived for a variety of different types of wave board motion (Schäffer, 1993). In addition to the well known transfer functions (Sand 1982, Barthel et al., 1983, Sand and Donslund, 1985, Sand and Mansard 1986a,b) some new terms appeared. These are related to the first order evanescent modes and accordingly they are significant when the wave board motion makes a poor fit to the velocity profile of the desired progressive wave component. This is for example the case for the high-frequency part of a primary wave spectrum when using a piston type wavemaker. The well known transfer functions are given in a relatively simple form by which the computational effort is reduced substantially. However, the new terms are represented by double infinite series of very slow convergence, which makes straightforward series summation extremely memory and time consuming. The computational effort can be reduced by orders of magnitude by use of the asymptotic expression for large j of the j'th term in the series. The paper focuses on this method for practical computation of the developed transfer functions.

1. INTRODUCTION

Second order wavemaker theory has been treated in a number of papers e.g. Sand (1982) and Barthel et al. (1983) for the subharmonic interaction and Mansard and Sand (1986a,b) for the superharmonics. The actual development of the theory is not the topic of the present paper and with regard to both introduction and theoretical derivation reference is made to Schäffer (1993), who rederived the full second order wavemaker theory in a unifying and compact form that includes both superharmonics and subharmonics and covers a variety of different types of wave board motion. This derivation includes the mutual interactions of evanescent modes which were overlooked in previous derivations. These terms which appear in the transfer functions as double infinite series of very slow convergence were also included in the analysis for regular waves given by Fontanet (1961) and Hudspeth and Sulisz (1991) and also recognized although not included by Suh and Dalrymple (1987).

---

† Danish Hydraulic Institute, Agern Allé 5, DK-2970 Hørsholm, Denmark

For irregular waves interactions between a large number of pairs of components at different frequencies must be evaluated and straightforward summation of the double infinite series becomes a formidable task. In the present paper it is demonstrated how the computational effort can be reduced by orders of magnitude using the asymptotic behaviour of the series terms.

In § 2 the wavemaker theory is summarized and the method of series summation is developed in § 3. Examples of application of this method are given in § 4 while § 5 is devoted to summary and conclusions.

2. WAVEMAKER THEORY

2.1. Types of wavemakers considered

Wavemakers of the piston and hinged type are consided and thus the position of the wave board can be written

$$X(z,t) = f(z)X_0(t) \tag{1}$$

where $X_0(t)$ is the position at still water level and $f(z)$ describes the type of wavemaker:

$$f(z) = \begin{cases} 1 + \dfrac{z}{h+\ell} & \text{for } -(h-d) \le z \le 0 \\ 0 & \text{for } -h \le z < -(h-d) \end{cases} \tag{2}$$

Here h is water depth, $z = -(h+\ell)$ gives the centre of rotation $(-h < \ell \le \infty)$ and $d \ge 0$ describes elevated wavemakers i.e. the last case in (2) is only relevant for $d > 0$. It turns out that the series that appear in the second order solution are only convergent for $\ell = -d$ or $d = 0$ for which $f(z)$ is continuous. However, the first order solution is valid for other cases as well, e.g. the piston type (i.e. $\ell = \infty$), where the piston does not reach the bottom (i.e. $d > 0$). It is emphasized that solutions for such discontinuous $f(z)$ should be used with extreme care due to problems of separated flow.

2.2. First order solution

At first order it suffices to analyse a monochromatic wave in the frequency domain since superposition and the inverse FFT can then be applied to obtain time series of paddle position for irregular waves.

The first order solution was obtained by Biésel (1951), Ursell et al. (1960), Flick and Guza (1980). Sand and Donslund (1985). and others. Here the solution will be given in a slightly more general version as regards the type of wave board motion, and in the compact notation provided by a complex representation.

Let the first order paddle position for each of the wavelets constituting the first order spectrum be given by

$$X_0^{(1)} = \frac{1}{2}\{-iX_a e^{i\omega t} + \text{c.c.}\} \tag{3}$$

where X_a is the constant complex first order wave board amplitude at still water level and where c.c. denotes the complex conjugate of the preceding term, then the first order surface elevation may be expressed as

$$\eta^{(1)} = \frac{1}{2}\left\{X_a \sum_{j=0}^{\infty} c_j e^{i(\omega t - k_j x)} + \text{c.c.}\right\} \tag{4}$$

which includes both the wanted progressive-wave term and the evanescent modes which are due to the mismatch between the shape of the progressive-wave velocity profile and the shape of the wave board. Here

$$\omega^2 = gk_j \tanh k_j h \tag{5}$$

which is the linear dispersion relation generalized to complex wavenumbers. It has one real solution, say k_0, and an infinity of purely imaginary solutions $(k_1, k_2, \ldots)$, where $ik_j > 0$, $j = 1, 2, \ldots$.

Furthermore

$$c_j = \sinh k_j h \frac{\Lambda_1(k_j)}{\Lambda_2(k_j)} \tag{6}$$

where

$$\Lambda_1(k_j) = \sinh k_j h - \frac{d+\ell}{h+\ell} \sinh k_j d + \frac{1}{h+\ell} \frac{\cosh k_j d - \cosh k_j h}{k_j} \tag{7}$$

$$\Lambda_2(k_j) = \frac{1}{2}(k_j h + \sinh k_j h \cosh k_j h) \tag{8}$$

For $j = 0$ this gives the real quantity c_0 which is known as the Biésel transfer function in terms of which the complex amplitude A of the progressive part of the first order wavefield in (4) may be related to the complex amplitude X_a of the first order paddle position through

$$A = c_0 X_a \tag{9}$$

the elevation for the progressive part of the first order waves being

$$\eta_0^{(1)} = \frac{1}{2}\{Ae^{i(\omega t - kx)} + \text{c.c.}\} \tag{10}$$

For $j = 1, 2, \ldots$ c_j is purely imaginary and $ic_j > 0$, $j = 1, 2, \ldots$.

With $A = a - ib$ (3) and (10) may be written

$$X_0^{(1)} = c_0^{-1}(a \sin \omega t - b \cos \omega t) \tag{11}$$

and

$$\eta_0^{(1)} = a \cos(\omega t - kx) + b \sin(\omega t - kx) \tag{12}$$

2.3. Second order solution

At second order interactions between two wavelets of generally different angular frequencies ω_n and ω_m constitute the basis of the second order spectrum. The interaction terms appear as subharmonics (difference frequencies) and superharmonics (sum frequencies) and thus bichromatic waves should be analysed in the frequency domain and superposition and the inverse FFT can then be applied to obtain second order time series of paddle position for irregular waves to be added to the corresponding first order time series.

The aim of second order wave generation is to get a second order bound surface elevation $\eta_0^{(21)\pm}$ without any spurious free wave generation. Let the summation indices for the series (4) giving the first order solution for ω_n and ω_m be denoted j and l, respectively, then $\eta_0^{(21)\pm}$ is given by

$$\eta_0^{(21)\pm} = \frac{1}{2}\{G_{nm}^{\pm} A_n A_m^{-:*} e^{i(\theta_{0n} \pm \theta_{0m})} + \text{c.c.}\} \tag{13}$$

where G_{nm}^{-} is the transfer function for subharmonic frequencies first derived by Ottesen–Hansen (1978) and G_{nm}^{+} is the equivalent for the superharmonics, see e.g. Schäffer (1993). Furthermore

$$\theta_{jn} \equiv \omega_n t - k_{jn} x; \qquad \theta_{lm} \equiv \omega_m t - k_{lm} x \tag{14}$$

and the symbol $^{-:*}$ introduced for brevity is defined by

$$Z^{-:*} = \begin{cases} Z & \text{for superharmonics} \\ Z^* & \text{for subharmonics} \end{cases} \tag{15}$$

where $*$ denotes complex conjugation, i.e. $^{-:*}$ is to be interpreted as complex conjugation in case of subharmonics, while being ignored for superharmonics.

The solution to second order wavemaker theory indicates how to move the wave board in order to obtain the desired waves, and the second order paddle position may be written as

$$X_0^{(2)\pm} = \frac{1}{2}\{ - i\mathcal{F}^{\pm} \frac{A_n A_m^{-:*}}{h} e^{i(\omega_n \pm \omega_m)t} + \text{c.c.}\} \tag{16}$$

where $\mathcal{F}^{\pm}$ is a complex transfer function derived in Schäffer (1993). It can be shown that the expression for $\mathcal{F}^{\pm}$ contains divergent series unless either $d = 0$ or $d + l = 0$ invalidating the solution for discontinuous shape functions $f(z)$, e.g. the elevated piston wavemaker. In case of $d = 0$ or $d + l = 0$ $\mathcal{F}^{\pm}$ may be recast in the form

$$\mathcal{F}^{\pm} = \frac{\delta_{nm} K_0^{\pm 2} h}{c_{0n} c_{0m} (\omega_n \pm \omega_m)^3 (1 + M_1(K_0^{\pm}))} \times$$
$$\Big\{ \mp \frac{g}{2\omega_n} \sum_{j=0}^{\infty} c_{jn} \frac{k_{jn}^2}{k_{jn}^2 - K_0^{\pm 2}} \left(\omega_n^2 - (\omega_n \pm \omega_m)^2 + M_2(k_{jn}, K_0^{\pm})\right) + \widehat{lmjn}$$
$$+ \sum_{j=0}^{\infty} \sum_{l=0}^{\infty} c_{jn} c_{lm}^{-:*} \frac{k_{jn} \pm k_{lm}^{-:*}}{(k_{jn} \pm k_{lm}^{-:*})^2 - K_0^{\pm 2}} H_{jnlm}^{\pm} \Big\} \tag{17}$$

where $\widehat{lmjn}$ means "the preceding term permuting l and j as well as m and n". Furthermore $K_0^{\pm}$ is the real solution to

$$(\omega_n \pm \omega_m)^2 = g K_0^{\pm} \tanh K_0^{\pm} h \tag{18}$$

and

$$\delta_{nm} \equiv \begin{cases} \frac{1}{2} & \text{for } n = m \\ 1 & \text{for } n \neq m \end{cases} \tag{19}$$

$$H^{\pm}_{jnlm} \equiv (\omega_n \pm \omega_m)\left(\pm\omega_n\omega_m - \frac{g^2 k_{jn} k^{-:*}_{lm}}{\omega_n\omega_m}\right) + \frac{\omega_n^3 \pm \omega_m^3}{2} - \frac{g^2}{2}\left(\frac{k^2_{jn}}{\omega_n} \pm \frac{k^2_{lm}}{\omega_m}\right) \tag{20}$$

$$M_1(K_0^{\pm}) \equiv \frac{1}{h+l}\frac{g}{(\omega_n \pm \omega_m)^2}\left(\frac{\cosh K_0^{\pm} d}{\cosh K_0^{\pm} h} - 1\right) \tag{21}$$

$$M_2(k_{jn}, K_0^{\pm}) \equiv -\frac{g}{h+l}\frac{K_0^{\pm}/k_{jn}}{k^2_{jn} - K_0^{\pm 2}}\left\{2k_{jn}K_0^{\pm}\left(1 - \frac{\cosh k_{jn} d}{\cosh k_{jn} h}\frac{\cosh K_0^{\pm} d}{\cosh K_0^{\pm} h}\right)\right.$$
$$\left. - (k^2_{jn} + K_0^{\pm 2})\left(\frac{\omega_n^2(\omega_n \pm \omega_m)^2}{g^2 k_{jn} K_0^{\pm}} - \frac{\sinh k_{jn} d}{\cosh k_{jn} h}\frac{\sinh K_0^{\pm} d}{\cosh K_0^{\pm} h}\right)\right\} \tag{22}$$

Both M_1 and M_2 vanish for a piston type wavemaker.

In a real representation with $A_n = a_n - ib_n$ and $A_m^{-:*} = a_m \mp ib_m$ (13) and (16) may be written

$$\eta_0^{(21)\pm} = G^{\pm}_{nm}\{(a_n a_m \mp b_n b_m)\cos(\theta_{0n} \pm \theta_{0m}) + (\pm a_n b_m + a_m b_n)\sin(\theta_{0n} \pm \theta_{0m})\} \tag{23}$$

and

$$X_0^{(2)\pm} = \frac{\Re\{\mathcal{F}^{\pm}\}}{h}\{(a_n a_m \mp b_n b_m)\sin(\omega_n \pm \omega_m)t + (\mp a_n b_m - a_m b_n)\cos(\omega_n \pm \omega_m)t\}$$
$$-\frac{\Im\{\mathcal{F}^{\pm}\}}{h}\{(\mp a_n b_m - a_m b_n)\sin(\omega_n \pm \omega_m)t - (a_n a_m \mp b_n b_m)\cos(\omega_n \pm \omega_m)t\} \tag{24}$$

3. ASYMPTOTIC BEHAVIOUR AND SERIES SUMMATION

The asymptotic behaviour of the infinite series in the transfer function is of interest for two reasons. First of all it reveals whether the series are convergent or not. As mentioned previously convergence can be shown to require a continuous shape function $f(z)$ (at least among the ones considered in (2)). Secondly it provides a method for evaluating the series with immense savings as regard to computational effort.

The key to the asymptotic analysis is the evanescent-mode dispersion relation (5) from which the following relations may be obtained

$$k_j h \asymp -\pi i j \tag{25}$$

$$\cosh k_j h \asymp (-1)^j \tag{26}$$

$$\sinh k_j h \asymp \frac{\omega^2 h}{g} \frac{i}{j\pi}(-1)^j \tag{27}$$

$$\cosh k_j d \asymp \cos j\pi \frac{d}{h} \tag{28}$$

$$\sinh k_j d \asymp -i \sin j\pi \frac{d}{h} \tag{29}$$

Here the symbol $\asymp$ denotes the asymtotic value for large j. In the practical computation it is important to precalculate and save (for each primary-wave frequency) quantities like $k_j h$ ($j = 0, \ldots, j_{max}$) needed in the computation of the transfer function (17). For memory reasons there should be as few of these quantities as possible with the constraint that different asymptotic behaviour should not be mixed, as will be explained below. With due regard to these principles the first order transfer function (6) is rewritten as

$$c_j = \left(\frac{\omega^2 h}{g} - \frac{h}{h+l}\right) \frac{1}{D_j(k_j)} + \frac{h}{h+l} \frac{1}{D_j(k_j)} \frac{\cosh k_j d}{\cosh k_j h} \tag{30}$$

where

$$D_j(k_j) \equiv \frac{k_j h}{2} \left(\frac{k_j h}{\sinh k_j h \cosh k_j h} + 1\right) \tag{31}$$

Here we have assumed $d + l = 0$ or $d = 0$, by which the second term in (7) vanishes. Since

$$\frac{1}{D_j(k_j)} \asymp \frac{-2i\frac{\omega^2 h}{g}}{\pi^3 j^3} \tag{32}$$

we get for the two terms in c_j

$$\left(\frac{\omega^2 h}{g} - \frac{h}{h+l}\right) \frac{1}{D_j(k_j)} \propto \frac{1}{j^3} \tag{33}$$

and

$$\frac{h}{h+l} \frac{1}{D_j(k_j)} \frac{\cosh k_j d}{\cosh k_j h} \propto \frac{(-1)^j}{j^3} \cos j\pi \frac{d}{h} \tag{34}$$

The transfer function (17) is already in a carefully chosen form, and is seen that in (17)–(22) and (30) the hyperbolic functions only appear in the context $\cosh k_j d / \cosh k_j h$ or $\sinh k_j d / \cosh k_j h$. Thus it suffices to precompute these together with $k_j h$ and $D_j(k_j)$. ($j = 0, \ldots, j_{max}$). For a piston type wavemaker only the two latter quantities are needed and c_j could be chosen instead of $D_j(k_j)$.

3.1. Piston type wavemaker

Since for a piston type wavemaker $M_2(k_{jn}, K_0^{\pm})$ vanishes and c_j reduces to

$$c_j = \frac{\omega^2 h}{g} \frac{1}{D_j(k_j)} \propto \frac{1}{j^3} \tag{35}$$

it appears that the leading term of the first series in (17) is $\propto 1/j^3$ for large j. The double infinite series in (17) is evaluated as

$$\sum_{j=0}^{\infty}\sum_{l=0}^{\infty} = \sum_{j=0}^{\infty}\sum_{l=j}^{\infty} + \sum_{l=0}^{\infty}\sum_{j=l+1}^{\infty} \tag{36}$$

and it follows from (17), (20), (25) and (35) that the leading term in $\sum_{l=j}^{\infty}$ and $\sum_{j=l+1}^{\infty}$ is $\propto 1/j^2$. Two exceptions are $\sum_{l=j}^{\infty}$ for $j = 0$ and $\sum_{j=l+1}^{\infty}$ for $l = 0$ where the imaginary part of the leading term is $\propto 1/j^3$. These exceptional series show up in the quantity $iF_{24}^{\pm}$ (see eq. (50) in Schäffer 1993) from which the quoted asymptotic behaviour may be shown.

Now let d_j represent the j'th term in a series and assume that

$$d_j \propto \frac{1}{j^p} \tag{37}$$

where $p \geq 2$ is an integer describing the asymptotic behaviour, then

$$\sum_{j=0}^{\infty} d_j = \sum_{j=0}^{J-1} d_j + d_J J^p \sum_{j=J}^{\infty} \frac{1}{j^p} + \epsilon_J \tag{38}$$

where

$$\epsilon_J = d_J J^p \sum_{j=J+1}^{\infty} \frac{1}{j^p}\left\{ \left(\frac{j}{J}\right)^p \frac{d_j}{d_J} - 1 \right\} \tag{39}$$

In (38) the infinite series can be summed up explicitly and ϵ_J gives the absolute error on the series evaluation when using the asymtotic approximation. In order to quantify ϵ_J we introduce the relative error ε_j on each term caused by using the asymtotic behaviour to extrapolate from the previous term instead of direct evaluation of d_j:

$$\varepsilon_j = \frac{d_j}{d_{j-1}}\left(\frac{j}{j-1}\right)^p - 1 \tag{40}$$

Now the last factor (in braces) of the first series term ($j = J+1$) in (39) equals ε_{J+1} and we can estimate the error by

$$|\epsilon_J| \approx |\varepsilon_{J+1} d_J| J^p \sum_{j=J+1}^{\infty} \frac{1}{j^p} \tag{41}$$

Assuming that d_j is only evaluated up to $j = J$ then this also applies to ε_j, and ε_{J+1} is conveniently substituted by ε_J to get the more conservative estimate

$$|\epsilon_J| \approx \left|\frac{d_J}{d_{J-1}}\left(\frac{J}{J-1}\right)^p - 1\right| |d_J| J^p \sum_{j=J+1}^{\infty} \frac{1}{j^p} \tag{42}$$

Unfortunately these estimates are not always reliable and they have to be supplemented by a criterion based on consecutive estimates of the infinite sum.

In order to reduce repetitious computations the following quantities are precalculated for $J = 2$ to $J = j_{max}$ where the choice of j_{max} depends on the required accuracy:

$$R(J) \equiv \left(\frac{J}{J-1}\right)^p \tag{43}$$

$$Res(J) \equiv J^p \sum_{j=J}^{\infty} \frac{1}{j^p} \tag{44}$$

where the latter is conveniently computed by the recurrence relation

$$Res(J) = R(J)(Res(J-1) - 1) \tag{45}$$

As initial value we have for the relevant values of p

$$Res(1) = \sum_{j=1}^{\infty} \frac{1}{j^p} = \begin{cases} \pi^2/6 & \text{for} \quad p = 2 \\ 1.20206\ldots & \text{for} \quad p = 3 \end{cases} \tag{46}$$

Actually $Res(1)$ may be recognized as the Riemann zeta function of argument p.

Neglecting ϵ_J (38) reads in terms of $Res(J)$

$$\sum_{j=0}^{\infty} d_j \approx \sum_{j=0}^{J-1} d_j + d_J Res(J) \tag{47}$$

and in terms of $R(J)$ and $Res(J)$ the error estimate (42) becomes

$$|\epsilon_J| \approx \left|\frac{d_J}{d_{J-1}} R(J) - 1\right| |d_J| Res(J) \tag{48}$$

In practice (47) is evaluated successively for $J = 2, 3, \ldots$ until the error (48) is acceptable and until consecutive estimates of the infinite sum show sufficiently small differences e.g. 5% of the error accepted on $|\epsilon_J|$.

If $Res(J)$ was equal to unity (47) would represent straightforward series summation up to $j = J$. With $Res(J)$ given by (44) the right hand side of (47) accounts for the infinity of terms, but for $j > J$ in an approximate way.

3.2. Hinged type wavemaker

For the hinged type wavemaker the series summation tecnique is somewhat more complicated. In addition to the type of terms (37) described in the previous section, the first infinite sum in (17) together with (30) and (34) shows that we also have to consider

$$e_j \propto \frac{(-1)^j}{j^p} \cos j\pi \frac{d}{h} \tag{49}$$

where e_j represents the j'th term in a series and $p = 3$. Fortunately $M_2(k_{jn}, K_0^{\pm})$ in (17) as given by (22) does not give rise to any leading terms as compared with the preceding ω-terms. In the double infinite sum in (17) terms like $c_j k_j$ are leading and thus the new terms to be considered are covered by (49) for $p = 2$. As for the piston–type case two exceptions apply where $p = 3$ instead of $p = 2$, see the text following (36).

Equivalent to (38) and (39) we can formally write

$$\sum_{j=0}^{\infty} e_j = \sum_{j=0}^{J-1} e_j + \frac{e_J J^p}{(-1)^J \cos J\pi\frac{d}{h}} \sum_{j=J}^{\infty} \frac{(-1)^j}{j^p} \cos j\pi\frac{d}{h} + \tilde{\epsilon}_J \tag{50}$$

where

$$\tilde{\epsilon}_J = \frac{e_J J^p}{(-1)^J \cos J\pi\frac{d}{h}} \sum_{j=J+1}^{\infty} \frac{(-1)^j \cos j\pi\frac{d}{h}}{j^p} \left\{ \left(\frac{j}{J}\right)^p \frac{e_j}{e_J} \frac{(-1)^J \cos J\pi\frac{d}{h}}{(-1)^j \cos j\pi\frac{d}{h}} - 1 \right\} \tag{51}$$

but the problem with this formulation is that $\cos J\pi(d/h)$ may be zero or close to zero by which $\tilde{\epsilon}_J$ becomes singular or very large. However, in this case e_J is very small due to the factor $\cosh k_j d / \cosh k_j h$ and the problem is solved using the slightly different form

$$\sum_{j=0}^{\infty} e_j = \sum_{j=0}^{J-1} e_j + d_J J^p \sum_{j=J}^{\infty} \frac{(-1)^j}{j^p} \cos j\pi\frac{d}{h} + \epsilon_J \tag{52}$$

where

$$\epsilon_J = \sum_{j=J}^{\infty} \left\{ e_j - d_J \left(\frac{J}{j}\right)^p (-1)^j \cos j\pi\frac{d}{h} \right\} \tag{53}$$

and the relation between e_j and d_j is

$$e_j = d_j \frac{\cosh k_j d}{\cosh k_j h} \tag{54}$$

consistent with (26), (28), (37) and (49). The problem in (50) is really that $\cosh k_j d / \cos j\pi(d/h)$ is not necessarily close to unity for large j even though $\cosh k_j d \asymp \cos j\pi(d/h)$. This problem is eliminated in (52), but the magnitude of the error term (53) is difficult to assess. However, due to the alternating sign of e_j the corresponding series will converge somewhat faster than the series of d_j terms (which has to be evaluated anyway), and the J-value for which the estimated error (48) is accepted will also be sufficient in the evaluation of (52).

For convenience we define

$$Resca(J) \equiv J^p \sum_{j=J}^{\infty} \frac{(-1)^j}{j^p} \cos j\pi\frac{d}{h} \tag{55}$$

which may be computed by the recurrence relation

$$Resca(J) = R(J)\Big(Res(J-1) + (-1)^j \cos(j-1)\pi\frac{d}{h}\Big) \quad (56)$$

where the initial value can be given analytically for $p = 2$

$$Resca(1) = \sum_{j=1}^{\infty} \frac{(-1)^j}{j^p} \cos j\pi\frac{d}{h} = \frac{\pi^2}{12}\left(3\left(\frac{d}{h}\right)^2 - 1\right) \quad \text{for} \quad p = 2 \quad (57)$$

whereas for $p = 3$ it must generally be computed numerically. Now (52) becomes

$$\sum_{j=0}^{\infty} e_j \approx \sum_{j=0}^{J-1} e_j + d_J Resca(J) \quad (8)$$

neglecting the error term ϵ_J.

As in (48) the last term in (58) accounts for an infinite number of terms in an approximate manner.

4. EXAMPLES OF APPLICATION

A variety of examples of the capability of the asymptotic summation method is shown in Fig. 1–3. Before commenting on each example the common features are explained. Each figure shows the estimate of an infinite sum using the asymptotic summation method as well as straightforward summation versus the maximum summation index J up to $J = 50$. For reference the correct value of the infinite sum based on $J = 500$ is also shown. The examples were drawn from the double infinite sum in (17) as the real part of the series $\sum_{l=j}^{\infty}$ for $j = 0$ (see the first term of the right hand side of (36)) which is typical for the series constituting the double infinite sum. No direct comparison with the magnitude of $\mathcal{F}^{\pm}$ can be made from the figures, since a factor was omitted in the examples given.

For three different wavemakers Fig. 1 shows results for the superharmonic with dimensionless first order frequencies $2\pi f_n\sqrt{h/g} = 2\pi f_m\sqrt{h/g} = 1.5$. In Fig. 1a a piston-type wavemaker ($d = 0, l = \infty$) is considered while Figs. 1b and 1c give results for the hinged type where the hinge is located half way from the surface to the bottom ($d = h/2, l = -h/2$) and at the bottom ($d = l = 0$), respectively.

In the example of Fig. 1a a maximum error of 1% requires $J = 6$ when using the asymptotic summation method, while straightforward summation requires no less than $J = 115$. If the maximum error accepted is relaxed to 3% then the numbers are $J = 4$ and $J = 32$, respectively. Generally the advantage of the asymptotic method is larger if higher acuracy is required, but typically a reduction in the computational effort of a factor 10 is obtained. Since this applies for both index l and index j in the double infinite series the resulting reduction is of the order of a factor 100.

Figure 2 shows results for the subharmonic with dimensionless first order frequencies $2\pi f_n\sqrt{h/g} = 1.5$ and $2\pi f_m\sqrt{h/g} = 2.0$. The three cases considered

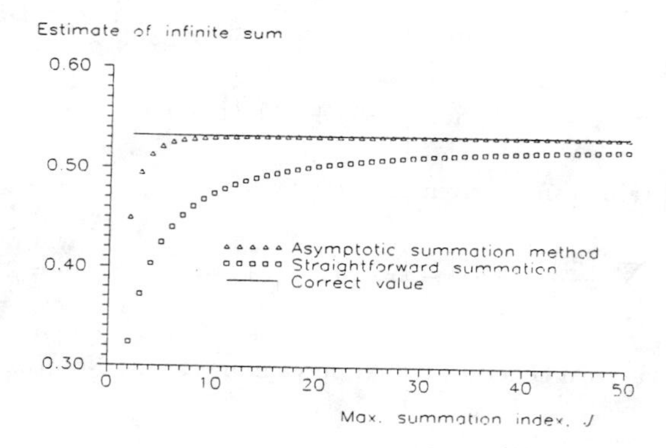

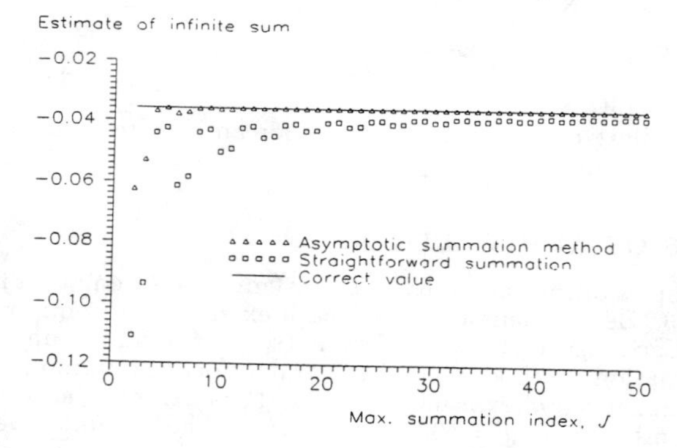

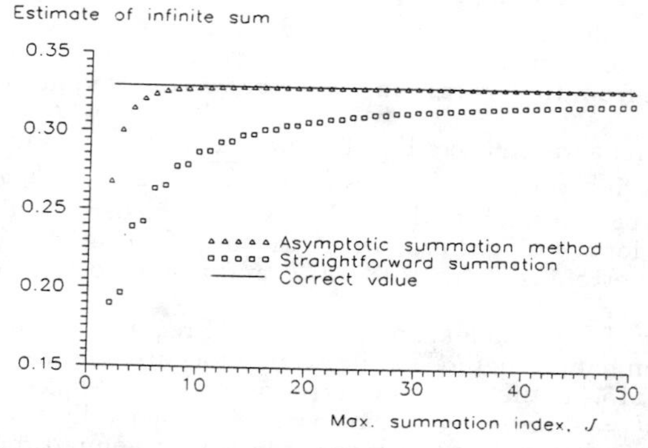

Figure 1 Typical examples of the capability of the asymptotic summation method for part of the superharmonic second order transfer function for a) a piston type wavemaker b) a hinged type wavemaker where the hinge is placed halfway from the surface to the bottom and c) a bottom hinged wavemaker.

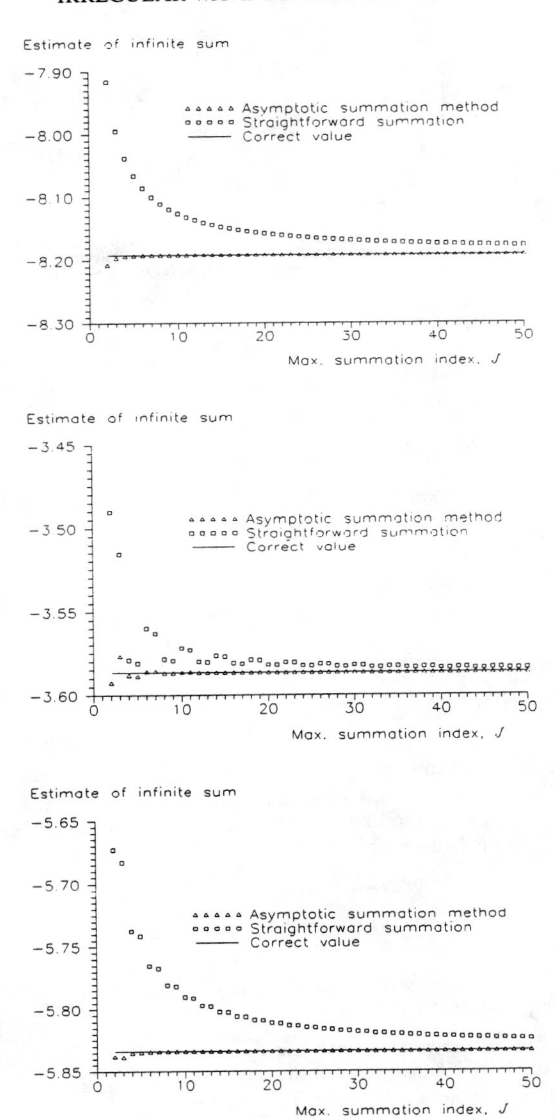

Figure 2 The same as Fig. 1 but for the subharmonic second order transfer function and for different primary-wave frequencies.

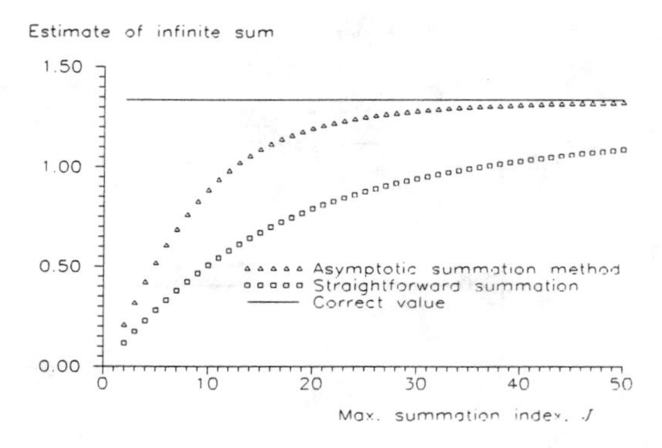

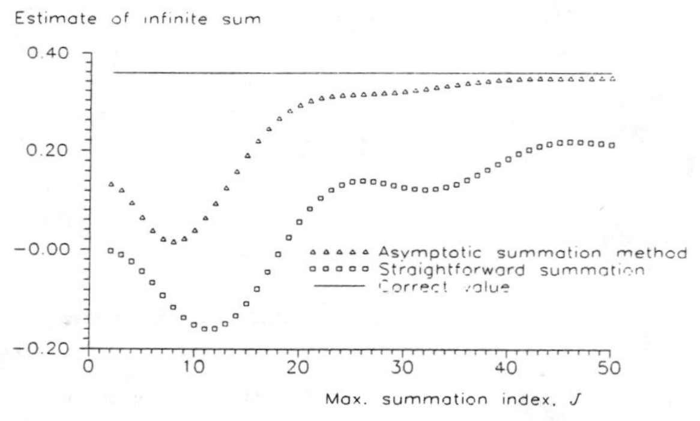

Figure 3 The same comparison as in Fig. 1 but for an extreme deep water case and for a) a piston type wavemaker and b) a hinged type wavemaker where the hinge is placed one tenth of the depth from the surface.

correspond to Fig. 1 as regard to the type of wavemaker. Again the improvement by the new technique is clearly seen.

Finally Fig. 3 shows two extreme deep water cases for which the rate of convergence of the series is very small. The results shown are for the superharmonic with dimensionless first order frequencies $2\pi f_n \sqrt{h/g} = 2\pi f_m \sqrt{h/g} = 4.0$ corresponding to $k_n h = k_m h = 16.0$. In Fig. 3a a piston-type wavemaker is considered and straightforward summation with $J = 50$ gives a 20% error while the error is only 1% using the asymptotic method. In Fig. 3b the wavemaker is of the hinged type with the hinge situated one tenth of the depth from the surface ($d = 0.9h, l = -0.9h$). Now straightforward summation for $J = 50$ gives an error of approximately 60% while the error using asymptotic summation is only 2%.

5. SUMMARY AND CONCLUSIONS

On the basis of the asymptotic values (25) of the solution to the evanescent-mode dispersion relation (5) a method (47) & (58) has been developed for approximate summation of the infinite series that appear in the transfer function for second order wave generation. The method is derived for different types of waveboard motion and it provides a reduction by orders of magnitude of the computational effort required for the evaluation of the transfer function in intermediate and deep water.

Acknowledgement. This work was partially funded by a research grant from the Danish Energy Agency under the EFP'93 program.

References

Barthel, V., Mansard, E.P.D. Sand, S.E., and Vis, F.C., 1983. Group bounded long waves in physical models. *Ocean Engrg.* **10**(4), 261–294.

Biésel, F., 1951. Etude théorique d'un type d'appareil à houle. *La Houille Blanche* **6**(2), 152–165.

Flick, R.E. and Guza, R.T., 1980. Paddle generated waves in laboratory channels. *J. WatWays, Port, Coastal Ocean Div., ASCE* **106**(WW1), 79–97.

Fontanet, P., 1961. Théorie de la génération de la houle cylindrique par un batteur plan. *La Huille Blanche* **16**(1), 3–31 (part 1) and 174–196 (part 2).

Hudspeth. R. T. and Sulisz, W., 1991. Stokes drift in two-dimensional wave flumes. *J. Fluid Mech.* **230**, 209–229.

Ottesen Hansen, N.-E., 1978. Long Period waves in natural wave trains. Prog. Rep. **46**, Inst. of Hydrodyn. and Hydraulic Engrg. (ISVA), Tech. Univ. of Denmark, 13–24.

Sand, S.E., 1982. Long wave problems in laboratory models. *J. WatWays, Port, Coastal Ocean Div., ASCE* **108**(WW4), 492–503.

Sand, S.E. and Donslund, B., 1985. Influence of wave board type on bounded long waves. *J. Hydraulic Res.* **23**(2), 147–163.

Sand, S.E. and Mansard, E.P.D., 1986a. Description and reproduction of higher harmonic waves. National Research Council of Canada, Hydraulics Laboratory Tech. Rep. TR-HY-012.

Sand, S.E. and Mansard, E.P.D., 1986b. Reproduction of higher harmonics in irregular waves. *Ocean Engrg.* **13**(1), 57–83.

Suh, K. and Dalrymple, R.A., 1987. Directional wavemaker theory: A spectral approach. Proc. IAHR-Seminar *Wave analysis and generation in laboratory basins*, Lausanne, Switzerland, 1987, 389–395.

Schäffer, H. A., 1993. Laboratory wave generation correct to second order. Proc. Int. Conf. *Wave kinematics and environmental forces*, London, 1993. Society for Underwater Technology, **29**.

Ursell, F., Dean, R.G., and Yu, Y.S., 1960. Forced small-amplitude water waves: A comparison of theory and experiment. *J. Fluid Mech.* **7**.

NUMERICAL MODEL OF THE LONGSHORE CURRENT ON A BARRED BEACH

Jane McKee Smith[1], Magnus Larson[2], and Nicholas C. Kraus[3]

ABSTRACT: The longshore current model of Larson and Kraus (1991) is modified to include a transport equation for the turbulent kinetic energy generated under the breaking wave roller. Predictions of wave height and current obtained with this five-equation model (equations for wave height, wave angle, turbulent kinetic energy transport, longshore current, and mean water surface elevation) are compared with field data for a barred beach. The wave input is provided by an individual wave height model run in random mode through Monte-Carlo selection. Additional verification of the model is made with Visser (1984) laboratory measurements of longshore current and mean water surface elevation generated by monochromatic waves arriving obliquely to a uniformly sloping beach. The five-equation model correctly reproduces observed distributions of longshore current on a barred beach and generally gives improved results for both random and monochromatic waves.

INTRODUCTION

Modeling of the longshore current has previously been based on four governing equations: (1) Wave shoaling, breaking, and decay to calculate wave height, (2) Snell's law to calculate wave angle, (3) Cross-shore momentum to calculate mean water level, and (4) Longshore momentum to calculate longshore current. The longshore current is driven by gradients in radiation stress arising from the cross-shore decay in wave height. Four-equation models have been relatively

---

1) Research Hydraulic Engineer, USAE Waterways Experiment Station, Coastal Engineering Research Center (CERC), 3909 Halls Ferry Rd., Vicksburg, MS 39180-6199, USA.
2) Associate Professor, Department of Water Resources Engineering, Institute of Technology, University of Lund, Box 118, Lund S-221 00, Sweden.
3) Director, Conrad Blucher Inst. for Surveying and Science, Texas A&M Univ.-Corpus Christi, 6300 Ocean Dr., Corpus Christi, TX 78412-5599, USA (formerly, CERC).

successful for estimating longshore current on plane beaches. Figure 1 shows wave height transformation, longshore current, and mean water level on a plane beach calculated with a four-equation model. The magnitude of the longshore current is greatest in the region of the maximum decay in wave height. Figure 2 shows the same quantities on a barred beach calculated with a four-equation model. Again, the maximum predicted longshore current velocities occur in regions of maximum wave height decay, i.e., over the bar and on the steep foreshore. Measurements of the longshore current on a barred beach obtained in the multi-institutional DELILAH field data collection project conducted in 1990 at the Corps of Engineers' Field Research Facility (FRF) in North Carolina revealed a persistent broad peak in the current velocity in the trough in between the nearshore bar and shore. Four-equation models predict a local minimum in the longshore current in the trough (Figure 2) because of the absence of a gradient in wave height as the waves reform in the trough. Present four-equation models cannot, therefore, predict the current peak in the trough observed in the field. In addition to disparities between predicted and measured longshore current on barred beaches, precise measurements of the setdown under monochromatic breaking waves on a uniformly sloping beach in the laboratory indicate that the point of maximum setdown lies shoreward of the break point, and the region of maximum setdown is broad and flat rather than a distinct, sharp minimum, as predicted by four-equation models.

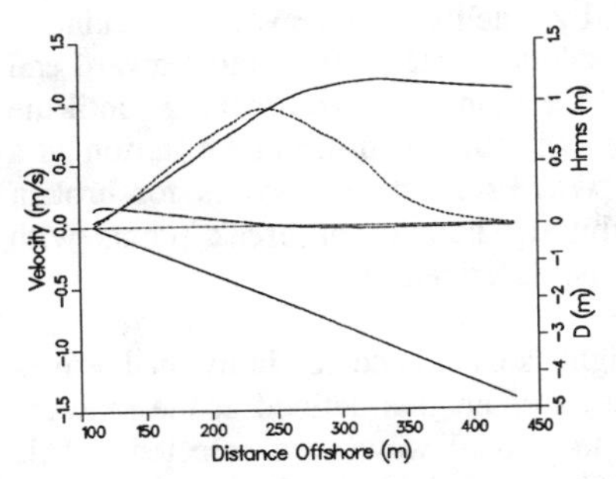

Figure 1. Four-equation model results for a plane beach (wave height - solid line, longshore current - dashed line, and mean water level - chain-dot line).

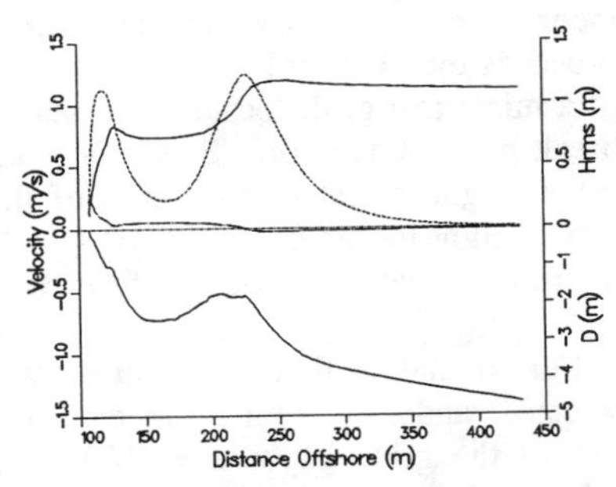

Figure 2. Four-equation model results for a barred beach (wave height - solid line, longshore current - dashed line, and mean water level - chain-dot line).

In this paper, we describe a model with an added term in the stress formulation, an extension of the approach of Roelvink and Stive (1989), to account for the momentum in the surface roller of breaking waves which is neglected in four-equation models. This addition improves model agreement with measurements on both barred and plane beaches. First, the numerical model formulation is described, introducing a fifth equation for the turbulent kinetic energy transport that is needed to obtain the additional stress term. Next, the field data from the DELILAH project are presented. These data showed a strong longshore current velocity in the trough and motivated this study. Results comparing model predictions with field and laboratory measurement are given, and conclusions are presented.

NUMERICAL MODEL

The four-equation wave transformation and longshore current model NMLONG (Larson and Kraus 1991) was extended with a fifth equation for the transport of turbulent kinetic energy (Smith et al. 1993). The fifth equation is used to calculate the Reynolds stresses associated with turbulence in the breaking wave surface roller which are included in the momentum equations. The principal assumptions in the model NMLONG are longshore uniformity in the waves, wind, and bottom topography; stationarity of wave conditions; applicability of linear-wave theory; and neglect of bottom friction in the cross-shore plane. The following sections describe wave transformation, turbulent kinetic energy transport, and longshore current and mean water surface elevation calculations in the five-equation NMLONG model.

Wave Transformation Model

The wave transformation model employs the Dally et al. (1985) model for wave decay and reformation after breaking, as modified by Larson and Kraus (1991) to describe oblique wave incidence. Wave energy flux is calculated with linear wave theory, and wave direction is determined by Snell's Law. Wave calculation proceeds by a forward-difference numerical solution scheme from the seaward end of a numerical grid, located well seaward of the point of wave breaking, until the break point is detected. Breaking is initiated with the depth-limited criterion of a wave-height-to-water-depth ratio of 0.78. Wave reformation occurs for broken wave-height-to-depth ratio of 0.4. An arbitrary number of break points with intermediate zones of wave reformation can be described.

For simulating field conditions, wave height is calculated for individual waves selected randomly from a Rayleigh distribution of heights defined at the seaward end of the grid. Typically, 100 or more individual waves are selected. This calculation procedure has replicated measured wave height distribution in the surf zone that deviate from the Rayleigh distribution because of depth-limited wave breaking over an irregular bottom (Dally 1990, Larson and Kraus 1991). Monochromatic wave conditions, such as produced in the laboratory, are simulated by input of a single wave height.

Turbulent Kinetic Energy Transport

The longshore current is generally calculated under the assumption that the driving force is the gradients in radiation stress; that is, as the waves break and create turbulence, the decrease in the wave moment flux is balanced by a shear stress associated with the generated longshore current. However, as pointed out by Roelvink and Stive (1989), a distinction should be made between the loss of organized wave energy and the dissipation of turbulent kinetic energy for determining nearshore currents and changes in mean water level. Wave-induced turbulence contributes to the onshore transport of momentum, implying that decay in radiation stress does not correspond to immediate dissipation by viscous forces, and that a shift in time and space accompanies the transport of turbulent kinetic energy (Svendsen 1987).

Svendsen (1987) showed through laboratory measurements that only a small portion of the energy in the breakers is dissipated below wave trough level (2 to 6 percent), also indicating that the vertical variation in the time-mean of the turbulent kinetic energy below trough level is small. Thus, most of the transport of turbulent momentum takes place above trough level. Our approach in describing the transport of momentum is to schematize the water column into two layers; one layer extending from the bottom up to trough level, and another layer above trough level with thickness equal to the wave height. Similar schematizations have previously been employed for calculating macro-features of surf zone hydrodynamics (Stive and Wind 1986, Thieke and Sobey 1990).

The cross-shore distribution of the mean turbulent kinetic energy above trough level, with a thickness assumed to be the wave height, is calculated in the present study through a general steady-state transport equation (McGuirk and Rodi 1987, Rastogi and Rodi 1978, Rodi 1980)

$$\frac{d}{dx}\left(\nu_t\, H \frac{dk}{dx}\right) + \frac{d}{dx}\left(k\, C\, H \cos\theta\right) + \frac{P}{\rho} - c_D\, k^{3/2} = 0 \qquad (1)$$

where x denotes the cross-shore coordinate, ν_t is kinematic eddy viscosity, H is wave height, k is turbulent kinetic energy, C is wave celerity, θ is wave angle relative to shore normal, P is production of turbulence per unit surface area, ρ is density of water, and c_D is an empirical coefficient. The first term in (1) describes diffusion of turbulent kinetic energy, for which the kinematic eddy viscosity is parameterized as (Rodi 1980)

$$\nu_t = c_\mu \sqrt{k}\, l \qquad (2)$$

where c_μ is an empirical coefficient, and the length scale l of the turbulent eddies is set to the root-mean-square (rms) wave height H_{rms}. The second term in (1) represents the advection of turbulent kinetic energy with a speed equal to the wave celerity. The third term is the production of turbulent kinetic energy where P is the cross-shore gradient in wave energy flux from the wave model. The last term in (1) is the energy dissipation integrated over the above-trough layer under the assumption of k independent of the vertical coordinate. The physical picture of turbulent kinetic energy uniform in the above-trough layer is intuitively reasonable for a well-mixed surf. In turbulence modeling, it is generally accepted that $c_\mu c_D$ = 0.08 (Launder and Spalding 1972, Rodi 1980), and this value was employed in the present study with c_μ = 0.5 and c_D = 0.16.

Equation (1), with the advection term represented by an upwind difference, is solved by a double-sweep method analogous to that used by Larson and Kraus (1991) for the longshore current velocity. At the seaward boundary, a condition of $k = 0$ was used. The shoreward boundary condition $dk/dx = 0$ is used, allowing advection of k into the swash zone.

Longshore Current and Mean Water Surface Elevation

Under the assumption of longshore uniformity, the vertically-integrated, time-averaged momentum equations for nearshore water motion (Mei 1983) are

$$g \frac{d\eta}{dx} = -\frac{1}{\rho\, d} \frac{dS_{xx}}{dx} - \frac{\rho_a C_W}{\rho\, d} |W|\, W \cos\phi \tag{3}$$

for the cross-shore (x) component, and

$$\frac{d}{dx}\left(\epsilon_L\, d\, \frac{dV}{dx}\right) - f_{by} = \frac{1}{\rho} \frac{dS_{xy}}{dx} - \frac{\rho_a C_W}{\rho} |W|\, W \sin\phi \tag{4}$$

for the longshore (y) component. In these equations, η is the mean water surface elevation, d is the total water depth, S_{xx} and S_{xy} are components of the excess momentum flux tensors, ρ_a is the density of air, C_W is the wind drag coefficient, W is the wind speed at 10-m elevation, ϕ is the wind direction defined as positive directed onshore, ϵ_L is the eddy viscosity controlling lateral diffusion of momentum below trough level, and f_{by} is the bottom friction stress. The wind drag coefficient given by the WAMDI Group (1988), derived from measurements made in intermediate-depth water, is adopted.

The excess momentum flux tensors each consist of two parts, one contribution from the transport of wave momentum (radiation stress) and another from the transport of turbulent momentum above trough level (Reynolds stress). If the two contributions are uncorrelated, they may be separated according to

$$S_{xx} = S_{xx}^{w} + S_{xx}^{t}$$
$$S_{xy} = S_{xy}^{w} + S_{xy}^{t} \tag{5}$$

where the superscripts w and t refer to wave and turbulence, respectively. In the present model, S_{xx}^w and S_{xy}^w are specified from Longuet-Higgins and Stewart (1964), and S_{xx}^t and S_{xy}^t are parameterized using k. Also, it is assumed that the turbulent momentum transport above trough level is dominant, and only this region is modeled in detail through the turbulent kinetic energy transport equation. The turbulent momentum exchange below trough level is schematically described through the mixing term in (4).

Equation (3) governs the mean water surface displacement η, and (4) gives the longshore current V, for which the driving forces are the radiation and Reynolds stresses. The effective eddy viscosity coefficient ϵ_L, which parameterizes the lateral mixing below trough level, is expressed according to the empirical formulation by Larson and Kraus (1991) as

$$\epsilon_L = \Lambda \, u_m \, H \tag{6}$$

where Λ is an empirical coefficient representing the lateral mixing strength, and u_m is the amplitude of the horizontal component of the wave orbital velocity at the bottom. Putrevu and Svendsen (1992) showed that the high tail in eddy viscosity, long known to be necessary for longshore current modeling (Kraus and Larson 1991), is produced by momentum carried by the undertow. Equation (6) is then viewed as representing an effective eddy viscosity. The second term on the left side of (4) represents the longshore component of the bottom friction stress, the major retarding force for the longshore current, and is expressed by a quadratic law in the total fluid velocity as

$$f_{by} = c_f \left\langle (u_m^2 \cos^2\tau + V^2 + 2\, u_m\, V \cos\tau \, \sin\theta)^{1/2} \, (V + u_m \cos\tau \, \sin\theta) \right\rangle \tag{7}$$

where c_f is the depth-dependent Manning friction coefficient (Smith et al. 1993); $\tau = 2\pi t/T$, where t denotes time and T denotes wave period; $u_m \cos \tau \sin \theta$ is the longshore component of the wave orbital velocity at the bottom; and triangular brackets denote time averaging over a wave period. In evaluation of (7), the closed-form approximation of Nishimura (1988) is used to explicitly calculate the time average, but iteration is required because of the quadratic dependence in V.

In application of the five-equation model to a field situation, a Rayleigh distribution of wave height based on a given rms wave height and fixed wave angle and period are input to the wave model on a wave-by-wave basis. The spectral peak period and mean direction are used, implying a narrow-band, single-peak spectrum. Turbulent energy production values associated with each wave are averaged to obtain an ensemble mean production across the surf zone. The ensemble production is input to the turbulent kinetic energy transport equation to solve for a mean k in the above-trough layer, from which the Reynolds stresses in the momentum equation are determined. The solution process is coupled between the mean water level calculation (3) and the wave energy balance, and typically five iterations are required to achieve convergence to less than 1 percent change in η anywhere on the grid. In the iteration, the same individual waves are selected in a Monte-Carlo process from the Rayleigh distribution. This iterative procedure is equivalent to simultaneous solution of the wave transformation, mean water level, and longshore current as an ensemble average, typically calculated from 100 or more individual waves. The numerical solution scheme for calculating the wave height, wave angle, mean water level, and longshore current velocity is described by Kraus and Larson (1991) and Larson and Kraus (1991).

DELILAH FIELD DATA COLLECTION PROJECT

The DELILAH (Duck Experiment on Low-frequency and Incident-band Longshore and Across-shore Hydrodynamics) field data collection project was conducted on the barred beach at the U.S. Army Engineer Waterways Experiment Station, Coastal Engineering Research Center, FRF located in Duck, North Carolina, in October 1990. The objectives of the DELILAH project were to measure the wave- and wind-forced 3-D nearshore dynamics (with emphasis on infragravity waves, shear waves, mean circulation, setup, runup, and wave transformation) and to monitor the bathymetric response to the hydrodynamic processes. A dense array of current meters and pressure gauges was deployed in the nearshore during the period October 1-19, 1990. The project provided detailed synoptic measurement of nearshore mean currents as well as the conditions generating the currents; nearshore waves, wind, tide, and bathymetry. The strong longshore currents observed in the trough at DELILAH motivated the research described in this paper.

Nearshore waves and currents were measured at nine cross-shore positions with pressure gauges and electromagnetic current meters in water depths of 4 m to less than 1 m, arranged from 250 m offshore to the shoreline. Offshore waves were measured with an array of pressure gauges at the 8-m depth contour. These directional wave data provide offshore boundary conditions for wave forcing of the numerical model. Over-water winds and tidal elevation were measured at the FRF pier. The bathymetry in the area of the nearshore current meter/pressure gauge array (340-m by 600-m area) was surveyed daily during DELILAH. The median grain size in the FRF nearshore varies from 0.7 mm on the steep foreshore to 0.2 mm on the bar, and 0.12 mm in the offshore (Howd and Birkemeier 1987).

For this study, we examine measurements taken on October 14. On this day the wave spectra measured at the 8-m depth were narrow banded in frequency with symmetric directional distributions about a mean oblique wave direction (Smith et al. 1993). The unimodal, swell wave trains are favorable for modeling wave transformation with linear wave propagation. On October 14, the significant wave height at the 8-m gauge depth was 1.1 to 1.5 m with a peak period of 10 to 12 s and a mean direction of 10 to 30 deg south of shore-normal. The bottom was nearly uniform alongshore, as shown in Figure 3.

MODEL RESULTS

DELILAH Data

Root-mean-square wave height, mean longshore current velocity, and mean water surface elevation were calculated for the October 14 DELILAH data. The input mean wave direction and peak wave period were taken from the 8-m array for eight cases corresponding to the eight collection times at the array. The input rms wave height was taken from the most seaward of the nine nearshore pressure gauges, and

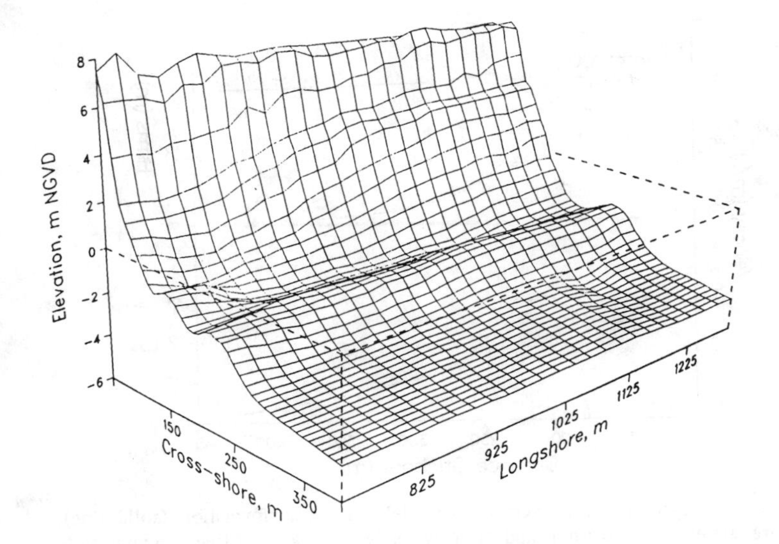

Figure 3. Nearshore bathymetry at the FRF on October 14, 1990.

the height was inversely refracted and shoaled to the 8-m depth. Input of the rms wave height measured at the 8-m array caused a 15 percent overprediction of wave height at the most seaward pressure gauge that may be attributable to use of linear refraction and shoaling in the model. The model grid consisted of 380 cross-shore cells with a spacing of 1 m. The mixing strength parameter Λ was set at 1.0, and the main fitting parameter was taken to be the Manning n, determined as 0.020 to give an optimal fit for all comparisons with the DELILAH data.

One hundred individual waves were randomly selected from a Rayleigh wave height distribution defined by the measured rms height to simulate the wave field. The nearshore wave and current measurements were averaged over the 2-hour interval concurrent with the 8-m array measurements. The input wave, wind, and tide elevation were averaged over the same interval. The wave field changed slowly through time, with the greatest nearshore variation caused by tide.

The five-equation model results for October 14 at 0100 and 0400 hours are given in Figures 4 and 5. The figures show the water depth along the profile (adjusted for the tide), the measured and calculated rms wave height, the measured and calculated mean longshore current, and the calculated mean water surface elevation. Results from the four-equation model (Manning n = 0.027) for the same cases are shown in Figures 6 and 7. The improvement in current prediction of the five-equation model over the four-equation model is striking. The wave transformations

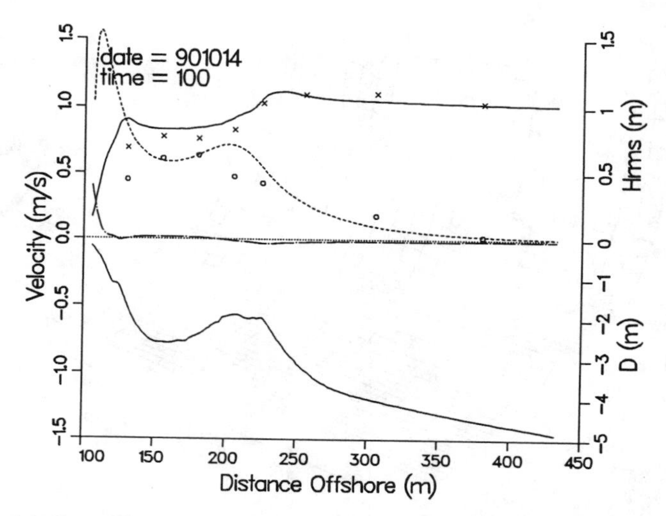

Figure 4. Comparison of five-equation model wave transformation (solid line), longshore current (dashed line), and mean water level (chain-dot line) calculation to DELILAH 10/14 at 0100 hours.

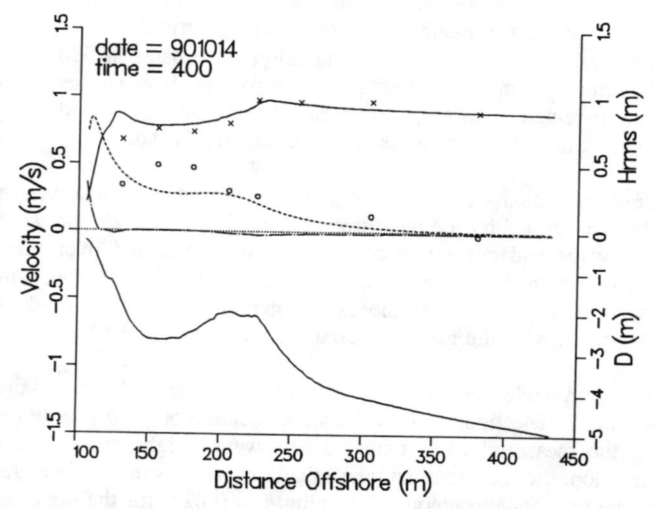

Figure 5. Comparison of five-equation model wave transformation (solid line), longshore current (dashed line), and mean water level (chain-dot line) calculation to DELILAH 10/14 at 0400 hours.

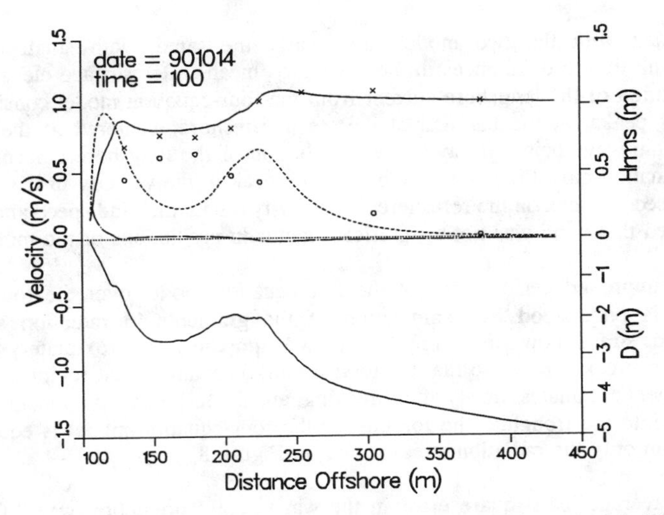

Figure 6. Comparison of four-equation model wave transformation (solid line), longshore current (dashed line), and mean water level (chain-dot line) calculation to DELILAH 10/14 at 0100 hours.

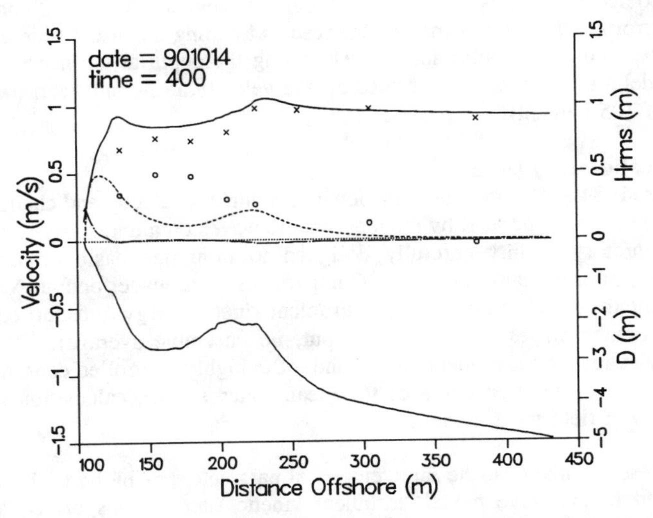

Figure 7. Comparison of four-equation model wave transformation (solid line), longshore current (dashed line), and mean water level (chain-dot line) calculation to DELILAH 10/14 at 0400 hours.

calculated with the two models are nearly the same, with small differences occurring due to differences in the calculated mean water surface elevation. The distribution of the longshore current from the four-equation model consists of two distinct peaks, at the bar and at the steep foreshore, centered at the points of maximum wave height decay. The distribution of the longshore current from the five-equation model is given as a broad, flat peak in the velocity in the bar trough with a second peak on the foreshore. Sensitivity tests with wind speed and direction indicated that wind was not a significant factor in predictions of the model.

The improved performance of the five-equation model over the four-equation model is understood by examination of the gradients in radiation stress and Reynolds stress shown in Figure 8. The two components approximately balance in the outer breaker zone, so that the total forcing is small in that region. The wave component dominates shortly after breaking, and the turbulent component dominates further into the trough. The forcing for the four-equation model is equivalent to that from only the radiation stress shown in Figure 8.

The average least-square error in the wave height prediction was 8.0 percent, ranging between 5.1 to 9.4 percent for the eight cases. The model tended to overpredict the measurements in the trough, but, overall, the model predicts the wave height well. The average least-square error in the longshore current was 55.9 percent, ranging between 36.0 to 90.2 percent for the eight cases. The average least-square error for the four-equation model was 67.2 percent. The greatest error tended to occur for cases with the maximum tide elevation (e.g., 0400 hours). The larger errors at these times may be the result of adding a constant tide elevation at each cross-shore grid point and overestimating the depth at the nearshore points. The model also consistently overpredicts the velocity at the most shoreward gauge (error of 125 percent).

Visser Laboratory Data

Visser (1984, 1991) measured the longshore current velocity and change in water surface elevation generated by monochromatic waves on a uniformly sloping beach in a laboratory facility carefully designed to minimize basin effects such as reflection and artificial circulation. Comparisons were made for four Visser cases to examine the effect of including the turbulent kinetic energy transport equation for monochromatic waves (single wave input, no ensemble average). Mean water surface elevation measurements made under the highly controlled conditions in the laboratory also allow checking of the mean water surface calculation that is not possible with field measurements.

For these calculations, the same empirical parameters as in the DELILAH cases were used in the wave model, turbulent kinetic energy transport equation, and longshore current model, except that the height-to-depth ratio at incipient breaking was set to the measured value (0.74 to 1.0) and the mixing strength parameter was set to $\Lambda = 0.15$ in both the four- and five-equations models. The optimal average

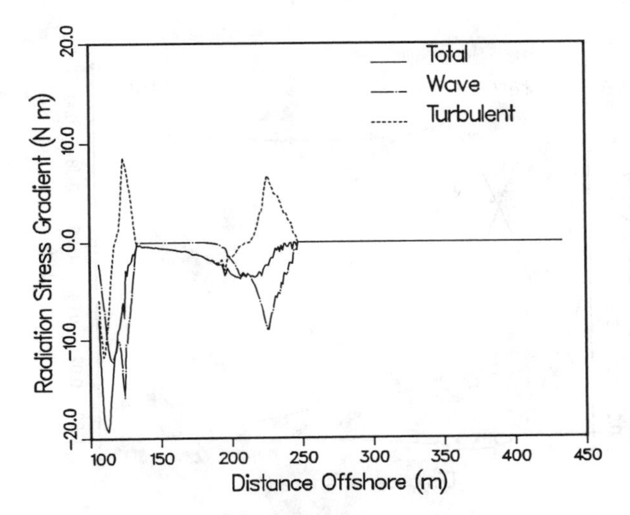

Figure 8. Gradient in total, radiation, and Reynolds stress across the beach profile (DELILAH 0100 hours).

Manning n was determined to be 0.015 and 0.016, respectively for the five- and four-equation models, for the smooth concrete bottom.

Comparisons of measured versus calculated wave height, longshore current, and mean water surface elevation for Visser Case 3 are given in Figure 9 for the five-equation model. Similar comparisons for the four-equation model are given in Figure 10. The figures show that the peak in the current distribution is better predicted by the five-equation model, as are the overall shapes of the distributions. The least-square error in wave height ranged from 8.7 to 35.6 percent, as compared to 9.0 to 36.7 percent for the four-equation model; the error in longshore velocity ranged from 13.0 to 20.5 percent, as compared to 27.9 to 48.5 percent for the four-equation model; and the error in water surface elevation ranged from 32.3 to 59.4 percent, as compared to 12.4 to 40.6 percent for the four-equation model. The errors in the five-equation model for predicting the longshore current were approximately half of those calculated for the four-equation model.

Errors in calculated mean water surface elevation were slightly larger for the five-equation model. However, the five-equation model gave improved qualitative agreement between calculated and measured mean water surface elevation in reproducing the smooth transition from point of setdown to start of setup, location of maximum setdown, and flatness in the region of setdown. Improved agreement in mean water surface elevation could be, in principle, obtained by tuning the wave transformation model, in particular, by adjustment to reproduce the location of the

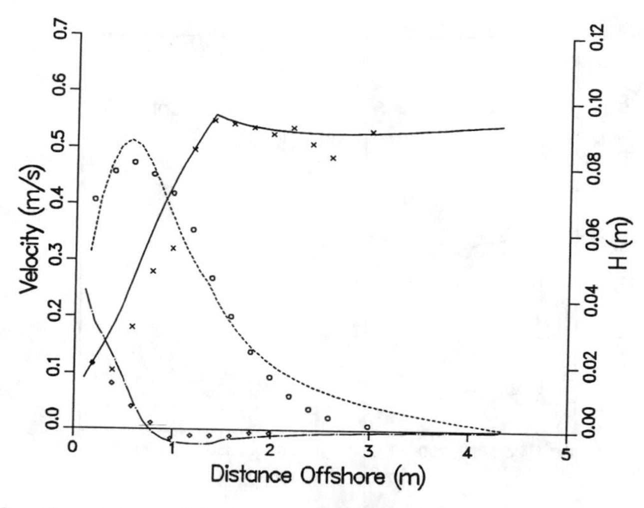

Figure 9. Comparison of five-equation model wave transformation (solid line), longshore current (dashed line), and mean water level (chain-dot line) calculation to Visser (1984) laboratory data.

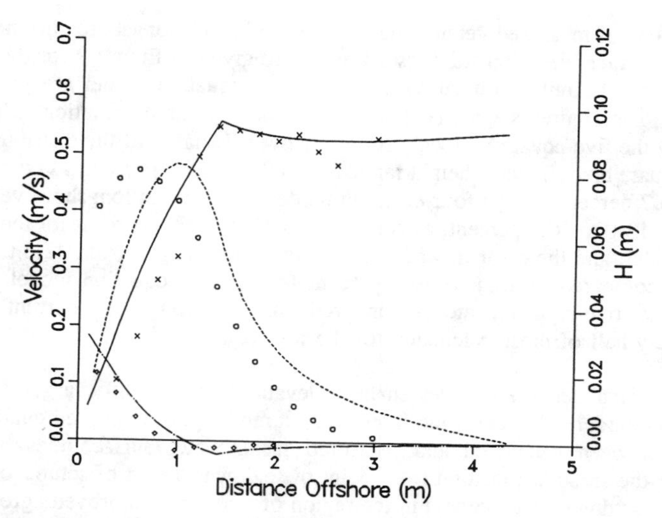

Figure 10. Comparison of four-equation model wave transformation (solid line), longshore current (dashed line), and mean water level (chain-dot line) calculation to Visser (1984) laboratory data.

measured point of wave breaking.

CONCLUSIONS

Measurements of the longshore current over the barred profile at Duck, North Carolina, showed a broad peak in the current distribution over the longshore trough that would be anomalous in traditional mathematical models of the current that compute the driving force only from wave height decay. Following and generalizing work of Roelvink and Stive (1989), a turbulent kinetic energy transport equation was introduced to compute the cross-shore distribution of turbulent kinetic energy from the turbulence production by wave breaking. Reynolds stress components associated with turbulent momentum were expressed in terms of turbulent kinetic energy. Use of the model, termed a five-equation model by the addition of the turbulent energy transport equation to the traditional model, gave improved agreement (20 to 50 percent reduction in least-square error) with the measured time-average longshore current distribution over the barred bottom field beach profile. The model also gave quantitatively improved agreement with high-quality laboratory data of the longshore current and qualitatively improved agreement for the associated mean water surface elevation generated by monochromatic waves obliquely incident to a uniformly sloping beach. In comparisons to field and laboratory measurements, the key factor in achieving improved agreement with measurements was the shoreward shift and smoothing of the energy dissipation provided by the turbulent kinetic energy transport equation.

The driving force for the random wave field comparisons was computed from an ensemble average of the turbulence production contributed from individual waves. The individual wave model does not depend on assumed forms for the wave height distribution in the surf zone, providing such distribution as part of the calculation, and it also allows ready calculation for laboratory situations involving monochromatic waves.

ACKNOWLEDGMENTS. Mr. William Birkemeier, Chief, FRF, and the FRF staff are acknowledged for their planning and execution of the DELILAH project. The authors thank Prof. Battjes for a terminology suggestion. This research was conducted under the Nearshore Waves and Currents work unit, Coastal Flooding and Storm Protection Program, and the Three-Dimensional Modeling of Wave-Induced Currents work unit, Laboratory Discretionary Research Program, by the USAE Waterways Experiment Station, Coastal Engineering Research Center. Permission to publish this paper was granted by the Office, Chief of Engineers.

REFERENCES

Dally, W. R. (1990). Random breaking waves: a closed-form solution for planar beaches, *Coastal Eng.*, 14, 233-263.

Dally, W. R., Dean, R. G., and Dalrymple, R. A. (1985). Wave height variation across beaches of arbitrary profile, *J. Geophys. Res.*, 90(C6), 11917-11927.

Howd, P. A., and Birkemeier, W. A. (1987). Beach and nearshore survey data: 1981-1984 CERC Field Research Facility, Tech Rep. CERC-87-9, USAE Waterways Experiment Station, Vicksburg, MS, 143 pp.

Kraus, N. C., and Larson, M. (1991). NMLONG: numerical model for simulating the longshore current, Tech. Rep. DRP-91-1, USAE Waterways Experiment Station, Vicksburg, MS, 166 pp.

Larson, M., and Kraus, N. C. (1991). Numerical model of longshore current over bar and trough beaches, *J. Wtrway., Port, Coast., and Oc. Engrg.*, ASCE, 117(4), 326-347.

Launder, B. E., and Spalding, D. B. (1972). *Mathematical Models of Turbulence*, Academic Press, San Diego, Calif, 169 pp.

Longuet-Higgins, M. S., and Stewart, R. W. (1964). Radiation stress in water waves; a physical discussion with applications, *Deep Sea Res.*, 11(4), 529-562.

McGuirk, J. J., and Rodi W. (1978). A depth-averaged mathematical model for the flow field of side discharges into open-channel flow, *J. Fluid Mech.*, 86, 761-781.

Mei, C. C. (1983). *The Applied Dynamics of Ocean Surface Waves*, John Wiley and Sons, New York, 740 pp.

Nishimura, H. (1988). Computation of nearshore current, in *Nearshore Dynamics and Coastal Processes*, Edited by K. Horikawa, Univ. Tokyo Press, Tokyo, Japan, 271-291.

Putrevu, U., and Svendsen, I. A. (1992). A mixing mechanism in the nearshore region, *Proc. 23rd Coastal Engrg. Conf.*, ASCE, 2758-2771.

Rastogi, A. K., and Rodi, W. (1978). Predictions of heat and mass transfer in open channels, *J. Hydraulics Div. ASCE*, HY3, 397-420.

Rodi, W. (1980). Turbulence models and their application in hydraulics - a state of the art review, *Sonderforshungsbereich 80*, Institut fur Hydromechanik, Univ. of Karlsruhe, Karlsruhe, Germany, 140 pp.

Roelvink, J. A., and Stive, M. J. F. (1989). Bar-generating cross-shore flow mechanisms on a beach, *J. Geophys. Res.*, 94(C4), 4785-4800.

Smith, J. M., Larson, M., and Kraus, N. C. (1993). Longshore current on a barred beach: field measurements and calculation, *J. Geophys. Res.*, in press.

Stive, M. J. F., and Wind, H. G. (1986). Cross-shore mean flow in the surf zone, *Coastal Eng.*, 10, 325-340.

Svendsen, I. A. (1987). Analysis of surf zone turbulence, *J. Geophys. Res.*, 92(C5), 5115-5124.

Thieke, R. J., and Sobey, R. J. (1990). Cross-shore wave transformation and mean flow circulation, *Coastal Eng.*, 14, 387-415.

Visser, P. J. (1984). Uniform longshore current measurements and calculations, *Proc. 19th Coastal Engrg. Conf.*, ASCE, NY, 1984.

__________. (1991). Laboratory measurements of uniform longshore currents, *Coastal Eng.*, 15, 563-593.

WAMDI Group (1988). The WAM model: a third generation ocean wave prediction model, *J. Phys. Oceanography*, Am. Met. Soc., Dec., 1775-1810.

SPECTRAL BOUSSINESQ MODELLING OF BREAKING WAVES

J.A. Battjes[1], Y. Eldeberky[2] and Y. Won[3]

Abstract

Random waves passing over a shallow bar are considered, in particular the amplification of bound harmonics in shoaling water, their subsequent release in deepening water, and the role of wave breaking. Calculations have been performed utilizing a set of coupled evolution equations for complex Fourier amplitudes based on ideal-fluid Boussinesq-type equations for the wave motion, supplemented with a quasi-linear dissipation term to account for wave breaking. This is used together with the assumption of random, independent initial phases to calculate the evolution of the energy spectrum of the random waves. The results show encouraging agreement with observed spectra both for nonbreaking waves and for breaking waves passing over a bar.

Introduction

The passage of nonbreaking and breaking waves over a shallow bar is accompanied by amplification and phase shifting of bound higher harmonics in the shoaling region, leading to asymmetric profiles, and by the partial release of the bound harmonics in the deepening region, leading to rapid profile deformations. The rapid growth of the harmonics can be ascribed to triad wave-wave interactions which become more nearly resonant in shallow water as the waves become more nearly non-dispersive. This phenomenon has been investigated experimentally and mathematically by several authors (e.g. Johnson et al. (1951), Jolas (1961), Byrne (1969), Drouin and Ouellet (1988), Young (1989) and Kojima et al. (1990)). Recent related work at the Delft University

---

1 2 3: Professor[1], Research Assistant[2] and former Post-Doc[3] at Delft University of Technology, Department of Civil Engineering, Section Hydraulics, Stevinweg 1, P.O. Box 5048, 2628 CN Delft, Netherlands.

of Technology has been reported by Battjes and Beji (1992) and Beji and Battjes (1993). The present contribution reports briefly on the results obtained in the continuation of this work.

The principal research questions were as follows:

(1) Can the observed phenomena, as described above, be modelled with Boussinesq-type equations?
(2) What is the influence of breaking?
(3) How can the influence of breaking be modelled?
(4) Can the results be cast in a form of an evolution equation for spectral energy density (so as to make them compatible with conventional numerical models for forecasting or hindcasting of wind-generated waves)?

Resume of previous results

The questions (1) and (2) were addressed in the references Beji and Battjes cited above. For the conditions of the observations, the following conclusions were obtained:

- The evolution of nonbreaking periodic or random waves over a bar can be modelled very well with the time-domain Boussinesq equations for sloping bottoms according to Peregrine (1967), extended for improved dispersion properties by Madsen et al. (1991) and Madsen and Sørensen (1992).
- The primary effect of wave breaking is to reduce the total energy, without affecting the evolution of the spectral shape significantly.

Based on these findings, it was suggested, in partial answer to question (3), to combine in some fashion the basically conservative (nondissipative) Boussinesq model, to simulate shoaling and nonlinear wave-wave interactions, with a conventional formulation for the rate of dissipation of the total energy (integrated spectrally), such that the latter would not affect the spectral shape (the relative values of spectral amplitudes or spectral energy density).

Spectral Boussinesq modelling of nonbreaking random waves

To implement this suggestion, and in view of the ultimate aim of a spectral energy model, it was decided to work with a frequency-domain version of the extended Boussinesq equations. For the one-dimensional propagation considered so far, the time (t) variation of the surface elevation (η) at each location (x) is expanded in a Fourier series as in

$$\eta(t;x) = \sum_{p=-\infty}^{\infty} A_p(x) \exp\{i(\omega_p t - \psi_p(x))\} \tag{1}$$

with A_p denoting a complex amplitude, p indicating the rank of the harmonic, $\omega_p = p\omega_1$, and $d\psi_p/dx = k_p$, the wave number corresponding to ω_p according to the dispersion equation for the linearized Boussinesq equations. By substituting (1) into the time-domain Boussinesq equations, and by neglecting certain higher-order terms on the assumption of a sufficiently gradual evolution of the wave field, Madsen and Sørensen (1993) develop a set of coupled evolution equations for the set of complex amplitudes A_p which in abbreviated form can be written as

$$\frac{dA_p}{dx} = L_p \frac{dh}{dx} A_p + \sum_{m=1}^{p-1} Q^+_{m,p} A_m A_{p-m} + \sum_{m=1}^{\infty} Q^-_{m,p} A^*_m A_{p+m} \tag{2}$$

The first term on the right represents linear shoaling, proportional to the bottom slope dh/dx, the second term the triad sum interactions and the third the triad difference interactions. Complete expressions for the coefficients L, Q^+ and Q^- can be found in Madsen and Sørensen (1993).

In applications of this coupled complex Fourier mode formulation to random waves with a given energy spectrum E(f) at an upwave location, this spectrum was discretized with bandwith Δf; initial amplitude values $|A_p|$ were set equal to $\{\frac{1}{2} E(f_p)\Delta f\}^{1/2}$ (the frequencies f are taken nonnegative). For each realization, a set of initial phases was drawn at random, assuming them to be mutually independent, each uniformly distributed over 2π. Following numerical integration of (2), raw values of spectral energy density at downwave locations were estimated as $2|A_p|^2/\Delta f$. These raw spectra appeared to vary between realizations, thus exhibiting influence of the particular set of initial phases. This was largely eliminated by suitable averaging over a sequence of realizations, each with a different set of initial phases, and/or over neighbouring frequency bands. In this manner the evolution of the spectra of the nonbreaking waves was simulated quite well (see Figure 1). Details of this work are described by Won and Battjes (1992).

Note that this formulation is basically deterministic. Attempts to eliminate *ab initio* the dependence on the initial phases so as to arrive at a formulation in terms of spectral energy were not (yet) successful. The approach taken by Abreu et al. (1992) for a spectral energy model including triad interactions in random waves, based on the nonlinear shallow-water equations (without frequency dispersion), may be promising in this respect.

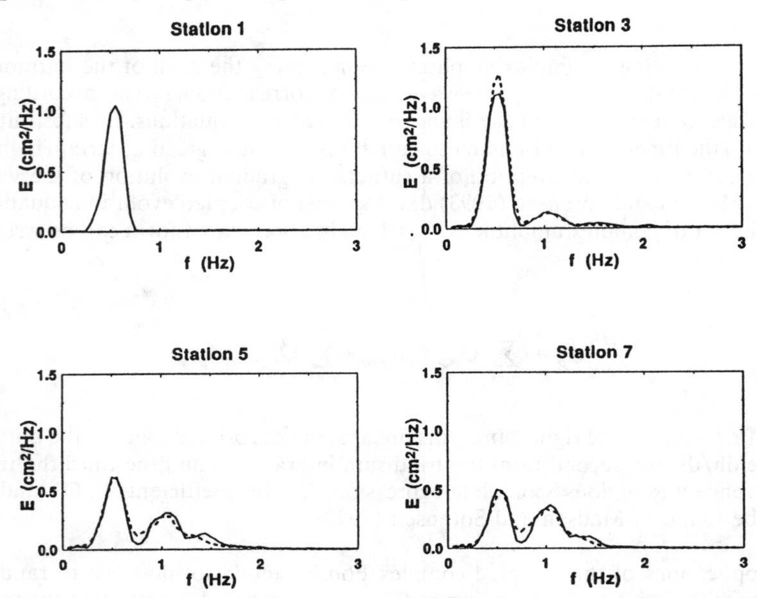

Figure 1: Energy spectra from experiments (——) and from Boussinesq model without breaking (---) for nonbreaking waves passing over a shallow bar. Propagation from station 1 (seaside) to station 7 (shoreside); stations 3 and 5 are over the bar crest. Measurements by Beji and Battjes (1993) for peak frequency $f_m = 0.50$ Hz, initial significant wave height H_s $(=4\sqrt{m_0}) = 0.020$ m, and minimum depth over the bar $h_{min} = 0.10$ m.

Inclusion of wave breaking

The effects of wave breaking can be accounted for along the lines sketched above by adding the following quasi-linear dissipation term (d_p) to the r.h.s. of (2):

$$d_p = -\frac{1}{2}\frac{D}{F}A_p \tag{3}$$

Here, D is the total local rate of random-wave energy dissipation per unit area due to breaking, and F is the total local rate of energy flux per unit width. The value of D is determined according to the model by Battjes and Janssen (1978) in which it depends mainly on the total energy (proportional to $\Sigma|A_p|^2$) and the local depth. The value of F is proportional to $\Sigma|A_p|^2c_{gp}$, with the group velocity c_{gp} determined from the linearised Boussinesq model (as in Madsen et al.,1991).

The dissipation term d_p is proportional to A_p, with a frequency-independent factor, so that this term reduces the spectral amplitudes in the same proportion; it does not affect the spectral shape. Note that the factor is chosen to be real; there is at present no basis for inclusion of certain phase shifts due to breaking.

Application of (2) including (3) to cases of random waves breaking over a bar showed rather good agreement with the observed spectral evolution, both for a mildly breaking case (Beji and Battjes, 1993), shown in Figure 2, and for a strongly breaking case for the same bar shape (measured by Delft Hydraulics), shown in Figure 3.

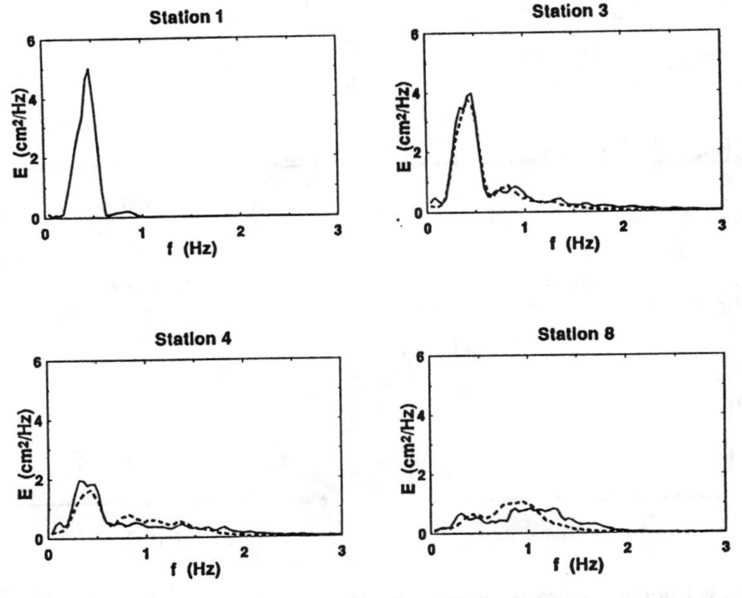

Figure 2: Energy spectra from experiments (——) and from Boussinesq model with breaking (---) for breaking waves passing over a shallow bar. Propagation from station 1 (seaside) to station 8 (shoreside); stations 3 and 4 are over the bar crest. Measurements by Beji and Battjes (1993) for peak frequency f_m =0.47 Hz, initial significant wave height H_s=0.043 m, and minimum depth over the bar h_{min}=0.10 m.

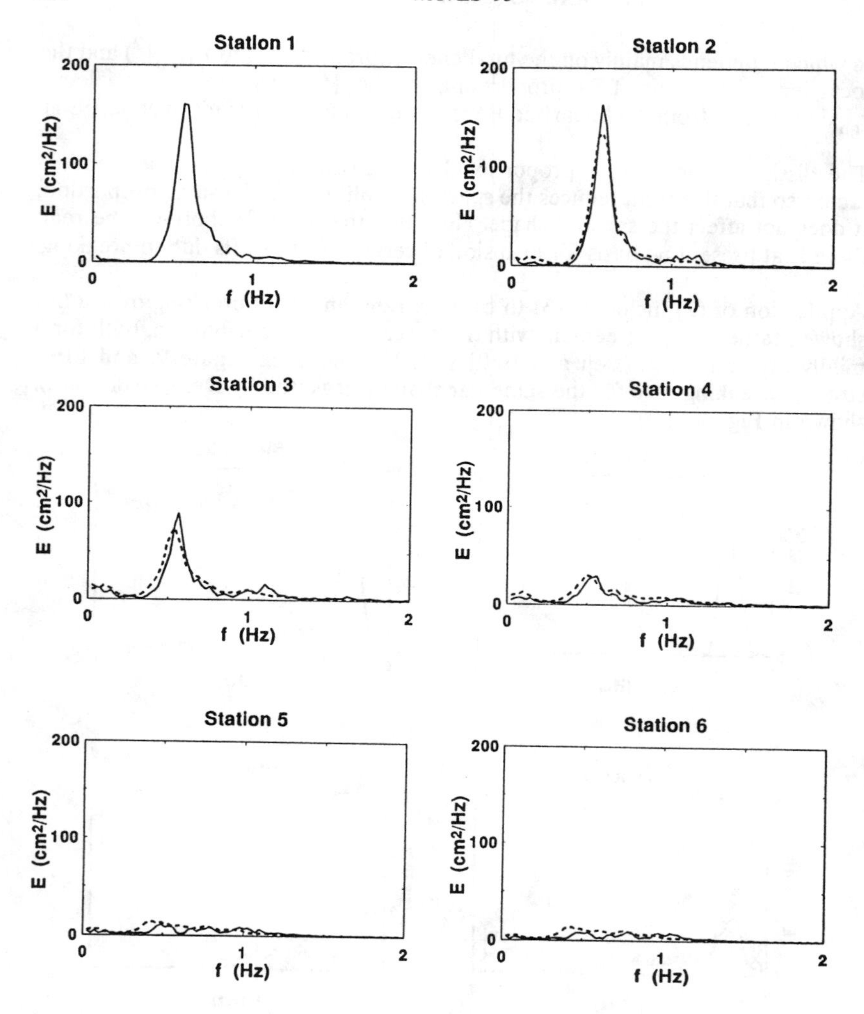

Figure 3: Energy spectra from experiments (——) and from Boussinesq model with breaking (---) for breaking waves passing over a shallow bar. Propagation from station 1 (seaside) to station 6 (shoreside); stations 3 and 4 are over the bar crest. Measurements by Delft Hydraulics for peak frequency $f_m = 0.55$ Hz, initial significant wave height $H_s = 0.22$ m, and minimum depth over the bar $h_{min} = 0.20$ m.

Conclusions

For the conditions of these investigations, the following conclusions can be drawn:

- Spectral evolution of nonbreaking random waves passing over a shallow bar is well predicted by a coupled Fourier mode formulation based on Boussinesq equations.
- The effect of breaking can be modelled by including a quasi-linear dissipation term in the evolution equations for the complex amplitudes, which reduces the total (spectrally integrated) energy without affecting the spectral shape. This formulation has been found to yield realistic predictions of the spectral evolution of random waves, breaking over a shallow bar.

Acknowledgements

The work presented here is part of an ongoing project at the Delft University of Technology which is financed in part by the Tidal Waters Division and the Road and Hydraulics Engineering Division of the Department of Public Works (Rijkswaterstaat). The J.M. Burgers Centre for Fluid Mechanics at the Delft University of Technology provided financial support for Y. Won. The experimental data shown in Figure 3 were kindly made available to the authors by Delft Hydraulics, through the cooperation of Dr. J. v.d. Meer. Thanks are due to Dr. Per Madsen of the Danish Hydraulics Institute for making manuscripts available to the authors prior to publication.

References

Abreu, M., Larraza, A., and Thornton, E.B., 1992. "Nonlinear transformation of directional wave spectra in shallow water". J. of Geophys. Res. 97 (C10), 15.579-15.589.

Battjes, J.A., and Beji, S., 1992. "Breaking waves propagating over a shoal". Proc. 23rd Intern. Conf. Coastal Eng., Venice, Vol.1, 42-50.

Battjes, J.A., and Won, Y., 1992. Spectral Boussinesq modelling of random waves. Report no. 93-2, Delft University of Technology, Delft, The Netherlands.

Beji, S., and Battjes, J.A., 1993. "Experimental investigation of wave propagation over a bar". Coastal Eng. 19, 151-162.

Byrne, R.J., 1969. "Field occurrences of induced multiple gravity waves". J. Geophys. Res., 74(10), 2590-2596.

Drouin, A., and Ouellet, Y., 1988. "Experimental study of immersed plates used as breakwaters". Proc. 21st. Intern. Conf. Coastal Eng., Malaga, Vol.3, 2272-2283.

Johnson, J.W., Fuchs, R.A., and Morison, J.R., 1951. "The damping action of submerged breakwaters". Trans. Amer. Geophys. Union, 32(5), 704-718.

Jolas, P., 1960. "Passage de la houle sur un seuil". La Houille Blanche, Vol.2, 148-152.

Kojima, H., Ijima, T., and Yoshida, A., 1990. "Decomposition and interception of long waves by a submerged horizontal plate". Proc. 22nd Intern. Coastal Eng., Delft, Vol.2, 1228-1241.

Madsen P.A., Murray, R., and Sørensen, O.R., 1991. "A new form of the Boussinesq equations with improved linear dispersion characteristics". Coastal Eng. 15, 371-388.

Madsen, P.A., and Sørensen, O.R., 1992. "A new form of the Boussinesq equations with improved linear dispersion characteristics, Part 2: A slowly-varying bathymetry". Coastal Eng. 18, 183-205.

Madsen, P.A., and Sørensen, O.R., 1993. "Bound waves and triad interactions in shallow water". J. Ocean Eng., 20 (4), 359-388.

Peregrine, D.H., 1967. "Long waves on a beach". J. Fluid Mech. 27, 815-827.

Young, I.R., 1989. "Wave transformation over coral reefs". J. Geophys. Res. 94, 9779-9789.

Kinematics of Wave Overtopping on Marine Structure

J.J. Lee[1], F. Zhuang[2], C. Chang[3]

ABSTRACT

Kinematics of transient wave overtopping on coastal breakwaters has been studied both experimentally and numerically. For the laboratory experiments, solitary waves with moderate wave amplitude are used as the incident waves. The wave profiles are obtained by resistance type wave gauge. The two dimensional water particle velocities are measured by a portable, four-beam, fiber optic Laser Doppler Velocimeter (LDV) system. For the numerical analysis, potential flow theory and Boundary Element Method (BEM) are used for analyzing the wave field induced by the coastal structure. It is found that the numerical results compared well with the experimental data in the cases where the numerical model is valid. Some interesting experimental observations are also presented.

1 Introduction

Overtopping of ocean waves over marine structures has been an important problem for coastal and ocean engineers due to the fact that it may severely damage the marine structure and it may transmit significant wave energy toward the coastal zone for which the marine structure is supposed to protect. Because of the complication of the physical process of overtopping and the lack of suitable instruments to properly measure the internal velocity and

[1]Professor of Civil Engineering, University of Southern California, Los Angeles, CA 90089-2531

[2]Graduate Research Assistant, Department of Civil Engineering, University of Southern California, Los Angeles, CA 90089

[3]Postdoctoral Research Associate, Department of Civil Engineering, University of Southern California, Los Angeles, CA 90089

acceleration fields, little progress has been achieved in obtaining the detailed kinematic structure in connection with the wave overtopping process, even though several investigations have been devoted to wave overtopping problem (see references [1],[5],[7]-[11]).

Coastal engineers are often faced with the problem determining the flow field and hence the loading condition on the structures. A breakwater structure may be constructed at a particular site for the non-overtopping condition. However, due to varying tide levels plus the unforeseen wave conditions, the submergence of the structure may be so altered to cause substantial overtopping of incident waves toward shoreward region. This experimental and numerical study has been directed to obtain the kinematic properties of overtopping nonlinear waves across a breakwater structure. For the laboratory experiments, solitary wave with moderate wave amplitude are used as the incident waves because stable wave profiles can be generated repeatedly. Wave profiles as a function of time are measured at various locations (upstream, downstream and above the breakwater region). Two dimensional water particle velocities are measured by a portable, four-beam fiber optic Laser Doppler Velocimeter (LDV) system equipped with frequency shifting. A numerical model has been developed, employing potential theory and Boundary Element Method (BEM), to simulate the wave field induced by the coastal structure as the incident wave interacting with the breakwater.

2 Experimental Equipment and Experimental Procedures

Experiments involving kinematic measurements of solitary waves overtopping various submerged breakwater configurations are conducted in a wave tank 15.2 meter long, 39.4 centimeter wide, and 61 centimeter deep. The sidewalls of the wave tank are made of glass and offer excellent transparency for laser beams. A programmable piston type wave generator is installed at one end of the tank and a sloping beach is installed at the other end of the wave tank to aid the wave dissipation thereby reducing the waiting time required between runs.

Two different breakwater configurations are used in the experiment. The first breakwater is 45 inch wide, 4.5 inch high and is made of plywood. The second breakwater is of one third in width with the same height and is made of lucite. The breakwater is sunk and fixed to the bottom of the wave tank by adding lead weight.

The wave generating device is a piston type wave generator. It is powered by a hydraulic piston whose motion is controlled by a micro computer. The computer determines a voltage time history which defines the trajectory of the wave plate through a hydraulic-servo system. The wave generation program used by the computer allows motions of the wave machine to be prescribed for generating small amplitude periodic waves, finite amplitude periodic waves and solitary waves.

Tap water is used to fill the wave tank to the desired depth. The depth is measured using a point gage which is mounted to a movable carriage traveling on a rail system installed at the top of both side walls of the wave tank. Resistance type wave gages are used to measure water surface elevations as a function of time. Three wave gages are installed at desired locations to make simultaneous wave profile measurements. The wave gages are connected to a Sanborn four channel oscillograph recorder which records the measurements on an oscillograph paper.

The water particle velocities are measured using a portable four-beam, two-component, fiber optic Laser Doppler Velocimeter (LDV) manufactured by TSI, Inc.. The LDV system used in the experiment consists of a 100 mW argon-ion laser, transmitting and receiving optics, a fiberoptic probe, frequency shifting and signal processing instruments. A multicolor beam separator separates the incident laser beam into four beams, two blue beams and two green beams. The four laser beams are focused into one point within the flow field to form a two-component system. Figure 1 provides a flow chart of the LDV system used for the present experiment. Two photomultipliers convert optical signals into electric signals. Two frequency shifters help attain accurate flow measurements in applications where high turbulence or flow reversals are anticipated. A fiberoptic probe which features focusing and receiving optics in one compact unit offer considerable ease for setting up the LDV system. Two IFA550 signal analysis systems are used for signal analysis. This is done by using a micro computer which manages the LDV measurements, stores velocity measurements, and displays the velocity-time history on the monitor. The whole LDV system is installed in a room close to the wave tank. When velocity measurements are needed, only the portable fiberoptic probe is moved to the measuring station and is mounted to a traversing mechanism which offers three dimensional positioning of the measuring point. Seeding is one of the key elements affecting the performance of the LDV measurements. Seeding particles must be small enough to move with the flow yet large enough to scatter sufficient light for ideal signal quality. Titanium Dioxide powder (TiO_2) is used in the experiment and is proved to be a good seeding agent for water in the wave tank used for the present experiments.

3 Numerical Analysis

For the numerical analysis, the problem of wave overtopping on marine structure is formulated as a two-dimensional boundary value problem. The fluid in the solution domain is assumed to be incompressible and the flow is assumed to be irrotational. Viscous force is also neglected. Application of potential theory leads to the governing Laplace's equation for the velocity potential function ϕ:

$$\nabla^2\phi(\mathbf{x},t) = 0 \qquad \mathbf{x} \in \Omega(t) \tag{1}$$

with the kinematic and the nonlinear dynamic free surface conditions as the boundary conditions on the free surface Γ_s:

$$\frac{D\mathbf{r}}{Dt} = (\frac{\partial}{\partial t} + \mathbf{u}\cdot\nabla)\mathbf{r} = \mathbf{u} = \nabla\phi \tag{2}$$

$$\frac{D\phi}{Dt} = -gy + \frac{1}{2}|\nabla\phi|^2 - \frac{p_a - p_o}{\rho} \tag{3}$$

where $\mathbf{r}$ is the position vector of a free surface fluid particle, g the acceleration due to gravity, y the vertical coordinate, p_a the pressure at the surface, p_o a reference pressure and ρ the fluid density.

The solution to the boundary value problem is expressed as a boundary integral using the free space Green's function $G(\mathbf{x}_i,\mathbf{x}_j) = -\frac{1}{2\pi}\log|\mathbf{x}_i - \mathbf{x}_j|$ and Green's theorem:

$$\alpha(\mathbf{x}_i)\phi(\mathbf{x}_i) = \int_{\Gamma(\mathbf{x})}\left[\frac{\partial\phi}{\partial n}G(\mathbf{x},\mathbf{x}_i) - \phi(\mathbf{x})\frac{\partial G(\mathbf{x},\mathbf{x}_i)}{\partial n}\right]d\Gamma(\mathbf{x}) \tag{4}$$

where $\mathbf{x}_i$ and $\mathbf{x}$ are position vectors for points on the boundary ($\mathbf{x}_i$ can also be any where within the domain), $\Gamma(\mathbf{x})$ is the boundary of the fluid domain Ω, $\mathbf{n}$ the unit outward normal vector and $\alpha(\mathbf{x}_i)$ a geometric coefficient. The boundary integral equation is then solved using the Boundary Element Method.

A time marching procedure which was first suggested by Dold and Peregrine (1986) has been used in the present analysis to update both the new position of the free surface and the potential value ϕ on the free surface at the next time step:

$$\mathbf{r}(t+\Delta t) = \mathbf{r}(t) + \sum_{k=1}^{n}\frac{(\Delta t)^k}{k!}\frac{D^k\mathbf{r}(t)}{Dt^k} + O[(\Delta t)^{n+1}] \tag{5}$$

$$\phi(\mathbf{r}(t+\Delta t), t+\Delta t) = \phi(\mathbf{r}(t),t) + \sum_{k=1}^{n}\frac{(\Delta t)^k}{k!}\frac{D^k\phi(\mathbf{r}(t),t)}{Dt^k} + O[(\Delta t)^{n+1}]. \tag{6}$$

The Boundary Element Method together with the time marching procedure form a unique solution technique for solving transient nonlinear wave problems. A detailed description of this solution method for the general wave-structure interaction problem can be found in Chang (1993).

Computation of wave profiles and the kinematic properties beneath the waves based on this method can be expected to work well for certain cases with good accuracy (when the water depth at the crest of the breakwater is not too shallow to induce wave breaking). However, for more complicated cases such as significant wave overtopping over very small water depth above the breakwater crest region, the numerical simulation will cease to be valid. Thus, major efforts are directed to experimental work to obtain the kinematic properties of the overtopping waves in the present study.

4 Presentation and Discussion of Results

When the breakwater is totally submerged in the water with sufficient depth above the crest of breakwater the numerical computation based on the Boundary Element Method outlined in previous section can be expected to work well for predicting the wave profiles and the accompanying kinematics of the water particle. An example of this condition is shown in Figure 2.

Figure 2 presents the comparison of water particle velocities measured and computed theoretically at three different locations for a solitary wave propagating over a submerged breakwater taken from Lee, Chang & Zhuang (1992). The crest of the breakwater is at one-half of the water depth. The wave height/water depth ratio is 0.2. The still water depth is 9 inches and the breakwater width (in the longitudinal direction) is ten times the breakwater height.

Included in Figure 2 are both the horizontal and vertical water particle velocities at the three different horizontal locations: five water depth upstream of the center of the breakwater, center of the breakwater, five water depth downstream of the breakwater. The vertical position of the measurement position is at 0.1 water depth below the still water level. In each of the three figures presented the horizontal velocities are significantly larger than the vertical velocities. Comparing the numerical results obtained by the boundary element model and experimental results one can see that the agreement is quite good even though there is some scattering when the velocity is very small. The good agreement between the theory and experiment for the two components of the water particle velocities provides a critical check of the reliability of the results from the numerical model under the given condition.

Figure 3 illustrates an experimental set up for a solitary wave propagating from the open sea toward the breakwater region. The crest of the breakwater is equal to the still water depth (the free board is zero). Thus, when an incident solitary wave strikes the breakwater from the left, the overtopping wave is propagating over the breakwater at zero water depth until it passes toward the shoreward side of the breakwater.

Figure 4 shows a solitary wave traveling in this zero free board situation. From Figure 4(A) it is clear that the majority of the first wave represents the incident solitary wave and the second wave represents the reflected wave from the upstream edge of the breakwater. The shape of the reflected wave is quite similar to the solitary wave also. The wave profile shown in Figure 4(B) represents the transmitted wave profile at 15" downstream (3.33 water depth) after the solitary wave has traveled above the breakwater crest region (at zero water depth). It shows that the primary wave is followed by a series of oscillatory tails. These oscillatory wave trains have been reformed into more regular oscillatory waves as they traveled further downstream (at 10 water depth downstream) as evidenced in the wave profile presented in Figure 4(C). It is interesting to observe the physical nature of the transmitted wave as it travels above the breakwater crest at zero depth. The jet like water mass is translated into the shoreward region of the breakwater. This water mass which is above the still water level then plunge into the shoreward region by the continuous effect of the gravity forces causing the water mass in the shoreward region to exhibit significant undulations. Comparing the wave profiles shown in Figure 4(A), 4(B) and 4(C), it is clear that the breakwater serves to break up the incident wave resulting in significant higher frequency wave components in the shoreward region as shown from the energy spectra in the right column of Figure 4. This physical phenomenon is significant for the assessment of basin response of the shoreward coastal region. Thus, the overtopping effect causes the generation of higher frequency waves toward the shoreward region. If the shoreward region is connected to a harbor, then the basin response could be significantly altered. The importance of this has been noted by Raichlen (1992).

Figure 5 presents u, v components of the water particle velocity at location A ($x = 2.22d$, $y = -0.1d$) obtained by LDV measurements. The maximum incident wave amplitude to depth ratio is 0.2. The ordinate is particle velocity normalized with respect to $\sqrt{gd}$ (the wave celerity for shallow water wave based on the upstream still water depth). The abscissa (time scale) is normalized as $t\sqrt{g/d}$. The starting point of the time scale is arbitrarily chosen which represents the time we start the LDV measurement. Looking at Figure 5 we can see that the horizontal water particle velocity is almost equal to the wave celerity. One can recall that the horizontal water particle velocity under an undisturbed solitary wave with 0.2 amplitude ratio at the location of 0.1 depth below the still water level would be approximately $0.2\sqrt{gd}$. Therefore, this is more than *four* times the water particle velocity at the similar location within the incident wave before the incident wave strikes the breakwater. Relatively large negative vertical component of velocity can also be seen from the lower graph of Figure 5 showing tremendous impact of the overtopping waves.

When the overtopping wave field are closely observed, it is found that a series of vortices is generated by the overtopping wave above the still water level which propagate downstream and modify drastically the velocity field in the shoreward vicinity of the breakwater. Figure 6 presents u, v components of the water particle velocity at location B ($x = 2.22d$, $y = -0.5d$), which is in the same vertical line with location A, but 0.4 depth lower. Comparing Figures 5 and 6, it is observed that when the wave is propagating across this vertical line, the horizontal velocity at location A is in the shoreward direction (positive), whereas at location B it is basically in the seaward direction (negative). This indicates a rotational motion of particles is generated in the shoreward vicinity of the breakwater. The overtopping wave induces a complex rotational velocity field. The flow field is quite similar to that found in backward facing step flow except that the present flow condition involves an unsteady flow condition and with a free surface. The sequence of the vortex motion can be seen in the experiment with different breakwater height/water depth ratio. When the advancing wave propagate further toward shoreward region the generated vortices move downstream and are disintegrated eventually. Such kinematic properties have direct impact on the stability of the breakwater units. Especially, as can be seen from Figure 6, a relatively large upward vertical velocity is generated near the shoreward breakwater, which could cause a lifting force on the breakwater units in the shoreward side of the breakwater.

5 Conclusions

The experimental and numerical study on kinematic properties of wave overtopping on marine structure has been presented. The major conclusions can be summarized as follows:

1. For incident solitary wave with moderate wave height and when the breakwater is deeply submerged the numerical results have been compared with the experimental data. It is shown that the agreement between the numerical results and the experimental data has been quite good.

2. A series of experimental data is presented for solitary wave interacting with the breakwater which has the same height as the still water depth. The experimental data show that a series of higher frequency oscillatory tails is generated in the shoreward region. This transmitted wave is reformed to become a series of well defined oscillatory wave as they propagate further away from the breakwater. This physical phenomenon is significant for the assessment of basin response of the shoreward coastal region.

3. In situations where breakwater height is about the same as the still water depth, when the incident solitary wave overtops the breakwater, the horizontal water particle velocity is nearly as large as the wave celerity, which indicates the wave motion is transformed temporarily into a bore-like flow by the breakwater in the local area. The large horizontal and vertical velocities produce significant impact on the breakwater due to the overtopping waves.

4. It is observed vortices are generated in the vicinity of the shoreward breakwater as the solitary wave propagates over the top of the breakwater. It is similar to the phenomenon which appears in the backward facing step flow. This flow is unsteady and with a free surface. The vortex motion can be seen in the experiments for various breakwater height/water depth ratio, ranging from one half to one. As the solitary wave propagating further away from the breakwater, the vortices continue to move downstream and are disintegrated eventually. The vortex motion will affect the pressure field near the surface of the shoreward breakwater. Also, because of the rotating direction of the vortex, there will be a relatively large upward velocity acting on the breakwater units, representing a lifting force. More attention to the induced flow field needs to be directed in the region.

6 Acknowledgment

This study is supported by USC Foundation for Cross-Connection Control and Hydraulic Research. The LDV System is supported by NSF under Grant No. 8906898. The authors are grateful for the generosity of Dr. Fredric Raichlen for permitting them to conduct the experiments at Caltech's W.M. Keck Laboratory of Hydraulics and Water Resources.

7 References

1. Ahrens, J.P. & Heimbaugh, M.S., "Seawall Overtopping Model", Proceedings of 20th Coastal Engineering Conference, pp.795-804, 1988.

2. Chang, Chun, "Simulation of Generation, Propagation and Wave-Structure Interaction of Transient Nonlinear Wave Using the Boundary Element Method", Ph.D. Thesis, University of Southern California, Los Angeles, CA, May 1993.

3. Dold, J.W. & Peregrine, D.H., "An Efficient Boundary Integral Method for Steep Unsteady Water Waves", Numerical Methods for Fluid Dynamics II (ed. K.W. Morton & M.J. Baines), pp.671-679, Clarendon Press, Oxford, 1986.

4. Grilli, S., Skourup, J. and Svendsen, I.A., "An Efficient Boundary Element Method for Nonlinear Water Waves", Engineering Analysis with Boundary Elements, 6(2), 97-107, 1989.

5. Kobayashi, N. and Wurjanto, A., " Wave Overtopping on Coastal Structures", J. Waterway, Port, Coastal and Ocean Engineering, Vol. 115, No. 2, pp235-251, March 1989.

6. Lee, J.J., Chang, C. and Zhuang, F., "Interaction of Nonlinear Waves with Coastal Structures", Proceedings of the Twenty-Third International Conference on Coastal Engineering, ASCE, Venice, Italy, Oct.4-9, 1992, pp.1327-1340.

7. Lee, J.J., Skjelbreia, J.E. & Raichlen, F., "Measurement of Velocities in Solitary Waves", J. Waterway, Port Costal and Ocean Division, ASCE, Vol. 108, No. WW2, May 1982.

8. Raichlen, F., "Armor Stability of Overtopped Breakwaters", J. Waterways Harbor Div. Proc. ASCE 98: 273-79, 1972.

9. Raichlen, F., Cox, J.C. and Ramsden, J.D., "Inner Harbor Wave Conditions due to Breakwater Overtopping", Proceedings Coastal Engineering Practice '92, ASCE, March 1992.

10. Saville, T. Jr., "Laboratory Data on Wave Runup and Overtopping on Shore Structures", TM-64 U.S. Army, Corps of Engineers, Beach Erosion Board, Washington, D.C., Oct. 1955.

11. Seelig, W., "Two-Dimensional Tests of Wave Transmission and Reflection Characteristics of Laboratory Breakwaters", TR80-1, Coastal Engineering Research Center, U.S. Army Engineer Waterways experiment Station, Vicksburg, Miss., June 1980.

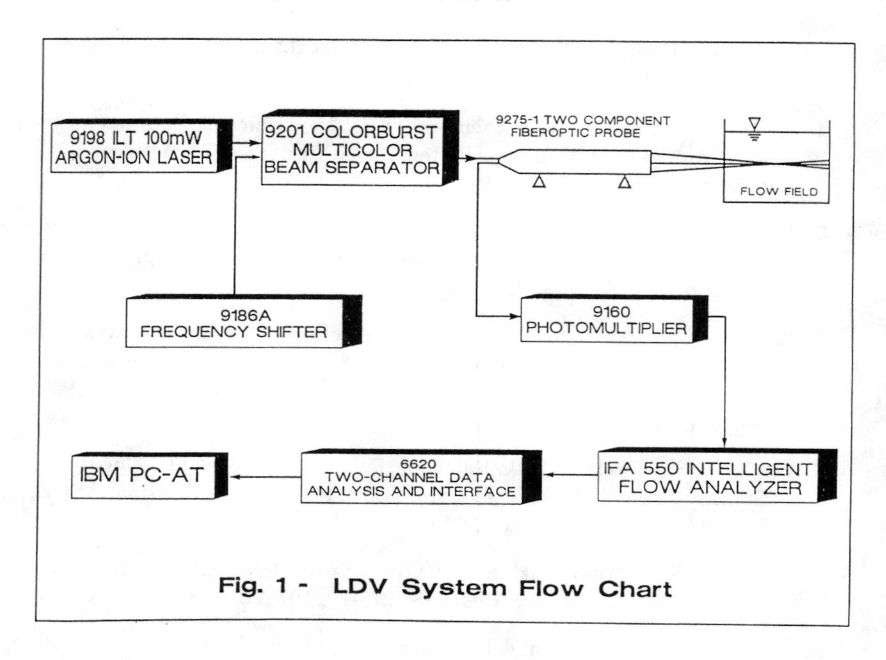

Fig. 1 - LDV System Flow Chart

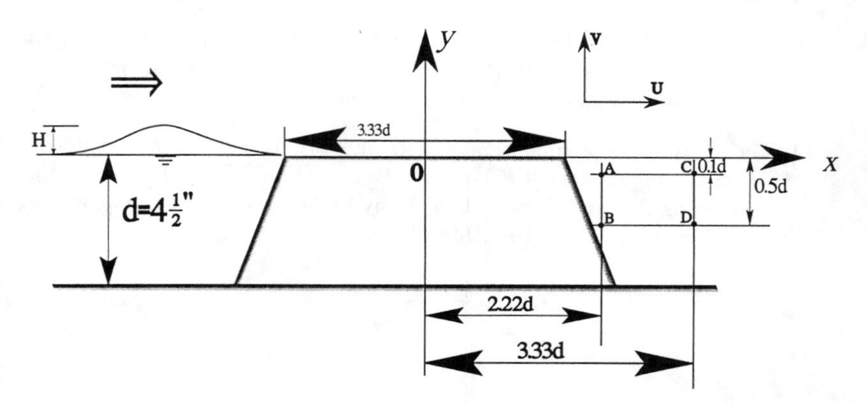

Figure 3 *Sketch of an experimental set up showing the breakwater height, still water depth and locations of LDV measurements A,B,C,D (not to scale)*

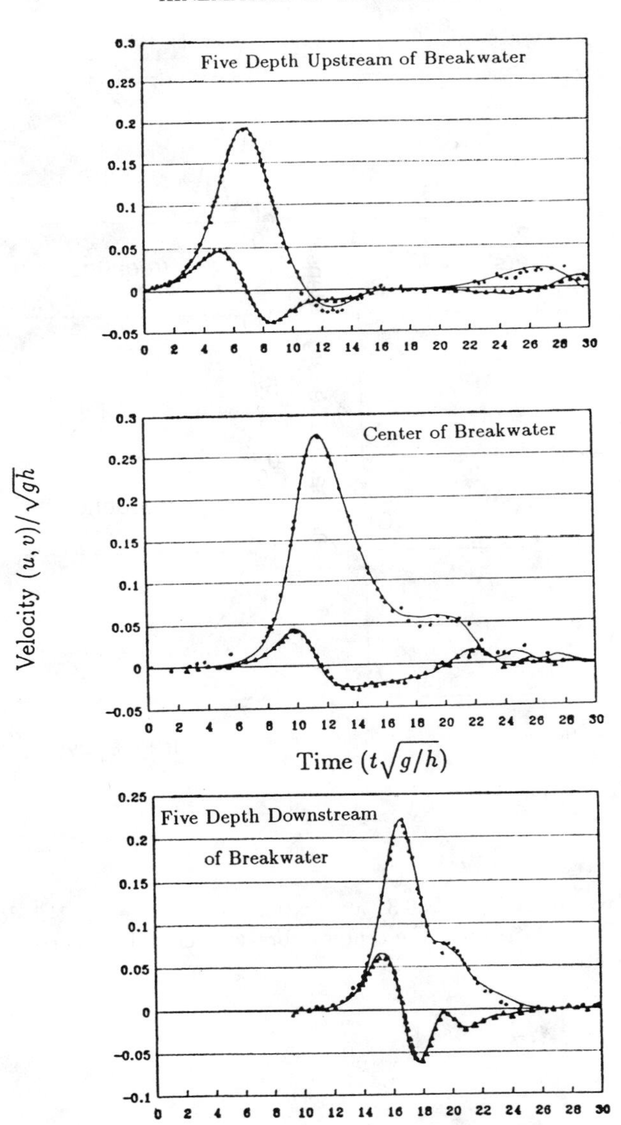

Figure 2: Comparison of water particle velocities at three different locations (— present theory; ··· experiment, u component; ▲▲▲ experiment, v component) (from Lee, Chang & Zhuang 1992)

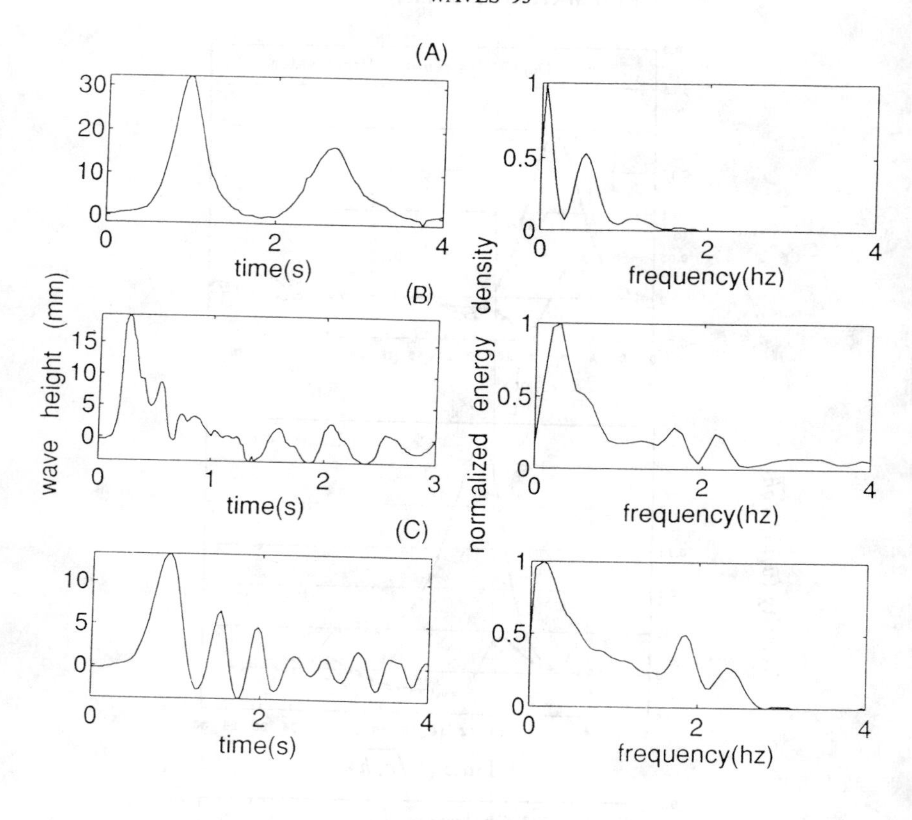

Figure 4: Wave records (H/d=0.3) at locations 45" upstream, 15" downstream and 45" downstream of the center of breakwater and the corresponding energy spectra.

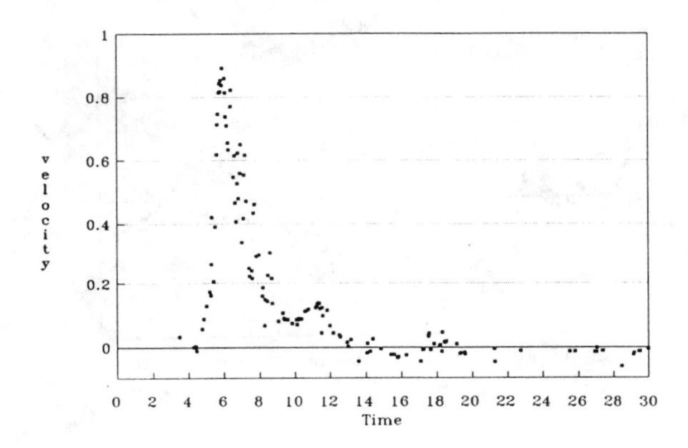

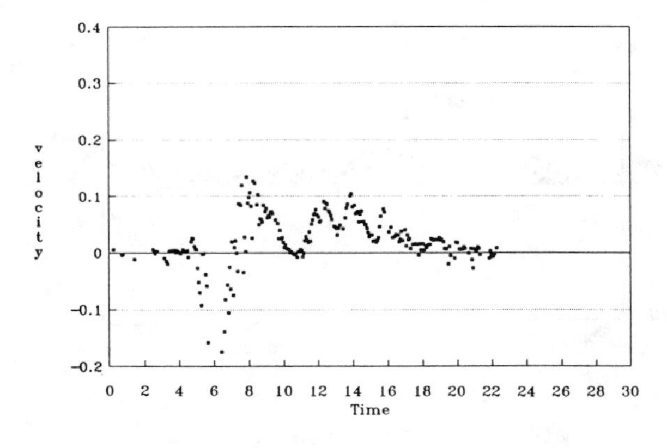

Figure 5: LDV measured u, v water particle velocity at location A downstream of the breakwater (x=2.22d, y=-0.1d)

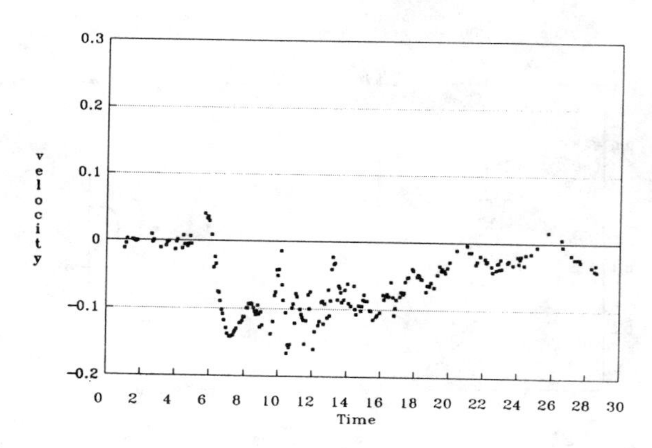

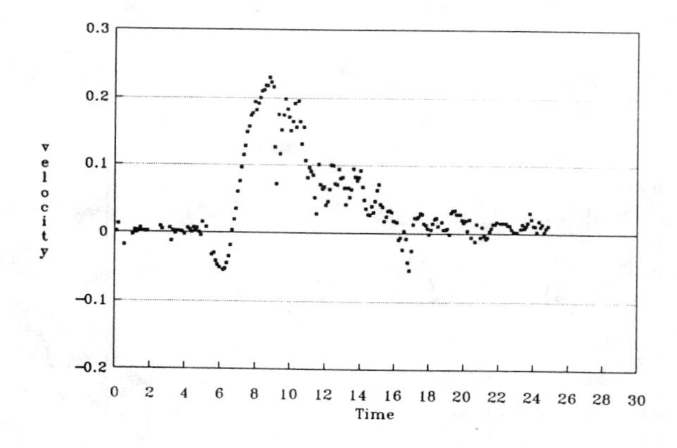

Figure 6: LDV measured u, v water particle velocity at location B downstream of the breakwater (x=2.22d, y=-0.5d)

WAVE SETUP NEAR A SEAWALL: EXPERIMENTS AND A NUMERICAL MODEL

K.A.Rakha[1], J.W.Kamphuis[2]

ABSTRACT

Most published studies on wave setup use data obtained in two-dimensional wave flumes with normal wave incidence. This paper presents a new data set of three-dimensional tests performed to study the effect of seawalls on an infinite beach. Setup values were measured every hour as the beach in front of the seawall eroded. The experiments were simulated with a numerical model, which included the effect of wave incidence and wave reflection on the wave setup. The numerical model simulated the measured wave setup adequately.

INTRODUCTION

Wave setup was identified in the field as early as 1938 on the east coast of North America, and its existence was confirmed in laboratory experiments by Savage (1957), Fairchild (1958), and Saville (1961). However, wave setup was not successfully explained theoretically until the early 1960s when the concept of radiation stress was developed by Longuet-Higgins and Stewart (1962). Radiation stress was applied to obtain analytical and numerical models for the prediction of wave setup (e.g. Bowen et al 1968, Svendsen 1984). There are several experimental and field investigations in which wave setup was measured. Wave setup was measured on fixed beaches, as well as on equilibrium profiles formed from beach materials of different permeabilities. The available setup data provide a considerable amount of information which could serve as a guide for the development of numerical models, but most of the collected data is restricted to normally incident waves on beaches. The effect of wave angle and of wave-structure interaction on wave setup has not received much

---

[1] Research Student, [2] Prof., Department of Civil Eng. Queen's University, Kingston, Ontario, Canada K7L 3N6.

attention.

Three-dimensional mobile bed hydraulic model tests on a beach backed by a seawall were performed at Queen's University. The purpose of the tests was to study the beach morphology in front of a seawall. The present paper analyses the wave setup data obtained during these tests. The measured wave setup data are compared with the results from a numerical model.

EXPERIMENTS

Six tests were carried out in the three-dimensional model basin at the Queen's University Coastal Engineering Laboratory shown in Figure 1. A complete description of the tests can be found in Kamphuis, Rakha, and Jui (1992). The case of a beach backed by a seawall was considered. A vertical seawall was constructed on a sandy beach of median grain size 0.11 mm. The test conditions are summarized in Table 1. Test G was a repetition of Test B. A time series of waves, simulating a Jonswap wave spectrum was used with a 10 degree incident wave angle, a peak period of 1.15 seconds and wave heights varying from 0.05 to 0.09 m.

A beach of initial slope 1:10 was first allowed to shape itself to a near equilibrium state. The offshore depth was chosen so that only about one percent of the waves impacted on the seawall when the beach reached equilibrium, thus simulating a seawall well back on the beach. The water level was then raised in some tests to simulate storm surge. This resulted in initial percentages of wave impact on the seawall as shown in Table 1. The test continued until a new equilibrium profile was approached. The wave heights were recorded using capacitance type wave gauges. One wave gauge was mounted offshore to record the incident wave height and a set of 15 wave gauges where mounted on a horizontal beam at regular intervals of 0.2 m through the breaking zone. Profiles of the beach in front of the seawall were measured along six parallel lines evenly spaced along the seawall. The profiler was mounted on a horizontal beam parallel to the beam carrying the wave gauges. The profiles presented here are the average of profiles B,C,D, and E (Figure 1).

Each test was performed as a series of hourly segments. Each hour the wave machine was stopped, the beach profiles were measured and the 15 wave gauges were read to obtain the Still Water Level (SWL). Shortly after the wave generator was started again and conditions had become steady, wave gauge readings were recorded over the full cycle of the wave time series. Averaging the readings of each wave gauge over the cycle resulted in the Mean Water Level (MWL) at each wave gauge location. Subtracting the SWL readings from the MWL provided the wave setup or setdown at 15 locations over the breaker zone.

TABLE 1
TEST SUMMARY

TEST	GENERATED WAVE CONDITIONS				t (hr)	% touch
	Hs (m)	Tp (sec)	α (deg)	d (m)		
A	0.07	1.15	10.0	0.55	32	1.0
B	0.07	1.15	10.0	0.57	18	10
D	0.07	1.15	10.0	0.57	12	20
E	0.09	1.15	10.0	0.54	14	10
F	0.05	1.15	10.0	0.59	22	10
G	0.07	1.15	10.0	0.57	18	10

The test durations varied from 12 to 32 hours. Thus, wave setup data for a large number of beach profiles was obtained. The profiles in front of the seawall varied from a natural beach to profiles greatly affected by the seawall. Sample profiles for Test A are shown in Figure 2. It is seen that the depth immediately in front of the seawall increases in time and that the material is placed offshore, forming a near-horizontal platform.

SETUP CALCULATION

The development of a numerical model to simulate the beach evolution in the vicinity of a seawall is in progress. The wave model was developed to account for combined shoaling, refraction, breaking, and reflection. Reflection was included in the model by using a mirror image technique. The details of the wave model will be published elsewhere. Wave setup was calculated by solving a simplified momentum equation in the offshore direction,

$$g\frac{\partial \eta}{\partial x} = -\frac{1}{\rho d}\frac{\partial S_{xx}}{\partial x} \tag{1}$$

where, η is the wave setup, x is the offshore direction, g is the acceleration of gravity, ρ is the fluid density, S_{xx} is the radiation stress in the x direction, and d is the total depth $(h+\eta)$, where, h is the still water depth.

Bottom friction was neglected in Eq.1, Stive and Wind (1982) mentioned that this can be justified.

Linear wave theory was used for these laboratory experiments, because Kamphuis (1991a) shows that wave transformation of irregular waves is adequately described by linear wave shoaling and refraction theory. Guza and

Thornton (1981) also mention that linear theory adequately relates local S_{xx} measurements from pressure, sea surface elevation, and velocity sensors to each other. They also find that linear wave theory predicts S_{xx} outside the surf zone adequately. Both studies performed by Kamphuis (1991a) and Guza and Thornton (1981) were based on irregular wave data. Longuet-Higgins (1964) used linear wave theory to introduce the following equations,

$$S_{xx} = [S_{xx}]_i + [S_{xx}]_r \tag{2}$$

$$S_{xx} = E[n(1+\cos^2\alpha) - \frac{1}{2}] \tag{3}$$

where the subscripts i and r denote the incident and the reflected waves respectively, α is the wave angle, E is the wave energy density:

$$E = \frac{1}{16}\rho g H_s^2 \tag{4}$$

and n is:

$$n = \frac{1}{2}\left(1 + \frac{2kd}{\sinh(2kd)}\right) \tag{5}$$

where k is the wave number, and H_s is the significant wave height. The radiation stress for the incident and the reflected waves are calculated from Eq.3 by substituting for the incident and reflected wave angles and wave heights respectively.

It is necessary to define where the waves break and how the wave energy is dissipated. Breaking is calculated in the present model by the incipient breaking criterion developed by Kamphuis (1991b).

$$H_{sb} = (0.095e^{4.0m})L_{pb}\tanh(2\pi d_b/L_{pb}) \tag{6}$$

where, L is the wave length, m is the beach slope, and the subscripts p and b denote peak period and breaker respectively. The total wave height was used in the above equation for the cases where wave reflection existed.

Wave energy dissipation is simulated using Dally, Dean, and Dalrymple's (1984) wave decay model, modified to take into account irregular waves as in Kamphuis (1993) and wave reflection. The modification of Dally et al's (1984) model was developed for the statistical wave height decay. This model is referred to as the Linear Time Domain Model (LTDM). Battjes and Janssen (1978) wave dissipation model was also modified to include reflection for the spectral wave decay. The Modified Battjes model is referred to as the Linear Frequency Domain Model (LFDM).

COMPARISON WITH REGULAR WAVE EXPERIMENTS

Stive and Wind (1982) obtained experimental results with regular waves and compared them with models based on James (1974), Cokelet (1977), and Linear wave theories. James wave theory is a nonlinear wave theory combining a third order Stokes and a Cnoidal theory. Figures 3 and 4 show the results of the two tests performed by Stive and Wind (1982). Stive and Wind concluded that non-linear wave theory is preferred. They mention that linear wave theory failed to provide a good fit unless the breaker index $(H/d)_b$ on a 1:40 slope was taken as 0.6 instead of the usual value of 0.8 derived from Solitary wave theory. They did not justify the value 0.6. Using Kamphuis' (1991b) simplified equation,

$$H_{sb} = (0.56 e^{3.5m}) d_b \tag{7}$$

Stive and Wind's 1:40 slope would yield a value of approximately 0.6 for the breaker index, which would justify Stive and Wind's assumption. However the measured breaker wave height is underpredicted in this case. Stive and Wind also mention that Linear theory can only yield a linear relationship between setup and depth of water in the surf zone and cannot predict the observed decrease in slope of the MWL in the surf zone.

The linear model developed in the present study is also shown in Figures 3 and 4. It can be seen to give results close to those achieved by Stive and Wind's model based on Cokelet wave theory. A curved relationship between MWL and depth resulted from using Dally et al's (1984) wave decay model.

It should be noted that Eq. 6 was developed to predict the significant wave height for irregular waves due to the large scatter in the data for regular waves. Since, only irregular wave data were used in Kamphuis (1991b), better results would be expected when the present numerical model is compared with tests using irregular waves. Goda (1970) breaker criterion was used for all other regular wave tests.

Hansen and Svendsen (1979) performed regular wave tests on a 1:35 slope. Figure 5 shows the results obtained for one of these tests, the present model is also shown in Figure 5. Linear wave theory failed to predict the wave heights just before wave breaking. Svendsen and Hansen (1977) obtained good results using Cnoidal wave theory. The present model was developed for irregular waves, which do not exhibit the sharp increase in significant wave height immediately before the breaker as shown by Kamphuis (1991a). In fact, Kamphuis found that Linear wave theory overpredicted significant wave heights for irregular waves

which is opposite to what is found for regular waves. To describe wave setup for regular waves Svendsen et al (1978) divide the surf zone into three regions:
i) The outer region where there is a rapid change in wave shape as the surface roller is developing. There is no significant energy dissipation, and wave setdown is constant in this region.
ii) Inner region where the surf roller is developed.
iii) Swash region where the uprush-backwash cycle occurs on the beach front. Dissipation of energy and setup begin at the transition between inner and outer regions.

Gourlay (1992) mentioned that with irregular waves the boundaries between the various regions of the surf zone will no longer be distinct, and setup will begin further offshore. The present model also does not distinguish between the outer and inner regions, because it is concerned with modelling irregular waves. In addition, because wave reflection had to be included it was necessary to use simple models.

Bowen et al (1968) performed some tests on a slope of 0.083. Their outer region was small as compared with Hansen and Svendsen (1979). This might be due to the steeper beach slope. Good results were obtained with the present numerical model using first order Cnoidal wave theory (Isobe, 1985), and Goda's (1970) breaker criterion..

COMPARISON WITH IRREGULAR WAVE FLUME EXPERIMENTS

Battjes and Janssen (1978) performed flume tests on a 1:20 slope for irregular waves. Figure 6 shows the results obtained for Run 3 in their study. The sharp increase in the wave height at the breaker shown for regular waves does not exist. This was true for all their published results, where a spectral wave definition was used. Both the LTDM and LFDM predicted the wave setup well, except for the outer region. Similar results were obtained for the barred beaches tested by Battjes and Janssen.

COMPARISON WITH IRREGULAR THREE-DIMENSIONAL EXPERIMENTS

The performance of the numerical model for cases with a non-zero incident wave angle and wave reflection is now tested by comparing with the experimental results obtained from the three-dimensional tests. Figures 7,and 8 represent the wave heights and wave setup at different stages of Tests A and E. The SWL reached the seawall at Hour 22 of Test A and Hour 7 of Test E. The tests performed indicated that wave reflection did not need to be included in the numerical model until the MWL touched the seawall. This condition was achieved at Hour 15 of Test A and Hour 6 of Test E. Similar results were obtained for the other tests. The LFDM however, overpredicted the spectral wave heights for the cases with reflection. The existence of an outer region was not clear from the tests performed. This could be due to the steep slope of the natural profiles at the breaker location, or due to the angle of incidence. A comparison with two-dimensional data for equilibrium profiles would be required to make any conclusion. Two-dimensional tests were performed at Queen's University, but only wave heights were measured. The predicted wave heights compared well with the those measured.

THE EFFECT OF THE SEAWALL ON WAVE SETUP

Wave setup is influenced by the beach profile. Figure 9 shows examples of the wave setup during Test A. It can be seen that the wave setup decreases as the beach in front of the seawall erodes. All the tests performed showed this trend. As the depth in front of the seawall increases the energy dissipation through breaking decreases causing wave setup to decrease. In the extreme case of complete reflection and no wave breaking, no wave setup will take place except that the MWL is slightly raised at the antinode and lowered at the nodes as mentioned by Longuet-Higgins (1964).

CONCLUSIONS

A numerical model for irregular waves based on Eq.1, using Linear theory, the Kamphuis (1991b) incipient breaking criterion, and a modified Dally, Dean, and Dalrymple's (1984) wave dissipation model yield adequate predictions of wave heights and wave setup along an infinite beach. Wave incidence is also included in the model.

To simulate the effect of a seawall wave reflection may be included by using a mirror image technique. It was found that reflection off the seawall can be ignored until the MWL touches the structure.

The model developed could predict wave heights and setup for regular waves as well. Because it is based on relationships for irregular waves, it has the following limitations. The Linear model underestimates the regular wave heights before the breaker, and may not estimate the breaker location correctly. Also, the outer region is not predicted by the present model.

ACKNOWLEDGMENTS

This study was carried out under the Strategic Grants Program of the Natural Sciences and Engineering Research Council (NSERC) of Canada. The careful hydraulic testing was carried out by J. Kooistra and J. Jui.

NOTATION

d = water depth to MWL;
d_b = water depth at breaking;
g = acceleration of gravity;
h = water depth to SWL;
H_s = significant wave height;
H_{sb} = significant breaking wave height;
k = wave number;
L_{pb} = wave length for peak period at breaker;
m = beach slope;
S_{xx} = radiation stress in offshore direction;
T_p = peak period;
x = offshore direction;
α = wave angle;
η = wave setup;
ρ = fluid density.

REFERENCES

Battjes, J.A. and Janssen, J.P.F.M. (1978). Energy loss and set-up due to breaking of random waves. Proc. 16th Int. Conf. Coastal Eng., ASCE, Vol.1, pp.569-589.

Bowen,A.J., Inman,D.L., and Simmons,V.P. (1968). Wave setdown and setup. J. of Geophys.Res.,73(8),pp.2569-2577.

Cokelet, E.D. (1977). Steep gravity waves in water of arbitrary uniform depth. Philos. Trans. R. Soc. London, Ser. A,286,pp.183-230

Dally, W.R., Dean, R.G., and Dalrymple, R.A. (1984). A model for breaker decay on beaches. Proc. 19th Int. Conf. Coastal Eng., ASCE, Vol.1, pp.82-98.

Fairchild, J.C. (1958). Model study of wave setup induced by hurricane waves at Narrangsett Pier. Rhode Island, Bull. 12, U.S. Army Corps of Eng., Washington D.C.

Goda, Y. (1970). A synthesis of breaker indices. Trans. of Japan Soc. of Civ. Eng., Vol.2, No., pp.227-230.

Gourlay, M.R. (1992). Wave set-up, wave run-up and beach water table: Interaction between surf zone hydraulics and ground water hydraulics. Coastal Eng., Vol.17, pp.93-144.

Guza, and Thornton, (1981). Wave set-up on a natural beach. J. Geophys. Res., Vol. 86, No., C5, pp.4133-4137.

Hansen, J.B., and Svendsen, I.A. (1979). Regular waves in shoaling water, experimental data. Inst. Hydrodyn. Hydr. Eng., Series Paper 21. Tech. Univ. of Denmark.

Isobe, M. (1985). Calculation and application of first order cnoidal wave theory. Coastal Eng., Vol.9, pp.309-325.

James, I.D. (1974). Non-linear waves in the nearshore region:shoaling and setup. Estuarine Coastal Mar. Sci.,2,pp.207-23.

Kamphuis, J.W. (1991a). Wave transformation. Coast. Eng., Vol.15, pp.173-184.

Kamphuis, J.W. (1991b). Incipient wave breaking. Coast. Eng., Vol.15, pp.185-203.

Kamphuis, J.W., Rakha, K.A., and Jui, J. (1992). Hydraulic model experiments on seawalls. Proc. 23rd Int. Conf. Coastal Eng., ASCE, Vol.2, PP.1272-1284.

Kamphuis, J.W. (1994). Wave heights from deep water through the breaking zone. Waterway, Port, Coastal, and Ocean Eng., ASCE, submitted for publication.

Longuet-Higgins, M.S., and Stewart, R.W. (1962). Radiation stress and mass transporting gravity waves with application to surf beats. J. Fluid Mech., 13, pp.481-504.

Longuet-Higgins, M.S. (1964). Radiation stresses in water waves; a physical discussion with applications. Deep-sea Research, Vol.11, pp.529-562.

Savage, R.P. (1957). Model tests for hurricane protection project, technical bulletin, U.S. Army Corps of Eng., Washington,D.C.

Saville, T. (1961). Experimental determination of wave setup, Proc. 2nd Tech Conf. on hurricanes. Beach erosion Bd., Miami Beach, Fla.

Stive, M.J.F., Wind, H.G. (1982). A study of radiation stress and setup in the nearshore region. Coastal Eng., Vol.6, pp.1-25.

Svendsen, I.A., and Hansen, J.B. (1977). The wave height variation for regular waves in shoaling water. Coast. Eng., Vol.1, pp.261-284.

Svendsen, I.A., Madsen, P.A., and Hansen, J.B. (1978). Wave characteristics in the surf zone. Proc. 16th Coast. Eng. Conf., pp.520-539.

Svendsen, I.A. (1984). Wave heights and setup in a surf zone. Coast. Eng., Vol.8, pp.303-329.

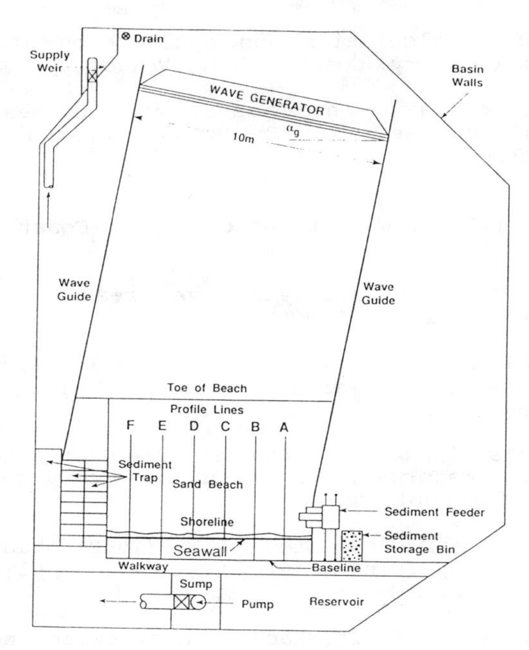

Figure 1: Wave Basin.

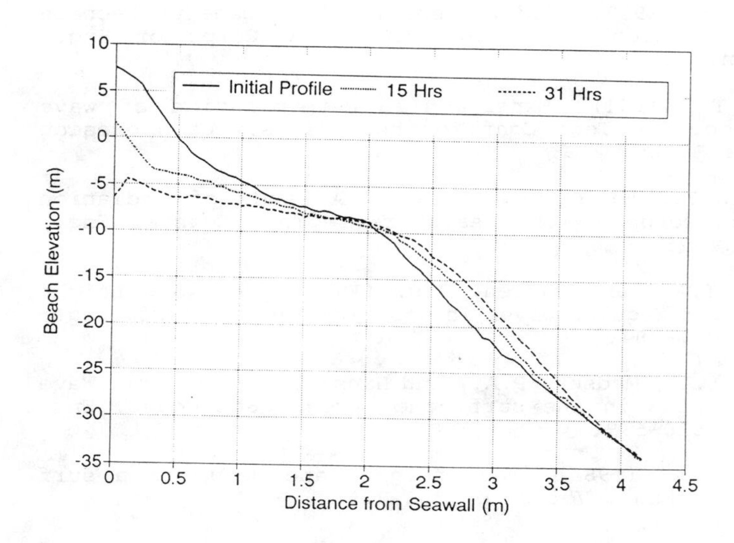

Figure 2: Beach Profiles for Test A.

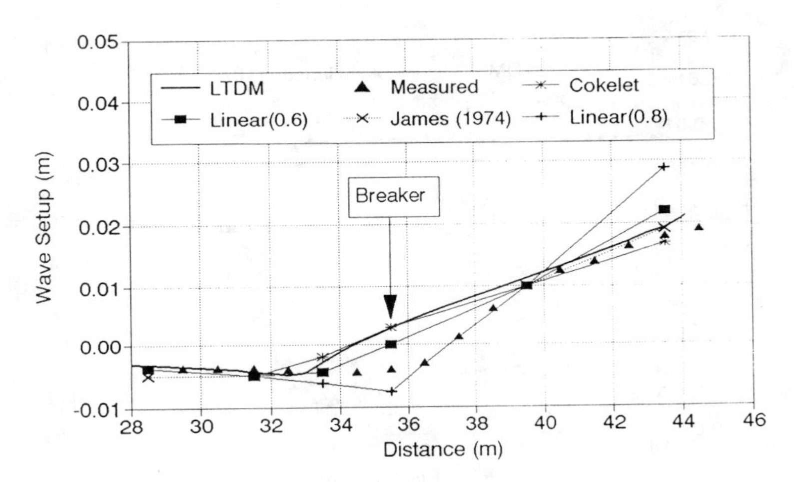

Figure 3: Stive and Wind (1982), Test 1.

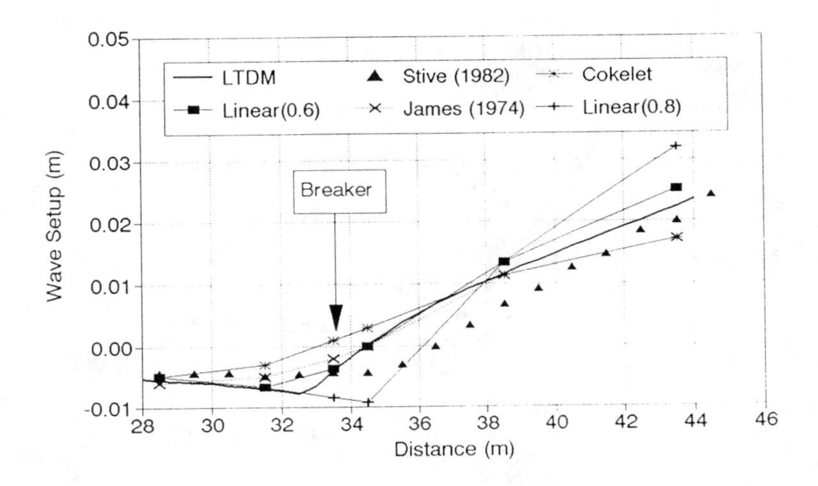

Figure 4: Stive and Wind (1982), Test 2.

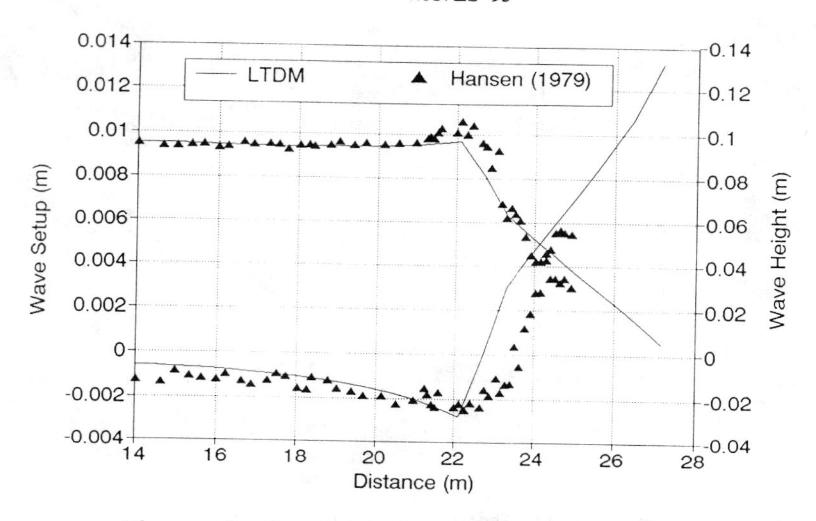

Figure 5: Hansen and Svendsen(1979), Test B.

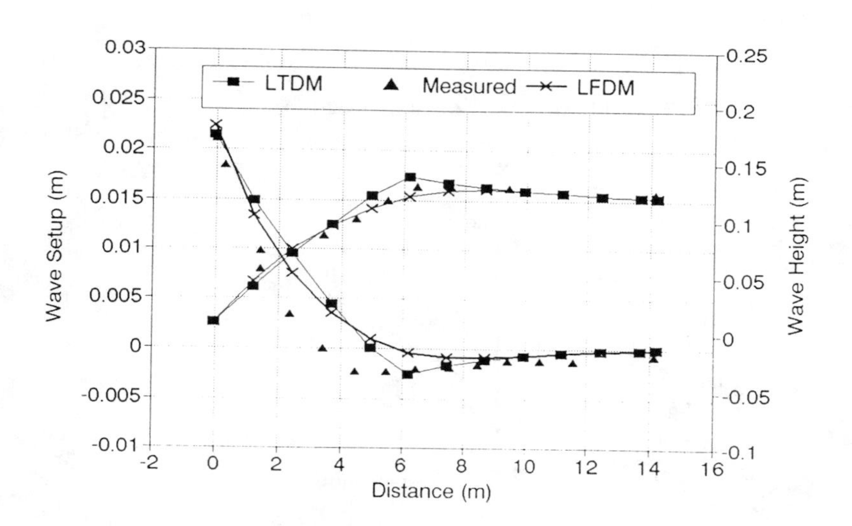

Figure 6: Battjes and Janssen (1978), Run 3.

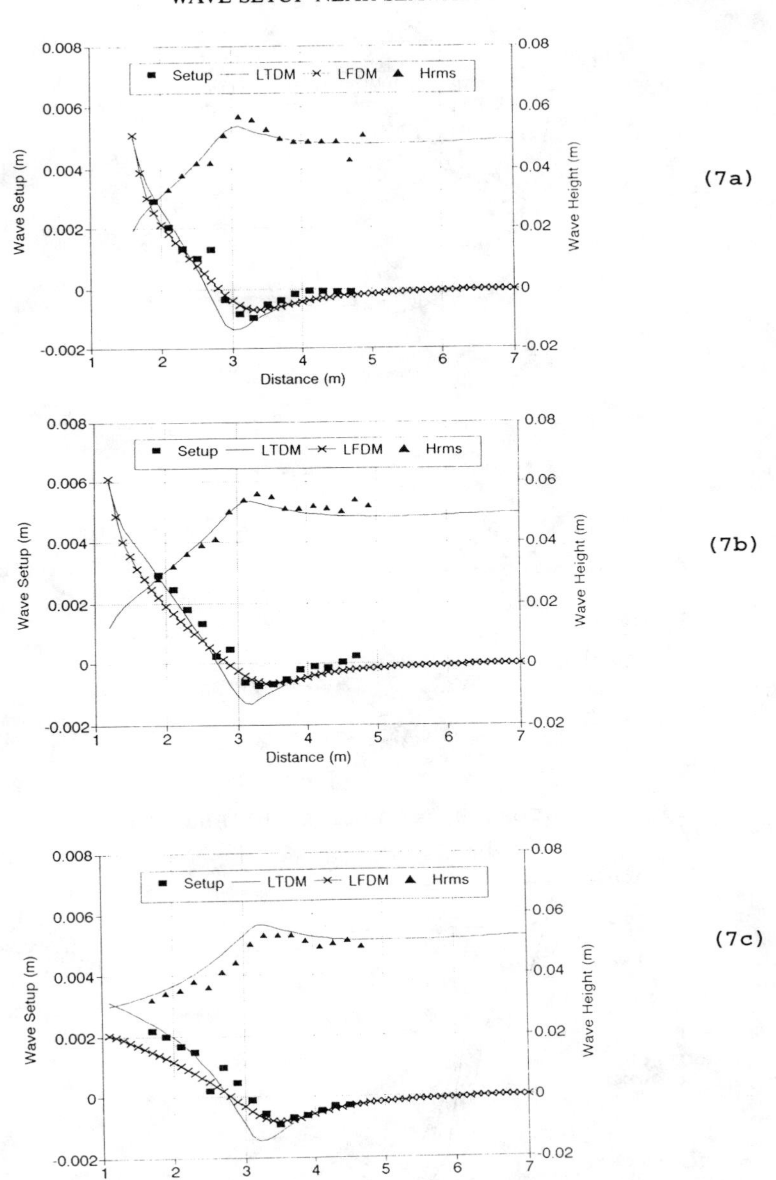

Figure 7: Test A; a) Hour 3, b) Hour 15, c) Hour 31.

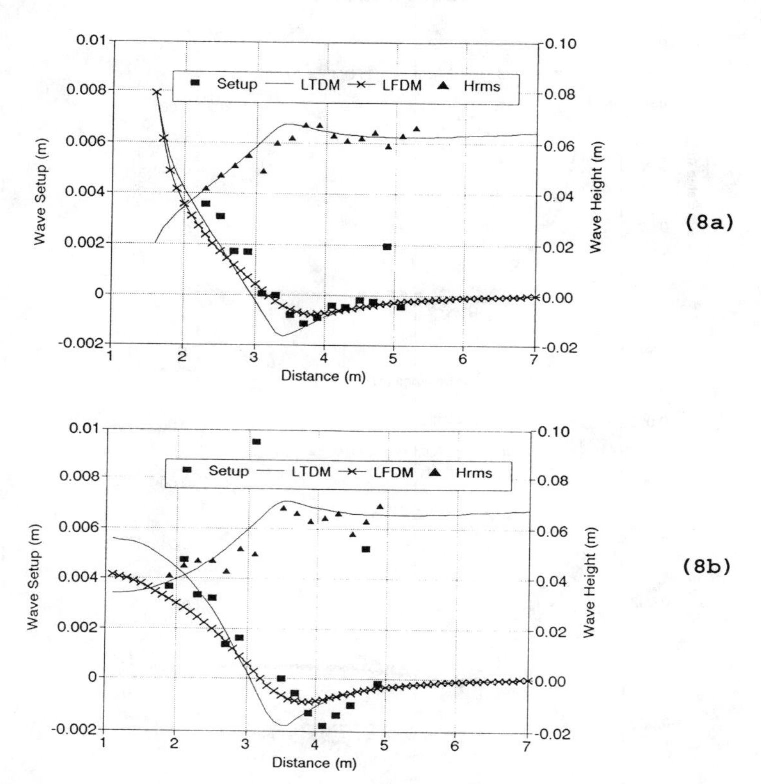

Figure 8: Test E; a) Hour 1, b) Hour 14.

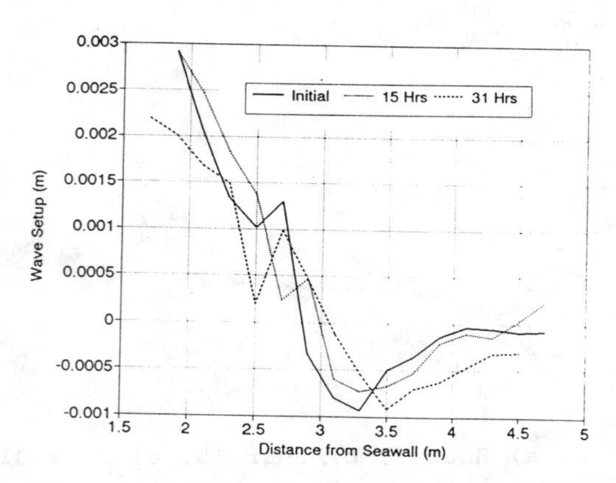

Figure 9: Wave Setup Change For Test A.

Evolution of Breaking Directional Spectral Waves in the Nearshore Zone

H. Tuba Özkan and James T. Kirby, M. ASCE [1]

Abstract

Methods used in parabolic refraction-diffraction models for computing the evolution of monochromatic waves in the nearshore zone are used to construct a model for spectral wave conditions. The two dimensional spectrum is divided into discrete wave components and the individual wave components are computed simultaneously in the domain using parabolic models for each wave component. Statistical quantities are computed at each forward step in the parabolic scheme and used to construct a statistical wave breaking model. The breaking model is built into the parabolic model equation using an additional breaking term. Model results are compared to available one- or two-dimensional data.

Introduction

The parabolic approximation has been shown to be a robust method for computing the evolution of monochromatic water waves in the nearshore zone (Kirby and Dalrymple, 1984). Recently, methods for computing a directional spectral sea state using the parabolic approximation have been developed. Panchang, et. al. (1990) and Grassa (1990) have each developed models which operate by first discretizing the spectrum into individual monochromatic directional components, and then running each component in a separate parabolic model. Results from the sequential runs are superimposed in order to obtain estimates of statistical wave heights. O'Reilly and Guza (1991) have further extended this formulation and have pointed out that the model could be used (in linear form) to develop a transfer function between onshore and offshore, after which any incident spectrum could be simply transformed using the computed transfer function.

In the nearshore region and especially near inlets, waves become strongly modified by significant variations in bathymetry and ambient current fields. These modifications lead, in turn, to depth-limited and current-limited breaking, which cause significant energy losses in the incident wave field and which are accompanied by strong nonlinear modifications of the wave field. In the models mentioned above, the assumptions made in model development or operation preclude an ability to predict local depth-limited breaking. The purpose of this study is to alleviate these restrictions and provide a platform for predicting strong breaking

[1]Center for Applied Coastal Research, Department of Civil Engineering, University of Delaware, Newark, DE 19716

energy losses in directional spectral seas in the vicinity of tidal inlets.

Description of the Modeling Scheme

The parabolic model for spectral wave conditions used here requires the input of a directional random sea at the offshore boundary. The random sea is represented by a two-dimensional spectrum in frequency and direction and is discretized resulting in wave components. The evolution of these wave components is computed simultaneously at each forward step in the parabolic scheme. Therefore, after each forward step in the shore normal direction it is possible to determine statistical properties on that row before taking another step forward. These statistical quantities are incorporated into a statistical wave breaking model.

In the following sections the wave field resulting from the discretization is described, the model equation for the individual wave components is stated and the incorporation of the breaking model is explained. Furthermore, an initial attempt to include significant effects of nonlinearity on the wave field is described.

Wave Climate

The discretization process of the two dimensional spectrum results in wave components of amplitude A with an associated frequency f and an angle of incidence θ to the assumed propagation direction. The water surface elevation can be described in terms of these discrete wave components. It is assumed that the water surface elevation η is periodic in time and that the spatial dependency can be split into a fast-varying phase and a slow-varying amplitude.

$$\eta(x,y,t) = \sum_{all f}\sum_{all \theta}\left\{\frac{A(x,y;f,\theta)}{2}e^{i\psi} + c.c.\right\} \tag{1}$$

where f is the frequency, θ is the direction of any individual wave component and

$$\psi = \int \mathbf{k}\cdot d\mathbf{x} - \omega t \tag{2}$$

In order to determine energy losses associated with randomly occuring wave breaking, it is necessary to have estimates of the statistical wave height at each point in the model grid as computation passes that point. Using the computed information about the spectral components at a location (x, y) the significant wave height can be computed as

$$H_{1/3}(x,y) = \sqrt{8\sum_{n=1}^{N} \mid A(x,y)_n \mid^2} \tag{3}$$

where N is the total number of wave components and $A(x,y)_n$ is the amplitude of the wave component n at location (x, y).

Wave Models for Individual Wave Components

The refraction, diffraction and shoaling of the discrete wave components is assumed to be governed by the parabolic approximation to the mild slope equation derived by Berkhoff (1972). To minimize the restrictions placed on the range of allowed wave angles with respect to the assumed wave direction, the procedure derived by Booij (1981) is used, enabling the model to handle wave direction up to about 45° from the shore normal, herein called x-direction. The model also has the ability to handle strong currents by using the formulation of the mild slope equation including the influence of currents derived by Kirby (1986).

The model uses the parabolic approximation including wave-current interactions and rederived for a wide angle approximation. The governing equation is:

$$
\begin{aligned}
& (C_{gn}+U)(A_n)_x - 2\Delta_1 V (A_n)_y + i(\bar{k}_n - a_0 k_n)(C_{gn}+U)A_n \\
+ & \left\{ \frac{\sigma_n}{2}\left(\frac{C_{gn}+U}{\sigma_n}\right)_x - \Delta_1 \sigma_n \left(\frac{V}{\sigma_n}\right)_y \right\} A_n + i\Delta_n' \left[\left((CC_g)_n - V^2\right)\left(\frac{A_n}{\sigma_n}\right)_y\right]_y \\
- & i\Delta_1 \left\{ \left[UV\left(\frac{A_n}{\sigma_n}\right)_y\right]_x + \left[UV\left(\frac{A_n}{\sigma_n}\right)_x\right]_y \right\} + \alpha A_n \\
+ & \frac{-b_1}{k_n}\left\{ \left[\left((CC_g)_n - V^2\right)\left(\frac{A_n}{\sigma_n}\right)_y\right]_{yx} + 2i\left(\sigma_n V\left(\frac{A_n}{\sigma_n}\right)_y\right)_x \right\} \\
+ & b_1\beta_n \left\{ 2i\omega_n U\left(\frac{A_n}{\sigma_n}\right)_x + 2i\sigma_n V\left(\frac{A_n}{\sigma_n}\right)_y - 2UV\left(\frac{A_n}{\sigma_n}\right)_{xy} \right. \\
+ & \left. \left[\left((CC_g)_n - V^2\right)\left(\frac{A_n}{\sigma_n}\right)_y\right]_y \right\} - \frac{i}{k_n} b_1 \left\{(\omega_n V)_y + 3(\omega_n U)_x\right\}\left(\frac{A_n}{\sigma_n}\right)_x \\
- & \Delta_2 \left\{ \omega_n U\left(\frac{A_n}{\sigma_n}\right)_x + \frac{1}{2}\omega_n U_x\left(\frac{A_n}{\sigma_n}\right) \right\} + ik\omega_n U(a_0 - 1)\left(\frac{A_n}{\sigma_n}\right) = 0
\end{aligned}
\tag{4}
$$

where

$$
\begin{aligned}
\sigma_n &= \omega_n - k_n U \\
\beta_n &= \frac{(k_n)_x}{k_n^2} + \frac{\left(k_n\left((CC_g)_n - U^2\right)\right)_x}{2k_n^2\left((CC_g)_n - U^2\right)} \\
\Delta_1 &= a_1 - b_1 \\
\Delta_2 &= 1 + 2a_1 - 2b_1 \\
\Delta_n' &= a_1 - b_1 \frac{\bar{k}_n}{k_n}
\end{aligned}
\tag{5}
$$

The model coefficients

$$
\begin{aligned}
a_0 &= 1 \\
a_1 &= -0.75 \\
b_1 &= -0.25
\end{aligned}
\tag{6}
$$

recover the approximation of Booij.

For more detailed information about the development of the wave model, the reader is referred to the documentation of the numerical model **REF/DIF S** by Kirby and Özkan (1992).

The Breaking Model

The statistical information obtained after each step in the parabolic scheme is used to construct a model for the dissipation of energy due to breaking. The wave breaking scheme involves a uniform dissipation of energy over the individual wave components. The amount of total energy dissipation is a function of the local statistical quantities and the local water depth. Therefore breaking of individual waves in a wave train is not considered, but a spectral approach to the wave breaking process is taken.

To determine the uniformly distributed energy dissipation, a simple model by Thornton and Guza (1980) is used. They showed that the energy dissipation can be expressed as

$$\frac{\partial E C_g}{\partial x} = -\epsilon_b \tag{7}$$

where the energy E and the bore dissipation ϵ_b can be expressed as

$$E = \frac{1}{8}\rho g H_{rms}^2 \tag{8}$$

$$\epsilon_b = \frac{3\sqrt{\pi}}{16}\frac{\rho g \bar{f} B^3}{\gamma^4 h^5} H_{rms}^7 \tag{9}$$

Here h is the local water depth and $\bar{f}$ is a representative frequency for the frequency spectrum and is chosen to the peak frequency. Following Mase and Kirby (1992), the constants B and γ are chosen to be equal to 1 and 0.6, respectively. The root-mean-square (rms) waveheight can be written as

$$H_{rms} = \frac{1}{\sqrt{2}} H_{1/3} \tag{10}$$

The energy dissipation model is built into the model equation using an additional breaking term. The model equation includes terms that can easily be related to the energy flux EC_g. In the same manner the new dissipation term will be related to the bore dissipation ϵ_b. Therefore the model equation includes terms like

$$C_g \frac{\partial A}{\partial x} = -\alpha A \tag{11}$$

The coefficient α is given by

$$\alpha = \frac{4\epsilon_b}{\rho g H_{rms}^2} = \frac{3\sqrt{\pi}}{4}\frac{\bar{f} B^3}{\gamma^4 h^5} H_{rms}^5 \tag{12}$$

The magnitude of the coefficient α will be infinitesimally small when breaking does not occur but grows to a significant value when breaking starts to occur. Furthermore, the presence of the breaking term in the equation at all times makes any criterion for turning breaking on or off unnecessary. It should be noted, though, that no modifications have been made to Thornton and Guza's dissipation model to account for directional effects.

Nonlinear Formulation for the Wave Speed

It is anticipated that nonlinear effects, present during wave breaking, will cause the phase speed of the shoaling and breaking waves to increase with respect to the unbroken waves in the domain. An effect like this would cause waves on top of a submerged shoal to speed up with respect to the waves at the sides of the shoal, acting to nullify any focusing tendency. Because this would change the spatial wave field significantly, an initial attempt has been made to take the effect of the nonlinearity on the wave speed into account. It should be noted that the construction of a full nonlinear model to simulate this or other nonlinear effects is not within the scope of this study.

Hedges (1976) developed an approximate relationship for nonlinear dispersion of monochromatic waves in shallow water. This formulation involved a simple modification to the linear dispersion relation to approximate nonlinear behavior. The modified dispersion relation by Hedges (1976) is given by

$$\sigma^2 = gk \tanh\left(kh(1 + |A|/h)\right) \tag{13}$$

This formulation can be altered by using the significant wave height instead of the wave amplitudes in shallow water in an attempt to include the effect of a random sea. The Hedges (1976) formulation would then be altered to

$$\sigma^2 = gk \tanh\left(kh(1 + H_{1/3}/2h)\right) \tag{14}$$

It should be noted that in shallow water and for small $|A|/h$, the Hedges (1976) relationship represents the propagation speed of a solitary wave.

$$C^2 = gh\,(1 + |A|/h) \tag{15}$$

It can be argued that a breaking wave propagates with the speed of a bore which would be given by

$$C^2 = gh\,(1 + 3|A|/h) \tag{16}$$

which is larger than the shallow water relationship given by Hedges (1976). The application of this relationship to the model would cause the breaking waves in the domain to travel faster.

Comparison to Data

Five example cases will be discussed making use of experimental data collected by Vincent and Briggs (1989) and by Mase and Kirby (1992).The test set-ups for both experiments will be summarized and results presented and dicussed in the following sections.

Waves Shoaling on a Plane Beach

Mase and Kirby (1992) conducted experiments using a Pierson-Moskowitz spectrum without directional spreading. The waves were generated in constant depth of 47 cm, and then shoaled and dissipated on a 1 : 20 beach. Wave gages were placed at several locations on the beach. Figure 1 shows the set-up of the experiment in a wave flume.

The comparisons between this data set and the model will be made using Case 1 in Mase and Kirby (1992). The available data are in terms of time series at the wave gage locations. The time series are used to compute the significant wave

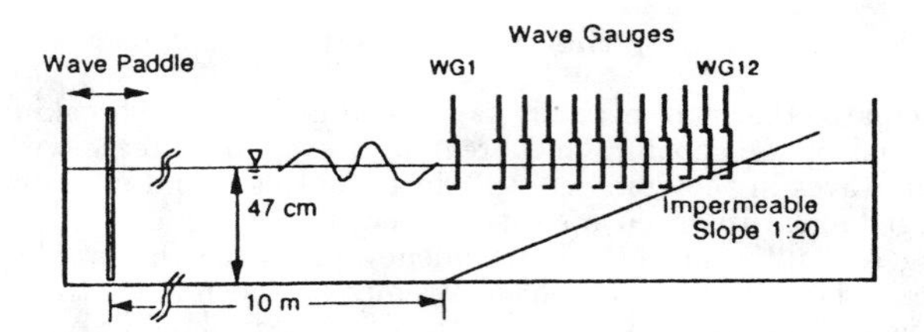

Figure 1: Experimental Setup (from Mase and Kirby (1992))

height for comparison with model results. Measured data at the offshore gage is used to create a smoothed incident frequency spectrum for the model input. The incident spectrum is divided into 50 discrete wave components. Figure 2 shows the incident frequency spectrum.

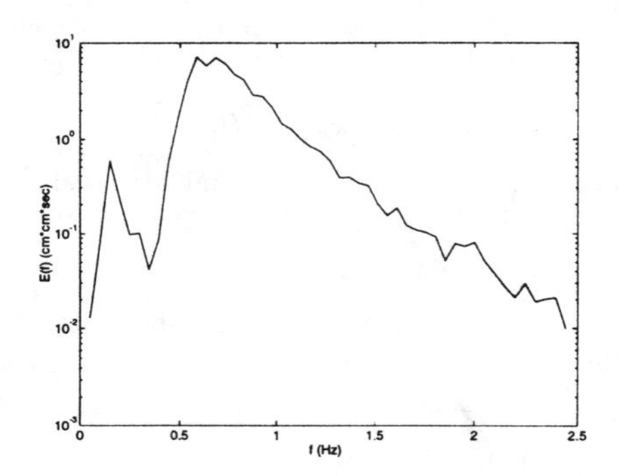

Figure 2: Incident Frequency spectrum

Results for this case are presented in Figure 3. It can be seen that the model shoals the waves up to the well predicted breaking point. The wave height decay after breaking starts is also very well predicted. The data points close to the shore show the presence of setup which the model cannot predict.

Waves over a Submerged Shoal

Vincent and Briggs (1989) conducted their experiments in a wave tank that is 35 m (114 ft) wide and 29 m (96 ft) long. The waves were generated by the directional spectral wave generator, which is 27.43 m (90 ft) long. The center of the shoal was located at $x = 6.10$ m and $y = 13.72$ m. The elliptical shoal had a major radius of $3.96m$, a minor radius of 3.05 m and a maximum height of 30.48 cm at the center. Maximum water depth was 47 cm. Expressions for the shoal perimeter and the elevation of each point on the shoal are given by Vincent and

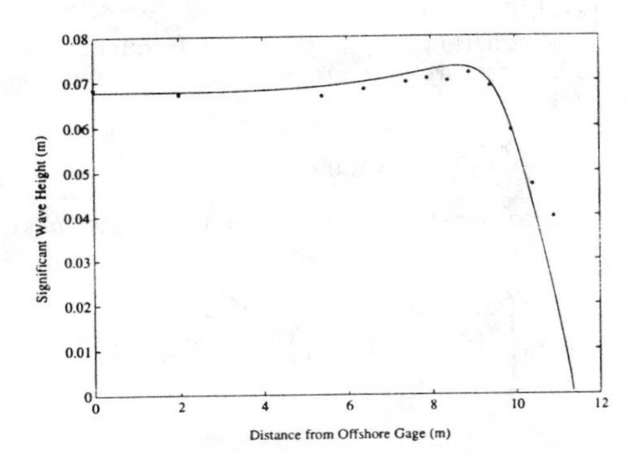

Figure 3: Waves Shoaling on a Beach: Significant Wave Height for Data and Model

Briggs (1989). Reference should be made to that document for further information about the domain.

Figure 4 shows the experimental set-up in the basin and the measurement transects 1–9. Data collection for the cases of interest to this study were only performed on transect 4.

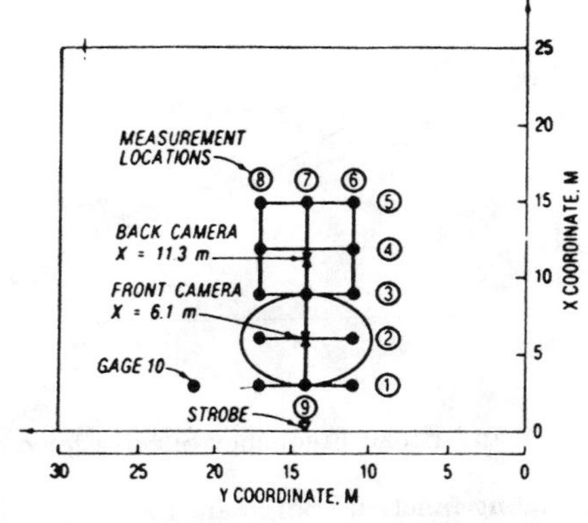

Figure 4: Experimental Setup (from Vincent and Briggs (1989))

In all the cases a TMA spectrum (Hughes, 1984) in conjunction with a directional spreading function was used to establish the target spectrum.

The TMA spectrum is given by the energy density $E(f)$ for frequency f

$$E(f) = \frac{\alpha g^2}{(2\pi)^4 f^5} \exp\left\{-1.25\left(\frac{f_m}{f}\right)^4 + (\ln\gamma)\exp\left[\frac{-(f-f_m)^2}{2\sigma^2 f_m^2}\right]\right\}\phi(f,h) \quad (17)$$

Özkan

where α is the Phillips' constant, f_m is the peak frequency and γ is the peak enhancement factor. Furthermore σ = shape parameter defined by

$$\sigma = \begin{cases} \sigma_a = 0.07 & \text{if } f < f_m \\ \sigma_b = 0.09 & \text{if } f \geq f_m \end{cases} \tag{18}$$

The factor $\phi(f,h)$ incorporates the effect of the depth h and is computed following Hughes (1984) by

$$\phi = \begin{cases} 0.5\,(\omega_h)^2 & \text{if } \omega_h < 1 \\ 1 - 0.5\,(2-\omega_h)^2 & \text{if } 1 \leq \omega_h \leq 2 \\ 1 & \text{if } \omega_h > 2 \end{cases} \tag{19}$$

where

$$\omega_h = 2\pi f \sqrt{\frac{h}{g}}$$

Narrow or broad frequency spectra used in the experiment can be obtained assigning the values 20 and 2 to the parameter γ, respectively. Representative plots of the narrow and broad frequency spectra are given in Figure 5.

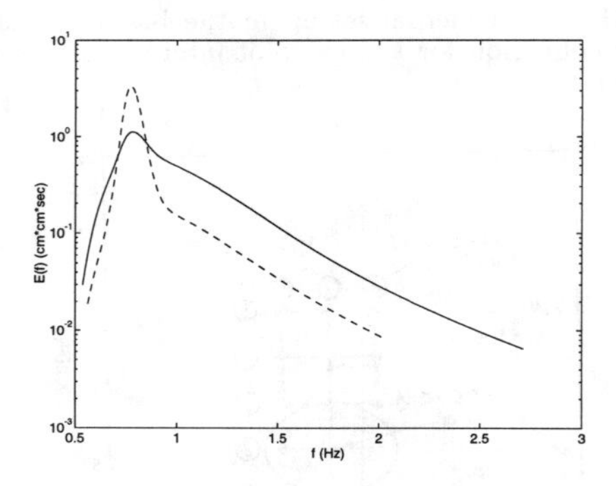

Figure 5: Narrow and Broad Frequency Spectra for $\alpha = 0.00047$

The directional spreading function (Borgman, 1985)is given by

$$D(\theta) = \frac{1}{2\pi} + \frac{1}{\pi}\sum_{j=1}^{J} \exp\left[-\frac{(j\sigma_m)^2}{2}\right] \cos j\,(\theta - \theta_m) \tag{20}$$

where

θ_m = mean wave direction = 0°.

J = number of terms in the series (chosen to be 20 in the numerical calculations).

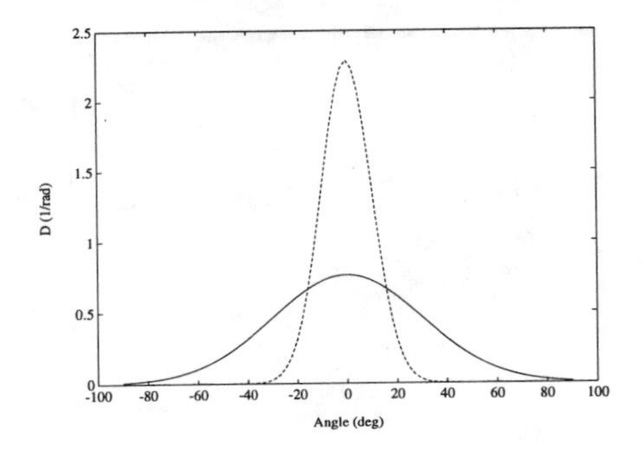

Figure 6: Narrow and Broad Directional Spreading Functions

σ_m is chosen to be 10° or 30° giving a narrow or broad directional spreading, respectively. Representative plots of the narrow and broad directional spreading functions are given in Figure 6.

The two-dimensional spectrum is then given by the product

$$S(f,\theta) = E(f)\,D(\theta) \tag{21}$$

For the simulation of these experiments, the target spectrum was divided into 10 frequency components and 20 directional components, yielding a total of 200 discrete wave components. Four cases out of this data set will be used for comparisons with the model. The first two cases do not involve breaking waves, whereas the next two cases involve intensive breaking of the waves over the submerged shoal.

Non-Breaking Cases

The first case is denoted as Case N4 by Vincent and Briggs (1989) and involves a narrow frequency spectrum used in conjunction with a narrow directional spreading. Another case (B4) involves a narrow frequency spectrum and a broad directional spreading. Neither of these cases involve breaking waves and are simulated to show that the breaking term that is always in effect does not induce an unrealistic wave height reduction.

Results for Case N4 are presented in Figures 7 and 8, where the first of these shows a contour plot of the computed normalized significant wave height and the second shows a comparison of model and data at Transect 4.

It can be seen that the general shape of the curve is well modeled. The fact that the peak of the wave height for this case with no wave breaking is well represented shows that the coefficient of the breaking term stays small and does not dissipate energy from the system.

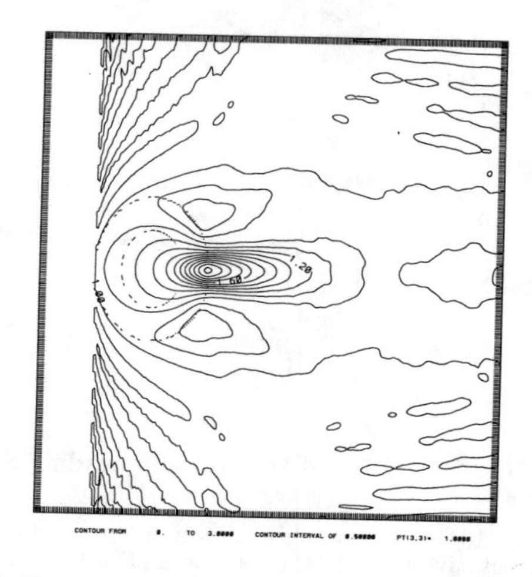

Figure 7: Case N4: Contour Plot of Normalized Significant Wave Height

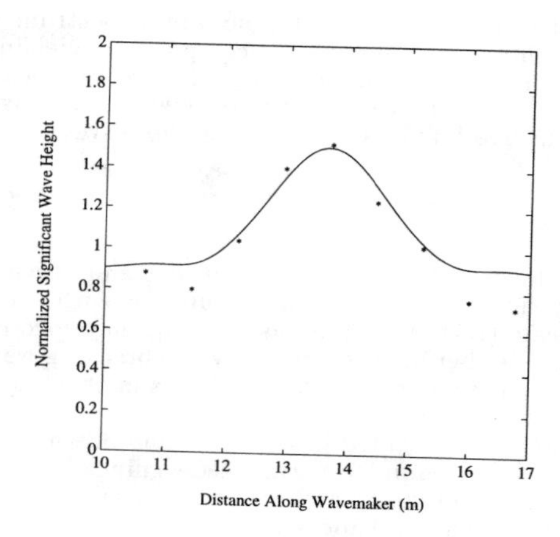

Figure 8: Case N4: Normalized Significant Wave Height at Transect at $x = 12.2$ m

Results for Case B4 are presented in Figures 9 and 10. These results also confirm that the effect of the breaking term cannot be seen in these cases without breaking.

Figure 9: Case B4: Contour Plot of Normalized Significant Wave Height

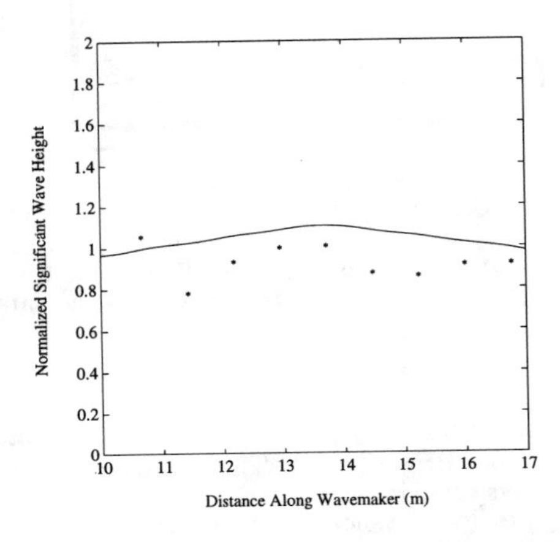

Figure 10: Case B4: Normalized Significant Wave Height at Transect at $x = 12.2$ m

Breaking Cases

The next two cases involve breaking waves and are denoted as Cases N5 and B5. Case N5 involves a narrow frequency spectrum as well as a narrow directional distribution, whereas Case B5 involves a broad frequency spectrum and a broad direction distribution. Details about the values of the free parameters for these cases can be obtained from Vincent and Briggs (1989).

Results for Case N5 are presented in Figures 11 and 12, where the first of these shows a contour plot of the computed normalized significant wave height and the second shows a comparison of model and data at Transect 4.

It can be seen that the experimental wave height shows a tendency to decrease around $y = 14$ m; the model does not predict this behavior. It is also seen that the data show a tendency to recover the initial significant wave height at $y = 10$ m and $y = 17$ m, whereas the model does not show this tendency.

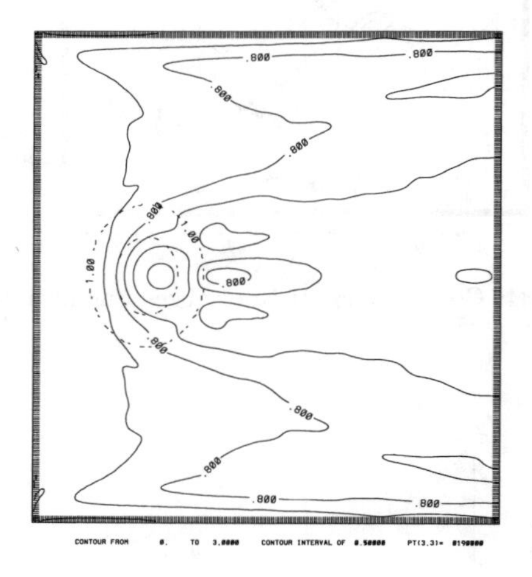

Figure 11: Case N5: Contour Plot of Normalized Significant Wave Height

Results for Case B5 are presented in Figures 13 and 14. Figure 14 shows that the model is predicting a fairly constant wave height, but the data show a decrease in the midsection.

Conclusions

Computed results involving cases without wave breaking agree well with laboratory data by Vincent and Briggs (1989). This demonstrates that the coefficient of the breaking term stays small in cases where wave breaking does not occur. The model also accurately predicts the shoaling and decay of statistical wave height for unidirectional random wave propagation onto a beach. However, the model does not accurately predict the shoaling and decay of the wave height for a multidirectional random sea. This behavior might be linked to the fact that no modifications have been made to the energy dissipation model to account for directional effects. Pronounced nonlinear effects may also have to be modeled more completely.

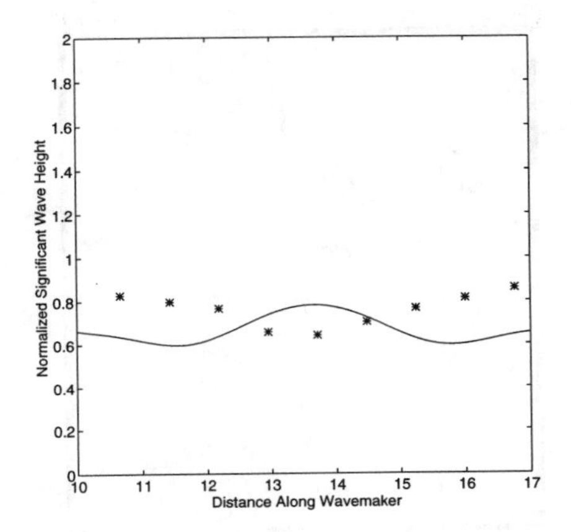

Figure 12: Case N5: Normalized Significant Wave Height at Transect at $x = 12.2$ m

In order to address these problems we need a more complete understanding of the physical processes causing the behavior observed in the two cases involving breaking waves in Vincent and Briggs' (1989) experiment. For this reason, an experiment involving waves on a submerged shoal has been designed with the intention of concentrating on directional effects on the energy dissipation and on nonlinear effects during breaking. The speed of the breaking waves will be observed and similarities to the speed of a bore will be investigated. The experiment is currently in progress and it is anticipated that the collected data will aid in the further development of the model.

Acknowledgement

This research has been sponsored by the U.S. Army Corps of Engineers, Coastal Engineering Research Center (Contract No. DACW 39-90-D-0006-D002) and by NOAA Office of Sea Grant, Department of Commerce, under Grant No. NA/6RG0162-01 (Project No. R/OE-9). The U.S. Government is authorized to produce and distribute reprints for governmental purposes, not withstanding any copyright notation that may appear hereon.

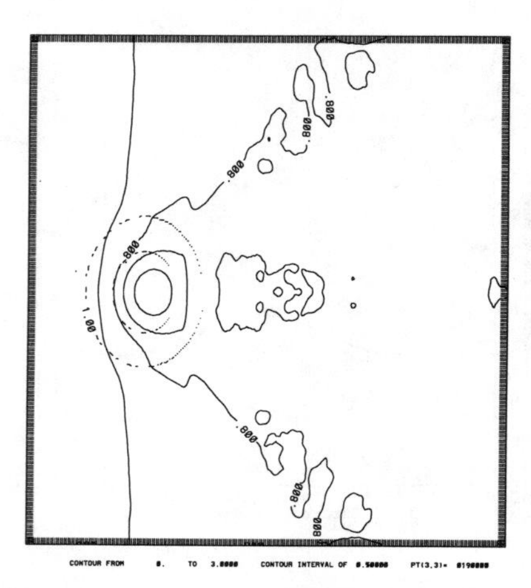

Figure 13: Case B5: Contour Plot of Normalized Significant Wave Height

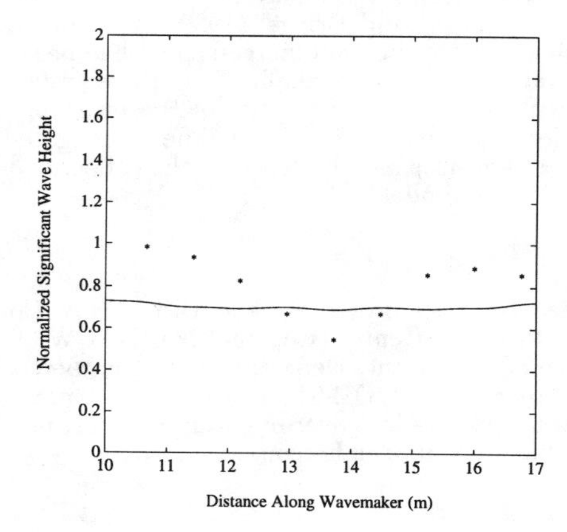

Figure 14: Case B5: Normalized Significant Wave Height at Transect at $x = 12.2$ m

References

Berkhoff, J. C. W., 1972, "Computation of combined refraction-diffraction", *Proc. 13th Intl. Conf, Coast. Engineering*, Vancouver, 471-490.

Booij, N., 1981, " Gravity Waves on Water with Non-uniform Depth and Current", Reprt No. 81-1, *Communication on Hydraulics*, Department of Civil Engineering, Delft University of Technology.

Borgman, L. E., 1985, "Directional spectrum estimation for the S_{xy} gauges," *Technical Report*, Coast. Engrg. Res. Center, Waterways Experiment Station, Vicksburg, MS, 1-104.

Grassa, J. M., 1990, "Directional random waves propagation on beaches", *Proc. 22nd Intl. Conf. Coastal Engrng.*, Delft, 798-811.

Hedges, T.S., 1976, "An empirical modification to linear wave theory," *Proc. Inst. Civ. Eng.*, 61, 575-579.

Hughes, S. A., 1984, "The TMA shallow-water spectrum description and applications," *Technical Report*, Coast. Engrg. Res. Center, Waterways Experiment Station, Vicksburg, MS.

Kirby, J.T., 1986, "Higher-order approximations in the parabolic equation method for water waves", *J. Geophys. Res.*, 91, 933-952.

Kirby, J. T. and Dalrymple, R. A., 1984, "Verification of a parabolic equation for propagation of weakly nonlinear waves", *Coastal Engrng.* 8, 219-232.

Kirby, J. T. and Dalrymple, R. A., 1986, "An approximate model for nonlinear dispersion in monochromatic wave propagation models", *Coast. Eng.*, **9**, 545-561.

Kirby, J. T. and Özkan, H. T., 1992, "REF/DIF S Version 1.0, Documentation and User's Manual", CACR 92-06, Coast. Engrg. Res. Center, Waterways Experiment Station, Vicksburg, Miss

Mase, H. and Kirby, J.T., 1992, "Modified frequency-domain KdV equation for random wave shoaling," *Proc. Intl. Conf. Coast. Engrg.*, Venice.

O'Reilly, W. C. and Guza, R. T., 1991, "Comparison of spectral refraction and refraction-diffraction wave models", *J. Waterway, Port, Coastal and Ocean Engrng.*, 117, 199-215.

Panchang, V. G., Pearce, B. R., Wei, G. and Cushman-Roisin, B., 1990, "Solution of the mild-slope wave problem by iteration", *Applied Ocean Res.*, 13, 187-199.

Thornton, E. B. and Guza, R. T., 1983, "Transformation of wave height distribution", *J. Geophys. Res.*, 88, 5925-5938.

Vincent, C. L. and Briggs, M. J., 1989, "Refraction and diffraction of irregular waves over a mound", *J. of Waterway, Port, Coastal, and Ocean Engineering*, 115, 269-284.

NEARSHORE BREAKING WAVES: A NUMERICAL MODEL

by Scott L. Douglass,[1] M. ASCE and Juan C. Ortiz,[1] A.M. ASCE

ABSTRACT

A numerical model is proposed for evaluating the breaking of individual waves on mild beach slopes. The model uses a boundary element method formulation of the mixed Eulerian-Lagrangian technique. The initial condition is a periodic, finite-amplitude, nearshore wave of permanent form and the bottom slope is simulated by invoking a "locally flat bottom" assumption and gradually reducing the depth in the computation. Results of the deformation of the water surface in time are presented for a series of different tests including a range of constant beach slopes from 1:20 to 1:40 and a range of initial wave steepnesses from 0.018 to 0.049. The overall breaker geometry results agree qualitatively with laboratory results and a quantitative breaker type parameter is proposed. The use of this numerical model to evaluate the phenomenon of nearshore breaking waves avoids some of the limitations inherent in laboratory work while introducing different limitations. The results indicate that the depth-induced breaking wave instability may not be very sensitive to bottom slope, wavelength or energy conservation in the final few wave periods approaching breaking but rather is most strongly controlled by the water depth.

INTRODUCTION

As waves propagate toward the beach, they eventually reach a depth in which they break. Depth-induced wave breaking defines the width of the surf zone and thus the region of most of the wave-driven currents and sediment transport on a beach. However, engineering and oceanographic

---

[1] Asst. Prof., Dept. of Civil Engrg., Univ. of South Alabama, Mobile, AL 36688 (205)460-6174.

models of the breaking process are at present, very limited. Munk (1949) postulated that within the surf zone each wave behaves independently and wrote that as waves progress into very shallow water and the water surface becomes non-sinusoidal with higher crests and flat troughs "the character of these isolated crests scarcely depends upon the distance L between the crests." Although laboratory investigations have provided useful information about the initiation of breaking, there are some significant limitations in their ability to adequately model the actual phenomenon of interest, wave breaking on prototype beaches. These limitations include scaling problems dependent on the specific tank geometry, such as long-period tank oscillations, bottom and sidewall friction, bottom response and percolation, surface contamination, and free wave harmonics. Another limitation of laboratory investigations is the constructability of realistic beach slopes. On sandy shorelines, beach slopes offshore of breaking of 1:30 to 1:100 are common. Very few lab flumes are long enough to test a range of realistically mild beach slopes. Also, many natural beach slopes are concave upward, a feature that is very difficult to construct in laboratory facilities of limited length. This paper presents an attempt to numerically model nearshore breaking waves on mild slopes.

Longuet-Higgins and Cokelet (1976) pioneered the use of a numerical, mixed Eulerian-Lagrangian, boundary integral technique for modeling breaking waves. The technique reduces the 2-dimensional problem to a 1-dimensional computation around the boundary by assuming the existence of a velocity potential. Other approaches that depend on solving the interior flow problem (Harlow and Welch, 1965; Nichols, et al. 1980) have been less successful at producing realistic solutions to the overturning wave problem. Vinje and Brevig (1980), New, et al. (1985), Dold and Peregrine (1984), Seo and Dalrymple (1990) and others have made adaptations to the boundary integral technique for solving problems with spatial periodicity. Dommermuth, et al. (1988) used the technique to model a laboratory experiment on deep-water wave breaking. A series of papers, Grilli, et al. (1988, 1989) and Grilli and Svendsen (1989, 1990), has attempted to develop a "numerical wave tank" for non-periodic depth-induced wave breaking. Romate (1990) and Xu and Yue (1992) have expanded the technique to the full 3-dimensional flow problem for deep-water wave breaking.

New, et al. (1985) used an initial condition of a wave of permanent form in finite-depth. Breaking was induced by instantaneously placing that wave in much shallower water. By varying the magnitude of this "depth perturbation" they were able to produce either small or large overhanging jets and tubes which correspond with spilling or plunging breakers. A parameter which New, et al. likened to a beach slope is the ratio of the vertical decrease in depth with the horizontal distance traveled before the wave face became vertical. The parameter showed some agreement with experimental relationships between beach slopes and $(H/d)_b$. However, one term in the "beach slope" ratio, the horizontal distance to overturning, is a function of the solution results and not an independent variable.

The numerical model presented in this paper is a different approach to applying the mixed Eulerian-Lagrangian technique to the phenomenon of interest, wave breaking on prototype, i.e. real world, beaches. This work uses the "locally flat" bottom assumption, which is common in analytical shoaling theories, to simulate breaking with a numerical scheme. The bottom depth is gradually decreased to simulate a given beach slope. The model formulation, solution, results, and limitations are presented below.

GOVERNING EQUATIONS

Fluid particle velocities, u and v, in the cartesian, x and y, directions of an inviscid, incompressible fluid in irrotational flow are described in terms of the velocity potential Φ satisfying Laplace's equation,

$$\nabla^2\Phi=0 \tag{1}$$

as,

$$u=\frac{\partial\Phi}{\partial x}\ ;\ v=\frac{\partial\Phi}{\partial y} \tag{2}$$

The boundary condition used in all of the computations in this paper at the bottom surface, Γ_b, is one of no flow normal to the surface,

$$\frac{\partial\Phi}{\partial n}=0 \qquad on\ \Gamma_b \tag{3}$$

where n is the outward pointing normal to the surface Γ bounding the domain of the flow which is taken as a single, spatially periodic wave as shown in Figure 1.

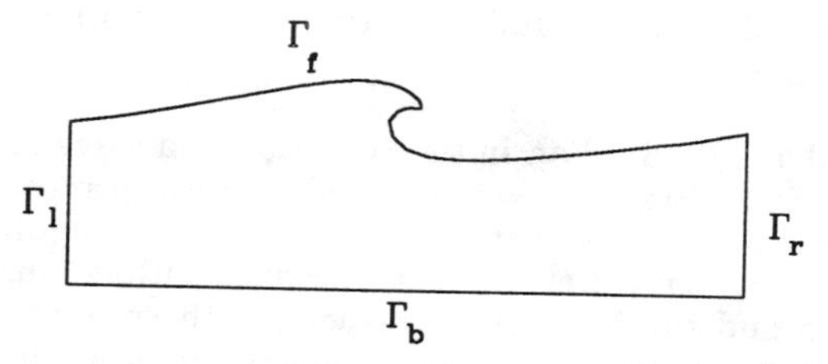

Figure 1. Computational domain and boundary definition sketch.

The spatial periodicity boundary conditions on the left and right sides, Γ_l and Γ_r, are

$$\Phi(\Gamma_l) = \Phi(\Gamma_r) \tag{4}$$

and

$$\frac{\partial\Phi(\Gamma_l)}{\partial n} = -\frac{\partial\Phi(\Gamma_r)}{\partial n} \tag{5}$$

On the free surface, or upper surface, the kinematic condition can be written in terms of the total or material derivative

$$\frac{Dx}{Dt} = u \qquad \frac{Dy}{Dt} = v \tag{6}$$

where t = time. The dynamic condition is the satisfaction of Bernoulli's Equation, by which the total derivative of the velocity potential is

$$\frac{D\Phi}{Dt} = \frac{1}{2}(u^2+v^2) - gy \tag{7}$$

where g = acceleration of gravity, and the pressure on the free surface is taken as zero.

APPROXIMATIONS

The general form of the algorithm (Ortiz and Douglass 1991, 1993a) used to follow the time-dependent motion of the water surface is that of Longuet-Higgins and Cokelet (1976). However, the solution of the governing equations is addressed in the physical plane in this model. Given an initial geometry of the free surface and a velocity potential Φ along the surface; specific locations, or fluid particles, are selected on the surface. The free surface boundary conditions are then used to estimate the motion of these particles and thus, the movement of the free surface. In the present work, the boundary integral equation form of Laplace's Equation is solved with a boundary element method (BEM) to estimate the velocity field on the free surface at each step in time.

BEM Solution of Laplace's Equation

The BEM uses Green's third formula to obtain a solution to Laplace's equation over a domain as a function of integrals along the boundary of the domain only (Gipson 1987)

$$C\phi(r_l)=\int_{\Gamma}\phi(r)\frac{\partial W}{\partial n}Wd\Gamma-\int_{\Gamma}\frac{\partial\phi(r_l)}{\partial n}Wd\Gamma \tag{8}$$

where C = a constant, $\mathbf{r}_l$ = the location of the source point, $\mathbf{r}$ = a dummy point over which the integrations are performed, and W is the Green's function corresponding to Laplace's equation in two dimensions:

$$W=\frac{1}{2\pi}\log|r-r_l| \tag{9}$$

By discretizing the boundary and evaluating Equation (8) at every node, a system of simultaneous equations of the form

$$[H]\{\phi\}=[G]\{\frac{\partial\phi}{\partial n}\} \tag{10}$$

is obtained, where *[H]* and *[G]* = the coefficient matrices, dependant only on the boundary geometry, and the { } represents evaluation at the nodes. With the boundary conditions defined in Equations (3)-(5), Equation (10) is rearranged into a standard system of linear equations which is then solved for the unknown values of potential and normal derivatives of potential. The only necessary values for the time-stepping scheme are the derivatives of the potential on the free surface of the wave which are used to march the surface in time.

Boundary Discretization

In the BEM system used in this work, the wave surface is discretized with cubic Overhauser spline elements. Ortiz and Douglass (1993a) show that use of the Overhauser element produced both more stable results than some other higher-order elements and more realistic water surface shapes in the vicinity of the overturning jet.

Time-stepping

The time-stepping of the free surface particle locations and Φ is accomplished with a Adams-Bashford-Moulton method applied to Equations (6) and (7). The fourth-order, predictor-corrector method uses three previous values of the individual particle velocity and potential. The model is initialized with a Runge-Kutta scheme. The time steps are automatically subdivided using a Courant number

$$C_0 = \frac{\Delta t'}{\Delta x'}$$

(where $\Delta t'$= the time step divided by the initial wave period and $\Delta x'$= the distance between the two closest nodes divided by the wave length) to keep the solution from becoming unstable as the nodes near the wave tip get closer and closer together. Throughout the time-stepping process whenever $C_0 > 0.4$, the time step is halved.

The numerical model does not require any smoothing, re-gridding, or post-blending of the results between time-steps to avoid numerical instabilities. The results presented below used an initial time step of 1/200th the wave period, 120 free surface particles, and took up to several hours of CPU time on a Cray X-MP supercomputer.

Tests of the Numerical Model

The numerical model has been tested with waves of permanent form and by comparing with published results (Ortiz and Douglass 1993a). The most basic test of the model was a check of its ability to model the propagation of wave of permanent form in finite-depth water. Input water surface location and potential for such waves were taken from Dean (1974). These waves moved through the model without change of height or form for several wave periods. The model was also tested by comparing with the results of Vinje and Brevig (1980). The water surfaces agree in time and location as the wave face becomes vertical, overturns, and the jet approaches the trough.

Gradually Decreasing Depth to Simulate Beach Slope

This research effort uses the BEM system to study the transformations experienced by a wave as the depth of the water gradually decreases. The initial free surface conditions are Dean's (1974) numerical solutions to the governing equations for one wave length of a symmetric wave form of permanent form for given d/L and H/L conditions. The initial depth is the depth from Dean (1974) at which that wave will propagate without change of form. The bottom slope is modeled by gradually decreasing the depth at each time step. The depth decreases at a rate controlled by the wave celerity to model any desired beach slope. The movement of the crest of the wave is used as the wave celerity.

RESULTS

The results from the numerical model include the full flow kinematics including water surface elevations and water particle velocities and accelerations. This presentation focusses on the water surface deformations. Model results simulating wave breaking on constant slope beaches are presented.

Transformation from Symmetric Form to Breaking

Figure 2 shows the time-evolution of the free surface computed by the model for the same initial wave condition on three different slopes.

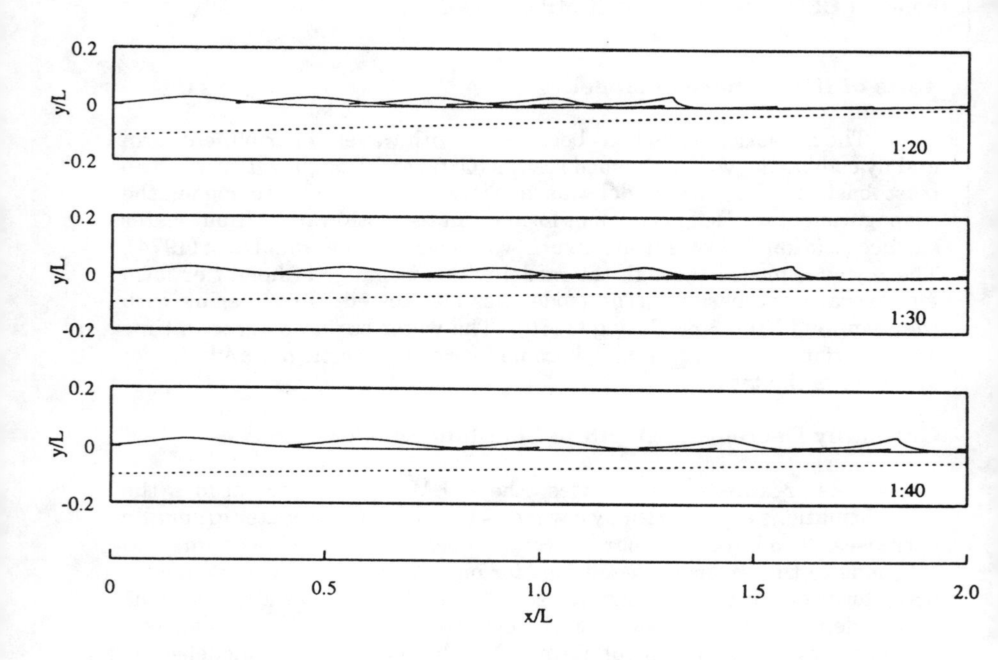

Figure 2. Water surface elevation results for wave propagation up simulated 1:20, 1:30, and 1:40 slopes. Initial condition is Dean's (1974) wave case 5-B.

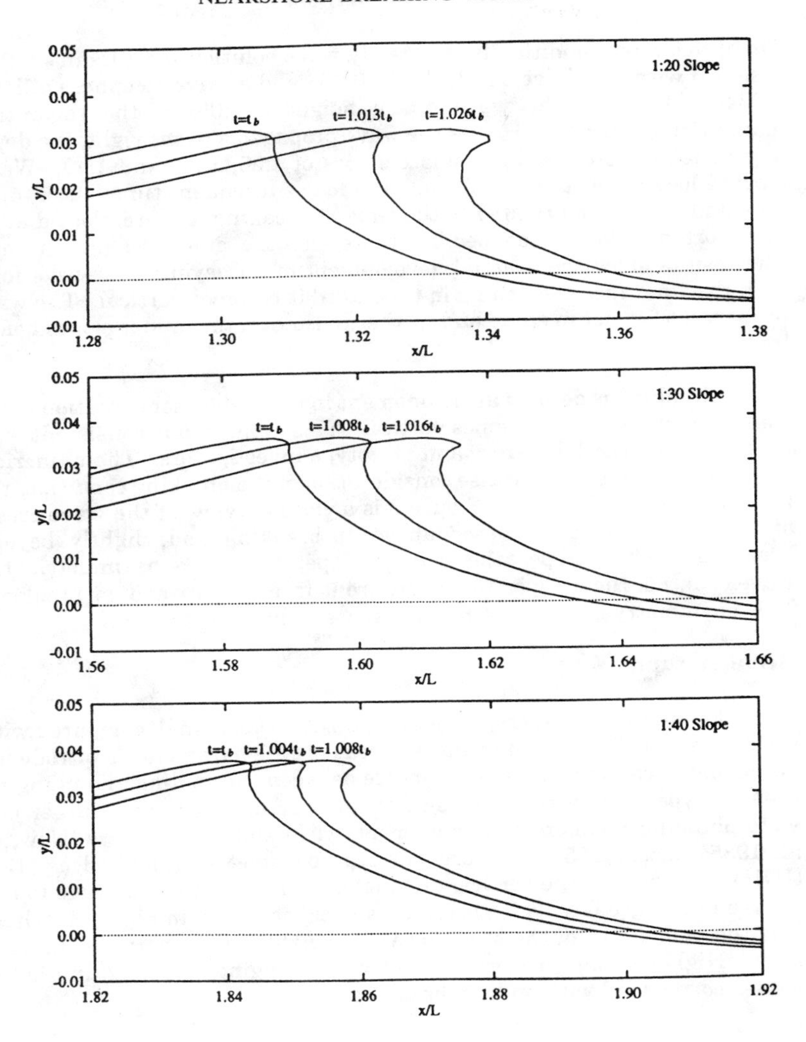

Figure 3. Close-up of the water surface elevation results at the wave crests at and beyond breaking for the waves in Figure 2.

The input wave condition is the steady wave solution from Dean's (1974) case 5-B with a relative depth of d/L=0.088 and a wave steepness of H/L= 0.0344. Initially, the wave has a height of 50% of the theoretical maximum, i.e. $H = 0.39\ H_b$. As the wave propagates to the right, the depth is gradually decreased to simulate slopes of 1:20, 1:30, and 1:40. Water surface location at the initial (t=0) and four subsequent times (25%, 50%, 75% and 100% of t_b) related to the time of breaking, t_b, are plotted at an undistorted scale. The dashed line represents the simulated bottom slope. The results show the gradual development of an asymmetric wave form with the front face steepening in time until it becomes vertical. The wave propagates farther in space and further in time on the milder slopes before breaking.

Breaking is defined as the moment in time when some portion of the front face of the wave becomes vertical. This definition is consistent with that used by most laboratory and prototype investigators. The numerical model allows for a very precise consideration of the breaking condition, the first occurrence of dx/dy=0. Figure 3 is a close-up view of the wave crests of the three waves discussed above at breaking and slightly beyond breaking. The steeper the bottom slope, the more asymmetric the waveform, i.e. the steeper the entire front face of the wave, at breaking. This asymmetry contributes to differences in breaker type.

Breaker type

For spilling breakers, the jet and tube are very small compared with the entire wave face. A quantitative breaker type parameter is introduced based on the concept that the difference between the spilling and plunging breaker types is related to the size of the overturning jet, i.e. larger jets yield plunging breakers and smaller jets yield spilling breakers (New, et al. 1985; Basco 1985). The breaker type parameter is defined as $B = (H_j/H)_b$, where H_j = the vertical distance from the crest of the wave to the point on the wave face which becomes vertical as shown in Figure 4. Thus, B is the percentage of the breaking wave height which overturns to form a jet. Higher values of B correspond with plunging breakers and lower values correspond with spilling breakers.

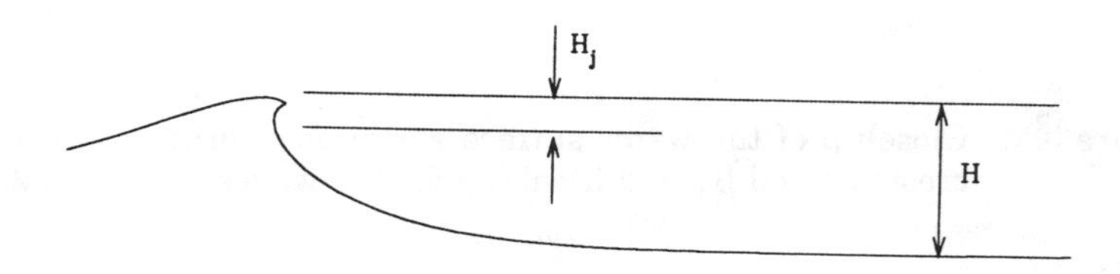

Figure 4. Definition sketch of the breaker type parameter, $B=(H_j/H)_b$.

The variation of B with beach slope for the three breakers discussed above are B=0.25, 0.16, and 0.12 for slopes of 1:20, 1:30, and 1:40 respectively. The milder slopes produce lower B values. Thus, the qualitative trends agree with expectations from laboratory results that steeper beaches tend toward plunging waves and milder beaches tend toward spilling waves. Since the difference between the breaker types is gradual, most laboratory investigators have allowed the precise demarcation between the types to be a subjective matter. The differences in subsequent vortex formation and wave energy decay are some of the important surf zone dynamics ramifications of breaker type. Thus, a wave which initially breaks with a small jet which touches down high on its own wave face has initial vortex formation that does not penetrate the water column and thus behaves as a spilling breaker. For a practical demarcation, B=0.20 is proposed here as a reasonable criterion. In other words, if less than one-fifth of the wave face is overturning in the jet, the wave is spilling.

The variation of B with wave steepness was investigated by choosing different initial wave steepnesses and propagating them up the same slope. Figure 5 shows the computed water surface elevations near breaking for three different initial wave steepnesses on a 1:20 slope. The input waves are Dean's (1974) cases 5-A, 5-B, and 5-C with wave steepnesses of H/L=0.018, 0.034, and 0.049; respectively and the same initial relative depth. The B values found are B = 0.43, 0.25, and 0.16 for the 5-A, 5-B, and 5-C waves respectively. The steeper waves produce lower B values. Thus, the qualitative trends agree with expectations from laboratory results that steeper waves tend toward spilling waves and less steep waves tend toward plunging waves.

The surf parameter, $\xi = m/(H/L)^{0.5}$, has been shown to describe the general form of the relationship between two independent parameters, wave steepness and beach slope, and a number of aspects of wave breaking including runup and breaker geometry in laboratory investigations (Battjes 1974). The effects of these independent parameters on the breaker type parameter were investigated with the present numerical model. Figure 6 shows the influence of wave steepness on B, the influence of beach slope on B, and the influence of ξ on B for nine different wave steepness and beach slope combinations. The breaker type parameter, B, varies linearly with beach slope in Figure 6a. The breaker type parameter, B, varies inversely at some root power of wave steepness in Figure 6b. The breaker type parameter, B, varies roughly linearly with ξ (computed with H_b) as shown in Figure 6c. The results indicate that ξ is a reasonable combination of the independent parameters and a reasonable model for B.

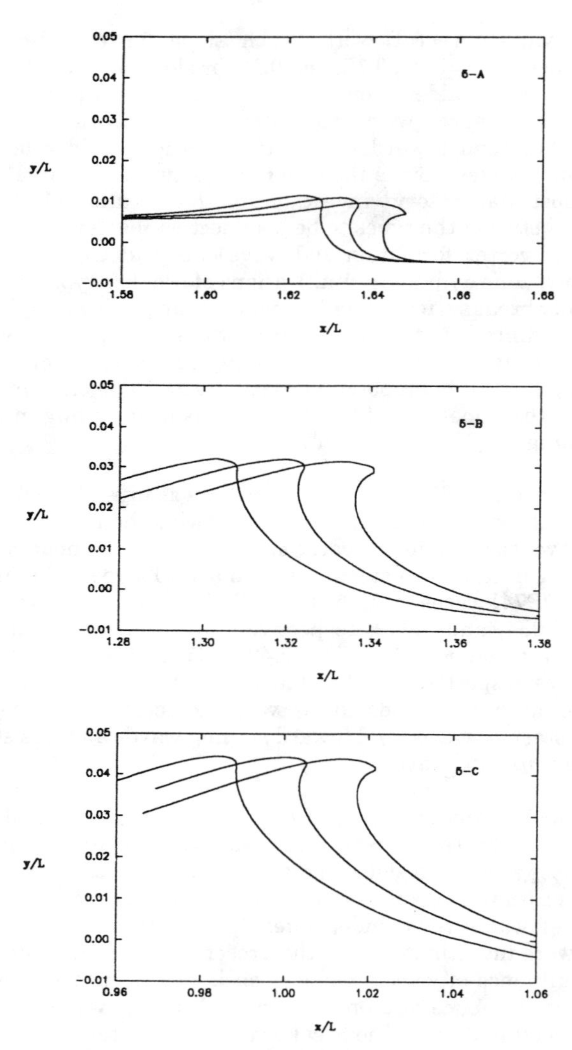

Figure 5. Water surface elevation results at the wave crests near breaking for three different initial wave steepnesses (Dean's 1974 wave cases 5-A, 5-B, and 5-C) on a simulated 1:20 slope.

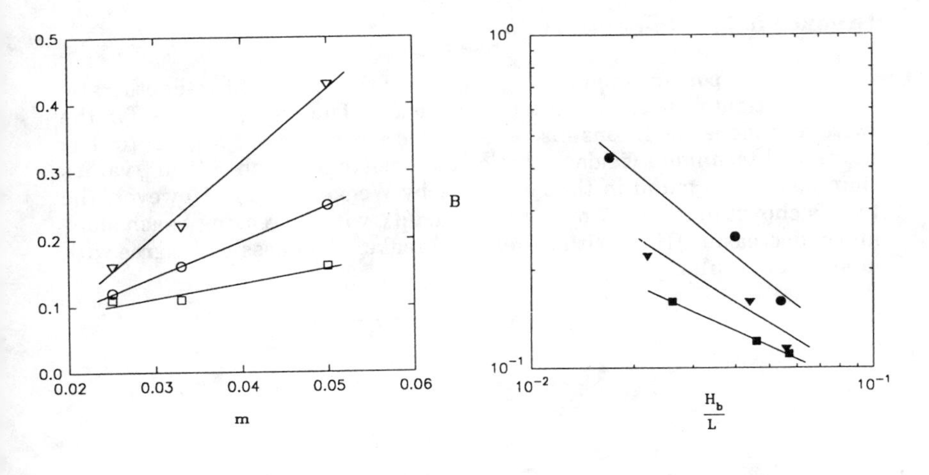

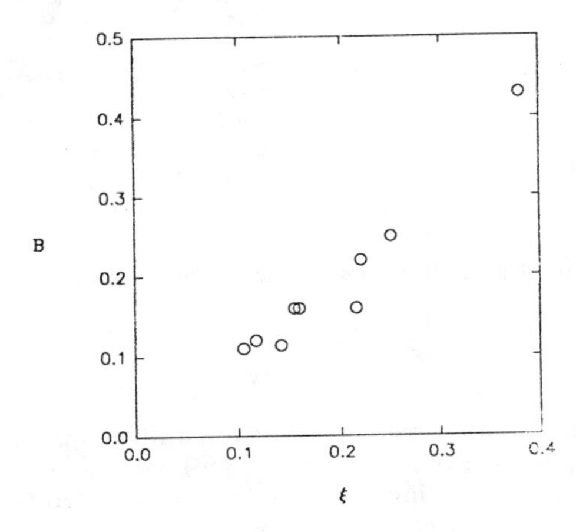

Figure 6. a) The influence of beach slope, m, on breaker type parameter, B, (▽ H_i/L=.018, ○ H_i/L=.034, □ H_i/L=.049); b) the influence of wave steepness, H_b/L, on breaker type parameter, B, (● m=.05, ▾ m=.033, ■ m=.025); c) the influence of surf parameter, ξ, on breaker type parameter, B.

Breaker height-to-depth ratio

The important engineering value of breaker height-to-depth ratio was investigated with the numerical model. The $(H/d)_b$ results for the wave and slope conditions discussed above are shown on Figure 7. The numerical technique yields realistic but consistently higher $(H/d)_b$ values than have been found in the laboratory by Weggel (1972). However, the trends shown in Figure 7 of increasing $(H/d)_b$ with increasing beach slope and of decreasing $(H/d)_b$ with increasing breaker steepness both agree with laboratory results.

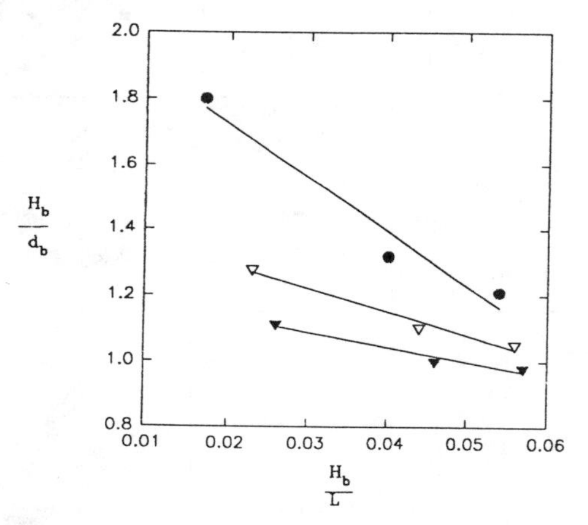

Figure 7. Breaker height-to-depth ratios (● m=.05, ▽m=.033, ▼m=.025)

CONCLUSIONS

A numerical model of wave breaking on mild slopes has been proposed. It is an application the mixed Eulerian-Lagrangian numerical technique to model wave breaking induced by gradual depth changes similar to those on many real coastal beaches. This type of modeling avoids some of the limitations inherent in laboratory modeling including wave field contamination, sidewall friction and limits on the constructability of mild and non-constant slopes.

Some disadvantages of the use of this technique to model the phenomenon of interest, depth-induced breaking, relate to numerical limitations and some relate to the theoretical limitations. By assuming that the bottom slope is locally flat, no effects due to the actual slope are present. However, considering the mildness of the slopes, such effects are probably small and the assumption has been used effectively elsewhere.

Because the technique uses a periodic wave form solution, the wavelength must be specified. In this work, the wavelength was fixed at the initial condition wavelength. Thus, the well-known phenomenon of decreasing wavelength with decreasing depth is not included. Given the shallowness of the initial conditions chosen, the wavelength would only be expected to decrease several percent prior to breaking. The wavelength could be reduced proportionally in the computation as predicted by linear or some other wave theory.

Another limitation of the technique is the lack of energy conservation due to the bottom slope simulation technique. As the bottom is gradually decreased, each depth reduction discards a very small slice of fluid from bottom of the domain. Each thin layer has some kinetic energy. For the waves discussed above the total energy loss varied from 10% to 30% prior to breaking.

Given these two theoretical shortcomings of the model, a specified wavelength and a non-conservative computation, the ability of the method to simulate the breaking process as well as it does is surprising. The method's results indicate that perhaps the wave instability of breaking is not very sensitive to wavelength or energy conservation in the final few wave periods approaching breaking. Similarly, breaking may not be sensitive to the bottom slope under an individual wavelength. Rather, the depth alone appears to be strongly controlling breaking.

The technique appears to be limited to slopes no steeper than 1:20. At steeper simulated slopes, too much of the flow field is discarded. At the other extreme, the model is able to simulate extremely mild slopes and is limited by computational resources and efficiency only. For beach slopes as mild as 1:100 the overall wave form approached breaking in a reasonable manner, but the code was limited in its ability to discern overturning jets by the resolution provided by the spacing of the surface particles. In order to adequately resolve the expected very small overturning jets, many more surface particles would be needed than were possible with the computational budget of this project.

Other limitations of the technique to model the phenomenon of interest, wave breaking in the prototype surf zone, relate to theoretical limitations and phenomenon which are excluded from the analysis. The fundamental assumption of the existence of a velocity potential assumes irrotational flow and negligible viscosity. At some point in the breaking process, flow becomes rotational, but, as results from the boundary integral technique have indicated, it might not occur until the actual touch-down of the jet on the forward face (Longuet-Higgins and Cokelet 1976). Surface tension is not included in the technique. Wave breaking is probably more sensitive to surface conditions and assumptions for waves with very small overturning jets, i.e. spilling breakers. Influences of local wind (Douglass 1990), other incident or reflected wave energy and existing turbulence are ignored in this approach.

This paper presents results on the water surface deformation only. The numerical technique can be used to investigate the full velocity and acceleration fields in shoaling and breaking waves (Ortiz and Douglass 1993b). Another straight-forward extension of this model is to non-constant beach slopes. In spite of the limitations discussed above, the numerical model has extended the work of previous numerical investigators to the problem of nearshore periodic wave breaking. The technique provides a different research approach to modeling the phenomenon.

ACKNOWLEDGEMENTS

Both authors gratefully acknowledge the support of Cray Research, Inc. and the Alabama Supercomputer Network Authority.

APPENDIX I. REFERENCES

Basco, D.R. (1985). "A qualitative description of wave breaking," *J. Wtrwy, Port, Coast. and Oc. Engrg.*, ASCE, 111:2, 171-188.

Battjes, J.A. (1974) "A computation of set-up, longshore currents, run-up and overtopping due to wind-generated waves," Ph.D. Dissertation, Delft Univ. of Tech., The Netherlands.

Dean, R.G. (1974) "Evaluation and comparison of engineering water wave theories," Spec. Rep. No. 1, US Army Coast. Engrg. Res. Ctr, Ft. Belvoir, Va.

Dold, J.W. and Peregrine, D.H. (1984). "Steep unsteady water waves: an efficient computational scheme," *Proc. 19th Coast. Engrg. Conf.*, Edge, B., ed., ASCE, 955-967.

Dommermuth, D.G., Yue, D.K.P., Lin, W.M., Rapp, R.J., Chan, E.S., and Melville, W.K. (1988). "Deep-water plunging breakers: a comparison between potential theory and experiments," *J. Fluid Mech.*, 180, 423-442.

Douglass, S.L. (1990). "Influence of wind on breaking waves," *Jour. Wtrwy, Port, Coast. & Oc. Engrg.* ASCE. WW6:116:651-663.

Gipson, G.S. (1987) *Boundary Element Fundamentals---Basic Concepts and Recent Developments in the Poisson Equation*, Computational Mechanics Publications, Southampton, U.K.

Grilli S.T, Skourup, J. and Svendsen, I.A. (1988) "The modeling of highly nonlinear water waves: a step toward a numerical wave tank," *Proc. 10th Inter. Conf. on Bound. Elements*, Springer-Verlag, 1:549-564.

Grilli S.T, Skourup, J. and Svendsen, I.A. (1989) "An efficient boundary element method for nonlinear water waves," *Engrg. Anal. Boundary*

Elements, Comp. Mech. Pub, 6:2:97-106.

Grilli, S.T. and Svendsen, I.A. (1989) "The modeling of highly nonlinear waves: some improvements to the numerical wave tank", *Boundary Elements XII,* Brebbia, C.A. and Conner, J.J., ed., Comp. Mech. Pub., Boston, 2:269-281.

Grilli, S.T. and Svendsen, I.A. (1990) "Corner problems and global accuracy in the boundary element solution of nonlinear wave flows," *Engrg. Anal. Boundary Elements*, Comp. Mech. Pub., 7:4:178-195.

Harlow, F.H. and Welch, J.E. (1965) "Numerical calculation of time-dependent viscous incompressible flow of fluid with free surface," *The Phys. of Fluids*, 8:12.

Longuet-Higgins, M.S. and Cokelet, E.D. (1976). "The deformation of steep surface waves on water, i. A numerical method of computation," *Proc. Roy. Soc. London*, 350A:1-26.

Munk, W.H. (1949). "The solitary wave theory and its application to surf problems," *Annals of the N.Y. Acad. Sciences*, 51, 376-462.

New, A.L., McIver, P. and Peregrine, D.H. (1985). "Computation of overturning waves," *J. Fluid Mech.*, 150:233-251.

Nichols, B.D., Hotchkiss, C.W. and Hurt, R.S. (1980). "SOLA-VOF: A solution algorithm for transient fluid flow with multiple free boundaries," Los Alamos Sci. Lab. Rep. No. LA-8355.

Ortiz, J.C. and Douglass, S.L. (1991). "Water wave modeling using a boundary element method with overhauser spline elements", in *Boundary Elements XIII*, Brebbia, C.A. and Gipson, G.S., eds., Elsevier Applied Sci., New York, 273-282.

Ortiz, J.C. and Douglass, S.L. (1993)a "Overhauser boundary elements solution for periodic water waves in the physical plane," *Engrg. Anal. Boundary Elements*, Comp. Mech. Pub., Southampton, 11:47-54.

Ortiz, J.C. and Douglass, S.L. (1993)b "A boundary element solution of water particle velocities of waves breaking on mild slopes", in *Boundary Elements XV*, Brebbia, C.A. and Rencis, J.J., eds., Elsevier Sci. Pub., New York, 1:221-232.

Romate, J.E. (1990) "The numerical simulation of nonlinear gravity waves," *Engrg. Anal. Boundary Elements*, Comp. Mech. Pub., 7:4:156-166.

Seo, S.N. and Dalrymple, R.A. (1990). "An efficient model for periodic overturning waves," *Engrg. Anal. Boundary Elements*, Comp. Mech. Pub., 7:4:196-204.

Vinje, T. and Brevig, P. (1980). "Numerical simulation of breaking waves," *Proc. 3rd Intl. Conf. on Finite Elements in Water Resources*, University of Mississippi, 5.196-5.210.

Weggel, J.R. (1972). "Maximum breaker height," *J. Wtrwys, Harbors and Coast. Engrg.*, ASCE, 98:4.

Xu, H.B. and Yue, D.K.P. (1992) "A numerical study of kinematics of nonlinear water waves in three dimensions," *Proc. 5th Conf. on Civ. Engrg. in the Oceans*, ASCE.

The Significance of Hydrodynamic Forces in Coastal Embankment Design

Bang-Fuh Chen[1], member

Abstract

The dynamic forces of sea water and backfill soil acting on coastal embankment are analyzed by using a new finite-difference method. The Euler's equation and continuity equation are used to solved the nonlinear hydrodynamic pressures of the sea water. The backfill soil at another side of the embankment is modeled as the fluid-filled solid mixture. The momentum equation and storage equation of the saturated elastic solid mixture are combined to derive the pressure wave equation of the backfill soil. The pressure wave equation of sea water was solved by a developed finite difference method, while the pressure wave equation of solid mixture are solved by an implicit finite difference method coupled with a fast Poisson solver. The hydrodynamic pressure of sea water on embankment face are significantly increasing as the slope of the sea bottom increasing. The pore pressure and the interaction force between soil solid and fluid can be augmented significantly when the backfill soil is compressed during earthquakes, and the augmented dynamic pressure may result in the liquefaction of the backfill soil.

Introduction

The hydrodynamic analysis of dam-reservoir system has been studied extensively during the last three decades. The first rigorous analysis of hydrodynamic forces on dam faces during earthquakes was done by Westergaard in 1933. At the same time, von Karman(1933) used a momentum balance method analyzed the same problem. In the followed fifty years, many researchers have studied the hydrodynamic pressures on dams using various methods. The compressibility effect (Hung and Wang 1987), flexibility of dam (Hung and Chen 1990), arbitrary reservoir shape (Chen 1993), dam-reservoir foundation interaction (Chopra et. al 1981), absorption of reservoir sediment (Fenves and Chopra 1983, Chen and Hung 1991), nonlinear free surface wave (Chwang 1983, Hung and Chen 1990),

---

[1] Associate Professor, Department of Marine Environment, National Sun Yat-Sen University, Kaohsiung, Taiwan, R.O.C.

and even the viscous effect (Yang and Chwang 1991, Chen 1993) were included in the analysis of dam hydrodynamics. The importance of the dynamic forces due to reservoir water and sediment has been recognized, and the dam hydrodynamic analysis has become a necessary analysis in a new dam-project and the safety-evaluation of the existing dam.

The coastal embankment is designed for preventing coastal line erosion, runup of the waves and sliding of the backfill soil. The embankment and sea water system can be regarded as a relatively small dam associated with a very large reservoir. At the another side, the embankment can be considered as a retaining wall for preventing sliding of the backfill soil. Traditionaly, the design loadings in embankment structural analysis are wave forces due to sea waves and tides, and static active and passive forces due to backfill soil. Extensive formulas for evaluating wave forces, in active way, on embankment have been reported, such as Sainflou (1928), Hiroi (1919) and Goda (1974) etc. The dynamic forces, generated by embankment motion which is in a passive way, of sea water and backfill soil during earthquakes are always not included in coastal embankment design. In the present analysis, a developed finite-difference model(1993) was extended to analyze coastal embankment problem. The non-horizontal sea bottom is considered in the analysis.

Basic Equations for Inviscid Flow

For the analysis of fluid motion near the embankment during earthquakes, the coordinate system is chosen to move with the ground motion. The momentum equations for a two-dimensional case can be written as

$$\frac{\partial(u+u_0)}{\partial t}+u\frac{\partial u}{\partial x}+v\frac{\partial u}{\partial y}=-\frac{1}{\rho}\frac{\partial p}{\partial x} \tag{1}$$

$$\frac{\partial(v+v_0)}{\partial t}+u\frac{\partial v}{\partial x}+v\frac{\partial v}{\partial y}=-g-\frac{1}{\rho}\frac{\partial p}{\partial y} \tag{2}$$

where u and v are, respectively, the velocity components of fluid in the $x-$ and $y-$ directions, u_0 and v_0 the corresponding ground velocity components, p the total pressure, ρ the fluid density, and g the gravitational acceleration. The general definition sketch of the system in present study is shown in Figure 1.

The continuity equation for the compressible flow can be written as

$$\frac{\partial u}{\partial x}+\frac{\partial v}{\partial y}=-\frac{1}{\rho c^2}\frac{\partial p}{\partial t} \tag{3}$$

where c is the sound speed in sea water. Taking partial differentiation of equations (1) and (2) with respective to x and y, respectively, and summing up the results, one can obtain

$$\left(\frac{\partial^2 p}{\partial x^2}+\frac{\partial^2 p}{\partial y^2}\right)=\frac{1}{c^2}\frac{\partial^2 p}{\partial t^2}-\rho\left(u\frac{\partial u}{\partial x}+v\frac{\partial u}{\partial y}\right)-\rho\left(u\frac{\partial v}{\partial x}+v\frac{\partial v}{\partial y}\right) \tag{4}$$

The first term of the right hand side of equation (4) accounts for the compressibility effect, and the other terms are due to convective acceleration.

Constitutive Equations for Fluid-Filled Mixture

The backfill-soil sediment is modeled as the fluid-filled mixture which is a two-phase system consisting of a solid and a fluid phases. Each phase is regarded as a continuum and follows its own motion. Based on Atkin and Craine's(1976) theory of continuum of mixture of fluid and elastic solid, for a two-dimensional analysis, the field equations of the fluid-filled elastic solid can be written as follow:
Momentum equations:

$$(1-n)\rho_s\frac{\partial(v_x+u_0)}{\partial t} = \frac{\partial\sigma_{xx}}{\partial x}+\frac{\partial\sigma_{xy}}{\partial y}-(1-n)\frac{\partial p}{\partial x}+\frac{n^2}{k}(u_x-v_x)+p\frac{\partial n}{\partial x} \quad (5)$$

$$(1-n)\rho_s\frac{\partial(v_y+v_0)}{\partial t} = \frac{\partial\sigma_{xy}}{\partial x}+\frac{\partial\sigma_{yy}}{\partial y}-(1-n)\frac{\partial p}{\partial y}+\frac{n^2}{k}(u_y-v_y)+p\frac{\partial n}{\partial y}+(1-n)\rho_s g \quad (6)$$

$$n\rho\frac{\partial(u_x+u_0)}{\partial t} = -n\frac{\partial p}{\partial x}-\frac{n^2}{k}(u_x-v_x)-p\frac{\partial n}{\partial x} \quad (7)$$

$$n\rho\frac{\partial(u_y+v_0)}{\partial t} = -n\frac{\partial p}{\partial y}-\frac{n^2}{k}(u_y-v_y)-p\frac{\partial n}{\partial y}+n\rho g \quad (8)$$

in which u_j and v_j are velocities of fluid and solid and the subscripts x and y indicate the velocity components in the horizontal and vertical directions, σ_{xx} and σ_{yy} the corresponding normal stresses, and σ_{xy} the shear stress, n is the total porosity which is the sum of the static porosity n_o and the perturbation n'; ρ_s is the density of solid grain which is assumed to be incompressible and ρ_w is the density of the water and k is the permeability.

Storage equation:

$$n(\frac{\partial u_x}{\partial x}+\frac{\partial u_y}{\partial y}-\frac{\partial v_x}{\partial x}-\frac{\partial v_y}{\partial y})+(\frac{\partial v_x}{\partial x}+\frac{\partial v_y}{\partial y}) = -\frac{n}{\beta}\frac{\partial p}{\partial t} \quad (9)$$

Linear constitutive equations:

$$\frac{\partial\sigma_{xx}}{\partial t} = (\lambda+2\mu)\frac{\partial v_x}{\partial x}+\lambda\frac{\partial v_y}{\partial y} \quad (10)$$

$$\frac{\partial\sigma_{xy}}{\partial t} = \mu(\frac{\partial v_x}{\partial y}+\frac{\partial v_y}{\partial x}) \quad (11)$$

$$\frac{\partial\sigma_{yy}}{\partial t} = (\lambda+2\mu)\frac{\partial v_y}{\partial y}+\lambda\frac{\partial v_x}{\partial x} \quad (12)$$

where λ and μ are the effective Lame's moduli.

In the proposed study, the pressure in backfill soil region will be obtained from both the mass balance and the momentum balance. Similar to the derivation of the pressure wave expression (equation (4)) for fluid flow in the sea field, the storage equation can be coupled with four momentum equations to form the pressure wave equation in the soil region:

$$\frac{\partial^2 p}{\partial x^2} + \frac{\partial^2 p}{\partial y^2} = f(x, y, t) \tag{13}$$

where

$$f(x,y,t) = \frac{\rho_s \rho}{(1-n)\rho + n\rho_s}\left(\frac{n}{\beta}\frac{\partial^2 p}{\partial t^2} + \frac{1}{\rho_s}\left(\frac{\partial^2 \sigma_{xx}}{\partial x^2} + 2\frac{\partial^2 \sigma_{xy}}{\partial x^2} + \frac{\partial^2 \sigma_{yy}}{\partial y^2}\right)\right.$$

$$\left. + \frac{\rho - \rho_s}{\rho \rho_s}\frac{n^2}{k}\left[\frac{\partial (u_x - v_x)}{\partial x} + \frac{\partial (u_y - v_y)}{\partial y}\right]\right) \tag{14}$$

Notice that the term, $\frac{n}{\beta}\frac{\partial^2 p}{\partial t^2}$, in $f(x, y, t)$ will vanish when β approaches infinity for incompressible fluid in backfill soil. The dilatational term, $p\nabla^2 n$, and the nonlinear terms associated with the gradient of the porosity are small; they are neglected in the present analysis.

In summary, the dynamic response of the backfill soil was calculated from equations (5) through (14). They were solved by an implicit finite difference method with an application of the fast Poisson solver (Strang 1970, Chen and Hung 1993).

Results and Discussion

In the present analysis, a rigid embankment motion is assumed. That is, the geometry of embankment face is not varied with time. The hydrodynamic pressure of sea water and the dynamic force of backfill soil can be evaluated separately. We will, first, discuss the results of hydrodynamic pressures of sea water on embankment.

A 0.5g horizontal acceleration is used as the ground excitation for calculating the hydrodynamic pressures on the vertical embankment face. In the present study, the sea bottom is consist of two parts, a sloped bottom and a horizontal bottom. Figure 2 shows the onset hydrodynamic pressure of sea water on embankment faces with various sea depthes, the associated sea bottom slope ϕ is equal to 15°. The results of $\phi = 30^0$ and 45^0 are shown in figure 3 and 4 respectively. The result of $\phi = 0^0$ is also shown in these figures. As shown in these figures, the hydrodynamic pressure is increasing as the slope region of the sea bottom, C_2, increasing. Note that, the factor C_2 can also be explained as the factor of the depth of horizontal sea bottom. Actually, the hydrodynamic pressure is generated by the embankment motion. If the slope region of sea bottom is combined with the real embankment as a part of the coastal embankment and call it as an *equivalent embankment*. The amount of sea water pushed by the equivalent embankment is actually by a larger region with a sea-water depth d_2 instead of d_0, the water depth at toe of the real embankment. Although, the increment of C_2 in Figs. 3 and 4, is the same ($= 0.25$), the gap between two consecutive curves reduces rapidly. For $\phi = 30^0$ and $C_2 > 0.75$ the hydrodynamic pressures on embankment face are equal to that of $C_2 = 0.75$ and bascially remaining constant.

Let's shfit our attention to the otherside of the embankemnt and discuss the dynamic response of the backfill soil. In the present study, the backfill soil is assumed to be uniform distributed coarse sand with a homogeneous porosity of 0.3. The depth of the backfill soil is 5 meters and the material properties slected are $\lambda = 290$ x $10^4 N/m^2$, $\mu = 145$ x $10^4 N/m^2$, $k = 0.09m^3 - sec/N$, and $\rho_s = 2.5\rho_w$. The same ground acceleration, $0.5g$, is used as the ground excitation for calculating dynamic pressures of soil on the embankment face. The ground water level is assumed to be at the same level of the free surfcae of the backfill soil, i.e. the soil is basically fully saturated.

The profiles of the onset dynamic pore water pressures for various effective bulk modulus β are shown in Figure 5. Also shown in the figure is the distribution of the onset hydrodynamic pressure of pure water, i.e. the backfill soil is replaced by the water. As indicated in the figure, the dynamic pore water distribution along the embankment face is remaining constant if the effective bulk modulus is greater than $10^{13}psf$, and is drastically reduced when β is less than $10^{12}psf$. For a sound velocity of $1473m/sec$ in water,the corresponding bulk modulus is about 5 x $10^9 psf$. As shown in the figure, the onset dynamic pore pressure acting on emebankment face is much less than the onset hydrodynamic pressure of pure water. When the constant ground acceleration is prolong to $0.005sec$, the distribution of the dynamic pore water pressure on embankment faces is plotted against the dimensionless ground displacement and shown in Fig. 6. As presented in the figure , significant augmentation of dynamic pore water pressure is noted and the dynamic pore pressure could be four times as that at onset. The increases are due to the compressibility of pore water, the reduction in porosity of soil and a rise of the Stokes drag.

Conclusions

A new finite difference method has been used for calculating the dynamic forces on coastal embankment. The significance of dynamic forces due to the hydrodynamic pressure of sea water and dynamic pore pressure of backfill soil has been noted during the analysis. Both the slope and the length of the sloped sea bottom will significantly affect the magnitude of the hydrodynamic pressure on embankment faces. The onset dynamic pore pressure of backfill soil is much less the onset hydrodynamic pressure of pure water. However, the dynamic pore pressure is augmented rapidly during a prolonged ground excitation. The profile of the increasing pressure will approach that of the hydrodynamic reponse of the heavier fluid for liquefied backfill soil.

Acknowledgement

The study is partially supported by the National Science Council under the grant NSC 81-0209-E-110-02.

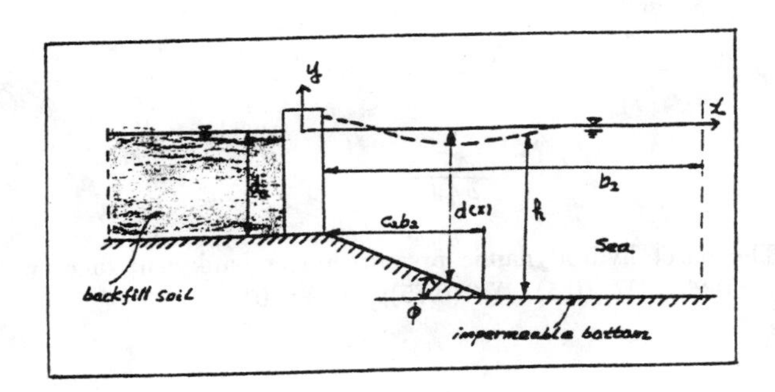

Figure 1: The definition sketch

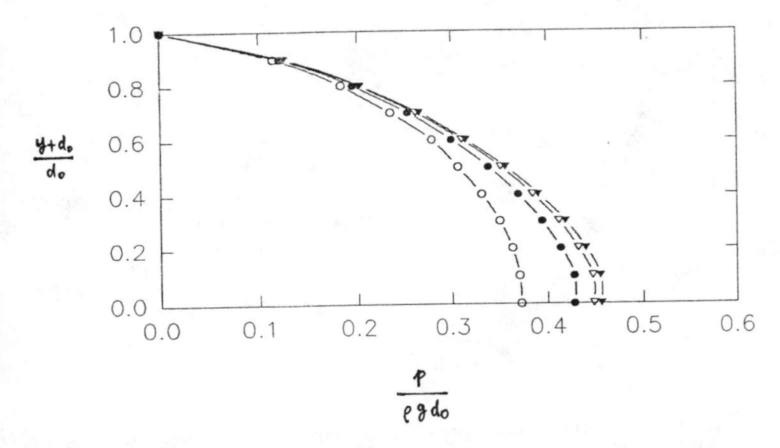

Figure 2: The onset hydrodynamic pressure on embankment faces with $\phi = 15^0$, $\bullet$ $(C_2 = 0.25)$, ∇ (0.5), ∇ (0.75); $\circ$ $\phi = 0^0$.

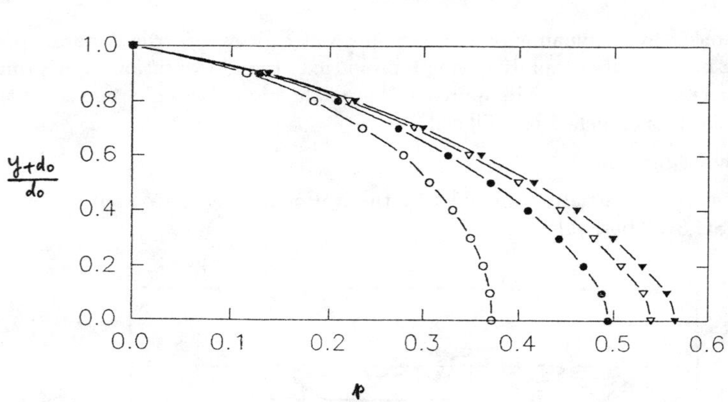

Figure 3: The onset hydrodynamic pressure on embankment faces with $\phi = 30^0$, $\bullet$ ($C_2 = 0.25$), ∇ (0.5), ∇ (0.75); $\circ$ $\phi = 0^0$.

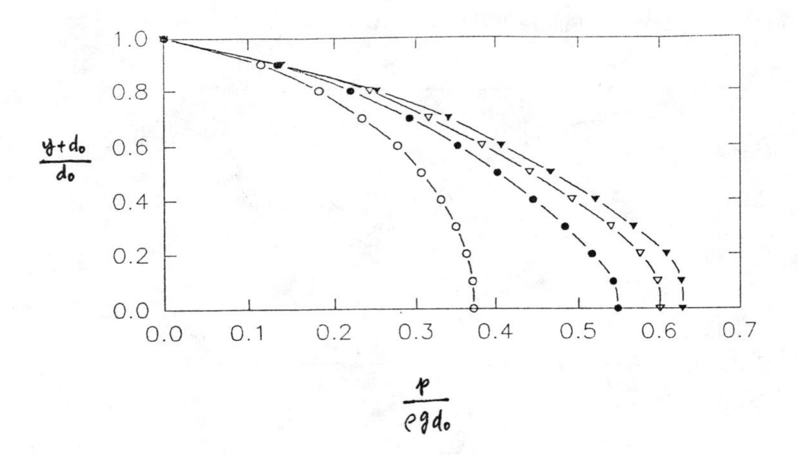

Figure 4: The onset hydrodynamic pressure on embankment faces with $\phi = 45^0$, $\bullet$ ($C_2 = 0.25$), ∇ (0.5), ∇ (0.75); $\circ$ $\phi = 0^0$.

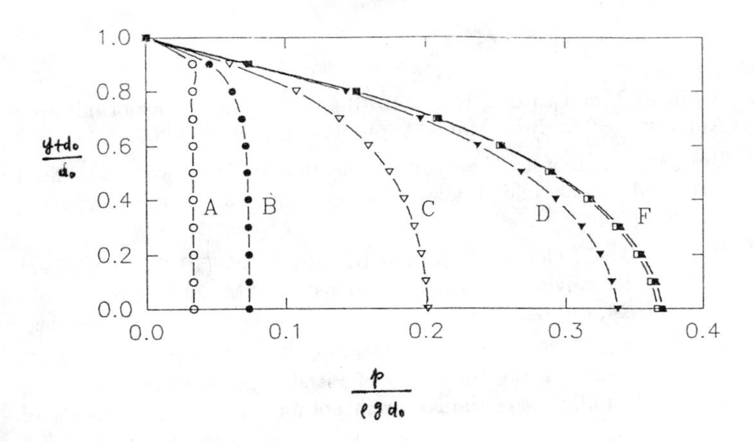

Figure 5: The onset dynamic pore pressures on vertical embankment face with $\beta = 10^8, 10^9, 10^{10}, 10^{11}, 10^{12} and 10^{13} psf$ for curves A to F

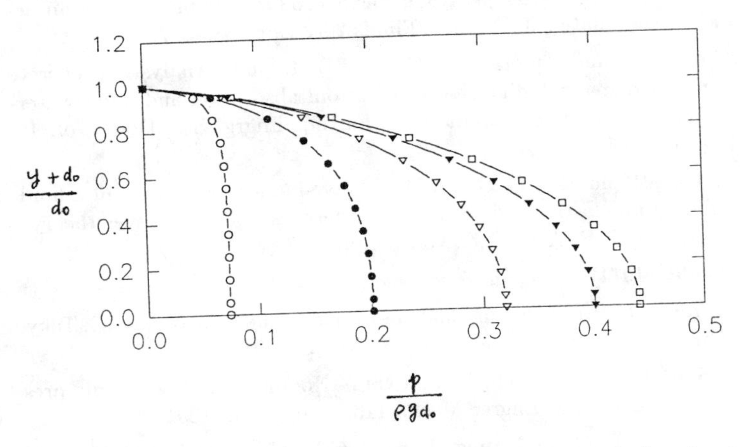

Figure 6: The increases of dynamic pore pressure on vertical embankemnt face with dimensionless timeε^2 for $\beta = 10^9 psf$.

References

1. Atkin R. J. and Craine, R. E. (1976) "Continuum theory of mixture: Applications", J. Inst. Maths Appls, Vol. 17, pp. 153-207.
2. Biot, M. A., (1956), "Theory of propagation of elastic waves in a fluid saturated porous solid. I. low-frequency range", J. Acoust. Soc. Am., Vol. 28, p 168-.
3. Buzbee, B. C., Golub, G. H. and Nielson C. W. (1970), "On direct methods for solving Poisson's equations", SIAM, J. Numer. Anal., Vol. 7, No.4, pp. 627-656.
4. Chen, B. F., (1989), "Dynamic forces on dams during earthquakes", thesis presented to the University of Pittsburgh, at Pittsburgh, PA., in partial fulfillment of the requirements for the degree of Doctor of Philosophy.
5. Chen, B. F. (1993), "A study of viscous effects on dam hydrodynamics", submitted to J. IAHR.
6. Cheng, A. H. D., (1986), "Effect of sediment on earthquake-induced reservoir hydrodynamic response", J. Engrg. Mech., ASCE, vol. 111, pp. 654-663.
7. Chwang, A. T., (1983). "Nonlinear hydrodynamic pressure on an accelerating plate", J. Physics. Fluids 26(2), pp. 383-387.
8. Fenves, G. and Chopra A. K. (1984), "Earthquake Analysis of concrete gravity dams including reservoir bottom absorption and dam -water-foundation reservoir interaction", Earthq. Engrg. Str. Dyn., Vol. 12, pp. 663-680.
9. Foda A., and Mei, C. C., (1981), "Wave-induced responses in a fluid-filled poro-elastic solid with afree surface: a boundary layer theory", Geophys, J. R. Astr. Soc., vol. 66, pp. 597-631.
10. Goda, (1974), in Japanese.
11. Hiroi, I., (1919), "Ona method of estimating the force of waves", Tokyo Univ. Engineering reports, Vol. X, No. 1, p. 19-.
12. Hung, T. K. and Chen, B. F.(1990), "Nonlinear hydrodynamic pressure on dams", J. Engrg. Mech., 116(6), pp. 1372-1391.
13. Macagno, E. O., and Hung, T. K. (1967), "Computational and experimental study of a captive annular eddy", J. Fluid Mech., 28(1), pp. 43-64.
14. Mei, C. C., Foda, M. A. and Tong, P. (1979), "Exact and hybrid-element solutions for the vibration of a thin elastic structure seated on the sea floor". , Appl. Ocean Res., 1(2), pp. 79-88.

15. Prevost, J. H. (1985) "Wave propagation in fluid-saturated porous media: an efficient finite element procedure", Int. J. Soil Dyn. Earthq. Engrg., vol. 4, No. 4, pp. 183-202.

16. Sainflou, G. (1928), "Essai sur les diques maritimmes verticales", Annales des Ponts et Chaussees, Vol. 98, No. 1, pp. 5-48.

17. Seed, H. (1976) "Evaluation of soil liquefaction effects on level ground during earthquakes", Liquefaction problems in geotechnical engineering, ASCE, Speciality session, Philadelphia, pp. 1-183.

18. Strang, G. (1986), "Introduction to applied mathematics", Wellesey-Cambridge Press., Wellesly, MA., pp. 39-62.

19. von Karman, T. (1933), "Discussion of water pressures on dam during earthquakes", by H. M., Westergarrd, Trans. , ASCE, 98, pp. 434-436.

20. Westergarrd, H. M., (1933), "Water pressures on dams during earthquakes", Trans. , ASCE, 98, pp. 418-433.

21. Yang, S. A. and Chwang, A. T., (1989), "Nonlinear viscous waves produced by an impulsively moving plate", IIHR Report No. 332, The Univ. of Iowa.

ASA.WAVES: An Interactive PC-based Wave Forecasting Tool

Stéphan T. Grilli[1], T. Opishinski [1], M.L. Spaulding[2] & T. Isaji [2]

Abstract : An interactive PC-based system for wave refraction-diffraction in coastal regions has been developed, based on a Mild-slope equation finite element wave model. The system includes an embedded Geographic Information System, and sophisticated graphics user interface. The model accounts for partial wave absorption at boundaries, and energy dissipation by bottom friction. Interactive graphic tools have been developed, for both boundary fitted grid generation (pre-processing), and for the analysis and verification of model predictions (pre- and post-processing).

Introduction

The design of coastal structures, the study of beach erosion and restoration, and the design and management of harbors and marinas, all require that wave refraction-diffraction be accurately calculated in coastal regions. Many factors control, and complicate this analysis including arbitrary bathymetry and coastline geometry, bottom friction, and wave reflection and absorption at the coastline and on structures.

In general, offshore wave climates are either directly measured or forecast based on hindcasting procedures or through the use of wind-wave models. The resulting directional wave spectrum gives the distribution of incident wave energy as a function of frequency, f, and direction, θ. Although, at a research level, nonlinear wave models are being developed for the study of wave shoaling and breaking in shallow water (e.g., Grilli, 1993), most state-of-the-art models used in design calculate wave refraction-diffraction in coastal regions, based on linear wave theory, and often assume the bottom slope is mild.

Among these methods, the Mild Slope Equation (MSE), first derived by Berkhoff (1972), has become increasingly popular because it is has the advantages of being two-dimensional (2D), and of transforming into a simple second-order elliptic equatio when expressed in the frequency domain. Although the MSE is in principle restricted to mild slopes, Booij (1983) compared MSE results with the full three-dimensional linear

[1]Assoc. Prof. and Graduate Res. Asst., Dept. of Ocean Engng., Univ. of Rhode Island, Narragansett, RI 02882. Tel.: (401) 792-6636; Fax.: (401) 792-6837

[2]Senior Scientists, *Applied Science Associate, Inc.* (ASA), 70 Dean Knauss Dr., Narragansett, RI 02882. Tel.: (401) 789-6224; Fax.: (401) 789-1932

solution and showed the MSE is accurate up to 1:3 bottom slopes in the direction of wave propagation, and up to 1:1 slopes in the direction perpendicular to wave propagation. Parabolic approximations to the MSE, less general than the full expression but more computationally efficient, have been derived (e.g., Kirby & Dalrymple, 1983) and are being successfully used in design. The MSE has also been extended to cases of waves with current, as well as to transient cases.

Numerical solutions to the MSE have been developed based primarily upon the Finite Element Method (FEM) (e.g., by Berkhoff 1972; Lejeune, *et al.*, 1982). Behrendt (1985) developed an efficient FEM model for the MSE based on the hybrid element method introduced by Chen & Mei (1974) (see Mei, 1989). In this method, only part of the horizontal plane is divided into finite elements. Truncated eigenfunction expansions are used to describe the wave field in the remaining part of the water covered area. In his model, Behrendt developed higher-order partial absorbing and reflecting boundary conditions, to model wave energy dissipation by friction and breaking, on lateral boundaries of the computational domain. These conditions have proved to be among the best available for linear wave refraction-diffraction problems and have successfully been implemented in other linear wave models (e.g., Grilli, 1986 : 3D Boundary Element Method; and Kirby, 1991 : Parabolic Wave Equation). Behrendt also introduced wave damping due to bottom friction in his model.

Whatever the selected modeling approach, however, efficient design requires that the modeling tools be flexible, easy to use and, particularly, that they be optimized at both the pre- and post-processing stages of the calculations. Whereas running a model can be fairly quick and cheap on todays high-end PC platforms, it is well known to modelers that the design of accurate computational grids (especially when using a FEM) is the most time consuming, and therefore the most expensive aspect, of an analysis. Closely following this in terms of cost, is the analysis and presentation of results, i.e., post-processing. Without easy access to graphic visualization tools, post-processing may become discouraging, or even prohibitive, and prevent designers from performing the "modeling iterations" necessary to approach optimal design.

In connection with various research and development projects, Applied Science Associates (ASA) has developed an innovative approach based on interfacing a custom designed Geographic Information Systems (GIS), ASA.MAP, and graphics user interface, with traditional environmental models (tidal and wind driven coastal circulation, ecosystem, pollutant transport). GIS's are interactive graphic display tools that give immediate access to various layers of information, for a selected geographic area (Fig. 1). With this new approach, it is possible for models to readily access geographically referenced data and environmental parameters for a selected area from which computational grids and boundary conditions for the models can be interactively generated (pre-processing). Models are then executed (processing), and computed results are stored as data bases of "simulated data" that can be visualized and processed, using various tools, through the graphics user interface (post-processing), along with any "observed data" available for the selected area. When installed on a powerful laptop PC computer, the combined system user-interface/GIS/models, provides a fast, interactive, and flexible engineering design tool.

Models for oil spills (OIL.MAP), and hydrodynamic and water quality simulation (WQ.MAP), have been developed. It is the purpose of this paper to describe and illustrate features of ASA.WAVES, an interactive PC-based wave forecasting tool developed using the same approach.

Description of the model

ASA.WAVES solves water wave refraction-diffraction problems, based on the MSE, using a FEM model similar to the hybrid element model by Behrendt (1985). A 2nd-order partial absorption boundary condition and energy damping due to bottom friction can be specified in the model. Values of the parameters in the absorption condition can be adjusted to account for energy loss by friction and breaking at structures and along coastlines. Behrendt's model was implemented by Losada *et al.* (1990) on a PC platform, following a more traditional "batch" approach, with the addition of a boundary fitted grid generation procedure. This software served as the initial basis for the interactive model developed under the present project. The three main functions in ASA.WAVES are detailed in the following.

Pre-processing.—A boundary fitted grid made of triangular finite elements is interactively generated based on user-specified control nodes, and on topographical information extracted from, or provided to, the GIS data bases. The grid generation method performs a curvilinear transformation between an orthogonal auxiliary space (I,J), and the physical space (x,y) (Figs. 3-6). Two windows and a set of menus are availables for specifying control nodes in both spaces, along with parameters for grid generation.

Main functions available in the grid generation are : (i) selection of the general area of interest for the wave calculation, in the GIS data bases (Figs. 1,2); (ii) selection of control nodes along the boundary, in both (x,y) and (I,J) (Figs. 3,4); (iii) definition of segments between control nodes (Fig. 4); (iv) specification of required resolution along boundary segments (Fig. 5); (v) specification of boundary conditions; (vi) grid generation (Fig. 6); (vii) verification of the grid, element distortions, resolution with respect to incident wave length, or local depth; (viii) specification of the depth (coordinate z) at each grid point based on user-specified parallel transects, or on information extracted from the GIS data bases. The grid can be interactively visualized and modified and coastal structures can be introduced at their existing or planned location (e.g., groins, breakwaters, wave barriers, etc). Lateral boundary conditions can be interactively specified, using menu options, along the model boundaries. A bottom friction coefficient can also be specified.

Processing.—The MSE is solved using a hybrid-FEM model, with linear triangular finite elements, for each specified deep water wave frequency, f, and direction, θ. This part of the software is based on a block-solver elimination procedure that uses

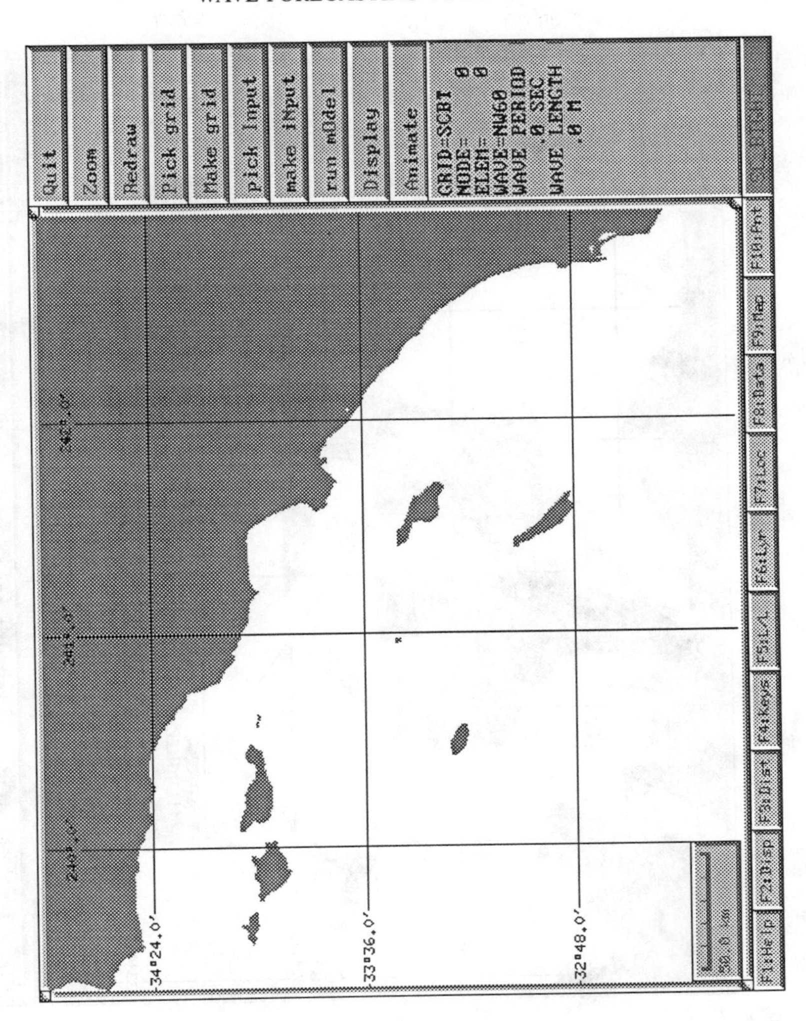

Figure 1: General geographic location for the problem (US West Coast).

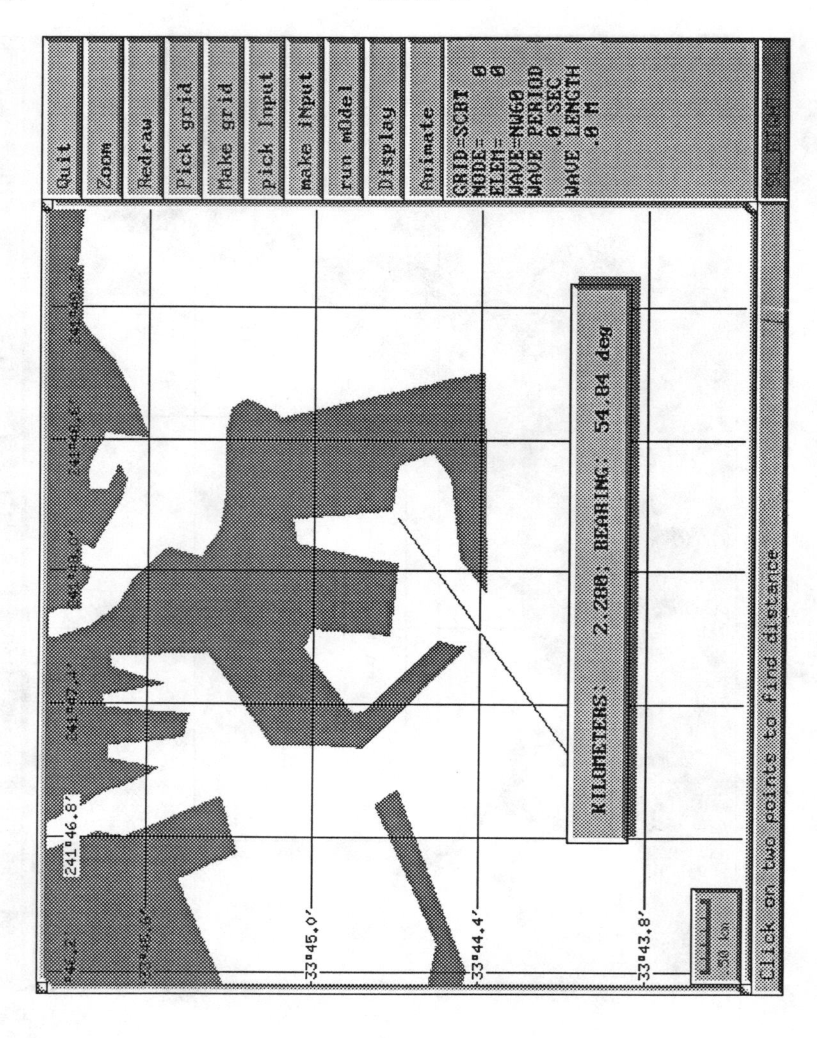

Figure 2: Zoom on the Long Beach, LA, area in Fig. 1, using the GIS graphics tools. Illustration of the distance measuring tool.

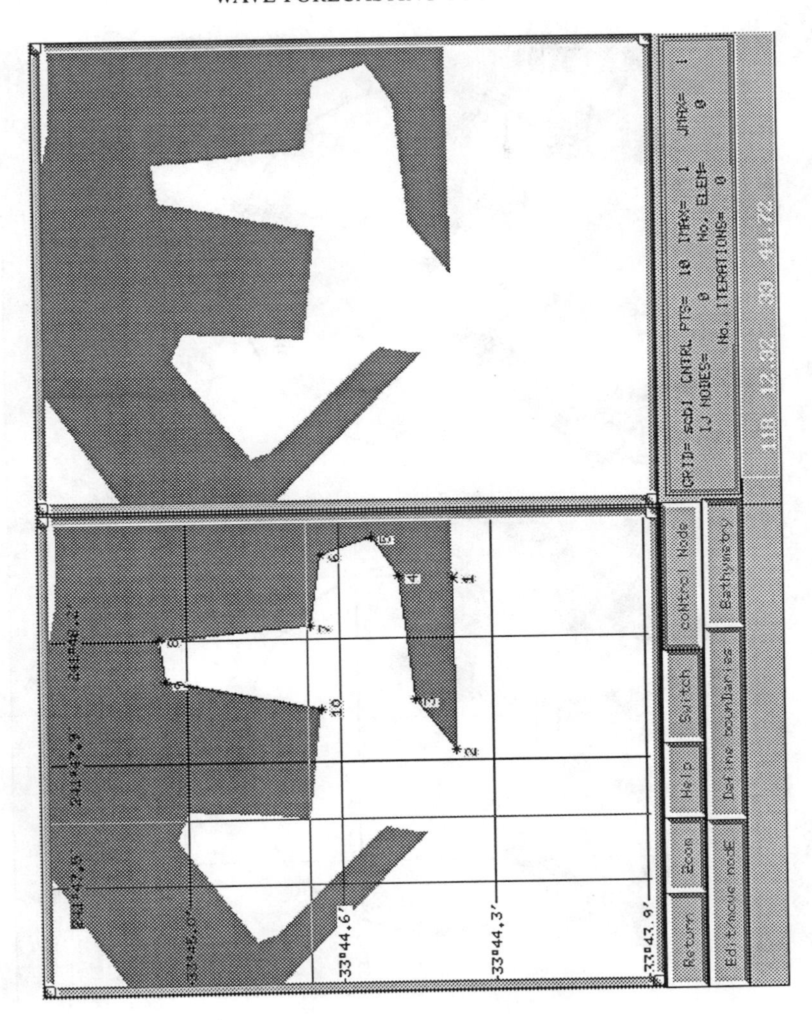

Figure 3: "Make grid" function in ASA.WAVES main menu (Fig. 2). Interactive specification of control nodes in (x, y).

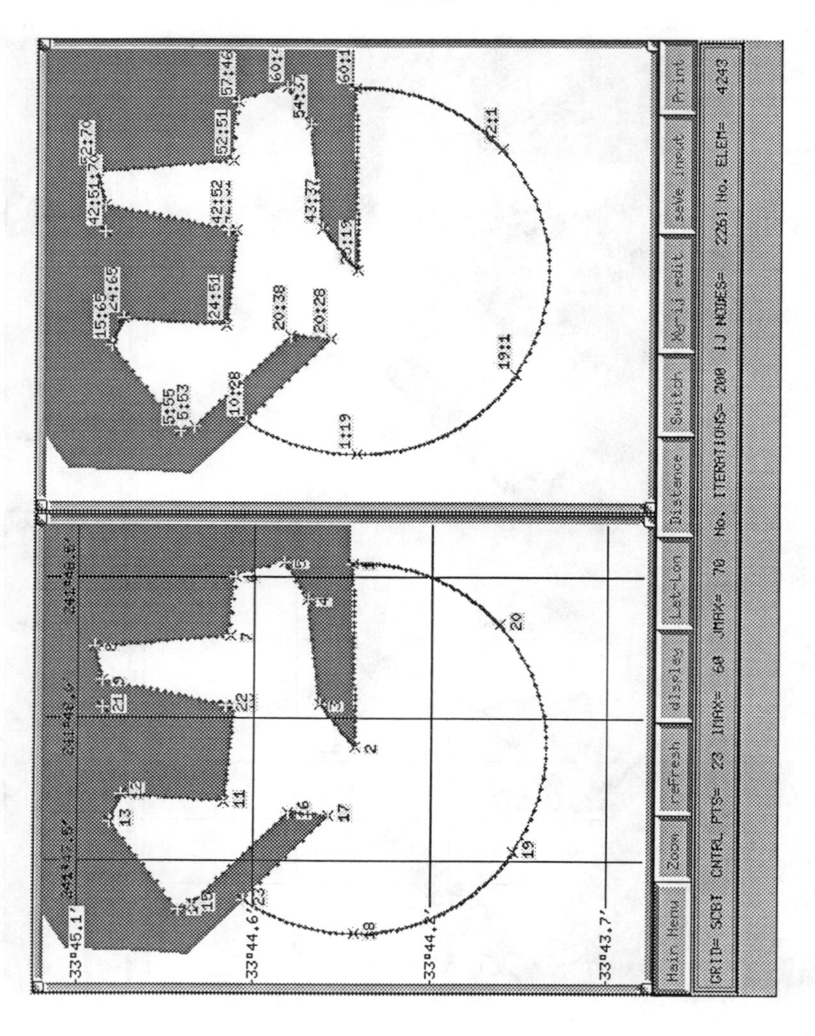

Figure 4: All 20 control nodes have been specified in both (x, y) and (I,J) planes ((I,J) coordinates are viewed in the xy-plane).

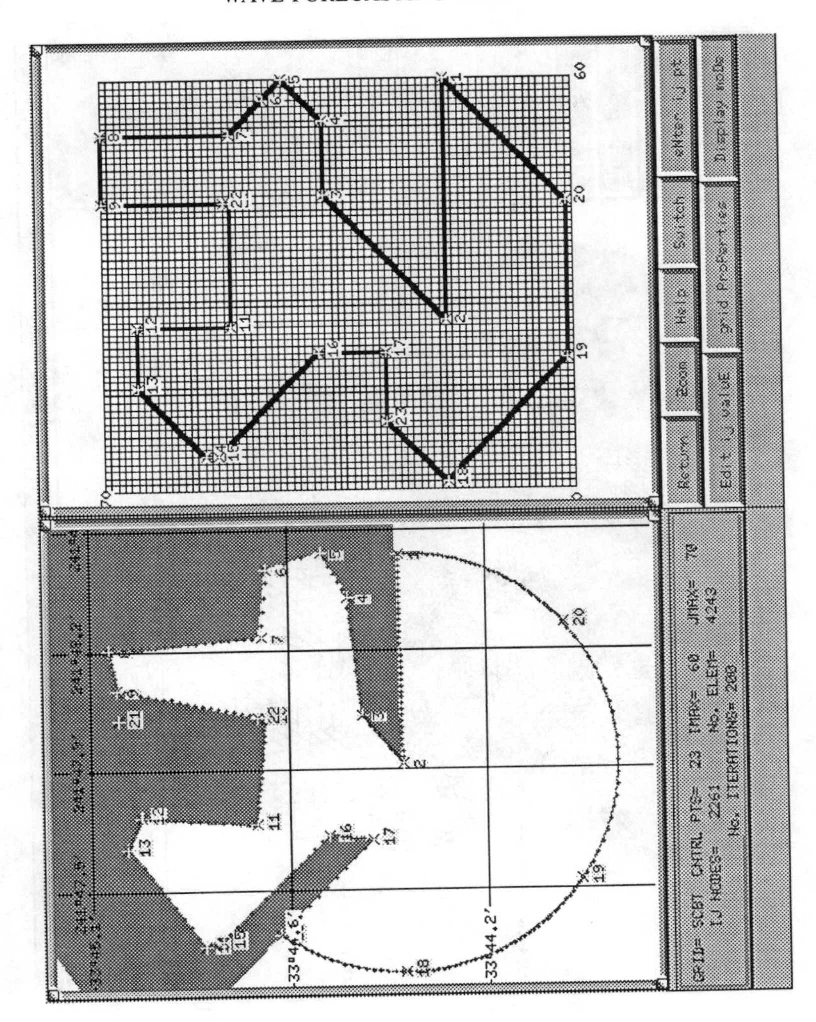

Figure 5: Same as Fig. 4, with (I,J) control nodes now viewed in the orthogonal plane.

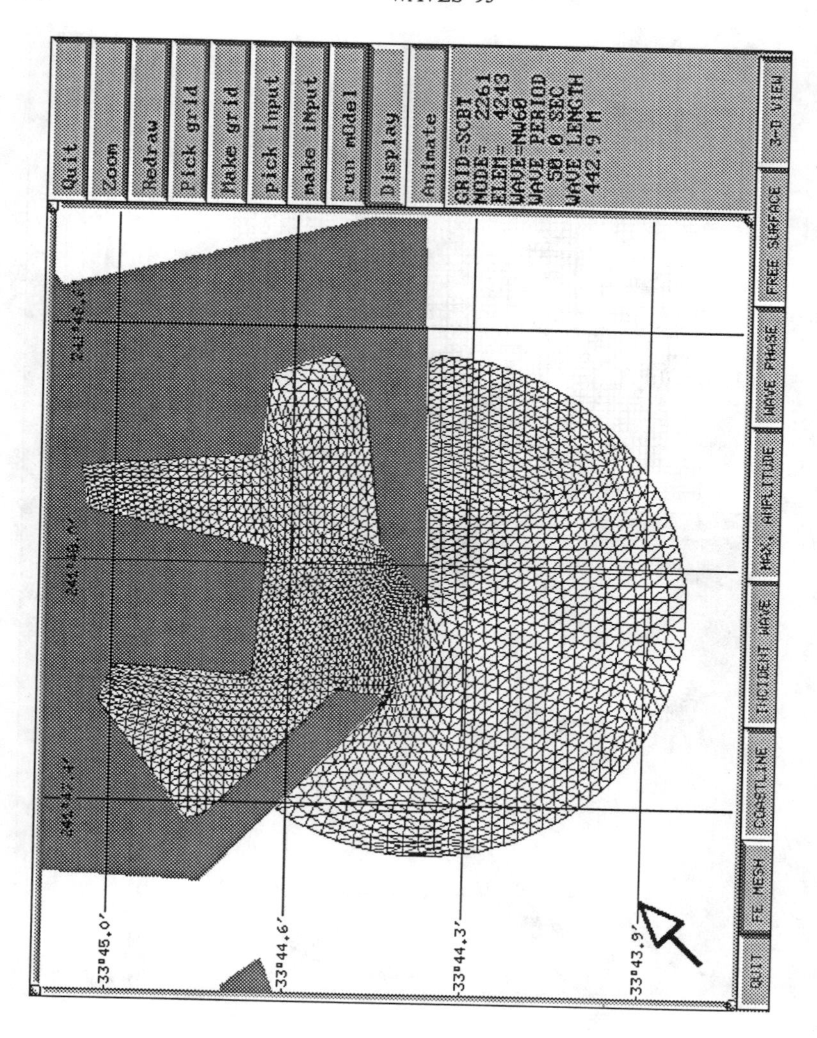

Figure 6: FEM grid generated based on the data in Figs. 4 and 5 (2261 nodes, 4243 triangular elements).

virtual memory storage on the PC hard disk when required by the size of the problem. Problems with discretizations on the order of 30,000 elements have been solved with reasonable response time (~30 min) on top-of-the-line 486-PC computers (66MHz, 24/200Mbytes). Under the present approach the maximum number of grid nodes, however, for a given RAM size, is only limited by the space available on the computer hard disk. The following results are provided at each grid node : (i) wave amplitude $A(x, y)$; (ii) wave phase $\varphi(x, y)$; and (iii) free surface elevation $\eta(x, y) = A \cos \varphi$.

Post-processing : Results are visualized and analyzed within the system. Predicted results, for each calculated incident wave set (f, θ), include : (i) Contour plots of $A(x, y)$ (Fig. 7); (ii) Contour plots of $\varphi(x, y)$; (iii) Contour plots of $\eta(x, y)$ (with animation) (Fig. 8); (iv) 3D perspective view of $\eta(x, y)$ (with animation) (Fig. 9).

Linear superposition of results for several wave frequencies and directions can be made and local (directional) spectrum calculated. Wave amplitude and currents can also be determined at each grid node, as a function of depth, and stored as multiple layers of information in the GIS data bases. Corresponding bottom shear stress is also available (e.g., required as input for wave induced sediment transport models). The model has been verified for selected case examples with simplified geometry (e.g., rectangular harbor of constant depth in Behrendt, 1985; Fig. 10). In the following, the model is applied to a typical case to predict harbor oscillations. This application illustrates the efficiency of the new approach for solving practical engineering problems.

Results and discussions

A complete calculation of harbor oscillations due to long waves is presented in the following, to illustrate the interactive graphic capabilities of ASA.WAVES.

Pre-processing.—Figs. 1-6 illustrate the pre-processing functions. In Figs. 1 and 2, a geographic working area (US West coast, Long Beach) is selected in an existing GIS coastline map. In Fig. 2, the GIS measuring tool (F3:Dist., in Fig. 1) is used to calculate a critical wave propagation distance (2.2km), and bearing (54.84deg.). In Figs. 3-5, the "Make grid" function of ASA.WAVES has been selected, and control nodes and grid generation commands are introduced. In Fig. 4, (I,J) coordinates are visualized in the (x, y) plane (right hand window) for each of the 20 control nodes defined in the physical plane (x, y) (in the left hand window). In Fig. 5, (I,J) nodes are seen in their corresponding orthogonal grid, with connecting segments, as in physical space. Fig. 6 shows the FEM grid generated based on the data in Figs. 4 and 5, with 2261 nodes, and 4243 triangular elements.

Processing.— Computations are carried out with the MSE program in the grid of Fig. 6 for an incident wave with $f = 0.02$Hz, $\theta = 60$deg., incident height 1m, and assuming a constant depth of 8m, both within and outside the harbor. The corresponding shallow water wave length is 443m. Computation time is on the order of 4min.

Post-processing.— Results are presented in Figs. 7-9, for this case. Wave amplitude

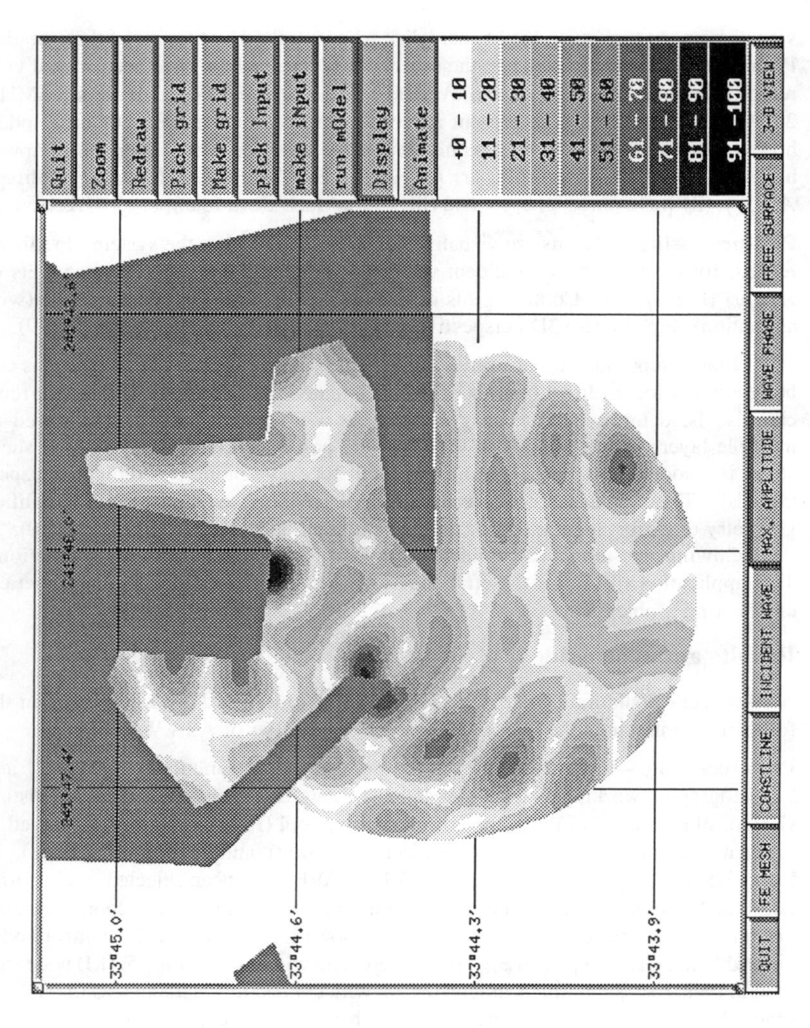

Figure 7: Contour lines of wave amplitude, A, calculated in the grid of Fig. 5, for $f = 0.02$Hz, $\theta = 60$deg, and 8m constant depth everywhere (scaled in percentage of the maximum amplitude, 7.24m).

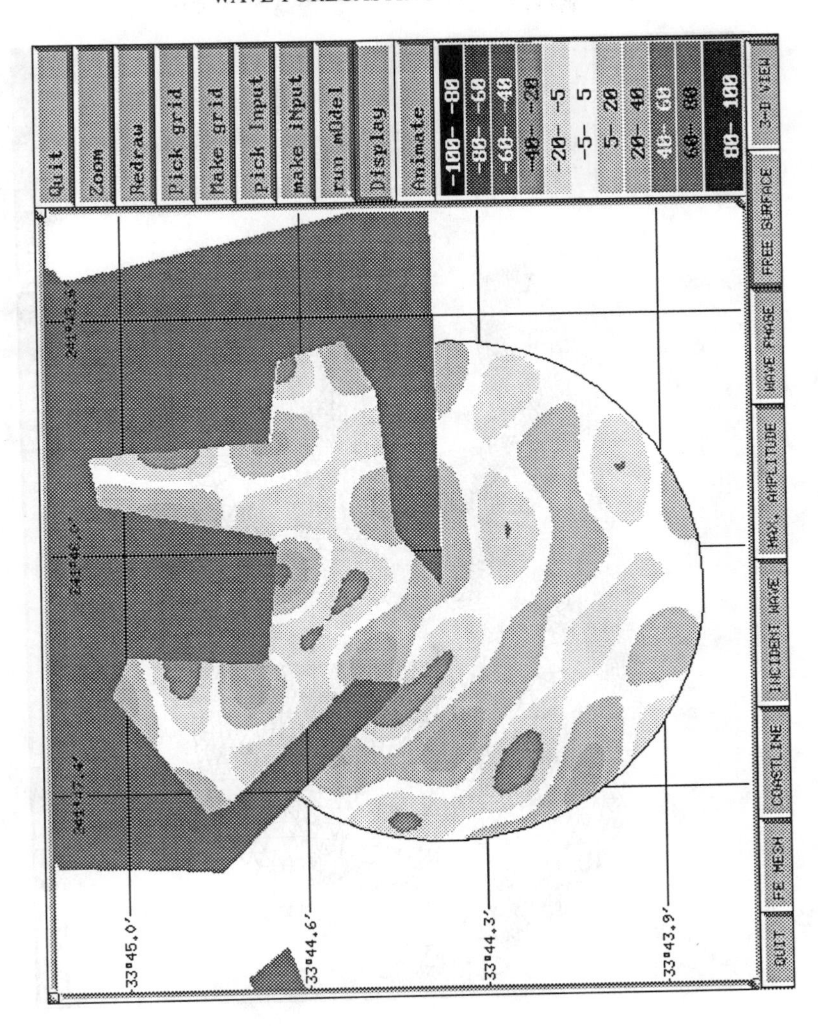

Figure 8: Contour lines of surface elevation, η, for the same case as in Fig. 7.

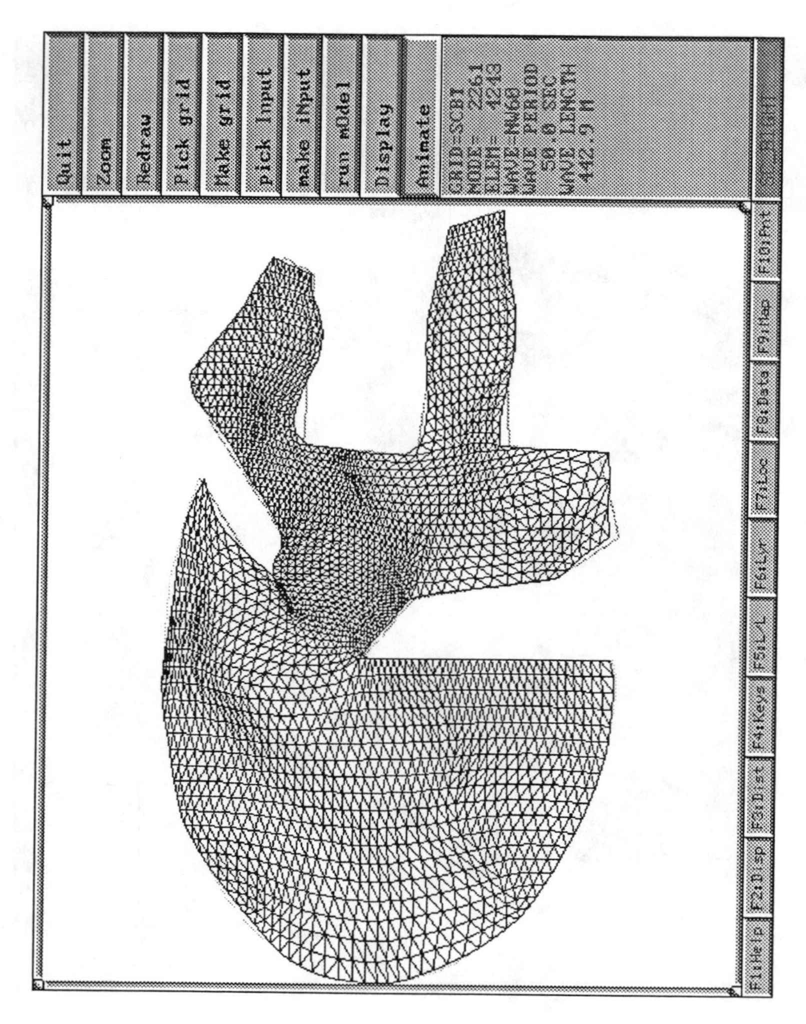

Figure 9: 3D view of wave elevation as in Fig. 7.

(a)

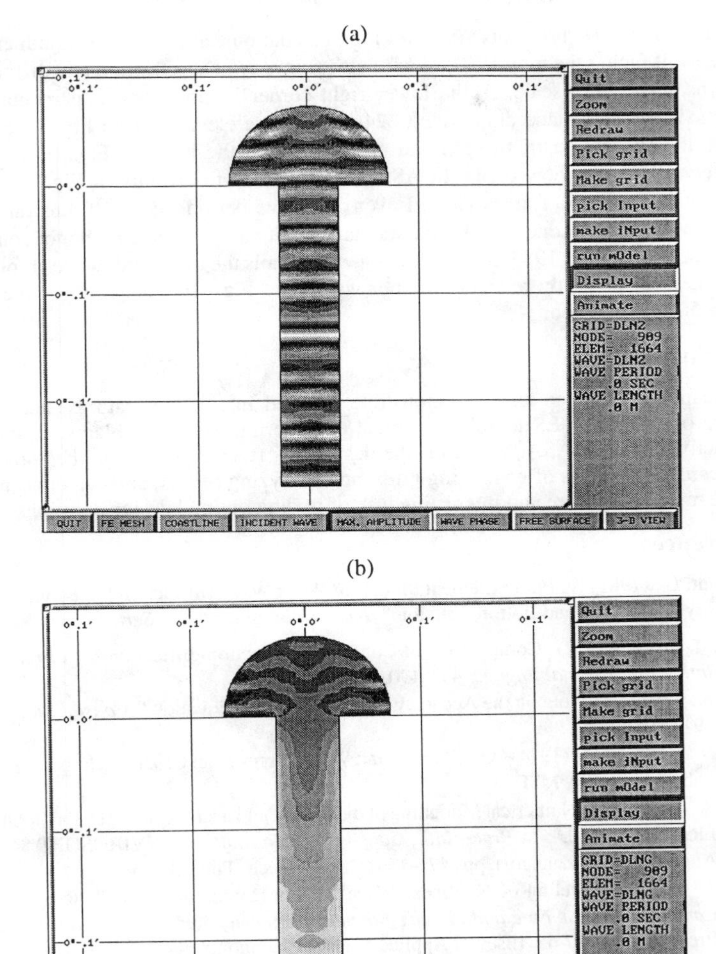

(b)

Figure 10: Contour lines of wave amplitude, A, for oscillations in a rectangular harbor with : (a) reflecting walls; (b) partially absorbing walls.

A in Fig. 7 shows "hot spots"(dark) of high oscillations, at the harbor mouth and on the seawall facing the mouth. The figure also shows typical reflection and diffraction patterns. The grey scale, in the lower right corner is graduated in percentage of the maximum calculated amplitude (7.24m). Surface elevation η in Fig. 8, clearly shows the succession of troughs and crests entering the harbor. Fig. 9 shows a very evocative 3D view of η. In ASA.WAVES, both the results in Fig. 8 and 9 are animated with 60 frames viewed over one wave period. Fig. 10 illustrates the absorbing boundary condition, for the standard case of a rectangular harbor of constant depth (see, Behrendt, 1985, for detail). Fig. 10a details the maximum wave amplitude calculated for perfectly reflecting harbor walls, and Fig. 10b shows the same case with a 50% absorption.

Conclusions

Applications presented above demonstrate selected interactive features, and capabilities of ASA.WAVES to solve practical engineering design problems. Using an interactive tool like ASA.WAVES's, the designer is freed from the cumbersome and time consuming tasks of generating grids, and analyzing results, and can concentrate on the more productive and interesting task of finding optimal design solutions.

References

Behrendt, L. (1985) "A Finite Element Model for Water Wave Diffraction Including Boundary Absorption and Bottom Friction." *Tech. Univ. of Denmark Series Paper No.* **37.**

Berkhoff, J.C.W (1972) "Computation of Combined Refraction-diffraction." In *Proc. 13th Intl. Conf. Coastal Engng.* 471-490, ASCE.

Booij, N. (1983) "A Note on the Accuracy of the Mild-slope Equation." *Coastal Engng.* **7,** 191-203.

Chen, N. & Mei, C.C. (1974) "Oscillations and Wave Forces in an Offshore Harbor." Parson Lab. Report **190**, MIT.

Grilli, S. (1986) "The Numerical Modelling of the Wave Field near the Shore by the Boundary Element Method." In *Proc. Intl. Conf. on Envir. Soft.* (ENVIROSOFT 86, Los Angeles)(ed. P. Zannetti), pp. 471-486. Comp. Mech. Pub., Boston.

Grilli, S. (1993) "Modeling of Nonlinear Wave Motion in Shallow Water." Chapter 3 in *Comp. Meth. for Free and Moving Boundary Problems* (eds. L.C. Wrobel & C.A. Brebbia), pps. 37-65, Elsevier Applied Sciences, London, UK.

Kirby, J.T. & Dalrymple, R.A. (1983) "A Parabolic Equation for the Combined Ref.-Dif. of Stokes Waves by Mildly Varying Topography." *J. Fluid Mech.* **136,** 453-466.

Kirby, J.T. (1989) "A Note on Parabolic Radiation Boundary Conditions for Elliptic Wave Calculation." *Coastal Engng.* **13,** 211-218.

Lejeune A., Marchal J., Hoffait, T., Grilli, S. & Sahloul, M. (1982) "Wave Actions on Floating Structures and Wave Propagation using Finite Element Method." In *Proc. 4th Intl. Conf. on Finite Elements in Water Res.* pp. 17.13-17.23. Springer-Verlag, Berlin.

Losada *et al.* (1990) "Descripcion Del Paquete de Programas de la Mild-slope". *Report,* Dept. de Ciencias y Tec. del Agua, Univ. of Cantabria, Spain.

Mei, C.C (1989) *The Applied Dynamics of Ocean Surface Waves (2nd ed.).* World Scientific.

JOINT PROBABILITY OF SUPERELEVATED WATER LEVELS AND WAVE HEIGHTS AT DUCK, NORTH CAROLINA

Douglas A. Gaffney[1]
Gregory L. Williams, P.E. Associate Member, ASCE[2]

ABSTRACT

Storm events are typically classified as a percent occurrence or return interval based on their peak storm stage elevation. The stage is the combination of storm surge, astronomical tide and other large spatial scale characteristics. Alternatively, wave heights are classified using the statistical probability of the entire range of wave conditions that one could expect to exist at a location. Storm waves are more locally affected and of a smaller spatial scale than the storm stage.

Intuitively, we know that storms, having superelevated water levels, occur simultaneously with increased wave heights, yet the joint occurrences of both parameters cannot be predicted with confidence based on empirical data. Wave forces and water levels are used in the design of coastal structures and the determination of the level of protection. However, the probabilistic combination of high energy waves and storm stage are not often considered jointly.

By considering the joint occurrence of both the water level and wave heights, a more complete idea of the actual recurrence interval of a storm

[1]USAE District, Philadelphia, 100 Penn Square East, Wanamaker Bldg., Philadelphia, PA 19107

[2]Coastal Engineering Research Center, USAE Waterways Experiment Station, 3909 Halls Ferry Rd., Vicksburg, MS 39180-6199

may be determined, as well as the level of protection of a shore protection project. Additionally, knowing the joint probability of occurrence can help define levels of protection in the case of emergency response measures in the aftermath of a storm. This paper addresses these concerns by investigating the joint probability of storm surge and wave heights.

INTRODUCTION

There are two major concerns which should be initially addressed relative to developing a robust joint probability distribution. First, which parameters are best used to describe a storm and secondly, once the parameters are chosen, what is their statistical dependency?

The desire to quantify the probability of a damaging storm event is a primary concern which is found to some extent in the literature, and is a concern in this paper. However, a simple, less practical search was carried out to find a possible relationship between two parameters which have apparent correlation, storm surge and wave height. These parameters can be thought of as representing the potential energy in the water during a storm.

One reason why larger waves would occur simultaneously with superelevated water is that deeper water can support higher waves. This is true at the nearshore. However, the research conducted for this paper used a wave record which is assumed not to be depth limited. The other, more general reason that larger waves accompany superelevated water is that wind causes setup and generates local wave conditions. Wind is therefore a potential causal mechanism which may imply statistical dependance of the observations.

Most of the literature which addresses the joint probability of these parameters focuses on evaluating the probability of storms which cause damage and how best to predict their destructive impact. An early investigation into the relationship of wave height and storm stage was conducted by Tancreto (1958). Before the advent of computers, Tancreto determined an insightful linear relationship between surge and wave height for Boston Harbor. This relationship was compared to an arbitrarily chosen month of data measured at the Field research Facility (FRF), Duck, NC. For the month of April 1988, a very high regression coefficient seemed to indicate that in fact, a linear relationship may exist. With this information in hand, a search for a joint probability distribution of wave heights and surge began.

The Dutch have been concerned with storm surge barriers for

centuries. In particular, dune erosion has been studied by Van de Graaff (1983 and 1986), who concluded that the dune erosion rate can be calculated using the following boundary conditions and determining factors. They are maximum storm stage (ie. astronomical tide plus storm surge), significant wave height, sediment particle diameter, shape of initial profile, storm stage duration, and squall oscillations and gust bumps. Of interest here is the use of the Weibull exceedance distribution for wave heights at any maximum stage level.

Van Aalst (1983) found maximum stage level versus significant wave height relationships along the Dutch coast. Van Aalst used methods in which the wind speed determined set-up as well as locally generated waves. Although nonlinear relationships can be derived, Van de Graaff (1986) realized that given a surge level, different significant wave heights can occur due to the compounding effects of wind direction and duration. Thus, variables which are seemingly dependant (due to wind speed) can also be statistically independent.

Vrijling and Bruinsma (1990) investigated the joint occurrences of waves and stage related to the design of storm surge barriers. That work proposed a three dimensional probability function of maximum storm stage, basin level behind the barrier, and wave energy. Statistical independence is assumed through the use of a conditional probability density function (PDF) of maximum storm stage level and local wind speed. The resulting PDF is simply the multiplication of two independent probabilities. Once again, however, the researchers found only loose correlation between storm stage and waves, presumably due to wave energy originating from two different sources.

The research which is presented in this paper had three main objectives. They were to 1) investigate the relationship between wave height and surge, 2) investigate the probability density functions of each variable, and 3) to determine if a robust joint probability distribution exists for surge and wave height at Duck, NC.

THE DATA

Water level and wave data were obtained from the Field Research Facility (FRF) of the US Army Engineer Waterways Experiment Station's (WES) Coastal Engineering Research Center (CERC) in Duck, NC. Water level data were obtained from the tide gage (National Ocean Service Station 865-1730) located at the seaward end of the 561 meter long FRF pier. This tide gage collects instantaneous water levels every six minutes. Wave data were obtained from the offshore wave rider buoy, FRF Gage 630, located six kilometers offshore in approximately 17 meters of depth. Gage 630 samples sea surface elevations at a frequency of 0.5 Hz for 34 minutes four times a day: 0100, 0700, 1300, and 1900 EST. During storm conditions, sampling is increased to every 3 hours. Wave heights used in this study were H_{m0}, which is an energy based statistic equal to four times the standard deviation of the sea surface elevations in each 34 minute record.

The measure of superelevated water to be used in this investigation is surge which is the difference between the expected or predicted water level and the actual measured water level. Using CERC's Automated Coastal Engineering System (ACES) software and standard tidal harmonics for Duck, NC, the predicted tidal curves were calculated. Surge was obtained by determining the difference between the predicted tidal record from the actual measurements from the NOS tide gage.

DATA EXPLORATION. H_{m0} was taken to be representative of the general wave climate and indicative of the total wave energy for each hour of record. As such, this parameter may contain energy from multiple sources such as seas or swell. H_{mo} was compared to the simultaneous superelevation of water. A time series plot was generated showing H_{m0} and surge for the entire month of April, 1988 as seen in Figure 1. This plot

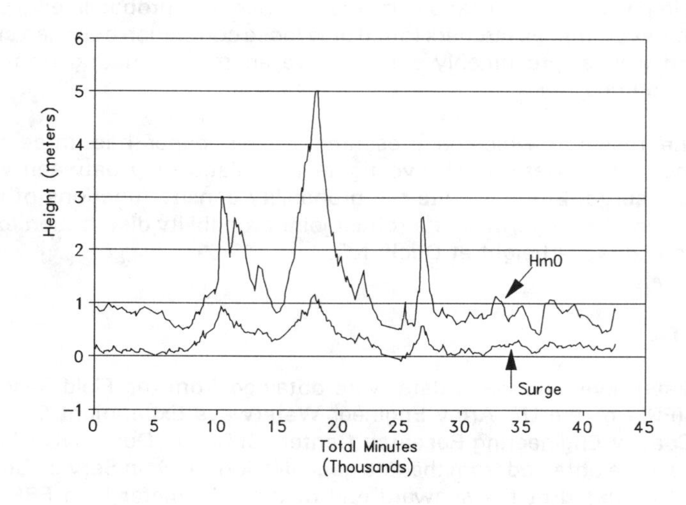

Figure 1

showed the apparent correlation of larger wave heights with superelevated water levels.

Linear Regression. The first data set consisted of hourly H_{m0} and surge for the month of April 1988. The data were highly correlated, and had an R^2 of 0.86 using linear regression (see Figure 2). The H_{m0} data fit most closely to the Rayleigh distribution while the surge data most closely resembled the normal distribution. The regression line results in the following equation in meters:

$$S=-0.028+0.255HmO \qquad \text{eq. (1)}$$

where S is the superelevation of water above the predicted astronomical tide and H_{m0} is the zeroeth moment of the wave time series record. There were 276 paired observations in this data set.

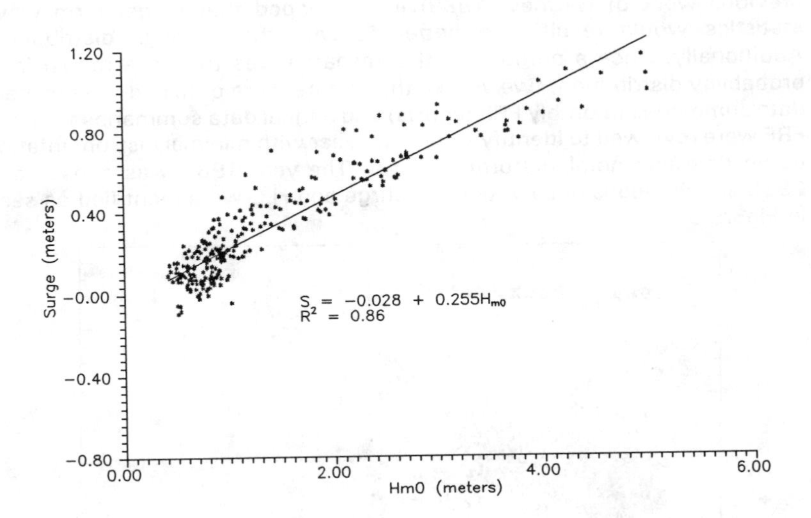

Figure 2

Comparing equation 1 to Tancreto (1958) which, in contrast, uses total water elevation during a storm, S, and computed significant wave height offshore, Hs, at the time of maximum surge at Boston, converted to meters:

$$S=0.073+0.11Hs \qquad \text{eq. (2)}$$

it can be seen that conceptually, and intuitively, wave heights increase as surge levels increase. Equation 2 is based on far fewer data points than equation 1. The y intercept is nearly zero in both equations. From Figure 2, it can be seen that the slope of the line would be smaller if non-storm observations were deleted. The good linear fit implies a dependence of H_{m0} on surge. The results of the contingency table Chi squared test however, indicate that independence of the parameters can not be rejected.

The wave height distribution of the April 1988 data resembled the Rayleigh curve but the goodness of fit was not irrefutable. From the previous work of Gaffney (1989) it was hoped that longer term wave statistics would result in a better fit with the Rayleigh distribution. Additionally, since a purpose of this research was to find a robust joint probability distribution, twelve months of data were obtained. Preliminary data summaries (monthly FRF reports) and annual data summaries from the FRF were reviewed to identify a calendar year with minimal instrumentation down time and notable storm activity. The year 1989 was chosen and 2938 simultaneous observations of surge and H_{mo} were identified as seen in Figure 3.

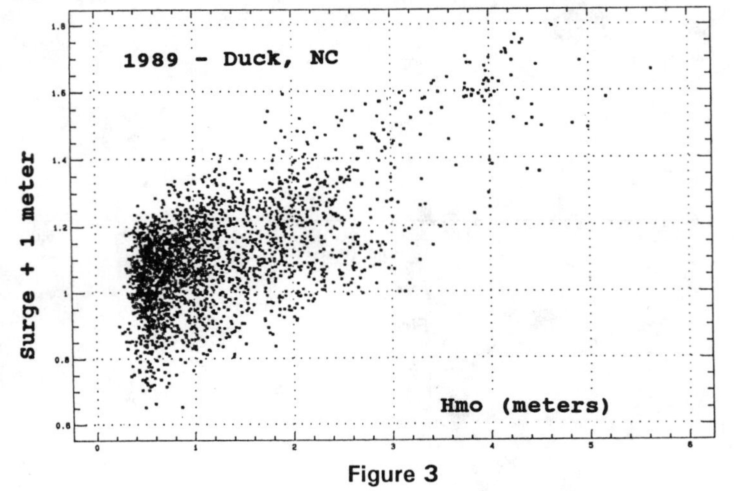

Figure 3

Since the water level is not always greater than the predicted astronomical tide, the surge value was sometimes negative. To facilitate statistical analysis, a new datum was arbitrarily chosen such that all surge values were positive. One meter was arbitrarily added to all surge values. The new surge value was matched with its corresponding H_{m0} to obtain a record of paired observations. Adding one meter to all surge values does not affect the relative distribution or correlation of parameters.

Correlation. From looking at the time series of wave height and surge, it was immediately obvious that this data (1989) showed less correlation than the month of April, 1988. The correlation coefficient for the April, 1988 data set was 0.93 while the correlation coefficient for the entire year of 1989 was 0.61.

Linear and 2nd Order Regression. The second data set had an R^2 of only 0.37 using linear regression, with a statistical significance level of zero. The results of second order polynomial regression were still disappointing with an R^2 of 0.46.

Based on the results thus far, storm events were separated from non-storm events for further analysis. Wave gage 630 records data every three hours when a different gage (625) at the end of the FRF pier measures a 2.0 meter H_{m0}. This definition of a storm event was used to separate the data. When all storm events were evaluated as a separate data set from all calm periods, no statistically significant relationship could be determined. However, individual storms were then evaluated with the finding that for the three storms with the highest H_{m0}, surge was highly correlated (R^2 of 0.86, 0.81 and 0.88). The slope of the regression line differed among the storms, possibly the function of a third storm parameter which was not accounted for, presumably duration.

PROBABILITY DISTRIBUTIONS

As stated earlier, it had been assumed that the data would fit reasonably well with established probability density functions. However, the 1989 H_{m0} did not satisfy a goodness of fit test for the Rayleigh or Weibull distributions and only fit the lognormal distribution for about half the surge categories. This was initially thought to be due to a second peak seen in the H_{m0} data between 3.75 and 4.25 meters which could imply a significant difference between storm and non-storm activity (see Figure 4).

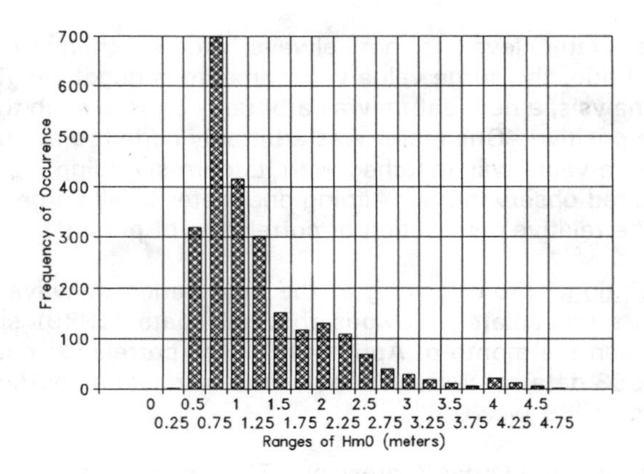

Figure 4

The surge data most closely resembled the normal distribution as seen in Figure 5, however did not pass the chi squared goodness of fit test. Similar to the wave record, the distribution showed a secondary peak near 0.6 meters. It should be noted that the range of surges in this study is relatively small. A longer period of record may have had different results.

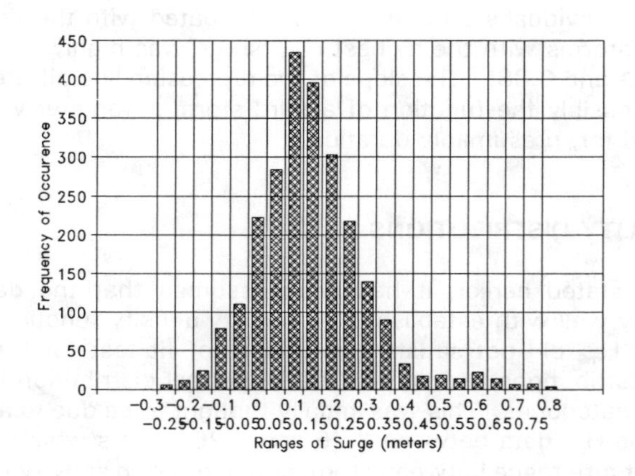

Figure 5

Since storm data is sampled twice as often as non-storm data, the possibility exists that the data set is skewed toward the higher surge and wave height values. Approximately 13.4 percent of the observations were considered storm conditions while 87.6 percent were non-storm. This may affect the significance of the small peaks in the tails of Figures 4 and 5.

A considerable body of research exists which provides rationale and data to support the use of several probability density functions for both waves and storm stage. Wave heights, especially significant wave height, is often said to follow the Rayleigh, Weibull, lognormal (St. Q. Isaacson and MacKenzie, 1981), and even Gaussian distributions. Even though the 1989 data did not fit any known distributions, including several others not mentioned, it could be *assumed* to fit based on earlier research if a valid joint probability distribution could be found. Therefore, the concept of determining a joint probability was still a reasonable goal.

JOINT OCCURRENCES

Storm stage recurrence is usually determined using a maximum exceedance distribution such as the Weibull and includes two independent occurrences, surge and astronomical tide. This research sought to remove the influence of tide, and focus on the superelevation of water caused by local climatic factors.

It would be counter-intuitive to assume that the joint occurrence of H_{m0} and surge is the simple combination of independently estimated values. This is because some dependance is assumed based on the physical environment. One method to determine the joint occurrence is to assume the extreme value of one parameter, and estimate the most probable condition associated with the other variable (Hogben, 1989).

APPLICABILITY. If the joint occurrence of the water level and wave height were quantifiable, the distribution could provide better predictability of recurrence intervals. In the past, the primary method of estimating the strength of a storm has been to examine the associated tidal records. The maximum tidal elevations are then compared against the historical stage frequency curve for the area to determine the recurrence probability. Measured wave heights, when available, are evaluated as an aside in the inevitable classification of the storm. However, throughout the tide cycle and associated storm surge, wave energy continues to impact the shoreline. High water level, when coupled with waves, has a very different damaging impact on structures and beaches than superelevated water alone. Therefore by jointly considering the probability of both, a truer estimation of storm's magnitude can be realized. For coastal structure design purposes, performance criteria can be better designated as well as verified.

CONDITIONAL PROBABILITY. The H_{m0} data were put into 23 categories of 0.25 meter increments from 0 to 5.75 meters. The surge data were put into 11 categories of 0.1 meter from 0.6 to 1.79 meters. The joint occurrence of H_{m0} and surge can be seen in Table 1 and Figure 6.

	SURGE GROUPS											
H_{m0} Groups	0.6-0.79	0.8-0.89	0.9-0.99	1.0-1.09	1.1-1.19	1.2-1.29	1.3-1.39	1.4-1.49	1.5-1.59	1.6-1.69	1.7-1.79	TOTAL
5.5-5.75										1		1
5.25-5.5										0		0
5.0-5.25									1	0		1
4.75-5.0								1	1	1		3
4.5-4.75								0	1	0		1
4.25-4.5							2	0	4	1	1	8
4.0-4.25							1	0	3	5	3	12
3.75-4.0						2	2	0	6	12	5	27
3.5-3.75						0	1	1	2	3	2	9
3.25-3.5					2	3	1	1	4	3		14
3.0-3.25				4	4	4	1	4	6			23
2.75-3.0				5	10	6	5	10	4			40
2.5-2.75			1	17	15	11	10	4	4			62
2.25-2.5			0	21	17	25	20	5	2			90
2.0-2.25			15	26	22	39	25	11	1			139
1.75-2.0		3	16	27	48	54	24	5	1			178
1.5-1.75		2	13	41	43	33	11	0	2			145
1.25-1.5		13	22	44	40	47	15	2				183
1.0-1.25	1	22	47	78	110	80	16	2				356
0.75-1.0	4	34	48	148	173	62	11	0				480
0.5-0.75	28	55	166	262	244	42	2	0				799
0.25-0.5	15	28	83	149	78	12		1				366
0.0-0.25			1									1
TOTAL	48	157	412	822	806	420	147	47	42	26	11	2938

Table 1

As an example, if the peak surge level is 0.7 meters (1.7 minus the added 1 meter), the conditional probability distribution tells us that the most probable H_{m0} is between 3.75 and 4.0 meters. The design engineer could then assume a PDF and arrive at a maximum wave height. Of course, a longer term data set would provide more information from which to extrapolate design parameters.

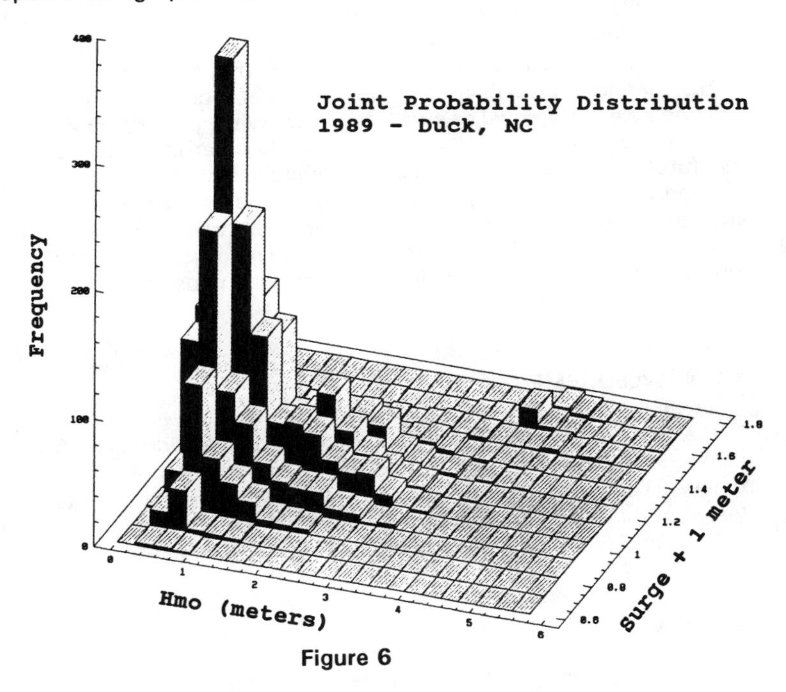

Figure 6

CONCLUSIONS

The data does not give one the sense that assumptions regarding commonly used probability distributions are valid for either wave heights or surge, based on one year of data. Marginal distributions did not show goodness of fit to standard distributions, therefore, the conditional probability approach seems to be the most valid method at this time to determine joint occurrences of superelevated water and wave heights.

Dependence of the paired variables, H_{m0} and surge for 1989, was not statistically shown. Additionally, independence of the variables was not demonstrated yet cannot be rejected. The joint probability distribution derived by multiplying independent probabilities for storm surge level, wave energy and basin level by Vrijling and Bruinsma (1990) was not thoroughly tested since statistically significant PDFs could not be found. There is an apparent lack of research aimed at addressing this problem on the U.S. East Coast.

While it did not appear that the combination of storm and non-storm data adversely impacted the success of the analysis, the inclusion of sea and swell inherent in H_{m0} may have been. Perhaps the fatal flaw in the use of H_{m0} for a joint probability distribution is that it combines the energy from sea and swell. Removal of swell, based on wave period alone, subsequent to the generation of the H_{m0} statistic, would be arbitrary at best. Perhaps the future direction of this research should include wind direction as a method to remove swell, as a third input to an energy based calculation, and as a potential causal mechanism between wave heights and superelevated water levels. Other potential research directions include the removal of negative surges since they may be caused by offshore winds, and the inclusion of storm duration.

ACKNOWLEDGEMENTS

The authors wish to acknowledge Mr. Bill Birkmeier, Chief of the Field Research Facility, for the data sets, technical support and his thorough review. Thanks to Dr. Ed Thompson for his helpful comments during the design of this research, and to Dr. Jan Van de Graaff of the Delft University of Technology for providing several valuable references.

REFERENCES

Aalst, W. van, "Wave Height-Surge Relationships for the Dutch Coast" WWKZ-83G.218; Rijkswaterstaat, the Hague, the Netherlands.

Gaffney, D.A., "A Method to Determine Optimal Locations for a Seawave Powered Desalinator," Master's Thesis, University of Delaware, 1989.

Graaff, J. van de, "Probabilistic Design of Dunes," Coastal Structures '83. 1983.

Graaff, J. van de, "Probabilistic Design of Dunes; An Example from the Netherlands," Coastal Engineering. 9, 1986.

Hogben, N., "Long Term Wave Statistics", Chapter for the Book Ocean Engineering Science: The Sea, Vol. 9. 1989.

St. Q. Isaacson, M. and N. G. MacKenzie, "Long Term Distributions of Ocean Waves: A Review", Journal of Waterway, Port, Coastal and Ocean Engineering, Vol. 107, No. WW2, May 1981.

Tancreto, A.E., "A Method for Forecasting the Maximum Surge at Boston Due to Extratropical Storms," Monthly Weather Review Vol. 86, Number 6, June 1958.

Vrijling, J.K. and Bruinsma, J., "Hydraulic Boundary Conditions Related to the Design of the Oosterschelde Storm Surge Barrier in the Netherlands: An Example of a Joint Distribution of Waves and Surges," Proceedings on Hydraulic Aspects of Coastal Structures Part I: Developments in Hydraulic Engineering, Delft University Press, Delft, 1990.

ESTIMATING LABORATORY WAVE REFLECTION USING LASER DOPPLER

Steven A. Hughes[1]

Abstract: Estimates of irregular wave reflection in two-dimensional laboratory wave tanks are typically accomplished using an array of spatially separated wave gauges. This paper presents a frequency domain analysis method that utilizes synoptic time series of either: (1) horizontal water velocity and sea surface elevation collected in a vertical array, or (2) horizontal and vertical water velocities collected at the same location in the water column.

The *"co-located gauge method"* was tested using laboratory measurements obtained using a two-component laser Doppler velocimeter. Estimates of incident and reflected wave spectra compared well to similar estimates derived from spatially separated wave gauges. However, reliable estimates are limited to the range of spectral frequencies having appreciable energy and exhibiting good coherence in the cross-spectrum of the time series signals. Outside this range, improbable results occur.

Also in this paper, the co-located gauge reflection analysis method is extended to the quasi-three-dimensional case of unidirectional, long-crested irregular waves being obliquely reflected by a straight reflective surface. This extension is based on the assumption that the oblique angle of wave incidence is known a priori.

1 Introduction

Waves in the nearshore region are reflected by beaches, coastal structures, and floating or submerged solid bodies. The reflected waves interact with the incident waves and contribute to the overall characteristics of the wave and flow fields. Physical models of the coastal region reproduce these wave reflections, with additional complications arising from model boundary reflections and re-reflection of waves by non-absorbing wave boards.

[1]Research Hydraulic Engineer, US Army Engineer Waterways Experiment Station, Coastal Engineering Research Center, 3909 Halls Ferry Road, Vicksburg, MS 39180-6199.

Field and laboratory techniques have been developed to resolve incident and reflected wave spectra using synoptic measurements from multiple gauges. These reflection analyses allow investigators to relate coastal process response to characteristics of the incident wave spectrum.

Generally, the reflection analysis methods utilize either spatially separated wave gauges, or combinations of wave gauge and current meter configured in a vertical array. Several of these methods for irregular waves are briefly summarized in the following sections.

1.1 Spatially Separated Gauge Methods

Goda and Suzuki (1976) introduced an irregular wave analysis technique that uses synoptic time series of sea surface elevations from two wave gauges separated by a short distance on a line parallel to the direction of wave travel. This frequency-domain, linear theory method assumes a horizontal bottom, and it is recommended that the wave gauge nearest to the reflective surface must be located at least one wavelength away from the surface in order to resolve incident and reflected spectra. In laboratory practice, it is best to stay at least several wavelengths from the structure and even farther away from nonabsorbing wave boards.

Mansard and Funke (1980) presented a linear theory, least-squares method for separating incident and reflected waves from synoptic time series recorded at three wave gauges with known spacing. In their frequency-domain method, auto- and cross-spectral analyses of gauge pairs result in estimates of incident and reflected wave energy and a noise signal. A least-squares procedure is used to minimize the noise signal for all three gauges. This method assumes the gauges are placed over a flat bottom, and it provides improved stability in the estimates and an increased range of frequency resolution with less sensitivity to gauge spacing.

Field application of Mansard and Funke's least-squares method was described by Nelson and Gonsalves (1990). They applied the technique to resolve incident and reflected infragravity wave spectra using wave data obtained in the surf zone.

Gaillard, et al. (1980) described a linear analysis technique that is similar to that of Mansard and Funke (1980), and Maoxiang and Zhenquan (1988) presented two methods for estimating second-order long waves in a wave tank.

Most laboratories presently employ either the *Goda and Suzuki Method* or the *Mansard and Funke Method* for estimating incident and reflected spectra in two-dimensional wave flumes, and the Mansard and Funke Method is considered to be a little more accurate.

1.2 Co-Located Gauge Methods

Several investigators have described techniques for resolving incident and reflected two-dimensional waves using a wave gauge (or pressure gauge) and current meter located on the same vertical line in the water column.

Guza, et al. (1984) developed a time domain procedure for resolving nondispersive **long wave** incident and reflected wave trains. Synoptic values of horizontal water velocity and sea surface elevation are combined to form new time series of incident and reflected long waves. These wave trains can then be Fourier transformed to give incident and reflected long-wave energy spectra.

Tatavarti, et al. (1988) developed a frequency domain analysis procedure that estimates the frequency dependent wave reflection coefficient in terms of the gain and the phase relationships between synoptic measurements of sea surface elevation and horizontal velocity time series collected from a vertical array.

Walton (1992) applied a frequency domain method to time series data obtained from co-located pressure gauge/current meter combinations located outside the surf zone at two natural beach sites. His results showed small reflection in the wind wave frequencies.

It appears that the co-located gauge methods for resolving incident and reflected irregular waves have not seen much application in the laboratory. This may stem from that fact that reliable low-cost current meters have only recently become available.

2 Reflection Analysis From Co-Located Gauges (2-D)

Hughes (1993) presented a frequency domain analysis for resolving incident and reflected irregular wave spectra using either (1) synoptic time series of sea surface elevation, $\eta(t)$, and horizontal velocity, $u(t)$, collected from a wave gauge and current meter located in a vertical array; or (2) synoptic time series of horizontal velocity, $u(t)$, and vertical velocity, $v(t)$, collected at the same point in the water column. This method most closely resembles the method of Goda and Suzuki (1976), although it is similar to the methods presented by Tatavarti, et al. (1988) and Walton (1992).

Derivation of the co-located gauge reflection analysis method was given by Hughes (1993) and is summarized below. The reflection analysis method applies irregular first-order wave theory to analyze two-dimensional, irregular, nonbreaking waves propagating over a horizontal bottom. The waves are assumed normally incident to the reflective structure (i.e., wave tank studies).

2.1 Basic Linear Equations

Based on the coordinate system shown in Figure 1, the first-order velocity potential for irregular, linearly superimposed reflected waves in two dimensions on a flat bottom can be written as

$$\phi(x,z,t) = \sum_{i=1}^{\infty} \frac{-a_{I_i} g}{\sigma_i} \frac{\cosh k_i(h+z)}{\cosh(k_i h)} \sin(k_i x - \sigma_i t + \epsilon_{I_i})$$

$$+ \sum_{i=1}^{\infty} \frac{a_{R_i} g}{\sigma_i} \frac{\cosh k_i(h+z)}{\cosh(k_i h)} \sin(k_i x + \sigma_i t + \epsilon_{R_i}) \quad (1)$$

where

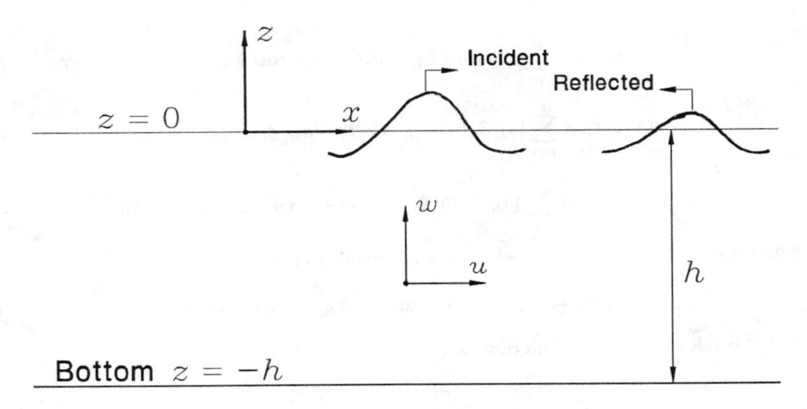

Figure 1: Coordinate system and gauge configuration for 2-d reflection analysis.

a_{I_i}	–	amplitude of the ith incident wave component
a_{R_i}	–	amplitude of the ith reflected wave component
k_i	–	wavenumber of the ith component
σ_i	–	angular wave frequency of the ith component [$= 2\pi/T_i$, where T_i is wave period of the ith component]
ϵ_{I_i}	–	phase angle of the ith incident wave component
ϵ_{R_i}	–	phase angle of the ith reflected wave component
x	–	horizontal coordinate
h	–	water depth
z	–	vertical coordinate with $z = 0$ at still water level and $z = -h$ at bottom
t	–	time
g	–	gravitational acceleration

The first term in Eqn. 1 represents the incident wave components propagating in the positive x-direction, and the second term represents the partially reflected wave components moving in the opposite direction.

Time series expressions for sea surface elevation (η), horizontal components of water velocity (u and v), and vertical component of water velocity (w) are found from the velocity potential as

$$\eta = \frac{1}{g}\left[\frac{\partial\phi}{\partial t}\right]_{z=0} \qquad u = -\frac{\partial\phi}{\partial x} \qquad v = -\frac{\partial\phi}{\partial y} \qquad w = -\frac{\partial\phi}{\partial z} \tag{2}$$

Substituting Eqn. 1 into the expressions of Eqn. 2 yields the following equations for incident and reflected irregular linear waves in two dimensions:

$$\eta(x,t) = \sum_{i=1}^{\infty} [a_{I_i} \cos(\Phi_{I_i} - \sigma_i t) + a_{R_i} \cos(\Phi_{R_i} + \sigma_i t)] \tag{3}$$

$$u(x,z,t) = \sum_{i=1}^{\infty} [a_{I_i} Z_i \cos(\Phi_{I_i} - \sigma_i t) - a_{R_i} Z_i \cos(\Phi_{R_i} + \sigma_i t)] \tag{4}$$

$$w(x,z,t) = \sum_{i=1}^{\infty} [a_{I_i} Y_i \sin(\Phi_{I_i} - \sigma_i t) - a_{R_i} Y_i \sin(\Phi_{R_i} + \sigma_i t)] \tag{5}$$

where the incident and reflected phases are defined as

$$\Phi_{I_i} = (k_i x + \epsilon_{I_i}) \quad \text{and} \quad \Phi_{R_i} = (k_i x + \epsilon_{R_i}) \tag{6}$$

and the velocity transfer functions are given by

$$Z_i = \frac{g k_i}{\sigma_i} \frac{\cosh k_i(h+z)}{\cosh(k_i h)} \tag{7}$$

$$Y_i = \frac{g k_i}{\sigma_i} \frac{\sinh k_i(h+z)}{\cosh(k_i h)} \tag{8}$$

2.2 Solution Using Wave Gauge and Horizontal Velocity

The time series representations for sea surface (Eqn. 3) and horizontal velocity (Eqn. 4) can be expanded and rearranged into the forms[2]

$$\eta(x,t) = \sum_{i=1}^{\infty} [A_i \cos(\sigma_i t) + B_i \sin(\sigma_i t)] \tag{9}$$

and

$$u(x,z,t) = \sum_{i=1}^{\infty} [C_i \cos(\sigma_i t) + D_i \sin(\sigma_i t)] \tag{10}$$

where the Fourier coefficients in Eqns. 9 and 10 are given by the expressions

$$A_i = a_{I_i} \cos(\Phi_{I_i}) + a_{R_i} \cos(\Phi_{R_i}) \tag{11}$$

$$B_i = a_{I_i} \sin(\Phi_{I_i}) - a_{R_i} \sin(\Phi_{R_i}) \tag{12}$$

$$\frac{C_i}{Z_i} = a_{I_i} \cos(\Phi_{I_i}) - a_{R_i} \cos(\Phi_{R_i}) \tag{13}$$

$$\frac{D_i}{Z_i} = a_{I_i} \sin(\Phi_{I_i}) + a_{R_i} \sin(\Phi_{R_i}) \tag{14}$$

and the velocity transfer function, Z_i, is given by Eqn. 7.

[2] Hughes(1993) also included a time lag in Eqn. 3 to allow for measurement systems that do not collect precisely synoptic samples from the two gauges, but that extension is not included here.

Equations 11 – 14 are solved for the four unknowns (a_{I_i}, a_{R_i}, Φ_{I_i}, and Φ_{R_i}) for each frequency component. These unknowns are expressed in terms of the Fourier coefficients (A_i, B_i, C_i, and D_i) obtained via fast-Fourier transform from the two time series represented by Eqns. 9 and 10. The solution is given as

$$a_{I_i} = \frac{1}{2}\sqrt{\left(A_i + \frac{C_i}{Z_i}\right)^2 + \left(B_i + \frac{D_i}{Z_i}\right)^2} \tag{15}$$

$$a_{R_i} = \frac{1}{2}\sqrt{\left(A_i - \frac{C_i}{Z_i}\right)^2 + \left(B_i - \frac{D_i}{Z_i}\right)^2} \tag{16}$$

$$\Phi_{I_i} = \tan^{-1}\left[\frac{\left(B_i + \frac{D_i}{Z_i}\right)}{\left(A_i + \frac{C_i}{Z_i}\right)}\right] \tag{17}$$

$$\Phi_{R_i} = \tan^{-1}\left[\frac{-\left(B_i - \frac{D_i}{Z_i}\right)}{\left(A_i - \frac{C_i}{Z_i}\right)}\right] \tag{18}$$

Equations 15 – 18 give the incident and reflected wave amplitude and phase spectral components at each frequency. Solving the set of equations over the frequency range of interest provides estimates of discrete incident and reflected wave amplitude and phase spectra.

2.3 Solution Using Horizontal and Vertical Velocity

A similar solution is found for the case where synoptic time series of horizontal and vertical velocity are collected at the same location. Expanding and rearranging the vertical velocity time series (Eqn. 5) yields the form

$$w(x, z, t) = \sum_{i=1}^{\infty} [E_i \cos(\sigma_i t) + F_i \sin(\sigma_i t)] \tag{19}$$

where the Fourier coefficients in Eqn. 19 are given by the expressions

$$\frac{E_i}{Y_i} = a_{I_i} \sin(\Phi_{I_i}) - a_{R_i} \sin(\Phi_{R_i}) \tag{20}$$

$$\frac{F_i}{Y_i} = -a_{I_i} \cos(\Phi_{I_i}) - a_{R_i} \cos(\Phi_{R_i}) \tag{21}$$

and the velocity transfer function, Y_i, is given by Eqn. 8.

Equations 13, 14, 20, and 21 can be solved for the four unknowns (a_{I_i}, a_{R_i}, Φ_{I_i}, and Φ_{R_i}) at each frequency component. These unknowns are expressed in terms of the Fourier coefficients (C_i, D_i, E_i, and F_i) obtained via fast-Fourier transform from the two time series represented by Eqns. 10 and 19. The solution is

$$a_{I_i} = \frac{1}{2}\sqrt{\left(\frac{D_i}{Z_i} + \frac{E_i}{Y_i}\right)^2 + \left(\frac{C_i}{Z_i} - \frac{F_i}{Y_i}\right)^2} \tag{22}$$

$$a_{R_i} = \frac{1}{2}\sqrt{\left(\frac{D_i}{Z_i} - \frac{E_i}{Y_i}\right)^2 + \left(\frac{C_i}{Z_i} + \frac{F_i}{Y_i}\right)^2} \tag{23}$$

$$\Phi_{I_i} = \tan^{-1}\left[\frac{\left(\frac{D_i}{Z_i} + \frac{E_i}{Y_i}\right)}{\left(\frac{C_i}{Z_i} - \frac{E_i}{Y_i}\right)}\right] \tag{24}$$

$$\Phi_{R_i} = \tan^{-1}\left[\frac{-\left(\frac{D_i}{Z_i} - \frac{E_i}{Y_i}\right)}{\left(\frac{C_i}{Z_i} + \frac{F_i}{Y_i}\right)}\right] \tag{25}$$

Estimates of incident and reflected wave amplitude spectra and phase spectra arise from solving Equations 22 – 25 over the frequency range of interest.

2.4 Application of the 2-D Co-Located Gauges Method

2.4.1 Limitations

There are a few practical limitations that must be considered when applying the co-located gauges method for resolving incident and reflected wave spectra:

1. The time series of velocities must be a strong signal that is well above the noise level for the instrument. Consequently, location of the velocity gauge has the following considerations:

    - The depth of the velocity gauge placement determines how much of the high frequency spectrum can be resolved.
    - The procedure that utilizes the vertical velocity component should not be applied when the velocity gauge is placed near the bottom because the vertical signal will be too weak. In this situation use the "$\eta - u$" gauge combination.
    - Also, resolution of long-wave frequencies can only be done using the $\eta - u$ gauge arrangement because vertical velocities are small in long waves.

2. The reflection analysis is only valid over the frequency range that contains appreciable variance in the spectra of the measured time series. Outside this frequency range, bizarre results can occur.

3. The coherence of the cross-spectrum of the two measured time series is a good measure of confidence. Reflection estimates for frequencies exhibiting high coherence are more reliable.

The limitations given by items 2 and 3 also apply to the spatially separated wave gauge methods, as noted by Mansard and Funke (1980) and Tatavarti, et al. (1988).

Failure to adhere to the above restrictions will result in erroneous estimates at some frequencies. These errors are sometimes evident from inspection of the resolved spectra or reflection coefficient as a function of frequency. However, errors in values of bulk parameters, such as incident and reflected significant wave height (H_{mo}) or overall reflection coefficient, may not be apparent (particularly if the analysis procedure is automated).

2.4.2 Example Application

The upper plot in Figure 2 shows horizontal and vertical velocity spectra calculated from time series collected in a laboratory wave flume with a laser Doppler velocimeter. The irregular waves were being reflected by a 1:30 impermeable slope, and velocity data were collected over the flat-bottom portion of the wave tank at approximately the same location as an array of three spatially separated wave gauges. Further details of the experiment setup and data collection are given in Sultan and Hughes (1993).

The velocity spectra indicate that there is minimal variance below 0.5 Hz or above 1.6 Hz. Cross-spectral analysis of the velocity time series produced good coherence (middle plot of Figure 2) between the frequencies 0.5–1.6 Hz, which corresponds approximately to the frequency range containing appreciable spectral energy. Therefore, we should expect reasonable estimates within this frequency range.

Calculated incident and reflected wave amplitude spectra are shown on the bottom plot of Figure 2 over the frequency range 0.0–2.5 Hz. The analysis procedure estimated significant energy for frequencies above 1.6 Hz, and these values are obviously suspect because no variance at these frequencies was evident in the measured velocity spectra. Therefore, estimated values above 1.6 Hz should not be included in any estimate of bulk parameters. Likewise, spectral estimates below 0.5 Hz also show energy that didn't exist in the original spectra, and thus, it is concluded that this is also an artifact of the analysis.

In this particular example an overall bulk reflection coefficient[3] of 11% was determined using only the frequency range between 0.5–1.6 Hz, whereas the coefficient increased to 63% when the entire frequency range was considered. Although this is an extreme example, it serves to illustrate the error that can arise from automated calculation of reflection without consideration of the coherence function.

Additional examples of the co-located velocities reflection analysis method are given in Hughes (1993) for the case of irregular wave reflection by a vertical wall. In that paper it was shown that reliable estimates could be made using velocity data collected quite close to the reflective surface (well within one wavelength).

2.4.3 Comparison to the Goda and Suzuki Method

Additional data from the mild-slope irregular wave tests reported in Sultan and Hughes (1993) were analyzed for incident and reflected spectra using the co-located gauge method and the method of Goda and Suzuki (1976). Figure 3 compares values of incident significant wave height for forty-one cases where velocity data were obtained at mid-depth in the vicinity of the three-gauge "Goda array." Equivalence is indicated by the solid line. The frequency range over which the co-located gauge analysis was performed was determined by inspection of the velocity spectra and the coherence function between velocity signals. The spatially-separated wave gauge analysis was also limited to this same frequency range. Generally, the comparison shown by Figure 3 is judged to be good with the Goda and Suzuki method giving slightly higher estimates of incident H_{mo} for the conditions tested.

[3]The bulk reflection coefficient is defined as $\sqrt{E_R/E_I}$, where E_I and E_R represent the total spectral energy contained in the incident and reflected spectra over the specified frequency range.

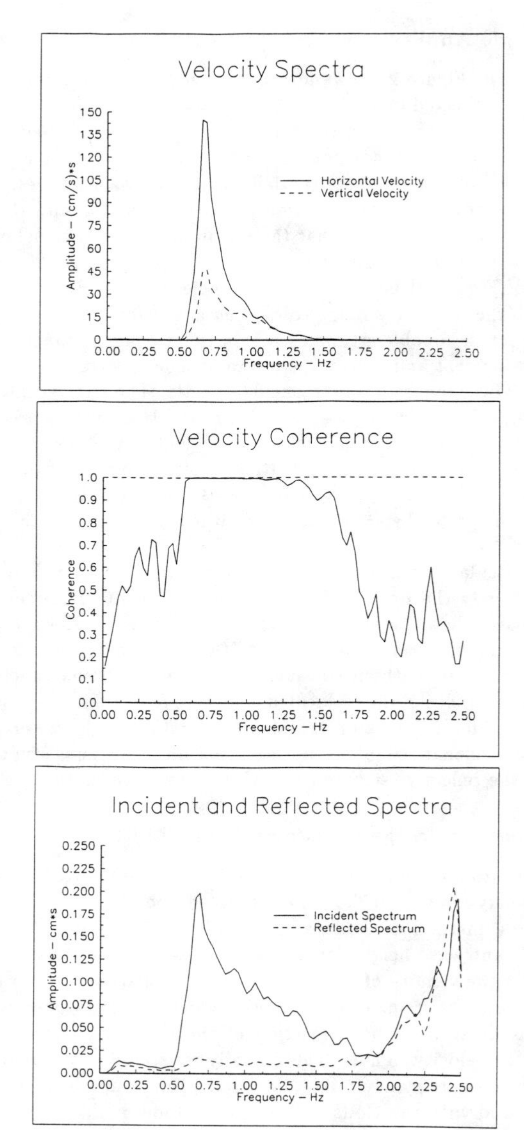

Figure 2: Reflection analysis of irregular waves on a mild slope.

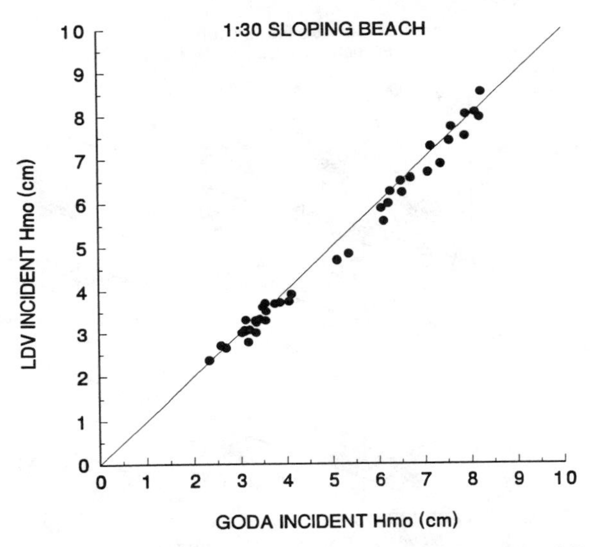

Figure 3: Comparison of co-Located velocities method and Goda and Suzuki method.

3 Oblique Reflection Analysis From Co-Located Gauges

The two-dimensional reflection analysis methods using co-located gauges can be extended to cover the case of irregular, unidirectional, long-crested waves being obliquely reflected by a straight structure as illustrated in Figure 4. Previously, Isaacson (1991) had outlined a procedure for applying a similar extension for the spatially-separated wave gauge methods.

The analysis of oblique wave reflection using co-located gauges can be performed using either:

1. Synoptic time series of sea surface elevation, $\eta(t)$, and the y-direction horizontal velocity, $v(t)$, collected in a vertical array.
2. Synoptic time series of the x-direction and y-direction horizontal velocities, $u(t)$ and $v(t)$, respectively, collected at the same point.
3. Synoptic time series of vertical velocity, $w(t)$, and the y-direction horizontal velocity, $v(t)$, collected at the same point.

Due to space limitations, only the first two cases are developed in this paper. Derivation of the third case proceeds along the same lines.

Development of the oblique reflection analysis method is based on irregular, first-order wave theory applied to nonbreaking, unidirectional, long-crested irregular waves propagating over a horizontal bottom. In addition, we assume that the angle of oblique wave incidence, θ, is known and the angle of reflection is the same as angle of incidence.

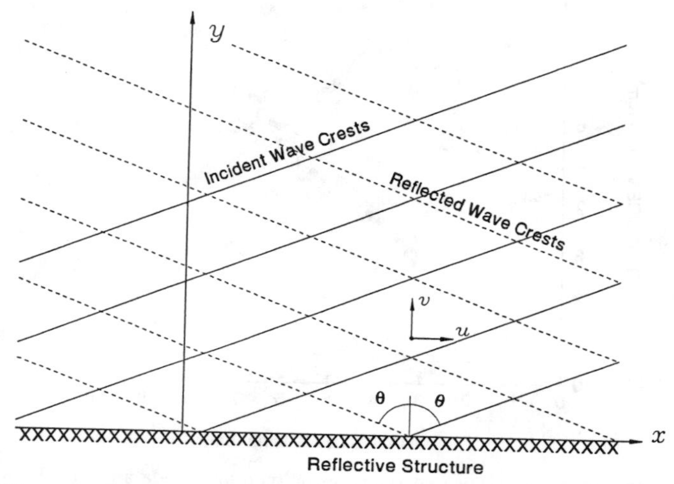

Figure 4: Coordinate system for oblique reflection of long-crested waves.

3.1 Basic Linear Equations

Using the coordinate system specified in Figure 4, the first-order velocity potential for irregular, oblique waves being reflected by a straight structure on a flat bottom is given as

$$\phi(x,y,z,t) = \sum_{i=1}^{\infty} \frac{-a_{I_i} g}{\sigma_i} \frac{\cosh k_i(h+z)}{\cosh(k_i h)} \sin(k_i x \cos\theta - k_i y \sin\theta - \sigma_i t + \epsilon_{I_i}) +$$

$$+ \sum_{i=1}^{\infty} \frac{-a_{R_i} g}{\sigma_i} \frac{\cosh k_i(h+z)}{\cosh(k_i h)} \sin(k_i x \cos\theta + k_i y \sin\theta - \sigma_i t + \epsilon_{R_i}) \qquad (26)$$

where

x – horizontal coordinate parallel to the reflective structure
y – horizontal coordinate perpendicular to the reflective structure
θ – angle of wave incidence [$\theta = 90^o$ is normal incidence]

and the other symbols were defined previously for Eqn. 1. The first term in Eqn. 26 represents the incident wave components propagating obliquely toward the reflective surface located on the x-axis, and the second term represents the partially reflected wave components moving obliquely away from the reflective surface at an angle equal to the angle of incidence.

Time series expressions for sea surface elevation (η) and horizontal components of water velocity (u and v) are found by substituting the Eqn. 26 velocity potential into

Eqn. 2 to yield the following first-order equations for obliquely incident and reflected irregular waves:

$$\eta(x,y,t) = \sum_{i=1}^{\infty} [a_{I_i} \cos(\Phi_{I_i} - \sigma_i t) + a_{R_i} \cos(\Phi_{R_i} - \sigma_i t)] \tag{27}$$

$$u(x,y,z,t) = \sum_{i=1}^{\infty} [(a_{I_i} Z_i \cos\theta) \cos(\Phi_{I_i} - \sigma_i t) + (a_{R_i} Z_i \cos\theta) \cos(\Phi_{R_i} - \sigma_i t)] \tag{28}$$

$$v(x,y,z,t) = \sum_{i=1}^{\infty} [-(a_{I_i} Z_i \sin\theta) \cos(\Phi_{I_i} - \sigma_i t) + (a_{R_i} Z_i \sin\theta) \cos(\Phi_{R_i} - \sigma_i t)] \tag{29}$$

where the incident and reflected phases are now defined as

$$\Phi_{I_i} = (k_i x \cos\theta - k_i y \sin\theta + \epsilon_{I_i}) \quad \text{and} \quad \Phi_{R_i} = (k_i x \cos\theta + k_i y \sin\theta + \epsilon_{R_i}) \tag{30}$$

and the velocity transfer function is given by Eqn. 7.

3.2 Solution Using Wave Gauge and Horizontal Velocity

The first-order solution for oblique reflection using a vertical array consisting of a wave gauge and a current meter aligned to measure velocity in the y-direction is determined as follows.

First, expand and rearrange Eqns. 27 and 29 into the forms given by

$$\eta(x,y,t) = \sum_{i=1}^{\infty} \left[\hat{A}_i \cos(\sigma_i t) + \hat{B}_i \sin(\sigma_i t)\right] \tag{31}$$

and

$$v(x,y,z,t) = \sum_{i=1}^{\infty} \left[\hat{C}_i \cos(\sigma_i t) + \hat{D}_i \sin(\sigma_i t)\right] \tag{32}$$

where the Fourier coefficients[4] in Eqns. 31 and 32 are given by the expressions

$$\hat{A}_i = a_{I_i} \cos(\Phi_{I_i}) + a_{R_i} \cos(\Phi_{R_i}) \tag{33}$$

$$\hat{B}_i = a_{I_i} \sin(\Phi_{I_i}) + a_{R_i} \sin(\Phi_{R_i}) \tag{34}$$

$$\frac{\hat{C}_i}{Z_i \sin\theta} = -\, a_{I_i} \cos(\Phi_{I_i}) + a_{R_i} \cos(\Phi_{R_i}) \tag{35}$$

$$\frac{\hat{D}_i}{Z_i \sin\theta} = -\, a_{I_i} \sin(\Phi_{I_i}) + a_{R_i} \sin(\Phi_{R_i}) \tag{36}$$

Equations 33 – 36 are solved for the four unknowns (a_{I_i}, a_{R_i}, Φ_{I_i}, and Φ_{R_i}) for each frequency component. These unknowns are expressed in terms of the angle of incidence,

[4] The "hat" symbol is used on the oblique solution Fourier coefficients to distinguish them from the Fourier coefficients given for the two-dimensional solution.

θ, and the Fourier coefficients ($\hat{A}_i$, $\hat{B}_i$, $\hat{C}_i$, $\hat{D}_i$) obtained via fast-Fourier transform from the two time series represented by Eqns. 31 – 32. The solution is

$$a_{I_i} = \frac{1}{2}\sqrt{\left(\hat{A}_i - \frac{\hat{C}_i}{Z_i \sin\theta}\right)^2 + \left(\hat{B}_i - \frac{\hat{D}_i}{Z_i \sin\theta}\right)^2} \tag{37}$$

$$a_{R_i} = \frac{1}{2}\sqrt{\left(\hat{A}_i + \frac{\hat{C}_i}{Z_i \sin\theta}\right)^2 + \left(\hat{B}_i + \frac{\hat{D}_i}{Z_i \sin\theta}\right)^2} \tag{38}$$

$$\Phi_{I_i} = \tan^{-1}\left[\frac{\left(\hat{B}_i - \frac{\hat{D}_i}{Z_i \sin\theta}\right)}{\left(\hat{A}_i - \frac{\hat{C}_i}{Z_i \sin\theta}\right)}\right] \tag{39}$$

$$\Phi_{R_i} = \tan^{-1}\left[\frac{\left(\hat{B}_i + \frac{\hat{D}_i}{Z_i \sin\theta}\right)}{\left(\hat{A}_i + \frac{\hat{C}_i}{Z_i \sin\theta}\right)}\right] \tag{40}$$

When the angle of incidence, θ, is 90°, the solution collapses to the two dimensional case of incident waves moving in the negative y-direction; and when $\theta \to 0$, the solution approaches a singularity as the horizontal component vanishes.

3.3 Solution Using Two Components of Horizontal Velocity

A similar first-order solution for oblique reflection can be obtained by using two time series of co-located components of horizontal velocity aligned as illustrated in Figure 4. For this case the Fourier series representation for the y-direction horizontal velocity (given by Eqn. 32) is combined with Eqn. 28, which can be expanded and rearranged into the form

$$u(x, y, z, t) = \sum_{i=1}^{\infty}\left[\hat{E}_i \cos(\sigma_i t) + \hat{F}_i \sin(\sigma_i t)\right] \tag{41}$$

where the Fourier coefficients are expressed as

$$\frac{\hat{E}_i}{Z_i \cos\theta} = a_{I_i} \cos(\Phi_{I_i}) + a_{R_i} \cos(\Phi_{R_i}) \tag{42}$$

$$\frac{\hat{F}_i}{Z_i \cos\theta} = a_{I_i} \sin(\Phi_{I_i}) + a_{R_i} \sin(\Phi_{R_i}) \tag{43}$$

Equations 35, 36, 42, and 43 are solved for the four unknowns (a_{I_i}, a_{R_i}, Φ_{I_i}, and Φ_{R_i}) for each frequency component. These unknowns are expressed in terms of the angle of incidence, θ, and the Fourier coefficients ($\hat{C}_i$, $\hat{D}_i$, $\hat{E}_i$, $\hat{F}_i$) obtained via fast-Fourier transform from the two time series represented by Eqns. 32 and 41. The solution is given by

$$a_{I_i} = \frac{1}{2}\sqrt{\left(\frac{\hat{E}_i}{Z_i \cos\theta} - \frac{\hat{C}_i}{Z_i \sin\theta}\right)^2 + \left(\frac{\hat{F}_i}{Z_i \cos\theta} - \frac{\hat{D}_i}{Z_i \sin\theta}\right)^2} \tag{44}$$

$$a_{R_i} = \frac{1}{2}\sqrt{\left(\frac{\hat{E}_i}{Z_i \cos\theta} + \frac{\hat{C}_i}{Z_i \sin\theta}\right)^2 + \left(\frac{\hat{F}_i}{Z_i \cos\theta} + \frac{\hat{D}_i}{Z_i \sin\theta}\right)^2} \tag{45}$$

$$\Phi_{I_i} = \tan^{-1}\left[\frac{\left(\frac{\hat{F}_i}{Z_i \cos\theta} - \frac{\hat{D}_i}{Z_i \sin\theta}\right)}{\left(\frac{\hat{E}_i}{Z_i \cos\theta} - \frac{\hat{C}_i}{Z_i \sin\theta}\right)}\right] \tag{46}$$

$$\Phi_{R_i} = \tan^{-1}\left[\frac{\left(\frac{\hat{F}_i}{Z_i \cos\theta} + \frac{\hat{D}_i}{Z_i \sin\theta}\right)}{\left(\frac{\hat{E}_i}{Z_i \cos\theta} + \frac{\hat{C}_i}{Z_i \sin\theta}\right)}\right] \tag{47}$$

Singularities occur if the angle of incidence approaches either 0° or 90°, which corresponds to one of the horizontal velocity components being perpendicular to the two-dimensional flow. Therefore, care must be taken not to apply this method under these circumstances.

The technique of estimating oblique wave reflection using the above methods has not yet been implemented in the laboratory, so no assessment can be made of its capability to resolve incident and reflected wave spectra. We do know, however, that the restrictions on the range of frequencies as noted for the two-dimensional case also apply to the oblique reflection case. Coherence between the time series signals may not be as high as obtained in the two-dimensional cases. Nevertheless, the method does open up the possibility of studying oblique reflection of waves by coastal structures.

4 Summary and Conclusions

This paper has summarized a frequency domain method for estimating incident and reflected irregular wave spectra using synoptic time series obtained from co-located gauges. The two-dimensional solution has been tested in a laboratory wave tank using velocity measurements recorded by a laser Doppler velocimeter (LDV), and results indicate the method gives reasonable estimates. Although the title of this paper might imply that a LDV is necessary for application of this method, that is not the case. Any reliable instrument capable of measuring oscillatory flow velocity can be used.

The co-located gauges method is probably slight less accurate than the method of Goda and Suzuki (1976). Both techniques invoke linear theory in their formulation, but the co-located gauge method must also assume that velocities beneath the waves are linearly related to the sea surface elevations. Therefore, it is best to avoid making velocity measurements in the upper portion of the water column where greater departure from linear wave theory velocities is known to occur.

The co-located gauges method has the advantages that it avoids the singularities that occur with the spatially separated gauge methods, and the co-located gauges can be deployed very close to a reflective structure and still give reliable estimates. (This was demonstrated by Hughes (1993) for the case of a vertical wall.) Furthermore, the co-located gauges methods have the potential for use on a mildly-sloping bottom, even though this is in violation of the theory assumptions. The spatially separated gauges methods must compensate for changes in wavelength with depth when deployed on a sloping bottom.

As the irregular waves become more nonlinear, coherence between the time series signals will decrease, and reflection estimates will become less reliable, as should be expected for any linear theory analysis technique.

Extension of the co-located gauges method to the case of oblique reflection of long-crested irregular waves with known angle of incidence appears to have potential application for laboratory basin studies. However, testing of the oblique reflection analysis method has not been completed as of this writing.

Finally, it must be stressed that reliable estimates of incident and reflected wave spectra occur only for those frequencies that contain appreciable energy and exhibit high coherence in the cross-spectrum of the measured time series. Application of the method outside this range of frequency will surely produce erroneous results.

Acknowledgements

The work described and the results presented herein, unless otherwise noted, were obtained from research conducted under the Scour Holes at Ends of Structures work unit of the U.S. Army Corps of Engineers, Coastal Engineering Research Center. Permission was granted by the Chief of Engineers to publish this information. Dr. J. Fowler, Messrs. N. Sultan, R. Herrington and D. Dailey, and Ms. D. Green are acknowledged for their valuable contributions to the laboratory experiments.

5 References

Gaillard, P, Gauthier, M., and Holly, F. (1980). "Method of Analysis of Random Wave Experiments with Reflecting Coastal Structures," *Proceedings of the 17th Coastal Engineering Conference*, ASCE, Vol 1, pp 201-220.

Goda, Y., and Suzuki, Y. (1976). "Estimation of Incident and Reflected Waves in Random Wave Experiments," *Proceedings of the 15th Coastal Engineering Conference*, ASCE, Vol 1, pp 828-845.

Guza, R. T., Thornton, E. B., and Holman, R. A. (1984). "Swash on Steep and Shallow Beaches," *Proceedings of the 19th Coastal Engineering Conference*, ASCE, Vol 1, pp 708-723.

Hughes, S. A. (1993). "Laboratory Wave Reflection Analysis Using Co-Located Gages," *Coastal Engineering*, Vol 20, pp 223-247.

Isaacson, M. (1991). "Measurement of Regular Wave Reflection," *Journal of Waterway, Port, Coastal, and Ocean Engineering*, American Society of Civil Engineers, Vol 117, No. 6, pp 553-569.

Mansard, E. P., and Funke, E. R. (1980). "The Measurement of Incident and Reflected Spectra Using a Least Squares Method," *Proceedings of the 17th Coastal Engineering Conference*, ASCE, Vol 1, pp 154-172.

Maoxiang, G., and Zhenquan, Z. (1988). "An Analysis of Second Order Waves in a Wave Tank by 3-Probe Correlation Analysis and by Square-Low Operation," *Ocean Engineering*, Vol 15, No. 4, pp 319-343.

Nelson, R. C., and Gonsalves, J. (1990). "A Field Study of Wave Reflections from an Exposed Dissipative Beach," *Coastal Engineering*, Vol 14, pp 457-477.

Sultan, N. J., and Hughes, S. A. (1993). "Irregular Wave-Induced Velocities in Shallow Water," *Journal of Waterways, Port, Coastal, and Ocean Engineering*, ASCE, Vol 119, No. 4, pp 429-447.

Tatavarti, R. V., Huntley, D. A., and Bowen, A. J. (1988). "Incoming and Outgoing Wave Interactions on Beaches," *Proceedings of the 21st Coastal Engineering Conference*, ASCE, Vol 1, pp 136-150.

Walton, T. L. (1992). "Wave Reflection from Natural Beaches," *Ocean Engineering*, Vol 19, No. 3, pp 239-258.

Observations of Wind Wave Growth by ERS-1 SAR

Chich Y. Peng[1]
Antony K. Liu[2]

Abstract

Spatial variations of ocean waves can be observed from space-borne synthetic aperture radar (SAR) images of ocean surface. For a steady offshore wind, the growth of wind wave from an upwind coastline depends primarily on the finite fetch distance. From the European Remote Sensing Satellite-1 SAR images, such a case of fetch-dependent growth of wind waves is identified and analyzed. The fetch-dependence of peak spectral density and peak frequency extraced from SAR images compares reasonbly well with the power laws obtained from the field measured data.

Introduction

The development of wind waves depends not only on the magnitude of wind speed, but also on the fetch distance and duration of the wind. If an upwind coastline exists and the wind field persists over a certain minimum duration, the growth of the wind waves depends primarily on the fetch distance. The first detailed field measurements of such fetch-limited growing wind waves have obtained during the Joint Sea Wave Project (Hasselmann, et al, 1973). The wave parameters such as wave height variance and peak frequency versus fetch can be expressed in power laws. Such power laws for fetch limited wind wave growth have also been found from

---

1 Science Systems & Applications Inc., 5900 Princess Garden Parkway, Lanham MD 20706
2 NASA/GSFC, Oceans and Ice Branch, Code 971, Greenbelt, Marylnad 20771

experiments in wave tanks (Plate, 1978) and from aircraft microwave radar measurements over ocean surface (Walsh, 1987). While many phenomena related to spatial variation of ocean waves over open oceans have been observed from SEASAT synthetic aperture radar images, the observations of the growth of ocean waves within a finite fetch from SEASAT SAR data have not yet been reported.

With a full resolution of about 25 m, it is hardly possible to observe the initial stage of the growth of wind waves from SAR data. But under favorable conditions, the intermediate and later stage of growth may be possibly observed from high resolution satellite SAR images. Since the launch of the First European Remote Sensing Satellite in 1991, many SAR images of high resolution have been obtained over the Bering Sea and the coastal area off Alaska. Through spectral analysis of some of these SAR images, we present a case study of fetch-dependent wind wave growth off the Alaska coast in January 17,1992. The analysis is based on the comparison of field measured fetch power laws for spectral peak frequency and peak spectral density with those extracted from SAR images.

SAR Data and Wind Conditions

In January 17, 1992, ERS-1 passed over the coastal area of the Alaska Peninsula along a ground track pointing 196 degree from the true north. A standard full-resulotion SAR images has 8192x8192 pixels with a 12.5 meter pixel spacing and covers an area of about 100x100square kilometers. The range resolution is 29 meters and azimuthal resolution is 27 meters. The location of such a standard SAR image to be analyzed is shown in shown in Figure 1. With center latitutde and longitude 57.256 N and 166.011 W respectively, the nearest distance of the covered area is about 300 km from the west coastline of the Alaska Peninsula.

About 600 km to the west of the scene, a NDBC buoy 46035 is located at 57.0N and 177.7W in the Bering Sea. The wind speed and direction during January 1992 obtained from the buoy are shown in Figure 2. We note that the mean wind direction has been from the west (about 270 degree from true north) during a period of about five days prior to January 17, 1992. Assuming a spatially uniform wind direction over a few hundred kilometers and considering the proximity of the coastline and the persistence of quite steady offshore wind direction, we might expect to observe the fetch-dependent growth of wind waves within an area covered by the SAR image. In the following we shall show this is the case by examining the fetch dependence of the wave parameters extracted from the SAR images.

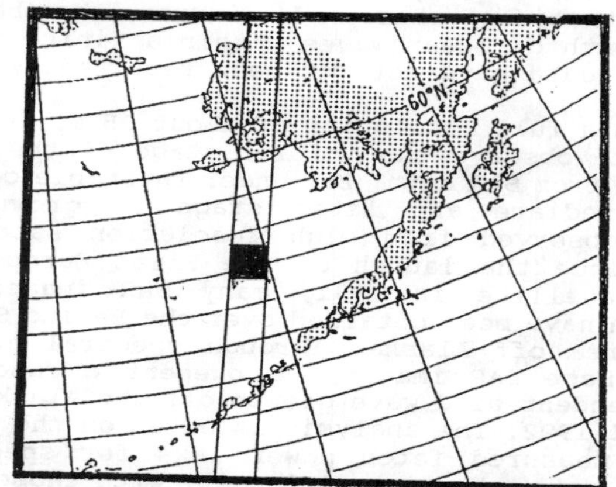

Figure 1. Location of the study area

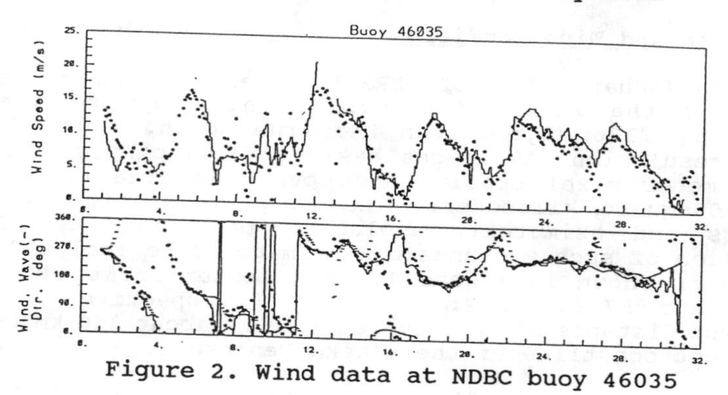

Figure 2. Wind data at NDBC buoy 46035

Results

The directional spectra are computed for a series of subscenes consisting of 512x512 pixels both along and perpendicular to the wind direction. An example of such a scene is shown in Figure 3. The procedures for computing the directional spectra from the SAR images are briefly outlined as follows: (1) estimate the fractional modulation intensity as (I-<I>)/<I>, where I is the intensity (amplitude squared) at each pixel and <I> is the average intensity of the subscene, (2) compute the two dimensional FFT of the estimated modulation intensity, (3) compute the squared modulas from the FFT result, (4) swap the modulas for quadrant one and three with those of quadrant two and four, and (5) smooth the sample spectrum with a Gaussian kernal.

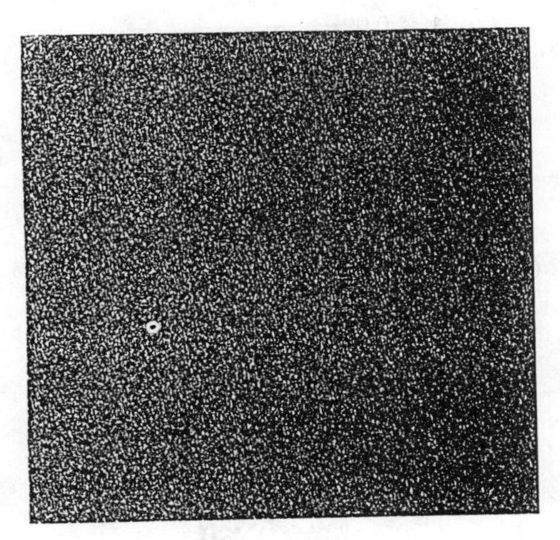

Figure 3. A subscene of the SAR image

The existing power laws of the wave parameters of the fetch-limited wave growth are expressed in terms of frequency spectra. To comapre, we need to convert the computed directional wavenumber spectra from SAR images into the directional frequency spectra by using the dispersion relation (Phillips, 1977):

$$< \xi^2 > = \iint S(f,\phi)\, df\, d\phi = \frac{2}{(2\pi)^4 g^2} \iint k^{\frac{3}{2}}(f)\, F(k_x, k_y)\, df\, d\phi \quad (1)$$

where $< \xi^2 >$ is the wave height variance, ϕ is the direction angle, f is the frequency, F is the directional wavenumber spectrum, and S is the directional frequency spectrum. Three such directional frequency spectra at a fetch of 300 km, 339 km and 378 km respectively are shown in Figure 4. We note that the spectra are bi-modal: one component is along the north-south direction, and the other along the east-west direction. As the local wind estimated from the NDBC buoy wind measurements is about 270 degree from the true north, this east-west component is identified as the local wind component. We shall focus on this component only and examine its spatial evolution along the wind direction along a distance of about 100 km across the whole SAR image.

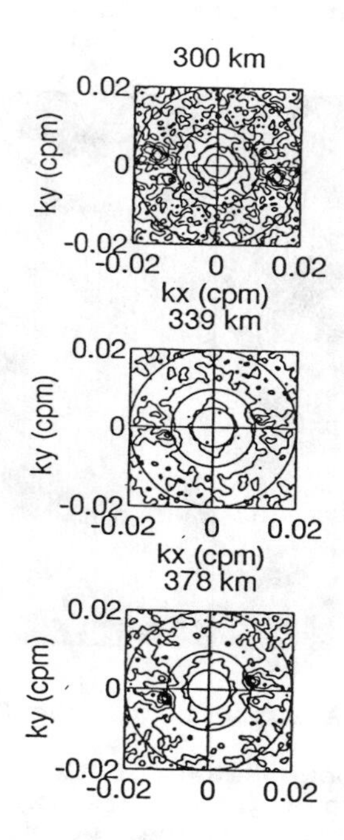

Figure 4. Directional spectra at different fetches

The spatial evolution of the two wave spectral parmeters to be considered are the peak frequency f_p and the peak spectral density. Their dependence on the fetch, F, in the fetch-limited case is as follows (Hasselmann et al, 1973):

$$f_p \sim F^{-0.33}$$

$$\Phi_p(f_p) \sim f^{-5} \sim F^{1.65} \quad (2)$$

Assuming the directional frequency spectrum can be represented as a product of a nondirectional spectrum,

Φ(f) and a directional spread factor which is a function of dircetion angle only as follows:

$$S(f,\phi) = \Phi(f)\,G(\phi) \tag{3}$$

we can then evaluate the peak spectral frequency and the peak spectral density from the maximum of the directional frequency spectra as follows:

$$\Phi_p(f_p) \sim Max\left[K^{3/2}F(k_x,k_y)\right] \tag{4}$$

These two parameters are extrated from 18 spectra for partially overlapped subscenes located at the same fetch and then averaged. This procedure is repeated for 12 different fetch distances. Figure 5 shows the peak frequency vs. fetch in a log-log plot. The solid line is the best least square fit. The slope of the line is found to be -0.27, which is close to the the exponent, -0.33, from the field measured power law for the peak spectral frequency. The peak period is about 7.56 seconds at a fetch of 300 km. Using the power law for the peak period from field measurements (Hasselmann et al,1973),

$$\frac{T_p\,g}{U_{10}} = 0.286\left(\frac{gF}{U_{10}^2}\right)^{0.33} \tag{5}$$

we estimate the minimum wind speed that is necessary to sustain the fetch-limted wind wave growth to be 6.6 m/s. This value is quite consistent with the averaged wind speed of about 7 m/s measured at the buoy (see Figure 2).

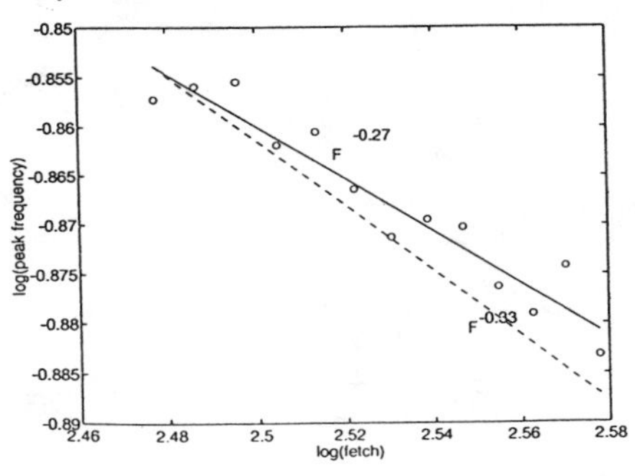

Figure 5. Peak spectral frequency versus fetch

While the frequency decreases with fetch, the peak spectral density increases with fetch. Figure 6 shows the variation of peak spectral density as a function of fetch. The unit for the spectral density is arbitrary by a constant factor and has not been calibrated. The least square fit yield the exponent of 2.05, versus 1.65 obtained from the field measured data. Finally, Figure 7 show the cuts of the frequency spectra along the direciton of the spectral peak of the wind wave component at three equally spaced fetch distances. The down-shift of the peak frequency and the growing of the peak density versus fetch are clearly evident. We note that the spectral peaks falls as f^{-5}, which is indicated by the dashed line.

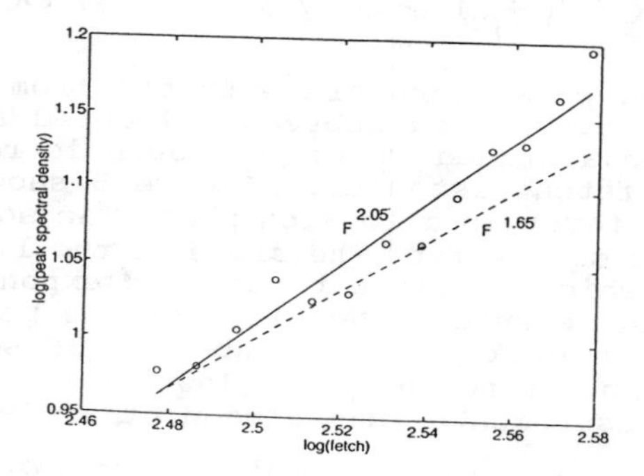

Figure 6. Peak spectral density versus fetch

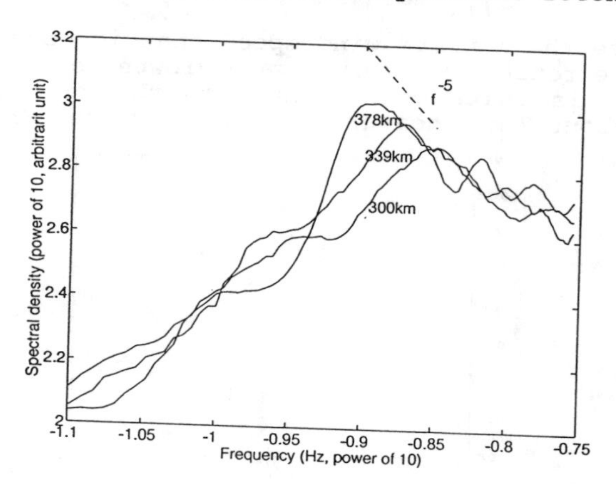

Figure 7. Evolution of frequecy spectra with fetch

Concluding Remarks

The fetch dependence of the wind wave growth have been identified form the ERS-1 SAR data. The down-shift of the peak frequency and growing of the peak density versus fetch have been demonstrated. Their fetch dependence appears to agree reasonbly well with the power laws obtained from the field measurements. We have not attempted to compare the vaiation of the wave energy by integrating the directainal spectra, because of the ditortion of the spectral shape by the SAR tranfer functions, and because of the diffuculties in isolating the wind wave component from the bi-modal spectra in integration.

We note that the fetch considered is more than 300 km. From the power law for the peak period (see Equation 5), a critical fetch may be shown to be proportional to approximately the third power of a given peak period and inversely proportional to a given wind speed. Thus for fetch-limited wind wave growing at a far fetch, it is necessary to have long wave period and low wind speed. These conditions appear to be satified in the present case.

Acknowledgements

The authors wish to thank Dr. H. S. Chen for providing the NDBC buoy wind data and helpful comments. GSFC work was supported by the National Aeronautics and Space Administration and Office of Naval Research.

References

Hasselmann,K. et al.,1973, Measurements of wind wave growth and swell decay during the Joint North Sea Wave Project (JONSWAP). Dtsch. Hydrogr. Z. A8 (No.12)

Longuet-Higgins, M. S., D. E. Cartwright, and N. D. Smith 1963, Observations of Directional spectrum of sea waves using the motion of a floating buoy. Ocean Wave Spectra, Prentice-Hall, Englewood Cliffs, N.J., pp 111-136.

Phillips, O. M., 1977, The Dynamics of the Upper Ocean. Cambridge unoversity Press, Cambridge, London, pp 366.

Plate, E. J., 1978, Wind-generated water surface waves: the laboratory evidence In Turbulent Fluxes through the SEA Surface, Wave Dynamics,and Prediction, Plenum Press,New York,pp385.

Walsh, E. J., Hancock, D. W., Hines, D. E., Swift, R. N. and Scott, J. F. 1989, An observation of directional spectrum evolution from shoreline to fully developed, J. Phys. Oceanogr., 19, 670-690.

Optical Remote Sensing of Wave Surface Kinematics

Stephen Riedl[1], Richard Seymour[2], member, ASCE, Jun Zhang[3], associate member, ASCE

Abstract

The measurement of the directional properties of surface slopes, velocities and accelerations is difficult, even in the laboratory, under conditions of steep, random and short-crested waves. A technique has been developed to use small surface marking floats, color-coded to facilitate identification, as targets for stereo video imaging of the kinematics of the wave surface. The images from successive frames are compared, using both stereo and standard image processing techniques for enhancement and target detection, to determine changes in target position in three dimensions.

This provides an unusual synoptic view of the wave surface that is comparable to having a dense array of wave probes with co-located current meters capable of operating in the uppermost water layer. Since it has been amply demonstrated that surface velocities of steep waves are not well predicted by any existing theory even under long-crested waves, it follows that accurate measurement of the kinematics of the surface of short-crested waves will provide valuable data against which to test improved models.

The paper describes a multi-year research program to develop a fully operational laboratory instrument. Results from the first year's work, utilizing 2-D waves initially, will be presented in detail.

---

1 Grad. Student, Ocean Engrg. Program, Dept. of Civ. Engr., Texas A&M Univ., College Station, TX 77845-3400
2 Dir., Offshore Tech. Res. Ctr., 1200 Mariner Dr., Texas A&M Univ., College Station, TX 77845-3400
3 Prof., Ocean Engrg. Program, Dept. of Civ. Engrg., Texas A&M Univ., College Station, TX 77843-3136

Introduction

Accurate measurement of the kinematics of laboratory waves has been made possible with recent advances in measurement techniques and laboratory equipment. Although these new techniques and equipment are accurate when measuring wave kinematics in a region below the wave surface, their design does not always permit them to accurately measure kinematics on the wave surface. The need for accurately measuring wave surface kinematics becomes apparent when considering the application of both linear and non-linear wave theory upon the surface of a wave.

Linear wave theory tells us that particle displacements are at a maximum on the wave surface and thus the particle velocities must also be at a maximum on the surface of a wave. This theory also makes the assumptions that waves are sinusoidal and that the water particles acted upon by the waves move in closed orbits. One knows that even when producing waves in a controlled environment, such as a wave basin, they are not symmetric about the still water level. The crests of these waves rise above the SWL by an elevation greater than one-half the wave height while the troughs are below the SWL by an elevation less than one-half the wave height. This non-symmetric pattern is predicted by non-linear wave theory. This effect increases the magnitude of the crest velocity and decreases the magnitude of the trough velocities, negating the closed orbit particle path assumption and introducing a net displacement in the direction of wave propagation. In order to measure these surface kinematics, a laboratory instrument has been developed that uses image processing techniques in conjunction with real-time video images of wave surfaces, to provide an accurate measure of surface particle displacements in the horizontal direction. This information can then be used to determine surface properties.

Experimental Setup and Procedures

The laboratory experimentation was conducted in a two dimensional, glass-walled wave tank (Figure 1) measuring 37 m long, 0.91 m wide and 1.22 m deep, using a water depth of 0.91 m. Wave generation was provided by an electrically driven single flap wave maker. Wave absorption in the tank is provided by a 1:3.5 slope artificial beach. A black sheet-metal liner was placed inside the tank in the area of interest. This liner provided a uniform background for the video images and prevented any light from entering the study area through the tanks glass walls. In order to reduce specular reflection off the water surface, the wave trains were filmed using Ultraviolet (black light) illumination. The particle/wave interactions were filmed using a Panasonic Super-VHS video cam-corder that was mounted directly above the tank at a distance of 1.82 m from the tanks still water level. The surface marking targets consisted of 0.6 cm diameter fluorescent paper, laminated for water resistance.

The video images of the particle oscillations were digitized and imported, frame by frame, into an IBM compatible 486/50 personal computer. A sampling rate of 30 Hz was used because it is the natural sampling rate of the video camera (30 frames/sec). These digitized images were then run through a series of enhancement and tracking routines. The enhancement routines filtered out any noise in the image that might interfere with particle identification, while the tracking routines determined the center point of each particle within each video frame. The particle locations over

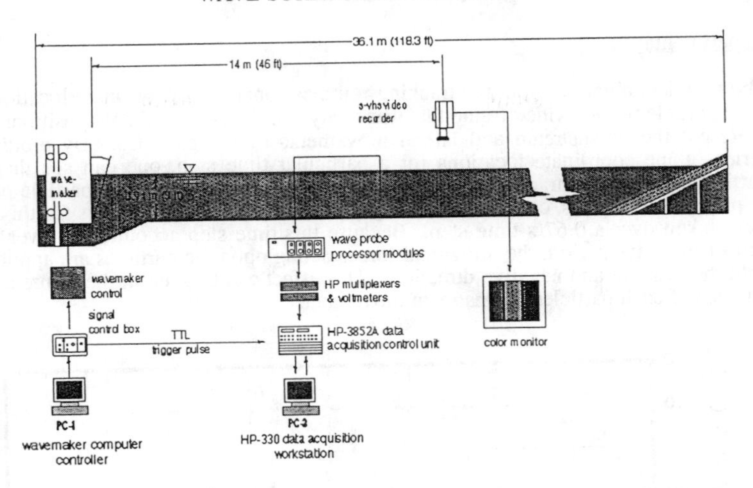

Figure 1. Schematic of laboratory wave tank and instrumentation

time were then used to study the surface kinematics of the observed waves. Kinematic properties that can be obtained using this method include particle drift velocities, wave surface particle velocities, the phase speed of the wave and certain characteristics of the wave tank.

Table 1. Wave Parameters

T (s)	H (cm)	H/L
0.8	4.39	0.044
0.8	6.14	0.061
1.0	9.66	0.062
1.0	12.95	0.083

This measurement method was used to analyze two sets of regular, deep water waves. Each wave set contained two different waves. The first wave set consisted of a 1.0 s period wave having heights of 9.66 cm and 12.95 cm, while the second wave set consisted of a 0.8 s period wave having wave heights of 4.39 cm and 6.14 cm. These wave parameters can be seen in Table 1. This table also shows wave steepness values for each of the waves. These steepness values (0.044 to 0.083) are small when compared to a maximum steepness value of 0.142 and therefore eliminates the possibility of the particles losing contact with the wave surface and "surfing" down the face of the wave. For each wave case, the input wave series was sinusoidal with the wave trains being attenuated in amplitude at the beginning of the series. The results presented here are from the 1.0 s, 9.66 cm wave.

Experimental Results

The raw data obtained from the tracking routines consists of coordinate locations of each particle in each video frame that was analyzed. These coordinate positions represent the downstream and lateral movements of the particles. By plotting a series of the coordinate locations for a particular time step, one can see that the particle paths can be tracked using this measurement technique. The particle paths, or particle streaklines, can be seen in Figure 2. The particle coordinates for this plot were taken over a 0.67 s time step. Because this time step accounts for over one-half of the wave period, the horizontal forces acting upon the particles are applied in both the positive and negative directions. This effect can be seen in the figure as the reversal of each particles downstream movement.

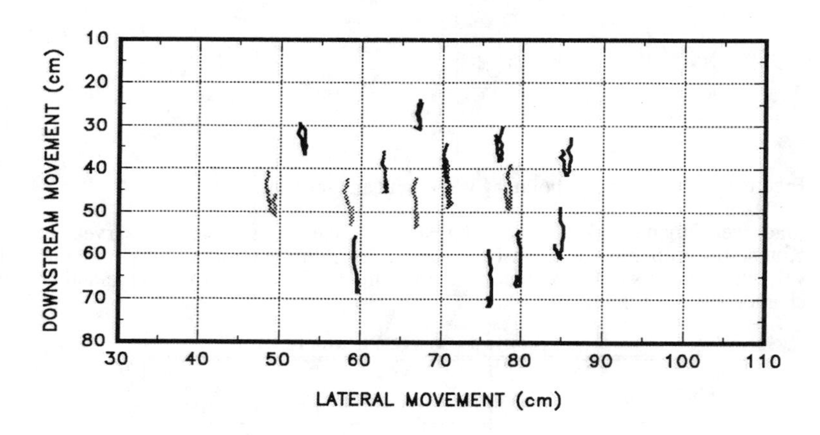

Figure 2. Particle streaklines for T=1.0 s, H=9.66 cm and a time step of 0.67 s

Figure 3 presents the horizontal displacement as a function of time for an individual particle. The distinct upward trend in the plot is the mass transport component of the wave train at the surface. A linear regression curve was fit to the data to determine the particles mean down-tank position. By calculating the slope of this regression curve, one can determine the particles drift velocity.

To obtain the horizontal particle velocities, the horizontal particle displacements were differentiated using a cubic spline interpolation method. Figure 4 shows a comparison of the measured horizontal velocities to the horizontal velocities as predicted by linear wave theory. The increase in magnitude of the crest velocity and the decrease in magnitude of the trough velocities of the wave is expected, as it is predicted using 2nd order wave theory.

In order to remove the down-tank drift component from the original particle displacements (Figure 3), the displacement data was detrended using the residuals from the regression curve plotted in Figure 3. The detrended particle displacements can be seen in Figure 5. The figure clearly shows the presence of a slight upward

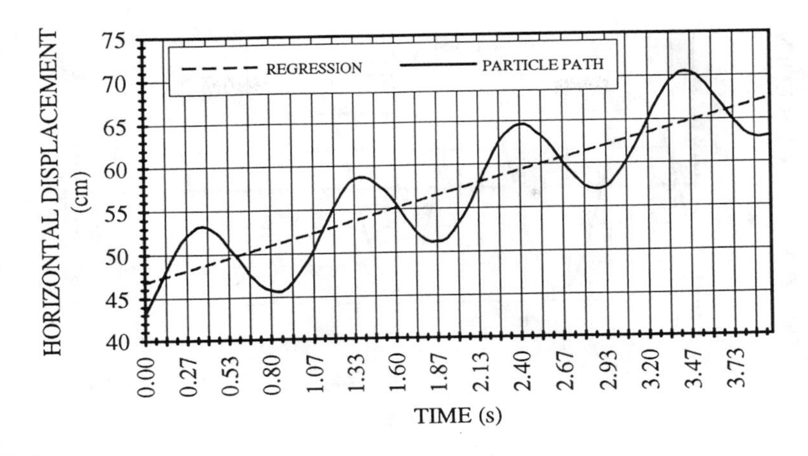

Figure 3. Horizontal displacements of a single particle as a function of time for T=1.0 s and H=9.66 cm

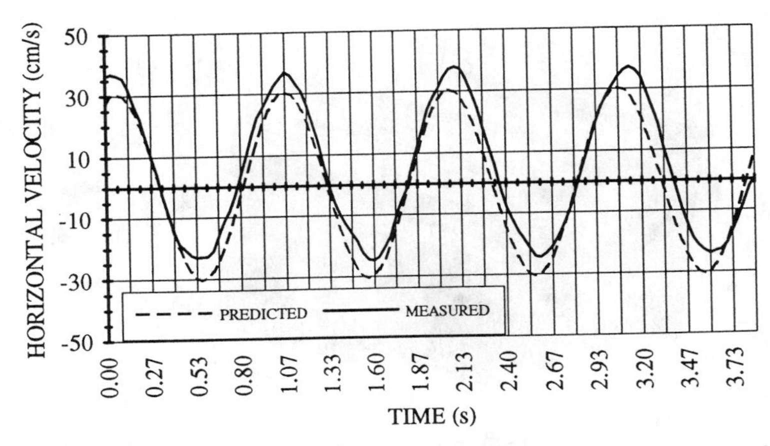

Figure 4. Horizontal velocities of a single particle as a function of time for T=1.0 s and H=9.66 cm

trend for this particular particle. This upward trend does not represent an increase in velocity because the magnitude of the displacement over one wave period remains constant. After examining displacement plots from particles located further down-tank it was found that this trend evolves into a downward trend as one moves down the tank through the particle cluster. Because of this transformation, the trend is believed to be the result of a low frequency disturbance in the wave basin.

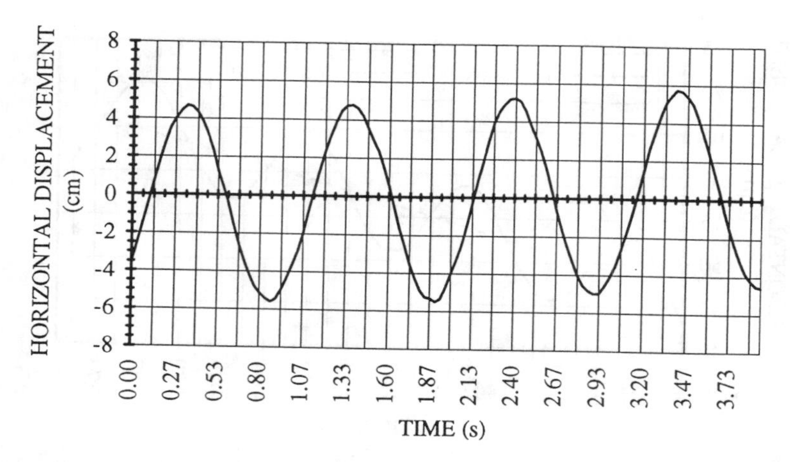

Figure 5. Detrended horizontal displacements for up-tank particle for T=1.0 s and H=9.66 cm

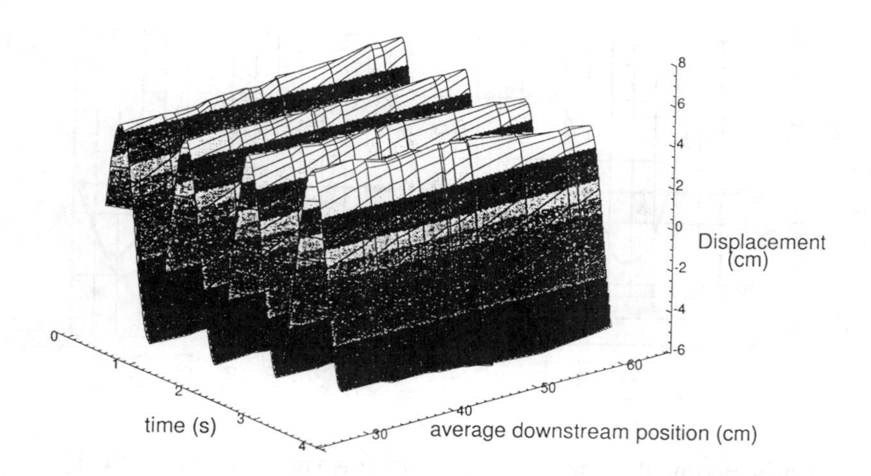

Figure 6. 3-D plot of global particle displacements

Combining the detrended horizontal displacement plots for each particle present in the wave set and normalizing them based on their mean down-tank positions, as determined in Figure 3, results in a 3-D plot of global particle displacements (Figure 6). The upward trend in particle displacements can be seen as the increase in peak height in the up-tank particles while the downward trend can be seen as the decrease in peak height in the down-tank particles. As previously mentioned, this is a characteristic of a low frequency disturbance and is believed to be a long wave present in the wave tank. If this assumption holds true, the period of this long wave can be determined by tracking the particles for an extended period of time and the wave length can be determined by introducing more particles further down the tank.

Figure 7 shows an overhead view of the global particle displacements plot as seen in Figure 6. The distinct angle of alignment of the peaks and valleys can be calculated to determine the phase speed of the wave. Using this method, the phase speed of the wave was determined to be 1.60 m/s which agrees quite nicely with a theoretical value of 1.56 m/s.

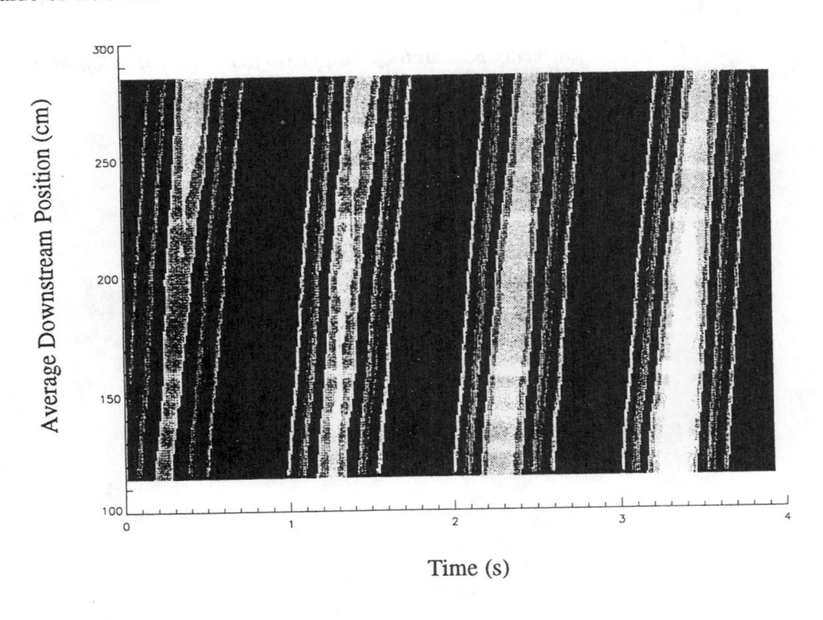

Figure 7. Overhead view of global particle displacements plot

Conclusion

Utilizing video imaging techniques, we have shown that it is possible to measure wave surface kinematics by tracking the movement of surface marking particles. The development of this technique into an operational laboratory instrument that is both fast and easy to use will provide not only a means to accurately measure wave surface kinematics but also data against which to test present and future models.

At present, this instrument utilizes 2-D processing techniques and can measure horizontal surface velocities, particle displacements, particle drift velocities and the phase speed of the wave. Plans are underway to expand this process to 3-D by using stereo image processing to track the horizontal and vertical displacements of the particles. This expansion will allow for the measurement of vertical velocities and surface slopes in addition to the previously available properties.

Acknowledgments

The research presented in this paper was supported in part by the Offshore Technology Research Center, National Science Foundation Engineering Research Centers program grant CDR-8721512.

Appendix I. References

U.S. Army, Coastal Engineering Research Center, (1977). *Shore Protection Manual*, Vol. I, U.S. Government Printing Office, Washington, D.C.

Breaking Wave Measurement by a Void Fraction Technique

Ming-Yang Su[1]
John Cartmill[2]

Abstract

Breaking of steep surface gravity waves in seas under moderate to strong wind forcing always involves the entrainment of air into the water column below the air-sea interface. As such, the presence of air in the near-surface zone may be a positive indicator for occurrence of wave breaking. This simple idea is the physical basis for our measurement of breaking waves by means of a new technique that directly measures the percentage of air content (void fraction) under a pre-fixed depth below the sea surface. The data from such void fraction measurements conducted in the open sea under moderate to high sea state were processed to obtain several breaking wave statistics. A comparison of these statistics with five previously-published statistics employing different techniques shows that the new void fraction technique is a better technique both in its easier execution and its more quantitative discrimination of various sizes of breaking waves. We shall also present the space-time variations in void fraction of a large breaking wave in a laboratory tank, acquired in comparison to other previous results.

Introduction

As a steep surface wave breaks, air is entrained into the water, forming clouds of bubbles. This highly nonlinear, complex, two-phase flow phenomenon has significant effects on near surface physics both above and below the water surface.

---

[1]Naval Research Laboratory, Stennis Space Center, MS 39529-5004
[2]Planning Systems, Inc., Slidell, LA 70458

Even though the importance of air entrainment due to wave breaking has been well recognized for quite some time, detailed laboratory and field measurements of this dynamic surface process are still lacking. This is, without question, due to the intrinsic difficulty involved in measuring the phenomena bordering the interface of two fluid media, with the interface itself undulating randomly and rapidly in three dimensions. Additional problems involve the development of sensors for the accurate and rapid measurement of the bubble size density and void fraction, which is the total volume of air contained in the water.

Surprisingly, up to now there has been only one such detailed laboratory measurement, performed by Rapp and Melville (1990) using mechanically generated breaking waves. Less detailed measurements of bubble distribution due to wave breaking in a laboratory under high wind speed (up to 16 m/sec) have been reported by Balldy (1988) and Hwang, et al. (1990). The purpose of this paper is two-fold: (1) to present some recent experimental results on the temporal and spatial variation of void fraction due to a large-scale breaking wave event in a fresh-water wave channel, and (2) to present the breaking wave statistics by this technique during the Surface Wave Dynamics Experiment in 1990.

The Void Fraction Meter - Theory and Calibration

The void fraction meter we employed consisted of two parallel electrodes, one inch square and separated by a one inch gap as sketched in Figure 1. bubbly water can flow freely between these two electrodes. The void fraction was computed from the increase in resistance between the two electrodes due to the presence of bubbles in comparison with bubble-free water. The theory for relating the change of resistance of the flow medium to the void fraction was developed by Maxwell (1891) a century ago. A short summary of his theoretical analysis of the principle is reproduced here:

Let the specific resistance of water alone be r_2 and the resultant void fraction be p. Then the specific resistance r of the bubbly flow mixture which contains small identical spheres of air bubbles, with their mean separation large compared with the bubble diameter, is given by

$$r = \frac{2r_1 + r_2 + p(r_1 - r_2)}{2r_1 + r_2 - 2p(r_1 - r_2)} r_2 \tag{1}$$

Now, the resistance of air is much greater than that of the water. If one assumes that the air resistance in the bubble is infinite, then the above expression is reduced to

$$r = \frac{2+p}{2-2p} r_2 \tag{2}$$

or

$$p = \frac{(r/r_2)-1}{(r/r_2)+1}.$$

Thus, the ratio of resistance of the bubbly water to the bubble-free water is directly related to the void fraction p. Since several assumptions are involved in Maxwell's derivation of equations (1) and (2), some empirical calibration must be done in order to check the ranges of validity. The calibration was performed by submerging the void fraction meter in a circular plexiglass tube containing water. Air bubbles were then introduced into the tube from below by means of an aquarium aerator as shown in the sketch in Figure 3(b). By direct measurement of height raise in the top surface of the water column in the tube due to the introduction of air bubbles, one can obtain the void fraction while, at the same time, the void fraction is computed based on equation (2). A typical plot of the comparison of the two void fraction values up to p = 27% is given in Figure 2. This graph shows a good agreement between the Maxwell theory and direct measurement, even though the assumptions on which his theory is based are rather far from the situation encountered at high void fractions (27% in particular).

A bridge circuit, shown in Figure 3, is used in order to expand the dynamic range of resistance ratio measurement. The bridge is excited by a 4 kHz sine wave and is "balanced" for the bubble free case. As the resistance changes due to air bubbles in the water, a voltage is measured between points A and B in Figure 5. The resistance can then be computed from:

$$V_{ab} = V_{Total}\left[\frac{R}{R+R_2} - \frac{R_3}{R_3+R_1}\right] \tag{3}$$

Void Fraction Measurement in a Laboratory Tank

Figure 4 is a sketch of the test section of the void fraction measurement under a breaking wave within an evolving wave group in a wave tank at the Oregon State University. Three void fraction meters were mounted on a surface following steel frame at depths of 0.30m, 0.76m, and 1.22m below the water surface, respectively.

With the three-second wave period and the corresponding wave length of about 15.25m we observed that the separation distance from the start of the wave breaking to its completion is about 18.3m as indicated in Figure 4. At the beginning, bubbles only penetrated to a very shallow depth. Similarly, near the end, the bubbles were detected only in a region close to the surface. For the beginning and ending conditions, the topmost sensor was located below the bubble layer. Based on these direct observations, we selected six stations (marked as X_0, X_1, X_2, X_3, X_4, and X_5 in Figure 4) to cover the horizontal variation of void fraction. Station X_0 was about 1.53m after the initiation of wave breaking, which shows the wave crest breaking down and air entrainment beginning. Station X_5 was about 3.05m ahead of the last trace of white capping. With the exception of the separation of 1.5m between station X_0 and X_1, the separation for all of the other adjacent stations was 3.05m.

A sampling rate of ten samples per second was used for each void fraction meter, this rate provides adequate representation of the temporal variations at the selected depths and horizontal stations.

The void fraction data were taken at each of six stations along the channel, X_0, X_1, ...,X_5, and three depths per station, for 40 seconds from the start time of the wave maker which generated a wave packet. The measurement at each location was repeated four times. These data records were first converted from their original resistance reading into corresponding void fraction using Equation (2) and then corrected for slight deviations from these theoretical values using the calibration curve for each particular void fraction gauge. The four time records for each location were then averaged. The three-dimensional bubble plumes were very similar in gross structure, but they were not identical from case to case, even with the identical paddle motion from the computer-controlled wave maker; consequently, some averaging of the data were required. This instability caused some degree of variation in both size and spacing of bubble plums despite the initial wave packet being identical for each case.

Three .8 averaged void fraction variations, in space and in time, are plotted in Figures 5 through 7, showing the short-term high void fraction (i.e., time shorter than one wave period after the passing of the breaking wave crest). In the void fraction variation at the first station, X_0 (Figure 5), note that for the top two depths (C_1=0.3m and C_2=0.76m; i.e., about 1/5 and 1/2 of the wave height, respectively, there were clearly two maxima with respect to time for each depth. The first maximum was less than 1% and the second maximum reached 40% for C_1 and about 13% for C_2. Only one maximum of about 10% occurred in the lowest depth, C_3=1.22m (about 4/5 of the wave height). The first maximum in both C_1 and C_2 appeared to be caused by the splash up in front of the main wave plunging that entrains most of the air into water forming bubble plumes.

The void fractions of the breaking waves as they continue to move forward are shown in Figures 5 through 7. We see that the maximum remains near 40% for C_1, but the duration for this high void fraction becomes shorter from X_0 to X_5. The two maxima in C_1 at X_0 are merged gradually into only one maximum, resulting in a skew distribution of void fraction with time. This disappearance of the leading smaller maximum is expected as the wave moves from X_0 to X_5, since the initial splash up decays in strength rather quickly. The e-folding time of void fraction with respect to its maximum is about one second; i.e., 1/3 of the wave period, which implies that the large bubbles with radii larger than, say, millimeters, which contribute most importantly to the high void fraction, rise quickly above the depth of 1/5 of the wave height.

Breaking Wave Statistics in SWADE

A modified version of the instrument frame as described above and used in the laboratory tanks was constructed for deployment in the sea in moderate to strong wind/sea situations. The system was tethered to the ship by a 150 meter cable for power and data transmission during the Surface Wave Dynamics Experiment (SWADE) conducted in 1990 off the east coast of the United States.

In order to obtain the breaking wave statistics from the directly-measured void fraction statistics, we proceed as follows: The time series of void fraction measurement at the depth of 0.25m is screened for a continuous segment that exceeds some threshold values. Each such segment is called a void fraction event, whose maximum value of void fraction is used to represent the first-order magnitude of breaking wave intensity.

Each "void fraction event" may be interpreted as one "breaking wave event" if we define a breaking wave to be a wave which entrains air into the water that becomes a plume of bubbles with a measurable void fraction to a depth of 0.25m below the surface. This definition will preclude all the micro-breaking by capillary waves and gravity-capillary waves with wave lengths less than about 2m. The cumulative average in percentages for four cases (of a total 13 hours) during SWADE are plotted in Figure 8 along with the five ranges of void fraction denoted by letters *a,b,c,d,* and *e*, respectively, and with the mean wind speeds of 10 and 15 m/s.

We shall next compare our field measurements with several published field measurements on breaking wave statistics by various investigators employing different techniques and definitions on what constitutes a "breaking wave."

In the paper by Holthuijsen and Herbers (1986), the observed occurrence percentage (fraction) of breaking waves from five different field measurements as a function of wind speed (U_{10}) are compared. Their Figure 7 is reproduced here in Figure 8 for comparison with our results.

Toba et al. (1971) and Holthuijsen and Herbers (1986) use the similar technique of coupling visual identification of white capping with wave height measurement. Even though this method might be the most positive way to say a wave is in the process of breaking, some degree of subjective judgement is still involved, particularly when the breaking area is small. Thorpe and Humphries (1980) use telephotography in place of direct visual observation. This method could miss counting smaller breaking waves in conterast to the former method. As such, the observed number of breaking waves should be less in the latter case, as exhibited in Figure 2. Weissman et al. (1984) employed the method of detecting the presence of high frequency components (5-50 Hz) in the waves as the definition of breaking waves. Their measurement at a low wind speed of about 6 m/s involved only very small breaking waves, which should agree better with those by direct visual observation (such as Toba, et al.). Longuet-Higgins and Smith (1983) employed a "jump meter" to measure the slope of large and/or dominant waves near a North Sea tower. Waves with the steep slope beyond some chosen threshold values are then considered to be breaking waves. Since the "jump meter" technique is designed for detecting larger waves, many small breaking waves are ignored. It is reasonable to suggest that small breaking waves occur more often than large breaking waves. Hence, Longuet-Higgins and Smith's count would be much lower than all of the other methods described earlier.

It is very interesting to note that curves *a* and *e* encompass all the earlier data points, and curves *b*, *c*, and *d* provide some dividing boundaries to data points from different techniques. The fact that *e* and *c* with

1%>VF>9.30% agrees more with Toba et al. (1971) might imply that Toba and his colleagues have been more careful than the others to pick up many more smaller breaking waves. The observed data by Holthuijsen and Hervers (1986) are bounded nicely by curves *d* and *c* with 2%>VF>1%. The data obtained by Thorpe and Humphries (1980) fall more into the bound defined by curves *b* and *c* with 5%>VF2%. The data presented by Longuet-Higgins and Smith (1983) seem to lie more below curve *b* with VF>5% and above curve *a* with VF>10%.

The surprisingly good overall match between the void fraction event statistics, interpreted as the equivalent breaking wave statistics, from the five previous field measurements, may strongly suggest that the quantitative measure of void fraction is physically better than other techniques for detecting breaking waves and their relative scale and intensity in the open ocean.

Summary

The void fraction technique based on the theory of J.C. Maxwell for detecting and measuring the strength (and thus size) of breaking waves both in the laboratory tank and in the sea under moderate to strong winds are shown to be a simple, but also highly effective methods that can provide more quantitative information than any other technique currently in use. The current limitation of this technique is that the lowest threshold in void fraction is about 0.1%.

References

1. Breitz, N. and H. Medwin. 1989. Instrumentation for In-Situ Acoustical Measurements of Bubble Spectra Under Breaking Waves. JASA 86(2). pp. 739-743.

2. Farmer, D.M. and S. Vagle. 1989. Wave Guide Propagation of Ambient Sound in the Open Surface Bubble Layer. JASA 86(5). pp. 1987-1908.

3. Holthuijsen, L.H. and T.H.C. Herbers. 1986. Statistics of Breaking Waves Observed as Whitecaps in the Open Sea. J. Phys. Oceano. 16. pp 290-297.

4. Longuet-Higgins, M.S. and N.D. Smith, 1983. Measurement of Breaking Waves by a Surface Jump Meter. J. Geophys. Res. 88-C14. pp. 9823-9831.

5. Maxwell, J.C. 1891. A Treatise on Electricity and Magnetism. Dover 1945. p. 314.

6. Rapp, J.J. and W.K. Melville. Laboratory Measurements of Deep-Water Breaking Waves. phil. trans. R. Soc. London, A331 (1990). pp. 735-800.

7. Su, M.Y,. S.C. Ling and J. Cartmill. 1988. Optical Microbubble Measurement in the North Sea in Sea Surface Sound, ed. by B.R. Kerman, pp. 221-224.

8. Thorpe, S.A. and P.N. Humphries. 1980. Bubbles and Breaking Waves. Nature 283. pp. 643-665.

9. Toba, Y., H. Kunishi, K. Hishi, S. Kawai, Y Shimada and N. Shibota. 1971. Study on Air-Sea Boundary Processes at the Shirahama Oceanographic Tower Station. Disaster Prevention Research Institute. Kyoto University Annals. 148. pp. 519-531. (In Japanese with English abstract).

10. Weissman, M.A., S.S. Atakturk and K.B. Katsaros. 1984. Detection of Breaking Events in a Wind-Generated Wave Field. J. Phys. Oceano. 14. pp. 1608-1619.

11. Weller, R.A., M.A. Donelan, M.G. Briscoe and N.E. Husug. 1991. Riding the Crest - A Tale of Two Wave Experiments. Bulletin Am. Met. Soc. 72(2). pp. 163-183.

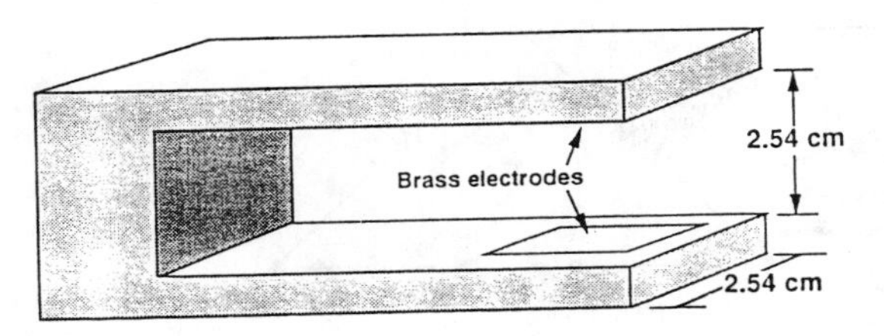

Figure 1. Diagram of void fraction resistance gauge.

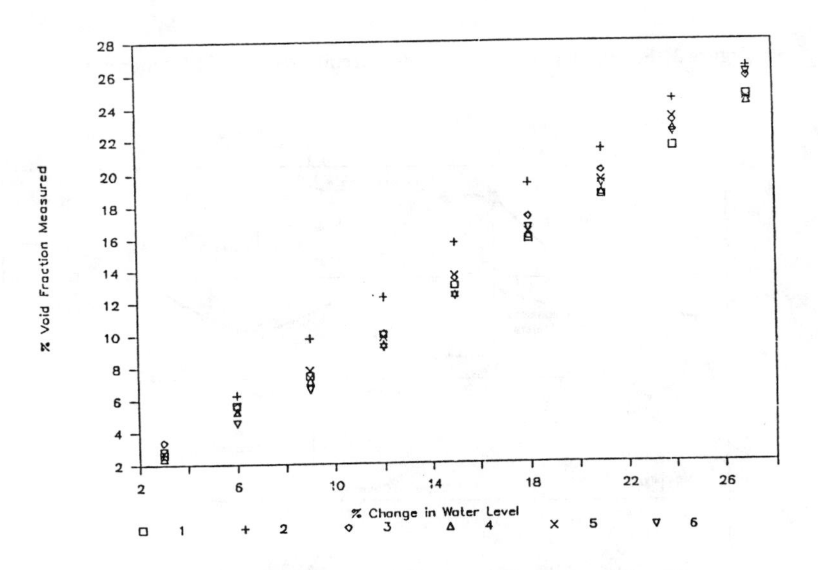

Figure 2. Calibration of 6 void fraction meters.

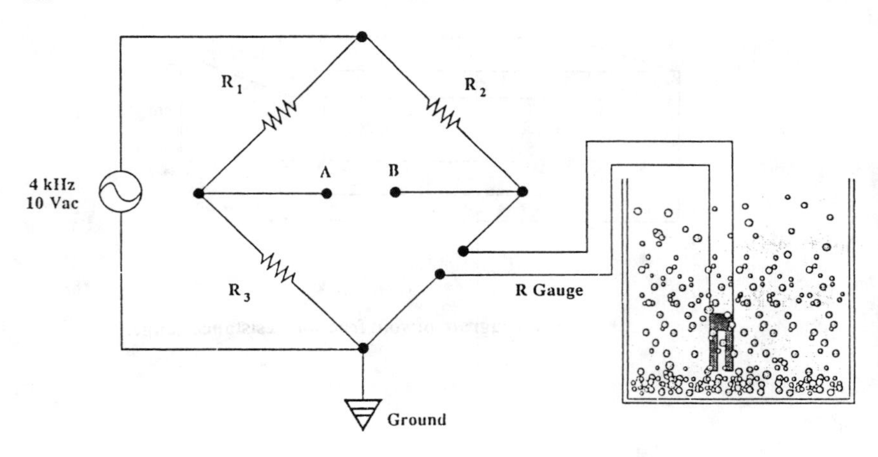

Figure 3. Schematic diagram of bridge circuit used in void fraction meters.

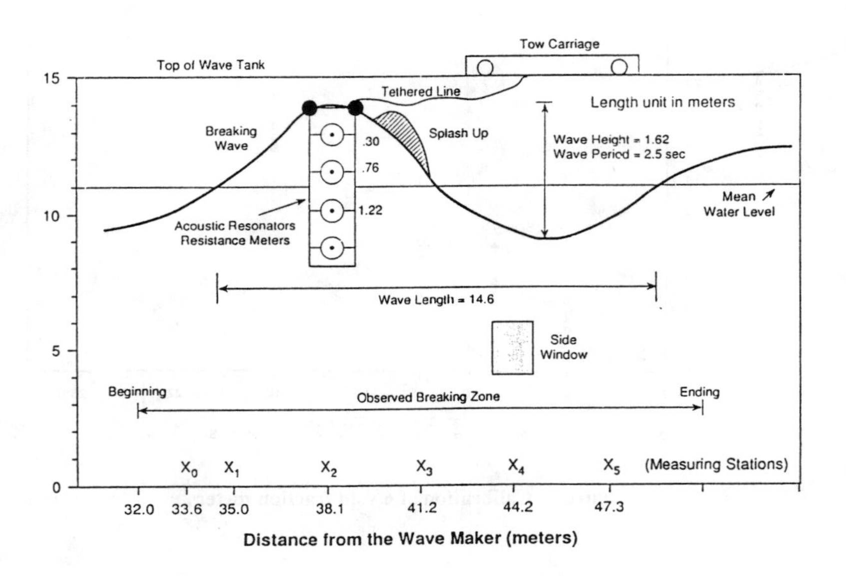

Figure 4. Diagram of Oregon State University Laboratory experiment.

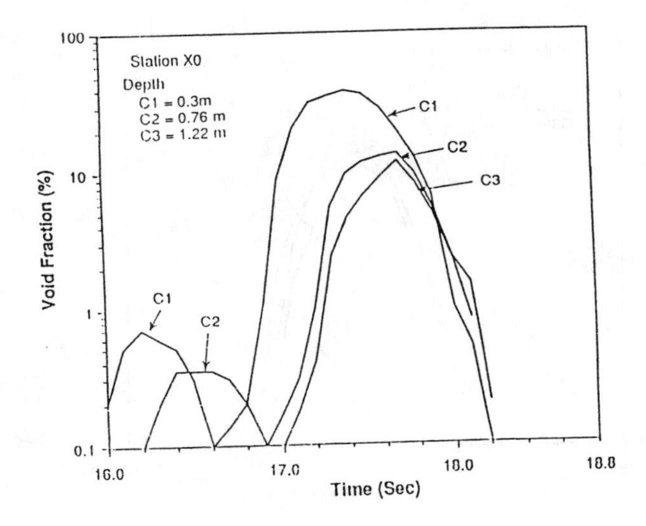

Figure 5. Void fraction versus time station X0.

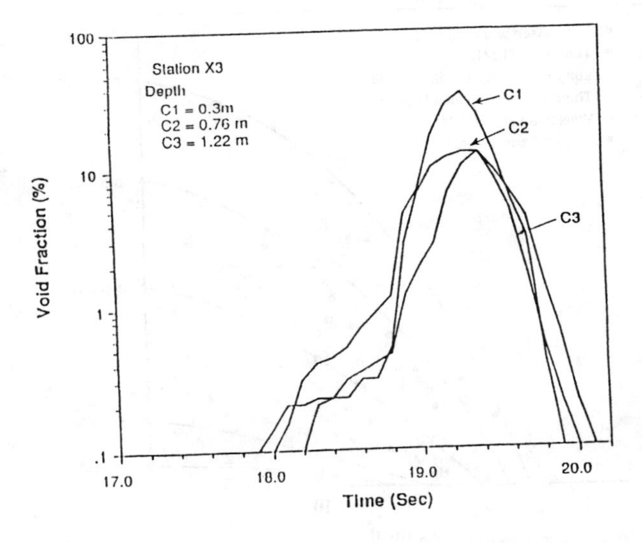

Figure 6. Void fraction versus time station X3.

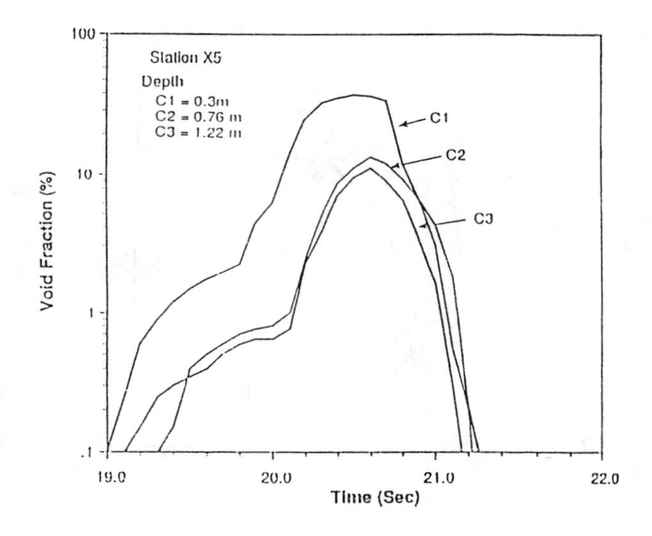

Figure 7. Void fraction versus time station X5.

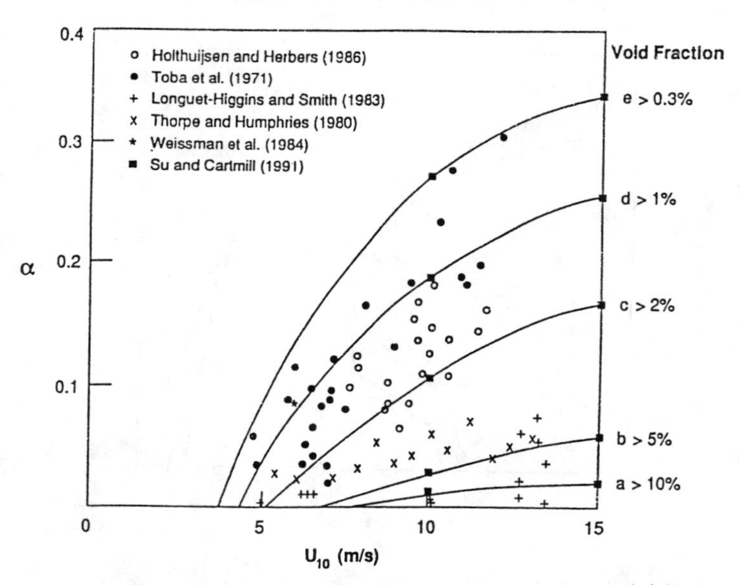

Figure 8. Breaking wave statistics. The probability occurrence(α) in terms of void fraction with comparison to previous field measurements.

INTERCOMPARISON OF EXTREMAL WAVE ANALYSIS METHODS USING NUMERICALLY SIMULATED DATA

Yoshimi Goda,* Peter Hawkes,† Etienne Mansard, ‡
Maria Jesus Martin § Martin Mathiesen, ¶ Eric Peltier, ||
Edward Thompson, ** Gerbrant van Vledder ††

Abstract

Several methods of extremal wave analysis were applied to 1000 samples of numerically simulated data for evaluation of their performance in the estimation of return wave heights. The Weibull distribution with the shape parameter $k = 1.4$ was selected as the parent population, and the FT-I, FT-II and Weibull distributions were fitted to the samples by the Methods of Moments, Least Squares, and Maximum Likelihood. The mean value of the estimated return wave heights was almost the same as the true value, but their statistical deviations were large owing to the sampling variability. For uncensored samples, the Maximum Likelihood Method performed well, but its performance for censored samples was not much different from the other methods.

Introduction

The statistical analysis of extreme storm wave data is an indispensable tool for the rational determination of design wave height for maritime structures. The method of extremal wave analysis is fundamentally the same as that used for storm wind speeds, flood discharges, and other rare events in the nature. The methodology for the statistical analysis of these extreme data, however, is quite diversified depending on the preference of analysts, and there seems to be no consensus as to the most reliable method of extremal data analysis. Extremal wave analysis is further handicapped by the lack of long records of storm wave data.

The Section of Maritime Hydraulics of the International Association for Hydraulic Research organized a Working Group on Extreme Wave Statistics in 1990 with the aim of reaching a mutual understanding of the merits and demerits of the methods used in extreme wave statistical analysis. Eight institutions in Canada, Europe, Japan, and USA joined the Working Group, and Mr. Martin Mathiesen of the Norwegian Hydrotechnical Laboratory has been serving as the Chairman.

*Prof., Yokohama National University, Dept. Civil Eng., Hodogaya-Ku, Yokohama 240, Japan
†HR Wallingford, U.K.
‡National Research Council, Canada
§Centro de Estudios de Puertos y Costas - CEDEX, Spain
¶SINTEF NHL, Norweginan Hydrotechnical Laboratory, Norway
||EDF Laboratoire National d'Hydraulique Chatou), France
**USAE Waterways Experiment Station, Coastal Engineering Research Center, U.S.A.
††Delft Hydraulics, The Netherlands

The first assignment of each member of the Working Group was to analyze two sets of field wave data and to predict the 100-year wave height, using the methods of their choice. The results of this comparative analysis are presented in this Symposium as a separate paper [Van Vledder *et al.*, 1993]. This joint work indicated the necessity for a close review of the merits and demerits of various methods of extremal wave analysis. It was decided to employ the numerically simulated samples of extreme data from a preselected distribution for intercomparison of various methods. The group members were asked to analyze the samples without knowledge of the true distribution and to examine how far the best-fitting distributions for individual samples may deviate from the true distribution. Both uncensored (complete) data and censored data were used, each with 500 samples. The methodology and results of this simulation study are presented in this paper.

Parent Distribution and Preparation of Data Set

The parent distribution was selected as the following Weibull distribution with the shape parameter $k = 1.4$, by referring to one of the field data sets analyzed in the first joint work:

$$F(x) = 1 - \exp\{-[(x - B)/A]^k\} \quad : B \leq x < \infty \tag{1}$$

where $F(x)$ denotes the distribution function of the extreme data (wave height) x. The scale parameter A and the location parameter B were so selected to yield the 1-year wave height $x_1 = 8.00$ m and the 100-year wave height $x_{100} = 13.00$ m at the mean rate of $\lambda = 5$ (five extreme data per year on the average). The parameters were thus given the following values:

$$k = 1.4, \quad A = 2.190484\,\mathrm{m}, \quad B = 4.922738\,\mathrm{m} \tag{2}$$

A sample of data belonging to the distribution of Eqs. 1 and 2 can be generated by Monte Carlo simulation technique. The variate R, uniformly distributed between 0 and 1, is generated by computer, the distribution function F is equated to R, and the extreme data x is obtained as F^{-1}. The uniform variate R was generated by the following pseudo random number generating algorithm:

$$Y = aX_i, \quad X_{i+1} = \mathrm{mod}\{Y, b\}, \quad R_{i+1} = X_{i+1}/q \tag{3}$$

where $\mathrm{mod}\{Y, b\}$ denotes the residue of Y divided by b, and the constants a, b, and q are given the following values:

$$a = 7909, \quad b = 2^{36}, \quad q = 2^{35} - 1 \tag{4}$$

An arbitrary integer (recommended to be an odd number) is input as X_1 to yield X_2, it is then used as the input for X_3, and the cycle is repeated as many times as necessary. If $X_{i+1} > 2^{35}$, then Eq. 3 is modified to use $(2^{36} - X_{i+1})$ as X_{i+1}. This algorithm has been proven to generate a good series of random numbers up to 2×10^7 cycles.

The sample size N, or the number of data per sample, was chosen as 100, which corresponds to the record duration of 20 years with the mean rate of $\lambda = 5$. By Monte Carlo simulation, one thousand samples were generated. The initial 500 samples were used without censoring, constituting the Uncensored Sample Set A. The remaining 500 samples were censored at the threshold level of $x_c = 6.50$ m, and these censored samples were called Set B. The number of data in a censored sample varied from sample to sample. The mean sample size was $\overline{N} = 53.25$ with the standard deviation of $\sigma(N) = 5.31$.

Statistical Properties of Sample Sets

A sample of extreme data drawn from its population exhibits statistical properties different from those of population. The mean and the standard deviation of data in a sample are different from sample to sample, for example. The Sample Sets A and B had the following means and standard deviations:

Uncensored Sample Set A:

Mean of data in a sample: $\overline{x} = 6.925 \pm 0.149\,\text{m}$

Standard deviation of data in a sample: $s_x = 1.441 \pm 0.142\,\text{m}$

Censored Sample Set B:

Mean of data in a sample: $\overline{x} = 7.933 \pm 0.179\,\text{m}$

Standard deviation of data in a sample: $s_x = 1.248 \pm 0.196\,\text{m}$

in which the numerals before and after the double sign ± indicate the mean and standard deviation of the statistic in concern calculated from 500 samples, respectively. The standard deviation s_x is calculated by the following formula:

$$s_x = \left[\frac{1}{N}\sum_{i=1}^{N}(x_i - \overline{x})^2\right]^{1/2} \tag{5}$$

The mean and the standard deviation of the uncensored population of the distribution Eqs. 1 and 2 are calculated as

$$\overline{x} = A\Gamma(1 + 1/k) + B = 6.919\,\text{m}$$

$$\sigma = A\left[\Gamma(1 + 2/k) - \Gamma^2(1 + 1/k)\right]^{1/2} = 1.445\,\text{m}$$

where Γ denotes the gamma function. The mean and the standard deviation of the uncensored samples conforms with the above population values.

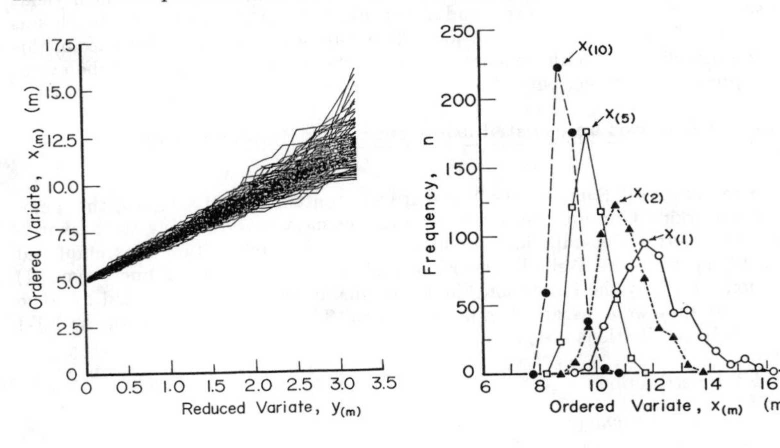

Fig. 1 Plot of 100 samples of simulated storm wave heights.

Fig. 2 Histograms of the largest, the second-largest, the fifth-largest, and the tenth-largest data.

The sampling flucutation of extreme data is demonstrated in Fig. 1, in which the data of the first 100 samples from the Uncensored Sample Set A are plotted together.

The extreme data have been sorted and arranged in descending order from the largest one $x_{(1)}$ to the smallest one $x_{(N)}$. The abscissa is the reduced variate $y_{(m)}$ for the m-th largest data $x_{(m)}$ which is defined by

$$y_{(m)} = \left[-\ln(1 - \hat{F}_m)\right]^{1/k} \tag{6}$$

The plotting position is estimated by

$$\hat{F}_m = 1 - \frac{m - (0.20 + 0.27/\sqrt{k})}{N + 0.20 + 0.23/\sqrt{k}} \tag{7}$$

which was originally given by Petruaskas and Aagaard (1970) as the unbiased plotting position formula for the Weibull distribution and later modified by Goda (1988). The data in the same sample is connected with straight lines. The crisscross of plotting lines indicates the degree of the spread of extreme data in various samples drawn from the same population.

Figure 2 supplies further evidence of sampling variability. The histograms of the largest, the second-largest, the fifth-largest, and the tenth-largest data from the 500 samples of the Uncensored Sample Set A are plotted there. The highest value $x_{(1)}$, for example, varies in the range from 9 to 16 m. The range of variation decreases as the descending order number increases. The lowest value $x_{(100)}$ is confined to the range from 4.9 to 5.2. If the parent distribution were the Fisher-Tippett type I (or Gumbel) distribution (hereinafter abbreviated as FT-I), the lowest values would have shown a much wider variation.

Such statistical variability of samples of extreme data has already been demonstrated by Benson (1952) for the FT-I distribution in the flood-frequency analysis. Though the sampling variability cannot be identified with the field data which yields only a single sample, it causes over- and under- estimations of return wave heights as well as fitting of samples to a distribution different from the true one. Because of this sampling variability, the confidence interval of any return wave height must be evaluated and presented together with the estimate of return height.

Methods of Analysis and Distribution Functions Employed by Group Members

Upon receiving the Sample Sets A and B each containing 500 samples, the members of the Working Group proceeded to perform the data analysis using their favorite methods and distribution functions for fitting. The distribution functions employed for data fitting were the Weibull, the FT-I, and the FT-II (Fisher-Tippett type II) distributions. Initially the Pareto and the log-normal distributions were tried by some members, but they were discarded in later analysis. The functional forms of the FT-I and the FT-II distributions are given by

FT-I distribution:

$$F(x) = \exp\{-\exp[-(x - B)/A]\} \quad : \; -\infty < x < \infty \tag{8}$$

FT-II distribution:

$$F(x) = \exp\left\{-[1 + (x - B)/kA]^{-k}\right\} \quad : \; B - kA \le x < \infty \tag{9}$$

In addition, the truncated Weibull distribution of the following form was used by one member for the analysis of censored samples:

Truncated Weibull distribution:

$$F(x) = 1 - \exp\left\{-[(x - B)/A]^k + [(x_c - B)/A]^k\right\} \quad : x_c \leq x < \infty \tag{10}$$

where x_c denotes the threshold level at which the distribution is truncated.

The methods of extremal analysis employed by the group members could be classified into several categories. Some methods were used by several members. There are ten different combinations of analysis method and distribution function for fitting. They are briefly introduced in the following:

(1) Case A : Least Squares Method 1 with FT-I, FT-II, and Weibull distributions

The Weibull (Eq. 1), the FT-I (Eq. 8), and the FT-II (Eq. 9) are used to fit a sample. The shape parameter k for the Weibull distribution is limited to four particular values, namely 0.75, 1.0, 1.4, and 2.0 as advocated by Goda (1988). The shape parameter k for the FT-II distribution is also limited to the four values of 2.5, 3.33, 5.0, and 10.0 (Goda and Kobune 1990). Thus, nine candidate distributions are computed for best-fitting to a sample of extreme data.

The scale and location parameters A and B are evaluated by the Least Squares Method applied between the ordered data of $x_{(m)}$ and its reduced variate $y_{(m)}$. The linear regression equation is

$$x_{(m)} = Ay_{(m)} + B \tag{11}$$

The plotting position for calculation of reduced variate is estimated with Eq. 7 for the Weibull, the formula by Gringorten (1963) for the FT-I, and an empirical formula by Goda and Kobune (1990) for the FT-II distribution. For a censored sample, the total size of sample N_T before censoring is used instead of the sample size N after censoring (Muir and El-Shaarawi 1986).

The selection of the best-fitting distribution is made with the MIR criterion, which is based on the correlation coefficient between $x_{(m)}$ and $y_{(m)}$. Instead of simply taking the largest correlation coefficient as representative of the best-fitting, the residues of correlation coefficients from 1 are calculated for all the candidate distributions and their ratios to the respective population values are compared. Then the distribution which shows the minimum ratio is selected as the best-fitting one.

The 90% confidence interval of return wave height is estimated as $\pm 1.64\sigma(x_R)$ by assuming the normal distribution of the sample estimates of return wave height. The standard deviation of return wave height $\sigma(x_R)$ is estimated with empirical formulas derived through Monte Carlo simulation studies (Goda 1988, Goda and Kobune 1990).

(2) Case B : Least Squares Method 1 with FT-I and Weibull distributions

The FT-II distribution is dropped as a candidate for data fitting. The FT-I distribution and the four Weibull distributions remain as the candidates. The rest is the same as Case A.

(3) Case C : Least Squares Method 1 with Weibull distributions

Only the Weibull distributions with four fixed values of the shape parameters are employed as the candidate distributions of data fitting. The rest is the same as described in Case A.

(4) Case D : Least Squares Method 2 with Weibull distributions

The MIR criterion for the selection of the best-fitting distribution was replaced by a simple method of choosing the distribution with the largest correlation coefficient between $x_{(m)}$ and $y_{(m)}$. The rest is the same as Case C.

(5) Case E : Least Squares Method 3 with FT-I and Weibull distributions

The FT-I and the Weibull distributions are employed as the candidate distributions. The Least Squares Method listed in Case A is modified for the Weibull distribution by one of the group members, by employing a finer resolution of the shape parameter. The empirical formulas for the goodness of fit MIR criterion and for the computation of the standard deviation of return wave heights have been extended by local interpolation and extrapolation. Similar technique was also adopted to compute the DOL and REC rejection criteria by Goda and Kobune (1990), and the all samples were subjected to the rejection test.

In assessing the plotting position for a censored sample, the actual sample size N after censoring is used as part of the investigation to evaluate the recommendation by Muir and El-Shaarawi (1986).

(6) Case F : Least Squares Method 4 and Weibull Distribution

The candidate distribution is the Weibull distribution only. In this case, the location parameter B is preset at some value and the shape parameter k and the scale parameter A are estimated by solving the following linear regression equation. The value of location parameter B is then varied and the process is repeated until the optimization is reached.

$$y_{(m)} = kX_{(m)} + C \tag{12}$$

where

$$X_{(m)} = \ln[x_{(m)} - B] \tag{13}$$

$$y_{(m)} = \ln[-\ln(1 - \hat{F}_m)] \tag{14}$$

and the scale parameter is retrieved as

$$A = \exp[-C/k] \tag{15}$$

The optimization of B is judged by the value of the correlation coefficient between $X_{(m)}$ and $y_{(m)}$.

The plotting position in Eq. 14 is estimated by the Weibull formula of

$$\hat{F}_m = 1 - \frac{m}{N+1} \tag{16}$$

The actual sample size N is used for the censored samples in Set B.

In Case F, the estimate of confidence interval has not been given.

(7) Case G : Method of Moments with Weibull Distribution

The mean $\overline{x}$, the standard deviation σ_x, and the skewness γ of a sample are calculated. The three parameters of A, B, and k of the Weibull distribution are then obtained by numerically solving the following nonlinear equation with an iterative procedure:

$$\overline{x} = A\Gamma(1 + 1/k) + B \tag{17}$$

$$\sigma_x = A\left[\Gamma(1+2/k) - \Gamma^2(1+1/k)\right]^{1/2} \quad (18)$$

$$\gamma = A^3\left[\Gamma(1+3/k) - 3\Gamma(1+1/k)\Gamma(1+2/k) + 2\Gamma^3(1+1/k)\right]/\sigma_x^3 \quad (19)$$

The Method of Moments was applied to the Uncensored Sample Set A only, because the bias correction for the standard deviation for censored samples is not known. The confidence interval of return wave heights was calculated for each sample through 1000 simulations by Monte Carlo technique by using the parameters estimated.

(8) Case H : Maximum Likelihood Method 1 with FT-I and Weibull distributions

The three-parameter Weibull distribution is converted to the two-parameter distribution by preselecting the location parameter B by inspection of data in a sample. It was set at the lowest value -0.1 m for all the samples in Sets A and B. The shape parameter k and the scale parameter A are so set to maximize the likelihood function or its logarithm. This condition is realized when the following two equations hold:

$$\frac{N}{k} + \sum_{i=1}^{N} \ln X_i - \frac{N\sum_{i=1}^{N} X_i^k \ln X_i}{\sum_{i=1}^{N} X_i^k} = 0 \quad (20)$$

$$A = \left[\frac{1}{N}\sum_{i=1}^{N} X_i^k\right]^{1/k} \quad (21)$$

where X_i stands for $x_i - B$. The shape parameter k is obtained by iteratively solving Eq. 20, and then the scale parameter A is estimated with Eq. 21.

The scale and location parameters of the FT-I distribution are obtained by solving similar equations for the maximization of the likelihood function. Between the FT-I and the Weibull distributions thus fitted to a sample, the one with the larger correlation coefficient between the extreme data and its prediction is chosen as the best-fitted distribution. Selection between the two distributions was also tried with the χ^2 value and the Kolmogorov- Smirnov test, but the results were almost identical with that by the correlation coefficient.

The 90% confidence interval of return wave height is estimated with the theoretical value of standard deviation by assuming the normal distribution of return wave heights.

(9) Case I : Maximum Likelihood Method 2 with Weibull distributions

This is almost the same as Case H except that the FT-I distribution is dropped from the candidate distributions for data fitting. The location parameter was set at B = 4.90 m for the Uncensored Sample Set A and B = 6.49 m for the Censored Sample Set B, both of which were the lowest value of 500 samples, respectively. The algorithm for solving Eq. 20 as well as the formula for the estimate of confidence interval are different from those of Case H.

(10) Case J : Maximum Likelihood Method 3 with truncated Weibull distributions

The truncated Weibull distribution of Eq. 10 is employed as the candidate distribution for data fitting for samples in Set B. The location parameter was set at B = 4.60 m and the threshold level was given as x_c = 6.49 m.

The nonlinear equations for the condition of the maximum likelihood are

$$\frac{N}{k} + \sum_{i=1}^{N} \ln X_i - \frac{N\left[\frac{1}{N}\sum_{i=1}^{N} X_i^k \ln X_i - X_c^k \ln X_c\right]}{\left[\frac{1}{N}\sum_{i=1}^{N} X_i^k - X_c^k\right]} = 0 \tag{22}$$

$$A = \left[\frac{1}{N}\sum_{i=1}^{N} X_i^k - X_c^k\right]^{1/k} \tag{23}$$

where X_i and X_c stand for $(x_i - B)$ and $(x_c - B)$, respectively.

The confidence interval of return wave heights was calculated for each sample through 1000 simulations by Monte Carlo technique by using the estimated values of parameters.

Results of Extremal Analysis of Simulated Wave Data

The group members analyzed the whole samples in the Uncensored Set A and the Censored Set B by using the methods described above. They were asked to report the following information:

1) The frequencies of extremal distributions judged as best-fitted to samples.
2) The estimates of the average and standard deviation of the 100-year return wave height in various classes of extremal distributions.
3) The half range of the confidence interval of the estimated 100-year return wave height in various classes of extremal distributions.
4) The number of samples for which the true 100-year return wave height of $x_{100} =$ 13.0 m fell within the range of the estimated 90% confidence interval.

The results of their reports are summarized in Tables A-1 to A-3 for the Uncensored Sample Set A and Tables B-1 to B-3 for the Censored Sample Set B. The Weibull distribution is classified by the value of shape parameter k with the class interval of $\Delta k = 0.2$; thus, the entry of $k = 1.2$ in Tables A-1 to B-3 stands for the range of $1.1 < k \leq 1.3$, for example. If the distributions fitted to individual samples gather closely around the true distribution of the Weibull with $k = 1.4$, then the method of analysis is a good one. If the fitted distributions show a wide scatter, then the method of analysis is an inferior one. Tables A-1 and B-1 list the frequencies of various classes of extremal distributions having been judged as best-fitted to samples.

The estimate of 100-year return wave height should be close to the true value as much as possible, but the sampling variability illustrated in Figs. 1 and 2 produces a certain amount of scatter in the estimate of return wave heights. The goodness of the method of extremal analysis is evaluated by the mean and the standard deviation of estimates of return wave heights. If the average value of estimates differs from the true value, the method is said to yield a bias. The standard deviation of estimates is a measure of the efficiency of the method of extremal analysis. The smaller the standard deviation is, the more efficient the method is. In Table A-2 and B-2, the average of estimated return heights is listed for each class as well as for all samples combined. The standard deviation is listed only for all samples combined. This overall value of standard deviation is calculated either by direct computation or by the following formula with the class-wise mean and standard deviation:

$$\sigma(x_R) = \left[\frac{1}{N}\sum_{i=1}^{m} n_i(\overline{x_i}^2 + \sigma_i^2) - \overline{x_R}^2\right]^{1/2} \tag{24}$$

in which m is the number of the class of distributions, n_i , $\overline{x_i}$, and σ_i denote the frequencies, mean, and standard deviation of return wave heights in the i-th class, respectively, and $\overline{x_R}$ is the mean of all samples.

Tables A-3 and B-3 list the half range of 90% confidence interval of the estimate of 100-year return wave height for each class of extremal distribution, the overall mean of half range, and the number of samples for which the true 100-year return wave height is located within the estimated range of 90% confidence interval. The overall mean of the half range of 90% confidence interval has been calculated as a measure of its magnitude for the purpose of intercomparison between various methods of extremal analysis. The calculation is made in analogy with the overall value of standard deviation by using the following formula:

$$\overline{Q}_{100} = \left[\frac{1}{N}\sum_{i=1}^{m} n_i(\overline{x_i}^2 + Q_i^2) - \overline{x_R}^2\right]^{1/2} \tag{25}$$

Because Sets A and B contain 500 samples each, about 450 samples from each set should produce the confidence intervals within which the true return wave height of x_{100} = 13.0 m falls. The number of samples less than about 450 implies a poor performance of the method of extremal analysis in predicting the confidence interval.

Discussion of The Results of Uncensored Samples

For uncensored (complete) samples, all methods perform quite well even though the true distribution is unknown. Inclusion of the FT-II distribution as the candidate (Case A), however, produces an appreciable amount of positive bias in the estimate of the 100-year return wave height. The confidence interval of the estimated return wave height for the samples fitted to the FT-II distribution with k = 3.33 or 5.0 is so large that the selection of a design wave height from such a wide interval would become very difficult.

Inappropriateness of the FT-II distribution as the candidate of fitting to samples is also apparent in the case of censored samples. The inappropriateness, however, should not be taken as the general rule. The present simulation analysis employs the Weibull distribution with k = 1.4 having a relatively narrow tail as the parent distribution. A possibility that the FT-II distribution constitutes the population distribution of storm waves in some ocean areas cannot be denied. Thom (1973a and 1973b) argued for the applicability of the FT-II distribution for the extreme winds and wave heights over oceans. There is also some evidence that the annual maxima of daily rainfall fit to the FT-II distribution with k = 10 better than to the FT-I or the Weibull distributions (Goda 1990).

Among Cases B to I, the Maximum Likelihood Method yielded the best performance. The method shows the least dispersion from the true distribution, the least amount of the standard deviation of return height estimates, and good prediction of confidence interval. This good performance would have been assisted by the nature of the Weibull distribution having the lower limit in variate ($x \geq B$). Because the lowest value in a sample drawn from the Weibull population exhibits only a small fluctuation, the location parameter B could be set rather accurately a priori. The other methods could not take advantage of this fact, and thus the estimated value of location parameter had a scatter around the true value of B = 4.92 m. Accordingly, the distributions fitted to samples showed wider spreading around the true one than did the Maximum Likelihood Method. A wider spreading in turn caused an increase in the standard deviation of return height estimates, decreasing the efficiency of the methods. A wider spreading also deteriorated the ability to predict the confidence interval of estimated return wave height in the cases using the Least Squares Method. It might be recommended for application of the Least Squares Method to estimate the shape parameter k and the scale parameter A after presetting the location parameter B slightly below the value of smallest data in a sample.

The bias or the deviation of the overall mean of return wave height from the true value is relatively small nevertheless, being about 1% to 2% except for Case A. The Least Squares Method has a tendency toward positive bias, while the Method of Moments and the Maximum Likelihood Method tend to give negative bias. The bias is insignificant for the samples fitted to the class of Weibull distribution with $k = 1.4$ which is the same as the true distribution. The bias originates from incorrect fitting of a sample to a distribution different from the true distribution. Therefore, it can be said that the method of analysis itself functions correctly but the lack of the information about the true distribution causes bias.

The standard deviation of the estimated 100-year wave height is 0.8 to 1.2 m except for Case A. The Maximum Likelihood Method yields a smaller amount of standard deviation than the other two methods, indicating its superior efficiency in the estimate of return value for uncensored data.

The confidence interval of estimated return wave height increases as the tail of the fitted distribution becomes elongated as seen in Table A-3; the smaller the k-value of the Weibull and FT-II distribution is, the longer the tail of the distribution becomes. The overall mean of estimated half range of the 90% confidence interval varies from 1.4 to 2.0 m. With the application of the Maximum Likelihood Method, approximately 90% of samples yielded an estimate of 100-year return wave height which is within the 90% confidence interval from the true value. The prediction of confidence interval by Monte Carlo simulation for the samples analyzed by the Method of Moments is acceptable, though the range of confidence interval itself is larger than those by other methods. The prediction of confidence interval by empirical formulas for the samples analyzed by the Least Squares Method is not so good; it is due to a wide spread of best-fitting distributons around the true distribution.

Discussion of The Results of Censored Samples

Compared with the case of uncensored samples, the censored data present much greater difficulty in obtaining a result close to the population values. As seen in Table B-1, the Least Squares Methods 3 and 4 (Cases E and F) as well as the Maximum Likelihood Methods 1 and 2 (Cases H and I) yield the prediction of best-fitting distribution skewed toward the functions with longer tails. The weighted mean of the shape parameter k for the samples fitted to the Weibull distribution in these four cases is about 1.2. The distortion in the functional shape seems to have originated from the use of the actual sample size N instead of the total sample size N_T in the assessment of plotting position in Cases E and F, which employ the Least Squares Method. By cutting the data below the threshold level by censoring and by neglecting the existence of such data, the histogram of sampled data is forced to have a sharp rise at the threshold level. Such a sharply-rising histogram is likely to be fitted to the Weibull distribution with a small k value. In Cases H and I which employ the Maximum Likelihood Method, setting the location parameter B slightly below the threshold level of $x_c = 6.5$ m (despite the true value of $B = 4.92$ m) has the same effect of functional distortion as the use of actual sample size N in the assessment of plotting position in the Least Squares Method. Development of a new algorithm for the Maximum Likelihood Method which takes account of data censoring in the analysis will be required for reliable estimation of distribution functions.

In Cases B to D which employ the total size of sample N_T in the assessment of plotting position, the weighted mean of the shape parameter for samples fitted to the Weibull distribution is close to the true value of 1.4. Thus the use of the total sample size N_T produces a better estimation of the shape parameter than the use of the actual sample size N. The application of the truncated Weibull distribution (Eq. 10) to the censored samples, as represented by Case J is effective in recovering the original shape of the distribution function, though it produces a slight bias toward a large k value and a tendency toward large spreading of the k value.

The bias of the overall mean of the estimated 100-year wave height in the censored samples is about 1% to 3% except for Case A which includes the FT-II distribution as the candidate for data fitting: the magnitude of bias is about the same as that for the uncensored samples. The Least Squares Method yields a tendency toward positive bias. The amount of bias associated with the Maximum Likelihood Method seems to depend on the algorithm of numerical solution, because Case I yields a positive bias but Cases H and J show a negative bias though small in quantity. The standard deviation of the estimated 100-year return wave height in the censored samples is 1.1 to 1.4 m, which is greater than that in the uncensored samples. The increase is large for the cases analyzed with the Maximum Likelihood Method and relatively small with the Least Squares Method. As a result, the amount of standard deviation is about the same regardless of the methodology of extremal analysis for the censored samples. Thus the difference in the efficiency of analytical methods diminishes for the censored samples.

The half range of the 90% confidence interval of 100-year return wave height for the censored samples is larger than that for the uncensored samples. The overall mean of the estimated half-range varies from 1.9 to 2.7 m depending on the methodology used. This is an increase of 25% to 75% from that of uncensored samples. Generally speaking, for the cases in which the overall mean of estimated half-range of confidence interval is large, the percentage of samples for which the estimated 100-year wave height remains within the 90% confidence interval around the true value becomes high. In comparison with the cases of uncensored samples listed in Table A-3, the difference between the Least Squares Method and the Maximum Likelihood Method with regard to the predictability of the confidence interval is rather small.

Among the results obtained by the Least Squares Method, a simple criterion of best-fitting using the absolute value of correlation coefficient produces slightly better results than the MIR criterion for both the uncensored and censored samples. Compared with Case C employing the MIR criterion, Case D with the simple criterion yields a narrower spread of best-fitting distributions, smaller bias, more efficiency (smaller standard deviation), and narrower range of 90% confidence interval. The advantage of the simple criterion over the MIR criterion is probably confined to the samples from populations with a short tail of distribution. Samples from the Weibull distribution with $k = 1.0$ or less would yield results of opposite tendency.

Concluding Remarks

It should be emphasized that the fluctuation of samples drawn from the same population of extreme storm waves is very large, and beyond the range ordinarily considered by people. The sample size employed in this simulation study is 100 for uncensored samples and 52.3 on average for censored ones, which are larger than those encountered in most real storm wave data sets. In spite of such a good sample size, difficulties have been observed in finding the true distribution function and in making reliable estimates of return wave heights. Difficulties will be reduced by employing samples of much larger size. Collection of storm wave data over a long time span would be necessary to increase the sample size. Use of many storm peaks, say 20 storms per year, is another approach, but care should be taken to ensure the storm wave data belongs to the same population.

Under the circumstance of large sampling variability and no knowledge of the true distribution function, all the methods of extremal wave analysis practiced by the member institutions yielded an estimate of the 100-year return wave height very close to the true value. The bias was up to about 3%. Individual estimates from single samples varied considerably from the mean with a standard deviation of about 6% to 12% depending on the methodology and the degree of data censoring.

For the analysis of uncensored samples, the Maximum Likelihood Method exhibited performance superior to other methods. For the analysis of censored samples, however, the Maximum Likelihood Method proved to be less efficient. In order to improve the reliability of return wave height estimates from the Maximum Likelihood Method, it would be neccessary to develop a scheme with due consideration for data censoring.

The conclusions drawn from the present simulation study are confined to the case of the population defined by a Weibull distribution with the shape parameter $k = 1.4$. Different conclusions should be expected when the true distribution differs from the one employed here. The most pressing question in extremal wave analysis is to determine the population distribution of extremal storm waves in various oceans. There is a possibility that the population distribution depends on the nature of storms which generate extreme waves and thus on the locality of oceans. Effort should be dedicated to the search and for clarification of the population distributions of extremal waves.

Provision of Data Set: Copies of the Sample Sets A and B can be obtained from any member of the Working Group for the purpose of examining and improving the methodology of extremal wave analysis.

References

Benson, M.A. [1952]: Characteristics of frequency curves based on theoretical 1000 years record, Reproduced in *Flood-Frequency Analysis*, U.S. Geological Survey Water Supply Paper 1543-A (1960).

Goda, Y. [1988]: On the methodology of selecting design wave height, *Proc. 21st Int. Conf. Coastal Engg.*, Malaga, pp.899-913.

Goda, Y. [1990]: Rejection criteria for outliers and their applications in extreme statistics, *Proc. Japan Soc. Civil Engrs.*, No.417/II-13, pp.245- 254 (*in Japanese*).

Goda, Y. and Kobune, K. [1990]: Distribution function fitting for storm wave data, *Proc. 22nd Int. Conf. Coastal Engg.*, Delft, pp.18-31.

Goda, Y. [1992]: Uncertainty in design parameter from the viewpoint of statistical variability, *J. Offshore Mechanics and Arctic Engg., Trans. ASME,* Vol.114, May, pp.76-82.

Gringorten, I.I. [1963]: A plotting rule for extreme probability paper, *J. Geophys. Res.*, Vol.68, No.3, pp.813-814.

Muir, L.R. and El-Shaarawi, A.H. [1986]: On the calculation of extreme wave heights: a review, *Ocean Engg.*, Vol.13, No.1, pp.93-118.

Petruaskas, C. and Aagaard, P.M. [1970]: Extrapolation of historical storm data for estimating design wave heights, *Prepr. 2nd Ann. Offshore Tech. Conf.*, OTC1190.

Thom, H.C.S. [1973a]: Distributions of extreme winds over oceans, *Proc. ASCE,* Vol.99, No.WW1, pp.1-17.

Thom, H.C.S. [1973b]: Extreme wave height distributions over oceans, *Proc. ASCE,* Vol.99, No.WW3, pp.355-374.

Van Vledder, G.Ph, Goda, Y., Hawkes, P., Mansard, E., Martin, M.J., Mathiesen, M., Peltier, E., and Thompson, E. [1993]: A case study of extreme wave analysis: a comparative analysis, *Proc. WAVES '93 Conf., 26-28 July, New Orleans, USA.*

Table A-1. Frequencies of Best-Fitted Distributions, n, for Uncensored Samples

Case	Method	FT-II with k-value of				FT-I	Weibull with representative k-value of									Total
		2.5	3.33	5.0	10.0		<0.9	1.0	1.2	1.4	1.6	1.8	2.0	2.2	>2.3	
A	LSQ1	0	1	27	138	37	0	18	-	235	-	-	45	-	-	500
B	LSQ1	-	-	-	-	72	5	93	-	285	-	-	45	-	-	500
C	LSQ1	-	-	-	-	-	5	99	-	348	-	-	48	-	-	500
D	LSQ2	-	-	-	-	-	0	65	-	370	-	-	65	-	-	500
E	LSQ3	-	-	-	-	37	13	40	164	114	94	30	6	0	0	498[a)]
F	LSQ4	-	-	-	-	-	1	29	129	161	125	38	15	2	0	500
G	MoM	-	-	-	-	-	1	20	95	142	138	69	28	6	1	500
H	MLM1	-	-	-	-	109	0	0	30	251	104	6	0	0	0	500
I	MLM2	-	-	-	-	-	0	0	45	312	133	10	0	0	0	500

Note: a) In this analysis, 2 samples were rejected for fitting to any distribution by the DOL and REC criteria.
b) The symbol "-" indicates that the samples were not fitted to this distribution.

Table A-2. Estimates of the Mean Value of 100 Year Wave Height, x_{100}, for Uncensored Samples

Case	Method	Mean Value of 100 Year Wave Height, x_{100} (m)														All Mean	Stand. Dev.
		FT-II with k-value of				FT-I	Weibull with representative k-value of										
		2.5	3.33	5.0	10.0		<0.9	1.0	1.2	1.4	1.6	1.8	2.0	2.2	>2.3	$\overline{x}_{100}$	$\sigma(x_{100})$
A	LSQ1	-	17.5	17.1	14.9	12.8	-	15.4	-	13.0	-	-	11.6	-	-	13.71	1.52
B	LSQ1	-	-	-	-	13.0	17.0	15.0	-	13.0	-	-	11.6	-	-	13.31	1.23
C	LSQ1	-	-	-	-	-	17.0	15.0	-	13.0	-	-	11.6	-	-	13.28	1.23
D	LSQ2	-	-	-	-	-	-	15.1	-	13.0	-	-	11.6	-	-	13.11	1.14
E	LSQ3	-	-	-	-	13.0	15.9	14.9	13.7	13.1	12.2	11.8	11.3	-	-	13.25	1.14
F	LSQ4	-	-	-	-	-	17.5	15.5	13.9	13.2	12.4	11.9	11.6	11.1	-	13.18	1.16
G	MoM	-	-	-	-	-	14.8	15.0	13.8	13.1	12.5	11.8	11.5	11.4	11.2	12.85	1.08
H	MLM1	-	-	-	-	12.8	-	-	13.9	13.1	12.1	11.8	-	-	-	12.86	0.77
I	MLM2	-	-	-	-	-	-	-	14.0	13.0	12.2	11.6	-	-	-	12.85	0.83

Note: 1) The standard deviation $\sigma(x_{100})$ is calculated for the estimated 100 year wave heights derived from the whole samples.
2) The symbol "-" indicates that the samples were not fitted to this distribution.

Table A-3. Estimates of the Mean Half-Range of Confidence Interval of 100 Year Wave Height, Q_{100}, for Uncensored Samples

Case	Method	Mean Half-Range of Cofidence Interval of 100 Year Wave Height, Q_{100} (m)														All Mean	Sample Nos. within Confid.
		FT-II with k-value of				FT-I	Weibull with representative k-value of										
		2.5	3.33	5.0	10.0		<0.9	1.0	1.2	1.4	1.6	1.8	2.0	2.2	>2.3	$\overline{Q}_{100}$	Interval
A	LSQ1	-	4.36	3.32	1.45	1.11	-	2.22	-	1.16	-	-	0.70	-	-	1.99	346 / 500
B	LSQ1	-	-	-	-	1.15	3.92	2.12	-	1.16	-	-	0.65	-	-	1.73	385 / 500
C	LSQ1	-	-	-	-	-	3.92	2.10	-	1.15	-	-	0.69	-	-	1.74	381 / 500
D	LSQ2	-	-	-	-	-	-	2.14	-	1.16	-	-	0.70	-	-	1.49	376 / 500
E	LSQ3	-	-	-	-	1.15	3.08	2.09	1.52	1.20	0.96	0.83	0.65	-	-	1.70	381 / 498
G	MoM	-	-	-	-	-	3.50	2.95	2.22	1.82	1.58	1.19	1.07	0.95	0.90	1.98	437 / 500
H	MLM1	-	-	-	-	0.93	-	-	1.80	1.50	1.20	1.04	-	-	-	1.43	446 / 500
I	MLM2	-	-	-	-	-	-	-	1.72	1.40	1.13	0.94	-	-	-	1.45	441 / 500

Note: 1) The all mean value of Q_{100} is calculated by taking the mean value of x_{100} for respective distribution into consideration in the manner similar as the standard deviation of x_{100}.
2) The symbol " - " indicates that the samples were not fitted to this distribution.

Table B-1. Frequencies of Best-Fitted Distributions, n, for Censored Samples

Case	Method	FT-II with k-value of				FT-I	Weibull with representative k-value of									Total
		2.5	3.33	5.0	10.0		<0.9	1.0	1.2	1.4	1.6	1.8	2.0	2.2	>2.3	
A	LSQ1	6	10	31	66	73	13	36	-	109	-	-	156	-	-	500
B	LSQ1	-	-	-	-	90	70	75	-	109	-	-	156	-	-	500
C	LSQ1	-	-	-	-	-	70	114	-	160	-	-	156	-	-	500
D	LSQ2	-	-	-	-	-	37	123	-	139	-	-	201	-	-	500
E	LSQ3	-	-	-	-	10	74	118	176	58	47	10	2	1	0	496[a]
F	LSQ4	-	-	-	-	-	39	111	151	117	57	19	5	1	0	500
H	MLM1	-	-	-	-	47	0	46	279	114	14	0	0	0	0	500
I	MLM2	-	-	-	-	-	6	177	258	55	4	0	0	0	0	500
J	MLM3	-	-	-	-	-	31	35	66	98	100	70	42	29	29	500

Note: a) In this analysis, 4 samples were rejected for fitting to any distribution by the DOL and REC criteria.
b) The symbol " - " indicates that the samples were not fitted to this distribution.

Table B-2. Estimates of the Mean Value of 100 Year Wave Height, x_{100}, for Censored Samples

Case	Method	Mean Value of 100 Year Wave Height, x_{100} (m)														All Mean	Stand. Dev.
		FT-II with k-value of				FT-I	Weibull with representative k-value of										
		2.5	3.33	5.0	10.0		<0.9	1.0	1.2	1.4	1.6	1.8	2.0	2.2	>2.3	$\overline{x}_{100}$	$\sigma(x_{100})$
A	LSQ1	17.9	17.5	15.6	14.5	13.4	15.9	14.1	-	13.2	-	-	12.2	-	-	13.52	1.63
B	LSQ1	-	-	-	-	13.3	15.3	14.2	-	13.2	-	-	12.2	-	-	13.31	1.39
C	LSQ1	-	-	-	-	-	15.3	14.0	-	13.1	-	-	12.2	-	-	13.28	1.39
D	LSQ2	-	-	-	-	-	15.6	14.0	-	13.2	-	-	12.3	-	-	13.19	1.34
E	LSQ3	-	-	-	-	11.8	15.3	13.8	13.1	12.5	12.0	11.5	12.2	11.9	-	13.36	1.34
F	LSQ4	-	-	-	-	-	16.5	14.2	13.4	12.6	12.1	11.8	11.4	11.5	-	13.39	1.55
H	MLM1	-	-	-	-	12.1	-	14.4	13.0	12.2	11.6	-	-	-	-	12.82	1.05
I	MLM2	-	-	-	-	-	15.3	14.2	13.1	11.9	11.4	-	-	-	-	13.44	1.25
J	MLM3	-	-	-	-	-	15.7	14.3	13.7	13.2	12.7	12.3	11.8	11.5	11.1	12.93	1.33

Note: 1) The standard deviation $\sigma(x_{100})$ is calculated for the estimated 100 year wave heights derived from the whole samples.
2) The symbol "-" indicates that the samples were not fitted to this distribution.

Table B-3. Estimates of the Mean Half-Range of Confidence Interval of 100 Year Wave Height, Q_{100}, for Censored Samples

Case	Method	Mean Half-Range of Cofidence Interval of 100 Year Wave Height, Q_{100} (m)														All Mean	Sample Nos. within Confid. Interval
		FT-II with k-value of				FT-I	Weibull with representative k-value of										
		2.5	3.33	5.0	10.0		<0.9	1.0	1.2	1.4	1.6	1.8	2.0	2.2	>2.3	$\overline{Q}_{100}$	
A	LSQ1	8.7	6.1	3.8	2.6	1.7	3.8	2.2	-	1.6	-	-	1.2	-	-	2.71	421 / 500
B	LSQ1	-	-	-	-	1.7	3.6	2.3	-	1.6	-	-	1.2	-	-	2.27	419 / 500
C	LSQ1	-	-	-	-	-	3.5	2.2	-	1.6	-	-	1.2	-	-	2.17	381 / 500
D	LSQ2	-	-	-	-	-	3.7	2.3	-	1.6	-	-	1.2	-	-	1.94	412 / 500
E	LSQ3	-	-	-	-	1.1	3.4	2.2	1.6	1.3	1.0	0.8	0.7	0.6	-	2.27	423 / 496
H	MLM1	-	-	-	-	1.0	-	2.8	2.0	1.5	1.2	-	-	-	-	2.01	429 / 500
I	MLM2	-	-	-	-	-	3.3	2.4	1.8	1.3	1.0	-	-	-	-	2.20	463 / 500
J	MLM3	-	-	-	-	-	5.1	3.4	2.6	2.0	1.8	1.4	1.4	1.1	0.9	2.53	427 / 500

Note: 1) The all mean value of Q_{100} is calculated by taking the mean value of Q_{100} for respective distribution into consideration in the manner similar as the standard deviation of x_{100}.
2) The symbol "-" indicates that the samples were not fitted to this distribution.

Case studies of extreme wave analysis: a comparative analysis

Gerbrant van Vledder[1], Yoshimi Goda[2], Peter Hawkes[3], Etienne Mansard[4], Maria Jesus Martin[5], Martin Mathiesen[6], Eric Peltier[7], Edward Thompson[8]

Abstract

Several methods of extremal wave analysis have been applied to two sets of deep water extreme wave data. One set consisted of three-hourly sea-state records collected during a nine year period with a wave buoy at Haltenbanken off the Norwegian coast. The other set comprised a 20-year period of storm peak wave heights, obtained via a numerical hindcast of historical storms. For both data sets, 100-year return wave heights with corresponding 90% confidence intervals were computed with a variety of extremal analysis techniques, including the Initial Distribution Method, the Annual Maxima Method, and the Peak over Threshold Method. FT-I, Weibull and other probability distributions were fitted to the data by the Method of Moments, Least Squares Method, and the Maximum Likelihood Method. The results of the study show that for both data sets the estimated 100-year return wave heights differ less than 10% of one another. It was also found that the effect of choices in the data selection, e.g. the choice of the threshold, has a significant effect on the estimated wave heights.

1 Introduction

The statistical analysis of extreme wave data is an important tool in the determination of the design wave for many coastal and offshore structures. A large number of techniques are presently in use to determine design wave height levels which will be exceeded with a certain probability in the life-time of a structure.

The section on Maritime Hydraulics of the International Association for Hydraulic Research (IAHR) organized a Working Group on Extreme Wave Statistics in 1990 with the aim to reach a mutual understanding of the merits and demerits of the methods used in the statistical analysis of extreme wave heights. Eight institutions in Canada, Europe, Japan and the USA joined the working group with Mr. Mathiesen acting as chairman. The objective of the working group was to identify methods and procedures that produce the best estimates of extreme wave heights, i.e. which produce minimum bias and scatter. In addition, a description was to be made of recommended procedures for the analysis of extreme wave heights as well as an identification of areas where more research would be needed.

---

1) Delft Hydraulics, P.O. Box 152, 8300 AD Emmeloord, The Netherlands, 2) Yokohama National University, Japan, 3) HR, Wallingford, UK, 4) National Research Council, Canada, 5) CEPYC - CEDEX, Spain, 6) SINTEF NHL, Norwegian Hydrotechnical Laboratory, Norway, 7) EDF, Laboratoire National d'Hydraulique (Chatou), France, 8) USAE Coastal Engineering Research Center, USA.

The first task of the working group was to perform a comparative analysis of presently used methods. To this end, two sets of deep water wave data were analyzed: one set of measured wave data collected near the west coast of Norway and one set of hindcast storm peak wave data from Kodiak, Alaska, USA. In the present analysis, each group member used their own preferred method. The results of this analysis are presented in this paper. Another task of the working group was to analyze a set of numerically simulated samples of extreme data to study in detail the merits and demerits of the various methods of extremal analysis, of which the results are presented in a separate paper (Goda et al., 1993).

The purpose of the analysis of the two data sets was to provide a basis for the comparison of different analysis techniques presently in use at the participating institutes. The techniques cover different data selection techniques, preferred candidate extreme probability distributions, fitting techniques, goodness-of-fit tests and rejection criteria. The goal of this study is to find a set of best-fitting distributions and corresponding estimates for the 100-year return wave height, i.e. the wave height which on average is exceeded once in a 100 years. In addition, the sensitivity of the computed 100-year wave height to the choice of the threshold value is addressed as well as its sensitivity to the minimum required duration between successive storm events. All wave heights mentioned in this paper are considered as significant wave heights.

The organization of the subsequent chapters of this paper is as follows: Chapter 2 provides a description of the data sets used. The applied analysis methods are dealt with in Chapter 3. Results of the analysis are presented in Chapter 4. The results are discussed in Chapter 5, and conclusions are presented in Chapter 6.

2 Description of data sets

2.1 The Haltenbanken data

Haltenbanken is located on the Norwegian continental shelf. It is part of the most exposed Atlantic coast of Norway (See Fig. 1). Wave data were collected with the Norwave directional buoy at the deep water locations 65°05'N, 7°34'E and 65°11'N, 7°15'E. The local water depth at these locations is 280 m and 290 m respectively. Wave data were collected at three-hourly intervals from March 1980 until October 1987 for the first location, and from November 1987 until March 1988 for the second location.

Missing or poor wave data have been supplemented by hindcast data enabling the composition of a continuous series of 3-hourly data for a 9 year period, starting on January 1, 1980 and ending on December 31, 1988. The hindcast data were obtained with the WINCH model (Eide et al., 1986) at 6-hourly intervals after which they were interpolated to obtain 3-hourly data. The Haltenbanken data set consists of 15,578 observations, 5,365 hindcast data including the largest wave height in the total series, and 5,361 interpolated data, thus making up a total of 26,304 data. The maximum wave height in this data set is 12.51 m.

2.2 The Kodiak data

The Kodiak data consist of hindcast results, obtained through the Wave Information Study (WIS) performed at CERC (cf. Andrew et al., 1985). Data have been retrieved from deep water grid point 17, located at 57°50'N, 148°78'W, and shown in Fig. 2. The data set consists of all peak storm wave heights with a significant wave height exceeding 6 m. In total 78 storms have been hindcast to generate a data set covering 20 years, from January 1, 1956 till December 31, 1975. The maximum wave height in this data set is 11.70 m.

3 Methods of analysis

3.1 Introduction

The Haltenbanken and Kodiak data sets have been analyzed by using several analysis methods. Despite the large number of differences all analysis methods have the following phases in common:

- pre-processing
- data selection
- candidate probability distributions
- fitting techniques
- outlier detection
- assessment of goodness-of-fit and choice of favourable fit
- computation of confidence bands
- computation of wave heights for different return periods

3.2 Pre-processing

The first step in the extremal analysis is the composition of a set of all relevant wave height maxima. To that end one should check that all significant storm events are considered in a selected period. This may be important, since loss of data is often associated with periods of bad weather and this cannot be ignored. A common method to supplement missing data is to add hindcast data. This technique has been used for the Haltenbanken data. Missing data in periods of "fine weather" need not always to be supplemented.

3.3 Data selection

The purpose of the data selection phase is to obtain a sample of mutually independent wave height maxima, all belonging to one parent population with an unknown distribution. Moreover, it is desired that this distribution can be described by a documented theoretical probability distribution. In practice two main methods are in use to select independent wave height maxima: namely Partial Duration Series (PDS) and Annual Maxima (AM).

The Partial Duration Series is used to detect significant storm periods. A common technique is to apply a fixed threshold value to identify storm periods comprising a sequence of sea-states with wave heights all exceeding the given threshold. It is then common practice to select the highest wave height in a storm period, which is denoted as the peak or extreme wave height. This method is also known as the Peak over Threshold (POT) method. The resulting sample of peak wave heights is used in the further statistical analysis. In the Annual Maxima method the highest wave height per year is used in the subsequent statistical analysis.

The duration of a storm and the shape of the storm profile, i.e. the time variation of wave height, is important for the determination of the statistics of the highest individual wave height or wave crest. Such statistics, however, are not the subject of this study and will not be addressed.

Various methods are in use to ensure independence between extreme wave events. In general, one should be aware that storm events may come from different populations. Resio (1978), for example, demonstrated the necessity of separate analyses of extreme waves off Cleveland in Lake Erie. A practical method of separating different populations is to use historical meteorological information (c.f. Cavaleri et al., 1986). Another common method is to impose a certain minimum time between the occurrence of successive local wave height maxima in a wave record. The optimum minimum time between successive storm events can be assessed by means of correlation studies. In addition, one may use directional wave information to separate different types of storms. This can be achieved by constructing polar plots of wave height versus wave direction (Van Vledder and Zitman, 1992). For the present study, the storm events in the Kodiak and Haltenbanken data sets are all considered as "winter" storms, i.e. each sample of resulting wave height maxima is considered to belong to one parent population.

The minimum duration between successive storm events and the threshold level determine the number of extremes that will be used in the subsequent statistical analysis. In practice an optimum must be found between maximizing the number of extremes to increase the statistical significance and minimizing the number of extremes to ensure that they all are independent samples from one population.

In the Initial Distribution Method (IDM) all available data are used, including relatively low wave heights. Individual data values can be given equal weight or the data values can be grouped in classes which are given equal weight before fitting a distribution to them. The Initial Distribution Method is sometimes used in combination with unequal weights to improve the quality of the fit, e.g. by assigning the higher data values a larger weight. The use of weighted data, however, is bound to be fairly arbitrary and it requires much experience.

3.4 Candidate probability distributions

There is no physical, theoretical, or empirical evidence for selecting a particular probability distribution function for extreme wave heights. As a consequence one should try several candidate distributions and select the one which gives the best fit to the sample distribution.

The following probability distributions have been tried out by the group members: Exponential, Fisher-Tippett type-I (Gumbel), Fisher-Tippett type-II (Fréchet), Log-Normal, Pareto, 2 and 3-parameter Weibull and truncated Weibull. Expressions for these distributions are given in the Appendix to this paper.

3.5 Fitting techniques

Three objective fitting techniques are used by the group members: Method of Moments (MOM), Least Squares Method (LSQ) and the Maximum Likelihood Method (MLM). Each of these methods tries to fit a target distribution function $P(H)$ with coefficients $\alpha_1,\ldots,\alpha_m$ to the observed sample of size N.

Method of Moments

The Method of Moments is the most widely used fitting technique, because of its simplicity. The method works by equating the first m statistical moments (mean value, standard deviation, skewness, etc...) of the target distribution to the moments derived from the observations. The number of statistical moments that needs to be used is equal to the number of parameters of the target distribution.

Least Squares Method

The Least Squares Method is easy to use for almost any data set and distribution function. Least Squares Methods are widely used for the fitting of two-parameter distributions to the observed data. To that end, the target distribution function $P(H;\alpha_1,\alpha_2)$ is rewritten in the form of a linear regression function

$$X(H_i) = a(\alpha_1,\alpha_2)\,Y\big(P(H_i)\big) + b(\alpha_1,\alpha_2) \tag{1}$$

where X and Y are functions which follow from the target distribution function. The regression coefficients a and b, and consequently the parameters α_1 and α_2 can then be computed using standard methods of linear regression. Three-parameter distributions can also be fitted with a linear regression method. To that end the third parameter is usually determined such that a goodness-of-fit criterion is optimized. Another technique is to use nonlinear regression methods to fit distributions with three or more parameters.

When Least Squares Methods are used, one is faced with the problem of assigning proper probability levels $P_i(H)$ to the observed data values H_i, i.e. choosing the plotting positions such that the bias in the parameter estimates is minimized. This is a vital problem, as an improper choice of the plotting position formula may lead to highly biased parameter estimates. The optimal choice of the plotting position formula is the subject of many papers (c.f. Petrauskas and Aagaard, 1971; Goda and Kobune, 1990; Mathiesen, 1991).

Maximum Likelihood Method

The Maximum Likelihood Method is based on the assertion that the parameters of the target probability density function $p(H;\alpha_1,..,\alpha_m)$ should be chosen such that the likelihood function

$$L(H_1,H_2,\ldots,H_N) = \prod_{i=1}^{N} p(H_i;\alpha_1,..,\alpha_m) \tag{2}$$

is maximized by the choice of the model parameters $\alpha_1,..,\alpha_m$. L can be considered as the probability of getting the particular set of data values H_i. The maximization of L is achieved by determining the m model parameters α_i for which $dL/d\alpha_i = 0$. In practice, it is more convenient to work with the logarithm of the likelihood L, which attains its maximum for the same values of the model parameters as L itself.

3.6 Outlier detection

An outlier is a suspected data value that deviates markedly from the other data values in a sample to which a distribution is fitted. An outlier may exist for various reasons:
- it was generated by another meteorological process, i.e. it belongs to another population,
- it may be a rare event from the same population,
- it may be an incorrect data value, e.g. it was generated by an error in the measurements or in the hindcast.

An objective criterion for the detection of outliers has been proposed by Goda and Kobune (1990). This DOL criterion (Deviation of OutLiers) applies Thompson's statistical test on the dimensionless deviation

$$\xi = (x_{max} - \bar{x})/s \tag{3}$$

in which x_{max}, $\bar{x}$ and s are the largest data value, the mean and standard deviation of the sample data, respectively. Goda and Kobune (1990) recommend that the ξ value of a given sample be compared with values of $\xi_{5\%}$ and $\xi_{95\%}$, which they have computed for various theoretical distributions. The 5% and 95% levels were tentatively chosen as the threshold values for rejection.

Another test for the detection of outliers is presented by Barnett and Lewis (1984, pp 144-150). They compute a test statistic for a single upper outlier x_{max} in a sample with an exponential distribution

$$t = x_{max} / \sum_{i=1}^{n} x_i \tag{4}$$

Critical values for t are tabulated in Barnett and Lewis (1984, pp 369-370). Outlier tests for other target distributions can also be tested with the above test t statistic by means of simple transformations.

When wave measurements are used, a suspicious data value can be tested by carefully scrutinizing the underlying physical processes (e.g. check wind and wave records for consistency). When ships observations are used, the wave data have to be checked with respect to wave steepness and to the relation between wind and wave direction. Outliers should be removed from the sample when they are incorrect or they should be replaced by a correct value.

3.7 Assessment of goodness-of-fit and choice of favourable fit

The rejection or acceptance of a fitted distribution is usually based on a qualitative (visual) inspection and quantitative judgements of the fits. Visual checks are easy to perform but they usually need an experienced eye. When large quantities of fits are to be judged, however, visual checks are less suited. A number of quantitative tests are in use by the various group members to quantify the goodness-of-fit, to reject or accept distributions and to choose between various fitted distributions. They are:

- Kolmogorov-Smirnov (KS)
- Anderson-Darling (AD)
- Chi-square (χ^2)
- Correlation coefficient (r)
- Residue of Correlation coefficient (REC)
- Minimum Ratio of residual correlation coefficient (MIR)

The first three tests are applied to examine the shape of the distribution. These tests are effective at detecting deviations in the middle of the distribution (c.f. Muir and El-Shaarawi, 1986).

The coefficient r of correlation between the observed wave heights and the best-fit wave heights, both with the same probability of exceedance, can also be used to judge the quality of the fit. The higher the correlation coefficient, the better the fit. In practice, simple subjective criteria are used to reject or accept a fit on the basis of the correlation coefficient.

Based on the correlation coefficient r and numerous numerical simulations, Goda and Kobune (1990) proposed the REC and MIR criteria to test the goodness-of-fit of a particular distribution. The REC criterion considers the residue of r from 1, i.e. $\Delta r = 1 - r$. The cumulative distribution of Δr has been obtained through extensive numerical simulations.

Goda and Kobune (1990) give parameterizations for the $\Delta r_{95\%}$ exceedance levels for a number of distribution functions. The MIR criterion can be used to discriminate between various fitted distributions, but the sole use of the correlation coefficient r is not sufficient to rank distributions. The shape of the distribution and the sample size N should also be accounted for. Based on these considerations, Goda and Kobune (1990) developed a procedure to assess the goodness-of-fit of a fitted distribution that is based on the ratio $\Delta r / \Delta r_{mean}$. The distribution with the lowest value of $\Delta r / \Delta r_{mean}$ is then considered as the optimum.

3.8 Confidence intervals

The reliability of extreme wave height estimates is normally indicated with confidence bands, computed through various methods. Most methods quantify the uncertainty in the estimated extreme wave heights due to sampling variability. This approach, however, does not address the uncertainties related to the accuracy of the original data or long-term non-stationarity. A common method to calculate confidence bands is to assume a normal distribution of possible extremes around the best-fit line. This method uses information on the model distribution, the estimation technique and the sample size. This method produces symmetric confidence intervals, More detailed estimates of the confidence intervals can be obtained with Monte Carlo simulations, which simulate the statistical uncertainty due to finite record lengths. Such techniques are able to represent the asymmetry of confidence intervals. Both of the above types of method have been used by the group members.

3.9 Computation of wave heights for different return periods

The computation of return periods for a specified peak sea state is based on the encounter frequency of extreme events and the exceedance probability of a given level in a single event. For POT methods the return period R, expressed in years, for a specific level of exceedance H is computed as:

$$R = \frac{1}{\lambda(1 - P(H))} \tag{5}$$

in which λ is the mean number of extreme events per year and $P(H)$ the probability of non-exceedance for a single extreme wave height. Based on (5) the wave height with a return period of R years is computed as:

$$H = P^{-1}\left(1 - \frac{1}{\lambda R}\right) \tag{6}$$

where P^{-1} denotes the inverse of $P(H)$. For AM methods the parameter λ is set to one.

4 Results of extremal analysis

The results of the extremal analysis are summarized in Table 4.1 for the Haltenbanken data and in Table 4.2 for the Kodiak data for each group member. The letters in the first column correspond to the individual group members. (Their order is not in agreement with the order of group members listed in the title.) For some group members the results of more than one analysis technique are presented, with the preferred or best fitting one on top. One group member has also applied the bootstrap resampling approach, as described by Andrew and Hemsley (1990). Although this approach does not include a fitting technique, it has been nevertheless included in this overview. In these tables the abbreviations W3 and TW refer to the 3-parameter and truncated Weibull distribution respectively. The preferred 100-year return wave heights with corresponding confidence intervals are summarized in Table 4.3.

A number of methods are in use to estimate the parameters of the 3-parameter Weibull distribution. For LSQ methods, a common method (Goda, 1988) is to fix the shape factor k to one of the values 0.75, 1.0, 1.4 and 2.0, and subsequently to estimate the scale factor σ and the location parameter x_0. For the MLM, another approach is necessary since the location parameter x_0 tends to the smallest value in the sample, giving an unrealistic likelihood value. Therefore, the location parameter x_0 is fixed to a value which is somewhat lower than the lowest value in the sample, followed by the estimation of the shape and scale factors. Application of the MLM in combination with the POT method usually yields a location parameter x_0 which is close to the threshold value. These problems may be avoided by using the (untested) Truncated Weibull distribution, since this distribution already contains the threshold level.

Some group members have performed additional studies and examined the data for their sensitivity of the estimated return values to the threshold level and the minimum separation time between different storms. The results of these analyses are presented in the Tables 4.4 and 4.5. Examples of fitted Weibull distributions are given in Figs. 3 and 4.

	Data selection method and threshold	Best fitting distribution and parameters				Fitting technique	H_{100} (90% Conf. Int.)	Goodness-of-fit criteria	Number of data
			σ	x_0	k				
A	POT (x_c=6) IDM (0.25)	TW W3	6.594 2.109	0.0 0.723	2.774 1.245	MLM MOM	14.3 (13.1-15.7) 16.7 (15.3-18.4)	KS, χ^2, DT KS, χ^2, DT	181 26,304
B	AM POT (6.0) POT (6.0)	FT-I W3 FT-I	0.663 1.971 1.113	11.166 5.849 7.027	- 1.30 -	LSQ LSQ LSQ	14.2 (13.0-16.5) 15.4 (-) 15.7 (-)	visual	9 134 134
C	POT (6.0)	W3	2.79	1.47	2	LSQ	15.3 (14.0-16.6)	DOL, REC, MIR	269
D	POT (8.7)	W3	1.380	8.6	1.225	MLM	14.7 (12.6-16.9)	KS, χ^2, r	46
E	POT (7.5)	W3	1.642	7.5	1.265	MLM	15.0 (13.3-16.7)	χ^2, AD	84
F	POT (6.0) AM	W3 FT-I	3.34 0.88	3.98 10.53	2 -	LSQ MOM	15.8 (13.2-18.4) 14.6 (12.2-17.0)	r	174 9
G	POT (6.0)	W3	5.203	0.047	2	LSQ	15.5 (14.3-16.8)	DOL, REC, MIR	144
H	POT (7.0)	FT-I W3	1.025 3.834	8.176 4.043	- 2	LSQ LSQ	15.4 (14.2-16.6) 14.7 (-)	DOL, REC, MIR	105
I	Bootstrap						15.3 (-)		

Table 4.1 Summary of the analysis of the Haltenbanken data.

	Data selection method and threshold	Best fitting distribution and parameter values				Fitting technique	H_{100} (90% Conf. Int.)	Goodness-of-fit criteria	Number of data
			σ	x_0	k				
A	POT (x_c=6)	TW	5.854	0.0	2.693	MLM	12.1 (10.8-13.5)	KS, χ^2,DT	78
B	AM	FT-I FT-I	1.040 1.129	8.050 8.061	- -	LSQ MOM	13.3 (11.7-15.1) 12.8 (-)	visual	20 20
C	POT (6.0)	W3	1.86	5.81	1.4	LSQ	12.5 (11.4-13.6)	DOL, REC, MIR	78
D	POT (6.7)	W3 FT-I	1.578 0.776	6.7 7.665	1.386 -	MLM	12.1 (10.7-13.5) 11.9 (-)	KS, χ^2, r	50 50
E	POT (5.9)	W3	1.74	5.9	1.324	MLM	12.6 (11.2-14.0)	χ^2, AD	78
F	POT (6.0) AM	W3 FT-I	1.86 1.06	5.81 8.02	1.4 -	LSQ MOM	13.0 (10.8-15.2) 12.9 (10.9-14.9)	r	78 19
G	POT (6.0)	W3	1.862	5.805	1.4	LSQ	12.5 (11.4-13.5)	DOL, REC, MIR	78
H	POT (7.0)	FT-I W3	0.876 3.188	7.616 4.145	- 2	LSQ LSQ	12.5 (11.2-13.8) 12.1 (-)	DOL, REC, MIR	46 46
I	Bootstrap						12.8 (-)		

Table 4.2 Summary of the analysis of the Kodiak data

Group member	Haltenbanken		Kodiak	
	H_{100}	half width of confidence interval	H_{100}	half width of confidence interval
A	14.3	1.6	12.1	1.4
B	14.2	1.8	12.8	1.7
C	15.3	1.3	12.5	1.1
D	14.7	2.2	12.1	1.4
E	15.0	1.7	12.6	1.4
F	15.8	2.6	13.0	2.2
G	15.5	1.3	12.5	1.1
H	15.4	1.2	12.5	1.3
Mean	15.0	1.7	12.5	1.5
Standard deviation	0.6	0.5	0.3	0.4

Table 4.3 Summary of preferred 100-year return wave heights with corresponding half-widths of 90% confidence interval.

Haltenbanken data, 3-parameter Weibull distribution, fitted with LSQ method					
Peak gap	Number of peaks	H_{100} (90% Conf. Int.)	σ	x_0	k
18 hours	228	15.2 (14.3 - 16.1)	5.182	-0.876	2
48 hours	167	15.5 (14.4 - 16.6)	5.267	-0.436	2
72 hours	144	15.5 (14.3 - 16.8)	5.203	0.047	2

Table 4.4 Sensitivity of the 100-year return wave height to the separation time between storm peaks. POT method with threshold at 6.0 m.

Haltenbanken data, 3-parameter Weibull distribution, fitted with LSQ method					
Threshold	Number of peaks	H_{100} (90% Conf. Int.)	σ	x_0	k
3.0	342	16.0 (15.1 - 16.9)	5.489	-0.054	2
6.0	134	15.4 (14.2 - 16.6)	5.026	0.699	2
9.0	30	14.5 (13.0 - 16.1)	3.799	4.050	2

Table 4.5 Sensitivity of the 100-year return wave height to the threshold level. POT method and minimum duration between peak events set to 4 days.

5 Discussion

The results presented in the Tables 4.1 through 4.3 indicate a remarkable consistency between the estimated 100-year return wave heights. For each data set the wave heights differ less than 10% of one another. For each data set the average half-width of the confidence intervals is about one eighth of their average return wave height.

Three different data selection methods have been applied. Most group members preferred the Peak over Threshold method, although the Annual Maxima and Initial Distribution method were also used. Least Squares methods were used by most members, although this does not mean that LSQ methods are the best methods. The Maximum Likelihood Method was applied by three of the eight group members. Least Squares methods are simple to use, but should be applied with care regarding the choice of the plotting position formula. A slight preference was shown for the 3-parameter Weibull distribution, followed by the Fisher-Tippett I distribution (Gumbel).

A closer look at the Haltenbanken data shows the following results. The IDM method provides by far the highest return wave height. This is probably due to the fact that all wave heights have been used in the distribution fitting. The lower wave height classes probably affect the fit in an inappropriate manner since they cannot be considered as extreme wave heights. The AM method has been used by two group members with almost similar results. The differences in their results are due to the different methods used to extract return values. The results of the POT method depend on the user, namely those who favour an FT-I distribution and those who favour the 3-parameter Weibull distribution. The two FT-I distributions that have been fitted with the LSQ method produce similar results. Their difference is due to the different threshold levels and thus to the different number of data values applied (134 versus 105). The results for the 3-parameter Weibull distributions have been obtained with MLM and LSQ fitting techniques. The analyses with the MLM produce similar results for the considered two threshold values, where the difference is due to the different number of data values used (46 versus 84). The resulting shape factors are close to 1.3. This value was also obtained by one group member who used a LSQ method to estimate all three parameters. Four group members applied the LSQ method with a 3-parameter Weibull distribution with the shape factor k set to 2. The corresponding results are close to one another, apart from the wider range of uncertainty provided by group member F. This difference is due to the introduction of a bias correction to account for the uncertainty in not knowing the true distribution. All results show a clear trend of an increasing return wave height with a decreasing threshold value. Based on a visual inspection of fitted distributions it was noted by one group member that the Haltenbanken data probably consist of two populations. This has not further been worked out in this study.

A closer look at the analysis of the Kodiak data shows similar results. Two group members applied the Annual Maxima method with nearly similar results. Group member B fitted the FT-I distribution to the data with the MOM and the LSQ method, with nearly identical results. This implies that the method of fitting has only a small effect on the estimated return values, at least for this example. The Peak-over-Threshold method was used in combination with LSQ and MLM techniques to fit the FT-I and 3-parameter Weibull distributions to the data. The most favourable distribution seems to be a 3-parameter Weibull with a shape factor of about 1.35. Again, it is noted that the results lie close to one another, with a clear dependence on the threshold level.

Since the Haltenbanken data consist of a continuous period of three hourly wave heights it was possible to investigate the sensitivity to the estimated return wave height to the minimum time interval between successive storm peaks. As can be seen in Table 4.4, the estimated values of the return wave heights show hardly any no dependency on the minimum separation time between storm events. It was found that an appropriate minimum time duration is 18 hours. Computations with larger minimum time durations showed no significant differences. This implies, at least for the Haltenbanken data, that the condition of independence is not very critical. It is worthwhile to investigate whether or not such a result is generally found in extreme wave data analysis.

In Peak over Threshold methods the choice of the threshold level strongly determines the estimated return wave height. For both data sets it is found that a lower threshold gives more peak wave data and an increase in the estimated return wave height. This can be seen in the Tables 4.1 and 4.2, and, more specifically, in Table 4.5. There is no ideal number of independent peak wave data, but a suitable number might be in the range from 50 to 75, to be chosen from a wave data record that is as long as possible. The inclusion of a few extra storms makes little difference provided that they are not the most severe in the data set. An objective method to determine the optimum number of data values may be by optimizing the goodness-of-fit for various threshold levels or by minimizing the width of the confidence interval.

Some members mentioned testing for outliers. Two objective tests are in use, none of which caused a data value to be discarded from the subsequent analysis. A number of goodness-of-fit tests are in use by the group members: correlation coefficient, Anderson-Darling test, Kolmogorov-Smirnov test, Chi-square test, REC and MIR tests. These tests have been used to reject or accept fitted distributions and to choose between the different accepted distributions. It is noted that these case studies are too limited to draw firm conclusions about the merits or demerits of goodness-of-fit tests.

Most members obtained narrow symmetrical 90% confidence bands (of about plus or minus 10%) for their extremes. The inclusion of these intervals is recommended since they give essential information about the reliability of the estimated return wave height. Clients do not generally ask for confidence limits, but they probably should be educated to expect them and to know what to do with them, for instance in the probabilistic design of structures.

It is noted that the present study does not give any judgement about the recommended estimated return wave height for either location, nor for the preferred data selection technique and fitting method.

6 Concluding remarks

Although a wide variety of methods to compute the "100-year" wave height was applied, only relatively small differences were found in the extreme wave height predictions. This is encouraging, since it is an indication of the robustness of each extremal analysis technique in use at the participating institutes.

Only small differences are found between the results obtained through the various fitting techniques, provided they use the same data sample. The differences found in this study are primarily due to the different methods of data selection, i.e. which peak wave data should be included in the statistical analysis.

In POT methods two criteria are essential to select peak wave heights: wave height threshold and minimum separation time between successive storm peaks. For the present study it was found that the threshold level has the most significant effect on the resulting estimated return wave heights. A lower threshold yields a higher estimated 100-year wave height.

Obtaining a good fit is easy compared to the decisions that need to be made concerning the techniques of pre-processing and data selection.

Relatively large differences are found in the width of the estimated 90% confidence intervals. Further comparative studies are recommended that may explain these differences.

Provision of data set: Copies of the Haltenbanken and Kodiak data sets can be obtained from any member of the Working Group for purposes of comparison.

References

Andrew, M., and J.M. Hemsley, 1990: Resampling approach to extreme wave-height analysis. Journal of Waterway, Port, Coastal, and Ocean Engineering, Vol. 116, No. 4, 444-458.

Andrew, M., O.P. Smith and J.M. Mckee, 1985: Extremal analysis of hindcast wind and wave data at Kodiak, Alaska, Techn. Report CERC-85-4, U.S. Army Corps of Engineers, Waterways Experiment Station, Vicksburg, Miss.

Barnett, V., and T. Lewis, 1984: Outliers in statistical data, Second Edition, John Wiley & Sons, 463 pp.

Cavaleri, L., P.L. De Filippi, G.F. Grancini, G.L. Iovenitti and R. Tosi., 1986: Extreme wave conditions in the Tyrrhenian sea. Ocean Engineering, Vol. 13, No. 2, 157-180.

Eide, L.I., M. Reistad and J. Guddal, 1986: A comparison of hindcast data studies with a) a coupled discrete wave model and b) a coupled hybrid wave model. Proc. Int. Workshop on Wave Hindcasting and Forecasting, Halifax, Nova Scotia, 153-159.

Goda, Y., 1988: On the methodology of selecting design wave height. Proc. 21st Int. Conf. on Coastal Engineering, Malaga, Spain, 899-913.

Goda, Y., and K. Kobune, 1990: Distribution function fitting for storm wave data. Proc. 21st Int. Conf. on Coastal Engineering, Delft, The Netherlands, 18-31.

Goda, Y., P. Hawkes, E.P.D. Mansard, M. Jesus Martin, M. Mathiesen, E. Peltier, E.F. Thompson and G. Ph. van Vledder, 1993: Intercomparison of extremal wave analysis methods using numerically simulated data, Proc. WAVES'93 Conf., 26-28 July, New Orleans, USA.

Mathiesen, M., 1991: Long-term wave and wind statistics. SINTEF NHL Report STF60 A91084, Norwegian Hydrotechnical Laboratory.

Muir, L.R., and A.H. El-Shaarawi, 1986: On the calculation of extreme wave heights: A review. Ocean Engineering, Vol. 13, No. 1, 93-118.

Petrauskas, C. and P.M. Aagaard, 1971: Extrapolation of historical storm data for estimating design-wave heights, Journal of the Soc. of Petr. Engineering, Vol. 11, 23-37.

Resio, D.T., 1978: Some aspects of extreme wave prediction related to climatic variations. OTC paper 3278.

Van Vledder, G.Ph., and T.J. Zitman, 1992: Design Waves: Statistics and Engineering Practice. Proc. 2nd Int. Offshore and Polar Engineering Conference, ISOPE, San Francisco, USA, Vol. 3, 170-178.

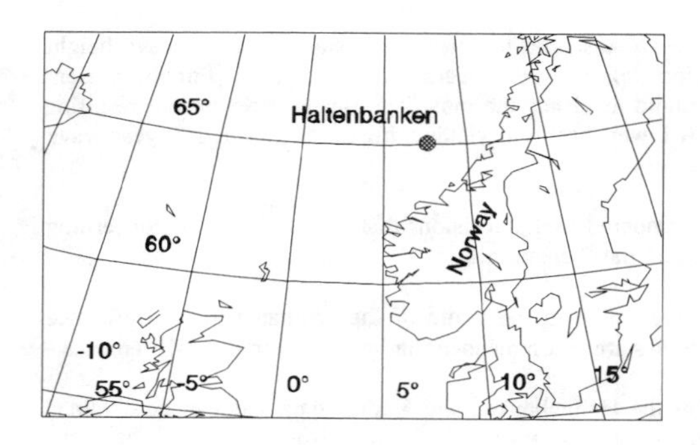

Figure 1: Geographic location of the Haltenbanken measurements

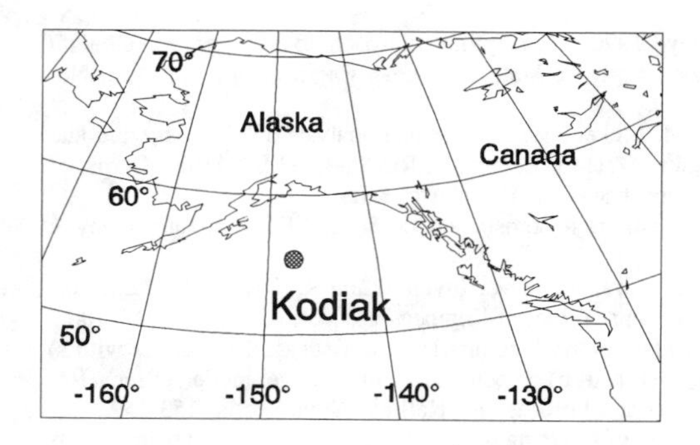

Figure 2: Geographic location of the Kodiak grid point.

Appendix: Candidate probability density functions

Exponential

$$F(x) = 1 - \exp\left(-\frac{x - x_0}{\sigma}\right)$$

Fisher-Tippett type I or Gumbel

$$F(x) = \exp\left\{-\exp\left(-\frac{x - x_0}{\sigma}\right)\right\}$$

Fisher-Tippett type II Fréchet

$$F(x) = \exp\left\{-\left(\frac{x - x_0}{\sigma}\right)^{-k}\right\}$$

Log-normal

$$F(x) = \frac{1}{k\sqrt{2\pi}} \int_0^x \frac{1}{t} \exp\left[-\frac{1}{2}\left(\frac{\ln(t) - \ln(\sigma)}{k}\right)^2\right] dt$$

Generalized Pareto

$$F(x) = 1 - \left[1 - \frac{(x - x_0)}{\sigma}\right]^{\frac{1}{k}}$$

Truncated Weibull

$$F(x) = 1 - \exp\left[-\left(\frac{x - x_0}{\sigma}\right)^k + \left(\frac{x_c - x_0}{\sigma}\right)^k\right] \quad \text{for } x > x_c$$

Weibull 2p

$$F(x) = 1 - \exp\left\{-\left(\frac{x}{\sigma}\right)^k\right\}$$

Weibull 3p

$$F(x) = 1 - \exp\left\{-\left(\frac{x - x_0}{\sigma}\right)^k\right\}$$

with σ the scale parameter, k the shape parameter and x_0 the location parameter. For the truncated Weibull distribution x_c is the threshold level at which the distribution is truncated.

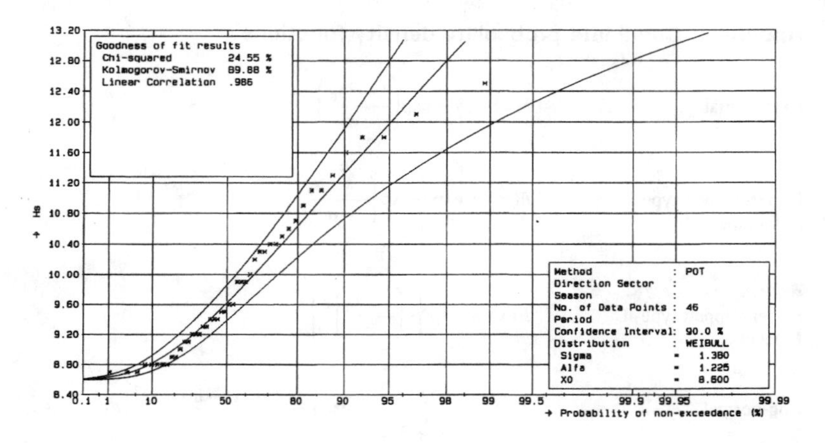

Figure 3: Example of a Weibull distribution fitted to the Haltenbanken data

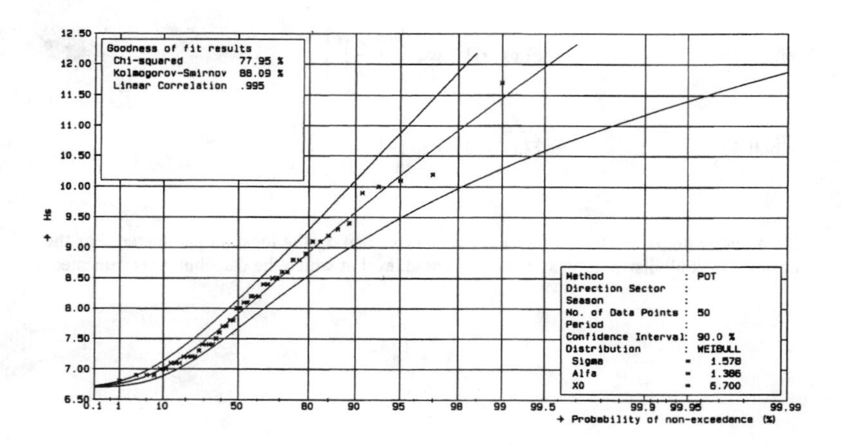

Figure 4: Example of a Weibull distribution fitted to the Kodiak data

Design Waves and Wave Spectra for Engineering Applications

Chung-Chu Teng[1], Gerald L. Timpe[1], Ian M. Palao[2], and David A. Brown[2]

Abstract

Thirteen years of hourly wave data measured at two of the National Data Buoy Center's (NDBC) buoy stations, one located in the northeastern Pacific and the other located in the northwestern Atlantic, were used to form a data base to develop design wave information. It was found that both the significant wave height and average wave period fit well with the log-normal distribution. The relationships between significant wave height and average wave period were established based on the lower limit and the average sense. Following the extreme value analysis, the design wave heights for both buoy stations were calculated using different extreme value distributions and parameter estimation techniques. It was also found that the Pierson-Moskowitz spectrum can be used to estimate climatic wave spectra for high sea states in the northeastern Pacific Ocean, while the JONSWAP spectrum is more appropriate for the northwestern Atlantic Ocean.

Introduction

The development of design waves and wave spectra, which is generally based on long-term wave statistics, is important in ocean and marine engineering design. Due to the limited amount of long-term measured wave data, most long-term wave statistics are based on either hindcast (Gu *et al.* 1992; Wyland & Thornton 1991) or visually measured data (Hogben *et al.* 1986). Although hindcast data are relatively easy to obtain and useful, it is desirable to have data measured by precise instrumentation (instead of visually measured data) to represent "live" long-term wave characteristics. Since 1972, the U.S. National Data Buoy Center (NDBC) has used moored data buoys to provide *in situ* wave measurements. This has created a large and expanding data base of wave measurements.

Thirteen years of hourly wave data measured at two NDBC buoy stations were used to examine characteristics of design waves and wave spectra in this study. These two buoy stations (both of them are in very deep water) are located in the northeastern Pacific Ocean and

[1] National Data Buoy Center, Stennis Space Center, MS 39529-6000 USA

[2] Computer Sciences Corporation, Stennis Space Center, MS 39529-6000 USA

in the northwest Atlantic Ocean. Thus, comparisons of design waves and wave spectra can be made between these two different geographic locations.

In this paper, the distributions of long-term significant wave heights and average wave periods for both stations are first examined. Then, the design waves for 50-year and 100-year return periods are computed based on extreme value analysis. Finally, the climatic (average) wave spectra for both high wave conditions and different wave height ranges are presented and compared to the theoretical spectra, namely the Pierson-Moskowitz and the Joint North Sea Wave Project (JONSWAP) spectra.

Buoy Wave Data

Data used in this study consisted of 13 years of hourly wave data from NDBC buoy stations 44004 and 46001. Station 44004 is located in the Atlantic Ocean off the New Jersey coast in 3,231 meters of water, while station 46001 is located in 4,115 meters of water in the Gulf of Alaska (Table 1).

The data buoys moored at these stations were equipped with an accelerometer for use in the *in situ* measurement of sea surface elevation. This information was collected for 20 minutes each hour and the data processed into wave spectral format. Then, the processed data were transmitted back to shore via satellite.

For both sites, the time period of the data extends from January 1980 through December 1992. Throughout these 13 years, there were periods of missing data in the wave records for both stations. These outages were due to sensor failure, payload failure, periodic loss of data during transmission, or buoys not being on station due to refurbishment or mooring failure. In this analysis, any effect of this lost data on the long-term wave distribution was assumed to be negligible. As presented in Table 1, 103,886 sets of hourly wave data were successfully collected at station 46001 during the 13 years. This represents 91.2% (the last column of the table) of the maximum possible hourly data during the time period. At station 44004, 82,396 hourly data points were collected during the same time span, which covers 72.4% of the 13 years.

Wave data used in this study include point wave spectra (from 0.04 Hz to 0.35 Hz in 0.01 Hz increments), significant wave heights (H_s), and average wave periods (T_z).

Significant Wave Heights and Average Wave Periods

Table 2 presents the maximum, mean, and standard deviations of significant wave heights measured during the 13 years for both buoy stations. The maximum and mean wave heights of the west coast buoy station (46001) (i.e., 13.3 and 2.77 meters) are much higher than those of the east coast station (44004) (i.e., 10.1 and 2.04 meters).

One of the basic tools for long-term wave statistics study is the scatter diagram, which shows the frequency of occurrence of wave height versus wave period. The scatter diagrams for buoy stations 46001 and 44004 are presented in Figures 1 and 2, respectively, in which the significant wave heights (H_s) and average wave periods (T_z) are used. Two important types of information are available from the diagrams: (1) the long-term distributions of significant wave heights and average wave periods, and (2) the relationship between wave heights and wave periods for both the lower limit case and the average case.

Figures 3 and 4 show the histograms of significant wave heights and average periods for the two stations. The associated log-normal distributions are also plotted in the figures. As discussed by Ochi (1975), it can be seen in Figure 3 that both the significant wave heights and average wave periods for station 46001 fit well with the log-normal distribution, although it is a little over-predicted at the peak. The fit for station 44004 is acceptable (see Fig. 4); however, it is not quite as good as the fit for station 46001, and the peak is under-predicted. One possible reason for the minor inferiority is that station 44004 has a higher percentage of missing data during the 13-year period.

From the scatter diagrams (Figs. 1 and 2), it can be observed that there is a lower limit on the average wave periods due to wave breaking. For example, there is no wave period less than 7 seconds for wave heights greater than 5.5 meters for either station. In this study, this lower limit is established by assuming an exponential relation between significant wave heights and average wave periods; i.e., $T_z=a{\cdot}H_s^{\,b}$. Regression analysis on the lowest wave height-period pairs (H_s, T_z) at each wave height range was conducted to obtain the two coefficients, a and b, in the exponential relationship. In addition to the relation for the lower limit, a similar relationship between significant wave heights and the mean average wave periods can be obtained based on the same exponential form. Table 3 shows the coefficients a and b, obtained from regression analysis, in the exponential relation between H_s and the corresponding T_z for both buoy stations. Since these coefficients are dimensional, the values displayed in the table are based on the SI system of units.

Battjes (1970) developed the following expression for the lower limit of average wave period (T_z):

$$T_z = (32\pi\, H_s/g)^{1/2}$$

where g is the gravitational acceleration. Using SI units, the above equation becomes $T_z \approx 3.2\, H_s^{1/2}$. From Table 3, it is clear that the coefficients for the lower limit for both stations are reasonably close to those given by Battjes (i.e., a=3.2 and b=0.5). Based on a very limited amount of extreme wave data, Buckley (1988) also determined the following relationship between wave heights and peak wave periods: $H_s/T_p^{\,2} = 0.00776{\cdot}g$. Using SI units, it becomes $T_p \approx 3.62{\cdot}H_s^{\,0.5}$. Note that the peak wave period was used in Buckley's equation. His result will be even closer to those derived from this study if the average wave period is used.

Design Wave Heights

Determination of design wave height is based on an extreme value analysis. Since the "true" extreme value distributions of wave height for the buoy stations are not known *a priori*, several theoretical distributions are fitted to the data. Then, various goodness-of-fit criteria are used to evaluate the fitted distributions. The design wave heights for a given return period are determined by extrapolation.

Extreme Value Distributions

Three extreme value distributions are used to fit the measured data in the study: the Fisher-Tippet 1 (FT-1) distribution, the log-normal distribution and the three-parameter Weibull distribution. Their simple cumulative distribution functions are as follows:

FT-1 distribution:

$$F_s(H_s \le x) = e^{-e^{-\frac{x-A}{B}}} \tag{1}$$

Log-normal distribution:

$$F_s(H_s \le x) = \frac{1}{\sqrt{2\pi}} \int_0^H \left(\frac{1}{Bx}\right) e^{-\frac{1}{2}\left(\frac{(\ln(x)-A)}{B}\right)^2} dx \tag{2}$$

Weibull distribution:

$$F_s(H_s \le x) = 1 - e^{-(\frac{x-A}{B})^C} \tag{3}$$

where A, B, and C are the location, scale, and shape parameters to be estimated, respectively.

Three methods are used in this study to estimate the parameters in the above distributions: the method of least squares, the method of maximum likelihood, and the method of moments. For the FT-1 and log-normal distributions, parameter estimation from the above mentioned methods is straightforward. For the three-parameter Weibull distribution, the third parameter is varied by trial and error until the best fit, as evaluated by the following goodness-of-fit criteria, is achieved. Details of these distributions and methods of parameter estimation can be found in Isaacson and Mackenzie (1981), Muir and El-Shaarawi (1986), and Tucker (1991).

Goodness-of-Fit Criteria

Once the parameters are estimated, the following four criteria are used in evaluating the goodness-of-fit of the distributions to the data:

(1) Coefficient of determination assesses the degree of linearity of the reduced variates and is defined as:

$$r^2 = \frac{\sum_{i=1}^{n} ((x_i - \mu_x)(y_i - \mu_y))^2}{\sum_{i=1}^{n} (x_i - \mu_x)^2 (y_i - \mu_y)^2} \tag{4}$$

where x and y represent the reduced variates of the abscissa and the ordinate, respectively. This statistic also explains the amount of variance of the dependent variable accounted for by the independent variable.

(2) Mean of the squared errors is a standard criteria used to assess the errors between the model and the observed data and is represented by

$$MSE = \frac{1}{n} \sum_{i=1}^{n} (model - data)^2 \tag{5}$$

This equation utilizes the probability domain and should not be used in the linear plotting domain, especially for the Weibull distribution.

(3) Cramer-von Mises statistic is also calculated in the probability domain and is defined as:

$$W_n = \sum_{i=1}^{n} [F_s(x_i) - \frac{i-0.5}{n}]^2 + \frac{1}{12n} \tag{6}$$

where i is the rank of the data point in ascending order.

(4) Kolmogorov-Smirnov statistic assesses the fit of the distribution to each observed data point and chooses the worst individual fit to be proportional to the statistic in question. Therefore, the goodness-of-fit of the distribution to a data set is dependent upon the worst fit data point. This statistic is defined as:

$$D_n = \max[\max(\frac{i}{n} - F_s(x_i)), \max(F_s(x_i) - \frac{i-1.0}{n})] \tag{7}$$

In general, values of MSE, W_n, and D_n close to 0 are desirable while values of r^2 close to ±1 indicate better fits. However, these criteria may not provide consistent results. Under this circumstance, the criterion used as the "major" criterion is mainly based on judgement and preference.

Data Censorship

There may be a temptation to use all the data samples to increase the size of the record. However, this approach ignores the statistical dependency that exists between consecutive data samples, and it also includes a large number of lower sea state values that may skew the final results toward the low end. Thus, it is appropriate to select the data before the analysis. One commonly used method is to review the wave record, selecting only the highest wave data in a given period of time (e.g., highest wave per week, highest wave per month, highest wave per year, etc.) and then fit the censored data to an extreme value distribution. Generally, the annual maxima is used by most studies. However, in the case of this study it does not provide enough data points to reliably fit to a distribution function; therefore, monthly maxima were used.

Since it is desirable to use the maximum number of data points possible while still excluding those lower wave heights which do not positively contribute to extreme wave height, threshold monthly maxima are used for this study. The first step for data censoring is to select monthly maximum significant wave heights from the hourly significant wave height observations. Two data selection criteria were used in the selection process. First, months with more than half of the data missing were excluded so that only statistically meaningful, monthly maximum significant wave heights would be used in the analyses. Second, since two monthly maximum heights caused by the same storm will bias the data set, independence of the data set to be analyzed was insured by imposing a 60-hour minimum time interval between monthly maxima as proposed by Rossouw (1988). After the data set for monthly maximums was built, a subset of the data was obtained by selecting the higher values to eliminate the low-value effect on the extreme distribution. Because the choice of threshold is actually an educated guess, the validity and reliability of the guess can deviate between analyses. This deviation can then influence the fit of the distribution, and therefore the design wave height. Due to the different wave climate between the two stations as shown in the previous section, it was decided that analysis of a percentage of total data points of each data set was more reliable than analyzing a fixed threshold value. In this study, the upper two-thirds of the monthly maximum significant wave height data are selected for analysis.

The computation of the plotting position in this study also differs from the technique commonly used. When computing Gringorten's plotting position, we used

$$P_i = \frac{i-0.44}{n+0.12} \tag{8}$$

instead of the traditional formula

$$P_i = \frac{i-0.44}{N+0.12} \tag{9}$$

where P_i is the plotting position, i is the rank of the data point in ascending order, n is the number of threshold monthly maxima, and N is the total number of monthly maxima. In other words, the fact that our set of censored points were taken from a much larger data sample is not taken into consideration, The plotting position values for the bulk of the data do not differ between the two methods. The main difference between the two methods is that the traditional method (Eq. 9) places value on data at the lower tail while the current method (Eq. 8) does not. Castillo and Sarabia (1992) also suggested that little or no weight should be placed on data near the lower tail (or upper tail, as is fitting to the particular study) for the plotting position. Our previous analyses showed that accounting for the much larger sample size in the plotting position does not greatly enhance the goodness-of-fit of the model to the data. However, it is recognized that our method may produce larger return period waves than those when the larger sample size is accounted for, as shown by Muir and El-Shaarawi (1986).

Computation of Design Wave Heights

Design wave heights for a specified return period were first calculated in this study using the traditional method, which associates a probability (F_r) to a particular return period (T_r):

$$F_r = 1.0 - \frac{R}{T_r} \tag{10}$$

in which R is the measurement interval associated with each data point. This method is based on the recurrence concept, which means that the event will occur on average once in a T_r. However, as discussed by Rossouw (1988), it may be more appropriate to compute design wave heights for a return period (which can be assumed to be the design life of the structure) with a risk level (k) which is the probability that the return wave height is going to be exceeded during the return period:

$$P^r_{T_r} = (1.0 - k)^{\frac{R}{T_r}} \tag{11}$$

The advantage of this method is that it lets designers or engineers control the acceptable risk for their designs, whereas the former method does not.

The results of design wave heights from different extreme value distributions, together with associated parameters and values of goodness-of-fit criteria, are tabulated in Table 4 for station 46001. It is clear that the Weibull distribution has the best fit as judged by the four criteria.

Figures 5(a-e) show the measured data set and the distribution fits plotted in probability scales for all three distributions examined in this study. For both the FT-I and the log-normal distributions, distributions obtained from three parameter estimate techniques are plotted in the same figure (see Figs. 5(a) and 5(b)). Since the probability scale for the Weibull distribution involves the location parameter and since different parameter estimate techniques could produce different location parameters, three figures are used to present results from the three parameter estimate techniques (see Figs. 5(c-e). The FT-1 distribution shows a "drooping" of the data at the extreme upper tail. Thus, the estimated return wave heights tend to be high, as shown in Table 4. From the table, it is clear that the design wave heights vary based on the probability distribution and fitting method used. For the log-normal distribution, the results are very consistent for the different parameter estimation techniques. With respect to the Weibull distribution, the least square and maximum likelihood methods show consistency in their fits and design wave estimates. However, the method of moments tends to neglect the lower part of the data (Fig. 5(e)). The design wave heights, based on the traditional method (Eq. 10), range from 14.0 to 14.7 meters for a 50-year return period and from 14.5 to 15.2 meters for a 100-year return period. With a 5% risk level (based on Eq. 11), the design wave heights range from 15.6 to 16.6 meters for a 30-year return period and from 15.9 to 17.0 meters for a 50-year return period.

In the preliminary stage of this study (Timpe & Teng 1992), only five years of wave data from several buoy stations in the northeastern Pacific were analyzed. The design wave results using the FT-1 distribution for station 46001 were significantly higher that those found in this study. The 50-year wave at 46001 calculated in the preliminary study was 17.8 meters, as compared to 15.5 meters computed in this study. Thus, it can be said that the design wave results are closely linked to the length of the data record and/or data censorship technique.

Similar to those presented for station 46001, the results for station 44004 are presented in Table 5 and Figure 6. Compared to Figure 5, data for station 44004 presented in Figure 6 show more variation in the probability plots. The FT-1 distribution does not fit as well as either the log-normal or the Weibull distributions; this is most probably due to a flattening out of the data points at the extreme upper tail. Both Weibull and log-normal fit the bulk of the data well. Although there are four criteria to evaluate goodness-of-fit of the extreme distributions, high emphasis is placed on the r^2 value since it inherently describes the degree of linearity the distribution has successfully imposed on the data. Based on the r^2 criterion, the results from the log-normal distribution were adopted. The design wave heights are about 12.2 and 12.7 meters for 50- and 100-year return periods, respectively. With a 5% risk level, the design wave heights for 30- and 50-year return periods are about 13.6 and 13.8 meters, respectively.

Wave Spectra

In addition to extreme wave heights and associated wave periods, it is also desirable to obtain wave information in a spectral form, especially for extreme sea states. The advantage of using spectra to model extreme seas is that it gives a statistical description of all the wave components in a given sea state. The spectra can also be easily used to analyze the response of ocean and marine structures.

Figure 7 shows the climatic wave spectrum for extreme sea states (i.e., for H_s greater than 10 meters) for station 46001 together with the corresponding Pierson-Moskowitz spectrum. Both the peak frequency and spectral shape for the extreme climatic wave spectrum at this station correspond well to those calculated using the Pierson-Moskowitz spectrum. From a design viewpoint, it can be said that the Pierson-Moskowitz spectrum can be used as the design wave spectra for severe sea states in the areas of the northeastern Pacific near 46001. However, as shown in Figure 8, the Pierson-Moskowitz spectrum underestimates the climatic spectrum for extreme sea states (i.e., H_s greater than 9.0 meters for this buoy station since there is only one H_s greater than 10 meters) for the east coast station 44004. Instead, as seen in Figure 8, the JONSWAP spectrum, with proper spectral parameters, better represents the climatic spectrum for high waves in the northwestern Atlantic near 44004.

Climatic wave spectra for each of the wave height ranges in Figures 1 and 2 were developed by averaging all the wave spectra in that range. Figures 9 and 10 show the climatic wave spectra for different wave ranges for stations 46001 and 44004, respectively. Comparing these two figures, it is clear that, for the same wave height range, both the peak wave frequency and the peak wave energy of the climatic wave spectrum for the west coast station (46001) are lower than those for the east coast station (44004).

Summary and Conclusions

By using 13 years of hourly wave data from two NDBC buoy stations, it was shown that long-term distributions of significant wave height and average wave period for both stations fit well with the log-normal distribution. The relationships between significant wave height and average wave period for both the lower limit of wave period and the average values were established for both stations from nonlinear regression analysis.

Based on the threshold monthly maxima, the design wave heights for both stations were computed from different extreme value distributions and different parameter fitting methods. The design wave heights for a 100-year return period ranges from 14.5 to 15.2 meters for station 46001 and is about 12.7 meters for station 44004. For a 50-year return period with 5% risk level, the design wave height ranges from 16.2 to 17.0 meters for 46001, and is approximately 13.8 meters for 44004.

In the northeastern Pacific, it was shown that the climatic wave spectra for high waves can be modelled by the Pierson-Moskowitz spectrum. However, in the northwestern Atlantic, the Pierson-Moskowitz spectrum is not an acceptable representation for high wave spectra; instead, the JONSWAP spectrum appears to be more representative.

References

Battjes, J.A. (1970). "Long-term wave height distribution at seven stations around the British Isles." N.I.O. Report No. A 44.

Buckley, W.H. (1988). "Extreme and climatic wave spectra for use in the structural design of ships." *Naval Engineers Journal*, 36-58.

Castillo, E., and Sarabia, J.M. (1992). "Engineering analysis of extreme value data: Selection of models." *J. of Waterway, Port, Coastal and Ocean Division*, ASCE, 118(WW2) 129-146.

Gu, G.Z., Berek, E.P., Dello Stritto, F.J., Szabo, D., and Leder, H.V. (1992). "Extremal analysis of Hibernia wave hindcast data." *Third International Workshop on Wave Hindcasting and Forecasting*, Montreal, Quebec, May 19-22, 283-308.

Hogben, N., Dacunha, N.M., and Olliver, G.F. (1986). *Global wave statistics.* Unwin Brothers.

Isaacson, M., and MacKenzie, N. (1981). "Long-Term distributions of ocean waves: A review." *J. of Waterway, Port, Coastal and Ocean Div.*, ASCE, 107(WW2), 93-109.

Muir, L.R. and El-Shaarawi, A.H. (1986). "On the calculation of extreme wave heights: A review." *J. of Ocean Engineering*, 13(1), 93-118.

Ochi, M.K. (1978). "Wave statistics for the design of ships and ocean structures." *SNAME Transactions*, 86, 47-76.

Rossouw, J. (1988). "Design waves and their probability density functions." *Proceedings ASCE Conference on Coastal Engineering*, 822-834.

Timpe, G.L., and Teng, C.C. (1992). "Developing design wave criteria for use in data buoy design," *Proceedings Oceans '92*, 764-769.

Tucker, M.J. (1991). *Waves in ocean engineering: Measurement, analysis, interpretation*. Ellis Horwood Ltd., England.

Wyland, R.M., and Thornton, E.B. (1991). "Extremal wave statistics using three hindcasts." *J. of Waterway, Port, Coastal and Ocean Div.*, ASCE, 117(WW1), 60-74.

Table 1. Buoy Locations and Data Points Used in This Study.

Station	Location	Depth (m)	No. of Data Points	%
46001	56.3° N., 148.3° W. Gulf of Alaska	4,115	103,886	91.2
44004	38.5° N., 70.7° W. New Jersey Coast	3,231	82,396	72.4

Table 2. Maximum, Mean, and Standard Deviation of Significant Wave Heights.

Station	Maximum Wave Height (m)	Mean (m)	Std (m)
46001	13.30 (12/1/81 0200 Z)	2.77	1.43
44004	10.07 (12/19/86 1900 Z)	2.04	1.25

Table 3. Relationship between Significant Wave Height and Average Wave Period.

Station	Lower Limit	Average
46001	$T = 3.23 \cdot H^{0.47}$	$T = 5.24 \cdot H^{0.30}$
44004	$T = 3.28 \cdot H^{0.43}$	$T = 5.24 \cdot H^{0.26}$

Table 4. Design Wave Results for Station 46001.

46001 Censored

									Standard		5% Risk	
		A	B	C	r^2	W_n	MSE	D_n	50	100	30	50
FT-1	LSM	7.465	1.264	-	0.980	0.103	1E-3	0.073	15.55	16.42	18.66	19.31
	MLM	7.432	1.316	-	0.980	0.071	7E-4	0.067	15.85	16.76	19.09	19.76
	Moment	7.467	1.249	-	0.980	0.061	1E-3	0.057	15.46	16.32	18.53	19.17
LOG-N	LSM	2.084	0.193	-	0.985	0.065	6E-4	0.059	14.18	14.76	15.82	15.98
	MLM	2.084	0.192	-	0.985	0.068	7E-4	0.060	14.14	14.72	15.78	15.94
	Moment	2.083	0.194	-	0.985	0.061	6E-4	0.057	14.20	14.79	15.85	16.01
WEI.	LSM	5.420	3.114	1.708	0.991	0.029	3E-4	0.049	14.65	15.22	16.59	16.96
	MLM	5.270	3.292	1.904	0.989	0.033	3E-4	0.051	14.00	14.48	15.62	15.93
	Moment	5.372	3.350	1.900	0.991	0.113	1E-3	0.076	14.27	14.76	15.93	16.25

Table 5. Design Wave Results for Station 44004.

44004 Censored

									Standard		5% Risk	
		A	B	C	r^2	W_n	MSE	D_n	50	100	30	50
FT-1	LSM	6.453	1.048	-	0.955	0.142	2E-3	0.106	13.16	13.88	15.74	16.27
	MLM	6.398	1.199	-	0.955	0.114	1E-3	0.095	14.07	14.90	17.02	17.63
	Moment	6.448	1.045	-	0.955	0.148	2E-3	0.108	13.13	13.86	15.70	16.24
LOG-N	LSM	1.935	0.193	-	0.978	0.062	8E-4	0.074	12.21	12.72	13.63	13.77
	MLM	1.935	0.193	-	0.978	0.063	8E-4	0.074	12.21	12.72	13.63	13.77
	Moment	1.935	0.188	-	0.978	0.065	8E-4	0.075	12.04	12.55	13.50	13.62
WEI.	LSM	3.785	3.654	2.700	0.961	0.056	7E-4	0.066	11.05	11.33	12.20	12.15
	MLM	3.750	3.719	2.676	0.960	0.054	7E-4	0.074	11.19	11.48	12.14	12.33
	Moment	3.942	3.660	2.700	0.961	0.144	2E-3	0.109	11.22	11.50	12.15	12.34

***** 46001 (1980 - 1992) 7/8/93

T avg (sec) / H(s)(meter)	0 - 2.0	2.0 - 3.0	3.0 - 4.0	4.0 - 5.0	5.0 - 6.0	6.0 - 7.0	7.0 - 8.0	8.0 - 9.0	9.0 - 10.0	10.0 - 11.0	11.0 - 12.0	12.0 - 13.0	13.0 - 14.0	14.0 - 15.0	15.0 - 16.0	16.0 - >	SUM	%
.0- .5	0	12	76	141	74	24	0	0	0	0	0	0	0	0	0	0	327	.31
.5- 1.0	0	5	345	1990	2157	966	238	41	11	0	0	0	0	0	0	0	5753	5.54
1.0- 1.5	0	0	131	3834	5942	2840	594	108	31	1	0	0	0	0	0	0	13481	12.98
1.5- 2.0	0	0	6	2033	7979	5119	1539	272	44	0	0	0	0	0	0	0	16992	16.36
2.0- 2.5	0	0	1	485	6411	6301	2687	635	109	10	0	0	0	0	0	0	16639	16.02
2.5- 3.0	0	0	0	22	2719	6513	3383	950	138	16	0	0	0	0	0	0	13741	13.23
3.0- 3.5	0	0	0	0	751	5132	3659	1046	219	19	1	0	0	0	0	0	10827	10.42
3.5- 4.0	0	0	0	0	85	2827	3469	1323	268	41	5	0	0	0	0	0	8018	7.72
4.0- 4.5	0	0	0	0	2	1073	3083	1392	325	48	5	3	0	0	0	0	5931	5.71
4.5- 5.0	0	0	0	0	0	252	2149	1300	386	67	11	4	1	0	0	0	4170	4.01
5.0- 5.5	0	0	0	0	0	29	1256	1138	432	79	7	0	0	0	0	0	2941	2.83
5.5- 6.0	0	0	0	0	0	0	408	946	346	68	10	0	0	0	0	0	1778	1.71
6.0- 6.5	0	0	0	0	0	0	87	728	293	75	14	0	0	0	0	0	1197	1.15
6.5- 7.0	0	0	0	0	0	0	12	454	275	76	15	0	0	0	0	0	832	.80
7.0- 7.5	0	0	0	0	0	0	0	199	227	54	15	1	0	0	0	0	496	.48
7.5- 8.0	0	0	0	0	0	0	0	68	184	52	12	3	0	0	0	0	319	.31
8.0- 8.5	0	0	0	0	0	0	0	20	122	47	11	0	0	0	0	0	200	.19
8.5- 9.0	0	0	0	0	0	0	0	1	73	32	12	3	0	0	0	0	121	.12
9.0- 9.5	0	0	0	0	0	0	0	0	22	29	9	3	1	0	0	0	64	.06
9.5-10:0	0	0	0	0	0	0	0	0	1	18	4	4	0	0	0	0	27	.03
10.0-10.5	0	0	0	0	0	0	0	0	1	6	4	6	0	0	0	0	17	.02
10.5-11.0	0	0	0	0	0	0	0	0	0	3	2	1	0	0	0	0	6	.01
11.0-11.5	0	0	0	0	0	0	0	0	0	2	1	0	2	0	0	0	5	.00
11.5-12.0	0	0	0	0	0	0	0	0	0	0	1	1	0	0	0	0	2	.00
12.0-12.5	0	0	0	0	0	0	0	0	0	0	1	0	0	0	0	0	1	.00
> 12.5	0	0	0	0	0	0	0	0	0	0	1	0	0	0	0	0	1	.00
SUM	0	17	559	8505	26120	31076	22564	10621	3507	743	141	29	4	0	0	0	103886	
%	.00	.02	.54	8.19	25.14	29.91	21.72	10.22	3.38	.72	.14	.03	.00	.00	.00	.00		

Figure 1. Scatter Diagram for Station 46001

***** 44004 (1980 - 1992)

T avg (sec) / H(s)(meter)	0 - 2.0	2.0 - 3.0	3.0 - 4.0	4.0 - 5.0	5.0 - 6.0	6.0 - 7.0	7.0 - 8.0	8.0 - 9.0	9.0 - 10.0	10.0 - 11.0	11.0 - 12.0	12.0 - 13.0	13.0 - 14.0	14.0 - 15.0	15.0 - 16.0	16.0 - >	SUM	%
.0- .5	0	1	65	406	345	135	10	0	0	0	0	0	0	0	0	0	962	1.17
.5- 1.0	0	0	716	6760	5009	1608	309	28	2	0	0	0	0	0	0	0	14432	17.52
1.0- 1.5	0	0	170	8537	8258	2255	609	114	25	1	0	0	0	0	0	0	19969	24.24
1.5- 2.0	0	0	18	3436	8788	2386	624	175	38	11	1	0	0	0	0	0	15477	18.78
2.0- 2.5	0	0	0	464	6126	2856	570	172	27	10	2	0	0	0	0	0	10227	12.41
2.5- 3.0	0	0	0	19	2819	2983	701	170	57	4	0	0	0	0	0	0	6753	8.20
3.0- 3.5	0	0	0	3	700	2953	775	152	73	10	0	0	0	0	0	0	4666	5.66
3.5- 4.0	0	0	0	5	55	2040	921	194	59	29	4	0	0	0	0	0	3307	4.01
4.0- 4.5	0	0	0	0	11	867	1122	189	32	18	9	0	0	0	0	0	2248	2.73
4.5- 5.0	0	0	0	0	3	185	1063	246	33	12	4	0	0	0	0	0	1546	1.88
5.0- 5.5	0	0	0	0	0	17	607	299	44	7	1	0	0	0	0	0	975	1.18
5.5- 6.0	0	0	0	0	0	0	269	345	63	4	3	0	0	0	0	0	684	.83
6.0- 6.5	0	0	0	0	0	0	97	313	66	6	3	0	0	0	0	0	485	.59
6.5- 7.0	0	0	0	0	0	0	9	205	66	3	1	1	0	0	0	0	285	.35
7.0- 7.5	0	0	0	0	0	0	2	105	65	1	1	0	0	0	0	0	174	.21
7.5- 8.0	0	0	0	0	0	0	0	35	55	3	0	0	0	0	0	0	93	.11
8.0- 8.5	0	0	0	0	0	0	0	9	37	6	0	0	0	0	0	0	52	.06
8.5- 9.0	0	0	0	0	0	0	0	0	25	6	0	0	0	0	0	0	31	.04
9.0- 9.5	0	0	0	0	0	0	0	0	13	9	0	0	0	0	0	0	22	.03
9.5-10:0	0	0	0	0	0	0	0	0	3	4	0	0	0	0	0	0	7	.01
10.0-10.5	0	0	0	0	0	0	0	0	0	1	0	0	0	0	0	0	1	.00
10.5-11.0	0	0	0	0	0	0	0	0	0	0	0	0	0	0	0	0	0	.00
11.0-11.5	0	0	0	0	0	0	0	0	0	0	0	0	0	0	0	0	0	.00
11.5-12.0	0	0	0	0	0	0	0	0	0	0	0	0	0	0	0	0	0	.00
12.0-12.5	0	0	0	0	0	0	0	0	0	0	0	0	0	0	0	0	0	.00
> 12.5	0	0	0	0	0	0	0	0	0	0	0	0	0	0	0	0	0	.00
SUM	0	1	969	19630	32114	18285	7688	2751	783	145	29	1	0	0	0	0	82396	
%	.00	.00	1.18	23.82	38.98	22.19	9.33	3.34	.95	.18	.04	.00	.00	.00	.00	.00		

Figure 2. Scatter Diagram for Station 44004

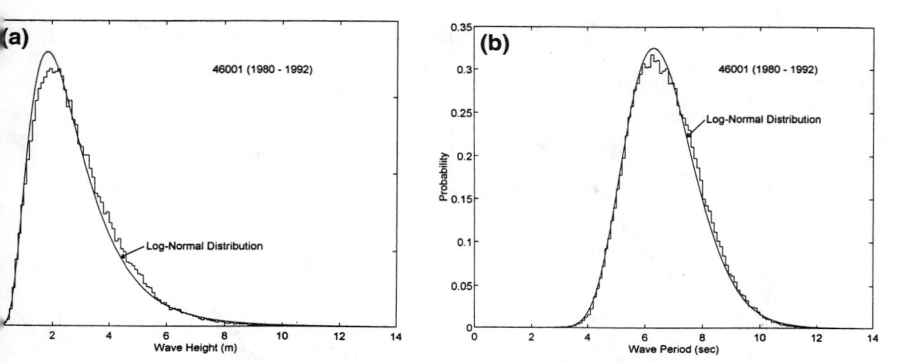

Figure 3. Histogram of Measured Data and associated Log-Normal Probability Density Function for Station 46001 for (a) Significant Wave Height and (b) Average Wave Period.

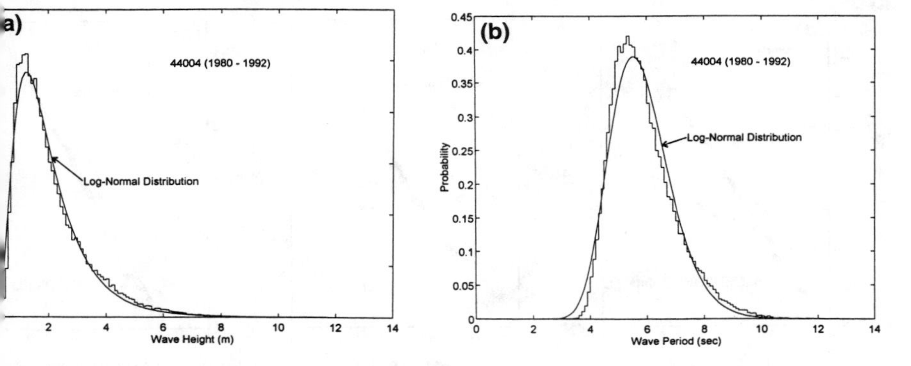

Figure 4. Histogram of Measured Data and associated Log-Normal Probability Density Function for Station 44004 for (a) Significant Wave Height and (b) Average Wave Period.

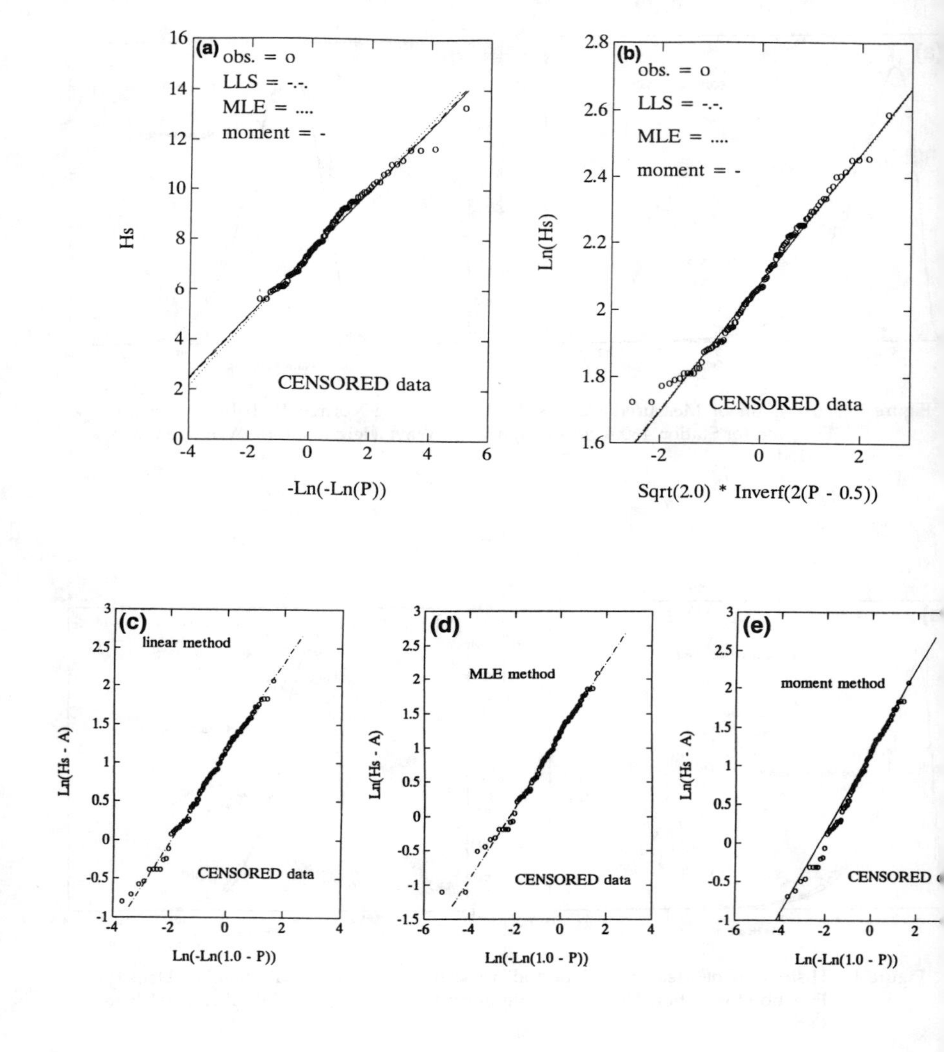

Figure 5. Extreme Value Distribution Fits for Station 46001 for (a) FT-1 Distribution, (b) Log-Normal Distribution, (c) Weibull Distribution - Least Squared, (d) Weibull - Maximum Likelihood, and (e) Weibull - Moment Method.

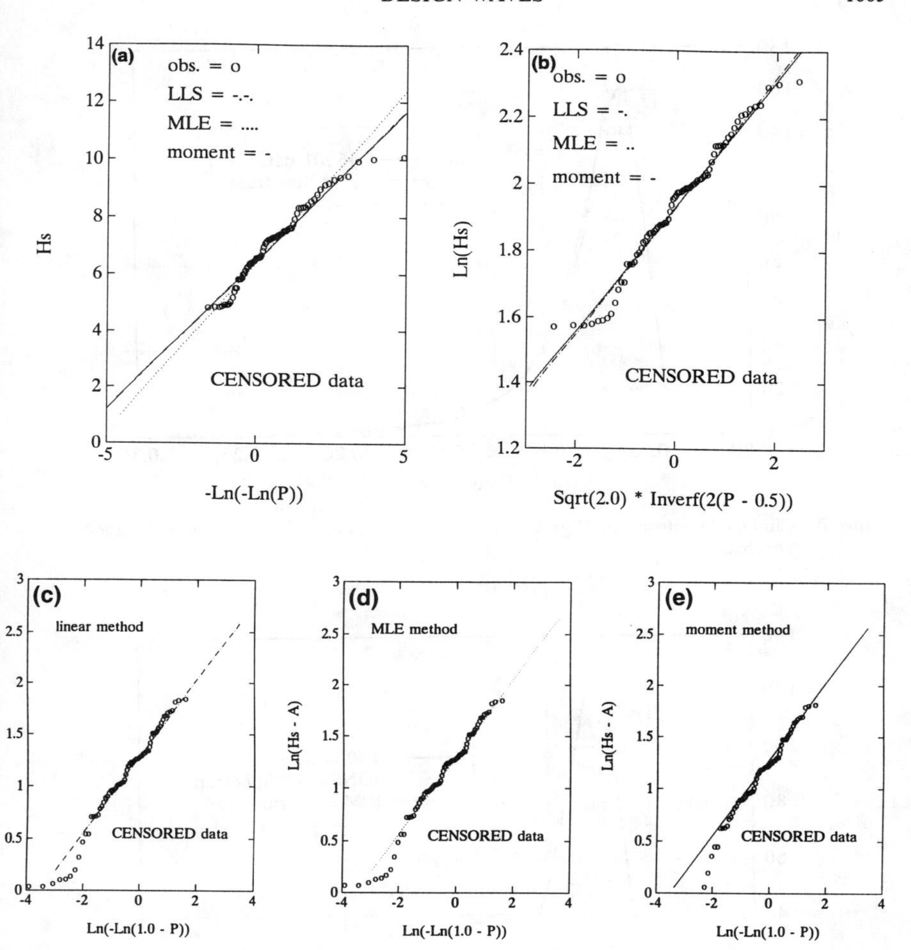

Figure 6. Extreme Value Distribution Fits for Station 44004 for (a) FT-1 Distribution, (b) Log-Normal Distribution, (c) Weibull Distribution - Least Squared, (d) Weibull - Maximum Likelihood, and (e) Weibull - Moment Method.

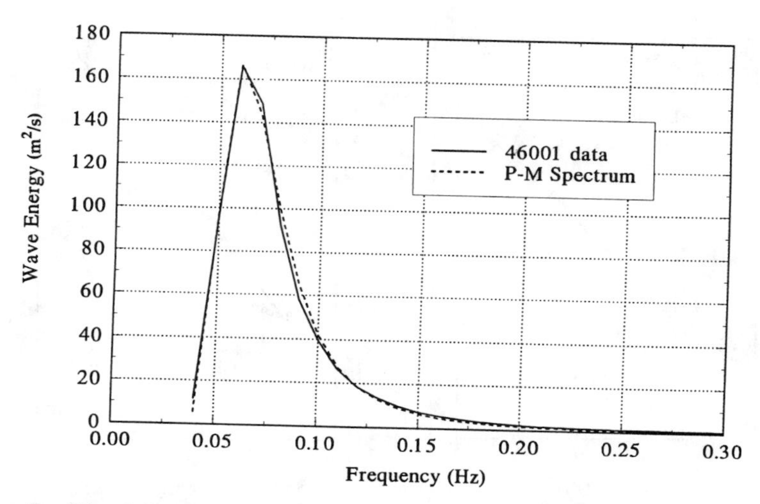

Figure 7. Climatic Spectrum for High Wave (> 10 m) for Station 46001 with associated P-M Spectrum.

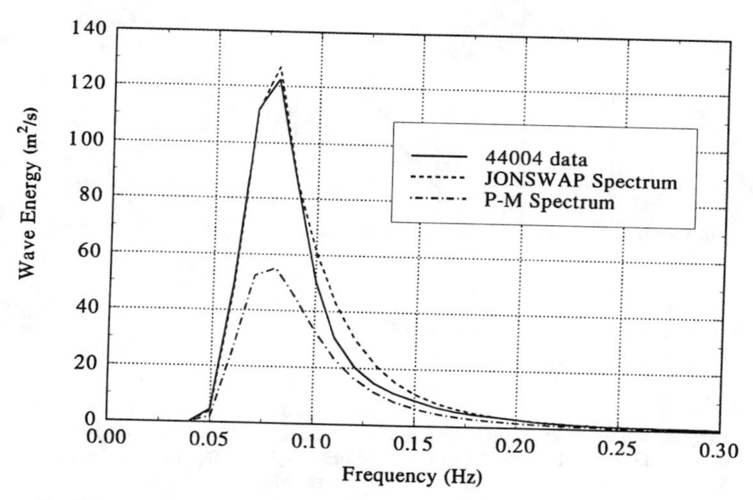

Figure 8. Climatic Spectrum for High Wave (> 9 m) for Station 44004 with associated P-M and JONSWAP Spectra.

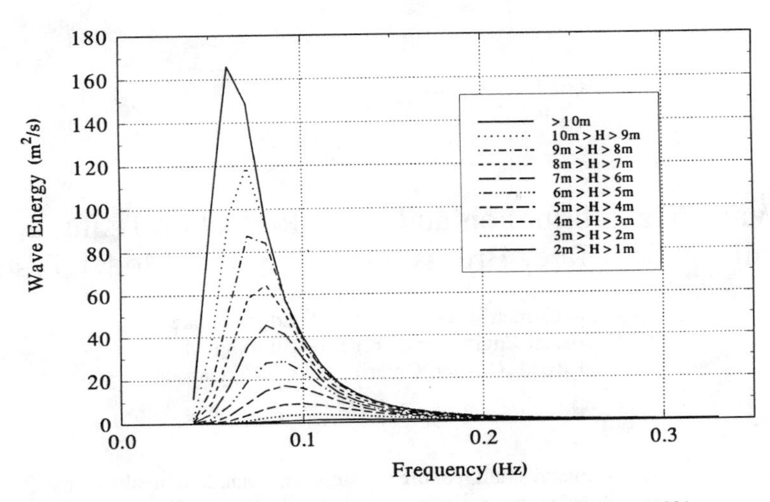

Figure 9. Climatic Spectra for Various Wave Height Ranges for Station 46001.

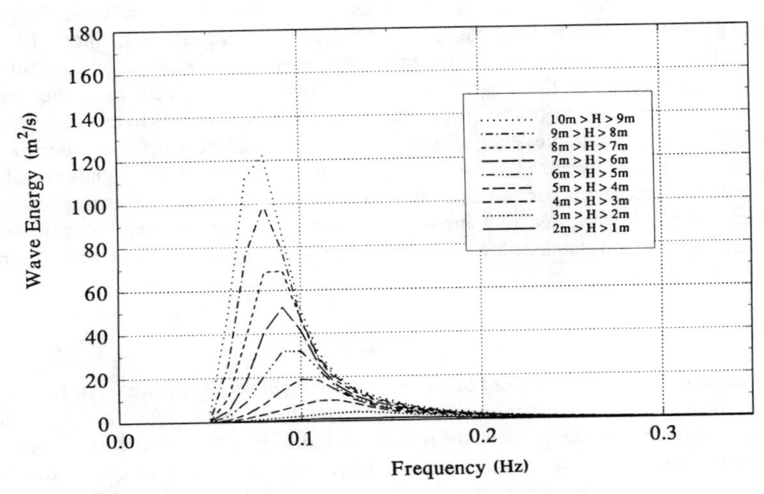

Figure 10. Climatic Spectra for Various Wave Height Ranges for Station 44004.

Wave Transformation and Load Reduction Using a Small Tandem Reef Breakwater - Physical Model Tests

Andrew Cornett, Etienne Mansard, Edgar Funke
NRC Coastal Engineering Program, Building M32
Montreal Road, Ottawa, Ontario, K1A 0R6, Canada

Abstract

The performance of several small reef breakwaters, and a tandem breakwater system (consisting of a reef breakwater positioned upwave from a surface-piercing main breakwater), are quantified through a limited set of physical model tests. Wave reflection and transmission coefficients for the reef breakwaters are computed from wave probe measurements in irregular waves. The character, and wave height distribution, of waves transmitted past the reef breakwaters are also examined. Fluid forces acting on the armour stones of the main breakwater are measured using a force panel, and compared to the characteristics of waves transmitted past the reef breakwaters, and to observations of armour stone damage. Reductions to extreme wave heights and extreme loading events, induced by several different reef breakwaters, are quantified and favourably correlated to improved armour stone stability. These results demonstrate the successful performance of a tandem breakwater system incorporating a relatively small, low-crested reef breakwater. Extrapolation and generalization of these results to different tandem breakwater systems and wave conditions will require further model testing.

Introduction

The tandem breakwater concept employs a submerged reef breakwater in front of a main surface piercing structure to protect the latter from severe wave action. The submerged reef breakwater serves to reflect and dissipate wave energy, thereby reducing the intensity of wave action on the main structure, which can therefore be designed using more economic materials. Cox and Clark (1992) describe a tandem breakwater system designed for Lake Michigan that could be built for $1 Million less than a single conventional structure designed to meet the same operating criteria. Efficient design of an effective tandem breakwater system relies on precise, quantitative information regarding the performance (stability) of reef breakwaters, and the character of the waves that are transmitted. Articles treating the transmission and reflection of wave energy from reef breakwaters include van der Meer and Angremond (1992),

Ahrens (1987), Adams and Sonu (1986), Seelig (1980), Dattatri, Raman and Shankar (1978) and Tanaka (1976). The stability of reef breakwaters is treated by Vidal and Mansard (1993), van der Meer and Pilarczyk (1990), and Ahrens (1989). In a survey of existing knowledge relating to the design and performance of reef breakwaters, Ahrens and Cox (1990) conclude that while reef breakwaters have considerable potential as an effective method of shore protection, there is still very little quantitative information on their design and performance characteristics.

This paper presents results from a short physical model test series designed to quantify the performance of a tandem breakwater system using three relatively low-crested reef breakwaters in a single wave environment. Making use of an existing model, it was possible to quantify characteristics of the waves transmitted past the submerged breakwater, as well as the reduction in wave loads acting on the armour of the main breakwater. Because of the opportunistic nature of this research, the data set is unfortunately restricted to a small number of tandem breakwater configurations and wave conditions. In consequence, it is difficult to draw general conclusions that can be applied to all situations. Nonetheless, the results of this research are a useful example of the successful application of the tandem breakwater concept.

Experiments

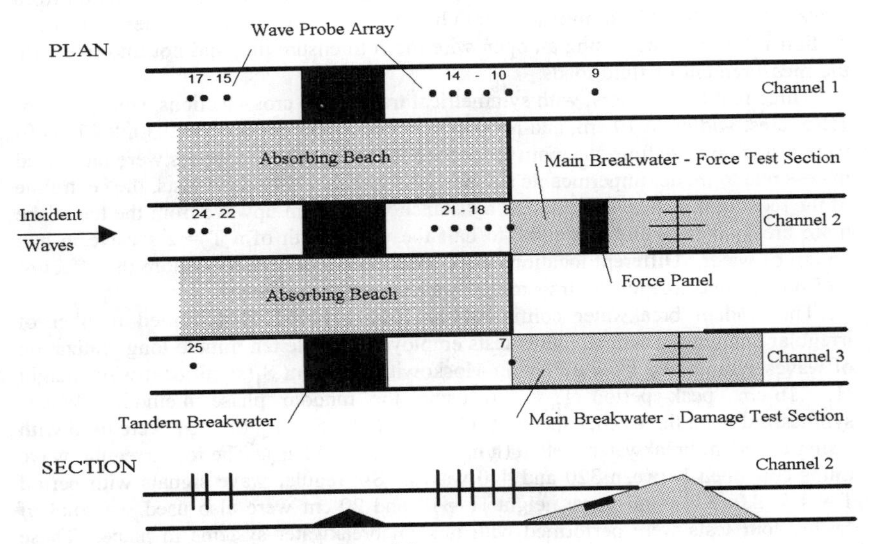

Figure 1. Sketch of the physical model.

Figure 1 shows a sketch of the physical model. Three separate test channels were constructed near the centre of a 14 m wide wave basin. The channels were calibrated to ensure very similar incident waves. Waves incident to the main breakwater were measured in channel 1 using two arrays of capacitance wire probes. Wave loads acting on the armour of the main breakwater were measured in channel 2 using a force panel. Channel 3 was used to monitor the extent of damage to (erosion of) armour stones. The downwave end of the basin was faced with a 1:15 porous gravel beach, characterized by a low reflection coefficient $C_r = 100 \cdot H_r / H_i \sim 5$ %. Similar beaches were installed between the three test channels. This setup allows the simultaneous measurement of incident waves, wave loads, and breakwater damage, relatively uncontaminated by unwanted wave reflections.

Wave loads were measured on a rigid, porous panel comprised of fifty irregularly shaped, aluminum model rocks, spot-welded together at points of contact. The panel represents a patch of surface layer armour stones with approximate overall dimensions 23 cm by 63 cm. Five degree-of-freedom measurements of the fluid force acting on the panel were made by a custom built load cell dynamometer. The panel was installed on the face of a breakwater test section, just below the still water level, and was isolated form the surrounding breakwater material by a thin gap that follows the irregular contours of the panel. Forces measured by the panel provide a steady, repeatable measure of the fluid forces acting on a patch of surface layer armour stones.

The main breakwater consisted of two layers of finely graded armour (D_{n50} = nominal diameter of armour stones = 4.2 cm, D_{85}/D_{15} = 1.1) placed at a constant slope of 1:1.75 over a permeable trapezoidal core (D_{n50} = 1.9 cm, D_{85}/D_{15} = 1.3). The force panel spanned the face of the breakwater test section from elevation z ~ 40 - 52 cm in water depth h = 55 cm. Loose armour stones on this test section were restrained using an open wire mesh to ensure they did not interfere with the measurement of fluid loads.

Three reef breakwaters, with symmetrical trapezoidal cross-sections, side slopes of 1:1.5, crest widths of 10 cm, and heights h_b = 17, 25, 33 cm (h/h_b = 0.30, 0.45, 0.60) were tested upwave from the main breakwater. These reef breakwaters were fabricated in concrete to model impermeable submerged structures. For most tests, the centreline of the reef breakwaters were located a distance x = 2.11 m upwave from the toe of the main breakwater, a distance equal to half the wavelength of a T = 2 s wave in h = 55 cm of water. Different locations were used in several tests to explore the effect of reef breakwater position on transmitted waves and armour loads.

The tandem breakwater configurations were exposed to a limited number of irregular and regular waves. Most tests employed a single ten minute long realization of waves synthesized from a Pierson-Moskowitz spectrum S_1(significant wave height H_s = 16 cm, peak period T_p = 2 s) using the random phase method. Waves synthesized from three other spectra S_2(14,2), S_3(14,1.67), S_4(14, 2.5), were used with a single tandem breakwater system (h/h_b = 0.6, x = 2.11 m). The four irregular wave trains contained between 320 and 440 waves. Six regular wave signals with period T = 1.5, 2.0, 3.0 s and wave height H = 12 and 20 cm were also used. A total of twenty-four tests were performed with tandem breakwater systems in place. These were supplemented by additional tests using identical waves without reef breakwaters.

Wave Transmission and Reflection

Submerged reef breakwaters can be used as part of a shore defence or protection scheme to moderate the waves that impinge on the shoreline or primary structure. The energy of the waves transmitted past a submerged breakwater is reduced corresponding to the energy that is reflected and dissipated through various mechanisms such as wave breaking, bottom friction, and the generation of turbulence. These processes are known to depend on characteristics of the submerged breakwater, such as its height, shape, permeability, and roughness, as well as on the height, and period of the incident waves, and on the water depth.

Van der Meer (1991) has reviewed experimental data on wave transmission at low-crested structures from six sources. The transmission coefficient $K_t = H_t/H_i$, where H_t is the transmitted wave height and H_i is the incident wave height, can be plotted somewhat satisfactorily as a function of the relative freeboard F/H_i, where $F = h_b - h$. For the range $-2 < F/H_i < -1$, the data surprisingly suggests that the transmission coefficient is a constant $K_t = 0.8$. This result is somewhat inconsistent with the known asymptote $K_t = 1.0$ that must hold at $F/H_i = -h/H_i$, corresponding to a submerged breakwater of zero height, and also represents a non-conservative design recommendation.

A three probe reflection analysis scheme, described by Mansard and Funke (1980), was used to separate the incident and reflected wave spectra measured by the two probe arrays upwave and downwave from the reef breakwater in channel 1 (see Figure 1). Reflected waves measured downwave from the submerged breakwater are produced by the 1:15 porous gravel beach at the rear of the basin, while reflected waves measured upwave from the structure are assumed to result from a combination of components produced by the submerged breakwater and the gravel beach. A transmission coefficient is defined as

$$K_t = \frac{H_{m0,i,u}}{H_{m0,i,d}} \tag{1}$$

where $H_{m0,i,u}$ is the spectral estimate of the significant wave height travelling in the incident wave direction measured upwave from the submerged breakwater, and $H_{m0,i,d}$ is the same parameter measured at the downwave array.

Transmission coefficients computed from twenty-three tests in irregular waves are plotted in Figure 2 as a function of relative freeboard. These results indicate that K_t approaches 1.0 as $F/H_{m0,i} \sim -2.5$. For the conditions tested, the transmission of wave energy was virtually unaffected by submerged breakwaters with freeboard larger than 2.5 times the significant wave height of the incident waves.

Three definitions of reflection coefficient K_r, representing the ratio of the reflected wave height to the incident wave height, are:

$$K_{r1} = \frac{H_{m0,r,u}}{H_{m0,i,u}} \tag{2}$$

$$K_{r2} = \frac{(H^2_{m0,r,u} - H^2_{m0,r,d})^{1/2}}{H_{m0,i,u}} \tag{3}$$

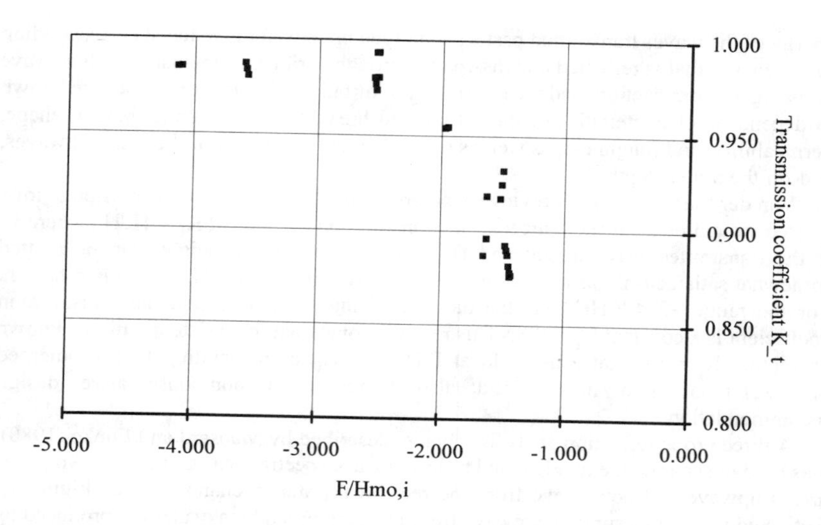

Figure 2. Transmission coefficients for low reef breakwaters.

$$K_{r3} = (1 - K_t^{\,2})^{1/2} \tag{4}$$

K_{r1} is based only on measurements at the upwave probes; K_{r2} is a modification to K_{r1} representing a correction for the energy reflected from the beach at the rear of the basin; and K_{r3} assumes that no energy dissipation takes place so that $K_t^{\,2} + K_{r3}^{\,2} = 1$. Reflection coefficients derived from the present tests with irregular waves are plotted in Figure 3 as a function of relative freeboard. Reflection coefficients very close to zero where recorded for submerged breakwaters with relative freeboard $F/Hi < -3$. Note that K_{r2} is not much different from K_{r1}, confirming the low level of wave energy reflected from the rear beach. The significant difference between K_{r1} or K_{r2} and K_{r3} indicates that energy dissipation should not be neglected.

Ahrens (1987) presents the following empirical formulae to estimate the wave transmission and reflection from reef breakwaters.

$$K_r = \exp\left[C_1\left(\frac{h}{Lp}\right)+C_2\left(\frac{h}{h_b}\right)+C_3\left(\frac{A_t}{h_b^{\,2}}\right)+C_4\left(\frac{F}{H_{m0}}\right)\right] \tag{5}$$

with $C_1 = -6.774$, $C_2 = -0.293$, $C_3 = -0.0860$, and $C_4 = 0.0833$; and

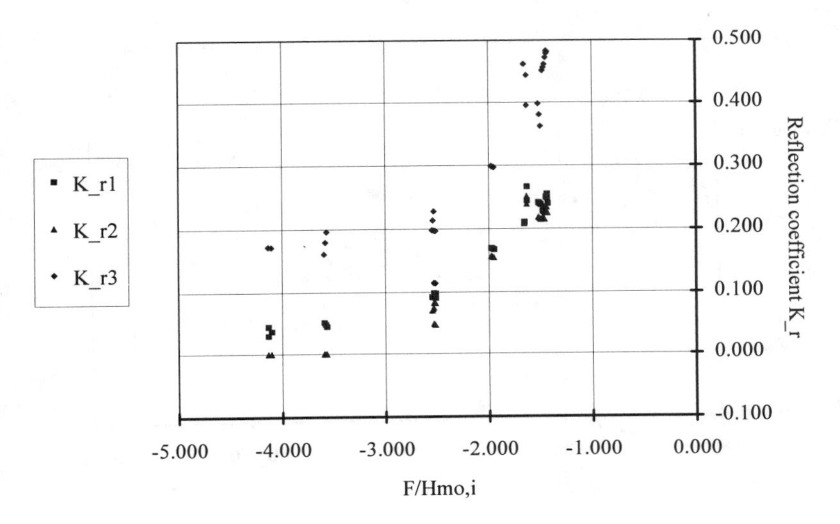

Figure 3. Reflection coefficients for low reef breakwaters.

$$K_t = \left\{ 1.0 + \left(\frac{h_b}{h}\right)^{C_1} \left(\frac{A_t}{hL_p}\right)^{C_2} \exp\left[C_3\left(\frac{F}{H_{m0}}\right) + C_4\left(\frac{A_t^{3/2}}{D_{n50}^2 L_p}\right)\right]\right\}^{-1} \quad , \quad \text{for} \quad \frac{F}{H_{m0}} < 1 \tag{6}$$

with $C_1 = 1.188$, $C_2 = 0.261$, $C_3 = 0.529$ and $C_4 = 0.00551$. In these formulae, L_p = wave length of peak period waves, A_t = average cross-sectional area of the reef breakwater. Figures 4 and 5 show a comparison between the present irregular wave test data and the predictions of K_r and K_t according to Ahrens' formulae. The agreement is generally good, supporting the use of Ahrens' formulae for design of wave transmission and reflection at low-crested reef breakwaters with $F/H_{m0,i} < -1.4$.

Character of Transmitted Waves

The character of the waves transmitted past a reef breakwater has considerable effect on local coastal processes, such as sediment transport and bedform geometry, and on the design and performance of additional protective structures, such as breakwaters or revetments.

Submerged sand bars, shoals, and reefs are known to trigger a shift of wave energy from fundamental to higher frequencies through "harmonic decoupling" of higher order wave components from their fundamental carriers. Beji and Battjes (1993) show that this effect is quite pronounced for a broad, shallow sloping shoal. Kojima, Ijima and Yoshida (1990) describe similar extensive transformations over long fixed horizontal

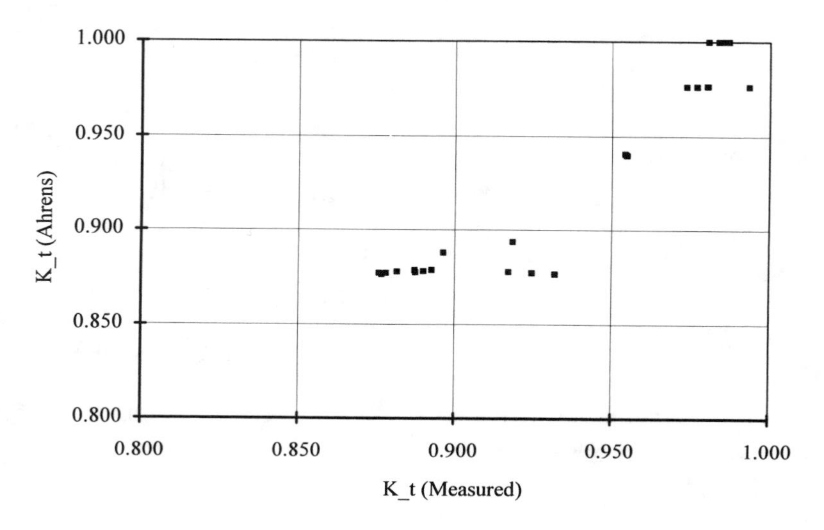

Figure 4. Comparison of measured and predicted transmission coefficients.

plates. Tests with regular waves indicate a strong spatial variation in wave characteristics downwave from a submerged structure, depending on the phasing of the free (free to propagate at their natural celerity) higher harmonics relative to the fundamental wave components. For irregular waves, the decomposition of longer waves and the release of higher harmonics as free wave components downwave from a submerged structure, produces a wave field with reduced significant wave period and wave heights that deviate from the Rayleigh distribution. The effect of these nonlinear processes on the spatial variability of random wave statistics is not known. Somewhat less dramatic transformations are expected for low-crested, steeply sloping, reef breakwaters that are not broad in comparison to either the water depth or incident wave length. Nonetheless, even subtle transformations to the higher frequency wave components, and thus to the distribution of wave heights, wave periods, crest and trough excursions, can have important effects on nearshore processes and the performance of shore protection structures.

Figure 6 depicts the incident and transmitted energy spectra for S_1 waves on a reef breakwater with $h_b/h = 0.6$ ($F/H_{mo,i} \sim -1.4$), which demonstrate an overall reduction of wave energy, and a typical shift to higher frequencies downwave of the submerged structure. Equivalent segments of time series of S_1 waves measured at probe 14 (downstream from the submerged breakwater), both with and without an $h_b/h = 0.6$ reef breakwater, are plotted in Figure 7. The submerged breakwater effects a consistent reduction in the trough excursions of the largest waves, however, the crest elevations are not necessarily reduced.

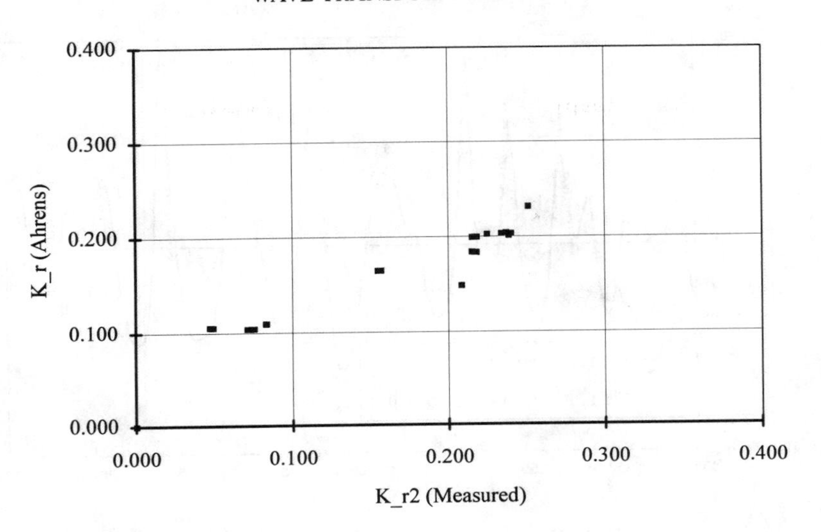

Figure 5. Comparison of measured and predicted reflection coefficients.

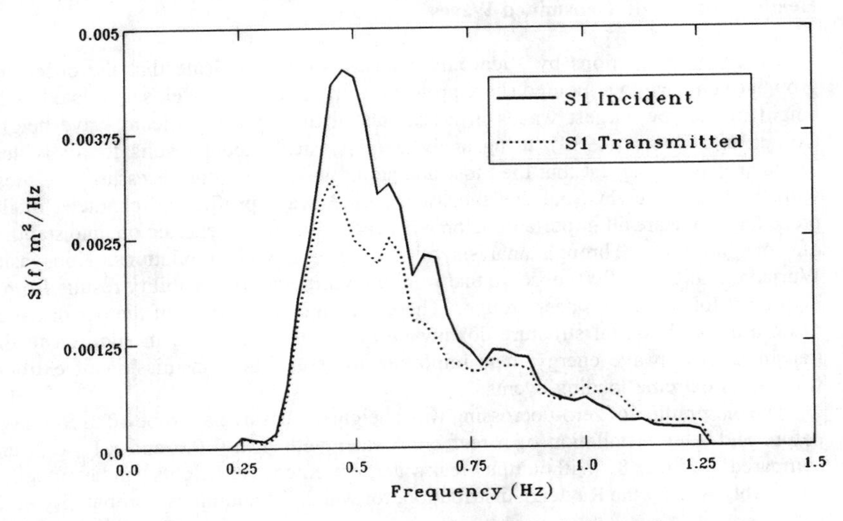

Figure 6. Incident and transmitted spectra for S1 seas, $h_b/h = 0.6$.

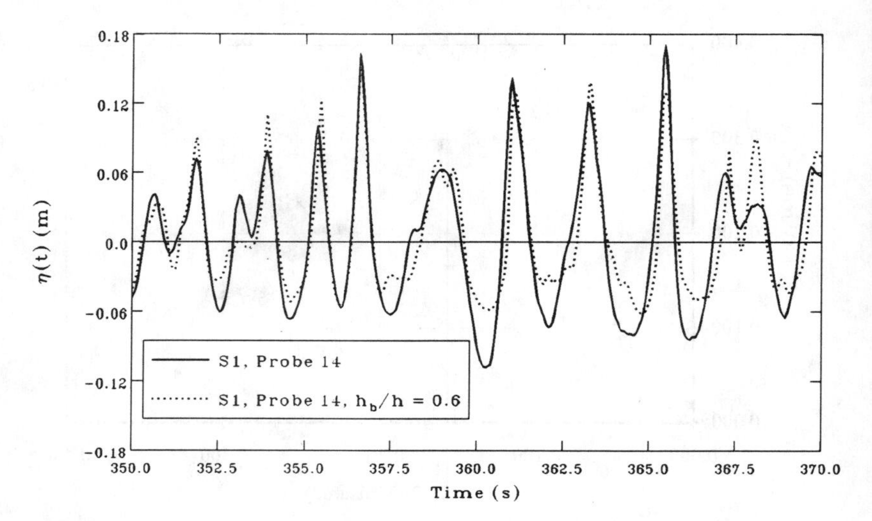

Figure 7. Effect of the $h_b/h = 0.6$ reef breakwater on S1 waves.

Height Statistics of Transmitted Waves

Recent investigations by Vidal and Mansard (1993) indicate that the onset and growth of damage to armoured shore protection structures is more closely related to the height of the few largest waves in a sea state than to the significant wave height. Cornett and Davies (1993), using analysis of physical model results for individual irregular waves, suggest that the most damaging waves may not necessarily be those with the largest heights; but that the wave period, wave profile and character of the preceding wave are all important factors influencing the forces exerted on, and stability of, rock armour. Through analysis of numerical model simulations, Kobayashi, Wurjanto and Cox (1990) observe that minimum armour stone stability results from a high crest followed by a deep trough. Therefore, in consideration of the performance of an armoured coastal structure downwave of a reef breakwater, in addition to the transmission of wave energy, it is important to assess the transmission of extreme waves and extreme loading events.

The distribution of zero-upcrossing wave heights, measured at probe 10 in S_1 waves before and after installation of a reef breakwater with $h_b/h = 0.6$ and $x/L_p = 2$, are compared in Figure 8. Without the breakwater, extreme wave heights can be modelled reasonably well by the Rayleigh distribution, for which the cumulative probability P(H) is

$$P(H) = 1 - \exp\left[-\left(\frac{H}{\tau}\right)^2\right] \tag{7}$$

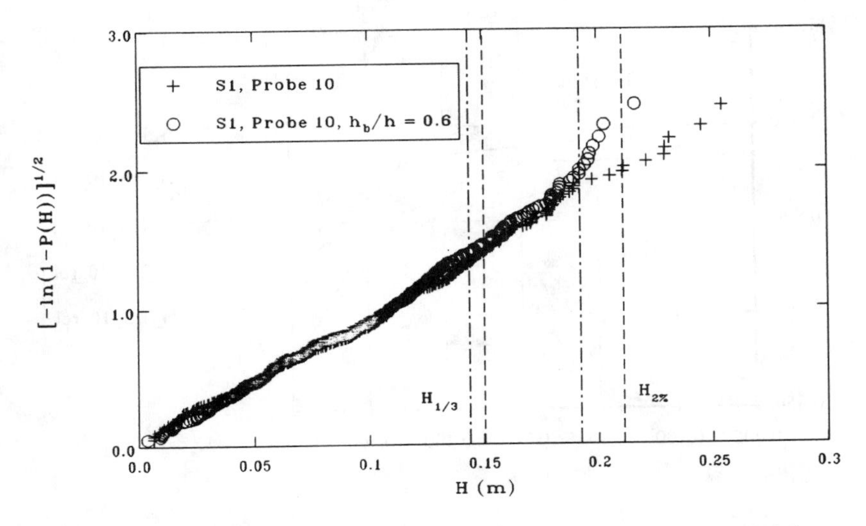

Figure 8. Effect of h_b/h = 0.6 reef breakwater on distribution of S1 wave heights.

where $\tau = 2\mu_H\pi^{-1/2}$ and μ_H = mean of H. However, behind the reef breakwater, extreme wave heights are reduced such that they no longer fit the Rayleigh model. Vertical lines corresponding to the 2 % threshold of wave height $H_{2\%}$ (height exceeded by 2 % of all waves), and significant wave height $H_{1/3}$ (average height of the largest third of all waves) are included in the Figure. In this case, the reef breakwater reduces $H_{2\%}$ by 9 % and $H_{1/3}$ by 5 % at probe 10. This reduction in extreme heights is not realized at all locations behind the submerged breakwater: - the reduction varies from probe to probe, approaching 0 % at probe 14. This suggests that the spacing between tandem breakwaters may influence the performance of the second structure, and that an optimal spacing may exist, dependent on wave conditions and the geometry of the tandem structures.

Figure 9 shows the effect of reef breakwater height on the change in extreme wave height statistics measured at probe 10. Each statistic of wave heights, measured with the reef breakwater in place, is divided by the equivalent statistic for the same waves without the structure. Several observations can be made: no significant change in height statistics is realized for $h_b/h < 0.6$; the $h_b/h = 0.6$ structure has a greater effect on the largest waves in the sea state; and the significant zero-crossing wave height $H_{1/3}$ is reduced less than the significant wave height derived from energy considerations H_{m0}.

Load Reduction

Fluid forces acting on armour stones located on the face of the main breakwater were measured using a force panel (see Figure 1). These measurements make it possible to quantify the effect of reef breakwaters on the armour of the main breakwater.

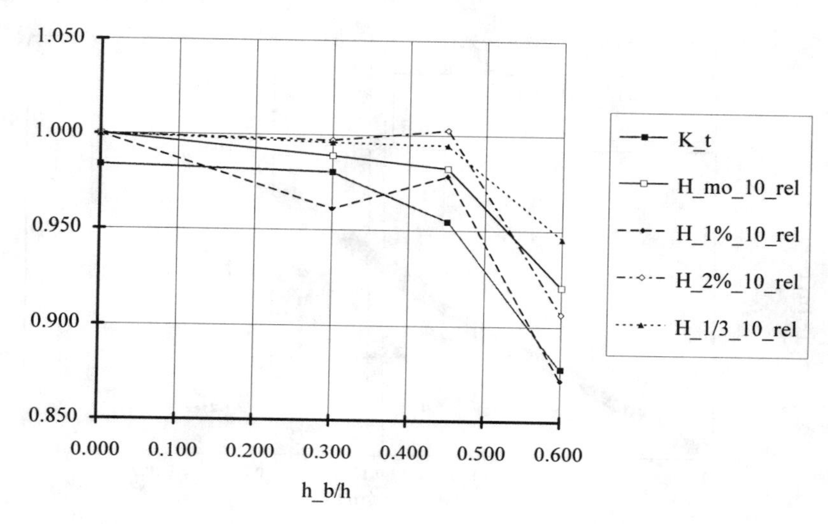

Figure 9. Reduction in transmitted wave height statistics with increasing reef height.

Kobayashi and Otta (1987) consider the hydraulic stability of armour units in several modes of failure, including lifting, sliding and rolling. They conclude that the rolling and sliding failure criteria are equivalent, and represent the minimum stability condition for realistic situations. The stability of armour stones can be expressed as a ratio R of the forces acting to bring about failure normalized by those acting to resist failure, as

$$R = \frac{\left|F_p - \sin\alpha(F_w - F_b)\right|}{\tan\phi\left[(F_w - F_b)\cos\alpha - F_n\right]} \tag{8}$$

where: F_p is the component of fluid force parallel to the surface (positive upslope); F_n is the component of fluid force normal to the surface (positive away from the structure); F_w is the weight of armour stones in air (positive downward); F_b is the buoyancy force (positive upward); ϕ is the internal friction angle for the armour stones; and α is the inclination of the armour layer (positive upward from the horizontal). R can be defined for a single armour stone, or a patch of armour stones such as the force panel. $R > 1$ indicates a potentially unstable loading condition, while $R < 1$ denotes greater armour stone stability. Another useful quantity is the magnitude of the force resultant F acting on the force panel, defined as

$$F = \left(F_p^2 + F_n^2\right)^{1/2} \tag{9}$$

which indicates the intensity of wave loading.

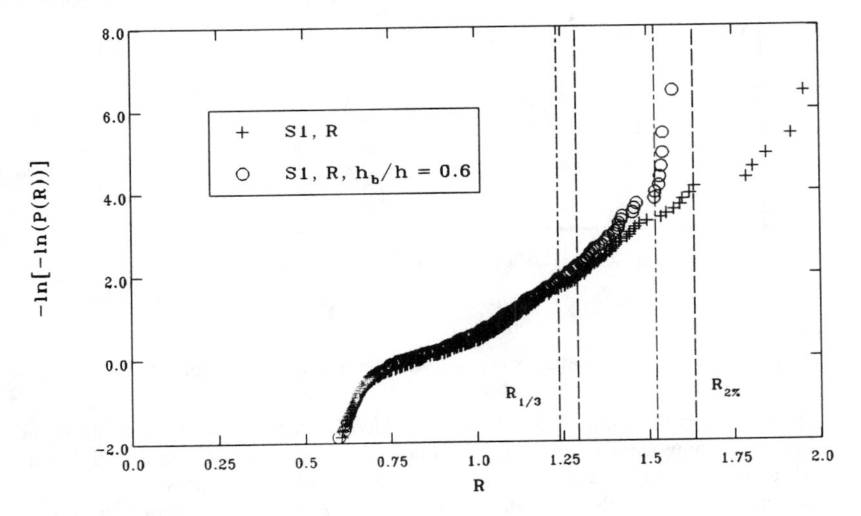

Figure 10. Effect of $h_b/h = 0.6$ reef breakwater on the distribution of loads for S1 seas.

Using force panel measurements, time series of F(t) and R(t) have been constructed for each tandem breakwater test. Maxima of F(t) and R(t) were identified for each wave cycle based on timing information from zero-upcrossing analysis of the water level measured on the face of the main breakwater. Only one maxima is identified for each full upcrossing cycle of water level on the breakwater face. Figure 10 shows the effect of the $h_b/h = 0.6$, $x/L_p = 2$ reef breakwater on the distribution of R maxima for S_1 waves. The maxima have been modelled using the Gumbel extreme value distribution, for which the cumulative probability is

$$P(R) = \exp\left\{-\exp\left[-\left(\frac{R-\varepsilon}{\delta}\right)\right]\right\} \tag{10}$$

where ε and δ are parameters of the distribution. The 4 % and 8 % reductions to $R_{1/3}$ and $R_{2\%}$ shown in Figure 10 are comparable to those realized during the same test for $H_{1/3}$ and $H_{2\%}$ (see Figure 8). Note that in this case, the reef breakwater is particularly effective at reducing the largest R maxima, and thus, reducing the potential for damage to the breakwater armour.

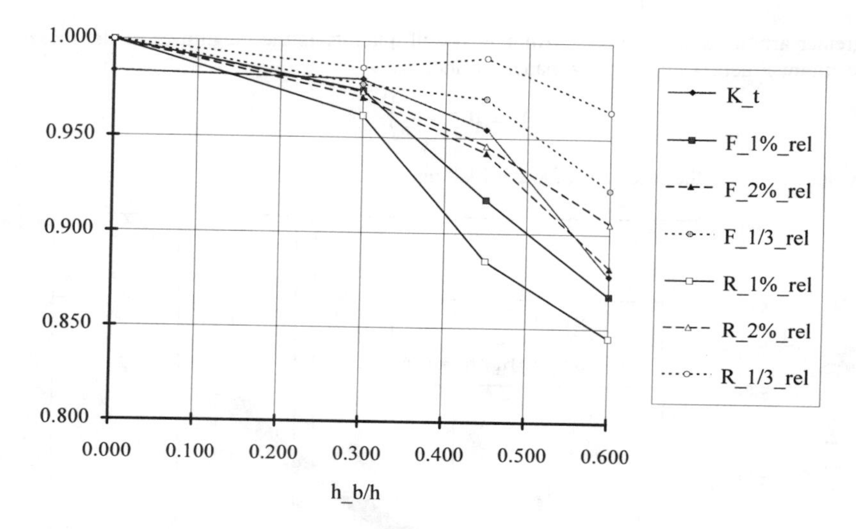

Figure 11. Reduction in loading statistics with increasing reef height.

The reduction in armour stone loads, provided by reef breakwaters of different heights in the same wave climate, are summarized in Figure 11. Several statistics of F and R maxima (1%, 2%, 1/3), measured with reef breakwaters in place, are normalized by the equivalent statistics obtained from tests without a reef breakwater. The $h_b/h = 0.45$ reef has very little effect on the significant level of armour stone loads, but provides roughly 10% reduction at the 1% load exceedence level. The $h_b/h = 0.6$ reef provides up to 15% reduction to armour stone loads at the 1% exceedence level.

This modest reduction in wave heights and load levels can have dramatic effect on the performance of rock armour. In conditions near the threshold of armour stone motion, such a reduction in loading can make the difference between extensive damage and no damage. Above the threshold of motion, numerous design equations (ie. van der Meer (1988)) relate armour damage to the cubic power of wave height, so that a 15% reduction in height translates to a significant 50 % reduction in damage. Armour damage can be quantified as $S = A/D_{n50}^2$, where A is the area of armour eroded in a cross-section and D_{n50} is the nominal diameter of armour stones. An alternative expression, in terms of the number of stones N_A eroded from a particular test section of width B, is

$$S = \frac{N_A D_{n50}}{B(1-p)} \tag{11}$$

where p is the porosity of the armour. Vidal and Mansard (1993) describe four distinct levels of damage and suggest corresponding threshold values of S:

(1) Initiation of damage, S=1;
(2) Iribarren's damage, S=2.5;
(3) Start of destruction, S=4;
(4) Destruction, S=9.

Damage results for the main breakwater, sheltered by tandem reefs of different heights exposed to approximately 3000 identical S_1 waves, are presented in Table 1. A trend towards lower damage with increased reef height is evident. In these tests, the h_b/h = 0.6 reef provides shelter sufficient to lower the observed damage from "start of destruction" to the threshold for "initiation of damage". This considerable abatement is achieved through only modest transformation of incident waves and peak loading events. Higher and/or broader reef breakwaters will likely cause more pronounced wave transformation, greater reductions to extreme loading events, and even more dramatic attenuation of damage. Such attenuation would allow more economic design of the main breakwater.

Table I. Damage after 3000 S_1 waves.

h_b/h	N_A	S
0	36	4.4
0.3	21	2.5
0.45	26	3.2
0.6	8	1.0

Summary

The performance of several small reef breakwaters, and a tandem breakwater system (consisting of a reef breakwater positioned upwave from a surface-piercing main breakwater), have been quantified through a limited set of physical model tests. Wave reflection and transmission coefficients for the reef breakwaters have been computed from wave probe measurements in irregular waves. The character, and wave height distribution, of waves transmitted past the reef breakwaters have also been examined. Fluid forces acting on a patch of surface-layer armour stones were measured in waves using a load panel installed on the face of the main breakwater. The reduction to extreme wave heights incident to the main breakwater, and extreme loading events, created by several different reef breakwaters, have been quantified. Moreover, reductions in extreme wave heights and loading events have been favourably correlated to improved armour stone stability. These results demonstrate the successful performance of a tandem breakwater system incorporating a relatively small, low-crested reef breakwater. Extrapolation and generalization of these results to different structures and wave conditions will require further model testing.

Specific conclusions to emerge from this work include the following.

• A low-crested reef breakwater, located upwave from a main breakwater, can reduce the loading on, and erosion of, rock armour on the face of the main breakwater. This represents a validation of the tandem breakwater concept.

• The transmission coefficient K_t approaches 1.0 for submerged breakwaters with relative freeboard $F/H_{m0,i} \leq -2.5$. Reflection coefficients K_r less than 0.1 were observed for such low-crested breakwaters.

• The empirical formulae for K_r and K_t of Ahrens (1987) provide a reasonable fit to the data over the encountered range of relative freeboard ($F/H_{m0,i} < -1.4$).

• A submerged breakwater with $h_b/h = 0.6$ reduces the energy of the transmitted waves, and shifts some wave energy to higher frequencies. Extreme wave statistics of the transmitted waves are spatially variable.

• A submerged breakwater with $h_b/h = 0.6$ provides greater wave height reduction to the largest waves in a sea state. The largest waves are modulated such that wave heights behind the submerged breakwater no longer satisfy the Rayleigh distribution.

• The largest peak wave loads on the armour of a main breakwater are significantly reduced by a $h_b/h \geq 0.6$ submerged reef breakwater located upwave.

• Modest reductions to transmitted wave heights and peak wave loading can lead to dramatically improved armour performance.

Considerably more research and testing of reef breakwaters and the tandem breakwater concept are required to develop a complete understanding of the transformations to transmitted waves, loading events, and the implications for the performance of diverse tandem breakwater systems. The effectiveness of the tandem breakwater concept has been quantified for a series of relatively low-crested reef breakwaters in one particular wave climate. These results should be treated as a case study that demonstrates the feasibility of using a submerged reef breakwater in front of a main breakwater to modify the transmitted waves and improve the performance of the primary structure. Waves realized from other spectra, and indeed, different realizations synthesized from the same spectrum, could give substantially different results. Mansard et al. (1988) document the extensive natural variability of wave statistics in shoaling water depths for multiple fixed-length realizations of a single sea state. Such inherent variability in shallow water wave statistics could influence the results presented here.

Wave period will influence the transmission of waves past a submerged reef breakwater, and can therefore be expected to affect the performance of a tandem breakwater system. Preliminary results in other wave conditions raise concerns about the general suitability of tandem breakwater systems using reef breakwaters with such low crests.

References

Adams, C.B. and Sonu, C.J. 1986. Wave Transmission Across Submerged Near-Surface Breakwaters. *Proc. 20th Int. Coastal Engineering Conference*, pp. 1729-1738.

Ahrens, J.P. 1987. Characteristics of Reef Breakwaters. *CERC Technical Report CERC-87-17*. U.S. Army Waterways Experiment Station, Vicksburg, Miss. U.S.A.

Ahrens, J.P. 1989. Stability of Reef Breakwaters. *ASCE Journal of Waterway, Port, Coastal and Ocean Engineering*, Vol. 115, No. 2. pp. 221-234.

Ahrens, J.P. and Cox, J. 1990. Design and Performance of Reef Breakwaters. *Journal of Coastal Research*, Special Issue No. 7. pp. 61-75.

Beji, S. and Battjes, J.A. 1993. Experimental Investigation of Wave Propagation Over a Bar. *Coastal Engineering*, 19. pp. 151-162.

Cornett, A. and Davies, M. 1993. Individual Irregular Wave Loads on Breakwater Armour. *Proc. Canadian Coastal Conference 1993*. Vol. 1., Vancouver, Canada.

Cox, J.C. and Clark, G.R. 1992. Design development of a tandem breakwater system for Hammond Indiana. *Coastal Structures and Breakwaters*. pp. 111-121. Thomas Telford, London, U.K.

Dattatri, J., Raman, H. and Shankar, N.J. 1978. Performance Characteristics of Submerged Breakwaters. *Proc. 16th Int. Coastal Engineering Conference*, pp. 2153-2171.

Kobayashi, N. and Otta, A.K. 1987. Hydraulic Stability Analysis of Armour Units. *ASCE Journal of Waterway, Port, Coastal and Ocean Engineering*. Vol. 113, No. 2. pp. 171-186.

Kobayashi, N., Wurjanto, A. and Cox, D.T. 1990. Irregular Waves on Rough Permeable Slopes. *Journal of Coastal Research*. Special Issue No. 7. pp. 167-184.

Kojima, H., Ijima, T. and Yoshida, A. 1990. Decomposition and Interception of Long Waves by a Submerged Horizontal Plate. *Proc. 22nd Int. Coastal Engineering Conference*, pp. 1228-1241.

Mansard, E.P.D., Funke, E.R., Readshaw, J.S. and Girard, R.K. 1988. On the Transformation of Wave Statistics due to Shoaling. *Proc. 21st Int. Coastal Engineering Conference*. pp. 106-120.

Mansard, E.P.D. and Funke, E.R. 1980. The Measurement of Incident and Reflected Spectra Using a Least Squares Method. *Proc. 17th Int. Coastal Engineering Conference*. pp. 154-172.

Seelig, W.N. 1980. Two-Dimensional Tests of Wave Transmission and Reflection Characteristics of Laboratory Breakwaters. *CERC Technical Report CERC-80-1*. U.S. Army Waterways Experiment Station, Vicksburg, Miss. U.S.A.

Tanaka, N. 1976. Effects of Submerged Rubble-Mound Breakwater on Wave Attenuation and Shoreline Stabilization. *Proc. Japanese Coastal Engineering* Conference.

Van der Meer, J.W. 1988. Rock Slopes and Gravel Beaches Under Wave Attack. *Doctoral Thesis, Delft University of Technology*. Also Delft Hydraulics Communication No. 396.

Van der Meer, J.W. and Pilarczyk, K.W. 1990. Stability of Low-Crested and Reef Breakwaters. *Proc. 22nd Int. Coastal Engineering Conference*, pp. 1375-1388.

Van der Meer, J.W. and Angremond, K. 1992. Wave Transmission at Low-Crested Structures. *Coastal Structures and Breakwaters*. pp. 25-54. Thomas Telford, London, UK.

Vidal, C., Losada, M.A. and Medina, R. 1991. Stability of Mound Breakwater's Head and Trunk. *ASCE Journal of Waterway, Port, Costal and Ocean Engineering*. Vol. 117. No. 6.

Vidal, C. and Mansard, E. 1993. On the Stability of Reef Breakwaters. *NRC Technical Report (in press)*. Ottawa, Canada.

The Evolution of Breakwater Design

by

Billy L. Edge[1], Orville T. Magoon[2] and William F. Baird[3]

Introduction

Since the earliest works of the Phoenicians, it has been recognized that one of the most important environmental effects on the construction and performance of a breakwater is the wave conditions. Understanding these conditions and the ensuing forces have taken centuries. The purpose of this paper is to present a view of the evolution of both rubble mound and vertical faced breakwaters from the earliest designs through the Iribarren and Hudson era. This presentation is not all inclusive but instead is focused on providing those events which the writers believe to have had a long lasting contribution to the field. Because of the significant and large number of achievements made in the later part of this century, only contributions up to the time of the extensive failures of breakwaters in the 1970's are included.

Earliest Known Designs

In describing some of the earliest documented efforts of port construction and coastal protection, Minikin (1954) noted the important contributions of Leonardo da Vinci as reported by d'Arrigo (1940). Leonardo da Vinci was not only an artist and philosopher, he also spent time studying the coast and its processes. In his studies of the Cesenatico coast on the Adriatic Sea north of Rimini he made numerous observations about the mechanics of wind waves, movement of sediment and detritus along the sea bed, wave breaking in shallow water and coastal defense. Because of his famous works, Francois le Premier commissioned Leonardo da Vinci to determine the protection works (breakwater) necessary to make the port of La Havre the "foremost port in the world."

Ancient Mediterranean harbors and protecting structures are also described in Inman (1974) and Mediterranean and northern European harbors and harbor works are described in Cornick (1958). The above referenced works demonstrate that perhaps as early as 7000 B.C. there was coastal trade and that early civilization had an understanding of waves and eventually built breakwaters to provide protection to their port and harbor areas. Inman (personal reference) has continued his research into ancient Mediterranean ports.

[1] Professor of Ocean and Civil Engineering, Ocean Engineering Program, Department of Civil Engineering, Texas A&M University, College Station, TX 77843-3136.

[2] Chairman, American Society of Civil Engineers Rubble Mound Structures Committee, Guenoc Winery, P.O. Box 279, Middletown, CA 95461-0279.

[3] President, Baird & Associates, 38 Antares Drive, Suite 150, Ottawa, Ontario, Canada, K2E 7V2.

Sir John Rennie (1854) prepared an extended treatment on the history of ancient harbors in his excellent two volume treatment of the history of breakwaters and harbor construction. With some admitted difficulty he put together much information from the historical literature given by such early writers as Herodotus, Polybius and Strabo about the earliest noted harbor construction works. He believed that the Phoenicians' knowledge and development of harbor construction, including that of harbor protection, was superior to that of the Romans and the Greeks.

In about 572 B.C. a breakwater (mole) was constructed at Tyre by loose blocks of stone as they were raised from the quarries and "thrown promiscuously into the sea to find their own base." There were many other Greek ports which had improvements commonly consisting of moles extending from the land. One of the most

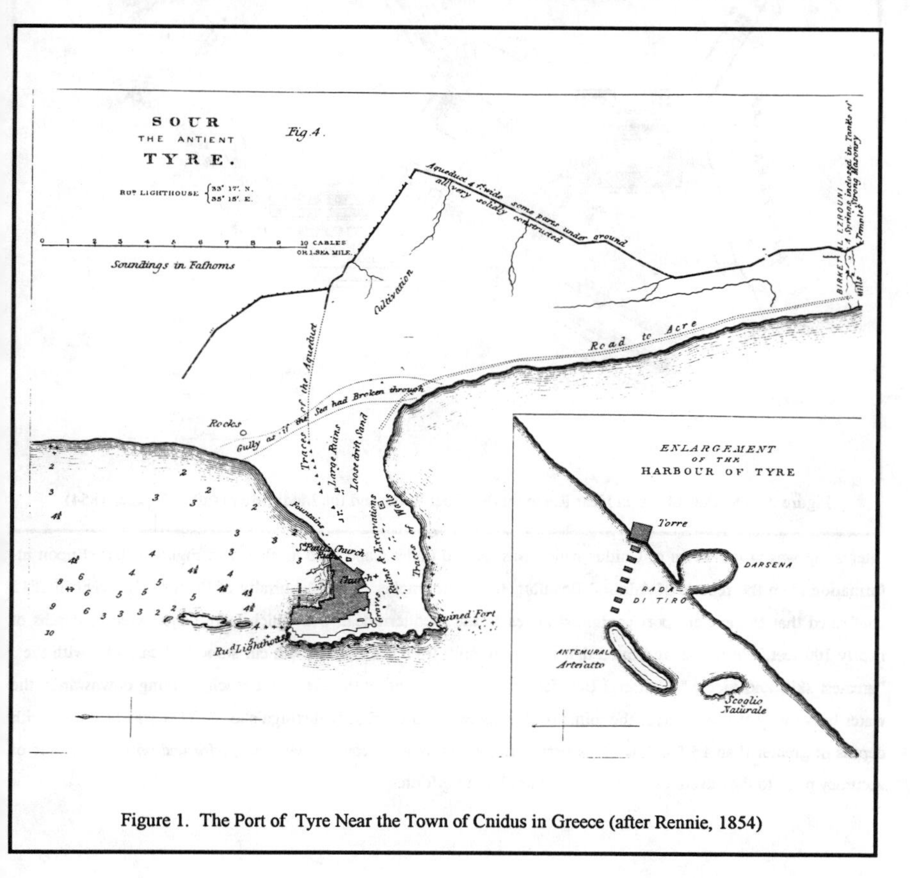

Figure 1. The Port of Tyre Near the Town of Cnidus in Greece (after Rennie, 1854)

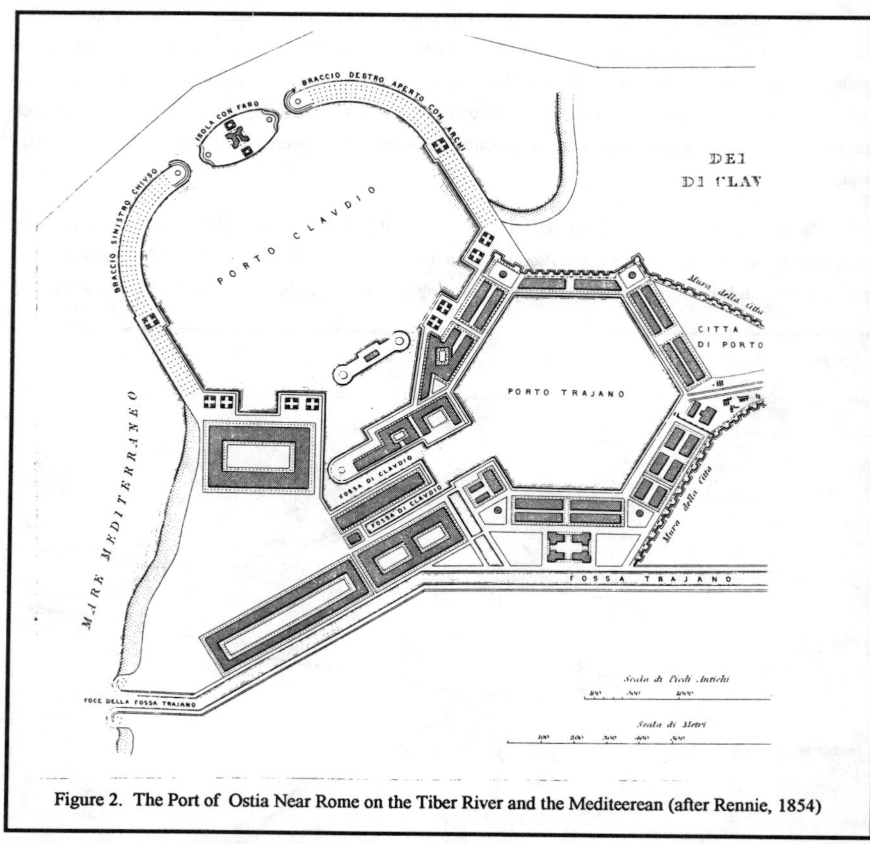

Figure 2. The Port of Ostia Near Rome on the Tiber River and the Mediteerean (after Rennie, 1854)

interesting was at the Town of Cnidus which was located under Cape Krio as shown in Figure 1, based upon information from the reports of Admiral Beaufort, the hydrographer to the Admiralty of Britain. His visit in 1812 confirmed that the ancient port contained moles in the Mediterranean Sea which were constructed to depths of nearly 100 feet. He further indicated that they were built from the bottom with cut stone laid carefully with the "greatest skill and nicely." Admiral Beaufort stated that "as far as the eye could reach, looking downwards, the water being very still and clear, the joints of the stones could be plainly distinguished." This was in waters with depths of greater than 15 fathoms. It is perplexing how this work could have been performed with any degree of accuracy prior to the advent of the diving bell or diving uniform.

Like the Greeks, the Romans had numerous improved harbors; however, Rennie (1854) indicated that there was little improvement over the original concepts used by the Phoenicians in developing these harbors. Ostia was the primary port of Rome at the mouth of the Tiber River. At the time of Julius Caesar, however, it was noted that significant shoaling limited the size of vessels which could be employed in the harbor. The Emperor Claudius constructed a new port independent of the Tiber River but connected to it. The plan consisted of two large moles each approximately 1900 feet long projecting nearly at right angles to the shore as shown in Figure 2. The top width of the breakwater was approximately 180 feet and the distance between the two seaward ends was approximately 1100 feet.

Between the two breakwaters an island was constructed leaving approximately 140 ft on either side for a dual entrance to the harbor. The breakwater was somewhat open to allow currents to circulate freely, preventing stagnation and subsequent deposition of material but sufficiently solid to reduce wave energy (Rennie, 1854). The breakwater foundation was constructed with large masses of rubble which was dumped into the sea and consolidated there by the waves. The breakwater also contained masonry of various kinds and it is suggested that it contained the beton or concrete system to join the masonry blocks. Considering the exaggeration of the historians at the time it is still nevertheless inspiring to see that the designers of ancient harbors had apparently studied the sciences well enough to create harbors which were able to last for centuries.

The modern breakwater at the Port of Algiers was initially constructed in 1529 to connect the four small islands to the mainland. The earliest known breakwater in the Port of Algiers connected the outer islands and it was called ICOSISIM by the Phoenicians and ICOSIUM by the Romans (Tabet-Aoul and Oumeraci, 1987). The earliest reported use of concrete armor units was at Algiers in 1834. Large concrete armor blocks were also cast at Cherbourg at about the same time (Bonnin, 1857).

There were several proposals for constructing a breakwater at Plymouth, England, including such unusual schemes as a wooden floating structure anchored in place (Rennie, 1848). The breakwater at Plymouth is one of the famous breakwaters in Europe which have survived over a century of high wave energy without catastrophic damage; over the life of this structure, only minor damage has been experienced, requiring some repair. The proposed detached breakwater design of Sir John Rennie was selected over a typical attached design. Prior to and during the construction extensive debate occurred about using caissons or a vertical parapet instead of the rubble mound construction. The final design, which was somewhat modified during construction employed randomly placed quarry stone "into the sea to enable them to form their own slope." Because Rennie recognized that most damage from storms is done between high and low water levels, he added a layer of fitted stone to provide a permanent layer. After the rubble had stabilized over time, a course of heavy squared granite and limestone was placed, with the blocks being dovetailed, bonded and doweled together. Sir John Rennie observed that the success of his design was attributed to the effectiveness of the "exterior berm or benching at the base of the main slope" to break the wave energy before it reached the intertidal part of the structure. Sir John Rennie (1848) felt that "the true elements of success consist in following out the laws of Nature." Another very important part of the design process was the planned continual evolution of the design as the work progressed and the effects of the sea were noted. This was obviously not a fixed price design or construction contract; in his report to the British Admiralty, Rennie noted "... in such an exposed situation as this is many unforeseen expenses may arise which no human foresight can calculate."

Various ideas were entertained for constructing the final design of the Plymouth breakwater. Rennie studied extensively all the different plans which had been adopted in similar works previously executed elsewhere, including the caissons, solid walls of masonry and the cones at Cherbourg,. Rennie (1854) contacted historians to research the earliest known works and he sent his engineers to visit and detail many breakwaters in the Mediterranean, along the coast of Britain and in the United States. These were all thoroughly considered in the design for Plymouth; but under all the circumstances, the orginial plan of dumping rubble, from half a ton to ten tons, as raised from the quarries, in the line of the intended breakwater, was considered the best and most advisable plan, and was accordingly adopted. These blocks of stone, Rennie reasoned, would naturally find their own position and slope according to the depth of the water, the wave climate and the specific gravity of the stone. He believed that after an initial period the waves would consolidate the quarry stone and a "great savings of time and costs would be effected in carrying on the work and giving the protection to the Sound." Figure 3 shows a sketch of the original proposed section of the breakwater and its final configuration after the waves had shaped the breakwater and the armor stone was added to the top of the section. Figure 4 is a lithograph[4] made at the time of construction showing the manner of devices used. The effects of the storms in shaping the breakwater and modifying the section with armor stone is described also by Baird and Hall (1983).

John Rennie (1848) summarized his reasons for selecting the rubble mound breakwater as follows:

> The principle of employing rough blocks of stone called rubble, as raised from the quarries and throwing them promiscuously into the sea to enable them to form their own slope or angle of repose, has been adopted as being the most effectual, and at the same time most economical mode of giving protection against the waves, as well as being accomplished in the shortest time period. If the vertical wall, composed of squared blocks of stone of masonry or caissons, had been adopted, both the time as well as the cost would have been so considerable as almost to have amounted to impracticability; whereas, by employing rubble, the greatest facility of construction has been obtained, and the storms themselves have been rendered instrumental in consolidating and perfecting the work.

Rennie also added that "the exterior berm or benching at the base of the main slope has been found of the greatest service in breaking and dispersing the main body of the waves before they arrive at the mass of the work and in thus materially diminishing their effect."

The breakwater at Cherbourg, France, was used by Rennie as an example in the design of the Plymouth breakwater. The Cherbourg breakwater was constructed (Bonnin, 1857) using two cones into which rubble was

[4] This lithograph is in the possession of Orville T. Magoon.

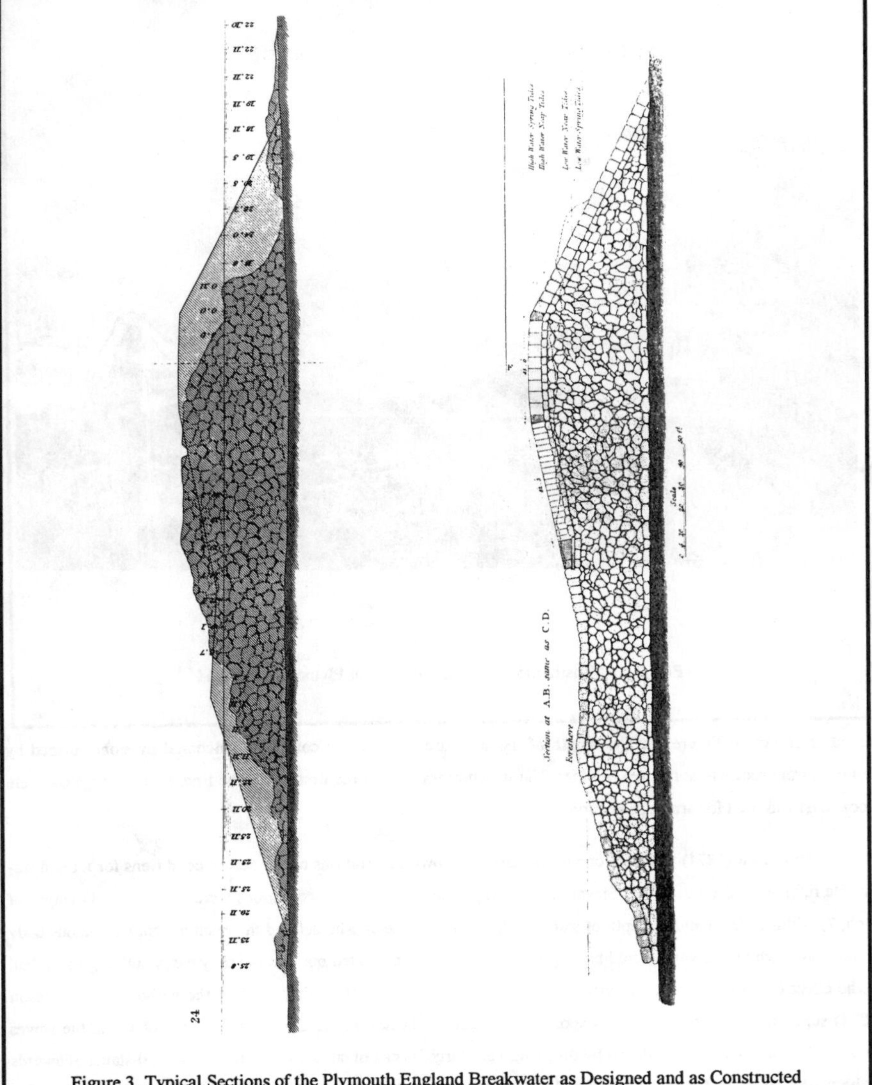

Figure 3. Typical Sections of the Plymouth England Breakwater as Designed and as Constructed (after Rennie, 1848)

Figure 4. Construction of the Breakwater at Plymouth, England

placed as shown in Figure 5. The length of the structure between the cones was anchored by stone, placed by dumping from scows tethered to a tie line. The structure was one of the first, if not the first, to have large concrete blocks cast and used for armor protection.

Stevenson (1874) provided one of the earliest known descriptions of the design conditions for a breakwater. He referenced the work of D. Stevenson (1838) and noted that wave conditions were based upon 1) length of fetch, 2) width of fetch and 3) depth of water in the area of the fetch; he defined the fetch as that continuous body of water over which the wind would blow generating waves. He pointed out that in many cases outlying rocks had "...the effect of tripping up the heavier seas, so as to destroy them before they reached the harbour..." Stevenson (1874) suggested "in cases of severe exposure, where it would not interfere with the passage of ships, the waves might to a certain extent be reduced by dropping very large blocks of stone or concrete at some distance seawards of the works, so as, by forming an artificial shoal, to cause the waves to crest and break outside." Observations by Scott Russell, and referenced by Stevenson, that the maximum wave height never exceeded the depth of water were

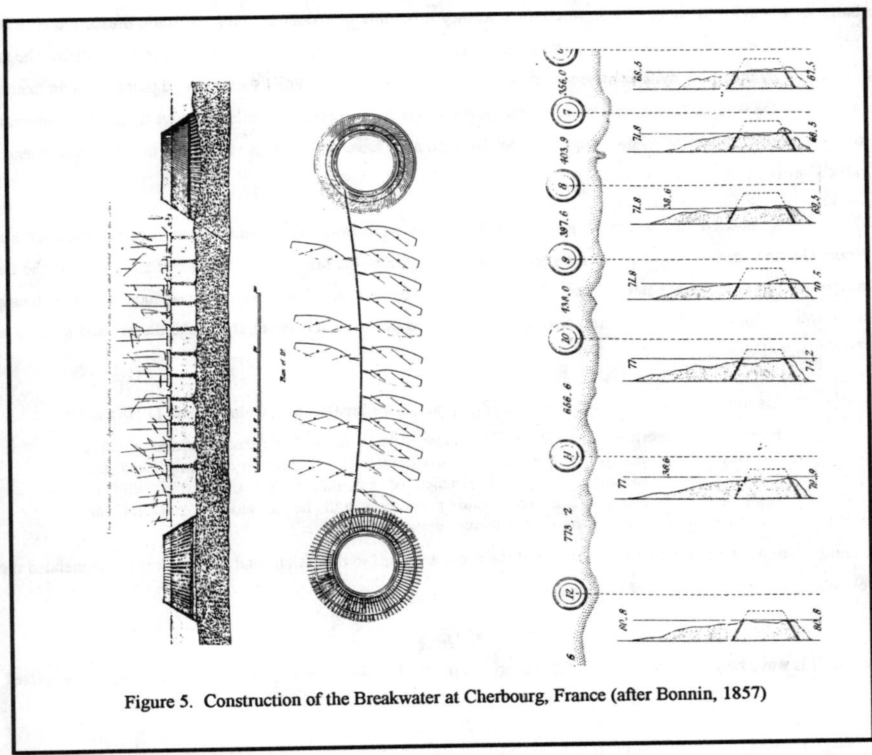

Figure 5. Construction of the Breakwater at Cherbourg, France (after Bonnin, 1857)

corroborated with measurements by Stevenson who measured breaking waves from the pier at Scarborough to be 5.25 feet with a depth below the trough of 10.25 feet. He then gave the relationship for the maximum height of wave that could exist in water of a specified depth as:

$$H = \frac{D}{2.5}$$

where

D = depth of water below the mean level

H = height of wave from hollow to crest

Stevenson (1874) provided considerable discussion about the ongoing controversy at the time between vertical-faced and rubble mound breakwaters. At the time there was keen competition between the designers which resulted in extensive hearings by the Institution of Civil Engineers. The proponents of the vertical-faced structures claimed that since waves in deep water are "purely oscillatory, and exert no impact against vertical barriers, which are therefore the most eligible, as they have only to encounter the hydrostatic pressure due to the height of the reflected billows, which are reflected without breaking." Stevenson attempted to better define the actual forces by installing dynamometers on an unfinished wall at Dunbar. His results showed that the waves exerted a force on the vertical face which was six times greater than the theoretical, oscillatory wave force. It is important to note that the important relationships provided by Stevenson have been improved through the extensive research which has occurred since his work.

The most curious design suggestions by Stevenson came from the recommendations of Scott Russell and Franz Gerstner in which it was recommended that the profile of the breakwater should be determined by the nature of the wave; in deep water with oscillatory waves, a convex shape should be employed such as at Cherbourg, as shown in Figure 5, and when translatory waves exist, a cycloidal (concave) shape should be used such as at Trinity, near Edinburgh .

In his introduction to the then current design practice for breakwaters, Cunningham (1918) stated:

> From an engineering point of view, we have little to do with abstract theories of wave formation. Mathematically, the subject is too obtuse for any but very accomplished and capable mathematicians, and the intricacies of calculation are interesting only as academical exercises... Thus no useful purpose would be served by pursuing an investigation into the laws and phenomena of water undulation.

Cunningham went on to note that the height of the wave is related to the "fetch" and he therefore recommended the following to obtain the height of the wave:

$$H = 1.5\sqrt{F}$$

where H is wave height in feet and F is fetch length in miles. For short fetches, of less than 30 miles, he suggested

$$H = 1.5F + \left(2.5 - F^{\frac{1}{4}}\right)$$

Cunningham cautioned that the waves cannot achieve their full height when the depth is not great enough to obtain the wave period. For deep water he suggested

$$\lambda = \frac{g}{2\pi}T^2$$

where λ is wave length, g is acceleration due to gravity and T is period. He determined that waves would break when the depth corresponded to $0.75H$ to $1.5H$. Cunningham provided a set of tables to determine wave velocity and pressure for breaking waves. He also provided a detailed comparison of the rubble and vertical faced breakwaters including, initial cost of construction, cost of maintenance, and efficiency without taking sides in the continuing controversy.

The Iribarren and Hudson Era

At the International Maritime Congress held at Milan in 1905, Signor Bernardini reported at length on the damages to the breakwater at Genoa from the Great Storm of 1898 (see Cunningham, 1918). Subsequently, extensive studies were begun in many countries to better define the relationship of design waves to the integrity of breakwaters. The major breakthroughs occurred during the period called the Iribarren and Hudson Era when a direct relationship was developed directly linking the size of armor stone to the size of the incident wave, the specific gravity of the stone and the specific gravity of the water. This was also an era in which hydraulic model studies were developed and then were used extensively in the design process. The ideal section for rubble mound structures and design of artificial armor units constituted the most extensive and thorough studies during the Iribarren and Hudson era.. This era concluded with the onset of extensive failures beginning in the mid 1970's which caused serious reflection of the direction the design process had taken during the Iribarren and Hudson Era.

The Iribarren formula initially presented in Iribarren (1938) is the earliest known analytical tool for determining the size of cover layer for rubble mound breakwaters. The equation developed by Iribarren is given as:

$$W = \frac{KH^3 S_r}{(\cos\theta - \sin\theta)^3 (S_r - 1)^3}$$

where

W = weight of armor stone

K = empirical coefficient

H = design wave height

S_r = specific gravity of stone (in metric tons per cubic meter)

θ = angle the armor face makes with the horizontal

This equation is still used and its applicability is still debated at the present time.

During the series of tests which were conducted by the Waterways Experiment Station for the United States Navy on the stability of rubble mound breakwaters of various designs it became evident that additional research needed to be done in this area. The report (U.S. Army Corps of Engineers, 1953) on the stability of rubble mound breakwaters was initially prompted by the interest of the US Navy in a large military facility at Roosevelt Roads, Puerto Rico, where a substantial breakwater would be required. Following the Second World War the importance of Roosevelt Roads became less significant and the study evolved into a general investigation of hydraulic stability of rubble mound breakwaters. The most significant findings of the study concerned the use of Iribarren's formula for the design of rubble mound breakwaters. The studies showed that the coefficient to be applied in Iribarren's formula varied significantly with the seaward slope. Another significant finding was that Froude-based models allowed adequate transfer from prototype conditions into the laboratory for all kinematic and dynamic wave conditions protecting the stability of rubble mound breakwaters. In other words, hydraulic models were determined to furnish accurate information on the design of these structures. They provided a set of relationships to be used with the Iribarren formula which would make it consistent with the results of the model study. Lastly, they determined that the masonry-type breakwaters using fitted stone were more resistant to storm wave action than ordinary dumped rubble mound yet they cautioned against this type of construction because of the potential damage from settlement and the resulting catastrophic failure.

The United States Army Corps of Engineers (1958) report (the famous Hudson report) presented the results of the extensive series of hydraulic model tests conducted over a thirteen year period. As a part of producing design information for rubble mound breakwaters the reported tests were to check the accuracy of the Iribarren and Epstein-Tyrrell formulas for stability. The Epstein-Tyrrell formula which was presented in the U.S. Army Corps of Engineers Report (1953) report, is essentially similar to the Iribarren formula. Results of the testing program indicated that the Iribarren formula was not adequate for designing armor layers for rubble mound breakwaters unless accurate values of the experimental coefficient, K', for each breakwater slope and shape of armor unit were available from individual hydraulic model tests. Since the Iribarren coefficient, K', was a function of the coefficient of friction, μ, it was important that the coefficient of friction be obtained. Unfortunately the methodology, equipment and method of placement of armor stone revealed erratic results in measurements of the friction coefficient. Moreover, when armor units other than quarry stone were used additional scatter was measured in the friction coefficient. Because of the difficulties in obtaining appropriate values for Iribarren's formula "a new, but similar, formula, derived on the basis of both theory and the results of tests completed to date, was adopted." The resulting equation was the famous Hudson equation. This equation has been used as a primary standard for breakwater design around the world since its introduction. The Hudson equation is given as:

$$W = \frac{w_r H^3}{K_D \cot\theta (S_r - 1)}$$

where

K_D = damage coefficient

w_r = unit weight of stone

The report further provided additional information on wave runup and thickness of the armor layer and required porosity for stability. It is important to note that the authors of the report identifed the problem of using essentially harmonic waves in the laboratory and transfering those results to the field where random events are the norm. The conclusions of the report noted that for the conditions tested, the stability of rubble mound breakwaters was not appreciably influenced by variation in the ratio of relative depth, D/λ, where λ is the wavelength, and wave steepness H/λ, and for a given shape of armor unit the experimental coefficient K_D is a function primarily of the amount of damage done to the layer of armor units for a given wave height. With regard to the actual wave height to be used for the design of a given structure, the authors confused the issue by leaving the selection of the design wave in the hands of the engineer since the harmonic waves in the laboratory do not closely simulate that which is found in nature. The use of a single harmonic design wave unfortunately left many design engineers confused about what wave should be used for the design of the armor layer: $H_{1/3}$, the significant wave height; $H_{10\%}$, the height of the wave which is exceeded only 10% of the time and $H_{1\%}$ the height of the wave which is exceeded only 1% of the time.

Following this report a long series of tests were conducted between 1955 and 1963 as reported by Jackson (1968) which included evaluation of the following types of armor units: smooth quarry stone, rough quarry stone, tetrapods, quadrapods, tribars, modified cubes, hexapods and modified tetrahedrons. The Hudson formula was used for each of these to determine a stability coefficient. The Corps subsequently produced a revised working draft of the Engineering Manual for the Design of Breakwaters and Jetties (U.S. Army Corps of Engineers, 1984). The purpose for the manual was to provide information on procedures accepted by the Corps of Engineers for the design of breakwaters and jetties. The Corps of Engineers at that time was involved with the design and construction of over 600 breakwaters and jetties some of which dated back to the mid-nineteenth century. They indicated that originally the design and construction was based upon trial and error.

The Corps of Engineers (1984) noted that by 1930, model tests had been introduced to aid in the design of breakwaters and jetties and to supplement the few analytical tools which existed. The manual described all known factors in the preliminary and final design for a breakwater or jetty. With adequate consideration for social and environmental effects many of the structures which were presented were no longer considered appropriate for general application such as pneumatic and floating tire breakwaters. The engineering manual suggested a design optimization procedure based on a defined economic design life of 50 years and it suggested a procedure for incorporating the frequency of damage based upon wave characteristics, cost of maintenance and initial construction cost. More specifically it proposed that the design elements to be considered in the economic optimization of a breakwater were as follows:

project economic life
construction costs for various design levels
maintenance costs for various design levels
replacement costs for various design levels
benefits for various design levels
probability for exceedance of design conditions

Shortly thereafter Merrifield and Zwamborn (1966) introduced the dolos armor unit which was reported to have much higher stability characteristics than any armor unit in existence. It should be noted that a shape similar to the dolos had been previously patented by the French Laboratory in Grenoble (NERPRIC). The Waterways Experiment Station proceeded to independently determine the stability characteristics of the dolos armor units in order to find a stable design for the Humboldt Jetties in Eureka, California. The test showed that with a two layer randomly placed armor layer of dolos with a porosity of 56%, a stability coefficient, K_D, of 31 could be obtained for "nonbreaking, nonovertopping waves." They were not able to verify that placement of the dolos in geometric patterns would increase stability over random placement. The tests indicated that dolos may be placed on slopes as steep as 1 to 1.5 if stability is indicated by site specific model tests. However, it was noted that when testing at a slope of 1 to 1.5, en masse sliding on the dolos layer was observed. Although this was assumed to be a Reynold's scale effect which could not be easily eliminated in the laboratory, Magoon et al. (1990) observed en masse sliding with dolos units on a 1 to 1.5 slope at the St. Thomas airport runway extension. The stability coefficient recommended by Carver and Davidson (1977) has been extensively discussed and subsequently refuted in additional tests using better information on statistical properties of waves and especially considering the strength of the dolos armor unit.

Whalin (1977) noted that hydraulic model tests are usually conducted to determine stable designs for typically a 1 and 20 year event. Subsequent tests of waves exceeding the noted damage design wave are performed in order to optimize the initial and probable maintenance costs. To identify properly a frequency of occurrence of damage from the model tests, Whalin indicated that additional extreme wave data are required including the joint probability distribution of waves and water level. This was also noted by Edge, Magoon and Sperling (1977) when they pointed out the limited data available in the North Atlantic.

Walton and Weggel (1981) offer one of several suggested modifications to the Hudson equation which would provide for wave steepness, and consequently wave period. They presented a stability number based on wave steepness including inertial effects. The results of their analysis is similar to that obtained by Raichlen (1975) but without the lift and inertia terms.

Lessons from Large Failures

Magoon et al. (1976) presented a summary of available information on the design, construction and maintenance of the Humboldt Jetties, one of the oldest breakwaters in the United States. The jetties are exposed to some of the highest wave energy of any structure on the Pacific coast. The construction was begun in 1888. Initially the jetties were built above the high tide to a depth of -18 feet, a distance of approximately 4000 feet offshore. By 1907 the head of the jetties had deteriorated significantly and were buried in sand. Initially there was little attention paid to the size of stone required. Efforts to stabilize the heads included construction of a concrete monolithic cap. By 1915 the design changed to include 6 to 20 ton armor stone on a 1 on 2 slope with a crest elevation of 19 feet. Because of construction limitations all repairs had to be conducted from the crest which significantly limited the size of armor stone that could be used. This basically changed the slope to 1 to 1.5. By 1932 efforts to cast and place 100 ton blocks were initiated to repair damaged areas with limited success -- many of the blocks broke during placement. Again in 1960-1961, 100 ton cubes were placed to repair damaged areas and lasted only until the winter storms of 1964-1965. The jetty heads were destroyed by 1970. It is not clear that the decision for using the 100 ton cubes were the result of the application of the Hudson equation or the limitations of the construction equipment.

Before the next rehabilitation, extensive studies were done to change the design. Using hydraulic model tests (Magoon, et al., 1976) 42 and 43 ton dolos units were selected for the head to be placed at a 1:5 slope which was dictated by the existing conditions of the previous works. In as much as the stable design tests showed that many dolos armor units were rocking, although not displaced, steel reinforcing was added to the dolos armor units. As difficulty in reaching stability was found at selected locations, high unit weight concrete was used, resulting in the heavier unit weight (43 ton) and increased stability (according to Hudson's formula).

An extensive summary was given by Takeyama and Nakayama (1975) of recorded disasters of caisson and rubble mound breakwaters between 1965 and 1972 in Japan. Their summary included 63 examples from 49 harbors where damage to breakwaters were noted. Diagrams given in the report indicated breakwater configurations before damage, the damage condition and the restored cross-section. Some analyses were attempted on the total number of breakwater damages observed in order to determine if any trends in breakwater damage was apparent. Although a large number of tables were summarized in the data, there was no synopsis which would lead to any trends that might be noted from the data of breakwater damages. The authors noted that it was evident from their data that rubble mound structures were weaker than block and caisson breakwaters; they referenced specific data from the damage at the Kagoshima Sangoro breakwater and the Wakayama main harbor breakwater to support their conclusions. They found equally good performance between the block and caisson breakwaters after forty to sixty years.

Since the beginning of the Iribarren and Hudson era, a number of significant if not catastrophic failures have occurred. Some of these are listed below:

1954	Kahului, HI
1958	Nawiliwili, HI
1960-1972	Humboldt, CA
1960-1973	Crescent City, CA
1976	Rosslyn Bay, Queensland
1977	Gansbaai, South Africa
1977	Bilbao, Spain
1978	Sines, Portugal
1980	Port dArzew El Djedid
1982	San Ciprian, Spain
1985	Cleveland, OH
1988	St. George, AK
1989	St. Thomas, VI

The actual cause of failure in each case is not fully known; however, it is known that in each case the wave conditions were a significant factor. In some cases the wave conditions were well beyond the expected or "design" conditions and in other cases, such as depth limited cases, the waves only reached their design condition before the structure failed. In response to these failures, a workshop was held at the Waterways Experiment

Station (U.S. Army Corps of Engineers, 1985) to discuss the current state-of-the-art of the structural strength of concrete armor units used in breakwaters.

Conclusions

The determination of the design wave and maximum single wave and the design structure life for rubble mound structures in deep water remains an unsolved problem to designers. Determining the design wave for a rubble mound structure sited in a water depth so that unbroken waves reach the structure is still an art in spite of the theoretical accomplishments made be engineers and statisticians. The derived maximum wave and the design waves (regardless of the spectral shape and distribution used) contain considerable uncertainties. Additionally, if concrete armor units are used in the structure, questions remain concerning whether or not a single large wave can cause damage of the armor units or armor layer and if so, how is that single largest wave to be calculated.

Although the design of breakwaters has advanced significantly from the earliest days of the Phoenicians, there are still questions which remain today and new questions which arise as engineers push the limits of design beyond the questions which have already been answered.

References

1. d'Arrigo, A. *Leonardo da Vinci e il Regime Della Spiaggia di Cesenatico*, Roma, 1940.

2. Baird, W.F. and K.R. Hall "The Design of Armour Systems for the Protection of Rubble Mound Breakwaters," *Proceedings of Conference on Breakwaters - Design & Construction*, ICE, London, 1983.

3. Baird, W.F. and B.L. Edge, O.T. Magoon and D.D. Treadwell, "Cyril E. King Airport Runway Extension and Vicinity at St. Thomas, U.S. Virgin Islands--Damage During Hurricane Hugo," *Journal of the American Shore & Beach Preservation Association,* Vol. 58, No. 4, Oct. 1990.

4. Bonnin, Par M. Joseph *Travaux D'Achevement de la Digue de Cherbourg, de 1830 a 1853*, precedes d'une Introduction Historique sur "Les Travaux Executes Depuis L'Origine Jusqu'en 1830" par Antoine-Elie de Lamblardie, 2 volumes, Victor Dalmont, Paris, 1857.

5. Carver, R.D. and D.D. Davidson, "Dolos Armor Units on Rubble-Mound Breakwater Trunks Subjected to Nonbreaking Waves with No Overtopping," Technical Report H-77-19, WES, Vicksburg, 1977.

6. Cornick, H.F. *Dock and Harbour Engineering,* vol. 1., Charles Griffin and Co. Ltd., London, 1958.

7. Cunningham, Brysson, *A Treatise on the Principles and Practice of Harbour Engineering*, 2nd Edition, Charles Griffin & Company, Limited, London, 377 p., 1918.

8. Edge, B.L., O.T. Magoon and P.H. Sperling, "A Compendium of Ocean Wave Data for North America," *Proceedings of Ports '77*, American Society of Civil Engineers, Long Beach, 1977.

9. Inman, D.L. "Ancient and Modern Harbors: A Repeating Phylogeny," *Proceedings of the 14th International Conference on Coastal Engineering*, American Society of Civil Engineers, Copenhagen, 1974.

10. Iribarren Cavanilles, R. "Una Formula para el Calculo de los Diques de Escollera," Revista de Obras Publicas, Madrid, Spain 1938. (Translation by D. Heinrick, University of California, Dept. of Civil Engineering TR-HE-116-295, Berkeley, CA, 1948.)

11. Jackson, R.A., "Design of Cover Layers for Rubble-Mound Breakwaters Subjected to Nonbreaking Waves," Research Report No. 2-11, WES Vicksburg, 1968.

12. Magoon, O.T., R.L. Sloan and N. Shimizu "Design and Construction of Humboldt Jetties," *Proceedings of the 15th ICCE,* American Society of Civil Engineers, Hawaii, 1976, pp. 2474-2498.

13. Merrifield, E.M. and J.A. Zwamborn, "The Economic Value of a New Breakwater Armor Unit 'Dolos'," *Proceedings of the 10th Coastal Engineering Conference*, American Society of Civil Engineers, Tokyo, 1966.

14. Minikin, R.R. "Fundamentals of Coast Erosion and Defence," *Proceedings of the 5th Conference on Coastal Engineering,* American Society of Civil Engineers, Grenoble, 1954.

15. Raichlen, F. "The Effect of Waves on Rubble-Mound Structures,: *Annual Review of Fluid Mechanics,* Vol. 7, pp. 327-356, 1975.

16. Rennie, John *An Historical, Practical and Theoretical Account of the Breakwater in Plymouth Sound,* Henry G. Bohn, London, 1848.

17. Rennie, Sir John, *The Theory, Formation and Construction of British and Foreign Harbours,* 2 volumes with 123 engravings, John Weale, London, 1854.

18. Stevenson, D., *Engineering of North America*, London, 1838.

19. Stevenson, Thomas, *The Design and Construction of Harbours*, 2nd Edition, Adam and Charles Black, Edinburgh, 272 p., 1874.

20. Tabet-Aoul, E.H. and H. Oumeraci, "Port of Algiers Extension: Optimum Design Lay-out", *Proceedings of Coastal & Port Engineering in Developing Countries*, China Ocean Press, p. 557-571, 1987.

21. Takeyama, H. and T. Nakayama, "Disasters of Breakwaters by Wave Action," Technical Notes No. 200 of the Port & Harbour Research Institute, Ministry of Transport, Tokyo, Japan, 255 p., 1975. Translation from the Japanese by LTRS, INC.

22. U.S. Army Corps of Engineers, "Stability of Rubble Mound Breakwaters - Hydraulic Model Investigation," TM No. 2-365, Waterways Experiment Station, 66 p., 1953.

23. U.S. Army Corps of Engineers, "Design of Quarry-Stone Cover Layers for Rubble-Mound Breakwaters," Research Report No. 2-2, WES, Vicksburg, 1958.

24. U.S. Army Corps of Engineers, "Engineering and Design - Design of Breakwaters and Jetties,: Engineering Manual No. 1110-2-2904, Department of the Army, 1984.

25. U.S. Army Corps of Engineers, *Proceedings of Workshop on Measurement and Analysis of Structural Response in Concrete Armor Units,* WES, Vicksburg, 1985.

26. Whalin, R.W. "Wave Data Needs for Model Studies and Construction Projects," prepared for the Ocean Wave Climate Symposium sponsored by National Ocean Survey, Dallas, 1977.

27. Walton, Jr., T.L. and J. R. Weggel, "Stability of Rubble Mound Breakwaters," *Journal of the Waterway, Port, Coastal and Ocean Division*, American Society of Civil Engineers, Vol. 107, pp. 195-201, August, 1981.

The Determination Of Typhoon Design Wave By Synthetic Probability Method

Nai Kuang Liang[1] and Hai Kuan Peng[1]

Abstract

In summer and autumn, typhoons always take place in the northwest pacific. In the northern hemisphere, the typhoon is a large, anticlockwise, atmospheric eddy. Due to the strong wind speed, huge waves are generated within the typhoon area. Hence, the design waves for maritime structures are determined by typhoon waves.

Bretschneider's hindcasting method for typhoon waves is the unique way to estimate typhoon waves by a single set of typhoon parameters. The parameters are the latitude and longitude of typhoon center, center atmospheric pressure, radius of typhoon wind field, moving speed and moving direction. If a location is specified, a wave height and period can be estimated. The typhoon parameters can be regarded as random variables. Some of the random variables can be regarded as being independent. Hence, the sample joint distribution of the typhoon parameters can be easily set up. Then a distribution of wave height is calculated. By introducing an average typhoon occurrence per year, the return period of a specified wave height is estimated.

1. Introduction

The wave is one of the most important natural forces in the ocean, which influence the design of maritime structures. Typhoon is a tropical cyclone, which may originate strong wind and huge waves. Typhoon wave determines the design wave height in Taiwan. In general, there are hardly long term measurements for any proposed sites. The design typhoon waves are determined usually by wave hindcasting techniques. In the past, there are two ways of determining a typhoon design wave, i.e. (1) A design typhoon is selected and a design wave is calculated

[1] Inst. of Oceanography, National Taiwan University

by a hindcasting technique. This design typhoon may not be a real typhoon, especially the route of the typhoon is artificial. The design wave of Taichung Harbour is an example. (2) All the past typhoon waves are hindcasted for a site. For each typhoon, the largest wave is reserved. For a N year period, N extreme wave heights are selected. By employing an extreme statistics theory, design waves pertaining to various return periods are estimated. For the former way, we may overestimate the design wave, because one usually creates a critical track for the design typhoon. For the latter one, if in the historical typhoon data the tracks of strong typhoons are far from a site and those of weak typhoons are close to the site, then the design typhoon wave may be underestimated, because in the future a strong typhoon may pass near the site.

Generally a typhoon wave hindcasting technique requires a series of typhoon data to calculate typhoon waves. However, Bretschneider's method (Bretschneider and Tamaye, 1976) needs only the typhoon parameters at a time. The hindcasted waves by Bretschneider's method were compared with the measured waves several times. The results are quite good. As a site is out of the radius of 10R, in which R is the radius of maximun wind speed, Bretschneider's method fails to estimate waves. One can employ Liang's typhoon swell prediction method (Liang, 1989,1990). Basically Liang's method takes advantage of the wave at radius R for a stationary typhoon then estimates a swell height and period.

Bretschneider's method needs the central atmospheric pressure, central latitude and longitude,radius of maximun wind speed R, moving speed and direction and the latitude & longitude of a site. The radius of maximun wind speed R is hard to estimate. We follow Wang's suggestion(Wang, 1978) i.e. R is equal to one tenth of the radius of Beaufort Scale 7 (R_7). The site latitude & longitude are constants. The 6 typhoon parameters are random variables. If we know the joint probability for the 6 typhoon parameters, each set of 6 typhoon parameters results to a wave height and a probability. Then we can obtain a wave height probability distribution function P (H≤Hs). The return period Tr, of which wave height is greater than Hs, is equal to $1/\alpha(1-P)$, where α is the average annual typhoon number.

2. Correlation Analysis of Typhoon Parameters

The correlation coefficient is the ratio of cross-covariance to the product of two standard deviations. Theoretically ρ_{xy} is between -1 and +1 ,and if ρ_{xy} is 1 or -1,it means that x is a linear function of y and vice versa. Collecting the typhoon data from 1945 till 1991, there are 215 typhoons. The moving speed and direction are calculated by the sucessive typhoon center latitudes and longitudes. The correlation coeffecients are shown in Table 2-1.

Table 2-1 Correlation Coefficients of Typhoon Parameters

	Longitude	Moving Speed	Moving Direction	Central Pressure	R_7
Altitude	- 0.208	0.083	- 0.372	- 0.007	0.197
Longitude		0.132	0.010	- 0.054	- 0.141
Moving Speed			- 0.029	0.054	0.021
Moving Direction				- 0.059	0.010
Central Pressur					- 0.259

3. Bretschneider's Typhoon Wave and Liang's Typhoon Swell Hindcasting Techniques

Bretschneider's technique can be found in Ref.(Bretschneider and Tamage,1976). The only difference is that after Liaw(1980),

$$\Delta P = (1000 + (1000 - P_0)/10 - P_0)/33.87 \qquad (3-1)$$

The propagation speed of swell is usually greater than the moving speed of typhoon. As the distance between a wave station and a typhoon center is larger than the radius of Beaufort Scale 7 (R_7), the station is considered as being outside of the typhoon. The procedure of Liang's typhoon swell hindcasting technique is as follows:

(1)Swell Period Formula

$$T_{1/3} = (0.00147 \cdot r \cdot R_7/T_R{}^4 + 0.2) \cdot U_{RS} \qquad (3-2)$$

in which

r: distance between station and typhoon center, in nautical mile.

T_R: wave period at R by Bretschneider's method, in second.

U_{RS}: 10-minute average wind speed at 10-meter height at R, in knots.

(2)Calculating Raw Swell Height

$$H_2 = 0.087 \cdot H_R\sqrt{R_7/r} \qquad (3-3)$$

in which

H_R: wave height at R by Bretschneider's method,in feet.

(3)Modification Factor due to the Movement of Typhoon

As a typhoon is approaching a station, the swell energy will be piled up, and as the typhoon is leaving the station, the swell energy will decay. The modification coefficient λ is as follows:

$$\lambda = (0.5/(0.5 + T_{lag1} - T_{lag2}))^{1/2} \tag{3-4}$$

in which

$T_{lag1} = r_1/(1.56 \cdot T_{1/3})$, $T_{lag2} = r_2/(1.56 \cdot T_{1/3})$

r_1: distance between station and typhoon center

r_2: distance between station and typhoon center 0.5 hour later

As shown in Fig.3-1, Q_1 is the present typhoon center and Q_2 is the typhoon center 0.5 hour later. θ is the typhoon moving direction. α is the angle between the east and the line from Q_1 to the station. Then by Cosine Law, we have

$$r_2^2 = r_1^2 + (0.5V)^2 - 2 \cdot (0.5V) \cdot r_1 \cdot \cos(\theta - \alpha) \tag{3-5}$$

in which V is the moving speed of typhoon.

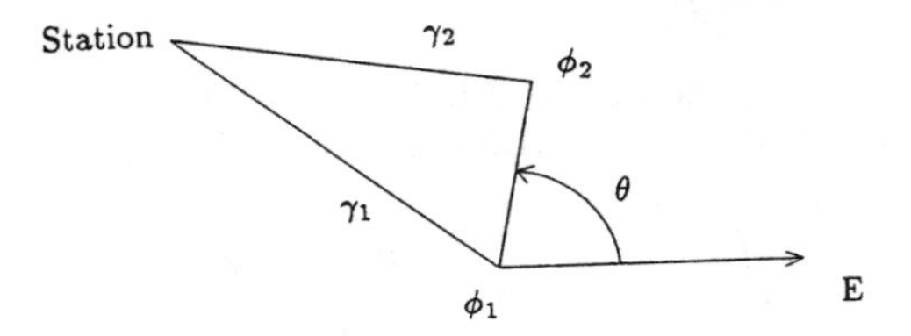

Fig.3-1 Successive Typhoon Centers

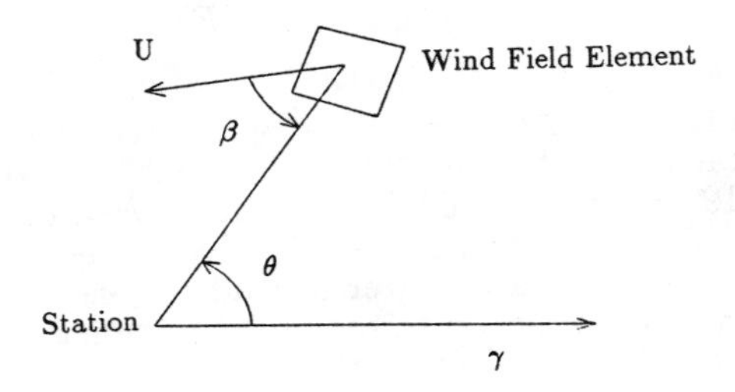

Fig.3-2 Elementatary Wave Model Scheme

(4)Swell Height Disregarding Land-Sheltering and Shoaling Effect

$$H_{1/3} = \lambda H_2 \qquad (3-6)$$

(5)Land-Sheltering Effect

There are two land-sheltering cases: One is the swell blocked by a land partially , another is partial wind field blowing on land. As shown in Fig.3-2, according to Elementary Wave Model(Liang, 1973), the wave energy flux per unit crest width is

$$dE = \xi U^2 \cdot \cos^2 \beta \cdot dA/r, |\beta| < 90 \qquad (3-7)$$

in which ξ is an empirical constant. For a whole wind field,

$$E = \int_A dE = \int_A \xi U^2 \cos^2 \beta \cdot dA/r \qquad (3-8)$$

For simplicity, we consider only the area from $R_7/10$ to R_7 radius. For the sheltered wind field,

$$E' = \int_{A'} dE = \int_{A'} \xi U^2 \cos^2 \beta \dot{d}A/r \qquad (3-9)$$

in which A' is the unsheltered wind field. Assuming that typhoon is far away, r can be assumed to be a constant and ξ is homogeneous. Then,

$$E = (\xi/r) \int_A U^2 \cos^2 \beta \cdot dA \qquad (3-10)$$

$$E' = (\xi/r) \int_{A'} U^2 \cos^2 \beta \cdot dA \qquad (3-11)$$

Supposing that the swell period is constant, the modified coefficient Ce is

$$Ce = (E'/E)^{1/2} \qquad (3-12)$$

4. Determination of Design Wave by Synthetic Probability Method

The station location is $22.397° N, 120.29° E$. All typhoon data from 1945 till 1991 are collected from Central Weather Bureau, ROC. The data from $20° - 25° N$ and $115° - 121° E$ are selected for calculating. The typhoon parameters are center latitude X, center longitude Y, center pressure P_0, moving speed V, moving direction θ and radius of Beaufort Scale 7 – R_7. Except X & Y, the other parameters are assumed to be independent. Then the sample joint probability is

$$F(X,Y,P_0,V,\theta,R_7) = F_1(X,Y) \cdot F_2(P_0) \cdot F_3(V) \cdot F_4(\theta) \cdot F_5(R_7) \quad (4-1)$$

where $F_1(X,Y), F_2(P_0), F_3(V), F_4(\theta), andF_5(R_7)$ are obtained from the selected typhoon data.

In regards of land sheltering calculation, it is impossible to calculate for each case, because the computation amount is too big. A representative typhoon is at first chosen. A system of modification coefficients for different positions of typhoon centers is found. The parameters of the representative typhoon are as follows:

$$P_0 = 980mb, V = 19km/hour, \theta = 180°, R_7 = 160km$$

The wave height prebability distribution and accumulation are shown in Fig.4-1 and Table 4-1. $F*C'$ and $SUM*C'$ mean those sheltered. There are 144 typhoons in 44 years. The design wave heights for different return periods are shown in Table 4-2.

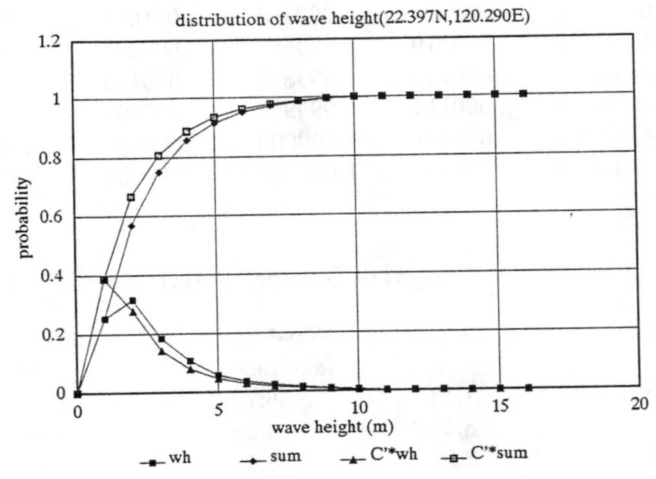

Fig.4-1 Wave Height Probability Distribution and Accumulation

Table 4-1 Wave Height Prabability Distribution and Accumulation
Measure Station : 22.397°N 120.290°E

Wave Height(M)	f	Sum	C'*f	C'*Sum
From 0 To 1	.252122	.252122	.386813	.386813
From 1 To 2	.315458	.567580	.279213	.666026
From 2 To 3	.182666	.750246	.141611	.807637
From 3 To 4	.105739	.855985	.078774	.886411
From 4 To 5	.056277	.912262	.044686	.931096
From 5 To 6	.035132	.947394	.027700	.958797
From 6 To 7	.022695	.970088	.017318	.976115
From 7 To 8	.014395	.984483	.010991	.987106
From 8 To 9	.009263	.993746	.007642	.994749
From 9 To 10	.004438	.998184	.003743	.998492
From 10 To 11	.001221	.999406	.001014	.999506
From 11 To 12	.000279	.999685	.000237	.999743
From 12 To 13	.000193	.999878	.000155	.999898
From 13 To 14	.000114	.999992	.000094	.999992
From 14 To 15	.000008	1.000000	.000008	1.000000
From 15 To 16	.000000	1.000000	.000000	1.000000

Table 4-2 Wave Height of Different Return Period

Return Period	Wave Height(m)	Wave Height(m) (sheltered)	Oceanweather
10	7.13	6.72	7.99
25	8.41	8.02	8.84
37	8.82	8.55	9.18
50	9.08	8.84	9.43
100	9.76	9.55	9.97
200	10.35	9.99	10.49

For larger return periods, the results are very close for Synthetic Probability Method and Oceanwether, which follows the old way, using Gumbel distribution (Oceanweather Inc.,1990).

5. Conclusion

Synthetic Probability Method is a simple and reliable tool to determine design typhoon wave in deep water.

6. References

Bretschneider, C.L. and E.E. Tamage(1976) Hurricane Wind and Wave Forecasting Techniques, Proceedings 15th International Coastal Engineering Conference, pp.202-237.

Liang, N.K.(1973) Elementary Wave Model and the Definition of "Fetch Area" in Wave Prediction, ACTA OCEANOGRAPHICA TAIWANICA, NO.3, pp.87-96.

Liang, N.K.(1989) A Revised Typhoon Swell Prediction Method, Harbour Technolog, Vol.4, 1-10.

Liang, N.K.(1990) A Study on Typhoon Swell Height Prediction, ACTA OCEANOGRAPHICA TAIWANICA, NO.25, PP.77-86.

Liaw (1980) Comment on Typhoon Prediction in Japan, Proceedings, Conference on Typhoon Prediction in 1980, R.O.C.(in Chinese).

Wang, G.C.Y.(1978) Sea-level Pressure Profile and Gusts within a Typhoon Circulation, Monthly Weather Review, Vol.106, pp. 954-961.

Oceanweather Inc.(1990) Extreme Meteorological and Oceanographic Data for Offshore Taiwan, Phase II Final Report, Vol.I.

Subject Index

Page number refers to first page of paper

Author Index

Page number refers to the first page of paper